# A FOUNDATION IN DIGITAL COMMUNICATION

This intuitive but rigorous introduction derives the core results and engineering schemes of digital communication from first principles. Theory, rather than industry standards, motivates the engineering approaches, and key results are stated with all the required assumptions.

The book emphasizes the geometric view, opening with the inner product, the matched filter for its computation, Parseval's theorem, the sampling theorem as an orthonormal expansion, the isometry between passband signals and their baseband representation, and the spectral-efficiency optimality of quadrature amplitude modulation (QAM). Subsequent chapters address noise, with a comprehensive study of hypothesis testing, Gaussian stochastic processes, the sufficiency of the matched filter outputs, and some coding theory.

New is a treatment of white noise without generalized functions and a presentation of the power spectral density without artificial random jitters and random phases in the analysis of QAM.

This systematic and insightful book – with over 300 exercises – is ideal for graduate courses in digital communication, and for anyone asking "why" and not just "how."

AMOS LAPIDOTH received his Ph.D. in electrical engineering from Stanford University. He was an assistant and associate professor at the Massachusetts Institute of Technology, and is currently Professor of Information Theory at ETH Zürich, the Swiss Federal Institute of Technology. He is a Fellow of the IEEE.

# A FOUNDATION IN DIGITAL COMMUNICATION

AMOS LAPIDOTH

*ETH Zürich, Swiss Federal Institute of Technology*

CAMBRIDGE UNIVERSITY PRESS
Cambridge, New York, Melbourne, Madrid, Cape Town, Singapore, São Paulo, Delhi

Cambridge University Press
The Edinburgh Building, Cambridge CB2 8RU, UK

Published in the United States of America by Cambridge University Press, New York

www.cambridge.org
Information on this title: www.cambridge.org/9780521193955

First published 2009

Printed in the United Kingdom at the University Press, Cambridge

*A catalogue record for this publication is available from the British Library*

ISBN 978-0-521-19395-5 hardback

Additional resources for this publication at www.cambridge.org/9780521193955

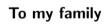

To my family

# Contents

# Preface

Claude Shannon, the father of Information Theory, described the fundamental problem of point-to-point communications in his classic 1948 paper as "that of reproducing at one point either exactly or approximately a message selected at another point." How engineers solve this problem is the subject of this book. But unlike Shannon's general problem, where the message can be an image, a sound clip, or a movie, here we restrict ourselves to bits. We thus envision that the original message is either a binary sequence to start with, or else that it was described using bits by a device outside our control and that our job is to reproduce the describing bits with high reliability. The issue of how images or text files are converted efficiently into bits is the subject of lossy and lossless data compression and is addressed in texts on information theory and on quantization.

The engineering solutions to the point-to-point communication problem greatly depend on the available resources and on the channel between the points. They typically bring together beautiful techniques from Fourier Analysis, Hilbert Spaces, Probability Theory, and Decision Theory. The purpose of this book is to introduce the reader to these techniques and to their interplay.

The book is intended for advanced undergraduates and beginning graduate students. The key prerequisites are basic courses in Calculus, Linear Algebra, and Probability Theory. A course in Linear Systems is a plus but not a must, because all the results from Linear Systems that are needed for this book are summarized in Chapters 5 and 6. But more importantly, the book requires a certain mathematical maturity and patience, because we begin with first principles and develop the theory before discussing its engineering applications. The book is for those who appreciate the views along the way as much as getting to the destination; who like to "stop and smell the roses;" and who prefer fundamentals to acronyms. I firmly believe that those with a sound foundation can easily pick up the acronyms and learn the jargon on the job, but that once one leaves the academic environment, one rarely has the time or peace of mind to study fundamentals.

In the early stages of the planning of this book I took a decision that greatly influenced the project. I decided that every key concept should be unambiguously defined; that every key result should be stated as a mathematical theorem; and that every mathematical theorem should be correct. This, I believe, makes for a solid foundation on which one can build with confidence. But it is also a tall order. It required that I scrutinize each "classical" result before I used it in order to be sure that I knew what the needed qualifiers were, and it forced me to include background material to which the reader may have already been exposed, because I needed the results "done right." Hence Chapters 5 and 6 on Linear Systems and Fourier Analysis. This is also partly the reason why the book is so long. When I started out my intention was to write a much shorter book. But I found that to do justice to the beautiful mathematics on which Digital Communications is based I had to expand the book.

Most physical layer communication problems are at their core of a continuous-time nature. The transmitted physical waveforms are functions of time and not sequences synchronized to a clock. But most solutions first reduce the problem to a discrete-time setting and then solve the problem in the discrete-time domain. The reduction to discrete-time often requires great ingenuity, which I try to describe. It is often taken for granted in courses that open with a discrete-time model from Lecture 1. I emphasize that most communication problems are of a continuous-time nature, and that the reduction to discrete-time is not always trivial or even possible. For example, it is extremely difficult to translate a peak-power constraint (stating that at no epoch is the magnitude of the transmitted waveform allowed to exceed a given constant) to a statement about the sequence that is used to represent the waveform. Similarly, in Wireless Communications it is often very difficult to reduce the received waveform to a sequence without any loss in performance.

The quest for mathematical precision can be demanding. I have therefore tried to precede the statement of every key theorem with its gist in plain English. Instructors may well choose to present the material in class with less rigor and direct the students to the book for a more mathematical approach. I would rather have textbooks be more mathematical than the lectures than the other way round. Having a rigorous textbook allows the instructor in class to discuss the intuition knowing that the students can obtain the technical details from the book at home.

The communication problem comes with a beautiful geometric picture that I try to emphasize. To appreciate this picture one needs the definition of the inner product between energy-limited signals and some of the geometry of the space of energy-limited signals. These are therefore introduced early on in Chapters 3 and 4. Chapters 5 and 6 cover standard material from Linear Systems. But note the early introduction of the matched filter as a mechanism for computing inner products in Section 5.8. Also key is Parseval's Theorem in Section 6.2.2 which relates the geometric pictures in the time domain and in the frequency domain.

Chapter 7 deals with passband signals and their baseband representation. We emphasize how the inner product between passband signals is related to the inner product between their baseband representations. This elegant geometric relationship is often lost in the haze of various trigonometric identities. While this topic is important in wireless applications, it is not always taught in a first course in Digital Communications. Instructors who prefer to discuss baseband communication only can skip Chapters 7, 9, 16, 17, 18, 24 27, and Sections 26.10 and 28.5. But it would be a shame.

Chapter 8 presents the celebrated Sampling Theorem from a geometric perspective. It is inessential to the rest of the book but is a striking example of the geometric approach. Chapter 9 discusses the Sampling Theorem for passband signals.

Chapter 10 discusses modulation. I have tried to motivate Linear Modulation and Pulse Amplitude Modulation and to minimize the use of the "that's just how it is done" argument. The use of the Matched Filter for detecting (here in the absence of noise) is emphasized. This also motivates the Nyquist Theory, which is treated in Chapter 11. I stress that the motivation for the Nyquist Theory is not to avoid inter-symbol interference at the sampling points but rather to guarantee the orthogonality of the time shifts of the pulse shape by integer multiples of the baud period. This ultimately makes more engineering sense and leads to cleaner mathematics: compare Theorem 11.3.2 with its corollary, Corollary 11.3.4.

The result of modulating random bits is a stochastic process, a concept which is first encountered in Chapter 10; formally defined in Chapter 12; and revisited in Chapters 13, 17, and 25. It is an important concept in Digital Communications, and I find it best to first introduce man-made synthesized stochastic processes (as the waveforms produced by an encoder when fed random bits) and only later to introduce the nature-made stochastic processes that model noise. Stationary discrete-time stochastic processes are introduced in Chapter 13 and their complex counterparts in Chapter 17. These are needed for the analysis in Chapter 14 of the power in Pulse Amplitude Modulation and for the analysis in Chapter 17 of the power in Quadrature Amplitude Modulation.

I emphasize that power is a physical quantity that is related to the time-averaged energy in the continuous-time transmitted power. Its relation to the power in the discrete-time modulating sequence is a nontrivial result. In deriving this relation I refrain from adding random timing jitters that are often poorly motivated and that turn out to be unnecessary. (The transmitted power does not depend on the realization of the fictitious jitter.) The Power Spectral Density in Pulse Amplitude Modulation and Quadrature Amplitude Modulation is discussed in Chapters 15 and 18. The discussion requires a definition for Power Spectral Density for non-stationary processes (Definitions 15.3.1 and 18.4.1) and a proof that this definition coincides with the classical definition when the process is wide-sense stationary (Theorem 25.14.3).

Chapter 19 opens the second part of the book, which deals with noise and detection. It introduces the univariate Gaussian distribution and some related distributions. The principles of Detection Theory are presented in Chapters 20–22. I emphasize the notion of Sufficient Statistics, which is central to Detection Theory. Building on Chapter 19, Chapter 23 introduces the all-important multivariate Gaussian distribution. Chapter 24 treats the complex case.

Chapter 25 deals with continuous-time stochastic processes with an emphasis on stationary Gaussian processes, which are often used to model the noise in Digital Communications. This chapter also introduces white Gaussian noise. My approach to this topic is perhaps new and is probably where this text differs the most from other textbooks on the subject.

I define **white Gaussian noise of double-sided power spectral density $N_0/2$ with respect to the bandwidth $W$** as any measurable,[1] stationary, Gaussian stochastic process whose power spectral density is a nonnegative, symmetric, inte-

---

[1]This book does not assume any Measure Theory and does not teach any Measure Theory. (I do define sets of Lebesgue measure zero in order to be able to state uniqueness theorems.) I

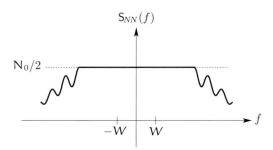

**Figure 1:** The power spectral density of a white Gaussian noise process of double-sided power spectral density $N_0/2$ with respect to the bandwidth $W$.

grable function of frequency that is equal to $N_0/2$ at all frequencies $f$ satisfying $|f| \leq W$. The power spectral density at other frequencies can be arbitrary. An example of the power spectral density of such a process is depicted in Figure 1. Adopting this definition has a number of advantages. The first is, of course, that such processes exist. One need not discuss "generalized processes," Gaussian processes with infinite variances (that, by definition, do not exist), or introduce the Itô calculus to study stochastic integrals. (Stochastic integrals with respect to the Brownian motion are mathematically intricate and physically unappealing. The idea of the noise having infinite power is ludicrous.) The above definition also frees me from discussing Dirac's Delta, and, in fact, Dirac's Delta is never used in this book. (A rigorous treatment of Generalized Functions is beyond the engineering curriculum in most schools, so using Dirac's Delta always gives the reader the unsettling feeling of being on unsure footing.)

The detection problem in white Gaussian noise is treated in Chapter 26. No course in Digital Communications should end without Theorem 26.4.1. Roughly speaking, this theorem states that if the mean-signals are bandlimited to $W$ Hz and if the noise is white Gaussian noise with respect to the bandwidth $W$, then the inner products between the received signal and the mean-signals form a sufficient statistic. Numerous examples as well as a treatment of colored noise are also discussed in this chapter. Extensions to noncoherent detection are addressed in Chapter 27 and implications for Pulse Amplitude Modulation and for Quadrature Amplitude Modulation in Chapter 28.

The book concludes with Chapter 29, which introduces Coding. It emphasizes how the code design influences the transmitted power, the transmitted power spectral density, the required bandwidth, and the probability of error. The construction of good codes is left to texts on Coding Theory.

use Measure Theory only in stating theorems that require measurability assumptions. This is in line with my attempt to state theorems together with all the assumptions that are required for their validity. I recommend that students ignore measurability issues and just make a mental note that whenever measurability is mentioned there is a minor technical condition lurking in the background.

## Basic Latin

Mathematics sometimes reads like a foreign language. I therefore include here a short glossary for such terms as "*i.e.*," "*that is*," "*in particular*," "*a fortiori*," "*for example*," and "*e.g.*," whose meaning in Mathematics is slightly different from the definition you will find in your English dictionary. In mathematical contexts these terms are actually logical statements that the reader should verify. Verifying these statements is an important way to make sure that you understand the math.

What are these logical statements? First note the synonym "*i.e.*" = "*that is*" and the synonym "*e.g.*" = "*for example*." Next note that the term "*that is*" often indicates that the statement following the term is equivalent to the one preceding it: "We next show that $p$ is a prime, *i.e.*, that $p$ is a positive integer that is not divisible by any number other than one and itself." The terms "*in particular*" or "*a fortiori*" indicate that the statement following them is implied by the one preceding them: "Since $g(\cdot)$ is differentiable and, *a fortiori*, continuous, it follows from the Mean Value Theorem that the integral of $g(\cdot)$ over the interval $[0,1]$ is equal to $g(\xi)$ for some $\xi \in [0,1]$." The term "*for example*" can have its regular day-to-day meaning but in mathematical writing it also sometimes indicates that the statement following it implies the one preceding it: "Suppose that the function $g(\cdot)$ is monotonically nondecreasing, *e.g.*, that it is differentiable with a nonnegative derivative."

Another important word to look out for is "*indeed*," which in this book typically signifies that the statement just made is about to be expanded upon and explained. So when you read something that is unclear to you, be sure to check whether the next sentence begins with the word "indeed" before you panic.

The Latin phrases "*a priori*" and "*a posteriori*" show up in Probability Theory. The former is usually associated with the unconditional probability of an event and the latter with the conditional. Thus, the "*a priori*" probability that the sun will shine this Sunday in Zurich is 25%, but now that I know that it is raining today, my outlook on life changes and I assign this event the *a posteriori* probability of 15%.

The phrase "*prima facie*" is roughly equivalent to the phrase "before any further mathematical arguments have been presented." For example, the definition of the projection of a signal $\mathbf{v}$ onto the signal $\mathbf{u}$ as the vector $\mathbf{w}$ that is collinear with $\mathbf{u}$ and for which $\mathbf{v} - \mathbf{w}$ is orthogonal to $\mathbf{u}$, may be followed by the sentence: "*Prima facie*, it is not clear that the projection always exists and that it is unique. Nevertheless, as we next show, this is the case."

## Syllabuses or Syllabi

The book can be used as a textbook for a number of different courses. For a course that focuses on deterministic signals one could use Chapters 1–9 & Chapter 11. A course that covers Stochastic Processes and Detection Theory could be based on Chapter 12 and Chapters 19–26 with or without discrete-time stochastic processes (Chapter 13) and with or without complex random variables and processes

(Chapters 17 & 24).

For a course on Digital Communications one could use the entire book or, if time does not permit it, discuss only baseband communication. In the latter case one could omit Chapters 7, 9, 16, 17, 18, 24, 27, and Section 28.5,

The dependencies between the chapters are depicted on Page xxiii.

A web page for this book can be found at

`www.afoundationindigitalcommunication.ethz.ch`

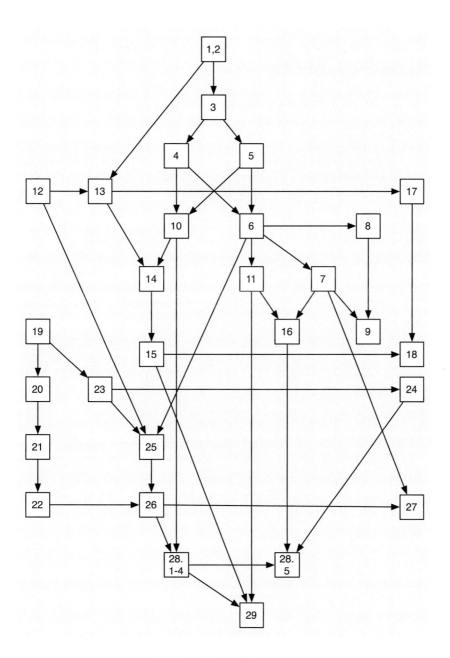

A Dependency Diagram.

# Acknowledgments

This book has a long history. Its origins are in a course entitled "Introduction to Digital Communication" that Bob Gallager and I developed at the Massachusetts Institute of Technology (MIT) in the years 1997 (course number 6.917) and 1998 (course number 6.401). Assisting us in these courses were Emre Koksal and Poompat Saengudomlert (Tengo) respectively. The course was first conceived as an advanced undergraduate course, but at MIT it has since evolved into a first-year graduate course leading to the publication of the textbook (Gallager, 2008). At ETH the course is still an advanced undergraduate course, and the lecture notes evolved into the present book. Assisting me at ETH were my former and current Ph.D. students Stefan Moser, Daniel Hösli, Natalia Miliou, Stephan Tinguely, Tobias Koch, Michèle Wigger, and Ligong Wang. I thank them all for their enormous help. Marion Brändle was also a great help.

I also thank Bixio Rimoldi for his comments on an earlier draft of this book, from which he taught at École Polytechnique Fédérale de Lausanne (EPFL) and Thomas Mittelholzer, who used a draft of this book to teach a course at ETH during my sabbatical.

Extremely helpful were discussions with Amir Dembo, Sanjoy Mitter, Alain-Sol Sznitman, and Ofer Zeitouni about some of the more mathematical aspects of this book. Discussions with Ezio Biglieri, Holger Boche, Stephen Boyd, Young-Han Kim, and Sergio Verdú are also gratefully acknowledged.

Special thanks are due to Bob Gallager and Dave Forney with whom I had endless discussions about the material in this book both while at MIT and afterwards at ETH. Their ideas have greatly influenced my thinking about how this course should be taught.

I thank Helmut Bölcskei, Andi Loeliger, and Nikolai Nefedov for having tolerated my endless ramblings regarding Digital Communications during our daily lunches. Jim Massey was a huge help in patiently answering my questions regarding English usage. I should have asked him much more!

A number of dear colleagues read parts of this manuscript. Their comments were extremely useful. These include Helmut Bölcskei, Moritz Borgmann, Samuel Braendle, Shraga Bross, Giuseppe Durisi, Yariv Ephraim, Minnie Ho, Young-Han Kim, Yiannis Kontoyiannis, Nick Laneman, Venya Morgenshtern, Prakash Narayan, Igal Sason, Brooke Shrader, Aslan Tchamkerten, Sergio Verdú, Pascal Vontobel, and Ofer Zeitouni. I am especially indebted to Emre Telatar for his enormous help in all aspects of this project.

I would like to express my sincere gratitude to the Rockefeller Foundation at whose Study and Conference Center in Bellagio, Italy, this all began.

Finally, I thank my wife, Danielle, for her encouragement, her tireless editing, and for making it possible for me to complete this project.

# Chapter 1

# Some Essential Notation

Reading a whole chapter about notation can be boring. We have thus chosen to collect here only the essentials and to introduce the rest when it is first used. The "List of Symbols" on Page 704 is more comprehensive.

We denote the set of complex numbers by $\mathbb{C}$, the set of real numbers by $\mathbb{R}$, the set of integers by $\mathbb{Z}$, and the set of natural numbers (positive integers) by $\mathbb{N}$. Thus,

$$\mathbb{N} = \{n \in \mathbb{Z} : n \geq 1\}.$$

The above equation is not meant to belabor the point. We use it to introduce the notation

$$\boxed{\{x \in \mathcal{A} : \text{statement}\}}$$

for the set consisting of all those elements of the set $\mathcal{A}$ for which "statement" holds.

In treating real numbers, we use the notation $(a, b)$, $[a, b)$, $[a, b]$, $(a, b]$ to denote open, half open on the right, closed, and half open on the left intervals of the real line. Thus, for example,

$$[a, b) = \{x \in \mathbb{R} : a \leq x < b\}.$$

A statement followed by a comma and a condition indicates that the statement holds whenever the condition is satisfied. For example,

$$|a_n - a| < \epsilon, \quad n \geq n_0$$

means that $|a_n - a| < \epsilon$ whenever $n \geq n_0$.

We use I{statement} to denote the indicator of the statement. It is equal to 1, if the statement is true, and it is equal to 0, if the statement is false. Thus

$$\boxed{\text{I\{statement\}} = \begin{cases} 1 & \text{if statement is true,} \\ 0 & \text{if statement is false.} \end{cases}}$$

1

In dealing with complex numbers we use i to denote the purely imaginary unit-magnitude complex number

$$i = \sqrt{-1}.$$

We use $z^*$ to denote the complex conjugate of $z$, we use $\mathrm{Re}(z)$ to denote the real part of $z$, we use $\mathrm{Im}(z)$ to denote the imaginary part of $z$, and we use $|z|$ to denote the absolute value (or "modulus", or "complex magnitude") of $z$. Thus, if $z = a + ib$, where $a, b \in \mathbb{R}$, then $z^* = a - ib$, $\mathrm{Re}(z) = a$, $\mathrm{Im}(z) = b$, and $|z| = \sqrt{a^2 + b^2}$.

The notation used to define functions is extremely important and is, alas, sometimes confusing to students, so please pay attention. A **function** or a **mapping** associates with each element in its **domain** a unique element in its **range**. If a function has a name, the name is often written in bold as in $\mathbf{u}$.[1] Alternatively, we sometimes denote a function $\mathbf{u}$ by $u(\cdot)$. The notation

$$\mathbf{u} \colon \mathcal{A} \to \mathcal{B}$$

indicates that $\mathbf{u}$ is a function of domain $\mathcal{A}$ and range $\mathcal{B}$. The rule specifying for each element of the domain the element in the range to which it is mapped is often written to the right or underneath. Thus, for example,

$$\mathbf{u} \colon \mathbb{R} \to (-5, \infty), \qquad t \mapsto t^2$$

indicates that the domain of the function $\mathbf{u}$ is the reals, that its range is the set of real numbers that exceed $-5$, and that $\mathbf{u}$ associates with $t$ the nonnegative number $t^2$. We write $u(t)$ for the result of applying the mapping $\mathbf{u}$ to $t$. The **image** of a mapping $\mathbf{u} \colon \mathcal{A} \to \mathcal{B}$ is the set of all elements of the range $\mathcal{B}$ to which at least one element in the domain is mapped by $\mathbf{u}$:

$$\Big(\text{image of } \mathbf{u} \colon \mathcal{A} \to \mathcal{B}\Big) = \big\{u(x) : x \in \mathcal{A}\big\}. \tag{1.1}$$

The image of a mapping is a subset of its range. In the above example, the image of the mapping is the set of nonnegative reals $[0, \infty)$. A mapping $\mathbf{u} \colon \mathcal{A} \to \mathcal{B}$ is said to be **onto** (or **surjective**) if its image is equal to its range. Thus, $\mathbf{u} \colon \mathcal{A} \to \mathcal{B}$ is onto if, and only if, for every $y \in \mathcal{B}$ there corresponds some $x \in \mathcal{A}$ (not necessarily unique) such that $u(x) = y$. If the image of $g(\cdot)$ is a subset of the domain of $h(\cdot)$, then the **composition** of $g(\cdot)$ and $h(\cdot)$ is the mapping $x \mapsto h\big(g(x)\big)$, which is denoted by $\mathbf{h} \circ \mathbf{g}$.

Sometimes we do not specify the domain and range of a function if they are clear from the context. Thus, we might write $\mathbf{u} \colon t \mapsto v(t)\cos(2\pi f_{\mathrm{c}} t)$ without making explicit what the domain and range of $\mathbf{u}$ are. In fact, if there is no need to give a function a name, then we will not. For example, we might write $t \mapsto v(t)\cos(2\pi f_{\mathrm{c}} t)$ to designate the unnamed function that maps $t$ to $v(t)\cos(2\pi f_{\mathrm{c}} t)$. (Here $v(\cdot)$ is some other function, which was presumably defined before.)

If the domain of a function $\mathbf{u}$ is $\mathbb{R}$ and if the range is $\mathbb{R}$, then we sometimes say that $\mathbf{u}$ is a **real-valued signal** or a **real signal**, especially if the argument of $\mathbf{u}$

---

[1] But some special functions such as the self-similarity function $\mathsf{R}_{\mathbf{gg}}$, the autocovariance function $\mathsf{K}_{XX}$, and the power spectral density $\mathsf{S}_{XX}$, which will be introduced in later chapters, are not in boldface.

stands for time. Similarly we shall sometimes refer to a function $\mathbf{u} \colon \mathbb{R} \to \mathbb{C}$ as a **complex-valued signal** or a **complex signal**. If we refer to $\mathbf{u}$ as a signal, then the question whether it is complex-valued or real-valued should be clear from the context, or else immaterial to the claim.

We caution the reader that, while $\mathbf{u}$ and $u(\cdot)$ denote functions, $u(t)$ denotes the result of applying $\mathbf{u}$ to $t$. If $\mathbf{u}$ is a real-valued signal then $u(t)$ is a real number!

Given two signals $\mathbf{u}$ and $\mathbf{v}$ we define their **superposition** or **sum** as the signal $t \mapsto u(t) + v(t)$. We denote this signal by $\mathbf{u} + \mathbf{v}$. Also, if $\alpha \in \mathbb{C}$ and $\mathbf{u}$ is any signal, then we define the **amplification** of $\mathbf{u}$ by $\alpha$ as the signal $t \mapsto \alpha u(t)$. We denote this signal by $\alpha \mathbf{u}$. Thus,

$$\alpha \mathbf{u} + \beta \mathbf{v}$$

is the signal

$$t \mapsto \alpha u(t) + \beta v(t).$$

We refer to the function that maps every element in its domain to zero as the **all-zero function** and we denote it by $\mathbf{0}$. The **all-zero signal** $\mathbf{0}$ maps every $t \in \mathbb{R}$ to zero. If $\mathbf{x} \colon \mathbb{R} \to \mathbb{C}$ is a signal that maps every $t \in \mathbb{R}$ to $x(t)$, then its **reflection** or **mirror image** is denoted by $\overleftarrow{\mathbf{x}}$ and is the signal that is defined by

$$\overleftarrow{\mathbf{x}} \colon t \mapsto x(-t).$$

Dirac's Delta (which will hardly be mentioned in this book) *is not a function*.

A probability space is defined as a triplet $(\Omega, \mathcal{F}, P)$, where the set $\Omega$ is the set of **experiment outcomes**, the elements of the set $\mathcal{F}$ are subsets of $\Omega$ and are called **events**, and where $P \colon \mathcal{F} \to [0, 1]$ assigns probabilities to the various events. It is assumed that $\mathcal{F}$ forms a $\sigma$-algebra, i.e., that $\Omega \in \mathcal{F}$; that if a set is in $\mathcal{F}$ then so is its complement (with respect to $\Omega$); and that every finite or countable union of elements of $\mathcal{F}$ is also an element of $\mathcal{F}$. A random variable $X$ is a mapping from $\Omega$ to $\mathbb{R}$ that satisfies the technical condition that

$$\{\omega \in \Omega : X(\omega) \leq \xi\} \in \mathcal{F}, \quad \xi \in \mathbb{R}. \tag{1.2}$$

This condition guarantees that it is always meaningful to evaluate the probability that the value of $X$ is smaller or equal to $\xi$.

# Chapter 2

# Signals, Integrals, and Sets of Measure Zero

## 2.1 Introduction

The purpose of this chapter is not to develop the Lebesgue theory of integration. Mastering this theory is not essential to understanding Digital Communications. But some concepts from this theory are needed in order to state the main results of Digital Communications in a mathematically rigorous way. In this chapter we introduce these required concepts and provide references to the mathematical literature that develops them.

The less mathematically-inclined may gloss over most of this chapter. Readers who interpret the integrals in this book as Riemann integrals; who interpret "measurable" as "satisfying a minor mathematical restriction"; who interpret "a set of Lebesgue measure zero" as "a set that is so small that integrals of functions are not sensitive to the value the integrand takes in this set"; and who swap orders of summations, expectations and integrations fearlessly will not miss any engineering insights.

But all readers should pay attention to the way the integral of complex-valued signals is defined (Section 2.3); to the basic inequality (2.13); and to the notation introduced in (2.6).

## 2.2 Integrals

Recall that a real-valued signal $\mathbf{u}$ is a function $\mathbf{u} \colon \mathbb{R} \to \mathbb{R}$. The integral of $\mathbf{u}$ is denoted by

$$\int_{-\infty}^{\infty} u(t)\,\mathrm{d}t. \tag{2.1}$$

For (2.1) to be meaningful some technical conditions must be met. (You may recall from your calculus studies, for example, that not every function is Riemann integrable.) In this book all integrals will be understood to be Lebesgue integrals, but nothing essential will be lost on readers who interpret them as Riemann integrals. For the Lebesgue integral to be defined the integrand $\mathbf{u}$ must be a **Lebesgue measurable function**. Again, do not worry if you have not studied the Lebesgue integral or the notion of measurable functions. We point this out merely to cover ourselves when we state various theorems. Also, for the integral in (2.1) to be defined we insist that

$$\int_{-\infty}^{\infty} |u(t)| \, dt < \infty. \tag{2.2}$$

(There are ways of defining the integral in (2.1) also when (2.2) is violated, but they lead to fragile expressions that are difficult to manipulate.)

A function $\mathbf{u} \colon \mathbb{R} \to \mathbb{R}$ which is Lebesgue measurable and which satisfies (2.2) is said to be **integrable**, and we denote the set of all such functions by $\mathcal{L}_1$. We shall refrain from integrating functions that are not elements of $\mathcal{L}_1$.

## 2.3   Integrating Complex-Valued Signals

This section should assuage your fear of integrating complex-valued signals. (Some of you may have a trauma from your Complex Analysis courses where you dealt with integrals of functions from the complex plane to the complex plane. Here things are much simpler because we are dealing only with integrals of functions from the real line to the complex plane.) We formally define the integral of a complex-valued function $\mathbf{u} \colon \mathbb{R} \to \mathbb{C}$ by

$$\int_{-\infty}^{\infty} u(t) \, dt \triangleq \int_{-\infty}^{\infty} \mathrm{Re}\big(u(t)\big) \, dt + \mathrm{i} \int_{-\infty}^{\infty} \mathrm{Im}\big(u(t)\big) \, dt. \tag{2.3}$$

For this to be meaningful, we require that the real functions $t \mapsto \mathrm{Re}\big(u(t)\big)$ and $t \mapsto \mathrm{Im}\big(u(t)\big)$ both be integrable real functions. That is, they should both be Lebesgue measurable and we should have

$$\int_{-\infty}^{\infty} \big|\mathrm{Re}\big(u(t)\big)\big| \, dt < \infty \quad \text{and} \quad \int_{-\infty}^{\infty} \big|\mathrm{Im}\big(u(t)\big)\big| \, dt < \infty. \tag{2.4}$$

It is not difficult to show that (2.4) is equivalent to the more compact condition

$$\int_{-\infty}^{\infty} |u(t)| \, dt < \infty. \tag{2.5}$$

We say that a complex signal $\mathbf{u} \colon \mathbb{R} \to \mathbb{C}$ is **Lebesgue measurable** if the mappings $t \mapsto \mathrm{Re}\big(u(t)\big)$ and $t \mapsto \mathrm{Im}\big(u(t)\big)$ are Lebesgue measurable real signals. We say that a function $\mathbf{u} \colon \mathbb{R} \to \mathbb{C}$ is **integrable** if it is Lebesgue measurable and (2.4) holds. The set of all Lebesgue measurable integrable complex signals is denoted by $\mathcal{L}_1$. Note that we use the same symbol $\mathcal{L}_1$ to denote both the set of integrable real signals and the set of integrable complex signals. To which of these two sets we refer should be clear from the context, or else immaterial.

For $\mathbf{u} \in \mathcal{L}_1$ we define $\|\mathbf{u}\|_1$ as

$$\|\mathbf{u}\|_1 \triangleq \int_{-\infty}^{\infty} |u(t)| \, dt. \tag{2.6}$$

Before summarizing the key properties of the integral of complex signals we remind the reader that if **u** and **v** are complex signals and if $\alpha, \beta$ are complex numbers, then the complex signal $\alpha\mathbf{u} + \beta\mathbf{v}$ is defined as the complex signal $t \mapsto \alpha u(t) + \beta v(t)$. The intuition for the following proposition comes from thinking about the integrals as Riemann integrals, which can be approximated by finite sums and by then invoking the analogous results about finite sums.

**Proposition 2.3.1 (Properties of Complex Integrals).** *Let the complex signals* **u**, **v** *be in* $\mathcal{L}_1$*, and let* $\alpha, \beta$ *be arbitrary complex numbers.*

*(i) Integration is linear in the sense that* $\alpha\mathbf{u} + \beta\mathbf{v} \in \mathcal{L}_1$ *and*

$$\int_{-\infty}^{\infty} \big(\alpha\,u(t) + \beta\,v(t)\big)\,\mathrm{d}t = \alpha \int_{-\infty}^{\infty} u(t)\,\mathrm{d}t + \beta \int_{-\infty}^{\infty} v(t)\,\mathrm{d}t. \qquad (2.7)$$

*(ii) Integration commutes with complex conjugation*

$$\int_{-\infty}^{\infty} u^*(t)\,\mathrm{d}t = \left( \int_{-\infty}^{\infty} u(t)\,\mathrm{d}t \right)^*. \qquad (2.8)$$

*(iii) Integration commutes with the operation of taking the real part*

$$\mathrm{Re}\left( \int_{-\infty}^{\infty} u(t)\,\mathrm{d}t \right) = \int_{-\infty}^{\infty} \mathrm{Re}\big(u(t)\big)\,\mathrm{d}t. \qquad (2.9)$$

*(iv) Integration commutes with the operation of taking the imaginary part*

$$\mathrm{Im}\left( \int_{-\infty}^{\infty} u(t)\,\mathrm{d}t \right) = \int_{-\infty}^{\infty} \mathrm{Im}\big(u(t)\big)\,\mathrm{d}t. \qquad (2.10)$$

**Proof.** For a proof of (i) see, for example, (Rudin, 1974, Theorem 1.32). The rest of the claims follow easily from the definition of the integral of a complex-valued signal (2.3). $\qquad\square$

## 2.4  An Inequality for Integrals

Probably the most important inequality for complex numbers is the **Triangle Inequality for Complex Numbers**

$$|w + z| \leq |w| + |z|, \quad w, z \in \mathbb{C}. \qquad (2.11)$$

This inequality extends by induction to finite sums:

$$\left| \sum_{j=1}^{n} z_j \right| \leq \sum_{j=1}^{n} |z_j|, \quad z_1, \ldots, z_n \in \mathbb{C}. \qquad (2.12)$$

The extension to integrals is the most important inequality for integrals:

**Proposition 2.4.1.** *For every complex-valued or real-valued signal* **u** *in* $\mathcal{L}_1$

$$\left| \int_{-\infty}^{\infty} u(t)\,\mathrm{d}t \right| \le \int_{-\infty}^{\infty} |u(t)|\,\mathrm{d}t. \tag{2.13}$$

**Proof.** See, for example, (Rudin, 1974, Theorem 1.33). □

Note that in (2.13) we should interpret $|\cdot|$ as the absolute-value function if **u** is a real signal, and as the modulus function if **u** is a complex signal.

Another simple but useful inequality is

$$\|\mathbf{u} + \mathbf{v}\|_1 \le \|\mathbf{u}\|_1 + \|\mathbf{v}\|_1, \quad \mathbf{u}, \mathbf{v} \in \mathcal{L}_1, \tag{2.14}$$

which can be proved using the calculation

$$
\begin{aligned}
\|\mathbf{u} + \mathbf{v}\|_1 &\triangleq \int_{-\infty}^{\infty} |u(t) + v(t)|\,\mathrm{d}t \\
&\le \int_{-\infty}^{\infty} \big(|u(t)| + |v(t)|\big)\,\mathrm{d}t \\
&= \int_{-\infty}^{\infty} |u(t)|\,\mathrm{d}t + \int_{-\infty}^{\infty} |v(t)|\,\mathrm{d}t \\
&= \|\mathbf{u}\|_1 + \|\mathbf{v}\|_1,
\end{aligned}
$$

where the inequality follows by applying the Triangle Inequality for Complex Numbers (2.11) with the substitution of $u(t)$ for $w$ and $v(t)$ for $z$.

## 2.5 Sets of Lebesgue Measure Zero

It is one of life's minor grievances that the integral of a nonnegative function can be zero even if the function is not identically zero. For example, $t \mapsto \mathrm{I}\{t = 17\}$ is a nonnegative function whose integral is zero and which is nonetheless not identically zero (it maps 17 to one). In this section we shall derive a necessary and sufficient condition for the integral of a nonzero function to be zero. This condition will allow us later to state conditions under which various integral inequalities hold with equality. It will give mathematical meaning to the physical intuition that if the waveform describing some physical phenomenon (such as voltage over a resistor) is nonnegative and integrates to zero then "for all practical purposes" the waveform is zero.

We shall define sets of Lebesgue measure zero and then show that a nonnegative function $\mathbf{u}\colon \mathbb{R} \to [0, \infty)$ integrates to zero if, and only if, the set $\{t \in \mathbb{R} : u(t) > 0\}$ is of Lebesgue measure zero. We shall then introduce the notation $\mathbf{u} \equiv \mathbf{v}$ to indicate that the set $\{t \in \mathbb{R} : u(t) \ne v(t)\}$ is of Lebesgue measure zero.

It should be noted that since the integral is unaltered when the integrand is changed at a finite (or countable) number of points, it follows that any nonnegative function that is zero except at a countable number of points integrates to zero. The reverse, however, is not true. One can find nonnegative functions that integrate to zero and that are nonzero on an uncountable set of points.

The less mathematically inclined readers may skip the mathematical definition of sets of measure zero and just think of a subset of the real line as being of Lebesgue measure zero if it is so "small" that the integral of any function is unaltered when the values it takes in the subset are altered. Such readers should then think of the statement $\mathbf{u} \equiv \mathbf{v}$ as indicating that $\mathbf{u} - \mathbf{v}$ is just the result of altering the all-zero signal $\mathbf{0}$ on a set of Lebesgue measure zero and that, consequently,

$$\int_{-\infty}^{\infty} |u(t) - v(t)| \, \mathrm{d}t = 0.$$

**Definition 2.5.1 (Sets of Lebesgue Measure Zero).** *We say that a subset $\mathcal{N}$ of the real line $\mathbb{R}$ is a **set of Lebesgue measure zero** (or a **Lebesgue null set**) if for every $\epsilon > 0$ we can find a sequence of intervals $[a_1, b_1]$, $[a_2, b_2], \ldots$ such that the total length of the intervals is smaller than or equal to $\epsilon$*

$$\sum_{j=1}^{\infty} (b_j - a_j) \leq \epsilon \tag{2.15a}$$

*and such that the union of the intervals cover the set $\mathcal{N}$*

$$\mathcal{N} \subseteq [a_1, b_1] \cup [a_2, b_2] \cup \cdots. \tag{2.15b}$$

As an example, note that the set $\{1\}$ is of Lebesgue measure zero. Indeed, it is covered by the single interval $[1 - \epsilon/2, 1 + \epsilon/2]$ whose length is $\epsilon$. Similarly, any finite set is of Lebesgue measure zero. Indeed, the set $\{\alpha_1, \ldots, \alpha_n\}$ can be covered by $n$ intervals of total length not exceeding $\epsilon$ as follows:

$$\{\alpha_1, \ldots, \alpha_n\} \subset \left[\alpha_1 - \epsilon/(2n), \alpha_1 + \epsilon/(2n)\right] \cup \cdots \cup \left[\alpha_n - \epsilon/(2n), \alpha_n + \epsilon/(2n)\right].$$

This argument can be also extended to show that any countable set is of Lebesgue measure zero. Indeed the countable set $\{\alpha_1, \alpha_2, \ldots\}$ can be covered as

$$\{\alpha_1, \alpha_2, \ldots\} \subseteq \bigcup_{j=1}^{\infty} \left[\alpha_j - 2^{-j-1}\epsilon, \alpha_j + 2^{-j-1}\epsilon\right]$$

where we note that the length of the interval $\left[\alpha_j - 2^{-j-1}\epsilon, \alpha_j + 2^{-j-1}\epsilon\right]$ is $2^{-j}\epsilon$, which when summed over $j$ yields $\epsilon$.

With a similar argument one can show that the union of a countable number of sets of Lebesgue measure zero is of Lebesgue measure zero.

The above examples notwithstanding, it should be emphasized that there exist sets of Lebesgue measure zero that are not countable.[1] Thus, the concept of a set of Lebesgue measure zero is different from the concept of a countable set.

Loosely speaking, we say that two signals are indistinguishable if they agree except possibly on a set of Lebesgue measure zero. We warn the reader, however, that this terminology is not standard.

---

[1]For example, the Cantor set is of Lebesgue measure zero and uncountable; see (Rudin, 1976, Section 11.11, Remark (f), p. 309).

**Definition 2.5.2 (Indistinguishable Functions).** *We say that the Lebesgue measurable functions* $\mathbf{u}, \mathbf{v}$ *from* $\mathbb{R}$ *to* $\mathbb{C}$ *(or to* $\mathbb{R}$*) are* ***indistinguishable*** *and write*

$$\mathbf{u} \equiv \mathbf{v}$$

*if the set* $\{t \in \mathbb{R} : u(t) \neq v(t)\}$ *is of Lebesgue measure zero.*

Note that $\mathbf{u} \equiv \mathbf{v}$ if, and only if, the signal $\mathbf{u} - \mathbf{v}$ is indistinguishable from the all-zero signal $\mathbf{0}$

$$\big(\mathbf{u} \equiv \mathbf{v}\big) \Leftrightarrow \big(\mathbf{u} - \mathbf{v} \equiv \mathbf{0}\big). \tag{2.16}$$

The main result of this section is the following:

**Proposition 2.5.3.**

(i) *A nonnegative Lebesgue measurable signal integrates to zero if, and only if, it is indistinguishable from the all-zero signal* $\mathbf{0}$.

(ii) *If* $\mathbf{u}, \mathbf{v}$ *are Lebesgue measurable functions from* $\mathbb{R}$ *to* $\mathbb{C}$ *(or to* $\mathbb{R}$*), then*

$$\left( \int_{-\infty}^{\infty} |u(t) - v(t)| \, \mathrm{d}t = 0 \right) \Leftrightarrow \big( \mathbf{u} \equiv \mathbf{v} \big) \tag{2.17}$$

*and*

$$\left( \int_{-\infty}^{\infty} |u(t) - v(t)|^2 \, \mathrm{d}t = 0 \right) \Leftrightarrow \big( \mathbf{u} \equiv \mathbf{v} \big). \tag{2.18}$$

(iii) *If* $\mathbf{u}$ *and* $\mathbf{v}$ *are integrable and indistinguishable, then their integrals are equal:*

$$\big( \mathbf{u} \equiv \mathbf{v} \big) \Rightarrow \left( \int_{-\infty}^{\infty} u(t) \, \mathrm{d}t = \int_{-\infty}^{\infty} v(t) \, \mathrm{d}t \right), \quad \mathbf{u}, \mathbf{v} \in \mathcal{L}_1. \tag{2.19}$$

**Proof.** The proof of (i) is not very difficult, but it requires more familiarity with Measure Theory than we are willing to assume. The interested reader is thus referred to (Rudin, 1974, Theorem 1.39).

The equivalence in (2.17) follows by applying Part (i) to the nonnegative function $t \mapsto |u(t) - v(t)|$. Similarly, (2.18) follows by applying Part (i) to the nonnegative function $t \mapsto |u(t) - v(t)|^2$ and by noting that the set of $t$'s for which $|u(t) - v(t)|^2 \neq 0$ is the same as the set of $t$'s for which $u(t) \neq v(t)$.

Part (iii) follows from (2.17) by noting that

$$\left| \int_{-\infty}^{\infty} u(t) \, \mathrm{d}t - \int_{-\infty}^{\infty} v(t) \, \mathrm{d}t \right| = \left| \int_{-\infty}^{\infty} \big( u(t) - v(t) \big) \, \mathrm{d}t \right|$$

$$\leq \int_{-\infty}^{\infty} \big| u(t) - v(t) \big| \, \mathrm{d}t,$$

where the first equality follows by the linearity of integration, and where the subsequent inequality follows from Proposition 2.4.1. $\qquad \square$

## 2.6   Swapping Integration, Summation, and Expectation

In numerous places in this text we shall swap the order of integration as in

$$\int_{-\infty}^{\infty} \left( \int_{-\infty}^{\infty} u(\alpha, \beta)\, d\alpha \right) d\beta = \int_{-\infty}^{\infty} \left( \int_{-\infty}^{\infty} u(\alpha, \beta)\, d\beta \right) d\alpha \qquad (2.20)$$

or the order of summation as in

$$\sum_{\nu=1}^{\infty} \left( \sum_{\eta=1}^{\infty} a_{\nu,\eta} \right) = \sum_{\eta=1}^{\infty} \left( \sum_{\nu=1}^{\infty} a_{\nu,\eta} \right) \qquad (2.21)$$

or the order of summation and integration as in

$$\int_{-\infty}^{\infty} \left( \sum_{\nu=1}^{\infty} a_\nu u_\nu(t) \right) dt = \sum_{\nu=1}^{\infty} \left( a_\nu \int_{-\infty}^{\infty} u_\nu(t)\, dt \right) \qquad (2.22)$$

or the order of integration and expectation as in

$$\mathsf{E} \left[ \int_{-\infty}^{\infty} X\, u(t)\, dt \right] = \int_{-\infty}^{\infty} \mathsf{E}[X u(t)]\, dt = \mathsf{E}[X] \int_{-\infty}^{\infty} u(t)\, dt.$$

These changes of order are usually justified using Fubini's Theorem, which states that these changes of order are permissible provided that a very technical measurability condition is satisfied and that, in addition, either the integrand is nonnegative or that in some order (and hence in all orders) the integrals/summation/expectation of the absolute value of the integrand is finite.

For example, to justify (2.20) it suffices to verify that the function $\mathbf{u} \colon \mathbb{R}^2 \to \mathbb{R}$ in (2.20) is Lebesgue measurable and that, in addition, it is either nonnegative or

$$\int_{-\infty}^{\infty} \left( \int_{-\infty}^{\infty} |u(\alpha, \beta)|\, d\alpha \right) d\beta < \infty$$

or

$$\int_{-\infty}^{\infty} \left( \int_{-\infty}^{\infty} |u(\alpha, \beta)|\, d\beta \right) d\alpha < \infty.$$

Similarly, to justify (2.21) it suffices to show that $a_{\nu,\eta} \geq 0$ or that

$$\sum_{\eta=1}^{\infty} \left( \sum_{\nu=1}^{\infty} |a_{\nu,\eta}| \right) < \infty$$

or that

$$\sum_{\nu=1}^{\infty} \left( \sum_{\eta=1}^{\infty} |a_{\nu,\eta}| \right) < \infty.$$

(No need to worry about measurability which is automatic in this setup.)

As a final example, to justify (2.22) it suffices that the functions $\{\mathbf{u}_\nu\}$ are all measurable and that either $a_\nu u_\nu(t)$ is nonnegative for all $\nu \in \mathbb{N}$ and $t \in \mathbb{R}$ or

$$\int_{-\infty}^{\infty} \left( \sum_{\nu=1}^{\infty} |a_\nu| \, |u_\nu(t)| \right) \mathrm{d}t < \infty$$

or

$$\sum_{\nu=1}^{\infty} |a_\nu| \left( \int_{-\infty}^{\infty} |u_\nu(t)| \, \mathrm{d}t \right) < \infty.$$

A precise statement of Fubini's Theorem requires some Measure Theory that is beyond the scope of this book. The reader is referred to (Rudin, 1974, Theorem 7.8) and (Billingsley, 1995, Chapter 3, Section 18) for such a statement and for a proof.

We shall frequently use the swapping-of-order argument to manipulate the square of a sum or the square of an integral.

**Proposition 2.6.1.**

(i) If $\sum_\nu |a_\nu| < \infty$ then

$$\left( \sum_{\nu=1}^{\infty} a_\nu \right)^2 = \sum_{\nu=1}^{\infty} \sum_{\nu'=1}^{\infty} a_\nu a_{\nu'}. \tag{2.23}$$

(ii) If $\mathbf{u}$ is an integrable real-valued or complex-valued signal, then

$$\left( \int_{-\infty}^{\infty} u(\alpha) \, \mathrm{d}\alpha \right)^2 = \int_{-\infty}^{\infty} \int_{-\infty}^{\infty} u(\alpha) \, u(\alpha') \, \mathrm{d}\alpha \, \mathrm{d}\alpha'. \tag{2.24}$$

**Proof.** The proof is a direct application of Fubini's Theorem. But ignoring the technicalities, the intuition is quite clear: it all boils down to the fact that $(a+b)^2$ can be written as $(a+b)(a+b)$, which can in turn be written as $aa+ab+ba+bb$. $\square$

## 2.7   Additional Reading

Numerous books cover the basics of Lebesgue integration. Classic examples are (Riesz and Sz.-Nagy, 1990), (Rudin, 1974) and (Royden, 1988). These texts also cover the notion of sets of Lebesgue measure zero, e.g., (Riesz and Sz.-Nagy, 1990, Chapter 1, Section 2). For the changing of order of Riemann integration see (Körner, 1988, Chapters 47 & 48).

## 2.8   Exercises

**Exercise 2.1 (Integrating an Exponential).** Show that

$$\int_0^{\infty} e^{-zt} \, \mathrm{d}t = \frac{1}{z}, \quad \mathrm{Re}(z) > 0.$$

**Exercise 2.2 (Triangle Inequality for Complex Numbers).** Prove the Triangle Inequality for complex numbers (2.11). Under what conditions does it hold with equality?

**Exercise 2.3 (When Are Complex Numbers Equal?).** Prove that if the complex numbers $w$ and $z$ are such that $\operatorname{Re}(\beta z) = \operatorname{Re}(\beta w)$ for all $\beta \in \mathbb{C}$, then $w = z$.

**Exercise 2.4 (An Integral Inequality).** Show that if $\mathbf{u}$, $\mathbf{v}$, and $\mathbf{w}$ are integrable signals, then

$$\int_{-\infty}^{\infty} \left| u(t) - w(t) \right| dt \leq \int_{-\infty}^{\infty} \left| u(t) - v(t) \right| dt + \int_{-\infty}^{\infty} \left| v(t) - w(t) \right| dt.$$

**Exercise 2.5 (An Integral to Note).** Given some $f \in \mathbb{R}$, compute the integral

$$\int_{-\infty}^{\infty} \mathrm{I}\{t = 17\} e^{-\mathrm{i}2\pi f t}\, dt.$$

**Exercise 2.6 (Subsets of Sets of Lebesgue Measure Zero).** Show that a subset of a set of Lebesgue measure zero must also be of Lebesgue measure zero.

**Exercise 2.7 (Nonuniqueness of the Probability Density Function).** We say that the random variable $X$ is of density $f_X(\cdot)$ if $f_X(\cdot)$ is a (Lebesgue measurable) nonnegative function such that

$$\Pr[X \leq x] = \int_{-\infty}^{x} f_X(\xi)\, d\xi, \quad x \in \mathbb{R}.$$

Show that if $X$ is of density $f_X(\cdot)$ and if $g(\cdot)$ is a nonnegative function that is indistinguishable from $f_X(\cdot)$, then $X$ is also of density $g(\cdot)$. (The reverse is also true: if $X$ is of density $g_1(\cdot)$ and also of density $g_2(\cdot)$, then $g_1(\cdot)$ and $g_2(\cdot)$ must be indistinguishable.)

**Exercise 2.8 (Indistinguishability).** Let $\psi \colon \mathbb{R}^2 \to \mathbb{R}$ satisfy $\psi(\alpha, \beta) \geq 0$, for all $\alpha, \beta \in \mathbb{R}$ with equality only if $\alpha = \beta$. Let $\mathbf{u}$ and $\mathbf{v}$ be Lebesgue measurable signals. Show that

$$\left( \int_{-\infty}^{\infty} \psi\big(u(t), v(t)\big)\, dt = 0 \right) \Rightarrow \left( \mathbf{v} \equiv \mathbf{u} \right).$$

**Exercise 2.9 (Indistinguishable Signals).** Show that if the Lebesgue measurable signals $\mathbf{g}$ and $\mathbf{h}$ are indistinguishable, then the set of epochs $t \in \mathbb{R}$ where the sums $\sum_{j=-\infty}^{\infty} g(t+j)$ and $\sum_{j=-\infty}^{\infty} h(t+j)$ are different (in the sense that they both converge but to different limits or that one converges but the other does not) is of Lebesgue measure zero.

**Exercise 2.10 (Continuous Nonnegative Functions).** A subset of $\mathbb{R}$ containing a nonempty open interval cannot be of Lebesgue measure zero. Use this fact to show that if a continuous function $\mathbf{g} \colon \mathbb{R} \to \mathbb{R}$ is nonnegative except perhaps on a set of Lebesgue measure zero, then the exception set is empty and the function is nonnegative.

**Exercise 2.11 (Order of Summation Sometimes Matters).** For every $\nu, \eta \in \mathbb{N}$ define

$$a_{\nu,\eta} = \begin{cases} 2 - 2^{-\nu} & \text{if } \nu = \eta \\ -2 + 2^{-\nu} & \text{if } \nu = \eta + 1 \\ 0 & \text{otherwise.} \end{cases}$$

Show that (2.21) is not satisfied. See (Royden, 1988, Chapter 12, Section 4, Exercise 24.).

**Exercise 2.12 (Using Fubini's Theorem).** Using the relation

$$\frac{1}{x} = \int_0^\infty e^{-xt}\, \mathrm{d}t, \quad x > 0$$

and Fubini's Theorem, show that

$$\lim_{\alpha \to \infty} \int_0^\alpha \frac{\sin x}{x}\, \mathrm{d}x = \frac{\pi}{2}.$$

See (Rudin, 1974, Chapter 7, Exercise 12).

*Hint: See also Problem 2.1.*

# Chapter 3

# The Inner Product

## 3.1 The Inner Product

The inner product is central to Digital Communications, so it is best to introduce it early. The motivation will have to wait.

Recall that $\mathbf{u}\colon \mathcal{A} \to \mathcal{B}$ indicates that $\mathbf{u}$ (sometimes denoted $u(\cdot)$) is a **function** (or **mapping**) that maps each element in its **domain** $\mathcal{A}$ to an element in its **range** $\mathcal{B}$. If both the domain and the range of $\mathbf{u}$ are the set of real numbers $\mathbb{R}$, then we sometimes refer to $\mathbf{u}$ as being a **real signal**, especially if the argument of $u(\cdot)$ stands for time. Similarly, if $\mathbf{u}\colon \mathbb{R} \to \mathbb{C}$ where $\mathbb{C}$ denotes the set of complex numbers and the argument of $u(\cdot)$ stands for time, then we sometimes refer to $\mathbf{u}$ as a **complex signal**.

The **inner product** between two real functions $\mathbf{u}\colon \mathbb{R} \to \mathbb{R}$ and $\mathbf{v}\colon \mathbb{R} \to \mathbb{R}$ is denoted by $\langle \mathbf{u}, \mathbf{v} \rangle$ and is defined as

$$\langle \mathbf{u}, \mathbf{v} \rangle \triangleq \int_{-\infty}^{\infty} u(t)v(t)\,\mathrm{d}t, \tag{3.1}$$

whenever the integral is defined. (In Section 3.2 we shall study conditions under which the integral is defined, i.e., conditions on the functions $\mathbf{u}$ and $\mathbf{v}$ that guarantee that the product function $t \mapsto u(t)v(t)$ is an integrable function.)

The signals that arise in our study of Digital Communications often represent electric fields or voltages over resistors. The energy required to generate them is thus proportional to the integral of their squared magnitude. This motivates us to define the **energy** of a Lebesgue measurable real-valued function $\mathbf{u}\colon \mathbb{R} \to \mathbb{R}$ as

$$\int_{-\infty}^{\infty} u^2(t)\,\mathrm{d}t.$$

(If this integral is not finite, then we say that $\mathbf{u}$ is of infinite energy.) We say that $\mathbf{u}\colon \mathbb{R} \to \mathbb{R}$ is of **finite energy** if it is Lebesgue measurable and if

$$\int_{-\infty}^{\infty} u^2(t)\,\mathrm{d}t < \infty.$$

The class of all finite-energy real-valued functions $\mathbf{u} \colon \mathbb{R} \to \mathbb{R}$ is denoted by $\mathcal{L}_2$.

Since the energy of $\mathbf{u} \colon \mathbb{R} \to \mathbb{R}$ is nonnegative, we can discuss its nonnegative square root, which we denote[1] by $\|\mathbf{u}\|_2$:

$$\|\mathbf{u}\|_2 \triangleq \sqrt{\int_{-\infty}^{\infty} u^2(t)\, \mathrm{d}t}. \tag{3.2}$$

(Throughout this book we denote by $\sqrt{\xi}$ the nonnegative square root of $\xi$ for every $\xi \geq 0$.) We can now express the energy in $\mathbf{u}$ using the inner product as

$$\begin{aligned}
\|\mathbf{u}\|_2^2 &= \int_{-\infty}^{\infty} u^2(t)\, \mathrm{d}t \\
&= \langle \mathbf{u}, \mathbf{u} \rangle.
\end{aligned} \tag{3.3}$$

In writing $\|\mathbf{u}\|_2^2$ above we used different fonts for the subscript and the superscript. The subscript is just a graphical character which is part of the notation $\|\cdot\|_2$. We could have replaced it with $\blacklozenge$ and designated the energy by $\|\mathbf{u}\|_\blacklozenge^2$ without any change in mathematical meaning.[2] The superscript, however, indicates that the quantity $\|\mathbf{u}\|_2$ is being squared.

For complex-valued functions $\mathbf{u} \colon \mathbb{R} \to \mathbb{C}$ and $\mathbf{v} \colon \mathbb{R} \to \mathbb{C}$ we define the inner product $\langle \mathbf{u}, \mathbf{v} \rangle$ by

$$\boxed{\langle \mathbf{u}, \mathbf{v} \rangle \triangleq \int_{-\infty}^{\infty} u(t)\, v^*(t)\, \mathrm{d}t,} \tag{3.4}$$

whenever the integral is defined. Here $v^*(t)$ denotes the complex conjugate of $v(t)$. The above integral in (3.4) is a complex integral, but that should not worry you: it can also be written as

$$\langle \mathbf{u}, \mathbf{v} \rangle = \int_{-\infty}^{\infty} \operatorname{Re}\big(u(t)\, v^*(t)\big)\, \mathrm{d}t + \mathrm{i} \int_{-\infty}^{\infty} \operatorname{Im}\big(u(t)\, v^*(t)\big)\, \mathrm{d}t, \tag{3.5}$$

where $\mathrm{i} = \sqrt{-1}$ and where $\operatorname{Re}(\cdot)$ and $\operatorname{Im}(\cdot)$ denote the functions that map a complex number to its real and imaginary parts: $\operatorname{Re}(a + \mathrm{i}b) = a$ and $\operatorname{Im}(a + \mathrm{i}b) = b$ whenever $a, b \in \mathbb{R}$. Each of the two integrals appearing in (3.5) is the integral of a real signal. See Section 2.3.

Note that (3.1) and (3.4) are in agreement in the sense that if $\mathbf{u}$ and $\mathbf{v}$ happen to take on only real values (i.e., satisfy that $u(t), v(t) \in \mathbb{R}$ for every $t \in \mathbb{R}$), then viewing them as real functions and thus using (3.1) would yield the same inner product as viewing them as (degenerate) complex functions and using (3.4). Note also that for complex functions $\mathbf{u}, \mathbf{v} \colon \mathbb{R} \to \mathbb{C}$ the inner product $\langle \mathbf{u}, \mathbf{v} \rangle$ is in general not the same as $\langle \mathbf{v}, \mathbf{u} \rangle$. One is the complex conjugate of the other.

---

[1]The subscript $2$ is here to distinguish $\|\mathbf{u}\|_2$ from $\|\mathbf{u}\|_1$, where the latter was defined in (2.6) as $\|\mathbf{u}\|_1 = \int_{-\infty}^{\infty} |u(t)|\, \mathrm{d}t$.

[2]We prefer $\|\cdot\|_2$ to $\|\cdot\|_\blacklozenge$ because it reminds us that in the definition (3.2) the integrand is raised to the second power. This should be contrasted with the symbol $\|\cdot\|_1$ where the integrand is raised to the first power (and where no square root is taken of the result); see (2.6).

Some of the properties of the inner product between complex-valued functions $\mathbf{u}, \mathbf{v} \colon \mathbb{R} \to \mathbb{C}$ are given below.

$$\langle \mathbf{u}, \mathbf{v} \rangle = \langle \mathbf{v}, \mathbf{u} \rangle^* \tag{3.6}$$

$$\langle \alpha \mathbf{u}, \mathbf{v} \rangle = \alpha \langle \mathbf{u}, \mathbf{v} \rangle, \quad \alpha \in \mathbb{C} \tag{3.7}$$

$$\langle \mathbf{u}, \alpha \mathbf{v} \rangle = \alpha^* \langle \mathbf{u}, \mathbf{v} \rangle, \quad \alpha \in \mathbb{C} \tag{3.8}$$

$$\langle \mathbf{u}_1 + \mathbf{u}_2, \mathbf{v} \rangle = \langle \mathbf{u}_1, \mathbf{v} \rangle + \langle \mathbf{u}_2, \mathbf{v} \rangle \tag{3.9}$$

$$\langle \mathbf{u}, \mathbf{v}_1 + \mathbf{v}_2 \rangle = \langle \mathbf{u}, \mathbf{v}_1 \rangle + \langle \mathbf{u}, \mathbf{v}_2 \rangle. \tag{3.10}$$

The above equalities hold whenever the inner products appearing on the right-hand side (RHS) are defined. The reader is encouraged to produce a similar list of properties for the inner product between real-valued functions $\mathbf{u}, \mathbf{v} \colon \mathbb{R} \to \mathbb{R}$.

The **energy** in a Lebesgue measurable complex-valued function $\mathbf{u} \colon \mathbb{R} \to \mathbb{C}$ is defined as

$$\int_{-\infty}^{\infty} |u(t)|^2 \, \mathrm{d}t,$$

where $|\cdot|$ denotes absolute value so $|a + ib| = \sqrt{a^2 + b^2}$ whenever $a, b \in \mathbb{R}$. This definition of energy might seem a bit contrived because there is no such thing as complex voltage, so *prima facie* it seems meaningless to define the energy of a complex signal. But this is not the case. Complex signals are used to represent real passband signals, and the representation is such that the energy in the real passband signal is proportional to the integral of the squared modulus of the complex-valued signal representing it; see Section 7.6 ahead.

**Definition 3.1.1 (Energy-Limited Signal).** *We say that* $\mathbf{u} \colon \mathbb{R} \to \mathbb{C}$ *is **energy-limited** or **of finite energy** if* $\mathbf{u}$ *is Lebesgue measurable and*

$$\int_{-\infty}^{\infty} |u(t)|^2 \, \mathrm{d}t < \infty.$$

The set of all energy-limited complex-valued functions $\mathbf{u} \colon \mathbb{R} \to \mathbb{C}$ is denoted by $\mathcal{L}_2$. Note that whether $\mathcal{L}_2$ stands for the class of energy-limited *complex*-valued or *real*-valued functions should be clear from the context, or else immaterial.

For every $\mathbf{u} \in \mathcal{L}_2$ we define $\|\mathbf{u}\|_2$ as the nonnegative square root of its energy

$$\|\mathbf{u}\|_2 \triangleq \sqrt{\langle \mathbf{u}, \mathbf{u} \rangle}, \tag{3.11}$$

so

$$\|\mathbf{u}\|_2 = \sqrt{\int_{-\infty}^{\infty} |u(t)|^2 \, \mathrm{d}t}. \tag{3.12}$$

Again (3.12) and (3.2) are in agreement in the sense that for every $\mathbf{u} \colon \mathbb{R} \to \mathbb{R}$, computing $\|\mathbf{u}\|_2$ via (3.2) yields the same result as if we viewed $\mathbf{u}$ as mapping from $\mathbb{R}$ to $\mathbb{C}$ and computed $\|\mathbf{u}\|_2$ via (3.12).

## 3.2 When Is the Inner Product Defined?

As noted in Section 2.2, in this book we shall only discuss the integral of **integrable** functions, where a function $\mathbf{u}\colon \mathbb{R} \to \mathbb{R}$ is integrable if it is Lebesgue measurable and if $\int_{-\infty}^{\infty} |u(t)| \, \mathrm{d}t < \infty$. (We shall sometimes make an exception for functions that take on only nonnegative values. If $\mathbf{u}\colon \mathbb{R} \to [0,\infty)$ is Lebesgue measurable and if $\int u(t) \, \mathrm{d}t$ is not finite, then we shall say that $\int u(t) \, \mathrm{d}t = +\infty$.)

Similarly, as in Section 2.3, in integrating complex signals $\mathbf{u}\colon \mathbb{R} \to \mathbb{C}$ we limit ourselves to signals that are **integrable** in the sense that both $t \mapsto \mathrm{Re}(u(t))$ and $t \mapsto \mathrm{Im}(u(t))$ are Lebesgue measurable real-valued signals and $\int_{-\infty}^{\infty} |u(t)| \, \mathrm{d}t < \infty$.

Consequently, we shall say that the inner product between $\mathbf{u}\colon \mathbb{R} \to \mathbb{C}$ and $\mathbf{v}\colon \mathbb{R} \to \mathbb{C}$ is well-defined only when they are both Lebesgue measurable (thus implying that $t \mapsto u(t)\, v^*(t)$ is Lebesgue measurable) and when

$$\int_{-\infty}^{\infty} \big|u(t)\, v(t)\big| \, \mathrm{d}t < \infty. \tag{3.13}$$

We next discuss conditions on the Lebesgue measurable complex signals $\mathbf{u}$ and $\mathbf{v}$ that guarantee that (3.13) holds. The simplest case is when one of the functions, say $\mathbf{u}$, is bounded and the other, say $\mathbf{v}$, is integrable. Indeed, if $\sigma_\infty \in \mathbb{R}$ is such that $|u(t)| \leq \sigma_\infty$ for all $t \in \mathbb{R}$, then $|u(t)\, v(t)| \leq \sigma_\infty |v(t)|$ and

$$\int_{-\infty}^{\infty} \big|u(t)\, v(t)\big| \, \mathrm{d}t \leq \sigma_\infty \int_{-\infty}^{\infty} \big|v(t)\big| \, \mathrm{d}t = \sigma_\infty \, \|\mathbf{v}\|_1 \,,$$

where the RHS is finite by our assumption that $\mathbf{v}$ is integrable.

Another case where the inner product is well-defined is when both $\mathbf{u}$ and $\mathbf{v}$ are of finite energy. To prove that in this case too the mapping $t \mapsto u(t)\, v(t)$ is integrable we need the inequality

$$\alpha\beta \leq \frac{1}{2}(\alpha^2 + \beta^2), \quad \alpha, \beta \in \mathbb{R}, \tag{3.14}$$

which follows directly from the inequality $(\alpha - \beta)^2 \geq 0$ by simple algebra:

$$0 \leq (\alpha - \beta)^2$$
$$= \alpha^2 + \beta^2 - 2\alpha\beta.$$

By substituting $|u(t)|$ for $\alpha$ and $|v(t)|$ for $\beta$ in (3.14) we obtain the inequality $|u(t)\, v(t)| \leq (|u(t)|^2 + |v(t)|^2)/2$ and hence

$$\int_{-\infty}^{\infty} \big|u(t)\, v(t)\big| \, \mathrm{d}t \leq \frac{1}{2} \int_{-\infty}^{\infty} \big|u(t)\big|^2 \, \mathrm{d}t + \frac{1}{2} \int_{-\infty}^{\infty} \big|v(t)\big|^2 \, \mathrm{d}t, \tag{3.15}$$

thus demonstrating that if both $\mathbf{u}$ and $\mathbf{v}$ are of finite energy (so the RHS is finite), then the inner product is well-defined, i.e., $t \mapsto u(t)v(t)$ is integrable.

As a by-product of this proof we can obtain an upper bound on the magnitude of the inner product in terms of the energies of $\mathbf{u}$ and $\mathbf{v}$. All we need is the inequality

$$\left| \int_{-\infty}^{\infty} f(\xi) \, \mathrm{d}\xi \right| \leq \int_{-\infty}^{\infty} |f(\xi)| \, \mathrm{d}\xi$$

(see Proposition 2.4.1) to conclude from (3.15) that

$$
\begin{aligned}
|\langle \mathbf{u}, \mathbf{v} \rangle| &= \left| \int_{-\infty}^{\infty} u(t)\, v^*(t)\, \mathrm{d}t \right| \\
&\leq \int_{-\infty}^{\infty} |u(t)|\, |v(t)|\, \mathrm{d}t \\
&\leq \frac{1}{2} \int_{-\infty}^{\infty} |u(t)|^2\, \mathrm{d}t + \frac{1}{2} \int_{-\infty}^{\infty} |v(t)|^2\, \mathrm{d}t \\
&= \frac{1}{2} \left( \|\mathbf{u}\|_2^2 + \|\mathbf{v}\|_2^2 \right).
\end{aligned}
\tag{3.16}
$$

This inequality will be improved in Theorem 3.3.1, which introduces the Cauchy-Schwarz Inequality.

We finally mention here, without proof, a third case where the inner product between the Lebesgue measurable signals $\mathbf{u}, \mathbf{v}$ is defined. The result here is that if for some numbers $1 < p, q < \infty$ satisfying $1/p + 1/q = 1$ we have that

$$
\int_{-\infty}^{\infty} |u(t)|^p\, \mathrm{d}t < \infty \quad \text{and} \quad \int_{-\infty}^{\infty} |v(t)|^q\, \mathrm{d}t < \infty,
$$

then $t \mapsto u(t)\, v(t)$ is integrable. The proof of this result follows from Hölder's Inequality; see Theorem 3.3.2. Notice that the second case we addressed (where $\mathbf{u}$ and $\mathbf{v}$ are both of finite energy) follows from this case by considering $p = q = 2$.

## 3.3 The Cauchy-Schwarz Inequality

The Cauchy-Schwarz Inequality is probably the most important inequality on the inner product. Its discrete version is attributed to Augustin-Louis Cauchy (1789–1857) and its integral form to Victor Yacovlevich Bunyakovsky (1804–1889) who studied with him in Paris. Its (double) integral form was derived independently by Hermann Amandus Schwarz (1843–1921). See (Steele, 2004, pp. 10–12) for more on the history of this inequality and on how inequalities get their names.

**Theorem 3.3.1 (Cauchy-Schwarz Inequality).** *If the functions* $\mathbf{u}, \mathbf{v} \colon \mathbb{R} \to \mathbb{C}$ *are of finite energy, then the mapping* $t \mapsto u(t)\, v^*(t)$ *is integrable and*

$$
\boxed{ |\langle \mathbf{u}, \mathbf{v} \rangle| \leq \|\mathbf{u}\|_2\, \|\mathbf{v}\|_2 . }
\tag{3.17}
$$

*That is,*

$$
\left| \int_{-\infty}^{\infty} u(t)\, v^*(t)\, \mathrm{d}t \right| \leq \sqrt{ \int_{-\infty}^{\infty} |u(t)|^2\, \mathrm{d}t } \, \sqrt{ \int_{-\infty}^{\infty} |v(t)|^2\, \mathrm{d}t }.
$$

Equality in the Cauchy-Schwarz Inequality is possible, e.g., if $\mathbf{u}$ is a scaled version of $\mathbf{v}$, i.e., if for some constant $\alpha$

$$
u(t) = \alpha v(t), \quad t \in \mathbb{R}.
$$

In fact, the Cauchy-Schwarz Inequality holds with equality if, and only if, either $v(t)$ is zero for all $t$ outside a set of Lebesgue measure zero or for some constant $\alpha$ we have $u(t) = \alpha v(t)$ for all $t$ outside a set of Lebesgue measure zero.

There are a number of different proofs of this important inequality. We shall focus here on one that is based on (3.16) because it demonstrates a general technique for improving inequalities. The idea is that once one obtains a certain inequality—in our case (3.16)—one can try to improve it by taking advantage of one's understanding of how the quantity in question is affected by various transformations. This technique is beautifully illustrated in (Steele, 2004).

**Proof.** The quantity in question is $|\langle \mathbf{u}, \mathbf{v} \rangle|$. We shall take advantage of our understanding of how this quantity behaves when we replace $\mathbf{u}$ with its scaled version $\alpha \mathbf{u}$ and when we replace $\mathbf{v}$ with its scaled version $\beta \mathbf{v}$. Here $\alpha, \beta \in \mathbb{C}$ are arbitrary. The quantity in question transforms as

$$|\langle \alpha \mathbf{u}, \beta \mathbf{v} \rangle| = |\alpha|\,|\beta|\,|\langle \mathbf{u}, \mathbf{v} \rangle|. \tag{3.18}$$

We now use (3.16) to upper-bound the left-hand side (LHS) of the above by substituting $\alpha \mathbf{u}$ and $\beta \mathbf{v}$ for $\mathbf{u}$ and $\mathbf{v}$ in (3.16) to obtain

$$|\alpha|\,|\beta|\,|\langle \mathbf{u}, \mathbf{v} \rangle| = |\langle \alpha \mathbf{u}, \beta \mathbf{v} \rangle|$$
$$\leq \frac{1}{2}|\alpha|^2\,\|\mathbf{u}\|_2^2 + \frac{1}{2}|\beta|^2\,\|\mathbf{v}\|_2^2, \quad \alpha, \beta \in \mathbb{C}. \tag{3.19}$$

If both $\|\mathbf{u}\|_2$ and $\|\mathbf{v}\|_2$ are positive, then (3.17) follows from (3.19) by choosing $\alpha = 1/\|\mathbf{u}\|_2$ and $\beta = 1/\|\mathbf{v}\|_2$. To conclude the proof it thus remains to show that (3.17) also holds when either $\|\mathbf{u}\|_2$ or $\|\mathbf{v}\|_2$ is zero so the RHS of (3.17) is zero. That is, we need to show that if either $\|\mathbf{u}\|_2$ or $\|\mathbf{v}\|_2$ is zero, then $\langle \mathbf{u}, \mathbf{v} \rangle$ must also be zero. To show this, suppose first that $\|\mathbf{u}\|_2$ is zero. By substituting $\alpha = 1$ in (3.19) we obtain in this case that

$$|\beta|\,|\langle \mathbf{u}, \mathbf{v} \rangle| \leq \frac{1}{2}|\beta|^2\,\|\mathbf{v}\|_2^2,$$

which, upon dividing by $|\beta|$, yields

$$|\langle \mathbf{u}, \mathbf{v} \rangle| \leq \frac{1}{2}|\beta|\,\|\mathbf{v}\|_2^2, \quad \beta \neq 0.$$

Upon letting $|\beta|$ tend to zero from above this demonstrates that $\langle \mathbf{u}, \mathbf{v} \rangle$ must be zero as we set out to prove. (As an alternative proof of this case one notes that $\|\mathbf{u}\|_2 = 0$ implies, by Proposition 2.5.3, that the set $\{t \in \mathbb{R} : u(t) \neq 0\}$ is of Lebesgue measure zero. Consequently, since every zero of $t \mapsto u(t)$ is also a zero of $t \mapsto u(t)\,v^*(t)$, it follows that $\{t \in \mathbb{R} : u(t)\,v^*(t) \neq 0\}$ is included in $\{t \in \mathbb{R} : u(t) \neq 0\}$, and must therefore also be of Lebesgue measure zero (Exercise 2.6). Consequently, by Proposition 2.5.3, $\int_{-\infty}^{\infty} |u(t)\,v^*(t)|\,\mathrm{d}t$ must be zero, which, by Proposition 2.4.1, implies that $|\langle \mathbf{u}, \mathbf{v} \rangle|$ must be zero.)

The case where $\|\mathbf{v}\|_2 = 0$ is very similar: by substituting $\beta = 1$ in (3.19) we obtain that (in this case)

$$|\langle \mathbf{u}, \mathbf{v} \rangle| \leq \frac{1}{2}|\alpha|\,\|\mathbf{u}\|_2^2, \quad \alpha \neq 0$$

and the result follows upon letting $|\alpha|$ tend to zero from above.                    $\square$

While we shall not use the following inequality in this book, it is sufficiently important that we mention it in passing.

**Theorem 3.3.2 (Hölder's Inequality).** *If* $\mathbf{u}\colon \mathbb{R} \to \mathbb{C}$ *and* $\mathbf{v}\colon \mathbb{R} \to \mathbb{C}$ *are Lebesgue measurable functions satisfying*

$$\int_{-\infty}^{\infty} |u(t)|^p \, \mathrm{d}t < \infty \quad and \quad \int_{-\infty}^{\infty} |v(t)|^q \, \mathrm{d}t < \infty$$

*for some* $1 < p, q < \infty$ *satisfying* $1/p + 1/q = 1$, *then the function* $t \mapsto u(t)\, v^*(t)$ *is integrable and*

$$\left| \int_{-\infty}^{\infty} u(t)\, v^*(t) \, \mathrm{d}t \right| \leq \left( \int_{-\infty}^{\infty} |u(t)|^p \, \mathrm{d}t \right)^{1/p} \left( \int_{-\infty}^{\infty} |v(t)|^q \, \mathrm{d}t \right)^{1/q}. \tag{3.20}$$

Note that the Cauchy-Schwarz Inequality corresponds to the case where $p = q = 2$.

**Proof.** See, for example, (Rudin, 1974, Theorem 3.5) or (Royden, 1988, Section 6.2).                                                                                      $\square$

## 3.4    Applications

There are numerous applications of the Cauchy-Schwarz Inequality. Here we only mention a few. The first relates the energy in the superposition of two signals to the energies of the individual signals. The result holds for both complex-valued and real-valued functions, and—as is our custom—we shall thus not make the range explicit.

**Proposition 3.4.1 (Triangle Inequality for $\mathcal{L}_2$).** *If* $\mathbf{u}$ *and* $\mathbf{v}$ *are in* $\mathcal{L}_2$, *then*

$$\|\mathbf{u} + \mathbf{v}\|_2 \leq \|\mathbf{u}\|_2 + \|\mathbf{v}\|_2. \tag{3.21}$$

**Proof.** The proof is a straightforward application of the Cauchy-Schwarz Inequality and the basic properties of the inner product (3.6)–(3.9):

$$\begin{aligned}
\|\mathbf{u} + \mathbf{v}\|_2^2 &= \langle \mathbf{u} + \mathbf{v}, \mathbf{u} + \mathbf{v} \rangle \\
&= \langle \mathbf{u}, \mathbf{u} \rangle + \langle \mathbf{v}, \mathbf{v} \rangle + \langle \mathbf{u}, \mathbf{v} \rangle + \langle \mathbf{v}, \mathbf{u} \rangle \\
&\leq \langle \mathbf{u}, \mathbf{u} \rangle + \langle \mathbf{v}, \mathbf{v} \rangle + |\langle \mathbf{u}, \mathbf{v} \rangle| + |\langle \mathbf{v}, \mathbf{u} \rangle| \\
&= \|\mathbf{u}\|_2^2 + \|\mathbf{v}\|_2^2 + 2|\langle \mathbf{u}, \mathbf{v} \rangle| \\
&\leq \|\mathbf{u}\|_2^2 + \|\mathbf{v}\|_2^2 + 2\|\mathbf{u}\|_2 \|\mathbf{v}\|_2 \\
&= \left( \|\mathbf{u}\|_2 + \|\mathbf{v}\|_2 \right)^2,
\end{aligned}$$

from which the result follows by taking square roots. Here the first line follows from the definition of $\|\cdot\|_2$ (3.11); the second by (3.9) & (3.10); the third by the Triangle Inequality for Complex Numbers (2.12); the fourth because, by (3.6), $\langle \mathbf{v}, \mathbf{u} \rangle$ is the complex conjugate of $\langle \mathbf{u}, \mathbf{v} \rangle$ and is hence of equal modulus; the fifth by the Cauchy-Schwarz Inequality; and the sixth by simple algebra.                    $\square$

Another important mathematical consequence of the Cauchy-Schwarz Inequality is the continuity of the inner product. To state the result we use the notation $a_n \to a$ to indicate that the sequence $a_1, a_2, \ldots$ converges to $a$, i.e., that $\lim_{n \to \infty} a_n = a$.

**Proposition 3.4.2 (Continuity of the Inner Product).** *Let* **u** *and* **v** *be in* $\mathcal{L}_2$. *If the sequence* $\mathbf{u}_1, \mathbf{u}_2, \ldots$ *of elements of* $\mathcal{L}_2$ *satisfies*

$$\|\mathbf{u}_n - \mathbf{u}\|_2 \to 0,$$

*and if the sequence* $\mathbf{v}_1, \mathbf{v}_2, \ldots$ *of elements of* $\mathcal{L}_2$ *satisfies*

$$\|\mathbf{v}_n - \mathbf{v}\|_2 \to 0,$$

*then*

$$\langle \mathbf{u}_n, \mathbf{v}_n \rangle \to \langle \mathbf{u}, \mathbf{v} \rangle.$$

**Proof.**

$$
\begin{aligned}
|\langle \mathbf{u}_n, \mathbf{v}_n \rangle &- \langle \mathbf{u}, \mathbf{v} \rangle| \\
&= |\langle \mathbf{u}_n - \mathbf{u}, \mathbf{v} \rangle + \langle \mathbf{u}_n - \mathbf{u}, \mathbf{v}_n - \mathbf{v} \rangle + \langle \mathbf{u}, \mathbf{v}_n - \mathbf{v} \rangle| \\
&\leq |\langle \mathbf{u}_n - \mathbf{u}, \mathbf{v} \rangle| + |\langle \mathbf{u}_n - \mathbf{u}, \mathbf{v}_n - \mathbf{v} \rangle| + |\langle \mathbf{u}, \mathbf{v}_n - \mathbf{v} \rangle| \\
&\leq \|\mathbf{u}_n - \mathbf{u}\|_2 \|\mathbf{v}\|_2 + \|\mathbf{u}_n - \mathbf{u}\|_2 \|\mathbf{v}_n - \mathbf{v}\|_2 + \|\mathbf{u}\|_2 \|\mathbf{v}_n - \mathbf{v}\|_2 \\
&\to 0,
\end{aligned}
$$

where the first equality follows from the basic properties of the inner product (3.6)–(3.10); the subsequent inequality by the Triangle Inequality for Complex Numbers (2.12); the subsequent inequality from the Cauchy-Schwarz Inequality; and where the final limit follows from the proposition's hypotheses. $\quad\square$

Another useful consequence of the Cauchy-Schwarz Inequality is in demonstrating that if a signal is energy-limited and is zero outside an interval, then it is also integrable.

**Proposition 3.4.3 (Finite-Energy Functions over Finite Intervals are Integrable).** *If for some real numbers* $a$ *and* $b$ *satisfying* $a \leq b$ *we have*

$$\int_a^b |x(\xi)|^2 \, d\xi < \infty,$$

*then*

$$\int_a^b |x(\xi)| \, d\xi \leq \sqrt{b - a} \sqrt{\int_a^b |x(\xi)|^2 \, d\xi},$$

*and, in particular,*

$$\int_a^b |x(\xi)| \, d\xi < \infty.$$

**Proof.**

$$\int_a^b \left| x(\xi) \right| \mathrm{d}t = \int_{-\infty}^{\infty} \mathrm{I}\{a \le \xi \le b\} \left| x(\xi) \right| \mathrm{d}\xi$$

$$= \int_{-\infty}^{\infty} \underbrace{\mathrm{I}\{a \le \xi \le b\}}_{u(\xi)} \underbrace{\mathrm{I}\{a \le \xi \le b\} \left| x(\xi) \right|}_{v(\xi)} \mathrm{d}\xi$$

$$\le \sqrt{b-a} \sqrt{\int_a^b \left| x(\xi) \right|^2 \mathrm{d}\xi},$$

where the inequality is just an application of the Cauchy-Schwarz Inequality to the function $\xi \mapsto \mathrm{I}\{a \le \xi \le b\} \left| x(\xi) \right|$ and the indicator function $\xi \mapsto \mathrm{I}\{a \le \xi \le b\}$.  □

Note that, in general, an energy-limited signal need not be integrable. For example, the real signal

$$t \mapsto \begin{cases} 0 & \text{if } t \le 1, \\ 1/t & \text{otherwise,} \end{cases} \tag{3.22}$$

is of finite energy but is not integrable.

The Cauchy-Schwarz Inequality demonstrates that if both $\mathbf{u}$ and $\mathbf{v}$ are of finite energy, then their inner product $\langle \mathbf{u}, \mathbf{v} \rangle$ is well-defined, i.e., the integrand in (3.4) is integrable. It can also be used in slightly more sophisticated ways. For example, it can be used to treat cases where one of the functions, say $\mathbf{u}$, is not of finite energy but where the second function decays to zero sufficiently quickly to compensate for that. For example:

**Proposition 3.4.4.** *If the Lebesgue measurable functions* $\mathbf{x} \colon \mathbb{R} \to \mathbb{C}$ *and* $\mathbf{y} \colon \mathbb{R} \to \mathbb{C}$ *satisfy*

$$\int_{-\infty}^{\infty} \frac{|x(t)|^2}{t^2 + 1} \, \mathrm{d}t < \infty$$

*and*

$$\int_{-\infty}^{\infty} |y(t)|^2 \, (t^2 + 1) \, \mathrm{d}t < \infty,$$

*then the function* $t \mapsto x(t) \, y^*(t)$ *is integrable and*

$$\left| \int_{-\infty}^{\infty} x(t) \, y^*(t) \, \mathrm{d}t \right| \le \sqrt{\int_{-\infty}^{\infty} \frac{|x(t)|^2}{t^2 + 1} \, \mathrm{d}t} \sqrt{\int_{-\infty}^{\infty} |y(t)|^2 \, (t^2 + 1) \, \mathrm{d}t}.$$

**Proof.** This is a simple application of the Cauchy-Schwarz Inequality to the functions $t \mapsto x(t)/\sqrt{t^2 + 1}$ and $t \mapsto y(t)\sqrt{t^2 + 1}$. Simply write

$$\int_{-\infty}^{\infty} x(t) \, y^*(t) \, \mathrm{d}t = \int_{-\infty}^{\infty} \underbrace{\frac{x(t)}{\sqrt{t^2 + 1}}}_{u(t)} \underbrace{\sqrt{t^2 + 1} \, y^*(t)}_{v^*(t)} \, \mathrm{d}t$$

and apply the Cauchy-Schwarz Inequality to the functions $u(\cdot)$ and $v(\cdot)$.  □

## 3.5   The Cauchy-Schwarz Inequality for Random Variables

There is also a version of the Cauchy-Schwarz Inequality for random variables. It is very similar to Theorem 3.3.1 but with time integrals replaced by expectations. We denote the expectation of the random variable $X$ by $\mathsf{E}[X]$ and remind the reader that the variance $\mathsf{Var}[X]$ of the random variable $X$ is defined by

$$\mathsf{Var}[X] = \mathsf{E}\big[(X - \mathsf{E}[X])^2\big]. \tag{3.23}$$

**Theorem 3.5.1 (Cauchy-Schwarz Inequality for Random Variables).** *Let the random variables $U$ and $V$ be of finite variance. Then*

$$\big|\mathsf{E}[UV]\big| \leq \sqrt{\mathsf{E}[U^2]}\sqrt{\mathsf{E}[V^2]}, \tag{3.24}$$

*with equality if, and only if, $\Pr[\alpha U = \beta V] = 1$ for some real $\alpha$ and $\beta$ that are not both equal to zero.*

**Proof.** Use the proof of Theorem 3.3.1 with all time integrals replaced with expectations. For a different proof and for the conditions for equality see (Grimmett and Stirzaker, 2001, Chapter 3, Section 3.5, Theorem 9).                    □

For the next corollary we need to recall that the covariance $\mathsf{Cov}[U, V]$ between the finite-variance random variables $U$, $V$ is defined by

$$\mathsf{Cov}[U, V] = \mathsf{E}\big[(U - \mathsf{E}[U])(V - \mathsf{E}[V])\big]. \tag{3.25}$$

**Corollary 3.5.2 (Covariance Inequality).** *If the random variables $U$ and $V$ are of finite variance $\mathsf{Var}[U]$ and $\mathsf{Var}[V]$, then*

$$\big|\mathsf{Cov}[U, V]\big| \leq \sqrt{\mathsf{Var}[U]}\sqrt{\mathsf{Var}[V]}. \tag{3.26}$$

**Proof.** Apply Theorem 3.5.1 to the random variables $U - \mathsf{E}[U]$ and $V - \mathsf{E}[V]$.   □

Corollary 3.5.2 shows that the **correlation coefficient**, which is defined for random variables $U$ and $V$ having strictly positive variances as

$$\rho = \frac{\mathsf{Cov}[U, V]}{\sqrt{\mathsf{Var}[U]}\sqrt{\mathsf{Var}[V]}}, \tag{3.27}$$

satisfies

$$-1 \leq \rho \leq +1. \tag{3.28}$$

## 3.6   Mathematical Comments

(i) Mathematicians typically consider $\langle \mathbf{u}, \mathbf{v} \rangle$ only when both $\mathbf{u}$ and $\mathbf{v}$ are of finite energy. We are more forgiving and simply require that the integral defining the inner product be well-defined, i.e., that the integrand be integrable.

(ii) Some refer to $\|\mathbf{u}\|_2$ as the "norm of $\mathbf{u}$" or the "$\mathcal{L}_2$ norm of $\mathbf{u}$." We shall refrain from this usage because mathematicians use the term "norm" very selectively. They require that no function other than the all-zero function be of zero norm, and this is not the case for $\|\cdot\|_2$. Indeed, any function $\mathbf{u}$ that is indistinguishable from the all-zero function satisfies $\|\mathbf{u}\|_2 = 0$, and there are many such functions (e.g., the function that is equal to one at rational times and that is equal to zero at all other times). This difficulty can be overcome by defining two functions to be the same if their difference is of zero energy. In this case $\|\cdot\|_2$ is a norm in the mathematical sense and is, in fact, what mathematicians call the $L_2$ norm. This issue is discussed in greater detail in Section 4.7. To stay out of trouble we shall refrain from giving $\|\cdot\|_2$ a name.

## 3.7 Exercises

**Exercise 3.1 (Manipulating Inner Products).** Show that if $\mathbf{u}$, $\mathbf{v}$, and $\mathbf{w}$ are energy-limited complex signals, then

$$\langle \mathbf{u} + \mathbf{v}, 3\mathbf{u} + \mathbf{v} + i\mathbf{w} \rangle = 3 \|\mathbf{u}\|_2^2 + \|\mathbf{v}\|_2^2 + \langle \mathbf{u}, \mathbf{v} \rangle + 3 \langle \mathbf{u}, \mathbf{v} \rangle^* - i \langle \mathbf{u}, \mathbf{w} \rangle - i \langle \mathbf{v}, \mathbf{w} \rangle.$$

**Exercise 3.2 (Orthogonality to All Signals).** Let $\mathbf{u}$ be an energy-limited signal. Show that

$$\left( \mathbf{u} \equiv \mathbf{0} \right) \Leftrightarrow \left( \langle \mathbf{u}, \mathbf{v} \rangle = 0, \quad \mathbf{v} \in \mathcal{L}_2 \right).$$

**Exercise 3.3 (Finite-Energy Signals).** Let $\mathbf{x}$ be an energy-limited signal.

(i) Show that, for every $t_0 \in \mathbb{R}$, the signal $t \mapsto x(t - t_0)$ must also be energy-limited.

(ii) Show that the reflection of $\mathbf{x}$ is also energy-limited. I.e., show that the signal $\bar{\mathbf{x}}$ that maps $t$ to $x(-t)$ is energy-limited.

(iii) How are the energies in $t \mapsto x(t)$, $t \mapsto x(t - t_0)$, and $t \mapsto x(-t)$ related?

**Exercise 3.4 (Inner Products of Mirror Images).** Express the inner product $\langle \bar{\mathbf{x}}, \bar{\mathbf{y}} \rangle$ in terms of the inner product $\langle \mathbf{x}, \mathbf{y} \rangle$.

**Exercise 3.5 (On the Cauchy-Schwarz Inequality).** Show that the bound obtained from the Cauchy-Schwarz Inequality is at least as tight as (3.16).

**Exercise 3.6 (Truncated Polynomials).** Consider the signals $\mathbf{u}\colon t \mapsto (t + 2)\,\mathrm{I}\{0 \le t \le 1\}$ and $\mathbf{v}\colon t \mapsto (t^2 - 2t - 3)\,\mathrm{I}\{0 \le t \le 1\}$. Compute the energies $\|\mathbf{u}\|_2^2$ & $\|\mathbf{v}\|_2^2$ and the inner product $\langle \mathbf{u}, \mathbf{v} \rangle$.

**Exercise 3.7 (Indistinguishability and Inner Products).** Let $\mathbf{u} \in \mathcal{L}_2$ be indistinguishable from $\mathbf{u}' \in \mathcal{L}_2$, and let $\mathbf{v} \in \mathcal{L}_2$ be indistinguishable from $\mathbf{v}' \in \mathcal{L}_2$. Show that the inner product $\langle \mathbf{u}', \mathbf{v}' \rangle$ is equal to the inner product $\langle \mathbf{u}, \mathbf{v} \rangle$.

**Exercise 3.8 (Finite Energy and Integrability).** Let $\mathbf{x}\colon \mathbb{R} \to \mathbb{C}$ be Lebesgue measurable.

(i) Show that the conditions that $\mathbf{x}$ is of finite energy and that the mapping $t \mapsto t\,x(t)$ is of finite energy are simultaneously met if, and only if,

$$\int_{-\infty}^{\infty} |x(t)|^2 \, (1 + t^2) \, \mathrm{d}t < \infty. \tag{3.29}$$

(ii) Show that (3.29) implies that $\mathbf{x}$ is integrable.

(iii) Give an example of an integrable signal that does not satisfy (3.29).

**Exercise 3.9 (The Cauchy-Schwarz Inequality for Sequences).**

(i) Let the complex sequences $a_1, a_2, \ldots$ and $b_1, b_2, \ldots$ satisfy

$$\sum_{\nu=1}^{\infty} |a_\nu|^2, \sum_{\nu=1}^{\infty} |b_\nu|^2 < \infty.$$

Show that

$$\left| \sum_{\nu=1}^{\infty} a_\nu b_\nu^* \right|^2 \leq \left( \sum_{\nu=1}^{\infty} |a_\nu|^2 \right) \left( \sum_{\nu=1}^{\infty} |b_\nu|^2 \right).$$

(ii) Derive the Cauchy-Schwarz Inequality for $d$-tuples:

$$\left| \sum_{\nu=1}^{d} a_\nu b_\nu^* \right|^2 \leq \left( \sum_{\nu=1}^{d} |a_\nu|^2 \right) \left( \sum_{\nu=1}^{d} |b_\nu|^2 \right).$$

**Exercise 3.10 (Summability and Square Summability).** Let $a_1, a_2, \ldots$ be a sequence of complex numbers. Show that

$$\left( \sum_{\nu=1}^{\infty} |a_\nu| < \infty \right) \Rightarrow \left( \sum_{\nu=1}^{\infty} |a_\nu|^2 < \infty \right).$$

**Exercise 3.11 (A Friendlier GPA).** Use the Cauchy-Schwarz Inequality for $d$-tuples (Problem 3.9) to show that for any positive integer $d$,

$$\frac{a_1 + \cdots + a_d}{d} \leq \sqrt{\frac{a_1^2 + \cdots + a_d^2}{d}}, \quad a_1, \ldots, a_d \in \mathbb{R}.$$

# Chapter 4

# The Space $\mathcal{L}_2$ of Energy-Limited Signals

## 4.1 Introduction

In this chapter we shall study the space $\mathcal{L}_2$ of energy-limited signals in greater detail. We shall show that its elements can be viewed as vectors in a vector space and begin developing a geometric intuition for understanding its structure. We shall focus on the case of complex-valued signals, but with some minor changes the results are also applicable to real-valued signals. (The main changes that are needed for translating the results to real-valued signals are replacing $\mathbb{C}$ with $\mathbb{R}$, ignoring the conjugation operation, and interpreting $|\cdot|$ as the absolute value function for real arguments as opposed to the modulus function.)

We remind the reader that the space $\mathcal{L}_2$ was defined in Definition 3.1.1 as the set of all Lebesgue measurable complex-valued signals $\mathbf{u} \colon \mathbb{R} \to \mathbb{C}$ satisfying

$$\int_{-\infty}^{\infty} |u(t)|^2 \, \mathrm{d}t < \infty, \tag{4.1}$$

and that in (3.12) we defined for every $\mathbf{u} \in \mathcal{L}_2$ the quantity $\|\mathbf{u}\|_2$ as

$$\|\mathbf{u}\|_2 = \sqrt{\int_{-\infty}^{\infty} |u(t)|^2 \, \mathrm{d}t}. \tag{4.2}$$

We refer to $\mathcal{L}_2$ as the space of energy-limited signals and to its elements as energy-limited signals or signals of finite energy.

## 4.2 $\mathcal{L}_2$ as a Vector Space

In this section we shall explain how to view the space $\mathcal{L}_2$ as a vector space over the complex field by thinking about signals in $\mathcal{L}_2$ as vectors, by interpreting the superposition $\mathbf{u} + \mathbf{v}$ of two signals as vector-addition, and by interpreting the amplification of $\mathbf{u}$ by $\alpha$ as the operation of multiplying the vector $\mathbf{u}$ by the scalar $\alpha \in \mathbb{C}$.

We begin by reminding the reader that the **superposition** of the two signals $\mathbf{u}$ and $\mathbf{v}$ is denoted by $\mathbf{u} + \mathbf{v}$ and is the signal that maps every $t \in \mathbb{R}$ to $u(t) + v(t)$.

26

The **amplification** of $\mathbf{u}$ by $\alpha$ is denoted by $\alpha\mathbf{u}$ and is the signal that maps every $t \in \mathbb{R}$ to $\alpha u(t)$. More generally, if $\mathbf{u}$ and $\mathbf{v}$ are signals and if $\alpha$ and $\beta$ are complex numbers, then $\alpha\mathbf{u} + \beta\mathbf{v}$ is the signal $t \mapsto \alpha u(t) + \beta v(t)$.

If $\mathbf{u} \in \mathcal{L}_2$ and $\alpha \in \mathbb{C}$, then $\alpha\mathbf{u}$ is also in $\mathcal{L}_2$. Indeed, the measurability of $\mathbf{u}$ implies the measurability of $\alpha\mathbf{u}$, and if $\mathbf{u}$ is of finite energy, then $\alpha\mathbf{u}$ is also of finite energy, because the energy in $\alpha\mathbf{u}$ is the product of $|\alpha|^2$ by the energy in $\mathbf{u}$. We thus see that the operation of amplification of $\mathbf{u}$ by $\alpha$ results in an element of $\mathcal{L}_2$ whenever $\mathbf{u} \in \mathcal{L}_2$ and $\alpha \in \mathbb{C}$.

We next show that if the signals $\mathbf{u}$ and $\mathbf{v}$ are in $\mathcal{L}_2$, then their superposition $\mathbf{u} + \mathbf{v}$ must also be in $\mathcal{L}_2$. This holds because a standard result in Measure Theory guarantees that the superposition of two Lebesgue measurable signals is a Lebesgue measurable signal and because Proposition 3.4.1 guarantees that if both $\mathbf{u}$ and $\mathbf{v}$ are of finite energy, then so is their superposition. Thus the superposition that maps $\mathbf{u}$ and $\mathbf{v}$ to $\mathbf{u} + \mathbf{v}$ results in an element of $\mathcal{L}_2$ whenever $\mathbf{u}, \mathbf{v} \in \mathcal{L}_2$.

It can be readily verified that the following properties hold:

(i) **commutativity**:
$$\mathbf{u} + \mathbf{v} = \mathbf{v} + \mathbf{u}, \quad \mathbf{u}, \mathbf{v} \in \mathcal{L}_2;$$

(ii) **associativity**:
$$(\mathbf{u} + \mathbf{v}) + \mathbf{w} = \mathbf{u} + (\mathbf{v} + \mathbf{w}), \quad \mathbf{u}, \mathbf{v}, \mathbf{w} \in \mathcal{L}_2,$$

$$(\alpha\beta)\mathbf{u} = \alpha(\beta\mathbf{u}), \quad \left(\alpha, \beta \in \mathbb{C}, \quad \mathbf{u} \in \mathcal{L}_2\right);$$

(iii) **additive identity**: the all-zero signal $\mathbf{0} \colon t \mapsto 0$ satisfies
$$\mathbf{0} + \mathbf{u} = \mathbf{u}, \quad \mathbf{u} \in \mathcal{L}_2;$$

(iv) **additive inverse**: to every $\mathbf{u} \in \mathcal{L}_2$ there corresponds a signal $\mathbf{w} \in \mathcal{L}_2$ (namely, the signal $t \mapsto -u(t)$) such that
$$\mathbf{u} + \mathbf{w} = \mathbf{0};$$

(v) **multiplicative identity**:
$$1\mathbf{u} = \mathbf{u}, \quad \mathbf{u} \in \mathcal{L}_2;$$

(vi) **distributive properties**:
$$\alpha(\mathbf{u} + \mathbf{v}) = \alpha\mathbf{u} + \alpha\mathbf{v}, \quad \left(\alpha \in \mathbb{C}, \quad \mathbf{u}, \mathbf{v} \in \mathcal{L}_2\right),$$

$$(\alpha + \beta)\mathbf{u} = \alpha\mathbf{u} + \beta\mathbf{u}, \quad \left(\alpha, \beta \in \mathbb{C}, \quad \mathbf{u} \in \mathcal{L}_2\right).$$

We conclude that with the operations of superposition and amplification the set $\mathcal{L}_2$ forms a **vector space** over the complex field (Axler, 1997, Chapter 1). This justifies referring to the elements of $\mathcal{L}_2$ as "vectors," to the operation of signal superposition as "vector addition," and to the operation of amplification of an element of $\mathcal{L}_2$ by a complex scalar as "scalar multiplication."

## 4.3 Subspace, Dimension, and Basis

Once we have noted that $\mathcal{L}_2$ together with the operations of superposition and amplification forms a vector space, we can borrow numerous definitions and results from the theory of vector spaces. Here we shall focus on the very basic ones.

A **linear subspace** (or just **subspace**) of $\mathcal{L}_2$ is a nonempty subset $\mathcal{U}$ of $\mathcal{L}_2$ that is closed under superposition

$$\mathbf{u}_1 + \mathbf{u}_2 \in \mathcal{U}, \quad \mathbf{u}_1, \mathbf{u}_2 \in \mathcal{U} \tag{4.3}$$

and under amplification

$$\alpha \mathbf{u} \in \mathcal{U}, \quad (\alpha \in \mathbb{C}, \quad \mathbf{u} \in \mathcal{U}). \tag{4.4}$$

**Example 4.3.1.** Consider the set of all functions of the form

$$t \mapsto p(t)\, e^{-|t|},$$

where $p(t)$ is any polynomial of degree no larger than 3. Thus, the set is the set of all functions of the form

$$t \mapsto \left(\alpha_0 + \alpha_1 t + \alpha_2 t^2 + \alpha_3 t^3\right) e^{-|t|}, \tag{4.5}$$

where $\alpha_0, \alpha_1, \alpha_2, \alpha_3$ are arbitrary complex numbers.

In spite of the polynomial growth of the pre-exponent, all such functions are in $\mathcal{L}_2$ because the exponential decay more than compensates for the polynomial growth. The above set is thus a subset of $\mathcal{L}_2$. Moreover, as we show next, this is a linear subspace of $\mathcal{L}_2$.

If $\mathbf{u}$ is of the form (4.5), then so is $\alpha \mathbf{u}$, because $\alpha \mathbf{u}$ is the mapping

$$t \mapsto \left(\alpha \alpha_0 + \alpha \alpha_1 t + \alpha \alpha_2 t^2 + \alpha \alpha_3 t^3\right) e^{-|t|},$$

which is of the same form.

Similarly, if $\mathbf{u}$ is as given in (4.5) and

$$\mathbf{v} \colon t \mapsto \left(\beta_0 + \beta_1 t + \beta_2 t^2 + \beta_3 t^3\right) e^{-|t|},$$

then $\mathbf{u} + \mathbf{v}$ is the mapping

$$t \mapsto \left((\alpha_0 + \beta_0) + (\alpha_1 + \beta_1)t + (\alpha_2 + \beta_2)t^2 + (\alpha_3 + \beta_3)t^3\right) e^{-|t|},$$

which is again of this form.

An **$n$-tuple** of vectors from $\mathcal{L}_2$ is a (possibly empty) ordered list of $n$ vectors from $\mathcal{L}_2$ separated by commas and enclosed in parentheses, e.g., $(\mathbf{v}_1, \ldots, \mathbf{v}_n)$. Here $n \geq 0$ can be any nonnegative integer, where the case $n = 0$ corresponds to the empty list.

A vector $\mathbf{v} \in \mathcal{L}_2$ is said to be a **linear combination** of the $n$-tuple $(\mathbf{v}_1, \ldots, \mathbf{v}_n)$ if it is equal to

$$\alpha_1 \mathbf{v}_1 + \cdots + \alpha_n \mathbf{v}_n, \tag{4.6}$$

which is written more succinctly as

$$\sum_{\nu=1}^{n} \alpha_\nu \mathbf{v}_\nu, \tag{4.7}$$

for some scalars $\alpha_1, \ldots, \alpha_n \in \mathbb{C}$. The all-zero signal is a linear combination of any $n$-tuple including the empty tuple.

The **span** of an $n$-tuple $(\mathbf{v}_1, \ldots, \mathbf{v}_n)$ of vectors in $\mathcal{L}_2$ is denoted by

$$\mathrm{span}(\mathbf{v}_1, \ldots, \mathbf{v}_n)$$

and is the set of all vectors in $\mathcal{L}_2$ that are linear combinations of $(\mathbf{v}_1, \ldots, \mathbf{v}_n)$:

$$\mathrm{span}(\mathbf{v}_1, \ldots, \mathbf{v}_n) \triangleq \{\alpha_1 \mathbf{v}_1 + \cdots + \alpha_n \mathbf{v}_n : \alpha_1, \ldots, \alpha_n \in \mathbb{C}\}. \tag{4.8}$$

(The span of the empty tuple is given by the one-element set $\{\mathbf{0}\}$ containing the all-zero signal only.)

Note that for any $n$-tuple of vectors $(\mathbf{v}_1, \ldots, \mathbf{v}_n)$ in $\mathcal{L}_2$ we have that $\mathrm{span}(\mathbf{v}_1, \ldots, \mathbf{v}_n)$ is a linear subspace of $\mathcal{L}_2$. Also, if $\mathcal{U}$ is a linear subspace of $\mathcal{L}_2$ and if the vectors $\mathbf{u}_1, \ldots, \mathbf{u}_n$ are in $\mathcal{U}$, then $\mathrm{span}(\mathbf{u}_1, \ldots, \mathbf{u}_n)$ is a linear subspace which is contained in $\mathcal{U}$. A subspace $\mathcal{U}$ of $\mathcal{L}_2$ is said to be **finite-dimensional** if there exists an $n$-tuple $(\mathbf{u}_1, \ldots, \mathbf{u}_n)$ of vectors in $\mathcal{U}$ such that $\mathrm{span}(\mathbf{u}_1, \ldots, \mathbf{u}_n) = \mathcal{U}$. Otherwise, we say that $\mathcal{U}$ is **infinite-dimensional**. For example, the space of all mappings of the form $t \mapsto p(t) e^{-|t|}$ for some polynomial $p(\cdot)$ can be shown to be infinite-dimensional, but under the restriction that $p(\cdot)$ be of degree smaller than 5, it is finite-dimensional. If $\mathcal{U}$ is a finite-dimensional subspace and if $\mathcal{U}'$ is a subspace contained in $\mathcal{U}$, then $\mathcal{U}'$ must also be finite-dimensional.

An $n$-tuple of signals $(\mathbf{v}_1, \ldots, \mathbf{v}_n)$ in $\mathcal{L}_2$ is said to be **linearly independent** if whenever the scalars $\alpha_1, \ldots, \alpha_n \in \mathbb{C}$ are such that $\alpha_1 \mathbf{v}_1 + \cdots \alpha_n \mathbf{v}_n = \mathbf{0}$, we have $\alpha_1 = \cdots = \alpha_n = 0$. I.e., if

$$\left(\sum_{\nu=1}^{n} \alpha_\nu \mathbf{v}_\nu = \mathbf{0}\right) \Rightarrow \left(\alpha_\nu = 0, \quad \nu = 1, \ldots, n\right). \tag{4.9}$$

(By convention, the empty tuple is linearly independent.) For example, the 3-tuple consisting of the signals $t \mapsto e^{-|t|}$, $t \mapsto t e^{-|t|}$, and $t \mapsto t^2 e^{-|t|}$ is linearly independent. If $(\mathbf{v}_1, \ldots, \mathbf{v}_n)$ is not linearly independent, then we say that it is **linearly dependent**. For example, the 3-tuple consisting of the signals $t \mapsto e^{-|t|}$, $t \mapsto t e^{-|t|}$, and $t \mapsto (2t + 1) e^{-|t|}$ is linearly dependent. The $n$-tuple $(\mathbf{v}_1, \ldots, \mathbf{v}_n)$ is linearly dependent if, and only if, (at least) one of the signals in the tuple can be written as a linear combination of the others.

The $d$-tuple $(\mathbf{u}_1, \ldots, \mathbf{u}_d)$ is said to form a **basis** for the linear subspace $\mathcal{U}$ if it is linearly independent and if $\mathrm{span}(\mathbf{u}_1, \ldots, \mathbf{u}_d) = \mathcal{U}$. The latter condition is equivalent to the requirement that every $\mathbf{u} \in \mathcal{U}$ can be represented as

$$\mathbf{u} = \alpha_1 \mathbf{u}_1 + \cdots + \alpha_d \mathbf{u}_d \tag{4.10}$$

for some $\alpha_1, \ldots, \alpha_d \in \mathbb{C}$. The former condition that the tuple $(\mathbf{u}_1, \ldots, \mathbf{u}_d)$ be linearly independent guarantees that if such a representation exists, then it is unique. Thus, $(\mathbf{u}_1, \ldots, \mathbf{u}_d)$ forms a basis for $\mathcal{U}$ if $\mathbf{u}_1, \ldots, \mathbf{u}_d \in \mathcal{U}$ (thus guaranteeing that span$(\mathbf{u}_1, \ldots, \mathbf{u}_d) \subseteq \mathcal{U})$ and if every $\mathbf{u} \in \mathcal{U}$ can be written *uniquely* as in (4.10).

Every finite-dimensional linear subspace $\mathcal{U}$ has a basis, and all bases for $\mathcal{U}$ have the same number of elements. This number is called the **dimension** of $\mathcal{U}$. Thus, if $\mathcal{U}$ is a finite-dimensional subspace and if both $(\mathbf{u}_1, \ldots, \mathbf{u}_d)$ and $(\mathbf{u}'_1, \ldots, \mathbf{u}'_{d'})$ form a basis for $\mathcal{U}$, then $d = d'$ and both are equal to the dimension of $\mathcal{U}$. The dimension of the subspace $\{\mathbf{0}\}$ is zero.

## 4.4    $\|\mathbf{u}\|_2$ as the "length" of the Signal $u(\cdot)$

Having presented the elements of $\mathcal{L}_2$ as vectors, we next propose to view $\|\mathbf{u}\|_2$ as the "length" of the vector $\mathbf{u} \in \mathcal{L}_2$. To motivate this view, we first present the key properties of $\|\cdot\|_2$.

**Proposition 4.4.1 (Properties of $\|\cdot\|_2$).** *Let $\mathbf{u}$ and $\mathbf{v}$ be elements of $\mathcal{L}_2$, and let $\alpha$ be some complex number. Then*

$$\|\alpha \mathbf{u}\|_2 = |\alpha| \, \|\mathbf{u}\|_2 \, , \tag{4.11}$$

$$\|\mathbf{u} + \mathbf{v}\|_2 \leq \|\mathbf{u}\|_2 + \|\mathbf{v}\|_2 \, , \tag{4.12}$$

*and*

$$\Big( \|\mathbf{u}\|_2 = 0 \Big) \Leftrightarrow \Big( \mathbf{u} \equiv \mathbf{0} \Big). \tag{4.13}$$

**Proof.** Identity (4.11) follows directly from the definition of $\|\cdot\|_2$; see (4.2). Inequality (4.12) is a restatement of Proposition 3.4.1. The equivalence of the condition $\|\mathbf{u}\|_2 = 0$ and the condition that $\mathbf{u}$ is indistinguishable from the all-zero signal $\mathbf{0}$ follows from Proposition 2.5.3. $\qquad\square$

Identity (4.11) is in agreement with our intuition that stretching a vector merely scales its length. Inequality (4.12) is sometimes called the **Triangle Inequality** because it is reminiscent of the theorem from planar geometry that states that the length of no side of a triangle can exceed the sum of the lengths of the others; see Figure 4.1.

Substituting $-\mathbf{y}$ for $\mathbf{u}$ and $\mathbf{x} + \mathbf{y}$ for $\mathbf{v}$ in (4.12) yields $\|\mathbf{x}\|_2 \leq \|\mathbf{y}\|_2 + \|\mathbf{x} + \mathbf{y}\|_2$, i.e., the inequality $\|\mathbf{x} + \mathbf{y}\|_2 \geq \|\mathbf{x}\|_2 - \|\mathbf{y}\|_2$. And substituting $-\mathbf{x}$ for $\mathbf{u}$ and $\mathbf{x} + \mathbf{y}$ for $\mathbf{v}$ in (4.12) yields the inequality $\|\mathbf{y}\|_2 \leq \|\mathbf{x}\|_2 + \|\mathbf{x} + \mathbf{y}\|_2$, i.e., the inequality $\|\mathbf{x} + \mathbf{y}\|_2 \geq \|\mathbf{y}\|_2 - \|\mathbf{x}\|_2$. Combining the two inequalities we obtain the inequality $\|\mathbf{x} + \mathbf{y}\|_2 \geq \big| \|\mathbf{x}\|_2 - \|\mathbf{y}\|_2 \big|$. This inequality can be combined with the inequality $\|\mathbf{x} + \mathbf{y}\|_2 \leq \|\mathbf{x}\|_2 + \|\mathbf{y}\|_2$ in the compact form of a double-sided inequality

$$\big| \|\mathbf{x}\|_2 - \|\mathbf{y}\|_2 \big| \leq \|\mathbf{x} + \mathbf{y}\|_2 \leq \|\mathbf{x}\|_2 + \|\mathbf{y}\|_2 \, , \quad \mathbf{x}, \mathbf{y} \in \mathcal{L}_2. \tag{4.14}$$

Finally, (4.13) "almost" supports the intuition that the only vector of length zero is the zero-vector. In our case, alas, we can only claim that if a vector is of zero length, then it is indistinguishable from the all-zero signal, i.e., that all $t$'s outside a set of Lebesgue measure zero are mapped by the signal to zero.

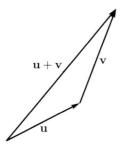

**Figure 4.1:** A geometric interpretation of the Triangle Inequality for energy-limited signals: $\|\mathbf{u} + \mathbf{v}\|_2 \leq \|\mathbf{u}\|_2 + \|\mathbf{v}\|_2$.

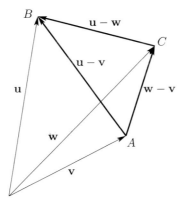

**Figure 4.2:** Illustration of the shortest path property in $\mathcal{L}_2$. The shortest path from A to B is no longer than the sum of the shortest path from A to C and the shortest path from C to B.

The Triangle Inequality (4.12) can also be stated slightly differently. In planar geometry the sum of the lengths of two sides of a triangle can never be smaller than the length of the remaining side. Thus, the shortest path from Point A to Point B cannot exceed the sum of the lengths of the shortest paths from Point A to Point C, and from Point C to Point B. By applying Inequality (4.12) to the signal $\mathbf{u} - \mathbf{w}$ and $\mathbf{w} - \mathbf{v}$ we obtain

$$\|\mathbf{u} - \mathbf{v}\|_2 \leq \|\mathbf{u} - \mathbf{w}\|_2 + \|\mathbf{w} - \mathbf{v}\|_2, \quad \mathbf{u}, \mathbf{v}, \mathbf{w} \in \mathcal{L}_2,$$

i.e., that the distance from $\mathbf{u}$ to $\mathbf{v}$ cannot exceed the sum of distances from $\mathbf{u}$ to $\mathbf{w}$ and from $\mathbf{w}$ to $\mathbf{v}$. See Figure 4.2.

## 4.5   Orthogonality and Inner Products

To further develop our geometric view of $\mathcal{L}_2$ we next discuss orthogonality. We shall motivate its definition with an attempt to generalize Pythagoras's Theorem to $\mathcal{L}_2$. As an initial attempt at defining orthogonality we might define two functions $\mathbf{u}, \mathbf{v} \in \mathcal{L}_2$ to be orthogonal if $\|\mathbf{u} + \mathbf{v}\|_2^2 = \|\mathbf{u}\|_2^2 + \|\mathbf{v}\|_2^2$. Recalling the definition of $\|\cdot\|_2$ (4.2) we obtain that this condition is equivalent to the condition $\mathrm{Re}\left(\int u(t)\,v^*(t)\,dt\right) = 0$, because

$$
\begin{aligned}
\|\mathbf{u} + \mathbf{v}\|_2^2 &= \int_{-\infty}^{\infty} |u(t) + v(t)|^2 \, dt \\
&= \int_{-\infty}^{\infty} \big(u(t) + v(t)\big)\big(u(t) + v(t)\big)^* \, dt \\
&= \int_{-\infty}^{\infty} \Big(|u(t)|^2 + |v(t)|^2 + 2\,\mathrm{Re}\big(u(t)\,v^*(t)\big)\Big) \, dt \\
&= \|\mathbf{u}\|_2^2 + \|\mathbf{v}\|_2^2 + 2\,\mathrm{Re}\left(\int_{-\infty}^{\infty} u(t)\,v^*(t)\,dt\right), \quad \mathbf{u}, \mathbf{v} \in \mathcal{L}_2, \qquad (4.15)
\end{aligned}
$$

where we have used the fact that integration commutes with the operation of taking the real part; see Proposition 2.3.1.

While this approach would work well for real-valued functions, it has some embarrassing consequences when it comes to complex-valued functions. It allows for the possibility that $\mathbf{u}$ is orthogonal to $\mathbf{v}$, but that its scaled version $\alpha \mathbf{u}$ is not. For example, with this definition, the function $t \mapsto \mathrm{i}\,\mathrm{I}\{|t| \leq 5\}$ is orthogonal to the function $t \mapsto \mathrm{I}\{|t| \leq 17\}$ but its scaled (by $\alpha = \mathrm{i}$) version $t \mapsto \mathrm{i}\,\mathrm{i}\,\mathrm{I}\{|t| \leq 5\} = -\mathrm{I}\{|t| \leq 5\}$ is not. To avoid this embarrassment, we define $\mathbf{u}$ to be orthogonal to $\mathbf{v}$ if

$$
\|\alpha \mathbf{u} + \mathbf{v}\|_2^2 = \|\alpha \mathbf{u}\|_2^2 + \|\mathbf{v}\|_2^2, \quad \alpha \in \mathbb{C}.
$$

This, by (4.15), is equivalent to

$$
\mathrm{Re}\left(\alpha \int_{-\infty}^{\infty} u(t)\,v^*(t)\,dt\right) = 0, \quad \alpha \in \mathbb{C},
$$

i.e., to the condition

$$
\int_{-\infty}^{\infty} u(t)\,v^*(t)\,dt = 0 \qquad (4.16)
$$

(because if $z \in \mathbb{C}$ is such that $\mathrm{Re}(\alpha z) = 0$ for all $\alpha \in \mathbb{C}$, then $z = 0$). Recalling the definition of the inner product $\langle \mathbf{u}, \mathbf{v} \rangle$ from (3.4)

$$
\langle \mathbf{u}, \mathbf{v} \rangle = \int_{-\infty}^{\infty} u(t)\,v^*(t)\,dt, \qquad (4.17)
$$

we conclude that (4.16) is equivalent to the condition $\langle \mathbf{u}, \mathbf{v} \rangle = 0$ or, equivalently (because by (3.6) $\langle \mathbf{u}, \mathbf{v} \rangle = \langle \mathbf{v}, \mathbf{u} \rangle^*$) to the condition $\langle \mathbf{v}, \mathbf{u} \rangle = 0$.

**Definition 4.5.1 (Orthogonal Signals in $\mathcal{L}_2$).** *The signals $\mathbf{u}, \mathbf{v} \in \mathcal{L}_2$ are said to be **orthogonal** if*

$$
\langle \mathbf{u}, \mathbf{v} \rangle = 0. \qquad (4.18)
$$

*The n-tuple* $(\mathbf{u}_1, \ldots, \mathbf{u}_n)$ *is said to be orthogonal if any two signals in the tuple are orthogonal*

$$\langle \mathbf{u}_\ell, \mathbf{u}_{\ell'} \rangle = 0, \quad \left( \ell \neq \ell', \quad \ell, \ell' \in \{1, \ldots, n\} \right). \tag{4.19}$$

The reader is encouraged to verify that if $\mathbf{u}$ is orthogonal to $\mathbf{v}$ then so is $\alpha\mathbf{u}$. Also, $\mathbf{u}$ is orthogonal to $\mathbf{v}$ if, and only if, $\mathbf{v}$ is orthogonal to $\mathbf{u}$. Finally every function is orthogonal to the all-zero function $\mathbf{0}$.

Having judiciously defined orthogonality in $\mathcal{L}_2$, we can now extend Pythagoras's Theorem.

**Theorem 4.5.2 (A Pythagorean Theorem).** *If the n-tuple of vectors* $(\mathbf{u}_1, \ldots, \mathbf{u}_n)$ *in* $\mathcal{L}_2$ *is orthogonal, then*

$$\|\mathbf{u}_1 + \cdots + \mathbf{u}_n\|_2^2 = \|\mathbf{u}_1\|_2^2 + \cdots + \|\mathbf{u}_n\|_2^2 \, .$$

**Proof.** This theorem can be proved by induction on $n$. The case $n = 2$ follows from (4.15) using Definition 4.5.1 and (4.17).

Assume now that the theorem holds for $n = \nu$, for some $\nu \geq 2$, i.e.,

$$\|\mathbf{u}_1 + \cdots + \mathbf{u}_\nu\|_2^2 = \|\mathbf{u}_1\|_2^2 + \cdots + \|\mathbf{u}_\nu\|_2^2 \, ,$$

and let us show that this implies that it also holds for $n = \nu + 1$, i.e., that

$$\|\mathbf{u}_1 + \cdots + \mathbf{u}_{\nu+1}\|_2^2 = \|\mathbf{u}_1\|_2^2 + \cdots + \|\mathbf{u}_{\nu+1}\|_2^2 \, .$$

To that end, let

$$\mathbf{v} = \mathbf{u}_1 + \cdots + \mathbf{u}_\nu. \tag{4.20}$$

Since the $\nu$-tuple $(\mathbf{u}_1, \ldots, \mathbf{u}_\nu)$ is orthogonal, our induction hypothesis guarantees that

$$\|\mathbf{v}\|_2^2 = \|\mathbf{u}_1\|_2^2 + \cdots + \|\mathbf{u}_\nu\|_2^2 \, . \tag{4.21}$$

Now $\mathbf{v}$ is orthogonal to $\mathbf{u}_{\nu+1}$ because

$$\begin{aligned}
\langle \mathbf{v}, \mathbf{u}_{\nu+1} \rangle &= \langle \mathbf{u}_1 + \cdots + \mathbf{u}_\nu, \mathbf{u}_{\nu+1} \rangle \\
&= \langle \mathbf{u}_1, \mathbf{u}_{\nu+1} \rangle + \cdots + \langle \mathbf{u}_\nu, \mathbf{u}_{\nu+1} \rangle \\
&= 0,
\end{aligned}$$

so by the $n = 2$ case

$$\|\mathbf{v} + \mathbf{u}_{\nu+1}\|_2^2 = \|\mathbf{v}\|_2^2 + \|\mathbf{u}_{\nu+1}\|_2^2 \, . \tag{4.22}$$

Combining (4.20), (4.21), and (4.22) we obtain

$$\begin{aligned}
\|\mathbf{u}_1 + \cdots + \mathbf{u}_{\nu+1}\|_2^2 &= \|\mathbf{v} + \mathbf{u}_{\nu+1}\|_2^2 \\
&= \|\mathbf{v}\|_2^2 + \|\mathbf{u}_{\nu+1}\|_2^2 \\
&= \|\mathbf{u}_1\|_2^2 + \cdots + \|\mathbf{u}_{\nu+1}\|_2^2 \, . \qquad \square
\end{aligned}$$

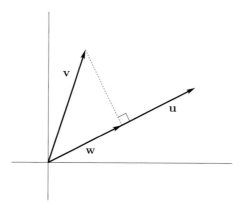

**Figure 4.3:** The projection **w** of the vector **v** onto **u**.

To derive a geometric interpretation for the inner product $\langle \mathbf{u}, \mathbf{v} \rangle$ we next extend to $\mathcal{L}_2$ the notion of the projection of a vector onto another. We first recall the definition for vectors in $\mathbb{R}^2$. Consider two nonzero vectors **u** and **v** in the real plane $\mathbb{R}^2$. The projection **w** of the vector **v** onto **u** is a scaled version of **u**. More specifically, it is a scaled version of **u** and its length is equal to the product of the length of **v** multiplied by the cosine of the angle between **v** and **u** (see Figure 4.3). More explicitly,

$$\mathbf{w} = (\text{length of } \mathbf{v}) \cos(\text{angle between } \mathbf{v} \text{ and } \mathbf{u}) \frac{\mathbf{u}}{\text{length of } \mathbf{u}}. \qquad (4.23)$$

This definition does not seem to have a natural extension to $\mathcal{L}_2$ because we have not defined the angle between two signals. An alternative definition of the projection, and one that is more amenable to extensions to $\mathcal{L}_2$, is the following. The vector **w** is the projection of the vector **v** onto **u**, if **w** is a scaled version of **u**, and if $\mathbf{v} - \mathbf{w}$ is orthogonal to **u**.

This definition makes perfect sense in $\mathcal{L}_2$ too, because we have already defined what we mean by "scaled version" (i.e., "amplification" or "scalar multiplication") and "orthogonality." We thus have:

**Definition 4.5.3 (Projection of a Signal in $\mathcal{L}_2$ onto another).** *Let* $\mathbf{u} \in \mathcal{L}_2$ *have positive energy. The **projection of the signal** $\mathbf{v} \in \mathcal{L}_2$ **onto the signal** $\mathbf{u} \in \mathcal{L}_2$ is the signal* **w** *that satisfies both of the following conditions:*

*1)* $\mathbf{w} = \alpha \mathbf{u}$ *for some* $\alpha \in \mathbb{C}$ *and*

*2)* $\mathbf{v} - \mathbf{w}$ *is orthogonal to* **u**.

Note that since $\mathcal{L}_2$ is closed with respect to scalar multiplication, Condition 1) guarantees that the projection **w** is in $\mathcal{L}_2$.

*Prima facie* it is not clear that a projection always exists and that it is unique. Nevertheless, this is the case. We prove this by finding an explicit expression for **w**. We need to find some $\alpha \in \mathbb{C}$ so that $\alpha \mathbf{u}$ will satisfy the requirements of the projection. The scalar $\alpha$ is chosen so as to guarantee that $\mathbf{v} - \mathbf{w}$ is orthogonal to $\mathbf{u}$. That is, we seek to solve for $\alpha \in \mathbb{C}$ satisfying

$$\langle \mathbf{v} - \alpha \mathbf{u}, \mathbf{u} \rangle = 0,$$

i.e.,

$$\langle \mathbf{v}, \mathbf{u} \rangle - \alpha \|\mathbf{u}\|_2^2 = 0.$$

Recalling our hypothesis that $\|\mathbf{u}\|_2 > 0$ (strictly), we conclude that $\alpha$ is uniquely given by

$$\alpha = \frac{\langle \mathbf{v}, \mathbf{u} \rangle}{\|\mathbf{u}\|_2^2},$$

and the projection $\mathbf{w}$ is thus unique and is given by

$$\mathbf{w} = \frac{\langle \mathbf{v}, \mathbf{u} \rangle}{\|\mathbf{u}\|_2^2} \mathbf{u}. \qquad (4.24)$$

Comparing (4.23) and (4.24) we can interpret

$$\frac{\langle \mathbf{v}, \mathbf{u} \rangle}{\|\mathbf{u}\|_2 \|\mathbf{v}\|_2} \qquad (4.25)$$

as the cosine of the angle between the function $\mathbf{v}$ and the function $\mathbf{u}$ (provided that neither $\mathbf{u}$ nor $\mathbf{v}$ is zero). If the inner product is zero, then we have said that $\mathbf{v}$ and $\mathbf{u}$ are orthogonal, which is consistent with the cosine of the angle between them being zero. Note, however, that this interpretation should be taken with a grain of salt because in the complex case the inner product in (4.25) is typically a complex number.

The interpretation of (4.25) as the cosine of the angle between $\mathbf{v}$ and $\mathbf{u}$ is further supported by noting that the magnitude of (4.25) is always in the range $[0, 1]$. This follows directly from the Cauchy-Schwarz Inequality (Theorem 3.3.1) to which we next give another (geometric) proof. Let $\mathbf{w}$ be the projection of $\mathbf{v}$ onto $\mathbf{u}$. Then starting from (4.24)

$$\begin{aligned}
\frac{|\langle \mathbf{v}, \mathbf{u} \rangle|^2}{\|\mathbf{u}\|_2^2} &= \|\mathbf{w}\|_2^2 \\
&\leq \|\mathbf{w}\|_2^2 + \|\mathbf{v} - \mathbf{w}\|_2^2 \\
&= \|\mathbf{w} + (\mathbf{v} - \mathbf{w})\|_2^2 \\
&= \|\mathbf{v}\|_2^2, \qquad (4.26)
\end{aligned}$$

where the first equality follows from (4.24); the subsequent inequality from the nonnegativity of $\|\cdot\|_2$; and the subsequent equality by the Pythagorean Theorem because, by its definition, the projection $\mathbf{w}$ of $\mathbf{v}$ onto $\mathbf{u}$ must satisfy that $\mathbf{v} - \mathbf{w}$ is orthogonal to $\mathbf{u}$ and hence also to $\mathbf{w}$, which is a scaled version of $\mathbf{u}$. The Cauchy-Schwarz Inequality now follows by taking the square root of both sides of (4.26).

## 4.6   Orthonormal Bases

We next consider orthonormal bases for finite-dimensional linear subspaces. These are special bases that are particularly useful for the calculation of projections and inner products.

### 4.6.1   Definition

**Definition 4.6.1 (Orthonormal Tuple).** *An n-tuple of signals in $\mathcal{L}_2$ is said to be* ***orthonormal*** *if it is orthogonal and if each of the signals in the tuple is of unit energy.*

Thus, the $n$-tuple $(\phi_1, \ldots, \phi_n)$ of signals in $\mathcal{L}_2$ is orthonormal, if

$$\langle \phi_\ell, \phi_{\ell'} \rangle = \begin{cases} 0 & \text{if } \ell \neq \ell', \\ 1 & \text{if } \ell = \ell', \end{cases} \quad \ell, \ell' \in \{1, \ldots, n\}. \tag{4.27}$$

Linearly independent tuples need not be orthonormal, but orthonormal tuples must be linearly independent:

**Proposition 4.6.2 (Orthonormal Tuples Are Linearly Independent).** *If a tuple of signals in $\mathcal{L}_2$ is orthonormal, then it must be linearly independent.*

**Proof.** Let the $n$-tuple $(\phi_1, \ldots, \phi_n)$ of signals in $\mathcal{L}_2$ be orthonormal, i.e., satisfy (4.27). We need to show that if

$$\sum_{\ell=1}^{n} \alpha_\ell \phi_\ell = \mathbf{0}, \tag{4.28}$$

then all the coefficients $\alpha_1, \ldots, \alpha_n$ must be zero. To that end, assume (4.28). It then follows that for every $\ell' \in \{1, \ldots, n\}$

$$\begin{aligned} 0 &= \langle \mathbf{0}, \phi_{\ell'} \rangle \\ &= \left\langle \sum_{\ell=1}^{n} \alpha_\ell \phi_\ell, \phi_{\ell'} \right\rangle \\ &= \sum_{\ell=1}^{n} \alpha_\ell \langle \phi_\ell, \phi_{\ell'} \rangle \\ &= \sum_{\ell=1}^{n} \alpha_\ell \, \mathrm{I}\{\ell = \ell'\} \\ &= \alpha_{\ell'}, \end{aligned}$$

thus demonstrating that (4.28) implies that $\alpha_{\ell'} = 0$ for every $\ell' \in \{1, \ldots, n\}$. Here the first equality follows because $\mathbf{0}$ is orthogonal to every energy-limited signal and, *a fortiori*, to $\phi_{\ell'}$; the second by (4.28); the third by the linearity of the inner product in its left argument (3.7) & (3.9); and the fourth by (4.27).  □

**Definition 4.6.3 (Orthonormal Basis).** *A d-tuple of signals in $\mathcal{L}_2$ is said to form an **orthonormal basis** for the linear subspace $\mathcal{U} \subset \mathcal{L}_2$ if it is orthonormal and its span is $\mathcal{U}$.*

### 4.6.2 Representing a Signal Using an Orthonormal Basis

Suppose that $(\phi_1, \ldots, \phi_d)$ is an orthonormal basis for $\mathcal{U} \subset \mathcal{L}_2$. The fact that $(\phi_1, \ldots, \phi_d)$ spans $\mathcal{U}$ guarantees that every $\mathbf{u} \in \mathcal{U}$ can be written as $\mathbf{u} = \sum_\ell \alpha_\ell \phi_\ell$ for some coefficients $\alpha_1, \ldots, \alpha_d \in \mathbb{C}$. The fact that $(\phi_1, \ldots, \phi_d)$ is orthonormal implies, by Proposition 4.6.2, that it is also linearly independent and hence that the coefficients $\{\alpha_\ell\}$ are unique. How does one go about finding these coefficients? We next show that the orthonormality of $(\phi_1, \ldots, \phi_d)$ also implies a very simple expression for $\alpha_\ell$ above. Indeed, as the next proposition demonstrates, $\alpha_\ell$ is given explicitly as $\langle \mathbf{u}, \phi_\ell \rangle$.

**Proposition 4.6.4 (Representing a Signal Using an Orthonormal Basis).**

*(i) If $(\phi_1, \ldots, \phi_d)$ is an orthonormal tuple of functions in $\mathcal{L}_2$ and if $\mathbf{u} \in \mathcal{L}_2$ can be written as $\mathbf{u} = \sum_{\ell=1}^{d} \alpha_\ell \phi_\ell$ for some complex numbers $\alpha_1, \ldots, \alpha_d$, then $\alpha_\ell = \langle \mathbf{u}, \phi_\ell \rangle$ for every $\ell \in \{1, \ldots, d\}$:*

$$\left( \mathbf{u} = \sum_{\ell=1}^{d} \alpha_\ell \phi_\ell \right) \Rightarrow \left( \alpha_\ell = \langle \mathbf{u}, \phi_\ell \rangle, \quad \ell \in \{1, \ldots, d\} \right),$$

$$\left( (\phi_1, \ldots, \phi_d) \text{ orthonormal} \right). \quad (4.29)$$

*(ii) If $(\phi_1, \ldots, \phi_d)$ is an orthonormal basis for the subspace $\mathcal{U} \subset \mathcal{L}_2$, then*

$$\mathbf{u} = \sum_{\ell=1}^{d} \langle \mathbf{u}, \phi_\ell \rangle \phi_\ell, \quad \mathbf{u} \in \mathcal{U}. \quad (4.30)$$

**Proof.** We begin by proving Part (i). If $\mathbf{u} = \sum_{\ell=1}^{d} \alpha_\ell \phi_\ell$, then for every $\ell' \in \{1, \ldots, d\}$

$$\langle \mathbf{u}, \phi_{\ell'} \rangle = \left\langle \sum_{\ell=1}^{d} \alpha_\ell \phi_\ell, \phi_{\ell'} \right\rangle$$

$$= \sum_{\ell=1}^{d} \alpha_\ell \langle \phi_\ell, \phi_{\ell'} \rangle$$

$$= \sum_{\ell=1}^{d} \alpha_\ell \, \mathrm{I}\{\ell = \ell'\}$$

$$= \alpha_{\ell'},$$

thus proving Part (i).

We next prove Part (ii). Let $\mathbf{u} \in \mathcal{U}$ be arbitrary. Since, by assumption, the tuple $(\boldsymbol{\phi}_1, \dots, \boldsymbol{\phi}_d)$ forms an orthonormal basis for $\mathcal{U}$ it follows *a fortiori* that its span is $\mathcal{U}$ and, consequently, that there exist coefficients $\alpha_1, \dots, \alpha_d \in \mathbb{C}$ such that

$$\mathbf{u} = \sum_{\ell=1}^{d} \alpha_\ell \boldsymbol{\phi}_\ell. \tag{4.31}$$

It now follows from Part (i) that for each $\ell \in \{1, \dots, d\}$ the coefficient $\alpha_\ell$ in (4.31) must be equal to $\langle \mathbf{u}, \boldsymbol{\phi}_\ell \rangle$, thus establishing (4.30). $\qquad\square$

This proposition shows that if $(\boldsymbol{\phi}_1, \dots, \boldsymbol{\phi}_d)$ is an orthonormal basis for the subspace $\mathcal{U}$ and if $\mathbf{u} \in \mathcal{U}$, then $\mathbf{u}$ is fully determined by the complex constants $\langle \mathbf{u}, \boldsymbol{\phi}_1 \rangle$, $\dots$, $\langle \mathbf{u}, \boldsymbol{\phi}_d \rangle$. Thus, any calculation involving $\mathbf{u}$ can be computed from these constants by first reconstructing $\mathbf{u}$ using the proposition. As we shall see in Proposition 4.6.9, calculations involving inner products and norms are, however, simpler than that.

### 4.6.3  Projection

We next discuss the projection of a signal $\mathbf{v} \in \mathcal{L}_2$ onto a finite-dimensional linear subspace $\mathcal{U}$ that has an orthonormal basis $(\boldsymbol{\phi}_1, \dots, \boldsymbol{\phi}_d)$.[1] To define the projection we shall extend the approach we adopted in Section 4.5 for the projection of the vector $\mathbf{v}$ onto the vector $\mathbf{u}$. Recall that in that section we defined the projection as the vector $\mathbf{w}$ that is a scaled version of $\mathbf{u}$ and that satisfies that $(\mathbf{v} - \mathbf{w})$ is orthogonal to $\mathbf{u}$. Of course, if $(\mathbf{v} - \mathbf{w})$ is orthogonal to $\mathbf{u}$, then it is orthogonal to any scaled version of $\mathbf{u}$, i.e., it is orthogonal to every signal in the space $\mathrm{span}(\mathbf{u})$.

We would like to adopt this approach and to define the projection of $\mathbf{v} \in \mathcal{L}_2$ onto $\mathcal{U}$ as the element $\mathbf{w}$ of $\mathcal{U}$ for which $(\mathbf{v} - \mathbf{w})$ is orthogonal to every signal in $\mathcal{U}$. Before we can adopt this definition, we must show that such an element of $\mathcal{U}$ always exists and that it is unique.

**Lemma 4.6.5.** *Let $(\boldsymbol{\phi}_1, \dots, \boldsymbol{\phi}_d)$ be an orthonormal basis for the linear subspace $\mathcal{U} \subset \mathcal{L}_2$. Let $\mathbf{v} \in \mathcal{L}_2$ be arbitrary.*

*(i) The signal $\mathbf{v} - \sum_{\ell=1}^{d} \langle \mathbf{v}, \boldsymbol{\phi}_\ell \rangle \boldsymbol{\phi}_\ell$ is orthogonal to every signal in $\mathcal{U}$:*

$$\left\langle \mathbf{v} - \sum_{\ell=1}^{d} \langle \mathbf{v}, \boldsymbol{\phi}_\ell \rangle \boldsymbol{\phi}_\ell, \mathbf{u} \right\rangle = 0, \quad \Big( \mathbf{v} \in \mathcal{L}_2, \quad \mathbf{u} \in \mathcal{U} \Big). \tag{4.32}$$

*(ii) If $\mathbf{w} \in \mathcal{U}$ is such that $\mathbf{v} - \mathbf{w}$ is orthogonal to every signal in $\mathcal{U}$, then*

$$\mathbf{w} = \sum_{\ell=1}^{d} \langle \mathbf{v}, \boldsymbol{\phi}_\ell \rangle \boldsymbol{\phi}_\ell. \tag{4.33}$$

---

[1] As we shall see in Section 4.6.5, not every finite-dimensional linear subspace of $\mathcal{L}_2$ has an orthonormal basis. Here we shall only discuss projections onto subspaces that do.

**Proof.** To prove (4.32) we first verify that it holds when $\mathbf{u} = \boldsymbol{\phi}_{\ell'}$, for some $\ell'$ in the set $\{1, \ldots, d\}$:

$$\left\langle \mathbf{v} - \sum_{\ell=1}^{d} \langle \mathbf{v}, \boldsymbol{\phi}_\ell \rangle \, \boldsymbol{\phi}_\ell, \boldsymbol{\phi}_{\ell'} \right\rangle = \langle \mathbf{v}, \boldsymbol{\phi}_{\ell'} \rangle - \left\langle \sum_{\ell=1}^{d} \langle \mathbf{v}, \boldsymbol{\phi}_\ell \rangle \, \boldsymbol{\phi}_\ell, \boldsymbol{\phi}_{\ell'} \right\rangle$$

$$= \langle \mathbf{v}, \boldsymbol{\phi}_{\ell'} \rangle - \sum_{\ell=1}^{d} \langle \mathbf{v}, \boldsymbol{\phi}_\ell \rangle \, \langle \boldsymbol{\phi}_\ell, \boldsymbol{\phi}_{\ell'} \rangle$$

$$= \langle \mathbf{v}, \boldsymbol{\phi}_{\ell'} \rangle - \sum_{\ell=1}^{d} \langle \mathbf{v}, \boldsymbol{\phi}_\ell \rangle \, \mathrm{I}\{\ell = \ell'\}$$

$$= \langle \mathbf{v}, \boldsymbol{\phi}_{\ell'} \rangle - \langle \mathbf{v}, \boldsymbol{\phi}_{\ell'} \rangle$$

$$= 0, \quad \ell' \in \{1, \ldots, d\}. \tag{4.34}$$

Having verified (4.32) for $\mathbf{u} = \boldsymbol{\phi}_{\ell'}$ we next verify that this implies that it holds for all $\mathbf{u} \in \mathcal{U}$. By Proposition 4.6.4 we obtain that any $\mathbf{u} \in \mathcal{U}$ can be written as $\mathbf{u} = \sum_{\ell'=1}^{d} \beta_{\ell'} \boldsymbol{\phi}_{\ell'}$, where $\beta_{\ell'} = \langle \mathbf{u}, \boldsymbol{\phi}_{\ell'} \rangle$. Consequently,

$$\left\langle \mathbf{v} - \sum_{\ell=1}^{d} \langle \mathbf{v}, \boldsymbol{\phi}_\ell \rangle \, \boldsymbol{\phi}_\ell, \mathbf{u} \right\rangle = \left\langle \mathbf{v} - \sum_{\ell=1}^{d} \langle \mathbf{v}, \boldsymbol{\phi}_\ell \rangle \, \boldsymbol{\phi}_\ell, \sum_{\ell'=1}^{d} \beta_{\ell'} \boldsymbol{\phi}_{\ell'} \right\rangle$$

$$= \sum_{\ell'=1}^{d} \beta_{\ell'}^* \left\langle \mathbf{v} - \sum_{\ell=1}^{d} \langle \mathbf{v}, \boldsymbol{\phi}_\ell \rangle \, \boldsymbol{\phi}_\ell, \boldsymbol{\phi}_{\ell'} \right\rangle$$

$$= \sum_{\ell'=1}^{d} \beta_{\ell'}^* \, 0$$

$$= 0, \quad \mathbf{u} \in \mathcal{U},$$

where the third equality follows from (4.34) and the basic properties of the inner product (3.6)–(3.10).

We next prove Part (ii) by showing that if $\mathbf{w}, \mathbf{w}' \in \mathcal{U}$ satisfy

$$\langle \mathbf{v} - \mathbf{w}, \mathbf{u} \rangle = 0, \quad \mathbf{u} \in \mathcal{U} \tag{4.35}$$

and

$$\langle \mathbf{v} - \mathbf{w}', \mathbf{u} \rangle = 0, \quad \mathbf{u} \in \mathcal{U}, \tag{4.36}$$

then $\mathbf{w} = \mathbf{w}'$.

This follows from the calculation:

$$\mathbf{w} - \mathbf{w}' = \sum_{\ell=1}^{d} \langle \mathbf{w}, \boldsymbol{\phi}_\ell \rangle \, \boldsymbol{\phi}_\ell - \sum_{\ell=1}^{d} \langle \mathbf{w}', \boldsymbol{\phi}_\ell \rangle \, \boldsymbol{\phi}_\ell$$

$$= \sum_{\ell=1}^{d} \langle \mathbf{w} - \mathbf{w}', \boldsymbol{\phi}_\ell \rangle \, \boldsymbol{\phi}_\ell$$

$$= \sum_{\ell=1}^{d} \left\langle (\mathbf{v} - \mathbf{w}') - (\mathbf{v} - \mathbf{w}), \boldsymbol{\phi}_\ell \right\rangle \boldsymbol{\phi}_\ell$$

$$= \sum_{\ell=1}^{d} \Big( \big\langle (\mathbf{v} - \mathbf{w}'), \boldsymbol{\phi}_\ell \big\rangle - \big\langle (\mathbf{v} - \mathbf{w}), \boldsymbol{\phi}_\ell \big\rangle \Big) \boldsymbol{\phi}_\ell$$

$$= \sum_{\ell=1}^{d} (0 - 0) \boldsymbol{\phi}_\ell$$

$$= \mathbf{0},$$

where the first equality follows from Proposition 4.6.4; the second by the linearity of the inner product in its left argument (3.9); the third by adding and subtracting $\mathbf{v}$; the fourth by the linearity of the inner product in its left argument (3.9); and the fifth equality from (4.35) & (4.36) applied by substituting $\boldsymbol{\phi}_\ell$ for $\mathbf{u}$.                    $\square$

With the aid of the above lemma we can now define the projection of a signal onto a finite-dimensional subspace that has an orthonormal basis.[2]

**Definition 4.6.6 (Projection of $\mathbf{v} \in \mathcal{L}_2$ onto $\mathcal{U}$).** *Let $\mathcal{U} \subset \mathcal{L}_2$ be a finite-dimensional linear subspace of $\mathcal{L}_2$ having an orthonormal basis. Let $\mathbf{v} \in \mathcal{L}_2$ be an arbitrary energy-limited signal. Then the **projection of $\mathbf{v}$ onto $\mathcal{U}$** is the unique element $\mathbf{w}$ of $\mathcal{U}$ such that*

$$\langle \mathbf{v} - \mathbf{w}, \mathbf{u} \rangle = 0, \quad \mathbf{u} \in \mathcal{U}. \tag{4.37}$$

**Note 4.6.7.** By Lemma 4.6.5 it follows that if $(\boldsymbol{\phi}_1, \ldots, \boldsymbol{\phi}_d)$ is an orthonormal basis for $\mathcal{U}$, then the projection of $\mathbf{v} \in \mathcal{L}_2$ onto $\mathcal{U}$ is given by

$$\sum_{\ell=1}^{d} \langle \mathbf{v}, \boldsymbol{\phi}_\ell \rangle \, \boldsymbol{\phi}_\ell. \tag{4.38}$$

To further develop the geometric picture of $\mathcal{L}_2$, we next show that, loosely speaking, the projection of $\mathbf{v} \in \mathcal{L}_2$ onto $\mathcal{U}$ is the element in $\mathcal{U}$ that is closest to $\mathbf{v}$. This result can also be viewed as an optimal approximation result: if we wish to approximate $\mathbf{v}$ by an element of $\mathcal{U}$, then the optimal approximation is the projection of $\mathbf{v}$ onto $\mathcal{U}$, provided that we measure the quality of our approximation using the energy in the error signal.

**Proposition 4.6.8 (Projection as Best Approximation).** *Let $\mathcal{U} \subset \mathcal{L}_2$ be a finite-dimensional subspace of $\mathcal{L}_2$ having an orthonormal basis $(\boldsymbol{\phi}_1, \ldots, \boldsymbol{\phi}_d)$. Let $\mathbf{v} \in \mathcal{L}_2$ be arbitrary. Then the projection of $\mathbf{v}$ onto $\mathcal{U}$ is the element $\mathbf{w} \in \mathcal{U}$ that, among all the elements of $\mathcal{U}$, is closest to $\mathbf{v}$ in the sense that*

$$\| \mathbf{v} - \mathbf{u} \|_2 \geq \| \mathbf{v} - \mathbf{w} \|_2, \quad \mathbf{u} \in \mathcal{U}. \tag{4.39}$$

**Proof.** Let $\mathbf{w}$ be the projection of $\mathbf{v}$ onto $\mathcal{U}$ and let $\mathbf{u}$ be an arbitrary signal in $\mathcal{U}$. Since, by the definition of projection, $\mathbf{w}$ is in $\mathcal{U}$ and since $\mathcal{U}$ is a linear subspace, it follows that $\mathbf{w} - \mathbf{u} \in \mathcal{U}$. Consequently, since by the definition of the projection

---

[2]A projection can also be defined if the subspace does not have an orthonormal basis, but in this case there is a uniqueness issue. There may be numerous vectors $\mathbf{w} \in \mathcal{U}$ such that $\mathbf{v} - \mathbf{w}$ is orthogonal to all vectors in $\mathcal{U}$. Fortunately, they are all indistinguishable.

$\mathbf{v} - \mathbf{w}$ is orthogonal to every element of $\mathcal{U}$, it follows that $\mathbf{v} - \mathbf{w}$ is *a fortiori* orthogonal to $\mathbf{w} - \mathbf{u}$. Thus

$$\|\mathbf{v} - \mathbf{u}\|_2^2 = \|(\mathbf{v} - \mathbf{w}) + (\mathbf{w} - \mathbf{u})\|_2^2$$
$$= \|\mathbf{v} - \mathbf{w}\|_2^2 + \|\mathbf{w} - \mathbf{u}\|_2^2 \tag{4.40}$$
$$\geq \|\mathbf{v} - \mathbf{w}\|_2^2 , \tag{4.41}$$

where the first equality follows by subtracting and adding $\mathbf{w}$, the second equality from the orthogonality of $(\mathbf{v} - \mathbf{w})$ and $(\mathbf{w} - \mathbf{u})$, and the final equality by the nonnegativity of $\|\cdot\|_2$. It follows from (4.41) that no signal in $\mathcal{U}$ is closer to $\mathbf{v}$ than $\mathbf{w}$ is. And it follows from (4.40) that if $\mathbf{u} \in \mathcal{U}$ is as close to $\mathbf{v}$ as $\mathbf{w}$ is, then $\mathbf{u} - \mathbf{w}$ must be an element of $\mathcal{U}$ that is of zero energy. We shall see in Proposition 4.6.10 that the hypothesis that $\mathcal{U}$ has an orthonormal basis implies that the only zero-energy element of $\mathcal{U}$ is $\mathbf{0}$. Thus $\mathbf{u}$ and $\mathbf{w}$ must be identical, and no other element of $\mathcal{U}$ is as close to $\mathbf{v}$ as $\mathbf{w}$ is. $\qquad\square$

### 4.6.4 Energy, Inner Products, and Orthonormal Bases

As demonstrated by Proposition 4.6.4, if $(\boldsymbol{\phi}_1, \ldots, \boldsymbol{\phi}_d)$ forms an orthonormal basis for the subspace $\mathcal{U} \subset \mathcal{L}_2$, then any signal $\mathbf{u} \in \mathcal{U}$ can be reconstructed from the $d$ numbers $\langle \mathbf{u}, \boldsymbol{\phi}_1 \rangle, \ldots, \langle \mathbf{u}, \boldsymbol{\phi}_d \rangle$. Any quantity that can be computed from $\mathbf{u}$ can thus be computed from $\langle \mathbf{u}, \boldsymbol{\phi}_1 \rangle, \ldots, \langle \mathbf{u}, \boldsymbol{\phi}_d \rangle$ by first reconstructing $\mathbf{u}$ and by then performing the calculation on $\mathbf{u}$. But some calculations involving $\mathbf{u}$ can be performed based on $\langle \mathbf{u}, \boldsymbol{\phi}_1 \rangle, \ldots, \langle \mathbf{u}, \boldsymbol{\phi}_d \rangle$ much more easily.

**Proposition 4.6.9.** *Let $(\boldsymbol{\phi}_1, \ldots, \boldsymbol{\phi}_d)$ be an orthonormal basis for the linear subspace $\mathcal{U} \subset \mathcal{L}_2$.*

(i) *The energy $\|\mathbf{u}\|_2^2$ of every $\mathbf{u} \in \mathcal{U}$ can be expressed in terms of the $d$ inner products $\langle \mathbf{u}, \boldsymbol{\phi}_1 \rangle, \ldots, \langle \mathbf{u}, \boldsymbol{\phi}_d \rangle$ as*

$$\|\mathbf{u}\|_2^2 = \sum_{\ell=1}^{d} |\langle \mathbf{u}, \boldsymbol{\phi}_\ell \rangle|^2. \tag{4.42}$$

(ii) *More generally, if $\mathbf{v} \in \mathcal{L}_2$ (not necessarily in $\mathcal{U}$), then*

$$\|\mathbf{v}\|_2^2 \geq \sum_{\ell=1}^{d} |\langle \mathbf{v}, \boldsymbol{\phi}_\ell \rangle|^2 \tag{4.43}$$

*with equality if, and only if, $\mathbf{v}$ is indistinguishable from some signal in $\mathcal{U}$.*

(iii) *The inner product between any $\mathbf{v} \in \mathcal{L}_2$ and any $\mathbf{u} \in \mathcal{U}$ can be expressed in terms of the inner products $\{\langle \mathbf{v}, \boldsymbol{\phi}_\ell \rangle\}$ and $\{\langle \mathbf{u}, \boldsymbol{\phi}_\ell \rangle\}$ as*

$$\langle \mathbf{v}, \mathbf{u} \rangle = \sum_{\ell=1}^{d} \langle \mathbf{v}, \boldsymbol{\phi}_\ell \rangle \langle \mathbf{u}, \boldsymbol{\phi}_\ell \rangle^*. \tag{4.44}$$

**Proof.** Part (i) follows directly from the Pythagorean Theorem (Theorem 4.5.2) applied to the $d$-tuple $(\langle \mathbf{u}, \boldsymbol{\phi}_1 \rangle \boldsymbol{\phi}_1, \ldots, \langle \mathbf{u}, \boldsymbol{\phi}_d \rangle \boldsymbol{\phi}_d)$.

To prove Part (ii) we expand the energy in $\mathbf{v}$ as

$$\|\mathbf{v}\|_2^2 = \left\| \left( \mathbf{v} - \sum_{\ell=1}^{d} \langle \mathbf{v}, \boldsymbol{\phi}_\ell \rangle \boldsymbol{\phi}_\ell \right) + \sum_{\ell=1}^{d} \langle \mathbf{v}, \boldsymbol{\phi}_\ell \rangle \boldsymbol{\phi}_\ell \right\|_2^2$$

$$= \left\| \mathbf{v} - \sum_{\ell=1}^{d} \langle \mathbf{v}, \boldsymbol{\phi}_\ell \rangle \boldsymbol{\phi}_\ell \right\|_2^2 + \left\| \sum_{\ell=1}^{d} \langle \mathbf{v}, \boldsymbol{\phi}_\ell \rangle \boldsymbol{\phi}_\ell \right\|_2^2$$

$$= \left\| \mathbf{v} - \sum_{\ell=1}^{d} \langle \mathbf{v}, \boldsymbol{\phi}_\ell \rangle \boldsymbol{\phi}_\ell \right\|_2^2 + \sum_{\ell=1}^{d} |\langle \mathbf{v}, \boldsymbol{\phi}_\ell \rangle|^2$$

$$\geq \sum_{\ell=1}^{d} |\langle \mathbf{v}, \boldsymbol{\phi}_\ell \rangle|^2, \tag{4.45}$$

where the first equality follows by subtracting and adding the projection of $\mathbf{v}$ onto $\mathcal{U}$; the second from the Pythagorean Theorem and by Lemma 4.6.5, which guarantees that the difference between $\mathbf{v}$ and its projection is orthogonal to any signal in $\mathcal{U}$ and hence *a fortiori* also to the projection itself; the third by Part (i) applied to the projection of $\mathbf{v}$ onto $\mathcal{U}$; and the final inequality by the nonnegativity of energy.

If Inequality (4.45) holds with equality, then the last inequality in its derivation must hold with equality, so $\left\| \mathbf{v} - \sum_{\ell=1}^{d} \langle \mathbf{v}, \boldsymbol{\phi}_\ell \rangle \boldsymbol{\phi}_\ell \right\|_2 = 0$ and hence $\mathbf{v}$ must be indistinguishable from the signal $\sum_{\ell=1}^{d} \langle \mathbf{v}, \boldsymbol{\phi}_\ell \rangle \boldsymbol{\phi}_\ell$, which is in $\mathcal{U}$.

Conversely, if $\mathbf{v}$ is indistinguishable from some $\mathbf{u}' \in \mathcal{U}$, then

$$\|\mathbf{v}\|_2^2 = \|(\mathbf{v} - \mathbf{u}') + \mathbf{u}'\|_2^2$$

$$= \|\mathbf{v} - \mathbf{u}'\|_2^2 + \|\mathbf{u}'\|_2^2$$

$$= \|\mathbf{u}'\|_2^2$$

$$= \sum_{\ell=1}^{d} |\langle \mathbf{u}', \boldsymbol{\phi}_\ell \rangle|^2$$

$$= \sum_{\ell=1}^{d} |\langle \mathbf{v}, \boldsymbol{\phi}_\ell \rangle + \langle \mathbf{u}' - \mathbf{v}, \boldsymbol{\phi}_\ell \rangle|^2$$

$$= \sum_{\ell=1}^{d} |\langle \mathbf{v}, \boldsymbol{\phi}_\ell \rangle|^2,$$

where the first equality follows by subtracting and adding $\mathbf{u}'$; the second follows from the Pythagorean Theorem because the fact that $\|\mathbf{v} - \mathbf{u}'\|_2 = 0$ implies that $\langle \mathbf{v} - \mathbf{u}', \mathbf{u}' \rangle = 0$ (as can be readily verified using the Cauchy-Schwarz Inequality $|\langle \mathbf{v} - \mathbf{u}', \mathbf{u}' \rangle| \leq \|\mathbf{v} - \mathbf{u}'\|_2 \|\mathbf{u}'\|_2$); the third from our assumption that $\mathbf{v}$ and $\mathbf{u}'$ are indistinguishable; the fourth from Part (i) applied to the function $\mathbf{u}'$ (which is in $\mathcal{U}$); the fifth by adding and subtracting $\mathbf{v}$; and where the final equality follows because

$\langle \mathbf{u}' - \mathbf{v}, \boldsymbol{\phi}_\ell \rangle = 0$ (as can be readily verified from the Cauchy Schwarz Inequality $|\langle \mathbf{u}' - \mathbf{v}, \boldsymbol{\phi}_\ell \rangle| \leq \|\mathbf{u}' - \mathbf{v}\|_2 \|\boldsymbol{\phi}_\ell\|_2$).

To prove Part (iii) we compute $\langle \mathbf{v}, \mathbf{u} \rangle$ as

$$\langle \mathbf{v}, \mathbf{u} \rangle = \left\langle \mathbf{v} - \sum_{\ell=1}^{d} \langle \mathbf{v}, \boldsymbol{\phi}_\ell \rangle \boldsymbol{\phi}_\ell + \sum_{\ell=1}^{d} \langle \mathbf{v}, \boldsymbol{\phi}_\ell \rangle \boldsymbol{\phi}_\ell, \mathbf{u} \right\rangle$$

$$= \left\langle \mathbf{v} - \sum_{\ell=1}^{d} \langle \mathbf{v}, \boldsymbol{\phi}_\ell \rangle \boldsymbol{\phi}_\ell, \mathbf{u} \right\rangle + \left\langle \sum_{\ell=1}^{d} \langle \mathbf{v}, \boldsymbol{\phi}_\ell \rangle \boldsymbol{\phi}_\ell, \mathbf{u} \right\rangle$$

$$= \left\langle \sum_{\ell=1}^{d} \langle \mathbf{v}, \boldsymbol{\phi}_\ell \rangle \boldsymbol{\phi}_\ell, \mathbf{u} \right\rangle$$

$$= \sum_{\ell=1}^{d} \langle \mathbf{v}, \boldsymbol{\phi}_\ell \rangle \langle \boldsymbol{\phi}_\ell, \mathbf{u} \rangle$$

$$= \sum_{\ell=1}^{d} \langle \mathbf{v}, \boldsymbol{\phi}_\ell \rangle \langle \mathbf{u}, \boldsymbol{\phi}_\ell \rangle^*,$$

where the first equality follows by subtracting and adding $\sum_{\ell=1}^{d} \langle \mathbf{v}, \boldsymbol{\phi}_\ell \rangle \boldsymbol{\phi}_\ell$; the second by the linearity of the inner product in its left argument (3.9); the third because, by Lemma 4.6.5, the signal $\mathbf{v} - \sum_{\ell=1}^{d} \langle \mathbf{v}, \boldsymbol{\phi}_\ell \rangle \boldsymbol{\phi}_\ell$ is orthogonal to any signal in $\mathcal{U}$ and *a fortiori* to $\mathbf{u}$; the fourth by the linearity of the inner product in its left argument (3.7) & (3.9); and the final equality by (3.6). $\square$

Proposition 4.6.9 has interesting consequences. It shows that if one thinks of $\langle \mathbf{u}, \boldsymbol{\phi}_\ell \rangle$ as the $\ell$-th coordinate of $\mathbf{u}$ (with respect to the orthonormal basis $(\boldsymbol{\phi}_1, \ldots, \boldsymbol{\phi}_d)$), then the energy in $\mathbf{u}$ is simply the sum of the squares of the coordinates, and the inner product between two functions is the sum of the products of each coordinate of $\mathbf{u}$ and the conjugate of the corresponding coordinate of $\mathbf{v}$.

We hope that the properties of orthonormal bases that we presented above have convinced the reader by now that there are certain advantages to describing functions using an orthonormal basis. A crucial question arises as to whether orthonormal bases always exist. This question is addressed next.

### 4.6.5 Does an Orthonormal Basis Exist?

Word on the street has it that every finite-dimensional subspace of $\mathcal{L}_2$ has an orthonormal basis, but this is not true. (It is true for the space $L_2$ that we shall encounter later.) For example, the set

$$\left\{ \mathbf{u} \in \mathcal{L}_2 : u(t) = 0 \quad \text{whenever } t \neq 17 \right\}$$

of all energy-limited signals that map $t$ to zero whenever $t \neq 17$ (with the value to which $t = 17$ is mapped being unspecified) is a one dimensional subspace of $\mathcal{L}_2$ that does not have an orthonormal basis. (All the signals in this subspace are of zero energy, so there are no unit-energy signals in it.)

**Proposition 4.6.10.** *If $\mathcal{U}$ is a finite-dimensional subspace of $\mathcal{L}_2$, then the following two statements are equivalent:*

*(a) $\mathcal{U}$ has an orthonormal basis.*

*(b) The only element of $\mathcal{U}$ of zero energy is the all-zero signal $\mathbf{0}$.*

**Proof.** The proof has two parts. The first consists of showing that (a) $\Rightarrow$ (b), i.e., that if $\mathcal{U}$ has an orthonormal basis and if $\mathbf{u} \in \mathcal{U}$ is of zero energy, then $\mathbf{u}$ must be the all-zero signal $\mathbf{0}$. The second part consists of showing that (b) $\Rightarrow$ (a), i.e., that if the only element of zero energy in $\mathcal{U}$ is the all-zero signal $\mathbf{0}$, then $\mathcal{U}$ has an orthonormal basis.

We begin with the first part, namely, (a) $\Rightarrow$ (b). We thus assume that $(\boldsymbol{\phi}_1, \ldots, \boldsymbol{\phi}_d)$ is an orthonormal basis for $\mathcal{U}$ and that $\mathbf{u} \in \mathcal{U}$ satisfies $\|\mathbf{u}\|_2 = 0$ and proceed to prove that $\mathbf{u} = \mathbf{0}$. We simply note that, by the Cauchy-Schwarz Inequality, $|\langle \mathbf{u}, \boldsymbol{\phi}_\ell \rangle| \leq \|\mathbf{u}\|_2 \|\boldsymbol{\phi}_\ell\|_2$ so the condition $\|\mathbf{u}\|_2 = 0$ implies

$$\langle \mathbf{u}, \boldsymbol{\phi}_\ell \rangle = 0, \quad \ell \in \{1, \ldots, d\}, \tag{4.46}$$

and hence, by Proposition 4.6.4, that $\mathbf{u} = \mathbf{0}$.

To show (b) $\Rightarrow$ (a) we need to show that if no signal in $\mathcal{U}$ other than $\mathbf{0}$ has zero energy, then $\mathcal{U}$ has an orthonormal basis. The proof is based on the Gram-Schmidt Procedure, which is presented next. As we shall prove, if the input to this procedure is a basis for $\mathcal{U}$ and if no element of $\mathcal{U}$ other than $\mathbf{0}$ is of energy zero, then the procedure produces an orthonormal basis for $\mathcal{U}$. The procedure is actually even more powerful. If it is fed a basis for a subspace that *does* contain an element other than $\mathbf{0}$ of zero-energy, then the procedure produces such an element and halts.

It should be emphasized that the Gram-Schmidt Procedure is not only useful for proving theorems; it can be quite useful for finding orthonormal bases for practical problems.[3]                                                                                          $\square$

## 4.6.6    The Gram-Schmidt Procedure

The Gram-Schmidt Procedure is named after the mathematicians Jørgen Pedersen Gram (1850–1916) and Erhard Schmidt (1876–1959). However, as pointed out in (Farebrother, 1988), this procedure was apparently already presented by Pierre-Simon Laplace (1749–1827) and was used by Augustin Louis Cauchy (1789–1857).

The input to the Gram-Schmidt Procedure is a basis $(\mathbf{u}_1, \ldots, \mathbf{u}_d)$ for a $d$-dimensional subspace $\mathcal{U} \subset \mathcal{L}_2$. We assume that $d \geq 1$. (The only 0-dimensional subspace of $\mathcal{L}_2$ is the subspace $\{\mathbf{0}\}$ containing the all-zero signal only, and for this subspace the empty tuple is an orthonormal basis; there is not much else to say here.) If $\mathcal{U}$ does not contain a signal of zero energy other than the all-zero signal $\mathbf{0}$, then the procedure runs in $d$ steps and produces an orthonormal basis for $\mathcal{U}$ (and thus also proves that $\mathcal{U}$ does not contain a zero-energy signal other than $\mathbf{0}$). Otherwise, the

---

[3]Numerically, however, it is unstable; see (Golub and van Loan, 1996).

procedure stops after $d$ or fewer steps and produces an element of $\mathcal{U}$ of zero energy other than $\mathbf{0}$.

**The Gram-Schmidt Procedure**:

> **Step 1**: If $\|\mathbf{u}_1\|_2 = 0$, then the procedure declares that there exists a zero-energy element of $\mathcal{U}$ other than $\mathbf{0}$, it produces $\mathbf{u}_1$ as proof, and it halts. Otherwise, it defines
>
> $$\boldsymbol{\phi}_1 = \frac{\mathbf{u}_1}{\|\mathbf{u}_1\|_2}$$
>
> and halts with the output $(\boldsymbol{\phi}_1)$ (if $d = 1$) or proceeds to Step 2 (if $d > 1$).
>
> Assuming that the procedure has run for $\nu - 1$ steps without halting and has defined the vectors $\boldsymbol{\phi}_1, \ldots, \boldsymbol{\phi}_{\nu-1}$, we next describe Step $\nu$.
>
> **Step $\nu$**: Consider the signal
>
> $$\tilde{\mathbf{u}}_\nu = \mathbf{u}_\nu - \sum_{\ell=1}^{\nu-1} \langle \mathbf{u}_\nu, \boldsymbol{\phi}_\ell \rangle \, \boldsymbol{\phi}_\ell. \tag{4.47}$$
>
> If $\|\tilde{\mathbf{u}}_\nu\|_2 = 0$, then the procedure declares that there exists a zero-energy element of $\mathcal{U}$ other than $\mathbf{0}$, it produces $\tilde{\mathbf{u}}_\nu$ as proof, and it halts. Otherwise, the procedure defines
>
> $$\boldsymbol{\phi}_\nu = \frac{\tilde{\mathbf{u}}_\nu}{\|\tilde{\mathbf{u}}_\nu\|_2} \tag{4.48}$$
>
> and halts with the output $(\boldsymbol{\phi}_1, \ldots, \boldsymbol{\phi}_d)$ (if $\nu$ is equal to $d$) or proceeds to Step $\nu + 1$ (if $\nu < d$).

We next prove that the procedure behaves as we claim.

**Proof.** To prove that the procedure behaves as we claim, we shall assume that the procedure performs Step $\nu$ (i.e., that it has not halted in the steps preceding $\nu$) and prove the following: if at Step $\nu$ the procedure declares that $\mathcal{U}$ contains a nonzero signal of zero-energy and produces $\tilde{\mathbf{u}}_\nu$ as proof, then this is indeed the case; otherwise, if it defines $\boldsymbol{\phi}_\nu$ as in (4.48), then $(\boldsymbol{\phi}_1, \ldots, \boldsymbol{\phi}_\nu)$ is an orthonormal basis for $\mathrm{span}(\mathbf{u}_1, \ldots, \mathbf{u}_\nu)$.

We prove this by induction on $\nu$. For $\nu = 1$ this can be verified as follows. If $\|\mathbf{u}_1\|_2 = 0$, then we need to show that $\mathbf{u}_1 \in \mathcal{U}$ and that it is not equal to $\mathbf{0}$. This follows from the assumption that the procedure's input $(\mathbf{u}_1, \ldots, \mathbf{u}_d)$ forms a basis for $\mathcal{U}$, so *a fortiori* the signals $\mathbf{u}_1, \ldots, \mathbf{u}_d$ must all be elements of $\mathcal{U}$ and neither of them can be the all-zero signal. If $\|\mathbf{u}_1\|_2 > 0$, then $\boldsymbol{\phi}_1$ is a unit-energy scaled version of $\mathbf{u}_1$ and thus $(\boldsymbol{\phi}_1)$ is an orthonormal basis for $\mathrm{span}(\mathbf{u}_1)$.

We now assume that our claim is true for $\nu - 1$ and proceed to prove that it is also true for $\nu$. We thus assume that Step $\nu$ is executed and that $(\boldsymbol{\phi}_1, \ldots, \boldsymbol{\phi}_{\nu-1})$ is an orthonormal basis for $\mathrm{span}(\mathbf{u}_1, \ldots, \mathbf{u}_{\nu-1})$:

$$\boldsymbol{\phi}_1, \ldots, \boldsymbol{\phi}_{\nu-1} \in \mathcal{U}; \tag{4.49}$$

$$\text{span}(\boldsymbol{\phi}_1, \ldots, \boldsymbol{\phi}_{\nu-1}) = \text{span}(\mathbf{u}_1, \ldots, \mathbf{u}_{\nu-1}); \qquad (4.50)$$

and

$$\langle \boldsymbol{\phi}_\ell, \boldsymbol{\phi}_{\ell'} \rangle = \mathrm{I}\{\ell = \ell'\}, \quad \ell, \ell' \in \{1, \ldots, \nu - 1\}. \qquad (4.51)$$

We need to prove that if $\tilde{\mathbf{u}}_\nu$ is of zero energy, then it is a nonzero element of $\mathcal{U}$ of zero energy, and that otherwise the $\nu$-tuple $(\boldsymbol{\phi}_1, \ldots, \boldsymbol{\phi}_\nu)$ is an orthonormal basis for $\text{span}(\mathbf{u}_1, \ldots, \mathbf{u}_\nu)$. To that end we first prove that

$$\tilde{\mathbf{u}}_\nu \in \mathcal{U} \qquad (4.52)$$

and that

$$\tilde{\mathbf{u}}_\nu \neq \mathbf{0}. \qquad (4.53)$$

We begin with a proof of (4.52). Since (4.47) expresses $\tilde{\mathbf{u}}_\nu$ as a linear combination of $(\boldsymbol{\phi}_1, \ldots, \boldsymbol{\phi}_{\nu-1}, \mathbf{u}_\nu)$, and since $\mathcal{U}$ is by assumption a linear subspace, it suffices to show that $\boldsymbol{\phi}_1, \ldots, \boldsymbol{\phi}_{\nu-1} \in \mathcal{U}$ and that $\mathbf{u}_\nu \in \mathcal{U}$. The former follows from (4.49) and the latter from our assumption that $(\mathbf{u}_1, \ldots, \mathbf{u}_d)$ forms a basis for $\mathcal{U}$.

We next prove (4.53). By (4.47) it suffices to show that $\mathbf{u}_\nu \notin \text{span}(\boldsymbol{\phi}_1, \ldots, \boldsymbol{\phi}_{\nu-1})$. By (4.50) this is equivalent to showing that $\mathbf{u}_\nu \notin \text{span}(\mathbf{u}_1, \ldots, \mathbf{u}_{\nu-1})$, which follows from our assumption that $(\mathbf{u}_1, \ldots, \mathbf{u}_d)$ is a basis for $\mathcal{U}$ and *a fortiori* linearly independent.

Having established (4.52) and (4.53) it follows that if $\|\tilde{\mathbf{u}}_\nu\|_2 = 0$, then $\tilde{\mathbf{u}}_\nu$ is a nonzero element of $\mathcal{U}$ which is of zero-energy as we had claimed.

To conclude the proof we now assume $\|\tilde{\mathbf{u}}_\nu\|_2 > 0$ and prove that $(\boldsymbol{\phi}_1, \ldots, \boldsymbol{\phi}_\nu)$ is an orthonormal basis for $\text{span}(\mathbf{u}_1, \ldots, \mathbf{u}_\nu)$. That $(\boldsymbol{\phi}_1, \ldots, \boldsymbol{\phi}_\nu)$ is orthonormal follows because (4.51) guarantees that $(\boldsymbol{\phi}_1, \ldots, \boldsymbol{\phi}_{\nu-1})$ is orthonormal; because (4.48) guarantees that $\boldsymbol{\phi}_\nu$ is of unit energy; and because Lemma 4.6.5 (applied to the linear subspace $\text{span}(\boldsymbol{\phi}_1, \ldots, \boldsymbol{\phi}_{\nu-1})$) guarantees that $\tilde{\mathbf{u}}_\nu$—and hence also its scaled version $\boldsymbol{\phi}_\nu$—is orthogonal to every element of $\text{span}(\boldsymbol{\phi}_1, \ldots, \boldsymbol{\phi}_{\nu-1})$ and in particular to $\boldsymbol{\phi}_1, \ldots, \boldsymbol{\phi}_{\nu-1}$. It thus only remains to show that $\text{span}(\boldsymbol{\phi}_1, \ldots, \boldsymbol{\phi}_\nu) = \text{span}(\mathbf{u}_1, \ldots, \mathbf{u}_\nu)$. We first show that $\text{span}(\boldsymbol{\phi}_1, \ldots, \boldsymbol{\phi}_\nu) \subseteq \text{span}(\mathbf{u}_1, \ldots, \mathbf{u}_\nu)$. This follows because (4.50) implies that

$$\boldsymbol{\phi}_1, \ldots, \boldsymbol{\phi}_{\nu-1} \in \text{span}(\mathbf{u}_1, \ldots, \mathbf{u}_{\nu-1}); \qquad (4.54)$$

because (4.54), (4.47) and (4.48) imply that

$$\boldsymbol{\phi}_\nu \in \text{span}(\mathbf{u}_1, \ldots, \mathbf{u}_\nu); \qquad (4.55)$$

and because (4.54) and (4.55) imply that $\boldsymbol{\phi}_1, \ldots, \boldsymbol{\phi}_\nu \in \text{span}(\mathbf{u}_1, \ldots, \mathbf{u}_\nu)$ and hence that $\text{span}(\boldsymbol{\phi}_1, \ldots, \boldsymbol{\phi}_\nu) \subseteq \text{span}(\mathbf{u}_1, \ldots, \mathbf{u}_\nu)$. The reverse inclusion can be argued very similarly: by (4.50)

$$\mathbf{u}_1, \ldots, \mathbf{u}_{\nu-1} \in \text{span}(\boldsymbol{\phi}_1, \ldots, \boldsymbol{\phi}_{\nu-1}); \qquad (4.56)$$

by (4.47) and (4.48) we can express $\mathbf{u}_\nu$ as a linear combination of $(\boldsymbol{\phi}_1, \ldots, \boldsymbol{\phi}_\nu)$

$$\mathbf{u}_\nu = \|\tilde{\mathbf{u}}_\nu\|_2 \, \boldsymbol{\phi}_\nu + \sum_{\ell=1}^{\nu-1} \langle \mathbf{u}_\nu, \boldsymbol{\phi}_\ell \rangle \, \boldsymbol{\phi}_\ell; \qquad (4.57)$$

and (4.56) & (4.57) combine to prove that $\mathbf{u}_1, \ldots, \mathbf{u}_\nu \in \text{span}(\boldsymbol{\phi}_1, \ldots, \boldsymbol{\phi}_\nu)$ and hence that $\text{span}(\mathbf{u}_1, \ldots, \mathbf{u}_\nu) \subseteq \text{span}(\boldsymbol{\phi}_1, \ldots, \boldsymbol{\phi}_\nu)$. $\qquad \square$

By far the more important scenario for us is when $\mathcal{U}$ does not contain a nonzero element of zero energy. This is because we shall mostly focus on signals that are bandlimited (see Chapter 6), and the only energy-limited signal that is bandlimited to $W$ Hz and that has zero-energy is the all-zero signal (Note 6.4.2). For subspaces not containing zero-energy signals other than $\mathbf{0}$ the key properties to note about the signals $\phi_1, \ldots, \phi_d$ produced by the Gram-Schmidt procedure are that they satisfy for each $\nu \in \{1, \ldots, d\}$

$$\text{span}(\mathbf{u}_1, \ldots, \mathbf{u}_\nu) = \text{span}(\phi_1, \ldots, \phi_\nu) \tag{4.58a}$$

and

$$\big(\phi_1, \ldots, \phi_\nu\big) \text{ is an orthonormal basis for } \text{span}(\mathbf{u}_1, \ldots, \mathbf{u}_\nu). \tag{4.58b}$$

These properties are, of course, of greatest importance when $\nu = d$.

We next provide an example of the Gram-Schmidt procedure.

**Example 4.6.11.** Consider the following three signals: $\mathbf{u}_1 \colon t \mapsto \mathrm{I}\{0 \leq t \leq 1\}$, $\mathbf{u}_2 \colon t \mapsto t\,\mathrm{I}\{0 \leq t \leq 1\}$, and $\mathbf{u}_3 \colon t \mapsto t^2\,\mathrm{I}\{0 \leq t \leq 1\}$. The tuple $(\mathbf{u}_1, \mathbf{u}_2, \mathbf{u}_3)$ forms a basis for the subspace of all signals of the form $t \mapsto p(t)\,\mathrm{I}\{0 \leq t \leq 1\}$, where $p(\cdot)$ is a polynomial of degree smaller than 3. To construct an orthonormal basis for this subspace with the Gram-Schmidt Procedure, we begin by normalizing $\mathbf{u}_1$. To that end, we compute

$$\|\mathbf{u}_1\|_2^2 = \int_{-\infty}^{\infty} |\mathrm{I}\{0 \leq t \leq 1\}|^2 \,\mathrm{d}t = 1$$

and set $\phi_1 = \mathbf{u}_1 / \|\mathbf{u}_1\|_2$, so

$$\phi_1 \colon t \mapsto \mathrm{I}\{0 \leq t \leq 1\}. \tag{4.59a}$$

The second function $\phi_2$ is now obtained by normalizing $\mathbf{u}_2 - \langle \mathbf{u}_2, \phi_1 \rangle \phi_1$. We first compute the inner product $\langle \mathbf{u}_2, \phi_1 \rangle$

$$\langle \mathbf{u}_2, \phi_1 \rangle = \int_{-\infty}^{\infty} \mathrm{I}\{0 \leq t \leq 1\}\, t\, \mathrm{I}\{0 \leq t \leq 1\} \,\mathrm{d}t = \int_0^1 t \,\mathrm{d}t = \frac{1}{2}$$

to obtain that $\mathbf{u}_2 - \langle \mathbf{u}_2, \phi_1 \rangle \phi_1 \colon t \mapsto (t - 1/2)\,\mathrm{I}\{0 \leq t \leq 1\}$, which is of energy

$$\|\mathbf{u}_2 - \langle \mathbf{u}_2, \phi_1 \rangle \phi_1\|_2^2 = \int_0^1 \left(t - \frac{1}{2}\right)^2 \mathrm{d}t = \frac{1}{12}.$$

Hence,

$$\phi_2 \colon t \mapsto \sqrt{12} \left(t - \frac{1}{2}\right) \mathrm{I}\{0 \leq t \leq 1\}. \tag{4.59b}$$

The third function $\phi_3$ is the normalized version of $\mathbf{u}_3 - \langle \mathbf{u}_3, \phi_1 \rangle \phi_1 - \langle \mathbf{u}_3, \phi_2 \rangle \phi_2$. The inner products $\langle \mathbf{u}_3, \phi_1 \rangle$ and $\langle \mathbf{u}_3, \phi_2 \rangle$ are respectively

$$\langle \mathbf{u}_3, \phi_1 \rangle = \int_0^1 t^2 \,\mathrm{d}t = \frac{1}{3},$$

$$\langle \mathbf{u}_3, \phi_2 \rangle = \int_0^1 t^2 \sqrt{12} \left(t - \frac{1}{2}\right) \mathrm{d}t = \frac{1}{\sqrt{12}}.$$

Consequently

$$\mathbf{u}_3 - \langle \mathbf{u}_3, \phi_1 \rangle \, \phi_1 - \langle \mathbf{u}_3, \phi_2 \rangle \, \phi_2 : t \mapsto \left( t^2 - \frac{1}{3} - \left( t - \frac{1}{2} \right) \right) \mathrm{I}\{0 \leq t \leq 1\}$$

with corresponding energy

$$\left\| \mathbf{u}_3 - \langle \mathbf{u}_3, \phi_1 \rangle \, \phi_1 - \langle \mathbf{u}_3, \phi_2 \rangle \, \phi_2 \right\|_2^2 = \int_0^1 \left( t^2 - t + \frac{1}{6} \right)^2 \mathrm{d}t = \frac{1}{180}.$$

Hence, the orthonormal basis is completed by the third function

$$\phi_3 : t \mapsto \sqrt{180} \left( t^2 - t + \frac{1}{6} \right) \mathrm{I}\{0 \leq t \leq 1\}. \tag{4.59c}$$

## 4.7   The Space $L_2$

Very informally one can describe the space $L_2$ as the space of all energy-limited complex-valued signals, where we think of two signals as being different only if they are distinguishable. This section defines $L_2$ more precisely. It can be skipped because we shall have only little to do with $L_2$. Understanding this space is, however, important for readers who wish to fully understand how the Fourier Transform is defined for energy-limited signals that are not integrable (Section 6.2.3). Readers who continue should recall from Section 2.5 that two energy-limited signals $\mathbf{u}$ and $\mathbf{v}$ are said to be **indistinguishable** if the set $\{t \in \mathbb{R} : u(t) \neq v(t)\}$ is of Lebesgue measure zero. We write $\mathbf{u} \equiv \mathbf{v}$ to indicate that $\mathbf{u}$ and $\mathbf{v}$ are indistinguishable. By Proposition 2.5.3, the condition $\mathbf{u} \equiv \mathbf{v}$ is equivalent to the condition $\| \mathbf{u} - \mathbf{v} \|_2 = 0$.

To motivate the definition of the space $L_2$, we begin by noting that the space $\mathcal{L}_2$ of energy-limited signals is "almost" an example of what mathematicians call an "inner product space," but it is not. The problem is that mathematicians insist that in an inner product space the only vector whose inner product with itself is zero be the zero vector. This is not the case in $\mathcal{L}_2$: it is possible that $\mathbf{u} \in \mathcal{L}_2$ satisfy $\langle \mathbf{u}, \mathbf{u} \rangle = 0$ (i.e., $\| \mathbf{u} \|_2 = 0$) and yet not be the all-zero signal $\mathbf{0}$. From the condition $\| \mathbf{u} \|_2 = 0$ we can only infer that $\mathbf{u}$ is indistinguishable from $\mathbf{0}$.

The fact that $\mathcal{L}_2$ is not an inner product space is an annoyance because it precludes us from borrowing from the vast literature on inner product spaces (and Hilbert spaces, which are special kinds of inner product spaces), and because it does not allow us to view some of the results about $\mathcal{L}_2$ as instances of more general principles. For this reason mathematicians prefer to study the space $L_2$, which *is* an inner product space (and which is, in fact, a Hilbert space) rather than $\mathcal{L}_2$. Unfortunately, for this luxury they pay a certain price that I am loath to pay. Consequently, in most of this book I have decided to stick to $\mathcal{L}_2$ even though this precludes me from using the standard results on inner product spaces. The price one pays for using $L_2$ will become apparent once we define it.

To understand how $L_2$ is constructed it is useful to note that the relation "$\mathbf{u} \equiv \mathbf{v}$", i.e., "$\mathbf{u}$ is indistinguishable from $\mathbf{v}$" is an **equivalence relation** on $\mathcal{L}_2$, i.e., it satisfies

$$\mathbf{u} \equiv \mathbf{u}, \quad \mathbf{u} \in \mathcal{L}_2; \tag{\textbf{reflexive}}$$

$$\Big(\mathbf{u} \equiv \mathbf{v}\Big) \Leftrightarrow \Big(\mathbf{v} \equiv \mathbf{u}\Big), \quad \mathbf{u}, \mathbf{v} \in \mathcal{L}_2; \qquad \textbf{(symmetric)}$$

and

$$\Big(\mathbf{u} \equiv \mathbf{v} \text{ and } \mathbf{v} \equiv \mathbf{w}\Big) \Rightarrow \Big(\mathbf{u} \equiv \mathbf{w}\Big), \quad \mathbf{u}, \mathbf{v}, \mathbf{w} \in \mathcal{L}_2. \qquad \textbf{(transitive)}$$

Using these properties one can verify that if for every $\mathbf{u} \in \mathcal{L}_2$ we define its **equivalence class** $[\mathbf{u}]$ as

$$[\mathbf{u}] \triangleq \{\tilde{\mathbf{u}} \in \mathcal{L}_2 : \tilde{\mathbf{u}} \equiv \mathbf{u}\}, \tag{4.60}$$

then two equivalence classes $[\mathbf{u}]$ and $[\mathbf{v}]$ must be either identical or disjoint. In fact, the sets $[\mathbf{u}] \subset \mathcal{L}_2$ and $[\mathbf{v}] \subset \mathcal{L}_2$ are identical if, and only if, $\mathbf{u}$ and $\mathbf{v}$ are indistinguishable

$$\Big([\mathbf{u}] = [\mathbf{v}]\Big) \Leftrightarrow \Big(\|\mathbf{u} - \mathbf{v}\|_2 = 0\Big), \quad \mathbf{u}, \mathbf{v} \in \mathcal{L}_2,$$

and they are disjoint if, and only if, $\mathbf{u}$ and $\mathbf{v}$ are distinguishable

$$\Big([\mathbf{u}] \cap [\mathbf{v}] = \emptyset\Big) \Leftrightarrow \Big(\|\mathbf{u} - \mathbf{v}\|_2 > 0\Big), \quad \mathbf{u}, \mathbf{v} \in \mathcal{L}_2.$$

We define $L_2$ as the set of all such equivalence classes

$$L_2 \triangleq \big\{[\mathbf{u}] : \mathbf{u} \in \mathcal{L}_2\big\}. \tag{4.61}$$

Thus, the elements of $L_2$ are not functions, but sets of functions. Each element of $L_2$ is an equivalence class, i.e., a set of the form $[\mathbf{u}]$ for some $\mathbf{u} \in \mathcal{L}_2$. And for each $\mathbf{u} \in \mathcal{L}_2$ the equivalence class $[\mathbf{u}]$ is an element of $L_2$.

As we next show, the space $L_2$ can also be viewed as a vector space. To this end we need to first define "amplification of an equivalence class by a scalar $\alpha \in \mathbb{C}$" and "superposition of two equivalence classes." How do we define the scaling-by-$\alpha$ of an equivalence class $\mathcal{S} \in L_2$? A natural approach is to find some function $\mathbf{u} \in \mathcal{L}_2$ such that $\mathcal{S}$ is its equivalence class (i.e., satisfying $\mathcal{S} = [\mathbf{u}]$), and to define the scaling-by-$\alpha$ of $\mathcal{S}$ as the equivalence class of $\alpha\mathbf{u}$, i.e., as $[\alpha\mathbf{u}]$. Thus we would define $\alpha\mathcal{S}$ as the equivalence class of the signal $t \mapsto \alpha u(t)$. While this turns out to be a good approach, the careful reader might be concerned by something. Suppose that $\mathcal{S} = [\mathbf{u}]$ but that also $\mathcal{S} = [\tilde{\mathbf{u}}]$. Should $\alpha\mathcal{S}$ be defined as the equivalence class of $t \mapsto \alpha u(t)$ or of $t \mapsto \alpha\tilde{u}(t)$? Fortunately, it does not matter because the two equivalence classes are the same! Indeed, if $[\mathbf{u}] = [\tilde{\mathbf{u}}]$, then the equivalence class of $t \mapsto \alpha u(t)$ is equal to the equivalence class of $t \mapsto \alpha\tilde{u}(t)$ (because $[\mathbf{u}] = [\tilde{\mathbf{u}}]$ implies that $\mathbf{u}$ and $\tilde{\mathbf{u}}$ agree except on a set of measure zero so $\alpha\mathbf{u}$ and $\alpha\tilde{\mathbf{u}}$ also agree except on a set of measure zero, which in turn implies that $[\alpha\mathbf{u}] = [\alpha\tilde{\mathbf{u}}]$).

Similarly, one can show that if $\mathcal{S}_1 \in L_2$ and $\mathcal{S}_2 \in L_2$ are two equivalence classes, then we can define their sum (or superposition) $\mathcal{S}_1 + \mathcal{S}_2$ as $[\mathbf{u}_1 + \mathbf{u}_2]$ where $\mathbf{u}_1$ is any function in $\mathcal{L}_2$ such that $\mathcal{S}_1 = [\mathbf{u}_1]$ and where $\mathbf{u}_2$ is any function in $\mathcal{L}_2$ such that $\mathcal{S}_2 = [\mathbf{u}_2]$. Again, to make sure that the result of the superposition of $\mathcal{S}_1$ and $\mathcal{S}_2$ does not depend on the choice of $\mathbf{u}_1$ and $\mathbf{u}_2$ we need to verify that if $\mathcal{S}_1 = [\mathbf{u}_1] = [\tilde{\mathbf{u}}_1]$ and if $\mathcal{S}_2 = [\mathbf{u}_2] = [\tilde{\mathbf{u}}_2]$ then $[\mathbf{u}_1 + \mathbf{u}_2] = [\tilde{\mathbf{u}}_1 + \tilde{\mathbf{u}}_2]$. This is not difficult but is omitted.

Using these definitions and by defining the zero vector to be the equivalence class $[\mathbf{0}]$, it is not difficult to show that $L_2$ forms a linear space over the complex field. To make it into an inner product space we need to define the inner product $\langle \mathcal{S}_1, \mathcal{S}_2 \rangle$ between two equivalence classes. If $\mathcal{S}_1 = [\mathbf{u}_1]$ and if $\mathcal{S}_2 = [\mathbf{u}_2]$ we define the inner product $\langle \mathcal{S}_1, \mathcal{S}_2 \rangle$ as the complex number $\langle \mathbf{u}_1, \mathbf{u}_2 \rangle$. Again, we have to show that our definition is good in the sense that it does not depend on the particular choice of $\mathbf{u}_1$ and $\mathbf{u}_2$. More specifically, we need to verify that if $\mathcal{S}_1 = [\mathbf{u}_1] = [\tilde{\mathbf{u}}_1]$ and if $\mathcal{S}_2 = [\mathbf{u}_2] = [\tilde{\mathbf{u}}_2]$ then $\langle \mathbf{u}_1, \mathbf{u}_2 \rangle = \langle \tilde{\mathbf{u}}_1, \tilde{\mathbf{u}}_2 \rangle$. This can be proved as follows:

$$
\begin{aligned}
\langle \mathbf{u}_1, \mathbf{u}_2 \rangle &= \langle \tilde{\mathbf{u}}_1 + (\mathbf{u}_1 - \tilde{\mathbf{u}}_1), \mathbf{u}_2 \rangle \\
&= \langle \tilde{\mathbf{u}}_1, \mathbf{u}_2 \rangle + \langle \mathbf{u}_1 - \tilde{\mathbf{u}}_1, \mathbf{u}_2 \rangle \\
&= \langle \tilde{\mathbf{u}}_1, \mathbf{u}_2 \rangle \\
&= \langle \tilde{\mathbf{u}}_1, \tilde{\mathbf{u}}_2 + (\mathbf{u}_2 - \tilde{\mathbf{u}}_2) \rangle \\
&= \langle \tilde{\mathbf{u}}_1, \tilde{\mathbf{u}}_2 \rangle + \langle \tilde{\mathbf{u}}_1, \mathbf{u}_2 - \tilde{\mathbf{u}}_2 \rangle \\
&= \langle \tilde{\mathbf{u}}_1, \tilde{\mathbf{u}}_2 \rangle,
\end{aligned}
$$

where the third equality follows because $[\mathbf{u}_1] = [\tilde{\mathbf{u}}_1]$ implies that $\|\mathbf{u}_1 - \tilde{\mathbf{u}}_1\|_2 = 0$ and hence that $\langle \mathbf{u}_1 - \tilde{\mathbf{u}}_1, \mathbf{u}_2 \rangle = 0$ (Cauchy-Schwarz Inequality), and where the last equality follows by a similar reasoning about $\mathbf{u}_2$ and $\tilde{\mathbf{u}}_2$. Using the above definition of the inner product between equivalence classes one can show that if for some equivalence class $\mathcal{S}$ we have $\langle \mathcal{S}, \mathcal{S} \rangle = 0$, then $\mathcal{S}$ is the zero vector, i.e., the equivalence class $[\mathbf{0}]$.

With these definitions of the scaling of an equivalence class by a scalar, the superposition of two equivalence classes, and the inner product between two equivalence classes, the space of equivalence classes $L_2$ becomes an inner product space in the sense that mathematicians like. In fact, it is a Hilbert space.

What is the price we have to pay for working in an inner product space? It is that the elements of $L_2$ are not functions but equivalence classes and that it is meaningless to talk about the value they take at a given time. For example, it is meaningless to discuss the supremum (or maximum) of an element of $L_2$.[4] To add to the confusion, mathematicians refer to elements of $L_2$ as "functions" (even though they are equivalence classes of functions), and they drop the square brackets. Things get even trickier when one deals with signals contaminated by noise. If one views the signals as elements of $L_2$, then the result of adding noise to them is not a stochastic process (Definition 12.2.1 ahead). We find this price too high, and in this book we shall mostly deal with $\mathcal{L}_2$.

## 4.8   Additional Reading

Most of the results of this chapter follows from basic results on inner product spaces and can be found, for example, in (Axler, 1997). However, since $\mathcal{L}_2$ is not an inner-product space, we had to introduce some slight modifications.

---

[4]To deal with this, mathematicians define the **essential supremum**.

More on the definition of the space $L_2$ can be found in most texts on analysis. See, for example, (Rudin, 1974, Chapter 3, Remark 3.10) and (Royden, 1988, Chapter 1 Section 7).

## 4.9  Exercises

**Exercise 4.1 (Linear Subspace).** Consider the set of signals $\mathbf{u}$ of the form $\mathbf{u}\colon t \mapsto e^{-t^2} p(t)$, where $p(\cdot)$ is a polynomial whose degree does not exceed $d$. Is this a linear subspace of $\mathcal{L}_2$? If yes, find a basis for this subspace.

**Exercise 4.2 (Characterizing Infinite-Dimensional Subspaces).** Recall that we say that a linear subspace is infinite dimensional if it is not of finite dimension. Show that a linear subspace $\mathcal{U}$ is infinite dimensional if, and only if, there exists a sequence $\mathbf{u}_1, \mathbf{u}_2, \ldots$ of elements of $\mathcal{U}$ such that for every $n \in \mathbb{N}$ the tuple $(\mathbf{u}_1, \ldots, \mathbf{u}_n)$ is linearly independent.

**Exercise 4.3 ($\mathcal{L}_2$ Is Infinite Dimensional).** Show that $\mathcal{L}_2$ is infinite dimensional.

*Hint: Exercises 4.1 and 4.2 may be useful.*

**Exercise 4.4 (Separation between Signals).** Given $\mathbf{u}_1, \mathbf{u}_2 \in \mathcal{L}_2$, let $\mathcal{V}$ be the set of all complex signals $\mathbf{v}$ that are equidistant to $\mathbf{u}_1$ and $\mathbf{u}_2$:

$$\mathcal{V} = \big\{ \mathbf{v} \in \mathcal{L}_2 : \| \mathbf{v} - \mathbf{u}_1 \|_2 = \| \mathbf{v} - \mathbf{u}_2 \|_2 \big\}.$$

(i)  Show that

$$\mathcal{V} = \left\{ \mathbf{v} \in \mathcal{L}_2 : \mathrm{Re}\big( \langle \mathbf{v}, \mathbf{u}_2 - \mathbf{u}_1 \rangle \big) = \frac{\| \mathbf{u}_2 \|_2^2 - \| \mathbf{u}_1 \|_2^2}{2} \right\}.$$

(ii)  Is $\mathcal{V}$ a linear subspace of $\mathcal{L}_2$?

(iii)  Show that $(\mathbf{u}_1 + \mathbf{u}_2)/2 \in \mathcal{V}$.

**Exercise 4.5 (Projecting a Signal).** Let $\mathbf{u} \in \mathcal{L}_2$ be of positive energy, and let $\mathbf{v} \in \mathcal{L}_2$ be arbitrary.

(i)  Show that Definitions 4.6.6 and 4.5.3 agree in the sense that the projection of $\mathbf{v}$ onto $\mathrm{span}(\mathbf{u})$ (according to Definition 4.6.6) is the same as the projection of $\mathbf{v}$ onto the signal $\mathbf{u}$ (according to Definition 4.5.3).

(ii)  Show that if the signal $\mathbf{u}$ is an element of a finite-dimensional subspace $\mathcal{U}$ having an orthonormal basis, then the projection of $\mathbf{u}$ onto $\mathcal{U}$ is given by $\mathbf{u}$.

**Exercise 4.6 (Orthogonal Subspace).** Given signals $\mathbf{v}_1, \ldots, \mathbf{v}_n \in \mathcal{L}_2$, define the set

$$\mathcal{U} = \big\{ \mathbf{u} \in \mathcal{L}_2 : \langle \mathbf{u}, \mathbf{v}_1 \rangle = \langle \mathbf{u}, \mathbf{v}_2 \rangle = \cdots = \langle \mathbf{u}, \mathbf{v}_n \rangle = 0 \big\}.$$

Show that $\mathcal{U}$ is a linear subspace of $\mathcal{L}_2$.

**Exercise 4.7 (Constructing an Orthonormal Basis).** Let $\mathsf{T}_s$ be a positive constant. Consider the signals $\mathbf{s}_1\colon t \mapsto \mathrm{I}\{0 \le t \le \mathsf{T}_s/2\} - \mathrm{I}\{\mathsf{T}_s/2 < t \le \mathsf{T}_s\}$; $\mathbf{s}_2\colon t \mapsto \mathrm{I}\{0 \le t \le \mathsf{T}_s\}$; $\mathbf{s}_3\colon t \mapsto \mathrm{I}\{0 \le t \le \mathsf{T}_s/4\} + \mathrm{I}\{3\mathsf{T}_s/4 \le t \le \mathsf{T}_s\}$; and $\mathbf{s}_4\colon t \mapsto \mathrm{I}\{0 \le t \le \mathsf{T}_s/4\} - \mathrm{I}\{3\mathsf{T}_s/4 \le t \le \mathsf{T}_s\}$.

   (i) Plot $\mathbf{s}_1$, $\mathbf{s}_2$, $\mathbf{s}_3$, and $\mathbf{s}_4$.

  (ii) Find an orthonormal basis for span $(\mathbf{s}_1, \mathbf{s}_2, \mathbf{s}_3, \mathbf{s}_4)$.

 (iii) Express each of the signals $\mathbf{s}_1$, $\mathbf{s}_2$, $\mathbf{s}_3$, and $\mathbf{s}_4$ as a linear combination of the basis vectors found in Part (ii).

**Exercise 4.8 (Is the $\mathcal{L}_2$-Limit Unique?).** Show that for signals $\boldsymbol{\zeta}, \mathbf{x}_1, \mathbf{x}_2, \ldots$ in $\mathcal{L}_2$ the statement

$$\lim_{n \to \infty} \|\mathbf{x}_n - \boldsymbol{\zeta}\|_2 = 0$$

is equivalent to the statement

$$\left( \lim_{n \to \infty} \|\mathbf{x}_n - \tilde{\boldsymbol{\zeta}}\|_2 = 0 \right) \Leftrightarrow \left( \tilde{\boldsymbol{\zeta}} \in [\boldsymbol{\zeta}] \right).$$

**Exercise 4.9 (Signals of Zero Energy).** Given $\mathbf{v}_1, \ldots, \mathbf{v}_n \in \mathcal{L}_2$, show that there exist integers $1 \leq \nu_1 < \nu_2 < \cdots < \nu_d \leq n$ such that the following three conditions hold: the $d$-tuple $\left( \mathbf{v}_{\nu_1}, \ldots, \mathbf{v}_{\nu_d} \right)$ is linearly independent; span$(\mathbf{v}_{\nu_1}, \ldots, \mathbf{v}_{\nu_d})$ contains no signal of zero energy other than the all-zero signal $\mathbf{0}$; and each element of span$(\mathbf{v}_1, \ldots, \mathbf{v}_n)$ is indistinguishable from some element of span$(\mathbf{v}_{\nu_1}, \ldots, \mathbf{v}_{\nu_d})$.

**Exercise 4.10 (Orthogonal Subspace).** Given $\mathbf{v}_1, \ldots, \mathbf{v}_n \in \mathcal{L}_2$, define the set

$$\mathcal{U} = \left\{ \mathbf{u} \in \mathcal{L}_2 : \langle \mathbf{u}, \mathbf{v}_1 \rangle = \langle \mathbf{u}, \mathbf{v}_2 \rangle = \cdots = \langle \mathbf{u}, \mathbf{v}_n \rangle = 0 \right\},$$

and the set of all energy-limited signals that are orthogonal to all the signals in $\mathcal{U}$:

$$\mathcal{U}^{\perp} = \left\{ \mathbf{w} \in \mathcal{L}_2 : \left( \langle \mathbf{w}, \mathbf{u} \rangle = 0, \ \mathbf{u} \in \mathcal{U} \right) \right\}.$$

   (i) Show that $\mathcal{U}^{\perp}$ is a linear subspace of $\mathcal{L}_2$.

  (ii) Show that an energy-limited signal is in $\mathcal{U}^{\perp}$ if, and only if, it is indistinguishable from some element of span$(\mathbf{v}_1, \ldots, \mathbf{v}_n)$.

*Hint: For Part (ii) you may find Exercise 4.9 useful.*

**Exercise 4.11 (More on Indistinguishability).** Given $\mathbf{v}_1, \ldots, \mathbf{v}_n \in \mathcal{L}_2$ and some $\mathbf{w} \in \mathcal{L}_2$, propose an algorithm to check whether there exists an element of span$(\mathbf{v}_1, \ldots, \mathbf{v}_n)$ that is indistinguishable from $\mathbf{w}$.

*Hint: Exercise 4.9 may be useful.*

# Chapter 5

# Convolutions and Filters

## 5.1  Introduction

Convolutions play a central role in the analysis of linear systems, and it is thus not surprising that they will appear repeatedly in this book. Most of the readers have probably seen the definition and key properties in an earlier course on linear systems, so this chapter can be viewed as a very short review. New perhaps is the following section on notation and the all-important Section 5.8 on the matched filter and its use in calculating inner products.

## 5.2  Time Shifts and Reflections

Suppose that $\mathbf{x} \colon \mathbb{R} \to \mathbb{R}$ is a real signal, where we think of the argument as being time. Such functions are typically plotted on paper with the time arrow pointing to the right. Take a moment to plot an example of such a function, and on the same coordinates plot the function

$$t \mapsto x(t - t_0),$$

which maps every $t \in \mathbb{R}$ to $x(t - t_0)$ for some positive $t_0$. Repeat with $t_0$ being negative. This may seem like a mindless exercise but there is a point to it. It will help you understand convolutions graphically and help you visualize mappings such as $t \mapsto \sum_\ell \alpha_\ell \, g(t - \ell T_s)$, which we will encounter later in our study of Pulse Amplitude Modulation (PAM). It will also help you visualize the matched filter.

Given a complex signal $\mathbf{x} \colon \mathbb{R} \to \mathbb{C}$, we denote its **reflection** or **mirror image** by $\vec{\mathbf{x}}$:

$$\vec{\mathbf{x}} \colon t \mapsto x(-t). \tag{5.1}$$

Its plot is the mirror image of the plot of $x(\cdot)$ about the vertical axis. The mirror image of the mirror image of $\mathbf{x}$ is $\mathbf{x}$.

## 5.3   The Convolution Expression

The convolution $\mathbf{x} \star \mathbf{h}$ between two complex signals $\mathbf{x} \colon \mathbb{R} \to \mathbb{C}$ and $\mathbf{h} \colon \mathbb{R} \to \mathbb{C}$ is formally defined as the complex signal whose time-$t$ value $(\mathbf{x} \star \mathbf{h})(t)$ is given by

$$(\mathbf{x} \star \mathbf{h})(t) = \int_{-\infty}^{\infty} x(\tau)\, h(t - \tau)\, \mathrm{d}\tau. \tag{5.2}$$

Note that the integrand in the above is complex. (See Section 2.3 for a discussion of such integrals.) This definition also holds for real signals.

We used the term "formally defined" because certain conditions need to be met for this integral to be defined. It is conceivable that for some $t \in \mathbb{R}$ the integrand $\tau \mapsto x(\tau)\, h(t - \tau)$ will not be integrable, so the integral will be undefined. (Recall that in this book we only allow integrals of the form $\int_{-\infty}^{\infty} g(t)\, \mathrm{d}t$ if the integrand $g(\cdot)$ is in $\mathcal{L}_1$ so $\int_{-\infty}^{\infty} |g(t)|\, \mathrm{d}t < \infty$. Otherwise, we say that the integral $\int_{-\infty}^{\infty} g(t)\, \mathrm{d}t$ is undefined.) We thus say that $\mathbf{x} \star \mathbf{h}$ is defined at $t \in \mathbb{R}$ if $\tau \mapsto x(\tau)\, h(t - \tau)$ is integrable.

While (5.2) does not make it apparent, the convolution is in fact symmetric in $\mathbf{x}$ and $\mathbf{h}$. Thus, the integral in (5.2) is defined for a given $t$ if, and only if, the integral

$$\int_{-\infty}^{\infty} h(\sigma)\, x(t - \sigma)\, \mathrm{d}\sigma \tag{5.3}$$

is defined. And if both are defined, then their values are identical. This follows directly by the change of variable $\sigma \triangleq t - \tau$.

## 5.4   Thinking About the Convolution

Depending on the application, we can think about the convolution operation in a number of different ways.

(i)  Especially when $h(\cdot)$ is nonnegative and integrates to one, one can think of the convolution as an averaging, or smoothing, operation. Thus, when $\mathbf{x}$ is convolved with $\mathbf{h}$ the result at time $t_0$ is not $x(t_0)$ but rather a smoothed version thereof, namely, $\int_{-\infty}^{\infty} x(t_0 - \tau)\, h(\tau)\, \mathrm{d}\tau$. For example, if $\mathbf{h}$ is the mapping $t \mapsto \mathrm{I}\{|t| \leq \mathsf{T}/2\}/\mathsf{T}$ for some $\mathsf{T} > 0$, then the convolution $\mathbf{x} \star \mathbf{h}$ at time $t_0$ is not $x(t_0)$ but rather

$$\frac{1}{\mathsf{T}} \int_{t_0 - \mathsf{T}/2}^{t_0 + \mathsf{T}/2} x(\tau)\, \mathrm{d}\tau.$$

Thus, in this example, we can think of $\mathbf{x} \star \mathbf{h}$ as being a "moving average," or a "sliding-window average" of $\mathbf{x}$.

(ii)  For energy-limited signals it is sometimes beneficial to think about $(\mathbf{x} \star \mathbf{h})(t_0)$ as the inner product between the functions $\tau \mapsto x(\tau)$ and $\tau \mapsto h^*(t_0 - \tau)$:

$$(\mathbf{x} \star \mathbf{h})(t_0) = \langle \tau \mapsto x(\tau),\, \tau \mapsto h^*(t_0 - \tau) \rangle. \tag{5.4}$$

(iii) Another useful informal way is to think about $\mathbf{x} \star \mathbf{h}$ as a limit of expressions of the form

$$\sum_j h(t_j) \, x(t - t_j), \tag{5.5}$$

i.e., as a limit of linear combinations of the time shifts of $\mathbf{x}$ where the coefficients are determined by $\mathbf{h}$.

## 5.5 When Is the Convolution Defined?

There are a number of useful theorems providing sufficient conditions for the convolution's existence. These theorems can be classified into two kinds: those that guarantee that the convolution $\mathbf{x} \star \mathbf{h}$ is defined at every epoch $t \in \mathbb{R}$ and those that only guarantee that the convolution is defined for all epochs $t$ outside a set of Lebesgue measure zero. Both types are useful. We begin with the former.

**Convolution defined for every** $t \in \mathbb{R}$:

(i) A particularly simple case where the convolution is defined at every time instant $t$ is when both $\mathbf{x}$ and $\mathbf{h}$ are energy-limited:

$$\mathbf{x}, \mathbf{h} \in \mathcal{L}_2. \tag{5.6a}$$

In this case we can use (5.4) and the Cauchy-Schwarz Inequality (Theorem 3.3.1) to conclude that the integral in (5.2) is defined for every $t \in \mathbb{R}$ and that $\mathbf{x} \star \mathbf{h}$ is a bounded function with

$$\left| (\mathbf{x} \star \mathbf{h})(t) \right| \leq \|\mathbf{x}\|_2 \, \|\mathbf{h}\|_2, \quad t \in \mathbb{R}. \tag{5.6b}$$

Indeed,

$$\begin{aligned}
\left| (\mathbf{x} \star \mathbf{h})(t) \right| &= \left| \langle \tau \mapsto x(\tau), \tau \mapsto h^*(t - \tau) \rangle \right| \\
&\leq \|\tau \mapsto x(\tau)\|_2 \, \|\tau \mapsto h^*(t - \tau)\|_2 \\
&= \|\mathbf{x}\|_2 \, \|\mathbf{h}\|_2 \,.
\end{aligned}$$

In fact, it can be shown that the result of convolving two energy-limited signals is not only bounded but also uniformly continuous.[1] (See, for example, (Adams and Fournier, 2003, Paragraph 2.23).)

Note that even if both $\mathbf{x}$ and $\mathbf{h}$ are of finite energy, the convolution $\mathbf{x} \star \mathbf{h}$ need not be. However, if $\mathbf{x}, \mathbf{h}$ are both of finite energy and if one of them is additionally also integrable, then the convolution $\mathbf{x} \star \mathbf{h}$ is a finite energy signal. Indeed,

$$\|\mathbf{x} \star \mathbf{h}\|_2 \leq \|\mathbf{h}\|_1 \, \|\mathbf{x}\|_2, \quad \mathbf{h} \in \mathcal{L}_1 \cap \mathcal{L}_2, \quad \mathbf{x} \in \mathcal{L}_2. \tag{5.7}$$

For a proof see, for example, (Rudin, 1974, Chapter 7, Exercise 4) or (Stein and Weiss, 1990, Chapter 1, Section 1, Theorem 1.3).

---

[1] A function $\mathbf{s} \colon \mathbb{R} \to \mathbb{C}$ is said to be **uniformly continuous** if for every $\epsilon > 0$ there corresponds some positive $\delta(\epsilon)$ such that $|s(\xi') - s(\xi'')|$ is smaller than $\epsilon$ whenever $\xi', \xi'' \in \mathbb{R}$ are such that $|\xi' - \xi''| < \delta(\epsilon)$.

(ii) Another simple case where the convolution is defined at every epoch $t \in \mathbb{R}$ is when one of the functions is measurable and bounded and when the other is integrable. For example, if

$$\mathbf{h} \in \mathcal{L}_1 \tag{5.8a}$$

and if $\mathbf{x}$ is a Lebesgue measurable function that is bounded in the sense that

$$|x(t)| \leq \sigma_\infty, \quad t \in \mathbb{R} \tag{5.8b}$$

for some constant $\sigma_\infty$, then for every $t \in \mathbb{R}$ the integrand in (5.3) is integrable because $|h(\sigma)x(t-\sigma)| \leq |h(\sigma)|\sigma_\infty$, with the latter being integrable by our assumption that $\mathbf{h}$ is integrable. The result of the convolution is a bounded function because

$$
\begin{aligned}
|(\mathbf{x} \star \mathbf{h})(t)| &= \left| \int_{-\infty}^{\infty} h(\tau)\, x(t-\tau)\, \mathrm{d}\tau \right| \\
&\leq \int_{-\infty}^{\infty} \left| h(\tau)\, x(t-\tau) \right| \mathrm{d}\tau \\
&\leq \sigma_\infty \, \|\mathbf{h}\|_1 \,, \quad t \in \mathbb{R},
\end{aligned}
\tag{5.8c}
$$

where the first inequality follows from Proposition 2.4.1, and where the second inequality follows from (5.8b).

For this case too one can show that the result of the convolution is not only bounded but also uniformly continuous.

(iii) Using Hölder's Inequality, we can generalize the above two cases to show that whenever $\mathbf{x}$ and $\mathbf{h}$ satisfy the assumptions of Hölder's Inequality, their convolution is defined at every epoch $t \in \mathbb{R}$ and is, in fact, a bounded uniformly continuous function. See, for example, (Adams and Fournier, 2003, Paragraph 2.23).

(iv) Another important case where the convolution is defined at every time instant will be discussed in Proposition 6.2.5. There it is shown that the convolution between an integrable function (of time) with the Inverse Fourier Transform of an integrable function (of frequency) is defined at every time instant and has a simple representation. This scenario is not as contrived as the reader might suspect. It arises quite naturally, for example, when discussing the lowpass filtering of an integrable signal (Section 6.4.2). The impulse response of an ideal lowpass filter (LPF) is not integrable, but it can be represented as the Inverse Fourier Transform of an integrable function; see (6.35).

Regarding theorems that guarantee that the convolution be defined for every $t$ outside a set of Lebesgue measure zero, we mention two.

**Convolution defined for $t$ outside a set of Lebesgue measure zero:**

(i) If both $\mathbf{x}$ and $\mathbf{h}$ are integrable, then one can show (see, for example, (Rudin, 1974, Theorem 7.14), (Katznelson, 1976, Section VI.1), or (Stein and Weiss,

1990, Chapter 1, Section 1, Theorem 1.3)) that, for all $t$ outside a set of Lebesgue measure zero, the mapping $\tau \mapsto x(\tau)h(t-\tau)$ is integrable, so for all such $t$ the function $(\mathbf{x} \star \mathbf{h})(t)$ is defined. Moreover, irrespective of how we define $(\mathbf{x} \star \mathbf{h})(t)$ for $t$ inside the set of Lebesgue measure zero

$$\|\mathbf{x} \star \mathbf{h}\|_1 \leq \|\mathbf{x}\|_1 \|\mathbf{h}\|_1, \quad \mathbf{x}, \mathbf{h} \in \mathcal{L}_1. \tag{5.9}$$

What is nice about this case is that the result of the convolution stays in the same class of integrable functions. This makes it meaningful to discuss associativity and other important properties of the convolution.

(ii) Another case where the convolution is defined for all $t$ outside a set of Lebesgue measure zero is when $\mathbf{h}$ is integrable and when $\mathbf{x}$ is a measurable function for which $\tau \mapsto |x(\tau)|^p$ is integrable for some $1 \leq p < \infty$. In this case we have (see, for example, (Rudin, 1974, Exercise 7.4) or (Stein and Weiss, 1990, Chapter 1, Section 1, Theorem 1.3)) that for all $t$ outside a set of Lebesgue measure zero the mapping $\tau \mapsto x(\tau)h(t-\tau)$ is integrable so for such $t$ the convolution $(\mathbf{x} \star \mathbf{h})(t)$ is well-defined. Moreover, irrespective of how we define $(\mathbf{x} \star \mathbf{h})(t)$ for $t$ inside the set of Lebesgue measure zero

$$\left( \int_{-\infty}^{\infty} |(\mathbf{x} \star \mathbf{h})(t)|^p \, dt \right)^{1/p} \leq \|\mathbf{h}\|_1 \left( \int_{-\infty}^{\infty} |x(t)|^p \, dt \right)^{1/p}. \tag{5.10}$$

This is written more compactly as

$$\|\mathbf{x} \star \mathbf{h}\|_p \leq \|\mathbf{h}\|_1 \|\mathbf{x}\|_p, \quad p \geq 1, \tag{5.11}$$

where we use the notation that for any measurable function $\mathbf{g}$ and $p > 0$

$$\|\mathbf{g}\|_p \triangleq \left( \int_{-\infty}^{\infty} |g(t)|^p \, dt \right)^{1/p}. \tag{5.12}$$

## 5.6 Basic Properties of the Convolution

The main properties of the convolution are summarized in the following theorem.

**Theorem 5.6.1 (Properties of the Convolution).** *The convolution is*

$$\mathbf{x} \star \mathbf{h} \equiv \mathbf{h} \star \mathbf{x}, \qquad \textbf{(commutative)}$$

$$(\mathbf{x} \star \mathbf{g}) \star \mathbf{h} \equiv \mathbf{x} \star (\mathbf{g} \star \mathbf{h}), \qquad \textbf{(associative)}$$

$$\mathbf{x} \star (\mathbf{g} + \mathbf{h}) \equiv \mathbf{x} \star \mathbf{g} + \mathbf{x} \star \mathbf{h}, \qquad \textbf{(distributive)}$$

*and linear in each of its arguments*

$$\mathbf{x} \star (\alpha \mathbf{g} + \beta \mathbf{h}) \equiv \alpha(\mathbf{x} \star \mathbf{g}) + \beta(\mathbf{x} \star \mathbf{h})$$
$$(\alpha \mathbf{g} + \beta \mathbf{h}) \star \mathbf{x} \equiv \alpha(\mathbf{g} \star \mathbf{x}) + \beta(\mathbf{h} \star \mathbf{x}),$$

*where the above hold for all* $\mathbf{g}, \mathbf{h}, \mathbf{x} \in \mathcal{L}_1$, *and* $\alpha, \beta \in \mathbb{C}$.

Some of these properties hold under more general or different sets of assumptions so the reader should focus here on the properties rather than on the restrictions.

## 5.7   Filters

A **filter of impulse response h** is a physical device that when fed the input waveform **x** produces the output waveform **h ⋆ x**. The impulse response **h** is assumed to be a real or complex signal, and it is tacitly assumed that we only feed the device with inputs **x** for which the convolution **x ⋆ h** is defined.[2]

**Definition 5.7.1 (Stable Filter).** *A filter is said to be **stable** if its impulse response is integrable.*

Stable filters are also called **bounded-input/bounded-output stable** or **BIBO stable**, because, as the next proposition shows, if such filters are fed a bounded signal, then their output is also a bounded signal.

**Proposition 5.7.2 (BIBO Stability).** *If **h** is integrable and if **x** is a bounded Lebesgue measurable signal, then the signal **x ⋆ h** is also bounded.*

**Proof.** If the impulse response **h** is integrable, and if the input **x** is bounded by some constant $\sigma_\infty$, then (5.8a) and (5.8b) are both satisfied, and the boundedness of the output then follows from (5.8c). □

**Definition 5.7.3 (Causal Filter).** *A filter of impulse response **h** is said to be **causal** or **nonanticipative** if **h** is zero at negative times, i.e., if*

$$h(t) = 0, \quad t < 0. \tag{5.13}$$

Causal filters play an important role in engineering because (5.13) guarantees that the present filter output be computable from the past filter inputs. Indeed, the time-$t$ filter output can be expressed in the form

$$(\mathbf{x} \star \mathbf{h})(t) = \int_{-\infty}^{\infty} x(\tau)\, h(t - \tau)\, \mathrm{d}\tau$$

$$= \int_{-\infty}^{t} x(\tau)\, h(t - \tau)\, \mathrm{d}\tau, \quad \mathbf{h} \text{ causal,}$$

where the calculation of the latter integral only requires knowledge of $x(\tau)$ for $\tau < t$. Here the first equality follows from the definition of the convolution (5.2), and the second equality follows from (5.13).

## 5.8   The Matched Filter

In Digital Communications inner products are often computed using a **matched filter**. In its definition we shall use the notation (5.1).

---

[2]This definition of a filter is reminiscent of the concept of a "linear time invariant system." Note, however, that since we do not deal with Dirac's Delta in this book, our definition is more restrictive. For example, a device that produces at its output a waveform that is identical to its input is excluded from our discussion here because we do not allow **h** to be Dirac's Delta.

**Definition 5.8.1 (The Matched Filter).** *The **matched filter** for the signal $\phi$ is a filter whose impulse response is $\overleftarrow{\phi}^*$, i.e., the mapping*

$$t \mapsto \phi^*(-t). \tag{5.14}$$

The main use of the matched filter is for computing inner products:

**Theorem 5.8.2 (Computing Inner Products with a Matched Filter).** *The inner product $\langle \mathbf{u}, \phi \rangle$ between the energy-limited signals $\mathbf{u}$ and $\phi$ is given by the output at time $t = 0$ of a matched filter for $\phi$ that is fed $\mathbf{u}$:*

$$\langle \mathbf{u}, \phi \rangle = \left( \mathbf{u} \star \overleftarrow{\phi}^* \right)(0), \quad \mathbf{u}, \phi \in \mathcal{L}_2. \tag{5.15}$$

*More generally, if $\mathbf{g} \colon t \mapsto \phi(t - t_0)$, then $\langle \mathbf{u}, \mathbf{g} \rangle$ is the time-$t_0$ output corresponding to feeding the waveform $\mathbf{u}$ to the matched filter for $\phi$:*

$$\int_{-\infty}^{\infty} u(t)\, \phi^*(t - t_0)\, \mathrm{d}t = \left( \mathbf{u} \star \overleftarrow{\phi}^* \right)(t_0). \tag{5.16}$$

**Proof.** We shall prove the second part of the theorem, i.e., (5.16); the first follows from the second by setting $t_0 = 0$. We express the time-$t_0$ output of the matched filter as:

$$\left( \mathbf{u} \star \overleftarrow{\phi}^* \right)(t_0) = \int_{-\infty}^{\infty} u(\tau)\, \overleftarrow{\phi}^*(t_0 - \tau)\, \mathrm{d}\tau$$

$$= \int_{-\infty}^{\infty} u(\tau)\, \phi^*(\tau - t_0)\, \mathrm{d}\tau,$$

where the first equality follows from the definition of convolution (5.2) and the second from the definition of $\overleftarrow{\phi}^*$ as the conjugated mirror image of $\phi$. $\qquad\square$

From the above theorem we see that if we wish to compute, say, the three inner products $\langle \mathbf{u}, \mathbf{g}_1 \rangle$, $\langle \mathbf{u}, \mathbf{g}_2 \rangle$, and $\langle \mathbf{u}, \mathbf{g}_3 \rangle$ in the very special case where the functions $\mathbf{g}_1, \mathbf{g}_2, \mathbf{g}_3$ are all time shifts of the same waveform $\phi$, i.e., when $\mathbf{g}_1 \colon t \mapsto \phi(t - t_1)$, $\mathbf{g}_2 \colon t \mapsto \phi(t - t_2)$, and $\mathbf{g}_3 \colon t \mapsto \phi(t - t_3)$, then we need only one filter, namely, the matched filter for $\phi$. Indeed, we can feed $\mathbf{u}$ to the matched filter for $\phi$ and the inner products $\langle \mathbf{u}, \mathbf{g}_1 \rangle$, $\langle \mathbf{u}, \mathbf{g}_2 \rangle$, and $\langle \mathbf{u}, \mathbf{g}_3 \rangle$ simply correspond to the filter's outputs at times $t_1$, $t_2$, and $t_3$. One circuit computes all three inner products. This is so exciting that it is worth repeating:

**Corollary 5.8.3 (Computing Many Inner Products using One Filter).** *If the energy-limited signals $\{\mathbf{g}_j\}_{j=1}^{J}$ are all time shifts of the same signal $\phi$ in the sense that*

$$\mathbf{g}_j \colon t \mapsto \phi(t - t_j), \quad j = 1, \ldots, J,$$

*and if $\mathbf{u}$ is any energy-limited signal, then all $J$ inner products*

$$\langle \mathbf{u}, \mathbf{g}_j \rangle, \quad j = 1, \ldots, J$$

*can be computed using one filter by feeding* $\mathbf{u}$ *to a matched filter for* $\phi$ *and sampling the output at the appropriate times* $t_1, \ldots, t_J$:

$$\langle \mathbf{u}, \mathbf{g}_j \rangle = (\mathbf{u} \star \overleftarrow{\phi}^*)(t_j), \quad j = 1, \ldots, J. \tag{5.17}$$

## 5.9   The Ideal Unit-Gain Lowpass Filter

The impulse response of the **ideal unit-gain lowpass filter** of cutoff frequency $W_c$ is denoted by $\mathrm{LPF}_{W_c}(\cdot)$ and is given for every $W_c > 0$ by[3]

$$\mathrm{LPF}_{W_c}(t) \triangleq \begin{cases} 2W_c \frac{\sin(2\pi W_c t)}{2\pi W_c t} & \text{if } t \neq 0, \\ 2W_c & \text{if } t = 0, \end{cases} \quad t \in \mathbb{R}. \tag{5.18}$$

This can be alternatively written as

$$\mathrm{LPF}_{W_c}(t) = 2W_c \operatorname{sinc}(2W_c t), \quad t \in \mathbb{R}, \tag{5.19}$$

where the function $\operatorname{sinc}(\cdot)$ is defined by[4]

$$\operatorname{sinc}(\xi) \triangleq \begin{cases} \frac{\sin(\pi\xi)}{\pi\xi} & \text{if } \xi \neq 0, \\ 1 & \text{if } \xi = 0, \end{cases} \quad \xi \in \mathbb{R}. \tag{5.20}$$

Notice that the definition of $\operatorname{sinc}(0)$ as being 1 makes sense because, for very small (but nonzero) values of $\xi$ the value of $\sin(\xi)/\xi$ is approximately 1. In fact, with this definition at zero the function is not only continuous at zero but also infinitely differentiable there. Indeed, the function from $\mathbb{C}$ to $\mathbb{C}$

$$z \longmapsto \begin{cases} \frac{\sin(\pi z)}{\pi z} & \text{if } z \neq 0, \\ 1 & \text{otherwise,} \end{cases}$$

is an entire function, i.e., an analytic function throughout the complex plane.

The importance of the ideal unit-gain lowpass filter will become clearer when we discuss the filter's frequency response in Section 6.3. It is thus named because the Fourier Transform of $\mathrm{LPF}_{W_c}(\cdot)$ is equal to 1 (hence "unit gain"), whenever $|f| \leq W_c$, and is equal to zero, whenever $|f| > W_c$. See (6.38) ahead.

From a mathematical point of view, working with the ideal unit-gain lowpass filter is tricky because the impulse response (5.18) is not an integrable function. (It decays like $1/t$, which does not have a finite integral from $t = 1$ to $t = \infty$.) This filter is thus not a stable filter. We shall revisit this issue in Section 6.4. Note, however, that the impulse response (5.18) is of finite energy. (The square of the impulse response decays like $1/t^2$ which does have a finite integral from one to infinity.) Consequently, the result of feeding an energy-limited signal to the ideal unit-gain lowpass filter is always well-defined.

Note also that the ideal unit-gain lowpass filter is not causal.

---

[3]For convenience we define the impulse response of the ideal unit-gain lowpass filter of cutoff frequency zero as the all zero signal. This is in agreement with (5.19).

[4]Some texts omit the $\pi$'s in (5.20) and define the $\operatorname{sinc}(\cdot)$ function as $\sin(\xi)/\xi$ for $\xi \neq 0$.

## 5.10 The Ideal Unit-Gain Bandpass Filter

The ideal unit-gain bandpass filter (BPF) of bandwidth $W$ around the carrier frequency $f_c$, where $f_c > W/2 > 0$ is a filter of impulse response $\mathrm{BPF}_{W,f_c}(\cdot)$, where

$$\mathrm{BPF}_{W,f_c}(t) \triangleq 2W \cos(2\pi f_c t) \operatorname{sinc}(Wt), \quad t \in \mathbb{R}. \tag{5.21}$$

This filter too is nonstable and noncausal. It derives its name from its frequency response (discussed in Section 6.3 ahead), which is equal to one at frequencies $f$ satisfying $\big||f| - f_c\big| \leq W/2$ and which is equal to zero at all other frequencies.

## 5.11 Young's Inequality

Many of the inequalities regarding convolutions are special cases of a result known as Young's Inequality. Recalling (5.12), we can state Young's Inequality as follows.

**Theorem 5.11.1 (Young's Inequality).** *Let $\mathbf{x}$ and $\mathbf{h}$ be measurable functions such that $\|\mathbf{x}\|_p, \|\mathbf{h}\|_q < \infty$ for some $1 \leq p, q < \infty$ satisfying $1/p + 1/q > 1$. Define $r$ through $1/p + 1/q = 1 + 1/r$. Then the convolution integral (5.2) is defined for all $t$ outside a set of Lebesgue measure zero; it is a measurable function; and*

$$\|\mathbf{x} \star \mathbf{h}\|_r \leq \mathsf{K} \|\mathbf{x}\|_p \|\mathbf{h}\|_q, \tag{5.22}$$

*where $\mathsf{K} < 1$ is some constant that depends only on $p$ and $q$.*

**Proof.** See (Adams and Fournier, 2003, Corollary 2.25). Alternatively, see (Stein and Weiss, 1990, Chapter 5, Section 1) where it is derived from the M. Riesz Convexity Theorem. □

## 5.12 Additional Reading

For some of the properties of the convolution and its use in the analysis of linear systems see (Oppenheim and Willsky, 1997) and (Kwakernaak and Sivan, 1991).

## 5.13 Exercises

**Exercise 5.1 (Convolution of Delayed Signals).** Let $\mathbf{x}$ and $\mathbf{h}$ be energy-limited signals. Let $\mathbf{x}_d : t \mapsto x(t - t_d)$ be the result of delaying $\mathbf{x}$ by some $t_d \in \mathbb{R}$. Show that

$$\big(\mathbf{x}_d \star \mathbf{h}\big)(t) = \big(\mathbf{x} \star \mathbf{h}\big)(t - t_d), \quad t \in \mathbb{R}.$$

**Exercise 5.2 (The Convolution of Reflections).** Let the signals $\mathbf{x}, \mathbf{y}$ be such that their convolution $(\mathbf{x} \star \mathbf{y})(t)$ is defined at every $t \in \mathbb{R}$. Show that the convolution of their reflections is also defined at every $t \in \mathbb{R}$ and that it is equal to the reflection of their convolution:

$$\big(\bar{\mathbf{x}} \star \bar{\mathbf{y}}\big)(t) = \big(\mathbf{x} \star \mathbf{y}\big)(-t), \quad t \in \mathbb{R}.$$

**Exercise 5.3 (Convolving Brickwall Functions).** For a given $a > 0$, compute the convolution of the signal $t \mapsto I\{|t| \le a\}$ with itself.

**Exercise 5.4 (The Convolution and Inner Products).** Let $\mathbf{y}$ and $\phi$ be energy-limited complex signals, and let $\mathbf{h}$ be an integrable complex signal. Argue that

$$\left\langle \mathbf{y}, \mathbf{h} \star \phi \right\rangle = \left\langle \mathbf{y} \star \bar{\mathbf{h}}^*, \phi \right\rangle.$$

**Exercise 5.5 (The Convolution's Derivative).** Let the signal $\mathbf{g} \colon \mathbb{R} \to \mathbb{C}$ be differentiable, and let $\mathbf{g}'$ denote its derivative. Let $\mathbf{h} \colon \mathbb{R} \to \mathbb{C}$ be another signal. Assume that $\mathbf{g}$, $\mathbf{g}'$, and $\mathbf{h}$ are all bounded, continuous, and integrable. Show that $\mathbf{g} \star \mathbf{h}$ is differentiable and that its derivative $(\mathbf{g} \star \mathbf{h})'$ is given by $\mathbf{g}' \star \mathbf{h}$.

See (Körner, 1988, Chapter 53, Theorem 53.1).

**Exercise 5.6 (Continuity of the Convolution).** Show that if the signals $\mathbf{x}$ and $\mathbf{y}$ are both in $\mathcal{L}_2$ then their convolution is a continuous function.

*Hint: Use the Cauchy-Schwarz Inequality and the fact that if $\mathbf{x} \in \mathcal{L}_2$ and if we define $\mathbf{x}_\delta \colon t \mapsto x(t - \delta)$, then $\lim_{\delta \to 0} \|\mathbf{x} - \mathbf{x}_\delta\|_2 = 0$.*

**Exercise 5.7 (More on the Continuity of the Convolution).** Let $\mathbf{x}$ and $\mathbf{y}$ be in $\mathcal{L}_2$. Let the sequence of energy-limited signals $\mathbf{x}_1, \mathbf{x}_2, \ldots$ converge to $\mathbf{x}$ in the sense that $\|\mathbf{x} - \mathbf{x}_n\|_2$ tends to zero as $n$ tends to infinity. Show that at every epoch $t \in \mathbb{R}$,

$$\lim_{n \to \infty} \left( \mathbf{x}_n \star \mathbf{y} \right)(t) = \left( \mathbf{x} \star \mathbf{y} \right)(t).$$

*Hint: Use the Cauchy-Schwarz Inequality*

**Exercise 5.8 (Convolving Bi-Infinite Sequences).** The convolution of the bi-infinite sequence $\ldots, a_{-1}, a_0, a_1 \ldots$ with the bi-infinite sequence $\ldots, b_{-1}, b_0, b_1 \ldots$ is the bi-infinite sequence $\ldots, c_{-1}, c_0, c_1 \ldots$ formally defined by

$$c_m = \sum_{\nu = -\infty}^{\infty} a_\nu b_{m-\nu}, \quad m \in \mathbb{Z}. \tag{5.23}$$

Show that if

$$\sum_{\nu = -\infty}^{\infty} |a_\nu|, \ \sum_{\nu = -\infty}^{\infty} |b_\nu| < \infty,$$

then the sum on the RHS of (5.23) converges for every integer $m$, and

$$\sum_{m = -\infty}^{\infty} |c_m| \le \left( \sum_{\nu = -\infty}^{\infty} |a_\nu| \right) \left( \sum_{\nu = -\infty}^{\infty} |b_\nu| \right).$$

*Hint: Recall Problems 3.10 & 3.9 and the Triangle Inequality for Complex Numbers.*

**Exercise 5.9 (Stability of the Matched Filter).** Let $\mathbf{g}$ be an energy-limited signal. Under what conditions is the matched filter for $\mathbf{g}$ stable?

**Exercise 5.10 (Causality of the Matched Filter).** Let $\mathbf{g}$ be an energy-limited signal.

(i) Under what conditions is the matched filter for $\mathbf{g}$ causal?

(ii) Under what conditions can you find a causal filter of impulse response $\mathbf{h}$ and a sampling time $t_0$ such that

$$(\mathbf{r} \star \mathbf{h})(t_0) = \langle \mathbf{r}, \mathbf{g} \rangle, \quad \mathbf{r} \in \mathcal{L}_2?$$

(iii) Show that for every $\delta > 0$ we can find a stable causal filter of impulse response $\mathbf{h}$ and a sampling epoch $t_0$ such that for every $\mathbf{r} \in \mathcal{L}_2$

$$\left| (\mathbf{r} \star \mathbf{h})(t_0) - \langle \mathbf{r}, \mathbf{g} \rangle \right| \leq \delta \left\| \mathbf{r} \right\|_2.$$

**Exercise 5.11 (The Output of the Matched Filter).** Compute and plot the output of the matched filter for the signal $t \mapsto e^{-t} \, \mathrm{I}\{t \geq 0\}$ when it is fed the input $t \mapsto \mathrm{I}\{|t| \leq 1/2\}$.

# Chapter 6

# The Frequency Response of Filters and Bandlimited Signals

## 6.1 Introduction

We begin this chapter with a review of the Fourier Transform and its key properties. We then use these properties to define the frequency response of filters, to discuss the ideal unit-gain lowpass filter, and to define bandlimited signals.

## 6.2 Review of the Fourier Transform

### 6.2.1 On Hats, $2\pi$'s, $\omega$'s, and $f$'s

We denote the **Fourier Transform** (FT) of a (possibly complex) signal $x(\cdot)$ by $\hat{x}(\cdot)$. Some other books denote it by $X(\cdot)$, but we prefer our notation because, where possible, we use lowercase letters for deterministic quantities and reserve uppercase letters for random quantities. In places where convention forces us to use uppercase letters for deterministic quantities, we try to use a special font, e.g., P for power, W for bandwidth, or A for a deterministic matrix.

More importantly, our definition of the Fourier Transform may be different from the one you are used to.

**Definition 6.2.1 (Fourier Transform).** *The **Fourier Transform** (or the $\mathcal{L}_1$-**Fourier Transform**) of an integrable signal $\mathbf{x}\colon \mathbb{R} \to \mathbb{C}$ is the mapping $\hat{\mathbf{x}}\colon \mathbb{R} \to \mathbb{C}$ defined by*

$$\hat{\mathbf{x}}\colon f \mapsto \int_{-\infty}^{\infty} x(t)\, e^{-\mathrm{i}2\pi ft}\, \mathrm{d}t. \tag{6.1}$$

(The FT can also be defined in more general settings. For example, in Section 6.2.3 it will be defined via a limiting argument for finite-energy signals that are not integrable.)

64

This definition should be contrasted with the definition

$$X(\mathrm{i}\omega) = \int_{-\infty}^{\infty} x(t)\, e^{-\mathrm{i}\omega t}\, \mathrm{d}t, \tag{6.2}$$

which you may have seen before. Note the $2\pi$, which appears in the exponent in our definition (6.1) and not in (6.2). We apologize to readers who are used to (6.2) for forcing a new definition, but we have some good reasons:

(i) With our definition, the transform and its inverse are very similar; see (6.1) and (6.4) below. If one uses the definition of (6.2), then the expression for the Inverse Fourier Transform requires scaling the integral by $1/(2\pi)$.

(ii) With our definition, the Fourier Transform and the Inverse Fourier Transform of a symmetric function are the same; see (6.6). This simplifies the memorization of some Fourier pairs.

(iii) As we shall state more precisely in Section 6.2.2 and Section 6.2.3, with our definition the Fourier Transform possesses an extremely important property: it preserves inner products

$$\langle \mathbf{u}, \mathbf{v} \rangle = \langle \hat{\mathbf{u}}, \hat{\mathbf{v}} \rangle \quad \text{(certain restrictions apply).}$$

Again, no $2\pi$'s.

(iv) If $x(\cdot)$ models a function of time, then $\hat{x}(\cdot)$ becomes a function of frequency. Thus, it is natural to use the generic argument $t$ for such signals $x(\cdot)$ and the generic argument $f$ for their transforms. It is more common these days to describe tones in terms of their frequencies (i.e., in Hz) and not in terms of their radial frequency (in radians per second).

(v) It seems that all books on communications use our definition, perhaps because people are used to setting their radios in Hz, kHz, or MHz.

Plotting the FT of a signal is tricky, because it is a *complex-valued* function. This is generally true even for real signals. However, for any integrable *real* signal $\mathbf{x}\colon \mathbb{R} \to \mathbb{R}$ the Fourier Transform $\hat{x}(\cdot)$ is **conjugate-symmetric**, i.e.,

$$\left(\hat{x}(-f) = \hat{x}^*(f), \quad f \in \mathbb{R}\right), \quad \mathbf{x} \in \mathcal{L}_1 \text{ is real-valued.} \tag{6.3}$$

Equivalently, the magnitude of the FT of an integrable real signal is symmetric, and the argument is anti-symmetric.[1] (The reverse statement is "essentially" correct. If $\hat{\mathbf{x}}$ is conjugate-symmetric then the set of epochs $t$ for which $x(t)$ is not real is of Lebesgue measure zero.) Consequently, when plotting the FT of a "generic" real signal we shall plot a symmetric function, but with solid lines for the positive frequencies and dashed lines for the negative frequencies. This is to remind the reader that the FT of a real signal is not symmetric but *conjugate* symmetric. See, for example, Figures 7.1 and 7.2 for plots of the Fourier Transforms of real signals.

---

[1] The argument of a nonzero complex number $z$ is defined as the element $\theta$ of $[-\pi, \pi)$ such that $z = |z|\, e^{\mathrm{i}\theta}$.

When plotting the FT of a complex-valued signal, we shall use a generic plot that is "highly asymmetric," using solid lines. See, for example, Figure 7.4 for the FT of a complex signal.

**Definition 6.2.2 (Inverse Fourier Transform).** *The **Inverse Fourier Transform** (IFT) of an integrable function* $\mathbf{g} \colon \mathbb{R} \to \mathbb{C}$ *is denoted by* $\check{\mathbf{g}}$ *and is defined by*

$$\check{\mathbf{g}} \colon t \mapsto \int_{-\infty}^{\infty} g(f)\, e^{\mathrm{i}2\pi f t}\, \mathrm{d}f. \tag{6.4}$$

We emphasize that the word "inverse" here is just part of the name of the transform. Applying the IFT to the FT of a signal does not always recover the signal.[2] (Conditions under which the IFT does recover the signal are explored in Theorem 6.2.13.) However, if one does not insist on using the IFT, then every integrable signal can be reconstructed to within indistinguishability from its FT; see Theorem 6.2.12.

**Proposition 6.2.3 (Some Properties of the Inverse Fourier Transform).**

   (i) *If* $\mathbf{g}$ *is integrable, then its IFT is the FT of its mirror image*

$$\check{\mathbf{g}} = \hat{\bar{\mathbf{g}}}, \quad \mathbf{g} \in \mathcal{L}_1. \tag{6.5}$$

   (ii) *If* $\mathbf{g}$ *is integrable and also symmetric in the sense that* $\bar{\mathbf{g}} = \mathbf{g}$, *then the IFT of* $\mathbf{g}$ *is equal to its FT*

$$\hat{\mathbf{g}} = \check{\mathbf{g}}, \quad \left( \mathbf{g} \in \mathcal{L}_1 \ and \ \bar{\mathbf{g}} = \mathbf{g} \right). \tag{6.6}$$

   (iii) *If* $\mathbf{g}$ *is integrable and* $\check{\mathbf{g}}$ *is also integrable, then*

$$\hat{\check{\mathbf{g}}} = \check{\hat{\mathbf{g}}}. \tag{6.7}$$

**Proof.** Part (i) follows by a simple change of integration variable:

$$\check{g}(\xi) = \int_{-\infty}^{\infty} g(\alpha)\, e^{\mathrm{i}2\pi\alpha\xi}\, \mathrm{d}\alpha = -\int_{\infty}^{-\infty} g(-\beta)\, e^{-\mathrm{i}2\pi\beta\xi}\, \mathrm{d}\beta$$

$$= \int_{-\infty}^{\infty} \bar{g}(\beta)\, e^{-\mathrm{i}2\pi\beta\xi}\, \mathrm{d}\beta$$

$$= \hat{\bar{g}}(\xi), \quad \xi \in \mathbb{R},$$

where we have changed the integration variable to $\beta \triangleq -\alpha$.

---

[2] This can be seen by considering the signal $t \mapsto \mathrm{I}\{t = 17\}$, which is zero everywhere except at 17 where it takes on the value 1. Its FT is zero at all frequencies, but if one applies the IFT to the all-zero function one obtains the all-zero function, which is not the function we started with. Things could be much worse. The FT of some integrable signals (such as the signal $t \mapsto \mathrm{I}\{|t| \leq 1\}$) is not integrable, so the IFT of their FT is not even defined.

Part (ii) is a special case of Part (i). To prove Part (iii) we compute

$$\hat{\hat{g}}(\xi) = \int_{-\infty}^{\infty} \left( \int_{-\infty}^{\infty} g(f)\, e^{i2\pi ft}\, df \right) e^{-i2\pi\xi t}\, dt$$

$$= \int_{-\infty}^{\infty} \hat{g}(-t)\, e^{-i2\pi\xi t}\, dt$$

$$= \int_{-\infty}^{\infty} \hat{g}(\tau)\, e^{i2\pi\xi\tau}\, d\tau$$

$$= \check{\hat{g}}(\xi), \quad \xi \in \mathbb{R},$$

where we have changed the integration variable to $\tau \triangleq -t$. $\qquad\square$

Identity (6.6) will be useful in Section 6.2.5 when we memorize the FT of the Brickwall function $\xi \mapsto \beta\, \mathrm{I}\{|\xi| \leq \gamma\}$, which is symmetric. Once we succeed we will also know its IFT.

Table 6.1 summarizes some of the properties of the FT. Note that some of these properties require additional technical assumptions.

| Property | Function | Fourier Transform |
|---|---|---|
| linearity | $\alpha\mathbf{x} + \beta\mathbf{y}$ | $\alpha\hat{\mathbf{x}} + \beta\hat{\mathbf{y}}$ |
| time shifting | $t \mapsto x(t - t_0)$ | $f \mapsto e^{-i2\pi f t_0}\, \hat{x}(f)$ |
| frequency shifting | $t \mapsto e^{i2\pi f_0 t}\, x(t)$ | $f \mapsto \hat{x}(f - f_0)$ |
| conjugation | $t \mapsto x^*(t)$ | $f \mapsto \hat{x}^*(-f)$ |
| stretching ($\alpha \in \mathbb{R}, \alpha \neq 0$) | $t \mapsto x(\alpha t)$ | $f \mapsto \frac{1}{|\alpha|}\, \hat{x}(\frac{f}{\alpha})$ |
| convolution in time | $\mathbf{x} \star \mathbf{y}$ | $f \mapsto \hat{x}(f)\, \hat{y}(f)$ |
| multiplication in time | $t \mapsto x(t)\, y(t)$ | $\hat{\mathbf{x}} \star \hat{\mathbf{y}}$ |
| real part | $t \mapsto \mathrm{Re}\big(x(t)\big)$ | $f \mapsto \frac{1}{2}\hat{x}(f) + \frac{1}{2}\hat{x}^*(-f)$ |
| time reflection | $\overleftarrow{\mathbf{x}}$ | $\check{\mathbf{x}}$ |
| transforming twice | $\hat{\hat{\mathbf{x}}}$ | $\overleftarrow{\mathbf{x}}$ |
| FT of IFT | $\check{\mathbf{x}}$ | $\mathbf{x}$ |

**Table 6.1:** Basic properties of the Fourier Transform. *Some restrictions apply!*

## 6.2.2 Parseval-like Theorems

A key result on the Fourier Transform is that, subject to some restrictions, it preserves inner products. Thus, if $\hat{\mathbf{x}}_1$ and $\hat{\mathbf{x}}_2$ are the Fourier Transforms of $\mathbf{x}_1$ and $\mathbf{x}_2$, then the inner product $\langle \mathbf{x}_1, \mathbf{x}_2 \rangle$ between $\mathbf{x}_1$ and $\mathbf{x}_2$ is typically equal to the inner product $\langle \hat{\mathbf{x}}_1, \hat{\mathbf{x}}_2 \rangle$ between their transforms. In this section we shall describe two scenarios where this holds. A third scenario, which is described in Theorem 6.2.9, will have to wait until we discuss the FT of signals that are energy-limited but not integrable.

To see how the next proposition is related to the preservation of the inner product under the Fourier Transform, think about $\mathbf{g}$ as being a function of frequency and of its IFT $\check{\mathbf{g}}$ as a function of time.

**Proposition 6.2.4.** *If* $\mathbf{g} \colon f \mapsto g(f)$ *and* $\mathbf{x} \colon t \mapsto x(t)$ *are integrable mappings from* $\mathbb{R}$ *to* $\mathbb{C}$, *then*

$$\int_{-\infty}^{\infty} x(t)\,\check{g}^*(t)\,\mathrm{d}t = \int_{-\infty}^{\infty} \hat{x}(f)\,g^*(f)\,\mathrm{d}f, \tag{6.8}$$

*i.e.,*

$$\langle \mathbf{x}, \check{\mathbf{g}} \rangle = \langle \hat{\mathbf{x}}, \mathbf{g} \rangle , \quad \mathbf{g}, \mathbf{x} \in \mathcal{L}_1. \tag{6.9}$$

**Proof.** The key to the proof is to use Fubini's Theorem to justify changing the order of integration in the following calculation:

$$\int_{-\infty}^{\infty} x(t)\,\check{g}^*(t)\,\mathrm{d}t = \int_{-\infty}^{\infty} x(t) \left( \int_{-\infty}^{\infty} g(f)\, e^{\mathrm{i}2\pi ft}\,\mathrm{d}f \right)^* \mathrm{d}t$$

$$= \int_{-\infty}^{\infty} x(t) \int_{-\infty}^{\infty} g^*(f)\, e^{-\mathrm{i}2\pi ft}\,\mathrm{d}f\,\mathrm{d}t$$

$$= \int_{-\infty}^{\infty} g^*(f) \int_{-\infty}^{\infty} x(t)\, e^{-\mathrm{i}2\pi ft}\,\mathrm{d}t\,\mathrm{d}f$$

$$= \int_{-\infty}^{\infty} g^*(f)\,\hat{x}(f)\,\mathrm{d}f,$$

where the first equality follows from the definition of $\check{\mathbf{g}}$; the second because the conjugation of an integral is accomplished by conjugating the integrand (Proposition 2.3.1); the third by changing the order of integration; and the final equality by the definition of the FT of $\mathbf{x}$.  $\square$

A related result is that the convolution of an integrable function with the IFT of an integrable function is always defined:

**Proposition 6.2.5.** *If the mappings* $\mathbf{x} \colon t \mapsto x(t)$ *and* $\mathbf{g} \colon f \mapsto g(f)$ *from* $\mathbb{R}$ *to* $\mathbb{C}$ *are both integrable, then the convolution* $\mathbf{x} \star \check{\mathbf{g}}$ *is defined at every epoch* $t \in \mathbb{R}$ *and*

$$\left(\mathbf{x} \star \check{\mathbf{g}}\right)(t) = \int_{-\infty}^{\infty} g(f)\,\hat{x}(f)\, e^{\mathrm{i}2\pi ft}\,\mathrm{d}f, \quad t \in \mathbb{R}. \tag{6.10}$$

**Proof.** Here too the key is in changing the order of integration:

$$\left(\mathbf{x} \star \check{\mathbf{g}}\right)(t) = \int_{-\infty}^{\infty} x(\tau)\,\check{g}(t - \tau)\,\mathrm{d}\tau$$

$$= \int_{-\infty}^{\infty} x(\tau) \int_{-\infty}^{\infty} e^{\mathrm{i}2\pi f(t-\tau)}\, g(f)\,\mathrm{d}f\,\mathrm{d}\tau$$

$$= \int_{-\infty}^{\infty} g(f)\, e^{\mathrm{i}2\pi ft} \int_{-\infty}^{\infty} x(\tau)\, e^{-\mathrm{i}2\pi f\tau}\,\mathrm{d}\tau\,\mathrm{d}f$$

$$= \int_{-\infty}^{\infty} g(f)\,\hat{x}(f)\, e^{\mathrm{i}2\pi ft}\,\mathrm{d}f,$$

where the first equality follows from the definition of the convolution; the second from the definition of the IFT; the third by changing the order of integration; and the final equality by the definition of the FT. The justification of the changing of the order of integration can be argued using Fubini's Theorem because, by assumption, both **g** and **x** are integrable. □

We next present another useful version of the preservation of inner products under the FT. It is useful for functions (of time) that are zero outside some interval $[-T, T]$ or for the IFT of functions (of frequency) that are zero outside an interval $[-W, W]$.

**Proposition 6.2.6 (A Mini Parseval Theorem).**

(i) Let the signals $\mathbf{x}_1$ and $\mathbf{x}_2$ be given by

$$x_\nu(t) = \int_{-\infty}^{\infty} g_\nu(f) e^{i2\pi ft} \, df, \quad \left( t \in \mathbb{R}, \, \nu = 1, 2 \right), \tag{6.11a}$$

where the functions $\mathbf{g}_\nu : f \mapsto g_\nu(f)$ satisfy

$$g_\nu(f) = 0, \quad \left( |f| > W, \, \nu = 1, 2 \right), \tag{6.11b}$$

for some $W \geq 0$, and

$$\int_{-\infty}^{\infty} |g_\nu(f)|^2 \, df < \infty, \quad \nu = 1, 2. \tag{6.11c}$$

Then

$$\langle \mathbf{x}_1, \mathbf{x}_2 \rangle = \langle \mathbf{g}_1, \mathbf{g}_2 \rangle. \tag{6.11d}$$

(ii) Let $\mathbf{g}_1$ and $\mathbf{g}_2$ be given by

$$g_\nu(f) = \int_{-\infty}^{\infty} x_\nu(t) e^{-i2\pi ft} \, dt, \quad \left( f \in \mathbb{R}, \, \nu = 1, 2 \right), \tag{6.12a}$$

where the signals $\mathbf{x}_1, \mathbf{x}_2 \in \mathcal{L}_2$ are such that for some $T \geq 0$

$$x_\nu(t) = 0, \quad \left( |t| > T, \, \nu = 1, 2 \right). \tag{6.12b}$$

Then

$$\langle \mathbf{x}_1, \mathbf{x}_2 \rangle = \langle \mathbf{g}_1, \mathbf{g}_2 \rangle. \tag{6.12c}$$

**Proof.** See the proof of Lemma A.3.6 on Page 693 and its corollary in the appendix. □

### 6.2.3   The $L_2$-Fourier Transform

To appreciate some of the mathematical subtleties of this section, the reader is encouraged to review Section 4.7 in order to recall the difference between the space $L_2$ and the space $\mathcal{L}_2$ and in order to recall the difference between an energy-limited signal $\mathbf{x} \in \mathcal{L}_2$ and the equivalence class $[\mathbf{x}] \in L_2$ to which it belongs. In this section we shall sketch how the Fourier Transform is defined for elements of $L_2$. This section can be skipped provided that you are willing to take on faith that such a transform exists and that, very roughly speaking, it has some of the same properties of the Fourier Transform of Definition 6.2.1. To differentiate between the transform of Definition 6.2.1 and the transform that we are about to define for elements of $L_2$, we shall refer in this section to the former as the $\mathcal{L}_1$-**Fourier Transform** and to the latter as the $L_2$-**Fourier Transform**. Both will be denoted by a "hat." In subsequent sections the Fourier Transform will be understood to be the $\mathcal{L}_1$-Fourier Transform unless explicitly otherwise specified.

Some readers may have already encountered the $L_2$-Fourier Transform without even being aware of it. For example, the $\text{sinc}(\cdot)$ function, which is defined in (5.20), is an energy-limited signal that is not integrable. Consequently, its $\mathcal{L}_1$-Fourier Transform is undefined. Nevertheless, you may have seen its Fourier Transform being given as the Brickwall function. As we shall see, this is somewhat in line with how the $L_2$-Fourier Transform of the $\text{sinc}(\cdot)$ is defined.[3] For more on the Fourier Transform of the $\text{sinc}(\cdot)$ see Section 6.2.5. Another example of an energy-limited signal that is not integrable is $t \mapsto 1/(1 + |t|)$.

We next sketch how the $L_2$-Fourier Transform is defined and explore some of its key properties. We begin with the bad news.

    (i) There is no explicit simple expression for the $L_2$-Fourier Transform.

    (ii) The result of applying the transform is not a function but an equivalence class of functions.

The $L_2$-Fourier Transform is a mapping

$$\hat{} : L_2 \to L_2$$

that maps elements of $L_2$ to elements of $L_2$. It thus maps equivalence classes to equivalence classes, not functions. As long as the operation we perform on the result of the $L_2$-Fourier Transform does not depend on which member of the equivalence class it is performed on, there is no need to worry about this issue. Otherwise, we can end up performing operations that are ill-defined. For example, an operation that is ill-defined is evaluating the result of the transform at a given frequency, say at $f = 17$.

An operation you cannot go wrong with is integration, because the integrals of two functions that differ on a set of measure zero are equal; see Proposition 2.5.3. Consequently, inner products, which are defined via integration, are fine too. In

---

[3]However, as we shall see, the result of the $L_2$-Fourier Transform is an element of $L_2$, i.e., an equivalence class, and not a function.

this book we shall therefore refrain from applying to the result of the $L_2$-Fourier Transform any operation other than integration (or related operations such as the computation of energy or inner product). In fact, since we find the notion of equivalence classes somewhat abstract we shall try to minimize its use.

Suppose that $\mathbf{x} \in \mathcal{L}_2$ is an energy-limited signal and that $[\mathbf{x}] \in L_2$ is its equivalence class. How do we define the $L_2$-Fourier Transform of $[\mathbf{x}]$? We first define for every positive integer $n$ the time-truncated function

$$\mathbf{x}_n : t \mapsto x(t)\, \mathrm{I}\{|t| \leq n\}$$

and note that, by Proposition 3.4.3, $\mathbf{x}_n$ is integrable. Consequently, its $\mathcal{L}_1$-Fourier Transform $\hat{\mathbf{x}}_n$ is well-defined and is given by

$$\hat{x}_n(f) = \int_{-n}^{n} x(t)\, e^{-i2\pi ft}\, \mathrm{d}t, \quad f \in \mathbb{R}.$$

We then note that $\|\mathbf{x} - \mathbf{x}_n\|_2$ tends to zero as $n$ tends to infinity, so for every $\epsilon > 0$ there exists some $\mathrm{L}(\epsilon)$ sufficiently large so that

$$\|\mathbf{x}_n - \mathbf{x}_m\|_2 < \epsilon, \quad n, m > \mathrm{L}(\epsilon). \tag{6.13}$$

Applying Proposition 6.2.6 (ii) with the substitution of $\max\{n, m\}$ for $\mathsf{T}$ and of $\mathbf{x}_n - \mathbf{x}_m$ for both $\mathbf{x}_1$ and $\mathbf{x}_2$, we obtain that (6.13) implies

$$\|\hat{\mathbf{x}}_n - \hat{\mathbf{x}}_m\|_2 < \epsilon, \quad n, m > \mathrm{L}(\epsilon). \tag{6.14}$$

Because the space of energy-limited signals is complete in the sense of Theorem 8.5.1 ahead, we may infer from (6.14) that there exists some function $\boldsymbol{\zeta} \in \mathcal{L}_2$ such that $\|\hat{\mathbf{x}}_n - \boldsymbol{\zeta}\|_2$ converges to zero.[4] We then define the $L_2$-Fourier Transform of the equivalence class $[\mathbf{x}]$ to be the equivalence class $[\boldsymbol{\zeta}]$. In view of Footnote 4 we can define the $L_2$-Fourier Transform as follows.

**Definition 6.2.7 ($L_2$-Fourier Transform).** *The $L_2$-Fourier Transform of the equivalence class $[\mathbf{x}] \in L_2$ is denoted by $\widehat{[\mathbf{x}]}$ and is given by*

$$\widehat{[\mathbf{x}]} \triangleq \left\{ \mathbf{g} \in \mathcal{L}_2 : \lim_{n \to \infty} \int_{-\infty}^{\infty} \left| g(f) - \int_{-n}^{n} x(t)\, e^{-i2\pi ft}\, \mathrm{d}t \right|^2 \mathrm{d}f = 0 \right\}.$$

The main properties of the $L_2$-Fourier Transform are summarized in the following theorem.

**Theorem 6.2.8 (Properties of the $L_2$-Fourier Transform).** *The $L_2$-Fourier Transform is a mapping from $L_2$ onto $L_2$ with the following properties:*

(i) *If $\mathbf{x} \in \mathcal{L}_2 \cap \mathcal{L}_1$, then the $L_2$-Fourier Transform of $[\mathbf{x}]$ is the equivalence class of the mapping*

$$f \mapsto \int_{-\infty}^{\infty} x(t)\, e^{-i2\pi ft}\, \mathrm{d}t.$$

---

[4]The function $\boldsymbol{\zeta}$ is not unique. If $\|\mathbf{x}_n - \boldsymbol{\zeta}\|_2 \to 0$, then also $\|\mathbf{x}_n - \tilde{\boldsymbol{\zeta}}\|_2 \to 0$ whenever $\tilde{\boldsymbol{\zeta}} \in [\boldsymbol{\zeta}]$. And conversely, if $\|\mathbf{x}_n - \boldsymbol{\zeta}\|_2 \to 0$ and $\|\mathbf{x}_n - \tilde{\boldsymbol{\zeta}}\|_2 \to 0$, then $\tilde{\boldsymbol{\zeta}}$ must be in $[\boldsymbol{\zeta}]$.

*(ii) The $L_2$-Fourier Transform is **linear** in the sense that*

$$\alpha \widehat{[\mathbf{x}_1] + \beta [\mathbf{x}_2]} = \alpha \widehat{[\mathbf{x}_1]} + \beta \widehat{[\mathbf{x}_2]}, \quad \left( \mathbf{x}_1, \mathbf{x}_2 \in \mathcal{L}_2, \quad \alpha, \beta \in \mathbb{C} \right).$$

*(iii) The $L_2$-Fourier Transform is **invertible** in the sense that to each $[\mathbf{g}] \in L_2$ there corresponds a unique equivalence class in $L_2$ whose $L_2$-Fourier Transform is $[\mathbf{g}]$. This equivalence class can be obtained by reflecting each of the elements of $[\mathbf{g}]$ to obtain the equivalence class $[\breve{\mathbf{g}}]$ of $\breve{\mathbf{g}}$, and by then applying the $L_2$-Fourier Transform to it. The result $\widehat{[\breve{\mathbf{g}}]}$ then satisfies*

$$\widehat{[\breve{\mathbf{g}}]} = [\mathbf{g}], \quad \mathbf{g} \in \mathcal{L}_2. \tag{6.15}$$

*(iv) Applying the $L_2$-Fourier Transform twice is equivalent to reflecting the elements of the equivalence class*

$$\widehat{\widehat{[\mathbf{x}]}} = [\breve{\mathbf{x}}], \quad \mathbf{x} \in \mathcal{L}_2. \tag{6.16}$$

*(v) The $L_2$-Fourier Transform **preserves energies**:[5]*

$$\left\| \widehat{[\mathbf{x}]} \right\|_2 = \left\| [\mathbf{x}] \right\|_2, \quad \mathbf{x} \in \mathcal{L}_2. \tag{6.17}$$

*(vi) The $L_2$-Fourier Transform **preserves inner products**:[6]*

$$\left\langle [\mathbf{x}], [\mathbf{y}] \right\rangle = \left\langle \widehat{[\mathbf{x}]}, \widehat{[\mathbf{y}]} \right\rangle, \quad \mathbf{x}, \mathbf{y} \in \mathcal{L}_2. \tag{6.18}$$

**Proof.** This theorem is a restatement of (Rudin, 1974, Chapter 9, Theorem 9.13). Identity (6.16) appears in this form in (Stein and Weiss, 1990, Chapter 1, Section 2, Theorem 2.4).                                                                                                                      □

The result that the $L_2$-Fourier Transform preserves energies is sometimes called **Plancherel's Theorem** and the result that it preserves inner products **Parseval's Theorem**. We shall use "Parseval's Theorem" for both. It is so important that we repeat it here in the form of a theorem. Following mathematical practice, we drop the square brackets in the theorem's statement.

**Theorem 6.2.9 (Parseval's Theorem).** *For any $\mathbf{x}, \mathbf{y} \in L_2$*

$$\boxed{\langle \mathbf{x}, \mathbf{y} \rangle = \langle \hat{\mathbf{x}}, \hat{\mathbf{y}} \rangle} \tag{6.19}$$

*and*

$$\boxed{\| \mathbf{x} \|_2 = \| \hat{\mathbf{x}} \|_2 .} \tag{6.20}$$

---

[5]The energy of an equivalence class was defined in Section 4.7.
[6]The inner product between equivalence classes was defined in Section 4.7.

As we mentioned earlier, there is no simple explicit expression for the $L_2$-Fourier Transform. The following proposition simplifies its calculation under certain assumptions that are, for example, satisfied by the $\text{sinc}(\cdot)$ function.

**Proposition 6.2.10.** *If* $\mathbf{x} = \check{\mathbf{g}}$ *for some* $\mathbf{g} \in \mathcal{L}_1 \cap \mathcal{L}_2$, *then:*

*(i)* $\mathbf{x} \in \mathcal{L}_2$.

*(ii)* $\|\mathbf{x}\|_2 = \|\mathbf{g}\|_2$.

*(iii) The* $L_2$-*Fourier Transform of* $[\mathbf{x}]$ *is the equivalence class* $[\mathbf{g}]$.

**Proof.** It suffices to prove Part (iii) because Parts (i) and (ii) will then follow from the preservation of energy under the $L_2$-Fourier Transform (Theorem 6.2.8 (v)). To prove Part (iii) we compute

$$[\mathbf{g}] = \widehat{[\check{\mathbf{g}}]}$$
$$= \widehat{[\hat{\check{\mathbf{g}}}]}$$
$$= \widehat{[\mathbf{x}]},$$

where the first equality follows from (6.15); the second from Theorem 6.2.8 (i) (because the hypothesis $\mathbf{g} \in \mathcal{L}_1 \cap \mathcal{L}_2$ implies that $\check{\mathbf{g}} \in \mathcal{L}_1 \cap \mathcal{L}_2$); and the final equality from Proposition 6.2.3 (i) and from the hypothesis that $\mathbf{x} = \check{\mathbf{g}}$. $\qquad \square$

## 6.2.4   More on the Fourier Transform

In this section we present additional results that shed some light on the problem of reconstructing a signal from its FT. The first is a continuity result, which may seem technical but which has some useful consequences. It can be used to show that the IFT (of an integrable function) always yields a continuous signal. Consequently, if one starts with a discontinuous function, takes its FT, and then the IFT, one does not obtain the original function. It can also be used—once we define the frequency response of a filter in Section 6.3—to show that *no stable filter can have a discontinuous frequency response.*

**Theorem 6.2.11 (Continuity and Boundedness of the Fourier Transform).**

*(i) If* $\mathbf{x}$ *is integrable, then its FT* $\hat{\mathbf{x}}$ *is a uniformly continuous function satisfying*

$$\left|\hat{x}(f)\right| \leq \int_{-\infty}^{\infty} |x(t)|\, \mathrm{d}t, \quad f \in \mathbb{R}, \tag{6.21}$$

*and*

$$\lim_{|f| \to \infty} \hat{x}(f) = 0. \tag{6.22}$$

*(ii) If $\mathbf{g}$ is integrable, then its IFT $\check{g}$ is a uniformly continuous function satisfying*

$$|\check{g}(t)| \le \int_{-\infty}^{\infty} |g(f)| \, \mathrm{d}f, \quad t \in \mathbb{R}. \tag{6.23}$$

**Proof.** We begin with Part (i). Inequality (6.21) follows directly from the definition of the FT and from Proposition 2.4.1. The proof of the uniform continuity of $\hat{x}$ is not very difficult but is omitted. See (Katznelson, 1976, Section VI.1, Theorem 1.2). A proof of (6.22) can be found in (Katznelson, 1976, Section VI.1, Theorem 1.7).

Part (ii) follows by substituting $\tilde{\mathbf{g}}$ for $\mathbf{x}$ in Part (i) because the IFT of $\mathbf{g}$ is the FT of its mirror image (6.5). $\qquad \square$

The second result we present is that every integrable signal can be reconstructed from its FT, but not necessarily via the IFT. The reconstruction formula in (6.25) ahead works even when the IFT does not do the job.

**Theorem 6.2.12 (Reconstructing a Signal from Its Fourier Transform).**

*(i) If two integrable signals have the same FT, then they are indistinguishable:*

$$\Big( \hat{x}_1(f) = \hat{x}_2(f), \quad f \in \mathbb{R} \Big) \Rightarrow \Big( \mathbf{x}_1 \equiv \mathbf{x}_2 \Big), \quad \mathbf{x}_1, \mathbf{x}_2 \in \mathcal{L}_1. \tag{6.24}$$

*(ii) Every integrable function $\mathbf{x}$ can be reconstructed from its FT in the sense that*

$$\lim_{\lambda \to \infty} \int_{-\infty}^{\infty} \left| x(t) - \int_{-\lambda}^{\lambda} \left( 1 - \frac{|f|}{\lambda} \right) \hat{x}(f) \, e^{\mathrm{i}2\pi ft} \, \mathrm{d}f \right| \mathrm{d}t = 0. \tag{6.25}$$

**Proof.** See (Katznelson, 1976, Section VI.1.10). $\qquad \square$

Conditions under which the IFT of the FT of a signal recovers the signal are given in the following theorem.

**Theorem 6.2.13 (The Inversion Theorem).**

*(i) Suppose that $\mathbf{x}$ is integrable and that its FT $\hat{x}$ is also integrable. Define*

$$\tilde{\mathbf{x}} = \check{\hat{\mathbf{x}}}. \tag{6.26}$$

*Then $\tilde{\mathbf{x}}$ is a continuous function with*

$$\lim_{|t| \to \infty} \tilde{x}(t) = 0, \tag{6.27}$$

*and the functions $\mathbf{x}$ and $\tilde{\mathbf{x}}$ agree except on a set of Lebesgue measure zero.*

*(ii) Suppose that $\mathbf{g}$ is integrable and that its IFT $\check{g}$ is also integrable. Define*

$$\tilde{\mathbf{g}} = \hat{\check{\mathbf{g}}}. \tag{6.28}$$

*Then $\tilde{\mathbf{g}}$ is a continuous function with*

$$\lim_{|f| \to \infty} \tilde{g}(f) = 0 \tag{6.29}$$

*and the functions $\mathbf{g}$ and $\tilde{\mathbf{g}}$ agree except on a set of Lebesgue measure zero.*

**Proof.** For a proof of Part (i) see (Rudin, 1974, Theorem 9.11). Part (ii) follows by substituting $\mathbf{g}$ for $\mathbf{x}$ in Part (i) and using Proposition 6.2.3 (iii). $\qquad\square$

**Corollary 6.2.14.**

(i) *If* $\mathbf{x}$ *is a continuous integrable signal whose FT is integrable, then*

$$\hat{\check{\mathbf{x}}} = \mathbf{x}. \tag{6.30}$$

(ii) *If* $\mathbf{g}$ *is continuous and integrable, and if* $\check{\mathbf{g}}$ *is also integrable, then*

$$\check{\hat{\mathbf{g}}} = \mathbf{g}. \tag{6.31}$$

**Proof.** Part (i) follows from Theorem 6.2.13 (i) by noting that if two continuous functions are equal outside a set of Lebesgue measure zero, then they are identical. Part (ii) follows similarly from Theorem 6.2.13 (ii). $\qquad\square$

## 6.2.5 On the Brickwall and the $\mathrm{sinc}(\cdot)$ Functions

We next discuss the FT and the IFT of the Brickwall function

$$\xi \mapsto \mathrm{I}\{|\xi| \le 1\}, \tag{6.32}$$

which derives its name from the shape of its plot. Since it is a symmetric function, it follows from (6.6) that its FT and IFT are identical. Both are equal to a properly stretched and scaled $\mathrm{sinc}(\cdot)$ function (5.20).

More generally, we offer the reader advice on how to remember that for $\alpha, \gamma > 0$,

$$t \mapsto \delta\,\mathrm{sinc}(\alpha t) \text{ is the IFT of } f \mapsto \beta\,\mathrm{I}\{|f| \le \gamma\} \tag{6.33}$$

if, and only if,

$$\delta = 2\gamma\beta \tag{6.34a}$$

and

$$\gamma\frac{1}{\alpha} = \frac{1}{2}. \tag{6.34b}$$

Condition (6.34a) is easily remembered because its LHS is the value at $t = 0$ of $\delta\,\mathrm{sinc}(\alpha t)$ and its RHS is the value at $t = 0$ of the IFT of $f \mapsto \beta\,\mathrm{I}\{|f| \le \gamma\}$:

$$\int_{-\infty}^{\infty} \beta\,\mathrm{I}\{|f| \le \gamma\}\,e^{\mathrm{i}2\pi ft}\,\mathrm{d}f\bigg|_{t=0} = \int_{-\infty}^{\infty} \beta\,\mathrm{I}\{|f| \le \gamma\}\,\mathrm{d}f = 2\gamma\beta.$$

Condition (6.34b) is intimately related to the Sampling Theorem that you may have already seen and that we shall discuss in Chapter 8. Indeed, in the Sampling Theorem (Theorem 8.4.3) the time between consecutive samples $\mathsf{T}$ and the bandwidth $\mathsf{W}$ satisfy

$$\mathsf{TW} = \frac{1}{2}.$$

(In this application $\alpha$ corresponds to $1/\mathsf{T}$ and $\gamma$ corresponds to the bandwidth $\mathsf{W}$.)

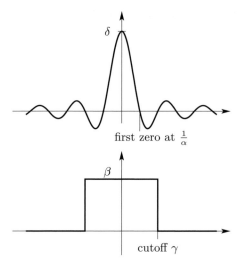

**Figure 6.1:** The stretched & scaled sinc(·) function and the stretched & scaled Brickwall function above are an $L_2$ Fourier pair if the value of the former at zero (i.e., $\delta$) is the integral of the latter (i.e., $2 \times \beta \times$ cutoff) and if the product of the location of the first zero of the former by the cutoff of the latter is $1/2$.

It is tempting to say that Conditions (6.34) also imply that the FT of the function $t \mapsto \delta \operatorname{sinc}(\alpha t)$ is the function $f \mapsto \beta \mathrm{I}\{|f| \le \gamma\}$, but there is a caveat. The signal $t \mapsto \delta \operatorname{sinc}(\alpha t)$ is not integrable. Consequently, its $\mathcal{L}_1$-Fourier Transform (Definition 6.2.1) is undefined. However, since it is energy-limited, its $L_2$-Fourier Transform is defined (Definition 6.2.7). Using Proposition 6.2.10 with the substitution of $f \mapsto \beta \mathrm{I}\{|f| \le \gamma\}$ for **g**, we obtain that, indeed, Conditions (6.34) imply that the $L_2$-Fourier Transform of the (equivalence class of the) function $t \mapsto \delta \operatorname{sinc}(\alpha t)$ is the (equivalence class of the) function $f \mapsto \beta \mathrm{I}\{|f| \le \gamma\}$.

The relation between the sinc(·) and the Brickwall functions is summarized in Figure 6.1.

The derivation of the result is straightforward: the IFT of the Brickwall function can be computed as

$$
\int_{-\infty}^{\infty} \beta \mathrm{I}\{|f| \le \gamma\} \, e^{i2\pi f t} \, \mathrm{d}f = \beta \int_{-\gamma}^{\gamma} e^{i2\pi f t} \, \mathrm{d}f
$$

$$
= \frac{\beta}{i2\pi t} e^{i2\pi f t} \Big|_{-\gamma}^{\gamma}
$$

$$
= \frac{\beta}{i2\pi t} \left( e^{i2\pi\gamma t} - e^{-i2\pi\gamma t} \right)
$$

$$
= \frac{\beta}{\pi t} \sin(2\pi\gamma t)
$$

$$
= 2\beta\gamma \operatorname{sinc}(2\gamma t). \tag{6.35}
$$

## 6.3 The Frequency Response of a Filter

Recall that in Section 5.7 we defined a filter of impulse response $\mathbf{h}$ to be a physical device that when fed the input $\mathbf{x}$ produces the output $\mathbf{x} \star \mathbf{h}$. Of course, this is only meaningful if the convolution is defined. Subject to some technical assumptions that are made precise in Theorem 6.3.2, the FT of the output waveform $\mathbf{x} \star \mathbf{h}$ is the product of the FT of the input waveform $\mathbf{x}$ by the FT of the impulse response $\mathbf{h}$. Consequently, we can think of a filter of impulse response $\mathbf{h}$ as a physical device that produces an output signal whose FT is the product of the FT of the input signal and the FT of the impulse response.

The FT of the impulse response is called the **frequency response of the filter**. If the filter is stable and its impulse response therefore integrable, then we define the filter's frequency response as the Fourier Transform of the impulse response using Definition 6.2.1 (the $\mathcal{L}_1$-Fourier Transform). If the impulse response is energy-limited but not integrable, then we define the frequency response as the Fourier Transform of the impulse response using the definition of the Fourier Transform for energy-limited signals that are not integrable as in Section 6.2.3 (the $L_2$-Fourier Transform).

**Definition 6.3.1 (Frequency Response).**

(i) The **frequency response** of a stable filter is the Fourier Transform of its impulse response as defined in Definition 6.2.1.

(ii) The **frequency response** of an unstable filter whose impulse response is energy-limited is the $L_2$-Fourier Transform of its impulse response as defined in Section 6.2.3.

As discussed in Section 5.5, if $\mathbf{x}, \mathbf{h}$ are both integrable, then $\mathbf{x} \star \mathbf{h}$ is defined at all epochs $t$ outside a set of Lebesgue measure zero, and $\mathbf{x} \star \mathbf{h}$ is integrable. In this case the FT of $\mathbf{x} \star \mathbf{h}$ is the mapping $f \mapsto \hat{x}(f)\,\hat{h}(f)$. If $\mathbf{x}$ is integrable and $\mathbf{h}$ is of finite energy, then $\mathbf{x} \star \mathbf{h}$ is also defined at all epochs $t$ outside a set of Lebesgue measure zero. But in this case the convolution is only guaranteed to be of finite energy; it need not be integrable. We can discuss its Fourier Transform using the definition of the $L_2$-Fourier Transform for energy-limited signals that are not integrable as in Section 6.2.3. In this case, again, the $L_2$-Fourier Transform of $\mathbf{x} \star \mathbf{h}$ is the (equivalence class of the) mapping $f \mapsto \hat{x}(f)\,\hat{h}(f)$:[7]

**Theorem 6.3.2 (The Fourier Transform of a Convolution).**

(i) If the signals $\mathbf{h}$ and $\mathbf{x}$ are both integrable, then the convolution $\mathbf{x} \star \mathbf{h}$ is defined for all $t$ outside a set of Lebesgue measure zero; it is integrable; and its $\mathcal{L}_1$-Fourier Transform $\widehat{\mathbf{x} \star \mathbf{h}}$ is given by

$$\boxed{\widehat{\mathbf{x} \star \mathbf{h}}(f) = \hat{x}(f)\,\hat{h}(f), \quad f \in \mathbb{R},} \tag{6.36}$$

---

[7]To be precise we should say that the $L_2$-Fourier Transform of $\mathbf{x} \star \mathbf{h}$ is the equivalence class of the product of the $\mathcal{L}_1$-Fourier Transform of $\mathbf{x}$ by any element in the equivalence class consisting of the $L_2$-Fourier Transform of $[\mathbf{h}]$.

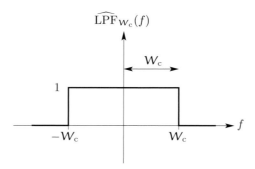

**Figure 6.2:** The frequency response of the ideal unit-gain lowpass filter of cutoff frequency $W_c$. Notice that $W_c$ is the length of the interval of positive frequencies where the gain is one.

*where $\hat{\mathbf{x}}$ and $\hat{\mathbf{h}}$ are the $\mathcal{L}_1$-Fourier Transforms of $\mathbf{x}$ and $\mathbf{h}$.*

*(ii) If the signal $\mathbf{x}$ is integrable and if $\mathbf{h}$ is of finite energy, then the convolution $\mathbf{x} \star \mathbf{h}$ is defined for all $t$ outside a set of Lebesgue measure zero; it is energy-limited; and its $L_2$-Fourier Transform $\widehat{\mathbf{x} \star \mathbf{h}}$ is also given by (6.36) with $\hat{\mathbf{x}}$, as before, being the $\mathcal{L}_1$-Fourier Transform of $\mathbf{x}$ but with $\hat{\mathbf{h}}$ now being the $L_2$-Fourier Transform of $\mathbf{h}$.*

**Proof.** For a proof of Part (i) see, for example, (Stein and Weiss, 1990, Chapter 1, Section 1, Theorem 1.4). For Part (ii) see (Stein and Weiss, 1990, Chapter 1, Section 2, Theorem 2.6). ☐

As an example, recall from Section 5.9 that the unit-gain ideal lowpass filter of cutoff frequency $W_c$ is a filter of impulse response

$$h(t) = 2W_c \operatorname{sinc}(2W_c t), \quad t \in \mathbb{R}. \tag{6.37}$$

This filter is not causal and not stable, but its impulse response is energy-limited. The filter's frequency response is the $L_2$-Fourier Transform of the impulse response (6.37), which, using the results from Section 6.2.5, is given by (the equivalence class of) the mapping

$$f \mapsto \mathrm{I}\{|f| \le W_c\}, \quad f \in \mathbb{R}. \tag{6.38}$$

This mapping maps all frequencies $f$ satisfying $|f| > W_c$ to 0 and all frequencies satisfying $|f| \le W_c$ to one. It is for this reason that we use the adjective "unit-gain" in describing this filter. We denote the mapping in (6.38) by $\widehat{\mathrm{LPF}}_{W_c}(\cdot)$ so

$$\widehat{\mathrm{LPF}}_{W_c}(f) \triangleq \mathrm{I}\{|f| \le W_c\}, \quad f \in \mathbb{R}. \tag{6.39}$$

This mapping is depicted in Figure 6.2. Note that $W_c$ is the length of the interval of positive frequencies where the response is one.

Turning to the ideal unit-gain bandpass filter of bandwidth $W$ around the carrier frequency $f_c$ satisfying $f_c \ge W/2$, we note that, by (5.21), its time-$t$ impulse

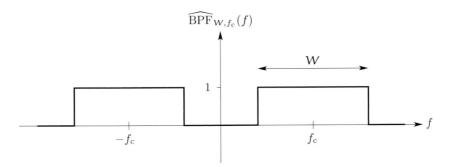

**Figure 6.3:** The frequency response of the ideal unit-gain bandpass filter of bandwidth $W$ around the carrier frequency $f_c$. Notice that, as for the lowpass filter, $W$ is the length of the interval of positive frequencies where the gain is one.

response $\mathrm{BPF}_{W,f_c}(t)$ is given by

$$\mathrm{BPF}_{W,f_c}(t) = 2W\cos(2\pi f_c t)\,\mathrm{sinc}(Wt)$$

$$= 2\,\mathrm{Re}\!\left(\mathrm{LPF}_{W/2}(t)\,e^{\mathrm{i}2\pi f_c t}\right). \tag{6.40}$$

This filter too is noncausal and nonstable. From (6.40) and (6.39) we obtain using Table 6.1 that its frequency response is (the equivalence class of) the mapping

$$f \mapsto \mathrm{I}\!\left\{ \big| |f| - f_c \big| \leq \frac{W}{2} \right\}.$$

We denote this mapping by $\widehat{\mathrm{BPF}}_{W,f_c}(\cdot)$ so

$$\widehat{\mathrm{BPF}}_{W,f_c}(f) \triangleq \mathrm{I}\!\left\{ \big| |f| - f_c \big| \leq \frac{W}{2} \right\}, \quad f \in \mathbb{R}. \tag{6.41}$$

This mapping is depicted in Figure 6.3. Note that, as for the lowpass filter, $W$ is the length of the interval of positive frequencies where the response is one.

## 6.4 Bandlimited Signals and Lowpass Filtering

In this section we define bandlimited signals and discuss lowpass filtering. We treat energy-limited signals and integrable signals separately. As we shall see, any integrable signal that is bandlimited to $W$ Hz is also an energy-limited signal that is bandlimited to $W$ Hz (Note 6.4.12).

### 6.4.1 Energy-Limited Signals

The main result of this section is that the following three statements are equivalent:

(a) The signal $\mathbf{x}$ is an energy-limited signal satisfying

$$(\mathbf{x} \star \mathrm{LPF}_W)(t) = x(t), \quad t \in \mathbb{R}. \tag{6.42}$$

(b) The signal $\mathbf{x}$ can be expressed in the form

$$x(t) = \int_{-W}^{W} g(f)\, e^{i2\pi ft}\, \mathrm{d}f, \quad t \in \mathbb{R}, \tag{6.43a}$$

for some measurable function $\mathbf{g}\colon f \mapsto g(f)$ satisfying

$$\int_{-W}^{W} |g(f)|^2\, \mathrm{d}f < \infty. \tag{6.43b}$$

(c) The signal $\mathbf{x}$ is a continuous energy-limited signal whose $L_2$-Fourier Transform $\hat{\mathbf{x}}$ satisfies

$$\int_{-\infty}^{\infty} |\hat{x}(f)|^2\, \mathrm{d}f = \int_{-W}^{W} |\hat{x}(f)|^2\, \mathrm{d}f. \tag{6.44}$$

We can thus define $\mathbf{x}$ to be an energy-limited signal that is bandlimited to $W$ Hz if one (and hence all) of the above conditions hold.

In deriving this result we shall take (a) as the definition. We shall then establish the equivalence (a) $\Leftrightarrow$ (b) in Proposition 6.4.5, which also establishes that the function $\mathbf{g}$ in (6.43a) can be taken as any element in the equivalence class of the $L_2$-Fourier Transform of $\mathbf{x}$, and that the LHS of (6.43b) is then $\|\mathbf{x}\|_2^2$. Finally, we shall establish the equivalence (a) $\Leftrightarrow$ (c) in Proposition 6.4.6.

We conclude the section with a summary of the key properties of the result of passing an energy-limited signal through an ideal unit-gain lowpass filter.

We begin by defining an energy-limited signal to be bandlimited to $W$ Hz if it is unaltered when it is lowpass filtered by an ideal unit-gain lowpass filter of cutoff frequency $W$. Recalling that we are denoting by $\mathrm{LPF}_W(t)$ the time-$t$ impulse response of an ideal unit-gain lowpass filter of cutoff frequency $W$ (see (5.19)), we have the following definition.[8]

**Definition 6.4.1 (Energy-Limited Bandlimited Signals).** *We say that the signal* $\mathbf{x}$ *is an* **energy-limited signal that is bandlimited to** $W$ Hz *if* $\mathbf{x}$ *is in* $\mathcal{L}_2$ *and*

$$(\mathbf{x} \star \mathrm{LPF}_W)(t) = x(t), \quad t \in \mathbb{R}. \tag{6.45}$$

**Note 6.4.2.** If an energy-limited signal that is bandlimited to $W$ Hz is of zero energy, then it is the all-zero signal $\mathbf{0}$.

**Proof.** Let $\mathbf{x}$ be an energy-limited signal that is bandlimited to $W$ Hz and that has zero energy. Then

$$\begin{aligned}
|x(t)| &= \left|(\mathbf{x} \star \mathrm{LPF}_W)(t)\right| \\
&\leq \|\mathbf{x}\|_2\, \|\mathrm{LPF}_W\|_2 \\
&= \|\mathbf{x}\|_2\, \sqrt{2W} \\
&= 0, \quad t \in \mathbb{R},
\end{aligned}$$

---

[8] Even though the ideal unit-gain lowpass filter of cutoff frequency $W$ is not stable, its impulse response $\mathrm{LPF}_W(\cdot)$ is of finite energy (because it decays like $1/t$ and the integral of $1/t^2$ from one to infinity is finite). Consequently, we can use the Cauchy-Schwarz Inequality to prove that if $\mathbf{x} \in \mathcal{L}_2$ then the mapping $\tau \mapsto x(\tau)\,\mathrm{LPF}_W(t - \tau)$ is integrable for every time instant $t \in \mathbb{R}$. Consequently, the convolution $\mathbf{x} \star \mathrm{LPF}_W$ is defined at every time instant $t$; see Section 5.5.

where the first equality follows because $\mathbf{x}$ is an energy-limited signal that is band-limited to $W$ Hz and is thus unaltered when it is lowpass filtered; the subsequent inequality follows from (5.6b); the subsequent equality by computing $\|\mathrm{LPF}_W\|_2$ using Parseval's Theorem and the explicit form of the frequency response of the ideal unit-gain lowpass filter of bandwidth $W$ (6.38); and where the final equality follows from the hypothesis that $\mathbf{x}$ is of zero energy. $\qquad\square$

Having defined what it means for an energy-limited signal to be bandlimited to $W$ Hz, we can now define its **bandwidth**.[9]

**Definition 6.4.3 (Bandwidth).** *The **bandwidth** of an energy-limited signal $\mathbf{x}$ is the smallest frequency $W$ to which $\mathbf{x}$ is bandlimited.*

The next lemma shows that the result of passing an energy-limited signal through an ideal unit-gain lowpass filter of cutoff frequency $W$ is an energy-limited signal that is bandlimited to $W$ Hz.

**Lemma 6.4.4.**

(i) *Let $\mathbf{y} = \mathbf{x} \star \mathrm{LPF}_W$ be the output of an ideal unit-gain lowpass filter of cutoff frequency $W$ that is fed the energy-limited input $\mathbf{x} \in \mathcal{L}_2$. Then $\mathbf{y} \in \mathcal{L}_2$;*

$$y(t) = \int_{-W}^{W} \hat{x}(f)\, e^{\mathrm{i}2\pi ft}\, \mathrm{d}f, \quad t \in \mathbb{R}; \tag{6.46}$$

*and the $\mathcal{L}_2$-Fourier Transform of $\mathbf{y}$ is the (equivalence class of the) mapping*

$$f \mapsto \hat{x}(f)\, \mathrm{I}\{|f| \le W\}. \tag{6.47}$$

(ii) *If $\mathbf{g}\colon f \mapsto g(f)$ is a bounded integrable function and if $\mathbf{x}$ is energy-limited, then $\mathbf{x} \star \check{\mathbf{g}}$ is in $\mathcal{L}_2$; it can be expressed as*

$$\left(\mathbf{x} \star \check{\mathbf{g}}\right)(t) = \int_{-\infty}^{\infty} \hat{x}(f)\, g(f)\, e^{\mathrm{i}2\pi ft}\, \mathrm{d}f, \quad t \in \mathbb{R}; \tag{6.48}$$

*and its $\mathcal{L}_2$-Fourier Transform is given by (the equivalence class of) the mapping $f \mapsto \hat{x}(f)\, g(f)$.*

**Proof.** Even though Part (i) is a special case of Part (ii) corresponding to $\mathbf{g}$ being the mapping $f \mapsto \mathrm{I}\{|f| \le W\}$, we shall prove the two parts separately. We begin with a proof of Part (i). The idea of the proof is to express for each $t \in \mathbb{R}$ the time-$t$ output $y(t)$ as an inner product and to then use Parseval's Theorem. Thus,

---

[9]To be more rigorous we should use in this definition the term "infimum" instead of "smallest," but it turns out that the infimum here is also a minimum.

(6.46) follows from the calculation

$$
\begin{aligned}
y(t) &= \big(\mathbf{x} \star \mathrm{LPF}_W\big)(t) \\
&= \int_{-\infty}^{\infty} x(\tau)\,\mathrm{LPF}_W(t - \tau)\,\mathrm{d}\tau \\
&= \big\langle \mathbf{x}, \tau \mapsto \mathrm{LPF}_W(t - \tau)\big\rangle \\
&= \big\langle \mathbf{x}, \tau \mapsto \mathrm{LPF}_W(\tau - t)\big\rangle \\
&= \big\langle \hat{\mathbf{x}}, f \mapsto e^{-\mathrm{i}2\pi ft}\,\widehat{\mathrm{LPF}}_W(f)\big\rangle \\
&= \big\langle \hat{\mathbf{x}}, f \mapsto e^{-\mathrm{i}2\pi ft}\,\mathrm{I}\{|f| \le W\}\big\rangle \\
&= \int_{-W}^{W} \hat{x}(f)\,e^{\mathrm{i}2\pi ft}\,\mathrm{d}f,
\end{aligned}
$$

where the fourth equality follows from the symmetry of the function $\mathrm{LPF}_W(\cdot)$, and where the fifth equality follows from Parseval's Theorem and the fact that delaying a function multiplies its FT by a complex exponential. Having established (6.46), Part (i) now follows from Proposition 6.2.10, because, by Parseval's Theorem, the mapping $f \mapsto \hat{x}(f)\,\mathrm{I}\{|f| \le W\}$ is of finite energy and hence, by Proposition 3.4.3, also integrable.

We next turn to Part (ii). We first note that the assumption that $\mathbf{g}$ is bounded and integrable implies that it is also energy-limited, because if $|g(f)| \le \sigma_\infty$ for all $f \in \mathbb{R}$, then $|g(f)|^2 \le \sigma_\infty |g(f)|$ and $\int |g(f)|^2\,\mathrm{d}f \le \sigma_\infty \int |g(f)|\,\mathrm{d}f$. Thus,

$$
\mathbf{g} \in \mathcal{L}_1 \cap \mathcal{L}_2. \tag{6.49}
$$

We next prove (6.48). To that end we express the convolution $\mathbf{x} \star \mathbf{\check{g}}$ at time $t$ as an inner product and then use Parseval's Theorem to obtain

$$
\begin{aligned}
\big(\mathbf{x} \star \mathbf{\check{g}}\big)(t) &= \int_{-\infty}^{\infty} x(\tau)\,\check{g}(t - \tau)\,\mathrm{d}\tau \\
&= \big\langle \mathbf{x}, \tau \mapsto \check{g}^*(t - \tau)\big\rangle \\
&= \big\langle \hat{\mathbf{x}}, f \mapsto e^{-\mathrm{i}2\pi ft}\,g^*(f)\big\rangle \\
&= \int_{-\infty}^{\infty} \hat{x}(f)\,g(f)\,e^{\mathrm{i}2\pi ft}\,\mathrm{d}f, \quad t \in \mathbb{R}, \tag{6.50}
\end{aligned}
$$

where the third equality follows from Parseval's Theorem and by noting that the $\mathcal{L}_2$-Fourier Transform of the mapping $\tau \mapsto \check{g}^*(t - \tau)$ is the equivalence class of the mapping $f \mapsto e^{-\mathrm{i}2\pi ft}\,g^*(f)$, as can be verified by expressing the mapping $\tau \mapsto \check{g}^*(t - \tau)$ as the IFT of the mapping $f \mapsto e^{-\mathrm{i}2\pi ft}\,g^*(f)$

$$
\begin{aligned}
\check{g}^*(t - \tau) &= \left(\int_{-\infty}^{\infty} g(f)\,e^{\mathrm{i}2\pi f(t - \tau)}\,\mathrm{d}f\right)^* \\
&= \int_{-\infty}^{\infty} g^*(f)\,e^{\mathrm{i}2\pi f(\tau - t)}\,\mathrm{d}f \\
&= \int_{-\infty}^{\infty} \big(g^*(f)\,e^{-\mathrm{i}2\pi ft}\big)\,e^{\mathrm{i}2\pi f\tau}\,\mathrm{d}f, \quad t, \tau \in \mathbb{R},
\end{aligned}
$$

and by then applying Proposition 6.2.10 to the mapping $f \mapsto g^*(f)\, e^{-\mathrm{i}2\pi ft}$, which is in $\mathcal{L}_1 \cap \mathcal{L}_2$ by (6.49).

Having established (6.48) we next examine the integrand in (6.48) and note that if $|g(f)|$ is upper-bounded by $\sigma_\infty$, then the modulus of the integrand is upper-bounded by $\sigma_\infty |\hat{x}(f)|$, so the assumption that $\mathbf{x} \in \mathcal{L}_2$ (and hence that $\hat{x}$ is of finite energy) guarantees that the integrand is square integrable. Also, by the Cauchy-Schwarz Inequality, the square integrability of $\mathbf{g}$ and of $\hat{x}$ implies that the integrand is integrable. Thus, the integrand is both square integrable and integrable so, by Proposition 6.2.10, the signal $\mathbf{x} \star \mathbf{g}$ is square integrable and its Fourier Transform is the (equivalence class of the) mapping $f \mapsto \hat{x}(f)\, g(f)$. $\qquad\square$

With the aid of the above lemma we can now give an equivalent definition for energy-limited signals that are bandlimited to $W$ Hz. This definition is popular among mathematicians, because it does not involve the $\mathcal{L}_2$-Fourier Transform and because the continuity of the signal is implied.

**Proposition 6.4.5 (On the Definition of Bandlimited Functions in $\mathcal{L}_2$).**

(i) *If $\mathbf{x}$ is an energy-limited signal that is bandlimited to $W$ Hz, then it can be expressed in the form*

$$x(t) = \int_{-W}^{W} g(f)\, e^{\mathrm{i}2\pi ft}\, \mathrm{d}f, \quad t \in \mathbb{R}, \tag{6.51}$$

*where $g(\cdot)$ satisfies*

$$\int_{-W}^{W} |g(f)|^2\, \mathrm{d}f < \infty \tag{6.52}$$

*and can be taken as (any function in the equivalence class of) $\hat{x}$.*

(ii) *If a signal $\mathbf{x}$ can be expressed as in (6.51) for some function $g(\cdot)$ satisfying (6.52), then $\mathbf{x}$ is an energy-limited signal that is bandlimited to $W$ Hz and $\hat{x}$ is (the equivalence class of) the mapping $f \mapsto g(f)\, \mathrm{I}\{|f| \le W\}$.*

**Proof.** We first prove Part (i). Let $\mathbf{x}$ be an energy-limited signal that is bandlimited to $W$ Hz. Then

$$x(t) = (\mathbf{x} \star \mathrm{LPF}_W)(t)$$
$$= \int_{-W}^{W} \hat{x}(f)\, e^{\mathrm{i}2\pi ft}\, \mathrm{d}f, \quad t \in \mathbb{R},$$

where the first equality follows from Definition 6.4.1, and where the second equality follows from Lemma 6.4.4 (i). Consequently, if we pick $\mathbf{g}$ as (any element of the equivalence class of) $f \mapsto \hat{x}(f)\, \mathrm{I}\{|f| \le W\}$, then (6.51) will be satisfied and (6.52) will follow from Parseval's Theorem.

To prove Part (ii) define $\tilde{\mathbf{g}} \colon f \mapsto g(f)\, \mathrm{I}\{|f| \le W\}$. From the assumption (6.52) and from Proposition 3.4.3 it then follows that $\tilde{\mathbf{g}} \in \mathcal{L}_1 \cap \mathcal{L}_2$. This and (6.51) imply that $\mathbf{x} \in \mathcal{L}_2$ and that the $\mathcal{L}_2$-Fourier Transform of (the equivalence class of) $\mathbf{x}$ is (the equivalence class of) $\tilde{\mathbf{g}}$; see Proposition 6.2.10. To complete the proof of Part (ii) it thus remains to show that $\mathbf{x} \star \mathrm{LPF}_W = \mathbf{x}$. This follows from the calculation:

$$
\begin{aligned}
(\mathbf{x} \star \mathrm{LPF}_W)(t) &= \int_{-W}^{W} \hat{x}(f)\, e^{i2\pi ft}\, \mathrm{d}f \\
&= \int_{-W}^{W} g(f)\, e^{i2\pi ft}\, \mathrm{d}f \\
&= x(t), \quad t \in \mathbb{R},
\end{aligned}
$$

where the first equality follows from Lemma 6.4.4 (i); the second because we have already established that the $L_2$-Fourier Transform of (the equivalence class of) $\mathbf{x}$ is (the equivalence class of) $f \mapsto g(f)\,\mathrm{I}\{|f| \leq W\}$; and where the last equality follows from (6.51). $\qquad\square$

In the engineering literature a function is often defined as bandlimited to $W$ Hz if its FT is zero for frequencies $f$ outside the interval $[-W, W]$. This definition is imprecise because the $L_2$-Fourier Transform of a signal is an equivalence class and its value at a given frequency is technically undefined. It would be better to define an energy-limited signal as bandlimited to $W$ Hz if $\|\mathbf{x}\|_2^2 = \int_{-W}^{W} |\hat{x}(f)|^2\, \mathrm{d}f$ so "all its energy is contained in the frequency band $[-W, W]$." However, this is not quite equivalent to our definition. For example, the $L_2$-Fourier Transform of the discontinuous signal

$$
x(t) = \begin{cases} 17 & \text{if } t = 0, \\ \operatorname{sinc} 2Wt & \text{otherwise,} \end{cases}
$$

is (the equivalence class of) the Brickwall (frequency domain) function

$$
\frac{1}{2W}\, \mathrm{I}\{|f| \leq W\}, \quad f \in \mathbb{R}
$$

(because the discontinuity at $t = 0$ does not influence the Fourier integral), but the signal *is* altered by the lowpass filter, which smooths it out to produce the continuous waveform $t \mapsto \operatorname{sinc}(2Wt)$. Readers who have already seen the Sampling Theorem will note that the above signal $x(\cdot)$ provides a counterexample to the Sampling Theorem as it is often imprecisely stated.

The following proposition clarifies the relationship between this definition and ours.

**Proposition 6.4.6 (More on the Definition of Bandlimited Functions in $\mathcal{L}_2$).**

(i) *If $\mathbf{x}$ is an energy-limited signal that is bandlimited to $W$ Hz, then $\mathbf{x}$ is a continuous function and all its energy is contained in the frequency interval $[-W, W]$ in the sense that its $L_2$-Fourier Transform $\hat{\mathbf{x}}$ satisfies*

$$
\int_{-\infty}^{\infty} |\hat{x}(f)|^2\, \mathrm{d}f = \int_{-W}^{W} |\hat{x}(f)|^2\, \mathrm{d}f. \tag{6.53}
$$

(ii) *If the signal* $\mathbf{x} \in \mathcal{L}_2$ *satisfies* (6.53), *then* $\mathbf{x}$ *is indistinguishable from the signal* $\mathbf{x} \star \mathrm{LPF}_\mathsf{W}$, *which is an energy-limited signal that is bandlimited to* $\mathsf{W}$ *Hz. If in addition to satisfying* (6.53) *the signal* $\mathbf{x}$ *is continuous, then* $\mathbf{x}$ *is an energy-limited signal that is bandlimited to* $\mathsf{W}$ *Hz.*

**Proof.** This proposition's claims are a subset of those of Proposition 6.4.7, which summarizes some of the results relating to lowpass filtering. The proof is therefore omitted. □

**Proposition 6.4.7.** *Let* $\mathbf{y} = \mathbf{x} \star \mathrm{LPF}_\mathsf{W}$ *be the result of feeding the signal* $\mathbf{x} \in \mathcal{L}_2$ *to an ideal unit-gain lowpass filter of cutoff frequency* $\mathsf{W}$. *Then:*

(i) $\mathbf{y}$ *is energy-limited with*

$$\|\mathbf{y}\|_2 \leq \|\mathbf{x}\|_2. \tag{6.54}$$

(ii) $\mathbf{y}$ *is an energy-limited signal that is bandlimited to* $\mathsf{W}$ *Hz.*

(iii) *Its* $\mathcal{L}_2$*-Fourier Transform* $\hat{\mathbf{y}}$ *is given by (the equivalence class of) the mapping* $f \mapsto \hat{x}(f)\,\mathrm{I}\{|f| \leq \mathsf{W}\}$.

(iv) *All the energy in* $\mathbf{y}$ *is concentrated in the frequency band* $[-\mathsf{W}, \mathsf{W}]$ *in the sense that:*

$$\int_{-\infty}^{\infty} |\hat{y}(f)|^2 \,\mathrm{d}f = \int_{-\mathsf{W}}^{\mathsf{W}} |\hat{y}(f)|^2 \,\mathrm{d}f.$$

(v) $\mathbf{y}$ *can be represented as*

$$y(t) = \int_{-\infty}^{\infty} \hat{y}(f)\,e^{\mathrm{i}2\pi ft}\,\mathrm{d}f, \quad t \in \mathbb{R} \tag{6.55}$$

$$= \int_{-\mathsf{W}}^{\mathsf{W}} \hat{x}(f)\,e^{\mathrm{i}2\pi ft}\,\mathrm{d}f, \quad t \in \mathbb{R}. \tag{6.56}$$

(vi) $\mathbf{y}$ *is uniformly continuous.*

(vii) *If* $\mathbf{x} \in \mathcal{L}_2$ *has all its energy concentrated in the frequency band* $[-\mathsf{W}, \mathsf{W}]$ *in the sense that*

$$\int_{-\infty}^{\infty} |\hat{x}(f)|^2 \,\mathrm{d}f = \int_{-\mathsf{W}}^{\mathsf{W}} |\hat{x}(f)|^2 \,\mathrm{d}f, \tag{6.57}$$

*then* $\mathbf{x}$ *is indistinguishable from the bandlimited signal* $\mathbf{x} \star \mathrm{LPF}_\mathsf{W}$.

(viii) $\mathbf{x}$ *is an energy-limited signal that is bandlimited to* $\mathsf{W}$ *if, and only if, it satisfies all three of the following conditions: it is in* $\mathcal{L}_2$; *it is continuous; and it satisfies* (6.57).

**Proof.** Part (i) follows from Lemma 6.4.4 (i), which demonstrates that $\hat{\mathbf{y}}$ is (the equivalence class of) the mapping $f \mapsto \hat{x}(f) \, \mathrm{I}\{|f| \le W\}$ so, by Parseval's Theorem,

$$\|\mathbf{y}\|_2^2 = \int_{-\infty}^{\infty} |\hat{y}(f)|^2 \, \mathrm{d}f$$

$$= \int_{-W}^{W} |\hat{x}(f)|^2 \, \mathrm{d}f$$

$$\le \int_{-\infty}^{\infty} |\hat{x}(f)|^2 \, \mathrm{d}f$$

$$= \|\mathbf{x}\|_2^2 \, .$$

Part (ii) follows because, by Lemma 6.4.4 (i), the signal $\mathbf{y}$ satisfies

$$y(t) = \int_{-W}^{W} \hat{x}(f) \, e^{\mathrm{i}2\pi f t} \, \mathrm{d}f$$

where

$$\int_{-W}^{W} |\hat{x}(f)|^2 \, \mathrm{d}f \le \int_{-\infty}^{\infty} |\hat{x}(f)|^2 \, \mathrm{d}f = \|\mathbf{x}\|_2^2 < \infty,$$

so, by Proposition 6.4.5, $\mathbf{y}$ is an energy-limited signal that is bandlimited to $W$ Hz.

Part (iii) follows directly from Lemma 6.4.4 (i). Part (iv) follows from Part (iii). Part (v) follows, again, directly from Lemma 6.4.4.

Part (vi) follows from the representation (6.56); from the fact that the IFT of integrable functions is uniformly continuous (Theorem 6.2.11); and because the condition $\|\mathbf{x}\|_2 < \infty$ implies, by Proposition 3.4.3, that $f \mapsto \hat{x}(f) \, \mathrm{I}\{|f| \le W\}$ is integrable.

To prove Part (vii) we note that by Part (ii) $\mathbf{x} \star \mathrm{LPF}_W$ is an energy-limited signal that is bandlimited to $W$ Hz, and we note that (6.57) implies that $\mathbf{x}$ is indistinguishable from $\mathbf{x} \star \mathrm{LPF}_W$ because

$$\|\mathbf{x} - \mathbf{x} \star \mathrm{LPF}_W\|_2^2 = \int_{-\infty}^{\infty} \left| \hat{x}(f) - \widehat{\mathbf{x} \star \mathrm{LPF}_W}(f) \right|^2 \, \mathrm{d}f$$

$$= \int_{-\infty}^{\infty} \left| \hat{x}(f) - \hat{x}(f) \, \mathrm{I}\{|f| \le W\} \right|^2 \, \mathrm{d}f$$

$$= \int_{|f|>W} \left| \hat{x}(f) \right|^2 \, \mathrm{d}f$$

$$= 0,$$

where the first equality follows from Parseval's Theorem; the second equality from Lemma 6.4.4 (i); the third equality because the integrand is zero for $|f| \le W$; and the final equality from (6.57).

To prove Part (viii) define $\mathbf{y} = \mathbf{x} \star \mathrm{LPF}_W$ and note that if $\mathbf{x}$ is an energy-limited signal that is bandlimited to $W$ Hz then, by Definition 6.4.1, $\mathbf{y} = \mathbf{x}$ so the continuity of $\mathbf{x}$ and the fact that its energy is concentrated in the interval $[-W, W]$ follow from Parts (iv) and (vi). In the other direction, if $\mathbf{x}$ satisfies (6.57) then by Part (vii) it is indistinguishable from the signal $\mathbf{y}$, which is continuous by Part (vi). If, additionally, $\mathbf{x}$ is continuous, then $\mathbf{x}$ must be identical to $\mathbf{y}$ because two continuous functions that are indistinguishable must be identical. $\qquad\square$

### 6.4.2   Integrable Signals

We next discuss what we mean when we say that $\mathbf{x}$ is an integrable signal that is bandlimited to $W$ Hz. Also important will be Note 6.4.11, which establishes that if $\mathbf{x}$ is such a signal, then $\mathbf{x}$ is equal to the IFT of its FT.

Even though the ideal unit-gain lowpass filter is unstable, its convolution with any integrable signal is well-defined. Denoting the cutoff frequency by $W_c$ we have:

**Proposition 6.4.8.** *For any* $\mathbf{x} \in \mathcal{L}_1$ *the convolution integral*

$$\int_{-\infty}^{\infty} x(\tau)\, \mathrm{LPF}_{W_c}(t - \tau)\, \mathrm{d}\tau$$

*is defined at every epoch* $t \in \mathbb{R}$ *and is given by*

$$\int_{-\infty}^{\infty} x(\tau)\, \mathrm{LPF}_{W_c}(t - \tau)\, \mathrm{d}\tau = \int_{-W_c}^{W_c} \hat{x}(f)\, e^{\mathrm{i}2\pi ft}\, \mathrm{d}f, \quad t \in \mathbb{R}. \tag{6.58}$$

*Moreover,* $\mathbf{x} \star \mathrm{LPF}_{W_c}$ *is an energy-limited function that is bandlimited to* $W_c$ *Hz. Its* $L_2$-*Fourier Transform is (the equivalence class of) the mapping*

$$f \mapsto \hat{x}(f)\, \mathrm{I}\{|f| \leq W_c\}.$$

**Proof.** The key to the proof is to note that, although the $\mathrm{sinc}(\cdot)$ function is not integrable, it follows from (6.35) that it can be represented as the Inverse Fourier Transform of an integrable function (of frequency). Consequently, the existence of the convolution and its representation as (6.58) follow directly from Proposition 6.2.5 and (6.35).

To prove the remaining assertions of the proposition we note that, since $\mathbf{x}$ is integrable, it follows from Theorem 6.2.11 that $|\hat{x}(f)| \leq \|\mathbf{x}\|_1$ and hence

$$\int_{-W_c}^{W_c} |\hat{x}(f)|^2\, \mathrm{d}f < \infty. \tag{6.59}$$

The result now follows from (6.58), (6.59), and Proposition 6.4.5. $\qquad\square$

With the aid of Proposition 6.4.8 we can now define bandlimited integrable signals:

**Definition 6.4.9 (Bandlimited Integrable Signals).** *We say that the signal* $\mathbf{x}$ *is an **integrable signal that is bandlimited to** $W$ Hz *if* $\mathbf{x}$ *is integrable and if it is unaltered when it is lowpass filtered by an ideal unit-gain lowpass filter of cutoff frequency* $W$:

$$x(t) = (\mathbf{x} \star \mathrm{LPF}_W)(t), \quad t \in \mathbb{R}.$$

**Proposition 6.4.10 (Characterizing Integrable Signals that Are Bandlimited to W Hz).** *If* $\mathbf{x}$ *is an integrable signal, then each of the following statements is equivalent to the statement that* $\mathbf{x}$ *is an integrable signal that is bandlimited to W Hz:*

(a) *The signal* $\mathbf{x}$ *is unaltered when it is lowpass filtered:*

$$x(t) = (\mathbf{x} \star \mathrm{LPF}_W)(t), \quad t \in \mathbb{R}. \tag{6.60}$$

(b) *The signal* $\mathbf{x}$ *can be expressed as*

$$x(t) = \int_{-W}^{W} \hat{x}(f)\, e^{\mathrm{i}2\pi ft}\, \mathrm{d}f, \quad t \in \mathbb{R}. \tag{6.61}$$

(c) *The signal* $\mathbf{x}$ *is continuous and*

$$\hat{x}(f) = 0, \quad |f| > W. \tag{6.62}$$

(d) *There exists an integrable function* $\mathbf{g}$ *such that*

$$x(t) = \int_{-W}^{W} g(f)\, e^{\mathrm{i}2\pi ft}\, \mathrm{d}f, \quad t \in \mathbb{R}. \tag{6.63}$$

**Proof.** Condition (a) is the condition given in Definition 6.4.9, so it only remains to show that the four conditions are equivalent. We proceed to do so by proving that (a) ⇔ (b); that (b) ⇒ (d); that (d) ⇒ (c); and that (c) ⇒ (b).

That (a) ⇔ (b) follows directly from Proposition 6.4.8 and, more specifically, from the representation (6.58). The implication (b) ⇒ (d) is obvious because nothing precludes us from picking $\mathbf{g}$ to be the mapping $f \mapsto \hat{x}(f)\, \mathrm{I}\{|f| \leq W\}$, which is integrable because $\hat{\mathbf{x}}$ is bounded by $\|\mathbf{x}\|_1$ (Theorem 6.2.11).

We next prove that (d) ⇒ (c). We thus assume that there exists an integrable function $\mathbf{g}$ such that (6.63) holds and proceed to prove that $\mathbf{x}$ is continuous and that (6.62) holds. To that end we first note that the integrability of $\mathbf{g}$ implies, by Theorem 6.2.11, that $\mathbf{x}$ ($= \check{\mathbf{g}}$) is continuous. It thus remains to prove that $\hat{\mathbf{x}}$ satisfies (6.62). Define $\mathbf{g}_0$ as the mapping $f \mapsto g(f)\, \mathrm{I}\{|f| \leq W\}$. By (6.63) it then follows that $\mathbf{x} = \check{\mathbf{g}}_0$. Consequently,

$$\hat{\mathbf{x}} = \hat{\check{\mathbf{g}}}_0. \tag{6.64}$$

Employing Theorem 6.2.13 (ii) we conclude that the RHS of (6.64) is equal to $\mathbf{g}_0$ outside a set of Lebesgue measure zero, so (6.64) implies that $\hat{\mathbf{x}}$ is indistinguishable from $\mathbf{g}_0$. Since both $\hat{\mathbf{x}}$ and $\mathbf{g}_0$ are continuous for $|f| > W$, this implies that $\hat{x}(f) = g_0(f)$ for all frequencies $|f| > W$. Since, by its definition, $g_0(f) = 0$ whenever $|f| > W$ we can conclude that (6.62) holds.

Finally (c) ⇒ (b) follows directly from Theorem 6.2.13 (i).  □

From Proposition 6.4.10 (*cf.* (b) and (c)) we obtain:

**Note 6.4.11.** If $\mathbf{x}$ is an integrable signal that is bandlimited to W Hz, then it is equal to the IFT of its FT.

By Proposition 6.4.10 it also follows that if $\mathbf{x}$ is an integrable signal that is bandlimited to $W$ Hz, then (6.61) is satisfied. Since the integrand in (6.61) is bounded (by $\|\mathbf{x}\|_1$) it follows that the integrand is square integrable over the interval $[-W, W]$. Consequently, by Proposition 6.4.5, $\mathbf{x}$ must be an energy-limited signal that is bandlimited to $W$ Hz. We have thus proved:

**Note 6.4.12.** An integrable signal that is bandlimited to $W$ Hz is also an energy-limited signal that is bandlimited to $W$ Hz.

The reverse statement is not true: the sinc$(\cdot)$ is an energy-limited signal that is bandlimited to $1/2$ Hz, but it is not integrable.

The definition of bandwidth for integrable signals is similar to Definition 6.4.3.[10]

**Definition 6.4.13 (Bandwidth).** *The **bandwidth** of an integrable signal is the smallest frequency $W$ to which it is bandlimited.*

## 6.5    Bandlimited Signals Through Stable Filters

In this section we discuss the result of feeding bandlimited signals to stable filters. We begin with energy-limited signals. In Theorem 6.3.2 we saw that the convolution of an integrable signal with an energy-limited signal is defined at all times outside a set of Lebesgue measure zero. The next proposition shows that if the energy-limited signal is bandlimited to $W$ Hz, then the convolution is defined at every time, and the result is an energy-limited signal that is bandlimited to $W$ Hz.

**Proposition 6.5.1.** *Let $\mathbf{x}$ be an energy-limited signal that is bandlimited to $W$ Hz and let $\mathbf{h}$ be integrable. Then $\mathbf{x} \star \mathbf{h}$ is defined for every $t \in \mathbb{R}$; it is an energy-limited signal that is bandlimited to $W$ Hz; and it can be represented as*

$$(\mathbf{x} \star \mathbf{h})(t) = \int_{-W}^{W} \hat{x}(f)\,\hat{h}(f)\,e^{i2\pi ft}\,\mathrm{d}f, \quad t \in \mathbb{R}. \tag{6.65}$$

**Proof.** Since $\mathbf{x}$ is an energy-limited signal that is bandlimited to $W$ Hz, it follows from Proposition 6.4.5 that

$$x(t) = \int_{-W}^{W} \hat{x}(f)\,e^{i2\pi ft}\,\mathrm{d}f, \quad t \in \mathbb{R}, \tag{6.66}$$

with the mapping $f \mapsto \hat{x}(f)\,\mathrm{I}\{|f| \leq W\}$ being square integrable and hence, by Proposition 3.4.3, also integrable. Thus the convolution $\mathbf{x} \star \mathbf{h}$ is the convolution between the IFT of the integrable mapping $f \mapsto \hat{x}(f)\,\mathrm{I}\{|f| \leq W\}$ and the integrable function $\mathbf{h}$. By Proposition 6.2.5 we thus obtain that the convolution $\mathbf{x} \star \mathbf{h}$ is defined at every time $t$ and has the representation (6.65). The proposition will now follow from (6.65) and Proposition 6.4.5 once we demonstrate that

$$\int_{-W}^{W} \left|\hat{x}(f)\,\hat{h}(f)\right|^2 \mathrm{d}f < \infty.$$

---

[10] Again, we omit the proof that the infimum is a minimum.

This can be proved by upper-bounding $|\hat{h}(f)|$ by $\|\mathbf{h}\|_1$ (Theorem 6.2.11) and by then using Parseval's Theorem.                                                                $\square$

We next turn to integrable signals passed through stable filters.

**Proposition 6.5.2 (Integrable Bandlimited Signals through Stable Filters).** *Let* $\mathbf{x}$ *be an integrable signal that is bandlimited to* $W$ *Hz, and let* $\mathbf{h}$ *be integrable. Then the convolution* $\mathbf{x} \star \mathbf{h}$ *is defined for every* $t \in \mathbb{R}$*; it is an integrable signal that is bandlimited to* $W$ *Hz; and it can be represented as*

$$\big(\mathbf{x} \star \mathbf{h}\big)(t) = \int_{-W}^{W} \hat{x}(f)\,\hat{h}(f)\,e^{i2\pi ft}\,\mathrm{d}f, \quad t \in \mathbb{R}. \tag{6.67}$$

**Proof.** Since every integrable signal that is bandlimited to $W$ Hz is also an energy-limited signal that is bandlimited to $W$ Hz, it follows from Proposition 6.5.1 that the convolution $\mathbf{x} \star \mathbf{h}$ is defined at every epoch and that it can be represented as (6.65). Alternatively, one can derive this representation from (6.61) and Proposition 6.2.5. It only remains to show that $\mathbf{x} \star \mathbf{h}$ is integrable, but this follows because the convolution of two integrable functions is integrable (5.9).                          $\square$

## 6.6    The Bandwidth of a Product of Two Signals

In this section we discuss the bandwidth of the product of two bandlimited signals. The result is a straightforward consequence of the fact that the FT of a product of two signals is the convolution of their FTs. We begin with the following result on the FT of a product of signals.

**Proposition 6.6.1 (The FT of a Product Is the Convolution of the FTs).** *If* $\mathbf{x}_1$ *and* $\mathbf{x}_2$ *are energy-limited signals, then their product*

$$t \mapsto x_1(t)\,x_2(t)$$

*is an integrable function whose FT is the mapping*

$$f \mapsto \big(\hat{\mathbf{x}}_1 \star \hat{\mathbf{x}}_2\big)(f).$$

**Proof.** Let $\mathbf{x}_1$ and $\mathbf{x}_2$ be energy-limited signals, and denote their product by $\mathbf{y}$:

$$y(t) = x_1(t)\,x_2(t), \quad t \in \mathbb{R}.$$

Since both $\mathbf{x}_1$ and $\mathbf{x}_2$ are square integrable, it follows from the Cauchy-Schwarz Inequality that their product $\mathbf{y}$ is integrable and that

$$\|\mathbf{y}\|_1 \leq \|\mathbf{x}_1\|_2\,\|\mathbf{x}_2\|_2. \tag{6.68}$$

Having established that the product is integrable, we next derive its FT and show that

$$\hat{y}(f) = (\hat{\mathbf{x}}_1 \star \hat{\mathbf{x}}_2)(f), \quad f \in \mathbb{R}. \tag{6.69}$$

This is done by expressing $\hat{y}(f)$ as an inner product between two finite-energy functions and by then using Parseval's Theorem:

$$
\begin{aligned}
\hat{y}(f) &= \int_{-\infty}^{\infty} y(t)\, e^{-i2\pi ft}\, \mathrm{d}t \\
&= \int_{-\infty}^{\infty} x_1(t)\, x_2(t)\, e^{-i2\pi ft}\, \mathrm{d}t \\
&= \left\langle t \mapsto x_1(t), t \mapsto x_2^*(t)\, e^{i2\pi ft} \right\rangle \\
&= \left\langle \tilde{f} \mapsto \hat{x}_1(\tilde{f}), \tilde{f} \mapsto \hat{x}_2^*(f - \tilde{f}) \right\rangle \\
&= \int_{-\infty}^{\infty} \hat{x}_1(\tilde{f})\, \hat{x}_2(f - \tilde{f})\, \mathrm{d}\tilde{f} \\
&= (\hat{\mathbf{x}}_1 \star \hat{\mathbf{x}}_2)(f), \quad f \in \mathbb{R}. \qquad \square
\end{aligned}
$$

**Proposition 6.6.2.** *Let* $\mathbf{x}_1$ *and* $\mathbf{x}_2$ *be energy-limited signals that are bandlimited to* $W_1$ *Hz and* $W_2$ *Hz respectively. Then their product is an energy-limited signal that is bandlimited to* $W_1 + W_2$ *Hz.*

**Proof.** We will show that

$$
x_1(t)x_2(t) = \int_{-(W_1+W_2)}^{W_1+W_2} g(f)\, e^{i2\pi ft}\, \mathrm{d}f, \quad t \in \mathbb{R}, \tag{6.70}
$$

where the function $g(\cdot)$ satisfies

$$
\int_{-(W_1+W_2)}^{W_1+W_2} |g(f)|^2\, \mathrm{d}f < \infty. \tag{6.71}
$$

The result will then follow from Proposition 6.4.5.

To establish (6.70) we begin by noting that since $\mathbf{x}_1$ is of finite energy and bandlimited to $W_1$ Hz we have by Proposition 6.4.5

$$
x_1(t) = \int_{-W_1}^{W_1} \hat{x}_1(f_1)\, e^{i2\pi f_1 t}\, \mathrm{d}f_1, \quad t \in \mathbb{R}.
$$

Similarly,

$$
x_2(t) = \int_{-W_2}^{W_2} \hat{x}_2(f_2)\, e^{i2\pi f_2 t}\, \mathrm{d}f_2, \quad t \in \mathbb{R}.
$$

Consequently,

$$
\begin{aligned}
x_1(t)\, x_2(t) &= \int_{-W_1}^{W_1} \hat{x}_1(f_1)\, e^{i2\pi f_1 t}\, \mathrm{d}f_1 \int_{-W_2}^{W_2} \hat{x}_2(f_2)\, e^{i2\pi f_2 t}\, \mathrm{d}f_2 \\
&= \int_{-W_1}^{W_1} \int_{-W_2}^{W_2} \hat{x}_1(f_1)\, \hat{x}_2(f_2)\, e^{i2\pi(f_1+f_2)t}\, \mathrm{d}f_1\, \mathrm{d}f_2 \\
&= \int_{-\infty}^{\infty} \int_{-\infty}^{\infty} \hat{x}_1(f_1)\, \hat{x}_2(f_2)\, e^{i2\pi(f_1+f_2)t}\, \mathrm{d}f_1\, \mathrm{d}f_2
\end{aligned}
$$

$$= \int_{-\infty}^{\infty} \int_{-\infty}^{\infty} \hat{x}_1(\tilde{f}) \, \hat{x}_2(f - \tilde{f}) \, e^{i2\pi ft} \, \mathrm{d}\tilde{f} \, \mathrm{d}f$$

$$= \int_{-\infty}^{\infty} e^{i2\pi ft} \left( \hat{\mathbf{x}}_1 \star \hat{\mathbf{x}}_2 \right)(f) \, \mathrm{d}f$$

$$= \int_{-\infty}^{\infty} e^{i2\pi ft} \, g(f) \, \mathrm{d}f, \quad t \in \mathbb{R}, \tag{6.72}$$

where

$$g(f) = \int_{-\infty}^{\infty} \hat{x}_1(\tilde{f}) \, \hat{x}_2(f - \tilde{f}) \, \mathrm{d}\tilde{f}, \quad f \in \mathbb{R}. \tag{6.73}$$

Here the second equality follows from Fubini's Theorem;[11] the third because $\mathbf{x}_1$ and $\mathbf{x}_2$ are bandlimited to $W_1$ and $W_2$ Hz respectively; and the fourth by introducing the variables $f \triangleq f_1 + f_2$ and $\tilde{f} \triangleq f_1$.

To establish (6.70) we now need to show that because $\mathbf{x}_1$ and $\mathbf{x}_2$ are bandlimited to $W_1$ and $W_2$ Hz respectively, it follows that

$$g(f) = 0, \quad |f| > W_1 + W_2. \tag{6.74}$$

To prove this we note that because $\mathbf{x}_1$ and $\mathbf{x}_2$ are bandlimited to $W_1$ Hz and $W_2$ Hz respectively, we can rewrite (6.73) as

$$g(f) = \int_{-\infty}^{\infty} \hat{x}_1(\tilde{f}) \, \mathrm{I}\big\{|\tilde{f}| \leq W_1\big\} \, \hat{x}_2(f - \tilde{f}) \, \mathrm{I}\big\{|f - \tilde{f}| \leq W_2\big\} \, \mathrm{d}\tilde{f}, \quad f \in \mathbb{R}, \tag{6.75}$$

and the product $\mathrm{I}\big\{|\tilde{f}| \leq W_1\big\} \mathrm{I}\big\{|f - \tilde{f}| \leq W_2\big\}$ is zero for all frequencies $\tilde{f}$ satisfying $|\tilde{f}| > W_1 + W_2$.

Having established (6.70) using (6.72) and (6.74), we now proceed to prove (6.71) by showing that the integrand in (6.71) is bounded. We do so by noting that the integrand in (6.71) is the convolution of two square-integrable functions ($\hat{\mathbf{x}}_1$ and $\hat{\mathbf{x}}_2$) so by (5.6b) (with the dummy variable now being $f$) we have

$$|g(f)| \leq \|\hat{\mathbf{x}}_1\|_2 \, \|\hat{\mathbf{x}}_2\|_2 = \|\mathbf{x}_1\|_2 \, \|\mathbf{x}_2\|_2 < \infty, \quad f \in \mathbb{R}. \qquad \square$$

## 6.7 Bernstein's Inequality

Bernstein's Inequality captures the engineering intuition that the rate at which a bandlimited signal can change is proportional to its bandwidth. The way the theorem is phrased makes it clear that it is applicable both to integrable signals that are bandlimited to $W$ Hz and to energy-limited signals that are bandlimited to $W$ Hz.

**Theorem 6.7.1 (Bernstein's Inequality).** *If* $\mathbf{x}$ *can be written as*

$$x(t) = \int_{-W}^{W} g(f) \, e^{i2\pi ft} \, \mathrm{d}f, \quad t \in \mathbb{R}$$

---

[11]The fact that $\int_{-W_1}^{W_1} |\hat{x}(f)| \, \mathrm{d}f$ is finite follows from the finiteness of $\int_{-W_1}^{W_1} |\hat{x}(f)|^2 \, \mathrm{d}f$ (which follows from Parseval's Theorem) and from Proposition 3.4.3. The same argument applies to $\mathbf{x}_2$.

*for some integrable function* **g**, *then*

$$\left| \frac{\mathrm{d}x(t)}{\mathrm{d}t} \right| \leq 4\pi W \sup_{\tau \in \mathbb{R}} |x(\tau)|, \quad t \in \mathbb{R}. \tag{6.76}$$

**Proof.** A proof of a slightly more general version of this theorem can be found in (Pinsky, 2002, Chapter 2, Section 2.3.8). □

## 6.8   Time-Limited and Bandlimited Signals

In this section we prove that no nonzero signal can be both time-limited and bandlimited. We shall present two proofs. The first is based on Theorem 6.8.1, which establishes a connection between bandlimited signals and entire functions. The second is based on the Fourier Series.

We remind the reader that a function $\boldsymbol{\xi} \colon \mathbb{C} \to \mathbb{C}$ is an **entire function** if it is analytic throughout the complex plane.

**Theorem 6.8.1.** *If* **x** *is an energy-limited signal that is bandlimited to* W *Hz, then there exists an entire function* $\boldsymbol{\xi} \colon \mathbb{C} \to \mathbb{C}$ *that agrees with* **x** *on the real axis*

$$\xi(t + \mathrm{i}0) = x(t), \quad t \in \mathbb{R} \tag{6.77}$$

*and that satisfies*

$$|\xi(z)| \leq \gamma \, e^{2\pi W |z|}, \quad z \in \mathbb{C}, \tag{6.78}$$

*where* $\gamma$ *is some constant that can be taken as* $\sqrt{2W} \, \|\mathbf{x}\|_2$.

**Proof.** Let **x** be an energy-limited signal that is bandlimited to W Hz. By Proposition 6.4.5 we can express **x** as

$$x(t) = \int_{-W}^{W} g(f) \, e^{\mathrm{i}2\pi f t} \, \mathrm{d}f, \quad t \in \mathbb{R} \tag{6.79}$$

for some square-integrable function **g** satisfying

$$\int_{-W}^{W} |g(f)|^2 \, \mathrm{d}f = \|\mathbf{x}\|_2^2. \tag{6.80}$$

Consider now the function $\boldsymbol{\xi} \colon \mathbb{C} \to \mathbb{C}$ defined by

$$\xi(z) = \int_{-W}^{W} g(f) \, e^{\mathrm{i}2\pi f z} \, \mathrm{d}f, \quad z \in \mathbb{C}. \tag{6.81}$$

This function is well-defined for every $z \in \mathbb{C}$ because in the region of integration the integrand can be bounded by

$$\begin{aligned} \left| g(f) \, e^{\mathrm{i}2\pi f z} \right| = |g(f)| \, e^{-2\pi f \, \mathrm{Im}(z)} \\ \leq |g(f)| \, e^{2\pi |f| \, |\mathrm{Im}(z)|} \\ \leq |g(f)| \, e^{2\pi W |z|}, \quad |f| \leq W, \end{aligned} \tag{6.82}$$

and the RHS of (6.82) is integrable over the interval $[-W, W]$ by (6.80) and Proposition 3.4.3.

By (6.79) and (6.81) it follows that $\boldsymbol{\xi}$ is an extension of the function $\mathbf{x}$ in the sense of (6.77). It is but a technical matter to prove that $\boldsymbol{\xi}$ is analytic. One approach is to prove that it is differentiable at every $z \in \mathbb{C}$ by verifying that the swapping of differentiation and integration, which leads to

$$\frac{\mathrm{d}\xi}{\mathrm{d}z}(z) = \int_{-W}^{W} g(f)\,(\mathrm{i}2\pi f)\,e^{\mathrm{i}2\pi f z}\,\mathrm{d}f, \quad z \in \mathbb{C}$$

is justified. See (Rudin, 1974, Section 19.1) for a different approach.

To prove (6.78) we compute

$$\begin{aligned}
|\xi(z)| &= \left| \int_{-W}^{W} g(f)\,e^{\mathrm{i}2\pi f z}\,\mathrm{d}f \right| \\
&\leq \int_{-W}^{W} \left| g(f)\,e^{\mathrm{i}2\pi f z} \right|\,\mathrm{d}f \\
&\leq e^{2\pi W |z|} \int_{-W}^{W} |g(f)|\,\mathrm{d}f \\
&\leq e^{2\pi W |z|}\,\sqrt{2W}\,\sqrt{\int_{-W}^{W} |g(f)|^2\,\mathrm{d}f} \\
&= \sqrt{2W}\,\|\mathbf{x}\|_2\,e^{2\pi W |z|},
\end{aligned}$$

where the inequality in the second line follows from Proposition 2.4.1; the inequality in the third line from (6.82); the inequality in the fourth line from Proposition 3.4.3; and the final equality from (6.80). $\qquad\square$


Using Theorem 6.8.1 we can now easily prove the main result of this section.

**Theorem 6.8.2.** *Let* $W$ *and* $T$ *be fixed nonnegative real numbers. If* $\mathbf{x}$ *is an energy-limited signal that is bandlimited to* $W$ *Hz and that is time-limited in the sense that it is zero for all* $t \notin [-T/2, T/2]$, *then* $x(t) = 0$ *for all* $t \in \mathbb{R}$.

By Note 6.4.12 this theorem also holds for integrable bandlimited signals.


**Proof.** By Theorem 6.8.1 $\mathbf{x}$ can be extended to an entire function $\boldsymbol{\xi}$. Since $\mathbf{x}$ has infinitely many zeros in a bounded interval (e.g., for all $t \in [T, 2T]$) and since $\boldsymbol{\xi}$ agrees with $\mathbf{x}$ on the real line, it follows that $\boldsymbol{\xi}$ also has infinitely many zeros in a bounded set (e.g., whenever $z \in \{w \in \mathbb{C} : \mathrm{Im}(w) = 0,\ \mathrm{Re}(w) \in [T, 2T]\}$). Consequently, $\boldsymbol{\xi}$ is an entire function that has infinitely many zeros in a bounded subset of the complex plane and is thus the all-zero function (Rudin, 1974, Theorem 10.18). But since $\mathbf{x}$ and $\boldsymbol{\xi}$ agree on the real line, it follows that $\mathbf{x}$ is also the all-zero function. $\qquad\square$

Another proof can be based on the Fourier Series, which is discussed in the appendix. Starting from (6.79) we obtain that the time-$\eta/(2W)$ sample of $x(\cdot)$ satisfies

$$\frac{1}{\sqrt{2W}} x\left(\frac{\eta}{2W}\right) = \int_{-W}^{W} g(f) \frac{1}{\sqrt{2W}} e^{i2\pi f \eta/(2W)} \, \mathrm{d}f, \quad \eta \in \mathbb{Z},$$

where we recognize the RHS of the above as the $\eta$-th Fourier Series Coefficient of the function $f \mapsto g(f) \, \mathrm{I}\{|f| \le W\}$ with respect to the interval $[-W, W)$ (Note A.3.5 on Page 693). But since $x(t) = 0$ whenever $|t| > T/2$, it follows that all but a finite number of these samples can be nonzero, thus leading us to conclude that all but a finite number of the Fourier Series Coefficients of $g(\cdot)$ are zero. By the uniqueness theorem for the Fourier Series (Theorem A.2.3) it follows that $g(\cdot)$ is equal to a trigonometric polynomial (except possibly on a set of measure zero). Thus,

$$g(f) = \sum_{\eta=-n}^{n} a_\eta \, e^{i2\pi\eta f/(2W)}, \quad f \in [-W, W] \setminus \mathcal{N}, \tag{6.83}$$

for some $n \in \mathbb{N}$; for some $2n + 1$ complex numbers $a_{-n}, \ldots, a_n$; and for some set $\mathcal{N} \subset [-W, W]$ of Lebesgue measure zero. Since the integral in (6.79) is insensitive to the behavior of $\mathbf{g}$ on the set $\mathcal{N}$, it follows from (6.79) and (6.83) that

$$x(t) = \int_{-W}^{W} e^{i2\pi f t} \sum_{\eta=-n}^{n} a_\eta \, e^{i2\pi\eta f/(2W)} \, \mathrm{d}f$$

$$= \sum_{\eta=-n}^{n} a_\eta \int_{-\infty}^{\infty} e^{i2\pi f \left(t + \frac{\eta}{2W}\right)} \mathrm{I}\{|f| \le W\} \, \mathrm{d}f$$

$$= 2W \sum_{\eta=-n}^{n} a_\eta \, \mathrm{sinc}(2Wt + \eta), \quad t \in \mathbb{R},$$

i.e., that $\mathbf{x}$ is a linear combination of a finite number of time-shifted $\mathrm{sinc}(\cdot)$ functions. It now remains to show that no linear combination of a finite number of time-shifted $\mathrm{sinc}(\cdot)$ functions can be zero for all $t \in [T, 2T]$ unless it is zero for all $t \in \mathbb{R}$. This can be established by extending the sincs to entire functions so that the linear combination of the time-shifted $\mathrm{sinc}(\cdot)$ functions is also an entire function and by then calling again on the theorem that an entire function that has infinitely many zeros in a bounded subset of the complex plane must be the all-zero function.

## 6.9  A Theorem by Paley and Wiener

The theorem of Paley and Wiener that we discuss next is important in the study of bandlimited functions, but it will not be used in this book.

Theorem 6.8.1 showed that every energy-limited signal $\mathbf{x}$ that is bandlimited to $W$ Hz can be extended to an entire function $\boldsymbol{\xi}$ satisfying (6.78) for some constant $\gamma$ by defining $\xi(z)$ as

$$\xi(z) = \int_{-W}^{W} \hat{x}(f) \, e^{i2\pi f z} \, \mathrm{d}f, \quad z \in \mathbb{C}. \tag{6.84}$$

The theorem of Paley and Wiener that we present next can be viewed as the reverse statement. It demonstrates that if $\boldsymbol{\xi} \colon \mathbb{C} \to \mathbb{C}$ is an entire function that satisfies (6.78) and whose restriction to the real axis is square integrable, then its restriction to the real axis is an energy-limited signal that is bandlimited to $W$ Hz and, moreover, if we denote this restriction by $\mathbf{x}$ so $x(t) = \xi(t + i0)$ for all $t \in \mathbb{R}$, then $\boldsymbol{\xi}$ is given by (6.84). This theorem demonstrates the close connection between entire functions satisfying (6.78)—functions that are called **entire functions of exponential type**—and energy-limited signals that are bandlimited to $W$ Hz.

**Theorem 6.9.1 (Paley-Wiener).** *If for some positive constants $W$ and $\gamma$ the entire function $\boldsymbol{\xi} \colon \mathbb{C} \to \mathbb{C}$ satisfies*

$$|\xi(z)| \leq \gamma \, e^{2\pi W |z|}, \quad z \in \mathbb{C} \tag{6.85}$$

*and if*

$$\int_{-\infty}^{\infty} |\xi(t + i0)|^2 \, \mathrm{d}t < \infty, \tag{6.86}$$

*then there exists an energy-limited function $\mathbf{g} \colon \mathbb{R} \to \mathbb{C}$ such that*

$$\xi(z) = \int_{-W}^{W} g(f) \, e^{i2\pi f z} \, \mathrm{d}f, \quad z \in \mathbb{C}. \tag{6.87}$$

**Proof.** See for example, (Rudin, 1974, Theorem 19.3) or (Katznelson, 1976, Chapter VI, Section 7) or (Dym and McKean, 1972, Section 3.3). □

## 6.10    Picket Fences and Poisson Summation

Engineering textbooks often contain a useful expression for the FT of an infinite series of equally-spaced Dirac's Deltas. Very roughly, the result is that the FT of the mapping

$$t \mapsto \sum_{j=-\infty}^{\infty} \delta\big(t + j\mathsf{T_s}\big)$$

is the mapping

$$f \mapsto \frac{1}{\mathsf{T_s}} \sum_{\eta=-\infty}^{\infty} \delta\Big(f + \frac{\eta}{\mathsf{T_s}}\Big),$$

where $\delta(\cdot)$ denotes Dirac's Delta. Needless to say, we are being extremely informal because we said nothing about convergence. This result is sometimes called the **picket-fence miracle**, because if we envision the plot of Dirac's Delta as an upward pointing bold arrow stemming from the origin, then the plot of a sum of shifted Delta's resembles a picket fence. The picket-fence miracle is that the FT of a picket fence is yet another scaled picket fence; see (Oppenheim and Willsky, 1997, Chapter 4, Example 4.8 and also Chapter 7, Section 7.1.1.) or (Kwakernaak and Sivan, 1991, Chapter 7, Example 7.4.19(c)).

In the mathematical literature, this result is called "the Poisson summation formula." It states that under certain conditions on the function $\psi \in \mathcal{L}_1$,

$$\sum_{j=-\infty}^{\infty} \psi(jT_s) = \frac{1}{T_s} \sum_{\eta=-\infty}^{\infty} \hat{\psi}\left(\frac{\eta}{T_s}\right). \tag{6.88}$$

To identify the roots of (6.88) define the mapping

$$\phi(t) = \sum_{j=-\infty}^{\infty} \psi(t + jT_s), \tag{6.89}$$

and note that this function is periodic in the sense that $\phi(t + T_s) = \phi(t)$ for every $t \in \mathbb{R}$. Consequently, it is instructive to study its Fourier Series on the interval $[-T_s/2, T_s/2]$ (Note A.3.5 in the appendix). Its $\eta$-th Fourier Series Coefficient with respect to the interval $[-T_s/2, T_s/2]$ is given by

$$\int_{-T_s/2}^{T_s/2} \phi(t) \frac{1}{\sqrt{T_s}} e^{-i2\pi\eta t/T_s} \, dt = \frac{1}{\sqrt{T_s}} \int_{-T_s/2}^{T_s/2} \sum_{j=-\infty}^{\infty} \psi(t + jT_s) e^{-i2\pi\eta t/T_s} \, dt$$

$$= \frac{1}{\sqrt{T_s}} \sum_{j=-\infty}^{\infty} \int_{-T_s/2+jT_s}^{T_s/2+jT_s} \psi(\tau) e^{-i2\pi\eta(\tau-jT_s)/T_s} \, d\tau$$

$$= \frac{1}{\sqrt{T_s}} \sum_{j=-\infty}^{\infty} \int_{-T_s/2+jT_s}^{T_s/2+jT_s} \psi(\tau) e^{-i2\pi\eta\tau/T_s} \, d\tau$$

$$= \frac{1}{\sqrt{T_s}} \int_{-\infty}^{\infty} \psi(\tau) e^{-i2\pi\eta\tau/T_s} \, d\tau$$

$$= \frac{1}{\sqrt{T_s}} \hat{\psi}\left(\frac{\eta}{T_s}\right), \quad \eta \in \mathbb{Z},$$

where the first equality follows from the definition of $\phi(\cdot)$ (6.89); the second by swapping the summation and the integration and by defining $\tau \triangleq t + jT_s$; the third by the periodicity of the complex exponential; the fourth because summing the integrals over disjoint intervals whose union is $\mathbb{R}$ is just the integral over $\mathbb{R}$; and the final equality from the definition of the FT.

We can thus interpret the RHS of (6.88) as the evaluation[12] at $t = 0$ of the Fourier Series of $\phi(\cdot)$ and the LHS as the evaluation of $\phi(\cdot)$ at $t = 0$. Having established the origin of the Poisson summation formula, we can now readily state conditions that guarantee that it holds. An example of a set of conditions that guarantees (6.88) is the following:

1) The function $\psi(\cdot)$ is integrable.

2) The RHS of (6.89) converges at $t = 0$.

3) The Fourier Series of $\phi(\cdot)$ converges at $t = 0$ to the value of $\phi(\cdot)$ at $t = 0$.

---

[12] At $t = 0$ the complex exponentials are all equal to one, and the Fourier Series is thus just the sum of the Fourier Series Coefficients.

We draw the reader's attention to the fact that it is not enough that both sides of (6.88) converge absolutely and that both $\psi(\cdot)$ and $\hat{\psi}(\cdot)$ be continuous; see (Katznelson, 1976, Chapter VI, Section 1, Exercise 15).

A setting where the above conditions are satisfied and where (6.88) thus holds is given in the following proposition.

**Proposition 6.10.1.** *Let $\psi(\cdot)$ be a continuous function satisfying*

$$\psi(t) = \begin{cases} 0 & \text{if } |t| \geq \mathsf{T}, \\ \int_{-\mathsf{T}}^{t} \xi(\tau)\,\mathrm{d}\tau & \text{otherwise,} \end{cases} \tag{6.90a}$$

*where*

$$\int_{-\mathsf{T}}^{\mathsf{T}} |\xi(\tau)|^2 \,\mathrm{d}\tau < \infty, \tag{6.90b}$$

*and where $\mathsf{T} > 0$ is some constant. Then for any $\mathsf{T_s} > 0$*

$$\sum_{j=-\infty}^{\infty} \psi(j\mathsf{T_s}) = \frac{1}{\mathsf{T_s}} \sum_{\eta=-\infty}^{\infty} \hat{\psi}\left(\frac{2\pi\eta}{\mathsf{T_s}}\right). \tag{6.90c}$$

**Proof.** The integrability of $\psi(\cdot)$ follows because $\psi(\cdot)$ is continuous and zero outside a finite interval. That the RHS of (6.89) converges at $t = 0$ follows because the fact that $\psi(\cdot)$ is zero outside the interval $[-\mathsf{T}, +\mathsf{T}]$ implies that only a finite number of terms contribute to the sum at $t = 0$. That the Fourier Series of $\phi(\cdot)$ converges at $t = 0$ to the value of $\phi(\cdot)$ at $t = 0$ follows from (Katznelson, 1976, Chapter 1, Section 6, Paragraph 6.2, Equation (6.2)) and from the corollary in (Katznelson, 1976, Chapter 1, Section 3, Paragraph 3.1). $\qquad\square$

## 6.11   Additional Reading

There are a number of excellent books on Fourier Analysis. We mention here (Katznelson, 1976), (Dym and McKean, 1972), (Pinsky, 2002), and (Körner, 1988). In particular, readers who would like to better understand how the FT is defined for energy-limited functions that are not integrable may wish to consult (Katznelson, 1976, Section VI 3.1) or (Dym and McKean, 1972, Sections 2.3–2.5). Numerous surprising applications of the FT can be found in (Körner, 1988).

Engineers often speak of the 2WT degrees of freedom that signals that are band-limited and time-limited have. A good starting point for the literature on this is (Slepian, 1976).

Bandlimited functions are intimately related to "entire functions of exponential type." For an accessible introduction to this concept see (Requicha, 1980); for a more mathematical approach see (Boas, 1954).

## 6.12   Exercises

**Exercise 6.1 (Symmetries of the FT).** Let $\mathbf{x}\colon \mathbb{R} \to \mathbb{C}$ be integrable, and let $\hat{\mathbf{x}}$ be its FT.

(i) Show that if $\mathbf{x}$ is a real signal, then $\hat{\mathbf{x}}$ is conjugate symmetric, i.e., $\hat{x}(-f) = \hat{x}^*(f)$, for every $f \in \mathbb{R}$.

(ii) Show that if $\mathbf{x}$ is purely imaginary (i.e., takes on only purely imaginary values), then $\hat{\mathbf{x}}$ is conjugate antisymmetric, i.e., $\hat{x}(-f) = -\hat{x}^*(f)$, for every $f \in \mathbb{R}$.

(iii) Show that $\hat{\mathbf{x}}$ can be written uniquely as the sum of a conjugate-symmetric function $\mathbf{g}_{\mathrm{cs}}$ and a conjugate-antisymmetric function $\mathbf{g}_{\mathrm{cas}}$. Express $\mathbf{g}_{\mathrm{cs}}$ & $\mathbf{g}_{\mathrm{cas}}$ in terms of $\hat{\mathbf{x}}$.

**Exercise 6.2 (Reconstructing a Function from Its IFT).** Formulate and prove a result analogous to Theorem 6.2.12 for the Inverse Fourier Transform.

**Exercise 6.3 (Eigenfunctions of the FT).** Show that if the energy-limited signal $\mathbf{x}$ satisfies $\hat{\mathbf{x}} = \lambda \mathbf{x}$ for some $\lambda \in \mathbb{C}$, then $\lambda$ can only be $\pm 1$ or $\pm i$. (The Hermite functions are such signals.)

**Exercise 6.4 (Existence of a Stable Filter (1)).** Let $W > 0$ be given. Does there exist a stable filter whose frequency response is zero for $|f| \leq W$ and is one for $W < f \leq 2W$?

**Exercise 6.5 (Existence of a Stable Filter (2)).** Let $W > 0$ be given. Does there exist a stable filter whose frequency response is given by $\cos(f)$ for all $|f| \geq W$?

**Exercise 6.6 (Existence of an Energy-Limited Signal).** Argue that there exists an energy-limited signal $\mathbf{x}$ whose FT is (the equivalence class of) the mapping $f \mapsto e^{-f}\, \mathrm{I}\{f \geq 0\}$. What is the energy in $\mathbf{x}$? What is the energy in the result of feeding $\mathbf{x}$ to an ideal unit-gain lowpass filter of cutoff frequency $W_{\mathrm{c}} = 1$?

**Exercise 6.7 (Passive Filters).** Let $\mathbf{h}$ be the impulse response of a stable filter. Show that the condition that "for every $\mathbf{x} \in \mathcal{L}_2$ the energy in $\mathbf{x} \star \mathbf{h}$ does not exceed the energy in $\mathbf{x}$" is equivalent to the condition

$$|\hat{h}(f)| \leq 1, \quad f \in \mathbb{R}.$$

**Exercise 6.8 (Real and Imaginary Parts of Bandlimited Signals).** Show that if $x(\cdot)$ is an integrable signal that is bandlimited to $W$ Hz, then its real and imaginary parts are also integrable signals that are bandlimited to $W$ Hz.

**Exercise 6.9 (Inner Products and Filtering).** Let $\mathbf{x}$ be an energy-limited signal that is bandlimited to $W$ Hz. Show that

$$\langle \mathbf{x}, \mathbf{y} \rangle = \langle \mathbf{x}, \mathbf{y} \star \mathrm{LPF}_W \rangle, \quad \mathbf{y} \in \mathcal{L}_2.$$

**Exercise 6.10 (Squaring a Signal).** Show that if $\mathbf{x}$ is an eneregy-limited signal that is bandlimited to $W$ Hz, then $t \mapsto x^2(t)$ is an integrable signal that is bandlimited to $2W$ Hz.

**Exercise 6.11 (Squared $\mathrm{sinc}(\cdot)$).** Find the FT and IFT of the mapping $t \mapsto \mathrm{sinc}^2(t)$.

**Exercise 6.12 (A Stable Filter).** Show that the IFT of the function

$$
\mathbf{g}_0 : f \mapsto
\begin{cases}
1 & \text{if } |f| \le a \\
\frac{b - |f|}{b - a} & \text{if } a < |f| < b \\
0 & \text{otherwise}
\end{cases}
$$

is given by

$$
\check{\mathbf{g}}_0 : t \mapsto \frac{1}{(\pi t)^2} \frac{\cos(2\pi a t) - \cos(2\pi b t)}{2(b - a)}
$$

and that this signal is integrable. Here $b > a > 0$.

**Exercise 6.13 (Multiplying Bandlimited Signals by a Carrier).** Let $\mathbf{x}$ be an integrable signal that is bandlimited to $W$ Hz.

(i) Show that if $f_c > W$, then

$$
\int_{-\infty}^{\infty} x(t) \cos(2\pi f_c t) \, \mathrm{d}t = \int_{-\infty}^{\infty} x(t) \sin(2\pi f_c t) \, \mathrm{d}t = 0.
$$

(ii) Show that if $f_c > W/2$, then

$$
\int_{-\infty}^{\infty} x(t) \cos^2(2\pi f_c t) \, \mathrm{d}t = \frac{1}{2} \int_{-\infty}^{\infty} x(t) \, \mathrm{d}t.
$$

**Exercise 6.14 (An Identity).** Prove that for every $W \in \mathbb{R}$

$$
\operatorname{sinc}(2Wt) \cos(2\pi Wt) = \operatorname{sinc}(4Wt), \quad t \in \mathbb{R}.
$$

Illustrate the identity in the frequency domain.

**Exercise 6.15 (Picket Fences).** If you are familiar with Dirac's Delta, explain how (6.88) is related to the heuristic statement that the FT of $\sum_{j \in \mathbb{Z}} \delta(t + j\mathsf{T_s})$ is $\mathsf{T_s}^{-1} \sum_{\eta \in \mathbb{Z}} \delta(f + \eta/\mathsf{T_s})$.

**Exercise 6.16 (Bounding the Derivative).** Show that if $\mathbf{x}$ is an energy-limited signal that is bandlimited to $W$ Hz, then its time-$t$ derivative $x'(t)$ satisfies

$$
|x'(t)| \le \sqrt{\frac{8}{3}} \, \pi W^{3/2} \, \|\mathbf{x}\|_2 \,, \quad t \in \mathbb{R}.
$$

*Hint: Use Proposition 6.4.5 and the Cauchy-Schwarz Inequality*

**Exercise 6.17 (Another Notion of Bandwidth).** Let $\mathcal{U}$ denote the set of all energy-limited signals $\mathbf{u}$ such that at least 90% of the energy of $\mathbf{u}$ is contained in the band $[-W, W]$. Is $\mathcal{U}$ a linear subspace of $\mathcal{L}_2$?

# Chapter 7

# Passband Signals and Their Representation

## 7.1 Introduction

The signals encountered in wireless communications are typically **real passband signals**. In this chapter we shall define such signals and define their **bandwidth around a carrier frequency**. We shall then explain how such signals can be represented using their complex **baseband representation**. We shall emphasize two relationships: that between the energy in the passband signal and in its baseband representation, and that between the bandwidth of the passband signal around the carrier frequency and the bandwidth of its baseband representation. We ask the reader to pay special attention to the fact that only *real* passband signals have a baseband representation.

Most of the chapter deals with the family of integrable passband signals. As we shall see in Corollary 7.2.4, an integrable passband signal must have finite energy, and this family is thus a subset of the family of energy-limited passband signals. Restricting ourselves to integrable signals—while reducing the generality of some of the results—simplifies the exposition because we can discuss the Fourier Transform without having to resort to the $L_2$-Fourier Transform, which requires all statements to be phrased in terms of equivalence classes. But most of the derived results will also hold for the more general family of energy-limited passband signals with only slight modifications. The required modifications are discussed in Section 7.7.

## 7.2 Baseband and Passband Signals

Integrable signals that are bandlimited to $W$ Hz were defined in Definition 6.4.9. By Proposition 6.4.10, an integrable signal $\mathbf{x}$ is bandlimited to $W$ Hz if it is continuous and if its FT is zero for all frequencies outside the band $[-W, W]$. The bandwidth of $\mathbf{x}$ is the smallest $W$ to which it is bandlimited (Definition 6.4.13). As an example, Figure 7.1 depicts the FT $\hat{\mathbf{x}}$ of a *real* signal $\mathbf{x}$, which is bandlimited to $W$ Hz. Since the signal $\mathbf{x}$ in this example is real, its FT is conjugate-symmetric, (i.e., $\hat{x}(-f) = \hat{x}^*(f)$ for all frequencies $f \in \mathbb{R}$). Thus, the magnitude of $\hat{\mathbf{x}}$ is symmetric (even), i.e., $|\hat{x}(f)| = |\hat{x}(-f)|$, but its phase is anti-symmetric (odd). In the figure dashed lines indicate this conjugate symmetry.

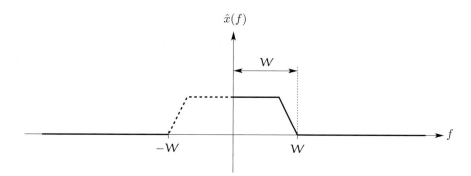

**Figure 7.1:** The FT $\hat{\mathbf{x}}$ of a real bandwidth-$W$ baseband signal $\mathbf{x}$.

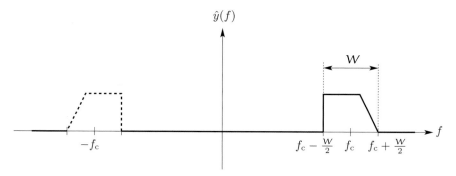

**Figure 7.2:** The FT $\hat{\mathbf{y}}$ of a real passband signal $\mathbf{y}$ that is bandlimited to $W$ Hz around the carrier frequency $f_{\mathrm{c}}$.

Consider now the real signal $\mathbf{y}$ whose FT $\hat{\mathbf{y}}$ is depicted in Figure 7.2. Again, since the signal is real, its FT is conjugate-symmetric, and hence the dashed lines. This signal (if continuous) is bandlimited to $f_{\mathrm{c}} + W/2$ Hz. But note that $\hat{\mathbf{y}}(f) = 0$ for all frequencies $f$ in the interval $|f| < f_{\mathrm{c}} - W/2$. Signals such as $\mathbf{y}$ are often encountered in wireless communication, because in a wireless channel the very-low frequencies often suffer severe attenuation and are therefore seldom used. Another reason is the concurrent use of the wireless spectrum by many systems. If all systems transmitted in the same frequency band, they would interfere with each other. Consequently, different systems are often assigned different carrier frequencies so that their transmitted signals will not overlap in frequency. This is why different radio stations transmit around different carrier frequencies.

## 7.2.1  Definition and Characterization

To describe signals such as $\mathbf{y}$ we use the following definition for passband signals. We ask the reader to recall the definition of the impulse response $\mathrm{BPF}_{W,f_{\mathrm{c}}}(\cdot)$ (see (5.21)) and of the frequency response $\widehat{\mathrm{BPF}}_{W,f_{\mathrm{c}}}(\cdot)$ (see (6.41)) of the ideal unit-gain bandpass filter of bandwidth $W$ around the carrier frequency $f_c$.

**Definition 7.2.1 (A Passband Signal).** *A signal $\mathbf{x}_{\mathrm{PB}}$ is said to be an **integrable passband signal that is bandlimited to** $W$ Hz **around the carrier frequency** $f_c$ if it is integrable*

$$\mathbf{x}_{\mathrm{PB}} \in \mathcal{L}_1; \tag{7.1a}$$

*the carrier frequency $f_c$ satisfies*

$$f_c > \frac{W}{2} > 0; \tag{7.1b}$$

*and if $\mathbf{x}_{\mathrm{PB}}$ is unaltered when it is fed to an ideal unit-gain bandpass filter of bandwidth $W$ around the carrier frequency $f_c$*

$$x_{\mathrm{PB}}(t) = \big(\mathbf{x}_{\mathrm{PB}} \star \mathrm{BPF}_{W,f_c}\big)(t), \quad t \in \mathbb{R}. \tag{7.1c}$$

*An **energy-limited passband signal that is bandlimited to** $W$ Hz **around the carrier frequency** $f_c$ is analogously defined but with (7.1a) replaced by the condition*

$$\mathbf{x}_{\mathrm{PB}} \in \mathcal{L}_2. \tag{7.1a'}$$

(That the convolution in (7.1c) is defined at every $t \in \mathbb{R}$ whenever $\mathbf{x}_{\mathrm{PB}}$ is integrable can be shown using Proposition 6.2.5 because $\mathrm{BPF}_{W,f_c}$ is the Inverse Fourier Transform of the integrable function $f \mapsto \mathrm{I}\{|\,|f| - f_c| \leq W/2\}$. That the convolution is defined at every $t \in \mathbb{R}$ also when $\mathbf{x}_{\mathrm{PB}}$ is of finite energy can be shown by noting that $\mathrm{BPF}_{W,f_c}$ is of finite energy, and the convolution of two finite-energy signals is defined at every time $t \in \mathbb{R}$; see Section 5.5.)

In analogy to Proposition 6.4.10 we have the following characterization:

**Proposition 7.2.2 (Characterizing Integrable Passband Signals).** *Let $f_c$ and $W$ satisfy $f_c > W/2 > 0$. If $\mathbf{x}_{\mathrm{PB}}$ is an integrable signal, then each of the following statements is equivalent to the statement that $\mathbf{x}_{\mathrm{PB}}$ is an integrable passband signal that is bandlimited to $W$ Hz around the carrier frequency $f_c$.*

(a) *The signal $\mathbf{x}_{\mathrm{PB}}$ is unaltered when it is bandpass filtered:*

$$x_{\mathrm{PB}}(t) = \big(\mathbf{x}_{\mathrm{PB}} \star \mathrm{BPF}_{W,f_c}\big)(t), \quad t \in \mathbb{R}. \tag{7.2}$$

(b) *The signal $\mathbf{x}_{\mathrm{PB}}$ can be expressed as*

$$x_{\mathrm{PB}}(t) = \int_{|\,|f|-f_c| \leq W/2} \hat{x}_{\mathrm{PB}}(f)\, e^{i2\pi f t}\, \mathrm{d}f, \quad t \in \mathbb{R}. \tag{7.3}$$

(c) *The signal $\mathbf{x}_{\mathrm{PB}}$ is continuous and*

$$\hat{x}_{\mathrm{PB}}(f) = 0, \quad |\,|f| - f_c| > \frac{W}{2}. \tag{7.4}$$

(d) *There exists an integrable function $\mathbf{g}$ such that*

$$x_{\mathrm{PB}}(t) = \int_{|\,|f|-f_c| \leq W/2} g(f)\, e^{i2\pi f t}\, \mathrm{d}f, \quad t \in \mathbb{R}. \tag{7.5}$$

**Proof.** The proof is similar to the proof of Proposition 6.4.10 and is omitted. $\square$

### 7.2.2  Important Properties

By comparing (7.4) with (6.62) we obtain:

**Corollary 7.2.3 (Passband Signals Are Bandlimited).** *If* $\mathbf{x}_{\mathrm{PB}}$ *is an integrable pass-band signal that is bandlimited to* W *Hz around the carrier frequency* $f_{\mathrm{c}}$*, then it is an integrable signal that is bandlimited to* $f_{\mathrm{c}} + W/2$ *Hz.*

Using Corollary 7.2.3 and Note 6.4.12 we obtain:

**Corollary 7.2.4 (Integrable Passband Signals Are of Finite Energy).** *Any inte-grable passband signal that is bandlimited to* W *Hz around the carrier frequency* $f_{\mathrm{c}}$ *is of finite energy.*

**Proposition 7.2.5 (Integrable Passband Signals through Stable Filters).** *If* $\mathbf{x}_{\mathrm{PB}}$ *is an integrable passband signal that is bandlimited to* W *Hz around the carrier frequency* $f_{\mathrm{c}}$*, and if* $\mathbf{h} \in \mathcal{L}_1$ *is the impulse response of a stable filter, then the convolution* $\mathbf{x}_{\mathrm{PB}} \star \mathbf{h}$ *is defined at every epoch; it is an integrable passband signal that is bandlimited to* W *Hz around the carrier frequency* $f_{\mathrm{c}}$*; and its FT is the mapping* $f \mapsto \hat{x}_{\mathrm{PB}}(f)\,\hat{h}(f)$*.*

**Proof.** The proof is similar to the proof of the analogous result for bandlimited signals (Proposition 6.5.2) and is omitted.                                              □

## 7.3  Bandwidth around a Carrier Frequency

**Definition 7.3.1 (The Bandwidth around a Carrier Frequency).** *The **bandwidth around the carrier*** $f_{\mathrm{c}}$ *of an integrable or energy-limited passband signal* $\mathbf{x}_{\mathrm{PB}}$ *is the smallest* W *for which both* (7.1b) *and* (7.1c) *hold.*

**Note 7.3.2 (The Carrier Frequency Is Critical).** The bandwidth of $\mathbf{x}_{\mathrm{PB}}$ around the carrier frequency $f_{\mathrm{c}}$ is determined not only by the FT of $\mathbf{x}_{\mathrm{PB}}$ but also by $f_{\mathrm{c}}$.

For example, the real passband signal whose FT is depicted in Figure 7.3 is of bandwidth W around the carrier frequency $f_{\mathrm{c}}$, but its bandwidth is smaller around a slightly higher carrier frequency.

At first it may seem that the definition of bandwidth for passband signals is incon-sistent with the definition for baseband signals. This, however, is not the case. A good way to remember the definitions is to focus on real signals. For such signals the bandwidth for both baseband and passband signals is defined as the length of an interval of *positive frequencies* where the FT of the signal may be nonzero. For baseband signals the bandwidth is the length of the smallest interval of positive frequencies of the form $[0, W]$ containing all positive frequencies where the FT may be nonzero. For passband signals it is the length of the smallest interval of positive frequencies that is *symmetric around the carrier frequency* $f_{\mathrm{c}}$ and that contains all positive frequencies where the signal may be nonzero. (For complex signals we have to allow for the fact that the zeros of the FT may not be symmetric sets around the origin.) See also Figures 6.2 and 6.3.

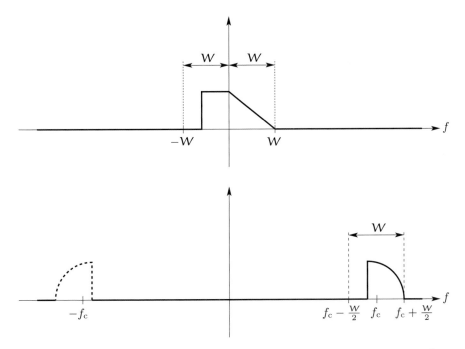

**Figure 7.3:** The FT of a complex baseband signal of bandwidth $W$ Hz (above) and of a real passband signal of bandwidth $W$ Hz around the carrier frequency $f_c$ (below).

We draw the reader's attention to an important consequence of our definition of bandwidth:

**Proposition 7.3.3 (Multiplication by a Carrier Doubles the Bandwidth).** *If* $\mathbf{x}$ *is an integrable signal of bandwidth* $W$ *Hz and if* $f_c > W$, *then* $t \mapsto x(t) \cos(2\pi f_c t)$ *is an integrable passband signal of bandwidth* $2W$ *around the carrier frequency* $f_c$.

**Proof.** Define $\mathbf{y} \colon t \mapsto x(t) \cos(2\pi f_c t)$. The proposition is a straightforward consequence of the definition of the bandwidth of $\mathbf{x}$ (Definition 6.4.13); the definition of the bandwidth of $\mathbf{y}$ around the carrier frequency $f_c$ (Definition 7.3.1); and the fact that if $\mathbf{x}$ is a continuous integrable signal of FT $\hat{x}$, then $\mathbf{y}$ is a continuous integrable signal of FT

$$\hat{y}(f) = \frac{1}{2}\big(\hat{x}(f - f_c) + \hat{x}(f + f_c)\big), \quad f \in \mathbb{R}, \tag{7.6}$$

where (7.6) follows from the calculation

$$\hat{y}(f) = \int_{-\infty}^{\infty} y(t)\, e^{-i2\pi f t}\, \mathrm{d}t$$

$$= \int_{-\infty}^{\infty} x(t) \cos(2\pi f_c t)\, e^{-i2\pi f t}\, \mathrm{d}t$$

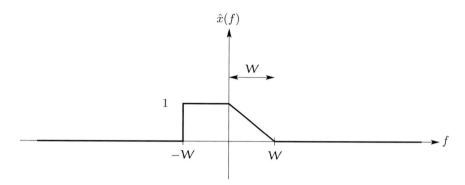

**Figure 7.4:** The FT of a complex baseband bandwidth-$W$ signal $\mathbf{x}$.

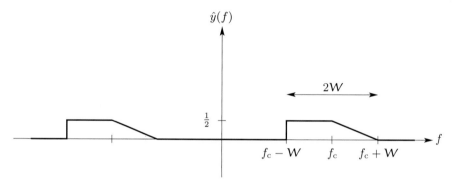

**Figure 7.5:** The FT of $\mathbf{y} \colon t \mapsto x(t) \cos(2\pi f_c t)$, where $\hat{x}$ is as depicted in Figure 7.4. Note that $\mathbf{x}$ is of bandwidth $W$ and that $\mathbf{y}$ is of bandwidth $2W$ around the carrier frequency $f_c$.

$$= \int_{-\infty}^{\infty} x(t) \frac{e^{\mathrm{i}2\pi f_c t} + e^{-\mathrm{i}2\pi f_c t}}{2} e^{-\mathrm{i}2\pi f t} \, \mathrm{d}t$$

$$= \frac{1}{2} \int_{-\infty}^{\infty} x(t) \, e^{-\mathrm{i}2\pi (f - f_c)t} \, \mathrm{d}t + \frac{1}{2} \int_{-\infty}^{\infty} x(t) \, e^{-\mathrm{i}2\pi (f + f_c)t} \, \mathrm{d}t$$

$$= \frac{1}{2} \bigl( \hat{x}(f - f_c) + \hat{x}(f + f_c) \bigr), \quad f \in \mathbb{R}.$$

As an illustration of the relation (7.6) note that if $\mathbf{x}$ is the complex bandwidth-$W$ signal whose FT is depicted in Figure 7.4, then the signal $\mathbf{y} \colon t \mapsto x(t) \cos(2\pi f_c t)$ is the complex passband signal of bandwidth $2W$ around $f_c$ whose FT is depicted in Figure 7.5.

Similarly, if $\mathbf{x}$ is the real baseband signal of bandwidth $W$ whose FT is depicted in Figure 7.6, then $\mathbf{y} \colon t \mapsto x(t) \cos(2\pi f_c t)$ is the real passband signal of bandwidth $2W$ around $f_c$ whose FT is depicted in Figure 7.7.     □

In wireless applications the bandwidth $W$ of the signals around the carrier frequency is typically much smaller than the carrier frequency $f_c$, but for most of our results

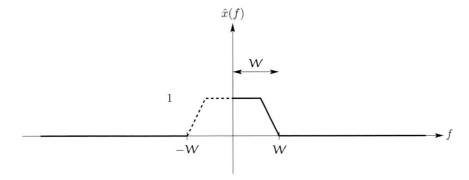

**Figure 7.6:** The FT of a real baseband bandwidth-$W$ signal $\mathbf{x}$.

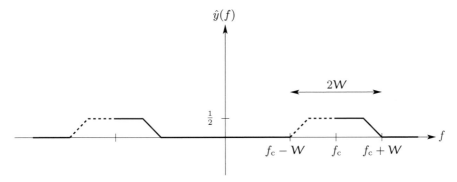

**Figure 7.7:** The FT of $\mathbf{y} \colon t \mapsto x(t) \cos\left(2\pi f_{\mathrm{c}} t\right)$, where $\hat{x}$ is as depicted in Figure 7.6. Here $\mathbf{x}$ is of bandwidth $W$ and $\mathbf{y}$ is of bandwidth $2W$ around the carrier frequency $f_{\mathrm{c}}$.

it suffices that (7.1b) hold.

The notion of a passband signal is also applied somewhat loosely in instances where the signals are not bandlimited. Engineers say that an energy-limited signal is a passband signal around the carrier frequency $f_{\mathrm{c}}$ if most of its energy is contained in frequencies that are close to $f_{\mathrm{c}}$ and $-f_{\mathrm{c}}$. Notice that in this "definition" we are relying heavily on Parseval's theorem. I.e., we think about the energy $\|\mathbf{x}\|_2^2$ of $\mathbf{x}$ as being computed in the frequency domain, i.e., by computing $\|\hat{\mathbf{x}}\|_2^2 = \int |\hat{x}(f)|^2 \, \mathrm{d}f$. By "most of the energy is contained in frequencies that are close to $f_{\mathrm{c}}$ and $-f_{\mathrm{c}}$" we thus mean that most of the contributions to this integral come from small frequency intervals around $f_{\mathrm{c}}$ and $-f_{\mathrm{c}}$. In other words, we say that $\mathbf{x}$ is a passband signal whose energy is mostly concentrated in a bandwidth $W$ around the carrier frequency $f_{\mathrm{c}}$ if

$$\int_{-\infty}^{\infty} |\hat{x}(f)|^2 \, \mathrm{d}f \approx \int_{||f| - f_{\mathrm{c}}| \leq W/2} |\hat{x}(f)|^2 \, \mathrm{d}f. \tag{7.7}$$

Similarly, a signal is approximately a baseband signal that is bandlimited to $W$ Hz if

$$\int_{-\infty}^{\infty} |\hat{x}(f)|^2 \, \mathrm{d}f \approx \int_{-W}^{W} |\hat{x}(f)|^2 \, \mathrm{d}f. \tag{7.8}$$

## 7.4   Real Passband Signals

Before discussing the baseband representation of real passband signals we emphasize the following.

(i) The passband signals transmitted and received in Digital Communications
are real.

(ii) Only real passband signals have a baseband representation.

(iii) The baseband representation of a real passband signal is typically a complex
signal.

(iv) While the FT of real signals is conjugate-symmetric (6.3), this does not imply
any symmetry with respect to the carrier frequency. Thus, the FT depicted
in Figure 7.2 and the one depicted in Figure 7.7 both correspond to real
passband signals. (The former is bandlimited to $W$ Hz around $f_c$ and the
latter to $2W$ around $f_c$.)

We also note that if $\mathbf{x}$ is a real integrable signal, then its FT must be conjugate-
symmetric. But if $\mathbf{g} \in \mathcal{L}_1$ is such that its IFT $\check{\mathbf{g}}$ is real, it does not follow that $\mathbf{g}$
must be conjugate-symmetric. For example, the conjugate symmetry could be
broken on a set of frequencies of Lebesgue measure zero, a set that does not influence the IFT. As the next proposition shows, this is the only way the conjugate
symmetry can be broken.

**Proposition 7.4.1.** *If* $\mathbf{x}$ *is a real signal and if* $\mathbf{x} = \check{\mathbf{g}}$ *for some integrable function*
$\mathbf{g} \colon f \mapsto g(f)$, *then:*

(i) *The signal* $\mathbf{x}$ *can be represented as the IFT of a conjugate-symmetric integrable function.*

(ii) *The function* $\mathbf{g}$ *and the conjugate-symmetric function* $f \mapsto \big(g(f)+g^*(-f)\big)/2$
*agree except on a set of frequencies of Lebesgue measure zero.*

**Proof.** Since $\mathbf{x}$ is real and since $\mathbf{x} = \check{\mathbf{g}}$ it follows that

$$
\begin{aligned}
x(t) &= \mathrm{Re}\big(x(t)\big) \\
&= \frac{1}{2}x(t) + \frac{1}{2}x^*(t) \\
&= \frac{1}{2}\int_{-\infty}^{\infty} g(f)\,e^{\mathrm{i}2\pi ft}\,\mathrm{d}f + \frac{1}{2}\left(\int_{-\infty}^{\infty} g(f)\,e^{\mathrm{i}2\pi ft}\,\mathrm{d}f\right)^* \\
&= \frac{1}{2}\int_{-\infty}^{\infty} g(f)\,e^{\mathrm{i}2\pi ft}\,\mathrm{d}f + \frac{1}{2}\int_{-\infty}^{\infty} g^*(f)\,e^{-\mathrm{i}2\pi ft}\,\mathrm{d}f
\end{aligned}
$$

$$= \frac{1}{2} \int_{-\infty}^{\infty} g(f) \, e^{\mathrm{i}2\pi ft} \, \mathrm{d}f + \frac{1}{2} \int_{-\infty}^{\infty} g^*(-\tilde{f}) \, e^{\mathrm{i}2\pi \tilde{f}t} \, \mathrm{d}\tilde{f}$$

$$= \int_{-\infty}^{\infty} \frac{g(f) + g^*(-f)}{2} \, e^{\mathrm{i}2\pi ft} \, \mathrm{d}f, \quad t \in \mathbb{R},$$

where the first equality follows from the hypothesis that $\mathbf{x}$ is a real signal; the second because for any $z \in \mathbb{C}$ we have $\mathrm{Re}(z) = (z + z^*)/2$; the third by the hypothesis that $\mathbf{x} = \check{\mathbf{g}}$; the fourth because conjugating a complex integral is tantamount to conjugating the integrand (Proposition 2.3.1 (ii)); the fifth by changing the integration variable in the second integral to $\tilde{f} \triangleq -f$; and the sixth by combining the integrals. Thus, $\mathbf{x}$ is the IFT of the conjugate-symmetric function defined by $f \mapsto \big(g(f) + g^*(-f)\big)/2$, and (i) is established.

As to (ii), since $\mathbf{x}$ is the IFT of both $\mathbf{g}$ and $f \mapsto \big(g(f) + g^*(-f)\big)/2$, it follows from the IFT analog of Theorem 6.2.12 that the two agree outside a set of Lebesgue measure zero. $\qquad\square$

## 7.5 The Analytic Signal

In this section we shall define the **analytic representation** of a real passband signal. This is also sometimes called the **analytic signal** associated with the signal. We shall use the two terms interchangeably. The analytic representation will serve as a steppingstone to the **baseband representation**, which is extremely important in Digital Communications. We emphasize that an analytic signal can only be associated with a *real* passband signal. The analytic signal itself, however, is complex-valued.

### 7.5.1 Definition and Characterization

Let $\mathbf{x}_{\mathrm{PB}}$ be a real integrable passband signal that is bandlimited to $W$ Hz around the carrier frequency $f_{\mathrm{c}}$. We would have liked to define its analytic representation as the complex signal $\mathbf{x}_{\mathrm{A}}$ whose FT is the mapping

$$f \mapsto \hat{x}_{\mathrm{PB}}(f) \, \mathrm{I}\{f \geq 0\}, \tag{7.9}$$

i.e., as the integrable signal whose FT is equal to zero at negative frequencies and to $\hat{x}_{\mathrm{PB}}(f)$ at nonnegative frequencies. While this is often the way we think about $\mathbf{x}_{\mathrm{A}}$, there are two problems with this definition: an existence problem and a uniqueness problem. It is not *prima facie* clear that there exists an integrable signal whose FT is the mapping (7.9). (We shall soon see that there does.) And, since two signals that differ on a set of Lebesgue measure zero have identical Fourier Transforms, the above definition would not fully specify $\mathbf{x}_{\mathrm{A}}$. This could be remedied by insisting that $\mathbf{x}_{\mathrm{A}}$ be continuous, but this would further exacerbate the existence issue. (We shall see that there does exist a unique integrable continuous signal whose FT is the mapping (7.9), but this requires proof.) Our approach is to define $\mathbf{x}_{\mathrm{A}}$ as the IFT of the mapping (7.9) and to then explore the properties of $\mathbf{x}_{\mathrm{A}}$.

**Definition 7.5.1 (Analytic Representation of a Real Passband Signal).** *The analytic representation of a real integrable passband signal* $\mathbf{x}_{\mathrm{PB}}$ *that is bandlimited to* W Hz *around the carrier frequency* $f_c$ *is the complex signal* $\mathbf{x}_{\mathrm{A}}$ *defined by*

$$x_{\mathrm{A}}(t) \triangleq \int_0^\infty \hat{x}_{\mathrm{PB}}(f)\, e^{\mathrm{i}2\pi ft}\, \mathrm{d}f, \quad t \in \mathbb{R}. \tag{7.10}$$

Note that, by Proposition 7.2.2, $\hat{x}_{\mathrm{PB}}(f)$ vanishes at frequencies $f$ that satisfy $\big||f| - f_c\big| > W/2$, so we can also write (7.10) as

$$x_{\mathrm{A}}(t) = \int_{f_c - \frac{W}{2}}^{f_c + \frac{W}{2}} \hat{x}_{\mathrm{PB}}(f)\, e^{\mathrm{i}2\pi ft}\, \mathrm{d}f, \quad t \in \mathbb{R}. \tag{7.11}$$

This latter expression has the advantage that it makes it clear that the integral is well-defined for every $t \in \mathbb{R}$, because the integrability of $\mathbf{x}_{\mathrm{PB}}$ implies that the integrand is bounded, i.e., that $\hat{x}_{\mathrm{PB}}(f) \leq \|\mathbf{x}_{\mathrm{PB}}\|_1$ for every $f \in \mathbb{R}$ (Theorem 6.2.11) and hence that the mapping $f \mapsto \hat{x}_{\mathrm{PB}}(f)\, \mathrm{I}\{|f - f_c| \leq W/2\}$ is integrable.

Also note that our definition of the analytic signal may be off by a factor of two or $\sqrt{2}$ from the one used in some textbooks. (Some textbooks introduce a factor of $\sqrt{2}$ in order to make the energy in the analytic signal equal that in the passband signal. We do not do so and hence end up with a factor of two in (7.23) ahead.)

We next show that the analytic signal $\mathbf{x}_{\mathrm{A}}$ is a continuous and integrable signal and that its FT is given by the mapping (7.9). In fact, we prove more.

**Proposition 7.5.2 (Characterizations of the Analytic Signal).** *Let* $\mathbf{x}_{\mathrm{PB}}$ *be a real integrable passband signal that is bandlimited to* W Hz *around the carrier frequency* $f_c$. *Then each of the following statements is equivalent to the statement that the complex-valued signal* $\mathbf{x}_{\mathrm{A}}$ *is its analytic representation.*

*(a) The signal* $\mathbf{x}_{\mathrm{A}}$ *is given by*

$$x_{\mathrm{A}}(t) = \int_{f_c - \frac{W}{2}}^{f_c + \frac{W}{2}} \hat{x}_{\mathrm{PB}}(f)\, e^{\mathrm{i}2\pi ft}\, \mathrm{d}f, \quad t \in \mathbb{R}. \tag{7.12}$$

*(b) The signal* $\mathbf{x}_{\mathrm{A}}$ *is a continuous integrable signal satisfying*

$$\hat{x}_{\mathrm{A}}(f) = \begin{cases} \hat{x}_{\mathrm{PB}}(f) & \text{if } f \geq 0, \\ 0 & \text{otherwise.} \end{cases} \tag{7.13}$$

*(c) The signal* $\mathbf{x}_{\mathrm{A}}$ *is an integrable passband signal that is bandlimited to* W Hz *around the carrier frequency* $f_c$ *and that satisfies (7.13).*

*(d) The signal* $\mathbf{x}_{\mathrm{A}}$ *is given by*

$$\mathbf{x}_{\mathrm{A}} = \mathbf{x}_{\mathrm{PB}} \star \check{\mathbf{g}} \tag{7.14a}$$

*for every integrable mapping* $\mathbf{g} \colon f \mapsto g(f)$ *satisfying*

$$g(f) = 1, \quad |f - f_c| \leq \frac{W}{2}, \tag{7.14b}$$

*and*

$$g(f) = 0, \quad |f + f_c| \leq \frac{W}{2} \tag{7.14c}$$

*(with $g(f)$ unspecified at other frequencies).*

**Proof.** That Condition (a) is equivalent to the statement that $\mathbf{x}_A$ is the analytic representation of $\mathbf{x}_{PB}$ is just a restatement of Definition 7.5.1. It thus only remains to show that Conditions (a), (b), (c), and (d) are equivalent. We shall do so by establishing that (a) $\Leftrightarrow$ (d); that (b) $\Leftrightarrow$ (c); that (b) $\Rightarrow$ (a); and that (d) $\Rightarrow$ (c).

To establish (a) $\Leftrightarrow$ (d) we use the integrability of $\mathbf{x}_{PB}$ and of $\mathbf{g}$ to compute $\mathbf{x}_{PB} \star \check{\mathbf{g}}$ using Proposition 6.2.5 as

$$
\begin{aligned}
\left(\mathbf{x}_{PB} \star \check{\mathbf{g}}\right)(t) &= \int_{-\infty}^{\infty} \hat{x}_{PB}(f)\, g(f)\, e^{i 2\pi f t}\, \mathrm{d}f \\
&= \int_{0}^{\infty} \hat{x}_{PB}(f)\, g(f)\, e^{i 2\pi f t}\, \mathrm{d}f \\
&= \int_{f_c - \frac{W}{2}}^{f_c + \frac{W}{2}} \hat{x}_{PB}(f)\, g(f)\, e^{i 2\pi f t}\, \mathrm{d}f \\
&= \int_{f_c - \frac{W}{2}}^{f_c + \frac{W}{2}} \hat{x}_{PB}(f)\, e^{i 2\pi f t}\, \mathrm{d}f, \quad t \in \mathbb{R},
\end{aligned}
$$

where the first equality follows from Proposition 6.2.5; the second because the assumption that $\mathbf{x}_{PB}$ is a passband signal implies, by Proposition 7.2.2 (*cf.* (c)), that the only negative frequencies $f < 0$ where $\hat{x}_{PB}(f)$ can be nonzero are those satisfying $|-f - f_c| \leq W/2$, and at those frequencies $\mathbf{g}$ is zero by (7.14c); the third by Proposition 7.2.2 (*cf.* (c)); and the fourth equality by (7.14b). This establishes that (a) $\Leftrightarrow$ (d).

The equivalence (b) $\Leftrightarrow$ (c) is an immediate consequence of Proposition 7.2.2. That (b) $\Rightarrow$ (a) can be proved using Corollary 6.2.14 as follows. If (b) holds, then $\mathbf{x}_A$ is a continuous integrable signal whose FT is given by the integrable function on the RHS of (7.13) and therefore, by Corollary 6.2.14, $\mathbf{x}_A$ is the IFT of the RHS of (7.13), thus establishing (a).

We now complete the proof by showing that (d) $\Rightarrow$ (c). To this end let $\mathbf{g} \colon f \mapsto g(f)$ be a continuous integrable function satisfying (7.14b) & (7.14c) and additionally satisfying that its IFT $\check{\mathbf{g}}$ is integrable. For example, $\mathbf{g}$ could be the function from $\mathbb{R}$ to $\mathbb{R}$ that is defined by

$$
g(f) = \begin{cases}
1 & \text{if } |f - f_c| \leq W/2, \\
0 & \text{if } |f - f_c| \geq W_c/2, \\
\frac{W_c - 2|f - f_c|}{W_c - W} & \text{otherwise,}
\end{cases} \tag{7.15}
$$

where $W_c$ can be chosen arbitrarily in the range

$$W < W_c < 2f_c. \tag{7.16}$$

This function is depicted in Figure 7.8. By direct calculation, it can be shown that its IFT is given by[1]

$$\check{g}(t) = e^{\mathrm{i}2\pi f_c t} \frac{1}{(\pi t)^2} \frac{\cos(\pi W t) - \cos(\pi W_c t)}{W_c - W}, \quad t \in \mathbb{R}, \tag{7.17}$$

which is integrable. Define now $\mathbf{h} = \check{\mathbf{g}}$ and note that, by Corollary 6.2.14, $\hat{\mathbf{h}} = \mathbf{g}$. If (d) holds, then

$$\mathbf{x}_A = \mathbf{x}_{PB} \star \check{\mathbf{g}}$$
$$= \mathbf{x}_{PB} \star \mathbf{h},$$

so $\mathbf{x}_A$ is the result of feeding an integrable passband signal that is bandlimited to $W$ Hz around the carrier frequency $f_c$ (the signal $\mathbf{x}_{PB}$) through a stable filter (of impulse response $\mathbf{h}$). Consequently, by Proposition 7.2.5, $\mathbf{x}_A$ is an integrable passband signal that is bandlimited to $W$ Hz around the carrier frequency $f_c$ and its FT is given by $f \mapsto \hat{x}_{PB}(f)\hat{h}(f)$. Thus, as we next justify,

$$\hat{x}_A(f) = \hat{x}_{PB}(f)\,\hat{h}(f)$$
$$= \hat{x}_{PB}(f)\,g(f)$$
$$= \hat{x}_{PB}(f)\,g(f)\,\mathrm{I}\{f \geq 0\}$$
$$= \hat{x}_{PB}(f)\,\mathrm{I}\{f \geq 0\}, \quad f \in \mathbb{R},$$

thus establishing (c). Here the third equality is justified by noting that the assumption that $\mathbf{x}_{PB}$ is a passband signal implies, by Proposition 7.2.2 (*cf.* (c)), that the only negative frequencies $f < 0$ where $\hat{x}_{PB}(f)$ can be nonzero are those satisfying $|-f - f_c| \leq W/2$, and at those frequencies $\mathbf{g}$ is zero by (7.15), (7.16), and (7.1b). The fourth equality follows by noting that the assumption that $\mathbf{x}_{PB}$ is a passband signal implies, by Proposition 7.2.2 (*cf.* (c)), that the only positive frequencies $f > 0$ where $\hat{x}_{PB}(f)$ can be nonzero are those satisfying $|f - f_c| \leq W/2$ and at those frequencies $g(f) = 1$ by (7.15). □

### 7.5.2   From $\mathbf{x}_A$ back to $\mathbf{x}_{PB}$

Proposition 7.5.2 describes the analytic representation $\mathbf{x}_A$ in terms of the real passband signal $\mathbf{x}_{PB}$. This representation would have been useless if we had not been able to recover $\mathbf{x}_{PB}$ from $\mathbf{x}_A$. Fortunately, we can. The key is that, because $\mathbf{x}_{PB}$ is real, its FT is conjugate-symmetric

$$\hat{x}_{PB}(-f) = \hat{x}_{PB}^*(f), \quad f \in \mathbb{R}. \tag{7.18}$$

Consequently, since the FT of $\mathbf{x}_A$ is equal to that of $\mathbf{x}_{PB}$ at the positive frequencies and to zero at the negative frequencies (7.13), we can add to $\hat{\mathbf{x}}_A$ its conjugated mirror-image to obtain $\hat{\mathbf{x}}_{PB}$:

$$\hat{x}_{PB}(f) = \hat{x}_A(f) + \hat{x}_A^*(-f), \quad f \in \mathbb{R}; \tag{7.19}$$

---

[1] At $t = 0$, the RHS of (7.17) should be interpreted as $(W + W_c)/2$.

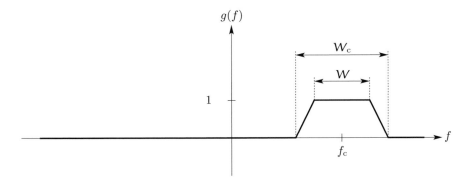

**Figure 7.8:** The function **g** of (7.15), which is used in the proof of Proposition 7.5.2.

see Figure 7.12 on Page 124. From here it is just a technicality to obtain the time-domain relationship

$$x_{\mathrm{PB}}(t) = 2\operatorname{Re}\big(x_{\mathrm{A}}(t)\big), \quad t \in \mathbb{R}. \tag{7.20}$$

These results are summarized in the following proposition.

**Proposition 7.5.3 (Recovering $x_{\mathrm{PB}}$ from $x_{\mathrm{A}}$).** *Let $x_{\mathrm{PB}}$ be a real integrable pass-band signal that is bandlimited to $W$ Hz around the carrier frequency $f_{\mathrm{c}}$, and let $x_{\mathrm{A}}$ be its analytic representation. Then,*

$$\hat{x}_{\mathrm{PB}}(f) = \hat{x}_{\mathrm{A}}(f) + \hat{x}_{\mathrm{A}}^*(-f), \quad f \in \mathbb{R}, \tag{7.21a}$$

*and*

$$x_{\mathrm{PB}}(t) = 2\operatorname{Re}\big(x_{\mathrm{A}}(t)\big), \quad t \in \mathbb{R}. \tag{7.21b}$$

**Proof.** The frequency relation (7.21a) is just a restatement of (7.19), whose derivation was rigorous. To prove (7.21b) we note that, by Proposition 7.2.2 (*cf.* (b) & (c)),

$$
\begin{aligned}
x_{\mathrm{PB}}(t) &= \int_{-\infty}^{\infty} \hat{x}_{\mathrm{PB}}(f)\, e^{\mathrm{i}2\pi ft}\, \mathrm{d}f \\
&= \int_{0}^{\infty} \hat{x}_{\mathrm{PB}}(f)\, e^{\mathrm{i}2\pi ft}\, \mathrm{d}f + \int_{-\infty}^{0} \hat{x}_{\mathrm{PB}}(f)\, e^{\mathrm{i}2\pi ft}\, \mathrm{d}f \\
&= x_{\mathrm{A}}(t) + \int_{-\infty}^{0} \hat{x}_{\mathrm{PB}}(f)\, e^{\mathrm{i}2\pi ft}\, \mathrm{d}f \\
&= x_{\mathrm{A}}(t) + \int_{0}^{\infty} \hat{x}_{\mathrm{PB}}(-\tilde{f})\, e^{-\mathrm{i}2\pi \tilde{f}t}\, \mathrm{d}\tilde{f} \\
&= x_{\mathrm{A}}(t) + \int_{0}^{\infty} \hat{x}_{\mathrm{PB}}^*(\tilde{f})\, e^{-\mathrm{i}2\pi \tilde{f}t}\, \mathrm{d}\tilde{f} \\
&= x_{\mathrm{A}}(t) + \left( \int_{0}^{\infty} \hat{x}_{\mathrm{PB}}(\tilde{f})\, e^{\mathrm{i}2\pi \tilde{f}t}\, \mathrm{d}\tilde{f} \right)^* \\
&= x_{\mathrm{A}}(t) + x_{\mathrm{A}}^*(t) \\
&= 2\operatorname{Re}\big(x_{\mathrm{A}}(t)\big), \quad t \in \mathbb{R},
\end{aligned}
$$

where in the second equality we broke the integral into two; in the third we used Definition 7.5.1; in the fourth we changed the integration variable to $\tilde{f} \triangleq -f$; in the fifth we used the conjugate symmetry of $\hat{x}_{\mathrm{PB}}$ (7.18); in the sixth we used the fact that conjugating the integrand results in the conjugation of the integral (Proposition 2.3.1); in the seventh we used the definition of the analytic signal; and in the last equality we used the fact that a complex number and its conjugate add up to twice its real part.                                                                    $\square$

### 7.5.3   Relating $\langle \mathbf{x}_{\mathrm{PB}}, \mathbf{y}_{\mathrm{PB}} \rangle$ to $\langle \mathbf{x}_{\mathrm{A}}, \mathbf{y}_{\mathrm{A}} \rangle$

We next relate the inner product between two real passband signals to the inner product between their analytic representations.

**Proposition 7.5.4 ($\langle \mathbf{x}_{\mathrm{PB}}, \mathbf{y}_{\mathrm{PB}} \rangle$ and $\langle \mathbf{x}_{\mathrm{A}}, \mathbf{y}_{\mathrm{A}} \rangle$).** *Let* $\mathbf{x}_{\mathrm{PB}}$ *and* $\mathbf{y}_{\mathrm{PB}}$ *be real integrable passband signals that are bandlimited to* W *Hz around the carrier frequency* $f_{\mathrm{c}}$, *and let* $\mathbf{x}_{\mathrm{A}}$ *and* $\mathbf{y}_{\mathrm{A}}$ *be their analytic representations. Then*

$$\langle \mathbf{x}_{\mathrm{PB}}, \mathbf{y}_{\mathrm{PB}} \rangle = 2\,\mathrm{Re}\big(\langle \mathbf{x}_{\mathrm{A}}, \mathbf{y}_{\mathrm{A}} \rangle\big), \tag{7.22}$$

*and*

$$\|\mathbf{x}_{\mathrm{PB}}\|_2^2 = 2\,\|\mathbf{x}_{\mathrm{A}}\|_2^2\,. \tag{7.23}$$

Note that in (7.22) the inner product appearing on the LHS is the inner product between *real* signals whereas the one appearing on the RHS is between *complex* signals.

**Proof.** We first note that the inner products and energies are well-defined because integrable passband signals are also energy-limited (Corollary 7.2.4). Next, even though (7.23) is a special case of (7.22), we first prove (7.23). The proof is a simple application of Parseval's Theorem. The intuition is as follows. Since $\mathbf{x}_{\mathrm{PB}}$ is real, it follows that its FT is conjugate-symmetric (7.18) so the magnitude of $\hat{x}_{\mathrm{PB}}$ is symmetric. Consequently, the positive frequencies and the negative frequencies of $\hat{x}_{\mathrm{PB}}$ contribute an equal share to the total energy in $\hat{x}_{\mathrm{PB}}$. And since the energy in the analytic representation is equal to the share corresponding to the positive frequencies only, its energy must be half the energy of $\mathbf{x}_{\mathrm{PB}}$.

This can be argued more formally as follows. Because $\mathbf{x}_{\mathrm{PB}}$ is real-valued, its FT $\hat{x}_{\mathrm{PB}}$ is conjugate-symmetric (7.18), so its magnitude is symmetric $|\hat{x}_{\mathrm{PB}}(f)| = |\hat{x}_{\mathrm{PB}}(-f)|$ for all $f \in \mathbb{R}$ and, *a fortiori*,

$$\int_0^\infty |\hat{x}_{\mathrm{PB}}(f)|^2\,\mathrm{d}f = \int_{-\infty}^0 |\hat{x}_{\mathrm{PB}}(f)|^2\,\mathrm{d}f. \tag{7.24}$$

Also, by Parseval's Theorem (applied to $\mathbf{x}_{\mathrm{PB}}$),

$$\int_0^\infty |\hat{x}_{\mathrm{PB}}(f)|^2\,\mathrm{d}f + \int_{-\infty}^0 |\hat{x}_{\mathrm{PB}}(f)|^2\,\mathrm{d}f = \|\mathbf{x}_{\mathrm{PB}}\|_2^2\,. \tag{7.25}$$

Consequently, by combining (7.24) and (7.25), we obtain

$$\int_0^\infty |\hat{x}_{\mathrm{PB}}(f)|^2\,\mathrm{d}f = \frac{1}{2}\,\|\mathbf{x}_{\mathrm{PB}}\|_2^2\,. \tag{7.26}$$

We can now establish (7.23) from (7.26) by using Parseval's Theorem (applied to $\mathbf{x}_A$) and (7.13) to obtain

$$\|\mathbf{x}_A\|_2^2 = \|\hat{\mathbf{x}}_A\|_2^2$$
$$= \int_{-\infty}^{\infty} |\hat{x}_A(f)|^2 \, \mathrm{d}f$$
$$= \int_0^{\infty} |\hat{x}_{PB}(f)|^2 \, \mathrm{d}f$$
$$= \frac{1}{2} \|\mathbf{x}_{PB}\|_2^2,$$

where the last equality follows from (7.26).

We next prove (7.22). We offer two proofs. The first is very similar to our proof of (7.23): we use Parseval's Theorem to express the inner products in the frequency domain, and then argue that the contribution of the negative frequencies to the inner product is the complex conjugate of the contribution of the positive frequencies. The second proof uses a trick to relate inner products and energies.

We begin with the first proof. Using Proposition 7.5.3 we have

$$\hat{x}_{PB}(f) = \hat{x}_A(f) + \hat{x}_A^*(-f), \quad f \in \mathbb{R},$$

$$\hat{y}_{PB}(f) = \hat{y}_A(f) + \hat{y}_A^*(-f), \quad f \in \mathbb{R}.$$

Using Parseval's Theorem we now have

$$\langle \mathbf{x}_{PB}, \mathbf{y}_{PB} \rangle = \langle \hat{\mathbf{x}}_{PB}, \hat{\mathbf{y}}_{PB} \rangle$$
$$= \int_{-\infty}^{\infty} \hat{x}_{PB}(f) \hat{y}_{PB}^*(f) \, \mathrm{d}f$$
$$= \int_{-\infty}^{\infty} \left( \hat{x}_A(f) + \hat{x}_A^*(-f) \right) \left( \hat{y}_A(f) + \hat{y}_A^*(-f) \right)^* \mathrm{d}f$$
$$= \int_{-\infty}^{\infty} \left( \hat{x}_A(f) + \hat{x}_A^*(-f) \right) \left( \hat{y}_A^*(f) + \hat{y}_A(-f) \right) \mathrm{d}f$$
$$= \int_{-\infty}^{\infty} \hat{x}_A(f) \hat{y}_A^*(f) \, \mathrm{d}f + \int_{-\infty}^{\infty} \hat{x}_A^*(-f) \hat{y}_A(-f) \, \mathrm{d}f$$
$$= \int_{-\infty}^{\infty} \hat{x}_A(f) \hat{y}_A^*(f) \, \mathrm{d}f + \left( \int_{-\infty}^{\infty} \hat{x}_A(-f) \hat{y}_A^*(-f) \, \mathrm{d}f \right)^*$$
$$= \int_{-\infty}^{\infty} \hat{x}_A(f) \hat{y}_A^*(f) \, \mathrm{d}f + \left( \int_{-\infty}^{\infty} \hat{x}_A(\tilde{f}) \hat{y}_A^*(\tilde{f}) \, \mathrm{d}\tilde{f} \right)^*$$
$$= \langle \hat{\mathbf{x}}_A, \hat{\mathbf{y}}_A \rangle + \langle \hat{\mathbf{x}}_A, \hat{\mathbf{y}}_A \rangle^*$$
$$= 2 \operatorname{Re}\left( \langle \hat{\mathbf{x}}_A, \hat{\mathbf{y}}_A \rangle \right)$$
$$= 2 \operatorname{Re}\left( \langle \mathbf{x}_A, \mathbf{y}_A \rangle \right),$$

where the fifth equality follows because at all frequencies $f \in \mathbb{R}$ the cross-terms $\hat{x}_A(f) \hat{y}_A(-f)$ and $\hat{x}_A^*(-f) \hat{y}_A^*(f)$ are zero, and where the last equality follows from Parseval's Theorem.

The second proof is based on (7.23) and on the identity

$$2\operatorname{Re}(\langle \mathbf{u}, \mathbf{v}\rangle) = \|\mathbf{u}+\mathbf{v}\|_2^2 - \|\mathbf{u}\|_2^2 - \|\mathbf{v}\|_2^2, \quad \mathbf{u}, \mathbf{v} \in \mathcal{L}_2, \tag{7.27}$$

which holds for both complex and real signals and which follows by expressing $\|\mathbf{u}+\mathbf{v}\|_2^2$ as

$$
\begin{aligned}
\|\mathbf{u}+\mathbf{v}\|_2^2 &= \langle \mathbf{u}+\mathbf{v}, \mathbf{u}+\mathbf{v}\rangle \\
&= \langle \mathbf{u}, \mathbf{u}\rangle + \langle \mathbf{u}, \mathbf{v}\rangle + \langle \mathbf{v}, \mathbf{u}\rangle + \langle \mathbf{v}, \mathbf{v}\rangle \\
&= \|\mathbf{u}\|_2^2 + \|\mathbf{v}\|_2^2 + \langle \mathbf{u}, \mathbf{v}\rangle + \langle \mathbf{u}, \mathbf{v}\rangle^* \\
&= \|\mathbf{u}\|_2^2 + \|\mathbf{v}\|_2^2 + 2\operatorname{Re}(\langle \mathbf{u}, \mathbf{v}\rangle).
\end{aligned}
$$

From Identity (7.27) and from (7.23) we have for the real signals $\mathbf{x}_{\mathrm{PB}}$ and $\mathbf{y}_{\mathrm{PB}}$

$$
\begin{aligned}
2\langle \mathbf{x}_{\mathrm{PB}}, \mathbf{y}_{\mathrm{PB}}\rangle &= 2\operatorname{Re}(\langle \mathbf{x}_{\mathrm{PB}}, \mathbf{y}_{\mathrm{PB}}\rangle) \\
&= \|\mathbf{x}_{\mathrm{PB}} + \mathbf{y}_{\mathrm{PB}}\|_2^2 - \|\mathbf{x}_{\mathrm{PB}}\|_2^2 - \|\mathbf{y}_{\mathrm{PB}}\|_2^2 \\
&= 2\left(\|\mathbf{x}_{\mathrm{A}} + \mathbf{y}_{\mathrm{A}}\|_2^2 - \|\mathbf{x}_{\mathrm{A}}\|_2^2 - \|\mathbf{y}_{\mathrm{A}}\|_2^2\right) \\
&= 4\operatorname{Re}(\langle \mathbf{x}_{\mathrm{A}}, \mathbf{y}_{\mathrm{A}}\rangle),
\end{aligned}
$$

where the first equality follows because the passband signals are real; the second from Identity (7.27) applied to the passband signals $\mathbf{x}_{\mathrm{PB}}$ and $\mathbf{y}_{\mathrm{PB}}$; the third from the second part of Proposition 7.5.4 and because the analytic representation of $\mathbf{x}_{\mathrm{PB}} + \mathbf{y}_{\mathrm{PB}}$ is $\mathbf{x}_{\mathrm{A}} + \mathbf{y}_{\mathrm{A}}$; and the final equality from Identity (7.27) applied to the analytic signals $\mathbf{x}_{\mathrm{A}}$ and $\mathbf{y}_{\mathrm{A}}$. $\qquad\square$

## 7.6   Baseband Representation of Real Passband Signals

Strictly speaking, the **baseband representation** $\mathbf{x}_{\mathrm{BB}}$ of a real passband signal $\mathbf{x}_{\mathrm{PB}}$ is not a "representation" because one cannot recover $\mathbf{x}_{\mathrm{PB}}$ from $\mathbf{x}_{\mathrm{BB}}$ alone; one also needs to know the carrier frequency $f_{\mathrm{c}}$. This may seem like a disadvantage, but engineers view this as an advantage. Indeed, in some cases, it may illuminate the fact that certain operations and results do not depend on the carrier frequency. This decoupling of various operations from the carrier frequency is very useful in hardware implementation of communication systems that need to work around selectable carrier frequencies. It allows for some of the processing to be done using carrier-independent hardware and for only a small part of the communication system to be tunable to the carrier frequency. Very loosely speaking, engineers think of $\mathbf{x}_{\mathrm{BB}}$ as everything about $\mathbf{x}_{\mathrm{PB}}$ that is not carrier-dependent. Thus, one does not usually expect the quantity $f_{\mathrm{c}}$ to show up in a formula for the baseband representation. Philosophical thoughts aside, the baseband representation has a straightforward definition.

### 7.6.1   Definition and Characterization

**Definition 7.6.1 (Baseband Representation).** *The **baseband representation** of a real integrable passband signal $\mathbf{x}_{\mathrm{PB}}$ that is bandlimited to* W *Hz around the carrier* *frequency* $f_c$ *is the complex signal*

$$x_{\mathrm{BB}}(t) \triangleq e^{-i2\pi f_c t}\, x_{\mathrm{A}}(t), \quad t \in \mathbb{R}, \tag{7.28}$$

*where* $\mathbf{x}_{\mathrm{A}}$ *is the analytic representation of* $\mathbf{x}_{\mathrm{PB}}$.

Note that, by (7.28), the magnitudes of $\mathbf{x}_{\mathrm{A}}$ and $\mathbf{x}_{\mathrm{BB}}$ are identical

$$\left| x_{\mathrm{BB}}(t) \right| = \left| x_{\mathrm{A}}(t) \right|, \quad t \in \mathbb{R}. \tag{7.29}$$

Consequently, since $\mathbf{x}_{\mathrm{A}}$ is integrable we also have:

**Proposition 7.6.2 (Integrability of $\mathbf{x}_{\mathrm{PB}}$ Implies Integrability of $\mathbf{x}_{\mathrm{BB}}$).** *The base-band representation of a real integrable passband signal that is bandlimited to* $W$ *Hz around the carrier frequency* $f_c$ *is integrable.*

By (7.28) and (7.13) we obtain that if $\mathbf{x}_{\mathrm{PB}}$ is a real integrable passband signal that is bandlimited to $W$ Hz around the carrier frequency $f_c$, then

$$\hat{x}_{\mathrm{BB}}(f) = \hat{x}_{\mathrm{A}}(f + f_c) = \begin{cases} \hat{x}_{\mathrm{PB}}(f + f_c) & \text{if } |f| \leq W/2, \\ 0 & \text{otherwise.} \end{cases} \tag{7.30}$$

Thus, the FT of $\mathbf{x}_{\mathrm{BB}}$ is the FT of $\mathbf{x}_{\mathrm{A}}$ but shifted to the left by the carrier frequency $f_c$. The relationship between the Fourier Transforms of $\mathbf{x}_{\mathrm{PB}}$, $\mathbf{x}_{\mathrm{A}}$, and $\mathbf{x}_{\mathrm{BB}}$ is depicted in Figure 7.9.

We have defined the baseband representation of a passband signal in terms of its analytic representation, but sometimes it is useful to define the baseband representation directly in terms of the passband signal. This is not very difficult. Rather than taking the passband signal and passing it through a filter of frequency response $\mathbf{g}$ satisfying (7.14) to obtain $\mathbf{x}_{\mathrm{A}}$ and then multiplying the result by $e^{-i2\pi f_c t}$ to obtain $\mathbf{x}_{\mathrm{BB}}$, we can multiply $\mathbf{x}_{\mathrm{PB}}$ by $t \mapsto e^{-i2\pi f_c t}$ and then filter the result to obtain the baseband representation. This procedure is depicted in the frequency domain in Figure 7.10 and is made precise in the following proposition.

**Proposition 7.6.3 (From $\mathbf{x}_{\mathrm{PB}}$ to $\mathbf{x}_{\mathrm{BB}}$ Directly).** *If* $\mathbf{x}_{\mathrm{PB}}$ *is a real integrable passband signal that is bandlimited to* $W$ *Hz around the carrier frequency* $f_c$, *then its baseband representation* $\mathbf{x}_{\mathrm{BB}}$ *is given by*

$$\mathbf{x}_{\mathrm{BB}} = \left( t \mapsto e^{-i2\pi f_c t}\, x_{\mathrm{PB}}(t) \right) \star \check{\mathbf{g}}_0, \tag{7.31a}$$

*where* $\mathbf{g}_0 \colon f \mapsto g_0(f)$ *is any integrable function satisfying*

$$g_0(f) = 1, \quad |f| \leq \frac{W}{2}, \tag{7.31b}$$

*and*

$$g_0(f) = 0, \quad |f + 2f_c| \leq \frac{W}{2}. \tag{7.31c}$$

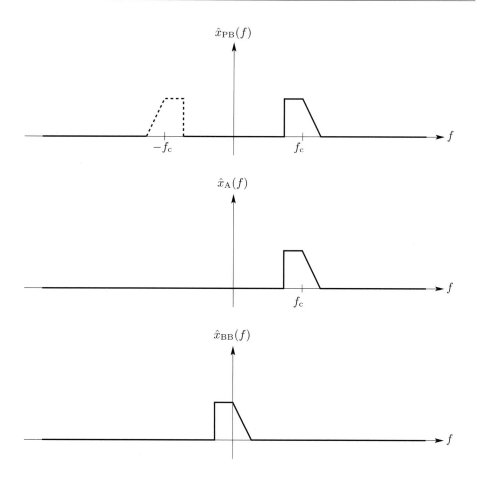

**Figure 7.9:** The Fourier Transforms of the analytic signal $\mathbf{x}_A$ and of the baseband representation $\mathbf{x}_{BB}$ of a real passband signal $\mathbf{x}_{PB}$.

**Proof.** The proof is all in Figure 7.10. For the pedantic reader we provide more details. By Definition 7.6.1 and by Proposition 7.5.2 (*cf.* (d)) we have for any integrable function $\mathbf{g} \colon f \mapsto g(f)$ satisfying (7.14b) & (7.14c)

$$
\begin{aligned}
x_{BB}(t) &= e^{-i2\pi f_c t} \big(\mathbf{x}_{PB} \star \check{\mathbf{g}}\big)(t) \\
&= e^{-i2\pi f_c t} \int_{-\infty}^{\infty} \hat{x}_{PB}(f)\, g(f)\, e^{i2\pi f t}\, \mathrm{d}f \\
&= \int_{-\infty}^{\infty} \hat{x}_{PB}(f)\, g(f)\, e^{i2\pi (f - f_c)t}\, \mathrm{d}f \\
&= \int_{-\infty}^{\infty} \hat{x}_{PB}(\tilde{f} + f_c)\, g(\tilde{f} + f_c)\, e^{i2\pi \tilde{f} t}\, \mathrm{d}\tilde{f} \\
&= \int_{-\infty}^{\infty} \hat{x}_{PB}(\tilde{f} + f_c)\, g_0(\tilde{f})\, e^{i2\pi \tilde{f} t}\, \mathrm{d}\tilde{f}
\end{aligned}
$$

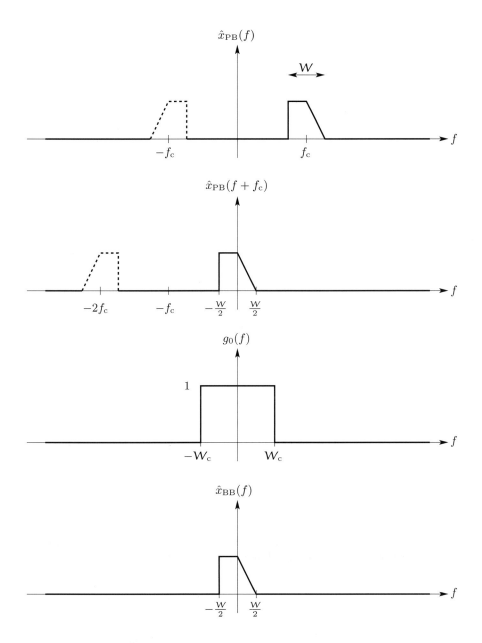

**Figure 7.10:** A frequency-domain description of the process for deriving $\mathbf{x}_{\mathrm{BB}}$ directly from $\mathbf{x}_{\mathrm{PB}}$. From top to bottom: $\hat{x}_{\mathrm{PB}}$; the FT of $t \mapsto e^{-\mathrm{i}2\pi f_c t} x_{\mathrm{PB}}(t)$; a function $\mathbf{g}_0$ satisfying (7.31b) & (7.31c); and $\hat{x}_{\mathrm{BB}}$.

$$= \Big( \big( t \mapsto e^{-i2\pi f_c t}\, x_{\mathrm{PB}}(t) \big) \star \check{g}_0 \Big)(t),$$

where we defined

$$g_0(f) = g(f + f_c), \quad f \in \mathbb{R}, \tag{7.32}$$

and where we use the following justification. The second equality follows from Proposition 6.2.5; the third by pulling the complex exponential into the integral; the fourth by the defining $\tilde{f} \triangleq f - f_c$; the fifth by defining the function $g_0$ as in (7.32); and the final equality by Proposition 6.2.5 using the fact that

$$\text{the FT of } t \mapsto e^{-i2\pi f_c t}\, x_{\mathrm{PB}}(t) \text{ is } f \mapsto \hat{x}_{\mathrm{PB}}(f + f_c). \tag{7.33}$$

The proposition now follows by noting that $\mathbf{g}$ satisfies (7.14b) & (7.14c) if, and only if, the mapping $g_0$ defined in (7.32) satisfies (7.31b) & (7.31c). $\qquad \square$

**Corollary 7.6.4.** *If $\mathbf{x}_{\mathrm{PB}}$ is a real integrable passband signal that is bandlimited to $\mathrm{W}$ Hz around the carrier frequency $f_c$, then its baseband representation $\mathbf{x}_{\mathrm{BB}}$ is given by*

$$\boxed{\mathbf{x}_{\mathrm{BB}} = \big( t \mapsto e^{-i2\pi f_c t}\, x_{\mathrm{PB}}(t) \big) \star \mathrm{LPF}_{\mathrm{W}_c},} \tag{7.34a}$$

*where the cutoff frequency $\mathrm{W}_c$ can be chosen arbitrarily in the range*

$$\boxed{\dfrac{\mathrm{W}}{2} \leq \mathrm{W}_c \leq 2f_c - \dfrac{\mathrm{W}}{2}.} \tag{7.34b}$$

**Proof.** Let $\mathrm{W}_c$ satisfy (7.34b) and define $g_0$ as follows: if $\mathrm{W}_c$ is strictly smaller than $2f_c - \mathrm{W}/2$, define $g_0(f) = \mathrm{I}\{|f| \leq \mathrm{W}_c\}$; otherwise define $g_0(f) = \mathrm{I}\{|f| < \mathrm{W}_c\}$. In both cases $g_0$ satisfies (7.31b) & (7.31c) and

$$\check{g}_0 = \mathrm{LPF}_{\mathrm{W}_c}. \tag{7.35}$$

The result now follows by applying Proposition 7.6.3 with this choice of $g_0$. $\qquad \square$

In analogy to Proposition 7.5.2, we can characterize the baseband representation of passband signals as follows.

**Proposition 7.6.5 (Characterizing the Baseband Representation).** *Let $\mathbf{x}_{\mathrm{PB}}$ be a real integrable passband signal that is bandlimited to $\mathrm{W}$ Hz around the carrier frequency $f_c$. Then each of the following statements is equivalent to the statement that the complex signal $\mathbf{x}_{\mathrm{BB}}$ is its baseband representation.*

*(a) The signal $\mathbf{x}_{\mathrm{BB}}$ is given by*

$$x_{\mathrm{BB}}(t) = \int_{-\mathrm{W}/2}^{\mathrm{W}/2} \hat{x}_{\mathrm{PB}}(f + f_c)\, e^{i2\pi f t}\, \mathrm{d}f, \quad t \in \mathbb{R}. \tag{7.36}$$

(b) *The signal* $\mathbf{x}_{\text{BB}}$ *is a continuous integrable signal satisfying*

$$\hat{x}_{\text{BB}}(f) = \hat{x}_{\text{PB}}(f + f_c) \, \text{I}\Big\{|f| \leq \frac{W}{2}\Big\}, \quad f \in \mathbb{R}. \tag{7.37}$$

(c) *The signal* $\mathbf{x}_{\text{BB}}$ *is an integrable signal that is bandlimited to* $W/2$ *Hz and that satisfies* (7.37).

(d) *The signal* $\mathbf{x}_{\text{BB}}$ *is given by* (7.31a) *for any* $\mathbf{g}_0 \colon f \mapsto g_0(f)$ *satisfying* (7.31b) *&* (7.31c).

**Proof.** Parts (a), (b), and (c) can be easily deduced from their counterparts in Proposition 7.5.2 using Definition 7.6.1 and the fact that (7.29) implies that the integrability of $\mathbf{x}_{\text{BB}}$ is equivalent to the integrability of $\mathbf{x}_{\text{A}}$. Part (d) is a restatement of Proposition 7.6.3. ☐

### 7.6.2  The In-Phase and Quadrature Components

The convolution in (7.34a) is a convolution between a complex signal (the signal $t \mapsto e^{-\mathrm{i}2\pi f_c t} \, x_{\text{PB}}(t)$) and a real signal (the signal $\text{LPF}_{W_c}$). This should not alarm you. The convolution of two complex signals evaluated at time $t$ is expressed as an integral (5.2), and in the case of complex signals this is an integral (over the real line) of a complex-valued integrand. Such integrals were addressed in Section 2.3. It should, however, be noted that since the definition of the convolution of two signals involves their products, *the real part of the convolution of two complex-valued signals is, in general, not equal to the convolution of their real parts.* However, as we next show, if one of the signals is real—as is the case in (7.34a)—then things become simpler: if $\mathbf{x}$ is a complex-valued function of time and if $\mathbf{h}$ is a real-valued function of time, then

$$\boxed{\text{Re}(\mathbf{x} \star \mathbf{h}) = \text{Re}(\mathbf{x}) \star \mathbf{h} \text{ and } \text{Im}(\mathbf{x} \star \mathbf{h}) = \text{Im}(\mathbf{x}) \star \mathbf{h}, \quad \mathbf{h} \text{ is real-valued.}} \tag{7.38}$$

This follows from the definition of the convolution,

$$(\mathbf{x} \star \mathbf{h})(t) = \int_{-\infty}^{\infty} x(\tau) \, h(t - \tau) \, \mathrm{d}\tau$$

and from the basic properties of complex integrals (Proposition 2.3.1) by noting that if $h(\cdot)$ is real-valued, then for all $t, \tau \in \mathbb{R}$,

$$\text{Re}\big(x(\tau) \, h(t - \tau)\big) = \text{Re}\big(x(\tau)\big) \, h(t - \tau),$$
$$\text{Im}\big(x(\tau) \, h(t - \tau)\big) = \text{Im}\big(x(\tau)\big) \, h(t - \tau).$$

We next use (7.38) to express the convolution in (7.31a) using real-number operations. To that end we first note that since $\mathbf{x}_{\text{PB}}$ is real, it follows from Euler's Identity

$$e^{\mathrm{i}\theta} = \cos\theta + \mathrm{i}\sin\theta, \quad \theta \in \mathbb{R} \tag{7.39}$$

that

$$\mathrm{Re}\big(x_{\mathrm{PB}}(t)\,e^{-\mathrm{i}2\pi f_c t}\big) = x_{\mathrm{PB}}(t)\cos(2\pi f_c t), \quad t \in \mathbb{R}, \tag{7.40a}$$

$$\mathrm{Im}\big(x_{\mathrm{PB}}(t)\,e^{-\mathrm{i}2\pi f_c t}\big) = -x_{\mathrm{PB}}(t)\sin(2\pi f_c t), \quad t \in \mathbb{R}, \tag{7.40b}$$

so by (7.34a), (7.38), and (7.40)

$$\mathrm{Re}(\mathbf{x}_{\mathrm{BB}}) = \Big(t \mapsto x_{\mathrm{PB}}(t)\cos(2\pi f_c t)\Big) \star \mathrm{LPF}_{\mathsf{W}_c/2}, \tag{7.41a}$$

$$\mathrm{Im}(\mathbf{x}_{\mathrm{BB}}) = -\Big(t \mapsto x_{\mathrm{PB}}(t)\sin(2\pi f_c t)\Big) \star \mathrm{LPF}_{\mathsf{W}_c/2}. \tag{7.41b}$$

It is common in the engineering literature to refer to the real part of $\mathbf{x}_{\mathrm{BB}}$ as the **in-phase component** of $\mathbf{x}_{\mathrm{PB}}$ and to the imaginary part as the **quadrature component** of $\mathbf{x}_{\mathrm{PB}}$.

**Definition 7.6.6 (In-Phase and Quadrature Components).** *The in-phase component of a real integrable passband signal $\mathbf{x}_{\mathrm{PB}}$ that is bandlimited to $\mathsf{W}$ Hz around the carrier frequency $f_c$ is the real part of its baseband representation, i.e.,*

$$\mathrm{Re}(\mathbf{x}_{\mathrm{BB}}) = \Big(t \mapsto x_{\mathrm{PB}}(t)\cos(2\pi f_c t)\Big) \star \mathrm{LPF}_{\mathsf{W}_c}. \quad \textbf{(In-Phase)}$$

*The **quadrature component** is the imaginary part of its baseband representation, i.e.,*

$$\mathrm{Im}(\mathbf{x}_{\mathrm{BB}}) = -\Big(t \mapsto x_{\mathrm{PB}}(t)\sin(2\pi f_c t)\Big) \star \mathrm{LPF}_{\mathsf{W}_c}. \quad \textbf{(Quadrature)}$$

*Here $\mathsf{W}_c$ is any cutoff frequency in the range $\mathsf{W}/2 \leq \mathsf{W}_c \leq 2f_c - \mathsf{W}/2$.*

Figure 7.11 depicts a block diagram of a circuit that produces the baseband representation of a real passband signal. This circuit will play an important role in Chapter 9 when we discuss the Sampling Theorem for passband signals and **complex sampling**.

### 7.6.3   Bandwidth Considerations

The following is a simple but exceedingly important observation regarding bandwidth. Recall that the bandwidth of $\mathbf{x}_{\mathrm{PB}}$ around the carrier frequency $f_c$ is defined in Definition 7.3.1 and that the bandwidth of the baseband signal $\mathbf{x}_{\mathrm{BB}}$ is defined in Definition 6.4.13.

**Proposition 7.6.7 ($\mathbf{x}_{\mathrm{PB}}$, $\mathbf{x}_{\mathrm{BB}}$, and Bandwidth).** *If the real integrable passband signal $\mathbf{x}_{\mathrm{PB}}$ is of bandwidth $\mathsf{W}$ Hz around the carrier frequency $f_c$, then its baseband representation $\mathbf{x}_{\mathrm{BB}}$ is an integrable signal of bandwidth $\mathsf{W}/2$ Hz.*

**Proof.** This can be seen graphically from Figure 7.9 or from Figure 7.10. It can be deduced analytically from (7.30). $\qquad\square$

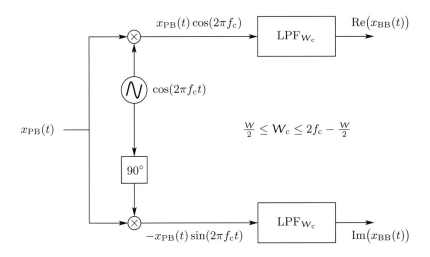

**Figure 7.11:** Obtaining the baseband representation of a real passband signal.

### 7.6.4 Recovering $\mathbf{x}_{\mathrm{PB}}$ from $\mathbf{x}_{\mathrm{BB}}$

Recovering a real passband signal $\mathbf{x}_{\mathrm{PB}}$ from its baseband representation $\mathbf{x}_{\mathrm{BB}}$ is conceptually simple. We can recover the analytic representation via (7.28) and then use Proposition 7.5.3 to recover $\mathbf{x}_{\mathrm{PB}}$:

**Proposition 7.6.8 (From $\mathbf{x}_{\mathrm{BB}}$ to $\mathbf{x}_{\mathrm{PB}}$).** *Let $\mathbf{x}_{\mathrm{PB}}$ be a real integrable passband signal that is bandlimited to $W$ Hz around the carrier frequency $f_{\mathrm{c}}$, and let $\mathbf{x}_{\mathrm{BB}}$ be its baseband representation. Then,*

$$\hat{x}_{\mathrm{PB}}(f) = \hat{x}_{\mathrm{BB}}(f - f_{\mathrm{c}}) + \hat{x}_{\mathrm{BB}}^*(-f - f_{\mathrm{c}}), \quad f \in \mathbb{R}, \tag{7.42a}$$

*and*

$$x_{\mathrm{PB}}(t) = 2\,\mathrm{Re}\big(x_{\mathrm{BB}}(t)\,e^{\mathrm{i}2\pi f_{\mathrm{c}}t}\big), \quad t \in \mathbb{R}. \tag{7.42b}$$

The process of recovering $\mathbf{x}_{\mathrm{PB}}$ from $\mathbf{x}_{\mathrm{BB}}$ is depicted in the frequency domain in Figure 7.12. It can, of course, also be carried out using real-number operations only by rewriting (7.42b) as

$$x_{\mathrm{PB}}(t) = 2\,\mathrm{Re}\big(x_{\mathrm{BB}}(t)\big)\cos(2\pi f_{\mathrm{c}}t) - 2\,\mathrm{Im}\big(x_{\mathrm{BB}}(t)\big)\sin(2\pi f_{\mathrm{c}}t), \quad t \in \mathbb{R}. \tag{7.43}$$

It should be emphasized that (7.42b) does not characterize the baseband representation of $\mathbf{x}_{\mathrm{PB}}$; it is possible that $x_{\mathrm{PB}}(t) = 2\,\mathrm{Re}\big(z(t)\,e^{\mathrm{i}2\pi f_{\mathrm{c}}t}\big)$ hold at every time $t$ and that $\mathbf{z}$ not be the baseband representation of $\mathbf{x}_{\mathrm{PB}}$. However, as the next proposition shows, this cannot happen if $\mathbf{z}$ is bandlimited to $W/2$ Hz.

**Proposition 7.6.9.** *Let $\mathbf{x}_{\mathrm{PB}}$ be a real integrable passband signal that is bandlimited to $W$ Hz around the carrier frequency $f_{\mathrm{c}}$. If the complex signal $\mathbf{z}$ satisfies*

$$x_{\mathrm{PB}}(t) = 2\,\mathrm{Re}\big(z(t)\,e^{\mathrm{i}2\pi f_{\mathrm{c}}t}\big), \quad t \in \mathbb{R}, \tag{7.44}$$

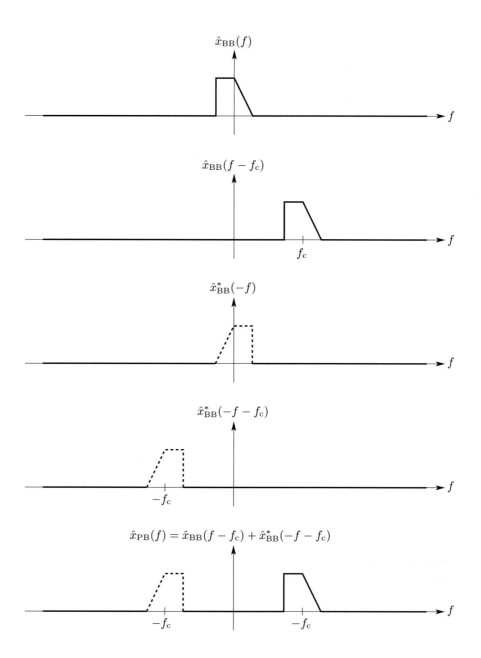

**Figure 7.12:** Recovering a passband signal from its baseband representation. Top plot of $\hat{\mathbf{x}}_{\mathrm{BB}}$ is the transform of $\mathbf{x}_{\mathrm{BB}}$; next is the transform of $t \mapsto x_{\mathrm{BB}}(t)\, e^{\mathrm{i}2\pi f_{\mathrm{c}}t}$; the transform of $x_{\mathrm{BB}}^{*}(t)$; the transform of $t \mapsto x_{\mathrm{BB}}^{*}(t)\, e^{-\mathrm{i}2\pi f_{\mathrm{c}}t}$; and finally the transform of $t \mapsto x_{\mathrm{BB}}(t)\, e^{\mathrm{i}2\pi f_{\mathrm{c}}t} + x_{\mathrm{BB}}^{*}(t)\, e^{-\mathrm{i}2\pi f_{\mathrm{c}}t} = 2\,\mathrm{Re}\big(x_{\mathrm{BB}}(t)\, e^{\mathrm{i}2\pi f_{\mathrm{c}}t}\big) = x_{\mathrm{PB}}(t)$.

*and is an integrable signal that is bandlimited to* $W/2$ *Hz, then* $\mathbf{z}$ *is the baseband representation of* $\mathbf{x}_{\mathrm{PB}}$.

**Proof.** Since $\mathbf{z}$ is bandlimited to $W/2$ Hz, it follows from Proposition 6.4.10 (*cf.* (c)) that $\mathbf{z}$ must be continuous and that its FT must vanish for $|f| > W/2$. Consequently, by Proposition 7.6.5 (*cf.* (b)), all that remains to show in order to establish that $\mathbf{z}$ is the baseband representation of $\mathbf{x}_{\mathrm{PB}}$ is that

$$\hat{z}(f) = \hat{x}_{\mathrm{PB}}(f + f_{\mathrm{c}}), \quad |f| \leq W/2, \tag{7.45}$$

and this is what we proceed to do. By taking the FT of both sides of (7.44) we obtain that

$$\hat{x}_{\mathrm{PB}}(f) = \hat{z}(f - f_{\mathrm{c}}) + \hat{z}^*(-f - f_{\mathrm{c}}), \quad f \in \mathbb{R}, \tag{7.46}$$

or, upon defining $\tilde{f} \triangleq f - f_{\mathrm{c}}$,

$$\hat{x}_{\mathrm{PB}}(\tilde{f} + f_{\mathrm{c}}) = \hat{z}(\tilde{f}) + \hat{z}^*(-\tilde{f} - 2f_{\mathrm{c}}), \quad \tilde{f} \in \mathbb{R}. \tag{7.47}$$

By recalling that $f_{\mathrm{c}} > W/2$ and that $\hat{z}$ is zero for frequencies $f$ satisfying $|f| > W/2$, we obtain that $\hat{z}^*(-\tilde{f} - 2f_{\mathrm{c}})$ is zero whenever $|\tilde{f}| \leq W/2$ so

$$\hat{z}(\tilde{f}) + \hat{z}^*(-\tilde{f} - 2f_{\mathrm{c}}) = \hat{z}(\tilde{f}), \quad |\tilde{f}| \leq W/2. \tag{7.48}$$

Combining (7.47) and (7.48) we obtain

$$\hat{x}_{\mathrm{PB}}(\tilde{f} + f_{\mathrm{c}}) = \hat{z}(\tilde{f}), \quad |\tilde{f}| \leq W/2,$$

thus establishing (7.45) and hence completing the proof. $\qquad \square$

Proposition 7.6.9 is more useful than its appearance may suggest. It provides an alternative way of computing the baseband representation of a signal. It demonstrates that if we can use algebra to express $\mathbf{x}_{\mathrm{PB}}$ in the form (7.44) for some signal $\mathbf{z}$, and if we can verify that $\mathbf{z}$ is bandlimited to $W/2$ Hz, then $\mathbf{z}$ must be the baseband representation of $\mathbf{x}_{\mathrm{PB}}$.

Note that the proof would also work if we replaced the assumption that $\mathbf{z}$ is an integrable signal that is bandlimited to $W/2$ Hz with the assumption that $\mathbf{z}$ is an integrable signal that is bandlimited to $f_{\mathrm{c}}$ Hz.

### 7.6.5 Relating $\langle \mathbf{x}_{\mathrm{PB}}, \mathbf{y}_{\mathrm{PB}} \rangle$ to $\langle \mathbf{x}_{\mathrm{BB}}, \mathbf{y}_{\mathrm{BB}} \rangle$

If $\mathbf{x}_{\mathrm{PB}}$ and $\mathbf{y}_{\mathrm{PB}}$ are integrable real passband signals that are bandlimited to $W$ Hz around the carrier frequency $f_{\mathrm{c}}$, and if $\mathbf{x}_{\mathrm{A}}$, $\mathbf{x}_{\mathrm{BB}}$, $\mathbf{y}_{\mathrm{A}}$, and $\mathbf{y}_{\mathrm{BB}}$ are their corresponding analytic and baseband representations, then, by (7.28),

$$\langle \mathbf{x}_{\mathrm{BB}}, \mathbf{y}_{\mathrm{BB}} \rangle = \langle \mathbf{x}_{\mathrm{A}}, \mathbf{y}_{\mathrm{A}} \rangle, \tag{7.49}$$

because

$$
\begin{aligned}
\langle \mathbf{x}_{\mathrm{BB}}, \mathbf{y}_{\mathrm{BB}} \rangle &= \int_{-\infty}^{\infty} x_{\mathrm{BB}}(t)\, y_{\mathrm{BB}}^{*}(t)\, \mathrm{d}t \\
&= \int_{-\infty}^{\infty} e^{-\mathrm{i}2\pi f_c t}\, x_{\mathrm{A}}(t) \big( e^{-\mathrm{i}2\pi f_c t}\, y_{\mathrm{A}}(t) \big)^{*} \mathrm{d}t \\
&= \int_{-\infty}^{\infty} e^{-\mathrm{i}2\pi f_c t}\, x_{\mathrm{A}}(t)\, e^{\mathrm{i}2\pi f_c t}\, y_{\mathrm{A}}^{*}(t)\, \mathrm{d}t \\
&= \langle \mathbf{x}_{\mathrm{A}}, \mathbf{y}_{\mathrm{A}} \rangle .
\end{aligned}
$$

Combining (7.49) with Proposition 7.5.4 we obtain the following relationship between the inner product between two real passband signals and the inner product between their corresponding complex baseband representations.

**Theorem 7.6.10 ($\langle \mathbf{x}_{\mathrm{PB}}, \mathbf{y}_{\mathrm{PB}} \rangle$ and $\langle \mathbf{x}_{\mathrm{BB}}, \mathbf{y}_{\mathrm{BB}} \rangle$).** *Let $\mathbf{x}_{\mathrm{PB}}$ and $\mathbf{y}_{\mathrm{PB}}$ be two real integrable passband signals that are bandlimited to $W$ Hz around the carrier frequency $f_c$, and let $\mathbf{x}_{\mathrm{BB}}$ and $\mathbf{y}_{\mathrm{BB}}$ be their corresponding baseband representations. Then*

$$
\boxed{\langle \mathbf{x}_{\mathrm{PB}}, \mathbf{y}_{\mathrm{PB}} \rangle = 2\,\mathrm{Re}\big( \langle \mathbf{x}_{\mathrm{BB}}, \mathbf{y}_{\mathrm{BB}} \rangle \big),} \tag{7.50}
$$

*and*

$$
\boxed{\|\mathbf{x}_{\mathrm{PB}}\|_2^2 = 2\,\|\mathbf{x}_{\mathrm{BB}}\|_2^2\,.} \tag{7.51}
$$

An extremely important corollary provides a necessary and sufficient condition for the inner product between two real passband signals to be zero, i.e., for two real passband signals to be orthogonal.

**Corollary 7.6.11 (Characterizing Orthogonal Real Passband Signals).** *Two integrable real passband signals $\mathbf{x}_{\mathrm{PB}}, \mathbf{y}_{\mathrm{PB}}$ that are bandlimited to $W$ Hz around the carrier frequency $f_c$ are orthogonal if, and only if, the inner product between their baseband representations is purely imaginary (i.e., of zero real part).*

Thus, for two such bandpass signals to be orthogonal their baseband representations need not be orthogonal. It suffices that their inner product be purely imaginary.

### 7.6.6 The Baseband Representation of $\mathbf{x}_{\mathrm{PB}} \star \mathbf{y}_{\mathrm{PB}}$

**Proposition 7.6.12 (The Baseband Representation of $\mathbf{x}_{\mathrm{PB}} \star \mathbf{y}_{\mathrm{PB}}$ Is $\mathbf{x}_{\mathrm{BB}} \star \mathbf{y}_{\mathrm{BB}}$).** *Let $\mathbf{x}_{\mathrm{PB}}$ and $\mathbf{y}_{\mathrm{PB}}$ be real integrable passband signals that are bandlimited to $W$ Hz around the carrier frequency $f_c$, and let $\mathbf{x}_{\mathrm{BB}}$ and $\mathbf{y}_{\mathrm{BB}}$ be their baseband representations. Then the convolution $\mathbf{x}_{\mathrm{PB}} \star \mathbf{y}_{\mathrm{PB}}$ is a real integrable passband signal that is bandlimited to $W$ Hz around the carrier frequency $f_c$ and whose baseband representation is $\mathbf{x}_{\mathrm{BB}} \star \mathbf{y}_{\mathrm{BB}}$.*

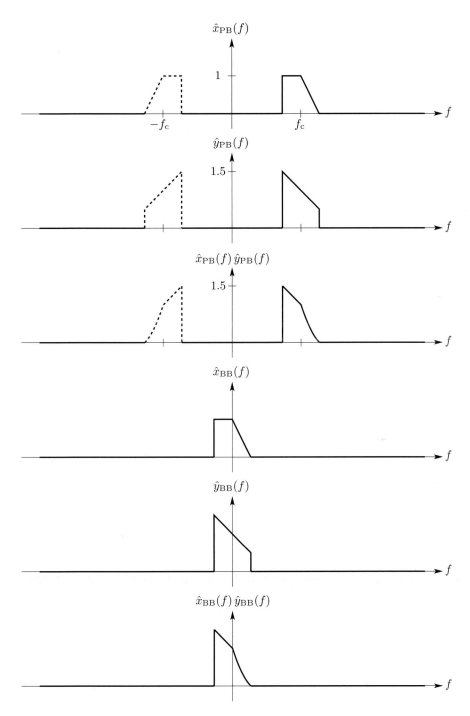

**Figure 7.13:** The convolution of two real passband signals and its baseband representation.

**Proof.** The proof is illustrated in Figure 7.13 on Page 127. All that remains is to add some technical details. We begin by defining

$$\mathbf{z} = \mathbf{x}_{\mathrm{PB}} \star \mathbf{y}_{\mathrm{PB}}$$

and by noting that, by Proposition 7.2.5, $\mathbf{z}$ is an integrable real passband signal that is bandlimited to $W$ Hz around the carrier frequency $f_c$ and that its FT is given by

$$\hat{z}(f) = \hat{x}_{\mathrm{PB}}(f)\,\hat{y}_{\mathrm{PB}}(f), \quad f \in \mathbb{R}. \tag{7.52}$$

Thus, it is at least meaningful to discuss the baseband representation of $\mathbf{x}_{\mathrm{PB}} \star \mathbf{y}_{\mathrm{PB}}$.

We next note that, by Proposition 7.6.5, both $\mathbf{x}_{\mathrm{BB}}$ and $\mathbf{y}_{\mathrm{BB}}$ are integrable signals that are bandlimited to $W/2$ Hz. Consequently, by Proposition 6.5.2, the convolution $\mathbf{u} = \mathbf{x}_{\mathrm{BB}} \star \mathbf{y}_{\mathrm{BB}}$ is defined at every epoch $t$ and is also an integrable signal that is bandlimited to $W/2$ Hz. Its FT is

$$\hat{u}(f) = \hat{x}_{\mathrm{BB}}(f)\,\hat{y}_{\mathrm{BB}}(f), \quad f \in \mathbb{R}. \tag{7.53}$$

From Proposition 7.6.5 we infer that to prove that $\mathbf{u}$ is the baseband representation of $\mathbf{z}$ it only remains to verify that $\hat{u}$ is the mapping $f \mapsto \hat{z}(f + f_c)\,\mathrm{I}\{|f| \le W/2\}$, which, in view of (7.52) and (7.53), is equivalent to showing that

$$\hat{x}_{\mathrm{BB}}(f)\,\hat{y}_{\mathrm{BB}}(f) = \hat{x}_{\mathrm{PB}}(f + f_c)\,\hat{y}_{\mathrm{PB}}(f + f_c)\,\mathrm{I}\{|f| \le W/2\}, \quad f \in \mathbb{R}. \tag{7.54}$$

But this follows because the fact that $\mathbf{x}_{\mathrm{BB}}$ and $\mathbf{y}_{\mathrm{BB}}$ are the baseband representations of $\mathbf{x}_{\mathrm{PB}}$ and $\mathbf{y}_{\mathrm{PB}}$ implies that

$$\hat{x}_{\mathrm{BB}}(f) = \hat{x}_{\mathrm{PB}}(f + f_c)\,\mathrm{I}\{|f| \le W/2\}, \quad f \in \mathbb{R},$$
$$\hat{y}_{\mathrm{BB}}(f) = \hat{y}_{\mathrm{PB}}(f + f_c)\,\mathrm{I}\{|f| \le W/2\}, \quad f \in \mathbb{R},$$

from which (7.54) follows. $\qquad\square$

### 7.6.7 The Baseband Representation of $\mathbf{x}_{\mathrm{PB}} \star \mathbf{h}$

We next study the result of passing a real integrable passband signal $\mathbf{x}_{\mathrm{PB}}$ that is bandlimited to $W$ Hz around the carrier frequency $f_c$ through a real stable filter of impulse response $\mathbf{h}$. Our focus is on the baseband representation of the result.

**Proposition 7.6.13 (Baseband Representation of $\mathbf{x}_{\mathrm{PB}} \star \mathbf{h}$).** *Let $\mathbf{x}_{\mathrm{PB}}$ be a real integrable passband signal that is bandlimited to $W$ Hz around the carrier frequency $f_c$, and let $\mathbf{h}$ be a real integrable signal. Then $\mathbf{x}_{\mathrm{PB}} \star \mathbf{h}$ is defined at every time instant; it is a real integrable passband signal that is bandlimited to $W$ Hz around the carrier frequency $f_c$; and its baseband representation is of FT*

$$f \mapsto \hat{x}_{\mathrm{BB}}(f)\,\hat{h}(f + f_c), \quad f \in \mathbb{R}, \tag{7.55}$$

*where $\mathbf{x}_{\mathrm{BB}}$ is the baseband representation of $\mathbf{x}_{\mathrm{PB}}$.*

**Proof.** That the convolution $\mathbf{x}_{\mathrm{PB}} \star \mathbf{h}$ is defined at every time instant follows from Proposition 7.2.5. Defining $\mathbf{y} = \mathbf{x}_{\mathrm{PB}} \star \mathbf{h}$ we have by the same proposition that $\mathbf{y}$ is a real integrable passband signal that is bandlimited to $\mathsf{W}$ Hz around the carrier frequency $f_{\mathrm{c}}$ and that its FT is given by

$$\hat{y}(f) = \hat{x}_{\mathrm{PB}}(f)\,\hat{h}(f), \quad f \in \mathbb{R}. \tag{7.56}$$

Applying Proposition 7.6.5 (*cf.* (b)) to the signal $\mathbf{y}$ we obtain that the baseband representation of $\mathbf{y}$ is of FT

$$f \mapsto \hat{x}_{\mathrm{PB}}(f + f_{\mathrm{c}})\,\hat{h}(f + f_{\mathrm{c}})\,\mathrm{I}\{|f| \leq \mathsf{W}/2\}, \quad f \in \mathbb{R}. \tag{7.57}$$

To conclude the proof it thus remains to establish that the mappings (7.57) and (7.55) are identical. But this follows because, by Proposition 7.6.5 (*cf.* (b)) applied to the signal $\mathbf{x}_{\mathrm{PB}}$,

$$\hat{x}_{\mathrm{BB}}(f) = \hat{x}_{\mathrm{PB}}(f + f_{\mathrm{c}})\,\mathrm{I}\left\{|f| \leq \frac{\mathsf{W}}{2}\right\}, \quad f \in \mathbb{R}. \qquad \square$$

Motivated by Proposition 7.6.13 we put forth the following definition.

**Definition 7.6.14 (Frequency Response with Respect to a Band).** *For a stable real filter of impulse response $\mathbf{h}$ we define the **frequency response with respect to the bandwidth $\mathsf{W}$ around the carrier frequency** $f_{\mathrm{c}}$ (satisfying $f_{\mathrm{c}} > \mathsf{W}/2$) as the mapping*

$$f \mapsto \hat{h}(f + f_{\mathrm{c}})\,\mathrm{I}\left\{|f| \leq \frac{\mathsf{W}}{2}\right\}. \tag{7.58}$$

Figure 7.14 illustrates the relationship between the frequency response of a real filter and its response with respect to the carrier frequency $f_{\mathrm{c}}$ and bandwidth $\mathsf{W}$. Heuristically, we can think of the frequency response with respect to the bandwidth $\mathsf{W}$ around the carrier frequency $f_{\mathrm{c}}$ of a filter of real impulse response $\mathbf{h}$ as the FT of the baseband representation of $\mathbf{h} \star \mathrm{BPF}_{\mathsf{W},f_{\mathrm{c}}}$.[2]

With the aid of Definition 7.6.14 we can restate Proposition 7.6.13 as stating that the baseband representation of the result of passing a real integrable passband signal that is bandlimited to $\mathsf{W}$ Hz around the carrier frequency $f_{\mathrm{c}}$ through a stable real filter is the product of the FT of the baseband representation of the signal by the frequency response with respect to the bandwidth $\mathsf{W}$ around the carrier frequency $f_{\mathrm{c}}$ of the filter. This relationship is illustrated in Figures 7.15 and 7.16. The former depicts the product of the FT of a real passband signal $\mathbf{x}_{\mathrm{PB}}$ and the frequency response of a real filter $\mathbf{h}$. The latter depicts the product of the baseband representation $\mathbf{x}_{\mathrm{BB}}$ of $\mathbf{x}_{\mathrm{PB}}$ by the frequency response of $\mathbf{h}$ with respect to the bandwidth $\mathsf{W}$ around the carrier frequency $f_{\mathrm{c}}$.

The relationship between some of the properties of $\mathbf{x}_{\mathrm{PB}}$, $\mathbf{x}_{\mathrm{A}}$, and $\mathbf{x}_{\mathrm{BB}}$ are summarized in Table 7.1 on Page 142.

---

[2]This is mathematically somewhat problematic because $\mathbf{h} \star \mathrm{BPF}_{\mathsf{W},f_{\mathrm{c}}}$ need not be an integrable signal. But this can be remedied because $\mathbf{h} \star \mathrm{BPF}_{\mathsf{W},f_{\mathrm{c}}}$ is an *energy-limited* passband signal that is bandlimited to $\mathsf{W}$ Hz around the carrier frequency, and, as such, also has a baseband representation; see Section 7.7.

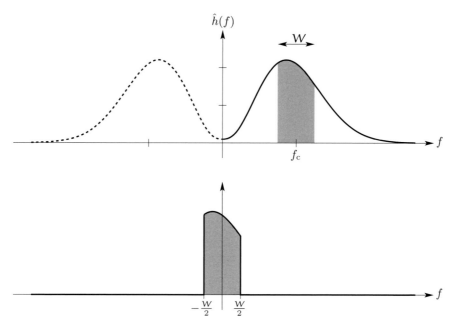

**Figure 7.14:** A real filter's frequency response (top) and its frequency response with respect to the bandwidth $W$ around the carrier frequency $f_c$ (bottom).

## 7.7 Energy-Limited Passband Signals

We next repeat the results of this chapter under the weaker assumption that the passband signal is energy-limited and not necessarily integrable. The key results require only minor adjustments, and most of the derivations are almost identical and are therefore omitted. The reader is encouraged to focus on the results and to read the proofs only if needed.

### 7.7.1 Characterization of Energy-Limited Passband Signals

Recall that energy-limited passband signals were defined in Definition 7.2.1 as energy-limited signals that are unaltered by bandpass filtering. In this subsection we shall describe alternative characterizations. Aiding us in the characterization is the following lemma, which can be viewed as the passband analog of Lemma 6.4.4 (i).

**Lemma 7.7.1.** *Let* $\mathbf{x}$ *be an energy-limited signal, and let* $f_c > W/2 > 0$ *be given. Then the signal* $\mathbf{x} \star \mathrm{BPF}_{W,f_c}$ *can be expressed as*

$$\left(\mathbf{x} \star \mathrm{BPF}_{W,f_c}\right)(t) = \int_{\left||f|-f_c\right| \leq W/2} \hat{x}(f)\, e^{i2\pi ft}\, \mathrm{d}f, \quad t \in \mathbb{R}; \qquad (7.59)$$

*it is of finite energy; and its* $L_2$*-Fourier Transform is (the equivalence class of) the mapping* $f \mapsto \hat{x}(f)\,\mathrm{I}\{\left||f| - f_c\right| \leq W/2\}$.

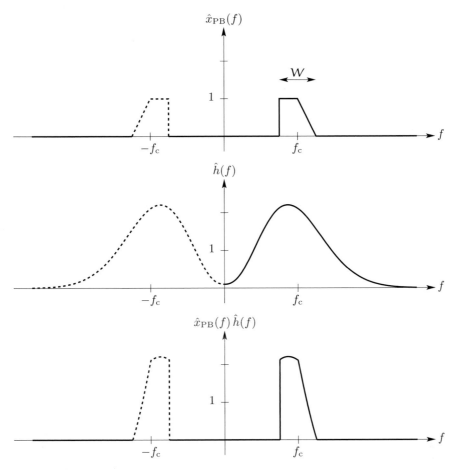

**Figure 7.15:** The FT of a passband signal (top); the frequency response of a real filter (middle); and their product (bottom).

**Proof.** The lemma follows from Lemma 6.4.4 (ii) by substituting for **g** the mapping $f \mapsto \mathrm{I}\{||f| - f_\mathrm{c}| \leq W/2\}$, whose IFT is $\mathrm{BPF}_{W,f_\mathrm{c}}$. $\qquad\square$

In analogy to Proposition 6.4.5 we can characterize energy-limited passband signals as follows.

**Proposition 7.7.2 (Characterizations of Passband Signals in $\mathcal{L}_2$).**

(i) *If* **x** *is an energy-limited passband signal that is bandlimited to* **W** *Hz around the carrier frequency* $f_\mathrm{c}$, *then it can be expressed in the form*

$$x(t) = \int_{||f|-f_\mathrm{c}| \leq W/2} g(f)\, e^{\mathrm{i}2\pi ft}\, \mathrm{d}f, \quad t \in \mathbb{R}, \tag{7.60}$$

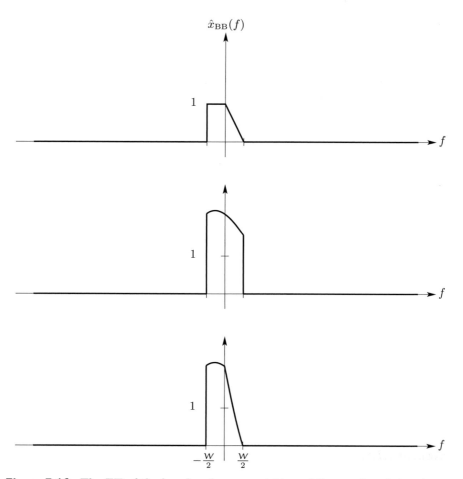

**Figure 7.16:** The FT of the baseband representation of the passband signal $\mathbf{x}_{\mathrm{PB}}$ of Figure 7.15 (top); the frequency response with respect to the bandwidth $W$ around the carrier frequency $f_{\mathrm{c}}$ of the filter of Figure 7.15 (middle); and their product (bottom).

*for some mapping* $\mathbf{g}\colon f \mapsto g(f)$ *satisfying*

$$\int_{\big||f|-f_c\big|\leq W/2} |g(f)|^2 \,\mathrm{d}f < \infty \qquad (7.61)$$

*that can be taken as (any function in the equivalence class of)* $\hat{\mathbf{x}}$.

(ii) *If a signal* $\mathbf{x}$ *can be expressed as in* (7.60) *for some function* $\mathbf{g}$ *satisfying* (7.61), *then* $\mathbf{x}$ *is an energy-limited passband signal that is bandlimited to* $W$ *Hz around the carrier frequency* $f_c$ *and its FT* $\hat{\mathbf{x}}$ *is (the equivalence class of) the mapping* $f \mapsto g(f)\,\mathrm{I}\big\{\big||f| - f_c\big| \leq W/2\big\}$.

**Proof.** The proof of Part (i) follows from Definition 7.2.1 and from Lemma 7.7.1 in very much the same way as Part (i) of Proposition 6.4.5 follows from Definition 6.4.1 and Lemma 6.4.4 (i).

The proof of Part (ii) is analogous to the proof of Part (ii) of Proposition 6.4.5. $\quad\square$

As a corollary we obtain the analog of Corollary 7.2.3:

**Corollary 7.7.3 (Passband Signals Are Bandlimited).** *If* $\mathbf{x}_{\mathrm{PB}}$ *is an energy-limited passband signal that is bandlimited to* $W$ *Hz around the carrier frequency* $f_c$, *then it is an energy-limited signal that is bandlimited to* $f_c + W/2$ *Hz.*

**Proof.** If $\mathbf{x}_{\mathrm{PB}}$ is an energy-limited passband signal that is bandlimited to $W$ Hz around the carrier frequency $f_c$, then, by Proposition 7.7.2 (i), there exists a function $\mathbf{g}\colon f \mapsto g(f)$ satisfying (7.61) such that $\mathbf{x}_{\mathrm{PB}}$ is given by (7.60). But this implies that the function $f \mapsto g(f)\,\mathrm{I}\big\{\big||f| - f_c\big| \leq W/2\big\}$ is an energy-limited function such that

$$x_{\mathrm{PB}}(t) = \int_{-f_c-W/2}^{f_c+W/2} g(f)\,\mathrm{I}\big\{\big||f| - f_c\big| \leq W/2\big\}\, e^{\mathrm{i}2\pi ft}\,\mathrm{d}f, \quad t \in \mathbb{R}, \qquad (7.62)$$

so, by Proposition 6.4.5 (ii), $\mathbf{x}_{\mathrm{PB}}$ is an energy-limited signal that is bandlimited to $f_c + W/2$ Hz. $\quad\square$

The following is the analog of Proposition 6.4.6.

**Proposition 7.7.4.**

(i) *If* $\mathbf{x}_{\mathrm{PB}}$ *is an energy-limited passband signal that is bandlimited to* $W$ *Hz around the carrier frequency* $f_c$, *then* $\mathbf{x}_{\mathrm{PB}}$ *is a continuous function and all its energy is contained in the frequencies* $f$ *satisfying* $\big||f| - f_c\big| \leq W/2$ *in the sense that*

$$\int_{-\infty}^{\infty} |\hat{x}_{\mathrm{PB}}(f)|^2 \,\mathrm{d}f = \int_{\big||f|-f_c\big|\leq W/2} |\hat{x}_{\mathrm{PB}}(f)|^2 \,\mathrm{d}f. \qquad (7.63)$$

(ii) *If* $\mathbf{x}_{\mathrm{PB}} \in \mathcal{L}_2$ *satisfies* (7.63), *then* $\mathbf{x}_{\mathrm{PB}}$ *is indistinguishable from the signal* $\mathbf{x}_{\mathrm{PB}} \star \mathrm{BPF}_{W,f_c}$, *which is an energy-limited passband signal that is bandlimited to* $W$ *Hz around* $f_c$. *If in addition to satisfying* (7.63) *the signal* $\mathbf{x}_{\mathrm{PB}}$ *is continuous, then* $\mathbf{x}_{\mathrm{PB}}$ *is an energy-limited passband signal that is bandlimited to* $W$ *Hz around the carrier frequency* $f_c$.

**Proof.** This proposition's claims are a subset of those of Proposition 7.7.5, which summarizes some of the results related to bandpass filtering.  □

**Proposition 7.7.5.** *Let* $\mathbf{y} = \mathbf{x} \star \mathrm{BPF}_{W, f_c}$ *be the result of feeding the signal* $\mathbf{x} \in \mathcal{L}_2$ *to an ideal unit-gain bandpass filter of bandwidth* $W$ *around the carrier frequency* $f_c$. *Assume* $f_c > W/2$. *Then:*

(i) $\mathbf{y}$ *is energy-limited with*

$$\|\mathbf{y}\|_2 \leq \|\mathbf{x}\|_2 . \tag{7.64}$$

(ii) $\mathbf{y}$ *is an energy-limited passband signal that is bandlimited to* $W$ *Hz around the carrier frequency* $f_c$.

(iii) *The* $\mathcal{L}_2$-*Fourier Transform of* $\mathbf{y}$ *is (the equivalence class of) the mapping* $f \mapsto \hat{x}(f) \, \mathrm{I}\{ \big||f| - f_c\big| \leq W/2 \}$.

(iv) *All the energy in* $\mathbf{y}$ *is concentrated in the frequencies* $\{ f : \big||f| - f_c\big| \leq W/2 \}$ *in the sense that*

$$\int_{-\infty}^{\infty} |\hat{y}(f)|^2 \, \mathrm{d}f = \int_{\big||f| - f_c\big| \leq W/2} |\hat{y}(f)|^2 \, \mathrm{d}f.$$

(v) $\mathbf{y}$ *can be represented as*

$$y(t) = \int_{-\infty}^{\infty} \hat{y}(f) \, e^{\mathrm{i}2\pi f t} \, \mathrm{d}f \tag{7.65}$$

$$= \int_{\big||f| - f_c\big| \leq W/2} \hat{x}(f) \, e^{\mathrm{i}2\pi f t} \, \mathrm{d}f, \quad t \in \mathbb{R}. \tag{7.66}$$

(vi) $\mathbf{y}$ *is uniformly continuous.*

(vii) *If all the energy of* $\mathbf{x}$ *is concentrated in the frequencies* $\{ f : \big||f| - f_c\big| \leq W/2 \}$ *in the sense that*

$$\int_{-\infty}^{\infty} |\hat{x}(f)|^2 \, \mathrm{d}f = \int_{\big||f| - f_c\big| \leq W/2} |\hat{x}(f)|^2 \, \mathrm{d}f, \tag{7.67}$$

*then* $\mathbf{x}$ *is indistinguishable from the passband signal* $\mathbf{x} \star \mathrm{BPF}_{W, f_c}$.

(viii) $\mathbf{z}$ *is an energy-limited passband signal that is bandlimited to* $W$ *Hz around the carrier frequency* $f_c$ *if, and only if, it satisfies all three of the following conditions: it is in* $\mathcal{L}_2$; *it is continuous; and all its energy is concentrated in the passband frequencies* $\{ f : \big||f| - f_c\big| \leq W/2 \}$.

**Proof.** The proof is very similar to the proof of Proposition 6.4.7 and is thus omitted.  □

### 7.7.2 The Analytic Representation

If $\mathbf{x}_{\text{PB}}$ is a real energy-limited passband signal that is bandlimited to $W$ Hz around the carrier frequency $f_{\text{c}}$, then we define its analytic representation via (7.11). (Since $\mathbf{x}_{\text{PB}} \in \mathcal{L}_2$, it follows from Parseval's Theorem that $\hat{x}_{\text{PB}}$ is energy-limited so, by Proposition 3.4.3, the mapping $f \mapsto \hat{x}_{\text{PB}}(f) \, \text{I}\{|f - f_{\text{c}}| \le W/2\}$ is integrable and the integral (7.11) is defined for every $t \in \mathbb{R}$. Also, the integral does not depend on which element of the equivalence class consisting of the $L_2$-Fourier Transform of $\mathbf{x}_{\text{PB}}$ it is applied to.)

In analogy to Proposition 7.5.2 we can characterize the analytic representation as follows.

**Proposition 7.7.6 (Characterizing the Analytic Representation of $\mathbf{x}_{\text{PB}} \in \mathcal{L}_2$).**
*Let $\mathbf{x}_{\text{PB}}$ be a real energy-limited passband signal that is bandlimited to $W$ Hz around the carrier frequency $f_{\text{c}}$. Then each of the following statements is equivalent to the statement that the complex signal $\mathbf{x}_{\text{A}}$ is the analytic representation of $\mathbf{x}_{\text{PB}}$:*

*(a) The signal $\mathbf{x}_{\text{A}}$ is given by*

$$x_{\text{A}}(t) = \int_{f_{\text{c}} - \frac{W}{2}}^{f_{\text{c}} + \frac{W}{2}} \hat{x}_{\text{PB}}(f) \, e^{i2\pi ft} \, \mathrm{d}f, \quad t \in \mathbb{R}. \tag{7.68}$$

*(b) The signal $\mathbf{x}_{\text{A}}$ is a continuous energy-limited signal whose $L_2$-Fourier Transform $\hat{x}_{\text{A}}$ is (the equivalence class of) the mapping*

$$f \mapsto \hat{x}_{\text{PB}}(f) \, \text{I}\{f \ge 0\}. \tag{7.69}$$

*(c) The signal $\mathbf{x}_{\text{A}}$ is an energy-limited passband signal that is bandlimited to $W$ Hz around the carrier frequency $f_{\text{c}}$ and whose $L_2$-Fourier Transform is (the equivalence class of) the mapping in (7.69).*

*(d) The signal $\mathbf{x}_{\text{A}}$ is given by*

$$\mathbf{x}_{\text{A}} = \mathbf{x}_{\text{PB}} \star \check{\mathbf{g}} \tag{7.70}$$

*where $\mathbf{g} \colon f \mapsto g(f)$ is any function in $\mathcal{L}_1 \cap \mathcal{L}_2$ satisfying*

$$g(f) = 1, \quad |f - f_{\text{c}}| \le W/2, \tag{7.71a}$$

*and*

$$g(f) = 0, \quad |f + f_{\text{c}}| \le W/2. \tag{7.71b}$$

**Proof.** The proof is not very difficult and is omitted. $\square$

We note that the reconstruction formula (7.21b) continues to hold also when $\mathbf{x}_{\text{PB}}$ is an energy-limited signal that is bandlimited to $W$ Hz around the carrier frequency $f_{\text{c}}$.

### 7.7.3 The Baseband Representation of $\mathbf{x}_{\mathrm{PB}} \in \mathcal{L}_2$

Having defined the analytic representation, we now use (7.28) to define the baseband representation.

As in Proposition 7.6.3, we can also describe a procedure for obtaining the baseband representation of a passband signal without having to go via the analytic representation.

**Proposition 7.7.7 (From $\mathbf{x}_{\mathrm{PB}} \in \mathcal{L}_2$ to $\mathbf{x}_{\mathrm{BB}}$ Directly).** *If $\mathbf{x}_{\mathrm{PB}}$ is a real energy-limited passband signal that is bandlimited to $\mathsf{W}$ Hz around the carrier frequency $f_{\mathrm{c}}$, then its baseband representation $\mathbf{x}_{\mathrm{BB}}$ is given by*

$$\mathbf{x}_{\mathrm{BB}} = \left(t \mapsto e^{-\mathrm{i}2\pi f_{\mathrm{c}} t}\, x_{\mathrm{PB}}(t)\right) \star \check{\mathbf{g}}_0, \tag{7.72}$$

*where $\mathbf{g}_0\colon f \mapsto g_0(f)$ is any function in $\mathcal{L}_1 \cap \mathcal{L}_2$ satisfying*

$$g_0(f) = 1, \quad |f| \leq \mathsf{W}/2, \tag{7.73a}$$

*and*

$$g_0(f) = 0, \quad |f + 2f_{\mathrm{c}}| \leq \mathsf{W}/2. \tag{7.73b}$$

**Proof.** The proof is very similar to the proof of Proposition 7.6.3 and is omitted. □

The following proposition, which is the analog of Proposition 7.6.5 characterizes the baseband representation of energy-limited passband signals.

**Proposition 7.7.8 (Characterizing the Baseband Representation of $\mathbf{x}_{\mathrm{PB}} \in \mathcal{L}_2$).** *Let $\mathbf{x}_{\mathrm{PB}}$ be a real energy-limited passband signal that is bandlimited to $\mathsf{W}$ Hz around the carrier frequency $f_{\mathrm{c}}$. Then each of the following statements is equivalent to the statement that the complex signal $\mathbf{x}_{\mathrm{BB}}$ is the baseband representation of $\mathbf{x}_{\mathrm{PB}}$.*

(a) *The signal $\mathbf{x}_{\mathrm{BB}}$ is given by*

$$x_{\mathrm{BB}}(t) = \int_{-\frac{\mathsf{W}}{2}}^{\frac{\mathsf{W}}{2}} \hat{x}_{\mathrm{PB}}(f + f_{\mathrm{c}})\, e^{\mathrm{i}2\pi f t}\, \mathrm{d}f, \quad t \in \mathbb{R}. \tag{7.74}$$

(b) *The signal $\mathbf{x}_{\mathrm{BB}}$ is a continuous energy-limited signal whose $\mathcal{L}_2$-Fourier Transform is (the equivalence class of) the mapping*

$$f \mapsto \hat{x}_{\mathrm{PB}}(f + f_{\mathrm{c}})\, \mathrm{I}\{|f| \leq \mathsf{W}/2\}. \tag{7.75}$$

(c) *The signal $\mathbf{x}_{\mathrm{BB}}$ is an energy-limited signal that is bandlimited to $\mathsf{W}/2$ Hz and whose $\mathcal{L}_2$-Fourier Transform is (the equivalence class of) the mapping (7.75).*

(d) *The signal $\mathbf{x}_{\mathrm{BB}}$ is given by (7.72) for any mapping $\mathbf{g}_0\colon f \mapsto g_0(f)$ satisfying (7.73).*

The **in-phase component** and the **quadrature component** of an energy-limited passband signal are defined, as in the integrable case, as the real and imaginary parts of its baseband representation.

Proposition 7.6.7, which asserts that the bandwidth of $\mathbf{x}_{BB}$ is half the bandwidth of $\mathbf{x}_{PB}$ continues to hold, as does the reconstruction formula (7.42b). Proposition 7.6.9 also extends to energy-limited signals. We repeat it (in a slightly more general way) for future reference.

**Proposition 7.7.9.**

(i) *If $\mathbf{z}$ is an energy-limited signal that is bandlimited to $W/2$ Hz, and if the signal $\mathbf{x}$ is given by*

$$x(t) = 2 \operatorname{Re}\big(z(t)\, e^{i2\pi f_c t}\big), \quad t \in \mathbb{R}, \tag{7.76}$$

*where $f_c > W/2$, then $\mathbf{x}$ is a real energy-limited passband signal that is bandlimited to $W$ Hz around $f_c$, and $\mathbf{z}$ is its baseband representation.*

(ii) *If $\mathbf{x}$ is an energy-limited passband signal that is bandlimited to $W$ Hz around the carrier frequency $f_c$ and if (7.76) holds for some energy-limited signal $\mathbf{z}$ that is bandlimited to $f_c$ Hz, then $\mathbf{z}$ is the baseband representation of $\mathbf{x}$ and is, in fact, bandlimited to $W/2$ Hz.*

**Proof.** Omitted. □

Identity (7.50) relating the inner products $\langle \mathbf{x}_{PB}, \mathbf{y}_{PB}\rangle$ and $\langle \mathbf{x}_{BB}, \mathbf{y}_{BB}\rangle$ continues to hold for energy-limited passband signals that are not necessarily integrable.

Proposition 7.6.12 does not hold for energy-limited signals, because the convolution of two energy-limited signals need not be energy-limited. But if we assume that at least one of the signals is also integrable, then things sail through. Consequently, using Corollary 7.2.4 we obtain:

**Proposition 7.7.10 (The Baseband Representation of $\mathbf{x}_{PB} \star \mathbf{y}_{PB}$ Is $\mathbf{x}_{BB} \star \mathbf{y}_{BB}$).**
*Let $\mathbf{x}_{PB}$ be a real integrable passband signal that is bandlimited to $W$ Hz around the carrier frequency $f_c$, and let $\mathbf{y}_{PB}$ be a real energy-limited passband signal that is bandlimited to $W$ Hz around the carrier frequency $f_c$. Let $\mathbf{x}_{BB}$ and $\mathbf{y}_{BB}$ be their corresponding baseband representations. Then $\mathbf{x}_{PB} \star \mathbf{y}_{PB}$ is a real energy-limited signal that is bandlimited to $W$ Hz around the carrier frequency $f_c$ and whose baseband representation is $\mathbf{x}_{BB} \star \mathbf{y}_{BB}$.*

Proposition 7.6.13 too requires only a slight modification to address energy-limited signals.

**Proposition 7.7.11 (Baseband Representation of $\mathbf{x}_{PB} \star \mathbf{h}$).** *Let $\mathbf{x}_{PB}$ be a real energy-limited passband signal that is bandlimited to $W$ Hz around the carrier frequency $f_c$, and let $\mathbf{h}$ be a real integrable signal. Then $\mathbf{x}_{PB} \star \mathbf{h}$ is defined at every time instant; it is a real energy-limited passband signal that is bandlimited to $W$ Hz around the carrier frequency $f_c$; and its baseband representation is given by*

$$\big(\mathbf{h} \star \mathbf{x}_{PB}\big)_{BB} = \mathbf{h}'_{BB} \star \mathbf{x}_{BB}, \tag{7.77}$$

where $\mathbf{h}'_{\mathrm{BB}}$ is the baseband representation of the energy-limited signal $\mathbf{h} \star \mathrm{BPF}_{W,f_c}$. The $L_2$-Fourier Transform of the baseband representation of $\mathbf{x}_{\mathrm{PB}} \star \mathbf{h}$ is (the equivalence class of) the mapping

$$f \mapsto \hat{x}_{\mathrm{BB}}(f)\,\hat{h}(f + f_c), \quad f \in \mathbb{R}, \tag{7.78}$$

where $\mathbf{x}_{\mathrm{BB}}$ is the baseband representation of $\mathbf{x}_{\mathrm{PB}}$.

The following theorem summarizes some of the properties of the baseband representation of energy-limited passband signals.

**Theorem 7.7.12 (Properties of the Baseband Representation).**

(i) The mapping $\mathbf{x}_{\mathrm{PB}} \mapsto \mathbf{x}_{\mathrm{BB}}$ that maps every real energy-limited passband signal that is bandlimited to $W$ Hz around the carrier frequency $f_c$ to its baseband representation is a one-to-one mapping onto the space of complex energy-limited signals that are bandlimited to $W/2$ Hz.

(ii) The mapping $\mathbf{x}_{\mathrm{PB}} \mapsto \mathbf{x}_{\mathrm{BB}}$ is linear in the sense that if $\mathbf{x}_{\mathrm{PB}}$ and $\mathbf{y}_{\mathrm{PB}}$ are real energy-limited passband signals that are bandlimited to $W$ Hz around the carrier frequency $f_c$, and if $\mathbf{x}_{\mathrm{BB}}$ and $\mathbf{y}_{\mathrm{BB}}$ are their corresponding baseband representations, then for every $\alpha, \beta \in \mathbb{R}$, the baseband representation of $\alpha\mathbf{x}_{\mathrm{PB}} + \beta\mathbf{y}_{\mathrm{PB}}$ is $\alpha\mathbf{x}_{\mathrm{BB}} + \beta\mathbf{y}_{\mathrm{BB}}$:

$$\left(\alpha\mathbf{x}_{\mathrm{PB}} + \beta\mathbf{y}_{\mathrm{PB}}\right)_{\mathrm{BB}} = \alpha\mathbf{x}_{\mathrm{BB}} + \beta\mathbf{y}_{\mathrm{BB}}, \quad \alpha, \beta \in \mathbb{R}. \tag{7.79}$$

(iii) The mapping $\mathbf{x}_{\mathrm{PB}} \mapsto \mathbf{x}_{\mathrm{BB}}$ is—to within a factor of two—energy preserving in the sense that

$$\|\mathbf{x}_{\mathrm{PB}}\|_2^2 = 2\,\|\mathbf{x}_{\mathrm{BB}}\|_2^2\,. \tag{7.80}$$

(iv) Inner products are related via

$$\langle \mathbf{x}_{\mathrm{PB}}, \mathbf{y}_{\mathrm{PB}} \rangle = 2\,\mathrm{Re}\big(\langle \mathbf{x}_{\mathrm{BB}}, \mathbf{y}_{\mathrm{BB}} \rangle\big), \tag{7.81}$$

for $\mathbf{x}_{\mathrm{PB}}$ and $\mathbf{y}_{\mathrm{PB}}$ as above.

(v) The (baseband) bandwidth of $\mathbf{x}_{\mathrm{BB}}$ is half the bandwidth of $\mathbf{x}_{\mathrm{PB}}$ around the carrier frequency $f_c$.

(vi) The baseband representation $\mathbf{x}_{\mathrm{BB}}$ can be expressed in terms of $\mathbf{x}_{\mathrm{PB}}$ as

$$\mathbf{x}_{\mathrm{BB}} = \big(t \mapsto e^{-\mathrm{i}2\pi f_c t}\, x_{\mathrm{PB}}(t)\big) \star \mathrm{LPF}_{W_c} \tag{7.82a}$$

where $W_c$ is any cutoff frequency satisfying

$$W/2 \le W_c \le 2f_c - W/2. \tag{7.82b}$$

(vii) The real passband signal $\mathbf{x}_{\mathrm{PB}}$ can be expressed in terms of its baseband representation $\mathbf{x}_{\mathrm{BB}}$ as

$$x_{\mathrm{PB}}(t) = 2\,\mathrm{Re}\big(x_{\mathrm{BB}}(t)\,e^{\mathrm{i}2\pi f_c t}\big), \quad t \in \mathbb{R}. \tag{7.83}$$

*(viii) If* **h** *is a real integrable signal, and if* $\mathbf{x}_{\mathrm{PB}}$ *is as above, then* $\mathbf{h} \star \mathbf{x}_{\mathrm{PB}}$ *is a real energy-limited passband signal that is bandlimited to* W Hz *around the carrier frequency* $f_{\mathrm{c}}$, *and its baseband representation is given by*

$$\left( \mathbf{h} \star \mathbf{x}_{\mathrm{PB}} \right)_{\mathrm{BB}} = \mathbf{h}'_{\mathrm{BB}} \star \mathbf{x}_{\mathrm{BB}}, \tag{7.84}$$

*where* $\mathbf{h}'_{\mathrm{BB}}$ *is the baseband representation of the energy-limited real signal* $\mathbf{h} \star \mathrm{BPF}_{\mathrm{W}, f_{\mathrm{c}}}$.

## 7.8 Shifting to Passband and Convolving

The following result is almost trivial if you think about its interpretation in the frequency domain. To that end, it is good to focus on the case where the signal **x** is a bandlimited baseband signal and where $f_{\mathrm{c}}$ is positive and large. In this case we can interpret the LHS of (7.85) as the result of taking the baseband signal **x**, up-converting it to passband by forming the signal $\tau \mapsto x(\tau) \, e^{\mathrm{i}2\pi f_{\mathrm{c}}\tau}$, and then convolving the result with **h**. The RHS corresponds to down-converting **h** to form the signal $\tau \mapsto e^{-\mathrm{i}2\pi f_{\mathrm{c}}\tau} h(\tau)$, then convolving this signal with **x**, and then up-converting the final result.

**Proposition 7.8.1.** *Suppose that* $f_{\mathrm{c}} \in \mathbb{R}$ *and that (at least) one of the following conditions holds:*

1) *The signal* **x** *is a measurable bounded signal and* $\mathbf{h} \in \mathcal{L}_1$.

2) *Both* **x** *and* **h** *are in* $\mathcal{L}_2$.

*Then, at every epoch* $t \in \mathbb{R}$,

$$\left( \left( \tau \mapsto x(\tau) \, e^{\mathrm{i}2\pi f_{\mathrm{c}}\tau} \right) \star \mathbf{h} \right)(t) = e^{\mathrm{i}2\pi f_{\mathrm{c}}t} \left( \mathbf{x} \star \left( \tau \mapsto e^{-\mathrm{i}2\pi f_{\mathrm{c}}\tau} h(\tau) \right) \right)(t). \tag{7.85}$$

**Proof.** We evaluate the LHS of (7.85) using the definition of the convolution:

$$\left( \left( \tau \mapsto x(\tau) \, e^{\mathrm{i}2\pi f_{\mathrm{c}}\tau} \right) \star \mathbf{h} \right)(t) = \int_{-\infty}^{\infty} x(\tau) \, e^{\mathrm{i}2\pi f_{\mathrm{c}}\tau} \, h(t-\tau) \, \mathrm{d}\tau$$

$$= e^{\mathrm{i}2\pi f_{\mathrm{c}}t} \, e^{-\mathrm{i}2\pi f_{\mathrm{c}}t} \int_{-\infty}^{\infty} x(\tau) \, e^{\mathrm{i}2\pi f_{\mathrm{c}}\tau} \, h(t-\tau) \, \mathrm{d}\tau$$

$$= e^{\mathrm{i}2\pi f_{\mathrm{c}}t} \int_{-\infty}^{\infty} x(\tau) \, e^{-\mathrm{i}2\pi f_{\mathrm{c}}(t-\tau)} \, h(t-\tau) \, \mathrm{d}\tau$$

$$= e^{\mathrm{i}2\pi f_{\mathrm{c}}t} \left( \mathbf{x} \star \left( \tau \mapsto e^{-\mathrm{i}2\pi f_{\mathrm{c}}\tau} h(\tau) \right) \right)(t). \qquad \square$$

## 7.9 Mathematical Comments

The analytic representation is related to the **Hilbert Transform**; see, for example, (Pinsky, 2002, Section 3.4). In our proof that $\mathbf{x}_{\mathrm{A}}$ is integrable whenever $\mathbf{x}_{\mathrm{PB}}$ is integrable we implicitly exploited the fact that the strict inequality $f_c > W/2$ implies that for the class of integrable passband signals that are bandlimited to $W$ Hz around the carrier frequency $f_c$ there exist Hilbert Transform kernels that are integrable. See, for example, (Logan, 1978, Section 2.5).

## 7.10    Exercises

**Exercise 7.1 (Purely Real and Purely Imaginary Baseband Representations).** Let $\mathbf{x}_{PB}$ be a real integrable passband signal that is bandlimited to $W$ Hz around the carrier frequency $f_c$, and let $\mathbf{x}_{BB}$ be its baseband representation.

(i) Show that $\mathbf{x}_{BB}$ is real if, and only if, $\hat{x}_{PB}$ satisfies

$$\hat{x}_{PB}(f_c - \delta) = \hat{x}_{PB}^*(f_c + \delta), \quad |\delta| \leq \frac{W}{2}.$$

(ii) Show that $\mathbf{x}_{BB}$ is imaginary if, and only if,

$$\hat{x}_{PB}(f_c - \delta) = -\hat{x}_{PB}^*(f_c + \delta), \quad |\delta| \leq \frac{W}{2}.$$

**Exercise 7.2 (Symmetry around the Carrier Frequency).** Let $\mathbf{x}_{PB}$ be a real integrable passband signal that is bandlimited to $W$ Hz around the carrier frequency $f_c$.

(i) Show that $\mathbf{x}_{PB}$ can be written in the form

$$x_{PB}(t) = w(t) \cos(2\pi f_c t)$$

where $w(\cdot)$ is a real integrable signal that is bandlimited to $W/2$ Hz if, and only if,

$$\hat{x}_{PB}(f_c + \delta) = \hat{x}_{PB}^*(f_c - \delta), \quad |\delta| \leq \frac{W}{2}.$$

(ii) Show that $\mathbf{x}_{PB}$ can be written in the form

$$x_{PB}(t) = w(t) \sin(2\pi f_c t), \quad t \in \mathbb{R}$$

for $w(\cdot)$ as above if, and only if,

$$\hat{x}_{PB}(f_c + \delta) = -\hat{x}_{PB}^*(f_c - \delta), \quad |\delta| \leq \frac{W}{2}.$$

**Exercise 7.3 (Viewing a Baseband Signal as a Passband Signal).** Let $\mathbf{x}$ be a real integrable signal that is bandlimited to $W$ Hz. Show that if we had informally allowed equality in (7.1b) and if we had allowed equality between $f_c$ and $W/2$ in (5.21), then we could have viewed $\mathbf{x}$ also as a real integrable passband signal that is bandlimited to $W$ Hz around the carrier frequency $f_c = W/2$. Viewed as such, what would have been its complex baseband representation?

**Exercise 7.4 (Bandwidth of the Product of Two Signals).** Let $\mathbf{x}$ be a real energy-limited signal that is bandlimited to $W_x$ Hz. Let $\mathbf{y}$ be a real energy-limited passband signal that is bandlimited to $W_y$ Hz around the carrier frequency $f_c$. Show that if $f_c > W_x + W_y/2$, then the signal $t \mapsto x(t)\, y(t)$ is a real integrable passband signal that is bandlimited to $2W_x + W_y$ Hz around the carrier frequency $f_c$.

**Exercise 7.5 (Phase Shift).** Let $\mathbf{x}$ be a real integrable signal that is bandlimited to $W$ Hz. Let $f_c$ be larger than $W$.

(i) Express the baseband representation of the real passband signal

$$z_{\text{PB}}(t) = x(t)\sin(2\pi f_c t + \phi), \quad t \in \mathbb{R}$$

in terms of $x(\cdot)$ and $\phi$.

(ii) Compute the Fourier Transform of $\mathbf{z}_{\text{PB}}$.

**Exercise 7.6 (Energy of a Passband Signal).** Let $\mathbf{x} \in \mathcal{L}_2$ be of energy $\|\mathbf{x}\|_2^2$.

(i) What is the approximate energy in $t \mapsto x(t)\cos(2\pi f_c t)$ if $f_c$ is very large?

(ii) Is your answer exact if $x(\cdot)$ is an energy-limited signal that is bandlimited to $W$ Hz, where $W < f_c$?

*Hint: In Part (i) approximate $\mathbf{x}$ as being constant over the periods of $t \mapsto \cos(2\pi f_c t)$. For Part (ii) see also Problem 6.13.*

**Exercise 7.7 (Differences in Passband).** Let $\mathbf{x}_{\text{PB}}$ and $\mathbf{y}_{\text{PB}}$ be real energy-limited passband signals that are bandlimited to $W$ Hz around the carrier frequency $f_c$. Let $\mathbf{x}_{\text{BB}}$ and $\mathbf{y}_{\text{BB}}$ be their baseband representations. Find the relationship between

$$\int_{-\infty}^{\infty} \big(x_{\text{PB}}(t) - y_{\text{PB}}(t)\big)^2 \, dt \quad \text{and} \quad \int_{-\infty}^{\infty} \big|x_{\text{BB}}(t) - y_{\text{BB}}(t)\big|^2 \, dt.$$

**Exercise 7.8 (Reflection of Passband Signal).** Let $\mathbf{x}_{\text{PB}}$ and $\mathbf{y}_{\text{PB}}$ be real integrable passband signals that are bandlimited to $W$ Hz around the carrier frequency $f_c$. Let $\mathbf{x}_{\text{BB}}$ and $\mathbf{y}_{\text{BB}}$ be their baseband representations.

(i) Express the baseband representation of $\tilde{\mathbf{x}}_{\text{PB}}$ in terms of $\mathbf{x}_{\text{BB}}$.

(ii) Express $\langle \mathbf{x}_{\text{PB}}, \tilde{\mathbf{y}}_{\text{PB}} \rangle$ in terms of $\mathbf{x}_{\text{BB}}$ and $\mathbf{y}_{\text{BB}}$.

**Exercise 7.9 (Deducing $\mathbf{x}_{\text{BB}}$).** Let $\mathbf{x}_{\text{PB}}$ be a real integrable passband signal that is bandlimited to $W$ Hz around the carrier frequency $f_c$. Show that it is possible that $x_{\text{PB}}(t)$ be given at every epoch $t \in \mathbb{R}$ by $2\operatorname{Re}\big(z(t)e^{i2\pi f_c t}\big)$ for some complex signal $z(t)$ and that $\mathbf{z}$ not be the baseband representation of $\mathbf{x}_{\text{PB}}$. Does this contradict Proposition 7.6.9?

| In terms of $\mathbf{x}_{\mathrm{PB}}$ | In terms of $\mathbf{x}_{\mathrm{A}}$ | In terms of $\mathbf{x}_{\mathrm{BB}}$ |
|---|---|---|
| $\mathbf{x}_{\mathrm{PB}}$ | $2\,\mathrm{Re}(\mathbf{x}_{\mathrm{A}})$ | $t \mapsto 2\,\mathrm{Re}\big(x_{\mathrm{BB}}(t)\,e^{\mathrm{j}2\pi f_c t}\big)$ |
| $\mathbf{x}_{\mathrm{PB}} \star \big(t \mapsto e^{\mathrm{j}2\pi f_c t}\,\mathrm{LPF}_{\mathsf{W}_c}(t)\big)$ | $\mathbf{x}_{\mathrm{A}}$ | $t \mapsto e^{\mathrm{j}2\pi f_c t}\,x_{\mathrm{BB}}(t)$ |
| $\big(t \mapsto e^{-\mathrm{j}2\pi f_c t}\,x_{\mathrm{PB}}(t)\big) \star \mathrm{LPF}_{\mathsf{W}_c}$ | $t \mapsto e^{-\mathrm{j}2\pi f_c t}\,x_{\mathrm{A}}(t)$ | $\mathbf{x}_{\mathrm{BB}}$ |
| $\hat{\mathbf{x}}_{\mathrm{PB}}$ | $f \mapsto \hat{x}_{\mathrm{A}}(f) + \hat{x}^*_{\mathrm{A}}(-f)$ | $f \mapsto \hat{x}_{\mathrm{BB}}(f - f_c) + \hat{x}^*_{\mathrm{BB}}(-f - f_c)$ |
| $f \mapsto \hat{x}_{\mathrm{PB}}(f)\,\mathrm{I}\{|f - f_c| \le \mathsf{W}_c\}$ | $f \mapsto \hat{x}_{\mathrm{A}}(f)$ | $f \mapsto \hat{x}_{\mathrm{BB}}(f - f_c)$ |
| $f \mapsto \hat{x}_{\mathrm{PB}}(f + f_c)\,\mathrm{I}\{|f| \le \mathsf{W}_c\}$ | $\hat{\mathbf{x}}_{\mathrm{A}}$ | $\hat{\mathbf{x}}_{\mathrm{BB}}$ |
| BW of $\mathbf{x}_{\mathrm{PB}}$ around $f_c$ | BW of $\mathbf{x}_{\mathrm{A}}$ around $f_c$ | $2 \times$ BW of $\mathbf{x}_{\mathrm{BB}}$ |
| $\frac{1}{2} \times$ BW of $\mathbf{x}_{\mathrm{PB}}$ around $f_c$ | $\frac{1}{2} \times$ BW of $\mathbf{x}_{\mathrm{A}}$ around $f_c$ | BW of $\mathbf{x}_{\mathrm{BB}}$ |
| $\|\mathbf{x}_{\mathrm{PB}}\|_2^2$ | $2\,\|\mathbf{x}_{\mathrm{A}}\|_2^2$ | $2\,\|\mathbf{x}_{\mathrm{BB}}\|_2^2$ |
| $\frac{1}{2}\,\|\mathbf{x}_{\mathrm{PB}}\|_2^2$ | $\|\mathbf{x}_{\mathrm{A}}\|_2^2$ | $\|\mathbf{x}_{\mathrm{BB}}\|_2^2$ |

**Table 7.1:** Table relating properties of a real integrable passband signal $\mathbf{x}_{\mathrm{PB}}$ that is bandlimited to $\mathsf{W}$ Hz around the carrier frequency $f_c$ to those of its analytic representation $\mathbf{x}_{\mathrm{A}}$ and its baseband representation $\mathbf{x}_{\mathrm{BB}}$. Same-row entries are equal. The cutoff frequency $\mathsf{W}_c$ is assumed to be in the range $\mathsf{W}/2 \le \mathsf{W}_c \le 2f_c - \mathsf{W}/2$, and BW stands for bandwidth. The transformation from $\mathbf{x}_{\mathrm{PB}}$ to $\mathbf{x}_{\mathrm{A}}$ is based on Proposition 7.5.2 with the function $\mathbf{g}$ in (d) being chosen as the mapping $f \mapsto \mathrm{I}\{|f - f_c| \le \mathsf{W}_c\}$.

# Chapter 8

# Complete Orthonormal Systems and the Sampling Theorem

## 8.1 Introduction

Like Chapter 4, this chapter deals with the geometry of the space $\mathcal{L}_2$ of energy-limited signals. Here, however, our focus is on infinite-dimensional linear subspaces of $\mathcal{L}_2$ and on the notion of a **complete orthonormal system** (CONS). As an application of this geometric picture, we shall present the Sampling Theorem as an orthonormal expansion with respect to a CONS for the space of energy-limited signals that are bandlimited to $W$ Hz.

## 8.2 Complete Orthonormal System

Recall that we denote by $\mathcal{L}_2$ the space of all measurable signals $\mathbf{u} \colon \mathbb{R} \to \mathbb{C}$ satisfying

$$\int_{-\infty}^{\infty} |u(t)|^2 \, \mathrm{d}t < \infty.$$

Also recall from Section 4.3 that a subset $\mathcal{U}$ of $\mathcal{L}_2$ is said to be a linear subspace of $\mathcal{L}_2$ if $\mathcal{U}$ is nonempty and if the signal $\alpha \mathbf{u}_1 + \beta \mathbf{u}_2$ is in $\mathcal{U}$ whenever $\mathbf{u}_1, \mathbf{u}_2 \in \mathcal{U}$ and $\alpha, \beta \in \mathbb{C}$. A linear subspace is said to be finite-dimensional if there exists a finite number of signals that span it; otherwise, it is said to be infinite-dimensional. The following are some examples of infinite-dimensional linear subspaces of $\mathcal{L}_2$.

(i) The set of all functions of the form $t \mapsto p(t) \, e^{-|t|}$, where $p(t)$ is any polynomial (of arbitrary degree).

(ii) The set of all energy-limited signals that vanish outside the interval $[-1, 1]$ (i.e., that map every $t$ outside this interval to zero).

(iii) The set of all energy-limited signals that vanish outside some *unspecified* finite interval (i.e., the set containing all signals $\mathbf{u}$ for which there exists some $a, b \in \mathbb{R}$ (depending on $\mathbf{u}$) such that $u(t) = 0$ whenever $t \notin [a, b]$).

143

(iv) The set of all energy-limited signals that are bandlimited to $W$ Hz.

While a basis for an infinite-dimensional subspace can be defined,[1] this notion does not turn out to be very useful for our purposes. Much more useful to us is the notion of a **complete orthonormal system**, which we shall define shortly.[2]

To motivate the definition, consider a bi-infinite sequence $\ldots, \phi_{-1}, \phi_0, \phi_1, \phi_2, \ldots$ in $\mathcal{L}_2$ satisfying the orthonormality condition

$$\langle \phi_\ell, \phi_{\ell'} \rangle = \mathrm{I}\{\ell = \ell'\}, \quad \ell, \ell' \in \mathbb{Z}, \tag{8.1}$$

and let $\mathbf{u}$ be an arbitrary element of $\mathcal{L}_2$. Define the signals

$$\mathbf{u}_\mathrm{L} \triangleq \sum_{\ell=-\mathrm{L}}^{\mathrm{L}} \langle \mathbf{u}, \phi_\ell \rangle \, \phi_\ell \quad \mathrm{L} = 1, 2, \ldots \tag{8.2}$$

By Note 4.6.7, $\mathbf{u}_\mathrm{L}$ is the projection of the vector $\mathbf{u}$ onto the subspace spanned by $(\phi_{-\mathrm{L}}, \ldots, \phi_\mathrm{L})$. By the orthonormality (8.1), the tuple $(\phi_{-\mathrm{L}}, \ldots, \phi_\mathrm{L})$ is an orthonormal basis for this subspace. Consequently, by Proposition 4.6.9,

$$\|\mathbf{u}\|_2^2 \geq \sum_{\ell=-\mathrm{L}}^{\mathrm{L}} |\langle \mathbf{u}, \phi_\ell \rangle|^2, \quad \mathrm{L} = 1, 2, \ldots, \tag{8.3}$$

with equality if, and only if, $\mathbf{u}$ is indistinguishable from some linear combination of $(\phi_{-\mathrm{L}}, \ldots, \phi_\mathrm{L})$. This motivates us to explore the situation where (8.3) holds with equality when $\mathrm{L} \to \infty$ and to hope that it corresponds to $\mathbf{u}$ being—in some sense that needs to be made precise—indistinguishable from a limit of finite linear combinations of $\ldots, \phi_{-1}, \phi_0, \phi_1, \ldots$

**Definition 8.2.1 (Complete Orthonormal System).** *A bi-infinite sequence of signals* $\ldots, \phi_{-1}, \phi_0, \phi_1, \ldots$ *is said to form a **complete orthonormal system** or a **CONS** for the linear subspace* $\mathcal{U}$ *of* $\mathcal{L}_2$ *if all three of the following conditions hold:*

*1) Each element of the sequence is in* $\mathcal{U}$

$$\phi_\ell \in \mathcal{U}, \quad \ell \in \mathbb{Z}. \tag{8.4}$$

*2) The sequence satisfies the orthonormality condition*

$$\langle \phi_\ell, \phi_{\ell'} \rangle = \mathrm{I}\{\ell = \ell'\}, \quad \ell, \ell' \in \mathbb{Z}. \tag{8.5}$$

*3) For every* $\mathbf{u} \in \mathcal{U}$ *we have*

$$\|\mathbf{u}\|_2^2 = \sum_{\ell=-\infty}^{\infty} |\langle \mathbf{u}, \phi_\ell \rangle|^2, \quad \mathbf{u} \in \mathcal{U}. \tag{8.6}$$

---

[1] A basis for a subspace is defined as a collection of functions such that any function in the subspace can be represented as a linear combination of a *finite* number of elements in the collection. More useful to us will be the notion of a complete orthonormal system. From a complete orthonormal system we only require that each function can be *approximated* by a linear combination of a finite number of functions in the system.

[2] Mathematicians usually define a CONS only for **closed** subspaces. Such subspaces are discussed in Section 8.5.

The following proposition considers equivalent definitions of a CONS and demonstrates that if $\{\phi_\ell\}$ is a CONS for $\mathcal{U}$, then, indeed, every element of $\mathcal{U}$ can be approximated by a finite linear combination of the functions $\{\phi_\ell\}$.

**Proposition 8.2.2.** *Let $\mathcal{U}$ be a subspace of $\mathcal{L}_2$ and let the bi-infinite sequence $\ldots, \phi_{-2}, \phi_{-1}, \phi_0, \phi_1, \ldots$ satisfy (8.4) & (8.5). Then each of the following conditions on $\{\phi_\ell\}$ is equivalent to the condition that $\{\phi_\ell\}$ forms a CONS for $\mathcal{U}$:*

(a) *For every $\mathbf{u} \in \mathcal{U}$ and every $\epsilon > 0$ there exists some positive integer $L(\epsilon)$ and coefficients $\alpha_{-L(\epsilon)}, \ldots, \alpha_{L(\epsilon)} \in \mathbb{C}$ such that*

$$\left\| \mathbf{u} - \sum_{\ell=-L(\epsilon)}^{L(\epsilon)} \alpha_\ell \phi_\ell \right\|_2 < \epsilon. \tag{8.7}$$

(b) *For every $\mathbf{u} \in \mathcal{U}$*

$$\lim_{L \to \infty} \left\| \mathbf{u} - \sum_{\ell=-L}^{L} \langle \mathbf{u}, \phi_\ell \rangle \phi_\ell \right\|_2 = 0. \tag{8.8}$$

(c) *For every $\mathbf{u} \in \mathcal{U}$*

$$\|\mathbf{u}\|_2^2 = \sum_{\ell=-\infty}^{\infty} |\langle \mathbf{u}, \phi_\ell \rangle|^2. \tag{8.9}$$

(d) *For every $\mathbf{u}, \mathbf{v} \in \mathcal{U}$*

$$\langle \mathbf{u}, \mathbf{v} \rangle = \sum_{\ell=-\infty}^{\infty} \langle \mathbf{u}, \phi_\ell \rangle \langle \mathbf{v}, \phi_\ell \rangle^*. \tag{8.10}$$

**Proof.** Since (8.4) & (8.5) hold (by hypothesis), it follows that the additional condition (c) is, by Definition 8.2.1, equivalent to $\{\phi_\ell\}$ being a CONS. It thus only remains to show that the four conditions are equivalent. We shall prove this by showing that (a) $\Leftrightarrow$ (b); that (b) $\Leftrightarrow$ (c); and that (c) $\Leftrightarrow$ (d).

That (b) implies (a) is obvious because nothing precludes us from choosing $\alpha_\ell$ in (8.7) to be $\langle \mathbf{u}, \phi_\ell \rangle$. That (a) implies (b) follows because, by Note 4.6.7, the signal

$$\sum_{\ell=-L}^{L} \langle \mathbf{u}, \phi_\ell \rangle \phi_\ell,$$

which we denoted in (8.2) by $\mathbf{u}_L$, is the projection of $\mathbf{u}$ onto the linear subspace spanned by $(\phi_{-L}, \ldots, \phi_L)$ and as such, by Proposition 4.6.8, best approximates $\mathbf{u}$ among all the signals in that subspace. Consequently, replacing $\alpha_\ell$ by $\langle \mathbf{u}, \phi_\ell \rangle$ can only reduce the LHS of (8.7).

To prove (b) $\Rightarrow$ (c) we first note that by letting $L$ tend to infinity in (8.3) it follows that

$$\|\mathbf{u}\|_2^2 \geq \sum_{\ell=-\infty}^{\infty} |\langle \mathbf{u}, \phi_\ell \rangle|^2, \quad \mathbf{u} \in \mathcal{L}_2, \tag{8.11}$$

so to establish (c) we only need to show that if $\mathbf{u}$ is in $\mathcal{U}$ then $\|\mathbf{u}\|_2^2$ is also upper-bounded by the RHS of (8.11). To that end we first upper-bound $\|\mathbf{u}\|_2$ as

$$
\begin{aligned}
\|\mathbf{u}\|_2 &= \left\| \left( \mathbf{u} - \sum_{\ell=-L}^{L} \langle \mathbf{u}, \boldsymbol{\phi}_\ell \rangle \, \boldsymbol{\phi}_\ell \right) + \sum_{\ell=-L}^{L} \langle \mathbf{u}, \boldsymbol{\phi}_\ell \rangle \, \boldsymbol{\phi}_\ell \right\|_2 \\
&\leq \left\| \mathbf{u} - \sum_{\ell=-L}^{L} \langle \mathbf{u}, \boldsymbol{\phi}_\ell \rangle \, \boldsymbol{\phi}_\ell \right\|_2 + \left\| \sum_{\ell=-L}^{L} \langle \mathbf{u}, \boldsymbol{\phi}_\ell \rangle \, \boldsymbol{\phi}_\ell \right\|_2 \\
&= \left\| \mathbf{u} - \sum_{\ell=-L}^{L} \langle \mathbf{u}, \boldsymbol{\phi}_\ell \rangle \, \boldsymbol{\phi}_\ell \right\|_2 + \left( \sum_{\ell=-L}^{L} |\langle \mathbf{u}, \boldsymbol{\phi}_\ell \rangle|^2 \right)^{1/2}, \quad \mathbf{u} \in \mathcal{L}_2, \qquad (8.12)
\end{aligned}
$$

where the first equality follows by adding and subtracting a term; the subsequent inequality by the Triangle Inequality (Proposition 3.4.1); and the final equality by the orthonormality assumption (8.5) and the Pythagorean Theorem (Theorem 4.5.2). If Condition (b) holds and if $\mathbf{u}$ is in $\mathcal{U}$, then the RHS of (8.12) converges to the square root of the infinite sum $\sum_{\ell \in \mathbb{Z}} |\langle \mathbf{u}, \boldsymbol{\phi}_\ell \rangle|^2$ and thus gives us the desired upper bound on $\|\mathbf{u}\|_2$.

We next prove (c) $\Rightarrow$ (b). We assume that (c) holds and that $\mathbf{u}$ is in $\mathcal{U}$ and set out to prove (8.8). To that end we first note that by the basic properties of the inner product (3.6)–(3.10) and by the orthonormality (8.1) it follows that

$$
\underbrace{\left\langle \mathbf{u} - \sum_{\ell=-L}^{L} \langle \mathbf{u}, \boldsymbol{\phi}_\ell \rangle \, \boldsymbol{\phi}_\ell, \boldsymbol{\phi}_{\ell'} \right\rangle}_{\mathbf{u}'} = \langle \mathbf{u}, \boldsymbol{\phi}_{\ell'} \rangle \, \mathrm{I}\{|\ell'| > L\}, \quad \left( \ell' \in \mathbb{Z}, \, \mathbf{u} \in \mathcal{L}_2 \right).
$$

Consequently, if we apply (c) to the under-braced signal $\mathbf{u}'$ (which for $\mathbf{u} \in \mathcal{U}$ is also in $\mathcal{U}$) we obtain that (c) implies

$$
\left\| \mathbf{u} - \sum_{\ell=-L}^{L} \langle \mathbf{u}, \boldsymbol{\phi}_\ell \rangle \, \boldsymbol{\phi}_\ell \right\|_2^2 = \sum_{|\ell| > L} |\langle \mathbf{u}, \boldsymbol{\phi}_\ell \rangle|^2, \quad \mathbf{u} \in \mathcal{U}.
$$

But by applying (c) to $\mathbf{u}$ we infer that the RHS of the above tends to zero as L tends to infinity, thus establishing (8.8) and hence (b).

We next prove (c) $\Leftrightarrow$ (d). The implication (d) $\Rightarrow$ (c) is obvious because we can always choose $\mathbf{v}$ to be equal to $\mathbf{u}$. We consequently focus on proving (c) $\Rightarrow$ (d). We do so by assuming that $\mathbf{u}, \mathbf{v} \in \mathcal{U}$ and calculating for every $\beta \in \mathbb{C}$

$$
\begin{aligned}
|\beta|^2 \, \|\mathbf{u}\|_2^2 &+ 2 \, \mathrm{Re}\big( \beta \langle \mathbf{u}, \mathbf{v} \rangle \big) + \|\mathbf{v}\|_2^2 \\
&= \|\beta \mathbf{u} + \mathbf{v}\|_2^2 \\
&= \sum_{\ell=-\infty}^{\infty} \big| \langle \beta \mathbf{u} + \mathbf{v}, \boldsymbol{\phi}_\ell \rangle \big|^2 \\
&= \sum_{\ell=-\infty}^{\infty} \big| \beta \langle \mathbf{u}, \boldsymbol{\phi}_\ell \rangle + \langle \mathbf{v}, \boldsymbol{\phi}_\ell \rangle \big|^2
\end{aligned}
$$

$$= |\beta|^2 \sum_{\ell=-\infty}^{\infty} |\langle \mathbf{u}, \boldsymbol{\phi}_\ell \rangle|^2 + 2 \operatorname{Re} \left( \beta \sum_{\ell=-\infty}^{\infty} \langle \mathbf{u}, \boldsymbol{\phi}_\ell \rangle \langle \mathbf{v}, \boldsymbol{\phi}_\ell \rangle^* \right)$$

$$+ \sum_{\ell=-\infty}^{\infty} |\langle \mathbf{v}, \boldsymbol{\phi}_\ell \rangle|^2, \quad \left( \mathbf{u}, \mathbf{v} \in \mathcal{U}, \ \beta \in \mathbb{C} \right), \tag{8.13}$$

where the first equality follows by writing $\|\beta \mathbf{u} + \mathbf{v}\|_2^2$ as $\langle \beta \mathbf{u} + \mathbf{v}, \beta \mathbf{u} + \mathbf{v} \rangle$ and using the basic properties of the inner product (3.6)–(3.10); the second by applying (c) to $\beta \mathbf{u} + \mathbf{v}$ (which for $\mathbf{u}, \mathbf{v} \in \mathcal{U}$ is also in $\mathcal{U}$); the third by the basic properties of the inner product; and the final equality by writing the squared magnitude of a complex number as its product by its conjugate. By applying (c) to $\mathbf{u}$ and by applying (c) to $\mathbf{v}$ we now obtain from (8.13) that

$$2 \operatorname{Re} \left( \beta \langle \mathbf{u}, \mathbf{v} \rangle \right) = 2 \operatorname{Re} \left( \beta \sum_{\ell=-\infty}^{\infty} \langle \mathbf{u}, \boldsymbol{\phi}_\ell \rangle \langle \mathbf{v}, \boldsymbol{\phi}_\ell \rangle^* \right), \quad \left( \mathbf{u}, \mathbf{v} \in \mathcal{U}, \ \beta \in \mathbb{C} \right),$$

which can only hold for all $\beta \in \mathbb{C}$ (and in particular for both $\beta = 1$ and $\beta = \mathrm{i}$) if

$$\langle \mathbf{u}, \mathbf{v} \rangle = \sum_{\ell=-\infty}^{\infty} \langle \mathbf{u}, \boldsymbol{\phi}_\ell \rangle \langle \mathbf{v}, \boldsymbol{\phi}_\ell \rangle^*, \quad \mathbf{u}, \mathbf{v} \in \mathcal{U},$$

thus establishing (d). $\qquad \square$

We next describe the two complete orthonormal systems that will be of most interest to us.

## 8.3 The Fourier Series

A CONS that you have probably already encountered is the one underlying the Fourier Series representation. You may have encountered the Fourier Series in the context of periodic functions, but we shall focus on a slightly different view.

**Proposition 8.3.1.** *For every* $\mathsf{T} > 0$, *the functions* $\{\boldsymbol{\phi}_\ell\}$ *defined for every integer* $\ell$ *by*

$$\boldsymbol{\phi}_\ell : t \mapsto \frac{1}{\sqrt{2\mathsf{T}}} e^{\mathrm{i}\pi \ell t / \mathsf{T}} \, \mathrm{I}\{|t| \leq \mathsf{T}\} \tag{8.14}$$

*form a CONS for the subspace*

$$\left\{ \mathbf{u} \in \mathcal{L}_2 : u(t) = 0 \ \text{whenever} \ |t| > \mathsf{T} \right\}$$

*of energy-limited signals that vanish outside the interval* $[-\mathsf{T}, \mathsf{T}]$.

**Proof.** Follows from Theorem A.3.3 in the appendix by substituting $2\mathsf{T}$ for $\mathsf{S}$. $\qquad \square$

Notice that in this case

$$\langle \mathbf{u}, \boldsymbol{\phi}_\ell \rangle = \frac{1}{\sqrt{2\mathsf{T}}} \int_{-\mathsf{T}}^{\mathsf{T}} u(t) \, e^{-\mathrm{i}\pi \ell t / \mathsf{T}} \, \mathrm{d}t \tag{8.15}$$

is the $\ell$-th Fourier Series Coefficient of $\mathbf{u}$; see Note A.3.5 in the appendix with 2T substituted for S.

**Note 8.3.2.** The dummy argument $t$ is immaterial in Proposition 8.3.1. Indeed, if we define for $W > 0$ the linear subspace

$$\mathcal{V} = \left\{ \mathbf{g} \in \mathcal{L}_2 : g(f) = 0 \text{ whenever } |f| > W \right\}, \tag{8.16}$$

then the functions defined for every integer $\ell$ by

$$f \mapsto \frac{1}{\sqrt{2W}} \, e^{i\pi \ell f / W} \, \mathrm{I}\{|f| \leq W\} \tag{8.17}$$

form a CONS for this subspace.

This note will be crucial when we next discuss a CONS for the space of energy-limited signals that are bandlimited to $W$ Hz.

## 8.4   The Sampling Theorem

We next provide a CONS for the space of energy-limited signals that are band-limited to $W$ Hz. Recall that if $\mathbf{x}$ is an energy-limited signal that is bandlimited to $W$ Hz, then there exists a measurable function[3] $\mathbf{g} \colon f \mapsto g(f)$ satisfying

$$g(f) = 0, \quad |f| > W \tag{8.18}$$

and

$$\int_{-W}^{W} |g(f)|^2 \, \mathrm{d}f < \infty, \tag{8.19}$$

such that

$$x(t) = \int_{-W}^{W} g(f) \, e^{i2\pi f t} \, \mathrm{d}f, \quad t \in \mathbb{R}. \tag{8.20}$$

Conversely, if $\mathbf{g}$ is any function satisfying (8.18) & (8.19), and if we define $\mathbf{x}$ via (8.20) as the Inverse Fourier Transform of $\mathbf{g}$, then $\mathbf{x}$ is an energy-limited signal that is bandlimited to $W$ Hz and its $L_2$-Fourier Transform $\hat{\mathbf{x}}$ is equal to (the equivalence class of) $\mathbf{g}$.

Thus, if, as in (8.16), we denote by $\mathcal{V}$ the set of all functions (of frequency) satisfying (8.18) & (8.19), then the set of all energy-limited signals that are bandlimited to $W$ Hz is just the image of $\mathcal{V}$ under the IFT, i.e., it is the set $\check{\mathcal{V}}$, where

$$\check{\mathcal{V}} \triangleq \{ \check{\mathbf{g}} : \mathbf{g} \in \mathcal{V} \}. \tag{8.21}$$

By the Mini Parseval Theorem (Proposition 6.2.6 (i)), if $\mathbf{x}_1$ and $\mathbf{x}_2$ are given by $\check{\mathbf{g}}_1$ and $\check{\mathbf{g}}_2$, where $\mathbf{g}_1, \mathbf{g}_2$ are in $\mathcal{V}$, then

$$\langle \mathbf{x}_1, \mathbf{x}_2 \rangle = \langle \mathbf{g}_1, \mathbf{g}_2 \rangle, \tag{8.22}$$

---

[3]Loosely speaking, this function is the Fourier Transform of $\mathbf{x}$. But since $\mathbf{x}$ is not necessarily integrable, its FT $\hat{\mathbf{x}}$ is an equivalence class of signals. Thus, more precisely, the equivalence class of $\mathbf{g}$ is the $L_2$-Fourier Transform of $\mathbf{x}$. Or, stated differently, $\mathbf{g}$ can be any one of the signals in the equivalence class of $\hat{\mathbf{x}}$ that is zero outside the interval $[-W, W]$.

i.e.,

$$\langle \check{\mathbf{g}}_1, \check{\mathbf{g}}_2 \rangle = \langle \mathbf{g}_1, \mathbf{g}_2 \rangle, \quad \mathbf{g}_1, \mathbf{g}_2 \in \mathcal{V}. \tag{8.23}$$

The following lemma is a simple but very useful consequence of (8.23).

**Lemma 8.4.1.** *If $\{\boldsymbol{\psi}_\ell\}$ is a CONS for the subspace $\mathcal{V}$, which is defined in (8.16), then $\{\check{\boldsymbol{\psi}}_\ell\}$ is a CONS for the subspace $\check{\mathcal{V}}$, which is defined in (8.21).*

**Proof.** Let $\{\boldsymbol{\psi}_\ell\}$ be a CONS for the subspace $\mathcal{V}$. By (8.23),

$$\langle \check{\boldsymbol{\psi}}_\ell, \check{\boldsymbol{\psi}}_{\ell'} \rangle = \langle \boldsymbol{\psi}_\ell, \boldsymbol{\psi}_{\ell'} \rangle, \quad \ell, \ell' \in \mathbb{Z},$$

so our assumption that $\{\boldsymbol{\psi}_\ell\}$ is a CONS for $\mathcal{V}$ (and hence that, *a fortiori*, it satisfies $\langle \boldsymbol{\psi}_\ell, \boldsymbol{\psi}_{\ell'} \rangle = \mathrm{I}\{\ell = \ell'\}$ for all $\ell, \ell' \in \mathbb{Z}$) implies that

$$\langle \check{\boldsymbol{\psi}}_\ell, \check{\boldsymbol{\psi}}_{\ell'} \rangle = \mathrm{I}\{\ell = \ell'\}, \quad \ell, \ell' \in \mathbb{Z}.$$

It remains to verify that for every $\mathbf{x} \in \check{\mathcal{V}}$

$$\sum_{\ell=-\infty}^{\infty} \left| \langle \mathbf{x}, \check{\boldsymbol{\psi}}_\ell \rangle \right|^2 = \|\mathbf{x}\|_2^2 \,.$$

Equivalently, since every $\mathbf{x} \in \check{\mathcal{V}}$ can be written as $\check{\mathbf{g}}$ for some $\mathbf{g} \in \mathcal{V}$, we need to show that

$$\sum_{\ell=-\infty}^{\infty} \left| \langle \check{\mathbf{g}}, \check{\boldsymbol{\psi}}_\ell \rangle \right|^2 = \|\check{\mathbf{g}}\|_2^2, \quad \mathbf{g} \in \mathcal{V}.$$

This follows from (8.23) and from our assumption that $\{\boldsymbol{\psi}_\ell\}$ is a CONS for $\mathcal{V}$ because

$$\sum_{\ell=-\infty}^{\infty} \left| \langle \check{\mathbf{g}}, \check{\boldsymbol{\psi}}_\ell \rangle \right|^2 = \sum_{\ell=-\infty}^{\infty} \left| \langle \mathbf{g}, \boldsymbol{\psi}_\ell \rangle \right|^2$$
$$= \|\mathbf{g}\|_2^2$$
$$= \|\check{\mathbf{g}}\|_2^2, \quad \mathbf{g} \in \mathcal{V},$$

where the first equality follows from (8.23) (by substituting $\mathbf{g}$ for $\mathbf{g}_1$ and by substituting $\boldsymbol{\psi}_\ell$ for $\mathbf{g}_2$); the second from the assumption that $\{\boldsymbol{\psi}_\ell\}$ is a CONS for $\mathcal{V}$; and the final equality from (8.23) (by substituting $\mathbf{g}$ for $\mathbf{g}_1$ and for $\mathbf{g}_2$). $\qquad \square$

Using this lemma and Note 8.3.2 we now derive a CONS for the subspace $\check{\mathcal{V}}$ of energy-limited signals that are bandlimited to $W$ Hz.

**Proposition 8.4.2 (A CONS for the Subspace of Energy-Limited Signals that Are Bandlimited to $W$ Hz).**

(i) *The sequence of signals that are defined for every integer $\ell$ by*

$$t \mapsto \sqrt{2W}\,\mathrm{sinc}(2Wt + \ell) \tag{8.24}$$

*forms a CONS for the space of energy-limited signals that are bandlimited to $W$ Hz.*

*(ii) If* **x** *is an energy-limited signal that is bandlimited to* W *Hz, then its inner product with the $\ell$-th signal is given by its scaled sample at time* $-\ell/(2W)$:

$$\left\langle \mathbf{x}, t \mapsto \sqrt{2W}\,\mathrm{sinc}(2Wt + \ell) \right\rangle = \frac{1}{\sqrt{2W}}\, x\!\left(-\frac{\ell}{2W}\right), \quad \ell \in \mathbb{Z}. \tag{8.25}$$

**Proof.** To prove Part (i) we recall that, by Note 8.3.2, the functions defined for every $\ell \in \mathbb{Z}$ by

$$\psi_\ell \colon f \mapsto \frac{1}{\sqrt{2W}}\, e^{i\pi\ell f/W}\, \mathrm{I}\{|f| \le W\} \tag{8.26}$$

form a CONS for the subspace $\mathcal{V}$. Consequently, by Lemma 8.4.1, their Inverse Fourier Transforms $\{\check{\psi}_\ell\}$ form a CONS for $\check{\mathcal{V}}$. It just remains to evaluate $\check{\psi}_\ell$ explicitly in order to verify that it is a scaled shifted $\mathrm{sinc}(\cdot)$:

$$\check{\psi}_\ell(t) = \int_{-\infty}^{\infty} \psi_\ell(f)\, e^{i2\pi ft}\, \mathrm{d}f$$

$$= \int_{-W}^{W} \frac{1}{\sqrt{2W}}\, e^{i\pi\ell f/W}\, e^{i2\pi ft}\, \mathrm{d}f \tag{8.27}$$

$$= \sqrt{2W}\,\mathrm{sinc}(2Wt + \ell), \tag{8.28}$$

where the last calculation can be verified by direct computation as in (6.35).

We next prove Part (ii). Since **x** is an energy-limited signal that is bandlimited to W Hz, it follows that there exists some $\mathbf{g} \in \mathcal{V}$ such that

$$\mathbf{x} = \check{\mathbf{g}}, \tag{8.29}$$

i.e.,

$$x(t) = \int_{-W}^{W} g(f)\, e^{i2\pi ft}\, \mathrm{d}f, \quad t \in \mathbb{R}. \tag{8.30}$$

Consequently,

$$\left\langle \mathbf{x}, t \mapsto \sqrt{2W}\,\mathrm{sinc}(2Wt + \ell) \right\rangle = \langle \mathbf{x}, \check{\psi}_\ell \rangle$$

$$= \langle \check{\mathbf{g}}, \check{\psi}_\ell \rangle$$

$$= \langle \mathbf{g}, \psi_\ell \rangle$$

$$= \int_{-W}^{W} g(f) \left(\frac{1}{\sqrt{2W}}\, e^{i\pi\ell f/W}\right)^{*} \mathrm{d}f$$

$$= \frac{1}{\sqrt{2W}} \int_{-W}^{W} g(f)\, e^{-i\pi\ell f/W}\, \mathrm{d}f$$

$$= \frac{1}{\sqrt{2W}}\, x\!\left(-\frac{\ell}{2W}\right), \quad \ell \in \mathbb{Z},$$

where the first equality follows from (8.28); the second by (8.29); the third by (8.23) (with the substitution of **g** for $\mathbf{g}_1$ and $\psi_\ell$ for $\mathbf{g}_2$); the fourth by the definition of the inner product and by (8.26); the fifth by conjugating the complex exponential; and the final equality by substituting $-\ell/(2W)$ for $t$ in (8.30).  □

Using Proposition 8.4.2 and Proposition 8.2.2 we obtain the following $\mathcal{L}_2$ version of the Sampling Theorem.

**Theorem 8.4.3 ($\mathcal{L}_2$-Sampling Theorem).** *Let* $\mathbf{x}$ *be an energy-limited signal that is bandlimited to* W *Hz, where* $W > 0$, *and let*

$$\mathsf{T} = \frac{1}{2W}. \tag{8.31}$$

(i) *The signal* $\mathbf{x}$ *can be reconstructed from the sequence* $\dots, x(-\mathsf{T}), x(0), x(\mathsf{T}), \dots$ *of its values at integer multiples of* $\mathsf{T}$ *in the sense that*

$$\lim_{L\to\infty} \int_{-\infty}^{\infty} \left| x(t) - \sum_{\ell=-L}^{L} x(-\ell\mathsf{T}) \operatorname{sinc}\left(\frac{t}{\mathsf{T}} + \ell\right) \right|^2 \mathrm{d}t = 0.$$

(ii) *The signal's energy can be reconstructed from its samples via the relation*

$$\int_{-\infty}^{\infty} |x(t)|^2 \,\mathrm{d}t = \mathsf{T} \sum_{\ell=-\infty}^{\infty} |x(\ell\mathsf{T})|^2.$$

(iii) *If* $\mathbf{y}$ *is another energy-limited signal that is bandlimited to* W *Hz, then*

$$\langle \mathbf{x}, \mathbf{y} \rangle = \mathsf{T} \sum_{\ell=-\infty}^{\infty} x(\ell\mathsf{T}) \, y^*(\ell\mathsf{T}).$$

**Note 8.4.4.** If $\mathsf{T} \leq 1/(2W)$, then any energy-limited signal $\mathbf{x}$ that is bandlimited to W Hz is also bandlimited to $1/(2\mathsf{T})$ Hz. Consequently, Theorem 8.4.3 continues to hold if we replace (8.31) with the condition

$$0 < \mathsf{T} \leq \frac{1}{2W}. \tag{8.32}$$

Table 8.1 highlights the duality between the Sampling Theorem and the Fourier Series.

We also mention here without proof a version of the Sampling Theorem that allows one to reconstruct the signal *pointwise*, i.e., at every epoch $t$. Thus, while Theorem 8.4.3 guarantees that, as more and more terms in the sum of the shifted sinc functions are added, the energy in the error function tends to zero, the following theorem demonstrates that at every fixed time $t$ the error tends to zero.

**Theorem 8.4.5 (Pointwise Sampling Theorem).** *If the signal* $\mathbf{x}$ *can be represented as*

$$x(t) = \int_{-W}^{W} g(f) \, e^{\mathrm{i}2\pi ft} \,\mathrm{d}f, \quad t \in \mathbb{R} \tag{8.33}$$

*for some function* $\mathbf{g}$ *satisfying*

$$\int_{-W}^{W} |g(f)| \,\mathrm{d}f < \infty, \tag{8.34}$$

*and if* $0 < \mathsf{T} \leq 1/(2W)$, *then for every* $t \in \mathbb{R}$

$$x(t) = \lim_{\mathsf{L} \to \infty} \sum_{\ell=-\mathsf{L}}^{\mathsf{L}} x(-\ell\mathsf{T}) \operatorname{sinc}\left(\frac{t}{\mathsf{T}} + \ell\right). \tag{8.35}$$

**Proof.** See (Pinsky, 2002, Chapter 4, Section 4.2.3, Theorem 4.2.13). □

The Sampling Theorem goes by various names. It is sometimes attributed to Claude Elwood Shannon (1916–2001), the founder of Information Theory. But it also appears in the works of Vladimir Aleksandrovich Kotelnikov (1908–2005), Harry Nyquist (1889–1976), and Edmund Taylor Whittaker (1873–1956). For further references regarding the history of this result and for a survey of many related results, see (Unser, 2000).

## 8.5   Closed Subspaces of $\mathcal{L}_2$

Our definition of a CONS for a subspace $\mathcal{U}$ is not quite standard, because we only assumed that $\mathcal{U}$ is a linear subspace; we did not assume that $\mathcal{U}$ is **closed**. In this section we shall define closed linear subspaces and derive a condition for a sequence $\{\phi_\ell\}$ to form a CONS for a closed subspace $\mathcal{U}$. (The set of energy-limited signals that vanish outside the interval $[-\mathsf{T}, \mathsf{T}]$ is closed, as is the class of energy-limited signals that are bandlimited to $W$ Hz.)

Before proceeding to define closed linear subspaces, we pause here to recall that the space $\mathcal{L}_2$ is **complete**.[4]

**Theorem 8.5.1 ($\mathcal{L}_2$ Is Complete).** *If the sequence* $\mathbf{u}_1, \mathbf{u}_2, \ldots$ *of signals in* $\mathcal{L}_2$ *is such that for any* $\epsilon > 0$ *there exists a positive integer* $\mathsf{L}(\epsilon)$ *such that*

$$\|\mathbf{u}_n - \mathbf{u}_m\|_2 < \epsilon, \quad n, m > \mathsf{L}(\epsilon),$$

*then there exists some function* $\mathbf{u} \in \mathcal{L}_2$ *such that*

$$\lim_{n \to \infty} \|\mathbf{u} - \mathbf{u}_n\|_2 = 0.$$

**Proof.** See, for example, (Rudin, 1974, Chapter 3, Theorem 3.11). □

**Definition 8.5.2 (Closed Subspace).** *A linear subspace* $\mathcal{U}$ *of* $\mathcal{L}_2$ *is said to be* ***closed*** *if for any sequence of signals* $\mathbf{u}_1, \mathbf{u}_2, \ldots$ *in* $\mathcal{U}$ *and any* $\mathbf{u} \in \mathcal{L}_2$, *the condition* $\|\mathbf{u} - \mathbf{u}_n\|_2 \to 0$ *implies that* $\mathbf{u}$ *is indistinguishable from some element of* $\mathcal{U}$.

Before stating the next theorem we remind the reader that a bi-infinite sequence of complex numbers $\ldots, \alpha_{-1}, \alpha_0, \alpha_1, \ldots$ is said to be **square summable** if

$$\sum_{\ell=-\infty}^{\infty} |\alpha_\ell|^2 < \infty.$$

---

[4]This property is usually stated about $L_2$ but we prefer to work with $\mathcal{L}_2$.

**Theorem 8.5.3 (Riesz-Fischer).** *Let $\mathcal{U}$ be a closed linear subspace of $\mathcal{L}_2$, and let the bi-infinite sequence $\ldots, \phi_{-1}, \phi_0, \phi_1, \ldots$ satisfy (8.4) & (8.5). Let the bi-infinite sequence of complex numbers $\ldots, \alpha_{-1}, \alpha_0, \alpha_1, \ldots$ be square summable. Then there exists an element $\mathbf{u}$ in $\mathcal{U}$ satisfying*

$$\lim_{\mathrm{L}\to\infty} \left\| \mathbf{u} - \sum_{\ell=-\mathrm{L}}^{\mathrm{L}} \alpha_\ell \phi_\ell \right\|_2 = 0; \tag{8.36a}$$

$$\langle \mathbf{u}, \phi_\ell \rangle = \alpha_\ell, \quad \ell \in \mathbb{Z}; \tag{8.36b}$$

*and*

$$\|\mathbf{u}\|_2^2 = \sum_{\ell=-\infty}^{\infty} |\alpha_\ell|^2. \tag{8.36c}$$

**Proof.** Define for every positive integer $\mathrm{L}$

$$\mathbf{u}_\mathrm{L} = \sum_{\ell=-\mathrm{L}}^{\mathrm{L}} \alpha_\ell \phi_\ell, \quad \mathrm{L} \in \mathbb{N}. \tag{8.37}$$

Since, by hypothesis, $\mathcal{U}$ is a linear subspace and the signals $\{\phi_\ell\}$ are all in $\mathcal{U}$, it follows that $\mathbf{u}_\mathrm{L} \in \mathcal{U}$. By the orthonormality assumption (8.5) and by the Pythagorean Theorem (Theorem 4.5.2), it follows that

$$\|\mathbf{u}_n - \mathbf{u}_m\|_2^2 = \sum_{\min\{m,n\}<|\ell|\le\max\{m,n\}} |\alpha_\ell|^2$$

$$\le \sum_{\min\{m,n\}<|\ell|<\infty} |\alpha_\ell|^2, \quad n, m \in \mathbb{N}.$$

From this and from the square summability of $\{\alpha_\ell\}$, it follows that for any $\epsilon > 0$ we have that $\|\mathbf{u}_n - \mathbf{u}_m\|_2$ is smaller than $\epsilon$ whenever both $n$ and $m$ are sufficiently large. By the completeness of $\mathcal{L}_2$ it thus follows that there exists some $\mathbf{u}' \in \mathcal{L}_2$ such that

$$\lim_{\mathrm{L}\to\infty} \|\mathbf{u}' - \mathbf{u}_\mathrm{L}\|_2 = 0. \tag{8.38}$$

Since $\mathcal{U}$ is closed, and since $\mathbf{u}_\mathrm{L}$ is in $\mathcal{U}$ for every $\mathrm{L} \in \mathbb{N}$, it follows from (8.38) that $\mathbf{u}'$ is indistinguishable from some element $\mathbf{u}$ of $\mathcal{U}$:

$$\|\mathbf{u} - \mathbf{u}'\|_2 = 0. \tag{8.39}$$

It now follows from (8.38) and (8.39) that

$$\lim_{\mathrm{L}\to\infty} \|\mathbf{u} - \mathbf{u}_\mathrm{L}\|_2 = 0, \tag{8.40}$$

as can be verified using (4.14) (with the substitution $(\mathbf{u}' - \mathbf{u}_\mathrm{L})$ for $\mathbf{x}$ and $(\mathbf{u} - \mathbf{u}')$ for $\mathbf{y}$). Combining (8.40) with (8.37) establishes (8.36a).

To establish (8.36b) we use (8.40) and the continuity of the inner product (Proposition 3.4.2) to calculate $\langle \mathbf{u}, \boldsymbol{\phi}_\ell \rangle$ for every fixed $\ell \in \mathbb{Z}$ as follows:

$$\langle \mathbf{u}, \boldsymbol{\phi}_\ell \rangle = \lim_{L \to \infty} \langle \mathbf{u}_L, \boldsymbol{\phi}_\ell \rangle$$

$$= \lim_{L \to \infty} \left\langle \sum_{\ell'=-L}^{L} \alpha_{\ell'} \boldsymbol{\phi}_{\ell'}, \boldsymbol{\phi}_\ell \right\rangle$$

$$= \lim_{L \to \infty} \alpha_\ell \, \mathrm{I}\{|\ell| \leq L\}$$

$$= \alpha_\ell, \quad \ell \in \mathbb{Z},$$

where the first equality follows from (8.40) and from the continuity of the inner product (Proposition 3.4.2); the second by (8.37); the third by the orthonormality (8.5); and the final equality because $\alpha_\ell \, \mathrm{I}\{|\ell| \leq L\}$ is equal to $\alpha_\ell$, whenever $L$ is large enough (i.e., exceeds $|\ell|$).

It remains to prove (8.36c). By the orthonormality of $\{\boldsymbol{\phi}_\ell\}$ and the Pythagorean Theorem (Theorem 4.5.2)

$$\|\mathbf{u}_L\|_2^2 = \sum_{\ell=-L}^{L} |\alpha_\ell|^2, \quad L \in \mathbb{N}. \tag{8.41}$$

Also, by (4.14) (with the substitution of $\mathbf{u}$ for $\mathbf{x}$ and of $(\mathbf{u}_L - \mathbf{u})$ for $\mathbf{y}$) we obtain

$$\|\mathbf{u}\|_2 - \|\mathbf{u} - \mathbf{u}_L\|_2 \leq \|\mathbf{u}_L\|_2 \leq \|\mathbf{u}\|_2 + \|\mathbf{u} - \mathbf{u}_L\|_2. \tag{8.42}$$

It now follows from (8.42), (8.40), and the Sandwich Theorem[5] that

$$\lim_{L \to \infty} \|\mathbf{u}_L\|_2 = \|\mathbf{u}\|_2, \tag{8.43}$$

which combines with (8.41) to prove (8.36c). $\qquad\qquad\qquad\qquad\qquad\qquad\Box$

By applying Theorem 8.5.3 to the space of energy-limited signals that are bandlimited to $W$ Hz and to the CONS that we derived for that space in Proposition 8.4.2 we obtain:

**Proposition 8.5.4.** *Any square-summable bi-infinite sequence of complex numbers corresponds to the samples at integer multiples of* $\mathsf{T}$ *of an energy-limited signal that is bandlimited to* $1/(2\mathsf{T})$ *Hz. Here* $\mathsf{T} > 0$ *is arbitrary.*

**Proof.** Let $\ldots, \beta_{-1}, \beta_0, \beta_1, \ldots$ be a square-summable bi-infinite sequence of complex numbers, and let $W = 1/(2\mathsf{T})$. We seek a signal $\mathbf{u}$ that is an energy-limited signal that is bandlimited to $W$ Hz and whose samples are given by $u(\ell\mathsf{T}) = \beta_\ell$, for every integer $\ell$. Since the set of all energy-limited signals that are bandlimited to $W$ Hz is a closed linear subspace of $\mathcal{L}_2$, and since the sequence $\{\boldsymbol{\check{\psi}}_\ell\}$ (given explicitly in (8.28) as $\boldsymbol{\check{\psi}}_\ell \colon t \mapsto \sqrt{2W}\,\mathrm{sinc}(2Wt+\ell)$) is an orthonormal sequence in that

---

[5]The Sandwich Theorem states that if the sequences of real number $\{a_n\}$, $\{b_n\}$ and $\{c_n\}$ are such that $b_n \leq a_n \leq c_n$ for every $n$, and if the sequences $\{b_n\}$ and $\{c_n\}$ converge to the same limit, then $\{a_n\}$ also converges to that limit.

subspace, it follows from Theorem 8.5.3 (with the substitution of $\check{\psi}_\ell$ for $\phi_\ell$ and of $\beta_{-\ell}/\sqrt{2W}$ for $\alpha_\ell$) that there exists an energy-limited signal $\mathbf{u}$ that is bandlimited to $W$ Hz and for which

$$\langle \mathbf{u}, \check{\psi}_\ell \rangle = \frac{1}{\sqrt{2W}} \beta_{-\ell}, \quad \ell \in \mathbb{Z}. \tag{8.44}$$

By Proposition 8.4.2,

$$\langle \mathbf{u}, \check{\psi}_\ell \rangle = \frac{1}{\sqrt{2W}} u(-\ell T), \quad \ell \in \mathbb{Z}, \tag{8.45}$$

so by (8.44) and (8.45)

$$u(-\ell T) = \beta_{-\ell}, \quad \ell \in \mathbb{Z}. \qquad \square$$

We now give an alternative characterization of a CONS for a closed subspace of $\mathcal{L}_2$. This result will not be used later in the book.

**Proposition 8.5.5 (Characterization of a CONS for a Closed Subspace).**

(i) *If the bi-infinite sequence $\{\phi_\ell\}$ is a CONS for the linear subspace $\mathcal{U} \subseteq \mathcal{L}_2$, then an element of $\mathcal{U}$ whose inner product with $\phi_\ell$ is zero for every integer $\ell$ must have zero energy:*

$$\Big( \langle \mathbf{u}, \phi_\ell \rangle = 0, \quad \ell \in \mathbb{Z} \Big) \Rightarrow \Big( \|\mathbf{u}\|_2 = 0 \Big), \quad \mathbf{u} \in \mathcal{U}. \tag{8.46}$$

(ii) *If $\mathcal{U}$ is a closed subspace of $\mathcal{L}_2$ and if the bi-infinite sequence $\{\phi_\ell\}$ satisfies (8.4) & (8.5), then Condition (8.46) is equivalent to the condition that $\{\phi_\ell\}$ forms a CONS for $\mathcal{U}$.*

**Proof.** We begin by proving Part (i). By definition, if $\{\phi_\ell\}$ is a CONS for $\mathcal{U}$, then (8.6) must hold for every every $\mathbf{u} \in \mathcal{U}$. Consequently, if for some $\mathbf{u} \in \mathcal{U}$ we have that $\langle \mathbf{u}, \phi_\ell \rangle$ is zero for all $\ell \in \mathbb{Z}$, then the RHS of (8.6) is zero and hence the LHS must also be zero, thus showing that $\mathbf{u}$ must be of zero energy.

We next turn to Part (ii) and assume that $\mathcal{U}$ is closed and that the bi-infinite sequence $\{\phi_\ell\}$ satisfies (8.4) & (8.5). That the condition that $\{\phi_\ell\}$ is a CONS implies Condition (8.46) follows from Part (i). It thus remains to show that if Condition (8.46) holds, then $\{\phi_\ell\}$ is a CONS. To prove this we now assume that $\mathcal{U}$ is a closed subspace; that $\{\phi_\ell\}$ satisfies (8.4) & (8.5); and that (8.46) holds and set out to prove that

$$\|\mathbf{u}\|_2^2 = \sum_{\ell=-\infty}^{\infty} \left| \langle \mathbf{u}, \phi_\ell \rangle \right|^2, \quad u \in \mathcal{U}. \tag{8.47}$$

To establish (8.47) fix some arbitrary $\mathbf{u} \in \mathcal{U}$. Since $\mathcal{U} \subseteq \mathcal{L}_2$, the fact that $\mathbf{u}$ is in $\mathcal{U}$ implies that it is of finite energy, which combines with (8.3) to imply that the bi-infinite sequence $\ldots, \langle \mathbf{u}, \phi_{-1} \rangle, \langle \mathbf{u}, \phi_0 \rangle, \langle \mathbf{u}, \phi_1 \rangle, \ldots$ is square summable. Since, by hypothesis, $\mathcal{U}$ is closed, this implies, by Theorem 8.5.3 (with the substitution of $\langle \mathbf{u}, \boldsymbol{\phi}_\ell \rangle$ for $\alpha_\ell$), that there exists some element $\tilde{\mathbf{u}} \in \mathcal{U}$ such that

$$\lim_{L \to \infty} \left\| \tilde{\mathbf{u}} - \sum_{\ell=-L}^{L} \langle \mathbf{u}, \boldsymbol{\phi}_\ell \rangle \, \boldsymbol{\phi}_\ell \right\|_2 = 0; \tag{8.48a}$$

$$\langle \tilde{\mathbf{u}}, \boldsymbol{\phi}_\ell \rangle = \langle \mathbf{u}, \boldsymbol{\phi}_\ell \rangle, \quad \ell \in \mathbb{Z}; \tag{8.48b}$$

and

$$\|\tilde{\mathbf{u}}\|_2^2 = \sum_{\ell=-\infty}^{\infty} \left| \langle \mathbf{u}, \boldsymbol{\phi}_\ell \rangle \right|^2. \tag{8.48c}$$

By (8.48b) it follows that the element $\mathbf{u} - \tilde{\mathbf{u}}$ of $\mathcal{U}$ satisfies

$$\langle \mathbf{u} - \tilde{\mathbf{u}}, \boldsymbol{\phi}_\ell \rangle = 0, \quad \ell \in \mathbb{Z},$$

and hence, by Condition (8.46), is of zero energy

$$\|\mathbf{u} - \tilde{\mathbf{u}}\|_2 = 0, \tag{8.49}$$

so $\mathbf{u}$ and $\tilde{\mathbf{u}}$ are indistinguishable and hence

$$\|\mathbf{u}\|_2 = \|\tilde{\mathbf{u}}\|_2 \, .$$

This combines with (8.48c) to prove (8.47).      $\square$

## 8.6   An Isomorphism

In this section we collect the results of Theorem 8.4.3 and Proposition 8.5.4 into a single theorem about the isomorphism between the space of energy-limited signals that are bandlimited to $W$ Hz and the space of square-summable sequences. This theorem is at the heart of quantization schemes for bandlimited signals. It demonstrates that to describe a bandlimited signal one can use discrete-time processing to quantize its samples and one can then map the quantized samples to a bandlimited signal. The energy in the error signal corresponding to the difference between the original signal and its description is then proportional to the sum of the squared differences between the samples of the original signal and the quantized version.

**Theorem 8.6.1 (Bandlimited Signals and Square-Summable Sequences).** *Let* $T = 1/(2W)$, *where* $W > 0$.

    *(i) If* $\mathbf{u}$ *is an energy-limited signal that is bandlimited to* $W$ *Hz, then the bi-infinite sequence*

$$\ldots, u(-T), u(0), u(T), u(2T), \ldots$$

    *consisting of its samples taken at integer multiples of* $T$ *is square summable and*

$$T \sum_{\ell=-\infty}^{\infty} \left| u(\ell T) \right|^2 = \|\mathbf{u}\|_2^2 \, .$$

(ii) *More generally, if* **u** *and* **v** *are energy-limited signals that are bandlimited to* W *Hz, then*

$$\mathsf{T} \sum_{\ell=-\infty}^{\infty} u(\ell\mathsf{T})\, v^*(\ell\mathsf{T}) = \langle \mathbf{u}, \mathbf{v} \rangle.$$

(iii) *If* $\{\alpha_\ell\}$ *is a bi-infinite square-summable sequence, then there exists an energy-limited signal* **u** *that is bandlimited to* W *Hz such that its samples are given by*

$$u(\ell\mathsf{T}) = \alpha_\ell, \quad \ell \in \mathbb{Z}.$$

(iv) *The mapping that maps every energy-limited signal that is bandlimited to* W *Hz to the square-summable sequence consisting of its samples is linear.*

## 8.7 Prolate Spheroidal Wave Functions

The following result, which is due to Slepian and Pollak, will not be used in this book; it is included for its sheer beauty.

**Theorem 8.7.1.** *Let the positive constants* $\mathsf{T} > 0$ *and* $\mathsf{W} > 0$ *be given. Then there exists a sequence of real functions* $\phi_1, \phi_2, \ldots$ *and a corresponding sequence of positive numbers* $\lambda_1 > \lambda_2 > \cdots$ *such that:*

(i) *The sequence* $\phi_1, \phi_2, \ldots$ *forms a CONS for the space of energy-limited signals that are bandlimited to* W *Hz, so, a fortiori,*

$$\int_{-\infty}^{\infty} \phi_\ell(t)\, \phi_{\ell'}(t)\, \mathrm{d}t = \mathrm{I}\{\ell = \ell'\}, \quad \ell, \ell' \in \mathbb{N}. \tag{8.50a}$$

(ii) *The sequence of scaled and time-windowed functions* $\tilde{\phi}_{1,\mathrm{w}}, \tilde{\phi}_{2,\mathrm{w}}, \ldots$ *defined at every* $t \in \mathbb{R}$ *by*

$$\tilde{\phi}_{\ell,\mathrm{w}}(t) = \frac{1}{\sqrt{\lambda_\ell}}\, \phi_\ell(t)\, \mathrm{I}\left\{|t| \leq \frac{\mathsf{T}}{2}\right\}, \quad \ell \in \mathbb{N} \tag{8.50b}$$

*forms a CONS for the subspace of* $\mathcal{L}_2$ *consisting of all energy-limited signals that vanish outside the interval* $[-\mathsf{T}/2, \mathsf{T}/2]$, *so, a fortiori,*

$$\int_{-\mathsf{T}/2}^{\mathsf{T}/2} \phi_\ell(t)\, \phi_{\ell'}(t)\, \mathrm{d}t = \lambda_\ell\, \mathrm{I}\{\ell = \ell'\}, \quad \ell, \ell' \in \mathbb{N}. \tag{8.50c}$$

(iii) *For every* $t \in \mathbb{R}$,

$$\int_{-\mathsf{T}/2}^{\mathsf{T}/2} \mathrm{LPF}_{\mathsf{W}}(t - \tau)\, \phi_\ell(\tau)\, \mathrm{d}\tau = \lambda_\ell\, \phi_\ell(t), \quad \ell \in \mathbb{N}. \tag{8.50d}$$

The above functions $\phi_1, \phi_2, \ldots$ are related to Prolate Spheroidal Wave Functions. For a discussion of this connection, a proof of this theorem, and numerous applications see (Slepian and Pollak, 1961) and (Slepian, 1976).

## 8.8   Exercises

**Exercise 8.1 (Expansion of a Function).** Expand the function $t \mapsto \mathrm{sinc}^2(t/2)$ as an orthonormal expansion in the functions

$$\ldots, t \mapsto \mathrm{sinc}(t+2), t \mapsto \mathrm{sinc}(t+1), t \mapsto \mathrm{sinc}(t), t \mapsto \mathrm{sinc}(t-1), t \mapsto \mathrm{sinc}(t-2), \ldots$$

**Exercise 8.2 (Inner Product with a Bandlimited Signal).** Show that if $\mathbf{x}$ is an energy-limited signal that is bandlimited to $W$ Hz, and if $\mathbf{y} \in \mathcal{L}_2$, then

$$\langle \mathbf{x}, \mathbf{y} \rangle = \mathsf{T_s} \sum_{\ell=-\infty}^{\infty} x(\ell \mathsf{T_s}) \, y^*_{\mathrm{LPF}}(\ell \mathsf{T_s}),$$

where $\mathbf{y}_{\mathrm{LPF}}$ is the result of passing $\mathbf{y}$ through an ideal unit-gain lowpass filter of bandwidth $W$ Hz, and where $\mathsf{T_s} = 1/(2W)$.

**Exercise 8.3 (Approximating a Sinc by Sincs).** Find the coefficients $\{\alpha_\ell\}$ that minimize the integral

$$\int_{-\infty}^{\infty} \left( \mathrm{sinc}(3t/2) - \sum_{\ell=-\infty}^{\infty} \alpha_\ell \, \mathrm{sinc}(t-\ell) \right)^2 \mathrm{d}t.$$

What is the value of this integral when the coefficients are chosen as you suggest?

**Exercise 8.4 (Integrability and Summability).** Show that if $\mathbf{x}$ is an integrable signal that is bandlimited to $W$ Hz and if $\mathsf{T_s} = 1/(2W)$, then

$$\sum_{\ell=-\infty}^{\infty} \left| x(\ell \mathsf{T_s}) \right| < \infty.$$

*Hint: Let $\mathbf{h}$ be the IFT of the mapping in (7.15) when we substitute $0$ for $f_c$; $2W$ for $W$; and $2W + \Delta$ for $W_c$, where $\Delta > 0$. Express $x(\ell \mathsf{T_s})$ as $(\mathbf{x} \star \mathbf{h})(\ell \mathsf{T_s})$; upper-bound the convolution integral using Proposition 2.4.1; and use Fubini's Theorem to swap the order of summation and integration.*

**Exercise 8.5 (Approximating an Integral by a Sum).** One often approximates an integral by a sum, e.g.,

$$\int_{-\infty}^{\infty} x(t) \, \mathrm{d}t \approx \delta \sum_{\ell=-\infty}^{\infty} x(\ell\delta).$$

(i) Show that if $\mathbf{u}$ is an energy-limited signal that is bandlimited to $W$ Hz, then, for every $0 < \delta \leq 1/(2W)$, the above approximation is exact when we substitute $|u(t)|^2$ for $x(t)$, that is,

$$\int_{-\infty}^{\infty} |u(t)|^2 \, \mathrm{d}t = \delta \sum_{\ell=-\infty}^{\infty} |u(\ell\delta)|^2.$$

(ii) Show that if $\mathbf{x}$ is an integrable signal that is bandlimited to $W$ Hz, then, for every $0 < \delta \leq 1/(2W)$,

$$\int_{-\infty}^{\infty} x(t) \, \mathrm{d}t = \delta \sum_{\ell=-\infty}^{\infty} x(\ell\delta).$$

(iii) Consider the signal $\mathbf{u} \colon t \mapsto \mathrm{sinc}(t)$. Compute $\|\mathbf{u}\|_2^2$ using Parseval's Theorem and use the result and Part (i) to show that

$$\sum_{m=0}^{\infty} \frac{1}{(2m+1)^2} = \frac{\pi^2}{8}.$$

**Exercise 8.6 (On the Pointwise Sampling Theorem).**

(i) Let the functions $\mathbf{g}, \mathbf{g}_0, \mathbf{g}_1, \ldots$ be elements of $\mathcal{L}_2$ that are zero outside the interval $[-W, W]$. Show that if $\|\mathbf{g} - \mathbf{g}_n\|_2 \to 0$, then for every $t \in \mathbb{R}$

$$\lim_{n \to \infty} \int_{-\infty}^{\infty} g_n(f)\, e^{\mathrm{i}2\pi f t}\, \mathrm{d}f = \int_{-\infty}^{\infty} g(f)\, e^{\mathrm{i}2\pi f t}\, \mathrm{d}f.$$

(ii) Use Part (i) to prove the Pointwise Sampling Theorem for energy-limited signals.

**Exercise 8.7 (Reconstructing from a Finite Number of Samples).** Show that there does not exist a universal positive integer L such that at $t = \mathsf{T}/2$

$$\left| x(t) - \sum_{\ell=-\mathsf{L}}^{\mathsf{L}} x(-\ell \mathsf{T})\, \mathrm{sinc}\!\left(\frac{t}{\mathsf{T}} + \ell\right) \right| < 0.1$$

for all energy-limited signals $\mathbf{x}$ that are bandlimited to $1/(2\mathsf{T})$ Hz.

**Exercise 8.8 (Inner Product between Passband Signals).** Let $\mathbf{x}_{\mathrm{PB}}$ and $\mathbf{y}_{\mathrm{PB}}$ be energy-limited passband signals that are bandlimited to $W$ Hz around the carrier frequency $f_{\mathrm{c}}$. Let $\mathbf{x}_{\mathrm{BB}}$ and $\mathbf{y}_{\mathrm{BB}}$ be their corresponding baseband representations. Let $\mathsf{T} = 1/W$. Show that

$$\langle \mathbf{x}_{\mathrm{PB}}, \mathbf{y}_{\mathrm{PB}} \rangle = 2\mathsf{T}\,\mathrm{Re}\!\left( \sum_{\ell=-\infty}^{\infty} x_{\mathrm{BB}}(\ell\mathsf{T})\, y_{\mathrm{BB}}^*(\ell\mathsf{T}) \right).$$

**Exercise 8.9 (Closed Subspaces).** Let $\mathcal{U}$ denote the set of energy-limited signals that vanish outside some interval. Thus, $\mathbf{u}$ is in $\mathcal{U}$ if, and only if, there exist $a, b \in \mathbb{R}$ (that may depend on $\mathbf{u}$) such that $u(t)$ is zero whenever $t \notin [a, b]$. Show that $\mathcal{U}$ is a linear subspace of $\mathcal{L}_2$, but that it is not closed.

**Exercise 8.10 (Projection onto an Infinite-Dimensional Subspace).**

(i) Let $\mathcal{U} \subset \mathcal{L}_2$ be the set of all elements of $\mathcal{L}_2$ that are zero outside the interval $[-1, +1]$. Given $\mathbf{v} \in \mathcal{L}_2$, let $\mathbf{w}$ be the signal $\mathbf{w} \colon t \mapsto v(t)\, \mathrm{I}\{|t| \le 1\}$. Show that $\mathbf{w}$ is in $\mathcal{U}$ and that $\mathbf{v} - \mathbf{w}$ is orthogonal to every signal in $\mathcal{U}$.

(ii) Let $\mathcal{U}$ be the subspace of energy-limited signals that are bandlimited to $W$ Hz. Given $\mathbf{v} \in \mathcal{L}_2$, define $\mathbf{w} = \mathbf{v} \star \mathrm{LPF}_W$. Show that $\mathbf{w}$ is in $\mathcal{U}$ and that $\mathbf{v} - \mathbf{w}$ is orthogonal to every signal in $\mathcal{U}$.

**Exercise 8.11 (A Maximization Problem).** Of all unit-energy real signals that are bandlimited to $W$ Hz, which one has the largest value at $t = 0$? What is its value at $t = 0$? Repeat for $t = 17$.

| $\tilde{\mathcal{V}}$ | $\mathcal{V}$ |
|---|---|
| energy-limited signals that are bandlimited to W Hz | energy-limited functions that vanish outside the interval $[-W, W)$ |
| generic element of $\tilde{\mathcal{V}}$<br>$\mathbf{x}: t \mapsto x(t)$ | generic element of $\mathcal{V}$<br>$\mathbf{g}: f \mapsto g(f)$ |
| a CONS<br>$\ldots, \tilde{\psi}_{-1}, \tilde{\psi}_0, \tilde{\psi}_1, \ldots$<br>$\tilde{\psi}_\ell(t) = \sqrt{2W}\,\mathrm{sinc}(2Wt + \ell)$ | a CONS<br>$\ldots, \psi_{-1}, \psi_0, \psi_1, \ldots$<br>$\psi_\ell(f) = \dfrac{1}{\sqrt{2W}}\, e^{i\pi \ell f/W}\,\mathrm{I}\{-W \le f < W\}$ |
| inner product<br>$\langle \mathbf{x}, \tilde{\psi}_\ell \rangle$<br>$\displaystyle\int_{-\infty}^{\infty} x(t)\sqrt{2W}\,\mathrm{sinc}(2Wt+\ell)\,dt$<br>$= \dfrac{1}{\sqrt{2W}}\, x\!\left(-\dfrac{\ell}{2W}\right)$ | inner product<br>$\langle \mathbf{g}, \psi_\ell \rangle$<br>$\displaystyle\int_{-W}^{W} g(f)\,\dfrac{1}{\sqrt{2W}}\, e^{-i\pi\ell f/W}\,df$<br>$= \mathbf{g}\text{'s }\ell\text{-th Fourier Series Coefficient } (\triangleq c_\ell)$ |
| Sampling Theorem<br>$\displaystyle\lim_{L\to\infty}\left\| \mathbf{x} - \sum_{\ell=-L}^{L} \langle \mathbf{x}, \tilde{\psi}_\ell \rangle\, \tilde{\psi}_\ell \right\|_2 = 0,$<br>i.e.,<br>$\displaystyle\int_{-\infty}^{\infty}\left| x(t) - \sum_{\ell=-L}^{L} x\!\left(-\dfrac{\ell}{2W}\right)\mathrm{sinc}(2Wt+\ell) \right|^2 dt \to 0$ | Fourier Series<br>$\displaystyle\lim_{L\to\infty}\left\| \mathbf{g} - \sum_{\ell=-L}^{L} \langle \mathbf{g}, \psi_\ell \rangle\, \psi_\ell \right\|_2 = 0,$<br>i.e.,<br>$\displaystyle\int_{-W}^{W}\left| g(f) - \sum_{\ell=-L}^{L} c_\ell \,\dfrac{1}{\sqrt{2W}}\, e^{i\pi\ell f/W} \right|^2 df \to 0$ |

**Table 8.1:** The duality between the Sampling Theorem and the Fourier Series Representation.

## Chapter 9

# Sampling Real Passband Signals

## 9.1   Introduction

In this chapter we present a procedure for representing a real energy-limited pass-band signal that is bandlimited to $W$ Hz around a carrier frequency $f_c$ using complex numbers that we accumulate at a rate of $W$ complex numbers per second. Alternatively, since we can represent every complex number as a pair of real numbers (its real and imaginary parts), we can view our procedure as allowing us to represent the signal using real numbers that we accumulate at a rate of $2W$ real numbers per second. Thus we propose to accumulate

$$\boxed{2W \text{ real samples per second,}}$$

or

$$\boxed{W \text{ complex samples per second.}}$$

Note that the carrier frequency $f_c$ plays no role here (provided, of course, that $f_c > W/2$): the rate at which we accumulate real numbers to describe the passband signal does not depend on $f_c$.[1]

For real baseband signals this feat is easily accomplished using the Sampling Theorem as follows. A real energy-limited baseband signal that is bandlimited to $W$ Hz can be reconstructed from its (real) samples that are taken $1/(2W)$ seconds apart (Theorem 8.4.3), so the signal can be reconstructed from real numbers (its samples) that are being accumulated at the rate of $2W$ real samples per second.

For passband signals we cannot achieve this feat by invoking the Sampling Theorem directly. Even though, by Corollary 7.7.3, every energy-limited passband signal $\mathbf{x}_{\mathrm{PB}}$ that is bandlimited to $W$ Hz around the center frequency $f_c$ is also an energy-limited bandlimited (baseband) signal, we are only guaranteed that $\mathbf{x}_{\mathrm{PB}}$ be bandlimited

---
[1] But the carrier frequency $f_c$ does play a role in the reconstruction.

to $f_c + W/2$ Hz. Consequently, if we were to apply the Sampling Theorem directly to $\mathbf{x}_{PB}$ we would have to sample $\mathbf{x}_{PB}$ every $1/(2f_c + W)$ seconds, i.e., we would have to accumulate $2f_c + W$ real numbers per second, which can be much higher than $2W$, especially in wireless communications where $f_c \gg W$.

Instead of applying the Sampling Theorem directly to $\mathbf{x}_{PB}$, the idea is to apply it to $\mathbf{x}_{PB}$'s baseband representation $\mathbf{x}_{BB}$. Suppose that $\mathbf{x}_{PB}$ is a real energy-limited passband signal that is bandlimited to $W$ Hz around the carrier frequency $f_c$. By Theorem 7.7.12 (vii), it can be represented using its baseband representation $\mathbf{x}_{BB}$, which is a complex baseband signal that is bandlimited to $W/2$ Hz (Theorem 7.7.12 (v)). Consequently, by the $\mathcal{L}_2$-Sampling Theorem (Theorem 8.4.3), $\mathbf{x}_{BB}$ can be described by sampling it at a rate of $W$ samples per second. Since the baseband signal is complex, its samples are also, in general, complex. Thus, in sampling $\mathbf{x}_{BB}$ every $1/W$ seconds we are accumulating one complex sample every $1/W$ seconds. Since we can recover $\mathbf{x}_{PB}$ from $\mathbf{x}_{BB}$ and $f_c$, it follows that, as we wanted, we have found a way to describe $\mathbf{x}_{PB}$ using complex numbers that are accumulated at a rate of $W$ complex numbers per second.

## 9.2 Complex Sampling

Recall from Section 7.7.3 (Theorem 7.7.12) that a real energy-limited passband signal $\mathbf{x}_{PB}$ that is bandlimited to $W$ Hz around a carrier frequency $f_c$ can be represented using its baseband representation $\mathbf{x}_{BB}$ as

$$x_{PB}(t) = 2\operatorname{Re}\big(e^{i2\pi f_c t}\, x_{BB}(t)\big), \quad t \in \mathbb{R}, \tag{9.1}$$

where $\mathbf{x}_{BB}$ is given by

$$\mathbf{x}_{BB} = \big(t \mapsto e^{-i2\pi f_c t}\, x_{PB}(t)\big) \star \mathrm{LPF}_{W_c}, \tag{9.2}$$

and where the cutoff frequency $W_c$ can be chosen arbitrarily in the range

$$\frac{W}{2} \leq W_c \leq 2f_c - \frac{W}{2}. \tag{9.3}$$

The signal $\mathbf{x}_{BB}$ is an energy-limited *complex* baseband signal that is bandlimited to $W/2$ Hz. Being bandlimited to $W/2$ Hz, it follows from the $\mathcal{L}_2$-Sampling Theorem that $\mathbf{x}_{BB}$ can be reconstructed from its samples taken $1/(2\,(W/2)) = 1/W$ seconds apart. We denote these samples by

$$x_{BB}\Big(\frac{\ell}{W}\Big), \quad \ell \in \mathbb{Z} \tag{9.4}$$

so, by (9.2),

$$x_{BB}\Big(\frac{\ell}{W}\Big) = \Big(\big(t \mapsto e^{-i2\pi f_c t}\, x_{PB}(t)\big) \star \mathrm{LPF}_{W_c}\Big)\Big(\frac{\ell}{W}\Big), \quad \ell \in \mathbb{Z}. \tag{9.5}$$

These samples are, in general, complex. Their real part corresponds to the samples of the in-phase component $\operatorname{Re}(\mathbf{x}_{BB})$, which, by (7.41a), is given by

$$\operatorname{Re}(\mathbf{x}_{BB}) = \big(t \mapsto x_{PB}(t)\cos(2\pi f_c t)\big) \star \mathrm{LPF}_{W_c} \tag{9.6}$$

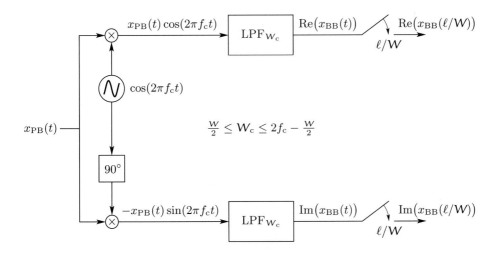

**Figure 9.1:** Sampling of a real passband signal $\mathbf{x}_{PB}$.

(for $W_c$ satisfying (9.3)) and their imaginary part corresponds to the samples of the quadrature-component $\mathrm{Im}(\mathbf{x}_{BB})$, which, by (7.41b), is given by

$$\mathrm{Im}(\mathbf{x}_{BB}) = -\big(t \mapsto x_{PB}(t)\sin(2\pi f_c t)\big) \star \mathrm{LPF}_{W_c}. \tag{9.7}$$

Thus,

$$x_{BB}\Big(\frac{\ell}{W}\Big) = \Big(\big(t \mapsto x_{PB}(t)\cos(2\pi f_c t)\big) \star \mathrm{LPF}_{W_c}\Big)\Big(\frac{\ell}{W}\Big)$$
$$- \mathrm{i}\Big(\big(t \mapsto x_{PB}(t)\sin(2\pi f_c t)\big) \star \mathrm{LPF}_{W_c}\Big)\Big(\frac{\ell}{W}\Big), \quad \ell \in \mathbb{Z}. \tag{9.8}$$

The procedure of taking a real passband signal $\mathbf{x}_{PB}$ and sampling its baseband representation to obtain the samples (9.8) is called **complex sampling**. It is depicted in Figure 9.1. The passband signal $\mathbf{x}_{PB}$ is first separately multiplied by $t \mapsto \cos(2\pi f_c t)$ and by $t \mapsto -\sin(2\pi f_c t)$, which are generated using a local oscillator and a $90°$-phase shifter. Each result is fed to a lowpass filter with cutoff frequency $W_c$ to produce the in-phase and quadrature component respectively. Each component is then sampled at a rate of $W$ real samples per second.

## 9.3 Reconstructing $x_{PB}$ from its Complex Samples

By the Pointwise Sampling Theorem (Theorem 8.4.5) applied to the energy-limited signal $\mathbf{x}_{BB}$ (which is bandlimited to $W/2$ Hz) we obtain

$$x_{BB}(t) = \sum_{\ell=-\infty}^{\infty} x_{BB}\Big(\frac{\ell}{W}\Big) \mathrm{sinc}(Wt - \ell), \quad t \in \mathbb{R}. \tag{9.9}$$

Consequently, by (9.1), $\mathbf{x}_{\mathrm{PB}}$ can be reconstructed from its complex samples as

$$x_{\mathrm{PB}}(t) = 2\,\mathrm{Re}\!\left( e^{\mathrm{i}2\pi f_c t} \sum_{\ell=-\infty}^{\infty} x_{\mathrm{BB}}\!\left(\frac{\ell}{W}\right) \mathrm{sinc}(Wt - \ell) \right), \quad t \in \mathbb{R}. \qquad (9.10\mathrm{a})$$

Since the $\mathrm{sinc}\,(\cdot)$ function is real, this can also be written as

$$x_{\mathrm{PB}}(t) = 2 \sum_{\ell=-\infty}^{\infty} \mathrm{Re}\!\left( e^{\mathrm{i}2\pi f_c t} x_{\mathrm{BB}}\!\left(\frac{\ell}{W}\right) \right) \mathrm{sinc}(Wt - \ell), \quad t \in \mathbb{R}, \qquad (9.10\mathrm{b})$$

or, using real operations, as

$$x_{\mathrm{PB}}(t) = 2 \sum_{\ell=-\infty}^{\infty} \mathrm{Re}\!\left( x_{\mathrm{BB}}\!\left(\frac{\ell}{W}\right) \right) \mathrm{sinc}(Wt - \ell) \cos(2\pi f_c t)$$
$$- 2 \sum_{\ell=-\infty}^{\infty} \mathrm{Im}\!\left( x_{\mathrm{BB}}\!\left(\frac{\ell}{W}\right) \right) \mathrm{sinc}(Wt - \ell) \sin(2\pi f_c t), \quad t \in \mathbb{R}. \qquad (9.10\mathrm{c})$$

As we next show, we can obtain another form of convergence using the $\mathcal{L}_2$-Sampling Theorem (Theorem 8.4.3). We first note that by that theorem

$$\lim_{\mathrm{L}\to\infty} \left\| t \mapsto x_{\mathrm{BB}}(t) - \sum_{\ell=-\mathrm{L}}^{\mathrm{L}} x_{\mathrm{BB}}\!\left(\frac{\ell}{W}\right) \mathrm{sinc}(Wt - \ell) \right\|_2^2 = 0. \qquad (9.11)$$

We next note that $\mathbf{x}_{\mathrm{BB}}$ is the baseband representation of $\mathbf{x}_{\mathrm{PB}}$ and that—as can be verified directly or by using Proposition 7.7.9—the mapping

$$t \mapsto x_{\mathrm{BB}}(\ell/W)\,\mathrm{sinc}(Wt - \ell)$$

is the baseband representation of the real passband signal

$$t \mapsto 2\,\mathrm{Re}\!\left( e^{\mathrm{i}2\pi f_c t} x_{\mathrm{BB}}\!\left(\frac{\ell}{W}\right) \mathrm{sinc}(Wt - \ell) \right).$$

Consequently, by linearity (Theorem 7.7.12 (ii)), the mapping

$$t \mapsto x_{\mathrm{BB}}(t) - \sum_{\ell=-\mathrm{L}}^{\mathrm{L}} x_{\mathrm{BB}}\!\left(\frac{\ell}{W}\right) \mathrm{sinc}(Wt - \ell)$$

is the baseband representation of the real passband signal

$$t \mapsto x_{\mathrm{PB}}(t) - 2\,\mathrm{Re}\!\left( e^{\mathrm{i}2\pi f_c t} \sum_{\ell=-\mathrm{L}}^{\mathrm{L}} x_{\mathrm{BB}}\!\left(\frac{\ell}{W}\right) \mathrm{sinc}(Wt - \ell) \right)$$

and hence, by Theorem 7.7.12 (iii),

$$\left\| t \mapsto x_{\mathrm{PB}}(t) - 2\,\mathrm{Re}\!\left( e^{\mathrm{i}2\pi f_c t} \sum_{\ell=-\mathrm{L}}^{\mathrm{L}} x_{\mathrm{BB}}\!\left(\frac{\ell}{W}\right) \mathrm{sinc}(Wt - \ell) \right) \right\|_2^2$$
$$= 2 \left\| t \mapsto x_{\mathrm{BB}}(t) - \sum_{\ell=-\mathrm{L}}^{\mathrm{L}} x_{\mathrm{BB}}\!\left(\frac{\ell}{W}\right) \mathrm{sinc}(Wt - \ell) \right\|_2^2. \qquad (9.12)$$

Combining (9.11) with (9.12) yields the $\mathcal{L}_2$ convergence

$$\lim_{L\to\infty}\left\| t \mapsto x_{PB}(t) - 2\operatorname{Re}\left( e^{i2\pi f_c t}\sum_{\ell=-L}^{L} x_{BB}\left(\frac{\ell}{W}\right)\operatorname{sinc}(Wt-\ell)\right)\right\|_2 = 0. \quad (9.13)$$

We summarize how a passband signal can be reconstructed from the samples of its baseband representation in the following theorem.

**Theorem 9.3.1 (The Sampling Theorem for Passband Signals).** *Let $\mathbf{x}_{PB}$ be a real energy-limited passband signal that is bandlimited to $W$ Hz around the carrier frequency $f_c$. For every integer $\ell$, let $x_{BB}(\ell/W)$ denote the time-$\ell/W$ sample of the baseband representation $\mathbf{x}_{BB}$ of $\mathbf{x}_{PB}$; see (9.5) and (9.8).*

*(i) $\mathbf{x}_{PB}$ can be pointwise reconstructed from the samples using the relation*

$$x_{PB}(t) = 2\operatorname{Re}\left( e^{i2\pi f_c t}\sum_{\ell=-\infty}^{\infty} x_{BB}\left(\frac{\ell}{W}\right)\operatorname{sinc}(Wt-\ell)\right), \quad t\in\mathbb{R}.$$

*(ii) $\mathbf{x}_{PB}$ can also be reconstructed from the samples in the $\mathcal{L}_2$ sense*

$$\lim_{L\to\infty}\int_{-\infty}^{\infty}\left( x_{PB}(t) - 2\operatorname{Re}\left( e^{i2\pi f_c t}\sum_{\ell=-L}^{L} x_{BB}\left(\frac{\ell}{W}\right)\operatorname{sinc}(Wt-\ell)\right)\right)^2 dt = 0.$$

*(iii) The energy in $\mathbf{x}_{PB}$ can be reconstructed from the sum of the squared magnitudes of the samples via*

$$\|\mathbf{x}_{PB}\|_2^2 = \frac{2}{W}\sum_{\ell=-\infty}^{\infty}\left| x_{BB}\left(\frac{\ell}{W}\right)\right|^2.$$

*(iv) If $\mathbf{y}_{PB}$ is another real energy-limited passband signal that is bandlimited to $W$ Hz around $f_c$, and if $\{y_{BB}(\ell/W)\}$ are the samples of its baseband representation, then*

$$\langle \mathbf{x}_{PB}, \mathbf{y}_{PB}\rangle = \frac{2}{W}\operatorname{Re}\left( \sum_{\ell=-\infty}^{\infty} x_{BB}\left(\frac{\ell}{W}\right) y_{BB}^*\left(\frac{\ell}{W}\right)\right).$$

**Proof.** Part (i) is just a restatement of (9.10b). Part (ii) is a restatement of (9.13). Part (iii) is a special case of Part (iv) corresponding to $\mathbf{y}_{PB}$ being equal to $\mathbf{x}_{PB}$. It thus only remains to prove Part (iv). This is done by noting that if $\mathbf{x}_{BB}$ and $\mathbf{y}_{BB}$ are the baseband representations of $\mathbf{x}_{PB}$ and $\mathbf{y}_{PB}$, then, by Theorem 7.7.12 (iv),

$$\langle \mathbf{x}_{PB}, \mathbf{y}_{PB}\rangle = 2\operatorname{Re}\left(\langle \mathbf{x}_{BB}, \mathbf{y}_{BB}\rangle\right)$$

$$= \frac{2}{W}\operatorname{Re}\left( \sum_{\ell=-\infty}^{\infty} x_{BB}\left(\frac{\ell}{W}\right) y_{BB}^*\left(\frac{\ell}{W}\right)\right),$$

where the second equality follows from Theorem 8.4.3 (iii). $\qquad\square$

Using the isomorphism between the family of complex square-summable sequences and the family of energy-limited signals that are bandlimited to $W$ Hz (Theorem 8.6.1), and using the relationship between real energy-limited passband signals and their baseband representation (Theorem 7.7.12), we can readily establish the following isomorphism between the family of complex square-summable sequences and the family of real energy-limited passband signals.

**Theorem 9.3.2 (Real Passband Signals and Square-Summable Sequences).** *Let $f_c$, $W$, and $T$ be constants satisfying*

$$f_c > W/2 > 0, \quad T = 1/W.$$

(i) *If $\mathbf{x}_{PB}$ is a real energy-limited passband signal that is bandlimited to $W$ Hz around $f_c$, and if $\mathbf{x}_{BB}$ is its baseband representation, then the bi-infinite sequence consisting of the samples of $\mathbf{x}_{BB}$ at integer multiples of $T$*

$$\ldots, x_{BB}(-T), x_{BB}(0), x_{BB}(T), x_{BB}(2T), \ldots$$

*is a square-summable sequence of complex numbers and*

$$2T \sum_{\ell=-\infty}^{\infty} \left| x_{BB}(\ell T) \right|^2 = \|\mathbf{x}_{PB}\|_2^2.$$

(ii) *More generally, if $\mathbf{x}_{PB}$ and $\mathbf{y}_{PB}$ are real energy-limited passband signals that are bandlimited to $W$ Hz around the carrier frequency $f_c$, and if $\mathbf{x}_{BB}$ and $\mathbf{y}_{BB}$ are their baseband representations, then*

$$2T \operatorname{Re}\left( \sum_{\ell=-\infty}^{\infty} x_{BB}(\ell T)\, y_{BB}^*(\ell T) \right) = \langle \mathbf{x}_{PB}, \mathbf{y}_{PB} \rangle.$$

(iii) *If $\ldots, \alpha_{-1}, \alpha_0, \alpha_1, \ldots$ is a square-summable bi-infinite sequence of complex numbers, then there exists a real energy-limited passband signal $\mathbf{x}_{PB}$ that is bandlimited to $W$ Hz around the carrier frequency $f_c$ such that the samples of its baseband representation $\mathbf{x}_{BB}$ are given by*

$$x_{BB}(\ell T) = \alpha_\ell, \quad \ell \in \mathbb{Z}.$$

(iv) *The mapping of every real energy-limited passband signal that is bandlimited to $W$ Hz around $f_c$ to the square-summable sequence consisting of the samples of its baseband representation is linear (over $\mathbb{R}$).*

## 9.4   Exercises

**Exercise 9.1 (A Specific Signal).** Let $\mathbf{x}$ be a real energy-limited passband signal that is bandlimited to $W$ Hz around the carrier frequency $f_c$. Suppose that all its complex samples are zero except for its zero-th complex sample, which is given by $1 + i$. What is $\mathbf{x}$?

**Exercise 9.2 (Real Passband Signals whose Complex Samples Are Real).** Characterize the Fourier Transforms of real energy-limited passband signals that are bandlimited to $W$ Hz around the carrier frequency $f_c$ and whose complex samples are real.

**Exercise 9.3 (Multiplying by a Carrier).** Let $\mathbf{x}$ be a real energy-limited signal that is bandlimited to $W/2$ Hz, and let $f_c$ be larger than $W/2$. Express the complex samples of $t \mapsto x(t) \cos(2\pi f_c t)$ in terms of $\mathbf{x}$. Repeat for $t \mapsto x(t) \sin(2\pi f_c t)$.

**Exercise 9.4 (Naively Sampling a Passband Signal).**

  (i) Consider the signal $\mathbf{x} \colon t \mapsto m(t) \sin(2\pi f_c t)$, where $m(\cdot)$ is an integrable signal that is bandlimited to 100 Hz and where $f_c = 100$ MHz. Can $\mathbf{x}$ be recovered from its samples $\dots, x(-\mathsf{T}), x(0), x(\mathsf{T}), \dots$ when $1/\mathsf{T} = 100$ MHz?

  (ii) Consider now the general case where $\mathbf{x}$ is an integrable real passband signal that is bandlimited to $W$ Hz around the carrier frequency $f_c$. Find conditions guaranteeing that $\mathbf{x}$ be reconstructible from its samples $\dots, x(-\mathsf{T}), x(0), x(\mathsf{T}), \dots$

**Exercise 9.5 (Orthogonal Passband Signals).** Let $\mathbf{x}_{\mathrm{PB}}$ and $\mathbf{y}_{\mathrm{PB}}$ be real energy-limited passband signals that are bandlimited to $W$ Hz around the carrier frequency $f_c$. Under what conditions on their complex samples are they orthogonal?

**Exercise 9.6 (Sampling a Baseband Signal As Though It Were a Passband Signal).** Recall that, ignoring some technicalities, a real baseband signal $\mathbf{x}$ of bandwidth $W$ Hz can be viewed as a real passband signal of bandwidth $W$ around the carrier frequency $f_c$, where $f_c = W/2$ (Problem 7.3). Compare the reconstruction formula for $\mathbf{x}$ from its samples to the reconstruction formula for $\mathbf{x}$ from its complex samples.

**Exercise 9.7 (Multiplying the Complex Samples).** Let $\mathbf{x}$ be a real energy-limited passband signal that is bandlimited to $W$ Hz around the carrier frequency $f_c$. Let $\dots, x_{-1}, x_0, x_1, \dots$ denote its complex samples taken $1/W$ second apart. Let $\mathbf{y}$ be a real energy-limited passband signal that is bandlimited to $W$ Hz around the carrier frequency $f_c$ and whose complex samples are like those of $\mathbf{x}$ but multiplied by i. Relate the FT of $\mathbf{y}$ to the FT of $\mathbf{x}$.

**Exercise 9.8 (Delayed Complex Sampling).** Let $\mathbf{x}$ and $\mathbf{y}$ be real energy-limited passband signals that are bandlimited to $W$ Hz around the carrier frequency $f_c$. Suppose that the complex samples of $\mathbf{y}$ are the same as those of $\mathbf{x}$, but delayed by one:

$$y_{\mathrm{BB}}\left(\frac{\ell}{W}\right) = x_{\mathrm{BB}}\left(\frac{\ell-1}{W}\right), \quad \ell \in \mathbb{Z}.$$

How are $\hat{\mathbf{x}}$ and $\hat{\mathbf{y}}$ related? Is $\mathbf{y}$ a delayed version of $\mathbf{x}$?

**Exercise 9.9 (On the Family of Real Passband Signals).** Is the set of all real energy-limited passband signals that are bandlimited to $W$ Hz around the carrier frequency $f_c$ a linear subspace of the set of all complex energy-limited signals?

**Exercise 9.10 (Complex Sampling and Inner Products).** Show that the $\ell$-th complex sample $x_{\mathrm{BB}}(\ell/W)$ of any real energy-limited passband signal that is bandlimited to $W$ Hz around the carrier frequency $f_c$ can be expressed as an inner product

$$x_{\mathrm{BB}}\left(\frac{\ell}{W}\right) = \langle \mathbf{x}, \boldsymbol{\phi}_\ell \rangle, \quad \ell \in \mathbb{Z},$$

where $\dots, \boldsymbol{\phi}_{-1}, \boldsymbol{\phi}_0, \boldsymbol{\phi}_1, \dots$ are orthogonal equi-energy complex signals. Is $\boldsymbol{\phi}_\ell$ in general a delayed version of $\boldsymbol{\phi}_0$?

**Exercise 9.11 (Absolute Summability of the Complex Samples).** Show that the complex samples of a real integrable passband signal that is bandlimited to $W$ Hz around the carrier frequency $f_c$ must be absolutely summable.

*Hint: See Exercise 8.4.*

**Exercise 9.12 (The Convolution Revisited).** Let $\mathbf{x}$ and $\mathbf{y}$ be real integrable passband signals that are bandlimited to $W$ Hz around the carrier frequency $f_c$. Express the complex samples of $\mathbf{x} \star \mathbf{y}$ in terms of those of $\mathbf{x}$ and $\mathbf{y}$.

**Exercise 9.13 (Complex Sampling and Filtering).** Let $\mathbf{x}$ be a real integrable passband signal that is bandlimited to $W$ Hz around the carrier frequency $f_c$, and let $\mathbf{h}$ be the impulse response of a real stable filter. Relate the complex samples of $\mathbf{x} \star \mathbf{h}$ to those of $\mathbf{x}$ and $\mathbf{h} \star \mathrm{BPF}_{W, f_c}$.

# Chapter 10

# Mapping Bits to Waveforms

## 10.1 What Is Modulation?

Data bits are mathematical entities that have no physical attributes. To send them over a channel, one needs to first map them into some physical signal, which is then "fed" into a channel to produce a physical signal at the channel's output. For example, when we send data over a telephone line, the data bits are first converted to an electrical signal, which then influences the voltage measured at the other end of the line. (We use the term "influences" because the signal measured at the other end of the line is usually not identical to the channel input: it is typically attenuated and also corrupted by thermal noise and other distortions introduced by various conversions in the telephone exchange system.) Similarly, in a wireless system, the data bits are mapped to an electromagnetic wave that then influences the electromagnetic field measured at the receiver antenna. In magnetic recording, data bits are written onto a magnetic medium by a mapping that maps them to a magnetization pattern, which is then measured (with some distortion and some noise) by the magnetic head at some later time when the data are read.

In the first example the bits are mapped to continuous-time waveforms corresponding to the voltage across an impedance, whereas in the last example the bits are mapped to a spatial waveform corresponding to different magnetizations at different locations across the magnetic medium. While some of the theory we shall develop holds for both cases, we shall focus here mainly on channels of the former type, where the channel input signal is some function of time rather than space.

We shall further focus on cases where the channel input corresponds to a time-varying voltage across a resistor, a time-varying current through a resistor, or a time-varying electric field, so the energy required to transmit the signal is proportional to the time integral of its square. Thus, if $x(t)$ denotes the channel input at time $t$, then we shall refer to $\int_t^{t+\Delta} x^2(\tau)\, d\tau$ as the transmitted energy during the time interval beginning at time $t$ and ending at time $t + \Delta$.

There are many mappings of bits to waveforms, and our goal is to find "good" ones. We will, of course, have to define some figures of merit to compare the quality of different mappings. We shall refer to the mapping of bits to a physical waveform as **modulation** and to the part of the system that performs the modulation as the

**modulator**.

Without going into too much detail, we can list a few qualitative requirements of a modulator. The modulation should be *robust* with respect to channel impairments, so that the receiver at the other end of the channel can reliably decode the data bits from the channel output. Also, the modulator should have *reasonable complexity*. Finally, in many applications we require that the transmitted signal be of *limited power* so as to preserve the battery. In wireless applications the transmitted signal may also be subject to *spectral restrictions* so as to not interfere with other systems.

## 10.2   Modulating One Bit

One does not typically expect to design a communication system in order to convey only one data bit. The purpose of the modulator is typically to map an entire bit stream to a waveform that extends over the entire life of the communication system. Nevertheless, for pedagogic reasons, it is good to first consider the simplest scenario of modulating a single bit. In this case the modulator is fully characterized by two functions $x_0(\cdot)$ and $x_1(\cdot)$ with the understanding that if the data bit $D$ is equal to zero, then the modulator produces the waveform $x_0(\cdot)$ and that otherwise it produces $x_1(\cdot)$. Thus, the signal produced by the modulator is given by

$$X(t) = \begin{cases} x_0(t) & \text{if } D = 0, \\ x_1(t) & \text{if } D = 1, \end{cases} \quad t \in \mathbb{R}. \tag{10.1}$$

For example, we could choose

$$x_0(t) = \begin{cases} A\, e^{-t/\mathsf{T}} & \text{if } t/\mathsf{T} \geq 0, \\ 0 & \text{otherwise,} \end{cases}, \quad t \in \mathbb{R},$$

and

$$x_1(t) = \begin{cases} A & \text{if } 0 \leq t/\mathsf{T} \leq 1, \\ 0 & \text{otherwise,} \end{cases}, \quad t \in \mathbb{R},$$

where $\mathsf{T} = 1$ sec and where $A$ is a constant such that $A^2$ has units of power.

This may seem like an odd way of writing these waveforms, but we have our reasons: we typically think of $t$ as having units of time, and we try to avoid applying transcendental functions (such as the exponential function) to quantities with units. Also, we think of the squared transmitted waveform as having units of power, whereas we think of the transcendental functions as returning unit-less arguments. Hence the introduction of the constant $A$ with the understanding that $A^2$ has units of power.

We denoted the bit to be sent by an uppercase letter ($D$) because we like to denote random quantities (such as random variables, random vectors, and stochastic processes) by uppercase letters, and we think of the transmitted bit as a random quantity. Indeed, if the transmitted bit were deterministic, there would be no need to transmit it! This may seem like a statement made in jest, but it is actually very important. In the first half of the twentieth century, engineers often analyzed the performance of (analog) communication systems by analyzing their performance in transmitting some particular signal, e.g., a sine wave. Nobody, of course, transmitted such "boring" signals, because those could always be produced at the receiver using a local oscillator. In the second half of the twentieth century, especially following the work of Claude Shannon, engineers realized that it is only meaningful to view the data to be transmitted as random, i.e., as quantities that are unknown at the receiver and also unknown to the system designer prior to the system's deployment. We thus view the bit to be sent $D$ as a random variable. Often we will assume that it takes on the values 0 and 1 equiprobably. This is a good assumption if prior to transmission a data compression algorithm is used.

By the same token, we view the transmitted signal as a random quantity, and hence the uppercase $X$. In fact, if we employ the above signaling scheme, then at every time instant $t' \in \mathbb{R}$ the value $X(t')$ of the transmitted waveform is a random variable. For example, at time $\mathsf{T}/2$ the value of the transmitted waveform is $X(\mathsf{T}/2)$, which is a random variable that takes on the values $\mathsf{A}\, e^{-1/2}$ and $\mathsf{A}$ equiprobably. Similarly, at time $2\mathsf{T}$ the value of the transmitted waveform is $X(2\mathsf{T})$, which is a random variable taking on the values $e^{-2}$ and 0 equiprobably. Mathematicians call such a waveform a **random process** or a **stochastic process** (SP). This will be defined formally in Section 12.2.

It is useful to think about a random process as a function of two arguments: time and "luck" or, more precisely, as a function of time and the result of all the random experiments in the system. For a fixed instant of time $t \in \mathbb{R}$, we have that $X(t)$ is a random variable, i.e., a real-valued function of the randomness in the system (in this case the realization of $D$). Alternatively, for a fixed realization of the randomness in the system, the random process is a deterministic function of time. These two views will be used interchangeably in this book.

## 10.3   From Bits to Real Numbers

Many of the popular modulation schemes can be viewed as operating in two stages. In the first stage the data bits are mapped to real numbers, and in the second stage the real numbers are mapped to a continuous-time waveform. If we denote by $k$ the number of data bits that will be transmitted by the system during its lifetime (or from the moment it is turned on until it is turned off), and if we denote the data bits by $D_1, D_2, \ldots, D_k$, then the first stage can be described as the application of a mapping $\varphi(\cdot)$ that maps length-$k$ sequences of bits to length-$n$ sequences of real numbers:

$$\varphi \colon \{0,1\}^k \to \mathbb{R}^n$$
$$(d_1, \ldots, d_k) \mapsto (x_1, \ldots, x_n).$$

From an engineering point of view, it makes little sense to allow for the encoding function to map two different binary $k$-tuples to the same real $n$-tuple, because this would result in the transmitted waveforms corresponding to the two $k$-tuples being identical. This may cause errors even in the absence of noise. We shall therefore assume throughout that the mapping $\varphi(\cdot)$ is **one-to-one** (injective) so no two distinct data $k$-tuples are mapped to the same $n$-tuple of real numbers.

An example of a mapping that maps bits to real numbers is the mapping that maps each data bit $D_j$ to the real number $X_j$ according to the rule

$$X_j = \begin{cases} +1 & \text{if } D_j = 0, \\ -1 & \text{if } D_j = 1, \end{cases} \quad j = 1, \ldots, k. \tag{10.2}$$

In this example one real symbol $X_j$ is produced for every data bit, so $n = k$. For this reason we say that this mapping has the rate of *one bit per real symbol*.

As another example consider the case where $k$ is even and the data bits $\{D_j\}$ are broken into pairs

$$(D_1, D_2), (D_3, D_4), \ldots, (D_{k-1}, D_k)$$

and each pair of data bits is then mapped to a single real number according to the rule

$$(D_{2j-1}, D_{2j}) \mapsto \begin{cases} +3 & \text{if } D_{2j-1} = D_{2j} = 0, \\ +1 & \text{if } D_{2j-1} = 0 \text{ and } D_{2j} = 1, \\ -3 & \text{if } D_{2j-1} = D_{2j} = 1, \\ -1 & \text{if } D_{2j-1} = 1 \text{ and } D_{2j} = 0, \end{cases} \quad j = 1, \ldots, k/2. \tag{10.3}$$

In this case $n = k/2$, and we say that the mapping has the rate of *two bits per real symbol*.

Note that the rate of the mapping could also be a fraction. Indeed, if each data bit $D_j$ produces two real numbers according to the repetition law

$$D_j \mapsto \begin{cases} (+1, +1) & \text{if } D_j = 0, \\ (-1, -1) & \text{if } D_j = 1, \end{cases} \quad j = 1, \ldots, k, \tag{10.4}$$

then $n = 2k$, and we say that the mapping is of rate *half a bit per real symbol*.

Since there is a natural correspondence between $\mathbb{R}^2$ and $\mathbb{C}$, i.e., between pairs of real numbers and complex numbers (where a pair of real numbers $(x, y)$ corresponds to the complex number $x + iy$), the rate of the above mapping (10.4) can also be stated as *one bit per complex symbol*. This may seem like an odd way of stating the rate, but it has some advantages that will become apparent later when we discuss the mapping of real (or complex) numbers to waveforms and the Nyquist Criterion.

## 10.4    Block-Mode Mapping of Bits to Real Numbers

The examples we gave in Section 10.3 of mappings $\varphi \colon \{0, 1\}^k \to \mathbb{R}^n$ have something in common. In each of those examples the mapping can be described as follows: the data bits $D_1, \ldots, D_k$ are first grouped into binary K-tuples; each K-tuple is then mapped to a real N-tuple by applying some mapping **enc**: $\{0, 1\}^K \to \mathbb{R}^N$; and the so-produced real N-tuples are then concatenated to form the sequence $X_1, \ldots, X_n$, where $n = (k/K)N$.

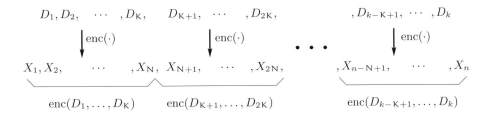

**Figure 10.1:** Block-mode encoding.

In the first example $K = N = 1$ and the mapping of K-tuples to N-tuples is the mapping (10.2). In the second example $K = 2$ and $N = 1$ with the mapping (10.3). And in the third example $K = 1$ and $N = 2$ with the repetition mapping (10.4).

To describe such mappings $\varphi\colon \{0,1\}^k \to \mathbb{R}^n$ more formally we need the notion of a binary-to-reals block encoder, which we define next.

**Definition 10.4.1 ($(K, N)$ Binary-to-Reals Block Encoder).** *A $(K, N)$ **binary-to-reals block encoder** is a one-to-one mapping from the set of binary K-tuples to the set of real N-tuples, where K and N are positive integers. The **rate** of a $(K, N)$ binary-to-reals block encoder is defined as*

$$\frac{K}{N} \left[ \frac{\text{bit}}{\text{real symbol}} \right].$$

Note that we shall sometimes omit the phrase "binary-to-reals" and refer to such an encoder as a $(K, N)$ **block encoder**. Also note that "one-to-one" means that no two distinct binary K-tuples may be mapped to the same real N-tuple.

We say that an encoder $\varphi\colon \{0,1\}^k \to \mathbb{R}^n$ **operates in block-mode using the $(K, N)$ binary-to-reals block encoder** $\mathrm{enc}(\cdot)$ if

1) $k$ is divisible by K;

2) $n$ is given by $(k/K)\,N$; and

3) $\varphi(\cdot)$ maps the binary sequence $D_1, \ldots, D_k$ to the sequence $X_1, \ldots, X_n$ by parsing the sequence $D_1, \ldots, D_k$ into consecutive length-K binary tuples and by then concatenating the results of applying $\mathrm{enc}(\cdot)$ to each such K-tuple as in Figure 10.1.

If $k$ is not divisible by K, we often introduce **zero padding**. In this case we choose $k'$ to be the smallest integer that is no smaller than $k$ and that is divisible by K, i.e.,

$$k' = \left\lceil \frac{k}{K} \right\rceil K,$$

(where for every $\xi \in \mathbb{R}$ we use $\lceil \xi \rceil$ to denote the smallest integer that is no smaller than $\xi$, e.g., $\lceil 1.24 \rceil = 2$) and map $D_1, \ldots, D_k$ to the sequence $X_1, \ldots, X_{n'}$ where

$$n' = \frac{k'}{K} N.$$

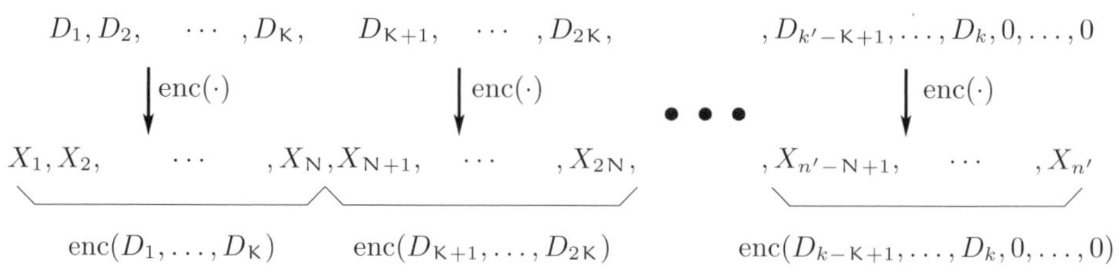

$$D_1, D_2, \quad \cdots \quad , D_K, \qquad D_{K+1}, \quad \cdots \quad , D_{2K}, \qquad\qquad , D_{k'-K+1}, \ldots, D_k, 0, \ldots, 0$$

$$\downarrow \text{enc}(\cdot) \qquad\qquad\qquad \downarrow \text{enc}(\cdot) \qquad\qquad \bullet\bullet\bullet \qquad\qquad \downarrow \text{enc}(\cdot)$$

$$X_1, X_2, \quad \cdots \quad , X_N, X_{N+1}, \quad \cdots \quad , X_{2N}, \qquad\qquad , X_{n'-N+1}, \quad \cdots \quad , X_{n'}$$

$$\underbrace{\qquad\qquad\qquad\qquad}_{\text{enc}(D_1,\ldots,D_K)} \quad \underbrace{\qquad\qquad\qquad\qquad}_{\text{enc}(D_{K+1},\ldots,D_{2K})} \qquad\qquad \underbrace{\qquad\qquad\qquad\qquad}_{\text{enc}(D_{k-K+1},\ldots,D_k,0,\ldots,0)}$$

**Figure 10.2:** Block-mode encoding with zero padding.

by applying the $(K, N)$ encoder in block-mode to the $k'$-length zero-padded binary tuple

$$D_1, \ldots, D_k, \underbrace{0, \ldots, 0}_{k' - k \text{ zeros}} \tag{10.5}$$

as in Figure 10.2.

## 10.5   From Real Numbers to Waveforms with Linear Modulation

There are numerous ways to map a sequence of real numbers $X_1, \ldots, X_n$ to a real-valued signal. Here we shall focus on mappings that have a linear structure. This additional structure simplifies the implementation of the modulator and demodulator. It will be described next.

Suppose we wish to modulate the $k$ data bits $D_1, \ldots, D_k$, and suppose that we have mapped these bits to the $n$ real numbers $X_1, \ldots, X_n$. Here $n$ can be smaller, equal, or greater than $k$. The transmitted waveform $X(\cdot)$ in a **linear modulation** scheme is then given by

$$X(t) = \mathsf{A} \sum_{\ell=1}^{n} X_\ell \, g_\ell(t), \quad t \in \mathbb{R}, \tag{10.6}$$

where the deterministic real waveforms $\mathbf{g}_1, \ldots, \mathbf{g}_n$ are specified in advance, and where $\mathsf{A} \geq 0$ is a scaling factor. The waveform $X(\cdot)$ can be thus viewed as a scaled-by-$\mathsf{A}$ linear combination of the tuple $(\mathbf{g}_1, \ldots, \mathbf{g}_n)$ with the coefficients $X_1, \ldots, X_n$:

$$\mathbf{X} = \mathsf{A} \sum_{\ell=1}^{n} X_\ell \, \mathbf{g}_\ell. \tag{10.7}$$

The transmitted energy is a random variable that is given by

$$\|\mathbf{X}\|_2^2 = \int_{-\infty}^{\infty} X^2(t) \, dt$$

$$= \int_{-\infty}^{\infty} \left( \mathsf{A} \sum_{\ell=1}^{n} X_\ell \, g_\ell(t) \right)^2 dt$$

$$= A^2 \sum_{\ell=1}^{n} \sum_{\ell'=1}^{n} X_\ell X_{\ell'} \int_{-\infty}^{\infty} g_\ell(t) \, g_{\ell'}(t) \, \mathrm{d}t$$

$$= A^2 \sum_{\ell=1}^{n} \sum_{\ell'=1}^{n} X_\ell X_{\ell'} \, \langle \mathbf{g}_\ell, \mathbf{g}_{\ell'} \rangle .$$

The transmitted energy takes on a particularly simple form if the waveforms $g_\ell(\cdot)$ are orthonormal, i.e., if

$$\langle \mathbf{g}_\ell, \mathbf{g}_{\ell'} \rangle = \mathrm{I}\{\ell = \ell'\}, \quad \ell, \ell' \in \{1, \dots, n\}, \tag{10.8}$$

in which case the energy is given by

$$\|\mathbf{X}\|_2^2 = A^2 \sum_{\ell=1}^{n} X_\ell^2, \quad \{\mathbf{g}_\ell\} \text{ orthonormal.} \tag{10.9}$$

As an exercise, the reader is encouraged to verify that there is no loss in generality in assuming that the waveforms $\{\mathbf{g}_\ell\}$ are orthonormal. More precisely:

**Theorem 10.5.1.** *Suppose that the waveform $X(\cdot)$ is generated from the binary $k$-tuple $D_1, \dots, D_k$ by applying the mapping $\varphi \colon \{0,1\}^k \to \mathbb{R}^n$ and by then linearly modulating the resulting $n$-tuple $\varphi(D_1, \dots, D_k)$ using the waveforms $\{\mathbf{g}_\ell\}_{\ell=1}^{n}$ as in (10.6).*

*Then there exist an integer $1 \leq n' \leq n$; a mapping $\varphi' \colon \{0,1\}^k \to \mathbb{R}^{n'}$; and $n'$ orthonormal signals $\{\phi_\ell\}_{\ell=1}^{n'}$ such that if $X'(\cdot)$ is generated from $D_1, \dots, D_k$ by applying linear modulation to $\varphi'(D_1, \dots, D_k)$ using the orthonormal waveforms $\{\phi_\ell\}_{\ell=1}^{n'}$, then $X'(\cdot)$ and $X(\cdot)$ are indistinguishable for every $k$-tuple $D_1, \dots, D_k$.*

**Proof.** The proof of this theorem is left as an exercise. □

Motivated by this theorem, we shall focus on linear modulation with orthonormal functions. But please note that even if the transmitted waveform satisfies (10.8), the received waveform might not. For example, the channel might consist of a linear filter that could destroy the orthogonality.

## 10.6 Recovering the Signal Coefficients with a Matched Filter

Suppose now that the binary $k$-tuple $(D_1, \dots, D_k)$ is mapped to the real $n$-tuple $(X_1, \dots, X_n)$ using the mapping

$$\varphi \colon \{0,1\}^k \to \mathbb{R}^n \tag{10.10}$$

and that the $n$-tuple $(X_1, \dots, X_n)$ is then mapped to the waveform

$$X(t) = A \sum_{\ell=1}^{n} X_\ell \, \phi_\ell(t), \quad t \in \mathbb{R}, \tag{10.11}$$

where $\phi_1, \ldots, \phi_n$ are orthonormal:

$$\langle \phi_\ell, \phi_{\ell'} \rangle = \mathrm{I}\{\ell = \ell'\}, \quad \ell, \ell' \in \{1, \ldots, n\}. \tag{10.12}$$

How can we recover the $k$-tuple $D_1, \ldots, D_k$ from $X(\cdot)$? The decoder's problem is, of course, harder, because the decoder usually does not have access to the transmitted waveform $X(\cdot)$ but only to the received waveform, which may be a noisy and distorted version of $X(\cdot)$. Nevertheless, it is instructive to consider the noiseless and distortionless problem first.

If we are able to recover the real numbers $\{X_\ell\}_{\ell=1}^n$ from the received signal $X(\cdot)$, and if the mapping $\varphi \colon \{0,1\}^k \to \mathbb{R}^n$ is one-to-one (as we assume), then the data bits $\{D_j\}_{j=1}^k$ can be reconstructed from $X(\cdot)$. Thus, the question is how to recover $\{X_\ell\}_{\ell=1}^n$ from $X(\cdot)$. But this is easy if the functions $\{\phi_\ell\}_{\ell=1}^n$ are orthonormal, because in this case, by Proposition 4.6.4 (i), $X_\ell$ is given by the scaled inner product between $\mathbf{X}$ and $\phi_\ell$:

$$X_\ell = \frac{1}{\mathsf{A}} \langle \mathbf{X}, \phi_\ell \rangle, \quad \ell = 1, \ldots, n. \tag{10.13}$$

Consequently, we can compute $X_\ell$ by feeding $\mathbf{X}$ to a matched filter for $\phi_\ell$ and scaling the time-0 output by $1/\mathsf{A}$ (Section 5.8). To recover $\{X_\ell\}_{\ell=1}^n$ we thus need $n$ matched filters, one matched to each of the waveforms $\{\phi_\ell\}$.

The implementation becomes much simpler if the functions $\{\phi_\ell\}$ have an additional structure, namely, if they are all time shifts of some function $\phi(\cdot)$:

$$\phi_\ell(t) = \phi(t - \ell \mathsf{T}_\mathrm{s}), \quad \left( \ell \in \{1, \ldots, n\}, \ t \in \mathbb{R} \right). \tag{10.14}$$

In this case it follows from Corollary 5.8.3 that we can compute all the inner products $\{\langle \mathbf{X}, \phi_\ell \rangle\}$ using one matched filter of impulse response $\overleftarrow{\phi}$ by feeding $\mathbf{X}$ to the filter and sampling its output at the appropriate times:

$$
\begin{aligned}
X_\ell &= \frac{1}{\mathsf{A}} \int_{-\infty}^{\infty} X(\tau)\, \phi_\ell(\tau)\, \mathrm{d}\tau \\
&= \frac{1}{\mathsf{A}} \int_{-\infty}^{\infty} X(\tau)\, \phi(\tau - \ell \mathsf{T}_\mathrm{s})\, \mathrm{d}\tau \\
&= \frac{1}{\mathsf{A}} \int_{-\infty}^{\infty} X(\tau)\, \overleftarrow{\phi}(\ell \mathsf{T}_\mathrm{s} - \tau)\, \mathrm{d}\tau \\
&= \frac{1}{\mathsf{A}} \left( X \star \overleftarrow{\phi} \right)(\ell \mathsf{T}_\mathrm{s}), \quad \ell = 1, \ldots, n.
\end{aligned}
\tag{10.15}
$$

Figure 10.3 demonstrates how the symbols $\{X_\ell\}$ can be recovered from $X(\cdot)$ using a single matched filter if the pulses $\{\phi_\ell\}$ satisfy (10.14).

## 10.7  Pulse Amplitude Modulation

Under Assumption (10.14), the transmitted signal $X(\cdot)$ in (10.11) is given by

$$X(t) = \mathsf{A} \sum_{\ell=1}^{n} X_\ell\, \phi(t - \ell \mathsf{T}_\mathrm{s}), \quad t \in \mathbb{R}, \tag{10.16}$$

**Figure 10.3:** Recovering the symbols from the transmitted waveform using a matched filter when (10.14) is satisfied.

which is a special case of **Pulse Amplitude Modulation** (PAM), which we describe next.

In PAM, the data bits $D_1, \ldots, D_k$ are mapped to real numbers $X_1, \ldots, X_n$, which are then mapped to the waveform

$$X(t) = \mathsf{A} \sum_{\ell=1}^{n} X_\ell \, g(t - \ell\mathsf{T_s}), \quad t \in \mathbb{R}, \tag{10.17}$$

for some scaling factor $\mathsf{A} \geq 0$, some function $\mathbf{g} \colon \mathbb{R} \to \mathbb{R}$, and some constant $\mathsf{T_s} > 0$. The function $\mathbf{g}$ (always assumed Borel measurable) is called the **pulse shape**; the constant $\mathsf{T_s}$ is called the **baud period**; and its reciprocal $1/\mathsf{T_s}$ is called the **baud rate**.[1] The units of $\mathsf{T_s}$ are seconds, and one often refers to the units of $1/\mathsf{T_s}$ as *real symbols per second*. PAM can thus be viewed as a special case of linear modulation (10.6) with $\mathbf{g}_\ell$ being given for every $\ell \in \{1, \ldots, n\}$ by the mapping $t \mapsto g(t - \ell\mathsf{T_s})$. The signal (10.16) can be viewed as a PAM signal where the pulse shape $\phi$ satisfies the orthonormality condition (10.14).

In this book we shall typically denote the PAM pulse shape by $\mathbf{g}$. But we shall use $\phi$ if we assume an additional orthonormality condition such as (10.12). In this case we shall refer to $1/\mathsf{T_s}$ as having units of *real dimensions per second*:

$$\frac{1}{\mathsf{T_s}} \left[ \frac{\text{real dimension}}{\text{sec}} \right], \quad \phi \text{ satisfies (10.12)}. \tag{10.18}$$

Note that according to Theorem 10.5.1 there is no loss in generality in assuming that the pulses $\{\phi_\ell\}$ are orthonormal. There is, however, a loss in generality in assuming that they satisfy (10.14).

## 10.8 Constellations

Recall that in PAM the data bits $D_1, \ldots, D_k$ are first mapped to the real $n$-tuple $X_1, \ldots, X_n$ using a one-to-one mapping $\varphi \colon \{0,1\}^k \to \mathbb{R}^n$, and that these real numbers are then mapped to the waveform $X(\cdot)$ via (10.17). Since there are only $2^k$ different binary $k$-tuples, it follows that each symbol $X_\ell$ can take on at most $2^k$ different values. The set of values that $X_\ell$ can take on may, in general, depend on $\ell$. The union of all these sets (over $\ell \in \{1, \ldots, n\}$) is called the **constellation** of

---

[1] These terms honor the French engineer J.M.E. Baudot (1845–1903) who invented a telegraph printing system.

the mapping $\varphi(\cdot)$. Denoting the constellation of $\varphi(\cdot)$ by $\mathcal{X}$, we thus have that a real number $x$ is in $\mathcal{X}$ if, and only if, for some choice of the binary $k$-tuple $(d_1, \ldots, d_k)$ and for some $\ell \in \{1, \ldots, n\}$ the $\ell$-th component of $\varphi\big((d_1, \ldots, d_k)\big)$ is equal to $x$.

For example, the constellation corresponding to the mapping (10.2) is the set $\{-1, +1\}$; the constellation corresponding to (10.3) is the set $\{-3, -1, +1, +3\}$; and the constellation corresponding to (10.4) is the set $\{-1, +1\}$. In all these examples, the constellation can be viewed as a special case of the constellation with $2\nu$ symbols

$$\big\{-(2\nu - 1), \ldots, -5, -3, -1, +1, +3, +5, \ldots, +(2\nu - 1)\big\} \tag{10.19}$$

for some positive integer $\nu$. A less prevalent constellation is the constellation

$$\{-2, -1, +1, +2\}. \tag{10.20}$$

The **number of points** in the constellation $\mathcal{X}$ is just $\#\mathcal{X}$, i.e., the number of elements (cardinality) of the set $\mathcal{X}$.

The **minimum distance** $\delta$ of a constellation is the Euclidean distance between the closest distinct elements in the constellation:

$$\delta \triangleq \min_{\substack{x, x' \in \mathcal{X} \\ x \neq x'}} |x - x'|. \tag{10.21}$$

The scaling of the constellation is arbitrary because of the scaling factor $\mathsf{A}$ in the signal's description. Thus, the signal $\mathsf{A} \sum_\ell X_\ell \, g(t - \ell \mathsf{T_s})$, where $X_\ell$ takes value in the set $\{\pm 1\}$ is of constellation $\{-1, +1\}$, but it can also be expressed in the form $\mathsf{A}' \sum_\ell X'_\ell \, g(t - \ell \mathsf{T_s})$, where $\mathsf{A}' = 2\mathsf{A}$ and $X'_\ell$ takes value in the set $\{-1/2, +1/2\}$, i.e., as a PAM signal of constellation $\{-1/2, +1/2\}$.

Different authors choose to normalize the constellation in different ways. One common normalization is to express the elements of the constellation as multiples of the minimum distance. Thus, we would represent the constellation $\{-1, +1\}$ as

$$\left\{ -\frac{1}{2}\delta, \; +\frac{1}{2}\delta \right\},$$

and the constellation $\{-3, -1, +1, +3\}$ as

$$\left\{ -\frac{3}{2}\delta, \; -\frac{1}{2}\delta, \; +\frac{1}{2}\delta, \; +\frac{3}{2}\delta \right\}.$$

The normalized version of the constellation (10.19) is

$$\left\{ \pm\frac{2\nu - 1}{2}\delta, \ldots, \pm\frac{5}{2}\delta, \; \pm\frac{3}{2}\delta, \; \pm\frac{1}{2}\delta \right\}. \tag{10.22}$$

The **second moment** of a constellation $\mathcal{X}$ is defined as

$$\frac{1}{\#\mathcal{X}} \sum_{x \in \mathcal{X}} x^2. \tag{10.23}$$

The second moment of the constellation in (10.22) is given by

$$\frac{1}{\#\,\mathcal{X}} \sum_{x \in \mathcal{X}} x^2 = \frac{1}{2\nu} 2 \sum_{\eta=1}^{\nu} (2\eta - 1)^2 \frac{\delta^2}{4}$$

$$= \frac{1}{3} \left( M^2 - 1 \right) \frac{\delta^2}{4}, \tag{10.24a}$$

where

$$M = 2\nu \tag{10.24b}$$

is the number of points in the constellation, and where (10.24a)–(10.24b) can be verified using the identity

$$\sum_{\eta=1}^{\nu} (2\eta - 1)^2 = \frac{1}{3}\nu(4\nu^2 - 1), \quad \nu = 1, 2, \ldots \tag{10.25}$$

## 10.9    Design Considerations

Designing a communication system employing PAM with a block encoder entails making choices. We need to choose the PAM parameters $A$, $T_s$, and $\mathbf{g}$, and we need to choose a $(K, N)$ block encoder $enc(\cdot)$. These choices greatly influence the overall system characteristics such as the transmitted power, bandwidth, and the performance of the system in the presence of noise. To design a system well, we must understand the effect of the design choices on the overall system at three levels. At the first level we must understand *which* design parameters influence *which* overall system characteristics. At the second level we must understand *how* the design parameters influence the system. And at the third level we must understand how to choose the design parameters so as to optimize the system characteristics subject to the given constraints.

In this book we focus on the first two levels. The third requires tools from Information Theory and from Coding Theory that are beyond the scope of this book. Here we offer a preview of the first level. We thus briefly and informally explain which design choices influence which overall system properties.

To simplify the preview, we shall assume in this section that the time shifts of the pulse shape by integer multiples of the baud period are orthonormal. Consequently, we shall denote the pulse shape by $\phi$ and assume that (10.12) holds. We shall also assume that $k$ and $n$ tend to infinity as in the bi-infinite block mode discussed in Section 14.5.2. Roughly speaking this assumption is tantamount to the assumption that the system has been running since time $-\infty$ and that it will continue running until time $+\infty$.

Our discussion is extremely informal, and we apologize to the reader for discussing concepts that we have not yet defined. Readers who are aggravated by this practice may choose to skip this section; the issues will be revisited in Chapter 29 after everything has been defined and all the claims proved.

The key observation we wish to highlight is that, to a great extent,

> the choice of the block encoder $\mathrm{enc}(\cdot)$ can be decoupled from the choice of the pulse shape. The bandwidth and power spectral density depend hardly at all on $\mathrm{enc}(\cdot)$ and very much on the pulse shape, whereas the probability of error on the white Gaussian noise channel depends very much on $\mathrm{enc}(\cdot)$ and not at all on the pulse shape $\phi$.

This observation greatly simplifies the design problem because it means that, rather than optimizing over $\phi$ and $\mathrm{enc}(\cdot)$ jointly, we can choose each of them separately.

We next briefly discuss the different overall system characteristics and which design choices influence them.

**Data Rate:** The data rate $R_b$ that the system supports is determined by the baud period $T_s$ and by the rate $K/N$ of the encoder. It is given by

$$R_b = \frac{1}{T_s} \frac{K}{N} \left[ \frac{\mathrm{bit}}{\mathrm{sec}} \right].$$

**Power:** The transmitted power does not depend on the pulse shape $\phi$ (Theorem 14.5.2). It is determined by the amplitude $A$, the baud period $T_s$, and by the block encoder $\mathrm{enc}(\cdot)$. In fact, if the block encoder $\mathrm{enc}(\cdot)$ is such that when it is fed the data bits it produces zero-mean symbols that are uniformly distributed over the constellation, then the transmitted power is determined by $A$, $T_s$, and the second moment of the constellation only.

**Power Spectral Density:** If the block encoder $\mathrm{enc}(\cdot)$ is such that when it is fed the data bits it produces zero-mean and uncorrelated symbols of equal variance, then the power spectral density is determined by $A$, $T_s$, and $\phi$ only; it is unaffected by $\mathrm{enc}(\cdot)$ (Section 15.4).

**Bandwidth:** The bandwidth of the transmitted waveform is equal to the bandwidth of the pulse shape $\phi$ (Theorem 15.4.1). We will see in Chapter 11 that for the orthonormality (10.12) to hold, the bandwidth $W$ of the pulse shape must satisfy

$$W \geq \frac{1}{2T_s}.$$

In Chapter 11 we shall also see how to design $\phi$ so as to satisfy (10.12) and so as to have its bandwidth as close as we wish to $1/(2T_s)$.[2]

**Probability of Error:** It is a remarkable fact that the pulse shape $\phi$ does not affect the performance of the system on the additive white Gaussian noise channel. Performance is determined only by $A$, $T_s$, and the block encoder $\mathrm{enc}(\cdot)$ (Section 26.5.2).

---

[2]Information-theoretic considerations suggest that this is a good approach.

The preceding discussion focused on PAM, but many of the results also hold for Quadrature Amplitude Modulation, which is discussed in Chapters 16, 18, and 28.

## 10.10   Some Implementation Considerations

It is instructive to consider some of the issues related to the generation of a PAM signal

$$X(t) = A \sum_{\ell=1}^{n} X_\ell \, g(t - \ell T_s), \quad t \in \mathbb{R}. \tag{10.26}$$

Here we focus on delay, causality, and digital implementation.

### 10.10.1   Delay

To illustrate the delay issue in PAM, suppose that the pulse shape $g(\cdot)$ is strictly positive. In this case we note that, irrespective of which epoch $t' \in \mathbb{R}$ we consider, the calculation of $X(t')$ requires knowledge of the entire $n$-tuple $X_1, \ldots, X_n$. Since the sequence $X_1, \ldots, X_n$ cannot typically be determined in its entirety unless the entire sequence $D_1, \ldots, D_k$ is determined first, it follows that, when $g(\cdot)$ is strictly positive, the modulator cannot produce $X(t')$ before observing the entire data sequence $D_1, \ldots, D_k$. And this is true for any $t' \in \mathbb{R}$! Since in the back of our minds we think about $D_1, \ldots, D_k$ as the data bits that will be sent during the entire life of the system or, at least, from the moment it is turned on until it is shut off, it is unrealistic to expect the modulator to observe the entire sequence $D_1, \ldots, D_k$ before producing any input to the channel.

The engineering solution to this problem is to find some positive integer L such that, for all practical purposes, $g(t)$ is zero whenever $|t| > L T_s$, i.e.,

$$g(t) \approx 0, \quad |t| > L T_s. \tag{10.27}$$

In this case we have that, irrespective of $t' \in \mathbb{R}$, only $2L + 1$ terms (approximately) determine $X(t')$. Indeed, if $\kappa$ is an integer such that

$$\kappa T_s \leq t' < (\kappa + 1) T_s, \tag{10.28}$$

then

$$X(t') \approx A \sum_{\ell = \max\{1, \kappa - L\}}^{\kappa + L} X_\ell \, g(t - \ell T_s), \quad \kappa T_s \leq t' < (\kappa + 1) T_s, \tag{10.29}$$

where the sum is assumed to be zero if $\kappa + L < 1$.

Thus, if (10.27) holds, then the approximate calculation of $X(t')$ can be performed without knowledge of the entire sequence $X_1, \ldots, X_n$ and the modulator can start producing the waveform $X(\cdot)$ as soon as it knows $X_1, \ldots, X_L$.

## 10.10.2   Causality

The reader may object to the fact that, even if (10.27) holds, the signal $X(\cdot)$ may be nonzero at negative times. It might therefore seem as though the transmitter needs to transmit a signal before the system has been turned on and that, worse still, this signal depends on the data bits that will be fed to the system in the future when the system is turned on. But this is not really an issue. It all has to do with how we define the epoch $t = 0$, i.e., to what physical time instant does $t = 0$ correspond. We never said it corresponded to the instant when the system was turned on and, in fact, there is no reason to set the time origin at that time instant or at the "Big Bang." For example, we can set the time origin at $\mathsf{L}\mathsf{T_s}$ seconds-past-system-turn-on, and the problem disappears. Similarly, if the transmitted waveform depends on $X_1, \ldots, X_\mathsf{L}$, and if these real numbers can only be computed once the data bits $D_1, \ldots, D_\kappa$ have been fed to the encoder, then it would make sense to set the time origin to the moment at which the last of these $\kappa$ data bits has been fed to the encoder.

Some problems in Digital Communications that appear like tough causality problems end up being easily solved by time delays and the redefinition of the time origin. Others can be much harder. It is sometimes difficult for the novice to determine which causality problem is of the former type and which of the latter. As a rule of thumb, you should be extra cautious when the system contains feedback loops.

## 10.10.3   Digital Implementation

Even when all the symbols among $X_1, \ldots, X_n$ that are relevant for the calculation of $X(t')$ are known, the actual computation may be tricky, particularly if the formula describing the pulse shape is difficult to implement in hardware. In such cases one may opt for a digital implementation using look-up tables. The idea is to compute only *samples* of $X(\cdot)$ and to then interpolate using a digital-to-analog (D/A) converter and an anti-aliasing filter. The samples must be computed at a rate determined by the Sampling Theorem, i.e., at least once every $1/(2W)$ seconds, where $W$ is the bandwidth of the pulse shape.

The computation of the values of $X(\cdot)$ at its samples can be done by choosing L sufficiently large so that (10.27) holds and by then approximating the sum (10.26) for $t'$ satisfying (10.28) by the sum (10.29). The samples of this latter sum can be computed with a digital computer or—as is more common if the symbols take on a finite (and small) number of values—using a pre-programmed look-up table. The size of the look-up table thus depends on two parameters: the number of samples one needs to compute every $\mathsf{T_s}$ seconds (determined via the bandwidth of $g(\cdot)$ and the Sampling Theorem), and the number of addresses needed (as determined by L and by the constellation size).

## 10.11   Exercises

**Exercise 10.1 (Exploiting Orthogonality).** Let the energy-limited real signals $\phi_1$ and $\phi_2$ be orthogonal, and let $A^{(1)}$ and $A^{(2)}$ be positive constants. Let the waveform $\mathbf{X}$ be given by

$$\mathbf{X} = \left(A^{(1)}X^{(1)} + A^{(2)}X^{(2)}\right)\phi_1 + \left(A^{(1)}X^{(1)} - A^{(2)}X^{(2)}\right)\phi_2,$$

where $X^{(1)}$ and $X^{(2)}$ are unknown real numbers. How can you recover $X^{(1)}$ and $X^{(2)}$ from $\mathbf{X}$?

**Exercise 10.2 (More Orthogonality).** Extend Exercise 10.1 to the case where $\phi_1, \ldots \phi_\eta$ are orthonormal;

$$\mathbf{X} = \left(a^{(1,1)}A^{(1)}X^{(1)} + \cdots + a^{(\eta,1)}A^{(\eta)}X^{(\eta)}\right)\phi_1 + \cdots$$
$$+ \left(a^{(1,\eta)}A^{(1)}X^{(1)} + \cdots + a^{(\eta,\eta)}A^{(\eta)}X^{(\eta)}\right)\phi_\eta;$$

and where the real numbers $a^{(\iota,\nu)}$ for $\iota, \nu \in \{1,\ldots,\eta\}$ satisfy the orthogonality condition

$$\sum_{\nu=1}^{\eta} a^{(\iota,\nu)}a^{(\iota',\nu)} = \begin{cases} \eta & \text{if } \iota = \iota', \\ 0 & \text{if } \iota \neq \iota', \end{cases} \quad \iota, \iota' \in \{1,\ldots,\eta\}.$$

**Exercise 10.3 (A Constellation and its Second Moment).** What is the constellation corresponding to the $(1,3)$ binary-to-reals block encoder that maps 0 to $(+1,+2,+2)$ and maps 1 to $(-1,-2,-2)$? What is its second moment? Let the real symbols $(X_\ell, \ell \in \mathbb{Z})$ be generated from IID random bits $(D_j, j \in \mathbb{Z})$ in block mode using this block encoder. Compute

$$\lim_{L\to\infty} \frac{1}{2L+1} \sum_{\ell=-L}^{L} \mathsf{E}\big[X_\ell^2\big].$$

**Exercise 10.4 (Orthonormal Signal Representation).** Prove Theorem 10.5.1.

*Hint: Recall the Gram-Schmidt procedure.*

**Exercise 10.5 (Unbounded PAM Signal).** Consider the formal expression

$$X(t) = \sum_{\ell=-\infty}^{\infty} X_\ell \operatorname{sinc}\left(\frac{t}{\mathsf{T_s}} - \ell\right), \quad t \in \mathbb{R}.$$

(i) Show that even if the $X_\ell$'s can only take on the values $\pm 1$, the value of $X(\mathsf{T_s}/2)$ can be arbitrarily high. That is, find a sequence $\{x_\ell\}_{-\infty}^{\infty}$ such that $x_\ell \in \{+1,-1\}$ for every $\ell \in \mathbb{Z}$ and

$$\lim_{L\to\infty} \sum_{\ell=-L}^{L} \operatorname{sinc}\left(\frac{1}{2} - \ell\right) = \infty.$$

(ii) Suppose now that $\mathbf{g}\colon \mathbb{R} \to \mathbb{R}$ satisfies

$$|g(t)| \leq \frac{\beta}{1 + |t/\mathsf{T_s}|^{1+\alpha}}, \quad t \in \mathbb{R}$$

for some $\alpha, \beta > 0$. Show that if for some $\gamma > 0$ we have $|x_\ell| \leq \gamma$ for all $\ell \in \mathbb{Z}$, then the sum

$$\sum_{\ell=-\infty}^{\infty} x_\ell\, g(t - \ell \mathsf{T_s})$$

converges at every $t$ and is a bounded function of $t$.

**Exercise 10.6 (Etymology).** Let $\mathbf{g}$ be an integrable real signal. Express the frequency response of the matched filter for $\mathbf{g}$ in terms of the FT of $\mathbf{g}$. Repeat when $\mathbf{g}$ is a complex signal. Can you guess the origin of the term "Matched Filter"?

*Hint: Recall the notion of a "matched impedance."*

**Exercise 10.7 (Recovering the Symbols from a Filtered PAM Signal).** Let $X(\cdot)$ be the PAM signal (10.17), where $\mathsf{A} > 0$, and where $g(t)$ is zero for $|t| \geq \mathsf{T_s}/2$ and positive for $|t| < \mathsf{T_s}/2$.

  (i) Suppose that $X(\cdot)$ is fed to a filter of impulse response $\mathbf{h}\colon t \mapsto \mathrm{I}\{|t| \leq \mathsf{T_s}/2\}$. Is it true that for every $\ell \in \{1, \ldots, n\}$ one can recover $X_\ell$ from the filter's output at time $\ell\mathsf{T_s}$? If so, how?

  (ii) Suppose now that the filter's impulse response is $\mathbf{h}\colon t \mapsto \mathrm{I}\{-\mathsf{T_s}/2 \leq t \leq 3\mathsf{T_s}/4\}$. Can one always recoever $X_\ell$ from the filter's output at time $\ell\mathsf{T_s}$? Can one recover the sequence $(X_1, \ldots, X_n)$ from the $n$ samples of the filter's output at the times $\mathsf{T_s}, \ldots, n\mathsf{T_s}$?

**Exercise 10.8 (Continuous Phase Modulation).** In Continuous Phase Modulation (CPM) the symbols $(X_\ell)$ are mapped to the waveform

$$X(t) = \mathsf{A} \cos\left(2\pi f_c t + 2\pi h \sum_{\ell=-\infty}^{\infty} X_\ell\, q(t - \ell\mathsf{T_s})\right), \quad t \in \mathbb{R},$$

where $f_c, h > 0$ are constants and $\mathbf{q}$ is a mapping from $\mathbb{R}$ to $\mathbb{R}$. Is CPM a special case of linear modulation?

# Chapter 11

# Nyquist's Criterion

## 11.1 Introduction

In Section 10.7 we discussed the benefit of choosing the pulse shape $\phi$ in Pulse Amplitude Modulation so that its time shifts by integer multiples of the baud period $T_s$ be orthonormal. We saw that if the real transmitted signal is given by

$$X(t) = A \sum_{\ell=1}^{n} X_\ell \, \phi(t - \ell T_s), \quad t \in \mathbb{R},$$

where for all integers $\ell, \ell' \in \{1, \ldots, n\}$

$$\int_{-\infty}^{\infty} \phi(t - \ell T_s) \, \phi(t - \ell' T_s) \, \mathrm{d}t = I\{\ell = \ell'\},$$

then

$$X_\ell = \frac{1}{A} \int_{-\infty}^{\infty} X(t) \, \phi(t - \ell T_s) \, \mathrm{d}t, \quad \ell = 1, \ldots, n,$$

and all the inner products

$$\int_{-\infty}^{\infty} X(t) \, \phi(t - \ell T_s) \, \mathrm{d}t, \quad \ell = 1, \ldots, n$$

can be computed using one circuit by feeding the signal $X(\cdot)$ to a matched filter of impulse response $\overleftarrow{\phi}$ and sampling the output at the times $t = \ell T_s$, for $\ell = 1, \ldots, n$. (In the complex case the matched filter is of impulse response $\overleftarrow{\phi^*}$.)

In this chapter we shall address the design of and the limitations on signals that are orthogonal to their time-shifts. While our focus so far has been on real functions $\phi$, for reasons that will become apparent in Chapter 16 when we discuss Quadrature Amplitude Modulation, we prefer to generalize the discussion and allow $\phi$ to be complex. The main results of this chapter are Corollary 11.3.4 and Corollary 11.3.5.

An obvious way of choosing a signal $\phi$ that is orthogonal to its time shifts by nonzero integer multiples of $T_s$ is by choosing a pulse that is zero outside some interval of length $T_s$, say $[-T_s/2, T_s/2)$. This guarantees that the pulse and its time shifts by nonzero integer multiples of $T_s$ do not overlap in time and that they are thus orthogonal. But this choice limits us to pulses of infinite bandwidth, because no nonzero bandlimited signal can vanish outside a finite (time) interval (Theorem 6.8.2).

Fortunately, as we shall see, there exist signals that are orthogonal to their time shifts and that are also bandlimited. This does not contradict Theorem 6.8.2 because these signals are not time-limited. They are orthogonal to their time shifts in spite of overlapping with them in time.

Since we have in mind using the pulse to send a very large number of symbols $n$ (where $n$ corresponds to the number of symbols sent during the lifetime of the system) we shall strengthen the orthonormality requirement to

$$\int_{-\infty}^{\infty} \phi(t - \ell T_s)\, \phi^*(t - \ell' T_s)\, dt = I\{\ell = \ell'\}, \quad \text{for all integers } \ell, \ell' \qquad (11.1)$$

and not only to those $\ell, \ell'$ in $\{1, \ldots, n\}$. We shall refer to Condition (11.1) as saying that "the time shifts of $\phi$ by integer multiples of $T_s$ are orthonormal."

Condition (11.1) can also be phrased as a condition on $\phi$'s self-similarity function, which we introduce next.

## 11.2   The Self-Similarity Function of Energy-Limited Signals

We next introduce the **self-similarity function** of energy-limited signals. This term is not standard; more common in the literature is the term "autocorrelation function." I prefer "self-similarity function," which was proposed to me by Jim Massey, because it reduces the risk of confusion with the autocovariance function and the autocorrelation function of stochastic processes. There is nothing random in our current setup.

**Definition 11.2.1 (Self-Similarity Function).** *The **self-similarity function** $R_{vv}$ of an energy-limited signal $\mathbf{v} \in \mathcal{L}_2$ is defined as the mapping*

$$R_{vv} : \tau \mapsto \int_{-\infty}^{\infty} v(t + \tau)\, v^*(t)\, dt, \quad \tau \in \mathbb{R}. \qquad (11.2)$$

If $\mathbf{v}$ is real, then the self-similarity function has a nice pictorial interpretation: one plots the original signal and the result of shifting the signal by $\tau$ on the same graph, and one then takes the pointwise product and integrates over time.

The main properties of the self-similarity function are summarized in the following proposition.

**Proposition 11.2.2 (Properties of the Self-Similarity Function).** *Let $R_{vv}$ be the self-similarity function of some energy-limited signal $\mathbf{v} \in \mathcal{L}_2$.*

*(i)* **Value at zero:**

$$R_{vv}(0) = \int_{-\infty}^{\infty} |v(t)|^2 \, dt. \qquad (11.3)$$

*(ii) Maximum at zero:*

$$|\mathsf{R_{vv}}(\tau)| \le \mathsf{R_{vv}}(0), \quad \tau \in \mathbb{R}. \tag{11.4}$$

*(iii) Conjugate symmetry:*

$$\mathsf{R_{vv}}(-\tau) = \mathsf{R_{vv}^*}(\tau), \quad \tau \in \mathbb{R}. \tag{11.5}$$

*(iv) Integral representation:*

$$\mathsf{R_{vv}}(\tau) = \int_{-\infty}^{\infty} |\hat{v}(f)|^2 \, e^{i2\pi f \tau} \, df, \quad \tau \in \mathbb{R}, \tag{11.6}$$

where $\hat{\mathbf{v}}$ is the $L_2$-Fourier Transform of $\mathbf{v}$.

*(v)* **Uniform Continuity:** $\mathsf{R_{vv}}$ is uniformly continuous.

*(vi)* **Convolution Representation:**

$$\mathsf{R_{vv}}(\tau) = (\mathbf{v} \star \tilde{\mathbf{v}}^*)(\tau), \quad \tau \in \mathbb{R}. \tag{11.7}$$

**Proof.** Part (i) follows by substituting $\tau = 0$ in (11.2).

Part (ii) follows by noting that $\mathsf{R_{vv}}(\tau)$ is the inner product between the mapping $t \mapsto v(t+\tau)$ and the mapping $t \mapsto v(t)$; by the Cauchy-Schwarz Inequality; and by noting that both of the above mappings have the same energy, namely, the energy of $\mathbf{v}$:

$$|\mathsf{R_{vv}}(\tau)| = \left| \int_{-\infty}^{\infty} v(t+\tau) \, v^*(t) \, dt \right|$$
$$\le \left( \int_{-\infty}^{\infty} |v(t+\tau)|^2 \, dt \right)^{1/2} \left( \int_{-\infty}^{\infty} |v^*(t)|^2 \, dt \right)^{1/2}$$
$$= \|\mathbf{v}\|_2^2$$
$$= \mathsf{R_{vv}}(0), \quad \tau \in \mathbb{R}.$$

Part (iii) follows from the substitution $s \triangleq t + \tau$ in the following:

$$\mathsf{R_{vv}}(\tau) = \int_{-\infty}^{\infty} v(t+\tau) \, v^*(t) \, dt$$
$$= \int_{-\infty}^{\infty} v(s) \, v^*(s-\tau) \, ds$$
$$= \left( \int_{-\infty}^{\infty} v(s-\tau) \, v^*(s) \, ds \right)^*$$
$$= \mathsf{R_{vv}^*}(-\tau), \quad \tau \in \mathbb{R}.$$

Part (iv) follows from the representation of $\mathsf{R_{vv}}(\tau)$ as the inner product between the mapping $t \mapsto v(t+\tau)$ and the mapping $t \mapsto v(t)$; by Parseval's Theorem;

and by noting that the $L_2$-Fourier Transform of the mapping $t \mapsto v(t + \tau)$ is the (equivalence class of the) mapping $f \mapsto e^{\mathrm{i}2\pi f\tau}\, \hat{v}(f)$:

$$
\begin{aligned}
\mathsf{R_{vv}}(\tau) &= \int_{-\infty}^{\infty} v(t + \tau)\, v^*(t)\, \mathrm{d}t \\
&= \big\langle t \mapsto v(t + \tau), t \mapsto v(t) \big\rangle \\
&= \big\langle f \mapsto e^{\mathrm{i}2\pi f\tau}\, \hat{v}(f), f \mapsto \hat{v}(f) \big\rangle \\
&= \int_{-\infty}^{\infty} e^{\mathrm{i}2\pi f\tau}\, |\hat{v}(f)|^2\, \mathrm{d}f, \quad \tau \in \mathbb{R}.
\end{aligned}
$$

Part (v) follows from the integral representation of Part (iv) and from the integrability of the function $f \mapsto |\hat{v}(f)|^2$. See, for example, the proof of (Katznelson, 1976, Section VI, Theorem 1.2).

Part (vi) follows from the substitution $s \triangleq t + \tau$ and by rearranging terms:

$$
\begin{aligned}
\mathsf{R_{vv}}(\tau) &= \int_{-\infty}^{\infty} v(t + \tau)\, v^*(t)\, \mathrm{d}t \\
&= \int_{-\infty}^{\infty} v(s)\, v^*(s - \tau)\, \mathrm{d}s \\
&= \int_{-\infty}^{\infty} v(s)\, \overleftarrow{v}^*(\tau - s)\, \mathrm{d}s \\
&= (\mathbf{v} \star \overleftarrow{\mathbf{v}}^*)(\tau).
\end{aligned}
$$
$\qquad\qquad\qquad\qquad\qquad\qquad\qquad\qquad\qquad\qquad\qquad\qquad\qquad\qquad\qquad\square$

With the above definition we can restate the orthonormality condition (11.1) in terms of the self-similarity function $\mathsf{R}_{\phi\phi}$ of $\phi$:

**Proposition 11.2.3 (Shift-Orthonormality and Self-Similarity).** *If $\phi$ is energy-limited, then the shift-orthonormality condition*

$$
\int_{-\infty}^{\infty} \phi(t - \ell\mathsf{T_s})\, \phi^*(t - \ell'\mathsf{T_s})\, \mathrm{d}t = \mathrm{I}\{\ell = \ell'\}, \quad \ell, \ell' \in \mathbb{Z} \tag{11.8}
$$

*is equivalent to the condition*

$$
\mathsf{R}_{\phi\phi}(\ell\mathsf{T_s}) = \mathrm{I}\{\ell = 0\}, \quad \ell \in \mathbb{Z}. \tag{11.9}
$$

**Proof.** The proposition follows by substituting $s \triangleq t - \ell'\mathsf{T_s}$ in the LHS of (11.8) to obtain

$$
\begin{aligned}
\int_{-\infty}^{\infty} \phi(t - \ell\mathsf{T_s})\, \phi^*(t - \ell'\mathsf{T_s})\, \mathrm{d}t &= \int_{-\infty}^{\infty} \phi\big(s + (\ell' - \ell)\mathsf{T_s}\big)\, \phi^*(s)\, \mathrm{d}s \\
&= \mathsf{R}_{\phi\phi}\big((\ell' - \ell)\mathsf{T_s}\big).
\end{aligned}
$$
$\qquad\qquad\qquad\qquad\qquad\qquad\qquad\qquad\qquad\qquad\qquad\qquad\qquad\qquad\qquad\square$

At this point, Proposition 11.2.3 does not seem particularly helpful because Condition (11.9) is not easy to verify. But, as we shall see in the next section, this condition can be phrased very elegantly in the frequency domain.

## 11.3   Nyquist's Criterion

**Definition 11.3.1 (Nyquist Pulse).** *We say that a complex signal* $\mathbf{v} \colon \mathbb{R} \mapsto \mathbb{C}$ *is a Nyquist Pulse of parameter* $\mathsf{T_s}$ *if*

$$v(\ell \mathsf{T_s}) = \mathrm{I}\{\ell = 0\}, \quad \ell \in \mathbb{Z}. \tag{11.10}$$

**Theorem 11.3.2 (Nyquist's Criterion).** *Let* $\mathsf{T_s} > 0$ *be given, and let the signal* $v(\cdot)$ *be given by*

$$v(t) = \int_{-\infty}^{\infty} g(f)\, e^{\mathrm{i}2\pi ft}\, \mathrm{d}f, \quad t \in \mathbb{R}, \tag{11.11}$$

*for some integrable function* $\mathbf{g} \colon f \mapsto g(f)$. *Then* $v(\cdot)$ *is a Nyquist Pulse of parameter* $\mathsf{T_s}$ *if, and only if,*

$$\lim_{\mathsf{J} \to \infty} \int_{-1/(2\mathsf{T_s})}^{1/(2\mathsf{T_s})} \left| \mathsf{T_s} - \sum_{j=-\mathsf{J}}^{\mathsf{J}} g\!\left(f + \frac{j}{\mathsf{T_s}}\right) \right| \mathrm{d}f = 0. \tag{11.12}$$

**Note 11.3.3.** Condition (11.12) is sometimes written imprecisely[1] in the form

$$\sum_{j=-\infty}^{\infty} g\!\left(f + \frac{j}{\mathsf{T_s}}\right) = \mathsf{T_s}, \quad -\frac{1}{2\mathsf{T_s}} \le f \le \frac{1}{2\mathsf{T_s}}, \tag{11.13}$$

or, in view of the periodicity of the LHS of (11.13), as

$$\sum_{j=-\infty}^{\infty} g\!\left(f + \frac{j}{\mathsf{T_s}}\right) = \mathsf{T_s}, \quad f \in \mathbb{R}. \tag{11.14}$$

Neither form is mathematically precise.

**Proof.** We will show that $v(-\ell \mathsf{T_s})$ is the $\ell$-th Fourier Series Coefficient of the function[2]

$$\frac{1}{\sqrt{\mathsf{T_s}}} \sum_{j=-\infty}^{\infty} g\!\left(f + \frac{j}{\mathsf{T_s}}\right), \quad -\frac{1}{2\mathsf{T_s}} \le f \le \frac{1}{2\mathsf{T_s}}. \tag{11.15}$$

It will then follow that the condition that $\mathbf{v}$ is a Nyquist Pulse of parameter $\mathsf{T_s}$ is equivalent to the condition that the function in (11.15) has Fourier Series Coefficients that are all zero except for the zeroth coefficient, which is one. The theorem will then follow by noting that a function is indistinguishable from a constant if, and only if, all but its zeroth Fourier Series Coefficient are zero. (This can be proved by applying Theorem A.2.3 with $\mathbf{g}_1$ chosen as the constant function.) The

---

[1] There is no guarantee that the sum converges at every frequency $f$.

[2] Since, by hypothesis, $\mathbf{g}$ is integrable, it follows that the sum in (11.15) converges in the $\mathcal{L}_1$ sense, i.e., that there exists some integrable function $\mathbf{s}_\infty$ such that

$$\lim_{\mathsf{J} \to \infty} \int_{-1/(2\mathsf{T_s})}^{1/(2\mathsf{T_s})} \left| s_\infty(f) - \sum_{j=-\mathsf{J}}^{\mathsf{J}} g(f + \frac{j}{\mathsf{T_s}}) \right| \mathrm{d}f = 0.$$

By writing $\sum_{j=-\infty}^{\infty} g(f + \frac{j}{\mathsf{T_s}})$ we are referring to this function $\mathbf{s}_\infty$.

value of the constant can be computed from the zeroth Fourier Series Coefficient. To conclude the proof we thus need to relate $v(-\ell T_s)$ to the $\ell$-th Fourier Series Coefficient of the function in (11.15). The calculation is straightforward: for every integer $\ell$,

$$
\begin{aligned}
v(-\ell T_s) &= \int_{-\infty}^{\infty} g(f)\, e^{-i2\pi f \ell T_s}\, df \\
&= \sum_{j=-\infty}^{\infty} \int_{\frac{j}{T_s} - \frac{1}{2T_s}}^{\frac{j}{T_s} + \frac{1}{2T_s}} g(f)\, e^{-i2\pi f \ell T_s}\, df \\
&= \sum_{j=-\infty}^{\infty} \int_{-\frac{1}{2T_s}}^{\frac{1}{2T_s}} g\!\left(\tilde{f} + \frac{j}{T_s}\right) e^{-i2\pi\left(\tilde{f} + \frac{j}{T_s}\right)\ell T_s}\, d\tilde{f} \\
&= \sum_{j=-\infty}^{\infty} \int_{-\frac{1}{2T_s}}^{\frac{1}{2T_s}} g\!\left(\tilde{f} + \frac{j}{T_s}\right) e^{-i2\pi \tilde{f} \ell T_s}\, d\tilde{f} \\
&= \int_{-\frac{1}{2T_s}}^{\frac{1}{2T_s}} \sum_{j=-\infty}^{\infty} g\!\left(\tilde{f} + \frac{j}{T_s}\right) e^{-i2\pi \tilde{f} \ell T_s}\, d\tilde{f} \\
&= \int_{-\frac{1}{2T_s}}^{\frac{1}{2T_s}} \left(\frac{1}{\sqrt{T_s}} \sum_{j=-\infty}^{\infty} g\!\left(\tilde{f} + \frac{j}{T_s}\right)\right) \sqrt{T_s}\, e^{-i2\pi \tilde{f} \ell T_s}\, d\tilde{f}, \qquad (11.16)
\end{aligned}
$$

which is the $\ell$-th Fourier Series Coefficient of the function in (11.15). Here the first equality follows by substituting $-\ell T_s$ for $t$ in (11.11); the second by partitioning the region of integration into intervals of length $\frac{1}{T_s}$; the third by the change of variable $\tilde{f} \triangleq f - \frac{j}{T_s}$; the fourth by the periodicity of the complex exponentials; the fifth by Fubini's Theorem, which allows us to swap the order summation and integration; and the final equality by multiplying and dividing by $\sqrt{T_s}$.                               $\square$

An example of a function $f \mapsto g(f)$ satisfying (11.12) is plotted in Figure 11.1.

**Corollary 11.3.4 (Characterization of Shift-Orthonormal Pulses).** *Let $\phi \colon \mathbb{R} \mapsto \mathbb{C}$ be energy-limited and let $T_s$ be positive. Then the condition*

$$
\int_{-\infty}^{\infty} \phi(t - \ell T_s)\, \phi^*(t - \ell' T_s)\, dt = \mathrm{I}\{\ell = \ell'\}, \quad \ell, \ell' \in \mathbb{Z} \qquad (11.17)
$$

*is equivalent to the condition*

$$
\sum_{j=-\infty}^{\infty} \left| \hat{\phi}\!\left(f + \frac{j}{T_s}\right) \right|^2 \equiv T_s, \qquad (11.18)
$$

*i.e., to the condition that the set of frequencies $f \in \mathbb{R}$ for which the LHS of (11.18) is not equal to $T_s$ is of Lebesgue measure zero.*[3]

---

[3]It is a simple technical matter to verify that the question as to whether or not (11.18) is satisfied outside a set of frequencies of Lebesgue measure zero does not depend on which element in the equivalence class of the $L_2$-Fourier Transform of $\phi$ is considered.

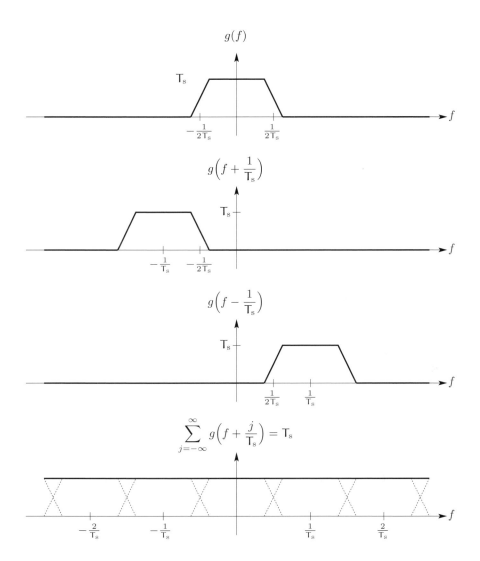

**Figure 11.1:** A function $g(\cdot)$ satisfying (11.12).

**Proof.** By Proposition 11.2.3, Condition (11.17) can be equivalently expressed in terms of the self-similarity function as

$$R_{\phi\phi}(mT_s) = I\{m = 0\}, \quad m \in \mathbb{Z}. \tag{11.19}$$

The result now follows from the integral representation of the self-similarity function $R_{\phi\phi}$ (Proposition 11.2.2 (iv)) and from Theorem 11.3.2 (with the additional simplification that for every $j \in \mathbb{Z}$ the function $f \mapsto |\hat{\phi}(f + \frac{j}{T_s})|^2$ is nonnegative, so the sum on the LHS of (11.18) converges (possibly to $+\infty$) for every $f \in \mathbb{R}$). $\qquad\square$

An extremely important consequence of Corollary 11.3.4 is the following corollary about the minimum bandwidth of a pulse $\phi$ satisfying the orthonormality condition (11.1).

**Corollary 11.3.5 (Minimum Bandwidth of Shift-Orthonormal Pulses).** *Let* $T_s > 0$ *be fixed, and let* $\phi$ *be an energy-limited signal that is bandlimited to* $W$ *Hz. If the time shifts of* $\phi$ *by integer multiples of* $T_s$ *are orthonormal, then*

$$W \geq \frac{1}{2T_s}. \tag{11.20}$$

*Equality is achieved if*

$$|\hat{\phi}(f)| = \sqrt{T_s}\, I\left\{|f| \leq \frac{1}{2T_s}\right\}, \quad f \in \mathbb{R} \tag{11.21}$$

*and, in particular, by the* $\mathrm{sinc}(\cdot)$ *pulse*

$$\phi(t) = \frac{1}{\sqrt{T_s}}\, \mathrm{sinc}\left(\frac{t}{T_s}\right), \quad t \in \mathbb{R} \tag{11.22}$$

*or any time-shift thereof.*

**Proof.** Figure 11.2 illustrates why $\phi$ cannot satisfy (11.18) if (11.20) is violated. The figure should also convince you of the conditions for equality in (11.20).

For the algebraically-inclined readers we prove the corollary by showing that if $W \leq 1/(2T_s)$, then (11.18) can only be satisfied if $\phi$ satisfies (11.21) (outside a set of frequencies of Lebesgue measure zero).[4] To see this, consider the sum

$$\sum_{j=-\infty}^{\infty} \left|\hat{\phi}\left(f + \frac{j}{T_s}\right)\right|^2 \tag{11.23}$$

for frequencies $f$ in the open interval $\left(-\frac{1}{2T_s}, +\frac{1}{2T_s}\right)$. The key observation in the proof is that for frequencies in this open interval, if $W \leq 1/(2T_s)$, then all the terms in the sum (11.23) are zero, except for the $j = 0$ term. That is,

$$\sum_{j=-\infty}^{\infty} \left|\hat{\phi}\left(f + \frac{j}{T_s}\right)\right|^2 = |\hat{\phi}(f)|^2, \quad \left(W \leq \frac{1}{2T_s}, f \in \left(-\frac{1}{2T_s}, +\frac{1}{2T_s}\right)\right). \tag{11.24}$$

---

[4]In the remainder of the proof we assume that $\hat{\phi}(f)$ is zero for frequencies $f$ satisfying $|f| > W$. The proof can be easily adjusted to account for the fact that, for frequencies $|f| > W$, it is possible that $\hat{\phi}(\cdot)$ be nonzero on a set of Lebesgue measure zero.

To convince yourself of (11.24), consider, for example, the term corresponding to $j = 1$, namely, $|\hat{\phi}(f + 1/\mathsf{T_s})|^2$. By the definition of bandwidth, it is zero whenever $|f + 1/\mathsf{T_s}| > W$, i.e., whenever $f > -1/\mathsf{T_s} + W$ or $f < -1/\mathsf{T_s} - W$. Since the former category $f > -1/\mathsf{T_s} + W$ includes—by our assumption that $W \leq 1/(2\mathsf{T_s})$— all frequencies $f > -1/(2\mathsf{T_s})$, we conclude that the term corresponding to $j = 1$ is zero for all the frequencies $f$ in the open interval $\left(-\frac{1}{2\mathsf{T_s}}, +\frac{1}{2\mathsf{T_s}}\right)$. More generally, the $j$-th term $|\hat{\phi}(f + j/\mathsf{T_s})|^2$ is zero for all frequencies $f$ satisfying the condition $|f + j/\mathsf{T_s}| > W$, a condition that is satisfied—assuming $j \neq 0$ and $W \leq 1/(2\mathsf{T_s})$—by the frequencies in the open interval that is of interest to us $\left(-\frac{1}{2\mathsf{T_s}}, +\frac{1}{2\mathsf{T_s}}\right)$.

For $W \leq 1/(2\mathsf{T_s})$ we thus obtain from (11.24) that the condition (11.18) implies (11.21), and, in particular, that $W = 1/(2\mathsf{T_s})$. $\qquad\square$

Functions satisfying (11.21) are seldom used in digital communication because they typically decay like $1/t$ so that even if the transmitted symbols $X_\ell$ are bounded, the signal $X(t)$ may take on very high values (albeit quite rarely). Consequently, the pulses $\phi$ that are used in practice have a larger bandwidth than $1/(2\mathsf{T_s})$.

This leads to the following definition.

**Definition 11.3.6 (Excess Bandwidth).** *The **excess bandwidth** in percent of a signal $\phi$ relative to $\mathsf{T_s} > 0$ is defined as*

$$100\% \left(\frac{\text{bandwidth of } \phi}{1/(2\mathsf{T_s})} - 1\right). \tag{11.25}$$

The following corollary to Corollary 11.3.4 is useful for the understanding of *real* signals of excess bandwidth smaller than 100%.

**Corollary 11.3.7 (Band-Edge Symmetry).** *Let $\mathsf{T_s}$ be positive, and let $\phi$ be a real energy-limited signal that is bandlimited to $W$ Hz, where $W < 1/\mathsf{T_s}$ so $\phi$ is of excess bandwidth smaller than 100%. Then the time shifts of $\phi$ by integer multiples of $\mathsf{T_s}$ are orthonormal if, and only if, $f \mapsto |\hat{\phi}(f)|^2$ satisfies the band-edge symmetry condition[5]*

$$\left|\hat{\phi}\left(\frac{1}{2\mathsf{T_s}} - f\right)\right|^2 + \left|\hat{\phi}\left(\frac{1}{2\mathsf{T_s}} + f\right)\right|^2 \equiv \mathsf{T_s}, \quad 0 < f \leq \frac{1}{2\mathsf{T_s}}. \tag{11.26}$$

**Proof.** We first note that, since we have assumed that $W < 1/\mathsf{T_s}$, only the terms corresponding to $j = -1$, $j = 0$, and $j = 1$ contribute to the sum on the LHS of (11.18) for $f \in \left(-\frac{1}{2\mathsf{T_s}}, +\frac{1}{2\mathsf{T_s}}\right)$. Moreover, since $\phi$ is by hypothesis real, it follows that $|\hat{\phi}(-f)| = |\hat{\phi}(f)|$, so the sum on the LHS of (11.18) is a symmetric function of $f$. Thus, the sum is equal to $\mathsf{T_s}$ on the interval $\left(-\frac{1}{2\mathsf{T_s}}, +\frac{1}{2\mathsf{T_s}}\right)$ if, and only if, it is equal to $\mathsf{T_s}$ on the interval $\left[0, +\frac{1}{2\mathsf{T_s}}\right)$. For frequencies in this shorter interval only two terms in the sum contribute: those corresponding to $j = 0$ and $j = -1$. We

---

[5]Condition (11.26) should be understood to indicate that the LHS and RHS of (11.26) are equal for all frequencies $0 \leq f \leq 1/(2\mathsf{T_s})$ outside a set of Lebesgue measure zero. Again, we ignore this issue in the proof and assume that $\hat{\phi}(f)$ is zero for all $|f| > W$.

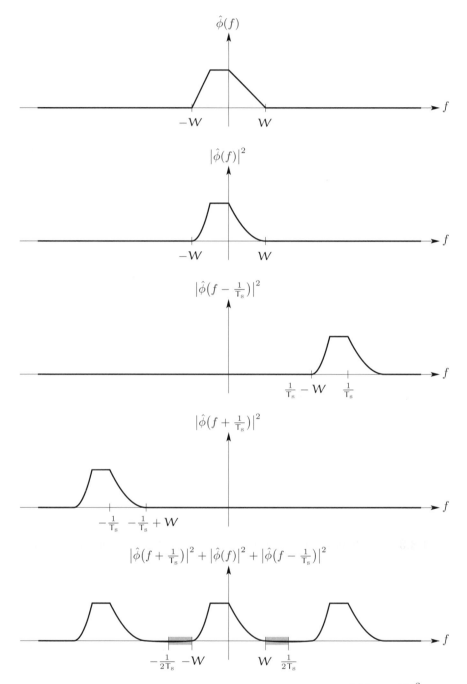

**Figure 11.2:** If $W < 1/(2T_s)$, then all the terms of the form $\left|\hat{\phi}\left(f + \frac{j}{T_s}\right)\right|^2$ are zero over the shaded frequencies $W < |f| < 1/(2T_s)$. Thus, for $W < 1/(2T_s)$ the sum $\sum_{j=-\infty}^{\infty}\left|\hat{\phi}\left(f + \frac{j}{T_s}\right)\right|^2$ cannot be equal to $T_s$ at any of the shaded frequencies.

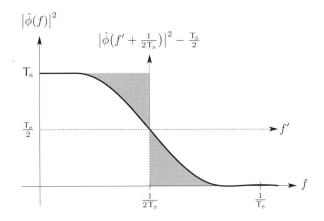

**Figure 11.3:** An example of a choice for $|\hat{\phi}(\cdot)|^2$ satisfying the band-edge symmetry condition (11.26).

thus conclude that, for real signals of excess bandwidth smaller than 100%, the condition (11.18) is equivalent to the condition

$$\left|\hat{\phi}(f)\right|^2 + \left|\hat{\phi}(f - 1/\mathsf{T_s})\right|^2 \equiv \mathsf{T_s}, \quad 0 \le f < \frac{1}{2\mathsf{T_s}}.$$

Substituting $f' \triangleq \frac{1}{2\mathsf{T_s}} - f$ in this condition leads to the condition

$$\left|\hat{\phi}\left(\frac{1}{2\mathsf{T_s}} - f'\right)\right|^2 + \left|\hat{\phi}\left(-f' - \frac{1}{2\mathsf{T_s}}\right)\right|^2 \equiv \mathsf{T_s}, \quad 0 < f' \le \frac{1}{2\mathsf{T_s}},$$

which, in view of the symmetry of $|\hat{\phi}(\cdot)|$, is equivalent to

$$\left|\hat{\phi}\left(\frac{1}{2\mathsf{T_s}} - f'\right)\right|^2 + \left|\hat{\phi}\left(f' + \frac{1}{2\mathsf{T_s}}\right)\right|^2 \equiv \mathsf{T_s}, \quad 0 < f' \le \frac{1}{2\mathsf{T_s}},$$

i.e., to (11.26).                                                                       $\square$

**Note 11.3.8.** The band-edge symmetry condition (11.26) has a nice geometric interpretation. This is best seen by rewriting the condition in the form

$$\underbrace{\left|\hat{\phi}\left(\frac{1}{2\mathsf{T_s}} - f'\right)\right|^2 - \frac{\mathsf{T_s}}{2}}_{=\tilde{g}(-f')} = -\underbrace{\left(\left|\hat{\phi}\left(\frac{1}{2\mathsf{T_s}} + f'\right)\right|^2 - \frac{\mathsf{T_s}}{2}\right)}_{=\tilde{g}(f')}, \quad 0 < f' \le \frac{1}{2\mathsf{T_s}}, \qquad (11.27)$$

which demonstrates that the band-edge condition is equivalent to the condition that the plot of $f \mapsto |\hat{\phi}(f)|^2$ in the interval $0 < f < 1/\mathsf{T_s}$ be invariant with respect to a 180°-rotation around the point $\left(\frac{1}{2\mathsf{T_s}}, \frac{\mathsf{T_s}}{2}\right)$. In other words, the function $\tilde{\mathbf{g}}\colon f' \mapsto \left|\hat{\phi}\left(\frac{1}{2\mathsf{T_s}} + f'\right)\right|^2 - \frac{\mathsf{T_s}}{2}$ should be anti-symmetric for $0 < f' \le \frac{1}{2\mathsf{T_s}}$. I.e., it should satisfy

$$\tilde{g}(-f') = -\tilde{g}(f'), \quad 0 < f' \le \frac{1}{2\mathsf{T_s}}.$$

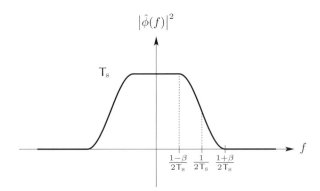

**Figure 11.4:** A plot of $f \mapsto |\hat{\phi}(f)|^2$ as given in (11.30) with $\beta = 0.5$.

Figure 11.3 is a plot over the interval $[0, 1/\mathsf{T}_\mathrm{s})$ of a mapping $f \mapsto |\hat{\phi}(f)|^2$ that satisfies the band-edge symmetry condition (11.26).

A popular choice of $\phi$ is based on the *raised-cosine* family of functions. For every $0 < \beta \le 1$ and every $\mathsf{T}_\mathrm{s} > 0$, the raised-cosine function is given by the mapping

$$
f \mapsto \begin{cases}
\mathsf{T}_\mathrm{s} & \text{if } 0 \le |f| \le \frac{1-\beta}{2\mathsf{T}_\mathrm{s}}, \\
\frac{\mathsf{T}_\mathrm{s}}{2}\left(1 + \cos\left(\frac{\pi \mathsf{T}_\mathrm{s}}{\beta}(|f| - \frac{1-\beta}{2\mathsf{T}_\mathrm{s}})\right)\right) & \text{if } \frac{1-\beta}{2\mathsf{T}_\mathrm{s}} < |f| \le \frac{1+\beta}{2\mathsf{T}_\mathrm{s}}, \\
0 & \text{if } |f| > \frac{1+\beta}{2\mathsf{T}_\mathrm{s}}.
\end{cases} \tag{11.28}
$$

Choosing $\phi$ so that its Fourier Transform is the square root of the raised-cosine mapping (11.28)

$$
\hat{\phi}(f) = \begin{cases}
\sqrt{\mathsf{T}_\mathrm{s}} & \text{if } 0 \le |f| \le \frac{1-\beta}{2\mathsf{T}_\mathrm{s}}, \\
\sqrt{\frac{\mathsf{T}_\mathrm{s}}{2}}\sqrt{1 + \cos\left(\frac{\pi \mathsf{T}_\mathrm{s}}{\beta}(|f| - \frac{1-\beta}{2\mathsf{T}_\mathrm{s}})\right)} & \text{if } \frac{1-\beta}{2\mathsf{T}_\mathrm{s}} < |f| \le \frac{1+\beta}{2\mathsf{T}_\mathrm{s}}, \\
0 & \text{if } |f| > \frac{1+\beta}{2\mathsf{T}_\mathrm{s}},
\end{cases} \tag{11.29}
$$

results in $\phi$ being real with

$$
|\hat{\phi}(f)|^2 = \begin{cases}
\mathsf{T}_\mathrm{s} & \text{if } 0 \le |f| \le \frac{1-\beta}{2\mathsf{T}_\mathrm{s}}, \\
\frac{\mathsf{T}_\mathrm{s}}{2}\left(1 + \cos\left(\frac{\pi \mathsf{T}_\mathrm{s}}{\beta}(|f| - \frac{1-\beta}{2\mathsf{T}_\mathrm{s}})\right)\right) & \text{if } \frac{1-\beta}{2\mathsf{T}_\mathrm{s}} < |f| \le \frac{1+\beta}{2\mathsf{T}_\mathrm{s}}, \\
0 & \text{if } |f| > \frac{1+\beta}{2\mathsf{T}_\mathrm{s}},
\end{cases} \tag{11.30}
$$

as depicted in Figure 11.4 for $\beta = 0.5$.

Using (11.29) and using the band-edge symmetry criterion (Corollary 11.3.7), it can be readily verified that the time shifts of $\phi$ by integer multiples of $\mathsf{T}_\mathrm{s}$ are orthonormal. Moreover, by (11.29), $\phi$ is bandlimited to $(1 + \beta)/(2\mathsf{T}_\mathrm{s})$ Hz. It is thus of excess bandwidth $\beta \times 100\%$. For every $0 < \beta \le 1$ we have thus found a pulse $\phi$ of excess bandwidth $\beta \times 100\%$ whose time shifts by integer multiples of $\mathsf{T}_\mathrm{s}$ are orthonormal.

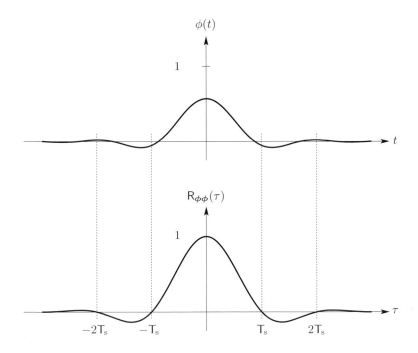

**Figure 11.5:** The pulse $\phi(\cdot)$ of (11.31) with $\beta = 0.5$ and its self-similarity function $R_{\phi\phi}(\cdot)$ of (11.32).

In the time domain

$$\phi(t) = \frac{2\beta}{\pi\sqrt{T_s}} \frac{\cos\left((1+\beta)\pi\frac{t}{T_s}\right) + \frac{\sin\left((1-\beta)\pi\frac{t}{T_s}\right)}{4\beta\frac{t}{T_s}}}{1 - (4\beta\frac{t}{T_s})^2}, \quad t \in \mathbb{R}, \tag{11.31}$$

with corresponding self-similarity function

$$R_{\phi\phi}(\tau) = \mathrm{sinc}\left(\frac{\tau}{T_s}\right)\frac{\cos(\pi\beta\tau/T_s)}{1 - 4\beta^2\tau^2/T_s^2}, \quad \tau \in \mathbb{R}. \tag{11.32}$$

The pulse $\phi$ of (11.31) is plotted in Figure 11.5 (top) for $\beta = 0.5$. Its self-similarity function (11.32) is plotted in the same figure (bottom). That the time shifts of $\phi$ by integer multiples of $T_s$ are orthonormal can be verified again by observing that $R_{\phi\phi}$ as given in (11.32) satisfies $R_{\phi\phi}(\ell T_s) = I\{\ell = 0\}$ for all $\ell \in \mathbb{Z}$.

Notice also that if $\phi(\cdot)$ is chosen as in (11.31), then for all $0 < \beta \leq 1$, the pulse $\phi(\cdot)$ decays like $1/t^2$. This decay property combined with the fact that the infinite sum $\sum_{\nu=1}^{\infty} \nu^{-2}$ converges (Rudin, 1976, Chapter 3, Theorem 3.28) will prove useful in Section 14.3 when we discuss the power in PAM.

## 11.4   The Self-Similarity Function of Integrable Signals

This section is a bit technical and can be omitted at first reading. In it we define the self-similarity function for integrable signals that are not necessarily energy-limited, and we then compute the Fourier Transform of the so-defined self-similarity function.

Recall that a Lebesgue measurable complex signal $\mathbf{v}\colon \mathbb{R} \to \mathbb{C}$ is integrable if $\int_{-\infty}^{\infty} |v(t)|\,\mathrm{d}t < \infty$ and that the class of integrable signal is denoted by $\mathcal{L}_1$. For such signals there may be $\tau$'s for which the integral in (11.2) is undefined. For example, if $\mathbf{v}$ is not energy-limited, then the integral in (11.2) will be infinite at $\tau = 0$. Nevertheless, we can discuss the self-similarity function of such signals by adopting the convolution representation of Proposition 11.2.2 as the definition. We thus define the self-similarity function $\mathsf{R_{vv}}$ of an integrable signal $\mathbf{v} \in \mathcal{L}_1$ as

$$\mathsf{R_{vv}} \triangleq \mathbf{v} \star \overleftarrow{\mathbf{v}}^*, \quad \mathbf{v} \in \mathcal{L}_1, \tag{11.33}$$

but we need some clarification. Since $\mathbf{v}$ is integrable, and since this implies that its reflected image $\overleftarrow{\mathbf{v}}$ is also integrable, it follows that the convolution in (11.33) is a convolution between two integrable signals. As such, we are guaranteed by the discussion leading to (5.9) that the integral

$$\int_{-\infty}^{\infty} v(\sigma)\,\overleftarrow{v}^*(\tau - \sigma)\,\mathrm{d}\sigma = \int_{-\infty}^{\infty} v(t + \tau)\,v^*(t)\,\mathrm{d}t$$

is defined for all $\tau$'s outside a set of Lebesgue measure zero. (This set of Lebesgue measure zero will include the point $\tau = 0$ if $\mathbf{v}$ is not of finite energy.) For $\tau$'s inside this set of measure zero we define the self-similarity function to be zero. The value zero is quite arbitrary because, irrespective of the value we choose for such $\tau$'s, we are guaranteed by (5.9) that the so-defined self-similarity function $\mathsf{R_{vv}}$ is integrable

$$\int_{-\infty}^{\infty} \left|\mathsf{R_{vv}}(\tau)\right|\,\mathrm{d}\tau \le \|\mathbf{v}\|_1^2, \quad \mathbf{v} \in \mathcal{L}_1, \tag{11.34}$$

and that its $\mathcal{L}_1$-Fourier Transform is given by the product of the $\mathcal{L}_1$-Fourier Transform of $\mathbf{v}$ and the $\mathcal{L}_1$-Fourier Transform of $\overleftarrow{\mathbf{v}}^*$, i.e.,

$$\boxed{\hat{\mathsf{R}}_{\mathsf{vv}}(f) = |\hat{v}(f)|^2, \quad \left(\mathbf{v} \in \mathcal{L}_1,\ f \in \mathbb{R}\right).} \tag{11.35}$$

## 11.5   Exercises

**Exercise 11.1 (Passband Signaling).** Let $f_0, \mathsf{T_s} > 0$ be fixed.

  (i) Show that a signal $\mathbf{x}$ is a Nyquist Pulse of parameter $\mathsf{T_s}$ if, and only if, the signal $t \mapsto e^{\mathrm{i}2\pi f_0 t}\,x(t)$ is such a pulse.

 (ii) Show that if $\mathbf{x}$ is a Nyquist Pulse of parameter $\mathsf{T_s}$, then so is $t \mapsto \cos(2\pi f_0 t)\,x(t)$.

(iii) If $t \mapsto \cos(2\pi f_0 t)\,x(t)$ is a Nyquist Pulse of parameter $\mathsf{T_s}$, must $\mathbf{x}$ also be one?

**Exercise 11.2 (The Self-Similarity Function of a Delayed Signal).** Let $\mathbf{u}$ be an energy-limited signal, and let the signal $\mathbf{v}$ be given by $\mathbf{v} \colon t \mapsto u(t - t_0)$. Express the self-similarity function of $\mathbf{v}$ in terms of the self-similarity of $\mathbf{u}$ and $t_0$.

**Exercise 11.3 (The Self-Similarity Function of a Frequency Shifted Signal).** Let $\mathbf{u}$ be an energy-limited complex signal, and let the signal $\mathbf{v}$ be given by $\mathbf{v} \colon t \mapsto u(t) \, e^{i 2\pi f_0 t}$ for some $f_0 \in \mathbb{R}$. Express the self-similarity function of $\mathbf{v}$ in terms of $f_0$ and the self-similarity function of $\mathbf{u}$.

**Exercise 11.4 (A Self-Similarity Function).** Compute and plot the self-similarity function of the signal $t \mapsto A \left( 1 - |t|/\mathsf{T} \right) \mathrm{I}\{|t| \leq \mathsf{T}\}$.

**Exercise 11.5 (Symmetry of the FT of the Self-Similarity Function of a Real Signal).** Show that if $\phi$ is an integrable real signal, then the FT of its self-similarity function is symmetric:

$$\left( \hat{R}_{\phi\phi}(f) = \hat{R}_{\phi\phi}(-f), \quad f \in \mathbb{R} \right), \quad \phi \in \mathcal{L}_1 \text{ is real.}$$

**Exercise 11.6 (The Self-Similarity Function is Positive Definite).** Show that if $\mathbf{v}$ is an energy-limited signal, $n$ is a positive integer, $\alpha_1, \ldots, \alpha_n \in \mathbb{C}$, and $t_1, \ldots, t_n \in \mathbb{R}$, then

$$\sum_{j=1}^{n} \sum_{\ell=1}^{n} \alpha_j \alpha_\ell^* \, R_{\mathbf{vv}}(t_j - t_\ell) \geq 0.$$

*Hint: Compute the energy in the signal $t \mapsto \sum_{j=1}^{n} \alpha_j \, v(t + t_j)$.*

**Exercise 11.7 (Relaxing the Orthonormality Condition).** What is the minimal bandwidth of an energy-limited signal whose time shifts by even multiples of $\mathsf{T}_s$ are orthonormal? What is the minimal bandwidth of an energy-limited signal whose time shifts by odd multiples of $\mathsf{T}_s$ are orthonormal?

**Exercise 11.8 (A Specific Signal).** Let $\mathbf{p}$ be the complex energy-limited bandlimited signal whose FT $\hat{p}$ is given by

$$\hat{p}(f) = \mathsf{T}_s \left( 1 - |\mathsf{T}_s f - 1| \right) \mathrm{I}\left\{ 0 \leq f \leq \frac{2}{\mathsf{T}_s} \right\}, \quad f \in \mathbb{R}.$$

(i) Plot $\hat{p}(\cdot)$.

(ii) Is $p(\cdot)$ a Nyquist Pulse of parameter $\mathsf{T}_s$?

(iii) Is the real part of $p(\cdot)$ a Nyquist Pulse of parameter $\mathsf{T}_s$?

(iv) What about the imaginary part of $p(\cdot)$?

**Exercise 11.9 (Nyquist's Third Criterion).** We say that an energy-limited signal $\psi(\cdot)$ satisfies Nyquist's Third Criterion if

$$\int_{(2\nu-1)\mathsf{T}_s/2}^{(2\nu+1)\mathsf{T}_s/2} \psi(t) \, \mathrm{d}t = \begin{cases} 1 & \text{if } \nu = 0, \\ 0 & \text{if } \nu \in \mathbb{Z} \setminus \{0\}. \end{cases} \tag{11.36}$$

(i) Express the LHS of (11.36) as an inner product between $\psi$ and some function $\mathbf{g}_\nu$.

(ii) Show that (11.36) is equivalent to

$$T_s \int_{-\infty}^{\infty} \hat{\psi}(f) \, e^{-i2\pi f \nu T_s} \, \mathrm{sinc}(T_s f) \, \mathrm{d}f = \begin{cases} 1 & \text{if } \nu = 0, \\ 0 & \text{if } \nu \in \mathbb{Z} \setminus \{0\}. \end{cases}$$

(iii) Show that, loosely speaking, $\psi$ satisfies Nyquist's Third Criterion if, and only if,

$$\sum_{j=-\infty}^{\infty} \hat{\psi}\left(f - \frac{j}{T_s}\right) \mathrm{sinc}(T_s f - j)$$

is indistinguishable from the all-one function. More precisely, if and only if,

$$\lim_{J \to \infty} \int_{-\frac{1}{2T_s}}^{\frac{1}{2T_s}} \left| 1 - \sum_{j=-J}^{J} \hat{\psi}\left(f - \frac{j}{T_s}\right) \mathrm{sinc}(T_s f - j) \right| \mathrm{d}f = 0.$$

(iv) What is the FT of the pulse of least bandwidth that satisfies Nyquist's Third Criterion with respect to the baud $T_s$? What is its bandwidth?

**Exercise 11.10 (Multiplication by a Carrier).**

(i) Let $\mathbf{u}$ be an energy-limited complex signal that is bandlimited to $W$ Hz, and let $f_0 > W$ be given. Let $\mathbf{v}$ be the signal $\mathbf{v} \colon t \mapsto u(t) \cos(2\pi f_0 t)$. Express the self-similarity function of $\mathbf{v}$ in terms of $f_0$ and the self-similarity function of $\mathbf{u}$.

(ii) Let the signal $\boldsymbol{\phi}$ be given by $\boldsymbol{\phi} \colon t \mapsto \sqrt{2} \cos(2\pi f_c t) \, \psi(t)$, where $f_c > W/2 > 0$; where $4 f_c T_s$ is an odd integer; and where $\psi$ is a real energy-limited signal that is bandlimited to $W/2$ Hz and whose time shifts by integer multiples of $(2T_s)$ are orthonormal. Show that the time shifts of $\boldsymbol{\phi}$ by integer multiples of $T_s$ are orthonormal.

**Exercise 11.11 (The Self-Similarity of a Convolution).** Let $\mathbf{p}$ and $\mathbf{q}$ be integrable signals of self-similarity functions $R_{pp}$ and $R_{qq}$. Show that the self-similarity function of their convolution $\mathbf{p} \star \mathbf{q}$ is indistinguishable from $R_{pp} \star R_{qq}$.

# Chapter 12

# Stochastic Processes: Definition

## 12.1 Introduction and Continuous-Time Heuristics

In this chapter we shall define stochastic processes. Our definition will be general so as to include the continuous-time stochastic processes of the type we encountered in Section 10.2 and also discrete-time processes.

In Section 10.2 we saw that since the data bits that we wish to communicate are random, the transmitted waveform is a stochastic process. But stochastic processes play an important role in Digital Communications not only in modeling the transmitted signals: they are also used to model the noise in the system and other sources of impairments.

The stochastic processes we encountered in Section 10.2 are continuous-time processes. We proposed that you think about such a process as a real-valued function of two variables: "time" and "luck." By "luck" we mean the realization of all the random components of the system, e.g., the bits to be sent, the realization of the noise processes (that we shall discuss later), or any other sources of randomness in the system.

Somewhat more precisely, recall that a probability space is defined as a triplet $(\Omega, \mathcal{F}, P)$, where the set $\Omega$ is the set of experiment outcomes, the set $\mathcal{F}$ is the set of events, and where $P(\cdot)$ assigns probabilities to the various events. A measurable real-valued function of the outcome is a random variable, and a function of time and the experiment outcome is a random process or a stochastic process. A continuous-time stochastic process $\mathbf{X}$ is thus a mapping

$$\mathbf{X} \colon \Omega \times \mathbb{R} \to \mathbb{R}$$
$$(\omega, t) \mapsto X(\omega, t).$$

If we fix some experiment outcome $\omega \in \Omega$, then the random process can be regarded as a function of one argument: time. This function is sometimes called a **sample-path**, **trajectory**, **sample-path realization**, or a **sample function**

$$X(\omega, \cdot) \colon \mathbb{R} \to \mathbb{R}$$
$$t \mapsto X(\omega, t).$$

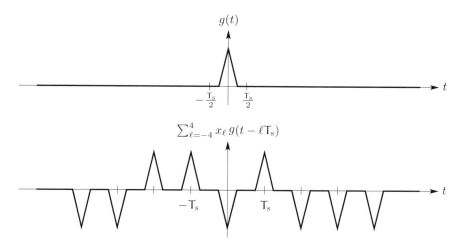

**Figure 12.1:** The pulse shape $\mathbf{g}\colon t \mapsto \left(1 - 4|t|/\mathsf{T_s}\right) \mathrm{I}\{|t| < \mathsf{T_s}/4\}$, and the sample function $t \mapsto \sum_{\ell=-4}^{4} x_\ell \, g(t - \ell\mathsf{T_s})$ when $\left(x_{-4}, x_{-3}, x_{-2}, x_{-1}, x_0, x_1, x_2, x_3, x_4\right) = (-1, -1, +1, +1, -1, +1, -1, -1, -1)$.

Similarly, if we fix an epoch $t \in \mathbb{R}$ and view the stochastic process as a function of "luck" only, we obtain a random variable:

$$X(\cdot, t)\colon \Omega \to \mathbb{R}$$
$$\omega \mapsto X(\omega, t).$$

This random variable is sometimes called **the value of the process at time** $t$ or **the time-$t$ sample of the process**.

Figure 12.1 shows the pulse shape $\mathbf{g}\colon t \mapsto \left(1 - 4|t|/\mathsf{T_s}\right) \mathrm{I}\{|t| < \mathsf{T_s}/4\}$ and a sample-path of the PAM signal

$$X(t) = \sum_{\ell=-4}^{4} X_\ell \, g(t - \ell\mathsf{T_s}) \tag{12.1}$$

with $\{X_\ell\}$ taking value in the set $\{-1, +1\}$. Notice that in this example the functions $t \mapsto g(t - \ell\mathsf{T_s})$ and $t \mapsto g(t - \ell'\mathsf{T_s})$ do not "overlap" if $\ell \neq \ell'$.

Figure 12.2 shows the pulse shape

$$\mathbf{g}\colon t \mapsto \begin{cases} 1 - \frac{4}{3\mathsf{T_s}}|t| & |t| \le \frac{3\mathsf{T_s}}{4}, \\ 0 & |t| > \frac{3\mathsf{T_s}}{4}, \end{cases} \quad t \in \mathbb{R} \tag{12.2}$$

and a sample-path of the PAM signal (12.1) for $\{X_\ell\}$ taking value in the set $\{-1, +1\}$. In this example the mappings $t \mapsto g(t - \ell\mathsf{T_s})$ and $t \mapsto g(t - \ell'\mathsf{T_s})$ do overlap (when $\ell' \in \{\ell - 1, \ell, \ell + 1\}$).

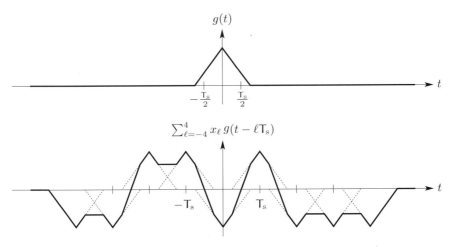

**Figure 12.2:** The pulse shape **g** of (12.2) and the trajectory $t \mapsto \sum_{\ell=-4}^{4} x_\ell\, g(t-\ell\mathsf{T_s})$ for $(x_{-4}, x_{-3}, x_{-2}, x_{-1}, x_0, x_1, x_2, x_3, x_4) = (-1,-1,+1,+1,-1,+1,-1,-1,-1)$.

## 12.2 A Formal Definition

We next give a formal definition of a **stochastic process**, which is also called a **random process**, or a **random function**.

**Definition 12.2.1 (Stochastic Process).** *A stochastic process $\big(X(t),\, t \in \mathcal{T}\big)$ is an indexed family of random variables that are defined on a common probability space $(\Omega, \mathcal{F}, P)$. Here $\mathcal{T}$ denotes the indexing set and $X(t)$ (or sometimes $X_t$) denotes the random variable indexed by $t$.*

Thus, $X(t)$ is the random variable to which $t \in \mathcal{T}$ is mapped. For each $t \in \mathcal{T}$ we have that $X(t)$ is a random variable, i.e., a measurable mapping from the experiment outcomes set $\Omega$ to the reals.[1]

A stochastic process $\big(X(t),\, t \in \mathcal{T}\big)$ is said to be **centered** or **of zero mean** if all the random variables in the family are of zero mean, i.e., if for every $t \in \mathcal{T}$ we have $\mathsf{E}[X(t)] = 0$. It is said to be **of finite variance** if all the random variables in the family are of finite variance, i.e., if $\mathsf{E}\big[X^2(t)\big] < \infty$ for all $t \in \mathcal{T}$.

The case where the indexing set $\mathcal{T}$ comprises only one element is not particularly exciting because in this case the stochastic process is just a random variable with fancy packaging. Similarly, when $\mathcal{T}$ is finite, the SP is just a random vector or a tuple of random variables in disguise. The cases that will be of most interest are enumerated below.

(i) When the indexing set $\mathcal{T}$ is the set of integers $\mathbb{Z}$, the stochastic process is said to be a **discrete-time stochastic process** and in this case it is simply

---

[1]Some authors, e.g., (Doob, 1990), allow for $X(t)$ to take on the values $\pm\infty$ provided that at each $t \in \mathcal{T}$ this occurs with zero probability, but we, following (Loève, 1963), insist that $X(t)$ only take on finite values.

a bi-infinite sequence of random variables

$$\ldots, X_{-2}, X_{-1}, X_0, X_1, X_2, \ldots$$

For discrete-time stochastic processes it is customary to denote the random variable to which $\nu \in \mathbb{Z}$ is mapped by $X_\nu$ rather than $X(\nu)$ and to refer to $X_\nu$ as the time-$\nu$ sample of the process $(X_\nu, \nu \in \mathbb{Z})$.

(ii) When the indexing set is the set of positive integers $\mathbb{N}$, the stochastic process is said to be a **one-sided discrete-time stochastic process** and it is simply a one-sided sequence of random variables

$$X_1, X_2, \ldots$$

Again, we refer to $X_\nu$ as the time-$\nu$ sample of $(X_\nu, \nu \in \mathbb{N})$.

(iii) When the indexing set $\mathcal{T}$ is the real line $\mathbb{R}$, the stochastic process is said to be a **continuous-time stochastic process** and the random variable $X(t)$ is the time-$t$ sample of $(X(t), t \in \mathbb{R})$.

In dealing with continuous-time stochastic processes we shall usually denote the process by $(X(t), t \in \mathbb{R})$, by $\mathbf{X}$, by $X(\cdot)$, or by $(X(t))$. The random variable to which $t$ is mapped, i.e., the time-$t$ sample of the process will be denoted by $X(t)$. Its realization will be denoted by $x(t)$, and the sample-path of the process by $\mathbf{x}$ or $x(\cdot)$.

Discrete-time processes will typically be denoted by $(X_\nu, \nu \in \mathbb{Z})$ or by $(X_\nu)$.

We shall need only a few results on discrete-time stochastic processes, and those will be presented in Chapter 13. Continuous-time stochastic processes will be discussed in Chapter 25.

## 12.3 Describing Stochastic Processes

The description of a continuous-time stochastic process in terms of a random variable (as in Section 10.2), in terms of a finite number of random variables (as in PAM signaling), or in terms of an infinite sequence of random variables (as in the transmission using PAM signaling of an infinite binary data stream) is particularly well suited for describing human-generated stochastic processes or stochastic processes that are generated using a mechanism that we fully understand. We simply describe how the stochastic process is *synthesized* from the random variables. The method is less useful when the stochastic process denotes a random signal (such as thermal noise or some other interference of unknown origin) that we observe rather than generate. In this case we can use measurements and statistical methods to *analyze* the process. Often, the best we can hope for is to be informed of the **finite-dimensional distributions** of the process, a concept that will be introduced in Section 25.2.

## 12.4    Additional Reading

Classic references on stochastic processes to which we shall frequently refer are (Doob, 1990) and (Loève, 1963). We also recommend (Gikhman and Skorokhod, 1996), (Cramér and Leadbetter, 2004), and (Grimmett and Stirzaker, 2001). For discrete-time stochastic processes, see (Pourahmadi, 2001) and (Porat, 2008).

## 12.5    Exercises

**Exercise 12.1 (Objects in a Basement).** Let $T_1, T_2, \ldots$ be a sequence of positive random variables, and let $N_1, N_2, \ldots$ be a sequence of random variables taking value in $\mathbb{N}$. Define

$$X(t) = \sum_{j=1}^{\infty} N_j \, \mathrm{I}\{t \geq T_j\}, \quad t \in \mathbb{R}.$$

Draw some sample paths of $\big(X(t),\ t \in \mathbb{R}\big)$. Assume that at time zero a basement is empty and that $N_j$ denotes the number of objects in the $j$-th box, which is brought down to the basement at time $T_j$. Explain why you can think of $X(t)$ as the number of objects in the basement at time $t$.

**Exercise 12.2 (A Queue).** Let $S_1, S_2, \ldots$ be a sequence of positive random variables. A system is turned on at time zero. The first customer arrives at the system at time $S_1$ and the next at time $S_1 + S_2$. More generally, Customer $\eta$ arrives $S_\eta$ minutes after Customer $(\eta - 1)$. The system serves one customer at a time. It takes the system one minute to serve each customer, and a customer leaves the system once it has been served. Let $X(t)$ denote the number of customers in the system at time $t$. Express $X(t)$ in terms of $S_1, S_2, \ldots$ Is $\big(X(t),\ t \in \mathbb{R}\big)$ a stochastic process? If so, draw a few of its sample paths. Compute $\Pr\big[X(0.5) > 0\big]$. Express your answer in terms of the distribution of $S_1, S_2, \ldots$

**Exercise 12.3 (A Continuous-Time Markov SP).** A particle is in State Zero at time $t = 0$. It stays in that state for $T_1^{(0)}$ seconds and then jumps to State One. It stays in State One for $T_1^{(1)}$ seconds and then jumps back to State Zero, where it stays for $T_2^{(0)}$ seconds. In general, $T_\nu^{(0)}$ is the duration of the particle's stay in State Zero on its $\nu$-th visit to that state. Similarly, $T_\nu^{(1)}$ is the duration of its stay in State One on its $\nu$-th visit. Assume that $T_1^{(0)}, T_1^{(1)}, T_2^{(0)}, T_2^{(1)}, T_3^{(0)}, T_3^{(1)}, \ldots$ are independent with $T_\nu^{(0)}$ being a mean-$\mu_0$ exponential and with $T_\nu^{(1)}$ being a mean-$\mu_1$ exponential for all $\nu \in \mathbb{N}$.

Let $X(t)$ be deterministically equal to zero for $t < 0$, and equal to the particle's state for $t \geq 0$.

(i) Plot some sample paths of $\big(X(t),\ t \in \mathbb{R}\big)$.

(ii) What is the probability that the sample path $t \mapsto X(\omega, t)$ is continuous in the interval $[0, t)$?

(iii) Conditional on $X(t) = 0$, where $t \geq 0$, what is the distribution of the remaining duration of the particle's stay in State Zero?

*Hint: An exponential RV $X$ has the memoryless property, i.e., that for every $s, t \geq 0$ we have $\Pr[X > s + t \,|\, X > t] = \Pr[X \geq s]$.*

**Exercise 12.4 (Peak Power).** Let the random variables $(D_j, \; j \in \mathbb{Z})$ be IID, each taking on the values 0 and 1 equiprobably. Let

$$X(t) = A \sum_{\ell=-\infty}^{\infty} \left(1 - 2D_\ell\right) g(t - \ell \mathsf{T_s}), \quad t \in \mathbb{R},$$

where $A, \mathsf{T_s} > 0$ and $\mathbf{g} \colon t \mapsto \mathrm{I}\{|t| \le 3\mathsf{T_s}/4\}$. Find the distribution of the random variable

$$\sup_{t \in \mathbb{R}} |X(t)|.$$

**Exercise 12.5 (Sample-Path Continuity).** Let the random variables $(D_j, \; j \in \mathbb{Z})$ be IID, each taking on the values 0 and 1 equiprobably. Let

$$X(t) = A \sum_{\ell=-\infty}^{\infty} \left(1 - 2D_\ell\right) g(t - \ell \mathsf{T_s}), \quad t \in \mathbb{R},$$

where $A, \mathsf{T_s} > 0$. Suppose that the function $\mathbf{g} \colon \mathbb{R} \to \mathbb{R}$ is continuous and is zero outside some interval, so $g(t) = 0$ whenever $|t| \ge \mathsf{T}$. Show that for every $\omega \in \Omega$, the sample-path $t \mapsto X(\omega, t)$ is a continuous function of time.

**Exercise 12.6 (Random Sampling Time).** Consider the setup of Exercise 12.5, with the pulse shape $\mathbf{g} \colon t \mapsto \left(1 - 2|t|/\mathsf{T_s}\right) \mathrm{I}\{|t| \le \mathsf{T_s}/2\}$. Further assume that the RV $T$ is independent of $(D_j, \; j \in \mathbb{Z})$ and uniformly distributed over the interval $[-\delta, \delta]$. Find the distribution of $X(k\mathsf{T_s} + T)$ for any integer $k$.

**Exercise 12.7 (A Strange SP).** Let $T$ be a mean-one exponential RV, and define the SP $(X(t), \; t \in \mathbb{R})$ by

$$X(t) = \begin{cases} 1 & \text{if } t = T, \\ 0 & \text{otherwise.} \end{cases}$$

Compute the distribution of $X(t_1)$ and the joint distribution of $X(t_1)$ and $X(t_2)$ for $t_1, t_2 \in \mathbb{R}$. What is the probability that the sample-path $t \mapsto X(\omega, t)$ is continuous at $t_1$? What is the probability that the sample-path is a continuous function (everwhere)?

**Exercise 12.8 (The Sum of Stochastic Processes: Formalities).** Let the stochastic processes $(X_1(t), \; t \in \mathbb{R})$ and $(X_2(t), \; t \in \mathbb{R})$ be defined on the same probability space $(\Omega, \mathcal{F}, P)$. Let $(Y(t), \; t \in \mathbb{R})$ be the SP corresponding to their sum. Express $\mathbf{Y}$ as a mapping from $\Omega \times \mathbb{R}$ to $\mathbb{R}$. What is $Y(\omega, t)$ for $(\omega, t) \in \Omega \times \mathbb{R}$?

**Exercise 12.9 (Independent Stochastic Processes).** Let the SP $(X_1(t), \; t \in \mathbb{R})$ be defined on the probability space $(\Omega_1, \mathcal{F}_1, P_1)$, and let $(X_2(t), \; t \in \mathbb{R})$ be defined on the space $(\Omega_2, \mathcal{F}_2, P_2)$. Define a new probability space $(\Omega, \mathcal{F}, P)$ with two stochastic processes $(\tilde{X}_1(t), \; t \in \mathbb{R})$ and $(\tilde{X}_2(t), \; t \in \mathbb{R})$ such that for every $\eta \in \mathbb{N}$ and epochs $t_1, \ldots, t_\eta \in \mathbb{R}$ the following three conditions hold:

1) The joint law of $\tilde{X}_1(t_1), \ldots, \tilde{X}_1(t_\eta)$ is the same as the joint law of $X_1(t_1), \ldots, X_1(t_\eta)$.

2) The joint law of $\tilde{X}_2(t_1), \ldots, \tilde{X}_2(t_\eta)$ is the same as the joint law of $X_2(t_1), \ldots, X_2(t_\eta)$.

3) The $\eta$-tuple $\tilde{X}_1(t_1), \ldots, \tilde{X}_1(t_\eta)$ is independent of the $\eta$-tuple $\tilde{X}_2(t_1), \ldots, \tilde{X}_2(t_\eta)$.

*Hint: Consider* $\Omega = \Omega_1 \times \Omega_2$.

**Exercise 12.10 (Pathwise Integration).** Let $\big(X_j,\ j \in \mathbb{Z}\big)$ be IID random variables defined over the probability space $(\Omega, \mathcal{F}, P)$, with $X_j$ taking on the values 0 and 1 equiprobably. Define the stochastic process $\big(X(t),\ t \in \mathbb{R}\big)$ as

$$X(t) = \sum_{j=-\infty}^{\infty} X_j\, \mathrm{I}\{j \le t < j+1\}, \quad t \in \mathbb{R}.$$

For a given $n \in \mathbb{N}$, compute the distribution of the random variable

$$\omega \mapsto \int_0^n X(\omega, t)\, \mathrm{d}t.$$

# Chapter 13

# Stationary Discrete-Time Stochastic Processes

## 13.1   Introduction

This chapter discusses some of the properties of real discrete-time stochastic processes. Extensions to complex discrete-time stochastic processes are discussed in Chapter 17.

## 13.2   Stationary Processes

A discrete-time stochastic process is said to be stationary if all equal-length tuples of consecutive samples have the same joint law. Thus:

**Definition 13.2.1 (Stationary Discrete-Time Processes).** *A discrete-time SP $(X_\nu)$ is said to be **stationary** or **strict sense stationary** or **strongly stationary** if for every $n \in \mathbb{N}$ and all integers $\eta, \eta'$ the joint distribution of the n-tuple $(X_\eta, \ldots X_{\eta+n-1})$ is identical to that of the n-tuple $(X_{\eta'}, \ldots, X_{\eta'+n-1})$:*

$$\left(X_\eta, \ldots X_{\eta+n-1}\right) \overset{\mathscr{L}}{=} \left(X_{\eta'}, \ldots X_{\eta'+n-1}\right). \tag{13.1}$$

Here $\overset{\mathscr{L}}{=}$ denotes equality of distribution (law) so $X \overset{\mathscr{L}}{=} Y$ indicates that the random variables $X$ and $Y$ have the same distribution; $(X, Y) \overset{\mathscr{L}}{=} (W, Z)$ indicates that the pair $(X, Y)$ and the pair $(W, Z)$ have the same joint distribution; and similarly for $n$-tuples.

By considering the case where $n = 1$ we obtain that if $(X_\nu)$ is stationary, then the distribution of $X_\eta$ is the same as the distribution of $X_{\eta'}$, for all $\eta, \eta' \in \mathbb{Z}$. That is, if $(X_\nu)$ is stationary, then all the random variables in the family $(X_\nu, \ \nu \in \mathbb{Z})$ have the same distribution: the random variable $X_1$ has the same distribution as the random variable $X_2$, etc. Thus,

$$\left((X_\nu, \ \nu \in \mathbb{Z}) \text{ stationary}\right) \Rightarrow \left(X_\nu \overset{\mathscr{L}}{=} X_1, \ \nu \in \mathbb{Z}\right). \tag{13.2}$$

By considering in the above definition the case where $n = 2$ we obtain that for a stationary process $(X_\nu)$ the joint distribution of $X_1, X_2$ is the same as the joint distribution of $X_\eta, X_{\eta+1}$ for any integer $\eta$. More, however, is true. If $(X_\nu)$ is stationary, then the joint distribution of $X_\nu, X_{\nu'}$ is the same as the joint distribution of $X_{\eta+\nu}, X_{\eta+\nu'}$:

$$\Big((X_\nu, \ \nu \in \mathbb{Z}) \text{ stationary}\Big) \Rightarrow \Big((X_\nu, X_{\nu'}) \stackrel{\mathscr{L}}{=} (X_{\eta+\nu}, X_{\eta+\nu'}), \ \nu, \nu', \eta \in \mathbb{Z}\Big). \quad (13.3)$$

To prove (13.3) first note that it suffices to treat the case where $\nu \geq \nu'$ because $(X, Y) \stackrel{\mathscr{L}}{=} (W, Z)$ if, and only if, $(Y, X) \stackrel{\mathscr{L}}{=} (Z, W)$. Next note that stationarity implies that

$$\big(X_{\nu'}, \ldots, X_\nu\big) \stackrel{\mathscr{L}}{=} \big(X_{\eta+\nu'}, \ldots, X_{\eta+\nu}\big) \quad (13.4)$$

because both are $(\nu - \nu' + 1)$-length tuples of consecutive samples of the process. Finally, (13.4) implies that the joint distribution of $(X_{\nu'}, X_\nu)$ is identical to the joint distribution of $(X_{\eta+\nu'}, X_{\eta+\nu})$ and (13.3) follows.

The above argument can be generalized to more samples. This yields the following proposition, which gives an alternative definition of stationarity, a definition that more easily generalizes to continuous-time stochastic processes.

**Proposition 13.2.2.** *A discrete-time SP $(X_\nu, \ \nu \in \mathbb{Z})$ is stationary if, and only if, for every $n \in \mathbb{N}$, all integers $\nu_1, \ldots, \nu_n \in \mathbb{Z}$, and every $\eta \in \mathbb{Z}$*

$$\big(X_{\nu_1}, \ldots, X_{\nu_n}\big) \stackrel{\mathscr{L}}{=} \big(X_{\eta+\nu_1}, \ldots, X_{\eta+\nu_n}\big). \quad (13.5)$$

**Proof.** One direction is trivial and simply follows by substituting consecutive integers for $\nu_1, \ldots, \nu_n$ in (13.5). The proof of the other direction is a straightforward extension of the argument we used to prove (13.3). $\qquad\square$

By noting that $(W_1, \ldots, W_n) \stackrel{\mathscr{L}}{=} (Z_1, \ldots, Z_n)$ if, and only if,[1] $\sum_j \alpha_j W_j \stackrel{\mathscr{L}}{=} \sum_j \alpha_j Z_j$ for all $\alpha_1, \ldots, \alpha_n \in \mathbb{R}$ we obtain the following equivalent characterization of stationary processes:

**Proposition 13.2.3.** *A discrete-time SP $(X_\nu)$ is stationary if, and only if, for every $n \in \mathbb{N}$, all $\eta, \nu_1, \ldots, \nu_n \in \mathbb{Z}$, and all $\alpha_1, \ldots, \alpha_n \in \mathbb{R}$*

$$\sum_{j=1}^{n} \alpha_j X_{\nu_j} \stackrel{\mathscr{L}}{=} \sum_{j=1}^{n} \alpha_j X_{\nu_j+\eta}. \quad (13.6)$$

## 13.3 Wide-Sense Stationary Stochastic Processes

**Definition 13.3.1 (Wide-Sense Stationary Discrete-Time SP).** *We say that a discrete-time SP $(X_\nu, \ \nu \in \mathbb{Z})$ is **wide-sense stationary** (WSS) or **weakly***

---

[1] This follows because the multivariate characteristic function determines the joint distribution (see Proposition 23.4.4 or (Dudley, 2003, Chapter 9, Section 5, Theorem 9.5.1)) and because the characteristic functions of all the linear combinations of the components of a random vector determine the multivariate characteristic function of the random vector (Feller, 1971, Chapter XV, Section 7).

*stationary* or ***covariance stationary*** or ***second-order stationary*** or **weak-sense stationary** *if the following three conditions are satisfied:*

1) *The random variables $X_\nu$, $\nu \in \mathbb{Z}$ are all of finite variance:*

$$\text{Var}[X_\nu] < \infty, \quad \nu \in \mathbb{Z}. \tag{13.7a}$$

2) *The random variables $X_\nu$, $\nu \in \mathbb{Z}$ have identical means:*

$$\mathsf{E}[X_\nu] = \mathsf{E}[X_1], \quad \nu \in \mathbb{Z}. \tag{13.7b}$$

3) *The quantity $\mathsf{E}[X_{\nu'} X_\nu]$ depends on $\nu'$ and $\nu$ only via $\nu - \nu'$:*

$$\mathsf{E}[X_{\nu'} X_\nu] = \mathsf{E}[X_{\eta+\nu'} X_{\eta+\nu}], \quad \nu, \nu', \eta \in \mathbb{Z}. \tag{13.7c}$$

**Note 13.3.2.** By considering (13.7c) when $\nu = \nu'$ we obtain that all the samples of a WSS SP have identical second moments. And since, by (13.7b), they also all have identical means, it follows that all the samples of a WSS SP have identical variances:

$$\Big( (X_\nu, \ \nu \in \mathbb{Z}) \ \text{WSS} \Big) \Rightarrow \Big( \text{Var}[X_\nu] = \text{Var}[X_1], \quad \nu \in \mathbb{Z} \Big). \tag{13.8}$$

An alternative definition of a WSS process in terms of the variance of linear functionals of the process is given below.

**Proposition 13.3.3.** *A finite-variance discrete-time SP $(X_\nu)$ is WSS if, and only if, for every $n \in \mathbb{N}$, every $\eta, \nu_1, \dots, \nu_n \in \mathbb{Z}$, and every $\alpha_1, \dots, \alpha_n \in \mathbb{R}$*

$$\sum_{j=1}^n \alpha_j X_{\nu_j} \text{ and } \sum_{j=1}^n \alpha_j X_{\nu_j + \eta} \quad \text{have the same mean \& variance.} \tag{13.9}$$

**Proof.** The proof is left as an exercise. Alternatively, see the proof of Proposition 17.5.5. ◻

## 13.4   Stationarity and Wide-Sense Stationarity

Comparing (13.9) with (13.6) we see that, for finite-variance stochastic processes, stationarity implies wide-sense stationarity, which is the content of the following proposition. This explains why stationary processes are sometimes called strong-sense stationary and why wide-sense stationary processes are sometimes called weak-sense stationary.

**Proposition 13.4.1 (Finite-Variance Stationary Stochastic Processes Are WSS).**
*Every finite-variance discrete-time stationary SP is WSS.*

**Proof.** While this is obvious from (13.9) and (13.6) we shall nevertheless give an alternative proof because the proof of Proposition 13.3.3 was left as an exercise. The proof is straightforward and follows directly from (13.2) and (13.3) by noting that if $X \stackrel{\mathscr{L}}{=} Y$, then $\mathsf{E}[X] = \mathsf{E}[Y]$ and that if $(X, Y) \stackrel{\mathscr{L}}{=} (W, Z)$, then $\mathsf{E}[XY] = \mathsf{E}[WZ]$. ◻

It is not surprising that not every WSS process is stationary. Indeed, the definition of WSS processes only involves means and covariances, so it cannot possibly say everything regarding the distribution. For example, the process whose samples are independent with the odd ones taking on the value $\pm 1$ equiprobably and with the even ones uniformly distributed over the interval $[-\sqrt{3}, +\sqrt{3}]$ is WSS but not stationary.

## 13.5 The Autocovariance Function

**Definition 13.5.1 (Autocovariance Function).** *The autocovariance function* $\mathsf{K}_{XX} \colon \mathbb{Z} \to \mathbb{R}$ *of a WSS discrete-time SP* $(X_\nu)$ *is defined by*

$$\mathsf{K}_{XX}(\eta) \triangleq \mathsf{Cov}[X_{\nu+\eta}, X_\nu], \quad \eta \in \mathbb{Z}. \tag{13.10}$$

Thus, the autocovariance function at $\eta$ is the covariance between two samples of the process taken $\eta$ units of time apart. Note that because $(X_\nu)$ is WSS, the RHS of (13.10) does not depend on $\nu$. Also, for WSS processes all samples are of equal mean (13.7b), so

$$
\begin{aligned}
\mathsf{K}_{XX}(\eta) &= \mathsf{Cov}[X_{\nu+\eta}, X_\nu] \\
&= \mathsf{E}[X_{\nu+\eta}X_\nu] - \mathsf{E}[X_{\nu+\eta}]\,\mathsf{E}[X_\nu] \\
&= \mathsf{E}[X_{\nu+\eta}X_\nu] - \big(\mathsf{E}[X_1]\big)^2, \quad \eta \in \mathbb{Z}.
\end{aligned}
$$

In some engineering texts the autocovariance function is called "autocorrelation function." We prefer the former because $\mathsf{K}_{XX}(\eta)$ does not measure the correlation coefficient between $X_\nu$ and $X_{\nu+\eta}$ but rather the covariance. These concepts are different also for zero-mean processes. Following (Grimmett and Stirzaker, 2001) we define the **autocorrelation function** of a WSS process of nonzero variance as

$$\rho_{XX}(\eta) \triangleq \frac{\mathsf{Cov}[X_{\nu+\eta}, X_\nu]}{\mathsf{Var}[X_1]}, \quad \eta \in \mathbb{Z}, \tag{13.11}$$

i.e., as the correlation coefficient between $X_{\nu+\eta}$ and $X_\nu$. (Recall that for a WSS process all samples are of the same variance (13.8), so for such a process the denominator in (13.11) is equal to $\sqrt{\mathsf{Var}[X_\nu]\,\mathsf{Var}[X_{\nu+\eta}]}$.)

Not every function from the integers to the reals is the autocovariance function of some WSS SP. For example, the autocovariance function must be symmetric in the sense that

$$\mathsf{K}_{XX}(-\eta) = \mathsf{K}_{XX}(\eta), \quad \eta \in \mathbb{Z}, \tag{13.12}$$

because, by (13.10),

$$
\begin{aligned}
\mathsf{K}_{XX}(\eta) &= \mathsf{Cov}[X_{\nu+\eta}, X_\nu] \\
&= \mathsf{Cov}[X_{\tilde{\nu}}, X_{\tilde{\nu}-\eta}] \\
&= \mathsf{Cov}[X_{\tilde{\nu}-\eta}, X_{\tilde{\nu}}] \\
&= \mathsf{K}_{XX}(-\eta), \quad \eta \in \mathbb{Z},
\end{aligned}
$$

where in the second equality we defined $\tilde{\nu} \triangleq \nu + \eta$, and where in the third equality we used the fact that for real random variables the covariance is symmetric: $\mathsf{Cov}[X, Y] = \mathsf{Cov}[Y, X]$.

Another property that the autocovariance function must satisfy is

$$\sum_{\nu=1}^{n} \sum_{\nu'=1}^{n} \alpha_\nu \alpha_{\nu'} \mathsf{K}_{XX}(\nu - \nu') \geq 0, \quad \alpha_1, \ldots, \alpha_n \in \mathbb{R}, \qquad (13.13)$$

because

$$\sum_{\nu=1}^{n} \sum_{\nu'=1}^{n} \alpha_\nu \alpha_{\nu'} \mathsf{K}_{XX}(\nu - \nu') = \sum_{\nu=1}^{n} \sum_{\nu'=1}^{n} \alpha_\nu \alpha_{\nu'} \mathsf{Cov}[X_\nu, X_{\nu'}]$$

$$= \mathsf{Cov}\left[\sum_{\nu=1}^{n} \alpha_\nu X_\nu, \sum_{\nu'=1}^{n} \alpha_{\nu'} X_{\nu'}\right]$$

$$= \mathsf{Var}\left[\sum_{\nu=1}^{n} \alpha_\nu X_\nu\right]$$

$$\geq 0.$$

It turns out that (13.12) and (13.13) fully characterize the autocovariance functions of discrete-time WSS stochastic processes in a sense that is made precise in the following theorem.

**Theorem 13.5.2 (Characterizing Autocovariance Functions).**

(i) *If* $\mathsf{K}_{XX}$ *is the autocovariance function of some discrete-time WSS SP* $(X_\nu)$, *then* $\mathsf{K}_{XX}$ *must satisfy (13.12) & (13.13).*

(ii) *If* $\mathsf{K} \colon \mathbb{Z} \to \mathbb{R}$ *is some function satisfying*

$$\mathsf{K}(-\eta) = \mathsf{K}(\eta), \quad \eta \in \mathbb{Z} \qquad (13.14)$$

*and*

$$\sum_{\nu=1}^{n} \sum_{\nu'=1}^{n} \alpha_\nu \alpha_{\nu'} \mathsf{K}(\nu - \nu') \geq 0, \quad \left(n \in \mathbb{N}, \ \alpha_1, \ldots, \alpha_n \in \mathbb{R}\right), \qquad (13.15)$$

*then there exists a discrete-time WSS SP* $(X_\nu)$ *whose autocovariance function* $\mathsf{K}_{XX}$ *is given by* $\mathsf{K}_{XX}(\eta) = \mathsf{K}(\eta)$ *for all* $\eta \in \mathbb{Z}$.

**Proof.** We have already proved Part (i). For a proof of Part (ii) see, for example, (Doob, 1990, Chapter X, § 3, Theorem 3.1) or (Pourahmadi, 2001, Theorem 5.1 in Section 5.1 and Section 9.7).[2]                                                                                      □

A function $\mathsf{K} \colon \mathbb{Z} \to \mathbb{R}$ satisfying (13.14) & (13.15) is called a **positive definite function**. Such functions have been extensively studied in the literature, and in Section 13.7 we shall give an alternative characterization of autocovariance functions based on these studies. But first we introduce the power spectral density.

---

[2]For the benefit of readers who have already encountered Gaussian stochastic processes, we mention here that if $\mathsf{K}(\cdot)$ satisfies (13.14) & (13.15) then we can even find a *Gaussian* SP whose autocovariance function is equal to $\mathsf{K}(\cdot)$.

## 13.6 The Power Spectral Density Function

Roughly speaking, the **power spectral density** (PSD) of a discrete-time WSS SP $(X_\nu)$ of autocovariance function $\mathsf{K}_{XX}$ is an integrable function on the interval $[-1/2, 1/2]$ whose $\eta$-th Fourier Series Coefficient is equal to $\mathsf{K}_{XX}(\eta)$. Such a function does not always exist. When it does, it is unique in the sense that any two such functions can only differ on a subset of the interval $[-1/2, 1/2]$ of Lebesgue measure zero. (This follows because integrable functions on the interval $[-1/2, 1/2]$ that have identical Fourier Series Coefficients can differ only on a subset of $[-1/2, 1/2]$ of Lebesgue measure zero; see Theorem A.2.3.) Consequently, we shall speak of "the" PSD but try to remember that this does not always exist and that, when it does, it is only unique in this restricted sense.

**Definition 13.6.1 (Power Spectral Density).** *We say that the discrete-time WSS SP $(X_\nu)$ is of **power spectral density** $\mathsf{S}_{XX}$ if $\mathsf{S}_{XX}$ is an integrable mapping from the interval $[-1/2, 1/2]$ to the reals such that*

$$\mathsf{K}_{XX}(\eta) = \int_{-1/2}^{1/2} \mathsf{S}_{XX}(\theta)\, e^{-\mathrm{i}2\pi\eta\theta}\, \mathrm{d}\theta, \quad \eta \in \mathbb{Z}. \tag{13.16}$$

*But see also Note 13.6.5 ahead.*

**Note 13.6.2.** We shall sometimes abuse notation and, rather than say that the *stochastic process* $(X_\nu, \ \nu \in \mathbb{Z})$ is of PSD $\mathsf{S}_{XX}$, we shall say that the *autocovariance function* $\mathsf{K}_{XX}$ is of PSD $\mathsf{S}_{XX}$.

By considering the special case of $\eta = 0$ in (13.16) we obtain that

$$\mathsf{Var}[X_\nu] = \mathsf{K}_{XX}(0)$$

$$= \int_{-1/2}^{1/2} \mathsf{S}_{XX}(\theta)\, \mathrm{d}\theta, \quad \nu \in \mathbb{Z}. \tag{13.17}$$

The main result of the following proposition is that power spectral densities are nonnegative (except possibly on a set of Lebesgue measure zero).

**Proposition 13.6.3 (PSDs Are Nonnegative and Symmetric).**

    (i) *If the WSS SP $(X_\nu, \ \nu \in \mathbb{Z})$ of autocovariance $\mathsf{K}_{XX}$ is of PSD $\mathsf{S}_{XX}$, then, except on subsets of $(-1/2, 1/2)$ of Lebesgue measure zero,*

$$\mathsf{S}_{XX}(\theta) \geq 0 \tag{13.18}$$

    *and*

$$\mathsf{S}_{XX}(\theta) = \mathsf{S}_{XX}(-\theta). \tag{13.19}$$

    (ii) *If the function $\mathsf{S} \colon [-1/2, 1/2] \to \mathbb{R}$ is integrable, nonnegative, and symmetric (in the sense that $\mathsf{S}(\theta) = \mathsf{S}(-\theta)$ for all $\theta \in (-1/2, 1/2)$), then there exists a WSS SP $(X_\nu)$ whose PSD $\mathsf{S}_{XX}$ is given by*

$$\mathsf{S}_{XX}(\theta) = \mathsf{S}(\theta), \quad \theta \in [-1/2, 1/2].$$

**Proof.** The nonnegativity of the PSD (13.18) will be established later in the more general setting of complex stochastic processes (Proposition 17.5.7 ahead). Here we only prove the symmetry (13.19) and establish the second half of the proposition.

That (13.19) holds (except on a set of Lebesgue measure zero) follows because $K_{XX}$ is symmetric. Indeed, for any $\eta \in \mathbb{Z}$ we have

$$\int_{-1/2}^{1/2} \big(S_{XX}(\theta) - S_{XX}(-\theta)\big) e^{-i2\pi\eta\theta} \, d\theta$$

$$= \int_{-1/2}^{1/2} S_{XX}(\theta) e^{-i2\pi\eta\theta} \, d\theta - \int_{-1/2}^{1/2} S_{XX}(-\theta) e^{-i2\pi\eta\theta} \, d\theta$$

$$= K_{XX}(\eta) - \int_{-1/2}^{1/2} S_{XX}(\tilde{\theta}) e^{-i2\pi(-\eta)\tilde{\theta}} \, d\tilde{\theta}$$

$$= K_{XX}(\eta) - K_{XX}(-\eta)$$

$$= 0, \quad \eta \in \mathbb{Z}. \tag{13.20}$$

Consequently, all the Fourier Series Coefficients of the function $\theta \mapsto S_{XX}(\theta) - S_{XX}(-\theta)$ are zero, thus establishing that this function is zero except on a set of Lebesgue measure zero (Theorem A.2.3).

We next prove that if the function $S\colon [-1/2, 1/2) \to \mathbb{R}$ is symmetric, nonnegative, and integrable, then it is the PSD of some WSS real SP. We cheat a bit because our proof relies on Theorem 13.5.2, which we never proved. From Theorem 13.5.2 it follows that it suffices to establish that the sequence $K\colon \mathbb{Z} \to \mathbb{R}$ defined by

$$K(\eta) = \int_{-1/2}^{1/2} S(\theta) e^{-i2\pi\eta\theta} \, d\theta, \quad \eta \in \mathbb{Z} \tag{13.21}$$

satisfies (13.14) & (13.15).

Verifying (13.14) is straightforward: by hypothesis, $S(\cdot)$ is symmetric so

$$K(-\eta) = \int_{-1/2}^{1/2} S(\theta) e^{-i2\pi(-\eta)\theta} \, d\theta$$

$$= \int_{-1/2}^{1/2} S(-\varphi) e^{-i2\pi\eta\varphi} \, d\varphi$$

$$= \int_{-1/2}^{1/2} S(\varphi) e^{-i2\pi\eta\varphi} \, d\varphi$$

$$= K(\eta), \quad \eta \in \mathbb{Z},$$

where the first equality follows from (13.21); the second from the change of variable $\varphi \triangleq -\theta$; the third from the symmetry of $S(\cdot)$, which implies that $S(-\varphi) = S(\varphi)$; and the last equality again from (13.21).

We next verify (13.15). To this end we fix arbitrary $\alpha_1, \ldots, \alpha_n \in \mathbb{R}$ and compute

$$\sum_{\nu=1}^{n} \sum_{\nu'=1}^{n} \alpha_\nu \alpha_{\nu'} K(\nu - \nu') = \sum_{\nu=1}^{n} \sum_{\nu'=1}^{n} \alpha_\nu \alpha_{\nu'} \int_{-1/2}^{1/2} S(\theta) e^{-i2\pi(\nu-\nu')\theta} \, d\theta$$

$$= \int_{-1/2}^{1/2} \mathsf{S}(\theta) \left( \sum_{\nu=1}^{n} \sum_{\nu'=1}^{n} \alpha_\nu \alpha_{\nu'} \, e^{-\mathrm{i}2\pi(\nu-\nu')\theta} \right) \mathrm{d}\theta$$

$$= \int_{-1/2}^{1/2} \mathsf{S}(\theta) \left( \sum_{\nu=1}^{n} \sum_{\nu'=1}^{n} \alpha_\nu \, e^{-\mathrm{i}2\pi\nu\theta} \, \alpha_{\nu'} \, e^{\mathrm{i}2\pi\nu'\theta} \right) \mathrm{d}\theta$$

$$= \int_{-1/2}^{1/2} \mathsf{S}(\theta) \left( \sum_{\nu=1}^{n} \alpha_\nu \, e^{-\mathrm{i}2\pi\nu\theta} \right) \left( \sum_{\nu'=1}^{n} \alpha_{\nu'} \, e^{-\mathrm{i}2\pi\nu'\theta} \right)^* \mathrm{d}\theta$$

$$= \int_{-1/2}^{1/2} \mathsf{S}(\theta) \left| \sum_{\nu=1}^{n} \alpha_\nu \, e^{-\mathrm{i}2\pi\nu\theta} \right|^2 \mathrm{d}\theta$$

$$\geq 0, \tag{13.22}$$

where the first equality follows from (13.21); the subsequent equalities by simple algebraic manipulation; and the final inequality from the nonnegativity of $\mathsf{S}(\cdot)$. □

**Corollary 13.6.4.** *If a discrete-time WSS SP $(X_\nu)$ has a PSD, then it also has a PSD $\mathsf{S}_{XX}$ for which (13.18) holds for every $\theta \in [-1/2, 1/2]$ and for which (13.19) holds for every $\theta \in (-1/2, 1/2)$ (and not only outside subsets of Lebesgue measure zero).*

**Proof.** Suppose that $(X_\nu)$ is of PSD $\mathsf{S}_{XX}$. Define the mapping $\mathsf{S} \colon [-1/2, 1/2] \to \mathbb{R}$ by[3]

$$\mathsf{S}(\theta) = \begin{cases} \frac{1}{2}\left(|\mathsf{S}_{XX}(\theta)| + |\mathsf{S}_{XX}(-\theta)|\right) & \text{if } \theta \in (-1/2, 1/2) \\ 1 & \text{if } \theta = -1/2. \end{cases} \tag{13.23}$$

By the proposition, $\mathsf{S}_{XX}$ and $\mathsf{S}(\cdot)$ differ only on a set of Lebesgue measure zero, so they must have identical Fourier Series Coefficients. Since the Fourier Series Coefficients of $\mathsf{S}_{XX}$ agree with $\mathsf{K}_{XX}$, it follows that so must those of $\mathsf{S}(\cdot)$. Thus, $\mathsf{S}(\cdot)$ is a PSD for $(X_\nu)$, and it is by (13.23) nonnegative on $[-1/2, 1/2]$ and symmetric on $(-1/2, 1/2)$. □

**Note 13.6.5.** In view of Corollary 13.6.4 we shall only say that $(X_\nu)$ is of PSD $\mathsf{S}_{XX}$ if the function $\mathsf{S}_{XX}$—in addition to being integrable and to satisfying (13.16)—is also nonnegative and symmetric.

As we have noted, not every WSS SP has a PSD. For example, the process defined by

$$X_\nu = X, \quad \nu \in \mathbb{Z},$$

where $X$ is some zero-mean unit-variance random variable has the all-one auto-covariance function $\mathsf{K}_{XX}(\eta) = 1$, $\eta \in \mathbb{Z}$, and this all-one sequence cannot be the Fourier Series Coefficients sequence of an integrable function because, by the Riemann-Lebesgue lemma (Theorem A.2.4), the Fourier Series Coefficients of an integrable function must converge to zero.[4]

---

[3]Our choice of $\mathsf{S}(-1/2)$ as 1 is arbitrary; any nonnegative value whould do.

[4]One could say that the PSD of this process is Dirac's Delta, but we shall refrain from doing so because we do not use Dirac's Delta in this book and because there is not much to be gained from this. (There exist processes that do not have a PSD even if one allows for Dirac's Deltas.)

In general, it is very difficult to characterize the autocovariance functions having a PSD. We know by the Riemann-Lebesgue lemma that such autocovariance functions must tend to zero, but this necessary condition is not sufficient. A very useful sufficient (but not necessary) condition is the following:

**Proposition 13.6.6 (PSD when $K_{XX}$ Is Absolutely Summable).** *If the autocovariance function $K_{XX}$ is absolutely summable, i.e.,*

$$\sum_{\eta=-\infty}^{\infty} \left| K_{XX}(\eta) \right| < \infty, \qquad (13.24)$$

*then the function*

$$S(\theta) = \sum_{\eta=-\infty}^{\infty} K_{XX}(\eta)\, e^{i2\pi\eta\theta}, \quad \theta \in [-1/2, 1/2] \qquad (13.25)$$

*is continuous, symmetric, nonnegative, and satisfies*

$$\int_{-1/2}^{1/2} S(\theta)\, e^{-i2\pi\eta\theta}\, d\theta = K_{XX}(\eta), \quad \eta \in \mathbb{Z}. \qquad (13.26)$$

*Consequently, $S(\cdot)$ is a PSD for $K_{XX}$.*

**Proof.** First note that because $|K_{XX}(\eta)\, e^{-i2\pi\theta\eta}| = |K_{XX}(\eta)|$, it follows that (13.24) guarantees that the sum in (13.25) converges uniformly and absolutely. And since each term in the sum is a continuous function, the uniform convergence of the sum guarantees that $S(\cdot)$ is continuous (Rudin, 1976, Chapter 7, Theorem 7.12). Consequently,

$$\int_{-1/2}^{1/2} |S(\theta)|\, d\theta < \infty, \qquad (13.27)$$

and it is meaningful to discuss the Fourier Series Coefficients of $S(\cdot)$.

We next prove that the Fourier Series Coefficients of $S(\cdot)$ are equal to $K_{XX}$, i.e., that (13.26) holds. This can be shown by swapping integration and summation and using the orthonormality property

$$\int_{-1/2}^{1/2} e^{i2\pi(\eta-\eta')\theta}\, d\theta = \mathrm{I}\{\eta = \eta'\}, \quad \eta, \eta' \in \mathbb{Z} \qquad (13.28)$$

as follows:

$$\int_{-1/2}^{1/2} S(\theta)\, e^{-i2\pi\eta\theta}\, d\theta = \int_{-1/2}^{1/2} \left( \sum_{\eta'=-\infty}^{\infty} K_{XX}(\eta')\, e^{i2\pi\eta'\theta} \right) e^{-i2\pi\eta\theta}\, d\theta$$

$$= \sum_{\eta'=-\infty}^{\infty} K_{XX}(\eta') \int_{-1/2}^{1/2} e^{i2\pi\eta'\theta}\, e^{-i2\pi\eta\theta}\, d\theta$$

$$= \sum_{\eta'=-\infty}^{\infty} K_{XX}(\eta') \int_{-1/2}^{1/2} e^{i2\pi(\eta'-\eta)\theta}\, d\theta$$

$$= \sum_{\eta'=-\infty}^{\infty} \mathsf{K}_{XX}(\eta')\, \mathrm{I}\{\eta' = \eta\}$$

$$= \mathsf{K}_{XX}(\eta), \quad \eta \in \mathbb{Z}.$$

It remains to show that $\mathsf{S}(\cdot)$ is symmetric, i.e., that $\mathsf{S}(\theta) = \mathsf{S}(-\theta)$, and that it is nonnegative. The symmetry of $\mathsf{S}(\cdot)$ follows directly from its definition (13.25) and from the fact that $\mathsf{K}_{XX}$, like every autocovariance function, is symmetric (Theorem 13.5.2 (i)).

We next prove that $\mathsf{S}(\cdot)$ is nonnegative. From (13.26) it follows that $\mathsf{S}(\cdot)$ can only be negative on a subset of the interval $[-1/2, 1/2)$ of Lebesgue measure zero (Proposition 13.6.3 (i)). And since $\mathsf{S}(\cdot)$ is continuous, this implies that $\mathsf{S}(\cdot)$ is nonnegative. $\qquad\square$

## 13.7 The Spectral Distribution Function

We next briefly discuss the case where $(X_\nu)$ does not necessarily have a power spectral density function. We shall see that in this case too we can express the autocovariance function as the Fourier Series of "something," but this "something" is not an integrable function. (It is, in fact, a measure.) The theorem will also yield a characterization of nonnegative definite functions. The proof, which is based on Herglotz's Theorem, is omitted. The results of this section will not be used in subsequent chapters.

Recall that a random variable taking value in the interval $[-\alpha, \alpha]$ is said to be **symmetric** (or to have a symmetric distribution) if $\Pr[X \leq -\xi] = \Pr[X \geq \xi]$ for all $\xi \in [-\alpha, \alpha]$.

**Theorem 13.7.1.** *A function $\boldsymbol{\rho}\colon \mathbb{Z} \to \mathbb{R}$ is the autocorrelation function of a real WSS SP if, and only if, there exists a symmetric random variable $\Theta$ taking value in the interval $[-1/2, 1/2]$ such that*

$$\rho(\eta) = \mathsf{E}\big[e^{-\mathrm{i}2\pi\eta\Theta}\big], \quad \eta \in \mathbb{Z}. \tag{13.29}$$

*The cumulative distribution function of $\Theta$ is fully determined by $\boldsymbol{\rho}$.*

**Proof.** See (Doob, 1990, Chapter X, § 3, Theorem 3.2), (Pourahmadi, 2001, Theorem 9.22), (Shiryaev, 1996, Chapter VI, § 1.1), or (Porat, 2008, Section 2.8). $\quad\square$

This theorem also characterizes autocovariance functions: a function $\mathsf{K}\colon \mathbb{Z} \to \mathbb{R}$ is the autocovariance function of a real WSS SP if, and only if, there exists a symmetric random variable $\Theta$ taking value in the interval $[-1/2, 1/2]$ and some constant $\alpha \geq 0$ such that

$$\mathsf{K}(\eta) = \alpha\, \mathsf{E}\big[e^{-\mathrm{i}2\pi\eta\Theta}\big], \quad \eta \in \mathbb{Z}. \tag{13.30}$$

(By equating (13.30) at $\eta = 0$ we obtain that $\alpha = \mathsf{K}(0)$, i.e., the variance of the stochastic process.)

Equivalently, we can state the theorem as follows. If $(X_\nu)$ is a real WSS SP, then its autocovariance function $\mathsf{K}_{XX}$ can be expressed as

$$\mathsf{K}_{XX}(\eta) = \mathsf{Var}[X_1]\, \mathsf{E}\left[e^{-\mathrm{i}2\pi\eta\Theta}\right], \quad \eta \in \mathbb{Z} \tag{13.31}$$

for some random variable $\Theta$ taking value in the interval $[-1/2, 1/2]$ according to some symmetric distribution. If, additionally, $\mathsf{Var}[X_1] > 0$, then the cumulative distribution function $F_\Theta(\cdot)$ of $\Theta$ is uniquely determined by $\mathsf{K}_{XX}$.

**Note 13.7.2.**

(i) If the random variable $\Theta$ above has a symmetric density $f_\Theta(\cdot)$, then the process is of PSD $\theta \mapsto \mathsf{Var}[X_1]\, f_\Theta(\theta)$. Indeed, by (13.31) we have for every integer $\eta$

$$\mathsf{K}_{XX}(\eta) = \mathsf{Var}[X_1]\, \mathsf{E}\left[e^{-\mathrm{i}2\pi\eta\Theta}\right]$$

$$= \mathsf{Var}[X_1] \int_{-1/2}^{1/2} f_\Theta(\theta)\, e^{-\mathrm{i}2\pi\eta\theta}\, \mathrm{d}\theta$$

$$= \int_{-1/2}^{1/2} \left(\mathsf{Var}[X_1]\, f_\Theta(\theta)\right) e^{-\mathrm{i}2\pi\eta\theta}\, \mathrm{d}\theta.$$

(ii) Some authors, e.g., (Grimmett and Stirzaker, 2001) refer to the cumulative distribution function $F_\Theta(\cdot)$ of $\Theta$, i.e., to the mapping $\theta \mapsto \Pr[\Theta \leq \theta]$, as the **Spectral Distribution Function** of $(X_\nu)$. This, however, is not standard. It is only in agreement with the more common usage in the case where $\mathsf{Var}[X_1] = 1$.[5]

## 13.8   Exercises

**Exercise 13.1 (Discrete-Time WSS Stochastic Processes).** Prove Proposition 13.3.3.

**Exercise 13.2 (Mapping a Discrete-Time Stationary SP).** Let $(X_\nu)$ be a stationary discrete-time SP, and let $\mathbf{g}\colon \mathbb{R} \to \mathbb{R}$ be some arbitrary (Borel measurable) function. For every $\nu \in \mathbb{Z}$, let $Y_\nu = g(X_\nu)$. Prove that the discrete-time SP $(Y_\nu)$ is stationary.

**Exercise 13.3 (Mapping a Discrete-Time WSS SP).** Let $(X_\nu)$ be a WSS discrete-time SP, and let $\mathbf{g}\colon \mathbb{R} \to \mathbb{R}$ be some arbitrary (Borel measurable) bounded function. For every $\nu \in \mathbb{Z}$, let $Y_\nu = g(X_\nu)$. Must the SP $(Y_\nu)$ be WSS?

**Exercise 13.4 (A Sliding-Window Mapping of a Stationary SP).** Let $(X_\nu)$ be a stationary discrete-time SP, and let $\mathbf{g}\colon \mathbb{R}^2 \to \mathbb{R}$ be some arbitrary (Borel measurable) function. For every $\nu \in \mathbb{Z}$ define $Y_\nu = g(X_{\nu-1}, X_\nu)$. Must $(Y_\nu)$ be stationary?

---

[5]The more common definition is that $\theta \mapsto \mathsf{Var}[X_1]\, \Pr[\Theta \leq \theta]$ is the spectral measure or spectral distribution function. But this is not a distribution function in the probabilistic sense because its value at $\theta = \infty$ is $\mathsf{Var}[X_1]$ which may be different from one.

**Exercise 13.5 (A Sliding-Window Mapping of a WSS SP).** Let $(X_\nu)$ be a WSS discrete-time SP, and let $\mathbf{g} \colon \mathbb{R}^2 \to \mathbb{R}$ be some arbitrary bounded (Borel measurable) function. For every $\nu \in \mathbb{Z}$ define $Y_\nu = g(X_{\nu-1}, X_\nu)$. Must $(Y_\nu)$ be WSS?

**Exercise 13.6 (Existence of a SP).** For which values of $\alpha, \beta \in \mathbb{R}$ is the function

$$\mathsf{K}_{XX}(m) = \begin{cases} 1 & \text{if } m = 0, \\ \alpha & \text{if } m = 1, \\ \beta & \text{if } m = -1, \\ 0 & \text{otherwise,} \end{cases} \quad m \in \mathbb{Z}$$

the autocovariance function of some WSS SP $(X_\nu, \ \nu \in \mathbb{Z})$?

**Exercise 13.7 (Dilating a Stationary SP).** Let $(X_\nu)$ be a stationary discrete-time SP, and define $Y_\nu = X_{2\nu}$ for every $\nu \in \mathbb{Z}$. Must $(Y_\nu)$ be stationary?

**Exercise 13.8 (Inserting Zeros Periodically).** Let $(X_\nu)$ be a stationary discrete-time SP, and let the RV $U$ be independent of it and take on the values 0 and 1 equiprobably. Define for every $\nu \in \mathbb{Z}$

$$Y_\nu = \begin{cases} 0 & \text{if } \nu \text{ is odd} \\ X_{\nu/2} & \text{if } \nu \text{ is even} \end{cases} \quad \text{and} \quad Z_\nu = Y_{\nu+U}. \tag{13.32}$$

Under what conditions is $(Y_\nu)$ stationary? Under what conditions is $(Z_\nu)$ stationary?

**Exercise 13.9 (The Autocovariance Function of a Dilated WSS SP).** Let $(X_\nu)$ be a WSS discrete-time SP of autocovariance function $\mathsf{K}_{XX}$. Define $Y_\nu = X_{2\nu}$ for every $\nu \in \mathbb{Z}$. Must $(Y_\nu)$ be WSS? If so, express its autocovariance function $\mathsf{K}_{YY}$ in terms of $\mathsf{K}_{XX}$.

**Exercise 13.10 (Inserting Zeros Periodically: the Autocovariance Function).** Let $(X_\nu)$ be a WSS discrete-time SP of autocovariance function $\mathsf{K}_{XX}$, and let the RV $U$ be independent of it and take on the values 0 and 1 equiprobably. Define $(Z_\nu)$ as in (13.32). Must $(Z_\nu)$ be WSS? If yes, express its autocovariance function in terms of $\mathsf{K}_{XX}$.

**Exercise 13.11 (Stationary But Not WSS).** Construct a discrete-time stationary SP that is not WSS.

**Exercise 13.12 (Complex Coefficients).** Show that (13.13) will hold for complex numbers $\alpha_1, \ldots, \alpha_n$ provided that we replace the product $\alpha_\nu \alpha_{\nu'}$ with $\alpha_\nu \alpha_{\nu'}^*$. That is, show that if $\mathsf{K}_{XX}$ is the autocovariance function of a real discrete-time WSS SP, then

$$\sum_{\nu=1}^{n} \sum_{\nu'=1}^{n} \alpha_\nu \alpha_{\nu'}^* \, \mathsf{K}_{XX}(\nu - \nu') \geq 0, \quad \alpha_1, \ldots, \alpha_n \in \mathbb{C}.$$

# Chapter 14

# Energy and Power in PAM

## 14.1 Introduction

Energy is an important resource in Digital Communications. The rate at which it is transmitted—the "transmit power"—is critical in battery-operated devices. In satellite applications it is a major consideration in determining the size of the required solar panels, and in wireless systems it influences the interference that one system causes to another. In this chapter we shall discuss the power in PAM signals. To define power we shall need some modeling trickery which will allow us to pretend that the system has been operating since "time $-\infty$" and that it will continue to operate indefinitely. Our definitions and derivations will be mathematically somewhat informal. A more formal account for readers with background in Measure Theory is provided in Section 14.6.

Before discussing power we begin with a discussion of the expected energy in transmitting a finite number of bits.

## 14.2 Energy in PAM

We begin with a seemingly completely artificial problem. Suppose that K independent data bits $D_1, \ldots, D_K$, each taking on the values 0 and 1 equiprobably, are mapped by a mapping $\mathbf{enc} \colon \{0,1\}^K \to \mathbb{R}^N$ to an N-tuple of real numbers $(X_1, \ldots, X_N)$, where $X_\ell$ is the $\ell$-th component of the N-tuple $\mathrm{enc}(D_1, \ldots, D_K)$. Suppose further that the symbols $X_1, \ldots, X_N$ are then mapped to the waveform

$$X(t) = A \sum_{\ell=1}^{N} X_\ell \, g(t - \ell T_\mathrm{s}), \quad t \in \mathbb{R}, \tag{14.1}$$

where $\mathbf{g} \in \mathcal{L}_2$ is an energy-limited real pulse shape, $A \geq 0$ is a scaling factor, and $T_\mathrm{s} > 0$ is the baud period. We seek the expected energy in the waveform $X(\cdot)$.

We assume that $X(\cdot)$ corresponds to the voltage across a unit-load or to the current through a unit-load, so the transmitted energy is the time integral of the mapping $t \mapsto X^2(t)$. Because the data bits are random variables, the signal $X(\cdot)$ is a stochastic process. Its energy $\int_{-\infty}^{\infty} X^2(t)\,dt$ is thus a random variable.[1] If $(\Omega, \mathcal{F}, P)$ is the probability space under consideration, then this RV is the mapping from $\Omega$ to $\mathbb{R}$ defined by

$$\omega \mapsto \int_{-\infty}^{\infty} X^2(\omega, t)\,dt.$$

This RV's expectation—the **expected energy**—is denoted by $\mathsf{E}$ and is given by

$$\mathsf{E} \triangleq \mathrm{E}\left[\int_{-\infty}^{\infty} X^2(t)\,dt\right]. \tag{14.2}$$

Note that even though we are considering the transmission of a finite number of symbols (N), the waveform $X(\cdot)$ may extend in time from $-\infty$ to $+\infty$.

We next derive an explicit expression for $\mathsf{E}$. Starting from (14.2) and using (14.1),

$$\begin{aligned}
\mathsf{E} &= \mathrm{E}\left[\int_{-\infty}^{\infty} X^2(t)\,dt\right] \\
&= \mathsf{A}^2 \mathrm{E}\left[\int_{-\infty}^{\infty} \left(\sum_{\ell=1}^{N} X_\ell\, g(t - \ell\mathsf{T_s})\right)^2 dt\right] \\
&= \mathsf{A}^2 \mathrm{E}\left[\int_{-\infty}^{\infty} \left(\sum_{\ell=1}^{N} X_\ell\, g(t - \ell\mathsf{T_s})\right)\left(\sum_{\ell'=1}^{N} X_{\ell'}\, g(t - \ell'\mathsf{T_s})\right) dt\right] \\
&= \mathsf{A}^2 \mathrm{E}\left[\int_{-\infty}^{\infty} \sum_{\ell=1}^{N} \sum_{\ell'=1}^{N} X_\ell X_{\ell'}\, g(t - \ell\mathsf{T_s})\, g(t - \ell'\mathsf{T_s})\,dt\right] \\
&= \mathsf{A}^2 \int_{-\infty}^{\infty} \sum_{\ell=1}^{N} \sum_{\ell'=1}^{N} \mathrm{E}[X_\ell X_{\ell'}]\, g(t - \ell\mathsf{T_s})\, g(t - \ell'\mathsf{T_s})\,dt \\
&= \mathsf{A}^2 \sum_{\ell=1}^{N} \sum_{\ell'=1}^{N} \mathrm{E}[X_\ell X_{\ell'}] \int_{-\infty}^{\infty} g(t - \ell\mathsf{T_s})\, g(t - \ell'\mathsf{T_s})\,dt \\
&= \mathsf{A}^2 \sum_{\ell=1}^{N} \sum_{\ell'=1}^{N} \mathrm{E}[X_\ell X_{\ell'}]\, \mathsf{R_{gg}}\big((\ell - \ell')\mathsf{T_s}\big), \tag{14.3}
\end{aligned}$$

where $\mathsf{R_{gg}}$ is the self-similarity function of the pulse $g(\cdot)$ (Section 11.2). Here the first equality follows from (14.2); the second from (14.1); the third by writing the square of a number as its product with itself ($\xi^2 = \xi\xi$); the fourth by writing the product of sums as the double sum of products; the fifth by swapping expectation with integration and by the linearity of expectation; the sixth by swapping integration and summation; and the final equality by the definition of the self-similarity function (Definition 11.2.1).

Using Proposition 11.2.2 (iv) we can also express $\mathsf{R_{gg}}$ as

$$\mathsf{R_{gg}}(\tau) = \int_{-\infty}^{\infty} |\hat{g}(f)|^2\, e^{\mathrm{i}2\pi f\tau}\,df, \quad \tau \in \mathbb{R} \tag{14.4}$$

---

[1] There are some slight measure-theoretic mathematical technicalities that we are sweeping under the rug. Those are resolved in Section 14.6.

and hence rewrite (14.3) as

$$E = A^2 \int_{-\infty}^{\infty} \sum_{\ell=1}^{N} \sum_{\ell'=1}^{N} E[X_\ell X_{\ell'}] \, e^{i2\pi f(\ell-\ell')T_s} \left| \hat{g}(f) \right|^2 df. \tag{14.5}$$

We define the **energy per bit** as

$$E_b \left[ \frac{\text{energy}}{\text{bit}} \right] \triangleq \frac{E}{K} \tag{14.6}$$

and the **energy per real symbol** as

$$E_s \left[ \frac{\text{energy}}{\text{real symbol}} \right] \triangleq \frac{E}{N}. \tag{14.7}$$

As we shall see in Section 14.5.2, if infinite data are transmitted using the binary-to-reals $(K, N)$ block encoder $enc(\cdot)$, then the resulting transmitted power $P$ is given by

$$\boxed{P = \frac{E_s}{T_s}.} \tag{14.8}$$

This result will be proved in Section 14.5.2 after we carefully define the average power. The units work out because if we think of $T_s$ as having units of seconds per real symbol then:

$$\frac{E_s \left[ \frac{\text{energy}}{\text{real symbol}} \right]}{T_s \left[ \frac{\text{second}}{\text{real symbol}} \right]} = \frac{E_s}{T_s} \left[ \frac{\text{energy}}{\text{second}} \right]. \tag{14.9}$$

Expression (14.3) for the expected energy $E$ is greatly simplified in two cases that we discuss next. The first is when the pulse shape $\mathbf{g}$ satisfies the orthogonality condition

$$\int_{-\infty}^{\infty} g(t) \, g(t - \kappa T_s) \, dt = \|\mathbf{g}\|_2^2 \, I\{\kappa = 0\}, \quad \kappa \in \{0, 1, \dots, N-1\}. \tag{14.10}$$

In this case (14.3) simplifies to

$$E = A^2 \|\mathbf{g}\|_2^2 \sum_{\ell=1}^{N} E[X_\ell^2], \quad \left( \{t \mapsto g(t - \ell T_s)\}_{\ell=0}^{N-1} \text{ orthogonal} \right). \tag{14.11}$$

(In this case one need not even go through the calculation leading to (14.3); the result simply follows from (14.1) and the Pythagorean Theorem (Theorem 4.5.2).)

The second case for which the computation of $E$ is simplified is when the distribution of $D_1, \dots, D_K$ and the mapping $enc(\cdot)$ result in the real symbols $X_1, \dots, X_N$ being of zero mean and uncorrelated:[2]

$$E[X_\ell] = 0, \quad \ell \in \{1, \dots, N\} \tag{14.12a}$$

---

[2] Actually, it suffices that (14.12b) hold; (14.12a) is not needed.

and

$$\mathsf{E}[X_\ell X_{\ell'}] = \mathsf{E}\big[X_\ell^2\big]\,\mathrm{I}\{\ell = \ell'\}, \quad \ell, \ell' \in \{1, \ldots, \mathsf{N}\}. \tag{14.12b}$$

In this case too (14.3) simplifies to

$$\mathsf{E} = \mathsf{A}^2 \,\|\mathbf{g}\|_2^2 \sum_{\ell=1}^{\mathsf{N}} \mathsf{E}\big[X_\ell^2\big], \quad \Big(\big(X_\ell,\ \ell \in \mathbb{Z}\big) \text{ zero-mean \& uncorrelated}\Big). \tag{14.13}$$

## 14.3  Defining the Power in PAM

If $\big(X(t),\ t \in \mathbb{R}\big)$ is a continuous-time stochastic process describing the voltage across a unit-load or the current through a unit-load, then it is reasonable to define the **power** $\mathsf{P}$ in $\big(X(t),\ t \in \mathbb{R}\big)$ as the limit

$$\mathsf{P} \triangleq \lim_{\mathsf{T} \to \infty} \frac{1}{2\mathsf{T}}\, \mathsf{E}\!\left[\int_{-\mathsf{T}}^{\mathsf{T}} X^2(t)\, \mathrm{d}t\right]. \tag{14.14}$$

But there is a problem. Over its lifetime, a communication system is only used to transmit a finite number of bits, and it only sends a finite amount of energy. Consequently, if $\big(X(t),\ t \in \mathbb{R}\big)$ corresponds to the transmitted waveform over the system's lifetime, then $\mathsf{P}$ as defined in (14.14) will always end up being zero. The definition in (14.14) is thus useless when discussing the transmission of a finite number of bits.

To define power in a useful way we need some modeling trickery. Instead of thinking about the encoder as producing a *finite* number of symbols, we should now pretend that the encoder produces an *infinite* sequence of symbols $\big(X_\ell,\ \ell \in \mathbb{Z}\big)$, which are then mapped to the infinite sum

$$X(t) = \mathsf{A} \sum_{\ell=-\infty}^{\infty} X_\ell\, g(t - \ell\mathsf{T}_\mathsf{s}), \quad t \in \mathbb{R}. \tag{14.15}$$

For the waveform in (14.15), the definition of $\mathsf{P}$ in (14.14) makes perfect sense. Philosophically speaking, the modeling trickery we employ corresponds to measuring power on a time scale much greater than the signaling period $\mathsf{T}_\mathsf{s}$ but much shorter than the system's lifetime.

But philosophy aside, there are still two problems we must address: how to model the generation of the infinite sequence $\big(X_\ell,\ \ell \in \mathbb{Z}\big)$, and how to guarantee that the sum in (14.15) converges for every $t \in \mathbb{R}$. We begin with the latter. If $\mathbf{g}$ is of finite duration, then at every epoch $t \in \mathbb{R}$ only a finite number of terms in (14.15) are nonzero and convergence is thus guaranteed. But we do not want to restrict ourselves to finite-duration pulse shapes because those, by Theorem 6.8.2, cannot be bandlimited. Instead, to guarantee convergence, we shall assume throughout that the following conditions both hold:

1) The symbols $\big(X_\ell,\ \ell \in \mathbb{Z}\big)$ are uniformly bounded in the sense that there exists some constant $\gamma$ such that

$$\big|X_\ell\big| \le \gamma, \quad \ell \in \mathbb{Z}. \tag{14.16}$$

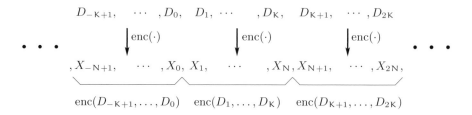

**Figure 14.1:** Bi-Infinite Block Encoding.

2) The pulse shape $t \mapsto g(t)$ decays faster than $1/t$ in the sense that there exist positive constants $\alpha, \beta > 0$ such that

$$|g(t)| \leq \frac{\beta}{1 + |t/T_\mathrm{s}|^{1+\alpha}}, \quad t \in \mathbb{R}. \tag{14.17}$$

Using the fact that the sum $\sum_{n \geq 1} n^{-(1+\alpha)}$ converges whenever $\alpha > 0$ (Rudin, 1976, Theorem 3.28), it is not difficult to show that if both (14.16) and (14.17) hold, then the infinite sum (14.15) converges at every epoch $t \in \mathbb{R}$.

As to the generation of $(X_\ell, \ \ell \in \mathbb{Z})$, we shall consider three scenarios. In the first, which we analyze in Section 14.5.1, we ignore this issue and simply assume that $(X_\ell, \ \ell \in \mathbb{Z})$ is a WSS discrete-time SP of a given autocovariance function. In the second scenario, which we analyze in Section 14.5.2, we tweak the block-encoding mode that we introduced in Section 10.4 to account for a bi-infinite data sequence. We call this tweaked mode **bi-infinite block encoding** and describe it more precisely in Section 14.5.2. It is illustrated in Figure 14.1. Finally, the third scenario, which we analyze in Section 14.5.3, is similar to the first except that we relax some of the statistical assumptions on $(X_\ell, \ \ell \in \mathbb{Z})$. But we only treat the case where the time shifts of the pulse shape by integer multiples of $T_\mathrm{s}$ are orthonormal.

Except in the third scenario, we shall only analyze the power in the stochastic process (14.15) assuming that the symbols $(X_\ell, \ \ell \in \mathbb{Z})$ are of zero mean

$$\mathsf{E}[X_\ell] = 0, \quad \ell \in \mathbb{Z}. \tag{14.18}$$

This not only simplifies the analysis but also makes engineering sense, because it guarantees that $(X(t), \ t \in \mathbb{R})$ is centered

$$\mathsf{E}[X(t)] = 0, \quad t \in \mathbb{R}, \tag{14.19}$$

and, for the reasons that we outline in Section 14.4, transmitting zero-mean waveforms is usually power efficient.

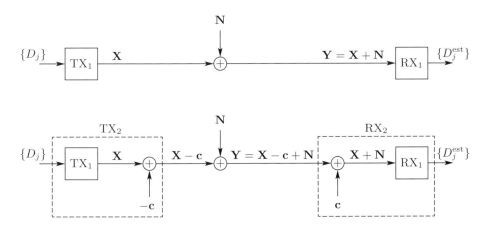

**Figure 14.2:** The above two systems have identical performance. In the former the transmitted power is the power in $t \mapsto X(t)$ whereas in the second it is the power in $t \mapsto X(t) - c(t)$.

## 14.4    On the Mean of Transmitted Waveforms

We next explain why the transmitted waveforms in digital communications are usually designed to be of zero mean.[3] We focus on the case where the transmitted signal suffers only from an additive disturbance. The key observation is that given any transmitter that transmits the SP $\big(X(t),\ t \in \mathbb{R}\big)$ and any receiver, we can design a new transmitter that transmits the waveform $t \mapsto X(t) - c(t)$ and a new receiver with identical performance. Here $c(\cdot)$ is any deterministic signal. Indeed, the new receiver can simply add $c(\cdot)$ to the received signal and then pass on the result to the old receiver. That the old and the new systems have identical performance follows by noting that if $\big(N(t),\ t \in \mathbb{R}\big)$ is the added disturbance, then the received signal on which the old receiver operates is given by $t \mapsto X(t) + N(t)$. And the received signal in the new system is $t \mapsto X(t) - c(t) + N(t)$, so after we add $c(\cdot)$ to this signal we obtain the signal $X(t) + N(t)$, which is equal the signal that the old receiver operated on. Thus, the performance of a system transmitting $X(\cdot)$ can be mimicked on a system transmitting $X(\cdot) - c(\cdot)$ by simply adding $c(\cdot)$ at the receiver. See Figure 14.2.

The addition at the receiver of $c(\cdot)$ entails no change in the transmitted power. Therefore, if a system transmits $X(\cdot)$, then we might be able to improve its power efficiency without hurting its performance by cleverly choosing $c(\cdot)$ so that the power in $X(\cdot) - c(\cdot)$ be smaller than the power in $X(\cdot)$ and by then transmitting $t \mapsto X(t) - c(t)$ instead of $t \mapsto X(t)$. The only additional change we would need to make is to add $c(\cdot)$ at the receiver.

How should we choose $c(\cdot)$? To answer this we shall need the following lemma.

---

[3]This, however, is not the case with some wireless systems that transmit training sequences to help the receiver learn the channel and acquire timing information.

**Lemma 14.4.1.** *If $W$ is a random variable of finite variance, then*

$$\mathsf{E}\left[(W - c)^2\right] \geq \mathsf{Var}[W], \quad c \in \mathbb{R} \tag{14.20}$$

*with equality if, and only if,*

$$c = \mathsf{E}[W]. \tag{14.21}$$

**Proof.**

$$\begin{aligned}
\mathsf{E}\left[(W - c)^2\right] &= \mathsf{E}\left[\left((W - \mathsf{E}[W]) + (\mathsf{E}[W] - c)\right)^2\right] \\
&= \mathsf{E}\left[(W - \mathsf{E}[W])^2\right] + 2\underbrace{\mathsf{E}[W - \mathsf{E}[W]]}_{0}(\mathsf{E}[W] - c) + (\mathsf{E}[W] - c)^2 \\
&= \mathsf{E}\left[(W - \mathsf{E}[W])^2\right] + (\mathsf{E}[W] - c)^2 \\
&\geq \mathsf{E}\left[(W - \mathsf{E}[W])^2\right] \\
&= \mathsf{Var}[W],
\end{aligned}$$

with equality if, and only if, $c = \mathsf{E}[W]$. $\qquad\qquad\square$

With the aid of Lemma 14.4.1 we can now choose $c(\cdot)$ to minimize the power in $t \mapsto X(t) - c(t)$ as follows. Keeping the definition of power (14.14) in mind, we study

$$\frac{1}{2\mathsf{T}}\int_{-\mathsf{T}}^{\mathsf{T}}\mathsf{E}\left[\left(X(t) - c(t)\right)^2\right]\,\mathrm{d}t$$

and note that this expression is minimized over all choices of the waveform $c(\cdot)$ by minimizing the integrand, i.e., by choosing at every epoch $t$ the value of $c(t)$ to be the one that mininimizes $\mathsf{E}\left[\left(X(t) - c(t)\right)^2\right]$. By Lemma 14.4.1 this corresponds to choosing $c(t)$ to be $\mathsf{E}[X(t)]$. It is thus optimal to choose $c(\cdot)$ as

$$c(t) = \mathsf{E}[X(t)], \quad t \in \mathbb{R}. \tag{14.22}$$

This choice results in the transmitted waveform being $t \mapsto X(t) - \mathsf{E}[X(t)]$, i.e., in the transmitted waveform being of zero mean.

Stated differently, if in a given system the transmitted waveform is not of zero mean, then a new system can be built that transmits a waveform of lower (or equal) average power and whose performance on any additive noise channel is identical.

## 14.5    Computing the Power in PAM

We proceed to compute the power in the signal

$$X(t) = \mathsf{A}\sum_{\ell=-\infty}^{\infty}X_\ell\, g(t - \ell\mathsf{T_s}), \quad t \in \mathbb{R} \tag{14.23}$$

under various assumptions on the bi-infinite random sequence $(X_\ell, \ \ell \in \mathbb{Z})$. We assume throughout that Conditions (14.16) & (14.17) are satisfied so the infinite sum converges at every epoch $t \in \mathbb{R}$. The power P is defined as in (14.14).[4]

## 14.5.1 $(X_\ell)$ Is Zero-Mean and WSS

Here we compute the power in the signal (14.23) when $(X_\ell, \ \ell \in \mathbb{Z})$ is a centered WSS SP of autocovariance function $\mathsf{K}_{XX}$:

$$\mathsf{E}[X_\ell] = 0, \quad \ell \in \mathbb{Z}, \tag{14.24a}$$

$$\mathsf{E}[X_\ell X_{\ell+m}] = \mathsf{K}_{XX}(m), \quad \ell, m \in \mathbb{Z}. \tag{14.24b}$$

We further assume that the pulse shape satisfies the decay condition (14.17) and that the process $(X_\ell, \ \ell \in \mathbb{Z})$ satisfies the boundedness condition (14.16).

We begin by calculating the expected energy of $X(\cdot)$ in a half-open interval $[\tau, \tau + \mathsf{T_s})$ of length $\mathsf{T_s}$ and in showing that this expected energy does not depend on $\tau$, i.e., that the expected energy in all intervals of length $\mathsf{T_s}$ are identical. We calculate the energy in the interval $[\tau, \tau + \mathsf{T_s})$ as follows:

$$
\begin{aligned}
\mathsf{E}&\left[ \int_\tau^{\tau+\mathsf{T_s}} X^2(t)\,\mathrm{d}t \right] \\
&= \mathsf{A}^2 \int_\tau^{\tau+\mathsf{T_s}} \mathsf{E}\left[ \left( \sum_{\ell=-\infty}^{\infty} X_\ell\, g(t - \ell\mathsf{T_s}) \right)^2 \right] \mathrm{d}t &&(14.25) \\
&= \mathsf{A}^2 \int_\tau^{\tau+\mathsf{T_s}} \mathsf{E}\left[ \sum_{\ell=-\infty}^{\infty} \sum_{\ell'=-\infty}^{\infty} X_\ell X_{\ell'}\, g(t - \ell\mathsf{T_s})\, g(t - \ell'\mathsf{T_s}) \right] \mathrm{d}t \\
&= \mathsf{A}^2 \int_\tau^{\tau+\mathsf{T_s}} \sum_{\ell=-\infty}^{\infty} \sum_{\ell'=-\infty}^{\infty} \mathsf{E}[X_\ell X_{\ell'}]\, g(t - \ell\mathsf{T_s})\, g(t - \ell'\mathsf{T_s})\,\mathrm{d}t \\
&= \mathsf{A}^2 \int_\tau^{\tau+\mathsf{T_s}} \sum_{\ell=-\infty}^{\infty} \sum_{m=-\infty}^{\infty} \mathsf{E}[X_\ell X_{\ell+m}]\, g(t - \ell\mathsf{T_s})\, g\big(t - (\ell+m)\mathsf{T_s}\big)\,\mathrm{d}t \\
&= \mathsf{A}^2 \int_\tau^{\tau+\mathsf{T_s}} \sum_{m=-\infty}^{\infty} \mathsf{K}_{XX}(m) \sum_{\ell=-\infty}^{\infty} g(t - \ell\mathsf{T_s})\, g\big(t - (\ell+m)\mathsf{T_s}\big)\,\mathrm{d}t \\
&= \mathsf{A}^2 \sum_{m=-\infty}^{\infty} \mathsf{K}_{XX}(m) \sum_{\ell=-\infty}^{\infty} \int_{\tau-\ell\mathsf{T_s}}^{\tau+\mathsf{T_s}-\ell\mathsf{T_s}} g(t')\, g(t' - m\mathsf{T_s})\,\mathrm{d}t' &&(14.26) \\
&= \mathsf{A}^2 \sum_{m=-\infty}^{\infty} \mathsf{K}_{XX}(m) \int_{-\infty}^{\infty} g(t')\, g(t' - m\mathsf{T_s})\,\mathrm{d}t' \\
&= \mathsf{A}^2 \sum_{m=-\infty}^{\infty} \mathsf{K}_{XX}(m)\, \mathsf{R_{gg}}(m\mathsf{T_s}), \quad \tau \in \mathbb{R}, &&(14.27)
\end{aligned}
$$

---

[4]A general mathematical definition of the power of a stochastic process is given in Definition 14.6.1 ahead.

where the first equality follows by the structure of $X(\cdot)$ (14.15); the second by writing $X^2(t)$ as $X(t)\,X(t)$ and rearranging terms; the third by the linearity of the expectation, which allows us to swap the double sum and the expectation and to take the deterministic term $g(t - \ell T_s)g(t - \ell' T_s)$ outside the expectation; the fourth by defining $m \triangleq \ell' - \ell$; the fifth by (14.24b); the sixth by defining $t' \triangleq t - \ell T_s$; the seventh by noting that the integrals of a function over all the intervals $[\tau - \ell T_s, \tau - \ell T_s + T_s)$ sum to the integral over the entire real line; and the final by the definition of the self-similarity function $R_{gg}$ (Section 11.2).

Note that, indeed, the RHS of (14.27) does not depend on the epoch $\tau$ at which the length-$T_s$ time interval starts. This observation will now help us to compute the power in $X(\cdot)$. Since the interval $[-T, +T]$ contains $\lfloor (2T)/T_s \rfloor$ disjoint intervals of the form $[\tau, \tau + T_s)$, and since it is contained in the union of $\lceil (2T)/T_s \rceil$ such intervals, it follows that

$$\left\lfloor \frac{2T}{T_s} \right\rfloor \mathsf{E}\left[\int_\tau^{\tau+T_s} X^2(t)\,\mathrm{d}t\right] \le \mathsf{E}\left[\int_{-T}^{T} X^2(t)\,\mathrm{d}t\right] \le \left\lceil \frac{2T}{T_s} \right\rceil \mathsf{E}\left[\int_\tau^{\tau+T_s} X^2(t)\,\mathrm{d}t\right], \quad (14.28)$$

where we use $\lfloor \xi \rfloor$ to denote the greatest integer smaller than or equal to $\xi$ (e.g., $\lfloor 4.2 \rfloor = 4$), and where we use $\lceil \xi \rceil$ to denote the smallest integer that is greater than or equal to $\xi$ (e.g., $\lceil 4.2 \rceil = 5$) so

$$\xi - 1 < \lfloor \xi \rfloor \le \lceil \xi \rceil < \xi + 1, \quad \xi \in \mathbb{R}. \quad (14.29)$$

Note that from (14.29) and the Sandwich Theorem it follows that

$$\lim_{T\to\infty} \frac{1}{2T} \left\lfloor \frac{2T}{T_s} \right\rfloor = \lim_{T\to\infty} \frac{1}{2T} \left\lceil \frac{2T}{T_s} \right\rceil = \frac{1}{T_s}, \quad T_s > 0. \quad (14.30)$$

Dividing (14.28) by 2T and using (14.30) we obtain that

$$\lim_{T\to\infty} \frac{1}{2T} \mathsf{E}\left[\int_{-T}^{T} X^2(t)\,\mathrm{d}t\right] = \frac{1}{T_s} \mathsf{E}\left[\int_\tau^{\tau+T_s} X^2(t)\,\mathrm{d}t\right],$$

which combines with (14.27) to yield

$$P = \frac{1}{T_s} A^2 \sum_{m=-\infty}^{\infty} \mathsf{K}_{XX}(m)\, \mathsf{R}_{gg}(m T_s). \quad (14.31)$$

The power P can be alternatively expressed in the frequency domain using (14.31) and (14.4) as

$$P = \frac{A^2}{T_s} \int_{-\infty}^{\infty} \sum_{m=-\infty}^{\infty} \mathsf{K}_{XX}(m)\, e^{i2\pi f m T_s}\, |\hat{g}(f)|^2\, \mathrm{d}f. \quad (14.32)$$

An important special case of (14.31) is when the symbols $(X_\ell)$ are zero-mean, uncorrelated, and of equal variance $\sigma_X^2$. In this case $\mathsf{K}_{XX}(m) = \sigma_X^2\, \mathsf{I}\{m = 0\}$, and the only nonzero term in (14.31) is the term corresponding to $m = 0$ so

$$P = \frac{1}{T_s} A^2\, \|\mathbf{g}\|_2^2\, \sigma_X^2, \quad \left((X_\ell) \text{ centered, variance } \sigma_X^2, \text{ uncorrelated}\right). \quad (14.33)$$

## 14.5.2 Bi-Infinite Block-Mode

The bi-infinite block-mode with a $(\mathsf{K}, \mathsf{N})$ binary-to-reals block encoder

$$\mathbf{enc} \colon \{0, 1\}^{\mathsf{K}} \to \mathbb{R}^{\mathsf{N}}$$

is depicted in Figure 14.1 and can be described as follows. A bi-infinite sequence of data bits $\left(D_j,\ j \in \mathbb{Z}\right)$ is fed to an encoder. The encoder parses this sequences into K-tuples and defines for every integer $\nu \in \mathbb{Z}$ the "$\nu$-th data block" $\mathbf{D}_\nu$

$$\mathbf{D}_\nu \triangleq \left(D_{\nu\mathsf{K}+1}, \ldots, D_{\nu\mathsf{K}+\mathsf{K}}\right), \quad \nu \in \mathbb{Z}. \tag{14.34}$$

Each data block $\mathbf{D}_\nu$ is then mapped by $\mathrm{enc}(\cdot)$ to a real N-tuple, which we denote by $\mathbf{X}_\nu$:

$$\mathbf{X}_\nu \triangleq \mathrm{enc}(\mathbf{D}_\nu), \quad \nu \in \mathbb{Z}. \tag{14.35}$$

The bi-infinite sequence $\left(X_\ell,\ \ell \in \mathbb{Z}\right)$ produced by the encoder is the concatenation of these N-tuples so

$$\left(X_{\nu\mathsf{N}+1}, \ldots, X_{\nu\mathsf{N}+\mathsf{N}}\right) = \mathbf{X}_\nu, \quad \nu \in \mathbb{Z}. \tag{14.36}$$

Stated differently, for every $\nu \in \mathbb{Z}$ and $\eta \in \{1, \ldots, \mathsf{N}\}$, the symbol $X_{\nu\mathsf{N}+\eta}$ is the $\eta$-th component of the N-tuple $\mathbf{X}_\nu$. The transmitted signal $X(\cdot)$ is as in (14.15) with the pulse shape $\mathbf{g}$ satisfying the decay condition (14.17) and with $\mathsf{T}_\mathrm{s} > 0$ being arbitrary. (The boundedness condition (14.16) is always guaranteed in bi-infinite block encoding.)

We next compute the power $\mathsf{P}$ in $X(\cdot)$ under the assumption that the data bits $\left(D_j,\ j \in \mathbb{Z}\right)$ are independent and identically distributed (IID) random bits, where we adopt the following definition.

**Definition 14.5.1 (IID Random Bits).** *We say that a collection of random variables are **IID random bits** if the random variables are independent and each of them takes on the values $0$ and $1$ equiprobably.*

The assumption that the bi-infinite data sequence $\left(D_j,\ j \in \mathbb{Z}\right)$ consists of IID random bits is equivalent to the assumption that the K-tuples $\left(\mathbf{D}_\nu,\ \nu \in \mathbb{Z}\right)$ are IID with $\mathbf{D}_\nu$ being uniformly distributed over the set of binary K-tuples $\{0, 1\}^{\mathsf{K}}$. We shall also assume that the real N-tuple $\mathrm{enc}(\mathbf{D})$ is of zero mean whenever the binary K-tuple is uniformly distributed over $\{0, 1\}^{\mathsf{K}}$. We will show that, subject to these assumptions,

$$\mathsf{P} = \frac{1}{\mathsf{N}\mathsf{T}_\mathrm{s}} \mathsf{E}\left[\int_{-\infty}^{\infty} \left(\mathsf{A}\sum_{\ell=1}^{\mathsf{N}} X_\ell\, g(t - \ell\mathsf{T}_\mathrm{s})\right)^2 \mathrm{d}t\right]. \tag{14.37}$$

This expression has an interesting interpretation. On the LHS is the power in the transmitted signal in bi-infinite block encoding using the $(\mathsf{K}, \mathsf{N})$ binary-to-reals block encoder $\mathrm{enc}(\cdot)$. On the RHS is the quantity $\mathsf{E}/(\mathsf{N}\mathsf{T}_\mathrm{s})$, where $\mathsf{E}$, as in (14.3), is the expected energy in the signal that results when only the K-tuple $(D_1, \ldots, D_\mathsf{K})$ is transmitted from time $-\infty$ to time $+\infty$. Using the definition of the energy per-symbol $E_s$ (14.7) we can also rewrite (14.37) as in (14.8). Thus, in bi-infinite block-mode, the transmitted power is the energy per real symbol $E_s$ normalized by the signaling period $T_s$. Also, by (14.5), we can rewrite (14.37) as

$$P = \frac{A^2}{NT_s} \int_{-\infty}^{\infty} \sum_{\ell=1}^{N} \sum_{\ell'=1}^{N} E[X_\ell X_{\ell'}] e^{i2\pi f(\ell-\ell')T_s} |\hat{g}(f)|^2 \, df. \qquad (14.38)$$

To derive (14.37) we first express the transmitted waveform $X(\cdot)$ as

$$X(t) = A \sum_{\ell=-\infty}^{\infty} X_\ell \, g(t - \ell T_s)$$

$$= A \sum_{\nu=-\infty}^{\infty} \sum_{\eta=1}^{N} X_{\nu N+\eta} \, g\big(t - (\nu N + \eta)T_s\big)$$

$$= A \sum_{\nu=-\infty}^{\infty} u\big(\mathbf{X}_\nu, t - \nu N T_s\big), \quad t \in \mathbb{R}, \qquad (14.39)$$

where the function $\mathbf{u} \colon \mathbb{R}^N \times \mathbb{R} \to \mathbb{R}$ is given by

$$\mathbf{u} \colon (x_1, \ldots, x_N, t) \mapsto \sum_{\eta=1}^{N} x_\eta \, g(t - \eta T_s). \qquad (14.40)$$

We now make three observations. The first is that because the law of $\mathbf{D}_\nu$ does not depend on $\nu$, neither does the law of $\mathbf{X}_\nu$ ($= \mathrm{enc}(\mathbf{D}_\nu)$):

$$\mathbf{X}_\nu \stackrel{\mathscr{L}}{=} \mathbf{X}_{\nu'}, \quad \nu, \nu' \in \mathbb{Z}. \qquad (14.41)$$

The second is that the assumption that $\mathrm{enc}(\mathbf{D})$ is of zero mean whenever $\mathbf{D}$ is uniformly distributed over $\{0,1\}^K$ implies by (14.40) that

$$E\big[u(\mathbf{X}_\nu, t)\big] = 0, \quad \big(\nu \in \mathbb{Z}, \, t \in \mathbb{R}\big). \qquad (14.42)$$

The third is that the hypothesis that the data bits $\big(D_j, \, j \in \mathbb{Z}\big)$ are IID implies that $\big(\mathbf{D}_\nu, \, \nu \in \mathbb{Z}\big)$ are IID and hence that $\big(\mathbf{X}_\nu, \, \nu \in \mathbb{Z}\big)$ are also IID. Consequently, since the independence of $\mathbf{X}_\nu$ and $\mathbf{X}_{\nu'}$ implies the independence of $u\big(\mathbf{X}_\nu, t\big)$ and $u\big(\mathbf{X}_{\nu'} t'\big)$, it follows from (14.42) that

$$E\big[u(\mathbf{X}_\nu, t)\, u(\mathbf{X}_{\nu'}, t')\big] = 0, \quad \big(t, t' \in \mathbb{R}, \, \nu \neq \nu', \, \nu, \nu' \in \mathbb{Z}\big). \qquad (14.43)$$

Using (14.39) and these three observations we can now compute for any epoch $\tau \in \mathbb{R}$ the expected energy in the time interval $[\tau, \tau + NT_s)$ as

$$\int_{\tau}^{\tau+NT_s} E\big[X^2(t)\big] \, dt$$

$$= \int_{\tau}^{\tau+NT_s} E\left[\left(A \sum_{\nu=-\infty}^{\infty} u\big(\mathbf{X}_\nu, t - \nu N T_s\big)\right)^2\right] dt$$

$$= A^2 \int_\tau^{\tau+NT_s} \sum_{\nu=-\infty}^{\infty} \sum_{\nu'=-\infty}^{\infty} \mathsf{E}\Big[u\big(\mathbf{X}_\nu, t - \nu NT_s\big)\, u\big(\mathbf{X}_{\nu'}, t - \nu' NT_s\big)\Big]\, dt$$

$$= A^2 \int_\tau^{\tau+NT_s} \sum_{\nu=-\infty}^{\infty} \mathsf{E}\big[u^2\big(\mathbf{X}_\nu, t - \nu NT_s\big)\big]\, dt$$

$$= A^2 \int_\tau^{\tau+NT_s} \sum_{\nu=-\infty}^{\infty} \mathsf{E}\big[u^2\big(\mathbf{X}_0, t - \nu NT_s\big)\big]\, dt$$

$$= A^2 \sum_{\nu=-\infty}^{\infty} \int_{\tau-\nu NT_s}^{\tau-(\nu-1)NT_s} \mathsf{E}\big[u^2\big(\mathbf{X}_0, t'\big)\big]\, dt'$$

$$= A^2 \int_{-\infty}^{\infty} \mathsf{E}\big[u^2\big(\mathbf{X}_0, t'\big)\big]\, dt'$$

$$= \mathsf{E}\left[\int_{-\infty}^{\infty} \left(A \sum_{\ell=1}^{N} X_\ell\, g(t' - \ell T_s)\right)^2 dt'\right], \quad \tau \in \mathbb{R}, \tag{14.44}$$

where the first equality follows from (14.39); the second by writing the square as a product and by using the linearity of expectation; the third from (14.43); the fourth because the law of $\mathbf{X}_\nu$ does not depend on $\nu$ (14.41); the fifth by changing the integration variable to $t' \triangleq t - NT_s$; the sixth because the sum of the integrals is equal to the integral over $\mathbb{R}$; and the seventh by (14.40).

Note that, indeed, the RHS of (14.44) does not depend on the starting epoch $\tau$ of the interval. Because there are $\lfloor 2T/(NT_s) \rfloor$ disjoint length-$NT_s$ half-open intervals contained in the interval $[-T, T)$ and because $\lceil 2T/(NT_s) \rceil$ such intervals suffice to cover the interval $[-T, T)$, it follows that

$$\left\lfloor \frac{2T}{NT_s} \right\rfloor \mathsf{E}\left[\int_{-\infty}^{\infty} \left(A \sum_{\ell=1}^{N} X_\ell\, g(t - \ell T_s)\right)^2 dt\right]$$

$$\leq \mathsf{E}\left[\int_T^T X^2(t)\, dt\right] \leq$$

$$\left\lceil \frac{2T}{NT_s} \right\rceil \mathsf{E}\left[\int_{-\infty}^{\infty} \left(A \sum_{\ell=1}^{N} X_\ell\, g(t - \ell T_s)\right)^2 dt\right].$$

Dividing by $2T$ and then letting $T$ tend to infinity establishes (14.37).


### 14.5.3   Time Shifts of Pulse Shape Are Orthonormal

We next consider the power in PAM when the time shifts of the real pulse shape by integer multiples of $T_s$ are orthonormal. To remind the reader of this assumption, we change notation and denote the pulse shape by $\phi(\cdot)$ and express the orthonormality condition as

$$\int_{-\infty}^{\infty} \phi(t - \ell T_s)\, \phi(t - \ell' T_s)\, dt = \mathsf{I}\{\ell = \ell'\}, \quad \ell, \ell' \in \mathbb{Z}. \tag{14.45}$$

The calculation of the power is a bit tricky because (14.45) only guarantees that the time shifts of the pulse shape are orthogonal over the interval $(-\infty, \infty)$; they need not be orthogonal over the interval $[-T, +T]$ (even for very large T). Nevertheless, intuition suggests that if $\ell T_s$ and $\ell' T_s$ are both much smaller than T, then the orthogonality of $t \mapsto \phi(t - \ell T_s)$ and $t \mapsto \phi(t - \ell' T_s)$ over the interval $(-\infty, \infty)$ should imply that they are nearly orthogonal over $[-T, T]$. Making this intuition rigorous is a bit tricky and the calculation of the energy in the interval $[-T, T]$ requires a fair number of approximations that must be justified.

To control these approximations we shall assume a decay condition on the pulse shape that is identical to (14.17). Thus, we shall assume that there exist positive constants $\alpha$ and $\beta$ such that

$$|\phi(t)| \leq \frac{\beta}{1 + |t/T_s|^{1+\alpha}}, \quad t \in \mathbb{R}. \tag{14.46}$$

(The pulse shapes used in practice, like those we encountered in (11.31), typically decay like $1/|t|^2$ so this is not a serious restriction.) We shall also continue to assume the boundedness condition (14.16) but otherwise make no statistical assumptions on the symbols $(X_\ell, \ell \in \mathbb{Z})$.

The main result of this section is the next theorem.

**Theorem 14.5.2.** *Let the continuous-time SP $(X(t), t \in \mathbb{R})$ be given by*

$$X(t) = A \sum_{\ell=-\infty}^{\infty} X_\ell \, \phi(t - \ell T_s), \quad t \in \mathbb{R}, \tag{14.47}$$

*where $A \geq 0$; $T_s > 0$; the pulse shape $\phi(\cdot)$ is a Borel measurable function satisfying the orthogonality condition (14.45) and the decay condition (14.46); and where the random sequence $(X_\ell, \ell \in \mathbb{Z})$ satisfies the boundedness condition (14.16). Then*

$$\lim_{T \to \infty} \frac{1}{2T} \, \mathsf{E}\!\left[ \int_{-T}^{T} X^2(t) \, dt \right] = \frac{A^2}{T_s} \lim_{L \to \infty} \frac{1}{2L+1} \sum_{\ell=-L}^{L} \mathsf{E}\!\left[ X_\ell^2 \right], \tag{14.48}$$

*whenever the limit on the RHS exists.*

**Proof.** The proof is somewhat technical and may be skipped. We begin by arguing that it suffices to prove the theorem for the case where $T_s = 1$. To see this, assume that $T_s > 0$ is not necessarily equal to 1. Define the function

$$\tilde{\phi}(t) = \sqrt{T_s} \, \phi(T_s t), \quad t \in \mathbb{R}, \tag{14.49}$$

and note that, by changing the integration variable to $\tau \triangleq t T_s$,

$$\int_{-\infty}^{\infty} \tilde{\phi}(t - \ell) \, \tilde{\phi}(t - \ell') \, dt = \int_{-\infty}^{\infty} \phi(\tau - \ell T_s) \, \phi(\tau - \ell' T_s) \, d\tau$$

$$= \mathrm{I}\{\ell = \ell'\}, \quad \ell, \ell' \in \mathbb{Z}, \tag{14.50a}$$

where the second equality follows from the theorem's assumption about the orthogonality of the time shifts of $\phi$ by integer multiples of $T_s$. Also, by (14.49) and (14.46) we obtain

$$
\begin{aligned}
|\tilde{\phi}(t)| &= \sqrt{T_s}\,|\phi(T_s t)| \\
&\leq \sqrt{T_s}\,\frac{\beta}{1+|t|^{1+\alpha}} \\
&= \frac{\beta'}{1+|t|^{1+\alpha}}, \quad t \in \mathbb{R},
\end{aligned}
\tag{14.50b}
$$

for some $\beta' > 0$ and $\alpha > 0$.

As to the power, by changing the integration variable to $\sigma \triangleq t/T_s$ we obtain

$$
\frac{1}{2T}\int_{-T}^{T}\left(\sum_{\ell\in\mathbb{Z}}X_\ell\,\phi(t-\ell T_s)\right)^2 dt = \frac{1}{T_s}\frac{1}{2(T/T_s)}\int_{-T/T_s}^{T/T_s}\left(\sum_{\ell\in\mathbb{Z}}X_\ell\,\tilde{\phi}(\sigma-\ell)\right)^2 d\sigma. \tag{14.50c}
$$

It now follows from (14.50a) & (14.50b) that if we prove the theorem for the pulse shape $\tilde{\phi}$ with $T_s = 1$, it will then follow that the power in $\sum X_\ell\,\tilde{\phi}(\sigma - \ell)$ is equal to $\lim_{L\to\infty}(2L+1)^{-1}\sum E[X_\ell^2]$ and that consequently, by (14.50c), the power in $\sum X_\ell\,\phi(t-\ell T_s)$ is equal $T_s^{-1}\lim_{L\to\infty}(2L+1)^{-1}\sum E[X_\ell^2]$. In the remainder of the proof we shall thus assume that $T_s = 1$ and express the decay condition (14.46) as

$$
|\phi(t)| \leq \frac{\beta}{1+|t|^{1+\alpha}}, \quad t \in \mathbb{R} \tag{14.51}
$$

for some $\beta, \alpha > 0$.

To further simplify notation we shall assume that $T$ is a positive integer. Indeed, if the limit is proved for positive integers, then the general result follows from the Sandwich Theorem by noting that for $T > 0$ (not necessarily an integer)

$$
\frac{\lfloor T\rfloor}{T}\frac{1}{\lfloor T\rfloor}\int_{-\lfloor T\rfloor}^{\lfloor T\rfloor}\left(\sum_{\ell\in\mathbb{Z}}X_\ell\,\phi(t-\ell)\right)^2 dt
$$

$$
\leq \frac{1}{T}\int_{-T}^{T}\left(\sum_{\ell\in\mathbb{Z}}X_\ell\,\phi(t-\ell)\right)^2 dt \leq
$$

$$
\frac{\lceil T\rceil}{T}\frac{1}{\lceil T\rceil}\int_{-\lceil T\rceil}^{\lceil T\rceil}\left(\sum_{\ell\in\mathbb{Z}}X_\ell\,\phi(t-\ell)\right)^2 dt \tag{14.52}
$$

and by noting that both $\lfloor T\rfloor/T$ and $\lceil T\rceil/T$ tend to 1, as $T\to\infty$.

We thus proceed to prove (14.48) for the case where $T_s = 1$ and where the limit $T\to\infty$ is only over positive integers. We also assume $A = 1$ because both sides of (14.48) scale like $A^2$. We begin by introducing some notation. For every integer $\ell$ we denote the mapping $t\mapsto\phi(t-\ell)$ by $\phi_\ell$, and for every positive integer $T$ we denote the windowed mapping $t\mapsto\phi(t-\ell)\,I\{|t|\leq T\}$ by $\phi_{\ell,w}$. Finally, we fix some (large) integer $\nu > 0$ and define for every $T > \nu$, the random processes

$$\mathbf{X}_0 = \sum_{|\ell| \leq T-\nu} X_\ell \, \phi_{\ell,\mathrm{w}}, \tag{14.53}$$

$$\mathbf{X}_1 = \sum_{T-\nu < |\ell| \leq T+\nu} X_\ell \, \phi_{\ell,\mathrm{w}}, \tag{14.54}$$

$$\mathbf{X}_2 = \sum_{T+\nu < |\ell| < \infty} X_\ell \, \phi_{\ell,\mathrm{w}}, \tag{14.55}$$

and the unwindowed version of $\mathbf{X}_0$

$$\mathbf{X}_0^{\mathrm{u}} = \sum_{|\ell| \leq T-\nu} X_\ell \, \phi_\ell \tag{14.56}$$

so

$$
\begin{aligned}
X(t)\, \mathrm{I}\{|t| \leq T\} &= X_0(t) + X_1(t) + X_2(t) \\
&= X_0^{\mathrm{u}} + \big(X_0(t) - X_0^{\mathrm{u}}(t)\big) + X_1(t) + X_2(t), \quad t \in \mathbb{R}. 
\end{aligned} \tag{14.57}
$$

Using arguments very similar to the ones leading to (4.14) (with integration replaced by integration and expectation) one can show that (14.57) leads to the bound

$$
\left( \sqrt{\mathsf{E}\Big[\|\mathbf{X}_0^{\mathrm{u}}\|_2^2\Big]} - \sqrt{\mathsf{E}\Big[\big\|(\mathbf{X}_0 - \mathbf{X}_0^{\mathrm{u}}) + \mathbf{X}_1 + \mathbf{X}_2\big\|_2^2\Big]} \right)^2
$$

$$
\leq \mathsf{E}\left[ \int_{-T}^{T} X^2(t)\, \mathrm{d}t \right] \leq
$$

$$
\left( \sqrt{\mathsf{E}\Big[\|\mathbf{X}_0^{\mathrm{u}}\|_2^2\Big]} + \sqrt{\mathsf{E}\Big[\big\|(\mathbf{X}_0 - \mathbf{X}_0^{\mathrm{u}}) + \mathbf{X}_1 + \mathbf{X}_2\big\|_2^2\Big]} \right)^2 . \tag{14.58}
$$

Note that, by the orthonormality assumption on the time shifts of $\phi$,

$$\|\mathbf{X}_0^{\mathrm{u}}\|_2^2 = \sum_{|\ell| \leq T-\nu} X_\ell^2$$

so

$$\lim_{T\to\infty} \frac{1}{2T}\, \mathsf{E}\Big[\|\mathbf{X}_0^{\mathrm{u}}\|_2^2\Big] = \lim_{L\to\infty} \frac{1}{2L+1} \sum_{|\ell| \leq L} \mathsf{E}\big[X_\ell^2\big] . \tag{14.59}$$

It follows from (14.58) and (14.59) that to conclude the proof of the theorem it suffices to show that for every fixed $\nu \geq 2$ we have for $T$ exceeding $\nu$

$$\lim_{T\to\infty} \frac{1}{2T}\mathsf{E}\Big[\|\mathbf{X}_1\|_2^2\Big] = 0, \tag{14.60}$$

$$\lim_{T\to\infty} \frac{1}{2T}\mathsf{E}\Big[\|\mathbf{X}_0 - \mathbf{X}_0^{\mathrm{u}}\|_2^2\Big] = 0, \tag{14.61}$$

and that

$$\lim_{\nu\to\infty} \overline{\lim_{T\to\infty}} \frac{1}{2T}\mathsf{E}\Big[\|\mathbf{X}_2\|_2^2\Big] = 0. \tag{14.62}$$

We begin with (14.60), which follows directly from the Triangle Inequality,

$$\|\mathbf{X}_1\|_2 \leq \sum_{\mathsf{T}-\nu < |\ell| \leq \mathsf{T}+\nu} |X_\ell| \|\boldsymbol{\phi}_{\ell,\mathrm{w}}\|_2$$

$$\leq 4\nu\gamma,$$

where the second inequality follows from the boundedness condition (14.16), from the fact that $\boldsymbol{\phi}_{\ell,\mathrm{w}}$ is a windowed version of the unit-energy signal $\boldsymbol{\phi}_\ell$ so $\|\boldsymbol{\phi}_{\ell,\mathrm{w}}\|_2 \leq \|\boldsymbol{\phi}\|_2 = 1$, and because there are $4\nu$ terms in the sum.

We next prove (14.62). To that end we upper-bound $|X_2(t)|$ for $|t| \leq \mathsf{T}$ as follows:

$$|X_2(t)| = \left| \sum_{\mathsf{T}+\nu < |\ell| < \infty} X_\ell\, \phi(t-\ell) \right|, \quad |t| \leq \mathsf{T}$$

$$\leq \gamma \sum_{\mathsf{T}+\nu < |\ell| < \infty} |\phi(t-\ell)|$$

$$\leq \gamma \sum_{\mathsf{T}+\nu < |\ell| < \infty} \frac{\beta}{|t-\ell|^{1+\alpha}}$$

$$\leq \gamma \sum_{\mathsf{T}+\nu < |\ell| < \infty} \frac{\beta}{\left||\ell| - |t|\right|^{1+\alpha}}$$

$$\leq \gamma \sum_{\mathsf{T}+\nu < |\ell| < \infty} \frac{\beta}{(|\ell| - \mathsf{T})^{1+\alpha}}, \quad |t| \leq \mathsf{T}$$

$$= 2\gamma\beta \sum_{\ell=\mathsf{T}+\nu+1}^{\infty} \frac{1}{(\ell-\mathsf{T})^{1+\alpha}}$$

$$= 2\gamma\beta \sum_{\tilde{\ell}=\nu+1}^{\infty} \frac{1}{\tilde{\ell}^{1+\alpha}}$$

$$\leq 2\gamma\beta \int_{\nu}^{\infty} \xi^{-1-\alpha}\, d\xi$$

$$= \frac{2\gamma\beta}{\alpha}\nu^{-\alpha}, \tag{14.63}$$

where the equality in the first line follows from the definition of $\mathbf{X}_2$ (14.55) by noting that for $|t| \leq \mathsf{T}$ we have $\phi_\ell(t) = \phi_{\ell,\mathrm{w}}(t)$); the inequality in the second line follows from the boundedness condition (14.16) and from the Triangle Inequality for Complex Numbers (2.12); the inequality in the third line from the decay condition (14.51); the inequality in the fourth line because $|\xi - \zeta| \geq \left||\xi| - |\zeta|\right|$ whenever $\xi, \zeta \in \mathbb{R}$; the inequality in the fifth line because we are only considering $|t| \leq \mathsf{T}$ and because over the range of this summation $|\ell| > \mathsf{T}+\nu$; the equality in the sixth line from the symmetry of the summand; the equality in the seventh line by defining $\tilde{\ell} \triangleq \ell - \mathsf{T}$; the inequality in the eighth line from the monotonicity of the function $\xi \mapsto \xi^{-1-\alpha}$, which implies that

$$\frac{1}{\tilde{\ell}^{1+\alpha}} \leq \int_{\tilde{\ell}-1}^{\tilde{\ell}} \frac{1}{\xi^{1+\alpha}}\, d\xi;$$

and where the final equality on the ninth line follows by computing the integral and by noting that for $t$ that does not satisfy $|t| \leq \mathsf{T}$ the LHS $|X_2(t)|$ is zero, so the inequality is trivial.

Using (14.63) and noting that $X_2(t)$ is zero for $|t| > \mathsf{T}$, we conclude that

$$\|\mathbf{X}_2\|_2^2 \leq 2\mathsf{T}\left(\frac{2\gamma\beta}{\alpha}\right)^2 \nu^{-2\alpha}, \tag{14.64}$$

from which (14.62) follows.

We next turn to proving (14.61). We begin by using the Triangle Inequality and the boundedness condition (14.16) to obtain

$$
\begin{aligned}
\|\mathbf{X}_0 - \mathbf{X}_0^{\mathrm{u}}\|_2^2 &= \left\| \sum_{|\ell| \leq \mathsf{T} - \nu} X_\ell\, \phi_{\ell,\mathrm{w}} - \sum_{|\ell| \leq \mathsf{T} - \nu} X_\ell\, \phi_\ell \right\|_2^2 \\
&= \left\| \sum_{|\ell| \leq \mathsf{T} - \nu} X_\ell\big(\phi_{\ell,\mathrm{w}} - \phi_\ell\big) \right\|_2^2 \\
&\leq \gamma^2 \left( \sum_{|\ell| \leq \mathsf{T} - \nu} \|\phi_{\ell,\mathrm{w}} - \phi_\ell\|_2 \right)^2.
\end{aligned}
\tag{14.65}
$$

We next proceed to upper-bound the RHS of (14.65) by first defining the function

$$\rho(\tau) = \sqrt{\int_{|t| > \tau} \phi^2(t)\, \mathrm{d}t} \tag{14.66}$$

and by then using this function to upper-bound $\|\phi_\ell - \phi_{\ell,\mathrm{w}}\|_2$ as

$$\|\phi_\ell - \phi_{\ell,\mathrm{w}}\|_2 \leq \rho(\mathsf{T} - |\ell|), \quad |\ell| \leq \mathsf{T}, \tag{14.67}$$

because

$$
\begin{aligned}
\|\phi_\ell - \phi_{\ell,\mathrm{w}}\|_2^2 &= \int_{-\infty}^{-\mathsf{T}} \phi^2(t - \ell)\, \mathrm{d}t + \int_{\mathsf{T}}^{\infty} \phi^2(t - \ell)\, \mathrm{d}t \\
&= \int_{-\infty}^{-\mathsf{T}-\ell} \phi^2(s)\, \mathrm{d}s + \int_{\mathsf{T}-\ell}^{\infty} \phi^2(s)\, \mathrm{d}s \\
&\leq \int_{-\infty}^{-\mathsf{T}+|\ell|} \phi^2(s)\, \mathrm{d}s + \int_{\mathsf{T}-|\ell|}^{\infty} \phi^2(s)\, \mathrm{d}s \\
&= \int_{|s| \geq \mathsf{T}-|\ell|} \phi^2(s)\, \mathrm{d}s, \quad |\ell| \leq \mathsf{T} \\
&= \rho^2(\mathsf{T} - |\ell|).
\end{aligned}
$$

It follows from (14.65) and (14.67) that

$$\|\mathbf{X}_0 - \mathbf{X}_0^{\mathrm{u}}\|_2^2 \leq \gamma^2 \left( \sum_{|\ell| \leq \mathsf{T} - \nu} \|\phi_{\ell,\mathrm{w}} - \phi_\ell\|_2 \right)^2$$

$$\leq \gamma^2 \left( \sum_{|\ell| \leq \mathsf{T} - \nu} \rho(\mathsf{T} - |\ell|) \right)^2$$

$$\leq \gamma^2 \left( 2 \sum_{0 \leq \ell \leq \mathsf{T} - \nu} \rho(\mathsf{T} - \ell) \right)^2$$

$$= 4\gamma^2 \left( \sum_{\eta = \nu}^{\mathsf{T}} \rho(\eta) \right)^2. \tag{14.68}$$

We next note that the decay condition (14.51) implies that

$$\rho(\tau) \leq \left( \frac{2\beta^2}{1 + 2\alpha} \right)^{1/2} \tau^{-\frac{1}{2} - \alpha}, \quad \tau > 0, \tag{14.69}$$

because for every $\tau > 0$,

$$\rho^2(\tau) = \int_{|t| > \tau} \phi^2(t) \, dt$$

$$\leq \int_{|t| > \tau} \frac{\beta^2}{|t|^{2 + 2\alpha}} \, dt$$

$$= 2\beta^2 \int_\tau^\infty t^{-2 - 2\alpha} \, dt$$

$$= \frac{2\beta^2}{1 + 2\alpha} \tau^{-1 - 2\alpha}.$$

It now follows from (14.69) that

$$\sum_{\eta = \nu}^{\mathsf{T}} \rho(\eta) \leq \left( \frac{2\beta^2}{1 + 2\alpha} \right)^{1/2} \sum_{\eta = \nu}^{\mathsf{T}} \eta^{-\frac{1}{2} - \alpha}$$

$$\leq \left( \frac{2\beta^2}{1 + 2\alpha} \right)^{1/2} \int_{\nu - 1}^{\mathsf{T}} \xi^{-\frac{1}{2} - \alpha} \, d\xi$$

and hence, by evaluating the integral explicitly, that

$$\lim_{\mathsf{T} \to \infty} \frac{1}{\mathsf{T}^{1/2}} \sum_{\eta = \nu}^{\mathsf{T}} \rho(\eta) = 0. \tag{14.70}$$

From (14.68) and (14.70) we thus obtain (14.61). $\qquad \square$

## 14.6 A More Formal Account

In this section we present a more formal definition of power and justify some of the mathematical steps that we took in deriving the power in PAM signals. This section is quite mathematical and is recommended for readers who have had some exposure to Measure Theory.

Let $\mathcal{R}$ denote the $\sigma$-algebra generated by the open sets in $\mathbb{R}$. A continuous-time stochastic process $\big(X(t)\big)$ defined over the probability space $(\Omega, \mathcal{F}, P)$ is said to be a **measurable stochastic process** if the mapping $(\omega, t) \mapsto X(\omega, t)$ from $\Omega \times \mathbb{R}$ to $\mathbb{R}$ is measurable when its range $\mathbb{R}$ is endowed with the $\sigma$-algebra $\mathcal{R}$ and when its domain $\Omega \times \mathbb{R}$ is endowed with the product $\sigma$-algebra $\mathcal{F} \times \mathcal{R}$. Thus, $\big(X(t),\ t \in \mathbb{R}\big)$ is measurable if the mapping $(\omega, t) \mapsto X(\omega, t)$ is $\mathcal{F} \times \mathcal{R}/\mathcal{R}$ measurable.[5]

From Fubini's Theorem it follows that if $\big(X(t),\ t \in \mathbb{R}\big)$ is measurable and if $\mathsf{T} > 0$ is deterministic, then:

(i) For every $\omega \in \Omega$, the mapping $t \mapsto X^2(\omega, t)$ is Borel measurable;

(ii) the mapping

$$\omega \mapsto \int_{-\mathsf{T}}^{\mathsf{T}} X^2(\omega, t)\,dt$$

is a random variable (i.e., $\mathcal{F}$ measurable) possibly taking on the value $+\infty$;

(iii) and

$$\mathsf{E}\left[\int_{-\mathsf{T}}^{\mathsf{T}} X^2(t)\,dt\right] = \int_{-\mathsf{T}}^{\mathsf{T}} \mathsf{E}\big[X^2(t)\big]\,dt, \quad \mathsf{T} \in \mathbb{R}. \tag{14.71}$$

**Definition 14.6.1 (Power of a Stochastic Process).** *We say that a measurable stochastic process $\big(X(t),\ t \in \mathbb{R}\big)$ is of **power** $\mathsf{P}$ if the limit*

$$\lim_{\mathsf{T} \to \infty} \frac{1}{2\mathsf{T}}\, \mathsf{E}\left[\int_{-\mathsf{T}}^{\mathsf{T}} X^2(t)\,dt\right] \tag{14.72}$$

*exists and is equal to $\mathsf{P}$.*

**Proposition 14.6.2.** *If the pulse shape $\mathbf{g}$ is a Borel measurable function satisfying the decay condition (14.17) for some positive $\alpha, \beta, \mathsf{T_s}$, and if the discrete-time SP $\big(X_\ell,\ \ell \in \mathbb{Z}\big)$ satisfies the boundedness condition (14.16) for some $\gamma \geq 0$, then the stochastic process*

$$\mathbf{X} \colon (\omega, t) \mapsto A \sum_{\ell = -\infty}^{\infty} X_\ell(\omega)\, g(t - \ell\mathsf{T_s}) \tag{14.73}$$

*is a measurable stochastic process.*

**Proof.** The mapping $(\omega, t) \mapsto X_\ell(\omega)$ is $\mathcal{F} \times \mathcal{R}/\mathcal{R}$ measurable because $X_\ell$ is a random variable, so the mapping $\omega \mapsto X_\ell(\omega)$ is $\mathcal{F}/\mathcal{R}$ measurable. The mapping $(\omega, t) \mapsto Ag(t - \ell\mathsf{T_s})$ is $\mathcal{F} \times \mathcal{R}/\mathcal{R}$ measurable because $\mathbf{g}$ is Borel measurable, so $t \mapsto g(t - \ell\mathsf{T_s})$ is $\mathcal{R}/\mathcal{R}$ measurable. Since the product of measurable functions is measurable (Rudin, 1974, Chapter 1, Section 1.9 (c)), it follows that the mapping

---

[5]See (Billingsley, 1995, Section 37, p. 503) or (Loève, 1963, Section 35) for the definition of a measurable stochastic process and see (Billingsley, 1995, Section 18) or (Loève, 1963, Section 8.2) or (Halmos, 1950, Chapter VII) for the definition of the product $\sigma$-algebra.

$(\omega, t) \mapsto A X_\ell(\omega) g(t - \ell T_s)$ is $\mathcal{F} \times \mathcal{R}/\mathcal{R}$ measurable. And since the sum of measurable functions is measurable (Rudin, 1974, Chapter 1, Section 1.9 (c)), it follows that for every positive integer $L \in \mathbb{Z}$, the mapping

$$(\omega, t) \mapsto A \sum_{\ell=-L}^{L} X_\ell(\omega) g(t - \ell T_s)$$

is $\mathcal{F} \times \mathcal{R}/\mathcal{R}$ measurable. The proposition now follows by recalling that the pointwise limit of every pointwise convergent sequence of measurable functions is measurable (Rudin, 1974, Theorem 1.14). □

Having established that the PAM signal (14.73) is a measurable stochastic process we would next like to justify the calculations leading to (14.31). To justify the swapping of integration and summations in (14.26) we shall need the following lemma, which also explains why the sum in (14.27) converges.

**Lemma 14.6.3.** *If $g(\cdot)$ is a Borel measurable function satisfying the decay condition*

$$|g(t)| \leq \frac{\beta}{1 + |t/T_s|^{1+\alpha}}, \quad t \in \mathbb{R} \tag{14.74}$$

*for some positive $\alpha$, $T_s$, and $\beta$, then*

$$\sum_{m=-\infty}^{\infty} \int_{-\infty}^{\infty} |g(t) \, g(t - mT_s)| \, dt < \infty. \tag{14.75}$$

**Proof.** The decay condition (14.74) guarantees that **g** is of finite energy. From the Cauchy-Schwarz Inequality it thus follows that the terms in (14.75) are all finite. Also, by symmetry, the term in (14.75) corresponding to $m$ is the same as the one corresponding to $-m$. Consequently, to establish (14.75), it suffices to prove

$$\sum_{m=2}^{\infty} \int_{-\infty}^{\infty} |g(t) \, g(t - mT_s)| \, dt < \infty. \tag{14.76}$$

Define the function

$$g_u(t) \triangleq \begin{cases} 1 & \text{if } |t| \leq 1, \\ |t|^{-1-\alpha} & \text{otherwise,} \end{cases} \quad t \in \mathbb{R}.$$

By (14.74) it follows that $|g(t)| \leq \beta \, g_u(t/T_s)$ for all $t \in \mathbb{R}$. Consequently,

$$\int_{-\infty}^{\infty} |g(t) \, g(t - mT_s)| \, dt \leq \beta^2 \int_{-\infty}^{\infty} g_u(t/T_s) \, g_u(t/T_s - m) \, dt$$

$$= \beta^2 T_s \int_{-\infty}^{\infty} g_u(\tau) \, g_u(\tau - m) \, d\tau,$$

and to establish (14.76) it thus suffices to prove

$$\sum_{m=2}^{\infty} \int_{-\infty}^{\infty} g_u(\tau) \, g_u(\tau - m) \, d\tau < \infty. \tag{14.77}$$

Since the integrand in (14.77) is symmetric around $\tau = m/2$, it follows that

$$\int_{-\infty}^{\infty} g_{\mathrm{u}}(\tau)\, g_{\mathrm{u}}(\tau - m)\, \mathrm{d}\tau = 2 \int_{m/2}^{\infty} g_{\mathrm{u}}(\tau)\, g_{\mathrm{u}}(\tau - m)\, \mathrm{d}\tau, \qquad (14.78)$$

and it thus suffices to establish

$$\sum_{m=2}^{\infty} \int_{m/2}^{\infty} g_{\mathrm{u}}(\tau)\, g_{\mathrm{u}}(\tau - m)\, \mathrm{d}\tau < \infty. \qquad (14.79)$$

We next upper-bound the integral in (14.79) for every $m \geq 2$ by first expressing it as

$$\int_{m/2}^{\infty} g_{\mathrm{u}}(\tau)\, g_{\mathrm{u}}(\tau - m)\, \mathrm{d}\tau = I_1 + I_2 + I_3,$$

where

$$I_1 \triangleq \int_{m/2}^{m-1} \frac{1}{\tau^{1+\alpha}} \frac{1}{(m - \tau)^{1+\alpha}}\, \mathrm{d}\tau,$$

$$I_2 \triangleq \int_{m-1}^{m+1} \frac{1}{\tau^{1+\alpha}}\, \mathrm{d}\tau,$$

$$I_3 \triangleq \int_{m+1}^{\infty} \frac{1}{\tau^{1+\alpha}} \frac{1}{(\tau - m)^{1+\alpha}}\, \mathrm{d}\tau.$$

We next upper-bound each of these terms for $m \geq 2$. Starting with $I_1$ we obtain upon defining $\xi \triangleq m - \tau$

$$\begin{aligned}
I_1 &= \int_{m/2}^{m-1} \frac{1}{\tau^{1+\alpha}} \frac{1}{(m - \tau)^{1+\alpha}}\, \mathrm{d}\tau \\
&= \int_{1}^{m/2} \frac{1}{(m - \xi)^{1+\alpha}} \frac{1}{\xi^{1+\alpha}}\, \mathrm{d}\xi \\
&\leq \int_{1}^{m/2} \frac{1}{(m/2)^{1+\alpha}} \frac{1}{\xi^{1+\alpha}}\, \mathrm{d}\xi \\
&= \frac{1}{\alpha}\, 2^{1+\alpha} \frac{1}{m^{1+\alpha}} \left(1 - \frac{2^{\alpha}}{m^{\alpha}}\right), \quad m \geq 2,
\end{aligned}$$

which is summable over $m$. As to $I_2$ we have

$$\begin{aligned}
I_2 &= \int_{m-1}^{m+1} \frac{1}{\tau^{1+\alpha}}\, \mathrm{d}\tau \\
&\leq \frac{2}{(m - 1)^{1+\alpha}}, \quad m \geq 2,
\end{aligned}$$

which is summable over $m$. Finally we upper-bound $I_3$ by defining $\xi \triangleq \tau - m$

$$\begin{aligned}
I_3 &= \int_{m+1}^{\infty} \frac{1}{\tau^{1+\alpha}} \frac{1}{(\tau - m)^{1+\alpha}}\, \mathrm{d}\tau \\
&= \int_{1}^{\infty} \frac{1}{(\xi + m)^{1+\alpha}} \frac{1}{\xi^{1+\alpha}}\, \mathrm{d}\xi
\end{aligned}$$

$$= \int_1^m \frac{1}{(\xi + m)^{1+\alpha}} \frac{1}{\xi^{1+\alpha}} \, d\xi + \int_m^\infty \frac{1}{(\xi + m)^{1+\alpha}} \frac{1}{\xi^{1+\alpha}} \, d\xi$$

$$\leq \frac{1}{m^{1+\alpha}} \int_1^m \frac{1}{\xi^{1+\alpha}} \, d\xi + \int_m^\infty \frac{1}{\xi^{1+\alpha}} \frac{1}{\xi^{1+\alpha}} \, d\xi$$

$$= \frac{1}{\alpha} \frac{1}{m^{1+\alpha}} \left(1 - \frac{1}{m^\alpha}\right) + \frac{1}{1 + 2\alpha} \frac{1}{m^{1+2\alpha}}, \quad m \geq 2,$$

which is summable over $m$. $\qquad\square$

We can now state (14.31) as a theorem.

**Theorem 14.6.4.** *Let the pulse shape* $\mathbf{g} \colon \mathbb{R} \to \mathbb{R}$ *be a Borel measurable function satisfying the decay condition (14.17) for some positive* $\alpha$, $\beta$, *and* $\mathsf{T_s}$. *Let* $\big(X_\ell, \ \ell \in \mathbb{Z}\big)$ *be a centered WSS SP of autocovariance function* $\mathsf{K}_{XX}$ *and satisfying the boundedness condition (14.16) for some* $\gamma \geq 0$. *Then the stochastic process (14.73) is measurable and is of the power* $\mathsf{P}$ *given in (14.31).*

**Proof.** The measurability of $\big(X(t), \ t \in \mathbb{R}\big)$ follows from Proposition 14.6.2. The power can be derived as in the derivation of (14.31) from (14.27) with the derivation of (14.27) now being justifiable by noting that (14.25) follows from (14.71) and by noting that (14.26) follows from Lemma 14.6.3 and Fubini's Theorem. $\qquad\square$

Similarly, we can state (14.37) as a theorem.

**Theorem 14.6.5 (Power in Bi-Infinite Block-Mode PAM).** *Let* $\big(D_j, \ j \in \mathbb{Z}\big)$ *be IID random bits. Let the* $(\mathsf{K}, \mathsf{N})$ *binary-to-reals encoder* $\mathbf{enc} \colon \{0, 1\}^{\mathsf{K}} \to \mathbb{R}^{\mathsf{N}}$ *be such that* $\mathrm{enc}(D_1, \ldots, D_{\mathsf{K}})$ *is of zero mean whenever the* $\mathsf{K}$-*tuple* $(D_1, \ldots, D_{\mathsf{K}})$ *is uniformly distributed over* $\{0, 1\}^{\mathsf{K}}$. *Let* $\big(X_\ell, \ \ell \in \mathbb{Z}\big)$ *be generated from* $\big(D_j, \ j \in \mathbb{Z}\big)$ *in bi-infinite block encoding mode using* $\mathrm{enc}(\cdot)$. *Assume that the pulse shape* $\mathbf{g}$ *is a Borel measurable function satisfying the decay condition (14.17) for some positive* $\alpha$, $\beta$, *and* $\mathsf{T_s}$. *Then the stochastic process (14.73) is measurable and is of the power* $\mathsf{P}$ *as given in (14.37).*

**Proof.** Measurability follows from Proposition 14.6.2. The derivation of (14.37) is justified using Fubini's Theorem. $\qquad\square$

## 14.7   Exercises

**Exercise 14.1 (Superimposing Independent Transmissions).** Let the two PAM signals $\big(X^{(1)}(t)\big)$ and $\big(X^{(2)}(t)\big)$ be given at every epoch $t \in \mathbb{R}$ by

$$X^{(1)}(t) = \mathsf{A}^{(1)} \sum_{\ell=-\infty}^\infty X_\ell^{(1)} g^{(1)}(t - \ell \mathsf{T_s}), \quad X^{(2)}(t) = \mathsf{A}^{(2)} \sum_{\ell=-\infty}^\infty X_\ell^{(2)} g^{(2)}(t - \ell \mathsf{T_s}),$$

where the zero-mean real symbols $\big(X_\ell^{(1)}\big)$ are generated from the data bits $\big(D_j^{(1)}\big)$ and the zero-mean real symbols $\big(X_\ell^{(2)}\big)$ from $\big(D_j^{(2)}\big)$. Assume that the bit streams $\big(D_j^{(1)}\big)$ and $\big(D_j^{(2)}\big)$ are independent and that $\big(X^{(1)}(t)\big)$ and $\big(X^{(1)}(t)\big)$ are of powers $\mathsf{P}^{(1)}$ and $\mathsf{P}^{(2)}$. Find the power in the sum of $\big(X^{(1)}(t)\big)$ and $\big(X^{(1)}(t)\big)$.

**Exercise 14.2 (The Minimum Distance of a Constellation and Power).** Consider the PAM signal (14.47) where the time shifts of the pulse shape $\phi$ by integer multiples of $\mathsf{T_s}$ are orthonormal, and where the symbols $(X_\ell)$ are IID and uniformly distributed over the set $\{\pm\frac{d}{2}, \pm\frac{3d}{2}, \ldots, \pm(2\nu-1)\frac{d}{2}\}$. Relate the power in $X(\cdot)$ to the minimum distance $d$ and the constant $\mathsf{A}$.

**Exercise 14.3 (PAM with Nonorthogonal Pulses).** Let the IID random bits $(D_j,\ j \in \mathbb{Z})$ be modulated using PAM with the pulse shape $\mathbf{g} \colon t \mapsto \mathrm{I}\{|t| \leq \mathsf{T_s}\}$ and the repetition block encoding map $0 \mapsto (+1,+1)$ and $1 \mapsto (-1,-1)$. Compute the average transmitted power.

**Exercise 14.4 (Non-IID Data Bits).** Expression (14.37) for the power in bi-infinite block mode was derived under the assumption that the data bits are IID. Show that it need not otherwise hold.

**Exercise 14.5 (The Power in Nonorthogonal PAM).** Consider the PAM signal (14.23) with the pulse shape $\mathbf{g} \colon t \mapsto \mathrm{I}\{|t| \leq \mathsf{T_s}\}$.

(i) Compute the power in $X(\cdot)$ when $(X_\ell)$ are IID of zero-mean and unit-variance.

(ii) Repeat when $(X_\ell)$ is a zero-mean WSS SP of autocovariance function

$$\mathsf{K}_{XX}(m) = \begin{cases} 1 & m = 0 \\ \frac{1}{2} & |m| = 1 \\ 0 & \text{otherwise} \end{cases}, \quad m \in \mathbb{Z}.$$

Note that in both parts $\mathsf{E}[X_\ell] = 0$ and $\mathsf{E}\left[X_\ell^2\right] = 1$.

**Exercise 14.6 (Pre-Encoding).** Rather than applying the mapping $\mathbf{enc} \colon \{0,1\}^\mathsf{K} \to \mathbb{R}^\mathsf{N}$ to the IID random bits $D_1, \ldots, D_\mathsf{K}$ directly, we first map the data bits using a one-to-one mapping $\phi \colon \{0,1\}^\mathsf{K} \to \{0,1\}^\mathsf{K}$ to $D'_1, \ldots, D'_\mathsf{K}$, and we then map $D'_1, \ldots, D'_\mathsf{K}$ using $\mathbf{enc}$ to $X_1, \ldots, X_\mathsf{N}$. Does this change the transmitted energy?

**Exercise 14.7 (Binary Linear Encoders Producing Pairwise-Independent Symbols).** Binary linear encoders with the antipodal mapping can be described as follows. Using a deterministic binary $\mathsf{K} \times \mathsf{N}$ matrix $\mathsf{G}$, the encoder first maps the row-vector $\mathbf{d} = (d_1, \ldots, d_\mathsf{K})$ to the row-vector $\mathbf{d}\mathsf{G}$, where $\mathbf{d}\mathsf{G}$ is computed using matrix multiplication over the binary field. (Recall that in the binary field multiplication is defined as $0 \cdot 0 = 0 \cdot 1 = 1 \cdot 0 = 0$, and $1 \cdot 1 = 1$; and addition is modulo 2, so $0 \oplus 0 = 1 \oplus 1 = 0$ and $0 \oplus 1 = 1 \oplus 0 = 1$). Thus, the $\ell$-th component $c_\ell$ of $\mathbf{d}\mathsf{G}$ is given by

$$c_\ell = d_1 \cdot g^{(1,\ell)} \oplus d_2 \cdot g^{(2,\ell)} \oplus \cdots \oplus d_\mathsf{K} \cdot g^{(\mathsf{K},\ell)}.$$

The real symbol $x_\ell$ is then computed according to the rule

$$x_\ell = \begin{cases} +1 & \text{if } c_\ell = 0, \\ -1 & \text{if } c_\ell = 1, \end{cases} \quad \ell = 1, \ldots, \mathsf{N}.$$

Let $X_1, X_2, \ldots, X_\mathsf{N}$ be the symbols produced by the encoder when it is fed IID random bits $D_1, D_2, \ldots, D_\mathsf{K}$. Show that:

(i) Unless all the entries in the $\ell$-th column of $\mathsf{G}$ are zero, $\mathsf{E}[X_\ell] = 0$.

(ii) $X_\ell$ is independent of $X_{\ell'}$ if, and only if, the $\ell$-th column and the $\ell'$-th column of $\mathsf{G}$ are not identical.

You may find it useful to first prove the following.

(i) If a RV $E$ takes value in the set $\{0, 1\}$, and if $F$ takes on the values 0 and 1 equiprobably and independently of $E$, then $E \oplus F$ is uniform on $\{0, 1\}$ and independent of $E$.

(ii) If $E_1$ and $E_2$ take value in $\{0, 1\}$, and if $F$ takes on the values 0 and 1 equiprobably and independently of $(E_1, E_2)$, then $E_1 \oplus F$ is independent of $E_2$.

**Exercise 14.8 (Zero-Mean Signals for Linearly Dispersive Channels).** Suppose that the transmitted signal $\mathbf{X}$ suffers not only from an additive random disturbance but also from a deterministic linear distortion. Thus, the received signal $\mathbf{Y}$ can be expressed as $\mathbf{Y} = \mathbf{X} \star \mathbf{h} + \mathbf{N}$, where $\mathbf{h}$ is a known (deterministic) impulse response, and where $\mathbf{N}$ is an unknown (random) additive disturbance. Show heuristically that transmitting signals of nonzero mean is power inefficient. How would you mimic the performance of a system transmitting $X(\cdot)$ using a system transmitting $X(\cdot) - c(\cdot)$?

**Exercise 14.9 (The Power in Orthogonal Code-Division Multi-Accessing).** Suppose that the data bits $\left(D_j^{(1)}\right)$ are mapped to the real symbols $\left(X_\ell^{(1)}\right)$ and that the data bits $\left(D_j^{(2)}\right)$ are mapped to $\left(X_\ell^{(2)}\right)$. Assume that

$$\frac{\left(\mathsf{A}^{(1)}\right)^2}{\mathsf{T_s}} \lim_{\mathsf{L} \to \infty} \frac{1}{2\mathsf{L}+1} \sum_{\ell=-\mathsf{L}}^{\mathsf{L}} \mathsf{E}\left[\left(X_\ell^{(1)}\right)^2\right] = \mathsf{P}^{(1)},$$

and similarly for $\mathsf{P}^{(2)}$. Further assume that the time shifts of $\phi$ by integer multiples of $\mathsf{T_s}$ are orthonormal and that $\phi$ satisfies the decay condition (14.46). Finally assume that $\left(X_\ell^{(1)}\right)$ and $\left(X_\ell^{(2)}\right)$ are bounded in the sense of (14.16). Compute the power in the signal

$$\sum_{\ell=-\infty}^{\infty} \left(\left(\mathsf{A}^{(1)} X_\ell^{(1)} + \mathsf{A}^{(2)} X_\ell^{(2)}\right) \phi\left(t - 2\ell\mathsf{T_s}\right) + \left(\mathsf{A}^{(1)} X_\ell^{(1)} - \mathsf{A}^{(2)} X_\ell^{(2)}\right) \phi\left(t - (2\ell+1)\mathsf{T_s}\right)\right).$$

**Exercise 14.10 (More on Orthogonal Code-Division Multi-Accessing).** Extend the result of Exercise 14.9 to the case with $\eta$ data streams, where the transmitted signal is given by

$$\sum_{\ell=-\infty}^{\infty} \left(\left(a^{(1,1)} \mathsf{A}^{(1)} X_\ell^{(1)} + \cdots + a^{(\eta,1)} \mathsf{A}^{(\eta)} X_\ell^{(\eta)}\right) \phi\left(t - \eta\ell\mathsf{T_s}\right)\right.$$
$$\left. + \cdots + \left(a^{(1,\eta)} \mathsf{A}^{(1)} X_\ell^{(1)} + \cdots + a^{(\eta,\eta)} \mathsf{A}^{(\eta)} X_\ell^{(\eta)}\right) \phi\left(t - (\eta\ell + \eta - 1)\mathsf{T_s}\right)\right)$$

and where the real numbers $a^{(\iota,\nu)}$ for $\iota, \nu \in \{1, \ldots, \eta\}$ satisfy the orthogonality condition

$$\sum_{\nu=1}^{\eta} a^{(\iota,\nu)} a^{(\iota',\nu)} = \begin{cases} \eta & \text{if } \iota = \iota', \\ 0 & \text{if } \iota \neq \iota', \end{cases} \quad \iota, \iota' \in \{1, \ldots, \eta\}.$$

The sequence $a^{(\iota,1)}, \ldots, a^{(\iota,\eta)}$ is sometimes called the **signature of the $\iota$-th stream**.

**Exercise 14.11 (The Samples of the Self-Similarity Function).** Let $\mathsf{g} \colon \mathbb{R} \to \mathbb{R}$ be of finite energy, and let $\mathsf{R_{gg}}$ be its self-similarity function.

(i) Show that there exists an integrable nonnegative function $G\colon [-1/2, 1/2) \to [0, \infty)$ such that

$$R_{gg}(m\,T_s) = \int_{-1/2}^{1/2} G(\theta)\, e^{-i2\pi m\theta}\, d\theta, \quad m \in \mathbb{Z},$$

and such that $G(-\theta) = G(\theta)$ for all $|\theta| < 1/2$. Express $G(\cdot)$ in terms of the FT of $\mathbf{g}$.

(ii) Show that if the samples of the self-similarity function are absolutely summable, i.e., if

$$\sum_{m \in \mathbb{Z}} \left| R_{gg}(m\,T_s) \right| < \infty,$$

then the function

$$\theta \mapsto \sum_{m=-\infty}^{\infty} R_{gg}(m\,T_s)\, e^{i2\pi m\theta}, \quad \theta \in [-1/2, 1/2),$$

is such a function, and it is continuous.

(iii) Show that if $(X_\ell)$ is of PSD $S_{XX}$, then the RHS of (14.31) can be expressed as

$$\frac{1}{T_s} A^2 \int_{-1/2}^{1/2} G(\theta)\, S_{XX}(\theta)\, d\theta.$$

**Exercise 14.12 (A Bound on the Power in PAM).** Let $G(\cdot)$ be as in Exercise 14.11.

(i) Show that if $(X_\ell)$ is of zero mean, of unit variance, and has a PSD, then the RHS of (14.31) is upper-bounded by

$$\frac{1}{T_s} A^2 \sup_{-1/2 \leq \theta < 1/2} G(\theta). \tag{14.80}$$

(ii) Suppose now that $G(\cdot)$ is continuous. Show that for every $\epsilon > 0$, there exists a zero-mean unit-variance SP $(X_\ell)$ with a PSD for which the RHS of (14.31) is within $\epsilon$ of (14.80).

# Chapter 15

# Operational Power Spectral Density

## 15.1 Introduction

The Power Spectral Density of a stochastic process tells us more about the SP than just its power. It tells us something about how this power is distributed among the different frequencies that the SP occupies. The purpose of this chapter is to clarify this statement and to derive the PSD of PAM signals. Most of this chapter is written informally with an emphasis on ideas and intuition as opposed to mathematical rigor. The mathematically-inclined readers will find precise statements of the key results of this chapter in Section 15.5. We emphasize that this chapter only deals with real continuous-time stochastic processes.

The classical definition of the PSD of continuous-time stochastic processes (Definition 25.7.2 ahead) is only applicable to wide-sense stationary stochastic processes, and PAM signals are not WSS.[1] Consequently, we shall have to introduce a new concept, which we call the **operational power spectral density**, or the **operational PSD** for short.[2] This new concept is applicable to a large family of stochastic processes that includes most WSS processes and most PAM signals. For WSS stochastic processes, the operational PSD and the classical PSD coincide (Section 25.14). In addition to being more general, the operational PSD is more intuitive in that it clarifies the origin of the words "power spectral density." Moreover, it gives an operational meaning to the concept.

## 15.2 Motivation

To motivate the new definition we shall first briefly discuss other "densities" such as charge density, mass density, and probability density.

In electromagnetism one encounters the concept of *charge density*, which is often denoted by $\varrho(\cdot)$. It measures the amount of charge per unit volume. Since the

---

[1] If the discrete-time symbol sequence is stationary then the PAM signal is **cyclostationary**. But this term will not be used in this book.

[2] These terms are not standard. Most of the literature does not seem to distinguish between the PSD in the sense of Definition 25.7.2 and what we call the operational PSD.

| function | quantity of interest | per unit of |
|---|---|---|
| charge (spatial) density | charge | space |
| mass (spatial) density | mass | space |
| mass line density | mass | length |
| probability (per unit of $X$) density | probability | unit of $X$ |
| power spectral density | power | spectrum (Hz) |

**Table 15.1:** Various densities and their units

charge need not be uniformly distributed, $\varrho(\cdot)$ is typically not constant so the charge density is a function of location. Thus, we usually write $\varrho(x, y, z)$ for the charge density at the location $(x, y, z)$. This can be defined differentially or integrally. The differential definition is

$$\varrho(x, y, z)$$
$$= \lim_{\Delta \downarrow 0} \frac{\text{Charge in Box} \left\{ (x', y', z') : |x - x'| \leq \frac{\Delta}{2}, |y - y'| \leq \frac{\Delta}{2}, |z - z'| \leq \frac{\Delta}{2} \right\}}{\text{Volume of Box} \left\{ (x', y', z') : |x - x'| \leq \frac{\Delta}{2}, |y - y'| \leq \frac{\Delta}{2}, |z - z'| \leq \frac{\Delta}{2} \right\}}$$
$$= \lim_{\Delta \downarrow 0} \frac{\text{Charge in box} \left\{ (x', y', z') : |x - x'| \leq \frac{\Delta}{2}, |y - y'| \leq \frac{\Delta}{2}, |z - z'| \leq \frac{\Delta}{2} \right\}}{\Delta^3},$$

and the integral definition is that a function $\varrho(\cdot)$ is the charge density if for every region $\mathcal{D} \subset \mathbb{R}^3$

$$\text{Charge in } \mathcal{D} = \int_{(x,y,z) \in \mathcal{D}} \varrho(x, y, z) \, dx \, dy \, dz, \quad \mathcal{D} \subset \mathbb{R}^3.$$

Ignoring some mathematical subtleties, the two definitions are equivalent. Perhaps a more appropriate name for charge density is "Charge Spatial Density," which makes it clear that the quantity of interest is charge and that we are interested in the way it is distributed in space. The units of $\varrho(x, y, z)$ are those of charge per unit volume.

*Mass density*—or as we would prefer to call it, "Mass Spatial Density"—is analogously defined. Either differentially, as

$$\varrho(x, y, z)$$
$$= \lim_{\Delta \downarrow 0} \frac{\text{Mass in Box} \left\{ (x', y', z') : |x - x'| \leq \frac{\Delta}{2}, |y - y'| \leq \frac{\Delta}{2}, |z - z'| \leq \frac{\Delta}{2} \right\}}{\text{Volume of Box} \left\{ (x', y', z') : |x - x'| \leq \frac{\Delta}{2}, |y - y'| \leq \frac{\Delta}{2}, |z - z'| \leq \frac{\Delta}{2} \right\}}$$
$$= \lim_{\Delta \downarrow 0} \frac{\text{Mass in box} \left\{ (x', y', z') : |x - x'| \leq \frac{\Delta}{2}, |y - y'| \leq \frac{\Delta}{2}, |z - z'| \leq \frac{\Delta}{2} \right\}}{\Delta^3},$$

or integrally as the function $\varrho(x, y, z)$ such that for every subset $\mathcal{D} \subset \mathbb{R}^3$

$$\text{Mass in } \mathcal{D} = \int_{(x,y,z) \in \mathcal{D}} \varrho(x, y, z) \, dx \, dy \, dz, \quad \mathcal{D} \subset \mathbb{R}^3.$$

The units are those of mass per unit volume. Since mass is nonnegative, the differential definition of mass density makes it clear that mass density must also be nonnegative. This is slightly less apparent from the integral definition, but (excluding subsets of $\mathbb{R}^3$ of measure zero) is true nonetheless. By convention, if one defines mass density integrally, then one typically insists that the density be nonnegative.

Similarly, in discussing *mass line density* one envisions a one-dimensional object, and its density with respect to unit length is defined differentially as

$$\varrho(x) = \lim_{\Delta \downarrow 0} \frac{\text{Mass in Interval} \left\{ x' : |x - x'| \leq \frac{\Delta}{2} \right\}}{\Delta},$$

or integrally as the nonnegative function $\varrho(\cdot)$ such that for every subset $\mathcal{D} \subset \mathbb{R}$ of the real line

$$\text{Mass in } \mathcal{D} = \int_{x \in \mathcal{D}} \varrho(x) \, dx, \quad \mathcal{D} \subset \mathbb{R}.$$

The units are units of mass per unit length.

In probability theory one encounters the *probability density function* of a random variable $X$. Here the quantity of interest is probability, and we are interested in how it is distributed on the real line. The units depend on the units of $X$. Thus, if $X$ measures the time in days until at least one piece in your new china set breaks, then the units of the probability density function $f_X(\cdot)$ of $X$ are those of probability (unit-less) per day. The probability density function can be defined differentially as

$$f_X(x) = \lim_{\Delta \downarrow 0} \frac{\Pr\left[ X \in \left( x - \frac{\Delta}{2}, x + \frac{\Delta}{2} \right) \right]}{\Delta}$$

or integrally by requiring that for every subset $\mathcal{E} \subset \mathbb{R}$

$$\Pr[X \in \mathcal{E}] = \int_{x \in \mathcal{E}} f_X(x) \, dx, \quad \mathcal{E} \subset \mathbb{R}. \tag{15.1}$$

Again, since probabilities are nonnegative, the differential definition makes it clear that the probability density function is nonnegative. In the integral definition we typically add the nonnegativity as a condition. That is, we say that $f_X(\cdot)$ is a density function for the random variable $X$ if $f_X(\cdot)$ is nonnegative and if (15.1) holds. (There is a technical uniqueness issue that we are sweeping under the rug here: if $f_X(\cdot)$ is a probability density function for $X$ and if $\xi(\cdot)$ is a nonnegative function that differs from $f_X(\cdot)$ only on a set of Lebesgue measure zero, then $\xi(\cdot)$ is also a probability density function for $X$.)

With these examples in mind, it is natural to interpret the power spectral density of a stochastic process $\left( X(t), \ t \in \mathbb{R} \right)$ as the distribution of the power of $X(\cdot)$ among the different frequencies. See Table 15.1 on Page 246. Heuristically, we would define the power spectral density $\mathsf{S}_{XX}$ at the frequency $f$ differentially as

$$\mathsf{S}_{XX}(f) = \lim_{\Delta \downarrow 0} \frac{\text{Power in the frequencies} \left[ f - \frac{\Delta}{2}, f + \frac{\Delta}{2} \right]}{\Delta}$$

or integrally by requiring that for any subset $\mathcal{D}$ of the spectrum

$$\text{Power of } \mathbf{X} \text{ in } \mathcal{D} = \int_{f \in \mathcal{D}} \mathsf{S}_{XX}(f) \, df, \quad \mathcal{D} \subset \mathbb{R}. \tag{15.2}$$

To make this meaningful we next explain what we mean by "the power of $\mathbf{X}$ in the frequencies $\mathcal{D}$." To that end it is best to envision a filter of impulse response $\mathbf{h}$ whose frequency response $\hat{\mathbf{h}}$ is given by

$$\hat{h}(f) = \begin{cases} 1 & \text{if } f \in \mathcal{D}, \\ 0 & \text{otherwise}, \end{cases} \tag{15.3}$$

and to think of the power of $X(\cdot)$ in the frequencies $\mathcal{D}$ as the average power at the output of that filter when it is fed $X(\cdot)$, i.e., the average power of the stochastic process $\mathbf{X} \star \mathbf{h}$.[3]

We are now almost ready to give a heuristic definition of the power spectral density. But there are three more points we would like to discuss first. The first is that (15.2) can also be rewritten as

$$\text{Power of } \mathbf{X} \text{ in } \mathcal{D} = \int_{\text{all frequencies}} \mathrm{I}\{f \in \mathcal{D}\} \, \mathsf{S}_{XX}(f) \, \mathrm{d}f, \quad \mathcal{D} \subset \mathbb{R}. \tag{15.4}$$

It turns out that if (15.2) holds for all sets $\mathcal{D} \subset \mathbb{R}$ of frequencies, then it also holds for all "nice" filters (of a frequency response that is not necessarily $\{0,1\}$ valued):

$$\text{Power of } \mathbf{X} \star \mathbf{h} = \int_{\text{all frequencies}} |\hat{h}(f)|^2 \, \mathsf{S}_{XX}(f) \, \mathrm{d}f, \quad \mathbf{h} \text{ "nice."} \tag{15.5}$$

That (15.4) typically implies (15.5) can be heuristically argued as follows. By (15.4) the set of frequency responses $\hat{\mathbf{h}}$ for which (15.5) holds includes all frequency responses of the form $\hat{h}(f) = \mathrm{I}\{f \in \mathcal{D}\}$. But if (15.5) holds for some frequency response $\hat{\mathbf{h}}$, then it must also hold for $\alpha \hat{\mathbf{h}}$, where $\alpha$ is any complex number, because scaling the frequency response by $\alpha$ merely multiplies the output power by $|\alpha|^2$. Also, if (15.5) holds for two responses $\hat{\mathbf{h}}_1$ and $\hat{\mathbf{h}}_2$ for which

$$\hat{h}_1(f) \, \hat{h}_2(f) = 0, \quad f \in \mathbb{R}, \tag{15.6}$$

then it must also hold for $\mathbf{h}_1 + \mathbf{h}_2$, because Parseval's Theorem and (15.6) imply that $\mathbf{X} \star \mathbf{h}_1$ and $\mathbf{X} \star \mathbf{h}_2$ must be orthogonal. Thus, (15.6) implies that the power in $\mathbf{X} \star (\mathbf{h}_1 + \mathbf{h}_2)$ is the sum of the power in $\mathbf{X} \star \mathbf{h}_1$ and the power in $\mathbf{X} \star \mathbf{h}_2$. It thus intuitively follows that if (15.4) holds for all subsets $\mathcal{D}$ of the spectrum, then it holds for all step functions $\hat{h}(f) = \sum_\nu \alpha_\nu \mathrm{I}\{f \in \mathcal{D}_\nu\}$, where $\{\mathcal{D}_\nu\}$ are disjoint. And since any "nice" frequency response $\hat{\mathbf{h}}$ can be arbitrarily well approximated by such step functions, we expect that (15.5) would hold for all "nice" responses.

Having heuristically established that (15.2) implies (15.5), we prefer to define the PSD as a function $\mathsf{S}_{XX}$ for which (15.5) holds, where "nice" will be taken to mean stable.

The second point we would like to make is regarding uniqueness. For real stochastic processes it is reasonable to require that (15.5) hold only for filters of *real* impulse response. Thus we would require

$$\text{Power of } \mathbf{X} \star \mathbf{h} = \int_{\text{all frequencies}} |\hat{h}(f)|^2 \, \mathsf{S}_{XX}(f) \, \mathrm{d}f, \quad \mathbf{h} \text{ real and "nice."} \tag{15.7a}$$

---

[3]We are ignoring the fact that the RHS of (15.3) is typically not the frequency response of a stable filter. A stable filter has a continuous frequency response (Theorem 6.2.11 (i)).

But since for filters of real impulse response the mapping $f \mapsto |\hat{h}(f)|^2$ is symmetric, (15.7a) can be rewritten as

$$\int_0^\infty |\hat{h}(f)|^2 \left(\mathsf{S}_{XX}(f) + \mathsf{S}_{XX}(-f)\right) \mathrm{d}f, \quad \mathbf{h} \text{ real and "nice."} \tag{15.7b}$$

This form makes it clear that for real stochastic processes, (15.7a) (or its equivalent form (15.7b)) can only specify the function $f \mapsto \mathsf{S}_{XX}(f) + \mathsf{S}_{XX}(-f)$; it cannot fully specify the mapping $f \mapsto \mathsf{S}_{XX}(f)$. For example, if a symmetric function $\mathsf{S}_{XX}$ satisfies (15.7a), then so does

$$f \mapsto \begin{cases} 2\mathsf{S}_{XX}(f) & \text{if } f > 0, \\ 0 & \text{otherwise,} \end{cases} \quad f \in \mathbb{R}.$$

In fact, if $\mathsf{S}_{XX}$ satisfies (15.7a), then so does any function $\tilde{\mathsf{S}}(\cdot)$ such that

$$\tilde{\mathsf{S}}(f) + \tilde{\mathsf{S}}(-f) = \mathsf{S}_{XX}(f) + \mathsf{S}_{XX}(-f), \quad f \in \mathbb{R}.$$

Thus, for the sake of uniqueness, we define the power spectral density $\mathsf{S}_{XX}$ to be a function of frequency that satisfies (15.7a) and that is additionally *symmetric*. It can be shown that this defines $\mathsf{S}_{XX}$ (to within indistinguishability) uniquely. In fact, once one has identified a nonnegative function $\mathsf{S}(\cdot)$ such that for any real impulse response $\mathbf{h}$ the integral

$$\int_{-\infty}^\infty \mathsf{S}(f)\, |\hat{h}(f)|^2 \,\mathrm{d}f$$

corresponds to the power in $\mathbf{X} \star \mathbf{h}$, then the PSD $\mathsf{S}_{XX}$ of $\mathbf{X}$ is given by the symmetrized version of $\mathsf{S}(\cdot)$, i.e.,

$$\mathsf{S}_{XX}(f) = \frac{1}{2}\left(\mathsf{S}(f) + \mathsf{S}(-f)\right), \quad f \in \mathbb{R}. \tag{15.8}$$

Note that the differential definition of the PSD would not have resolved the uniqueness issue because a filter of frequency response $f \mapsto \mathrm{I}\{f \in [f_0 - \frac{\Delta}{2}, f_0 + \frac{\Delta}{2}]\}$ is not real.

The final point we would like to make is regarding additivity. Apart from some mathematical details, what makes the definition of charge density possible is the fact that the total charge in the union of two disjoint regions in space is the sum of charges in the individual regions. The same holds for mass. For the probability densities the crucial property is that the probability of the union of two disjoint events is the sum of the probabilities. Consequently, if $\mathcal{D}_1$ and $\mathcal{D}_2$ are disjoint subsets of $\mathbb{R}$, then $\Pr[X \in \mathcal{D}_1 \cup \mathcal{D}_2] = \Pr[X \in \mathcal{D}_1] + \Pr[X \in \mathcal{D}_2]$. Does this hold for power? In general the power in the sum of two signals is *not* the sum of the individual powers. But if the signals are orthogonal, then their powers do add. Thus, while Parseval's theorem will not appear explicitly in our analysis of the PSD, it is really what makes it all possible. It demonstrates that if $\mathcal{D}_1, \mathcal{D}_2 \subset \mathbb{R}$ are disjoint frequency bands, then the signals $\mathbf{X} \star \mathbf{h}_1$ and $\mathbf{X} \star \mathbf{h}_2$ that result when $\mathbf{X}$ is passed through the filters of frequency response $\hat{h}_1(f) = \mathrm{I}\{f \in \mathcal{D}_1\}$ and $\hat{h}_2(f) = \mathrm{I}\{f \in \mathcal{D}_2\}$ are orthogonal, so their powers add. We will not bother to formulate this result precisely, because it does not show up in our analysis explicitly, but it is this result that allows us to define the power spectral density.

## 15.3   Defining the Operational PSD

Recall that in (14.14) we defined the power P in a SP $(Y(t), t \in \mathbb{R})$ as

$$P = \lim_{T \to \infty} \frac{1}{2T} \, \mathsf{E}\left[ \int_{-T}^{T} Y^2(t) \, \mathrm{d}t \right]$$

whenever the limit exists. Thus, the power is the limit, as T tends to infinity, of the ratio of the expected energy in the interval $[-T, T]$ to the interval's duration 2T. We define the operational power spectral density of a stochastic process as follows.

**Definition 15.3.1 (Operational PSD of a Real SP).** *We say that the continuous-time real stochastic process* $(X(t), t \in \mathbb{R})$ *is of **operational power spectral density** $\mathsf{S}_{XX}$ if $(X(t), t \in \mathbb{R})$ is a measurable SP; the mapping $\mathsf{S}_{XX} : \mathbb{R} \to \mathbb{R}$ is integrable and symmetric; and for every stable real filter of impulse response $\mathbf{h} \in \mathcal{L}_1$ the average power at the filter's output when it is fed $(X(t), t \in \mathbb{R})$ is given by*

$$\text{Power in } \mathbf{X} \star \mathbf{h} = \int_{-\infty}^{\infty} \mathsf{S}_{XX}(f) \, |\hat{h}(f)|^2 \, \mathrm{d}f.$$

We chose our words very carefully in the above definition, and, in doing so, we avoided two issues. The first is whether every SP is of some operational PSD. The answer to that is "no." (But most stochastic processes encountered in Digital Communications are.) The second issue we avoided is the uniqueness issue. Our wording did not indicate whether a SP could be of two different operational PSDs. It turns out that if a SP is of two different operational PSDs, then the two are equivalent in the sense that they agree except possibly on a set of frequencies of Lebesgue measure zero. Consequently, somewhat loosely, we shall speak of *the* operational power spectral density of $(X(t), t \in \mathbb{R})$ even though the uniqueness is only to within indistinguishability. The uniqueness is a corollary to the following somewhat technical lemma.

**Lemma 15.3.2.**

(i) *If* $\mathbf{s}$ *is an integrable function such that*

$$\int_{-\infty}^{\infty} s(f) \, |\hat{h}(f)|^2 \, \mathrm{d}f = 0 \tag{15.9}$$

*for every integrable **complex** function* $\mathbf{h} \colon \mathbb{R} \to \mathbb{C}$, *then $s(f)$ is zero for all frequencies outside a set of Lebesgue measure zero.*

(ii) *If* $\mathbf{s}$ *is a **symmetric** function such that* (15.9) *holds for every integrable **real** function* $\mathbf{h} \colon \mathbb{R} \to \mathbb{R}$, *then $s(f)$ is zero for all frequencies outside a set of Lebesgue measure zero.*

**Proof.** We begin with a proof of Part (i). For any $\lambda > 0$ and $f_0 \in \mathbb{R}$ define the function $\mathbf{h} : \mathbb{R} \to \mathbb{C}$ by

$$h(t) = \frac{1}{\sqrt{\lambda}} \, \mathrm{I}\left\{ |t| \leq \frac{\lambda}{2} \right\} e^{\mathrm{i} 2\pi f_0 t}, \quad t \in \mathbb{R}. \tag{15.10}$$

This function is in both $\mathcal{L}_1$ and $\mathcal{L}_2$. Since it is in $\mathcal{L}_2$, its self-similarity function $\mathsf{R_{hh}}(\tau)$ is defined at every $\tau \in \mathbb{R}$. In fact,

$$\mathsf{R_{hh}}(\tau) = \left(1 - \frac{|\tau|}{\lambda}\right) \mathrm{I}\{|\tau| \leq \lambda\} \, e^{\mathrm{i}2\pi f_0 \tau}, \quad \tau \in \mathbb{R}. \tag{15.11}$$

And since $\mathbf{h} \in \mathcal{L}_1$, it follows from (11.35) that the Fourier Transform of $\mathsf{R_{hh}}$ is the mapping $f \mapsto |\hat{h}(f)|^2$. Consequently, by Proposition 6.2.3 (i) (with the substitution $\check{\mathsf{R}}_{\mathsf{hh}}$ for $\mathbf{g}$), the mapping $f \mapsto |\hat{h}(f)|^2$ can be expressed as the Inverse Fourier Transform of $\check{\mathsf{R}}_{\mathsf{hh}}$. Thus, by (6.9) (with the substitutions of $\mathbf{s}$ for $\mathbf{x}$ and $\check{\mathsf{R}}_{\mathsf{hh}}$ for $\mathbf{g}$),

$$\int_{-\infty}^{\infty} s(f) \, |\hat{h}(f)|^2 \, \mathrm{d}f = \int_{-\infty}^{\infty} \hat{s}(f) \, \check{\mathsf{R}}_{\mathsf{hh}}^*(f) \, \mathrm{d}f. \tag{15.12}$$

It now follows from (15.9), (15.12), and (15.11) that

$$\int_{-\lambda}^{\lambda} \left(1 - \frac{|f|}{\lambda}\right) \hat{s}(f) \, e^{\mathrm{i}2\pi f_0 f} \, \mathrm{d}f = 0, \quad \lambda > 0, \; f_0 \in \mathbb{R}. \tag{15.13}$$

Part (i) now follows from (15.13) and from Theorem 6.2.12 (ii) (with the substitution of $\mathbf{s}$ for $\mathbf{x}$ and with the substitution of $f_0$ for $t$).

We next turn to Part (ii). For any integrable complex function $\mathbf{h} \colon \mathbb{R} \to \mathbb{C}$, define $\mathbf{h_R} \triangleq \mathrm{Re}(\mathbf{h})$ and $\mathbf{h_I} \triangleq \mathrm{Im}(\mathbf{h})$ so

$$\hat{h}_{\mathrm{R}}(f) = \frac{\hat{h}(f) + \hat{h}^*(-f)}{2}, \quad f \in \mathbb{R},$$

$$\hat{h}_{\mathrm{I}}(f) = \frac{\hat{h}(f) - \hat{h}^*(-f)}{2\mathrm{i}}, \quad f \in \mathbb{R}.$$

Consequently,

$$\left|\hat{h}_{\mathrm{R}}(f)\right|^2 = \frac{1}{4}\left(\left|\hat{h}(f)\right|^2 + \left|\hat{h}(-f)\right|^2 + 2\,\mathrm{Re}\!\left(\hat{h}(f)\,\hat{h}(-f)\right)\right), \quad f \in \mathbb{R}$$

$$\left|\hat{h}_{\mathrm{I}}(f)\right|^2 = \frac{1}{4}\left(\left|\hat{h}(f)\right|^2 + \left|\hat{h}(-f)\right|^2 - 2\,\mathrm{Re}\!\left(\hat{h}(f)\,\hat{h}(-f)\right)\right), \quad f \in \mathbb{R},$$

and

$$\left|\hat{h}_{\mathrm{R}}(f)\right|^2 + \left|\hat{h}_{\mathrm{I}}(f)\right|^2 = \frac{1}{2}\left(\left|\hat{h}(f)\right|^2 + \left|\hat{h}(-f)\right|^2\right), \quad f \in \mathbb{R}. \tag{15.14}$$

Applying the lemma's hypothesis to the real functions $\mathbf{h_R}$ and $\mathbf{h_I}$ we obtain

$$0 = \int_{-\infty}^{\infty} s(f) \left|\hat{h}_{\mathrm{R}}(f)\right|^2 \mathrm{d}f,$$

$$0 = \int_{-\infty}^{\infty} s(f) \left|\hat{h}_{\mathrm{I}}(f)\right|^2 \mathrm{d}f,$$

and thus, upon adding the equations,

$$0 = \int_{-\infty}^{\infty} s(f) \left( \left| \hat{h}_{\mathrm{R}}(f) \right|^2 + \left| \hat{h}_{\mathrm{I}}(f) \right|^2 \right) \mathrm{d}f$$

$$= \frac{1}{2} \int_{-\infty}^{\infty} s(f) \left( \left| \hat{h}(f) \right|^2 + \left| \hat{h}(-f) \right|^2 \right) \mathrm{d}f$$

$$= \int_{-\infty}^{\infty} \frac{s(f) + s(-f)}{2} \left| \hat{h}(f) \right|^2 \mathrm{d}f$$

$$= \int_{-\infty}^{\infty} s(f) \left| \hat{h}(f) \right|^2 \mathrm{d}f, \tag{15.15}$$

where the second equality follows from (15.14); the third by writing the integral of the sum as a sum of integrals and by changing the integration variable in the integral involving $\hat{h}(-f)$; and the last equality from the hypothesis that $\mathbf{s}$ is symmetric. Since we have established (15.15) for every complex $\mathbf{h} \colon \mathbb{R} \to \mathbb{C}$, we can now apply Part (i) to conclude that $\mathbf{s}$ is zero at all frequencies outside a set of Lebesgue measure zero. $\qquad\square$

**Corollary 15.3.3 (Uniqueness of PSD).** *If both* $\mathsf{S}_{XX}$ *and* $\mathsf{S}'_{XX}(\cdot)$ *are operational PSDs for the real SP* $\big( X(t),\ t \in \mathbb{R} \big)$*, then the set of frequencies at which they differ is of Lebesgue measure zero.*

**Proof.** Apply Lemma 15.3.2 (ii) to the function $\mathbf{s} \colon f \mapsto \mathsf{S}_{XX}(f) - \mathsf{S}'_{XX}(f)$. $\qquad\square$

As noted above, we make here no general claims about the existence of operational PSDs. Under certain restrictions that are made precise in Section 15.5, the operational PSD is defined for PAM signals. And by Theorem 25.13.2, the operational PSD always exists for measurable, centered, WSS, stochastic processes of integrable autocovariance functions.

**Definition 15.3.4 (Bandlimited Stochastic Processes).** *We say that a stochastic process* $\big( X(t),\ t \in \mathbb{R} \big)$ *of operational PSD* $\mathsf{S}_{XX}$ *is* **bandlimited to** $\mathsf{W}$ *Hz if, except on a set of frequencies of Lebesgue measure zero,* $\mathsf{S}_{XX}(f)$ *is zero for all frequencies* $f$ *satisfying* $|f| > \mathsf{W}$*.*

*The smallest* $\mathsf{W}$ *to which* $\big( X(t),\ t \in \mathbb{R} \big)$ *is limited is called the* **bandwidth** *of* $\big( X(t),\ t \in \mathbb{R} \big)$*.*

## 15.4 The Operational PSD of Real PAM Signals

Computing the operational PSD of PAM signals is much easier than you might expect. This is because, as we next show, passing a PAM signal of pulse shape $\mathbf{g}$ through a stable filter of impulse response $\mathbf{h}$ is tantamount to changing its pulse shape from $\mathbf{g}$ to $\mathbf{g} \star \mathbf{h}$:

$$\left( \left( \sigma \mapsto \mathsf{A} \sum_{\ell} X_{\ell} g(\sigma - \ell \mathsf{T_s}) \right) \star \mathbf{h} \right)(t) = \mathsf{A} \sum_{\ell} X_{\ell} (\mathbf{g} \star \mathbf{h})(t - \ell \mathsf{T_s}), \quad t \in \mathbb{R}. \tag{15.16}$$

(For a formal statement of this result, see Corollary 18.6.2, which also addresses the difficulty that arises when the sum is infinite.) Consequently, if one can compute the power in a PAM signal of arbitrary pulse shape (as explained in Chapter 14), then one can also compute the power in a filtered PAM signal.

That filtering a PAM signal is tantamount to convolving its pulse shape with the impulse response follows from two properties of the convolution: that it is linear

$$(\alpha \mathbf{u} + \beta \mathbf{v}) \star \mathbf{h} = \alpha \mathbf{u} \star \mathbf{h} + \beta \mathbf{v} \star \mathbf{h}$$

and that convolving a delayed version of a signal with $\mathbf{h}$ is equivalent to convolving the original signal and delaying the result

$$\Big( \big( \sigma \mapsto u(\sigma - t_0) \big) \star \mathbf{h} \Big)(t) = (\mathbf{u} \star \mathbf{h})(t - t_0), \quad t, t_0 \in \mathbb{R}.$$

Indeed, if $\mathbf{X}$ is the PAM signal

$$X(t) = \mathsf{A} \sum_{\ell = -\infty}^{\infty} X_\ell \, g(t - \ell \mathsf{T_s}), \tag{15.17}$$

then (15.16) follows from the calculation

$$(\mathbf{X} \star \mathbf{h})(t) = \left( \Big( \sigma \mapsto \mathsf{A} \sum_{\ell = -\infty}^{\infty} X_\ell \, g(\sigma - \ell \mathsf{T_s}) \Big) \star \mathbf{h} \right)(t)$$

$$= \mathsf{A} \sum_{\ell = -\infty}^{\infty} X_\ell \int_{-\infty}^{\infty} h(s) \, g(t - s - \ell \mathsf{T_s}) \, \mathrm{d}s$$

$$= \mathsf{A} \sum_{\ell = -\infty}^{\infty} X_\ell \, (\mathbf{g} \star \mathbf{h})(t - \ell \mathsf{T_s}), \quad t \in \mathbb{R}. \tag{15.18}$$

We are now ready to apply the results of Chapter 14 on the power in PAM signals to study the power in filtered PAM signals and hence to derive the operational PSD of PAM signals. We will not treat the case discussed in Section 14.5.3 where the only assumption is that the time shifts of the pulse shape by integer multiples of $\mathsf{T_s}$ are orthonormal, because this orthonomality is typically lost under filtering.

### 15.4.1 $(X_\ell, \ell \in \mathbb{Z})$ Are Centered, Uncorrelated, and of Equal Variance

We begin with the case where the symbols $(X_\ell, \ell \in \mathbb{Z})$ are of zero mean, uncorrelated, and of equal variance $\sigma_X^2$. As in (15.17) we denote the PAM signal by $(X(t), t \in \mathbb{R})$ and study its operational PSD by studying the power in $\mathbf{X} \star \mathbf{h}$. Using (15.18) we obtain that $\mathbf{X} \star \mathbf{h}$ is the PAM signal $\mathbf{X}$ but with the pulse shape $\mathbf{g}$ replaced by $\mathbf{g} \star \mathbf{h}$. Consequently, using Expression (14.33) for the power in PAM with zero-mean, uncorrelated, variance-$\sigma_X^2$ symbols, we obtain that the power in

$\mathbf{X} \star \mathbf{h}$ is given by

$$\text{Power in } \mathbf{X} \star \mathbf{h} = \frac{\mathsf{A}^2}{\mathsf{T}_\mathrm{s}} \sigma_X^2 \, \|\mathbf{g} \star \mathbf{h}\|_2^2$$

$$= \frac{\mathsf{A}^2 \sigma_X^2}{\mathsf{T}_\mathrm{s}} \int_{-\infty}^{\infty} |\hat{g}(f)|^2 \, |\hat{h}(f)|^2 \, \mathrm{d}f$$

$$= \int_{-\infty}^{\infty} \underbrace{\left( \frac{\mathsf{A}^2 \sigma_X^2}{\mathsf{T}_\mathrm{s}} |\hat{g}(f)|^2 \right)}_{\mathsf{S}_{XX}(f)} |\hat{h}(f)|^2 \, \mathrm{d}f, \qquad (15.19)$$

where the first equality follows from (14.33) applied to the PAM signal of pulse shape $\mathbf{g} \star \mathbf{h}$; the second follows from Parseval's Theorem by noting that the Fourier Transform of a convolution of two signals is the product of their Fourier Transforms; and where the third equality follows by rearranging terms. From (15.19) and from the fact that $f \mapsto |\hat{g}(f)|^2$ is a symmetric function (because $\mathbf{g}$ is real), it follows that the operational PSD of the PAM signal $\big(X(t), \, t \in \mathbb{R}\big)$ when $\big(X_\ell, \, \ell \in \mathbb{Z}\big)$ are zero-mean, uncorrelated, and of variance $\sigma_X^2$ is given by

$$\boxed{\mathsf{S}_{XX}(f) = \frac{\mathsf{A}^2 \sigma_X^2}{\mathsf{T}_\mathrm{s}} |\hat{g}(f)|^2, \quad f \in \mathbb{R}.} \qquad (15.20)$$

### 15.4.2 $\big(X_\ell\big)$ Is Centered and WSS

The more general case where the symbols $\big(X_\ell, \, \ell \in \mathbb{Z}\big)$ are not necessarily uncorrelated but form a centered, WSS, discrete-time SP can be treated with the same ease via (14.31) or (14.32). As above, passing $\mathbf{X}$ through a filter of impulse response $\mathbf{h}$ results in a PAM signal with identical symbols but with pulse shape $\mathbf{g} \star \mathbf{h}$. Consequently, the resulting power can be computed by substituting $\mathbf{g} \star \mathbf{h}$ for $\mathbf{g}$ in (14.32) to obtain that the power in $\mathbf{X} \star \mathbf{h}$ is given by

$$\text{Power in } \mathbf{X} \star \mathbf{h} = \int_{-\infty}^{\infty} \underbrace{\left( \frac{\mathsf{A}^2}{\mathsf{T}_\mathrm{s}} \sum_{m=-\infty}^{\infty} \mathsf{K}_{XX}(m) \, e^{\mathrm{i} 2\pi f m \mathsf{T}_\mathrm{s}} |\hat{g}(f)|^2 \right)}_{\mathsf{S}_{XX}(f)} |\hat{h}(f)|^2 \, \mathrm{d}f,$$

where again we are using the fact that the FT of $\mathbf{g} \star \mathbf{h}$ is $f \mapsto \hat{g}(f)\,\hat{h}(f)$. The operational PSD is thus

$$\boxed{\mathsf{S}_{XX}(f) = \frac{\mathsf{A}^2}{\mathsf{T}_\mathrm{s}} \sum_{m=-\infty}^{\infty} \mathsf{K}_{XX}(m) \, e^{\mathrm{i} 2\pi f m \mathsf{T}_\mathrm{s}} |\hat{g}(f)|^2, \quad f \in \mathbb{R},} \qquad (15.21)$$

because, as we next argue, the RHS of the above is a symmetric function of $f$. This symmetry follows from the symmetry of $|\hat{g}(\cdot)|$ (because the pulse shape $\mathbf{g}$ is real) and from the symmetry of the autocovariance function $\mathsf{K}_{XX}$ (because the symbols $\big(X_\ell, \, \ell \in \mathbb{Z}\big)$ are real; see (13.12)). Note that (15.21) reduces to (15.20) if $\mathsf{K}_{XX}(m) = \sigma_X^2 \, \mathrm{I}\{m = 0\}$.

### 15.4.3 The Operational PSD in Bi-Infinite Block-Mode

We now assume, as in Section 14.5.2, that the $(K, N)$ binary-to-reals block encoder $\mathbf{enc}\colon \{0,1\}^K \to \mathbb{R}^N$ is used in bi-infinite block encoding mode to map the bi-infinite IID random bits $(D_j, \ j \in \mathbb{Z})$ to the bi-infinite sequence of real numbers $(X_\ell, \ \ell \in \mathbb{Z})$, and that the transmitted signal is

$$X(t) = A \sum_{\ell=-\infty}^{\infty} X_\ell \, g(t - \ell \mathsf{T_s}), \tag{15.22}$$

where $\mathsf{T_s} > 0$ is the baud, and where $g(\cdot)$ is a pulse shape satisfying the decay condition (14.17). We do not assume that the time-shifts of $g(\cdot)$ by integer multiples of $\mathsf{T_s}$ are orthogonal, or that the symbols $(X_\ell, \ \ell \in \mathbb{Z})$ are uncorrelated. We do, however, continue to assume that the N-tuple $\mathbf{enc}(D_1, \dots, D_K)$ is of zero mean whenever $D_1, \dots, D_K$ are IID random bits.

We shall determine the operational PSD of $\mathbf{X}$ by computing the power of the signal that results when $\mathbf{X}$ is fed to a stable filter of impulse response $\mathbf{h}$. As before, we note that feeding $\mathbf{X}$ through a filter of impulse response $\mathbf{h}$ is tantamount to replacing its pulse shape $\mathbf{g}$ by $\mathbf{g} \star \mathbf{h}$. The power of this output signal can be thus computed from our expression for the power in bi-infinite block encoding with PAM signaling (14.38) but with the pulse shape being $\mathbf{g} \star \mathbf{h}$ and hence of FT $f \mapsto \hat{g}(f)\,\hat{h}(f)$:

$$\text{Power in } \mathbf{X} \star \mathbf{h} = \int_{-\infty}^{\infty} \underbrace{\left( \frac{A^2}{N\mathsf{T_s}} \sum_{\ell=1}^{N} \sum_{\ell'=1}^{N} \mathsf{E}[X_\ell X_{\ell'}] \, e^{i2\pi f(\ell-\ell')\mathsf{T_s}} \, |\hat{g}(f)|^2 \right)}_{\mathsf{S}_{XX}(f)} |\hat{h}(f)|^2 \, \mathrm{d}f.$$

As we next show, the underbraced term is a symmetric function of $f$, and we thus conclude that the PSD of $\mathbf{X}$ is:

$$\boxed{\mathsf{S}_{XX}(f) = \frac{A^2}{N\mathsf{T_s}} \sum_{\ell=1}^{N} \sum_{\ell'=1}^{N} \mathsf{E}[X_\ell X_{\ell'}] \, e^{i2\pi f(\ell-\ell')\mathsf{T_s}} \, |\hat{g}(f)|^2, \quad f \in \mathbb{R}.} \tag{15.23}$$

To see that the RHS of (15.23) is a symmetric function of $f$, use the identities

$$\sum_{\ell=1}^{N} \sum_{\ell'=1}^{N} a_{\ell,\ell'} = \sum_{\ell=1}^{N} a_{\ell,\ell} + \sum_{\ell=1}^{N} \sum_{\ell'=1}^{\ell-1} (a_{\ell,\ell'} + a_{\ell',\ell})$$

and $\mathsf{E}[X_\ell X_{\ell'}] = \mathsf{E}[X_{\ell'} X_\ell]$ to rewrite the RHS of (15.23) in the symmetric form

$$\frac{A^2}{N\mathsf{T_s}} \left( \sum_{\ell=1}^{N} \mathsf{E}[X_\ell^2] + \sum_{\ell=1}^{N} \sum_{\ell'=1}^{\ell-1} 2\,\mathsf{E}[X_\ell X_{\ell'}] \cos\big(2\pi f(\ell - \ell')\mathsf{T_s}\big) \right) |\hat{g}(f)|^2.$$

From (15.23) we obtain:

**Theorem 15.4.1 (The Bandwidth of PAM Is that of the Pulse Shape).** *Suppose that the operational PSD in bi-infinite block-mode of a PAM signal $(X(t))$ is as* *given in (15.23), e.g., that the conditions of Theorem 15.5.2 ahead are satisfied. Further assume*

$$A > 0, \quad \sum_{\ell=1}^{N} \mathsf{E}[X_\ell^2] > 0, \tag{15.24}$$

*e.g., that $(X(t))$ is not deterministically zero. Then the bandwidth of the SP $(X(t))$ is equal to the bandwidth of the pulse shape $\mathbf{g}$.*

**Proof.** If $\mathbf{g}$ is bandlimited to $W$ Hz, then so is $(X(t))$, because, by (15.23),

$$\left(\hat{g}(f) = 0\right) \Rightarrow \left(\mathsf{S}_{XX}(f) = 0\right).$$

We next complete the proof by showing that there are at most a countable number of frequencies $f$ such that $\mathsf{S}_{XX}(f) = 0$ but $\hat{g}(f) \neq 0$. From (15.23) it follows that to show this it suffices to show that there are at most a countable number of frequencies $f$ such that $\sigma(f) = 0$, where

$$
\begin{aligned}
\sigma(f) &\triangleq \frac{A^2}{NT_s} \sum_{\ell=1}^{N} \sum_{\ell'=1}^{N} \mathsf{E}[X_\ell X_{\ell'}] e^{i2\pi f(\ell-\ell')T_s} \\
&= \sum_{m=-N+1}^{N-1} \gamma_m e^{i2\pi f m T_s} \\
&= \sum_{m=-N+1}^{N-1} \gamma_m z^m \bigg|_{z=e^{i2\pi f T_s}},
\end{aligned}
\tag{15.25}
$$

and

$$\gamma_m = \frac{A^2}{NT_s} \sum_{\ell=\max\{1, m+1\}}^{\min\{N, N+m\}} \mathsf{E}[X_\ell X_{\ell-m}], \quad m \in \{-N+1, \ldots, N-1\}. \tag{15.26}$$

It follows from (15.25) that $\sigma(f)$ is zero if, and only if, $e^{i2\pi f T_s}$ is a root of the mapping

$$z \mapsto \sum_{m=-N+1}^{N-1} \gamma_m z^m.$$

Since $e^{i2\pi f T_s}$ is of unit magnitude, it follows that $\sigma(f)$ is zero if, and only if, $e^{i2\pi f T_s}$ is a root of the polynomial

$$z \mapsto \sum_{\nu=0}^{2N-2} \gamma_{\nu-N+1} z^\nu. \tag{15.27}$$

From (15.26) and (15.24) it follows that $\gamma_0 > 0$, so the polynomial in (15.27) is not zero. Consequently, since it is of degree $2N - 2$, it has at most $2N - 2$ distinct roots and, *a fortiori*, at most $2N - 2$ distinct roots of unit magnitude. Denote these roots by

$$e^{i\theta_1}, \ldots, e^{i\theta_d},$$

where $d \leq 2N - 2$ and $\theta_1, \ldots, \theta_d \in [-\pi, \pi)$. Since $f$ satisfies $e^{i2\pi f T_s} = e^{i\theta}$ if, and only if,

$$f = \frac{\theta}{2\pi T_s} + \frac{\eta}{T_s}$$

for some $\eta \in \mathbb{Z}$, we conclude that the set of frequencies $f$ satisfying $\sigma(f) = 0$ is the set

$$\left\{ \frac{\theta_1}{2\pi T_s} + \frac{\eta}{T_s} : \eta \in \mathbb{Z} \right\} \cup \cdots \cup \left\{ \frac{\theta_d}{2\pi T_s} + \frac{\eta}{T_s} : \eta \in \mathbb{Z} \right\},$$

and is thus countable. (The union of a finite (or countable) number of countable sets is countable.) $\square$

## 15.5 A More Formal Account

In this section we shall give a more formal account of the power at the output of a stable filter that is fed a PAM signal. There are two approaches to this. The first is based on carefully justifying the steps in our informal derivation.[4] This approach is pursued in Section 18.6.5, where the results are generalized to complex pulse shapes and complex symbols. The second approach is to convert the problem into one about WSS stochastic processes and to then rely heavily on Sections 25.13 and 25.14 on the filtering of WSS stochastic processes and, in particular, on the Wiener-Khinchin Theorem (Theorem 25.14.1). For the benefit of readers who have already encountered the Wiener-Khinchin Theorem we follow this latter approach here. We ask the readers to note that the Wiener-Khinchin Theorem is not directly applicable here because the PAM signal is not WSS. A "stationarization argument" is thus needed.

The key results of this section are the following two theorems.

**Theorem 15.5.1.** *Consider the setup of Theorem 14.6.4 with the additional assumption that the autocovariance function $K_{XX}$ of $(X_\ell)$ is absolutely summable:*

$$\sum_{m=-\infty}^{\infty} |K_{XX}(m)| < \infty. \tag{15.28}$$

*Let $\mathbf{h} \in \mathcal{L}_1$ be the impulse response of a stable real filter. Then:*

*(i) The PAM signal*

$$\mathbf{X} \colon (\omega, t) \mapsto A \sum_{\ell=-\infty}^{\infty} X_\ell(\omega) g(t - \ell T_s) \tag{15.29}$$

*is bounded in the sense that there exists a constant $\Gamma$ such that*

$$|X(\omega, t)| < \Gamma, \quad \left( \omega \in \Omega, \ t \in \mathbb{R} \right). \tag{15.30}$$

---

[4]The main difficulties in the justification are in making (15.16) rigorous and in controlling the decay of $\mathbf{g} \star \mathbf{h}$ for arbitrary $\mathbf{h} \in \mathcal{L}_1$.

(ii) *For every $\omega \in \Omega$ the convolution of the sample-path $t \mapsto X(\omega, t)$ with $\mathbf{h}$ is defined at every epoch.*

(iii) *The stochastic process*

$$(\omega, t) \mapsto \int_{-\infty}^{\infty} x(\omega, \sigma)\, h(t - \sigma)\, \mathrm{d}\sigma, \quad \left( \omega \in \Omega,\ t \in \mathbb{R} \right) \tag{15.31}$$

*that results when the sample-paths of $\mathbf{X}$ are convolved with $\mathbf{h}$ is a measurable stochastic process of power*

$$\mathsf{P} = \int_{-\infty}^{\infty} \left( \frac{\mathsf{A}^2}{\mathsf{T_s}} \sum_{m=-\infty}^{\infty} \mathsf{K}_{XX}(m)\, e^{\mathrm{i}2\pi f m \mathsf{T_s}}\, |\hat{g}(f)|^2 \right) |\hat{h}(f)|^2 \, \mathrm{d}f. \tag{15.32}$$

**Theorem 15.5.2.** *Consider the setup of Theorem 14.6.5. Let $\mathbf{h} \in \mathcal{L}_1$ be the impulse response of a real stable filter. Then:*

(i) *The sample-paths of the PAM stochastic process*

$$\mathbf{X} \colon (\omega, t) \mapsto \mathsf{A} \sum_{\ell=-\infty}^{\infty} X_\ell(\omega)\, g(t - \ell\mathsf{T_s}) \tag{15.33}$$

*are bounded in the sense of (15.30).*

(ii) *For every $\omega \in \Omega$ the convolution of the sample-path $t \mapsto X(\omega, t)$ and $\mathbf{h}$ is defined at every epoch.*

(iii) *The stochastic process $\left( X(t),\ t \in \mathbb{R} \right) \star \mathbf{h}$ that results when the sample-paths of $\mathbf{X}$ are convolved with $\mathbf{h}$ is a measurable stochastic process of power*

$$\mathsf{P} = \int_{-\infty}^{\infty} \left( \frac{\mathsf{A}^2}{\mathsf{N}\mathsf{T_s}} \sum_{\ell=1}^{\mathsf{N}} \sum_{\ell'=1}^{\mathsf{N}} \mathsf{E}[X_\ell X_{\ell'}]\, e^{\mathrm{i}2\pi f(\ell-\ell')\mathsf{T_s}}\, |\hat{g}(f)|^2 \right) |\hat{h}(f)|^2 \, \mathrm{d}f, \tag{15.34}$$

*where $\left( X_1, \ldots, X_\mathsf{N} \right) = \mathrm{enc}\left( D_1, \ldots, D_\mathsf{K} \right)$, and where $D_1, \ldots, D_\mathsf{K}$ are IID random bits.*

**Proof of Theorem 15.5.1.** Part (i) is a consequence of the assumption that $(X_\ell)$ is bounded in the sense of (14.16) and that the pulse shape $\mathbf{g}$ decays faster than $1/t$ in the sense of (14.17).

Part (ii) is a consequence of the fact that the convolution of a bounded function with an integrable function is defined at every epoch; see Section 5.5.

We next turn to Part (iii). The proof of the measurability of the convolution of $\left( X(t),\ t \in \mathbb{R} \right)$ with $\mathbf{h}$ is a bit technical. It is very similar to the proof of Theorem 25.13.2 (i). As in that proof, we first note that it suffices to prove the result for functions $\mathbf{h}$ that are Borel measurable; the extension to Lebesgue measurable functions will then follow by approximating $\mathbf{h}$ by a Borel measurable function that differs from it on a set of Lebesgue measure zero (Rudin, 1974, Chapter 7, Lemma 1) and by then noting that the convolution of $t \mapsto X(\omega, t)$ with $\mathbf{h}$ is unaltered when $\mathbf{h}$

is replaced by a function that differs from it on a set of Lebesgue measure zero. We thus assume that **h** is Borel measurable. Consequently, the mapping from $\mathbb{R}^2$ to $\mathbb{R}$ defined by $(t, \sigma) \mapsto h(t - \sigma)$ is also Borel measurable, because it is the composition of the continuous (and hence Borel measurable) mapping $(t, \sigma) \mapsto t - \sigma$ with the Borel measurable mapping $t \mapsto h(t)$.

As in the proof of Theorem 25.13.2, we prove the measurability of the convolution of $(X(t),\ t \in \mathbb{R})$ with **h** by proving the measurability of the mapping defined by $(\omega, t) \mapsto (1 + t^2)^{-1} \int_{-\infty}^{\infty} X(\omega, \sigma)\, h(t - \sigma)\, d\sigma$. To this end we study the function

$$((\omega, t), \sigma) \mapsto \frac{X(\omega, \sigma)\, h(t - \sigma)}{1 + t^2}, \quad \Big((\omega, t) \in \Omega \times \mathbb{R},\ \sigma \in \mathbb{R}\Big). \tag{15.35}$$

This function is measurable because, as noted above, $(t, \sigma) \mapsto h(t - \sigma)$ is measurable; because, by Proposition 14.6.2, $(X(t),\ t \in \mathbb{R})$ is measurable; and because the product of Borel measurable functions is Borel measurable (Rudin, 1974, Chapter 1, Section 1.9 (c)). Moreover, using (15.30) and Fubini's Theorem it can be readily verified that this function is integrable. Using Fubini's Theorem again, we conclude that the function

$$(\omega, t) \mapsto \frac{1}{1 + t^2} \int_{-\infty}^{\infty} X(\omega, \sigma)\, h(t - \sigma)\, d\sigma$$

is measurable. Consequently, so is $\mathbf{X} \star \mathbf{h}$.

To conclude the proof we now need to compute the power in the measurable (non-stationary) SP $\mathbf{X} \star \mathbf{h}$. This will be done in a roundabout way. We shall first define a new SP $\mathbf{X}'$. This SP is centered, measurable, and WSS so the power in $\mathbf{X}' \star \mathbf{h}$ can be computed using Theorem 25.14.1. We shall then show that the powers of $\mathbf{X} \star \mathbf{h}$ and $\mathbf{X}' \star \mathbf{h}$ are equal and hence that from the power in $\mathbf{X}' \star \mathbf{h}$ we can immediately obtain the power in $\mathbf{X} \star \mathbf{h}$.

We begin by defining the SP $(X'(t),\ t \in \mathbb{R})$ as

$$X'(t) = X(t + S), \quad t \in \mathbb{R}, \tag{15.36a}$$

where $S$ is independent of $(X(t))$ and uniformly distributed over the interval $[0, \mathsf{T_s}]$,

$$S \sim \mathcal{U}\left([0, \mathsf{T_s}]\right). \tag{15.36b}$$

That $(X'(t))$ is centered follows from the calculation

$$\begin{aligned}
\mathsf{E}[X'(t)] &= \mathsf{E}[X(t + S)] \\
&= \int_0^{\mathsf{T_s}} \frac{1}{\mathsf{T_s}} \mathsf{E}[X(t + s)]\, ds \\
&= 0,
\end{aligned}$$

where the first equality follows from the definition of $(X'(t))$; the second from the independence of $(X(t))$ and $S$ and from the specific form of the density of $S$; and the third because $(X(t))$ is centered. That $(X'(t))$ is measurable follows because the mapping $((\omega, s), t) \mapsto X(\omega, t + s)$ can be written as the composition of the mapping $\big((\omega, s), t\big) \mapsto (\omega, t + s)$ with the mapping $(\omega, t) \mapsto X(\omega, t)$. And that it is WSS follows from the calculation

$$
\begin{aligned}
\mathsf{E}&[X'(t)\,X'(t+\tau)] \\
&= \mathsf{E}[X(t+S)\,X(t+S+\tau)] \\
&= \frac{1}{\mathsf{T_s}} \int_0^{\mathsf{T_s}} \mathsf{E}[X(t+s)\,X(t+s+\tau)]\,\mathrm{d}s \\
&= \frac{1}{\mathsf{T_s}} \mathsf{A}^2 \int_0^{\mathsf{T_s}} \mathsf{E}\Bigg[ \sum_{\ell=-\infty}^{\infty} X_\ell\, g(t+s-\ell\mathsf{T_s}) \sum_{\ell'=-\infty}^{\infty} X_{\ell'}\, g(t+s+\tau-\ell'\mathsf{T_s}) \Bigg]\,\mathrm{d}s \\
&= \frac{1}{\mathsf{T_s}} \mathsf{A}^2 \sum_{\ell=-\infty}^{\infty} \sum_{\ell'=-\infty}^{\infty} \mathsf{E}[X_\ell X_{\ell'}] \int_0^{\mathsf{T_s}} g(t+s-\ell\mathsf{T_s})\, g(t+s+\tau-\ell'\mathsf{T_s})\,\mathrm{d}s \\
&= \frac{1}{\mathsf{T_s}} \mathsf{A}^2 \sum_{\ell=-\infty}^{\infty} \sum_{\ell'=-\infty}^{\infty} \mathsf{K}_{XX}(\ell-\ell') \int_0^{\mathsf{T_s}} g(t+s-\ell\mathsf{T_s})\, g(t+s+\tau-\ell'\mathsf{T_s})\,\mathrm{d}s \\
&= \frac{1}{\mathsf{T_s}} \mathsf{A}^2 \sum_{\ell=-\infty}^{\infty} \sum_{m=-\infty}^{\infty} \mathsf{K}_{XX}(m) \int_0^{\mathsf{T_s}} g\big(t+s-\ell\mathsf{T_s}\big)\, g\big(t+s+\tau-(\ell-m)\mathsf{T_s}\big)\,\mathrm{d}s \\
&= \frac{1}{\mathsf{T_s}} \mathsf{A}^2 \sum_{m=-\infty}^{\infty} \mathsf{K}_{XX}(m) \sum_{\ell=-\infty}^{\infty} \int_{-\ell\mathsf{T_s}+t}^{-\ell\mathsf{T_s}+\mathsf{T_s}+t} g(\xi)\, g(\xi+\tau+m\mathsf{T_s})\,\mathrm{d}\xi \\
&= \frac{1}{\mathsf{T_s}} \mathsf{A}^2 \sum_{m=-\infty}^{\infty} \mathsf{K}_{XX}(m) \int_{-\infty}^{\infty} g(\xi)\, g(\xi+\tau+m\mathsf{T_s})\,\mathrm{d}\xi \\
&= \frac{1}{\mathsf{T_s}} \mathsf{A}^2 \sum_{m=-\infty}^{\infty} \mathsf{K}_{XX}(m)\, \mathsf{R_{gg}}(m\mathsf{T_s}+\tau), \quad \tau, t \in \mathbb{R}.
\end{aligned}
\tag{15.37}
$$

Note that (15.37) also shows that $\big(X'(t)\big)$ is of PSD (as defined in Definition 25.7.2)

$$
\mathsf{S}_{X'X'}(f) = \frac{\mathsf{A}^2}{\mathsf{T_s}} \sum_{m=-\infty}^{\infty} \mathsf{K}_{XX}(m)\, e^{\mathrm{i}2\pi f m \mathsf{T_s}}\, |\hat{g}(f)|^2, \quad f \in \mathbb{R},
\tag{15.38}
$$

which is integrable by the absolute summability of $\mathsf{K}_{XX}$.

Defining $\big(Y'(t),\ t \in \mathbb{R}\big)$ to be $\big(X'(t),\ t \in \mathbb{R}\big) \star \mathbf{h}$ we can now use Theorem 25.14.1 to compute the power in $\big(Y'(t),\ t \in \mathbb{R}\big)$:

$$
\lim_{\mathsf{T} \to \infty} \frac{1}{2\mathsf{T}} \mathsf{E}\left[ \int_{-\mathsf{T}}^{\mathsf{T}} \big(Y'(t)\big)^2\,\mathrm{d}t \right] = \int_{-\infty}^{\infty} \left( \frac{\mathsf{A}^2}{\mathsf{T_s}} \sum_{m=-\infty}^{\infty} \mathsf{K}_{XX}(m)\, e^{\mathrm{i}2\pi f m \mathsf{T_s}}\, |\hat{g}(f)|^2 \right) |\hat{h}(f)|^2\,\mathrm{d}f.
$$

To conclude the proof we next show that the power in $\mathbf{Y}$ is the same as the power in $\mathbf{Y}'$. To that end we first note that from (15.36a) it follows that

$$
\big(\mathbf{X}' \star \mathbf{h}\big)\big((\omega, s), t\big) = \big(\mathbf{X} \star \mathbf{h}\big)(\omega, t+s), \quad \big(\omega \in \Omega,\ 0 \le s \le \mathsf{T_s},\ t \in \mathbb{R}\big),
$$

i.e., that

$$
Y'\big((\omega, s), t\big) = Y(\omega, t+s), \quad \big(\omega \in \Omega,\ 0 \le s \le \mathsf{T_s},\ t \in \mathbb{R}\big).
\tag{15.39}
$$

It thus follows that

$$\int_{-T}^{T} Y^2(\omega, t)\, \mathrm{d}t \leq \int_{-T-T_s}^{T} \big(Y'((\omega, s), t)\big)^2 \, \mathrm{d}t, \quad \Big(\omega \in \Omega,\ 0 \leq s \leq T_s,\ t \in \mathbb{R}\Big),$$

(15.40)

because

$$\int_{-T-T_s}^{T} \big(Y'((\omega, s), t)\big)^2 \, \mathrm{d}t = \int_{-T-T_s}^{T} Y^2(\omega, t+s)\, \mathrm{d}t$$

$$= \int_{-T-T_s+s}^{T+s} Y^2(\omega, \sigma)\, \mathrm{d}\sigma$$

$$\geq \int_{-T}^{T} Y^2(\omega, \sigma)\, \mathrm{d}\sigma, \quad 0 \leq s \leq T_s,$$

where the equality in the first line follows from (15.39); the equality in the second line from the substitution $\sigma \triangleq t+s$; and the final inequality from the nonnegativity of the integrand and because $0 \leq s \leq T_s$.

Similarly,

$$\int_{-T}^{T} Y^2(\omega, t)\, \mathrm{d}t \geq \int_{-T}^{T-T_s} \big(Y'((\omega, s), t)\big)^2 \, \mathrm{d}t, \quad \Big(\omega \in \Omega,\ 0 \leq s \leq T_s,\ t \in \mathbb{R}\Big),$$ (15.41)

because

$$\int_{-T}^{T-T_s} \big(Y'((\omega, s), t)\big)^2 \, \mathrm{d}t = \int_{-T}^{T-T_s} Y^2(\omega, t+s)\, \mathrm{d}t$$

$$= \int_{-T+s}^{T-T_s+s} Y^2(\omega, \sigma)\, \mathrm{d}\sigma$$

$$\leq \int_{-T}^{T} Y^2(\omega, \sigma)\, \mathrm{d}\sigma, \quad 0 \leq s \leq T_s.$$

Combining (15.40) and (15.41) and using the nonnegativity of the integrand we obtain that for every $\omega \in \Omega$ and $s \in [0, T_s]$

$$\int_{-T+T_s}^{T-T_s} \big(Y'((\omega, s), t)\big)^2 \, \mathrm{d}t \leq \int_{-T}^{T} Y^2(\omega, \sigma)\, \mathrm{d}\sigma \leq \int_{-T-T_s}^{T+T_s} \big(Y'((\omega, s), t)\big)^2 \, \mathrm{d}t. \quad (15.42)$$

Dividing by $2T$ and taking expectations we obtain

$$\frac{2T - 2T_s}{2T} \frac{1}{2T - 2T_s} \mathsf{E}\left[\int_{-T+T_s}^{T-T_s} \big(Y'(t)\big)^2 \, \mathrm{d}t\right]$$

$$\leq \frac{1}{2T} \mathsf{E}\left[\int_{-T}^{T} Y^2(\sigma)\, \mathrm{d}\sigma\right] \leq$$

$$\frac{2T + 2T_s}{2T} \frac{1}{2T + 2T_s} \mathsf{E}\left[\int_{-T-T_s}^{T+T_s} \big(Y'(t)\big)^2 \, \mathrm{d}t\right], \quad (15.43)$$

from which the equality between the power in $\mathbf{Y}'$ and in $\mathbf{Y}$ follows by letting $\mathsf{T}$ tend to infinity and using the Sandwich Theorem. □

**Proof of Theorem 15.5.2.** The proof of Theorem 15.5.2 is very similar to the proof of Theorem 15.5.1, so most of the details will be omitted. The main difference is that the process $\big(X'(t),\ t \in \mathbb{R}\big)$ is now defined as

$$X'(t) = X(t + S)$$

where the random variable $S$ is now uniformly distributed over the interval $[0, \mathrm{NT_s}]$,

$$S \sim \mathcal{U}\big([0, \mathrm{NT_s}]\big).$$

With this definition, the autocovariance of $\big(X'(t),\ t \in \mathbb{R}\big)$ can be computed as

$$
\begin{aligned}
\mathsf{K}_{X'X'}&(\tau) \\
&= \mathsf{E}\big[X(t + S)\,X(t + \tau + S)\big] \\
&= \frac{1}{\mathrm{NT_s}} \int_0^{\mathrm{NT_s}} \mathsf{E}\big[X(t + s)\,X(t + \tau + s)\big]\,\mathrm{d}s \\
&= \frac{A^2}{\mathrm{NT_s}} \mathsf{E}\left[\int_0^{\mathrm{NT_s}} \left(\sum_{\nu=-\infty}^{\infty} u(\mathbf{X}_\nu, t + s - \nu\mathrm{NT_s}) \sum_{\nu'=-\infty}^{\infty} u(\mathbf{X}_{\nu'}, t + \tau + s - \nu'\mathrm{NT_s})\right)\mathrm{d}s\right] \\
&= \frac{A^2}{\mathrm{NT_s}} \int_0^{\mathrm{NT_s}} \sum_{\nu=-\infty}^{\infty} \sum_{\nu'=-\infty}^{\infty} \mathsf{E}\big[u(\mathbf{X}_\nu, t + s - \nu\mathrm{NT_s})\,u(\mathbf{X}_{\nu'}, t + \tau + s - \nu'\mathrm{NT_s})\big]\,\mathrm{d}s \\
&= \frac{A^2}{\mathrm{NT_s}} \int_0^{\mathrm{NT_s}} \sum_{\nu=-\infty}^{\infty} \mathsf{E}\big[u(\mathbf{X}_\nu, t + s - \nu\mathrm{NT_s})\,u(\mathbf{X}_\nu, t + \tau + s - \nu\mathrm{NT_s})\big]\,\mathrm{d}s \\
&= \frac{A^2}{\mathrm{NT_s}} \int_0^{\mathrm{NT_s}} \sum_{\nu=-\infty}^{\infty} \mathsf{E}\big[u(\mathbf{X}_0, t + s - \nu\mathrm{NT_s})\,u(\mathbf{X}_0, t + \tau + s - \nu\mathrm{NT_s})\big]\,\mathrm{d}s \\
&= \frac{A^2}{\mathrm{NT_s}} \int_{-\infty}^{\infty} \mathsf{E}\big[u(\mathbf{X}_0, \xi)\,u(\mathbf{X}_0, \xi + \tau)\big]\,\mathrm{d}\xi \\
&= \frac{A^2}{\mathrm{NT_s}} \int_{-\infty}^{\infty} \mathsf{E}\left[\sum_{\eta=1}^{N} X_\eta\,g(\xi - \eta\mathrm{T_s}) \sum_{\eta'=1}^{N} X_{\eta'}\,g(\xi + \tau - \eta'\mathrm{T_s})\right]\mathrm{d}\xi \\
&= \frac{A^2}{\mathrm{NT_s}} \sum_{\eta=1}^{N} \sum_{\eta'=1}^{N} \mathsf{E}\big[X_\eta X_{\eta'}\big]\,\mathsf{R}_{\mathbf{gg}}\big(\tau + (\eta - \eta')\big), \quad t, \tau \in \mathbb{R},
\end{aligned}
$$

where the third equality follows from (14.36), (14.39), and (14.40); the fifth follows from (14.43); the sixth because the N-tuples $(\mathbf{X}_\eta,\ \eta \in \mathbb{Z})$ are IID; the seventh by defining $\xi = t + s$; the eighth by the definition (14.40) of the function $u(\cdot)$; and the final equality by swapping the summations and the expectation.

The process $\big(X'(t)\big)$ is thus a WSS process of PSD (as defined in Definition 25.7.2)

$$\mathsf{S}_{X'X'}(f) = \frac{A^2}{\mathrm{NT_s}} \sum_{\ell=1}^{N} \sum_{\ell'=1}^{N} \mathsf{E}\big[X_\ell X_{\ell'}\big]\,e^{i2\pi f(\ell - \ell')\mathrm{T_s}}\,|\hat{g}(f)|^2. \tag{15.44}$$

The proof proceeds now along the same lines as the proof of Theorem 15.5.1. $\qquad\square$

## 15.6 Exercises

**Exercise 15.1 (Scaling a SP).** Let $\big(Y(t)\big)$ be the result of scaling the SP $\big(X(t)\big)$ by the real number $\alpha$. Thus, $Y(t) = \alpha X(t)$ for every epoch $t \in \mathbb{R}$. Show that if $\big(X(t)\big)$ is of operational PSD $\mathsf{S}_{XX}$, then $\big(Y(t)\big)$ is of operational PSD $f \mapsto \alpha^2 \, \mathsf{S}_{XX}(f)$.

**Exercise 15.2 (The Operational PSD of a Sum of Independent SPs).** Intuition suggests that if $\big(X(t)\big)$ and $\big(Y(t)\big)$ are centered independent stochastic processes of operational PSDs $\mathsf{S}_{XX}$ and $\mathsf{S}_{YY}$, then their sum should be of operational PSD $f \mapsto \mathsf{S}_{XX}(f) + \mathsf{S}_{YY}(f)$. Explain why.

**Exercise 15.3 (Operational PSD of a Deterministic SP).** Let $\big(X(t)\big)$ be deterministically equal to the energy-limited signal $\mathbf{g} \colon \mathbb{R} \to \mathbb{R}$ in the sense that, at every epoch $t \in \mathbb{R}$, the RV $X(t)$ is deterministically equal to $g(t)$. Find the operational PSD of $\big(X(t)\big)$.

**Exercise 15.4 (Stretching Time).** Let $\big(X(t)\big)$ be of operational PSD $\mathsf{S}_{XX}$, and let $a > 0$ be fixed. Define the SP $\big(Y(t)\big)$ at every epoch $t \in \mathbb{R}$ as $Y(t) = X(t/a)$. Show that $\big(Y(t)\big)$ is of operational PSD $f \mapsto a\,\mathsf{S}_{XX}(af)$.

**Exercise 15.5 (The Operational PSD is Nonnegative).** Show that if $\big(X(t),\, t \in \mathbb{R}\big)$ is of operational PSD $\mathsf{S}_{XX}$, then $\mathsf{S}_{XX}(f)$ must be nonnegative outside a set of frequencies of Lebesgue measure zero. Would this also have been true if we had not insisted that the operational PSD be symmetric?

*Hint: Proceed along the lines of the proof of Lemma 15.3.2.*

**Exercise 15.6 (Operational PSD of PAM).** Let $\big(X_\ell,\, \ell \in \mathbb{Z}\big)$ be IID with $X_\ell$ taking on the values $\pm 1$ equiprobably. Let

$$g(t) = \mathrm{I}\Big\{|t| \leq \frac{\mathsf{T}_\mathrm{s}}{2}\Big\}, \quad t \in \mathbb{R},$$

$$X_1(t) = \mathsf{A} \sum_{\ell=-\infty}^{\infty} X_\ell\, g(t - \ell \mathsf{T}_\mathrm{s}), \quad t \in \mathbb{R},$$

where $\mathsf{A}, \mathsf{T}_\mathrm{s} > 0$ are deterministic.

  (i) Plot a sample function of $\mathbf{X}_1$ for a realization of $\big(X_\ell,\, \ell \in \mathbb{Z}\big)$ of your choice.

 (ii) Compute the operational PSD of $\mathbf{X}_1$.

(iii) Repeat Parts (i) and (ii) for

$$X_2(t) = \mathsf{A} \sum_{\ell=-\infty}^{\infty} X_\ell\, g(t - 2\ell \mathsf{T}_\mathrm{s}), \quad t \in \mathbb{R}.$$

(iv) How do the operational PSDs of $\mathbf{X}_1$ and $\mathbf{X}_2$ compare?

**Exercise 15.7 (Spectral Shaping via Precoding).** Let $\big(X_\ell,\, \ell \in \mathbb{Z}\big)$ be IID with $X_\ell$ taking on the values $\pm 1$ equiprobably. Let $\tilde{X}_\ell = X_\ell + X_{\ell-1}$ for every $\ell \in \mathbb{Z}$.

  (i) Compute the operational PSD of the PAM signal

$$X_1(t) = \sum_{\ell=-\infty}^{\infty} \tilde{X}_\ell\, g(t - \ell \mathsf{T}_\mathrm{s}), \quad t \in \mathbb{R}$$

for $g(\cdot)$ decaying to zero sufficiently fast as $|t| \to \infty$, e.g., satsifying (14.17).

(ii) Throw mathematical caution to the wind and evaluate your answer for the pulse shape whose FT is

$$\hat{g}(f) = \mathrm{I}\left\{|f| \le \frac{1}{2\mathsf{T_s}}\right\}, \quad f \in \mathbb{R}.$$

(Ignore the fact that this pulse shape does not satisfy (14.17).) Plot your answer and compare it to the operational PSD of the PAM signal

$$X_2(t) = \sum_{\ell=-\infty}^{\infty} X_\ell \, g(t - \ell \mathsf{T_s}), \quad t \in \mathbb{R}.$$

(iii) Show that $\mathbf{X}_1$ can also be written as a PAM signal with IID symbols but with a different pulse shape. That is,

$$X_1(t) = \sum_{\ell=-\infty}^{\infty} X_\ell \, h(t - \ell \mathsf{T_s}),$$

$$\mathbf{h} \colon t \mapsto g(t) + g(t - \mathsf{T_s}).$$

**Exercise 15.8 (The Operational PSD and Block Codes).** PAM is used in block-mode in conjunction with the $(1, 2)$ binary-to-reals block encoder

$$0 \mapsto (+1, -1), \quad 1 \mapsto (-1, +1)$$

to transmit IID random bits. The pulse shape $g(\cdot)$ satisfies the decay condition (14.17). Compute the power and operational PSD of the signal.

**Exercise 15.9 (Repetitions and the Operational PSD).** Let $\big(X(t)\big)$ be the signal (15.22) that results when the $(1, 2)$ binary-to-reals block-encoder (10.4) is used in bi-infinite block-mode. Find the operational PSD of $\big(X(t)\big)$.

**Exercise 15.10 (Direct-Sequence Spread-Spectrum Communications).** This problem is motivated by uncoded Direct-Sequence Spread-Spectrum communications with processing gain $\mathsf{N}$. Let the $(1, \mathsf{N})$ binary-to-reals block encoder map 0 to the sequence $a_1, \ldots, a_\mathsf{N}$ and 1 to $-a_1, \ldots, -a_\mathsf{N}$. Consider PAM with bi-infinite block encoding with this mapping. Express the operational PSD of the resulting PAM signal in terms of the sequence $a_1, \ldots, a_\mathsf{N}$ and the pulse shape $\mathbf{g}$. Calculate explicitly when the pulse shape is the mapping $t \mapsto \mathrm{I}\{|t| \le \mathsf{T_s}/2\}$ for two cases: when the sequence $a_1, \ldots, a_\mathsf{N}$ is the Barker-7 code $(+1, +1, +1, -1, -1, +1, -1)$ and when it is the sequence $(+1, +1, +1, +1, +1, +1, +1)$. Compare the latter case with the case where the mapping is the antipodal mapping $0 \mapsto +1$, and $1 \mapsto -1$, the baud period $7\mathsf{T_s}$, and the pulse shape is $t \mapsto \mathrm{I}\{|t| \le 7\mathsf{T_s}/2\}$

# Chapter 16

# Quadrature Amplitude Modulation

## 16.1 Introduction

We next discuss linear modulation in passband. We envision being allocated bandwidth $W$ around the carrier frequency $f_c$, so we can only send real signals whose Fourier Transform is zero at frequencies $f$ satisfying $\left| |f| - f_c \right| > W/2$. That is, the FT of the transmitted signal is allowed to be nonzero only in the frequency interval $[f_c - W/2, f_c + W/2]$ and in its negative frequency counterpart $[-f_c - W/2, -f_c + W/2]$ (Definition 7.3.1). We assume throughout this chapter that

$$f_c > \frac{W}{2}. \tag{16.1}$$

There are numerous ways to communicate in passband and, to complicate things further, sometimes seemingly different approaches lead to identical signals. Thus, while we would like to motivate the scheme we shall focus on—Quadrature Amplitude Modulation (QAM)—we cannot prove or claim that it is the only "optimal" solution.[1] Nevertheless, we shall try to motivate it by discussing some features that one would typically like to have and by then showing that QAM has these features.

From our studies of PAM we recall that if we are allocated (baseband) bandwidth $W$ Hz and if $T_s \geq 1/(2W)$, then we can find a bandwidth-$W$ pulse shape whose time shifts by integer multiples of $T_s$ are orthonormal. If $T_s = 1/(2W)$, then such a pulse is the bandwidth-$W$ unit-energy pulse $t \mapsto \sqrt{2W}\operatorname{sinc}(2Wt)$. (You may recall that such pulses are rarely used because they decay to zero too slowly over time, thus rendering the computation of the PAM signal unstable and the resulting peak power unbounded.) And if $T_s < 1/(2W)$, then no such pulse shape exists. (Corollary 11.3.5.)

From a somewhat more abstract perspective, PAM with the above pulse shape (or with the square root of a raised-cosine pulse shape (11.29) with very small excess

---

[1]There are information theoretic considerations that show that QAM can achieve the capacity of the bandlimited passband additive white Gaussian noise channel.

bandwidth) allows us to send symbols arriving at rate

$$R_s \left[ \frac{\text{real symbol}}{\text{second}} \right]$$

as the coefficients in a linear combination of orthonormal signals whose bandwidth does not exceed (or only slightly exceeds)

$$\frac{R_s}{2} \, [\text{Hz}] \, .$$

That is, for each spectral sliver of 1 Hz at baseband we obtain 2 real dimensions per second, i.e., we can communicate at spectral efficiency

$$2 \, \frac{[\text{real dimension/sec}]}{[\text{baseband Hz}]} \, .$$

This is an achievement that we would like to replicate for passband signaling:

**First Objective:** Find a way to transmit real symbols arriving at rate $R_s$ real symbols per second as the coefficients in a linear combination of orthonormal passband signals occupying a (passband) bandwidth of $W$ Hz around the carrier frequency $f_c$, where the bandwidth $W$ is equal to (or only slightly exceeds) $R_s/2$. That is, we would like to find a communication scheme that would allow us to communicate at

$$2 \, \frac{[\text{real dimension/sec}]}{[\text{passband Hz}]} \, .$$

Equivalently, since any stream of real symbols arriving at rate $R_s$ real symbols per second can be viewed as a stream of complex symbols arriving at rate $R_s/2$ complex symbols per second (simply by pairing tuples $(a, b)$ of real numbers $a, b \in \mathbb{R}$ into single complex numbers $a + ib$), we can restate our objective as follows: find a way to transmit complex symbols arriving at rate $R_s/2$ complex symbols per second as the coefficients in a linear combination of orthonormal passband signals occupying a (passband) bandwidth of $W$ Hz around the carrier frequency $f_c$, where the bandwidth $W$ is equal to, or only slightly exceeds $R_s/2$. That is, we would like to find a communication scheme that would allow us to communicate at

$$\boxed{1 \, \frac{[\text{complex dimension/sec}]}{[\text{passband Hz}]} \, .} \tag{16.2}$$

In addition, we would like our modulation scheme to be of reasonable complexity. One of the benefits of the baseband PAM scheme is that we can compute all the inner products required to reconstruct the coefficients (symbols) using the matched filter by feeding it with the transmitted signal and sampling its output at the appropriate times.

A naive approach that does *not* achieve our objective is to use real baseband PAM of the type we studied in Chapter 10 and to up-convert the PAM signal to passband by multiplying it by the mapping $t \mapsto \cos(2\pi f_c t)$. The problem with this approach is that the up-conversion doubles the bandwidth (Proposition 7.3.3).

## 16.2 PAM for Passband?

A natural approach to passband signaling might be to consider PAM directly without any up-conversion. We merely have to look for a pulse shape $\phi$ whose Fourier Transform is zero outside the band $\big||f| - f_c\big| \leq W/2$ and whose self-similarity function $\mathsf{R}_{\phi\phi}$ is a Nyquist Pulse. It turns out that with this approach we can only achieve our objective if $4f_c\mathsf{T}_s$ is an odd integer. Indeed, the reader is encouraged to use Corollary 11.3.4 to verify that if a pulse $\phi$ is an energy-limited passband signal that is bandlimited to $W$ Hz around the carrier frequency $f_c$, and if its time shifts by integer multiples of $\mathsf{T}_s$ are orthonormal, then

$$\mathsf{T}_s \geq \frac{1}{2W}$$

with equality being achievable only if *both*

$$|\hat{\phi}(f)|^2 = \mathsf{T}_s\, \mathrm{I}\big\{\big||f| - f_c\big| \leq W/2\big\}$$

(for all frequencies $f \in \mathbb{R}$ outside a set of Lebesgue measure zero) *and*

$$4f_c\mathsf{T}_s \quad \text{is an odd integer.} \tag{16.3}$$

In fact, it can be shown that if (16.3) is satisfied and if $\psi$ is any energy-limited signal that is bandlimited to $W/2$ Hz and whose time shifts by integer multiples of $2\mathsf{T}_s$ are orthonormal, then the passband signal

$$\phi(t) = \sqrt{2}\cos(2\pi f_c t)\, \psi(t), \quad t \in \mathbb{R}$$

is an energy-limited passband signal that is bandlimited to $W$ Hz around the carrier frequency $f_c$, and its time shifts by integer multiples of $\mathsf{T}_s$ are orthonormal.

It would thus seem that if (16.3) is satisfied, then PAM would be a viable solution to our problem. Nevertheless, this is not the standard solution. The reason may have to do with implementation. If the above approach is used, then the carrier frequency influences the choice of the pulse shape. Thus, a radio with a selectable carrier frequency would require a different pulse shape for each frequency! Moreover, the implementation of the modulator becomes carrier-dependent and fairly complex. This discussion motivates our second objective:

**Second Objective:** To allow for flexibility in the choice of the carrier, it is desirable to decouple the pulse shape selection from the carrier frequency.

## 16.3 The QAM Signal

Quadrature Amplitude Modulation achieves both our objectives. It achieves our desired spectral efficiency (16.2) and also decouples the signal design from the carrier frequency. It is easiest to describe QAM by describing the baseband representation $x_{\mathrm{BB}}(\cdot)$ of the transmitted passband signal $x_{\mathrm{PB}}(\cdot)$. Indeed, the baseband representation of the transmitted signal has the structure of PAM but with one important difference: we allow for complex symbols and for complex pulse shapes.[2]

---

[2]Allowing complex pulse shapes is not critical. Crucial is that we allow complex symbols.

In QAM the encoder

$$\varphi \colon \{0,1\}^k \to \mathbb{C}^n \tag{16.4}$$

maps $k$-tuples of data bits $(D_1, \ldots, D_k)$ to $n$-tuples of *complex* symbols $(C_1, \ldots, C_n)$, and the baseband representation of the transmitted signal is

$$X_{\mathrm{BB}}(t) = \mathsf{A} \sum_{\ell=1}^{n} C_\ell\, g(t - \ell \mathsf{T_s}), \quad t \in \mathbb{R}, \tag{16.5a}$$

where the  pulse shape $g(\cdot)$ may be complex (though it is often chosen to be real), $\mathsf{A} \geq 0$ is a real constant, $\mathsf{T_s} > 0$ is the baud period, and $1/\mathsf{T_s}$ is the baud rate. The rate of the encoder is given by

$$\frac{k}{n} \left[ \frac{\text{bit}}{\text{complex symbol}} \right], \tag{16.5b}$$

and the transmitted real passband QAM signal $X_{\mathrm{PB}}(\cdot)$ is given by

$$X_{\mathrm{PB}}(t) = 2 \operatorname{Re}\!\big(X_{\mathrm{BB}}(t)\, e^{\mathrm{i}2\pi f_c t}\big), \quad t \in \mathbb{R}. \tag{16.5c}$$

Using (16.5a) & (16.5c) we can also express the QAM signal as

$$\boxed{X_{\mathrm{PB}}(t) = 2 \operatorname{Re}\!\left( \mathsf{A} \sum_{\ell=1}^{n} C_\ell\, g(t - \ell \mathsf{T_s})\, e^{\mathrm{i}2\pi f_c t} \right), \quad t \in \mathbb{R}.} \tag{16.6}$$

Alternatively, we can use the identities

$$\operatorname{Re}(wz) = \operatorname{Re}(w)\operatorname{Re}(z) - \operatorname{Im}(w)\operatorname{Im}(z), \quad w, z \in \mathbb{C},$$

$$\operatorname{Im}(z) = -\operatorname{Re}(\mathrm{i}z), \quad z \in \mathbb{C}$$

to express the QAM signal as

$$X_{\mathrm{PB}}(t) = \sqrt{2}\mathsf{A} \sum_{\ell=1}^{n} \operatorname{Re}(C_\ell) \overbrace{2 \operatorname{Re}\!\left( \underbrace{\frac{1}{\sqrt{2}}\, g(t - \ell \mathsf{T_s})}_{g_{\mathrm{I},\ell,\mathrm{BB}}(t)}\, e^{\mathrm{i}2\pi f_c t} \right)}^{g_{\mathrm{I},\ell}(t)}$$

$$+ \sqrt{2}\mathsf{A} \sum_{\ell=1}^{n} \operatorname{Im}(C_\ell) \overbrace{2 \operatorname{Re}\!\left( \underbrace{\mathrm{i}\frac{1}{\sqrt{2}}\, g(t - \ell \mathsf{T_s})}_{g_{\mathrm{Q},\ell,\mathrm{BB}}(t)}\, e^{\mathrm{i}2\pi f_c t} \right)}^{g_{\mathrm{Q},\ell}(t)}, \quad t \in \mathbb{R}, \tag{16.7}$$

where we define

$$g_{\mathrm{I},\ell}(t) \triangleq 2 \operatorname{Re}\!\left( \frac{1}{\sqrt{2}}\, g(t - \ell \mathsf{T_s})\, e^{\mathrm{i}2\pi f_c t} \right) \tag{16.8a}$$

$$= 2 \operatorname{Re}\!\big(g_{\mathrm{I},\ell,\mathrm{BB}}(t)\, e^{\mathrm{i}2\pi f_c t}\big), \quad t \in \mathbb{R},$$

and

$$g_{Q,\ell}(t) \triangleq 2\,\mathrm{Re}\left(i\frac{1}{\sqrt{2}}\,g(t - \ell T_s)\,e^{i2\pi f_c t}\right) \tag{16.8b}$$

$$= 2\,\mathrm{Re}\left(g_{Q,\ell,\mathrm{BB}}(t)\,e^{i2\pi f_c t}\right), \quad t \in \mathbb{R},$$

with corresponding baseband representations:

$$g_{I,\ell,\mathrm{BB}}(t) \triangleq \frac{1}{\sqrt{2}}\,g(t - \ell T_s), \quad t \in \mathbb{R}, \tag{16.9a}$$

$$g_{Q,\ell,\mathrm{BB}}(t) \triangleq i\frac{1}{\sqrt{2}}\,g(t - \ell T_s), \quad t \in \mathbb{R}. \tag{16.9b}$$

Some comments about the QAM signal:

(i) The representation (16.7) demonstrates that the QAM signal is a linear combination of the waveforms $\{\mathbf{g}_{I,\ell}\}$ and $\{\mathbf{g}_{Q,\ell}\}$, where the coefficients are proportional to the real parts and the imaginary parts of the symbols $\{C_\ell\}$.

(ii) The normalization factor of $1/\sqrt{2}$ in the definition of the functions $\{\mathbf{g}_{I,\ell}\}$ and $\{\mathbf{g}_{Q,\ell}\}$ is for convenience only. Its role will become clearer in Section 16.5, where the pulse shape is chosen to be of unit energy. In this case the factor of $1/\sqrt{2}$ guarantees that the functions $\{\mathbf{g}_{I,\ell}\}$ and $\{\mathbf{g}_{Q,\ell}\}$ are also of unit energy.

(iii) We could also view QAM slightly differently as a modulation scheme where data bits $D_1, \ldots, D_k$ are mapped to $2n$ real numbers $X_1, \ldots, X_{2n}$, which are then grouped in pairs to form the $n$ complex numbers $C_\ell = X_{2\ell-1} + iX_{2\ell}$ for $\ell = 1, \ldots, n$ and where these complex numbers are then mapped into the passband signal whose baseband representation is given in (16.5a). The two views are, of course, completely equivalent.

The expression for the QAM signal $X_{\mathrm{PB}}(\cdot)$ is simplified if the pulse shape $\mathbf{g}$ is real. In this case we obtain from (16.6) for every $t \in \mathbb{R}$

$$X_{\mathrm{PB}}(t) = 2A \sum_{\ell=1}^{n} \mathrm{Re}(C_\ell)\,g(t - \ell T_s)\cos(2\pi f_c t)$$

$$- 2A \sum_{\ell=1}^{n} \mathrm{Im}(C_\ell)\,g(t - \ell T_s)\sin(2\pi f_c t), \quad \mathbf{g}\ \text{real}. \tag{16.10}$$

Thus, if the pulse shape $\mathbf{g}$ is real, then the QAM signal can be viewed as the sum of two signals: the first is the result of feeding $\{\mathrm{Re}(C_\ell)\}$ to a baseband PAM modulator of pulse shape $\mathbf{g}$ and multiplying the result by $\cos(2\pi f_c t)$, and the second is the result of feeding $\{\mathrm{Im}(C_\ell)\}$ to a baseband PAM modulator of pulse shape $\mathbf{g}$ and multiplying the result by $-\sin(2\pi f_c t)$. Figure 16.1 illustrates the generation of the QAM signal when the pulse shape $\mathbf{g}$ is real.

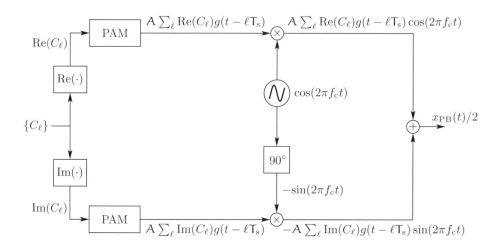

**Figure 16.1:** Generating a QAM signal when the pulse shape **g** is real.

## 16.4 Bandwidth Considerations

Recalling that the bandwidth of a passband signal around the carrier frequency is twice the bandwidth of its baseband representation (Proposition 7.6.7 and Theorem 7.7.12 (i)) we conclude:

**Note 16.4.1.** If the pulse shape **g** is bandlimited to $W/2$ Hz, then the QAM signal (16.6) is bandlimited to $W$ Hz around the carrier frequency $f_c$.

If the pulse shape **g** is real, then these bandwidth considerations can also be explained in another way. We note that if $g(\cdot)$ is bandlimited to $W/2$ Hz then the signal $\sum_\ell \mathrm{Re}(C_\ell)\, g(t - \ell T_s)$ is also bandlimited to $W/2$ Hz, so when it is upconverted by multiplication by $\cos(2\pi f_c t)$ the resulting signal is bandlimited to $W$ Hz around the carrier frequency $f_c$ (Proposition 7.3.3). A similar argument holds for the signal that is multiplied by $-\sin(2\pi f_c t)$.

## 16.5 Orthogonality Considerations

We next study the consequences of choosing the pulse shape $g(\cdot)$ so that its time shifts by integer multiples of $T_s$ be orthonormal. As in our treatment of PAM, we change notation and denote the pulse shape in this case by $\phi(\cdot)$. The orthonormality condition is thus

$$\int_{-\infty}^{\infty} \phi(t - \ell T_s)\, \phi^*(t - \ell' T_s)\, \mathrm{d}t = \mathrm{I}\{\ell = \ell'\}, \quad \ell, \ell' \in \mathbb{Z}. \tag{16.11}$$

By Corollary 11.3.4, this is equivalent to requiring that

$$\sum_{\ell=-\infty}^{\infty} \left| \hat{\phi}\left(f + \frac{\ell}{\mathsf{T_s}}\right)\right|^2 = \mathsf{T_s}, \tag{16.12}$$

for all frequencies $f$ outside a set of Lebesgue measure zero.

When the pulse shape satisfies the orthogonality condition (16.11) we refer to $1/\mathsf{T_s}$ as having units of complex dimensions per second. In analogy to Definition 11.3.6, we define the **excess bandwidth** as

$$100\% \left(\frac{\text{bandwidth of } \phi}{1/(2\mathsf{T_s})} - 1\right). \tag{16.13}$$

**Proposition 16.5.1.** *If the energy-limited pulse shape $\phi$ satisfies (16.11), then the QAM signal $X_{\mathrm{PB}}(\cdot)$ can be expressed as*

$$\mathbf{X}_{\mathrm{PB}} = \sqrt{2}A \sum_{\ell=1}^{n} \mathrm{Re}(C_\ell)\,\boldsymbol{\psi}_{\mathrm{I},\ell} + \sqrt{2}A \sum_{\ell=1}^{n} \mathrm{Im}(C_\ell)\,\boldsymbol{\psi}_{\mathrm{Q},\ell} \tag{16.14}$$

*where*

$$\ldots, \boldsymbol{\psi}_{\mathrm{I},-1}, \boldsymbol{\psi}_{\mathrm{Q},-1}, \boldsymbol{\psi}_{\mathrm{I},0}, \boldsymbol{\psi}_{\mathrm{Q},0}, \boldsymbol{\psi}_{\mathrm{I},1}, \boldsymbol{\psi}_{\mathrm{Q},1}, \ldots$$

*are orthonormal functions that are given by*

$$\boldsymbol{\psi}_{\mathrm{I},\ell} \colon t \mapsto 2\,\mathrm{Re}\left(\frac{1}{\sqrt{2}}\,\phi(t - \ell\mathsf{T_s})\,e^{i2\pi f_c t}\right), \quad \ell \in \mathbb{Z} \tag{16.15a}$$

$$\boldsymbol{\psi}_{\mathrm{Q},\ell} \colon t \mapsto 2\,\mathrm{Re}\left(i\frac{1}{\sqrt{2}}\,\phi(t - \ell\mathsf{T_s})\,e^{i2\pi f_c t}\right), \quad \ell \in \mathbb{Z}. \tag{16.15b}$$

**Proof.** Substituting $\phi$ for $\mathbf{g}$ in (16.7) we obtain

$$X_{\mathrm{PB}}(t) = \sqrt{2}A \sum_{\ell=1}^{n} \mathrm{Re}(C_\ell) \overbrace{2\,\mathrm{Re}\left(\underbrace{\frac{1}{\sqrt{2}}\,\phi(t - \ell\mathsf{T_s})\,e^{i2\pi f_c t}}_{\psi_{\mathrm{I},\ell,\mathrm{BB}}(t)}\right)}^{\psi_{\mathrm{I},\ell}(t)}$$

$$+ \sqrt{2}A \sum_{\ell=1}^{n} \mathrm{Im}(C_\ell) \overbrace{2\,\mathrm{Re}\left(\underbrace{i\frac{1}{\sqrt{2}}\,\phi(t - \ell\mathsf{T_s})\,e^{i2\pi f_c t}}_{\psi_{\mathrm{Q},\ell,\mathrm{BB}}(t)}\right)}^{\psi_{\mathrm{Q},\ell}(t)}, \quad t \in \mathbb{R},$$

where for every $t \in \mathbb{R}$

$$\psi_{\mathrm{I},\ell}(t) \triangleq 2\,\mathrm{Re}\left(\frac{1}{\sqrt{2}}\,\phi(t - \ell\mathsf{T_s})\,e^{i2\pi f_c t}\right) \tag{16.16a}$$

$$= 2\,\mathrm{Re}\left(\psi_{\mathrm{I},\ell,\mathrm{BB}}(t)\,e^{i2\pi f_c t}\right),$$

$$\psi_{\mathrm{Q},\ell}(t) \triangleq 2 \operatorname{Re}\left(\mathrm{i}\frac{1}{\sqrt{2}}\,\phi(t - \ell \mathsf{T_s})\, e^{\mathrm{i}2\pi f_c t}\right) \tag{16.16b}$$
$$= 2 \operatorname{Re}\left(\psi_{\mathrm{Q},\ell,\mathrm{BB}}(t)\, e^{\mathrm{i}2\pi f_c t}\right),$$

and the baseband representations are given by

$$\psi_{\mathrm{I},\ell,\mathrm{BB}}(t) \triangleq \frac{1}{\sqrt{2}}\,\phi(t - \ell \mathsf{T_s}) \tag{16.17a}$$

and

$$\psi_{\mathrm{Q},\ell,\mathrm{BB}}(t) \triangleq \mathrm{i}\frac{1}{\sqrt{2}}\,\phi(t - \ell \mathsf{T_s}). \tag{16.17b}$$

We next verify that, when $\phi$ satisfies (16.11), the functions

$$\dots, \boldsymbol{\psi}_{\mathrm{I},-1}, \boldsymbol{\psi}_{\mathrm{Q},-1}, \boldsymbol{\psi}_{\mathrm{I},0}, \boldsymbol{\psi}_{\mathrm{Q},0}, \boldsymbol{\psi}_{\mathrm{I},1}, \boldsymbol{\psi}_{\mathrm{Q},1}, \dots$$

are orthonormal. To this end we recall that the inner product between two real passband signals is twice the real part of the inner product between their baseband representations (Theorem 7.6.10). For $\ell \neq \ell'$ we thus have by (16.11)

$$\langle \boldsymbol{\psi}_{\mathrm{I},\ell}, \boldsymbol{\psi}_{\mathrm{I},\ell'} \rangle = 2 \operatorname{Re}\left(\langle \boldsymbol{\psi}_{\mathrm{I},\ell,\mathrm{BB}}, \boldsymbol{\psi}_{\mathrm{I},\ell',\mathrm{BB}} \rangle\right)$$
$$= 2 \operatorname{Re}\left(\left\langle t \mapsto \frac{1}{\sqrt{2}}\,\phi(t - \ell \mathsf{T_s}), t \mapsto \frac{1}{\sqrt{2}}\,\phi(t - \ell' \mathsf{T_s})\right\rangle\right)$$
$$= 0,$$

$$\langle \boldsymbol{\psi}_{\mathrm{Q},\ell}, \boldsymbol{\psi}_{\mathrm{Q},\ell'} \rangle = 2 \operatorname{Re}\left(\langle \boldsymbol{\psi}_{\mathrm{Q},\ell,\mathrm{BB}}, \boldsymbol{\psi}_{\mathrm{Q},\ell',\mathrm{BB}} \rangle\right)$$
$$= 2 \operatorname{Re}\left(\left\langle t \mapsto \mathrm{i}\frac{1}{\sqrt{2}}\,\phi(t - \ell \mathsf{T_s}), t \mapsto \mathrm{i}\frac{1}{\sqrt{2}}\,\phi(t - \ell' \mathsf{T_s})\right\rangle\right)$$
$$= 0,$$

and

$$\langle \boldsymbol{\psi}_{\mathrm{I},\ell}, \boldsymbol{\psi}_{\mathrm{Q},\ell'} \rangle = 2 \operatorname{Re}\left(\left\langle t \mapsto \frac{1}{\sqrt{2}}\,\phi(t - \ell \mathsf{T_s}), t \mapsto \mathrm{i}\frac{1}{\sqrt{2}}\,\phi(t - \ell' \mathsf{T_s})\right\rangle\right)$$
$$= 0.$$

And for $\ell = \ell'$ we have, again by (16.11),

$$\langle \boldsymbol{\psi}_{\mathrm{I},\ell}, \boldsymbol{\psi}_{\mathrm{I},\ell} \rangle = 2 \operatorname{Re}\left(\left\langle t \mapsto \frac{1}{\sqrt{2}}\,\phi(t - \ell \mathsf{T_s}), t \mapsto \frac{1}{\sqrt{2}}\,\phi(t - \ell \mathsf{T_s})\right\rangle\right)$$
$$= 1,$$

$$\langle \boldsymbol{\psi}_{\mathrm{I},\ell}, \boldsymbol{\psi}_{\mathrm{Q},\ell} \rangle = 2 \operatorname{Re}\left(\left\langle t \mapsto \frac{1}{\sqrt{2}}\,\phi(t - \ell \mathsf{T_s}), t \mapsto \mathrm{i}\frac{1}{\sqrt{2}}\,\phi(t - \ell \mathsf{T_s})\right\rangle\right)$$
$$= \operatorname{Re}\left(-\mathrm{i}\,\|\phi\|_2^2\right)$$
$$= 0,$$

and

$$\langle \pmb{\psi}_{\mathrm{Q},\ell}, \pmb{\psi}_{\mathrm{Q},\ell} \rangle = 2 \operatorname{Re}\left(\left\langle t \mapsto \mathrm{i}\frac{1}{\sqrt{2}}\,\phi(t - \ell \mathsf{T}_{\mathrm{s}}), t \mapsto \mathrm{i}\frac{1}{\sqrt{2}}\,\phi(t - \ell \mathsf{T}_{\mathrm{s}}) \right\rangle\right)$$

$$= 1. \qquad \qquad \square$$

Notice that (16.14)–(16.15) can be simplified when $\phi$ is real:

**Corollary 16.5.2.** *If, in addition to the assumptions of Proposition 16.5.1, we also assume that the pulse shape $\phi$ is real, then the QAM signal can be written as*

$$X_{\mathrm{PB}}(t) = \sqrt{2}\mathsf{A} \sum_{\ell=1}^{n} \operatorname{Re}(C_\ell)\,\sqrt{2}\,\phi(t - \ell \mathsf{T}_{\mathrm{s}}) \cos(2\pi f_{\mathrm{c}} t)$$

$$- \sqrt{2}\mathsf{A} \sum_{\ell=1}^{n} \operatorname{Im}(C_\ell)\,\sqrt{2}\,\phi(t - \ell \mathsf{T}_{\mathrm{s}}) \sin(2\pi f_{\mathrm{c}} t), \quad t \in \mathbb{R}, \quad (16.18)$$

*and*

$$\left\{ t \mapsto \sqrt{2}\,\phi(t - \ell \mathsf{T}_{\mathrm{s}}) \cos(2\pi f_{\mathrm{c}} t) \right\}_{\ell=-\infty}^{\infty}, \quad \left\{ t \mapsto \sqrt{2}\,\phi(t - \ell \mathsf{T}_{\mathrm{s}}) \sin(2\pi f_{\mathrm{c}} t) \right\}_{\ell=-\infty}^{\infty}$$

*are orthonormal.*

## 16.6   Spectral Efficiency

We next show that QAM achieves our spectral efficiency objective. We assume that we are only allowed to transmit signals of bandwidth $W$ around the carrier frequency $f_{\mathrm{c}}$, so the transmitted signal can only occupy the frequencies $f$ satisfying

$$\big||f| - f_{\mathrm{c}}\big| \le W/2.$$

In order for the QAM signal to meet this constraint, we choose a pulse shape $\phi$ that is bandlimited to $W/2$ Hz, because the up-conversion doubles the bandwidth (Note 16.4.1). Thus, by Corollary 11.3.5, the orthogonality (16.11) can only hold if the baud period $\mathsf{T}_{\mathrm{s}}$ satisfies $\mathsf{T}_{\mathrm{s}} \ge 1/(2 \times W/2)$ or

$$\mathsf{T}_{\mathrm{s}} \ge \frac{1}{W},$$

with the RHS being achievable by choosing $\phi$ to be the bandwidth-$W/2$ unit-energy signal $t \mapsto \sqrt{W}\operatorname{sinc}(Wt)$.

If we choose $\mathsf{T}_{\mathrm{s}}$ equal to $1/W$ (or only slightly larger than that), then our modulation will support the transmission of complex symbols arriving at a rate of $1/\mathsf{T}_{\mathrm{s}} \approx W$ complex symbols per second. And since our QAM signal only occupies $W$ Hz around the carrier frequency, our scheme achieves a spectral efficiency of 1 [complex dimension per second] per Hz. QAM thus achieves our spectral efficiency objective. This is so exciting that we highlight the achievement:

> QAM with the bandwidth-$W/2$ unit-energy pulse shape given by
> $t \mapsto \sqrt{W}\,\mathrm{sinc}(Wt)$ transmits a sequence of real symbols arriving at
> a rate of $2W$ real symbols per second as the coefficients in a linear
> combination of orthogonal signals, with the resulting waveform
> being bandlimited to $W$ Hz around the carrier frequency $f_c$. It
> thus achieves a spectral efficiency of
>
> $$2\,\frac{[\text{real dimension/sec}]}{[\text{passband Hz}]} = 1\,\frac{[\text{complex dimension/sec}]}{[\text{passband Hz}]}.$$

## 16.7   QAM Constellations

In analogy to the definition of the constellation of a PAM scheme (Section 10.8),
we define the **constellation** of a QAM scheme (or, perhaps more appropriately, of
the mapping $\varphi(\cdot)$ in (16.4)) as the smallest subset of $\mathbb{C}$ of which $C_\ell$ is an element
for every $\ell \in \{1,\ldots,n\}$ and for every realization of the data bits. We denote
the constellation by $\mathcal{C}$. The **number of points** in the constellation $\mathcal{C}$ is just the
number of elements of $\mathcal{C}$.

Important constellations include the square 4-QAM constellation (also knows as
QPSK)

$$\{+1 + i, -1 + i, -1 - i, +1 - i\},$$

the square QAM constellation with $(2\nu) \times (2\nu)$ points

$$\Big\{a + ib : a, b \in \big\{-(2\nu - 1), \ldots, -3, -1, +1, +3, \ldots, (2\nu - 1)\big\}\Big\}, \qquad (16.19)$$

and the M-PSK (M-ary Phase Shift Keying) constellation comprising the M complex numbers on the unit circle whose M-th power is one, i.e.,

$$\Big\{1, e^{i2\pi/M}, e^{i4\pi/M}, e^{i6\pi/M}, \ldots, e^{i(M-1)2\pi/M}\Big\}.$$

See Figure 16.2 for some common QAM constellations. Please note that the square
16-QAM and the 16-PSK are just two of many possible constellations with 16
points. However, some engineers omit the word "square" and write 4-QAM, 16-
QAM, 64-QAM, etc. for the respective square constellations.

We can also define the **minimum distance** $\delta$ of a constellation $\mathcal{C}$ in analogy to
(10.21) as

$$\delta \triangleq \min_{\substack{c,c' \in \mathcal{C} \\ c \neq c'}} |c - c'|. \qquad (16.20)$$

In analogy to (10.23), we define the **second moment** of a constellation $\mathcal{C}$ as

$$\frac{1}{\#\mathcal{C}} \sum_{c \in \mathcal{C}} |c|^2. \qquad (16.21)$$

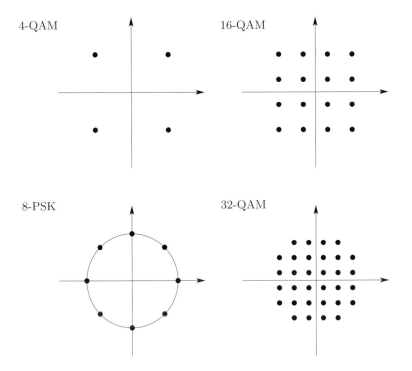

**Figure 16.2:** Some QAM constellations (drawn to no particular scale).

## 16.8 Recovering the Complex Symbols via Inner Products

Recall that, by Proposition 16.5.1, if the time shifts of $\phi$ by integer multiples of $\mathsf{T}_\mathrm{s}$ are orthonormal, then the QAM signal can be written as

$$\mathbf{X}_{\mathrm{PB}} = \sqrt{2}\mathsf{A} \sum_{\ell=1}^{n} \mathrm{Re}(C_\ell)\, \boldsymbol{\psi}_{\mathrm{I},\ell} + \sqrt{2}\mathsf{A} \sum_{\ell=1}^{n} \mathrm{Im}(C_\ell)\, \boldsymbol{\psi}_{\mathrm{Q},\ell},$$

where the signals $\ldots, \boldsymbol{\psi}_{\mathrm{I},-1}, \boldsymbol{\psi}_{\mathrm{Q},-1}, \boldsymbol{\psi}_{\mathrm{I},0}, \boldsymbol{\psi}_{\mathrm{Q},0}, \boldsymbol{\psi}_{\mathrm{I},1}, \boldsymbol{\psi}_{\mathrm{Q},1}, \ldots$, which are given in (16.15), are orthonormal. Consequently, the complex symbols can be recovered from the QAM signal (in the absence of noise) using the inner product:

$$\mathrm{Re}(C_\ell) = \frac{1}{\sqrt{2}\mathsf{A}} \langle \mathbf{X}_{\mathrm{PB}}, \boldsymbol{\psi}_{\mathrm{I},\ell} \rangle, \quad \ell \in \{1, \ldots, n\}, \tag{16.22a}$$

$$\mathrm{Im}(C_\ell) = \frac{1}{\sqrt{2}\mathsf{A}} \langle \mathbf{X}_{\mathrm{PB}}, \boldsymbol{\psi}_{\mathrm{Q},\ell} \rangle, \quad \ell \in \{1, \ldots, n\}. \tag{16.22b}$$

We next describe circuits to compute these inner products. With a view to future chapters where noise will be present, we shall describe more general circuits that compute the inner products $\langle \mathbf{r}, \boldsymbol{\psi}_{\mathrm{I},\ell} \rangle$ and $\langle \mathbf{r}, \boldsymbol{\psi}_{\mathrm{Q},\ell} \rangle$ for an arbitrary (not necessarily QAM) energy-limited signal $\mathbf{r}$. Moreover, since the calculation of the inner products will not exploit the orthogonality condition (16.11), we shall describe the more general setting where the pulse shape is arbitrary and refer to the notation of (16.7). Thus, we shall present circuits to compute

$$\langle \mathbf{r}, \mathbf{g}_{\mathrm{I},\ell} \rangle, \langle \mathbf{r}, \mathbf{g}_{\mathrm{Q},\ell} \rangle,$$

where $\mathbf{g}_{\mathrm{I},\ell}$ and $\mathbf{g}_{\mathrm{Q},\ell}$ and their baseband representations are given in (16.8) and (16.9). Here $\mathbf{r}$ is an arbitrary energy-limited signal. We present two approaches: an approach based on baseband conversion and a direct approach.

### 16.8.1 Inner Products via Baseband Conversion

We begin by noting that if the pulse shape $\mathbf{g}$ is bandlimited to $W/2$ Hz then both $\mathbf{g}_{\mathrm{I},\ell}$ and $\mathbf{g}_{\mathrm{Q},\ell}$ are bandlimited to $W$ Hz around the carrier frequency $f_{\mathrm{c}}$. Consequently, since they contain no energy outside the bands $[f_{\mathrm{c}} - W/2, f_{\mathrm{c}} + W/2]$ and $[-f_{\mathrm{c}} - W/2, -f_{\mathrm{c}} + W/2]$, it follows from Parseval's Theorem that the Fourier Transform of $\mathbf{r}$ outside these bands does not influence the value of the inner products. Thus, if $\mathbf{s}$ is the result of passing $\mathbf{r}$ through an ideal unit-gain bandpass filter of bandwidth $W$ around the carrier frequency $f_{\mathrm{c}}$, i.e.,

$$\mathbf{s} = \mathbf{r} \star \mathrm{BPF}_{W, f_{\mathrm{c}}}, \tag{16.23}$$

then

$$\langle \mathbf{r}, \mathbf{g}_{\mathrm{I},\ell} \rangle = \langle \mathbf{s}, \mathbf{g}_{\mathrm{I},\ell} \rangle, \tag{16.24a}$$

$$\langle \mathbf{r}, \mathbf{g}_{\mathrm{Q},\ell} \rangle = \langle \mathbf{s}, \mathbf{g}_{\mathrm{Q},\ell} \rangle. \tag{16.24b}$$

If we denote the baseband representation of $\mathbf{s}$ by $\mathbf{s}_{\mathrm{BB}}$, then

$$
\begin{aligned}
\langle \mathbf{r}, \mathbf{g}_{\mathrm{I},\ell} \rangle &= \langle \mathbf{s}, \mathbf{g}_{\mathrm{I},\ell} \rangle \\
&= 2 \operatorname{Re}\big( \langle \mathbf{s}_{\mathrm{BB}}, \mathbf{g}_{\mathrm{I},\ell,\mathrm{BB}} \rangle \big) \\
&= \sqrt{2} \operatorname{Re}\big( \langle \mathbf{s}_{\mathrm{BB}}, t \mapsto g(t - \ell \mathsf{T}_{\mathrm{s}}) \rangle \big),
\end{aligned}
\tag{16.25a}
$$

where the first equality follows from (16.24a); the second from Theorem 7.6.10; and the final equality from (16.9a). Similarly,

$$
\begin{aligned}
\langle \mathbf{r}, \mathbf{g}_{\mathrm{Q},\ell} \rangle &= \langle \mathbf{s}, \mathbf{g}_{\mathrm{Q},\ell} \rangle \\
&= 2 \operatorname{Re}\big( \langle \mathbf{s}_{\mathrm{BB}}, \mathbf{g}_{\mathrm{Q},\ell,\mathrm{BB}} \rangle \big) \\
&= \sqrt{2} \operatorname{Re}\big( \langle \mathbf{s}_{\mathrm{BB}}, t \mapsto i\, g(t - \ell \mathsf{T}_{\mathrm{s}}) \rangle \big) \\
&= \sqrt{2} \operatorname{Im}\big( \langle \mathbf{s}_{\mathrm{BB}}, t \mapsto g(t - \ell \mathsf{T}_{\mathrm{s}}) \rangle \big).
\end{aligned}
\tag{16.25b}
$$

We next describe circuits to compute the RHS of (16.25a) & (16.25b). The circuit to produce $\mathbf{s}_{\mathrm{BB}}$ from $\mathbf{s}$ was already discussed in Section 7.6 on the baseband representation of passband signals (Figure 7.11). One multiplies $s(t)$ by $e^{-i2\pi f_{\mathrm{c}} t}$ and then passes the result through a lowpass filter whose cutoff frequency $W_{\mathrm{c}}$ satisfies

$$\frac{W}{2} \leq W_{\mathrm{c}} \leq 2f_{\mathrm{c}} - \frac{W}{2},$$

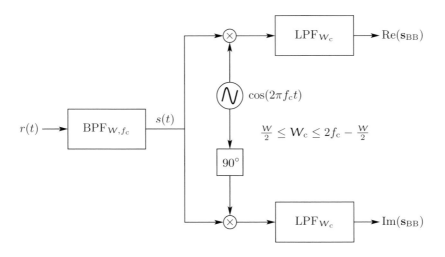

**Figure 16.3:** QAM demodulation: the front-end.

i.e.,

$$\mathbf{s}_{\mathrm{BB}} = \left(t \mapsto s(t)\, e^{-\mathrm{i}2\pi f_c t}\right) \star \mathrm{LPF}_{\mathsf{W}_c},$$

or, in terms of real operations:

$$\mathrm{Re}(\mathbf{s}_{\mathrm{BB}}) = \left(t \mapsto s(t)\cos(2\pi f_c t)\right) \star \mathrm{LPF}_{\mathsf{W}_c},$$

$$\mathrm{Im}(\mathbf{s}_{\mathrm{BB}}) = -\left(t \mapsto s(t)\sin(2\pi f_c t)\right) \star \mathrm{LPF}_{\mathsf{W}_c}.$$

This circuit is depicted in Figure 16.3. Notice that this circuit depends only on the carrier frequency $f_c$ and on the bandwidth $\mathsf{W}$; it does not depend on the pulse shape.

Once $\mathbf{s}_{\mathrm{BB}}$ has been computed, the calculation of the inner products on the RHS of (16.25a) & (16.25b) is straightforward. For example, to compute the inner product on the RHS of (16.25a) we note that from (16.25a)

$$\langle \mathbf{r}, \mathbf{g}_{\mathrm{I},\ell} \rangle = \sqrt{2}\,\mathrm{Re}\left( \int_{-\infty}^{\infty} s_{\mathrm{BB}}(t)\, g^*(t - \ell\mathsf{T}_s)\, \mathrm{d}t \right)$$

$$= \sqrt{2} \int_{-\infty}^{\infty} \mathrm{Re}(s_{\mathrm{BB}}(t))\, \mathrm{Re}(g(t - \ell\mathsf{T}_s))\, \mathrm{d}t$$

$$+ \sqrt{2} \int_{-\infty}^{\infty} \mathrm{Im}(s_{\mathrm{BB}}(t))\, \mathrm{Im}(g(t - \ell\mathsf{T}_s))\, \mathrm{d}t, \qquad (16.26)$$

where the terms on the RHS can be computed by feeding $\mathrm{Re}(\mathbf{s}_{\mathrm{BB}})$ to a matched filter matched to $\mathrm{Re}(\mathbf{g})$ and sampling the filter's output at time $\ell\mathsf{T}_s$

$$\int_{-\infty}^{\infty} \mathrm{Re}(s_{\mathrm{BB}}(t))\, \mathrm{Re}(g(t - \ell\mathsf{T}_s))\, \mathrm{d}t = \left(\mathrm{Re}(\mathbf{s}_{\mathrm{BB}}) \star \mathrm{Re}(\overleftarrow{\mathbf{g}})\right)(\ell\mathsf{T}_s), \qquad (16.27)$$

and by feeding $\mathrm{Im}(\mathbf{s}_{\mathrm{BB}})$ to a matched filter matched to $\mathrm{Im}(\mathbf{g})$ and sampling the filter's output at time $\ell T_{\mathrm{s}}$

$$\int_{-\infty}^{\infty} \mathrm{Im}\big(s_{\mathrm{BB}}(t)\big)\,\mathrm{Im}\big(g(t-\ell T_{\mathrm{s}})\big)\,\mathrm{d}t = \big(\mathrm{Im}(\mathbf{s}_{\mathrm{BB}}) \star \mathrm{Im}(\tilde{\mathbf{g}})\big)(\ell T_{\mathrm{s}}). \tag{16.28}$$

Similarly, to compute the inner product on the RHS of (16.25b) we note that from (16.25b)

$$\langle \mathbf{r}, \mathbf{g}_{\mathrm{Q},\ell} \rangle = \sqrt{2}\,\mathrm{Im}\left( \int_{-\infty}^{\infty} s_{\mathrm{BB}}(t)\,g^*(t-\ell T_{\mathrm{s}})\,\mathrm{d}t \right)$$

$$= \sqrt{2} \int_{-\infty}^{\infty} \mathrm{Im}\big(s_{\mathrm{BB}}(t)\big)\,\mathrm{Re}\big(g(t-\ell T_{\mathrm{s}})\big)\,\mathrm{d}t$$

$$- \sqrt{2} \int_{-\infty}^{\infty} \mathrm{Re}\big(s_{\mathrm{BB}}(t)\big)\,\mathrm{Im}\big(g(t-\ell T_{\mathrm{s}})\big)\,\mathrm{d}t, \tag{16.29}$$

where the inner products can be computed again using a matched filter:

$$\int_{-\infty}^{\infty} \mathrm{Im}\big(s_{\mathrm{BB}}(t)\big)\,\mathrm{Re}\big(g(t-\ell T_{\mathrm{s}})\big)\,\mathrm{d}t = \big(\mathrm{Im}(\mathbf{s}_{\mathrm{BB}}) \star \mathrm{Re}(\tilde{\mathbf{g}})\big)(\ell T_{\mathrm{s}}),$$

$$\int_{-\infty}^{\infty} \mathrm{Re}\big(s_{\mathrm{BB}}(t)\big)\,\mathrm{Im}\big(g(t-\ell T_{\mathrm{s}})\big)\,\mathrm{d}t = \big(\mathrm{Re}(\mathbf{s}_{\mathrm{BB}}) \star \mathrm{Im}(\tilde{\mathbf{g}})\big)(\ell T_{\mathrm{s}}).$$

Things become simpler when the pulse shape $\mathbf{g}$ is real. In this case (16.26) and (16.29) simplify to

$$\langle \mathbf{r}, \mathbf{g}_{\mathrm{I},\ell} \rangle = \sqrt{2} \int \mathrm{Re}\big(s_{\mathrm{BB}}(t)\big)\,g(t-\ell T_{\mathrm{s}})\,\mathrm{d}t, \quad \mathbf{g} \text{ real}, \tag{16.30a}$$

$$\langle \mathbf{r}, \mathbf{g}_{\mathrm{Q},\ell} \rangle = \sqrt{2} \int \mathrm{Im}\big(s_{\mathrm{BB}}(t)\big)\,g(t-\ell T_{\mathrm{s}})\,\mathrm{d}t, \quad \mathbf{g} \text{ real}. \tag{16.30b}$$

Diagrams demonstrating how these inner products are computed are given in Figures 16.3 and 16.4. We have already discussed the first diagram, which includes the front-end bandpass filter and the circuit for producing $\mathbf{s}_{\mathrm{BB}}$. The second diagram includes the matched filtering needed to compute the RHS of (16.30a) and the RHS of (16.30b). Notice that we have accomplished our second objective in that the first circuit depends only on the carrier frequency $f_{\mathrm{c}}$ (and the bandwidth $\mathsf{W}$) and the second circuit depends on the pulse shape but not on the carrier frequency.

## 16.8.2  Computing Inner Products Directly

The astute reader may have noticed that neither the bandpass filtering of the signal $\mathbf{r}$ nor the image rejection filters that produce $\mathbf{s}_{\mathrm{BB}}$ are needed for the computation of the inner products. Indeed, starting from (16.8a)

$$\langle \mathbf{r}, \mathbf{g}_{\mathrm{I},\ell} \rangle = \big\langle \mathbf{r}, t \mapsto 2\,\mathrm{Re}\big(g_{\mathrm{I},\ell,\mathrm{BB}}(t)\,e^{\mathrm{i}2\pi f_{\mathrm{c}}t}\big) \big\rangle$$

$$= 2\,\mathrm{Re}\big(\big\langle \mathbf{r}, t \mapsto g_{\mathrm{I},\ell,\mathrm{BB}}(t)\,e^{\mathrm{i}2\pi f_{\mathrm{c}}t} \big\rangle\big)$$

$$= 2\,\mathrm{Re}\big(\big\langle t \mapsto r(t)\,e^{-\mathrm{i}2\pi f_{\mathrm{c}}t}, \mathbf{g}_{\mathrm{I},\ell,\mathrm{BB}} \big\rangle\big)$$

$$= \sqrt{2}\,\mathrm{Re}\big(\big\langle t \mapsto r(t)\,e^{-\mathrm{i}2\pi f_{\mathrm{c}}t}, t \mapsto g(t-\ell T_{\mathrm{s}}) \big\rangle\big), \tag{16.31a}$$

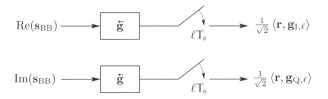

**Figure 16.4:** QAM demodulation: matched filtering (**g** real).

where the second equality follows because **r** is real and the last equality from (16.9a). Similarly, starting from (16.8b)

$$
\begin{aligned}
\langle \mathbf{r}, \mathbf{g}_{Q,\ell} \rangle &= \big\langle \mathbf{r}, t \mapsto 2\operatorname{Re}\big(g_{Q,\ell,\mathrm{BB}}(t)\, e^{\mathrm{i}2\pi f_c t}\big)\big\rangle \\
&= 2\operatorname{Re}\big(\big\langle \mathbf{r}, t \mapsto g_{Q,\ell,\mathrm{BB}}(t)\, e^{\mathrm{i}2\pi f_c t}\big\rangle\big) \\
&= 2\operatorname{Re}\big(\big\langle t \mapsto r(t)\, e^{-\mathrm{i}2\pi f_c t}, \mathbf{g}_{Q,\ell,\mathrm{BB}}\big\rangle\big) \\
&= \sqrt{2}\operatorname{Re}\big(\big\langle t \mapsto r(t)\, e^{-\mathrm{i}2\pi f_c t}, t \mapsto \mathrm{i}\,g(t - \ell T_s)\big\rangle\big) \\
&= \sqrt{2}\operatorname{Im}\big(\big\langle t \mapsto r(t)\, e^{-\mathrm{i}2\pi f_c t}, t \mapsto g(t - \ell T_s)\big\rangle\big),
\end{aligned}
\tag{16.31b}
$$

where the fourth equality follows from (16.9b). Notice that the RHS of (16.31a) and the RHS of (16.31b) do not involve any filtering. To see how to implement them with real operations we can write them more explicitly as:

$$
\langle \mathbf{r}, \mathbf{g}_{I,\ell} \rangle = \sqrt{2}\operatorname{Re}\left(\int_{-\infty}^{\infty} r(t)\, e^{-\mathrm{i}2\pi f_c t}\, g^*(t - \ell T_s)\, \mathrm{d}t\right),
$$

$$
\langle \mathbf{r}, \mathbf{g}_{Q,\ell} \rangle = \sqrt{2}\operatorname{Im}\left(\int_{-\infty}^{\infty} r(t)\, e^{-\mathrm{i}2\pi f_c t}\, g^*(t - \ell T_s)\, \mathrm{d}t\right),
$$

or even more explicitly in terms of real operations as:

$$
\langle \mathbf{r}, \mathbf{g}_{I,\ell} \rangle = \sqrt{2}\int_{-\infty}^{\infty} r(t)\cos(2\pi f_c t)\operatorname{Re}\big(g(t - \ell T_s)\big)\, \mathrm{d}t
$$
$$
- \sqrt{2}\int_{-\infty}^{\infty} r(t)\sin(2\pi f_c t)\operatorname{Im}\big(g(t - \ell T_s)\big)\, \mathrm{d}t,
\tag{16.32a}
$$

$$
\langle \mathbf{r}, \mathbf{g}_{Q,\ell} \rangle = -\sqrt{2}\int_{-\infty}^{\infty} r(t)\cos(2\pi f_c t)\operatorname{Im}\big(g(t - \ell T_s)\big)\, \mathrm{d}t
$$
$$
- \sqrt{2}\int_{-\infty}^{\infty} r(t)\sin(2\pi f_c t)\operatorname{Re}\big(g(t - \ell T_s)\big)\, \mathrm{d}t.
\tag{16.32b}
$$

The two approaches we discussed for computing the inner products are, of course, mathematically equivalent. The former makes more engineering sense, because the bandpass filter typically guarantees that the energy in **s** is significantly smaller than in **r**, thus reducing the dynamic range required from the rest of the receiver.

The latter approach is mathematically cleaner because it requires less mathematical justification. One need not check that the various filters satisfy the required integrability conditions. Moreover, this approach is more useful when $\mathbf{r}$ is not energy-limited and when this is compensated for by the fast decay of the pulse shape. (See, for example, the situation addressed by Proposition 3.4.4.)

## 16.9 Exercises

**Exercise 16.1 (Nyquist's Criterion and Passband Signals).** Corollary 11.3.4 provides conditions under which the time shifts of a signal by integer multiples of $T_s$ are orthonormal. Discuss how these conditions apply to real passband signals of bandwidth $W$ around the carrier frequency $f_c$. Specifically:

(i) Plot the function

$$ f \mapsto \sum_{\ell=-\infty}^{\infty} \left| \hat{y}\left( f + \frac{\ell}{T_s} \right) \right|^2 $$

for the passband signal $\mathbf{y}$ of Figure 7.2. Pay attention to how the sum at positive frequencies is influenced by the signal's FT at negative frequencies.

(ii) Show that there exists a passband signal $\phi(\cdot)$ whose bandwidth $W$ around the carrier frequency $f_c$ is $1/(2T_s)$ and whose time shifts by integer multiples of $T_s$ are orthonormal if, and only if, $4T_s f_c$ is an odd integer. Show that such a signal must satisfy (outside a set of frequencies of Lebesgue measure zero)

$$ |\hat{\phi}(f)| = \sqrt{T_s}\, I\left\{ ||f| - f_c| \le \frac{1}{4T_s} \right\}, \quad f \in \mathbb{R}. $$

(iii) Let $\phi$ be an energy-limited baseband signal of bandwidth $W/2$ whose FT is a symmetric function of frequency and whose time shifts by integer multiples of $(2T_s)$ are orthonormal. Let the carrier frequency $f_c$ be larger than $W/2$ and satisfy that $4T_s f_c$ is an odd integer. Show that the (possibly complex) passband signal $t \mapsto \sqrt{2}\cos(2\pi f_c t)\,\phi(t)$ is of bandwidth $W$ around the carrier $f_c$, and its time shifts by integer multiples of $T_s$ are orthonormal.

**Exercise 16.2 (How General is QAM?).** Under what conditions on $A$, $f_c$, $\phi$, $W$, and $T_s$ can we view the signal

$$ t \mapsto A \operatorname{Re}\left( e^{i(2\pi f_c t + \phi)} \sum_{\ell=1}^{n} C_\ell \operatorname{sinc}(Wt - \ell T_s) \right) $$

as a QAM signal?

**Exercise 16.3 (M-PSK).** Consider a QAM signal $\mathbf{X}_{\mathrm{PB}}$ of the form (16.6) with the pulse shape $\mathbf{g}\colon t \mapsto I\{-T_s/2 \le t < T_s/2\}$ and symbols $(C_\ell)$ that are IID and uniformly distributed over the set

$$ \{ e^{i2\pi/8}, e^{2i2\pi/8}, \ldots, e^{7i2\pi/8}, 1 \}. $$

(i) Plot a sample function of $(X_{\mathrm{PB}}(t),\ t \in \mathbb{R})$.

(ii) Are the sample paths continuous?

(iii) Express $X_{\mathrm{PB}}(t)$ in the form $2A\cos(2\pi f_c t + \Phi(t))$ and describe $\Phi(t)$. Plot a sample path of $(\Phi(t))$.

**Exercise 16.4 (Transmission Rate, Encoder Rate, and Bandwidth).** Data bits are to be transmitted at rate $R_b$ bits per second using QAM with a pulse shape $\phi$ satisfying the orthonormality condition (16.11).

(i) Let $W$ be the allotted bandwidth around the carrier frequency. What is the minimal constellation size required for the data bits to be reliably communicated in the absence of noise?

(ii) Repeat Part (i) if you are required to use a pulse shape of excess-bandwidth of $\beta = 15\%$ or more.

**Exercise 16.5 (Synthesis of 16-QAM).** Let $X_1(\cdot)$ and $X_2(\cdot)$ be 4-QAM (QPSK) signals that are given for every $t \in \mathbb{R}$ by

$$X_\nu(t) = 2A \operatorname{Re}\left( \sum_{\ell=1}^{n} C_\ell^{(\nu)} g(t - \ell T_s) e^{i2\pi f_c t} \right), \quad \nu = 1, 2,$$

where the symbols $\left(C_\ell^{(\nu)}\right)$ take on the values $\pm 1 \pm i$. Show that for the right choice of the constant $\alpha \in \mathbb{R}$, the signal

$$X(t) = \alpha X_1(t) + X_2(t), \quad t \in \mathbb{R}$$

can be viewed as a 16-QAM signal with a square constellation.

**Exercise 16.6 (Orthogonality of the In-Phase and Quadrature Components).** Let the pulse shape $\mathbf{g}$ be a *real* integrable signal that is bandlimited to $W/2$ Hz, and let the carrier frequency $f_c$ be larger than $W/2$. Show that, even if the time shifts of $\mathbf{g}$ by integer multiples of $T_s$ are not orthonormal, the signals

$$t \mapsto g(t - \ell T_s) \cos(2\pi f_c t + \varphi) \text{ and } t \mapsto g(t - \ell' T_s) \sin(2\pi f_c t + \varphi)$$

are orthogonal for all integers $\ell, \ell'$ (not necessarily distinct). Here $\varphi \in [-\pi, \pi)$ is arbitrary.

**Exercise 16.7 (The Importance of the Phase).** Let $\mathbf{x}$ and $\mathbf{y}$ be real integrable signals that are bandlimited to $W/2$ Hz. Let the transmitted signal $\mathbf{s}$ be

$$s(t) = \operatorname{Re}\left( (x(t) + iy(t)) e^{i(2\pi f_c t + \phi_T)} \right)$$
$$= x(t) \cos(2\pi f_c t + \phi_T) - y(t) \sin(2\pi f_c t + \phi_T), \quad t \in \mathbb{R},$$

where $f_c > W/2$, and where $\phi_T$ denotes the phase of the transmitted carrier. The receiver multiplies $s(t)$ by $2\cos(2\pi f_c t + \phi_R)$ (where $\phi_R$ denotes the phase of the receiver's oscillator) and passes the resulting product through a lowpass filter of cutoff frequency $W/2$ to produce the signal $\tilde{\mathbf{x}}$:

$$\tilde{x}(t) = \left( (\tau \mapsto s(\tau) 2\cos(2\pi f_c \tau + \phi_R)) \star \mathrm{LPF}_W \right)(t), \quad t \in \mathbb{R}.$$

Express $\tilde{x}(\cdot)$ in terms of $x(\cdot)$, $y(\cdot)$, $\phi_T$ and $\phi_R$. Evaluate your expression in the following cases: $\phi_T = \phi_R$, $\phi_T - \phi_R = \pi$, $\phi_T - \phi_R = \pi/2$, and $\phi_T - \phi_R = \pi/4$.

**Exercise 16.8 (Phase Imprecision).** Consider QAM with a real pulse shape and a receiver that performs a conversion to baseband followed by matched filtering (Section 16.8.1). Write an expression for the output of the receiver if its oscillator is at the right frequency but lags the phase of the transmitter's oscillator by $\Delta\phi$.

**Exercise 16.9 (Rotating a QAM Constellation).** Show that rotating a QAM constellation changes neither its second moment nor its minimum distance.

**Exercise 16.10 (Optimal Rectangular Constellation).** Consider all rectangular constellations of the form

$$\{a + ib, a - ib, -a + ib, -a - ib\},$$

where $a$ and $b$ are real. Which of these constellations whose second moment is one has the largest minimum distance?

# Chapter 17

# Complex Random Variables and Processes

## 17.1  Introduction

We first encountered complex random variables in Chapter 16 on QAM. There we considered an encoder that maps $k$-tuples of bits into $n$-tuples of complex numbers, and we then considered the result of applying this encoder to random bits. The resulting symbols were therefore random and were taking value in the complex field, i.e., they were complex random variables. Complex random variables are functions that map "luck" into the complex field: they map every outcome of the experiment $\omega \in \Omega$ to a complex number. Thus, they are very much like regular random variables, except that they take value in the complex field. They can always be considered as pairs of real variables: their real and imaginary parts.

It is perfectly meaningful to discuss their expectation and variance. If $C$ is a complex random variable, then

$$\mathsf{E}[C] = \mathsf{E}[\mathrm{Re}(C)] + \mathsf{i}\,\mathsf{E}[\mathrm{Im}(C)],$$

$$\mathsf{E}\big[|C|^2\big] = \mathsf{E}\big[(\mathrm{Re}(C))^2\big] + \mathsf{E}\big[(\mathrm{Im}(C))^2\big],$$

and

$$\mathsf{Var}[C] = \mathsf{E}\big[\big|C - \mathsf{E}[C]\big|^2\big]$$

$$= \mathsf{E}\big[|C|^2\big] - \big|\mathsf{E}[C]\big|^2.$$

In this chapter we shall make the above definition of complex random variables more formal and also discuss complex random vectors and complex stochastic processes.

Complex random variables can be avoided if one treats such variables as pairs of real variables. However, we do not recommend this approach. Many of the complex variables and processes encountered in Digital Communications possess additional properties that simplify their manipulation, and complex variables are better suited to take advantage of these simplifications.

We begin this chapter with some notation followed by some basic definitions for complex random variables. We next introduce a property that simplifies their manipulation: properness. (Another such property, circular symmetry, is described in Chapter 24.) Finally, we extend the discussion to complex random vectors and conclude with a discussion of complex stochastic processes.

## 17.2   Notation

The notation we use in this chapter is fairly standard. The only issue that may need clarification is the difference between three matrix/vector operations: transposition, conjugation, and Hermitian conjugation. These operations are described next.

All vectors in this chapter are column vectors. Thus, a vector $\mathbf{a}$ whose components are $a^{(1)}, \ldots, a^{(n)}$ is the column vector

$$\mathbf{a} = \begin{pmatrix} a^{(1)} \\ a^{(2)} \\ \vdots \\ a^{(n)} \end{pmatrix}. \tag{17.1}$$

We shall sometimes refer to such a vector $\mathbf{a}$ as an $n$-vector to make the number of its components explicit. For typesetting reasons, we shall usually use the notation

$$\mathbf{a} = \left( a^{(1)}, \ldots, a^{(n)} \right)^{\mathsf{T}}, \tag{17.2}$$

which is more space efficient. Here the operator $(\cdot)^{\mathsf{T}}$ denotes the **matrix transpose**. Thus if we think of $(a^{(1)}, \ldots a^{(n)})$ as a $1 \times n$ matrix, then $(a^{(1)}, \ldots a^{(n)})^{\mathsf{T}}$ is this matrix's transpose, i.e., an $n \times 1$ matrix, or a vector. More generally, if $\mathsf{A}$ is an $n \times m$ matrix, then $\mathsf{A}^{\mathsf{T}}$ is an $m \times n$ matrix whose Row-$j$ Column-$\ell$ component is the Row-$\ell$ Column-$j$ component of $\mathsf{A}$. We say that $\mathsf{A}$ is **symmetric** if $\mathsf{A}^{\mathsf{T}} = \mathsf{A}$.

We use $(\cdot)^*$ to denote **componentwise complex conjugation**. Thus, if $\mathbf{a}$ is as in (17.1), then

$$\mathbf{a}^* = \begin{pmatrix} \left( a^{(1)} \right)^* \\ \left( a^{(2)} \right)^* \\ \vdots \\ \left( a^{(n)} \right)^* \end{pmatrix}. \tag{17.3}$$

We use $(\cdot)^{\dagger}$ to denote **Hermitian conjugation**, i.e., the componentwise conjugate of the transposed matrix. Thus, if $\mathbf{a}$ is as in (17.1), then $\mathbf{a}^{\dagger}$ is the $1 \times n$ matrix

$$\mathbf{a}^{\dagger} = \left( \left( a^{(1)} \right)^*, \ldots, \left( a^{(n)} \right)^* \right). \tag{17.4}$$

The Hermitian conjugate $\mathsf{A}^{\dagger}$ of an $n \times m$ matrix $\mathsf{A}$ is an $m \times n$ matrix whose Row-$j$ Column-$\ell$ component is the complex conjugate of the Row-$\ell$ Column-$j$ component of the matrix $\mathsf{A}$. We say that a matrix $\mathsf{A}$ is **conjugate-symmetric** or **self-adjoint** or **Hermitian** if $\mathsf{A}^{\dagger} = \mathsf{A}$.

Note that if $\mathbf{a}$ and $\mathbf{b}$ are $n$-vectors, then $\mathbf{a}^\mathsf{T}\mathbf{b}$ is a scalar

$$\mathbf{a}^\mathsf{T}\mathbf{b} = \sum_{j=1}^{n} a^{(j)}b^{(j)}, \tag{17.5}$$

whereas $\mathbf{a}\mathbf{b}^\mathsf{T}$ is the $n \times n$ matrix

$$\mathbf{a}\mathbf{b}^\mathsf{T} = \begin{pmatrix} a^{(1)}b^{(1)} & a^{(1)}b^{(2)} & \cdots & a^{(1)}b^{(n)} \\ a^{(2)}b^{(1)} & a^{(2)}b^{(2)} & \cdots & a^{(2)}b^{(n)} \\ \vdots & \vdots & \vdots & \vdots \\ \vdots & \vdots & \vdots & \vdots \\ a^{(n)}b^{(1)} & a^{(n)}b^{(2)} & \cdots & a^{(n)}b^{(n)} \end{pmatrix}.$$

## 17.3   Complex Random Variables

We say that $C$ is a **complex random variable** (CRV) on the probability space $(\Omega, \mathcal{F}, P)$ if $C\colon \Omega \to \mathbb{C}$ is a mapping from $\Omega$ to the complex field $\mathbb{C}$ such that both $\mathrm{Re}(C)$ and $\mathrm{Im}(C)$ are random variables on $(\Omega, \mathcal{F}, P)$.

Any CRV $Z$ can be written in the form $Z = X + iY$, where $X$ and $Y$ are real random variables. But there are some advantages to studying complex random variables over pairs of real random variables. Those will become apparent when we discuss analytic functions of complex random variables and when we discuss complex random variables that have special properties such as that of being "proper" or that of being "circularly-symmetric."

Many of the definitions related to complex random variables are similar to the analogous definitions for pairs of real random variables, but some are not. We shall try to emphasize the latter.

### 17.3.1   Distribution and Density

Since it makes no sense to say that one complex number is smaller than another, we cannot define the cumulative distribution function (CDF) of a CRV as in the real case: an expression like "$\mathrm{Pr}[Z \le 1 + i]$" is meaningless. We can, however, discuss the joint distribution function of the real and imaginary parts of a CRV, which specifies $\mathrm{Pr}[\mathrm{Re}(Z) \le x,\ \mathrm{Im}(Z) \le y]$ for all $x, y \in \mathbb{R}$. We say that two complex random variables $W$ and $Z$ are of **equal law** (or have the same distribution) and write $W \overset{\mathscr{L}}{=} Z$, if the joint distribution of the pair $(\mathrm{Re}(W), \mathrm{Im}(W))$ is identical to the joint distribution of the pair $(\mathrm{Re}(Z), \mathrm{Im}(Z))$:

$$\left(W \overset{\mathscr{L}}{=} Z\right) \Leftrightarrow$$
$$\left(\mathrm{Pr}\big[\mathrm{Re}(W) \le x, \mathrm{Im}(W) \le y\big] = \mathrm{Pr}\big[\mathrm{Re}(Z) \le x, \mathrm{Im}(Z) \le y\big],\ x, y \in \mathbb{R}\right). \tag{17.6}$$

Similarly, we can define the **density function** $f_Z(\cdot)$ (if it exists) of a CRV $Z$ at the point $z \in \mathbb{C}$ as the joint density of the real pair $(\mathrm{Re}(Z), \mathrm{Im}(Z))$ at $(\mathrm{Re}(z), \mathrm{Im}(z))$:

$$f_Z(z) \triangleq f_{\mathrm{Re}(Z),\mathrm{Im}(Z)}\big(\mathrm{Re}(z), \mathrm{Im}(z)\big), \quad z \in \mathbb{C}, \tag{17.7}$$

which can also be written as

$$f_Z(z) = \left. \frac{\partial^2}{\partial x\, \partial y} \Pr\big[\mathrm{Re}(Z) \le x, \mathrm{Im}(Z) \le y\big] \right|_{x=\mathrm{Re}(z), y=\mathrm{Im}(z)}, \quad z \in \mathbb{C}. \tag{17.8}$$

The notions of distribution function and density of a CRV extend immediately to pairs of complex variables and, more generally, to $n$-tuples.

## 17.3.2   The Expectation

The **expectation** of a CRV can be defined in terms of the expectations of its real and imaginary parts:

$$\mathsf{E}[Z] = \mathsf{E}[\mathrm{Re}(Z)] + \mathrm{i}\, \mathsf{E}[\mathrm{Im}(Z)], \tag{17.9}$$

provided that the two real expectations $\mathsf{E}[\mathrm{Re}(Z)]$ and $\mathsf{E}[\mathrm{Im}(Z)]$ are finite. With this definition one can readily verify that, whenever $\mathsf{E}[Z]$ is defined, conjugation and expectation commute

$$\mathsf{E}[Z^*] = (\mathsf{E}[Z])^*, \tag{17.10}$$

and

$$\mathrm{Re}\big(\mathsf{E}[Z]\big) = \mathsf{E}\big[\mathrm{Re}(Z)\big], \tag{17.11a}$$

$$\mathrm{Im}\big(\mathsf{E}[Z]\big) = \mathsf{E}\big[\mathrm{Im}(Z)\big]. \tag{17.11b}$$

If the CRV $Z$ has a density $f_Z(\cdot)$, then the expectation $\mathsf{E}[g(Z)]$ for some measurable function $\mathbf{g} \colon \mathbb{C} \to \mathbb{C}$ can be formally written as

$$\mathsf{E}\big[g(Z)\big] = \int_{z \in \mathbb{C}} f_Z(z) g(z)\, \mathrm{d}z \tag{17.12}$$

or, in terms of real integrals, as

$$\mathsf{E}\big[g(Z)\big] = \int_{-\infty}^{\infty} \int_{-\infty}^{\infty} f_Z(x + \mathrm{i}y)\, \mathrm{Re}\big(g(x + \mathrm{i}y)\big)\, \mathrm{d}x\, \mathrm{d}y$$

$$+ \mathrm{i} \int_{-\infty}^{\infty} \int_{-\infty}^{\infty} f_Z(x + \mathrm{i}y)\, \mathrm{Im}\big(g(x + \mathrm{i}y)\big)\, \mathrm{d}x\, \mathrm{d}y. \tag{17.13}$$

Thus, rather than computing the distribution of $g(Z)$ and of then computing the expectations of its real and imaginary parts, one can use (17.12).

### 17.3.3   The Variance

The definition of the variance of a CRV is not consistent with viewing the CRV as a pair of real random variables. The **variance** $\mathsf{Var}[Z]$ of a CRV $Z$ is defined as

$$\mathsf{Var}[Z] \triangleq \mathsf{E}\big[|Z - \mathsf{E}[Z]|^2\big] \tag{17.14a}$$

$$= \mathsf{E}\big[|Z|^2\big] - |\mathsf{E}[Z]|^2 \tag{17.14b}$$

$$= \mathsf{Var}\big[\mathrm{Re}(Z)\big] + \mathsf{Var}\big[\mathrm{Im}(Z)\big]. \tag{17.14c}$$

This definition should be contrasted with the definition of the covariance *matrix* of the pair $(\mathrm{Re}(Z), \mathrm{Im}(Z))$

$$\begin{pmatrix} \mathsf{Var}\big[\mathrm{Re}(Z)\big] & \mathsf{Cov}\big[\mathrm{Re}(Z), \mathrm{Im}(Z)\big] \\ \mathsf{Cov}\big[\mathrm{Re}(Z), \mathrm{Im}(Z)\big] & \mathsf{Var}\big[\mathrm{Im}(Z)\big] \end{pmatrix}.$$

One can compute the variance of $Z$ from the covariance matrix of $(\mathrm{Re}(Z), \mathrm{Im}(Z))$, but not the other way around. Indeed, the variance of $Z$ is just the trace of the covariance matrix of $(\mathrm{Re}(Z), \mathrm{Im}(Z))$.

To derive (17.14b) from (17.14a) we note that

$$\begin{aligned} \mathsf{E}\big[|Z - \mathsf{E}[Z]|^2\big] &= \mathsf{E}\big[(Z - \mathsf{E}[Z])(Z - \mathsf{E}[Z])^*\big] \\ &= \mathsf{E}\big[(Z - \mathsf{E}[Z])(Z^* - \mathsf{E}[Z^*])\big] \\ &= \mathsf{E}\big[(Z - \mathsf{E}[Z])Z^*\big] - \mathsf{E}\big[(Z - \mathsf{E}[Z])\big]\,\mathsf{E}[Z^*] \\ &= \mathsf{E}\big[(Z - \mathsf{E}[Z])Z^*\big] \\ &= \mathsf{E}[ZZ^*] - \mathsf{E}[Z]\,\mathsf{E}[Z^*] \\ &= \mathsf{E}\big[|Z|^2\big] - |\mathsf{E}[Z]|^2, \end{aligned}$$

where we only used the linearity of expectation and (17.10). Here the first equality follows by writing $|w|^2$ as $ww^*$; the second by (17.10); the third by simple algebra; the fourth because the expectation of $Z - \mathsf{E}[Z]$ is zero; and the final by (17.10).

To derive (17.14c) from (17.14b) we write $\mathsf{E}\big[|Z|^2\big]$ as $\mathsf{E}\big[(\mathrm{Re}(Z))^2 + (\mathrm{Im}(Z))^2\big]$ and express $|\mathsf{E}[Z]|^2$ using (17.9) as $\mathsf{E}[\mathrm{Re}(Z)]^2 + \mathsf{E}[\mathrm{Im}(Z)]^2$.

### 17.3.4   Proper Complex Random Variables

Many of the complex random variables that appear in Digital Communications are **proper**. This is a concept that has no natural counterpart for real random variables.

**Definition 17.3.1 (Proper CRV).** *We say that the CRV $Z$ is **proper** if the following three conditions are all satisfied: it is of zero-mean; it is of finite-variance; and*

$$\mathsf{E}\big[Z^2\big] = 0. \tag{17.15}$$

Notice that the LHS of (17.15) is, in general, a complex number, so (17.15) is equivalent to two real equations:

$$\mathsf{E}\big[\mathrm{Re}(Z)^2\big] = \mathsf{E}\big[\mathrm{Im}(Z)^2\big] \tag{17.16a}$$

and
$$E\big[\operatorname{Re}(Z)\operatorname{Im}(Z)\big] = 0. \tag{17.16b}$$

This leads to the following characterization of proper complex random variables.

**Proposition 17.3.2.** *A CRV $Z$ is proper if, and only if, all three of the following conditions are satisfied: $Z$ is of zero mean; $\operatorname{Re}(Z)$ & $\operatorname{Im}(Z)$ have the same finite variance; and $\operatorname{Re}(Z)$ & $\operatorname{Im}(Z)$ are uncorrelated.*

An example of a proper CRV is one taking on the four values $\{\pm 1, \pm i\}$ equiprobably.

We mentioned earlier in Section 17.3.3 that the variance of a CRV is not the same as the covariance matrix of the tuple consisting of its real and imaginary parts. While the covariance matrix determines the variance, the variance does not uniquely determine the covariance matrix. However, if a CRV is proper, then its variance uniquely determines the covariance matrix of its real and imaginary parts. Indeed, by Proposition 17.3.2, a zero-mean finite-variance CRV is proper if, and only if, the covariance matrix of the pair $(\operatorname{Re}(Z), \operatorname{Im}(Z))$ is given by

$$\begin{pmatrix} \frac{1}{2}\operatorname{Var}[Z] & 0 \\ 0 & \frac{1}{2}\operatorname{Var}[Z] \end{pmatrix}.$$

## 17.3.5   The Covariance

The **covariance** $\operatorname{Cov}[Z, W]$ between the complex random variables $Z$ and $W$ is defined by

$$\operatorname{Cov}[Z, W] \triangleq E\Big[\big(Z - E[Z]\big)\big(W - E[W]\big)^*\Big]. \tag{17.17}$$

Again, this definition is different from the one for pairs of real random variables: the covariance between two pairs of real random variables is a real matrix, whereas the covariance between two CRVs is a complex scalar.

Some of the key properties of the covariance are listed next. They hold whenever the $\alpha$'s and $\beta$'s are deterministic complex numbers and the covariances on the RHS are defined.

 (i) Conjugate Symmetry:
$$\operatorname{Cov}[Z, W] = \big(\operatorname{Cov}[W, Z]\big)^*. \tag{17.18}$$

(ii) Sesquilinearity:
$$\operatorname{Cov}[\alpha Z, W] = \alpha \operatorname{Cov}[Z, W], \tag{17.19}$$
$$\operatorname{Cov}[Z_1 + Z_2, W] = \operatorname{Cov}[Z_1, W] + \operatorname{Cov}[Z_2, W], \tag{17.20}$$
$$\operatorname{Cov}[Z, \beta W] = \beta^* \operatorname{Cov}[Z, W], \tag{17.21}$$
$$\operatorname{Cov}[Z, W_1 + W_2] = \operatorname{Cov}[Z, W_1] + \operatorname{Cov}[Z, W_2], \tag{17.22}$$

and, more generally,
$$\operatorname{Cov}\left[\sum_{j=1}^{n} \alpha_j Z_j, \sum_{j'=1}^{n'} \beta_{j'} W_{j'}\right] = \sum_{j=1}^{n} \sum_{j'=1}^{n'} \alpha_j \beta_{j'}^* \operatorname{Cov}[Z_j, W_{j'}]. \tag{17.23}$$

(iii) Relation with Variance:

$$\mathsf{Var}[Z] = \mathsf{Cov}[Z, Z].\tag{17.24}$$

(iv) Variance of Linear Functionals:

$$\mathsf{Var}\left[\sum_{j=1}^{n}\alpha_j Z_j\right] = \sum_{j=1}^{n}\sum_{j'=1}^{n}\alpha_j \alpha_{j'}^{*}\,\mathsf{Cov}[Z_j, Z_{j'}].\tag{17.25}$$

### 17.3.6 The Characteristic Function

The definition of the characteristic function of a CRV is consistent with viewing it as a pair of real random variables. Recall that the characteristic function $\Phi_X\colon \mathbb{R} \to \mathbb{C}$ of a real random variable $X$ is defined by

$$\Phi_X\colon \varpi \mapsto \mathsf{E}\big[e^{i\varpi X}\big], \quad \varpi \in \mathbb{R}.\tag{17.26}$$

For a pair of real random variables $X, Y$ the joint characteristic function is the mapping $\Phi_{X,Y}\colon \mathbb{R}^2 \to \mathbb{C}$ defined by

$$\Phi_{X,Y}\colon (\varpi_1, \varpi_2) \mapsto \mathsf{E}\big[e^{i(\varpi_1 X + \varpi_2 Y)}\big], \quad \varpi_1, \varpi_2 \in \mathbb{R}.\tag{17.27}$$

Note that the expectations in (17.26) and (17.27) are always defined, because the argument to the expectation operator is of modulus one ($|e^{ir}| = 1$, whenever $r$ is real). This motivates us to define the characteristic function for a complex random variable as follows.

**Definition 17.3.3 (Characteristic Function of a CRV).** *The **characteristic function** $\Phi_Z\colon \mathbb{C} \to \mathbb{C}$ of a complex random variable $Z$ is defined as*

$$\Phi_Z(\varpi) \triangleq \mathsf{E}\big[e^{i\,\mathrm{Re}(\varpi^{*}Z)}\big], \quad \varpi \in \mathbb{C}$$
$$= \mathsf{E}\big[e^{i\left(\mathrm{Re}(\varpi)\,\mathrm{Re}(Z) + \mathrm{Im}(\varpi)\,\mathrm{Im}(Z)\right)}\big], \quad \varpi \in \mathbb{C}.$$

Here we can think of $\mathrm{Re}(\varpi)$ and $\mathrm{Im}(\varpi)$ as playing the role of $\varpi_1$ and $\varpi_2$ in (17.27).

### 17.3.7 Transforming Complex Variables

We next calculate the density of the result of applying a (deterministic) transformation to a CRV. The key to the calculation is to treat the CRV as a pair of real random variables and to then apply the analogous result regarding the transformation of a random real tuple. To that end we recall the following basic theorem regarding the transformation of real random vectors. In the theorem's statement we encounter the notion of an open subset of $\mathbb{R}^n$. Loosely speaking, $\mathcal{D} \subseteq \mathbb{R}^n$ is an open subset of $\mathbb{R}^n$ if to each $\mathbf{x} \in \mathcal{D}$ there corresponds some $\epsilon > 0$ such that the ball of radius $\epsilon$ and center $\mathbf{x}$ is fully contained in $\mathcal{D}$.[1]

---

[1]Thus, $\mathcal{D}$ is an open subset of $\mathbb{R}^n$ if $\mathcal{D} \subseteq \mathbb{R}^n$ and if to each $\mathbf{x} \in \mathcal{D}$ there corresponds some $\epsilon > 0$ such that each $\mathbf{y} \in \mathbb{R}^n$ satisfying $(\mathbf{x} - \mathbf{y})^{\mathsf{T}}(\mathbf{x} - \mathbf{y}) \leq \epsilon^2$ is in $\mathcal{D}$.

**Theorem 17.3.4 (Transforming Real Random Vectors).** *Let* $\mathbf{g} \colon \mathcal{D} \to \mathcal{R}$ *be a one-to-one mapping from an open subset* $\mathcal{D}$ *of* $\mathbb{R}^n$ *onto a subset* $\mathcal{R}$ *of* $\mathbb{R}^n$. *Assume that* $\mathbf{g}$ *has continuous partial derivatives in* $\mathcal{D}$ *and that the Jacobian determinant* $\det\left(\partial g(\mathbf{x})/\partial\mathbf{x}\right)$ *is at no point of* $\mathcal{D}$ *zero. Let the real random n-vector* $\mathbf{X}$ *have the density function* $f_{\mathbf{X}}(\cdot)$ *and satisfy* $\Pr[\mathbf{X} \in \mathcal{D}] = 1$. *Then the random n-vector* $\mathbf{Y} = g(\mathbf{X})$ *is of density*

$$f_{\mathbf{Y}}(\mathbf{y}) = \frac{f_{\mathbf{X}}(\mathbf{x})}{\left|\det \frac{\partial g(\mathbf{x})}{\partial \mathbf{x}}\right|}\Bigg|_{\mathbf{x}=g^{-1}(\mathbf{y})} \cdot \mathrm{I}\{\mathbf{y} \in \mathcal{R}\}. \tag{17.28}$$

Using Theorem 17.3.4 we can relate the density of a CRV $Z$ and the joint distribution of its phase and magnitude.

**Lemma 17.3.5 (The Joint Density of the Magnitude and Phase of a CRV).** *Let* $Z$ *be a CRV of density* $f_Z(\cdot)$, *and let* $R = |Z|$ *and* $\Theta \in [-\pi, \pi)$ *be the magnitude and argument of* $Z$:

$$Z = R\,e^{\mathrm{i}\Theta}, \quad Z \geq 0, \ \Theta \in [-\pi, \pi).$$

*Then the joint distribution of the pair* $(R, \Theta)$ *is of density*

$$f_{R,\Theta}(r, \theta) = r f_Z\big(r\,e^{\mathrm{i}\theta}\big), \quad r > 0, \ \theta \in [-\pi, \pi). \tag{17.29}$$

**Proof.** This result follows directly from Theorem 17.3.4 by computing the absolute value of the Jacobian determinant of the transformation[2] $(x, y) \mapsto (r, \theta)$ where $r = \sqrt{x^2 + y^2}$ and $\theta = \tan^{-1}(y/x)$:

$$\left|\det \begin{pmatrix} \frac{\partial r}{\partial x} & \frac{\partial r}{\partial y} \\ \frac{\partial \theta}{\partial x} & \frac{\partial \theta}{\partial y} \end{pmatrix}\right| = \frac{1}{\sqrt{x^2 + y^2}}$$

$$= \frac{1}{r}. \qquad \square$$

For the next change-of-variables result we recall some basic concepts from Complex Analysis. Given some $z_0 \in \mathbb{C}$ and some nonnegative real number $r \geq 0$, we denote by $\mathcal{D}(z_0, r)$ the disc of radius $r$ that is centered at $z_0$:

$$\mathcal{D}(z_0, r) \triangleq \{z \in \mathbb{C} : |z - z_0| < r\}.$$

We say that a subset $\mathcal{D}$ of the complex plane is **open** if to each $z \in \mathcal{D}$ there corresponds some $\epsilon > 0$ such that $\mathcal{D}(z_0, \epsilon) \subseteq \mathcal{D}$. Let $\mathbf{g} \colon \mathcal{D} \to \mathbb{C}$ be some function from an open set $\mathcal{D} \subseteq \mathbb{C}$ to $\mathbb{C}$. Let $z_0$ be in $\mathcal{D}$. We say that $g(\cdot)$ is **differentiable** at $z_0 \in \mathcal{D}$ and that its derivative at $z_0$ is the complex number $g'(z_0)$, if for every $\epsilon > 0$ there exists some $\delta > 0$ such that

$$\left|\frac{g(z_0 + h) - g(z_0)}{h} - g'(z_0)\right| \leq \epsilon, \tag{17.30}$$

---

[2]Here $\mathcal{D}$ is the set $\mathbb{R}^2$ without the origin.

whenever the complex number $h \in \mathbb{C}$ satisfies $0 < |h| \leq \delta$. It is important to note that here $h$ is complex. If $\mathbf{g}$ is differentiable at every $z \in \mathcal{D}$, then we say that $\mathbf{g}$ is **holomorphic** or **analytic** in $\mathcal{D}$.[3]

Define the mappings

$$\mathbf{u}, \mathbf{v} \colon \{x, y \in \mathbb{R} : x + \mathrm{i}y \in \mathcal{D}\} \to \mathbb{R} \qquad (17.31\mathrm{a})$$

by

$$u(x, y) = \mathrm{Re}\big(g(x + \mathrm{i}y)\big), \qquad (17.31\mathrm{b})$$

and

$$v(x, y) = \mathrm{Im}\big(g(x + \mathrm{i}y)\big). \qquad (17.31\mathrm{c})$$

**Proposition 17.3.6 (The Cauchy-Riemann Equations).** *Let $\mathcal{D} \subseteq \mathbb{C}$ be open and let $\mathbf{g} \colon \mathcal{D} \to \mathbb{C}$ be analytic in $\mathcal{D}$. Let $\mathbf{u}, \mathbf{v}$ be defined by (17.31). Then $\mathbf{u}$ and $\mathbf{v}$ satisfy the Cauchy-Riemann equations*

$$\frac{\partial u(x, y)}{\partial x} = \frac{\partial v(x, y)}{\partial y}, \qquad (17.32\mathrm{a})$$

$$\frac{\partial u(x, y)}{\partial y} = -\frac{\partial v(x, y)}{\partial x} \qquad (17.32\mathrm{b})$$

*at every $x, y \in \mathbb{R}$ such that $x + \mathrm{i}y \in \mathcal{D}$, and*

$$g'(z) = \left. \left( \frac{\partial u(x, y)}{\partial x} + \mathrm{i}\frac{\partial v(x, y)}{\partial x} \right) \right|_{(x,y)=\left( \mathrm{Re}(z), \mathrm{Im}(z) \right)}, \qquad z \in \mathcal{D}. \qquad (17.33)$$

*Moreover, the partial derivatives in (17.32) are continuous in the subset of $\mathbb{R}^2$ defined by $\{x, y \in \mathbb{R} : x + \mathrm{i}y \in \mathcal{D}\}$.*

**Proof.** See (Rudin, 1974, Chapter 11, Theorem 11.2 & Theorem 11.4) or (Nehari, 1975, Chapter II, Section 5 & Chapter III, Section 3). $\qquad \square$

We can now state the change-of-variables theorem for CRVs.

**Theorem 17.3.7 (Transforming Complex Random Variables).** *Let $\mathbf{g} \colon \mathcal{D} \to \mathcal{R}$ be a one-to-one mapping from an open subset $\mathcal{D}$ of $\mathbb{C}$ onto a subset $\mathcal{R}$ of $\mathbb{C}$. Assume that $\mathbf{g}$ is analytic in $\mathcal{D}$ and that at no point of $\mathcal{D}$ is the derivative of $\mathbf{g}$ zero. Let the CRV have the density function $f_Z(\cdot)$ and satisfy $\mathrm{Pr}[Z \in \mathcal{D}] = 1$. Then the CRV defined by $W = g(Z)$ is of density*

$$f_W(w) = \left. \frac{f_Z(z)}{|g'(z)|^2} \right|_{z=g^{-1}(w)} \mathrm{I}\{w \in \mathcal{R}\}. \qquad (17.34)$$

*Here $g^{-1}(w)$ denotes the point in $\mathcal{D}$ that is mapped by $\mathbf{g}$ to $w$.*

---

[3]There is some confusion in the literature about the terms **analytic**, **holomorphic**, and **regular**. We are following here (Rudin, 1974).

**Note 17.3.8.** The square in (17.34) does not appear in dealing with real random variables. It appears here because a mapping of complex numbers is essentially two-dimensional: scaling by $\alpha \in \mathbb{C}$ translates to a scaling of area by $|\alpha|^2$.

**Proof.** To prove (17.34) we begin by expressing the function $g(\cdot)$ as

$$g(x + iy) = u(x, y) + iv(x, y), \quad \Big(x, y \in \mathbb{R}, \ x + iy \in \mathcal{D}\Big),$$

where $u(x, y) = \mathrm{Re}(g(x + iy))$ and $v(x, y) = \mathrm{Im}(g(x + iy))$ are defined in (17.31b) and (17.31c). The density of $g(Z)$ is, by definition, the joint density of the pair $u(\mathrm{Re}(Z), \mathrm{Im}(Z)), v(\mathrm{Re}(Z), \mathrm{Im}(Z))$. And the joint density of the pair $(\mathrm{Re}(Z), \mathrm{Im}(Z))$ is just the density of $Z$. Thus, if we could relate the joint density of the pair $u(\mathrm{Re}(Z), \mathrm{Im}(Z)), v(\mathrm{Re}(Z), \mathrm{Im}(Z))$ to the joint density of the pair $(\mathrm{Re}(Z), \mathrm{Im}(Z))$, then we could relate the density of $g(Z)$ to the density of $Z$.

To relate the joint density of the pair $u(\mathrm{Re}(Z), \mathrm{Im}(Z)), v(\mathrm{Re}(Z), \mathrm{Im}(Z))$ to the joint density of the pair $(\mathrm{Re}(Z), \mathrm{Im}(Z))$ we employ Theorem 17.3.4. To that end we need to compute the absolute value of the Jacobian determinant. This we do as follows:

$$\left| \det \begin{pmatrix} \frac{\partial u}{\partial x} & \frac{\partial u}{\partial y} \\ \frac{\partial v}{\partial x} & \frac{\partial v}{\partial y} \end{pmatrix} \right| = \left| \det \begin{pmatrix} \frac{\partial u}{\partial x} & -\frac{\partial v}{\partial x} \\ \frac{\partial v}{\partial x} & \frac{\partial u}{\partial x} \end{pmatrix} \right|$$

$$= \left( \frac{\partial u}{\partial x} \right)^2 + \left( \frac{\partial v}{\partial x} \right)^2$$

$$= |g'(x + iy)|^2, \tag{17.35}$$

where the first equality follows from the Cauchy-Riemann equations (17.32); the second from a direct calculation of the determinant of a $2 \times 2$ matrix; and where the last equality follows from (17.33). The theorem now follows from (17.35) and Theorem 17.3.4. $\qquad\square$

## 17.4   Complex Random Vectors

We say that $\mathbf{Z} = (Z^{(1)}, \ldots, Z^{(n)})^\mathsf{T}$ is a complex random vector on the probability space $(\Omega, \mathcal{F}, P)$ if it is a mapping from the outcome set $\Omega$ to $\mathbb{C}^n$ such that the real vector

$$\Big( \mathrm{Re}\big(Z^{(1)}\big), \mathrm{Im}\big(Z^{(1)}\big), \ldots, \mathrm{Re}\big(Z^{(n)}\big), \mathrm{Im}\big(Z^{(n)}\big) \Big)^\mathsf{T}$$

comprising the real and imaginary parts of its components is a real random vector on $(\Omega, \mathcal{F}, P)$, i.e., if each of the components of $\mathbf{Z}$ is a CRV.

We say that the complex random vector $\mathbf{Z} = (Z^{(1)}, \ldots, Z^{(n)})^\mathsf{T}$ and the complex random vector $\mathbf{W} = (W^{(1)}, \ldots, W^{(n)})^\mathsf{T}$ are of **equal law** (or have the same distribution) and write $\mathbf{Z} \overset{\mathscr{L}}{=} \mathbf{W}$, if the real vector taking value in $\mathbb{R}^{2n}$ whose components are the real and imaginary parts of the components of $\mathbf{Z}$ has the same distribution as the analogous vector for $\mathbf{W}$, i.e., if for all $x_1, \ldots, x_n, y_1, \ldots, y_n \in \mathbb{R}$

$$\Pr\Big[\mathrm{Re}\big(Z^{(1)}\big) \leq x_1, \mathrm{Im}\big(Z^{(1)}\big) \leq y_1, \ldots, \mathrm{Re}\big(Z^{(n)}\big) \leq x_n, \mathrm{Im}\big(Z^{(n)}\big) \leq y_n\Big]$$
$$= \Pr\Big[\mathrm{Re}\big(W^{(1)}\big) \leq x_1, \mathrm{Im}\big(W^{(1)}\big) \leq y_1, \ldots, \mathrm{Re}\big(W^{(n)}\big) \leq x_n, \mathrm{Im}\big(W^{(n)}\big) \leq y_n\Big].$$

The **expectation** of a complex random vector is the vector consisting of the expectation of each of its components. We say that a complex random vector is of **finite variance** if each of its components is a CRV of finite variance.

### 17.4.1 The Covariance Matrix

The discussion in Section 17.3.5 can be generalized to random complex vectors. The **covariance matrix** $\mathsf{K}_{\mathbf{ZZ}}$ of a finite-variance complex random $n$-vector $\mathbf{Z}$ is defined as the conjugate-symmetric $n \times n$ matrix

$$\mathsf{K}_{\mathbf{ZZ}} \triangleq \mathsf{E}\big[(\mathbf{Z} - \mathsf{E}[\mathbf{Z}])(\mathbf{Z} - \mathsf{E}[\mathbf{Z}])^\dagger\big]. \qquad (17.36)$$

Once again, this definition is not consistent with viewing the random complex vector as a vector of length $2n$ of real random variables. The latter would have a real symmetric $2n \times 2n$ covariance matrix.

The reader may wonder why we have chosen to define the covariance and the covariance matrix with the conjugation sign. Why not look at $\mathsf{E}\big[(\mathbf{Z} - \mathsf{E}[\mathbf{Z}])(\mathbf{Z} - \mathsf{E}[\mathbf{Z}])^\top\big]$? The reason is that (17.36) is simply much more useful in applications. For example, for any deterministic $\alpha_1, \ldots, \alpha_n \in \mathbb{C}$ the variance of $\sum_{j=1}^{n} \alpha_j Z_j$ can be computed from $\mathsf{K}_{\mathbf{ZZ}}$ (using (17.25)) but not from $\mathsf{E}\big[(\mathbf{Z} - \mathsf{E}[\mathbf{Z}])(\mathbf{Z} - \mathsf{E}[\mathbf{Z}])^\top\big]$.

### 17.4.2 Proper Complex Random Vectors

The notion of proper random variables extends to vectors:

**Definition 17.4.1 (Proper Complex Random Vector).** *A complex random vector* $\mathbf{Z}$ *is said to be* **proper** *if the following three conditions are all met: it is of zero mean; it is of finite variance; and*

$$\mathsf{E}\big[\mathbf{Z}\mathbf{Z}^\top\big] = 0. \qquad (17.37)$$

An alternative definition can be given based on linear functionals:

**Proposition 17.4.2.** *The complex random $n$-vector* $\mathbf{Z}$ *is proper if, and only if, for every deterministic vector* $\boldsymbol{\alpha} \in \mathbb{C}^n$ *the CRV* $\boldsymbol{\alpha}^\top \mathbf{Z}$ *is proper.*

**Proof.** We begin by noting that $\mathbf{Z}$ is of zero mean if, and only if, $\boldsymbol{\alpha}^\top \mathbf{Z}$ is of zero mean for all $\boldsymbol{\alpha} \in \mathbb{C}^n$. This can be seen from the relation

$$\mathsf{E}\big[\boldsymbol{\alpha}^\top \mathbf{Z}\big] = \boldsymbol{\alpha}^\top \mathsf{E}[\mathbf{Z}], \quad \boldsymbol{\alpha} \in \mathbb{C}^n. \qquad (17.38)$$

Indeed, (17.38) demonstrates that if $\mathbf{Z}$ is of zero mean then so is $\boldsymbol{\alpha}^\top \mathbf{Z}$ for every $\boldsymbol{\alpha} \in \mathbb{C}^n$. Conversely, if $\boldsymbol{\alpha}^\top \mathbf{Z}$ is of zero mean for all $\boldsymbol{\alpha} \in \mathbb{C}^n$, then, *a fortiori*, it must also be of zero mean for the choice of $\alpha = \mathsf{E}[\mathbf{Z}]^*$, which yields that $0 = \mathsf{E}[\mathbf{Z}]^\dagger \mathsf{E}[\mathbf{Z}]$ and hence that $\mathsf{E}[\mathbf{Z}]$ must be zero (because $\mathsf{E}[\mathbf{Z}]^\dagger \mathsf{E}[\mathbf{Z}]$ is the sum of the squared magnitudes of the components of $\mathsf{E}[\mathbf{Z}]$).

We next note that $\mathbf{Z}$ is of finite variance if, and only if, $\alpha^\mathsf{T}\mathbf{Z}$ is of finite variance for every $\alpha \in \mathbb{C}^n$. The proof is not difficult and is omitted.

We thus continue with the proof under the assumption that $\mathbf{Z}$ is of zero mean and of finite variance. We note that for any deterministic complex vector $\alpha \in \mathbb{C}^n$

$$\begin{aligned}
\mathsf{E}\big[(\alpha^\mathsf{T}\mathbf{Z})^2\big] &= \mathsf{E}\big[(\alpha^\mathsf{T}\mathbf{Z})(\alpha^\mathsf{T}\mathbf{Z})\big] \\
&= \mathsf{E}\big[(\alpha^\mathsf{T}\mathbf{Z})(\alpha^\mathsf{T}\mathbf{Z})^\mathsf{T}\big] \\
&= \mathsf{E}\big[\alpha^\mathsf{T}\mathbf{Z}\mathbf{Z}^\mathsf{T}\alpha\big] \\
&= \alpha^\mathsf{T}\mathsf{E}\big[\mathbf{Z}\mathbf{Z}^\mathsf{T}\big]\alpha, \quad \alpha \in \mathbb{C}^n,
\end{aligned} \tag{17.39}$$

where the first equality follows by writing the square of a random variable as the product of the variable by itself; the second because the transpose of a scalar is the original scalar; the third by the transpose rule

$$(\mathsf{AB})^\mathsf{T} = \mathsf{B}^\mathsf{T}\mathsf{A}^\mathsf{T}, \tag{17.40}$$

and the final equality because $\alpha$ is deterministic.

From (17.39) it follows that if $\mathbf{Z}$ is proper, then so is $\alpha^\mathsf{T}\mathbf{Z}$ for all $\alpha \in \mathbb{C}^n$. Actually, (17.39) also proves the reverse implication by substituting $\mathsf{A} = \mathsf{E}\big[\mathbf{Z}\mathbf{Z}^\mathsf{T}\big]$ in the following fact from Matrix Theory:

$$\big(\alpha^\mathsf{T}\mathsf{A}\alpha = 0, \ \alpha \in \mathbb{C}^n\big) \Rightarrow \big(\mathsf{A} = 0\big), \quad \mathsf{A} \text{ symmetric}. \tag{17.41}$$

To prove this fact from Matrix Theory assume that $\mathsf{A}$ is symmetric, i.e., that

$$a^{(j,\ell)} = a^{(\ell,j)}, \quad j,\ell \in \{1,\ldots,n\}. \tag{17.42}$$

Let $\alpha = \mathbf{e}_\ell$ where $\mathbf{e}_\ell$ is all-zero except for its $\ell$-th component, which is one. The equality $\mathbf{e}_\ell^\mathsf{T}\mathsf{A}\mathbf{e}_\ell = 0$ for every $\ell \in \{1,\ldots,n\}$ is equivalent to

$$a^{(\ell,\ell)} = 0, \quad \ell \in \{1,\ldots,n\}. \tag{17.43}$$

Next choose $\alpha = \mathbf{e}_j + \mathbf{e}_\ell$. The equality

$$(\mathbf{e}_j + \mathbf{e}_\ell)^\mathsf{T}\mathsf{A}(\mathbf{e}_j + \mathbf{e}_\ell) = 0$$

for every $j,\ell \in \{1,\ldots,n\}$ is then equivalent to

$$a^{(j,\ell)} + a^{(j,j)} + a^{(\ell,j)} + a^{(\ell,\ell)} = 0, \quad j,\ell \in \{1,\ldots,n\}. \tag{17.44}$$

Equations (17.42), (17.43), and (17.44) guarantee that the matrix $\mathsf{A}$ is all-zero. $\square$

An important observation regarding complex random vectors is that a linearly-transformed proper vector is also proper:

**Proposition 17.4.3 (Linear Transformation of a Proper Random Vector).** *If the complex random n-vector* $\mathbf{Z}$ *is proper, then so is the complex random m-vector* $\mathsf{A}\mathbf{Z}$ *for every deterministic* $m \times n$ *complex matrix* $\mathsf{A}$.

**Proof.** We leave it to the reader to verify that the hypothesis that $\mathbf{Z}$ is proper implies that $\mathsf{A}\mathbf{Z}$ must be of zero mean and of finite variance. To show that $\mathsf{A}\mathbf{Z}$ is proper, it thus remains to show that $\mathsf{E}\big[(\mathsf{A}\mathbf{Z})(\mathsf{A}\mathbf{Z})^{\mathsf{T}}\big] = 0$. This we do by direct calculation:

$$
\begin{aligned}
\mathsf{E}\big[(\mathsf{A}\mathbf{Z})(\mathsf{A}\mathbf{Z})^{\mathsf{T}}\big] &= \mathsf{E}\big[\mathsf{A}\mathbf{Z}\mathbf{Z}^{\mathsf{T}}\mathsf{A}^{\mathsf{T}}\big] \\
&= \mathsf{A}\mathsf{E}\big[\mathbf{Z}\mathbf{Z}^{\mathsf{T}}\big]\mathsf{A}^{\mathsf{T}} \\
&= 0,
\end{aligned}
$$

where the first equality follows from the rule for the transpose of a product, namely, $(\mathsf{A}\mathsf{B})^{\mathsf{T}} = \mathsf{B}^{\mathsf{T}}\mathsf{A}^{\mathsf{T}}$; the second because $\mathsf{A}$ is deterministic; and the last from the hypothesis that $\mathbf{Z}$ is proper, so $\mathsf{E}\big[\mathbf{Z}\mathbf{Z}^{\mathsf{T}}\big] = 0$. $\qquad\square$

### 17.4.3   The Characteristic Function

The definition we gave in Section 17.3.6 for the characteristic function of a CRV extends naturally to vectors: the characteristic function $\Phi_{\mathbf{Z}}\colon \mathbb{C}^n \to \mathbb{C}$ of a complex random n-vector $\mathbf{Z}$ is defined as

$$
\Phi_{\mathbf{Z}}(\boldsymbol{\varpi}) \triangleq \mathsf{E}\Big[e^{\mathrm{i}\,\mathrm{Re}(\boldsymbol{\varpi}^{\dagger}\mathbf{Z})}\Big], \quad \boldsymbol{\varpi} \in \mathbb{C}^n.
$$

Invoking the analogous result for tuples of real random variables we have:

**Theorem 17.4.4.** *The complex random vectors* $\mathbf{Z}$ *and* $\mathbf{W}$ *are of equal law if, and only if, their characteristic functions are identical:*

$$
\Big(\mathbf{Z} \stackrel{\mathscr{L}}{=} \mathbf{W}\Big) \Leftrightarrow \Big(\Phi_{\mathbf{Z}}(\boldsymbol{\varpi}) = \Phi_{\mathbf{W}}(\boldsymbol{\varpi}),\ \boldsymbol{\varpi} \in \mathbb{C}^n\Big). \tag{17.45}
$$

**Corollary 17.4.5.** *The complex random n-vectors* $\mathbf{Z}$ *and* $\mathbf{W}$ *are of equal law if, and only if, for every deterministic vector* $\boldsymbol{\alpha} \in \mathbb{C}^n$ *the complex random variables* $\boldsymbol{\alpha}^{\mathsf{T}}\mathbf{Z}$ *and* $\boldsymbol{\alpha}^{\mathsf{T}}\mathbf{W}$ *are of equal law:*

$$
\Big(\mathbf{Z} \stackrel{\mathscr{L}}{=} \mathbf{W}\Big) \Leftrightarrow \Big(\boldsymbol{\alpha}^{\mathsf{T}}\mathbf{Z} \stackrel{\mathscr{L}}{=} \boldsymbol{\alpha}^{\mathsf{T}}\mathbf{W},\quad \boldsymbol{\alpha} \in \mathbb{C}^n\Big). \tag{17.46}
$$

**Proof.** The direction that needs proof is that equality in law of all linear combinations implies equality in law between the vectors. But this readily follows from the theorem because equality in law of the linear combinations implies that the law of $\boldsymbol{\varpi}^{\dagger}\mathbf{Z}$ is equal to the law of $\boldsymbol{\varpi}^{\dagger}\mathbf{W}$ for every $\boldsymbol{\varpi} \in \mathbb{C}^n$. This in turn implies $e^{\mathrm{i}\,\mathrm{Re}(\boldsymbol{\varpi}^{\dagger}\mathbf{Z})} \stackrel{\mathscr{L}}{=} e^{\mathrm{i}\,\mathrm{Re}(\boldsymbol{\varpi}^{\dagger}\mathbf{W})}$, from which, upon taking expectations, we obtain that $\mathbf{Z}$ and $\mathbf{W}$ have identical characteristic functions. Thus, by the theorem, they are equal in law. $\qquad\square$

### 17.4.4   Transforming Complex Random Vectors

The change of density rule (17.34) can be generalized to analytic multi-variable mappings (Exercise 17.6). But here we shall only present a version of this result for linear mappings:

**Lemma 17.4.6 (Linearly Transforming Complex Random Vectors).** *Let the complex random $n$-vector $\mathbf{W}$ be given by*

$$\mathbf{W} = \mathsf{A}\mathbf{Z},$$

*where $\mathsf{A}$ is a nonsingular deterministic complex $n \times n$ matrix, and where the complex random $n$-vector $\mathbf{Z}$ has the density $f_{\mathbf{Z}}(\cdot)$. Then $\mathbf{W}$ is of density*

$$f_{\mathbf{W}}(\mathbf{w}) = \frac{1}{|\det \mathsf{A}|^2} f_{\mathbf{Z}}(\mathsf{A}^{-1}\mathbf{w}), \quad \mathbf{w} \in \mathbb{C}^n. \tag{17.47}$$

**Proof.** The proof is based on viewing the complex $n \times n$ linear transformation from $\mathbf{Z}$ to $\mathbf{W}$ as a $2n \times 2n$ real transformation, and on then applying Theorem 17.3.4.

Stack the real parts of the components of $\mathbf{Z}$ on top of the imaginary parts in a real random $2n$-vector $\mathbf{S}$:

$$\mathbf{S} = \left( \mathrm{Re}\big(Z^{(1)}\big), \ldots, \mathrm{Re}\big(Z^{(n)}\big), \mathrm{Im}\big(Z^{(1)}\big), \ldots, \mathrm{Im}\big(Z^{(n)}\big) \right)^{\mathsf{T}}. \tag{17.48}$$

Similarly, stack the real parts of the components of $\mathbf{W}$ on top of the imaginary parts in a real random $2n$-vector $\mathbf{T}$:

$$\mathbf{T} = \left( \mathrm{Re}\big(W^{(1)}\big), \ldots, \mathrm{Re}\big(W^{(n)}\big), \mathrm{Im}\big(W^{(1)}\big), \ldots, \mathrm{Im}\big(W^{(n)}\big) \right)^{\mathsf{T}}.$$

We can then express $\mathbf{T}$ as the result of multiplying the random vector $\mathbf{S}$ by a $2n \times 2n$ real matrix:

$$\mathbf{T} = \begin{pmatrix} \mathrm{Re}(\mathsf{A}) & -\mathrm{Im}(\mathsf{A}) \\ \mathrm{Im}(\mathsf{A}) & \mathrm{Re}(\mathsf{A}) \end{pmatrix} \mathbf{S},$$

where $\mathrm{Re}(\mathsf{A})$ and $\mathrm{Im}(\mathsf{A})$ denote the componentwise real and imaginary parts of $\mathsf{A}$.

The result will follow from Theorem 17.3.4 once we show that the absolute value of the Jacobian determinant of this transformation is $|\det \mathsf{A}|^2$. Using elementary row and column operations we compute:

$$\begin{aligned} \det \begin{pmatrix} \mathrm{Re}(\mathsf{A}) & -\mathrm{Im}(\mathsf{A}) \\ \mathrm{Im}(\mathsf{A}) & \mathrm{Re}(\mathsf{A}) \end{pmatrix} &= \det \begin{pmatrix} \mathsf{A} & -\mathrm{Im}(\mathsf{A}) \\ -i\mathsf{A} & \mathrm{Re}(\mathsf{A}) \end{pmatrix} \\ &= \det \begin{pmatrix} \mathsf{A} & -\mathrm{Im}(\mathsf{A}) \\ 0 & \mathsf{A}^* \end{pmatrix} \\ &= (\det \mathsf{A})(\det \mathsf{A}^*) \\ &= |\det \mathsf{A}|^2, \end{aligned}$$

where the first equality follows by the elementary column operations of multiplying the right columns by $(-i)$ and adding the result to the left columns; the second from the elementary row operations of multiplying the top rows by i and adding the result to the bottom rows; the third from the identity

$$\det \begin{pmatrix} \mathsf{B} & \mathsf{C} \\ 0 & \mathsf{D} \end{pmatrix} = (\det \mathsf{B})\,(\det \mathsf{D});$$

and the last by noting that for any square matrix $\mathsf{B}$

$$\det(\mathsf{B}^*) = (\det \mathsf{B})^*. \qquad \square$$

## 17.5   Discrete-Time Complex Stochastic Processes

Definition 12.2.1 of a real stochastic process extends to the complex case as follows.

**Definition 17.5.1 (Complex Stochastic Process).** *A complex stochastic process (CSP) $\big(Z(t),\ t \in \mathcal{T}\big)$ is a collection of complex random variables that are defined on a common probability space $(\Omega, \mathcal{F}, P)$ and that are indexed by some set $\mathcal{T}$.*

A CSP $\big(Z(t),\ t \in \mathcal{T}\big)$ is said to be **centered** if for each $t \in \mathcal{T}$ the CRV $Z(t)$ is of zero mean. Similarly, the CSP is said to be of **finite variance** if for each $t \in \mathcal{T}$ the CRV $Z(t)$ is of finite variance. A **discrete-time** CSP corresponds to the case where the index set $\mathcal{T}$ is the set of integers $\mathbb{Z}$. Discrete-time complex stochastic processes are not very different from the real-valued ones we encountered in Chapter 13. Consequently, we shall present the main definitions and results succinctly with an emphasis on the issues where the complex and real processes differ. As in Chapter 13, when dealing with a discrete-time CSP we shall use subscripts to index the complex random variables and denote the process by $\big(Z_\nu,\ \nu \in \mathbb{Z}\big)$ or, more succinctly, by $\big(Z_\nu\big)$.

A discrete-time CSP $\big(Z_\nu,\ \nu \in \mathbb{Z}\big)$ is said to be **stationary**, or **strict-sense stationary**, or **strongly stationary** if for every positive integer $n$ and for every $\eta, \eta' \in \mathbb{Z}$, the joint distribution of the $n$-tuple $(Z_\eta, \ldots Z_{\eta+n-1})$ is identical to the joint distribution of the $n$-tuple $(Z_{\eta'}, \ldots, Z_{\eta'+n-1})$. This definition is essentially identical to the analogous definition for real processes (Definition 13.2.1). Similarly, Proposition 13.2.2 holds verbatim also for complex stochastic processes. Proposition 13.2.3 also holds for complex stochastic processes with the slight modification that the deterministic coefficients $\alpha_1, \ldots, \alpha_n$ are now allowed to be arbitrary *complex* numbers:

**Proposition 17.5.2.** *A discrete-time CSP $\big(Z_\nu\big)$ is stationary if, and only if, for every $n \in \mathbb{N}$, all $\eta, \nu_1, \ldots, \nu_n \in \mathbb{Z}$, and all $\alpha_1, \ldots, \alpha_n \in \mathbb{C}$,*

$$\sum_{j=1}^{n} \alpha_j Z_{\nu_j} \overset{\mathscr{L}}{=} \sum_{j=1}^{n} \alpha_j Z_{\nu_j + \eta}. \qquad (17.49)$$

The definition of a **wide-sense stationary** CSP is very similar to the analogous definition for real processes (Definition 13.3.1).

**Definition 17.5.3 (Wide-Sense Stationary Discrete-Time CSP).** *We say that a discrete-time CSP $(Z_\nu)$ is* **wide-sense stationary** *or* **weakly stationary** *or* **covariance stationary** *if the following three conditions all hold:*

1) *For every $\nu \in \mathbb{Z}$ the CRV $Z_\nu$ is of finite variance.*

2) *The mean of $Z_\nu$ does not depend on $\nu$.*

3) *The expectation $\mathsf{E}[Z_\nu Z_{\nu'}^*]$ depends on $\nu'$ and $\nu$ only via their difference $\nu - \nu'$:*

$$\mathsf{E}[Z_\nu Z_{\nu'}^*] = \mathsf{E}\left[Z_{\nu+\eta} Z_{\nu'+\eta}^*\right], \quad \nu, \nu', \eta \in \mathbb{Z}. \tag{17.50}$$

Note the conjugation in (17.50). We do not require that $\mathsf{E}[Z_{\nu'} Z_\nu]$ be computable from $\nu - \nu'$; it may or may not be. Thus, we do not require that the matrix

$$\begin{pmatrix} \mathsf{E}[\mathrm{Re}(Z_{\nu'})\,\mathrm{Re}(Z_\nu)] & \mathsf{E}[\mathrm{Re}(Z_{\nu'})\,\mathrm{Im}(Z_\nu)] \\ \mathsf{E}[\mathrm{Im}(Z_{\nu'})\,\mathrm{Re}(Z_\nu)] & \mathsf{E}[\mathrm{Im}(Z_{\nu'})\,\mathrm{Im}(Z_\nu)] \end{pmatrix}$$

be computable from $\nu - \nu'$. This matrix is, however, computable from $\nu - \nu'$ if the process is proper:

**Definition 17.5.4 (Proper CSP).** *A discrete-time CSP $(Z_\nu)$ is said to be* **proper** *if the following three conditions all hold: it is centered; it is of finite variance; and*

$$\mathsf{E}[Z_\nu Z_{\nu'}] = 0, \quad \nu, \nu' \in \mathbb{Z}. \tag{17.51}$$

Equivalently, a discrete-time CSP $(Z_\nu)$ is proper if, and only if, for every positive integer $n$ and all $\nu_1, \ldots, \nu_n \in \mathbb{Z}$ the complex random vector $(Z_{\nu_1}, \ldots, Z_{\nu_n})^\mathsf{T}$ is proper. Equivalently, $(Z_\nu)$ is proper if, and only if, for every positive integer $n$, all $\alpha_1, \ldots, \alpha_n \in \mathbb{C}$, and all $\nu_1, \ldots, \nu_n \in \mathbb{Z}$

$$\sum_{j=1}^{n} \alpha_j Z_{\nu_j} \text{ is proper} \tag{17.52}$$

(Proposition 17.4.2).

The alternative definition of WSS real processes in terms of the variance of linear functionals of the process (Proposition 13.3.3) requires little change:

**Proposition 17.5.5.** *A finite-variance discrete-time CSP $(Z_\nu)$ is WSS if, and only if, for every $n \in \mathbb{N}$, all $\eta, \nu_1, \ldots, \nu_n \in \mathbb{Z}$, and all $\alpha_1, \ldots, \alpha_n \in \mathbb{C}$*

$$\sum_{j=1}^{n} \alpha_j Z_{\nu_j} \text{ and } \sum_{j=1}^{n} \alpha_j Z_{\nu_j+\eta} \quad \text{have the same mean \& variance.} \tag{17.53}$$

**Proof.** We begin by assuming that $(Z_\nu)$ is WSS and prove (17.53). The equality of expectations follows directly from the linearity of expectation and from the fact that because $(Z_\nu)$ is WSS the mean of $Z_\nu$ does not depend on $\nu$. In proving the equality of the variances we use (17.25):

$$\text{Var}\left[\sum_{j=1}^{n} \alpha_j Z_{\nu_j+\eta}\right] = \sum_{j=1}^{n}\sum_{j'=1}^{n} \alpha_j \alpha_{j'}^* \text{Cov}\left[Z_{\nu_j+\eta}, Z_{\nu_{j'}+\eta}\right]$$

$$= \sum_{j=1}^{n}\sum_{j'=1}^{n} \alpha_j \alpha_{j'}^* \text{Cov}\left[Z_{\nu_j}, Z_{\nu_{j'}}\right]$$

$$= \text{Var}\left[\sum_{j=1}^{n} \alpha_j Z_{\nu_j}\right],$$

where the second equality follows from the wide-sense stationarity of $(Z_\nu)$ and the last equality again from (17.25).

We next turn to proving that (17.53) implies that $(Z_\nu)$ is WSS. Choosing $n = 1$ and $\alpha_1 = 1$ we obtain, by considering the equality of the means, that $\mathsf{E}[Z_\nu] = \mathsf{E}[Z_{\nu+\eta}]$ for all $\eta \in \mathbb{Z}$, i.e., that the mean of the process is constant. And, by considering the equality of the variances, we obtain that the random variables $(Z_\nu)$ all have the same variance

$$\text{Var}[Z_\nu] = \text{Var}[Z_{\nu+\eta}], \quad \nu, \eta \in \mathbb{Z}. \tag{17.54}$$

Choosing $n = 2$ and $\alpha_1 = \alpha_2 = 1$ we obtain from the equality of the variances

$$\text{Var}[Z_{\nu_1} + Z_{\nu_2}] = \text{Var}[Z_{\nu_1+\eta} + Z_{\nu_2+\eta}]. \tag{17.55}$$

But, by (17.25) and (17.54),

$$\text{Var}[Z_{\nu_1} + Z_{\nu_2}] = 2\text{Var}[Z_1] + 2\,\text{Re}\big(\text{Cov}[Z_{\nu_1}, Z_{\nu_2}]\big) \tag{17.56}$$

and similarly

$$\text{Var}[Z_{\nu_1+\eta} + Z_{\nu_2+\eta}] = 2\text{Var}[Z_1] + 2\,\text{Re}\big(\text{Cov}[Z_{\nu_1+\eta}, Z_{\nu_2+\eta}]\big). \tag{17.57}$$

By (17.55), (17.56), and (17.57)

$$\text{Re}\big(\text{Cov}[Z_{\nu_1+\eta}, Z_{\nu_2+\eta}]\big) = \text{Re}\big(\text{Cov}[Z_{\nu_1}, Z_{\nu_2}]\big), \quad \eta, \nu_1, \nu_2 \in \mathbb{Z}. \tag{17.58}$$

We now repeat the argument with $\alpha_1 = 1$ and $\alpha_2 = \mathsf{i}$:

$$\text{Var}[Z_{\nu_1} + \mathsf{i}\,Z_{\nu_2}] = \text{Var}[Z_{\nu_1}] + \text{Var}[Z_{\nu_2}] + 2\,\text{Re}\big(\text{Cov}[Z_{\nu_1}, \mathsf{i}\,Z_{\nu_2}]\big)$$

$$= 2\text{Var}[Z_1] + 2\,\text{Im}\big(\text{Cov}[Z_{\nu_1}, Z_{\nu_2}]\big)$$

and similarly

$$\text{Var}[Z_{\nu_1+\eta} + \mathsf{i}\,Z_{\nu_2+\eta}] = 2\text{Var}[Z_1] + 2\,\text{Im}\big(\text{Cov}[Z_{\nu_1+\eta}, Z_{\nu_2+\eta}]\big),$$

so the equality of the variances implies

$$\text{Im}\big(\text{Cov}[Z_{\nu_1+\eta}, Z_{\nu_2+\eta}]\big) = \text{Im}\big(\text{Cov}[Z_{\nu_1}, Z_{\nu_2}]\big), \quad \eta, \nu_1, \nu_2 \in \mathbb{Z},$$

which combines with (17.58) to prove $\text{Cov}[Z_{\nu_1+\eta}, Z_{\nu_2+\eta}] = \text{Cov}[Z_{\nu_1}, Z_{\nu_2}]$. $\qquad \square$

As with real processes, a comparison of Propositions 17.5.5 and 17.5.2 yields that *any finite-variance stationary CSP is also WSS.* The reverse is not true.

**Definition 17.5.6 (Autocovariance Function).** *We define the autocovariance function* $K_{ZZ} \colon \mathbb{Z} \to \mathbb{C}$ *of a discrete-time WSS CSP* $(Z_\nu)$ *as*[4]

$$K_{ZZ}(\eta) \triangleq \mathsf{Cov}[Z_{\nu+\eta}, Z_\nu] \tag{17.59}$$
$$= \mathsf{E}\Big[\big(Z_{\nu+\eta} - \mathsf{E}[Z_1]\big)\big(Z_\nu - \mathsf{E}[Z_1]\big)^*\Big], \quad \eta \in \mathbb{Z}.$$

By mimicking the derivations of (13.12) (taking into account the conjugate symmetry (17.18)) we obtain that the autocovariance function $K_{ZZ}$ of every discrete-time WSS CSP $(Z_\nu)$ satisfies the conjugate-symmetry condition

$$K_{ZZ}(-\eta) = K_{ZZ}^*(\eta), \quad \eta \in \mathbb{Z}. \tag{17.60}$$

Similarly, by mimicking the derivation of (13.13) (i.e., from the nonnegativity of the variance and from (17.25)), we obtain that the autocovariance function of such a process satisfies

$$\sum_{\nu=1}^{n} \sum_{\nu'=1}^{n} \alpha_\nu \alpha_{\nu'}^* \, K_{ZZ}(\nu - \nu') \geq 0, \quad \alpha_1, \ldots, \alpha_n \in \mathbb{C}. \tag{17.61}$$

In analogy to the real case, (17.60) and (17.61) fully characterize the possible autocovariance functions in the sense that any function $K \colon \mathbb{Z} \to \mathbb{C}$ satisfying

$$K(-\eta) = K^*(\eta), \quad \eta \in \mathbb{Z} \tag{17.62}$$

and

$$\sum_{\nu=1}^{n} \sum_{\nu'=1}^{n} \alpha_\nu \alpha_{\nu'}^* K(\nu - \nu') \geq 0, \quad \alpha_1, \ldots, \alpha_n \in \mathbb{C} \tag{17.63}$$

is the autocovariance function of some discrete-time WSS CSP.[5] If $K \colon \mathbb{Z} \to \mathbb{C}$ satisfies (17.62) and (17.63), then we say that $K(\cdot)$ is a positive definite function from the integers to the complex field.

Definition 13.16 of the **power spectral density** $S_{ZZ}$ requires no change. We require that $S_{ZZ}$ be integrable on the interval $[-1/2, 1/2]$ and that

$$K_{ZZ}(\eta) = \int_{-1/2}^{1/2} S_{ZZ}(\theta) \, e^{-\mathrm{i}2\pi\eta\theta} \, \mathrm{d}\theta, \quad \eta \in \mathbb{Z}. \tag{17.64}$$

Proposition 13.6.3 does require some alteration. Indeed, for complex stochastic processes the PSD need not be a symmetric function. However, the main result (that the PSD is real and nonnegative) remains true:

---

[4]Some authors, e.g., (Grimmett and Stirzaker, 2001), define $K_{ZZ}(m)$ as $\mathsf{Cov}[Z_\nu, Z_{\nu+m}]$. Our definition follows (Doob, 1990).

[5]In fact, it is the autocovariance function of some *proper Gaussian* stochastic process. Complex Gaussian random processes will be discussed in Chapter 24.

**Proposition 17.5.7 (PSDs of Complex Processes Are Nonnegative).**

(i) *If the discrete-time WSS CSP $(Z_\nu)$ is of PSD $\mathsf{S}_{ZZ}$, then*

$$\mathsf{S}_{ZZ}(\theta) \geq 0, \tag{17.65}$$

*except possibly on a subset of the interval $[-1/2, 1/2)$ of Lebesgue measure zero.*

(ii) *If a function $\mathsf{S}\colon [-1/2, 1/2) \to \mathbb{R}$ is integrable and nonnegative, then there exists a proper discrete-time WSS CSP[6] $(Z_\nu)$ whose PSD $\mathsf{S}_{ZZ}$ is given by*

$$\mathsf{S}_{ZZ}(\theta) = \mathsf{S}(\theta), \quad \theta \in [-1/2, 1/2).$$

As in the real case, by possibly changing the value of $\mathsf{S}_{ZZ}$ on the set of Lebesgue measure zero where (17.65) is violated, we can obtain a power spectral density that is nonnegative for *all* $\theta \in [-1/2, 1/2)$. Consequently, we shall always assume that the PSD, if it exists, is nonnegative for all $\theta \in [-1/2, 1/2)$.

**Proof.** We begin with Part (i) where we need to prove the nonnegativity of the PSD. We shall only sketch the proof. We recommend reading the appendix through Theorem A.2.2 before reading this proof.

Let $\mathsf{K}_{ZZ}$ denote the autocovariance function of the WSS CSP $(Z_\nu)$. Applying (17.61) with

$$\alpha_\nu = e^{i2\pi\nu\theta}, \quad \nu \in \{1, \ldots, n\}$$

and thus

$$\alpha_\nu \alpha_{\nu'}^* = e^{i2\pi(\nu-\nu')\theta}, \quad \nu, \nu' \in \{1, \ldots, n\},$$

we obtain

$$0 \leq \sum_{\nu=1}^{n} \sum_{\nu'=1}^{n} \alpha_\nu \alpha_{\nu'}^* \, \mathsf{K}_{ZZ}(\nu - \nu')$$

$$= \sum_{\nu=1}^{n} \sum_{\nu'=1}^{n} e^{i2\pi(\nu-\nu')\theta} \, \mathsf{K}_{ZZ}(\nu - \nu')$$

$$= \sum_{\eta=-(n-1)}^{n-1} \left(n - |\eta|\right) e^{i2\pi\eta\theta} \, \mathsf{K}_{ZZ}(\eta), \quad \theta \in [-1/2, 1/2).$$

Dividing by $n$ we obtain

$$0 \leq \sum_{\eta=-(n-1)}^{n-1} \left(1 - \frac{|\eta|}{n}\right) e^{i2\pi\eta\theta} \, \mathsf{K}_{ZZ}(\eta)$$

$$= \sum_{\eta=-(n-1)}^{n-1} \left(1 - \frac{|\eta|}{n}\right) e^{i2\pi\eta\theta} \, \hat{\mathsf{S}}_{ZZ}(\eta)$$

$$= \left(\mathbf{k}_{n-1} \star \mathsf{S}_{ZZ}\right)(\theta), \quad \theta \in [-1/2, 1/2),$$

---

[6]The process can be taken to be Gaussian; see Chapter 24.

where in the equality on the second line $\hat{S}_{ZZ}(\eta)$ denotes the $\eta$-th Fourier Series Coefficient of $S_{ZZ}$ and we use (17.64); and in the subsequent equality on the third line $\mathbf{k}_n$ denotes the degree-$n$ Fejér kernel (Definition A.1.3).

We have thus established that $\mathbf{k}_{n-1} \star S_{ZZ}$ is nonnegative. The result now follows from Theorem A.2.2 which guarantees that

$$\lim_{n \to \infty} \int_{-1/2}^{1/2} \left| S_{ZZ}(\theta) - \left(\mathbf{k}_n \star S_{ZZ}\right)(\theta) \right| \, d\theta = 0.$$

The proof of Part (ii) is very similar to the proof of the analogous result for real processes. As in (13.21), we define

$$K(\eta) \triangleq \int_{-1/2}^{1/2} S(\theta) \, e^{-i2\pi\eta\theta} \, d\theta, \quad \eta \in \mathbb{Z}, \tag{17.66}$$

and we prove that this function satisfies (17.62) and (17.63). To prove (17.62) we compute

$$K(-\eta) = \int_{-1/2}^{1/2} S(\theta) \, e^{-i2\pi(-\eta)\theta} \, d\theta$$

$$= \int_{-1/2}^{1/2} S^*(\theta) \, e^{i2\pi\eta\theta} \, d\theta$$

$$= \left( \int_{-1/2}^{1/2} S(\theta) \, e^{-i2\pi\eta\theta} \, d\theta \right)^*$$

$$= K^*(\eta), \quad \eta \in \mathbb{Z},$$

where the first equality follows from the definition of $K(\cdot)$ (17.66); the second because $S(\cdot)$ is, by assumption, real; the third because conjugating the integrand is equivalent to conjugating the integral; and the final equality again by (17.66).

To prove (17.63) we mimic the derivation of (13.22) with the constants $\alpha_1, \ldots, \alpha_n$ now being complex:

$$\sum_{\nu=1}^{n} \sum_{\nu'=1}^{n} \alpha_\nu \alpha_{\nu'}^* K(\nu - \nu') = \sum_{\nu=1}^{n} \sum_{\nu'=1}^{n} \alpha_\nu \alpha_{\nu'}^* \int_{-1/2}^{1/2} S(\theta) \, e^{-i2\pi(\nu-\nu')\theta} \, d\theta$$

$$= \int_{-1/2}^{1/2} S(\theta) \left( \sum_{\nu=1}^{n} \sum_{\nu'=1}^{n} \alpha_\nu \alpha_{\nu'}^* \, e^{-i2\pi(\nu-\nu')\theta} \right) d\theta$$

$$= \int_{-1/2}^{1/2} S(\theta) \left( \sum_{\nu=1}^{n} \sum_{\nu'=1}^{n} \alpha_\nu \, e^{-i2\pi\nu\theta} \, \alpha_{\nu'}^* \, e^{i2\pi\nu'\theta} \right) d\theta$$

$$= \int_{-1/2}^{1/2} S(\theta) \left( \sum_{\nu=1}^{n} \alpha_\nu \, e^{-i2\pi\nu\theta} \right) \left( \sum_{\nu'=1}^{n} \alpha_{\nu'} \, e^{-i2\pi\nu'\theta} \right)^* d\theta$$

$$= \int_{-1/2}^{1/2} S(\theta) \left| \sum_{\nu=1}^{n} \alpha_\nu \, e^{-i2\pi\nu\theta} \right|^2 d\theta$$

$$\geq 0. \qquad \qquad \square$$

Proposition 13.6.6 needs very little alteration. We only need to drop the symmetry property:

**Proposition 17.5.8 (PSD when $K_{ZZ}$ Is Absolutely Summable).** *If the autocovariance function $K_{ZZ}$ of a discrete-time WSS CSP is absolutely summable, i.e.,*

$$\sum_{\eta=-\infty}^{\infty} \left| K_{ZZ}(\eta) \right| < \infty, \tag{17.67}$$

*then the function*

$$S(\theta) = \sum_{\eta=-\infty}^{\infty} K_{ZZ}(\eta) \, e^{i2\pi\eta\theta}, \quad \theta \in [-1/2, 1/2] \tag{17.68}$$

*is continuous, nonnegative, and satisfies*

$$\int_{-1/2}^{1/2} S(\theta) \, e^{-i2\pi\eta\theta} \, d\theta = K_{ZZ}(\eta), \quad \eta \in \mathbb{Z}. \tag{17.69}$$

The **Spectral Distribution Function** that we encountered in Section 13.7 has a natural extension to discrete-time WSS CSPs:

**Theorem 17.5.9.**

(i) *If $(Z_\nu)$ is a WSS CSP of autocovariance function $K_{ZZ}$, then*

$$K_{ZZ}(\eta) = K_{ZZ}(0) \, \mathsf{E}\left[e^{i2\pi\eta\Theta}\right], \quad \eta \in \mathbb{Z}, \tag{17.70}$$

*for some random variable $\Theta$ taking value in the interval $[-1/2, 1/2)$. In the nontrivial case where $K_{ZZ}(0) > 0$ the distribution function of $\Theta$ is fully specified by $K_{ZZ}$.*

(ii) *If $\Theta$ is any random variable taking value in $[-1/2, 1/2)$ and if $\alpha > 0$, then there exists a proper discrete-time WSS CSP $(Z_\nu)$ whose autocovariance function $K_{ZZ}$ is given by*

$$K_{ZZ}(\eta) = \alpha \, \mathsf{E}\left[e^{i2\pi\eta\Theta}\right], \quad \eta \in \mathbb{Z} \tag{17.71}$$

*and whose variance is consequently given by $K_{ZZ}(0) = \alpha$.*

**Proof.** See (Shiryaev, 1996, Chapter VI, Section § 1 Theorem 3), (Doob, 1990, Chapter X § 3 Theorem 3.2), or (Feller, 1971, Chapter XIX, Section 6, Theorem 3). □

Some authors refer to the mapping $\theta \mapsto \Pr[\Theta \leq \theta]$ as the spectral distribution function of $(Z_\nu)$, but others refer to $\theta \mapsto K_{ZZ}(0) \, \Pr[\Theta \leq \theta]$ as the spectral distribution function. The latter is more common.

## 17.6 On the Eigenvalues of Large Toeplitz Matrices

Although it will not be used in this book, we cannot resist stating the following classic result, which is sometimes called "Szegő's Theorem." Let the function $s \colon [-1/2, 1/2] \to [0, \infty)$ be Lebesgue integrable. Define

$$c_\eta = \int_{-1/2}^{1/2} s(\theta) \, e^{-i2\pi\eta\theta} \, d\theta, \quad \eta \in \mathbb{Z}. \tag{17.72}$$

(In some applications $s(\cdot)$ is the PSD of a discrete-time real or complex stochastic process and $c_\eta$ is the value of the corresponding autocovariance function at $\eta$.)

The $n \times n$ matrix

$$\begin{pmatrix} c_0 & c_1 & \cdots & c_{n-1} \\ c_{-1} & c_0 & \cdots & c_{n-2} \\ \vdots & \vdots & \ddots & \vdots \\ c_{-n+1} & \cdots & \cdots & c_0 \end{pmatrix}$$

is positive semidefinite and conjugate-symmetric. Consequently, is has $n$ nonnegative eigenvalues (counting multiplicity), which we denote by

$$\lambda_n^{(1)} \leq \lambda_n^{(2)} \leq \cdots \leq \lambda_n^{(n)}. \tag{17.73}$$

As $n$ increases (with $s(\cdot)$ fixed), the number of eigenvalues increases. It turns out that we can say something quite precise about the distribution of these eigenvalues.

**Theorem 17.6.1.** *Let $s \colon [-1/2, 1/2] \to [0, \infty)$ be integrable, and let $\lambda_n^{(j)}$ be as in (17.73). Let $g \colon [0, \infty) \to \mathbb{R}$ be a continuous function such that the limit $\lim_{\xi \to \infty} \frac{g(\xi)}{\xi}$ exists and is finite. Then*

$$\lim_{n \to \infty} \frac{1}{n} \sum_{j=1}^n g\big(\lambda_n^{(j)}\big) = \int_{-1/2}^{1/2} g\big(s(\theta)\big) \, d\theta. \tag{17.74}$$

**Proof.** For a proof of a more general statement of this theorem see (Simon, 2005, Chapter 2, Section 7, Theorem 2.7.13). □

## 17.7 Exercises

**Exercise 17.1 (The Distribution of $\mathrm{Re}(Z)$ and $|Z|$).** Let the CRV $Z$ be uniformly distributed over the unit disc $\{z \in \mathbb{C} : |z| \leq 1\}$.

(i) What is the density of its real part $\mathrm{Re}(Z)$?

(ii) What is the density of its magnitude $|Z|$?

**Exercise 17.2 (The Density of $Z^2$).** Let $Z$ be a CRV of density $f_Z(\cdot)$. Express the density of $Z^2$ in terms of $f_Z(\cdot)$.

**Exercise 17.3 (The Conjugate of a Proper CRV).** Must the complex conjugate of a proper CRV be proper?

**Exercise 17.4 (Product of Proper CRVs).** Show that the product of independent proper complex random variables is proper. Is the assumption of independence essential?

**Exercise 17.5 (Sums of Proper CRVs).** Show that the sum of independent proper complex random variables is proper. Is the assumption of independence essential?

**Exercise 17.6 (Transforming Complex Random Vectors).** Let $\mathbf{Z}$ be a complex $n$-vector of PDF $f_{\mathbf{Z}}(\cdot)$. Let $\mathbf{W} = g(\mathbf{Z})$, where $\mathbf{g} \colon \mathcal{D} \to \mathcal{R}$ is a one-to-one function from an open subset $\mathcal{D}$ of $\mathbb{C}^n$ to $\mathcal{R} \subseteq \mathbb{C}^n$. Let the mappings $\mathbf{u}, \mathbf{v} \colon \mathbb{R}^{2n} \to \mathbb{R}^n$ be defined for $\mathbf{x}, \mathbf{y} \in \mathbb{R}^n$ as

$$\mathbf{u} \colon (\mathbf{x}, \mathbf{y}) \mapsto \mathrm{Re}\big(g(\mathbf{x} + \mathrm{i}\mathbf{y})\big) \quad \text{and} \quad \mathbf{v} \colon (\mathbf{x}, \mathbf{y}) \mapsto \mathrm{Im}\big(g(\mathbf{x} + \mathrm{i}\mathbf{y})\big).$$

Assume that $\mathbf{g}$ is differentiable in $\mathcal{D}$ in the sense that for all $j, \ell \in \{1, \ldots, n\}$ the partial derivatives

$$\frac{\partial u^{(j)}(\mathbf{x}, \mathbf{y})}{\partial x^{(\ell)}}, \frac{\partial u^{(j)}(\mathbf{x}, \mathbf{y})}{\partial y^{(\ell)}}, \frac{\partial v^{(j)}(\mathbf{x}, \mathbf{y})}{\partial x^{(\ell)}}, \frac{\partial v^{(j)}(\mathbf{x}, \mathbf{y})}{\partial y^{(\ell)}}$$

exist and are continuous in $\mathcal{D}$, and that they satisfy

$$\frac{\partial u^{(j)}(\mathbf{x}, \mathbf{y})}{\partial x^{(\ell)}} = \frac{\partial v^{(j)}(\mathbf{x}, \mathbf{y})}{\partial y^{(\ell)}} \quad \text{and} \quad \frac{\partial u^{(j)}(\mathbf{x}, \mathbf{y})}{\partial y^{(\ell)}} = -\frac{\partial v^{(j)}(\mathbf{x}, \mathbf{y})}{\partial x^{(\ell)}},$$

where $a^{(j)}$ denotes the $j$-th component of the vector $\mathbf{a}$. Further assume that the determinant of the Jacobian matrix

$$\det g'(\mathbf{z}) = \det \begin{pmatrix} \dfrac{\partial u^{(1)}(\mathbf{x}, \mathbf{y})}{\partial x^{(1)}} + \mathrm{i}\dfrac{\partial v^{(1)}(\mathbf{x}, \mathbf{y})}{\partial x^{(1)}} & \cdots & \dfrac{\partial u^{(1)}(\mathbf{x}, \mathbf{y})}{\partial x^{(n)}} + \mathrm{i}\dfrac{\partial v^{(1)}(\mathbf{x}, \mathbf{y})}{\partial x^{(n)}} \\ \vdots & \ddots & \vdots \\ \dfrac{\partial u^{(n)}(\mathbf{x}, \mathbf{y})}{\partial x^{(1)}} + \mathrm{i}\dfrac{\partial v^{(n)}(\mathbf{x}, \mathbf{y})}{\partial x^{(1)}} & \cdots & \dfrac{\partial u^{(n)}(\mathbf{x}, \mathbf{y})}{\partial x^{(n)}} + \mathrm{i}\dfrac{\partial v^{(n)}(\mathbf{x}, \mathbf{y})}{\partial x^{(n)}} \end{pmatrix}$$

is at no point in $\mathcal{D}$ zero. Show that the density $f_{\mathbf{W}}(\cdot)$ of $\mathbf{W}$ is given by

$$f_{\mathbf{W}}(\mathbf{w}) = \frac{f_{\mathbf{Z}}(\mathbf{z})}{|\det g'(\mathbf{z})|^2}\bigg|_{\mathbf{z} = g^{-1}(\mathbf{w})} \cdot \mathrm{I}\{\mathbf{w} \in \mathcal{R}\}.$$

**Exercise 17.7 (The Cauchy-Schwarz Inequality Revisited).** Let $(Z_\ell)$ be a discrete-time WSS CSP. Show that (17.61) implies

$$\big|\mathsf{Cov}[Z_\ell, Z_{\ell'}]\big| \leq \mathsf{Var}[Z_1], \quad \ell, \ell' \in \mathbb{Z}.$$

**Exercise 17.8 (On the Autocovariance Function of a Discrete-Time CSP).** Show that if $\mathsf{K}_{ZZ}$ is the autocovariance function of a discrete-time WSS CSP, then for every $n \in \mathbb{N}$, the matrix

$$\begin{pmatrix} \mathsf{K}_{ZZ}(0) & \mathsf{K}_{ZZ}(1) & \cdots & \mathsf{K}_{ZZ}(n-1) \\ \mathsf{K}_{ZZ}(-1) & \mathsf{K}_{ZZ}(0) & \cdots & \mathsf{K}_{ZZ}(n-2) \\ \vdots & \vdots & \ddots & \vdots \\ \mathsf{K}_{ZZ}(-n+1) & \mathsf{K}_{ZZ}(-n+2) & \cdots & \mathsf{K}_{ZZ}(0) \end{pmatrix}$$

is positive semidefinite.

**Exercise 17.9 (Reversing the Direction of Time).** Let $\mathsf{K}_{ZZ}$ be the autocovariance function of some discrete-time WSS CSP $(Z_\nu)$. For every $\nu \in \mathbb{Z}$ define $Y_\nu = Z_{-\nu}$. Show that the time-reversed CSP $(Y_\nu)$ is also a WSS CSP, and express its autocovariance function $\mathsf{K}_{YY}$ in terms of $\mathsf{K}_{ZZ}$.

**Exercise 17.10 (The Sum of Autocovariance Functions).** Show that the sum of the autocovariance functions of two discrete-time WSS complex stochastic processes is the autocovariance function of some discrete-time WSS CSP.

**Exercise 17.11 (The Real Part of an Autocovariance Function).** Let $\mathsf{K}_{ZZ}$ be the autocovariance function of some discrete-time WSS CSP $(Z_\nu)$. Show that the mapping $m \mapsto \mathrm{Re}\big(\mathsf{K}_{ZZ}(m)\big)$ is the autocovariance function of some real SP. Is this also true for the mapping $m \mapsto \mathrm{Im}\big(\mathsf{K}_{ZZ}(m)\big)$?

**Exercise 17.12 (Rotating a WSS CSP).** Let $(Z_\ell)$ be a zero-mean WSS discrete-time CSP, and let $\alpha \in \mathbb{C}$ be fixed. Define the new CSP $(W_\ell)$ as $W_\ell = \alpha^\ell Z_\ell$ for every $\ell \in \mathbb{Z}$.

  (i) Show that if $|\alpha| = 1$ then $(W_\ell)$ is WSS. Compute its autocovariance function.

  (ii) Does your answer change if $\alpha$ is not of unit magnitude?

# Chapter 18

# Energy, Power, and PSD in QAM

## 18.1  Introduction

The calculations of the power and of the operational power spectral density in QAM are not just repetitions of the analogous PAM calculations with complex notation. They contain two new elements that we shall try to highlight. The first is the relationship between the power (as opposed to energy) in passband and baseband, and the second is the fact that the energy and power in transmitting the complex symbols $\{C_\ell\}$ are only related to expectations of the form $\mathsf{E}[C_\ell C_{\ell'}^*]$; they are uninfluenced by those of the form $\mathsf{E}[C_\ell C_{\ell'}]$.

The signal $\big(X(t),\ t \in \mathbb{R}\big)$ (or $\mathbf{X}$ for short) that we consider is given by

$$X(t) = 2\operatorname{Re}\big(X_{\mathrm{BB}}(t)\, e^{\mathrm{i}2\pi f_{\mathrm{c}} t}\big), \quad t \in \mathbb{R}, \tag{18.1}$$

where

$$X_{\mathrm{BB}}(t) = \mathsf{A} \sum_\ell C_\ell\, g(t - \ell\mathsf{T_s}), \quad t \in \mathbb{R}. \tag{18.2}$$

Here $\mathsf{A} > 0$ is real; the symbols $\{C_\ell\}$ are complex random variables; the pulse shape $\mathbf{g}$ is an integrable complex function that is bandlimited to $W/2$ Hz; $\mathsf{T_s}$ is positive; and $f_{\mathrm{c}} > W/2$. The range of the summation will depend on the modes we discuss.

Our focus in this chapter is on $\mathbf{X}$'s energy, power, and operational PSD. These quantities are studied in Sections 18.2–18.4, albeit without all the fine mathematical details. Those are provided in Sections 18.5 & 18.6, which are recommended for the more mathematical readers. The definition of the operational PSD of complex stochastic processes is very similar to the one of real stochastic processes (Definition 15.3.1). It is given in Section 18.4 (Definition 18.4.1).

## 18.2  The Energy in QAM

As in our treatment in Chapter 14 of PAM, we begin with an analysis of the energy in transmitting K IID random bits $D_1, \ldots, D_\mathsf{K}$. We assume that the data bits are mapped to $N$ complex symbols $C_1, \ldots, C_N$ using a $(K, N)$ binary-to-complex block-encoder

$$\mathbf{enc} \colon \{0, 1\}^K \to \mathbb{C}^N \tag{18.3}$$

of rate

$$\frac{K}{N} \quad \left[\frac{\text{bit}}{\text{complex symbol}}\right].$$

The transmitted signal is then:

$$X(t) = 2 \operatorname{Re}\left(X_{\mathrm{BB}}(t)\, e^{\mathrm{i} 2\pi f_c t}\right) \tag{18.4}$$

$$= 2 \operatorname{Re}\left(A \sum_{\ell=1}^{N} C_\ell\, g(t - \ell \mathsf{T}_s)\, e^{\mathrm{i} 2\pi f_c t}\right), \quad t \in \mathbb{R}, \tag{18.5}$$

where the baseband representation of the transmitted signal is

$$X_{\mathrm{BB}}(t) = A \sum_{\ell=1}^{N} C_\ell\, g(t - \ell \mathsf{T}_s), \quad t \in \mathbb{R}. \tag{18.6}$$

Our interest is in the energy $\mathsf{E}$ in $\mathbf{X}$, which is defined by

$$\mathsf{E} \triangleq \mathsf{E}\left[\int_{-\infty}^{\infty} X^2(t)\, \mathrm{d}t\right]. \tag{18.7}$$

Our assumption that the pulse shape $\mathbf{g}$ is bandlimited to $W/2$ Hz implies that for every realization of the symbols $\{C_\ell\}$, the signal $X_{\mathrm{BB}}(\cdot)$ is also bandlimited to $W/2$ Hz. And since we assume that $f_c > W/2$, it follows from Theorem 7.6.10 that the energy in the passband signal $X(\cdot)$ is twice the energy in its baseband representation $X_{\mathrm{BB}}(\cdot)$, i.e.,

$$\mathsf{E} = 2\mathsf{E}\left[\int_{-\infty}^{\infty} \left|X_{\mathrm{BB}}(t)\right|^2 \mathrm{d}t\right]. \tag{18.8}$$

We can thus compute the energy in $X(\cdot)$ by computing the energy in $X_{\mathrm{BB}}(\cdot)$ and doubling the result. The energy of the baseband signal can be computed in much the same way that the energy was computed in Section 14.2 for PAM. The only difference is that the baseband signal is now complex:

$$\mathsf{E}\left[\int_{-\infty}^{\infty} \left|X_{\mathrm{BB}}(t)\right|^2 \mathrm{d}t\right]$$

$$= \int_{-\infty}^{\infty} \mathsf{E}\left[\left|A \sum_{\ell=1}^{N} C_\ell\, g(t - \ell \mathsf{T}_s)\right|^2\right] \mathrm{d}t$$

$$= \int_{-\infty}^{\infty} \mathsf{E}\left[\left(A \sum_{\ell=1}^{N} C_\ell\, g(t - \ell \mathsf{T}_s)\right)\left(A \sum_{\ell'=1}^{N} C_{\ell'}\, g(t - \ell' \mathsf{T}_s)\right)^*\right] \mathrm{d}t$$

$$= A^2 \sum_{\ell=1}^{N} \sum_{\ell'=1}^{N} \mathsf{E}[C_\ell C_{\ell'}^*] \int_{-\infty}^{\infty} g(t - \ell \mathsf{T}_s)\, g^*(t - \ell' \mathsf{T}_s)\, \mathrm{d}t$$

$$= A^2 \sum_{\ell=1}^{N} \sum_{\ell'=1}^{N} \mathsf{E}[C_\ell C_{\ell'}^*]\, \mathsf{R}_{\mathbf{gg}}\big((\ell' - \ell)\mathsf{T}_s\big), \tag{18.9}$$

where $\mathsf{R_{gg}}$ is the self-similarity function of the pulse shape $\mathbf{g}$ (Definition 11.2.1), i.e.,

$$\mathsf{R_{gg}}(\tau) = \int_{-\infty}^{\infty} g(t+\tau)\, g^*(t)\, \mathrm{d}t, \quad \tau \in \mathbb{R}. \tag{18.10}$$

This expression for the energy in $X_{\mathrm{BB}}(\cdot)$ is greatly simplified if the symbols $\{C_\ell\}$ are of zero mean and uncorrelated:

$$\mathsf{E}\left[\int_{-\infty}^{\infty}\left|X_{\mathrm{BB}}(t)\right|^2 \mathrm{d}t\right] = \mathsf{A}^2\, \|\mathbf{g}\|_2^2 \sum_{\ell=1}^{\mathsf{N}} \mathsf{E}\left[|C_\ell|^2\right],$$

$$\left(\mathsf{E}[C_\ell C_{\ell'}^*] = \mathsf{E}\left[|C_\ell|^2\right] \mathrm{I}\{\ell = \ell'\}, \quad \ell, \ell' \in \{1,\ldots,\mathsf{N}\}\right), \tag{18.11}$$

or if the time shifts of the pulse shape by integer multiples of $\mathsf{T_s}$ are orthonormal

$$\mathsf{E}\left[\int_{-\infty}^{\infty}\left|X_{\mathrm{BB}}(t)\right|^2 \mathrm{d}t\right] = \mathsf{A}^2 \sum_{\ell=1}^{\mathsf{N}} \mathsf{E}\left[|C_\ell|^2\right],$$

$$\left(\int_{-\infty}^{\infty} g(t - \ell\mathsf{T_s})g^*(t - \ell'\mathsf{T_s})\, \mathrm{d}t = \mathrm{I}\{\ell = \ell'\}, \quad \ell, \ell' \in \{1,\ldots,\mathsf{N}\}\right). \tag{18.12}$$

Since $\mathbf{g}$ is an integrable function that is bandlimited to $\mathsf{W}/2$ Hz, it is also energy-limited (Note 6.4.12). Consequently, by Proposition 11.2.2 (iv), we can express the self-similarity function $\mathsf{R_{gg}}$ in (18.9) as the Inverse Fourier Transform of the mapping $f \mapsto |\hat{g}(f)|^2$:

$$\mathsf{R_{gg}}(\tau) = \int_{-\infty}^{\infty} |\hat{g}(f)|^2\, e^{\mathrm{i}2\pi f\tau}\, \mathrm{d}f, \quad \tau \in \mathbb{R}. \tag{18.13}$$

With this representation of $\mathsf{R_{gg}}$ we obtain from (18.9) an equivalent representation of the energy as

$$\mathsf{E}\left[\int_{-\infty}^{\infty}\left|X_{\mathrm{BB}}(t)\right|^2 \mathrm{d}t\right] = \mathsf{A}^2 \int_{-\infty}^{\infty} \sum_{\ell=1}^{\mathsf{N}} \sum_{\ell'=1}^{\mathsf{N}} \mathsf{E}[C_\ell C_{\ell'}^*]\, e^{\mathrm{i}2\pi f(\ell'-\ell)\mathsf{T_s}}\, |\hat{g}(f)|^2\, \mathrm{d}f. \tag{18.14}$$

Using (18.8), (18.9), and (18.14) we obtain:

**Theorem 18.2.1 (Energy in QAM).** *Assume that* $\mathsf{A} \geq 0$, *that* $\mathsf{T_s} > 0$, *that* $\mathbf{g}\colon \mathbb{R} \to \mathbb{C}$ *is an integrable signal that is bandlimited to* $\mathsf{W}/2$ Hz, *and that* $f_{\mathrm{c}} > \mathsf{W}/2$. *Then the energy* $\mathsf{E}$ *in the QAM signal* $X(\cdot)$ *of* (18.5) *is given by*

$$\mathsf{E} = 2\mathsf{A}^2 \sum_{\ell=1}^{\mathsf{N}} \sum_{\ell'=1}^{\mathsf{N}} \mathsf{E}[C_\ell C_{\ell'}^*]\, \mathsf{R_{gg}}\big((\ell'-\ell)\mathsf{T_s}\big) \tag{18.15}$$

$$= 2\mathsf{A}^2 \int_{-\infty}^{\infty} \sum_{\ell=1}^{\mathsf{N}} \sum_{\ell'=1}^{\mathsf{N}} \mathsf{E}[C_\ell C_{\ell'}^*]\, e^{\mathrm{i}2\pi f(\ell'-\ell)\mathsf{T_s}}\, |\hat{g}(f)|^2\, \mathrm{d}f, \tag{18.16}$$

*whenever all the complex random variables* $C_1, \ldots, C_{\mathsf{N}}$ *are of finite variance*

$$\mathsf{E}\left[|C_\ell|^2\right] < \infty, \quad \ell = 1, \ldots, \mathsf{N}. \tag{18.17}$$

In analogy to PAM, we define the **energy per bit** $E_b$ by

$$E_b \triangleq \frac{E}{K} \tag{18.18}$$

and the **energy per complex symbol** $E_s$ by

$$E_s \triangleq \frac{E}{N}. \tag{18.19}$$

Using Theorem 18.2.1, we obtain

$$E_s = \frac{2}{N} A^2 \sum_{\ell=1}^{N} \sum_{\ell'=1}^{N} E[C_\ell C_{\ell'}^*] \, R_{gg}\big((\ell' - \ell)T_s\big) \tag{18.20}$$

$$= \frac{2}{N} A^2 \int_{-\infty}^{\infty} \sum_{\ell=1}^{N} \sum_{\ell'=1}^{N} E[C_\ell C_{\ell'}^*] \, e^{i2\pi f(\ell' - \ell)T_s} \, |\hat{g}(f)|^2 \, df. \tag{18.21}$$

Notice that, as promised, only terms of the form $E[C_\ell C_{\ell'}^*]$ influence the energy; terms of the form $E[C_\ell C_{\ell'}]$ do not appear in this analysis.

## 18.3 The Power in QAM

In order to discuss the power in QAM we must consider the transmission of an infinite sequence of complex symbols $(C_\ell)$. To guarantee convergence, we shall assume that the pulse shape **g**—in addition to being an integrable signal that is bandlimited to $W/2$ Hz—also satisfies the decay condition

$$|g(t)| \leq \frac{\beta}{1 + |t/T_s|^{1+\alpha}}, \quad t \in \mathbb{R} \tag{18.22}$$

for some $\alpha, \beta > 0$. Also, we shall only consider the transmission of bi-infinite sequences $(C_\ell)$ that are bounded in the sense that there exists some $\gamma > 0$ such that every realization of $(C_\ell)$ satisfies

$$|C_\ell| \leq \gamma, \quad \ell \in \mathbb{Z}. \tag{18.23}$$

As for PAM, we shall treat three different scenarios for the generation of $(C_\ell)$. In the first, we simply ignore the mechanism by which the sequence $(C_\ell)$ is generated and assume that it forms a wide-sense stationary complex stochastic process. In the second, we assume bi-infinite block encoding. And in the third we relax the statistical assumptions and consider the case where the time shifts of **g** by integer multiples of $T_s$ are orthonormal. In all these cases the transmitted waveform is given by

$$X(t) = 2\operatorname{Re}\big(X_{\mathrm{BB}}(t)\, e^{i2\pi f_c t}\big), \quad t \in \mathbb{R}, \tag{18.24}$$

where

$$X_{\mathrm{BB}}(t) = A \sum_{\ell=-\infty}^{\infty} C_\ell \, g(t - \ell T_s), \quad t \in \mathbb{R}. \tag{18.25}$$

It is tempting to derive the power in $X(\cdot)$ by using the complex version of the PAM results of Section 14.5 to compute the power in $X_{BB}(\cdot)$ and then doubling the result. This turns out to be a valid approach, but its justification requires some work. The difficulty is that the powers are defined as

$$\lim_{T\to\infty} \frac{1}{2T} E\left[\int_{-T}^{T} X^2(t)\,dt\right]$$

and

$$\lim_{T\to\infty} \frac{1}{2T} E\left[\int_{-T}^{T} |X_{BB}(t)|^2\,dt\right],$$

and—Theorem 7.6.10 notwithstanding—

$$\frac{1}{2T} E\left[\int_{-T}^{T} X^2(t)\,dt\right] \neq 2\,\frac{1}{2T} E\left[\int_{-T}^{T} |X_{BB}(t)|^2\,dt\right]. \tag{18.26}$$

The reason we cannot claim equality in (18.26) is that $t \mapsto X(t)\,I\{|t| \leq T\}$ is not bandlimited around $f_c$, so Theorem 7.6.10, which relates energies in passband and baseband, is not applicable. Nevertheless, it turns out that the limits as $T \to \infty$ of the RHS and the LHS of (18.26) do agree:

$$\lim_{T\to\infty} \frac{1}{2T} E\left[\int_{-T}^{T} X^2(t)\,dt\right] = 2 \lim_{T\to\infty} \frac{1}{2T} E\left[\int_{-T}^{T} |X_{BB}(t)|^2\,dt\right]. \tag{18.27}$$

Thus, the power in a QAM signal is, indeed, twice the power in its baseband representation. This is stated more precisely in Theorem 18.5.2 and is proved in Section 18.5. With the aid of (18.27) we can now readily compute the power in QAM.

## 18.3.1 $(C_\ell)$ Is Zero-Mean and WSS

We next ignore the mechanism by which the symbols $(C_\ell)$ are generated and merely assume that they form a zero-mean WSS discrete-time CSP of autocovariance function $K_{CC}$:

$$E[C_\ell] = 0, \quad \ell \in \mathbb{Z}, \tag{18.28a}$$

$$E[C_{\ell+m}C_\ell^*] = K_{CC}(m), \quad m, \ell \in \mathbb{Z}. \tag{18.28b}$$

The calculation of the RHS of (18.27) is very similar to the analogous computation in Section 14.5.1 for PAM. The only difference is that here $X_{BB}(\cdot)$ is complex. As in Section 14.5.1, we begin by computing the energy in a length-$T_s$ interval:

$$E\left[\int_{\tau}^{\tau+T_s} |X_{BB}(t)|^2\,dt\right]$$

$$= A^2 \int_{\tau}^{\tau+T_s} E\left[\left|\sum_{\ell=-\infty}^{\infty} C_\ell\, g(t - \ell T_s)\right|^2\right]\,dt$$

$$= A^2 \int_\tau^{\tau+T_s} \mathsf{E}\left[ \sum_{\ell=-\infty}^{\infty} \sum_{\ell'=-\infty}^{\infty} C_\ell C_{\ell'}^* \, g(t - \ell T_s) \, g^*(t - \ell' T_s) \right] dt$$

$$= A^2 \int_\tau^{\tau+T_s} \sum_{\ell=-\infty}^{\infty} \sum_{\ell'=-\infty}^{\infty} \mathsf{E}[C_\ell C_{\ell'}^*] \, g(t - \ell T_s) \, g^*(t - \ell' T_s) \, dt$$

$$= A^2 \int_\tau^{\tau+T_s} \sum_{m=-\infty}^{\infty} \sum_{\ell'=-\infty}^{\infty} \mathsf{E}[C_{\ell'+m} C_{\ell'}^*] \, g\big(t - (\ell' + m) T_s\big) \, g^*(t - \ell' T_s) \, dt$$

$$= A^2 \int_\tau^{\tau+T_s} \sum_{m=-\infty}^{\infty} \mathsf{K}_{CC}(m) \sum_{\ell'=-\infty}^{\infty} g\big(t - (\ell' + m) T_s\big) \, g^*(t - \ell' T_s) \, dt$$

$$= A^2 \sum_{m=-\infty}^{\infty} \mathsf{K}_{CC}(m) \sum_{\ell'=-\infty}^{\infty} \int_{\tau-\ell'T_s}^{\tau+T_s-\ell'T_s} g(t' - m T_s) \, g^*(t') \, dt'$$

$$= A^2 \sum_{m=-\infty}^{\infty} \mathsf{K}_{CC}(m) \int_{-\infty}^{\infty} g^*(t') \, g(t' - m T_s) \, dt'$$

$$= A^2 \sum_{m=-\infty}^{\infty} \mathsf{K}_{CC}(m) \, \mathsf{R}_{\mathsf{gg}}^*(m T_s), \tag{18.29}$$

where we have substituted $\ell' + m$ for $\ell$ (fourth equality) and $t'$ for $t - \ell' T_s$ (sixth equality).

As in the analogous analysis for real PAM signals, we lower-bound the energy of $X_{\mathrm{BB}}(\cdot)$ in the interval $[-T, +T]$ by

$$\left\lfloor \frac{2T}{T_s} \right\rfloor \mathsf{E}\left[ \int_\tau^{\tau+T_s} |X_{\mathrm{BB}}(t)|^2 \, dt \right]$$

and upper-bound it by

$$\left\lceil \frac{2T}{T_s} \right\rceil \mathsf{E}\left[ \int_\tau^{\tau+T_s} |X_{\mathrm{BB}}(t)|^2 \, dt \right],$$

so, by the Sandwich Theorem,

$$\lim_{T \to \infty} \frac{1}{2T} \mathsf{E}\left[ \int_{-T}^{+T} |X_{\mathrm{BB}}(t)|^2 \, dt \right] = \frac{1}{T_s} \mathsf{E}\left[ \int_\tau^{\tau+T_s} |X_{\mathrm{BB}}(t)|^2 \, dt \right]. \tag{18.30}$$

It thus follows from (18.30) and (18.29) that the power $\mathsf{P}_{\mathrm{BB}}$ in $X_{\mathrm{BB}}(\cdot)$ is

$$\mathsf{P}_{\mathrm{BB}} = \frac{A^2}{T_s} \sum_{m=-\infty}^{\infty} \mathsf{K}_{CC}(m) \, \mathsf{R}_{\mathsf{gg}}^*(m T_s) \tag{18.31}$$

$$= \frac{A^2}{T_s} \int_{-\infty}^{\infty} \sum_{m=-\infty}^{\infty} \mathsf{K}_{CC}(m) \, e^{-i 2\pi f m T_s} \, |\hat{g}(f)|^2 \, df, \tag{18.32}$$

where the second equality follows from (18.13).

Since the power in passband is twice the power in baseband, we conclude:

**Theorem 18.3.1.** *Let the QAM SP $\left(X(t)\right)$ be given by (18.24) & (18.25), where A, $T_s$, $\mathbf{g}$, $\mathbf{W}$, and $f_c$ are as in Theorem 18.2.1. Further assume that $\mathbf{g}$ satisfies the decay condition (18.22) and that the discrete-time SP $\left(C_\ell\right)$ is bounded in the sense of (18.23). If $\left(C_\ell\right)$ satisfies (18.28), then $\left(X(t)\right)$ is a measurable SP,*

$$\boxed{\lim_{T\to\infty} \frac{1}{2T}\mathsf{E}\left[\int_{-T}^{T} X^2(t)\,\mathrm{d}t\right] = \frac{2A^2}{T_s}\sum_{m=-\infty}^{\infty} \mathsf{K}_{CC}(m)\,\mathsf{R}_{\mathbf{gg}}^*(mT_s),} \tag{18.33}$$

*and*

$$\boxed{\lim_{T\to\infty} \frac{1}{2T}\mathsf{E}\left[\int_{-T}^{T} X^2(t)\,\mathrm{d}t\right] = \frac{2A^2}{T_s}\int_{-\infty}^{\infty}\sum_{m=-\infty}^{\infty} \mathsf{K}_{CC}(m)\,e^{-\mathrm{i}2\pi f m T_s}\,|\hat{g}(f)|^2\,\mathrm{d}f.}$$

$$\tag{18.34}$$

**Proof.** Follows by combining (18.27) (Theorem 18.5.2) and Theorem 14.6.4 (which extends to the case where the pulse shape and the symbols are complex). □

## 18.3.2   Bi-Infinite Block-Mode

The second scenario we consider is when $\left(C_\ell\right)$ is generated, as in Section 14.5.2, by applying a binary-to-complex block-encoder $\mathbf{enc}\colon \{0,1\}^K \to \mathbb{C}^N$ to bi-infinite IID random bits $\left(D_j\right)$. As in Section 14.5.2, we assume that the encoder, when fed IID random bits, produces symbols of zero mean.

By extending the results of Section 14.5.2 to complex pulse shapes and complex symbols, we obtain that the power in $X_{\mathrm{BB}}(\cdot)$ is given by:

$$P_{\mathrm{BB}} = \frac{1}{NT_s}\mathsf{E}\left[\left|A\sum_{\ell=1}^{N} C_\ell g(t-\ell T_s)\right|^2\right] \tag{18.35}$$

$$= \frac{A^2}{N}\int_{-\infty}^{\infty}\sum_{\ell=1}^{N}\sum_{\ell'=1}^{N} \mathsf{E}[C_\ell C_{\ell'}^*]\,e^{\mathrm{i}2\pi f(\ell'-\ell)T_s}\,|\hat{g}(f)|^2\,\mathrm{d}f. \tag{18.36}$$

Using the relationship between power in baseband and passband (18.27) and using the definitions of $\mathsf{E}$ (18.8) and of $\mathsf{E}_s$ (18.19), we obtain:

**Theorem 18.3.2.** *Under the assumptions of Theorem 18.3.1, if the symbols $\left(C_\ell\right)$ are generated from IID random bits $\left(D_j\right)$ in bi-infinite block-mode using the encoder $\mathrm{enc}(\cdot)$, where $\mathrm{enc}(\cdot)$ produces zero-mean symbols when fed IID random bits, then $\left(X(t)\right)$ is a measurable SP, and*

$$\boxed{\lim_{T\to\infty} \frac{1}{2T}\,\mathsf{E}\left[\int_{-T}^{T} X^2(t)\,\mathrm{d}t\right] = \frac{\mathsf{E}_s}{T_s},} \tag{18.37}$$

*where the energy per symbol $\mathsf{E}_s$ is defined in (18.19) and is given by (18.20) or (18.21).*

**Proof.** Follows from Theorem 18.5.2 and by noting that Theorem 14.6.5 also extends to the case where the pulse shape and the symbols are complex.     $\Box$

### 18.3.3   Time Shifts of Pulse Shape Are Orthonormal

We finally address the third scenario where the time shifts of the pulse shape by integer multiples of $T_s$ are orthonormal. This situation is very prevalent in Digital Communications and allows for significant simplifications. In this setting we denote the pulse shape by $\phi(\cdot)$ and state the orthonormality as

$$\int_{-\infty}^{\infty} \phi(t - \ell T_s)\, \phi^*(t - \ell' T_s)\, dt = I\{\ell = \ell'\}, \quad \ell, \ell' \in \mathbb{Z}. \tag{18.38}$$

The transmitted signal $\big(X(t),\ t \in \mathbb{R}\big)$ is thus given as in (18.24) but with

$$X_{\mathrm{BB}}(t) = A \sum_{\ell=-\infty}^{\infty} C_\ell\, \phi(t - \ell T_s), \quad t \in \mathbb{R}, \tag{18.39}$$

where we assume that the discrete-time CSP $\big(C_\ell\big)$ satisfies the boundedness condition (18.23) and that the complex pulse shape $\phi(\cdot)$ satisfies the orthogonality condition (18.38) and the decay condition

$$|\phi(t)| \leq \frac{\beta}{1 + |t/T_s|^{1+\alpha}}, \quad t \in \mathbb{R}, \tag{18.40}$$

for some $\alpha, \beta > 0$.

Computing the power in $\big(X_{\mathrm{BB}}(t),\ t \in \mathbb{R}\big)$ using Theorem 14.5.2, which easily extends to the complex case, we obtain from (18.27):

**Theorem 18.3.3.** *Let the SP $\big(X(t),\ t \in \mathbb{R}\big)$ be given by*

$$X(t) = 2\,\mathrm{Re}\left( A \sum_{\ell=-\infty}^{\infty} C_\ell\, \phi(t - \ell T_s)\, e^{i2\pi f_c t} \right), \quad t \in \mathbb{R}, \tag{18.41}$$

*where $A \geq 0$; $T_s > 0$; the pulse shape $\phi \colon \mathbb{R} \to \mathbb{C}$ is an integrable function that is bandlimited to $W/2$ Hz, is Borel measurable, satisfies the orthogonality condition (18.38), and satisfies the decay condition (18.40); the carrier frequency $f_c$ satisfies $f_c > W/2 > 0$; and where the CSP $\big(C_\ell\big)$ satisfies the boundedness condition (18.23). Then $\big(X(t),\ t \in \mathbb{R}\big)$ is a measurable stochastic process, and*

$$\lim_{T \to \infty} \frac{1}{2T} E\left[ \int_{-T}^{T} X^2(t)\, dt \right] = \frac{2A^2}{T_s} \lim_{L \to \infty} \frac{1}{2L+1} \sum_{\ell=-L}^{L} E\big[|C_\ell|^2\big], \tag{18.42}$$

*whenever the limit on the RHS exists.*

## 18.4   The Operational PSD of QAM Signals

We shall compute the operational PSD of the QAM signal $(X(t),\ t \in \mathbb{R})$ (18.24) by relating it to the operational PSD of the complex signal $(X_{\mathrm{BB}}(t),\ t \in \mathbb{R})$ (18.25) and by then computing the operational PSD of the latter using techniques similar to the ones we employed in Chapter 15 in our study of the operational PSD of real PAM signals. But first we must define the operational PSD of complex stochastic processes. The definition is very similar to that for real stochastic processes (Definition 15.3.1), but there are two issues to note. The first is that we do not require that the operational PSD be a symmetric function, and the second is that we allow for filters of complex impulse response.

**Definition 18.4.1 (Operational PSD of a CSP).** *We say that a CSP* $(Z(t),\ t \in \mathbb{R})$ *is of **operational power spectral density** $\mathsf{S}_{ZZ}$ if $(Z(t),\ t \in \mathbb{R})$ is a measurable CSP;[1] the mapping $\mathsf{S}_{ZZ} \colon \mathbb{R} \to \mathbb{R}$ is integrable; and for every integrable complex-valued function $\mathbf{h} \colon \mathbb{R} \to \mathbb{C}$ the average power of the convolution of $(Z(t),\ t \in \mathbb{R})$ and $\mathbf{h}$ is given by*

$$\text{Power in } \mathbf{Z} \star \mathbf{h} = \int_{-\infty}^{\infty} \mathsf{S}_{ZZ}(f)\, |\hat{h}(f)|^2 \, \mathrm{d}f. \tag{18.43}$$

By Lemma 15.3.2 (i) the PSD is unique:

**Note 18.4.2 (The Operational PSD Is Unique).** The operational PSD of a CSP is unique in the sense that if a CSP is of two different operational power spectral densities, then the two must be indistinguishable.

The relationship between the operational PSD of the real QAM signal $(X(t))$ (18.24) and of the CSP $(X_{\mathrm{BB}}(t))$ (18.25) turns out to be very simple. Indeed, subject to the conditions that are made precise in Theorem 18.6.6, if the baseband CSP $(X_{\mathrm{BB}}(t))$ is of operational PSD $\mathsf{S}_{\mathrm{BB}}$, then the real QAM SP $(X(t))$ is of operational PSD $\mathsf{S}_{XX}$, where

$$\boxed{\mathsf{S}_{XX}(f) = \mathsf{S}_{\mathrm{BB}}(|f| - f_{\mathrm{c}}), \quad f \in \mathbb{R}.} \tag{18.44}$$

This result is proved in Section 18.6 and relies heavily on the fact that $\mathbf{g}$ is bandlimited to $W/2$ Hz and that $f_{\mathrm{c}} > W/2$. Here we shall only derive it heuristically and then see how to apply it.

Recalling the definition of the operational PSD of a real SP (Definition 15.3.1), we note that in order to derive (18.44) we need to show that its RHS is an integrable symmetric function and that

$$\text{Power in } \mathbf{X} \star \mathbf{h} = \int_{-\infty}^{\infty} |\hat{h}(f)|^2\, \mathsf{S}_{\mathrm{BB}}(|f| - f_{\mathrm{c}})\, \mathrm{d}f, \tag{18.45}$$

---

[1] A complex stochastic processes is said to be measurable if its real and imaginary parts are measurable real stochastic processes.

whenever $\mathbf{h}\colon \mathbb{R} \to \mathbb{R}$ is integrable. The integrability of $f \mapsto \mathsf{S}_{\mathrm{BB}}(|f| - f_c)$ follows directly from the integrability of $\mathsf{S}_{\mathrm{BB}}(\cdot)$. The symmetry is obvious because the RHS of (18.44) depends on $f$ only via $|f|$. Our plan for computing the power in $\mathbf{X} \star \mathbf{h}$ is to first use the results of Section 7.6.7 to express the baseband representation of $\mathbf{X} \star \mathbf{h}$ in the form $\mathbf{X}_{\mathrm{BB}} \star \mathbf{h}'_{\mathrm{BB}}$, where $\mathbf{h}'_{\mathrm{BB}}$ is the baseband representation of the result of passing $\mathbf{h}$ through a unit-gain bandpass filter of bandwidth $W$ around the carrier frequency $f_c$. Using the relationship between power in passband and baseband, this will allow us to express the power in $\mathbf{X} \star \mathbf{h}$ as twice the power in $\mathbf{X}_{\mathrm{BB}} \star \mathbf{h}'_{\mathrm{BB}}$. Expressing the power in the latter using the operational PSD $\mathsf{S}_{\mathrm{BB}}(\cdot)$ of $\mathbf{X}_{\mathrm{BB}}$ will allow us to complete the calculation of the power in $\mathbf{X} \star \mathbf{h}$.

Before executing this plan, we pause here to heuristically argue that, loosely speaking, the condition that $\mathbf{g}$ is bandlimited to $W/2$ Hz implies that we may assume that

$$\mathsf{S}_{\mathrm{BB}}(f) = 0, \quad |f| > \frac{W}{2}. \tag{18.46}$$

For a precise statement of this result, see Proposition 18.6.3 in Section 18.6.2. The intuition behind this statement is that, since $\mathbf{g}$ is bandlimited to $W/2$ Hz, in some loose sense, all the power of the signal $\mathbf{X}_{\mathrm{BB}}$ is contained in the band $|f| \leq W/2$. To heuristically justify (18.46), we shall show that if $\mathsf{S}_{\mathrm{BB}}(\cdot)$ is an operational PSD for $\big(X_{\mathrm{BB}}(t)\big)$, then so is the mapping $f \mapsto \mathsf{S}_{\mathrm{BB}}(f)\,\mathrm{I}\{|f| \leq W/2\}$. This follows by noting that for every $\mathbf{h}\colon \mathbb{R} \to \mathbb{C}$ in $\mathcal{L}_1$

$$\text{Power in } \mathbf{X}_{\mathrm{BB}} \star \mathbf{h} = \text{Power in } \left(t \mapsto A \sum_{\ell \in \mathbb{Z}} C_\ell\, g(t - \ell \mathsf{T}_{\mathrm{s}})\right) \star \mathbf{h}$$

$$= \text{Power in } t \mapsto A \sum_{\ell \in \mathbb{Z}} C_\ell\, (\mathbf{g} \star \mathbf{h})(t - \ell \mathsf{T}_{\mathrm{s}})$$

$$= \text{Power in } t \mapsto A \sum_{\ell \in \mathbb{Z}} C_\ell\, \big((\mathbf{g} \star \mathrm{LPF}_{W/2}) \star \mathbf{h}\big)(t - \ell \mathsf{T}_{\mathrm{s}})$$

$$= \text{Power in } t \mapsto A \sum_{\ell \in \mathbb{Z}} C_\ell\, \big(\mathbf{g} \star (\mathbf{h} \star \mathrm{LPF}_{W/2})\big)(t - \ell \mathsf{T}_{\mathrm{s}})$$

$$= \text{Power in } \left(t \mapsto A \sum_{\ell \in \mathbb{Z}} C_\ell\, g(t - \ell \mathsf{T}_{\mathrm{s}})\right) \star (\mathbf{h} \star \mathrm{LPF}_{W/2})$$

$$= \int_{-\infty}^{\infty} \mathsf{S}_{\mathrm{BB}}(f)\, \big|\hat{h}(f)\,\mathrm{I}\{|f| \leq W/2\}\big|^2\, \mathrm{d}f$$

$$= \int_{-\infty}^{\infty} \big(\mathsf{S}_{\mathrm{BB}}(f)\,\mathrm{I}\{|f| \leq W/2\}\big)\, |\hat{h}(f)|^2\, \mathrm{d}f,$$

from which the result follows from the uniqueness (to within indistinguishability) of the operational PSD (Note 18.4.2). Here the first equality follows from the definition of $\mathbf{X}_{\mathrm{BB}}$ (18.25); the second because convolving a PAM signal of pulse shape $\mathbf{g}$ (in our case complex) with $\mathbf{h}$ is tantamount to replacing the pulse shape $\mathbf{g}$ with the new pulse shape $\mathbf{g} \star \mathbf{h}$ (see the derivation of (15.16) in Section 15.4 which extends verbatim to the complex case); the third because, by assumption, $\mathbf{g}$ is bandlimited to $W/2$ Hz; the fourth by the associativity of convolution (see Theorem 5.6.1, which, strictly speaking, is not applicable here because $\mathrm{LPF}_{W/2}$ is not integrable); the fifth because replacing the pulse shape $\mathbf{g}$ by $\mathbf{g} \star \big(\mathbf{h} \star \mathrm{LPF}_{W/2}\big)$ is tantamount to convolving the PAM signal with $(\mathbf{h} \star \mathrm{LPF}_{W/2})$; the sixth from our assumption that $\mathsf{S}_{\mathrm{BB}}(\cdot)$ is an operational PSD for $\mathbf{X}_{\mathrm{BB}}$ (and by ignoring the fact that $\mathbf{h} \star \mathrm{LPF}_{W/2}$ need not be integrable); and the seventh by trivial algebra.

Having established (18.46), we are now ready to compute the power in $\mathbf{X} \star \mathbf{h}$. Using the results of Section 7.6.7 we obtain that for every integrable $\mathbf{h} \colon \mathbb{R} \to \mathbb{R}$, the baseband representation of $\mathbf{X} \star \mathbf{h}$ is given by $\mathbf{X}_{\mathrm{BB}} \star \mathbf{h}'_{\mathrm{BB}}$ where $\mathbf{h}'_{\mathrm{BB}} \colon \mathbb{R} \to \mathbb{C}$ is the baseband representation of the result of passing $\mathbf{h}$ through a unit-gain bandpass filter of bandwidth $W$ around the carrier frequency $f_{\mathrm{c}}$:

$$\hat{h}'_{\mathrm{BB}}(f) = \hat{h}(f + f_{\mathrm{c}}) \, \mathrm{I}\{|f| \leq W/2\}, \quad f \in \mathbb{R}. \tag{18.47}$$

And since the power in passband is twice the power in baseband, we conclude that

$$
\begin{aligned}
\text{Power in } \mathbf{X} \star \mathbf{h} &= 2 \, \text{Power in } \mathbf{X}_{\mathrm{BB}} \star \mathbf{h}'_{\mathrm{BB}} \\
&= 2 \int_{-\infty}^{\infty} \mathsf{S}_{\mathrm{BB}}(f) \left| \hat{h}'_{\mathrm{BB}}(f) \right|^2 \mathrm{d}f \\
&= 2 \int_{-\infty}^{\infty} \mathsf{S}_{\mathrm{BB}}(f) \left| \hat{h}(f + f_{\mathrm{c}}) \right|^2 \mathrm{I}\{|f| \leq W/2\} \, \mathrm{d}f \\
&= 2 \int_{-\infty}^{\infty} \mathsf{S}_{\mathrm{BB}}(f) \left| \hat{h}(f + f_{\mathrm{c}}) \right|^2 \mathrm{d}f \\
&= 2 \int_{-\infty}^{\infty} \mathsf{S}_{\mathrm{BB}}(\tilde{f} - f_{\mathrm{c}}) \left| \hat{h}(\tilde{f}) \right|^2 \mathrm{d}\tilde{f} \\
&= \int_{-\infty}^{\infty} \mathsf{S}_{\mathrm{BB}}(\tilde{f} - f_{\mathrm{c}}) \left| \hat{h}(\tilde{f}) \right|^2 \mathrm{d}\tilde{f} + \int_{-\infty}^{\infty} \mathsf{S}_{\mathrm{BB}}(\tilde{f} - f_{\mathrm{c}}) \left| \hat{h}(-\tilde{f}) \right|^2 \mathrm{d}\tilde{f} \\
&= \int_{-\infty}^{\infty} \mathsf{S}_{\mathrm{BB}}(\tilde{f} - f_{\mathrm{c}}) \left| \hat{h}(\tilde{f}) \right|^2 \mathrm{d}\tilde{f} + \int_{-\infty}^{\infty} \mathsf{S}_{\mathrm{BB}}(-f' - f_{\mathrm{c}}) \left| \hat{h}(f') \right|^2 \mathrm{d}f' \\
&= \int_{-\infty}^{\infty} \left( \mathsf{S}_{\mathrm{BB}}(f - f_{\mathrm{c}}) + \mathsf{S}_{\mathrm{BB}}(-f - f_{\mathrm{c}}) \right) \left| \hat{h}(f) \right|^2 \mathrm{d}f \\
&= \int_{-\infty}^{\infty} \mathsf{S}_{\mathrm{BB}}(|f| - f_{\mathrm{c}}) \left| \hat{h}(f) \right|^2 \mathrm{d}f,
\end{aligned}
$$

where the first equality follows because the power in passband is twice the power in baseband; the second because $\mathbf{X}_{\mathrm{BB}}$ is of operational PSD $\mathsf{S}_{\mathrm{BB}}(\cdot)$; the third by (18.47); the fourth by (18.46); the fifth by changing the integration variable to $\tilde{f} \triangleq f + f_{\mathrm{c}}$; the sixth because $\mathbf{h}$ is real so its Fourier Transform must be conjugate-symmetric; the seventh by changing the integration variable in the second integral to $f' \triangleq -\tilde{f}$; the eighth by the linearity of integration; and the final equality by (18.46) and the assumption that $f_{\mathrm{c}} > W/2$. This establishes (18.45) and thus concludes the proof of (18.44).

We next apply (18.44) to calculate the operational PSD of QAM in two scenarios: when the complex symbols $(C_\ell)$ form a bounded, zero-mean, WSS, CSP and when they are generated in bi-infinite block-mode.

### 18.4.1  $(C_\ell)$ Zero-Mean WSS and Bounded

We next use (18.44) to derive the operational PSD of QAM when the discrete-time CSP $(C_\ell)$ is of zero mean and of autocovariance function $\mathsf{K}_{CC}$; see (18.28). To use (18.44) we first need to compute the operational PSD of the CSP $\mathbf{X}_{\mathrm{BB}}$. This is straightforward. As in Section 15.4.2, we note that $\mathbf{X}_{\mathrm{BB}} \star \mathbf{h}$ has the same form as (18.25) with the pulse shape $\mathbf{g}$ replaced by $\mathbf{g} \star \mathbf{h}$. Consequently, by substituting the FT of $\mathbf{g} \star \mathbf{h}$ for the FT of $\mathbf{g}$ in (18.32),[2] we obtain that

$$\text{Power in } \mathbf{X}_{\mathrm{BB}} \star \mathbf{h} = \frac{\mathsf{A}^2}{\mathsf{T}_{\mathrm{s}}} \int_{-\infty}^{\infty} \sum_{m=-\infty}^{\infty} \mathsf{K}_{CC}(m)\, e^{-\mathrm{i}2\pi f m \mathsf{T}_{\mathrm{s}}} |\hat{g}(f)|^2\, |\hat{h}(f)|^2\, \mathrm{d}f \quad (18.48)$$

and the operational PSD of $\mathbf{X}_{\mathrm{BB}}$ is thus

$$\mathsf{S}_{\mathrm{BB}}(f) = \frac{\mathsf{A}^2}{\mathsf{T}_{\mathrm{s}}} \sum_{m=-\infty}^{\infty} \mathsf{K}_{CC}(m)\, e^{-\mathrm{i}2\pi f m \mathsf{T}_{\mathrm{s}}} |\hat{g}(f)|^2, \quad f \in \mathbb{R}. \quad (18.49)$$

This is the complex analog of (15.21). From (18.49) and (18.44) we now obtain:

**Theorem 18.4.3.** *Under the assumptions of Theorem 18.3.1, the operational PSD of the QAM signal $\big(X(t),\, t \in \mathbb{R}\big)$ is given by*

$$\boxed{\mathsf{S}_{XX}(f) = \frac{\mathsf{A}^2}{\mathsf{T}_{\mathrm{s}}} \sum_{m=-\infty}^{\infty} \mathsf{K}_{CC}(m)\, e^{\mathrm{i}2\pi(|f|-f_{\mathrm{c}})m\mathsf{T}_{\mathrm{s}}} \big|\hat{g}\big(|f| - f_{\mathrm{c}}\big)\big|^2, \quad f \in \mathbb{R}.} \quad (18.50)$$

**Proof.** The justification of (18.44) is in Theorem 18.6.6. A formal derivation of the operational PSD of $\big(X_{\mathrm{BB}}(t),\, t \in \mathbb{R}\big)$ can be found in Section 18.6.5. We draw the reader's attention to the fact that the proof that we gave for the real case in Section 15.5 is not directly applicable to the complex case because that proof relied on Theorem 25.14.1 (Wiener-Khinchin), which we prove in Section 25.14 only for *real* WSS stochastic processes.[3]                                                             $\square$

Figure 18.1 depicts the relationship between the pulse shape $\mathbf{g}$ and the operational PSD of the QAM signal for the case where $\mathsf{K}_{CC}(m) = \mathrm{I}\{m = 0\}$ for every $m \in \mathbb{Z}$.

### 18.4.2  The Operational PSD of QAM in Bi-Infinite Block-Mode

The operational PSD of QAM in bi-infinite block-mode can also be computed using (18.44). All we need is the operational PSD of $\big(X_{\mathrm{BB}}(t)\big)$, which can be computed from (18.36) as follows. As in Section 15.4.2, we note that $\mathbf{X}_{\mathrm{BB}} \star \mathbf{h}$ has the same form as (18.25) with the pulse shape $\mathbf{g}$ replaced by $\mathbf{g} \star \mathbf{h}$. Consequently,

---

[2]We are ignoring here the fact that $\mathbf{g} \star \mathbf{h}$ need not satisfy the required decay condition.
[3]The extension to the complex case is not as trivial as one might think because the real and imaginary parts of a WSS complex SP need not be WSS.

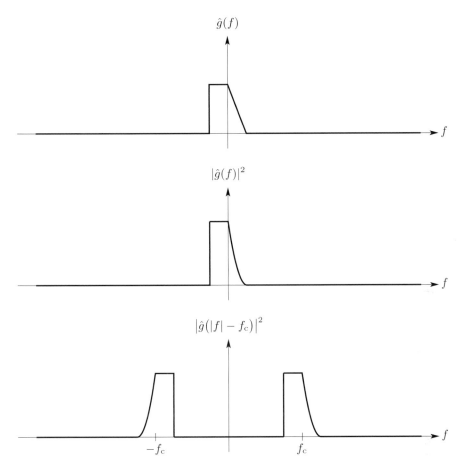

**Figure 18.1:** The relationship between the Fourier Transform of the pulse shape $g(\cdot)$ and the operational PSD of a QAM signal. The symbols $(C_\ell)$ are assumed to be of zero mean and uncorrelated.

by substituting the FT of $\mathbf{g} \star \mathbf{h}$ for the FT of $\mathbf{g}$ in (18.36), we obtain that

Power in $\mathbf{X}_{\mathrm{BB}} \star \mathbf{h}$

$$= \int_{-\infty}^{\infty} \left( \frac{A^2}{N} \sum_{\ell=1}^{N} \sum_{\ell'=1}^{N} \mathsf{E}[C_\ell C_{\ell'}^*] \, e^{\mathrm{i}2\pi f(\ell'-\ell)\mathsf{T_s}} \big|\hat{g}(f)\big|^2 \right) \big|\hat{h}(f)\big|^2 \, \mathrm{d}f, \quad (18.51)$$

and the operational PSD of $\mathbf{X}_{\mathrm{BB}}$ is thus

$$\mathsf{S}_{\mathrm{BB}}(f) = \frac{A^2}{N} \sum_{\ell=1}^{N} \sum_{\ell'=1}^{N} \mathsf{E}[C_\ell C_{\ell'}^*] \, e^{\mathrm{i}2\pi f(\ell'-\ell)\mathsf{T_s}} \big|\hat{g}(f)\big|^2, \quad f \in \mathbb{R}. \quad (18.52)$$

This is the complex analog of (15.23). (But note that, in our present case, $\mathsf{S}_{\mathrm{BB}}(\cdot)$ need not be a symmetric function.) From (18.52) and (18.44) we now obtain:

**Theorem 18.4.4 (Operational PSD of QAM in Bi-Infinite Block-Mode).** *Under the assumptions of Theorem 18.3.2, the operational PSD $\mathsf{S}_{XX}$ of the QAM signal $\big(X(t),\ t \in \mathbb{R}\big)$ is given for every $f \in \mathbb{R}$ by*

$$\mathsf{S}_{XX}(f) = \frac{\mathsf{A}^2}{N\mathsf{T}_\mathrm{s}} \sum_{\ell=1}^{N} \sum_{\ell'=1}^{N} \mathsf{E}[C_\ell C_{\ell'}^*]\, e^{\mathrm{i}2\pi(|f|-f_\mathrm{c})(\ell-\ell')\mathsf{T}_\mathrm{s}}\, \big|\hat{g}\big(|f| - f_\mathrm{c}\big)\big|^2. \qquad (18.53)$$

**Proof.** The justification of (18.44) is in Theorem 18.6.6, and a formal derivation of the operational PSD of $\big(X_{\mathrm{BB}}(t)\big)$ is given in Section 18.6.5. □

## 18.5　A Formal Account of Power in Passband and Baseband

In this section we formulate conditions under which (18.27) holds, i.e., under which the power in passband is twice the power in baseband. We first extend the Triangle Inequality (4.14) to stochastic processes.

**Proposition 18.5.1 (Triangle Inequality for Stochastic Processes).** *Let $\big(X(t)\big)$ and $\big(Y(t)\big)$ be (real or complex) measurable stochastic processes, and let $a < b$ be arbitrary real numbers. Suppose further that*

$$\mathsf{E}\left[\int_a^b |X(t)|^2\, \mathrm{d}t\right], \mathsf{E}\left[\int_a^b |Y(t)|^2\, \mathrm{d}t\right] < \infty. \qquad (18.54)$$

*Then*

$$\left(\sqrt{\mathsf{E}\left[\int_a^b |X(t)|^2\, \mathrm{d}t\right]} - \sqrt{\mathsf{E}\left[\int_a^b |Y(t)|^2\, \mathrm{d}t\right]}\right)^2$$

$$\leq \mathsf{E}\left[\int_a^b |X(t) + Y(t)|^2\, \mathrm{d}t\right] \leq$$

$$\left(\sqrt{\mathsf{E}\left[\int_a^b |X(t)|^2\, \mathrm{d}t\right]} + \sqrt{\mathsf{E}\left[\int_a^b |Y(t)|^2\, \mathrm{d}t\right]}\right)^2. \qquad (18.55)$$

*This also holds when $a$ is replaced with $-\infty$ and/or $b$ is replaced with $+\infty$.*

**Proof.** Replace all integrals in the proof of (4.14) with expectations of integrals. □

We can now state the main result of this section relating power in passband and baseband.

**Theorem 18.5.2.** *Let* $\mathsf{T_s}$, $\mathbf{g}$, $\mathsf{W}$, *and* $f_c$ *be as in Theorem 18.2.1 and, addition-ally, assume that* $\mathbf{g}$ *satisfies the decay condition* (18.22) *and that the CSP* $(C_\ell)$ *is bounded in the sense of* (18.23). *Then the condition*

$$\lim_{\mathsf{T}\to\infty} \frac{1}{2\mathsf{T}} \, \mathsf{E}\left[\int_{-\mathsf{T}}^{\mathsf{T}} \left|\sum_{\ell\in\mathbb{Z}} C_\ell \, g(t - \ell\mathsf{T_s})\right|^2 dt\right] = P \tag{18.56}$$

*is equivalent to the condition*

$$\lim_{\mathsf{T}\to\infty} \frac{1}{2\mathsf{T}} \, \mathsf{E}\left[\int_{-\mathsf{T}}^{\mathsf{T}} \left(2\operatorname{Re}\left(\sum_{\ell\in\mathbb{Z}} C_\ell \, g(t - \ell\mathsf{T_s}) \, e^{i2\pi f_c t}\right)\right)^2 dt\right] = 2P. \tag{18.57}$$

The rest of this section is dedicated to proving this theorem. To simplify the notation we begin by showing that it suffices to prove the result for the case where $\mathsf{T_s} = 1$. If $\mathsf{T_s} > 0$ is not necessarily equal to 1, then we define for every $t \in \mathbb{R}$,

$$\tilde{g}(t) = g(t\mathsf{T_s}),$$
$$\tilde{\mathsf{W}} = \mathsf{W}\mathsf{T_s},$$
$$\tilde{f}_c = f_c\mathsf{T_s},$$

and note that $\mathbf{g}$ is bandlimited to $\mathsf{W}/2$ Hz if, and only if, $\tilde{\mathbf{g}}$ is bandlimited to $\mathsf{W}\mathsf{T_s}/2$ Hz; that

$$(f_c \geq \mathsf{W}/2) \Leftrightarrow (\tilde{f}_c \geq \tilde{\mathsf{W}}/2);$$

and that $\mathbf{g}$ satisfies the decay condition (18.22) if, and only if,

$$|\tilde{g}(t)| \leq \frac{\beta}{1 + |t|^{1+\alpha}}, \quad t \in \mathbb{R}.$$

By defining $\tau \triangleq t/\mathsf{T_s}$ we obtain that

$$\frac{1}{2\mathsf{T}} \int_{-\mathsf{T}}^{\mathsf{T}} \left|\sum_{\ell\in\mathbb{Z}} C_\ell \, g(t - \ell\mathsf{T_s})\right|^2 dt = \frac{1}{2(\mathsf{T}/\mathsf{T_s})} \int_{-\mathsf{T}/\mathsf{T_s}}^{\mathsf{T}/\mathsf{T_s}} \left|\sum_{\ell\in\mathbb{Z}} C_\ell \, \tilde{g}(\tau - \ell)\right|^2 d\tau$$

so the power in the mapping $t \mapsto \sum C_\ell \, g(t - \ell\mathsf{T_s})$ is the same as in the mapping $\tau \mapsto \sum C_\ell \, \tilde{g}(\tau - \ell)$. Similarly,

$$\frac{1}{2\mathsf{T}} \int_{-\mathsf{T}}^{\mathsf{T}} \left(2\operatorname{Re}\left(\sum_\ell C_\ell \, g(t - \ell\mathsf{T_s}) \, e^{i2\pi f_c t}\right)\right)^2 dt$$

$$= \frac{1}{2(\mathsf{T}/\mathsf{T_s})} \int_{-\mathsf{T}/\mathsf{T_s}}^{\mathsf{T}/\mathsf{T_s}} \left(2\operatorname{Re}\left(\sum_\ell C_\ell \, \tilde{g}(\tau - \ell) \, e^{i2\pi \tilde{f}_c \tau}\right)\right)^2 d\tau$$

so the power in the mapping $t \mapsto 2\operatorname{Re}\left(\sum_\ell C_\ell \, g(t - \ell\mathsf{T_s}) \, e^{i2\pi f_c t}\right)$ is the same as in the mapping $\tau \mapsto 2\operatorname{Re}\left(\sum_\ell C_\ell \, \tilde{g}(t - \ell) \, e^{i2\pi \tilde{f}_c \tau}\right)$. Thus, if we establish that the inequality $\tilde{f}_c > \tilde{\mathsf{W}}/2$ implies that the power in the baseband signal $\tau \mapsto \sum C_\ell \, \tilde{g}(\tau - \ell)$ is equal to half the power in $\tau \mapsto 2\operatorname{Re}\left(\sum_\ell C_\ell \, \tilde{g}(t - \ell) \, e^{i2\pi \tilde{f}_c \tau}\right)$, then it will also follow that the inequality $f_c > W/2$ implies that the power in $t \mapsto \sum C_\ell g(t - \ell T_s)$ is equal to half the power in $t \mapsto \mathrm{Re}\big(\sum_\ell C_\ell g(t - \ell T_s) e^{i2\pi f_c t}\big)$.

Having established that it suffices to prove the theorem for $T_s = 1$, we assume for the remainder of this section that $T_s = 1$, so the decay condition (18.22) can be rewritten as

$$|g(t)| \leq \frac{\beta}{1 + |t|^{1+\alpha}}, \quad t \in \mathbb{R}. \tag{18.58}$$

As in the proof of Theorem 14.5.2, we shall simplify notation and assume that—in calculating power as the limiting ratio of the energy in the interval $[-T, T]$ to the length of the interval—$T$ is restricted to the positive integers. The justification is identical to the one we gave in proving Theorem 14.5.2; see (14.52).

We shall find it convenient to introduce an additional subscript "w" to indicate "windowing." Thus, if we define $X_{\mathrm{BB}}(\cdot)$ as

$$X_{\mathrm{BB}}(t) = \sum_{\ell \in \mathbb{Z}} C_\ell g(t - \ell), \quad t \in \mathbb{R},$$

then its windowed version $X_{\mathrm{BB,w}}(\cdot)$ is given by

$$X_{\mathrm{BB,w}}(t) = \sum_{\ell \in \mathbb{Z}} C_\ell g(t - \ell) \, \mathrm{I}\{|t| \leq \mathsf{T}\}, \quad t \in \mathbb{R}.$$

Similarly $X_{\mathrm{PB,w}}(\cdot)$ is the windowed version of the SP

$$X_{\mathrm{PB}}(t) = 2 \, \mathrm{Re}\left( \sum_{\ell \in \mathbb{Z}} C_\ell g(t - \ell) e^{i2\pi f_c t} \right), \quad t \in \mathbb{R},$$

and $\mathbf{g}_{\ell,\mathrm{w}}$ is the windowed version of

$$\mathbf{g}_\ell : t \mapsto g(t - \ell), \quad \ell \in \mathbb{Z}. \tag{18.59}$$

We can now express the power in baseband as the limit, as $\mathsf{T}$ tends to infinity, of $\mathsf{E}\Big[\|\mathbf{X}_{\mathrm{BB,w}}\|_2^2\Big] / (2\mathsf{T})$, and the power in passband as the limit of $\mathsf{E}\Big[\|\mathbf{X}_{\mathrm{PB,w}}\|_2^2\Big] / (2\mathsf{T})$. Note that, since the function $\mathrm{I}\{\cdot\}$ is real-valued,

$$X_{\mathrm{PB,w}}(t) = 2 \, \mathrm{Re}\big(X_{\mathrm{BB,w}}(t) e^{i2\pi f_c t}\big), \quad t \in \mathbb{R}. \tag{18.60}$$

But (18.60) notwithstanding, the energy in $\mathbf{X}_{\mathrm{PB,w}}$ need not be twice the energy in $\mathbf{X}_{\mathrm{BB,w}}$ because the signal $\mathbf{X}_{\mathrm{BB,w}}$—unlike its unwindowed version $\mathbf{X}_{\mathrm{BB}}$—is not bandlimited. It is time-limited, and as such cannot be bandlimited (Theorem 6.8.2).

The difficulty in proving the theorem is in relating the energy in $\mathbf{X}_{\mathrm{PB,w}}$ to the energy in $\mathbf{X}_{\mathrm{BB,w}}$ and, specifically, in showing that the difference between half the energy in $\mathbf{X}_{\mathrm{PB,w}}$ and the energy in $\mathbf{X}_{\mathrm{BB,w}}$, when normalized by $2\mathsf{T}$, tends to zero. Aiding us in this is the following lemma relating the energy in passband to the energy in baseband for signals that are not bandlimited.

**Lemma 18.5.3.** *Let* $\mathbf{z}$ *be a complex energy-limited signal that is not necessarily bandlimited, and consider the real signal* $\mathbf{x} \colon t \mapsto 2 \operatorname{Re}\!\left( z(t)\, e^{\mathrm{i} 2\pi f_{\mathrm{c}} t} \right)$, *where* $f_{\mathrm{c}} > 0$ *is arbitrary. Then,*

$$\left( \|\mathbf{z}\|_2 - \sqrt{2}\epsilon \right)^2 \leq \frac{1}{2} \|\mathbf{x}\|_2^2 \leq \left( \|\mathbf{z}\|_2 + \sqrt{2}\epsilon \right)^2, \tag{18.61}$$

*where*

$$\epsilon^2 = \int_{-\infty}^{-f_{\mathrm{c}}} |\hat{z}(f)|^2 \, \mathrm{d}f. \tag{18.62}$$

**Proof.** Expressing the FT of $\mathbf{x}$ in terms of the FT of $\mathbf{z}$, we obtain that for every $f \in \mathbb{R}$ outside a set of frequencies of Lebesgue measure zero,

$$\begin{aligned}
\hat{x}(f)\, \mathrm{I}\{f \geq 0\} \\
&= \hat{z}(f - f_{\mathrm{c}})\, \mathrm{I}\{f \geq 0\} + \hat{z}^*(-f - f_{\mathrm{c}})\, \mathrm{I}\{f \geq 0\} \\
&= \hat{z}(f - f_{\mathrm{c}}) + \hat{z}^*(-f - f_{\mathrm{c}})\, \mathrm{I}\{f \geq 0\} - \hat{z}(f - f_{\mathrm{c}})\, \mathrm{I}\{f < 0\}. \tag{18.63}
\end{aligned}$$

We next consider the integral over $f$ of the squared magnitude of the LHS and of the RHS of (18.63). Since $\mathbf{x}$ is real, its FT is conjugate-symmetric so, by Parseval's Theorem, the integral of the squared magnitude of the LHS of (18.63) is $\frac{1}{2} \|\mathbf{x}\|_2^2$. The integral of the squared magnitude of the first term on the RHS of (18.63) is given by $\|\mathbf{z}\|_2^2$. Finally, the integral of the squared magnitude of each of the last two terms on the RHS of (18.63) is $\epsilon^2$ and, since they are orthogonal, the integral of the squared magnitude of their sum is $2\epsilon^2$. The result now follows from the Triangle Inequality (4.14). $\qquad\square$

Applying Lemma 18.5.3 with the substitution of $\mathbf{x}_{\mathrm{BB,w}}$ for $\mathbf{z}$ and of $\mathbf{x}_{\mathrm{PB,w}}$ for $\mathbf{x}$ we obtain upon noting that $f_{\mathrm{c}} > W/2$ that, in order to establish the theorem, it suffices to show that the "out-of-band energy" term

$$e^2 \triangleq \int_{|f| \geq W/2} \left| \hat{x}_{\mathrm{BB,w}}(f) \right|^2 \mathrm{d}f \tag{18.64}$$

satisfies

$$\lim_{\mathsf{T} \to \infty} \frac{1}{\mathsf{T}} e^2 = 0, \tag{18.65}$$

with the convergence being uniform. That is, we need to show that $e^2/\mathsf{T}$ is upper-bounded by some function of $\alpha$, $\beta$, $\gamma$, and $\mathsf{T}$ that converges to zero as $\mathsf{T}$ tends to infinity with $\alpha$, $\beta$, $\gamma$ held fixed. Aiding us in the calculation of the out-of-band energy is the following lemma.

**Lemma 18.5.4.** *Let* $\mathbf{x}$ *be an energy-limited signal and let* $W \geq 0$.

(i) *If* $\mathbf{u}$ *is any energy-limited signal that is bandlimited to* $W/2$ *Hz, then*

$$\int_{|f| \geq W/2} |\hat{x}(f)|^2 \, \mathrm{d}f \leq \|\mathbf{x} - \mathbf{u}\|_2^2. \tag{18.66}$$

*(ii) In particular,*

$$\int_{|f|\geq W/2} |\hat{x}(f)|^2 \, \mathrm{d}f \leq \|\mathbf{x}\|_2^2 \, . \tag{18.67}$$

**Proof.** Part (ii) follows from Parseval's Theorem. Part (i) follows by noting that if $\mathbf{u}$ is an energy-limited signal that is bandlimited to $W/2$ Hz, then the Fourier Transforms of $\mathbf{x}$ and $\mathbf{x} - \mathbf{u}$ are indistinguishable for frequencies $f$ that satisfy $|f| \geq W/2$. Consequently,

$$\int_{|f|\geq W/2} |\hat{x}(f)|^2 \, \mathrm{d}f = \int_{|f|\geq W/2} |\hat{x}(f) - \hat{u}(f)|^2 \, \mathrm{d}f$$
$$\leq \|\mathbf{x} - \mathbf{u}\|_2^2 \, ,$$

where the inequality follows by applying Part (ii) to the signal $\mathbf{x} - \mathbf{u}$. $\qquad\square$

To prove (18.65) fix some integer $\nu \geq 2$ and express $\mathbf{x}_{\mathrm{BB,w}}$ as

$$\mathbf{x}_{\mathrm{BB,w}} = \mathbf{s}_{0,\mathrm{w}} + \mathbf{s}_{1,\mathrm{w}} + \mathbf{s}_{2,\mathrm{w}}, \tag{18.68}$$

where

$$\mathbf{s}_{0,\mathrm{w}} = \sum_{0\leq|\ell|\leq\mathsf{T}-\nu} c_\ell \, \mathbf{g}_{\ell,\mathrm{w}}, \tag{18.69}$$

$$\mathbf{s}_{1,\mathrm{w}} = \sum_{\mathsf{T}-\nu<|\ell|\leq\mathsf{T}+\nu} c_\ell \, \mathbf{g}_{\ell,\mathrm{w}}, \tag{18.70}$$

$$\mathbf{s}_{2,\mathrm{w}} = \sum_{\mathsf{T}+\nu<|\ell|<\infty} c_\ell \, \mathbf{g}_{\ell,\mathrm{w}}, \tag{18.71}$$

are of corresponding out-of-band energies

$$e_\kappa^2 = \int_{|f|\geq W/2} \left|\hat{s}_{\kappa,\mathrm{w}}(f)\right|^2 \mathrm{d}f, \quad \kappa = 0, 1, 2. \tag{18.72}$$

Note that by (18.64), (18.68), and the Triangle Inequality

$$e^2 \leq \left(e_0 + e_1 + e_2\right)^2. \tag{18.73}$$

Since the integer $\nu \geq 2$ is arbitrary, it follows from (18.73) that, to establish (18.65) and to thus complete the proof of the theorem, it suffices to show that for every fixed integer $\nu \geq 2$,

$$\lim_{\mathsf{T}\to\infty} \frac{1}{\mathsf{T}} e_0^2 = 0, \tag{18.74}$$

$$\lim_{\mathsf{T}\to\infty} \frac{1}{\mathsf{T}} e_1^2 = 0, \tag{18.75}$$

and that

$$\lim_{\nu\to\infty} \left(\overline{\lim_{\mathsf{T}\to\infty}} \, \frac{1}{\mathsf{T}} e_2^2\right) = 0. \tag{18.76}$$

We thus conclude the theorem's proof by establishing (18.74), (18.75), and (18.76).

We begin with the easiest, namely (18.75). To establish (18.75) we recall the definition of $e_1$ (18.72) & (18.70) and use the Triangle Inequality to obtain

$$
\begin{aligned}
e_1 &\leq \sum_{\mathsf{T}-\nu<|\ell|\leq\mathsf{T}+\nu} \left( \int_{|f|\geq\mathsf{W}/2} \left| c_\ell\, \hat{g}_{\ell,\mathrm{w}}(f) \right|^2 \mathrm{d}f \right)^{1/2} \\
&\leq \gamma \sum_{\mathsf{T}-\nu<|\ell|\leq\mathsf{T}+\nu} \|\mathbf{g}_{\ell,\mathrm{w}}\|_2 \\
&\leq 4\gamma\nu \|\mathbf{g}\|_2 ,
\end{aligned}
\tag{18.77}
$$

where the second inequality follows from (18.23) and from Lemma 18.5.4 (ii), and where the final inequality follows because windowing can only reduce energy so $\|\mathbf{g}_{\ell,\mathrm{w}}\|_2 \leq \|\mathbf{g}_\ell\|_2 = \|\mathbf{g}\|_2$. Inequality (18.77) establishes (18.75).

Having established (18.75), we next turn to proving (18.74). The proof is quite similar except that, instead of using Part (ii) of Lemma 18.5.4, we use Part (i) with the substitutions of $\mathbf{g}_{\ell,\mathrm{w}}$ for $\mathbf{x}$ and of $\mathbf{g}_\ell$ for $\mathbf{u}$ to obtain

$$
\int_{|f|\geq\mathsf{W}/2} \left| \hat{g}_{\ell,\mathrm{w}}(f) \right|^2 \mathrm{d}f \leq \|\mathbf{g}_{\ell,\mathrm{w}} - \mathbf{g}_\ell\|_2^2 , \quad \ell \in \mathbb{Z}.
\tag{18.78}
$$

We further upper-bound the RHS of (18.78) using the decay condition (18.58) as

$$
\begin{aligned}
\|\mathbf{g}_{\ell,\mathrm{w}} - \mathbf{g}_\ell\|_2^2 &= \int_{-\infty}^{\infty} \left| g_\ell(f) \right|^2 \mathrm{I}\{|t| > \mathsf{T}\}\, \mathrm{d}t \\
&= \int_{-\infty}^{-\mathsf{T}} |g(t-\ell)|^2\, \mathrm{d}t + \int_{\mathsf{T}}^{\infty} |g(t-\ell)|^2\, \mathrm{d}t \\
&= \int_{-\infty}^{-\mathsf{T}-\ell} |g(\tau)|^2\, \mathrm{d}t + \int_{\mathsf{T}-\ell}^{\infty} |g(\tau)|^2\, \mathrm{d}\tau \\
&\leq \int_{-\infty}^{-\mathsf{T}-\ell} \frac{\beta^2}{|\tau|^{2+2\alpha}}\, \mathrm{d}\tau + \int_{\mathsf{T}-\ell}^{\infty} \frac{\beta^2}{|\tau|^{2+2\alpha}}\, \mathrm{d}\tau \\
&\leq 2 \int_{\mathsf{T}-|\ell|}^{\infty} \frac{\beta^2}{|\tau|^{2+2\alpha}}\, \mathrm{d}\tau \\
&= \frac{2\beta^2}{1+2\alpha} \frac{1}{(\mathsf{T}-|\ell|)^{1+2\alpha}}, \quad |\ell| < \mathsf{T},
\end{aligned}
$$

to obtain

$$
\left( \int_{|f|\geq\mathsf{W}/2} \left| \hat{g}_{\ell,\mathrm{w}}(f) \right|^2 \mathrm{d}f \right)^{1/2} \leq \sqrt{\frac{2\beta^2}{1+2\alpha}} \frac{1}{(\mathsf{T}-|\ell|)^{1/2+\alpha}}, \quad |\ell| < \mathsf{T}.
\tag{18.79}
$$

Using (18.72), (18.69), (18.79), (18.23), and the Triangle Inequality we thus obtain

$$
\begin{aligned}
e_0 &\leq \sum_{0\leq|\ell|\leq\mathsf{T}-\nu} |c_\ell| \left( \int_{|f|\geq\mathsf{W}/2} \left| \hat{g}_{\ell,\mathrm{w}}(f) \right|^2 \mathrm{d}f \right)^{1/2} \\
&\leq \sqrt{\frac{2\gamma^2\beta^2}{1+2\alpha}} \sum_{0\leq|\ell|\leq\mathsf{T}-\nu} \frac{1}{(\mathsf{T}-|\ell|)^{1/2+\alpha}}
\end{aligned}
$$

$$
\leq 2\sqrt{\frac{2\gamma^2\beta^2}{1+2\alpha}} \sum_{\ell=0}^{\mathsf{T}-\nu} \frac{1}{(\mathsf{T}-\ell)^{1/2+\alpha}}
$$

$$
= 2\sqrt{\frac{2\gamma^2\beta^2}{1+2\alpha}} \sum_{\tilde{\ell}=\nu}^{\mathsf{T}} \frac{1}{\tilde{\ell}^{1/2+\alpha}}
$$

$$
\leq 2\sqrt{\frac{2\gamma^2\beta^2}{1+2\alpha}} \int_{\nu-1}^{\mathsf{T}} \frac{1}{\xi^{1/2+\alpha}}\,\mathrm{d}\xi
$$

$$
= \begin{cases} \sqrt{\frac{\gamma^2\beta^2}{1+2\alpha}}\,\frac{4}{1-2\alpha}\left(\mathsf{T}^{1/2-\alpha}-(\nu-1)^{1/2-\alpha}\right) & \text{if } \alpha \neq 1/2 \\ 2\gamma\beta\big(\ln\mathsf{T}-\ln(\nu-1)\big) & \text{if } \alpha = 1/2 \end{cases},
\tag{18.80}
$$

where the inequality in the first line follows from (18.72) and from the Triangle Inequality; the inequality in the second line from (18.79); the inequality in the third line by counting the term $\ell = 0$ twice; the equality in the fourth line by changing the summation variable to $\tilde{\ell} \triangleq \mathsf{T}-\ell$; the inequality in the fifth line from the monotonicity of the function $\xi \mapsto \xi^{-1/2-\alpha}$, which implies that

$$
\tilde{\ell}^{-1/2-\alpha} \leq \int_{\tilde{\ell}-1}^{\tilde{\ell}} \frac{1}{\xi^{1/2+\alpha}}\,\mathrm{d}\xi;
$$

and where the final equality on the sixth line follows by direct calculation. Inequality (18.80) combines with our assumption that $\alpha$ is positive to prove (18.74).

We now conclude the proof of the theorem by establishing (18.76). To that end, we begin by using Lemma 18.5.4 (ii) and the fact that $s_{2,\mathrm{w}}$ is zero outside the interval $[-\mathsf{T},\mathsf{T}]$ to obtain

$$
e_2^2 \leq \int_{-\mathsf{T}}^{\mathsf{T}} \big|s_{2,\mathrm{w}}(t)\big|^2\,\mathrm{d}t.
\tag{18.81}
$$

We next upper-bound the RHS of (18.81) using the boundedness of the symbols (18.23) and the decay condition (18.58):

$$
\big|s_{2,\mathrm{w}}(t)\big| = \left| \sum_{\mathsf{T}+\nu<|\ell|<\infty} c_\ell\, g_{\ell,\mathrm{w}}(t) \right|
$$

$$
\leq \gamma \sum_{\mathsf{T}+\nu<|\ell|<\infty} |g(t-\ell)|\, \mathrm{I}\{|t|\leq\mathsf{T}\}
$$

$$
\leq \gamma \sum_{\mathsf{T}+\nu<|\ell|<\infty} \frac{\beta}{|t-\ell|^{1+\alpha}}\, \mathrm{I}\{|t|\leq\mathsf{T}\}
$$

$$
\leq \gamma \sum_{\mathsf{T}+\nu<|\ell|<\infty} \frac{\beta}{\big||\ell|-|t|\big|^{1+\alpha}}\, \mathrm{I}\{|t|\leq\mathsf{T}\}
$$

$$
\leq \gamma \sum_{\mathsf{T}+\nu<|\ell|<\infty} \frac{\beta}{(|\ell|-\mathsf{T})^{1+\alpha}}
$$

$$
= 2\gamma\beta \sum_{\ell=\mathsf{T}+\nu+1}^{\infty} \frac{1}{(\ell-\mathsf{T})^{1+\alpha}}
$$

$$= 2\gamma\beta \sum_{\tilde{\ell}=\nu+1}^{\infty} \frac{1}{\tilde{\ell}^{1+\alpha}}$$

$$\leq 2\gamma\beta \int_{\nu}^{\infty} \xi^{-1-\alpha}\, d\xi$$

$$= \frac{2\gamma\beta}{\alpha} \nu^{-\alpha}, \tag{18.82}$$

where the equality in the first line follows from the definition of $\mathbf{s}_{2,\mathrm{w}}$ (18.71); the inequality in the second line from the Triangle Inequality for Complex Numbers (2.12), the boundedness of $(C_\ell)$ (18.23), and from the definition of $\mathbf{g}_\ell$ (18.59); the inequality in the third line from (18.58); the inequality in the fourth line because $|\xi - \zeta| \geq ||\xi| - |\zeta||$ whenever $\xi, \zeta \in \mathbb{R}$; the inequality in the fifth line because for $|t| > \mathsf{T}$ the LHS is zero and the RHS is positive, and because for $|t| \leq \mathsf{T}$ we have that $|\ell| - |t| \geq |\ell| - \mathsf{T}$ throughout the range of summation; the equality in the sixth line from the symmetry of the summand and from the assumption that $\mathsf{T}$ is an integer; the equality in the seventh line by changing the summation variable to $\tilde{\ell} = \ell - \mathsf{T}$; the inequality in the eighth line from the monotonicity of the function $\xi \mapsto \xi^{-1-\alpha}$, which implies that

$$\frac{1}{\tilde{\ell}^{1+\alpha}} \leq \int_{\tilde{\ell}-1}^{\tilde{\ell}} \frac{1}{\xi^{1+\alpha}}\, d\xi;$$

and the final equality in the ninth line by evaluating the integral.

It follows from (18.82) and (18.81) that

$$e_2^2 \leq 2\mathsf{T} \frac{4\gamma^2\beta^2}{\alpha^2} \nu^{-2\alpha} \tag{18.83}$$

and hence that

$$\overline{\lim_{\mathsf{T}\to\infty}} \frac{1}{\mathsf{T}} e_2^2 \leq \frac{8\gamma^2\beta^2}{\alpha^2} \nu^{-2\alpha}, \quad \nu \geq 2,$$

which proves (18.76).

## 18.6    A Formal Account of the PSD in Baseband and Passband

In this section we justify the derivations of Section 18.4.

### 18.6.1    On Limits of Convolutions

We begin with a lemma that justifies the swapping of infinite summation and convolution. As a corollary we establish conditions under which feeding a (real or complex) PAM signal of pulse shape $\mathbf{g}$ to a stable filter of impulse response $\mathbf{h}$ is tantamount to replacing its pulse shape $\mathbf{g}$ with the new pulse shape $\mathbf{g} \star \mathbf{h}$.

**Lemma 18.6.1.** *Let* $\mathbf{s}_1, \mathbf{s}_2, \ldots$ *be a sequence of measurable functions from* $\mathbb{R}$ *to* $\mathbb{C}$ *satisfying the following two conditions:*

1) *The sequence is uniformly bounded in the sense that there exists some positive number $\sigma_\infty$ such that*

$$\left|s_\ell(t)\right| \leq \sigma_\infty, \quad \left(t \in \mathbb{R}, \quad \ell = 1, 2, \ldots\right). \tag{18.84}$$

2) *The sequence converges to some function* **s** *uniformly over compact sets in the sense that for every fixed $\xi > 0$*

$$\lim_{\ell \to \infty} \sup_{|t| \leq \xi} \left|s(t) - s_\ell(t)\right| = 0. \tag{18.85}$$

*Then for every* $\mathbf{h} \in \mathcal{L}_1$,

$$\lim_{\ell \to \infty} \left(\mathbf{s}_\ell \star \mathbf{h}\right)(t) = \left(\mathbf{s} \star \mathbf{h}\right)(t), \quad t \in \mathbb{R}. \tag{18.86}$$

**Proof.** Fix some epoch $t_0 \in \mathbb{R}$ and some $\mathbf{h} \in \mathcal{L}_1$. We will show that for every $\epsilon > 0$ there exists some $L_0 \in \mathbb{N}$ (depending on $\epsilon$) such that

$$\left|\left(\mathbf{s}_\ell \star \mathbf{h}\right)(t_0) - \left(\mathbf{s} \star \mathbf{h}\right)(t_0)\right| < \epsilon, \quad \ell \geq L_0. \tag{18.87}$$

To that end note that our assumption that **h** is integrable implies that there exists some $\xi > 0$ such that

$$\int_{|\tau| \geq \xi} |h(\tau)| \, d\tau < \frac{\epsilon}{3\sigma_\infty}. \tag{18.88}$$

And when we apply our assumption that the sequence $\mathbf{s}_1, \mathbf{s}_2, \ldots$ converges to **s** uniformly over compact sets to the compact interval $[t_0 - \xi, t_0 + \xi]$, we obtain that there exists some $L_0$ (depending on $\epsilon$, $t_0$, and $\xi$) such that

$$\|\mathbf{h}\|_1 \sup_{t_0 - \xi \leq \tau \leq t_0 + \xi} \left|s(\tau) - s_\ell(\tau)\right| < \frac{\epsilon}{3}, \quad \ell \geq L_0. \tag{18.89}$$

We can now derive (18.87) as follows:

$$\left|\left(\mathbf{s}_\ell \star \mathbf{h}\right)(t_0) - \left(\mathbf{s} \star \mathbf{h}\right)(t_0)\right|$$

$$= \left|\int_{-\infty}^{\infty} s_\ell(t_0 - \tau) h(\tau) \, d\tau - \int_{-\infty}^{\infty} s(t_0 - \tau) h(\tau) \, d\tau\right|$$

$$\leq \left|\int_{-\xi}^{\xi} s_\ell(t_0 - \tau) h(\tau) \, d\tau - \int_{-\xi}^{\xi} s(t_0 - \tau) h(\tau) \, d\tau\right|$$

$$+ \left|\int_{|\tau| > \xi} s(t_0 - \tau) h(\tau) \, d\tau\right| + \left|\int_{|\tau| > \xi} s_\ell(t_0 - \tau) h(\tau) \, d\tau\right|$$

$$\leq \int_{-\xi}^{\xi} \left|s_\ell(t_0 - \tau) - s(t_0 - \tau)\right| |h(\tau)| \, d\tau$$

$$+ \int_{|\tau| > \xi} \left|s(t_0 - \tau) h(\tau)\right| d\tau + \int_{|\tau| > \xi} \left|s_\ell(t_0 - \tau) h(\tau)\right| d\tau$$

$$\leq \|\mathbf{h}\|_1 \left(\sup_{t_0 - \xi \leq \tau \leq t_0 + \xi} \left|s(\tau) - s_\ell(\tau)\right|\right) + 2\sigma_\infty \int_{|\tau| > \xi} |h(\tau)| \, d\tau$$

$$< \epsilon,$$

where the last equality follows from (18.88) and (18.89). $\qquad\square$

**Corollary 18.6.2.** *If the sequence* $(C_\ell)$ *is bounded in the sense of* (18.23) *and if the measurable function* **g** *satisfies the decay condition* (18.22), *then for every* $\mathbf{h} \in \mathcal{L}_1$ *and every epoch* $t_0 \in \mathbb{R}$

$$\left( \left( t \mapsto \sum_{\ell \in \mathbb{Z}} C_\ell\, g(t - \ell\mathsf{T}_\mathsf{s}) \right) \star \mathbf{h} \right)(t_0) = \sum_{\ell \in \mathbb{Z}} C_\ell (\mathbf{g} \star \mathbf{h})(t_0 - \ell\mathsf{T}_\mathsf{s}). \qquad (18.90)$$

**Proof.** Follows by applying Lemma 18.6.1 to the functions

$$\mathsf{s}_\mathsf{L} : t \mapsto \sum_{\ell = -\mathsf{L}}^{\mathsf{L}} C_\ell\, g(t - \ell\mathsf{T}_\mathsf{s}), \quad \mathsf{L} = 1, 2, \dots \qquad \square$$

## 18.6.2 On the Support of the Operational PSD of $\mathbf{X}_{\mathrm{BB}}$

We next prove that if the pulse shape **g** is bandlimited to $W/2$ Hz, then the operational PSD of $\mathbf{X}_{\mathrm{BB}}$ is zero at frequencies outside the band $[-W/2, W/2]$. That is, we justify (18.46).

**Proposition 18.6.3.** *Assume that* A, $\mathsf{T}_\mathsf{s}$, **g**, $W$, *and* $f_c$ *are as in Theorem 18.2.1 and, additionally, that* **g** *satisfies the decay condition* (18.22) *and that the CSP* $(C_\ell)$ *is bounded in the sense of* (18.23). *If the CSP* $(X_{\mathrm{BB}}(t),\ t \in \mathbb{R})$ *of* (18.25) *is of operational PSD* $\mathsf{S}_{\mathrm{BB}}(\cdot)$, *then* $\mathsf{S}_{\mathrm{BB}}(f)$ *is zero for all* $|f| > W/2$ *outside a set of Lebesgue measure zero, and consequently*

$$f \mapsto \mathsf{S}_{\mathrm{BB}}(f)\, \mathrm{I}\left\{ |f| \leq \frac{W}{2} \right\}$$

*is also an operational PSD for* $(X_{\mathrm{BB}}(t),\ t \in \mathbb{R})$.

**Proof.** We shall show that the proposition's hypotheses imply that if $\mathbf{h} \in \mathcal{L}_1$ is such that $\hat{h}(f) = 0$ at all frequencies $f$ satisfying $|f| \leq W/2$, then the power in $\mathbf{X}_{\mathrm{BB}} \star \mathbf{h}$ is zero, irrespective of the values of $\hat{h}(f)$ at other frequencies. That is, we shall show that

$$\left( \hat{h}(f) = 0, \quad |f| \leq W/2 \right) \Rightarrow \left( \text{Power in } \mathbf{X}_{\mathrm{BB}} \star \mathbf{h} = 0 \right), \quad \mathbf{h} \in \mathcal{L}_1. \qquad (18.91)$$

Since $\mathbf{X}_{\mathrm{BB}}$ is, by assumption, of operational PSD $\mathsf{S}_{\mathrm{BB}}(\cdot)$, it will then follow from (18.91) that

$$\left( \hat{h}(f) = 0, \quad |f| \leq W/2 \right) \Rightarrow \left( \int_{-\infty}^{\infty} \mathsf{S}_{\mathrm{BB}}(f)\, |\hat{h}(f)|^2 \, \mathrm{d}f = 0 \right), \quad \mathbf{h} \in \mathcal{L}_1. \qquad (18.92)$$

From (18.92) it is just a technicality to show that the nonnegative function $\mathsf{S}_{\mathrm{BB}}(\cdot)$ must be zero at all frequencies $|f| > W/2$ outside a set of Lebesgue measure zero. Indeed, if, in order to reach a contradiction, we assume that $\mathsf{S}_{\mathrm{BB}}(\cdot)$ is not indistinguishable from the all-zero function in some interval $[a, b]$, where $a$ and $b$ are such that $W/2 < a < b$, then picking **h** as an integrable function such that $\hat{h}(f)$ is zero for $|f| \leq W/2$ and such that $\hat{h}(f) = 1$ for $a \leq f \leq b$ would yield a contradiction to (18.92). (An example of such a function $\mathbf{h}$ is the IFT of the shifted-trapezoid mapping

$$
f \mapsto \begin{cases} 1 & \text{if } a \leq f \leq b, \\ 0 & \text{if } f \leq W/2 \text{ or } f \geq b + (a - W/2), \quad f \in \mathbb{R}, \\ 1 - \frac{|f - (a+b)/2| - (b-a)/2}{a - W/2} & \text{otherwise}, \end{cases}
$$

which is a frequency shifted version of the function we encountered in (7.15) and (7.17).) The assumption that $\mathsf{S}_{\mathrm{BB}}(\cdot)$ is not indistinguishable from the all-zero function in some interval $[a, b]$ where $a < b < -W/2$ can be similarly contradicted.

To complete the proof we thus need to justify (18.91). This follows from two observations. The first is that, by Corollary 18.6.2, for every $\mathbf{h} \in \mathcal{L}_1$

$$
\text{Power in } \mathbf{X}_{\mathrm{BB}} \star \mathbf{h} = \text{Power in } t \mapsto A \sum_{\ell \in \mathbb{Z}} C_\ell \left( \mathbf{g} \star \mathbf{h} \right) (t - \ell \mathsf{T}_{\mathrm{s}}). \tag{18.93}
$$

The second is that, because $\mathbf{g}$ is an integrable function that is bandlimited to $W/2$ Hz, it follows from Proposition 6.5.2 that

$$
\left( \mathbf{g} \star \mathbf{h} \right)(t) = \int_{-W/2}^{W/2} \hat{g}(f) \, \hat{h}(f) \, e^{\mathrm{i} 2\pi f t} \, \mathrm{d}f, \quad t \in \mathbb{R}
$$

and, in particular,

$$
\left( \hat{h}(f) = 0, \quad |f| \leq W/2 \right) \Rightarrow \left( \mathbf{g} \star \mathbf{h} = \mathbf{0} \right), \quad \mathbf{h} \in \mathcal{L}_1. \tag{18.94}
$$

Combining (18.93) and (18.94) establishes (18.91). $\qquad\square$

### 18.6.3  On the Definition of the Operational PSD

In order to demonstrate that $\left( Z(t), \; t \in \mathbb{R} \right)$ is of operational PSD $\mathsf{S}_{ZZ}$, one has to show that (18.43) holds for every function $\mathbf{h} \colon \mathbb{R} \to \mathbb{C}$ in $\mathcal{L}_1$ (Definition 18.4.1). It turns out that it suffices to establish (18.43) only for functions that are in a subset of $\mathcal{L}_1$, provided that the subset is sufficiently rich. This result will allow us to consider only functions $\mathbf{h}$ of compact support. To make this result precise we need the following definition. We say that the set $\mathcal{H}$ is a **dense subset of** $\mathcal{L}_1$ if $\mathcal{H}$ is a subset of $\mathcal{L}_1$ such that for every $\mathbf{h} \in \mathcal{L}_1$ there exists a sequence $\mathbf{h}_1, \mathbf{h}_2, \ldots$ of elements of $\mathcal{H}$ such that $\lim_{\nu \to \infty} \| \mathbf{h} - \mathbf{h}_\nu \|_1 = 0$. An example of a dense subset of $\mathcal{L}_1$ is the subset of functions of compact support, where a function $\mathbf{h} \colon \mathbb{R} \to \mathbb{C}$ is said to be of **compact support** if there exists some $\Delta > 0$ such that

$$
h(t) = 0, \quad |t| \geq \Delta. \tag{18.95}
$$

**Lemma 18.6.4 (On Functions of Compact Support).**

(i) *The set of integrable functions of compact support is a dense subset of $\mathcal{L}_1$.*

(ii) *If $\mathbf{h}$ is of compact support and if $\mathbf{g}$ satisfies the decay condition (18.22) with parameters $\alpha, \beta, \mathsf{T}_{\mathrm{s}} > 0$, then $\mathbf{g} \star \mathbf{h}$ also satisfies this decay condition with the same parameters $\alpha$ and $\mathsf{T}_{\mathrm{s}}$ but with a possibly different parameter $\beta'$.*

**Proof.** We begin with Part (i). Given any integrable function $\mathbf{h}$ (not necessarily of compact support) we define the sequence of integrable functions of compact support $\mathbf{h}_1, \mathbf{h}_2, \ldots$ by $\mathbf{h}_\nu : t \mapsto h(t) \, \mathrm{I}\{|t| \leq \nu\}$ for every $\nu \in \mathbb{N}$. It is then just a technicality to show that $\|\mathbf{h} - \mathbf{h}_\nu\|_1$ converges to zero. (This can be shown using the Dominated Convergence Theorem because $|h_\nu(t)| \leq |h(t)|$ for all $t \in \mathbb{R}$ and because $\mathbf{h}$ is integrable.)

We next prove Part (ii). Let $\mathbf{g}$ satisfy the decay condition (18.22) with the positive parameters $\alpha, \beta, \mathsf{T_s}$, and let $\Delta > 0$ be such that (18.95) is satisfied. We shall prove the lemma by showing that

$$\left|(\mathbf{g} \star \mathbf{h})(t)\right| \leq \frac{\beta'}{1 + (|t|/\mathsf{T_s})^{1+\alpha}}, \quad t \in \mathbb{R}, \tag{18.96}$$

where

$$\beta' = \beta \, \|\mathbf{h}\|_1 \, 2^{1+\alpha} \left(1 + (2\Delta/\mathsf{T_s})^{1+\alpha}\right). \tag{18.97}$$

To that end we shall first show that

$$\left|(\mathbf{g} \star \mathbf{h})(t)\right| \leq \beta \, \|\mathbf{h}\|_1, \quad t \in \mathbb{R} \tag{18.98}$$

and

$$\left|(\mathbf{g} \star \mathbf{h})(t)\right| \leq \beta \, \|\mathbf{h}\|_1 \, 2^{1+\alpha} \frac{1}{1 + (|t|/\mathsf{T_s})^{1+\alpha}}, \quad |t| \geq 2\Delta. \tag{18.99}$$

We shall then proceed to show that the RHS of (18.96) is larger than the RHS of (18.98) for $|t| \leq 2\Delta$ and that it is larger than the RHS of (18.99) for $|t| > 2\Delta$.

Both (18.98) and (18.99) follow from the bound

$$
\begin{aligned}
\left|(\mathbf{g} \star \mathbf{h})(t)\right| &= \left| \int_{t-\Delta}^{t+\Delta} g(\tau) h(t - \tau) \, \mathrm{d}\tau \right| \\
&\leq \int_{t-\Delta}^{t+\Delta} |g(\tau)| \, |h(t - \tau)| \, \mathrm{d}\tau \\
&\leq \int_{t-\Delta}^{t+\Delta} \left( \sup_{t-\Delta \leq \sigma \leq t+\Delta} |g(\sigma)| \right) |h(t - \tau)| \, \mathrm{d}\tau \\
&= \|\mathbf{h}\|_1 \sup_{t-\Delta \leq \sigma \leq t+\Delta} |g(\sigma)|
\end{aligned}
$$

as follows. Bound (18.98) simply follows by using (18.22) to upper-bound $|g(t)|$ by $\beta$. And Bound (18.99) follows by using (18.22) to upper-bound $|g(t)|$ for $|t| \geq \Delta$ by $\beta/\left(1 + ((|t| - \Delta)/\mathsf{T_s})^{1+\alpha}\right)$, and by then upper-bounding this latter expression in the range $|t| > 2\Delta$ by $\beta 2^{1+\alpha}/\left(1 + (|t|/\mathsf{T_s})^{1+\alpha}\right)$ because in this range

$$
\begin{aligned}
1 + \left((|t| - \Delta)/\mathsf{T_s}\right)^{1+\alpha} &= 1 + \left(\frac{|t|}{\mathsf{T_s}}\right)^{1+\alpha} \left(\frac{|t| - \Delta}{|t|}\right)^{1+\alpha} \\
&\geq 1 + \left(\frac{|t|}{\mathsf{T_s}}\right)^{1+\alpha} \left(\frac{1}{2}\right)^{1+\alpha} \\
&\geq 2^{-(1+\alpha)} + 2^{-(1+\alpha)} \left(\frac{|t|}{\mathsf{T_s}}\right)^{1+\alpha}, \quad |t| \geq 2\Delta.
\end{aligned}
$$

Having established (18.98) and (18.99) we now complete the proof by showing that the RHS of (18.96) upper-bounds the RHS of (18.98) whenever $|t| \leq 2\Delta$, and that it upper-bounds the RHS of (18.99) for $|t| \geq 2\Delta$. That the RHS of (18.96) upper-bounds the RHS of (18.98) whenever $|t| \leq 2\Delta$ follows because

$$\frac{\beta \|\mathbf{h}\|_1 \, 2^{1+\alpha} \left(1 + (2\Delta/\mathsf{T}_\mathrm{s})^{1+\alpha}\right)}{1 + (|t|/\mathsf{T}_\mathrm{s})^{1+\alpha}} \geq \beta \|\mathbf{h}\|_1 \, 2^{1+\alpha} \geq \beta \|\mathbf{h}\|_1, \quad |t| \leq 2\Delta.$$

And that the RHS of (18.96) upper-bounds the RHS of (18.99) whenever $|t| > 2\Delta$ follows because the term $1 + (2\Delta/\mathsf{T}_\mathrm{s})^{1+\alpha}$ is larger than one. $\qquad\square$

**Proposition 18.6.5.** *Assume that $\mathcal{H}$ is a dense subset of $\mathcal{L}_1$ and that the (real or complex) measurable stochastic process $\big(Z(t),\ t \in \mathbb{R}\big)$ is bounded in the sense that for some $\sigma_\infty$*

$$|Z(t)| \leq \sigma_\infty, \quad t \in \mathbb{R}. \tag{18.100}$$

*If $\mathsf{S}(\cdot)$ is a nonnegative integrable function such that the relation*

$$\text{Power in } \mathbf{Z} \star \mathbf{h} = \int_{-\infty}^{\infty} \mathsf{S}(f) \, |\hat{h}(f)|^2 \, \mathrm{d}f \tag{18.101}$$

*holds for every $\mathbf{h} \in \mathcal{H}$, then it holds for all $\mathbf{h} \in \mathcal{L}_1$.*

**Proof.** Let $\mathbf{h}$ be an element of $\mathcal{L}_1$ (but not necessarily of $\mathcal{H}$) for which we would like to prove (18.101). Since $\mathcal{H}$ is a dense subset of $\mathcal{L}_1$, there exists a sequence $\mathbf{h}_1, \mathbf{h}_2, \ldots$ of elements of $\mathcal{H}$

$$\mathbf{h}_\nu \in \mathcal{H}, \quad \nu = 1, 2, \ldots \tag{18.102}$$

such that

$$\lim_{\nu \to \infty} \|\mathbf{h} - \mathbf{h}_\nu\|_1 = 0. \tag{18.103}$$

We shall prove that (18.101) holds for $\mathbf{h}$ by justifying the calculation

$$\text{Power in } \mathbf{Z} \star \mathbf{h} = \lim_{\nu \to \infty} \text{Power in } \mathbf{Z} \star \mathbf{h}_\nu \tag{18.104}$$

$$= \lim_{\nu \to \infty} \int_{-\infty}^{\infty} \mathsf{S}(f) \, |\hat{h}_\nu(f)|^2 \, \mathrm{d}f \tag{18.105}$$

$$= \int_{-\infty}^{\infty} \mathsf{S}(f) \, |\hat{h}(f)|^2 \, \mathrm{d}f. \tag{18.106}$$

The justification of (18.105) is that, by (18.102), each of the functions $\mathbf{h}_\nu$ is in $\mathcal{H}$, and the proposition's hypothesis guarantees that (18.101) holds for such functions.

The justification of (18.106) is a bit technical. It is based on noting that (18.103) implies (by Theorem 6.2.11 (i) with the substitution of $\mathbf{h} - \mathbf{h}_\nu$ for $\mathbf{x}$) that

$$\lim_{\nu \to \infty} \hat{h}_\nu(f) = \hat{h}(f), \quad f \in \mathbb{R} \tag{18.107}$$

and by then using the Dominated Convergence Theorem to justify the swapping of the limit and integral. Indeed, (by Theorem 6.2.11 (i)) for every $\nu \in \mathbb{N}$, the function

$f \mapsto S(f)\hat{h}_\nu(f)$ is bounded by the function $f \mapsto \left(\sup_\nu \|\mathbf{h}_\nu\|_1\right) S(f)$, which is integrable because $S(\cdot)$ is integrable (by the proposition's hypothesis) and because the integrability of $\mathbf{h}$ and (18.103) imply that the supremum is finite as can be verified using the Triangle Inequality by writing $\mathbf{h}_\nu$ as $\mathbf{h} - (\mathbf{h} - \mathbf{h}_\nu)$.

We now complete the proof by justifying (18.104). Since $\mathbf{Z} \star \mathbf{h}_\nu = \mathbf{Z} \star \mathbf{h} - \mathbf{Z} \star (\mathbf{h} - \mathbf{h}_\nu)$, it follows from the Triangle Inequality for Stochastic Processes (Proposition 18.5.1) that for every $\mathsf{T} > 0$

$$\left| \sqrt{\mathsf{E}\left[ \int_{-\mathsf{T}}^{\mathsf{T}} |\mathbf{Z} \star \mathbf{h}_\nu(t)|^2 \, dt \right]} - \sqrt{\mathsf{E}\left[ \int_{-\mathsf{T}}^{\mathsf{T}} |\mathbf{Z} \star \mathbf{h}(t)|^2 \, dt \right]} \right|$$

$$\leq \sqrt{\mathsf{E}\left[ \int_{-\mathsf{T}}^{\mathsf{T}} |(\mathbf{Z} \star (\mathbf{h} - \mathbf{h}_\nu))(t)|^2 \, dt \right]}$$

$$\leq \sqrt{2\mathsf{T}} \sigma_\infty \|\mathbf{h} - \mathbf{h}_\nu\|_1, \qquad (18.108)$$

where the second inequality follows from (18.100) using (5.8c). Upon dividing by $\sqrt{2\mathsf{T}}$ and taking the limit of $\mathsf{T} \to \infty$, it now follows from (18.108) that

$$\left| \sqrt{\text{Power in } \mathbf{Z} \star \mathbf{h}_\nu} - \sqrt{\text{Power in } \mathbf{Z} \star \mathbf{h}} \right| \leq \sigma_\infty \|\mathbf{h} - \mathbf{h}_\nu\|_1,$$

from which (18.104) follows by (18.103). $\qquad \square$

### 18.6.4 Relating the Operational PSD in Passband and Baseband

We next make the relationship (18.44) between the operational PSD of $\mathbf{X}$ and the operational PSD of $\mathbf{X}_{\text{BB}}$ formal.

**Theorem 18.6.6.** *Under the assumptions of Proposition 18.6.3, if the complex stochastic process $\left(X_{\text{BB}}(t), \ t \in \mathbb{R}\right)$ of (18.25) is of operational PSD $\mathsf{S}_{\text{BB}}(\cdot)$ in the sense that $\mathsf{S}_{\text{BB}}(\cdot)$ is an integrable function satisfying that for every complex $\mathbf{h}_{\text{c}} \in \mathcal{L}_1$,*

$$\lim_{\mathsf{T} \to \infty} \frac{1}{2\mathsf{T}} \mathsf{E}\left[ \int_{-\mathsf{T}}^{\mathsf{T}} |(\mathbf{X}_{\text{BB}} \star \mathbf{h}_{\text{c}})(t)|^2 \, dt \right] = \int_{-\infty}^{\infty} \mathsf{S}_{\text{BB}}(f) \, |\hat{h}_{\text{c}}(f)|^2 \, df, \qquad (18.109)$$

*then the QAM real SP $\left(X(t), \ t \in \mathbb{R}\right)$ of (18.24) is of operational PSD*

$$\mathsf{S}_{\text{PB}}(f) \triangleq \mathsf{S}_{\text{BB}}(f - f_{\text{c}}) + \mathsf{S}_{\text{BB}}(-f - f_{\text{c}}), \quad f \in \mathbb{R} \qquad (18.110)$$

*in the sense that $\mathsf{S}_{\text{PB}}(\cdot)$ is an integrable symmetric function such that for every real $\mathbf{h}_{\text{r}} \in \mathcal{L}_1$*

$$\lim_{\mathsf{T} \to \infty} \frac{1}{2\mathsf{T}} \mathsf{E}\left[ \int_{-\mathsf{T}}^{\mathsf{T}} |(\mathbf{X} \star \mathbf{h}_{\text{r}})(t)|^2 \, dt \right] = \int_{-\infty}^{\infty} \mathsf{S}_{\text{PB}}(f) \, |\hat{h}_{\text{r}}(f)|^2 \, df. \qquad (18.111)$$

**Proof.** The hypothesis that $\mathsf{S}_{\text{BB}}(\cdot)$ is integrable clearly implies that $\mathsf{S}_{\text{PB}}(\cdot)$, as defined in (18.110), is integrable and symmetric. It remains to show that if (18.109) holds for every complex $\mathbf{h}_c \in \mathcal{L}_1$, then (18.111) must hold for every real $\mathbf{h}_r \in \mathcal{L}_1$. Since the set of integrable functions of compact support is a dense subset of $\mathcal{L}_1$ (Lemma 18.6.4 (i)), it follows from Proposition 18.6.5 that it suffices to establish (18.111) for real functions $\mathbf{h}_r$ that are of compact support. Let $\mathbf{h}_r$ be such a function. The following calculation demonstrates that passing the QAM signal $\mathbf{X}$ through a filter of impulse response $\mathbf{h}_r$ is tantamount to replacing its pulse shape $\mathbf{g}$ with the pulse shape consisting of the convolution of $\mathbf{g}$ with the complex signal $\tau \mapsto e^{-i2\pi f_c \tau} h_r(\tau)$:

$$
\begin{aligned}
(\mathbf{X} \star \mathbf{h}_r)(t) &= \left( \left( \tau \mapsto 2\operatorname{Re}\left(\mathbf{X}_{\mathrm{BB}}(\tau)\, e^{i2\pi f_c \tau}\right) \right) \star \mathbf{h}_r \right)(t) \\
&= 2\operatorname{Re}\left( \left( \left( \tau \mapsto \mathbf{X}_{\mathrm{BB}}(\tau)\, e^{i2\pi f_c \tau} \right) \star \mathbf{h}_r \right)(t) \right) \\
&= 2\operatorname{Re}\left( e^{i2\pi f_c t} \left( \mathbf{X}_{\mathrm{BB}} \star \left( \tau \mapsto e^{-i2\pi f_c \tau} h_r(\tau) \right) \right)(t) \right) \\
&= 2\operatorname{Re}\left( e^{i2\pi f_c t} A \sum_{\ell=-\infty}^{\infty} C_\ell \left( \mathbf{g} \star \left( \tau \mapsto e^{-i2\pi f_c \tau} h_r(\tau) \right) \right)(t - \ell T_s) \right) \\
&= 2\operatorname{Re}\left( A \sum_{\ell=-\infty}^{\infty} C_\ell \left( \mathbf{g} \star \mathbf{h}_c \right)(t - \ell T_s)\, e^{i2\pi f_c t} \right),
\end{aligned}
\tag{18.112}
$$

where the first equality follows from the definition of $\mathbf{X}$ in terms of $\mathbf{X}_{\mathrm{BB}}$; the second because $\mathbf{h}_r$ is real (see (7.38) on the convolution between a real and a complex signal); the third from Proposition 7.8.1; the fourth from Corollary 18.6.2; and where the fifth equality follows by defining the mapping

$$
\mathbf{h}_c : t \mapsto e^{-i2\pi f_c t} h_r(t).
\tag{18.113}
$$

Note that by (18.113)

$$
\hat{h}_c(f) = \hat{h}_r(f + f_c), \quad f \in \mathbb{R}.
\tag{18.114}
$$

It follows from (18.112) that $\mathbf{X} \star \mathbf{h}_r$ has the form of a QAM signal with pulse shape $\mathbf{g} \star \mathbf{h}_c$. We note that, because $\mathbf{g}$ (by hypothesis) satisfies the decay condition (18.22) and because the fact that $\mathbf{h}_r$ is of compact support implies by (18.113) that $\mathbf{h}_c$ is also of compact support, it follows from Lemma 18.6.4 (ii) that the pulse shape $\mathbf{g} \star \mathbf{h}_c$ satisfies the decay condition

$$
\left| (\mathbf{g} \star \mathbf{h}_c)(t) \right| \leq \frac{\beta'}{1 + (|t|/T_s)^{1+\alpha}}, \quad t \in \mathbb{R}
\tag{18.115}
$$

for some positive $\beta'$. Consequently, we can apply Theorem 18.5.2 to obtain that the power of $\mathbf{X} \star \mathbf{h}_r$ is given by

$$
\begin{aligned}
\text{Power in } \mathbf{X} \star \mathbf{h}_r &= 2\, \text{Power in } t \mapsto A \sum_{\ell=-\infty}^{\infty} C_\ell \left( \mathbf{g} \star \mathbf{h}_c \right)(t - \ell T_s) \\
&= 2\, \text{Power in } \left( t \mapsto A \sum_{\ell=-\infty}^{\infty} C_\ell\, g(t - \ell T_s) \right) \star \mathbf{h}_c
\end{aligned}
$$

$$= 2 \text{ Power in } (\mathbf{X}_{\mathrm{BB}} \star \mathbf{h}_{\mathrm{c}})$$

$$= 2 \int_{-\infty}^{\infty} \mathsf{S}_{\mathrm{BB}}(f) \left| \hat{h}_{\mathrm{c}}(f) \right|^2 \mathrm{d}f$$

$$= 2 \int_{-\infty}^{\infty} \mathsf{S}_{\mathrm{BB}}(f) \left| \hat{h}_{\mathrm{r}}(f + f_{\mathrm{c}}) \right|^2 \mathrm{d}f$$

$$= 2 \int_{-\infty}^{\infty} \mathsf{S}_{\mathrm{BB}}(\tilde{f} - f_{\mathrm{c}}) \left| \hat{h}_{\mathrm{r}}(\tilde{f}) \right|^2 \mathrm{d}\tilde{f}$$

$$= \int_{-\infty}^{\infty} \left( \mathsf{S}_{\mathrm{BB}}(\tilde{f} - f_{\mathrm{c}}) + \mathsf{S}_{\mathrm{BB}}(-\tilde{f} - f_{\mathrm{c}}) \right) \left| \hat{h}_{\mathrm{r}}(\tilde{f}) \right|^2 \mathrm{d}\tilde{f}, \qquad (18.116)$$

where the second equality follows from Corollary 18.6.2; the third by the definition of $\mathbf{X}_{\mathrm{BB}}$; the fourth because, by hypothesis, $\mathbf{X}_{\mathrm{BB}}$ is of operational PSD $\mathsf{S}_{\mathrm{BB}}(\cdot)$; the fifth from (18.114); the sixth by changing the integration variable to $\tilde{f} \triangleq f + f_{\mathrm{c}}$; and the seventh from the conjugate symmetry of $\hat{h}_{\mathrm{r}}(\cdot)$.

Since $\mathbf{h}_{\mathrm{r}}$ was an arbitrary integrable real function of compact support, (18.116) establishes (18.111) for all such functions. $\qquad\square$

**Corollary 18.6.7.** *Under the assumptions of Theorem 18.6.6, the QAM signal* $\big( X(t),\ t \in \mathbb{R} \big)$ *is of operational PSD*

$$\mathsf{S}_{XX}(f) = \mathsf{S}_{\mathrm{BB}}\big( |f| - f_{\mathrm{c}} \big), \quad f \in \mathbb{R}. \qquad (18.117)$$

**Proof.** Follows from the theorem by noting that, by Proposition 18.6.3 and by the assumption that $f_{\mathrm{c}} > \mathsf{W}/2$,

$$\mathsf{S}_{\mathrm{BB}}\big( f - f_{\mathrm{c}} \big) + \mathsf{S}_{\mathrm{BB}}\big( -f - f_{\mathrm{c}} \big) = \mathsf{S}_{\mathrm{BB}}\big( |f| - f_{\mathrm{c}} \big)$$

at all frequencies $f$ outside a set of frequencies of Lebesgue measure zero. $\qquad\square$

### 18.6.5   On the Operational PSD in Baseband

In the calculation of the operational PSD of the QAM signal $\big( X(t) \big)$ via (18.44) (which is formally stated as Corollary 18.6.7) we needed the operational PSD of the CSP $\big( X_{\mathrm{BB}}(t) \big)$ of (18.25). In this section we justify the calculations of this operational PSD that lead to Theorems 18.4.3 and 18.4.4. Specifically, we show:

**Proposition 18.6.8 (Operational PSD of a Complex PAM Signal).** *Let the CSP* $\big( X_{\mathrm{BB}}(t),\ t \in \mathbb{R} \big)$ *be given by (18.25), where* $\mathsf{A} \geq 0$, $\mathsf{T}_{\mathrm{s}} > 0$, *and where* $\mathbf{g}$ *is a complex Borel measurable function satisfying the decay condition (18.22) for some constants* $\alpha, \beta > 0$.

(i) *If* $\big( C_\ell \big)$ *is a bounded, zero-mean, WSS CSP of autocovariance function* $\mathsf{K}_{CC}$, *i.e., if it satisfies (18.23) and (18.28), then the CSP* $\big( X_{\mathrm{BB}}(t),\ t \in \mathbb{R} \big)$ *is of operational PSD* $\mathsf{S}_{\mathrm{BB}}(\cdot)$ *as given in (18.49).*

(ii) *If* $\big( C_\ell \big)$ *is produced in bi-infinite block-mode from IID random bits using an encoder* $\mathbf{enc}\colon \{0,1\}^{\mathsf{K}} \to \mathbb{C}^{\mathsf{N}}$ *that produces zero-mean symbols from IID random bits, then* $\big( X_{\mathrm{BB}}(t),\ t \in \mathbb{R} \big)$ *is of operational PSD* $\mathsf{S}_{\mathrm{BB}}(\cdot)$ *as given in (18.52).*

**Proof.** We have all the ingredients that are needed to justify our derivations of (18.49) and (18.52). All that remains is to piece them together. Let $\mathbf{h}$ be any complex integrable function of compact support. Then

$$\text{Power in } \mathbf{X}_{\text{BB}} \star \mathbf{h} = \text{Power in } \left( \left( t \mapsto A \sum_{\ell \in \mathbb{Z}} C_\ell \, g(t - \ell T_s) \right) \star \mathbf{h} \right)$$

$$= \text{Power in } t \mapsto A \sum_{\ell \in \mathbb{Z}} C_\ell \, (\mathbf{g} \star \mathbf{h})(t - \ell T_s), \qquad (18.118)$$

where the first equality follows from the definition of $\mathbf{X}_{\text{BB}}$ (18.25), and where the second equality follows from Corollary 18.6.2. Note that by Lemma 18.6.4 (ii) the function $\mathbf{g} \star \mathbf{h}$ satisfies the decay condition (18.96) for some $\beta' > 0$.

To prove Part (i) we now employ Theorem 14.6.4 (which extends to the case where the pulse shape and the symbols are complex) with the pulse shape $\mathbf{g} \star \mathbf{h}$ to obtain from (18.118) that

$$\text{Power in } \mathbf{X}_{\text{BB}} \star \mathbf{h} = \frac{A^2}{T_s} \int_{-\infty}^{\infty} \sum_{m=-\infty}^{\infty} \mathsf{K}_{CC}(m) \, e^{-i2\pi f m T_s} \, |\hat{g}(f)|^2 |\hat{h}(f)|^2 \, df, \quad (18.119)$$

for every integrable complex $\mathbf{h}$ of compact support. It follows from the fact that the set of integrable functions of compact support is a dense subset of $\mathcal{L}_1$ (Lemma 18.6.4 (i)) and from Proposition 18.6.5 that (18.119) must hold for all integrable functions. Recalling the definition of the operational PSD (Definition 18.4.1), it follows that $(X_{\text{BB}}(t), \ t \in \mathbb{R})$ is of operational PSD $\mathsf{S}_{\text{BB}}(\cdot)$ as given in (18.49).

The proof of Part (ii) is very similar except that we compute the RHS of (18.118) using (18.36) with the substitution of $\mathbf{g} \star \mathbf{h}$ for the pulse shape.  □

## 18.7   Exercises

**Exercise 18.1 (The Second Moment of the Square QAM Constellation).**

(i) Show that picking $X$ and $Y$ IID uniformly over the set in (10.19) results in $X + iY$ being uniformly distributed over the set in (16.19).

(ii) Compute the second moment of the square $2\nu \times 2\nu$ QAM constellation (16.19).

**Exercise 18.2 (Optimal Constellations).** Let $\mathcal{C}$ denote a QAM constellation, and define for every $z \in \mathbb{C}$ the constellation $\mathcal{C}' = \{c - z : c \in \mathcal{C}\}$.

(i) Relate the minimum distance of $\mathcal{C}'$ to that of $\mathcal{C}$.

(ii) Relate the second moment of $\mathcal{C}'$ to that of $\mathcal{C}$.

(iii) How would you choose $z$ to minimize the second moment of $\mathcal{C}'$?

**Exercise 18.3 (The Power in Baseband Is Real).** Show that the RHS of (18.29) is real. Which properties of the autocovariance function $\mathsf{K}_{CC}$ and of the self-similarity function $\mathsf{R}_{\mathbf{gg}}$ are you exploiting?

**Exercise 18.4 ($\pi/4$-QPSK).** In QPSK or 4-QAM the data bits are mapped to complex symbols $(C_\ell)$ which take value in set $\{\pm 1 \pm i\}$ and which are then transmitted using the signal $(X(t))$ defined in (18.24). Consider now $\pi/4$-QPSK where, prior to transmission, the complex symbols $(C_\ell)$ are rotated to form the complex symbols

$$\tilde{C}_\ell = \alpha^\ell C_\ell, \quad \ell \in \mathbb{Z},$$

where $\alpha = e^{i\pi/4}$. The transmitted signal is then

$$2A \operatorname{Re}\left( \sum_{\ell=-\infty}^\infty \tilde{C}_\ell \, g(t - \ell T_s) \, e^{i2\pi f_c t} \right), \quad t \in \mathbb{R}.$$

Compute the power and the operational PSD of the $\pi/4$-QPSK signal when $(C_\ell)$ is a zero-mean WSS CSP of autocovariance function $\mathsf{K}_{CC}$. Compare the power and operational PSD of $\pi/4$-QPSK with those of QPSK. How do they compare when the symbols $(C_\ell)$ are IID?

*Hint: See Exercise 17.12.*

**Exercise 18.5 (The Bandwidth of the QAM Signal).** Formulate and prove a result analogous to Theorem 15.4.1 for QAM.

**Exercise 18.6 (Bandwidth and Power in PAM and QAM).** Data bits $(D_j)$ are generated at rate $\mathsf{R}_b$ bits per second.

 (i) The bits are mapped to real symbols using a $(\mathsf{K}, \mathsf{N})$ binary-to-reals block-encoder of rate $\mathsf{K}/\mathsf{N}$ bits per real symbol. The symbols are mapped to a PAM signal of pulse shape $\phi$ whose time shifts by integer multiples of $T_s$ are orthonormal and whose excess bandwidth is $\eta$. Find the bandwidth of the transmitted signal (Definition 15.3.4).

 (ii) Repeat for the bandwidth around the carrier frequency $f_c$ in QAM when the bits are mapped to complex symbols using a $(\mathsf{K}, \mathsf{N})$ binary-to-complex block-encoder of rate $\mathsf{K}/\mathsf{N}$ bits per complex symbol. (As in Part (i), the pulse shape is of excess bandwidth $\eta$.)

 (iii) Show that if we express the rate $\rho$ of the block-encoder in both cases in bits per complex symbol, then in the former case $\rho = 2\mathsf{K}/\mathsf{N}$; in the latter case $\rho = \mathsf{K}/\mathsf{N}$; and in both cases the bandwidth can be expressed as the same function of $\mathsf{R}_b$, $\rho$, and $\eta$.

 (iv) Show that for both PAM and QAM the transmitted power is given by

$$\mathsf{P} = \frac{\mathsf{E}_s \mathsf{R}_b}{\rho}$$

provided that the energy per symbol $\mathsf{E}_s$ and the rate $\rho$ are computed in both cases per complex symbol.

*Hint: Exercise 18.5 is useful for Part (ii).*

**Exercise 18.7 (Operational PSD of Differential PSK).** Let the bi-infinite sequence of IID random bits $(D_j, \; j \in \mathbb{Z})$ be mapped to the complex symbols $(C_\ell, \; \ell \in \mathbb{Z})$ as follows:

$$C_{\ell+1} = C_\ell \exp\left( i\frac{2\pi}{8}(4D_{3\ell} + 2D_{3\ell+1} + D_{3\ell+2}) \right), \qquad \ell = 0, 1, 2, \ldots$$

$$C_\ell = C_{\ell+1} \exp\left( -i\frac{2\pi}{8}(4D_{3\ell} + 2D_{3\ell+1} + D_{3\ell+2}) \right), \quad \ell = \ldots, -2, -1,$$

where $C_0$ is independent of $(D_j)$ and uniformly distributed over the set

$$\mathcal{C} = \left\{ 1, e^{i\frac{2\pi}{8}}, e^{2i\frac{2\pi}{8}}, e^{3i\frac{2\pi}{8}}, \ldots, e^{7i\frac{2\pi}{8}} \right\}.$$

Find the operational PSD of the QAM signal under the assumptions of Section 18.3 on the pulse shape.

**Exercise 18.8 (PAM/QAM).** Let $D_1, \ldots, D_k$ be IID random bits. These bits are mapped by a mapping $\boldsymbol{\varphi}_{\mathrm{QAM}} \colon \{0, 1\}^k \to \mathbb{C}^n$ to the complex symbols $C_1, \ldots, C_n$, which are then mapped to the QAM signal

$$X_{\mathrm{QAM}}(t; D_1, \ldots, D_k) = 2A \operatorname{Re}\left( \sum_{\ell=1}^n C_\ell \, \phi_{\mathrm{QAM}}\left(t - \ell T_{\mathrm{s,QAM}}\right) e^{i2\pi f_c t} \right), \quad t \in \mathbb{R},$$

where the time shifts of $\phi_{\mathrm{QAM}}$ by integer multiples of $T_{\mathrm{s,QAM}}$ are orthonormal.

Define the real symbols $X_1, \ldots, X_{2n}$ by

$$X_{2\ell-1} = \operatorname{Re}(C_\ell), \quad X_{2\ell} = \operatorname{Im}(C_\ell), \quad \ell \in \{1, \ldots, n\}$$

and the corresponding PAM signal

$$X_{\mathrm{PAM}}(t; D_1, \ldots, D_k) = A \sum_{\ell=1}^{2n} X_\ell \, \phi_{\mathrm{PAM}}\left(t - \ell T_{\mathrm{s,PAM}}\right), \quad t \in \mathbb{R},$$

where $\phi_{\mathrm{PAM}}$ is real and its time shifts by integer multiples of $T_{\mathrm{s,PAM}}$ are orthonormal.

(i) Relate the expected energy in $\mathbf{X}_{\mathrm{QAM}}$ to that in $\mathbf{X}_{\mathrm{PAM}}$.

(ii) Relate the minimum squared distance

$$\min_{(d_1, \ldots, d_k) \neq (d_1', \ldots, d_k')} \int_{-\infty}^{\infty} \left( X_{\mathrm{QAM}}\left(t; d_1, \ldots, d_k\right) - X_{\mathrm{QAM}}\left(t; d_1', \ldots, d_k'\right) \right)^2 dt,$$

to

$$\min_{(d_1, \ldots, d_k) \neq (d_1', \ldots, d_k')} \int_{-\infty}^{\infty} \left( X_{\mathrm{PAM}}\left(t; d_1, \ldots, d_k\right) - X_{\mathrm{PAM}}\left(t; d_1', \ldots, d_k'\right) \right)^2 dt.$$

**Exercise 18.9 (The Operational PSD is Nonnegative).** Show that if the CSP $(Z(t), \, t \in \mathbb{R})$ is of operational PSD $\mathsf{S}_{ZZ}$, then $\mathsf{S}_{ZZ}(f)$ must be nonnegative outside a set of frequencies of Lebesgue measure zero.

*Hint: See Exercise 15.5.*

# Chapter 19

# The Univariate Gaussian Distribution

## 19.1 Introduction

In many communication scenarios the noise is modeled as a Gaussian stochastic process. This is sometimes justified by invoking a Central Limit Theorem, which demonstrates that many small independent disturbances add up to a stochastic process that is approximately Gaussian. Another justification is mathematical convenience: while Gaussian processes may seem daunting at first, they are actually well understood and often amenable to analysis. Finally, particularly in wireline communications, the Gaussian model is justified because it leads to robust results and to good engineering design. For other scenarios, e.g., fast-moving wireless mobile communications, more intricate models are needed.

Rather than starting immediately with the definition and analysis of Gaussian stochastic processes, we shall take the more moderate approach and start by first discussing Gaussian random variables. Building on that, we shall later discuss Gaussian random vectors in Chapter 23, and only then introduce continuous-time Gaussian stochastic processes in Chapter 25.

## 19.2 Standard Gaussian Random Variables

We begin with a special kind of Gaussian: the standard Gaussian.

**Definition 19.2.1 (Standard Gaussian).** *We say that the random variable $W$ is a* **standard Gaussian** *or that it has a* **standard Gaussian distribution***, if its density function $f_W(\cdot)$ is given by*

$$f_W(w) = \frac{1}{\sqrt{2\pi}} e^{-\frac{w^2}{2}}, \quad w \in \mathbb{R}. \tag{19.1}$$

This density is depicted in Figure 19.1. For this definition to be meaningful, the RHS of (19.1) had better be a valid density function, i.e., be nonnegative and integrate to one. This is indeed the case. In fact, the RHS of (19.1) is positive,

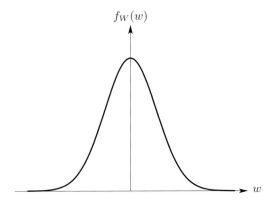

**Figure 19.1:** The standard Gaussian density function

and it integrates to one because, as we next show,

$$\int_{-\infty}^{\infty} e^{-w^2/2} \, dw = \sqrt{2\pi}. \tag{19.2}$$

This integral can be verified by computing its square as follows:

$$\left( \int_{-\infty}^{\infty} e^{-\frac{w^2}{2}} \, dw \right)^2 = \int_{-\infty}^{\infty} e^{-\frac{w^2}{2}} \, dw \int_{-\infty}^{\infty} e^{-\frac{v^2}{2}} \, dv$$

$$= \int_{-\infty}^{\infty} \int_{-\infty}^{\infty} e^{-\frac{w^2+v^2}{2}} \, dw \, dv$$

$$= \int_{0}^{\infty} \int_{-\pi}^{\pi} r \, e^{-\frac{r^2}{2}} \, d\varphi \, dr$$

$$= 2\pi \int_{0}^{\infty} r \, e^{-\frac{r^2}{2}} \, dr$$

$$= 2\pi \left( -e^{-r^2/2} \right) \Big|_{0}^{\infty}$$

$$= 2\pi,$$

where the first equality follows by writing $a^2$ as $a$ times $a$; the second by writing the product of the integrals as a double integral over $\mathbb{R}^2$; the third by changing from Cartesian to polar coordinates:

$$w = r \cos \varphi, \quad v = r \sin \varphi, \quad r \geq 0, \quad -\pi \leq \varphi < \pi,$$

$$dw \, dv = r \, dr \, d\varphi;$$

the fourth because the integrand does not depend on $\varphi$; the fifth because the derivative of $-e^{-r^2/2}$ is $r \, e^{-r^2/2}$; and where the final equality follows by direct evaluation.

Note that the density of a standard Gaussian random variable is symmetric (19.1). Consequently, if $W$ is a standard Gaussian, then so is $-W$. This symmetry also establishes that the expectation of a standard Gaussian is zero. The variance of a standard Gaussian can be computed using integration by parts:

$$
\begin{aligned}
\int_{-\infty}^{\infty} w^2 \frac{1}{\sqrt{2\pi}} e^{-\frac{w^2}{2}} \, dw &= \frac{1}{\sqrt{2\pi}} \int_{-\infty}^{\infty} w \left( -\frac{d}{dw} e^{-\frac{w^2}{2}} \right) dw \\
&= \frac{1}{\sqrt{2\pi}} \left( -w e^{-\frac{w^2}{2}} \Big|_{-\infty}^{\infty} + \int_{-\infty}^{\infty} e^{-\frac{w^2}{2}} \, dw \right) \\
&= \frac{1}{\sqrt{2\pi}} \int_{-\infty}^{\infty} e^{-\frac{w^2}{2}} \, dw \\
&= 1,
\end{aligned}
$$

where the last equality follows from (19.2).

## 19.3  Gaussian Random Variables

We next define a Gaussian (not necessarily standard) random variable as the result of applying an affine transformation to a standard Gaussian.

**Definition 19.3.1 (Centered Gaussians and Gaussians).** *We say that a random variable $X$ is a **centered Gaussian** or that it has a **centered Gaussian distribution** if it can be written in the form*

$$
X = aW \tag{19.3}
$$

*for some deterministic $a \in \mathbb{R}$ and for some standard Gaussian $W$. We say that the random variable $X$ is **Gaussian** or that it has a **Gaussian distribution** if*

$$
X = aW + b \tag{19.4}
$$

*for some deterministic $a, b \in \mathbb{R}$ and for some standard Gaussian $W$.*

**Note 19.3.2.** We do not preclude $a$ from being zero. The case $a = 0$ leads to $X$ being deterministically equal to $b$. We thus include the deterministic random variables in the family of Gaussian random variables.

**Note 19.3.3.** The family of Gaussian random variables is closed with respect to affine transformations: if $X$ is Gaussian and $\alpha, \beta \in \mathbb{R}$ are deterministic, then $\alpha X + \beta$ is also Gaussian.

**Proof.** Since $X$ is Gaussian, it can be written as $X = aW + b$, where $W$ is a standard Gaussian. Consequently

$$
\begin{aligned}
\alpha X + \beta &= \alpha(aW + b) + \beta \\
&= (\alpha a)W + (\alpha b + \beta),
\end{aligned}
$$

which has the form $a'W + b'$ for some deterministic $a', b' \in \mathbb{R}$. $\qquad\square$

If (19.4) holds, then the random variables on its RHS and LHS must have the same mean. The mean of a standard Gaussian is zero, so the mean of the RHS of (19.4) is $b$. The LHS is of mean $\mathsf{E}[X]$, and we thus conclude that in the representation (19.4) the deterministic constant $b$ is uniquely determined by the mean of $X$, and in fact,

$$b = \mathsf{E}[X].$$

Similarly, since the variance of a standard Gaussian is one, the variance of the RHS of (19.4) is $a^2$. And since the variance of the LHS is $\mathsf{Var}[X]$, we conclude that

$$a^2 = \mathsf{Var}[X].$$

Up to its sign, the deterministic constant $a$ in the representation (19.4) is thus also unique.

Based on the above, one might mistakenly think that for any given mean $\mu$ and variance $\sigma^2$ there are two different Gaussian distributions corresponding to

$$\sigma W + \mu, \quad \text{and} \quad -\sigma W + \mu, \tag{19.5}$$

where $W$ is a standard Gaussian. This, however, is not the case:

**Note 19.3.4.** There is only one Gaussian distribution of a given mean and variance.

**Proof.** This can be seen in two different ways. The first is to note that the two representations in (19.5) lead to the same distribution, because the standard Gaussian $W$ has a symmetric distribution, so $\sigma W$ and $-\sigma W$ have the same distribution. The second is based on computing the density of $\sigma W + \mu$ and showing that it is a symmetric function of $\sigma$; see (19.6) ahead.                                              □

Having established that there is only one Gaussian distribution of a given mean $\mu$ and variance $\sigma^2$, we denote it by

$$\mathcal{N}(\mu, \sigma^2)$$

and set out to study its density. Since the distribution does not depend on the sign of $\sigma$, it is customary to require that $\sigma$ be nonnegative and to refer to it as the **standard deviation**. Thus, $\sigma^2$ is the variance and $\sigma$ is the standard deviation. If $\sigma^2 = 0$, then the Gaussian distribution is deterministic with mean $\mu$ and has no density.[1] If $\sigma^2 > 0$, then the density can be computed from the density of the standard Gaussian distribution as follows. If $X \sim \mathcal{N}(\mu, \sigma^2)$, then $X$ has the same distribution as $\mu + \sigma W$, where $W$ is a standard Gaussian, because both $X$ and $\mu + \sigma W$ are of mean $\mu$ and variance $\sigma^2$ ($W$ is zero-mean and unit-variance); both are Gaussian (Note 19.3.3); and Gaussians of identical means and variances have identical distributions (Note 19.3.4). The density of $X$ is thus identical to the density of $\mu + \sigma W$. The density of the latter can be computed from the density of $W$ (19.1) to obtain that the density of a $\mathcal{N}(\mu, \sigma^2)$ Gaussian random variable of positive variance is

$$\frac{1}{\sqrt{2\pi\sigma^2}} e^{-\frac{(x-\mu)^2}{2\sigma^2}}, \quad x \in \mathbb{R}. \tag{19.6}$$

This density is depicted in Figure 19.2. To derive the density of $\mu + \sigma W$ from

---

[1] Some would say that the density of a deterministic random variable is given by Dirac's Delta, but we prefer not to use generalized functions in this book.

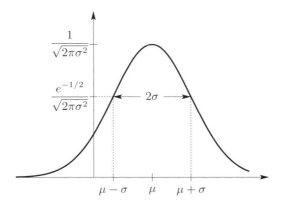

**Figure 19.2:** The Gaussian density function with mean $\mu$ and variance $\sigma^2$.

that of $W$, we have used the fact that if $X = g(W)$, where $g(\cdot)$ is a deterministic continuously differentiable function whose derivative never vanishes (in our case $g(w) = \mu + \sigma w$) and where $W$ is of density $f_W(\cdot)$ (in our case (19.1)), then the density $f_X(\cdot)$ of $X$ is given by:

$$f_X(x) = \begin{cases} 0 & \text{if for no } \xi \text{ is } x = g(\xi), \\ \frac{1}{|g'(\xi)|} f_W(\xi) & \text{if } \xi \text{ satisfies } x = g(\xi), \end{cases} \tag{19.7}$$

where $g'(\xi)$ denotes the derivative of $g(\cdot)$ at $\xi$. (For a more formal multivariate version of this fact see Theorem 17.3.4.)

Since the family of Gaussian random variables is closed under deterministic affine transformations (Note 19.3.3), it follows that if $X \sim \mathcal{N}(\mu, \sigma^2)$ with $\sigma^2 > 0$, then $(X - \mu)/\sigma$ is also a Gaussian random variable. Since it is of zero mean and of unit variance, it follows that it must be a standard Gaussian, because there is only one Gaussian distribution of zero mean and unit variance (Note 19.3.4). We thus conclude that for $\sigma^2 > 0$ and arbitrary $\mu \in \mathbb{R}$,

$$\boxed{\left( X \sim \mathcal{N}(\mu, \sigma^2) \right) \Rightarrow \left( \frac{X - \mu}{\sigma} \sim \mathcal{N}(0, 1) \right).} \tag{19.8}$$

Recall that the **Cumulative Distribution Function** $F_X(\cdot)$ of a RV $X$ is defined for $x \in \mathbb{R}$ as

$$F_X(x) = \Pr[X \le x],$$
$$= \int_{-\infty}^{x} f_X(\xi) \, \mathrm{d}\xi,$$

where the second equality holds if $X$ has a density function $f_X(\cdot)$. If $W$ is a standard Gaussian, then its CDF is thus given by

$$F_W(w) = \int_{-\infty}^{w} \frac{1}{\sqrt{2\pi}} e^{-\frac{\xi^2}{2}} \, \mathrm{d}\xi, \quad w \in \mathbb{R}.$$

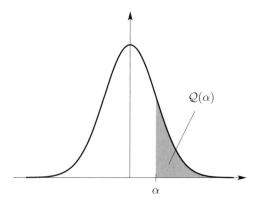

**Figure 19.3:** $\mathcal{Q}(\alpha)$ is the area to the right of $\alpha$ under the standard Gaussian density plot. Here it is represented by the shaded area.

There is, alas, no closed-form expression for this integral. To handle such expressions we next introduce the $\mathcal{Q}$-function.

## 19.4   The $\mathcal{Q}$-Function

The $\mathcal{Q}$-function maps every $\alpha \in \mathbb{R}$ to the probability that a standard Gaussian exceeds it:

**Definition 19.4.1 (The $\mathcal{Q}$-Function).** *The $\mathcal{Q}$-function is defined by*

$$\mathcal{Q}(\alpha) \triangleq \frac{1}{\sqrt{2\pi}} \int_{\alpha}^{\infty} e^{-\xi^2/2} \, d\xi, \quad \alpha \in \mathbb{R}. \tag{19.9}$$

For a graphical interpretation of this integral see Figure 19.3.

Since the $\mathcal{Q}$-function is a well-tabulated function, we are usually happy when we can express answers to various questions using this function. The CDF of a standard Gaussian $W$ can be expressed using the $\mathcal{Q}$-function as follows:

$$\begin{aligned} F_W(w) &= \Pr[W \leq w] \\ &= 1 - \Pr[W \geq w] \\ &= 1 - \mathcal{Q}(w), \quad w \in \mathbb{R}, \end{aligned} \tag{19.10}$$

where the second equality follows because the standard Gaussian has a density, so $\Pr[W = w] = 0$. Similarly, with the aid of the $\mathcal{Q}$-function we can express the probability that a standard Gaussian $W$ lies in some given interval $[a, b]$:

$$\begin{aligned} \Pr[a \leq W \leq b] &= \Pr[W \geq a] - \Pr[W \geq b] \\ &= \mathcal{Q}(a) - \mathcal{Q}(b), \quad a \leq b. \end{aligned}$$

More generally, if $X \sim \mathcal{N}(\mu, \sigma^2)$ with $\sigma > 0$, then

$$\Pr[a \leq X \leq b] = \Pr[X \geq a] - \Pr[X \geq b], \quad a \leq b$$

$$= \Pr\left[\frac{X - \mu}{\sigma} \geq \frac{a - \mu}{\sigma}\right] - \Pr\left[\frac{X - \mu}{\sigma} \geq \frac{b - \mu}{\sigma}\right], \quad \sigma > 0$$

$$= \mathcal{Q}\left(\frac{a - \mu}{\sigma}\right) - \mathcal{Q}\left(\frac{b - \mu}{\sigma}\right), \quad \left(a \leq b, \ \sigma > 0\right), \tag{19.11}$$

where the last equality follows because $(X - \mu)/\sigma$ is a standard Gaussian; see (19.8). Letting $b$ tend to $+\infty$ in (19.11), we obtain the probability of a half ray:

$$\Pr[X \geq a] = \mathcal{Q}\left(\frac{a - \mu}{\sigma}\right), \quad \sigma > 0. \tag{19.12a}$$

And letting $a$ tend to $-\infty$ we obtain

$$\Pr[X \leq b] = 1 - \mathcal{Q}\left(\frac{b - \mu}{\sigma}\right), \quad \sigma > 0. \tag{19.12b}$$

The $\mathcal{Q}$-function is usually only tabulated for nonnegative arguments, because the standard Gaussian density (19.1) is symmetric: if $W \sim \mathcal{N}(0, 1)$ then, by the symmetry of its density,

$$\Pr[W \geq -\alpha] = \Pr[W \leq \alpha]$$

$$= 1 - \Pr[W \geq \alpha], \quad \alpha \in \mathbb{R}.$$

Consequently, as illustrated in Figure 19.4,

$$\mathcal{Q}(\alpha) + \mathcal{Q}(-\alpha) = 1, \quad \alpha \in \mathbb{R}, \tag{19.13}$$

and it suffices to tabulate the $\mathcal{Q}$-function for nonnegative arguments. Note that, by (19.13),

$$\mathcal{Q}(0) = \frac{1}{2}. \tag{19.14}$$

An alternative expression for the $\mathcal{Q}$-function as an integral with fixed integration limits is known as Craig's formula:

$$\mathcal{Q}(\alpha) = \frac{1}{\pi} \int_0^{\pi/2} e^{-\frac{\alpha^2}{2\sin^2 \varphi}} \, d\varphi, \quad \alpha \geq 0. \tag{19.15}$$

This expression can be derived by computing a two-dimensional integral in two different ways as follows. Let $X \sim \mathcal{N}(0, 1)$ and $Y \sim \mathcal{N}(0, 1)$ be independent. Consider the probability of the event "$X \geq 0$ and $Y \geq \alpha$" where $\alpha \geq 0$. Since the two random variables are independent, it follows that

$$\Pr[X \geq 0 \text{ and } Y \geq \alpha] = \Pr[X \geq 0] \Pr[Y \geq \alpha]$$

$$= \frac{1}{2} \mathcal{Q}(\alpha), \tag{19.16}$$

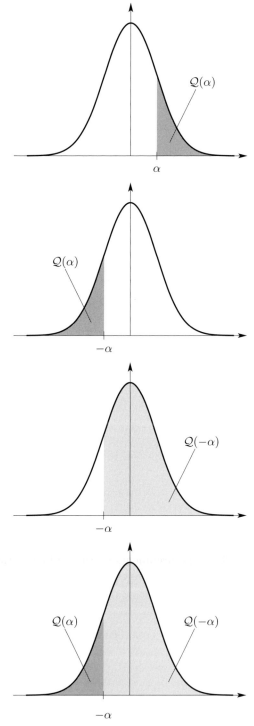

**Figure 19.4:** The identity $\mathcal{Q}(\alpha) + \mathcal{Q}(-\alpha) = 1$.

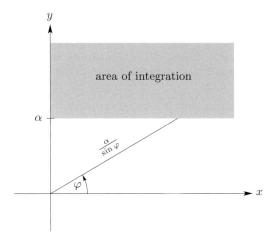

**Figure 19.5:** Use of polar coordinates to compute $\frac{1}{2}\mathcal{Q}(\alpha)$.

where the second equality follows from (19.14). We now proceed to compute the LHS of the above in polar coordinates centered at the origin (Figure 19.5):

$$
\begin{aligned}
\Pr[X \geq 0 \text{ and } Y \geq \alpha] &= \int_0^\infty \int_\alpha^\infty \frac{1}{2\pi} e^{-\frac{x^2+y^2}{2}} \, dy \, dx \\
&= \int_0^{\pi/2} \int_{\frac{\alpha}{\sin\varphi}}^\infty \frac{1}{2\pi} e^{-r^2/2} \, r \, dr \, d\varphi, \quad \alpha \geq 0 \\
&= \frac{1}{2\pi} \int_0^{\pi/2} \int_{\frac{\alpha^2}{2\sin^2\varphi}}^\infty e^{-t} \, dt \, d\varphi \\
&= \frac{1}{2\pi} \int_0^{\pi/2} e^{-\frac{\alpha^2}{2\sin^2\varphi}} \, d\varphi, \quad \alpha \geq 0, \quad\quad (19.17)
\end{aligned}
$$

where we have performed the change of variable $t \triangleq r^2/2$. The integral representation (19.15) now follows from (19.16) & (19.17).

We next describe various approximations for the $\mathcal{Q}$-function. We are particularly interested in its value for large arguments.[2] Since $\mathcal{Q}(\alpha)$ is the probability that a standard Gaussian exceeds $\alpha$, it follows that $\lim_{\alpha\to\infty} \mathcal{Q}(\alpha) = 0$. Thus, large arguments to the $\mathcal{Q}$-function correspond to small values of the $\mathcal{Q}$-function. The following bounds justify the approximation

$$
\mathcal{Q}(\alpha) \approx \frac{1}{\sqrt{2\pi\alpha^2}} e^{-\frac{\alpha^2}{2}}, \quad \alpha \gg 1. \quad\quad (19.18)
$$

**Proposition 19.4.2 (Estimates for the $\mathcal{Q}$-function).** *The $\mathcal{Q}$-function is bounded by*

$$
\frac{1}{\sqrt{2\pi\alpha^2}} e^{-\alpha^2/2} \left(1 - \frac{1}{\alpha^2}\right) < \mathcal{Q}(\alpha) < \frac{1}{\sqrt{2\pi\alpha^2}} e^{-\alpha^2/2}, \quad \alpha > 0 \quad\quad (19.19)
$$

---

[2]In Digital Communications this corresponds to scenarios with low probability of error.

*and*

$$Q(\alpha) \leq \frac{1}{2} e^{-\alpha^2/2}, \quad \alpha \geq 0. \tag{19.20}$$

**Proof.** The proof of (19.19) is omitted (but see Exercise 19.3). Inequality (19.20) is proved by replacing the integrand in (19.15) with its maximal value, namely, its value at $\varphi = \pi/2$. We shall see an alternative proof in Section 20.10. $\square$

## 19.5   Integrals of Exponentiated Quadratics

The fact that (19.6) is a density and hence integrates to one, i.e.,

$$\int_{-\infty}^{\infty} \frac{1}{\sqrt{2\pi\sigma^2}} e^{-\frac{(x-\mu)^2}{2\sigma^2}} \, \mathrm{d}x = 1, \tag{19.21}$$

can be used to compute seemingly complicated integrals. Here we shall show how (19.21) can be used to derive the identity

$$\boxed{\int_{-\infty}^{\infty} e^{-\alpha x^2 \pm \beta x} \, \mathrm{d}x = \sqrt{\frac{\pi}{\alpha}} \, e^{\frac{\beta^2}{4\alpha}}, \quad \beta \in \mathbb{R}, \ \alpha > 0.} \tag{19.22}$$

Note that this identity is meaningless when $\alpha \leq 0$, because in this case the integrand is not integrable. For exmples, if $\alpha < 0$, then the integrand tends to infinity as $|x|$ tends to $\infty$. If $\alpha = 0$ and $\beta \neq 0$, then the integrand tends to infinity either as $x$ tends to $+\infty$ or as $x$ tends to $-\infty$ (depending on the sign of $\beta$). Finally, if both $\alpha$ and $\beta$ are zero, then the integrand is 1, which is not integrable. Note also that, by considering the change of variable $u \triangleq -x$, one can verify that the sign of $\beta$ on the LHS of this identity is immaterial.

The trick to deriving (19.22) is to complete the exponent to a square and to then apply (19.21):

$$\int_{-\infty}^{\infty} e^{-\alpha x^2 + \beta x} \, \mathrm{d}x = \int_{-\infty}^{\infty} \exp\left(-\frac{x^2 - \frac{\beta}{\alpha}x}{2(1/\sqrt{2\alpha})^2}\right) \mathrm{d}x$$

$$= \int_{-\infty}^{\infty} \exp\left(-\frac{\left(x - \frac{\beta}{2\alpha}\right)^2}{2(1/\sqrt{2\alpha})^2} + \frac{\beta^2}{4\alpha}\right) \mathrm{d}x$$

$$= e^{\frac{\beta^2}{4\alpha}} \int_{-\infty}^{\infty} \exp\left(-\frac{\left(x - \frac{\beta}{2\alpha}\right)^2}{2(1/\sqrt{2\alpha})^2}\right) \mathrm{d}x$$

$$= e^{\frac{\beta^2}{4\alpha}} \sqrt{2\pi\left(1/\sqrt{2\alpha}\right)^2} \int_{-\infty}^{\infty} \frac{1}{\sqrt{2\pi\left(1/\sqrt{2\alpha}\right)^2}} \exp\left(-\frac{\left(x - \frac{\beta}{2\alpha}\right)^2}{2(1/\sqrt{2\alpha})^2}\right) \mathrm{d}x$$

$$= e^{\frac{\beta^2}{4\alpha}} \sqrt{2\pi\left(1/\sqrt{2\alpha}\right)^2}$$

$$= \sqrt{\frac{\pi}{\alpha}} \, e^{\frac{\beta^2}{4\alpha}},$$

where the first equality follows by rewriting the integrand so that the term $x^2$ in the numerator is of coefficient one and so that the denominator has the form $2\sigma^2$ for $\sigma$ which turns out here to be given by $\sigma \triangleq 1/\sqrt{2\alpha}$; the second follows by completing the square; the third by taking the multiplicative constant out of the integral; the fourth by multiplying and dividing the integral by $\sqrt{2\pi\sigma^2}$ so as to bring the integrand to the form of the density of a Gaussian; the fifth by (19.21); and the sixth equality by trivial algebra.

## 19.6   The Moment Generating Function

As an application of (19.22) we next derive the **Moment Generating Function** (MGF) of a Gaussian RV. Recall that the MGF of a RV $X$ is denoted by $M_X(\cdot)$ and is given by

$$M_X(\theta) \triangleq \mathsf{E}\big[e^{\theta X}\big] \tag{19.23}$$

for all $\theta \in \mathbb{R}$ for which this expectation is finite. If $X$ has density $f_X(\cdot)$, then its MGF can be written as

$$M_X(\theta) = \int_{-\infty}^{\infty} f_X(x)\, e^{\theta x}\, \mathrm{d}x, \tag{19.24}$$

thus highlighting the connection between the MGF of $X$ and the double-sided Laplace Transform of its density.

If $X \sim \mathcal{N}(\mu, \sigma^2)$ where $\sigma^2 > 0$, then

$$
\begin{aligned}
M_X(\theta) &= \int_{-\infty}^{\infty} f_X(x)\, e^{\theta x}\, \mathrm{d}x \\
&= \int_{-\infty}^{\infty} \frac{1}{\sqrt{2\pi\sigma^2}}\, e^{-\frac{(x-\mu)^2}{2\sigma^2}}\, e^{\theta x}\, \mathrm{d}x \\
&= \int_{-\infty}^{\infty} \frac{1}{\sqrt{2\pi\sigma^2}}\, e^{-\frac{\xi^2}{2\sigma^2}}\, e^{\theta(\xi+\mu)}\, \mathrm{d}\xi \\
&= e^{\theta\mu}\, \frac{1}{\sqrt{2\pi\sigma^2}} \int_{-\infty}^{\infty} e^{-\frac{\xi^2}{2\sigma^2} + \theta\xi}\, \mathrm{d}\xi \\
&= e^{\theta\mu}\, \frac{1}{\sqrt{2\pi\sigma^2}} \sqrt{\frac{\pi}{1/(2\sigma^2)}}\, e^{\frac{\theta^2}{4/(2\sigma^2)}} \\
&= e^{\theta\mu + \frac{1}{2}\theta^2\sigma^2}, \quad \theta \in \mathbb{R},
\end{aligned}
$$

where the first equality follows from (19.24); the second from (19.6); the third by changing the integration variable to $\xi \triangleq x - \mu$; the fourth by rearranging terms; the fifth from (19.22) with the substitution of $1/(2\sigma^2)$ for $\alpha$ and of $\theta$ for $\beta$; and the final by simple algebra. This can be verified to hold also when $\sigma^2 = 0$. Thus,

$$\Big(X \sim \mathcal{N}(\mu, \sigma^2)\Big) \Rightarrow \Big(M_X(\theta) = e^{\theta\mu + \frac{1}{2}\theta^2\sigma^2}, \quad \theta \in \mathbb{R}\Big). \tag{19.25}$$

## 19.7   The Characteristic Function of Gaussians

### 19.7.1   The Characteristic Function

Recall that the **Characteristic Function** $\Phi_X(\cdot)$ of a random variable $X$ is defined for every $\varpi \in \mathbb{R}$ by

$$\Phi_X(\varpi) = \mathsf{E}\left[e^{\mathrm{i}\varpi X}\right]$$
$$= \int_{-\infty}^{\infty} f_X(x) \, e^{\mathrm{i}\varpi x} \, \mathrm{d}x,$$

where the second equality holds if $X$ has density $f_X(\cdot)$. The second equality demonstrates that the characteristic function is related to the Fourier Transform of the density function but, by convention, there are no $2\pi$'s, and the complex exponential is not conjugated. If we allow for complex arguments to the MGF (by performing an analytic continuation), then the characteristic function can be viewed as the MGF evaluated on the imaginary axis:

$$\Phi_X(\varpi) = M_X(\mathrm{i}\varpi), \quad \varpi \in \mathbb{R}. \tag{19.26}$$

Some of the properties of the characteristic function are summarized next.

**Proposition 19.7.1 (On the Characteristic Function).** *Let $X$ be a random variable of characteristic function $\Phi_X(\cdot)$.*

(i) *If $\mathsf{E}[X^n] < \infty$ for some $n \in \mathbb{N}$, then $\Phi_X(\cdot)$ is differentiable $n$ times and the $\nu$-th moment of $X$ is related to the $\nu$-th derivative of $\Phi_X(\cdot)$ at zero via the relation*

$$\mathsf{E}[X^\nu] = \frac{1}{\mathrm{i}^\nu} \left. \frac{\mathrm{d}^\nu \Phi_X(\varpi)}{\mathrm{d}\varpi^\nu} \right|_{\varpi=0}, \quad \nu = 1, \ldots, n. \tag{19.27}$$

(ii) *Two random variables of identical characteristic functions must have the same distribution.*

(iii) *If $X$ and $Y$ are independent random variables of characteristic functions $\Phi_X(\cdot)$ and $\Phi_Y(\cdot)$, then the characteristic function $\Phi_{X+Y}(\cdot)$ of their sum is given by the product of their characteristic functions:*

$$\Big(X \ \& \ Y \text{ independent}\Big) \Rightarrow \Big(\Phi_{X+Y}(\varpi) = \Phi_X(\varpi)\,\Phi_Y(\varpi), \quad \varpi \in \mathbb{R}\Big). \tag{19.28}$$

**Proof.** For a proof of Part (i) see (Shiryaev, 1996, Chapter II, § 12.3, Theorem 1). For Part (ii) see (Shiryaev, 1996, Chapter II, § 12.4, Theorem 2). For Part (iii) see (Shiryaev, 1996, Chapter II, § 12.5, Theorem 4).                                          □

For $X \sim \mathcal{N}(\mu, \sigma^2)$ we obtain from (19.26) and (19.25) that[3]

$$\left( X \sim \mathcal{N}(\mu, \sigma^2) \right) \Rightarrow \left( \Phi_X(\varpi) = e^{i\varpi\mu - \frac{1}{2}\varpi^2\sigma^2}, \quad \varpi \in \mathbb{R} \right). \tag{19.29}$$

### 19.7.2 Moments

Since the standard Gaussian density decays faster than exponentially, it possesses moments of all orders. Those can be computed from the characteristic function (19.29) using Proposition 19.7.1 (i) by repeated differentiation. Using this approach we obtain that the moments of a standard Gaussian are

$$\mathsf{E}[W^\nu] = \begin{cases} 1 \times 3 \times \cdots \times (\nu - 1) & \text{if } \nu \text{ is even,} \\ 0 & \text{if } \nu \text{ is odd,} \end{cases} \quad W \sim \mathcal{N}(0,1). \tag{19.30}$$

We mention here in passing that[4]

$$\mathsf{E}[|W|^\nu] = \begin{cases} 1 \times 3 \times \cdots \times (\nu - 1) & \text{if } \nu \text{ is even,} \\ \sqrt{\frac{2}{\pi}} \, 2^{(\nu-1)/2} \left( \frac{\nu-1}{2} \right)! & \text{if } \nu \text{ is odd,} \end{cases} \quad W \sim \mathcal{N}(0,1) \tag{19.31}$$

(Johnson, Kotz, and Balakrishnan, 1994a, Chapter 18, Section 3, Equation (18.13)).

### 19.7.3 Sums of Independent Gaussians

Using the characteristic function we next show:

**Proposition 19.7.2 (The Sum of Two Independent Gaussians Is Gaussian).** *The sum of two independent Gaussian random variables is a Gaussian RV.*[5]

**Proof.** Let $X \sim \mathcal{N}(\mu_x, \sigma_x^2)$ and $Y \sim \mathcal{N}(\mu_y, \sigma_y^2)$ be independent. By (19.29),

$$\Phi_X(\varpi) = e^{i\varpi\mu_x - \frac{1}{2}\varpi^2\sigma_x^2}, \quad \varpi \in \mathbb{R},$$

$$\Phi_Y(\varpi) = e^{i\varpi\mu_y - \frac{1}{2}\varpi^2\sigma_y^2}, \quad \varpi \in \mathbb{R}.$$

---

[3]It does require a (small) leap of faith to accept that (19.25) also holds for *complex* $\theta$. This can be justified using analytic continuation. But there are also direct ways of deriving (19.29); see, for example, (Williams, 1991, Chapter E, Exercise E16.4) or (Shiryaev, 1996, Chapter II, Section 12, Paragraph 2, Example 2). Another approach is to express $\mathrm{d}\Phi_X(\varpi)/\mathrm{d}\varpi$ as $\mathsf{E}[iX\,e^{i\varpi X}]$ and to use integration by parts to verify that the latter's expectation is equal to $-\varpi\Phi_X(\varpi)$ and to then solve the differential equation $\mathrm{d}\Phi_X(\varpi)/\mathrm{d}\varpi = -\varpi\Phi_X(\varpi)$ with the condition $\Phi_X(0) = 1$ to obtain that $\ln\Phi_X(\varpi) = -\frac{1}{2}\varpi^2$.

[4]The distribution of $|W|$ is sometimes called **half-normal**. It is the positive square root of the central chi-squared distribution with one degree of freedom.

[5]More generally, as we shall see in Chapter 23, $X + Y$ is Gaussian whenever $X$ and $Y$ are jointly Gaussian. And independent Gaussians are jointly Gaussian.

Since the characteristic function of the sum of two independent random variables is equal to the product of their characteristic functions (19.28),

$$\Phi_{X+Y}(\varpi) = \Phi_X(\varpi)\,\Phi_Y(\varpi)$$
$$= e^{i\varpi\mu_x - \frac{1}{2}\varpi^2\sigma_x^2}\, e^{i\varpi\mu_y - \frac{1}{2}\varpi^2\sigma_y^2}$$
$$= e^{i\varpi(\mu_x+\mu_y) - \frac{1}{2}\varpi^2(\sigma_x^2+\sigma_y^2)}, \quad \varpi \in \mathbb{R}.$$

By (19.29), this is also the characteristic function of a $\mathcal{N}\!\left(\mu_x + \mu_y, \sigma_x^2 + \sigma_y^2\right)$ RV. Since the characteristic function of a random variable fully determines its law (Proposition 19.7.1 (ii)), $X + Y$ must be $\mathcal{N}\!\left(\mu_x + \mu_y, \sigma_x^2 + \sigma_y^2\right)$. □

Using induction one can generalize this proposition to any finite number of random variables: if $X_1, \ldots, X_n$ are independent Gaussian random variables, then their sum is Gaussian. Applying this to $\alpha_1 X_1, \ldots, \alpha_n X_n$, which are independent Gaussians whenever $X_1, \ldots, X_n$ are independent Gaussians, we obtain:

**Proposition 19.7.3 (Linear Combinations of Independent Gaussians).** *If the random variables $X_1, \ldots, X_n$ are independent Gaussians, and if $\alpha_1, \ldots, \alpha_n \in \mathbb{R}$ are deterministic, then the RV $Y = \sum_{\ell=1}^{n} \alpha_\ell X_\ell$ is Gaussian with mean and variance*

$$\mathsf{E}[Y] = \sum_{\ell=1}^{n} \alpha_\ell \, \mathsf{E}[X_\ell],$$

$$\mathsf{Var}[Y] = \sum_{\ell=1}^{n} \alpha_\ell^2 \, \mathsf{Var}[X_\ell].$$

## 19.8 Central and Noncentral Chi-Square Random Variables

We summarize here some of the definitions and main properties of the central and noncentral $\chi^2$ distributions and of some related distributions. We shall only use three results from this section: that the sum of the squares of two independent $\mathcal{N}(0,1)$ random variables has a mean-2 exponential distribution; that the distribution of the sum of the squares of $n$ independent Gaussian random variables of unit-variance and possibly different means depends only on $n$ and on the sum of the squared means; and that the MGF of this latter sum has a simple explicit form.

These results can be derived quite easily from the MGF of a squared Gaussian RV, an MGF which, using (19.22), can be shown to be given by

$$\left(X \sim \mathcal{N}\!\left(\mu, \sigma^2\right)\right) \Rightarrow \left(M_{X^2}(\theta) = \frac{1}{\sqrt{1 - 2\sigma^2\theta}}\, e^{-\frac{\mu^2}{2\sigma^2}}\, e^{\frac{\mu^2}{2\sigma^2(1 - 2\sigma^2\theta)}}, \ \theta < \frac{1}{2\sigma^2}\right). \quad (19.32)$$

With a small leap of faith we can assume that (19.32) also holds for complex arguments whose real part is smaller than $1/(2\sigma^2)$ so that upon substituting $i\varpi$ for $\theta$ we can obtain the characteristic function

$$\left(X \sim \mathcal{N}\!\left(\mu, \sigma^2\right)\right) \Rightarrow \left(\Phi_{X^2}(\varpi) = \frac{1}{\sqrt{1 - i2\sigma^2\varpi}}\, e^{-\frac{\mu^2}{2\sigma^2}}\, e^{\frac{\mu^2}{2\sigma^2(1 - i2\sigma^2\varpi)}}, \ \varpi \in \mathbb{R}\right). \quad (19.33)$$

### 19.8.1   The Central $\chi^2$ Distribution and Related Distributions

The **central $\chi^2$ distribution with $n$ degrees of freedom** is denoted by $\chi_n^2$ and is defined as the distribution of the sum of the squares of $n$ IID zero-mean unit-variance Gaussian random variables:

$$\left( X_1, \ldots, X_n \sim \text{IID } \mathcal{N}(0,1) \right) \Rightarrow \left( \sum_{j=1}^{n} X_j^2 \sim \chi_n^2 \right). \tag{19.34}$$

Using the fact that the MGF of the sum of independent random variables is the product of their MGFs and using (19.32) with $\mu = 0$ and $\sigma^2 = 1$, we obtain that the MGF of the central $\chi^2$ distribution with $n$ degrees of freedom is given by

$$\mathsf{E}\left[ e^{\theta \chi_n^2} \right] = \frac{1}{(1 - 2\theta)^{n/2}}, \quad \theta < \frac{1}{2}. \tag{19.35}$$

Similarly, by (19.33) and the fact that the characteristic function of the sum of independent random variables is the product of their characteristic functions, (or by substituting $i\varpi$ for $\theta$ in (19.35)), we obtain that the characteristic function of the central $\chi^2$ distribution with $n$ degrees of freedom is given by

$$\mathsf{E}\left[ e^{i\varpi \chi_n^2} \right] = \frac{1}{(1 - 2i\varpi)^{n/2}}, \quad \varpi \in \mathbb{R}. \tag{19.36}$$

Notice that for $n = 2$ this characteristic function is given by $\varpi \mapsto 1/(1 - i2\varpi)$, which is the characteristic function of the mean-2 exponential density

$$\frac{1}{2} e^{-x/2} \, \mathrm{I}\{x > 0\}, \quad x \in \mathbb{R}.$$

Since two random variables of identical characteristic functions must be of equal law (Proposition 19.7.1 (ii)), we conclude:

**Note 19.8.1.** The central $\chi^2$ distribution with two degrees of freedom $\chi_2^2$ is the mean-2 exponential distribution.

From (19.36) and the relationship between the moments of a distribution and the derivatives at zero of its characteristic function (19.27), one can verify that the $\nu$-th moment of a $\chi_n^2$ RV is given by

$$\mathsf{E}\left[ (\chi_n^2)^\nu \right] = n \times (n+2) \times \cdots \times \left( n + 2(\nu - 1) \right), \quad \nu \in \mathbb{N}, \tag{19.37}$$

so the mean is $n$; the second moment is $n(n+2)$; and the variance is $2n$.

Since the sum of the squares of random variables must be nonnegative, the density of the $\chi_n^2$ distribution is zero on the negative numbers. It is given by

$$f_{\chi_n^2}(x) = \frac{1}{2^{n/2}\, \Gamma(n/2)} \, e^{-x/2} \, x^{(n/2)-1} \, \mathrm{I}\{x > 0\}, \tag{19.38}$$

where $\Gamma(\cdot)$ is the **Gamma function**, which is defined by

$$\Gamma(\xi) \triangleq \int_0^\infty e^{-t}\, t^{\xi-1} \, \mathrm{d}t, \quad \xi > 0. \tag{19.39}$$

If the number of degrees of freedom is even, then the density has a particularly simple form:

$$f_{\chi^2_{2k}}(x) = \frac{1}{2^k(k-1)!} e^{-x/2} x^{k-1} \, \mathrm{I}\{x > 0\}, \quad k \in \mathbb{N}, \tag{19.40}$$

thus demonstrating again that when the number of degrees of freedom is two, the central $\chi^2$ distribution is the mean-2 exponential distribution (Note 19.8.1).

A related distribution is the **generalized Rayleigh distribution**, which is the distribution of the square root of a random variable having a $\chi^2_n$ distribution. The density of the generalized Rayleigh distribution is given by

$$f_{\sqrt{\chi^2_n}}(x) = \frac{2}{2^{n/2}\,\Gamma(n/2)} x^{n-1}\, e^{-x^2/2} \, \mathrm{I}\{x > 0\}, \tag{19.41}$$

and its moments by

$$\mathsf{E}\left[\left(\sqrt{\chi^2_n}\right)^\nu\right] = \frac{2^{\nu/2}\,\Gamma\big((n+\nu)/2\big)}{\Gamma(n/2)}, \quad \nu \in \mathbb{N}. \tag{19.42}$$

The **Rayleigh distribution** is the distribution of the square root of a $\chi^2_2$ random variable, i.e., the distribution of the square root of a mean-2 exponential random variable. The density of the Rayleigh distribution is obtained by setting $n = 2$ in (19.41):

$$f_{\sqrt{\chi^2_2}}(x) = x\, e^{-x^2/2} \, \mathrm{I}\{x > 0\}. \tag{19.43}$$

## 19.8.2 The Noncentral $\chi^2$ Distribution and Related Distributions

Using (19.32) and the fact that the MGF of the sum of independent random variables is the product of their MGFs, we obtain that if $X_1, \ldots, X_n$ are independent with $X_j \sim \mathcal{N}(\mu_j, \sigma^2)$, then the MGF of $\sum_j X_j^2$ is given by

$$\left(\frac{1}{\sqrt{1 - 2\sigma^2\theta}}\right)^n e^{-\frac{\sum_{j=1}^n \mu_j^2}{2\sigma^2}} e^{\frac{\sum_{j=1}^n \mu_j^2}{2\sigma^2(1 - 2\sigma^2\theta)}}, \quad \theta < \frac{1}{2\sigma^2}. \tag{19.44}$$

Noting that this MGF depends on the individual means $\mu_1, \ldots, \mu_n$ only via the sum of their squares $\sum \mu_j^2$, we obtain:

**Note 19.8.2.** The distribution of the sum of the squares of independent equivariance Gaussians is determined by their number, their common variance, and by the sum of the squares of their means.

The distribution of the sum of $n$ independent unit-variance Gaussians whose squared means sum to $\lambda$ is called the **noncentral $\chi^2$ distribution with $n$ degrees of freedom and noncentrality parameter $\lambda$**. This distribution is denoted by $\chi^2_{n,\lambda}$. Substituting $\sigma^2 = 1$ in (19.44) we obtain that the MGF of the $\chi^2_{n,\lambda}$ distribution is

$$\mathsf{E}\left[e^{\theta \chi^2_{n,\lambda}}\right] = \left(\frac{1}{\sqrt{1 - 2\theta}}\right)^n e^{-\frac{\lambda}{2}} e^{\frac{\lambda}{2(1 - 2\theta)}}, \quad \theta < \frac{1}{2}. \tag{19.45}$$

A special case of this distribution is the central $\chi^2$ distribution, which corresponds to the case where the noncentrality parameter $\lambda$ is zero.

Explicit expressions for the density of the noncentral $\chi^2$ distribution can be found in (Johnson, Kotz, and Balakrishnan, 1994b, Chapter 29, Equation (29.4)) and in (Simon, 2002, Chapter 2). An interesting representation of this density in terms of the density $f_{\chi^2_{\nu,0}}$ of the *central* $\chi^2$ distribution is:

$$f_{\chi^2_{n,\lambda}}(x) = \sum_{j=0}^{\infty} \left( \frac{(\frac{1}{2}\lambda)^j}{j!} e^{-\lambda/2} \right) f_{\chi^2_{n+2j,0}}(x), \quad x \in \mathbb{R}. \tag{19.46}$$

It demonstrates that a $\chi^2_{n,\lambda}$ random variable $X$ can be generated by picking a random integer $j$ according to the Poisson distribution of parameter $\lambda/2$ and by then generating a central $\chi^2$ random variable of $n + 2j$ degrees of freedom. That is, to generate a $\chi^2_{n,\lambda}$ random variable $X$, generate some random variable $J$ taking value in the nonnegative integers according to the law

$$\Pr[J = j] = e^{-\lambda/2} \frac{(\lambda/2)^j}{j!}, \quad j = 0, 1, \ldots \tag{19.47}$$

and then generate $X$ according the central $\chi^2$ distribution with $n + 2j$ degrees of freedom, where $j$ is the outcome of $J$.

The density of the $\chi^2_{2,\lambda}$ distribution is

$$f_{\chi^2_{2,\lambda}}(x) = \frac{1}{2} e^{-(\lambda+x)/2} \mathrm{I}_0\left(\sqrt{\lambda x}\right) \mathrm{I}\{x > 0\}, \tag{19.48}$$

where $\mathrm{I}_0(\cdot)$ is the modified zeroth-order Bessel function, which is defined in (27.47) ahead.

The **generalized Rice distribution** corresponds to the distribution of the square root of a noncentral $\chi^2$ distribution with $n$ degrees of freedom and noncentrality parameter $\lambda$. The case $n = 2$ is called the **Rice distribution**. The Rice distribution is thus the distribution of the square root of a random variable having the noncentral $\chi^2$ distribution with 2 degrees of freedom and noncentrality parameter $\lambda$. The density of the Rice distribution is

$$f_{\sqrt{\chi^2_{2,\lambda}}}(x) = x\, e^{-(x^2+\lambda)/2} \mathrm{I}_0\left(x\sqrt{\lambda}\right) \mathrm{I}\{x > 0\}. \tag{19.49}$$

The following property of the noncentral $\chi^2$ is useful in detection theory. In the statistics literature this property is called the **Monotone Likelihood Ratio** property (Lehmann and Romano, 2005, Section 3.4). Alternatively, it is called the **Total Positivity of Order 2** of the function $(x, \lambda) \mapsto f_{\chi^2_{n,\lambda}}(x)$.

**Proposition 19.8.3 (The Noncentral $\chi^2$ Family Has Monotone Likelihood Ratio).**
*Let $f_{\chi^2_{n,\lambda}}(\xi)$ denote the density at $\xi$ of the noncentral $\chi^2$ distribution with $n$ degrees of freedom and noncentrality parameter $\lambda \geq 0$; see (19.46). Then for $\xi_1, \xi_2 > 0$ and $\lambda_1, \lambda_2 \geq 0$ we have*

$$\left( \xi_0 < \xi_1 \text{ and } \lambda_0 < \lambda_1 \right) \Rightarrow \left( f_{\chi^2_{n,\lambda_1}}(\xi_0)\, f_{\chi^2_{n,\lambda_0}}(\xi_1) \leq f_{\chi^2_{n,\lambda_0}}(\xi_0)\, f_{\chi^2_{n,\lambda_1}}(\xi_1) \right), \tag{19.50}$$

*i.e.*,

$$\left(\lambda_1 > \lambda_0\right) \Rightarrow \left(\xi \mapsto \frac{f_{\chi^2_{n,\lambda_1}}(\xi)}{f_{\chi^2_{n,\lambda_0}}(\xi)} \text{ is nondecreasing in } \xi > 0\right). \tag{19.51}$$

**Proof.** See, for example, (Finner and Roters, 1997, Proposition 3.8). □

## 19.9   The Limit of Gaussians Is Gaussian

There are a number of useful definitions of convergence for sequences of random variables. Here we briefly mention a few and show that, under each of these definitions, the convergence of a sequence of Gaussian random variables to a random variable $X$ implies that $X$ is Gaussian.

Let the random variables $X, X_1, X_2, \ldots$ be defined over a common probability space $(\Omega, \mathcal{F}, P)$. We say that the sequence $X_1, X_2, \ldots$ converges to $X$ **with probability one** or **almost surely** if

$$\Pr\left(\left\{\omega \in \Omega : \lim_{n \to \infty} X_n(\omega) = X(\omega)\right\}\right) = 1. \tag{19.52}$$

Thus, the sequence $X_1, X_2, \ldots$ converges to $X$ almost surely if there exists an event $\mathcal{N} \in \mathcal{F}$ of probability zero such that for every $\omega \notin \mathcal{N}$ the sequence of real numbers $X_1(\omega), X_2(\omega), \ldots$ converges to the real number $X(\omega)$.

The sequence $X_1, X_2, \ldots$ converges to $X$ **in probability** if

$$\lim_{n \to \infty} \Pr\left[|X_n - X| \geq \epsilon\right] = 0, \quad \epsilon > 0. \tag{19.53}$$

The sequence $X_1, X_2, \ldots$ converges to $X$ **in mean square** if

$$\lim_{n \to \infty} \mathsf{E}\left[\left(X_n - X\right)^2\right] = 0. \tag{19.54}$$

We refer the reader to (Shiryaev, 1996, Ch. II, Section 10, Theorem 2) for a proof that convergence in mean-square implies convergence in probability and for a proof that almost-sure convergence implies convergence in probability. Also, if a sequence converges in probability to $X$, then it has a subsequence that converges to $X$ with probability one (Shiryaev, 1996, Ch. II, Section 10, Theorem 5).

**Theorem 19.9.1.** *Let the random variables* $X, X_1, X_2, \ldots$ *be defined over a common probability space* $(\Omega, \mathcal{F}, P)$. *Assume that each of the random variables* $X_1, X_2, \ldots$ *is Gaussian. If the sequence* $X_1, X_2, \ldots$ *converges to* $X$ *in the sense of* (19.52) *or* (19.53) *or* (19.54), *then* $X$ *must also be Gaussian.*

**Proof.** Since both mean-square convergence and almost-sure convergence imply convergence in probability, it suffices to prove the theorem in the case where the sequence $X_1, X_2, \ldots$ converges to $X$ in probability. And since every sequence converging to $X$ in probability has a subsequence converging to $X$ almost surely, it suffices to prove the theorem for almost sure convergence. Our proof for this case follows (Shiryaev, 1996, Ch. II, Section 13, Paragraph 5).

Since the random variables $X_1, X_2, \ldots$ are all Gaussian, it follows from (19.29) that

$$\mathsf{E}\left[e^{i\varpi X_n}\right] = e^{i\varpi\mu_n - \frac{1}{2}\varpi^2\sigma_n^2}, \quad \varpi \in \mathbb{R}, \tag{19.55}$$

where $\mu_n$ and $\sigma_n^2$ are the mean and variance of $X_n$. By the Dominated Convergence Theorem it follows that the almost sure convergence of $X_1, X_2, \ldots$ to $X$ implies that

$$\lim_{n\to\infty} \mathsf{E}\left[e^{i\varpi X_n}\right] = \mathsf{E}\left[e^{i\varpi X}\right], \quad \varpi \in \mathbb{R}. \tag{19.56}$$

It follows from (19.55) and (19.56) that

$$\lim_{n\to\infty} e^{i\varpi\mu_n - \frac{1}{2}\varpi^2\sigma_n^2} = \mathsf{E}\left[e^{i\varpi X}\right], \quad \varpi \in \mathbb{R}. \tag{19.57}$$

The limit in (19.57) can exist for every $\varpi \in \mathbb{R}$ only if there exist $\mu, \sigma^2$ such that $\mu_n \to \mu$ and $\sigma_n^2 \to \sigma^2$. And in this case, by (19.57),

$$\mathsf{E}\left[e^{i\varpi X}\right] = e^{i\varpi\mu - \frac{1}{2}\varpi^2\sigma^2}, \quad \varpi \in \mathbb{R},$$

so, by Proposition 19.7.1 (ii) and by (19.29), $X$ is $\mathcal{N}(\mu, \sigma^2)$. $\qquad\square$

Another type of convergence is **convergence in distribution** or **weak convergence**, which is defined as follows. Let $F_1, F_2, \ldots$ denote the cumulative distribution functions of the sequence of random variables $X_1, X_2, \ldots$ We say that the sequence $F_1, F_2, \ldots$ (or sometimes $X_1, X_2, \ldots$) converges in distribution to the cumulative distribution function $F(\cdot)$ if $F_n(\xi)$ converges to $F(\xi)$ at every point $\xi \in \mathbb{R}$ at which $F(\cdot)$ is continuous. That is,

$$\Big(F_n(\xi) \to F(\xi)\Big), \quad \Big(F(\cdot) \text{ is continuous at } \xi\Big). \tag{19.58}$$

**Theorem 19.9.2.** *Let the sequence of random variables $X_1, X_2, \ldots$ be such that $X_n \sim \mathcal{N}(\mu_n, \sigma_n^2)$, for every $n \in \mathbb{N}$. Then the sequence converges in distribution to some limiting distribution if, and only if, there exist some $\mu$ and $\sigma^2$ such that*

$$\mu_n \to \mu \text{ and } \sigma_n^2 \to \sigma^2. \tag{19.59}$$

*And if the sequence does converge in distribution, then it converges to the mean-$\mu$ variance-$\sigma^2$ Gaussian distribution.*

**Proof.** See (Gikhman and Skorokhod, 1996, Chapter I, Section 3, Theorem 4) where this statement is proved in the multivariate case. $\qquad\square$

For extensions of Theorems 19.9.1 & 19.9.2 to random vectors, see Theorems 23.9.1 & 23.9.2 in Section 23.9.

## 19.10 Additional Reading

The Gaussian distribution, its characteristic function, and its moment generating function appear in almost every basic book on Probability Theory. For more on the $Q$-function see (Verdú, 1998, Section 3.3) and (Simon, 2002). For more on distributions related to the Gaussian distribution see (Simon, 2002), (Johnson, Kotz, and Balakrishnan, 1994a), and (Johnson, Kotz, and Balakrishnan, 1994b). For more on the central $\chi^2$ distribution see (Johnson, Kotz, and Balakrishnan, 1994a, Chapter 18) and (Simon, 2002, Chapter 2). For more on the noncentral $\chi^2$ distribution see (Johnson, Kotz, and Balakrishnan, 1994b, Chapter 29) and (Simon, 2002, Chapter 2). Various characterizations of the Gaussian distribution can be found in (Bryc, 1995) and (Bogachev, 1998).

## 19.11 Exercises

**Exercise 19.1 (Sums of Independent Gaussians).** Let $X_1 \sim \mathcal{N}(0, \sigma_1^2)$ and $X_2 \sim \mathcal{N}(0, \sigma_2^2)$ be independent. Convolve their densities to show that $X_1 + X_2$ is Gaussian.

**Exercise 19.2 (Computing Probabilities).** Let $X \sim \mathcal{N}(1, 3)$ and $Y \sim \mathcal{N}(-2, 4)$ be independent. Express the probabilities $\Pr[X \leq 2]$ and $\Pr[2X + 3Y > -2]$ using the $Q$-function with nonnegative arguments.

**Exercise 19.3 (Bounds on the $Q$-function).** Prove (19.19). We suggest changing the integration variable in (19.9) to $\zeta \triangleq \xi - \alpha$ and then proving and using the inequality

$$1 - \frac{\zeta^2}{2} \leq \exp\left(-\frac{\zeta^2}{2}\right) \leq 1, \quad \xi \in \mathbb{R}.$$

**Exercise 19.4 (An Application of Craig's Formula).** Let the random variables $Z \sim \mathcal{N}(0, 1)$ and $A$ be independent, where $A^2$ is of MGF $M_{A^2}(\cdot)$. Show that

$$\Pr[Z \geq |A|] = \frac{1}{\pi} \int_0^{\pi/2} M_{A^2}\left(-\frac{1}{2\sin^2 \varphi}\right) d\varphi.$$

**Exercise 19.5 (An Expression for $Q^2(\alpha)$).** In analogy to (19.15), derive the identity

$$Q^2(\alpha) = \frac{1}{\pi} \int_0^{\pi/4} e^{-\frac{\alpha^2}{2\sin^2 \varphi}} d\varphi, \quad \alpha \geq 0.$$

**Exercise 19.6 (Expectation of $Q(X)$).** Show that for any RV $X$

$$\mathsf{E}[Q(X)] = \frac{1}{\sqrt{2\pi}} \int_{-\infty}^{\infty} \Pr[X \leq \xi] e^{-\xi^2/2} d\xi.$$

(See (Verdú, 1998, Chapter 3, Section 3.3, Eq. (3.57)).)

**Exercise 19.7 (Generating Gaussians from Uniform RVs).**

(i) Let $W_1$ and $W_2$ be IID $\mathcal{N}(0,1)$, and let $R = \sqrt{W_1^2 + W_2^2}$. Show that $R$ has a Rayleigh distribution, i.e., that its density $f_R(r)$ is given for every $r \in \mathbb{R}$ by $re^{-\frac{r^2}{2}} \mathrm{I}\{r \geq 0\}$. What is the CDF $F_R(\cdot)$ of $R$?

(ii) Prove that if a RV $X$ is of density $f_X(\cdot)$ and of CDF $F_X(\cdot)$, then $F_X(X) \sim \mathcal{U}(0,1)$.

(iii) Show that if $U_1$ and $U_2$ are IID $\mathcal{U}(0,1)$ and if we define $R = \sqrt{\ln \frac{1}{U_1}}$ and $\Theta = 2\pi U_2$, then $R\cos\Theta$ and $R\sin\Theta$ are IID $\mathcal{N}(0,1/2)$.

**Exercise 19.8 (Infinite Divisibility).** Show that for any $\mu \in \mathbb{R}$ and $\sigma^2 \geq 0$ there exist IID RVs $X$ and $Y$ such that $X + Y \sim \mathcal{N}(\mu, \sigma^2)$.

**Exercise 19.9 (MGF of the Square of a Gaussian).** Derive (19.32).

**Exercise 19.10 (The Distribution of the Magnitude).** Show that if a random variable $X$ is of density $f_X(\cdot)$ and if $Y = |X|$, then the density $f_Y(\cdot)$ of $Y$ is

$$f_Y(y) = \big(f_X(y) + f_X(-y)\big)\,\mathrm{I}\{y \geq 0\}, \quad y \in \mathbb{R}.$$

**Exercise 19.11 (Uniformly Distributed Random Variables).** Suppose that $X \sim \mathcal{U}([0,1])$.

(i) Find the characteristic function $\Phi_X(\cdot)$ of $X$.

(ii) Show that if $X$ and $Y$ are independent with $X$ as above, then $X+Y$ is not Gaussian.

**Exercise 19.12 (Sums and Differences of IID RVs).** Let $X$ and $Y$ be IID random variables with finite variances. Show that if $X + Y$ and $X - Y$ are independent, then $X$ and $Y$ are Gaussian.

(See (Feller, 1971, Chapter III, Section 4).)

# Chapter 20

# Binary Hypothesis Testing

## 20.1 Introduction

In Digital Communications the task of the receiver is to observe the channel outputs and to use these observations to accurately guess the data bits that were sent by the transmitter, i.e., the data bits that were fed to the modulator. Ideally, the guessing would be perfect, i.e., the receiver would make no errors. This, alas, is typically impossible because of the distortions and noise that the channel introduces. Indeed, while one can usually recover the data bits from the *transmitted* waveform (provided that the modulator is a one-to-one mapping), the receiver has no access to the transmitted waveform but only to the *received* waveform. And since the latter is typically a noisy version of the former, some errors are usually unavoidable.

In this chapter we shall begin our study of how to guess intelligently, i.e., how, given the channel output, one should guess the data bits with as low a probability of error as possible. This study will help us not only in the design of receivers but also in the design of modulators that allow for reliable decoding from the channel's output.

In the engineering literature the process of guessing the data bits based on the channel output is called "decoding." In the statistics literature this process is called "hypothesis testing." We like "guessing" because it demystifies the process.

In most applications the channel output is a continuous-time waveform and we seek to decode a large number of bits. Nevertheless, for pedagogical reasons, we shall begin our study with the simpler case where we wish to decode only a single data bit. This corresponds in the statistics literature to "binary hypothesis testing," where the term "binary" reminds us that in this guessing problem there are only two alternatives. Moreover, we shall assume that the observation, rather than being a continuous-time waveform, is a vector or a scalar. In fact, we shall begin our study with the simplest case where there are no observations at all.

## 20.2    Problem Formulation

In choosing a guessing strategy to minimize the probability of error, the labels of the two alternatives are immaterial. The principles that guide us in guessing the outcome of a fair coin toss (where the labels are "heads" or "tails") are the same as for guessing the value of a random variable that takes on the values $+1$ and $-1$ equiprobably. (These are, of course, extremely simple cases that can be handled with common sense.) Statisticians typically denote the two alternatives by $\mathcal{H}_0$ and $\mathcal{H}_1$ and call them "hypotheses." We shall denote the two alternatives by 0 and 1. We thus envision guessing the value of a random variable $H$ taking value in the set $\{0, 1\}$ with probabilities

$$\pi_0 = \Pr[H = 0], \qquad \pi_1 = \Pr[H = 1]. \tag{20.1}$$

The **prior** is the distribution of $H$ or the pair $(\pi_0, \pi_1)$. It reflects the state of our knowledge about $H$ before having made any observations. We say that the prior is **nondegenerate** if

$$\pi_0, \pi_1 > 0. \tag{20.2}$$

(If the prior is degenerate, then $H$ is deterministic and we can determine its value without any observation. For example if $\pi_0 = 0$ we always guess 1 and never err.) The prior is **uniform** if $\pi_0 = \pi_1 = 1/2$.

Aiding us in the guess work is the **observation Y**, which is a random vector taking value in $\mathbb{R}^d$. (When $d = 1$ the observation is a random variable and we denote it by $Y$.) We assume that $\mathbf{Y}$ is a column vector, so, using the notation of Section 17.2,

$$\mathbf{Y} = \left(Y^{(1)}, \ldots, Y^{(d)}\right)^{\mathsf{T}}.$$

Typically there is some statistical dependence between $\mathbf{Y}$ and $H$; otherwise, $\mathbf{Y}$ would be useless. If the dependence is so strong that from $\mathbf{Y}$ one can deduce $H$, then our guess work is very easy: we simply compute from $\mathbf{Y}$ the value of $H$ and declare the result as our guess; we never err. The cases of most interest to us are therefore those where $\mathbf{Y}$ neither determines $H$ nor is statistically independent of $H$. Unless otherwise specified, we shall assume that, conditional on $H = 0$, the observation $\mathbf{Y}$ is of density $f_{\mathbf{Y}|H=0}(\cdot)$ and that, conditional on $H = 1$, it is of density $f_{\mathbf{Y}|H=1}(\cdot)$. Here $f_{\mathbf{Y}|H=0}(\cdot)$ and $f_{\mathbf{Y}|H=1}(\cdot)$ are nonnegative Borel measurable functions from $\mathbb{R}^d$ to $\mathbb{R}$ that integrate to one.[1]

Our problem is how to use the observation $\mathbf{Y}$ to intelligently guess the value of $H$. At first we shall limit ourselves to deterministic guessing rules. Later we shall show that no randomized guessing rule can outperform an optimal deterministic rule. A deterministic **guessing rule** (or **decision rule** , or **decoding rule**) for guessing $H$ based on $\mathbf{Y}$ is a (Borel measurable) mapping from the set of possible observations $\mathbb{R}^d$ to the set $\{0, 1\}$. We denote such a mapping by

$$\phi_{\mathrm{Guess}} \colon \mathbb{R}^d \to \{0, 1\} \tag{20.3}$$

---

[1] Readers who are familiar with Measure Theory should note that these are densities with respect to the Lebesgue measure on $\mathbb{R}^d$, but that the reference measure is inessential to our analysis. We could have also chosen as our reference measure the sum of the probability measures on $\mathbb{R}^d$ corresponding to $H = 0$ and to $H = 1$. This would have guaranteed the existence of the densities.

and say that $\phi_{\text{Guess}}(\mathbf{y}_{\text{obs}})$ is the guess we make after having observed that $\mathbf{Y} = \mathbf{y}_{\text{obs}}$. The **probability of error** associated with the guessing rule $\phi_{\text{Guess}}(\cdot)$ is

$$\Pr(\text{error}) \triangleq \Pr[\phi_{\text{Guess}}(\mathbf{Y}) \neq H]. \tag{20.4}$$

Note that two sources of randomness determine whether the guessing rule $\phi_{\text{Guess}}(\cdot)$ errs or not: the realization of $H$ and the generation of $\mathbf{Y}$ conditional on that realization. We say that a guessing rule is **optimal** if no other guessing rule attains a smaller probability of error. (We shall later see that there always exists an optimal guessing rule.[2]) In general, there may be a number of different optimal guessing rules. We shall therefore try to refrain from speaking of *the* optimal guessing rule. We apologize if this results in cumbersome writing. The probability of error associated with optimal guessing rules is the **optimal probability of error** and is denoted throughout by

$$p^*(\text{error}).$$

## 20.3 Guessing in the Absence of Observables

We begin with the simplest case where there are no observables. Common sense dictates that in this case we should base our guess on the prior $(\pi_0, \pi_1)$ as follows. If $\pi_0 > \pi_1$, then we should guess that the value of $H$ is 0; if $\pi_0 < \pi_1$, then we should guess the value 1; and if $\pi_0 = \pi_1 = 1/2$, then it does not really matter what we guess: the probability of error will be either way $1/2$.

To verify that this intuition is correct note that, since there are no observables, there are only two guessing rules: the rule "guess 0" and the rule "guess 1." The former results in the probability of error $\pi_1$ (it is in error whenever $H = 1$, which happens with probability $\pi_1$), and the latter results in the probability of error $\pi_0$. Hence the former rule is optimal if $\pi_0 \geq \pi_1$ and the latter is optimal when $\pi_1 \geq \pi_0$. When $\pi_0 = \pi_1$ both rules are optimal and we can use either one.

We summarize that, in the absence of observations, an optimal guessing rule is:

$$\phi_{\text{Guess}}^* = \begin{cases} 0 & \text{if } \Pr[H = 0] \geq \Pr[H = 1], \\ 1 & \text{otherwise.} \end{cases} \tag{20.5}$$

(Here we guess 0 also when $\Pr[H = 0] = \Pr[H = 1]$. An equally good rule would guess 1 in this case.)

As we next show, the error probability $p^*(\text{error})$ of this rule is

$$p^*(\text{error}) = \min\{\Pr[H = 0], \Pr[H = 1]\}. \tag{20.6}$$

This can be verified by considering the case where $\Pr[H = 0] \geq \Pr[H = 1]$ and the case where $\Pr[H = 0] < \Pr[H = 1]$ separately. By (20.5), in the former case our

---

[2]Thus, while there is no such thing as "smallest strictly positive number," i.e., a positive number that is smaller-or-equal to any other positive number, we shall see that there always exists a guessing rule that no other guessing rule can outperform. Mathematicians paraphrase this by saying that "the infimum of the probability of error over all the guessing rules is achievable, i.e., is a minimum."

guess is 0 with the associated probability of error $\Pr[H = 1]$, whereas in the latter case our guess is 1 with the associated probability of error $\Pr[H = 0]$. In either case the probability of error is given by the RHS of (20.6).

## 20.4 The Joint Law of $H$ and $Y$

Before we can extend the results of Section 20.3 to the more interesting case where we guess $H$ after observing $\mathbf{Y}$, we pause to discuss the joint distribution of $H$ and $\mathbf{Y}$. This joint distribution is needed in order to derive an optimal decision rule and in order to analyze its performance. Some care must be exercised in describing this law because $H$ is discrete (binary) and $\mathbf{Y}$ has a density. It is usually simplest to describe the joint law by describing the prior (the distribution of $H$), and by then describing the conditional law of $\mathbf{Y}$ given $H = 0$ and the conditional law of $\mathbf{Y}$ given $H = 1$.

If, conditional on $H = 0$, the distribution of $\mathbf{Y}$ has the density $f_{\mathbf{Y}|H=0}(\cdot)$ and if, conditional on $H = 1$, the distribution of $\mathbf{Y}$ has the density $f_{\mathbf{Y}|H=1}(\cdot)$, then the joint distribution of $H$ and $\mathbf{Y}$ can be described using the prior $(\pi_0, \pi_1)$ (20.1) and the conditional densities

$$f_{\mathbf{Y}|H=0}(\cdot) \quad \text{and} \quad f_{\mathbf{Y}|H=1}(\cdot). \tag{20.7}$$

From the prior $(\pi_0, \pi_1)$ and the conditional densities $f_{\mathbf{Y}|H=0}(\cdot), f_{\mathbf{Y}|H=1}(\cdot)$ we can compute the (unconditional) density of $\mathbf{Y}$:

$$f_{\mathbf{Y}}(\mathbf{y}) = \pi_0 f_{\mathbf{Y}|H=0}(\mathbf{y}) + \pi_1 f_{\mathbf{Y}|H=1}(\mathbf{y}), \quad \mathbf{y} \in \mathbb{R}^d. \tag{20.8}$$

The conditional distribution of $H$ given $\mathbf{Y} = \mathbf{y}_{\mathrm{obs}}$ is a bit more tricky because the probability of $\mathbf{Y}$ taking on the value $\mathbf{y}_{\mathrm{obs}}$ (exactly) is zero. There are two approaches to defining $\Pr[H = 0 | \mathbf{Y} = \mathbf{y}_{\mathrm{obs}}]$ in this case: the heuristic one that is usually used in a first course on probability theory and the measure-theoretic one that was pioneered by Kolmogorov. Our approach is to define this quantity in a way that will be palatable to both mathematicians and engineers and to then give a heuristic justification for our definition.

We define the conditional probability that $H = 0$ given $\mathbf{Y} = \mathbf{y}_{\mathrm{obs}}$ as

$$\Pr\big[H = 0 \,\big|\, \mathbf{Y} = \mathbf{y}_{\mathrm{obs}}\big] \triangleq \begin{cases} \frac{\pi_0 f_{\mathbf{Y}|H=0}(\mathbf{y}_{\mathrm{obs}})}{f_{\mathbf{Y}}(\mathbf{y}_{\mathrm{obs}})} & \text{if } f_{\mathbf{Y}}(\mathbf{y}_{\mathrm{obs}}) > 0, \\ \frac{1}{2} & \text{otherwise,} \end{cases} \tag{20.9a}$$

where $f_{\mathbf{Y}}(\cdot)$ is given in (20.8), and analogously

$$\Pr\big[H = 1 \,\big|\, \mathbf{Y} = \mathbf{y}_{\mathrm{obs}}\big] \triangleq \begin{cases} \frac{\pi_1 f_{\mathbf{Y}|H=1}(\mathbf{y}_{\mathrm{obs}})}{f_{\mathbf{Y}}(\mathbf{y}_{\mathrm{obs}})} & \text{if } f_{\mathbf{Y}}(\mathbf{y}_{\mathrm{obs}}) > 0, \\ \frac{1}{2} & \text{otherwise.} \end{cases} \tag{20.9b}$$

Notice that our definition is meaningful in the sense that the values we assign to $\Pr[H = 0 | \mathbf{Y} = \mathbf{y}_{\mathrm{obs}}]$ and $\Pr[H = 1 | \mathbf{Y} = \mathbf{y}_{\mathrm{obs}}]$ are nonnegative and sum to one:

$$\Pr\big[H = 0 \,\big|\, \mathbf{Y} = \mathbf{y}_{\mathrm{obs}}\big] + \Pr\big[H = 1 \,\big|\, \mathbf{Y} = \mathbf{y}_{\mathrm{obs}}\big] = 1, \quad \mathbf{y}_{\mathrm{obs}} \in \mathbb{R}^d. \tag{20.10}$$

Also note that our definition of $\Pr[H = 0 \,|\, \mathbf{Y} = \mathbf{y}_{\text{obs}}]$ and $\Pr[H = 1 \,|\, \mathbf{Y} = \mathbf{y}_{\text{obs}}]$ for those $\mathbf{y}_{\text{obs}} \in \mathbb{R}^d$ for which $f_{\mathbf{Y}}(\mathbf{y}_{\text{obs}}) = 0$ is quite arbitrary; we chose $1/2$ just for concreteness.[3] Indeed, it is not difficult to verify that the probability that $\mathbf{y}_{\text{obs}}$ satisfies $\pi_0 f_{\mathbf{Y}|H=0}(\mathbf{y}_{\text{obs}}) + \pi_1 f_{\mathbf{Y}|H=1}(\mathbf{y}_{\text{obs}}) = 0$ is zero, and hence our definitions in this eventuality are not important; see (20.12) ahead.

If $d = 1$, then the observation is a random variable $Y$ and a heuristic way to motivate (20.9a) is to consider the limit

$$\lim_{\delta \downarrow 0} \frac{\Pr\big[H = 0, Y \in \big(y_{\text{obs}} - \delta, y_{\text{obs}} + \delta\big)\big]}{\Pr\big[Y \in \big(y_{\text{obs}} - \delta, y_{\text{obs}} + \delta\big)\big]}. \tag{20.11}$$

Assuming some regularity of the conditional densities (e.g., continuity) we can use the approximations

$$\Pr\big[H = 0, Y \in (y_{\text{obs}} - \delta, y_{\text{obs}} + \delta)\big] = \pi_0 \int_{y_{\text{obs}} - \delta}^{y_{\text{obs}} + \delta} f_{Y|H=0}(y) \, dy$$

$$\approx 2\pi_0 \delta f_{Y|H=0}(y_{\text{obs}}), \quad \delta \ll 1,$$

$$\Pr\big[Y \in (y_{\text{obs}} - \delta, y_{\text{obs}} + \delta)\big] = \int_{y_{\text{obs}} - \delta}^{y_{\text{obs}} + \delta} f_Y(y) \, dy$$

$$\approx 2\delta f_Y(y_{\text{obs}}), \quad \delta \ll 1,$$

to argue that, under suitable regularity conditions, (20.11) agrees with the RHS of (20.9a) when $f_Y(y_{\text{obs}}) > 0$. A similar calculation can be carried out in the vector case where $d > 1$.

We next remark on observations $\mathbf{y}_{\text{obs}}$ at which the density of $\mathbf{Y}$ is zero. Accounting for such observations makes the writing a bit cumbersome as in (20.9). Fortunately, the probability of such observations is zero:

**Note 20.4.1.** Let $H$ be drawn according to the prior $(\pi_0, \pi_1)$, and let the conditional densities of $\mathbf{Y}$ given $H$ be $f_{\mathbf{Y}|H=0}(\cdot)$ and $f_{\mathbf{Y}|H=1}(\cdot)$ with $f_{\mathbf{Y}}(\cdot)$ given in (20.8). Then

$$\Pr\big[\mathbf{Y} \in \{\tilde{\mathbf{y}} \in \mathbb{R}^d : f_{\mathbf{Y}}(\tilde{\mathbf{y}}) = 0\}\big] = 0. \tag{20.12}$$

**Proof.**

$$\Pr\big[\mathbf{Y} \in \{\tilde{\mathbf{y}} \in \mathbb{R}^d : f_{\mathbf{Y}}(\tilde{\mathbf{y}}) = 0\}\big] = \int_{\{\tilde{\mathbf{y}} \in \mathbb{R}^d : f_{\mathbf{Y}}(\tilde{\mathbf{y}}) = 0\}} f_{\mathbf{Y}}(\mathbf{y}) \, d\mathbf{y}$$

$$= \int_{\{\tilde{\mathbf{y}} \in \mathbb{R}^d : f_{\mathbf{Y}}(\tilde{\mathbf{y}}) = 0\}} 0 \, d\mathbf{y}$$

$$= 0,$$

where the second equality follows because the integrand is zero over the range of integration.   □

---

[3]In the measure-theoretic probability literature our definition is just a "version" (among many others) of the conditional probabilities of the event $H = 0$ (respectively $H = 1$), conditional on the $\sigma$-algebra generated by the random vector $\mathbf{Y}$ (Billingsley, 1995, Section 33), (Williams, 1991, Chapter 9).

We conclude this section with two technical remarks which are trivial if you ignore observations where $f_{\mathbf{Y}}(\cdot)$ is zero:

**Note 20.4.2.** Consider the setup of Note 20.4.1.

(i) For every $\mathbf{y} \in \mathbb{R}^d$

$$\min\{\pi_0 f_{\mathbf{Y}|H=0}(\mathbf{y}), \pi_1 f_{\mathbf{Y}|H=1}(\mathbf{y})\}$$
$$= \min\{\Pr[H = 0 | \mathbf{Y} = \mathbf{y}], \Pr[H = 1 | \mathbf{Y} = \mathbf{y}]\} f_{\mathbf{Y}}(\mathbf{y}). \quad (20.13)$$

(ii) For every $\mathbf{y} \in \mathbb{R}^d$

$$\left(\pi_0 f_{\mathbf{Y}|H=0}(\mathbf{y}) \geq \pi_1 f_{\mathbf{Y}|H=1}(\mathbf{y})\right)$$
$$\Leftrightarrow \left(\Pr[H = 0 | \mathbf{Y} = \mathbf{y}] \geq \Pr[H = 1 | \mathbf{Y} = \mathbf{y}]\right). \quad (20.14)$$

**Proof.** Identity (20.13) can be proved using (20.9) and (20.8) by separately considering the case $f_{\mathbf{Y}}(\mathbf{y}) > 0$ and the case $f_{\mathbf{Y}}(\mathbf{y}) = 0$ (where the latter is equivalent, by (20.8), to $\pi_0 f_{\mathbf{Y}|H=0}(\mathbf{y})$ and $\pi_1 f_{\mathbf{Y}|H=1}(\mathbf{y})$ both being zero).

To prove (20.14) we also separately consider the case $f_{\mathbf{Y}}(\mathbf{y}) > 0$ and the case $f_{\mathbf{Y}}(\mathbf{y}) = 0$. In the former case we note that for $c > 0$ the condition $a \geq b$ is equivalent to the condition $a/c \geq b/c$ so for $f_{\mathbf{Y}}(\mathbf{y}_{\text{obs}}) > 0$

$$\left(\pi_0 f_{\mathbf{Y}|H=0}(\mathbf{y}) \geq \pi_1 f_{\mathbf{Y}|H=1}(\mathbf{y})\right) \Leftrightarrow \left(\underbrace{\frac{\pi_0 f_{\mathbf{Y}|H=0}(\mathbf{y})}{f_{\mathbf{Y}}(\mathbf{y})}}_{\Pr[H=0|\mathbf{Y}=\mathbf{y}]} \geq \underbrace{\frac{\pi_1 f_{\mathbf{Y}|H=1}(\mathbf{y})}{f_{\mathbf{Y}}(\mathbf{y})}}_{\Pr[H=1|\mathbf{Y}=\mathbf{y}]}\right).$$

In the latter case where $f_{\mathbf{Y}}(\mathbf{y}) = 0$ we note that, by (20.8), both $\pi_0 f_{\mathbf{Y}|H=0}(\mathbf{y})$ and $\pi_1 f_{\mathbf{Y}|H=1}(\mathbf{y})$ are zero, so the condition on the LHS of (20.14) is true ($0 \geq 0$). Fortunately, when $f_{\mathbf{Y}}(\mathbf{y}) = 0$ the condition on the RHS of (20.14) is also true, because in this case (20.9) implies that $\Pr[H = 0 | \mathbf{Y} = \mathbf{y}]$ and $\Pr[H = 1 | \mathbf{Y} = \mathbf{y}]$ are both equal to $1/2$ (and $1/2 \geq 1/2$). $\qquad\square$

## 20.5  Guessing after Observing Y

We next derive an optimal rule for guessing $H$ after observing that $\mathbf{Y} = \mathbf{y}_{\text{obs}}$. We begin with a heuristic argument. Having observed that $\mathbf{Y} = \mathbf{y}_{\text{obs}}$, there are only two possible decision rules: to guess 0 or guess 1. Which should we choose? The answer now depends on the *a posteriori* distribution of $H$. Once it has been revealed to us that $\mathbf{Y} = \mathbf{y}_{\text{obs}}$, our outlook changes and we now assign the event $H = 0$ the *a posteriori* probability $\Pr[H = 0 | \mathbf{Y} = \mathbf{y}_{\text{obs}}]$ and the event $H = 1$ the complementary probability $\Pr[H = 1 | \mathbf{Y} = \mathbf{y}_{\text{obs}}]$. If the former is greater than the latter, then we should guess 0, and otherwise we should guess 1. Thus, after it has been revealed to us that $\mathbf{Y} = \mathbf{y}_{\text{obs}}$ the situation is equivalent to one in which we need to guess $H$ without any observables and where our distribution on $H$ is not its *a priori* distribution (prior) but its *a posteriori* distribution. Using our analysis from Section 20.3 we conclude that the guessing rule

$$\phi^*_{\text{Guess}}(\mathbf{y}_{\text{obs}}) = \begin{cases} 0 & \text{if } \Pr[H = 0 \,|\, \mathbf{Y} = \mathbf{y}_{\text{obs}}] \geq \Pr[H = 1 \,|\, \mathbf{Y} = \mathbf{y}_{\text{obs}}], \\ 1 & \text{otherwise,} \end{cases} \tag{20.15}$$

is optimal. Once again, the way we resolve ties is arbitrary: if the observation $\mathbf{Y} = \mathbf{y}_{\text{obs}}$ results in the *a posteriori* distribution of $H$ being uniform, that is, if $\Pr[H = 0 \,|\, \mathbf{Y} = \mathbf{y}_{\text{obs}}] = \Pr[H = 1 \,|\, \mathbf{Y} = \mathbf{y}_{\text{obs}}] = 1/2$, then either guess is optimal. Using Note 20.4.2 (ii) we can also express the decision rule (20.15) as

$$\phi^*_{\text{Guess}}(\mathbf{y}_{\text{obs}}) = \begin{cases} 0 & \text{if } \pi_0 f_{\mathbf{Y}|H=0}(\mathbf{y}_{\text{obs}}) \geq \pi_1 f_{\mathbf{Y}|H=1}(\mathbf{y}_{\text{obs}}), \\ 1 & \text{otherwise.} \end{cases} \tag{20.16}$$

Conditional on $\mathbf{Y} = \mathbf{y}_{\text{obs}}$, the probability of error of the optimal decision rule is, in analogy to (20.6), given by

$$p^*(\text{error}|\mathbf{Y} = \mathbf{y}_{\text{obs}}) = \min\{\Pr[H = 0 \,|\, \mathbf{Y} = \mathbf{y}_{\text{obs}}], \Pr[H = 1 \,|\, \mathbf{Y} = \mathbf{y}_{\text{obs}}]\}, \quad (20.17)$$

as can be seen by treating the case $\Pr[H = 0 \,|\, \mathbf{Y} = \mathbf{y}_{\text{obs}}] \geq \Pr[H = 1 \,|\, \mathbf{Y} = \mathbf{y}_{\text{obs}}]$ and the complementary case $\Pr[H = 0 \,|\, \mathbf{Y} = \mathbf{y}_{\text{obs}}] < \Pr[H = 1 \,|\, \mathbf{Y} = \mathbf{y}_{\text{obs}}]$ separately.

The unconditional probability of error associated with the rule (20.15) is thus

$$p^*(\text{error}) = \mathsf{E}\big[\min\{\Pr[H = 0 \,|\, \mathbf{Y}], \Pr[H = 1 \,|\, \mathbf{Y}]\}\big] \tag{20.18}$$

$$= \int_{\mathbb{R}^d} \min\{\Pr[H = 0 \,|\, \mathbf{Y} = \mathbf{y}], \Pr[H = 1 \,|\, \mathbf{Y} = \mathbf{y}]\} f_{\mathbf{Y}}(\mathbf{y}) \, \mathrm{d}\mathbf{y} \tag{20.19}$$

$$= \int_{\mathbb{R}^d} \min\{\pi_0 f_{\mathbf{Y}|H=0}(\mathbf{y}), \pi_1 f_{\mathbf{Y}|H=1}(\mathbf{y})\} \, \mathrm{d}\mathbf{y}, \tag{20.20}$$

where the last equality follows from Note 20.4.2 (i).

Before summarizing these conclusions in a theorem, we present the following simple lemma on the probabilities of error associated with general decision rules.

**Lemma 20.5.1.** *Consider the setup of Note 20.4.1. Let $\phi_{\text{Guess}}(\cdot)$ be an arbitrary guessing rule as in (20.3). Then the probabilities of error $p(\text{error}|H = 0)$, $p(\text{error}|H = 1)$, and $p(\text{error})$ associated with $\phi_{\text{Guess}}(\cdot)$ are given by*

$$p(\text{error}|H = 0) = \int_{\mathbf{y} \notin \mathcal{D}} f_{\mathbf{Y}|H=0}(\mathbf{y}) \, \mathrm{d}\mathbf{y}, \tag{20.21}$$

$$p(\text{error}|H = 1) = \int_{\mathbf{y} \in \mathcal{D}} f_{\mathbf{Y}|H=1}(\mathbf{y}) \, \mathrm{d}\mathbf{y}, \tag{20.22}$$

*and*

$$p(\text{error}) = \int_{\mathbb{R}^d} \big(\pi_0 f_{\mathbf{Y}|H=0}(\mathbf{y}) \, \mathrm{I}\{\mathbf{y} \notin \mathcal{D}\} + \pi_1 f_{\mathbf{Y}|H=1}(\mathbf{y}) \, \mathrm{I}\{\mathbf{y} \in \mathcal{D}\}\big) \, \mathrm{d}\mathbf{y}, \tag{20.23}$$

*where*

$$\mathcal{D} = \{\mathbf{y} \in \mathbb{R}^d : \phi_{\text{Guess}}(\mathbf{y}) = 0\}. \tag{20.24}$$

**Proof.** Conditional on $H = 0$ the guessing rule makes an error only if $\mathbf{Y}$ does not fall in the set of observations for which $\phi_{\text{Guess}}(\cdot)$ produces the guess "$H = 0$." This establishes (20.21). A similar argument proves (20.22). Finally, (20.23) follows from (20.21) & (20.22) using the identity

$$p(\text{error}) = \pi_0\, p(\text{error}|H = 0) + \pi_1\, p(\text{error}|H = 1). \qquad \square$$

We next state the key result about binary hypothesis testing. The statement is a bit cumbersome because, in general, there may be many observations that result in $H$ being *a posteriori* uniformly distributed, and an optimal decision rule can map each such observation to a different guess and still be optimal.

**Theorem 20.5.2 (Optimal Binary Hypothesis Testing).** *Suppose that a guessing rule* $\phi^*_{\text{Guess}} \colon \mathbb{R}^d \to \{0, 1\}$ *produces the guess "$H = 0$" only when* $\mathbf{y}_{\text{obs}}$ *is such that* $\pi_0 f_{\mathbf{Y}|H=0}(\mathbf{y}_{\text{obs}}) \geq \pi_1 f_{\mathbf{Y}|H=1}(\mathbf{y}_{\text{obs}})$, *i.e.,*

$$\left(\phi^*_{\text{Guess}}(\mathbf{y}_{\text{obs}}) = 0\right) \Rightarrow \left(\pi_0 f_{\mathbf{Y}|H=0}(\mathbf{y}_{\text{obs}}) \geq \pi_1 f_{\mathbf{Y}|H=1}(\mathbf{y}_{\text{obs}})\right), \qquad (20.25a)$$

*and produces the guess "$H = 1$" only when* $\pi_1 f_{\mathbf{Y}|H=1}(\mathbf{y}_{\text{obs}}) \geq \pi_0 f_{\mathbf{Y}|H=0}(\mathbf{y}_{\text{obs}})$, *i.e.,*

$$\left(\phi^*_{\text{Guess}}(\mathbf{y}_{\text{obs}}) = 1\right) \Rightarrow \left(\pi_1 f_{\mathbf{Y}|H=1}(\mathbf{y}_{\text{obs}}) \geq \pi_0 f_{\mathbf{Y}|H=0}(\mathbf{y}_{\text{obs}})\right). \qquad (20.25b)$$

*Then no other guessing rule has a smaller probability of error, and*

$$\Pr\left[\phi^*_{\text{Guess}}(\mathbf{Y}) \neq H\right] = \int_{\mathbb{R}^d} \min\left\{\pi_0 f_{\mathbf{Y}|H=0}(\mathbf{y}), \pi_1 f_{\mathbf{Y}|H=1}(\mathbf{y})\right\} \mathrm{d}\mathbf{y}. \qquad (20.26)$$

**Proof.** Let $\phi_{\text{Guess}} \colon \mathbb{R}^d \to \{0, 1\}$ be any guessing rule, and let

$$\mathcal{D} = \{\mathbf{y} \in \mathbb{R}^d : \phi_{\text{Guess}}(\mathbf{y}) = 0\} \qquad (20.27)$$

be the set of observations that result in $\phi_{\text{Guess}}(\cdot)$ producing the guess "$H = 0$." Then the probability of error associated with $\phi_{\text{Guess}}(\cdot)$ can be lower-bounded by

$$\Pr\left[\phi_{\text{Guess}}(\mathbf{Y}) \neq H\right] = \int_{\mathbb{R}^d} \left(\pi_0 f_{\mathbf{Y}|H=0}(\mathbf{y})\, \mathrm{I}\{\mathbf{y} \notin \mathcal{D}\} + \pi_1 f_{\mathbf{Y}|H=1}(\mathbf{y})\, \mathrm{I}\{\mathbf{y} \in \mathcal{D}\}\right) \mathrm{d}\mathbf{y}$$

$$\geq \int_{\mathbb{R}^d} \min\left\{\pi_0 f_{\mathbf{Y}|H=0}(\mathbf{y}), \pi_1 f_{\mathbf{Y}|H=1}(\mathbf{y})\right\} \mathrm{d}\mathbf{y}, \qquad (20.28)$$

where the equality follows from Lemma 20.5.1 and where the inequality follows because for every value of $\mathbf{y} \in \mathbb{R}^d$

$$\pi_0 f_{\mathbf{Y}|H=0}(\mathbf{y})\, \mathrm{I}\{\mathbf{y} \notin \mathcal{D}\} + \pi_1 f_{\mathbf{Y}|H=1}(\mathbf{y})\, \mathrm{I}\{\mathbf{y} \in \mathcal{D}\}$$
$$\geq \min\left\{\pi_0 f_{\mathbf{Y}|H=0}(\mathbf{y}), \pi_1 f_{\mathbf{Y}|H=1}(\mathbf{y})\right\}, \qquad (20.29)$$

as can be verified by noting that, irrespective of the set $\mathcal{D}$, one of the two terms $\mathrm{I}\{\mathbf{y} \in \mathcal{D}\}$ and $\mathrm{I}\{\mathbf{y} \notin \mathcal{D}\}$ is equal to one and the other is equal to zero, so the LHS of

(20.29) is either equal to $\pi_0 f_{\mathbf{Y}|H=0}(\mathbf{y})$ or to $\pi_1 f_{\mathbf{Y}|H=1}(\mathbf{y})$ and hence lower-bounded by $\min\{\pi_0 f_{\mathbf{Y}|H=0}(\mathbf{y}), \pi_1 f_{\mathbf{Y}|H=1}(\mathbf{y})\}$.

We prove the optimality of $\phi^*_{\text{Guess}}(\cdot)$ by next showing that the probability of error associated with $\phi^*_{\text{Guess}}(\cdot)$ is equal to the RHS of (20.28). To this end we define

$$\mathcal{D}^* = \{\mathbf{y} \in \mathbb{R}^d : \phi^*_{\text{Guess}}(\mathbf{y}) = 0\} \tag{20.30}$$

and note that if both (20.25a) and (20.25b) hold, then

$$\pi_0 f_{\mathbf{Y}|H=0}(\mathbf{y}) \, \mathrm{I}\{\mathbf{y} \notin \mathcal{D}^*\} + \pi_1 f_{\mathbf{Y}|H=1}(\mathbf{y}) \, \mathrm{I}\{\mathbf{y} \in \mathcal{D}^*\}$$
$$= \min\{\pi_0 f_{\mathbf{Y}|H=0}(\mathbf{y}), \pi_1 f_{\mathbf{Y}|H=1}(\mathbf{y})\}, \quad \mathbf{y} \in \mathbb{R}^d. \tag{20.31}$$

Applying Lemma 20.5.1 to the decoder $\phi^*_{\text{Guess}}(\cdot)$ we obtain

$$\Pr\left[\phi^*_{\text{Guess}}(\mathbf{Y}) \neq H\right] = \int_{\mathbb{R}^d} \left(\pi_0 f_{\mathbf{Y}|H=0}(\mathbf{y}) \, \mathrm{I}\{\mathbf{y} \notin \mathcal{D}^*\} + \pi_1 f_{\mathbf{Y}|H=1}(\mathbf{y}) \, \mathrm{I}\{\mathbf{y} \in \mathcal{D}^*\}\right) \mathrm{d}\mathbf{y}$$
$$= \int_{\mathbb{R}^d} \min\{\pi_0 f_{\mathbf{Y}|H=0}(\mathbf{y}), \pi_1 f_{\mathbf{Y}|H=1}(\mathbf{y})\} \, \mathrm{d}\mathbf{y}, \tag{20.32}$$

where the second equality follows from (20.31). The theorem now follows from (20.28) and (20.32).                                                                    □

Referring to a situation where the observation results in the *a posteriori* distribution of $H$ being uniform as a **tie** we have:

**Note 20.5.3.** The fact that both conditional on $H = 0$ and conditional on $H = 1$ the observation $\mathbf{Y}$ has a density does not imply that the probability of a tie is zero.

For example, if $H$ takes value in $\{0, 1\}$ equiprobably, and if the observation $Y$ is given by $Y = H + U$, where $U$ is uniformly distributed over the interval $[-2, 2]$ independently of $H$, then the *a posteriori* distribution of $H$ is uniform whenever $Y \in [-1, 2]$, and this occurs with probability $3/4$.

## 20.6    Randomized Decision Rules

So far we have restricted ourselves to deterministic decision rules, where the guess is a deterministic function of the observation. We next remove this restriction and allow for some randomization in the decision rule. As we shall see in this section and in greater generality in Section 20.11, when properly defined, randomization does not help: the lowest probability of error that is achievable with randomized decision rules can also be achieved with deterministic decision rules.

By a randomized decision rule we mean that, after observing that $\mathbf{Y} = \mathbf{y}_{\text{obs}}$, the guesser chooses some bias $b(\mathbf{y}_{\text{obs}}) \in [0, 1]$ and then tosses a coin of that bias. If the result is "heads" it guesses 0 and otherwise it guesses 1. Note that the deterministic rules we have considered before are special cases of the randomized ones: any deterministic decision rule can be viewed as a randomized decision rule where, depending on $\mathbf{y}_{\text{obs}}$, the bias $b(\mathbf{y}_{\text{obs}})$ is either zero or one.

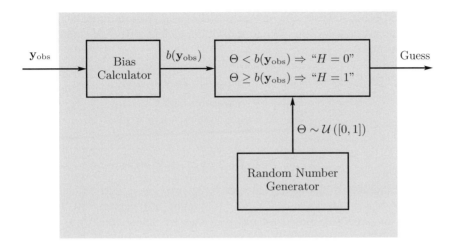

**Figure 20.1:** A block diagram of a randomized decision rule.

Some care must be exercised in defining the joint distribution of the coin toss with the other variables $(H, \mathbf{Y})$. We do not want to allow for "telepathic coins." That is, we want to make sure that once $\mathbf{Y} = \mathbf{y}_{\mathrm{obs}}$ has been observed and the bias $b(\mathbf{y}_{\mathrm{obs}})$ has been accordingly computed, the outcome of the coin toss is random, i.e., has nothing to do with $H$. Probabilists would say that we require that, conditional on $\mathbf{Y} = \mathbf{y}_{\mathrm{obs}}$, the outcome of the coin toss be independent of $H$. (We shall discuss conditional independence in Section 20.11.) We can clarify the setting as follows. Upon observing the outcome $\mathbf{Y} = \mathbf{y}_{\mathrm{obs}}$, the guesser computes the bias $b(\mathbf{y}_{\mathrm{obs}})$. Using a local random number generator the guesser then draws a random variable $\Theta$ uniformly over the interval $[0, 1]$, independently of the pair $(H, \mathbf{Y})$. If the outcome $\theta$ is smaller than $b(\mathbf{y}_{\mathrm{obs}})$, then it guesses "$H = 0$," and otherwise it guesses "$H = 1$." A randomized decision rule is depicted in Figure 20.1.

We offer two proofs that randomized decision rules cannot outperform the best deterministic ones. The first is by straightforward calculation. Conditional on $\mathbf{Y} = \mathbf{y}_{\mathrm{obs}}$, the randomized guesser makes an error either if $\Theta \leq b(\mathbf{y}_{\mathrm{obs}})$ (resulting in the guess "$H = 0$") while $H = 1$, or if $\Theta > b(\mathbf{y}_{\mathrm{obs}})$ (resulting in the guess "$H = 1$") while $H = 0$. Consequently,

$$\Pr\big(\mathrm{error} \,\big|\, \mathbf{Y} = \mathbf{y}_{\mathrm{obs}}\big)$$
$$= b(\mathbf{y}_{\mathrm{obs}}) \Pr\big[H = 1 \,\big|\, \mathbf{Y} = \mathbf{y}_{\mathrm{obs}}\big] + \big(1 - b(\mathbf{y}_{\mathrm{obs}})\big) \Pr\big[H = 0 \,\big|\, \mathbf{Y} = \mathbf{y}_{\mathrm{obs}}\big]. \quad (20.33)$$

Thus, $\Pr(\mathrm{error} | \mathbf{Y} = \mathbf{y}_{\mathrm{obs}})$ is a weighted average of $\Pr[H = 0 | \mathbf{Y} = \mathbf{y}_{\mathrm{obs}}]$ and $\Pr[H = 1 | \mathbf{Y} = \mathbf{y}_{\mathrm{obs}}]$. As such, irrespective of the weights, it cannot be smaller than the minimum of the two. But, by (20.17), the optimal deterministic decision rule (20.15) achieves just this minimum. We conclude that, irrespective of the bias, for each outcome $\mathbf{Y} = \mathbf{y}_{\mathrm{obs}}$ the conditional probability of error of the randomized decoder is lower-bounded by that of the optimal deterministic decoder (20.15).

Since this is the case for every outcome, it must also be the case when we average over the outcomes. This concludes the first proof.

In the second proof we view the outcome of the local random number generator $\Theta$ as an additional observation. Since it is independent of $(H, \mathbf{Y})$ and since it is uniform over $[0, 1]$,

$$\begin{aligned} f_{\mathbf{Y}, \Theta | H=0}(\mathbf{y}, \theta) &= f_{\mathbf{Y} | H=0}(\mathbf{y}) \, f_{\Theta | \mathbf{Y}=\mathbf{y}, H=0}(\theta) \\ &= f_{\mathbf{Y} | H=0}(\mathbf{y}) \, f_\Theta(\theta) \\ &= f_{\mathbf{Y} | H=0}(\mathbf{y}) \, \mathrm{I}\{0 \le \theta \le 1\}, \end{aligned} \tag{20.34a}$$

and similarly

$$f_{\mathbf{Y}, \Theta | H=1}(\mathbf{y}, \theta) = f_{\mathbf{Y} | H=1}(\mathbf{y}) \, \mathrm{I}\{0 \le \theta \le 1\}. \tag{20.34b}$$

Since the randomized decision rule can be viewed as a deterministic decision rule that is based on the pair $(\mathbf{Y}, \Theta)$, it cannot outperform any optimal deterministic guessing rule based on $(\mathbf{Y}, \Theta)$. But by Theorem 20.5.2 and (20.34) it follows that the deterministic decision rule that guesses "$H = 0$" whenever $\pi_0 f_{\mathbf{Y} | H=0}(\mathbf{y}) \ge \pi_1 f_{\mathbf{Y} | H=1}(\mathbf{y})$ is optimal not only for guessing $H$ based on $\mathbf{Y}$ but also for guessing $H$ based on $(\mathbf{Y}, \Theta)$, because it produces the guess "$H = 0$" only when $\pi_0 f_{\mathbf{Y}, \Theta | H=0}(\mathbf{y}, \theta) \ge \pi_1 f_{\mathbf{Y}, \Theta | H=1}(\mathbf{y}, \theta)$ and it produces the guess "$H = 1$" only when $\pi_1 f_{\mathbf{Y}, \Theta | H=1}(\mathbf{y}, \theta) \ge \pi_0 f_{\mathbf{Y}, \Theta | H=0}(\mathbf{y}, \theta)$. This concludes the second proof.

Even though randomized decision rules cannot outperform the best deterministic rules, they may have other advantages. For example, they allow for more symmetric ways of resolving ties. Suppose, for example, that we have no observations and that the prior is uniform. In this case guessing "$H = 0$" will give rise to a probability of error of $1/2$, with an error occurring whenever $H = 1$. Similarly guessing "$H = 1$" will also result in a probability of error of $1/2$, this time with an error occurring whenever $H = 0$. If we think about $H$ as being an information bit, then the former rule makes sending 0 less error prone than sending 1. A randomized test that flips a fair coin and guesses 0 if "heads" and 1 if "tails" gives rise to the same average probability of error (i.e., $1/2$) and makes sending 0 and sending 1 equally (highly) error prone.

If $\mathbf{Y} = \mathbf{y}_{\mathrm{obs}}$ results in a tie, i.e., if it yields a uniform *a posteriori* distribution on $H$,

$$\Pr[H = 0 \,|\, \mathbf{Y} = \mathbf{y}_{\mathrm{obs}}] = \Pr[H = 1 \,|\, \mathbf{Y} = \mathbf{y}_{\mathrm{obs}}] = \frac{1}{2},$$

then the probability of error of the randomized decoder (20.33) does not depend on the bias. In this case there is thus no loss in optimality in choosing $b(\mathbf{y}_{\mathrm{obs}}) = 1/2$, i.e., by employing a fair coin. This makes for a symmetric way of resolving the tie in the *a posteriori* distribution of $H$.

## 20.7 The MAP Decision Rule

In Section 20.5 we presented an optimal decision rule (20.15). A slight variation on that decoder is the **Maximum A Posteriori** (MAP) decision rule. The MAP rule is identical to (20.15) except in how it resolves ties. Unlike (20.15), which resolves ties by guessing "$H = 0$," the MAP rule resolves ties by flipping a fair coin. It can thus be summarized as follows:

$$\phi_{\text{MAP}}(\mathbf{y}_{\text{obs}}) \triangleq \begin{cases} 0 & \text{if } \Pr[H = 0 \,|\, \mathbf{Y} = \mathbf{y}_{\text{obs}}] > \Pr[H = 1 \,|\, \mathbf{Y} = \mathbf{y}_{\text{obs}}], \\ 1 & \text{if } \Pr[H = 0 \,|\, \mathbf{Y} = \mathbf{y}_{\text{obs}}] < \Pr[H = 1 \,|\, \mathbf{Y} = \mathbf{y}_{\text{obs}}], \\ \mathcal{U}(\{0, 1\}) & \text{if } \Pr[H = 0 \,|\, \mathbf{Y} = \mathbf{y}_{\text{obs}}] = \Pr[H = 1 \,|\, \mathbf{Y} = \mathbf{y}_{\text{obs}}], \end{cases}$$
$$(20.35)$$

where we use "$\mathcal{U}(\{0, 1\})$" to indicate that we guess the outcome uniformly at random.

Note that, like the rule in (20.15), the MAP rule is optimal. This follows because the way ties are resolved does not influence the probability of error, and because the MAP rule agrees with the rule (20.15) for all observations which do not result in a tie.

**Theorem 20.7.1 (The MAP Rule Is Optimal).** *The Maximum A Posteriori decision rule* (20.35) *is optimal.*

Since the MAP decoder is optimal,

$$p^*(\text{error}) = \pi_0 \, p_{\text{MAP}}(\text{error} | H = 0) + \pi_1 \, p_{\text{MAP}}(\text{error} | H = 1), \qquad (20.36)$$

where $p_{\text{MAP}}(\text{error} | H = 0)$ and $p_{\text{MAP}}(\text{error} | H = 1)$ denote the conditional probabilities of error for the MAP decoder. Note that one can easily find guessing rules (such as the rule "always guess 0") that yield a conditional probability of error smaller than $p_{\text{MAP}}(\text{error} | H = 0)$, but one cannot find a rule whose average probability of error outperforms the RHS of (20.36).

Using Note 20.4.2 (ii) we can express the MAP rule in terms of the densities and the prior as

$$\phi_{\text{MAP}}(\mathbf{y}_{\text{obs}}) = \begin{cases} 0 & \text{if } \pi_0 f_{\mathbf{Y}|H=0}(\mathbf{y}_{\text{obs}}) > \pi_1 f_{\mathbf{Y}|H=1}(\mathbf{y}_{\text{obs}}), \\ 1 & \text{if } \pi_0 f_{\mathbf{Y}|H=0}(\mathbf{y}_{\text{obs}}) < \pi_1 f_{\mathbf{Y}|H=1}(\mathbf{y}_{\text{obs}}), \\ \mathcal{U}(\{0, 1\}) & \text{if } \pi_0 f_{\mathbf{Y}|H=0}(\mathbf{y}_{\text{obs}}) = \pi_1 f_{\mathbf{Y}|H=1}(\mathbf{y}_{\text{obs}}). \end{cases} \qquad (20.37)$$

Alternatively, the MAP decision rule can be described using the **likelihood-ratio function** $\text{LR}(\cdot)$, which is defined by

$$\text{LR}(\mathbf{y}) \triangleq \frac{f_{\mathbf{Y}|H=0}(\mathbf{y})}{f_{\mathbf{Y}|H=1}(\mathbf{y})}, \quad \mathbf{y} \in \mathbb{R}^d \qquad (20.38)$$

using the convention

$$\left( \frac{\alpha}{0} = \infty, \quad \alpha > 0 \right) \quad \text{and} \quad \frac{0}{0} = 1. \qquad (20.39)$$

Since densities are nonnegative, and since we are defining the likelihood-ratio function using the convention (20.39), the range of $\text{LR}(\cdot)$ is the set $[0, \infty]$ consisting of the nonnegative reals and the special symbol $\infty$:

$$\text{LR}: \mathbb{R}^d \to [0, \infty].$$

Using the likelihood-ratio function and (20.37), we can rewrite the MAP rule for the case where the prior is nondegenerate (20.2) and where the observation $\mathbf{y}_{\mathrm{obs}}$ is such that $f_{\mathbf{Y}}(\mathbf{y}_{\mathrm{obs}}) > 0$ as

$$\phi_{\mathrm{MAP}}(\mathbf{y}_{\mathrm{obs}}) = \begin{cases} 0 & \text{if } \mathrm{LR}(\mathbf{y}_{\mathrm{obs}}) > \frac{\pi_1}{\pi_0}, \\ 1 & \text{if } \mathrm{LR}(\mathbf{y}_{\mathrm{obs}}) < \frac{\pi_1}{\pi_0}, \\ \mathcal{U}(\{0,1\}) & \text{if } \mathrm{LR}(\mathbf{y}_{\mathrm{obs}}) = \frac{\pi_1}{\pi_0}, \end{cases} \quad (\pi_0, \pi_1 > 0, \quad f_{\mathbf{Y}}(\mathbf{y}_{\mathrm{obs}}) > 0).$$
(20.40)

Since many of the densities that are of interest to us have an exponential form, it is sometimes more convenient to describe the MAP rule using the **log likelihood-ratio function** $\mathrm{LLR} \colon \mathbb{R}^d \to [-\infty, \infty]$, which is defined by

$$\mathrm{LLR}(\mathbf{y}) \triangleq \ln \frac{f_{\mathbf{Y}|H=0}(\mathbf{y})}{f_{\mathbf{Y}|H=1}(\mathbf{y})}, \quad \mathbf{y} \in \mathbb{R}^d, \tag{20.41}$$

using the convention

$$\left( \ln \frac{\alpha}{0} = +\infty, \ \ln \frac{0}{\alpha} = -\infty, \ \alpha > 0 \right) \quad \text{and} \quad \ln \frac{0}{0} = 0, \tag{20.42}$$

where $\ln(\cdot)$ denotes natural logarithm.

Using the log likelihood-ratio function $\mathrm{LLR}(\cdot)$ and the monotonicity of the logarithmic function

$$(a > b) \Leftrightarrow (\ln a > \ln b), \quad a, b > 0, \tag{20.43}$$

we can express the MAP rule (20.40) as

$$\phi_{\mathrm{MAP}}(\mathbf{y}_{\mathrm{obs}}) = \begin{cases} 0 & \text{if } \mathrm{LLR}(\mathbf{y}_{\mathrm{obs}}) > \ln \frac{\pi_1}{\pi_0}, \\ 1 & \text{if } \mathrm{LLR}(\mathbf{y}_{\mathrm{obs}}) < \ln \frac{\pi_1}{\pi_0}, \\ \mathcal{U}(\{0,1\}) & \text{if } \mathrm{LLR}(\mathbf{y}_{\mathrm{obs}}) = \ln \frac{\pi_1}{\pi_0}, \end{cases} \quad (\pi_0, \pi_1 > 0, \quad f_{\mathbf{Y}}(\mathbf{y}_{\mathrm{obs}}) > 0).$$
(20.44)

## 20.8   The ML Decision Rule

A different decision rule, which is typically suboptimal unless $H$ is *a priori* uniform, is the **Maximum-Likelihood** (ML) decision rule. Its structure is similar to that of the MAP rule except that it ignores the prior. In fact, if $\pi_0 = \pi_1$, then the two rules are identical. The ML rule is thus given by

$$\phi_{\mathrm{ML}}(\mathbf{y}_{\mathrm{obs}}) \triangleq \begin{cases} 0 & \text{if } f_{\mathbf{Y}|H=0}(\mathbf{y}_{\mathrm{obs}}) > f_{\mathbf{Y}|H=1}(\mathbf{y}_{\mathrm{obs}}), \\ 1 & \text{if } f_{\mathbf{Y}|H=0}(\mathbf{y}_{\mathrm{obs}}) < f_{\mathbf{Y}|H=1}(\mathbf{y}_{\mathrm{obs}}), \\ \mathcal{U}(\{0,1\}) & \text{if } f_{\mathbf{Y}|H=0}(\mathbf{y}_{\mathrm{obs}}) = f_{\mathbf{Y}|H=1}(\mathbf{y}_{\mathrm{obs}}). \end{cases} \tag{20.45}$$

The ML decision rule can be alternatively described using the likelihood-ratio function $\mathrm{LR}(\cdot)$ (20.38) as

$$\phi_{\mathrm{ML}}(\mathbf{y}_{\mathrm{obs}}) = \begin{cases} 0 & \text{if } \mathrm{LR}(\mathbf{y}_{\mathrm{obs}}) > 1, \\ 1 & \text{if } \mathrm{LR}(\mathbf{y}_{\mathrm{obs}}) < 1, \\ \mathcal{U}(\{0,1\}) & \text{if } \mathrm{LR}(\mathbf{y}_{\mathrm{obs}}) = 1. \end{cases} \tag{20.46}$$

Alternatively, using the log likelihood-ratio function $\mathrm{LLR}(\cdot)$ (20.41):

$$\phi_{\mathrm{ML}}(\mathbf{y}_{\mathrm{obs}}) = \begin{cases} 0 & \text{if } \mathrm{LLR}(\mathbf{y}_{\mathrm{obs}}) > 0, \\ 1 & \text{if } \mathrm{LLR}(\mathbf{y}_{\mathrm{obs}}) < 0, \\ \mathcal{U}(\{0,1\}) & \text{if } \mathrm{LLR}(\mathbf{y}_{\mathrm{obs}}) = 0. \end{cases} \tag{20.47}$$

## 20.9 Performance Analysis: the Bhattacharyya Bound

We next derive the **Bhattacharyya Bound**, which is a useful upper bound on the optimal probability of error $p^*(\text{error})$.

Starting with the exact expression (20.20) we obtain:

$$\begin{aligned} p^*(\text{error}) &= \int_{\mathbb{R}^d} \min\left\{ \pi_0 f_{\mathbf{Y}|H=0}(\mathbf{y}), \pi_1 f_{\mathbf{Y}|H=1}(\mathbf{y}) \right\} \, d\mathbf{y} \\ &\leq \int_{\mathbb{R}^d} \sqrt{\pi_0 f_{\mathbf{Y}|H=0}(\mathbf{y}) \pi_1 f_{\mathbf{Y}|H=1}(\mathbf{y})} \, d\mathbf{y} \\ &= \sqrt{\pi_0 \pi_1} \int_{\mathbb{R}^d} \sqrt{f_{\mathbf{Y}|H=0}(\mathbf{y}) f_{\mathbf{Y}|H=1}(\mathbf{y})} \, d\mathbf{y} \\ &\leq \frac{1}{2} \int_{\mathbb{R}^d} \sqrt{f_{\mathbf{Y}|H=0}(\mathbf{y}) f_{\mathbf{Y}|H=1}(\mathbf{y})} \, d\mathbf{y}, \end{aligned}$$

where the equality in the first line follows from (20.20); the inequality in the second line from the inequality

$$\min\{a,b\} \leq \sqrt{ab}, \quad a,b \geq 0, \tag{20.48}$$

(which can be easily verified by treating the case $a \geq b$ and the case $a < b$ separately); the equality in the third line by trivial algebra; and where the inequality in the fourth line follows by noting that if $c,d \geq 0$, then their geometric mean $\sqrt{cd}$ cannot exceed their arithmetic mean $(c+d)/2$, i.e.,

$$\sqrt{cd} \leq \frac{c+d}{2}, \quad c,d \geq 0, \tag{20.49}$$

and because in our case $c = \pi_0$ and $d = \pi_1$, so $c + d = 1$.

We have thus established the bound

$$p^*(\text{error}) \leq \frac{1}{2} \int_{\mathbf{y} \in \mathbb{R}^d} \sqrt{f_{\mathbf{Y}|H=0}(\mathbf{y}) f_{\mathbf{Y}|H=1}(\mathbf{y})} \, d\mathbf{y}, \tag{20.50}$$

which is known as the **Bhattacharyya Bound**.

## 20.10 Example

Consider the problem of guessing $H$ based on the observation $Y$, where $H$ takes on the values 0 and 1 equiprobably and where the conditional densities of $Y$ given

$H = 0$ and $H = 1$ are

$$f_{Y|H=0}(y) = \frac{1}{\sqrt{2\pi\sigma^2}}\, e^{-(y-\mathsf{A})^2/(2\sigma^2)}, \quad y \in \mathbb{R}, \tag{20.51a}$$

$$f_{Y|H=1}(y) = \frac{1}{\sqrt{2\pi\sigma^2}}\, e^{-(y+\mathsf{A})^2/(2\sigma^2)}, \quad y \in \mathbb{R} \tag{20.51b}$$

for some deterministic $\mathsf{A}, \sigma > 0$. Here the observable is a RV, so $d = 1$.

For these conditional densities the likelihood-ratio function (20.38) is given by:

$$
\begin{aligned}
\mathrm{LR}(y) &= \frac{f_{Y|H=0}(y)}{f_{Y|H=1}(y)} \\
&= \frac{\frac{1}{\sqrt{2\pi\sigma^2}}\, e^{-(y-\mathsf{A})^2/(2\sigma^2)}}{\frac{1}{\sqrt{2\pi\sigma^2}}\, e^{-(y+\mathsf{A})^2/(2\sigma^2)}} \\
&= e^{4y\mathsf{A}/(2\sigma^2)}, \quad y \in \mathbb{R}.
\end{aligned}
$$

Since the two hypotheses are *a priori* equally likely, the MAP rule is equivalent to the ML rule and both rules guess "$H = 0$" or "$H = 1$" depending on whether the likelihood-ratio $\mathrm{LR}(\mathbf{y}_{\mathrm{obs}})$ is greater or smaller than one. And since

$$
\begin{aligned}
\mathrm{LR}(y_{\mathrm{obs}}) > 1 &\Leftrightarrow e^{4y_{\mathrm{obs}}\mathsf{A}/(2\sigma^2)} > 1 \\
&\Leftrightarrow \ln\left(e^{4y_{\mathrm{obs}}\mathsf{A}/(2\sigma^2)}\right) > \ln 1 \\
&\Leftrightarrow 4y_{\mathrm{obs}}\mathsf{A}/(2\sigma^2) > 0 \\
&\Leftrightarrow y_{\mathrm{obs}} > 0,
\end{aligned}
$$

and

$$
\begin{aligned}
\mathrm{LR}(y_{\mathrm{obs}}) < 1 &\Leftrightarrow e^{4y_{\mathrm{obs}}\mathsf{A}/(2\sigma^2)} < 1 \\
&\Leftrightarrow \ln\left(e^{4y_{\mathrm{obs}}\mathsf{A}/(2\sigma^2)}\right) < \ln 1 \\
&\Leftrightarrow 4y_{\mathrm{obs}}\mathsf{A}/(2\sigma^2) < 0 \\
&\Leftrightarrow y_{\mathrm{obs}} < 0,
\end{aligned}
$$

it follows that the MAP decision rule guesses "$H = 0$," if $y_{\mathrm{obs}} > 0$; guesses "$H = 1$," if $y_{\mathrm{obs}} < 0$; and guesses "$H = 0$" or "$H = 1$" equiprobably, if $y_{\mathrm{obs}} = 0$ (i.e., in the case of a tie).

Note that in this example the probability of a tie is zero. Indeed, under both hypotheses, the probability that the observed variable $Y$ is exactly equal to zero is zero:

$$\Pr\big[Y = 0 \,\big|\, H = 0\big] = \Pr\big[Y = 0 \,\big|\, H = 1\big] = \Pr\big[Y = 0\big] = 0. \tag{20.52}$$

Consequently, the way ties are resolved is immaterial.

We next compute the probability of error of the MAP decoder. To this end, let $p_{\mathrm{MAP}}(\mathrm{error}|H = 0)$ and $p_{\mathrm{MAP}}(\mathrm{error}|H = 1)$ denote its conditional probabilities of error. Its (unconditional) probability of error, which is also the optimal probability of error, can be expressed as

$$p^*(\text{error}) = \pi_0 \, p_{\text{MAP}}(\text{error}|H = 0) + \pi_1 \, p_{\text{MAP}}(\text{error}|H = 1). \tag{20.53}$$

We proceed to compute the required terms on the RHS. Starting with the term $p_{\text{MAP}}(\text{error}|H = 0)$, we note that $p_{\text{MAP}}(\text{error}|H = 0)$ corresponds to the conditional probability that $Y$ is negative or that $Y$ is equal to zero and the coin toss that the MAP decoder uses to resolve the tie causes the guess to be "$H = 1$." By (20.52), the conditional probability of a tie is zero, so $p_{\text{MAP}}(\text{error}|H = 0)$ is, in fact, just the conditional probability that $Y$ is negative:

$$p_{\text{MAP}}(\text{error}|H = 0) = \Pr[Y < 0 | H = 0]$$

$$= \mathcal{Q}\left(\frac{A}{\sigma}\right), \tag{20.54}$$

where the second equality follows because, conditional on $H = 0$, the random variable $Y$ is $\mathcal{N}(A, \sigma^2)$, and the probability that it is smaller than zero can be thus computed using the $\mathcal{Q}$-function as in (19.12b). Similarly,

$$p_{\text{MAP}}(\text{error}|H = 1) = \Pr[Y > 0 | H = 1]$$

$$= \mathcal{Q}\left(\frac{A}{\sigma}\right). \tag{20.55}$$

Note that in this example the MAP rule is "fair" in the sense that the conditional probability of error given $H = 0$ is the same as given $H = 1$. This is a coincidence (that results from the symmetry in the problem). In general, the MAP rule need not be fair.

We conclude from (20.53), (20.54), and (20.55) that

$$p^*(\text{error}) = \mathcal{Q}\left(\frac{A}{\sigma}\right). \tag{20.56}$$

Figure 20.2 depicts the conditional densities of $y$ given $H = 0$ and given $H = 1$ and the decision regions of the MAP decision rule $\phi_{\text{MAP}}(\cdot)$. The area of the shaded region is the probability of an error conditioned on $H = 0$.

Note that the optimal decision rule for this example is not unique. Another optimal decision rule is to guess "$H = 0$" if $\mathbf{y}_{\text{obs}}$ is positive but not equal to 17, and to guess "$H = 1$" otherwise.

Even though we have an exact expression for the probability of error (20.56) it is instructive to compute the Bhattacharyya Bound too:

$$p^*(\text{error}) \le \frac{1}{2} \int_{-\infty}^{\infty} \sqrt{f_{Y|H=0}(y) f_{Y|H=1}(y)} \, dy$$

$$= \frac{1}{2} \int_{-\infty}^{\infty} \sqrt{\frac{1}{\sqrt{2\pi\sigma^2}} e^{-(y-A)^2/(2\sigma^2)} \frac{1}{\sqrt{2\pi\sigma^2}} e^{-(y+A)^2/(2\sigma^2)}} \, dy$$

$$= \frac{1}{2} e^{-A^2/2\sigma^2} \int_{-\infty}^{\infty} \frac{1}{\sqrt{2\pi\sigma^2}} e^{-y^2/2\sigma^2} \, dy$$

$$= \frac{1}{2} e^{-A^2/2\sigma^2}, \tag{20.57}$$

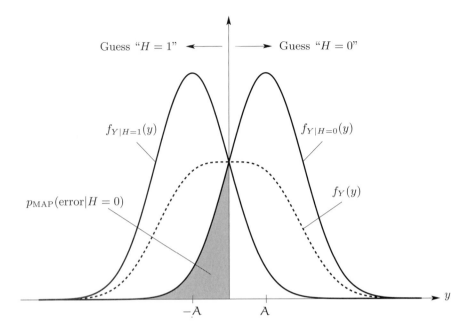

**Figure 20.2:** Binary hypothesis testing with a uniform prior. Conditional on $H = 0$ the observable $Y$ is $\mathcal{N}(A, \sigma^2)$ and conditional on $H = 1$ it is $\mathcal{N}(-A, \sigma^2)$. The area of the shaded region is the probability of error of the MAP rule conditional on $H = 0$.

where the first line follows from (20.50); the second from (20.51); the third by simple algebra; and the final equality because the Gaussian density (like all densities) integrates to one.

As an aside, we have from (20.57) and (20.56) the bound

$$Q(\alpha) \leq \frac{1}{2} e^{-\alpha^2/2}, \quad \alpha \geq 0, \tag{20.58}$$

which we encountered in Proposition 19.4.2.

## 20.11  (Nontelepathic) Processing

To further emphasize the optimality of the Maximum A Posteriori decision rule, and for ulterior motives that have to do with the introduction of conditional independence, we shall next show that no processing of the observables can reduce the probability of a guessing error. To that end we shall have to properly define what we mean by "processing."

The first thing that comes to mind is to consider processing as the application of some deterministic mapping. I.e., we think of mapping the observation $\mathbf{y}_{\mathrm{obs}}$ using

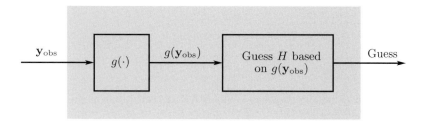

**Figure 20.3:** No decision rule based on $g(\mathbf{y}_{\mathrm{obs}})$ can outperform an optimal decision rule based on $\mathbf{y}_{\mathrm{obs}}$, because computing $g(\mathbf{y}_{\mathrm{obs}})$ and then forming the decision based on the answer can be viewed as a special case of guessing based on $\mathbf{y}_{\mathrm{obs}}$.

some deterministic function $g(\cdot)$ to $g(\mathbf{y}_{\mathrm{obs}})$ and then guessing $H$ based on $g(\mathbf{y}_{\mathrm{obs}})$. That this cannot reduce the probability of error is clear from Figure 20.3, which demonstrates that mapping $\mathbf{y}_{\mathrm{obs}}$ to $g(\mathbf{y}_{\mathrm{obs}})$ and then guessing $H$ based on $g(\mathbf{y}_{\mathrm{obs}})$ can be viewed as a special case of guessing $H$ based on $\mathbf{y}_{\mathrm{obs}}$ and, as such, cannot outperform the MAP decision rule, which is optimal among *all* decision rules based on $\mathbf{y}_{\mathrm{obs}}$.

A more general kind of processing involves randomization, or "dithering." Here we envision the processor as using a local random number generator to generate a random variable $\Theta$ and then producing an output of the form $g(\mathbf{y}_{\mathrm{obs}}, \theta_{\mathrm{obs}})$, where $\theta_{\mathrm{obs}}$ is the outcome of $\Theta$, and where $g(\cdot)$ is some deterministic function. Here $\Theta$ is assumed to be independent of the pair $(H, \mathbf{Y})$, so the processor can generate it using a local random number generator.

An argument very similar to the one we used in Section 20.6 (in the second proof of the claim that randomized decision rules cannot outperform optimal deterministic rules) can be used to show that this type of processing cannot improve our guessing. The argument is as follows. We view the application of the function $g(\cdot)$ to the pair $(\mathbf{Y}, \Theta)$ as deterministic processing of the pair $(\mathbf{Y}, \Theta)$, so no decision rule based on $g(\mathbf{Y}, \Theta)$ can outperform a decision rule that is optimal for guessing $H$ based on $(\mathbf{Y}, \Theta)$. It thus remains to show that the decision rule 'Guess "$H = 0$" if $\pi_0 f_{\mathbf{Y}|H=0}(\mathbf{y}_{\mathrm{obs}}) \geq \pi_1 f_{\mathbf{Y}|H=1}(\mathbf{y}_{\mathrm{obs}})$' is also optimal when observing $(\mathbf{Y}, \Theta)$ and not only $\mathbf{Y}$. This follows from Theorem 20.5.2 by noting that the independence of $\Theta$ and $(H, \mathbf{Y})$, implies that

$$f_{\mathbf{Y}, \Theta|H=0}(\mathbf{y}_{\mathrm{obs}}, \theta_{\mathrm{obs}}) = f_{\mathbf{Y}|H=0}(\mathbf{y}_{\mathrm{obs}})\, f_{\Theta}(\theta_{\mathrm{obs}}),$$

$$f_{\mathbf{Y}, \Theta|H=1}(\mathbf{y}_{\mathrm{obs}}, \theta_{\mathrm{obs}}) = f_{\mathbf{Y}|H=1}(\mathbf{y}_{\mathrm{obs}})\, f_{\Theta}(\theta_{\mathrm{obs}}),$$

and hence that this rule guesses "$H = 0$" only when $\mathbf{y}_{\mathrm{obs}}$ and $\theta_{\mathrm{obs}}$ are such that $\pi_0 f_{\mathbf{Y}, \Theta|H=0}(\mathbf{y}_{\mathrm{obs}}, \theta_{\mathrm{obs}}) \geq \pi_1 f_{\mathbf{Y}, \Theta|H=1}(\mathbf{y}_{\mathrm{obs}}, \theta_{\mathrm{obs}})$ and guesses "$H = 1$" only when $\pi_1 f_{\mathbf{Y}, \Theta|H=1}(\mathbf{y}_{\mathrm{obs}}, \theta_{\mathrm{obs}}) \geq \pi_0 f_{\mathbf{Y}, \Theta|H=0}(\mathbf{y}_{\mathrm{obs}}, \theta_{\mathrm{obs}})$.

Fearless readers who are not afraid to divide by zero should note that

$$\mathrm{LR}(\mathbf{y}_{\mathrm{obs}}, \theta_{\mathrm{obs}}) = \frac{f_{\mathbf{Y}, \Theta|H=0}(\mathbf{y}_{\mathrm{obs}}, \theta_{\mathrm{obs}})}{f_{\mathbf{Y}, \Theta|H=1}(\mathbf{y}_{\mathrm{obs}}, \theta_{\mathrm{obs}})}$$

$$= \frac{f_{\mathbf{Y}|H=0}(\mathbf{y}_{\mathrm{obs}})\, f_{\Theta}(\theta_{\mathrm{obs}})}{f_{\mathbf{Y}|H=1}(\mathbf{y}_{\mathrm{obs}})\, f_{\Theta}(\theta_{\mathrm{obs}})}$$

$$= \frac{f_{\mathbf{Y}|H=0}(\mathbf{y}_{\mathrm{obs}})}{f_{\mathbf{Y}|H=1}(\mathbf{y}_{\mathrm{obs}})}, \quad f_{\Theta}(\theta_{\mathrm{obs}}) \neq 0$$

$$= \mathrm{LR}(\mathbf{y}_{\mathrm{obs}}), \quad f_{\Theta}(\theta_{\mathrm{obs}}) \neq 0,$$

so (ignoring some technical issues) the MAP detector based on $(\mathbf{y}_{\mathrm{obs}}, \theta_{\mathrm{obs}})$ ignores $\theta_{\mathrm{obs}}$ and is identical to the MAP detector based on $\mathbf{y}_{\mathrm{obs}}$ only.[4]

Ostensibly more general is processing $\mathbf{Y}$ by mapping it to $g(\mathbf{Y}, \Theta)$, where the distribution of $\Theta$ is allowed to depend on $\mathbf{y}_{\mathrm{obs}}$. This motivates us to further extend the notion of processing. The cleanest way to define processing is to define its outcome rather than the way it is generated.

Before defining processing we remind the reader of the notion of conditional independence. But first we recall the definition of (unconditional) independence. We do so for discrete random variables using their Probability Mass Function (PMF). The extension to random variables with a joint density is straightforward. For the definition of independence in more general scenarios see, for example, (Billingsley, 1995, Section 20) or (Loève, 1963, Section 15) or (Williams, 1991, Chapter 4).

**Definition 20.11.1 (Independent Discrete Random Variables).** *We say that the discrete random variables $X$ and $Y$ of joint PMF $P_{X,Y}(\cdot, \cdot)$ and marginal PMFs $P_X(\cdot)$ and $P_Y(\cdot)$ are **independent** if $P_{X,Y}(\cdot, \cdot)$ factors as*

$$P_{X,Y}(x,y) = P_X(x)\, P_Y(y). \tag{20.59}$$

Equivalently, $X$ and $Y$ are independent if, for every outcome $y$ such that $P_Y(y) > 0$, the conditional distribution of $X$ given $Y = y$ is the same as its unconditional distribution:

$$P_{X|Y}(x|y) = P_X(x), \quad P_Y(y) > 0. \tag{20.60}$$

Equivalently, $X$ and $Y$ are independent if, for every outcome $x$ such that $P_X(x) > 0$, the conditional distribution of $Y$ given $X = x$ is the same as its unconditional distribution:

$$P_{Y|X}(y|x) = P_Y(y), \quad P_X(x) > 0. \tag{20.61}$$

The equivalence of (20.59) and (20.60) follows because, by the definition of the conditional probability mass function,

$$P_{X|Y}(x|y) = \frac{P_{X,Y}(x,y)}{P_Y(y)}, \quad P_Y(y) > 0.$$

Similarly, the equivalence of (20.59) and (20.61) follows from

$$P_{Y|X}(y|x) = \frac{P_{X,Y}(x,y)}{P_X(x)}, \quad P_X(x) > 0.$$

The beauty of (20.59) is that it is symmetric in $X, Y$. It makes it clear that $X$ and $Y$ are independent if, and only if, $Y$ and $X$ are independent. This is not obvious from (20.60) or (20.61).

---

[4]Technical issues arise when the outcome of $\Theta$, namely $\theta_{\mathrm{obs}}$, is such that $f_{\Theta}(\theta_{\mathrm{obs}}) = 0$.

The definition of the conditional independence of $X$ and $Y$ given $Z$ is similar, except that we condition everywhere on $Z$. Again we only consider the discrete case and refer the reader to (Loève, 1963, Section 25.3) and (Chung, 2001, Section 9.2) for the general case.

**Definition 20.11.2 (Conditionally Independent Discrete Random Variables).** *Let the discrete random variables $X, Y, Z$ have a joint PMF $P_{X,Y,Z}(\cdot, \cdot, \cdot)$. We say that $X$ and $Y$ are **conditionally independent** given $Z$ and write*

$$X\!-\!\circ\!-\!Z\!-\!\circ\!-\!Y$$

*if*

$$P_{X,Y|Z}(x, y|z) = P_{X|Z}(x|z)P_{Y|Z}(y|z), \quad P_Z(z) > 0. \tag{20.62}$$

Equivalently, $X$ and $Y$ are conditionally independent given $Z$ if, for any outcome $y, z$ with $P_{Y,Z}(y, z) > 0$, the conditional distribution of $X$ given that $Y = y$ and $Z = z$ is the same as the distribution of $X$ when conditioned on $Z = z$ only:

$$P_{X|Y,Z}(x|y, z) = P_{X|Z}(x|z), \quad P_{Y,Z}(y, z) > 0. \tag{20.63}$$

Or, equivalently, $X$ and $Y$ are conditionally independent given $Z$ if

$$P_{Y|X,Z}(y|x, z) = P_{Y|Z}(y|z), \quad P_{X,Z}(x, z) > 0. \tag{20.64}$$

The equivalence of (20.62) and (20.63) follows because, by the definition of the conditional probability mass function,

$$
\begin{aligned}
P_{X|Y,Z}(x|y, z) &= \frac{P_{X,Y,Z}(x, y, z)}{P_{Y,Z}(y, z)} \\
&= \frac{P_{X,Y|Z}(x, y|z)P_Z(z)}{P_{Y|Z}(y|z)P_Z(z)} \\
&= \frac{P_{X,Y|Z}(x, y|z)}{P_{Y|Z}(y|z)}, \quad P_{Y,Z}(y, z) > 0,
\end{aligned}
$$

and similarly the equivalence of (20.62) and (20.64) follows from

$$P_{Y|X,Z}(y|x, z) = \frac{P_{X,Y|Z}(x, y|z)}{P_{X|Z}(x|z)}, \quad P_{X,Z}(x, z) > 0.$$

Again, the beauty of (20.62) is that it is symmetric in $X, Y$. Thus $X\!-\!\circ\!-\!Z\!-\!\circ\!-\!Y$ if, and only if, $Y\!-\!\circ\!-\!Z\!-\!\circ\!-\!X$. When $X$ and $Y$ are conditionally independent given $Z$ we sometimes say that $X\!-\!\circ\!-\!Z\!-\!\circ\!-\!Y$ **forms a Markov chain**.

The equivalence between the different definitions of conditional independence continues to hold in the general case where the random variables are not necessarily discrete. We only reluctantly state this as a theorem, because we never defined conditional independence in nondiscrete settings.

**Theorem 20.11.3 (Equivalent Definition for Conditional Independence).** *Let $\mathbf{X}$, $\mathbf{Y}$, and $\mathbf{Z}$ be random vectors. Then the following statements are equivalent:*

(a) $\mathbf{X}$ and $\mathbf{Y}$ are conditionally independent given $\mathbf{Z}$.

(b) The conditional distribution of $\mathbf{Y}$ given $(\mathbf{X}, \mathbf{Z})$ is equal to its conditional distribution given $\mathbf{Z}$.

(c) The conditional distribution of $\mathbf{X}$ given $(\mathbf{Z}, \mathbf{Y})$ is equal to its conditional distribution given $\mathbf{Z}$.

**Proof.** For a precise definition of concepts appearing in this theorem and for a proof of the equivalence between the statements see (Loève, 1963, Section 25.3) and particularly Theorem 25.3A therein. $\qquad\square$

We are now ready to define the processing of the observation $\mathbf{Y}$ with respect to the hypothesis $H$.

**Definition 20.11.4 (Processing).** We say that $\mathbf{Z}$ is the result of **processing** $\mathbf{Y}$ **with respect to** $H$ if $H$ and $\mathbf{Z}$ are conditionally independent given $\mathbf{Y}$.

As we next show, this definition of processing extends the previous ones. We first show that if $\mathbf{Z} = g(\mathbf{Y})$ for some deterministic Borel measurable function $g(\cdot)$ then $H\!-\!\!\circ\!\!-\!\mathbf{Y}\!-\!\!\circ\!\!-\!g(\mathbf{Y})$. This follows by noting that, conditional on $\mathbf{Y}$, the random variable $g(\mathbf{Y})$ is deterministic and hence independent of everything and a fortiori of $H$.

We next show that if $\Theta$ is independent of $(H, \mathbf{Y})$, then $H\!-\!\!\circ\!\!-\!\mathbf{Y}\!-\!\!\circ\!\!-\!g(\mathbf{Y}, \Theta)$. Indeed, if $\mathbf{Z} = g(\mathbf{Y}, \Theta)$ with $\Theta$ being independent of $(\mathbf{Y}, H)$, then, conditionally on $\mathbf{Y} = \mathbf{y}$, the distribution of $\mathbf{Z}$ is simply the distribution of $g(\mathbf{y}, \Theta)$ so (under this conditioning) $\mathbf{Z}$ is independent of $H$.

We next show that processing the observables cannot help decrease the probability of error. The proof is conceptually very simple; the neat part is in the definition.

**Theorem 20.11.5 (Processing Is Futile).** If $\mathbf{Z}$ is the result of processing $\mathbf{Y}$ with respect to $H$, then no rule for guessing $H$ based on $\mathbf{Z}$ can outperform an optimal guessing rule based on $\mathbf{Y}$.

**Proof.** Surely no decision rule that guesses $H$ based on $\mathbf{Z}$ can outperform an optimal decision rule based on $\mathbf{Z}$, let alone outperform a decision rule that is optimal for guessing $H$ based on $\mathbf{Z}$ and $\mathbf{Y}$. But an optimal decision rule based on the pair $(\mathbf{Z}, \mathbf{Y})$ is the MAP rule, which compares

$$\Pr[H = 0 \,|\, \mathbf{Y} = \mathbf{y}, \mathbf{Z} = \mathbf{z}] \quad \text{and} \quad \Pr[H = 1 \,|\, \mathbf{Y} = \mathbf{y}, \mathbf{Z} = \mathbf{z}].$$

And, because $H\!-\!\!\circ\!\!-\!\mathbf{Y}\!-\!\!\circ\!\!-\!\mathbf{Z}$, it follows from Theorem 20.11.3 that this is equivalent to comparing

$$\Pr[H = 0 \,|\, \mathbf{Y} = \mathbf{y}] \quad \text{and} \quad \Pr[H = 1 \,|\, \mathbf{Y} = \mathbf{y}]$$

i.e., to an optimal (MAP) decision rule based on $\mathbf{Y}$ only. $\qquad\square$

The above theorem is more powerful than it seems. To demonstrate its strength, we next use it to show that in testing for a signal in Gaussian noise—irrespective of the prior—the optimal probability of error is monotonically nondecreasing in the noise variance. The setup we consider is one where $H$ is of prior $(\pi_0, \pi_1)$ and aiding us in guessing $H$ is the observable $Y$, which, conditional on $H = m$, is $\mathcal{N}(\alpha_m, \sigma^2)$ for $m \in \{0, 1\}$. We shall argue that, irrespective of the prior $(\pi_0, \pi_1)$, the optimal probability of error is monotonically nondecreasing in $\sigma^2$.

The beauty of the argument is that it allows us to prove the monotonicity result without having to calculate the optimal probability of error explicitly (as we did in Section 20.10 for the case of a uniform prior with $\alpha_0 = A$ and $\alpha_1 = -A$). While we could also compute the optimal probability of error for this more general setup and then use calculus to derive the monotonicity result, the argument we present instead has the advantage of also being applicable to multi-dimensional multi-hypothesis testing scenarios, where there is typically no closed-form expression for the optimal probability of error.

To prove this result, let $p_e^*(\sigma^2)$ denote the optimal probability of error as a function of $\sigma^2$. We need to show that $p_e^*(\sigma^2) \leq p_e^*(\sigma^2 + \delta^2)$, for all $\delta \in \mathbb{R}$. Consider the low-noise case where the conditional law of $Y$ given $H$ is $\mathcal{N}(\alpha_m, \sigma^2)$. Suppose that the receiver generates $W \sim \mathcal{N}(0, \delta^2)$ independently of $(H, Y)$ and adds $W$ to $Y$ to form $Z = Y + W$. Since $Z$ is the result of processing $Y$ with respect to $H$, it follows that the optimal probability of error based on $Y$, namely $p_e^*(\sigma^2)$, is at least as good as the optimal probability of error based on $Z$ (Theorem 20.11.5). We now complete the argument by showing that the optimal probability of error based on $Z$ is $p_e^*(\sigma^2 + \delta^2)$. This follows because, by Proposition 19.7.2, the conditional law of $Z$ given $H$ is $\mathcal{N}(\alpha_m, \sigma^2 + \delta^2)$.

Stated differently, since using a local random number generator the receiver can produce from an observation $Y$ of conditional law $\mathcal{N}(\alpha_m, \sigma^2)$ a random variable $Z$ whose conditional law is $\mathcal{N}(\alpha_m, \sigma^2 + \delta^2)$, the minimal probability of error based on an observation having conditional law $\mathcal{N}(\alpha_m, \sigma^2)$ cannot be larger than the optimal probability of error achievable based on an observation having conditional law $\mathcal{N}(\alpha_m, \sigma^2 + \delta^2)$. See Figure 20.4 for an illustration of this argument.

## 20.12 Sufficient Statistics

This section affords a first glance at the notion of sufficient statistics, which will be studied in greater depth and generality in Chapter 22. We begin with the following example. Consider the hypothesis testing problem with a uniform prior, where the observation is a tuple of real numbers $(Y_1, Y_2)$. Conditional on $H = 0$, the random variables $Y_1, Y_2$ are IID $\mathcal{N}(0, \sigma_0^2)$, whereas conditional on $H = 1$ they are IID $\mathcal{N}(0, \sigma_1^2)$, where

$$\sigma_0 > \sigma_1 > 0. \tag{20.65}$$

(If $\sigma_0^2 = \sigma_1^2$, then the problem is boring in that the conditional law of the observable given $H = 0$ is the same as given $H = 1$, so the two hypotheses cannot be differentiated. For $\sigma_0^2 \neq \sigma_1^2$ there is no loss in generality in assuming $\sigma_0 > \sigma_1$ because we can always relabel the hypotheses. And if $\sigma_0 > \sigma_1 = 0$, then the problem is trivial: we guess "$H = 1$" only if $Y_1 = Y_2 = 0$.) Thus, the observation space is the two-dimensional Euclidean space $\mathbb{R}^2$ and, using the explicit form of the Gaussian

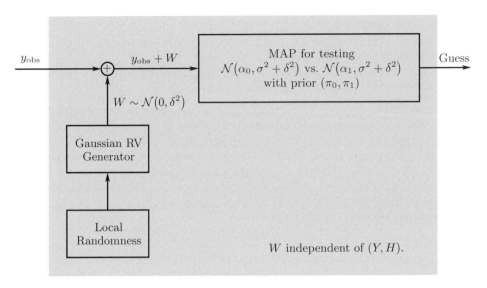

**Figure 20.4:** A suboptimal guessing rule (with randomization) for testing $\mathcal{N}(\alpha_0, \sigma^2)$ vs. $\mathcal{N}(\alpha_1, \sigma^2)$ with the given prior $(\pi_0, \pi_1)$. It attains the optimal probability of error for guessing $\mathcal{N}(\alpha_0, \sigma^2 + \delta^2)$ vs. $\mathcal{N}(\alpha_1, \sigma^2 + \delta^2)$ (with the given prior).

density (19.6),

$$f_{Y_1,Y_2|H=0}(y_1, y_2) = \frac{1}{2\pi\sigma_0^2} \exp\left(-\frac{1}{2\sigma_0^2}(y_1^2 + y_2^2)\right), \quad y_1, y_2 \in \mathbb{R}, \qquad (20.66\text{a})$$

$$f_{Y_1,Y_2|H=1}(y_1, y_2) = \frac{1}{2\pi\sigma_1^2} \exp\left(-\frac{1}{2\sigma_1^2}(y_1^2 + y_2^2)\right), \quad y_1, y_2 \in \mathbb{R}. \qquad (20.66\text{b})$$

Since we assumed a uniform prior, the ML decoding rule for guessing $H$ based on the tuple $(Y_1, Y_2)$ is optimal. To derive the ML rule explicitly, we compute the likelihood-ratio function

$$\begin{aligned}
\mathrm{LR}(y_1, y_2) &= \frac{f_{Y_1,Y_2|H=0}(y_1, y_2)}{f_{Y_1,Y_2|H=1}(y_1, y_2)} \\
&= \frac{\frac{1}{2\pi\sigma_0^2} \exp\left(-\frac{1}{2\sigma_0^2}(y_1^2 + y_2^2)\right)}{\frac{1}{2\pi\sigma_1^2} \exp\left(-\frac{1}{2\sigma_1^2}(y_1^2 + y_2^2)\right)} \\
&= \frac{\sigma_1^2}{\sigma_0^2} \exp\left(\frac{1}{2}\left(\frac{1}{\sigma_1^2} - \frac{1}{\sigma_0^2}\right)(y_1^2 + y_2^2)\right), \quad y_1, y_2 \in \mathbb{R}. \qquad (20.67)
\end{aligned}$$

Thus,

$$\begin{aligned}
\mathrm{LR}(y_1, y_2) > 1 &\Leftrightarrow \exp\left(\frac{1}{2}\left(\frac{1}{\sigma_1^2} - \frac{1}{\sigma_0^2}\right)(y_1^2 + y_2^2)\right) > \frac{\sigma_0^2}{\sigma_1^2} \\
&\Leftrightarrow \frac{1}{2}\left(\frac{1}{\sigma_1^2} - \frac{1}{\sigma_0^2}\right)(y_1^2 + y_2^2) > \ln\frac{\sigma_0^2}{\sigma_1^2}
\end{aligned}$$

$$\Leftrightarrow \frac{\sigma_0^2 - \sigma_1^2}{2\sigma_0^2\sigma_1^2}(y_1^2 + y_2^2) > \ln\frac{\sigma_0^2}{\sigma_1^2}$$

$$\Leftrightarrow y_1^2 + y_2^2 > \frac{2\sigma_0^2\sigma_1^2}{\sigma_0^2 - \sigma_1^2}\ln\frac{\sigma_0^2}{\sigma_1^2}, \tag{20.68}$$

where the second equivalence follows from the monotonicity of the logarithm function (20.43); and where the last equivalence follows by multiplying both sides of the inequality by the constant $2\sigma_0^2\sigma_1^2/(\sigma_0^2 - \sigma_1^2)$ (without the need to change the inequality direction because this constant is by (20.65) positive).

It follows from (20.68) that the ML decision rule for guessing $H$ based on $(Y_1, Y_2)$ computes $Y_1^2 + Y_2^2$ and then compares the result to a threshold. It is interesting to note that to implement this decision rule one need not observe $Y_1$ and $Y_2$ directly; it suffices to observe the sum of their squares

$$T \triangleq Y_1^2 + Y_2^2. \tag{20.69}$$

Of course, being the result of processing $(Y_1, Y_2)$ with respect to $H$, no guess of $H$ based on $T$ can outperform an optimal guess based on $(Y_1, Y_2)$ (Section 20.11). But what is interesting about this example is that, even though one cannot recover $(Y_1, Y_2)$ from $T$ (so there are some decision rules based on $(Y_1, Y_2)$ that cannot be implemented if one only knows $T$), the ML rule based on $(Y_1, Y_2)$ only requires knowledge of $T$. Thus, in this example, even though pre-processing the observations to produce $T = Y_1^2 + Y_2^2$ is not reversible, basing one's decision on $T$ incurs no loss in optimality. An optimal decision rule based on $T$ is just as good as an optimal rule based on $(Y_1, Y_2)$.

The reason for this can be traced to the fact that, in this example, to compute the likelihood-ratio $\mathrm{LR}(y_1, y_2)$ one need not know the pair $(y_1, y_2)$; it suffices that one know the sum of their squares $y_1^2 + y_2^2$; see (20.67). In this sense $T = Y_1^2 + Y_2^2$ forms a **sufficient statistic** for guessing $H$ from $(Y_1, Y_2)$, as we next define.

We would like to define a mapping $T(\cdot)$ from the observation space $\mathbb{R}^d$ to $\mathbb{R}^{d'}$ as being sufficient for the densities $f_{\mathbf{Y}|H=0}(\cdot)$ and $f_{\mathbf{Y}|H=1}(\cdot)$ if the likelihood-ratio $\mathrm{LR}(\mathbf{y}_{\mathrm{obs}})$ can be computed from $T(\mathbf{y}_{\mathrm{obs}})$ for *every* $\mathbf{y}_{\mathrm{obs}}$ in $\mathbb{R}^d$. However, for technical reasons, we require slightly less: we only require that $\mathrm{LR}(\mathbf{y}_{\mathrm{obs}})$ be computable from $T(\mathbf{y}_{\mathrm{obs}})$ for those observations $\mathbf{y}_{\mathrm{obs}}$ for which at least one of the densities is positive (so the likelihood-ratio is not of the form $0/0$) and that additionally lie outside some prespecified set $\mathcal{Y}_0 \subset \mathbb{R}^d$ of Lebesgue measure zero.[5] Thus, we shall require that there exist a set $\mathcal{Y}_0 \subset \mathbb{R}^d$ of Lebesgue measure zero and a function $\zeta \colon \mathbb{R}^{d'} \to [0, \infty]$ such that $\zeta\big(T(\mathbf{y}_{\mathrm{obs}})\big)$ is equal to $\mathrm{LR}(\mathbf{y}_{\mathrm{obs}})$ whenever

$$\mathbf{y}_{\mathrm{obs}} \notin \mathcal{Y}_0 \quad \text{and} \quad f_{\mathbf{Y}|H=0}(\mathbf{y}_{\mathrm{obs}}) + f_{\mathbf{Y}|H=1}(\mathbf{y}_{\mathrm{obs}}) > 0. \tag{20.70}$$

Note that the fact that $\mathcal{Y}_0$ is of Lebesgue measure zero implies that

$$\Pr[\mathbf{Y} \in \mathcal{Y}_0 \,|\, H = 0] = \Pr[\mathbf{Y} \in \mathcal{Y}_0 \,|\, H = 1] = 0. \tag{20.71}$$

---

[5]We allow this exception set so that the question of whether $T(\cdot)$ forms a sufficient statistic or not will not depend on our choice of the density function of the conditional distribution of the observable. (Recall that if a RV has a probability density function, then it has infinitely many different probability density functions, every two of which differ on a set of Lebesgue measure zero.)

To convince the reader that this really is only "slightly" less, we note:

**Note 20.12.1.** Both conditional on $H = 0$ and conditional on $H = 1$, the probability that the observable violates (20.70) is zero.

**Proof.** We shall show that conditional on $H = 0$, the probability that the observable violates (20.70) is zero. The conditional probability given $H = 1$ can be analogously shown to be zero. The condition that (20.70) is violated is equivalent to the condition that either $\mathbf{y}_{\mathrm{obs}} \in \mathcal{Y}_0$ or $f_{\mathbf{Y}|H=0}(\mathbf{y}_{\mathrm{obs}}) + f_{\mathbf{Y}|H=1}(\mathbf{y}_{\mathrm{obs}}) = 0$. By (20.71), $\Pr[\mathbf{Y} \in \mathcal{Y}_0 \,|\, H = 0] = 0$. And, by the nonnegativity of the densities,

$$\Pr\big[f_{\mathbf{Y}|H=0}(\mathbf{Y}) + f_{\mathbf{Y}|H=1}(\mathbf{Y}) = 0 \,\big|\, H = 0\big] \leq \Pr\big[f_{\mathbf{Y}|H=0}(\mathbf{Y}) = 0 \,\big|\, H = 0\big]$$

$$= \int_{\{\tilde{\mathbf{y}} \in \mathbb{R}^d : f_{\mathbf{Y}|H=0}(\tilde{\mathbf{y}})=0\}} f_{\mathbf{Y}|H=0}(\mathbf{y})\,\mathrm{d}\mathbf{y}$$

$$= \int_{\{\tilde{\mathbf{y}} \in \mathbb{R}^d : f_{\mathbf{Y}|H=0}(\tilde{\mathbf{y}})=0\}} 0\,\mathrm{d}\mathbf{y}$$

$$= 0.$$

Conditionally on $H = 0$, the probability of the observable violating (20.70) is thus the probability of the union of two events, each of which is of zero probability, and is thus of zero probability; see Corollary 21.5.2 ahead. $\qquad\square$

**Definition 20.12.2 (Sufficient Statistic for Two Densities).** *We say that a mapping* $T \colon \mathbb{R}^d \to \mathbb{R}^{d'}$ *forms a **sufficient statistic** for the density functions* $f_{\mathbf{Y}|H=0}(\cdot)$ *and* $f_{\mathbf{Y}|H=1}(\cdot)$ *on* $\mathbb{R}^d$ *if it is Borel measurable[6] and if there exists a set* $\mathcal{Y}_0 \subset \mathbb{R}^d$ *of Lebesgue measure zero and a Borel measurable function* $\boldsymbol{\zeta} \colon \mathbb{R}^{d'} \to [0, \infty]$ *such that for all* $\mathbf{y}_{\mathrm{obs}} \in \mathbb{R}^d$ *satisfying* (20.70)

$$\frac{f_{\mathbf{Y}|H=0}(\mathbf{y}_{\mathrm{obs}})}{f_{\mathbf{Y}|H=1}(\mathbf{y}_{\mathrm{obs}})} = \zeta\big(T(\mathbf{y}_{\mathrm{obs}})\big), \tag{20.72}$$

*where on the LHS of* (20.72) *we define* $a/0$ *to be* $+\infty$ *whenever* $a > 0$.

In our example the observation $(Y_1, Y_2)$ takes value in $\mathbb{R}^2$ so $d = 2$; the mapping $T \colon (y_1, y_2) \mapsto y_1^2 + y_2^2$ is a mapping from $\mathbb{R}^2$ to $\mathbb{R}$ so $d' = 1$; and by (20.67),

$$\zeta \colon t \mapsto \frac{\sigma_1^2}{\sigma_0^2} \exp\left(\frac{1}{2}\left(\frac{1}{\sigma_1^2} - \frac{1}{\sigma_0^2}\right)t\right).$$

---

[6]The technical condition that $T(\cdot)$ is Borel measurable guarantees that $T(\mathbf{Y})$ is a random vector. See for example (Billingsley, 1995, Theorem 13.1(ii)) for a discussion of this technical issue. The issue is best seen in the scalar case. Suppose that $Y$ is a RV defined over the probability space $(\Omega, \mathcal{F}, P)$. If $T(\cdot)$ is any function, then $T(Y)$ is a mapping from $\Omega$ to the $\mathbb{R}$, but we are not guaranteed that it be a RV, because for $T(Y)$ to be a RV we must have that, for every $\xi \in \mathbb{R}$, the set $\{\omega \in \Omega : T(Y(\omega)) \leq \xi\}$ be in $\mathcal{F}$, and this is, in general, not true. However, if $T(\cdot)$ is Borel measurable, then the above cited theorem guarantees that $T(X)$ is, indeed, a RV. Note that any continuous function is Borel measurable (Billingsley, 1995, Theorem 13.2). In practice, one never encounters functions that are not Borel measurable; In fact, it is hard work to construct one.

Here we can take $\mathcal{Y}_0$ to be the empty set.[7]

We next show that if $T(\cdot)$ is a sufficient statistic, then there is no loss in optimality in considering decision rules that base their decision on $T(\mathbf{Y})$. This result is almost obvious, because the MAP decision rule is optimal (Theorem 20.7.1); because it can be expressed in terms of the likelihood-ratio function (20.40); and because the sufficiency of $T(\cdot)$ implies that the likelihood-ratio function $\mathrm{LR}(\mathbf{y}_{\mathrm{obs}})$ is computable from $T(\mathbf{y}_{\mathrm{obs}})$. Nevertheless, we provide a formal proof because the result is important.

**Proposition 20.12.3.** *If $T\colon \mathbb{R}^d \to \mathbb{R}^{d'}$ is a sufficient statistic for the densities $f_{\mathbf{Y}|H=0}(\cdot)$ and $f_{\mathbf{Y}|H=1}(\cdot)$, then, irrespective of the prior of $H$, there exists a decision rule that guesses $H$ based on $T(\mathbf{Y})$ and which is as good as any optimal guessing rule based on $\mathbf{Y}$.*

**Proof.** We need to show that if $\phi^*_{\mathrm{Guess}}(\cdot)$ is an optimal decision rule for guessing $H$ based on $\mathbf{Y}$, then there exists a guessing rule based on $T(\mathbf{Y})$ that has the same probability of error. We note that it is enough to prove this result for a nondegenerate prior (20.2), because for degenerate priors one can achieve zero probability of error even without looking at $T(\mathbf{Y})$: if $\Pr[H = 0] = 1$ guess "$H = 0$," and if $\Pr[H = 1] = 1$ guess "$H = 1$." We thus proceed to assume a nondegenerate prior (20.2).

Let $\phi_{\mathrm{MAP}}(\cdot)$ be the MAP rule for guessing $H$ based on $\mathbf{Y}$. Since this rule is optimal, it suffices to exhibit a decoding rule $\phi_T(\cdot)$ based on $T(\mathbf{Y})$ of equal performance. Since $T(\cdot)$ is sufficient, it follows that there exists a set of Lebesgue measure zero $\mathcal{Y}_0$ and a Borel measurable function $\zeta(\cdot)$ such that $\zeta\big(T(\mathbf{y}_{\mathrm{obs}})\big) = \mathrm{LR}(\mathbf{y}_{\mathrm{obs}})$, whenever (20.70) holds. Based upon the observation $T(\mathbf{Y}) = T(\mathbf{y}_{\mathrm{obs}})$, the desired rule is to guess

$$\phi_T\big(T(\mathbf{y}_{\mathrm{obs}})\big) = \begin{cases} 0 & \text{if } \zeta\big(T(\mathbf{y}_{\mathrm{obs}})\big) > \frac{\pi_1}{\pi_0}, \\ 1 & \text{if } \zeta\big(T(\mathbf{y}_{\mathrm{obs}})\big) < \frac{\pi_1}{\pi_0}, \\ \mathcal{U}(\{0,1\}) & \text{if } \zeta\big(T(\mathbf{y}_{\mathrm{obs}})\big) = \frac{\pi_1}{\pi_0}. \end{cases} \tag{20.73}$$

That $\phi_T(\cdot)$ has the same performance as $\phi_{\mathrm{MAP}}(\cdot)$ now follows by noting that, by (20.72), the two decoding rules are in agreement except perhaps for observations $\mathbf{y}_{\mathrm{obs}}$ violating (20.70), but those, by Note 20.12.1, occur with probability zero. The performance of $\phi_{\mathrm{MAP}}(\cdot)$ (which is optimal based on $\mathbf{Y}$) and of $\phi_T(\cdot)$ (which is based on $T(\mathbf{Y})$) are thus identical. $\qquad\square$

Definition 20.12.2 is intuitive in that it demonstrates how one typically goes about identifying a sufficient statistic: one computes the likelihood-ratio and checks what it depends on. This definition, however, becomes a bit cumbersome in multi-hypothesis testing, which we shall discuss in Chapter 21. A definition that is more appropriate for that setting is given in Chapter 22 in terms of the computability of the *a posteriori* probabilities from $T(\mathbf{y}_{\mathrm{obs}})$ (Definition 22.2.1). The purpose of the next proposition is to show that the two definitions coincide in the binary case: ignoring sets of Lebesgue measure zero, the likelihood-ratio can be computed from

---

[7]We would have needed to choose a nontrivial set $\mathcal{Y}_0$ if we had changed the densities (20.66) at a finite number of points.

$T(\mathbf{y}_{\mathrm{obs}})$ (whenever the ratio is not $0/0$), if, and only if, for any prior $(\pi_0, \pi_1)$ one can compute the *a posteriori* distribution of $H$ from $T(\mathbf{y}_{\mathrm{obs}})$ (whenever $f_{\mathbf{Y}}(\mathbf{y}_{\mathrm{obs}}) > 0$).

We draw the reader's attention to the following subtle issue. Definition 20.12.2 makes it clear that the sufficiency of $T(\cdot)$ has nothing to do with the prior; it only depends on the densities $f_{\mathbf{Y}|H=0}(\cdot)$ and $f_{\mathbf{Y}|H=1}(\cdot)$. The equivalent definition of sufficient statistics in terms of the computability of the *a posteriori* distribution ostensibly depends also on the prior, because it is only meaningful to discuss the *a posteriori* distribution if $H$ has a prior. Nevertheless, the definitions are equivalent because in the latter definition we require that the *a posteriori* distribution be computable from $T(\mathbf{Y})$ *for every prior*, and not just for the prior given in the problem's formulation.

**Proposition 20.12.4 (Computability of the a Posteriori Distribution).** *Let the mapping $T \colon \mathbb{R}^d \to \mathbb{R}^{d'}$ be Borel measurable, and let $f_{\mathbf{Y}|H=0}(\cdot)$ and $f_{\mathbf{Y}|H=1}(\cdot)$ be densities on $\mathbb{R}^d$. Then the following two conditions are equivalent:*

*(a) $T(\cdot)$ forms a sufficient statistic for the densities $f_{\mathbf{Y}|H=0}(\cdot)$ and $f_{\mathbf{Y}|H=1}(\cdot)$.*

*(b) For some set $\mathcal{Y}_0 \subset \mathbb{R}^d$ of Lebesgue measure zero we have that for every prior $(\pi_0, \pi_1)$ there exist Borel measurable functions from $\mathbb{R}^{d'}$ to $[0, 1]$*

$$\mathbf{t} \mapsto \psi_m(\pi_0, \pi_1, \mathbf{t}), \quad m = 0, 1,$$

*such that the vector*

$$\left( \psi_0(\pi_0, \pi_1, T(\mathbf{y}_{\mathrm{obs}})), \ \psi_1(\pi_0, \pi_1, T(\mathbf{y}_{\mathrm{obs}})) \right)^{\mathsf{T}}$$

*is a probability vector, and this probability vector is equal to the vector*

$$\left( \Pr[H = 0 \,|\, \mathbf{Y} = \mathbf{y}_{\mathrm{obs}}], \ \Pr[H = 1 \,|\, \mathbf{Y} = \mathbf{y}_{\mathrm{obs}}] \right)^{\mathsf{T}}, \tag{20.74}$$

*whenever both the condition $\mathbf{y}_{\mathrm{obs}} \notin \mathcal{Y}_0$, and the condition*

$$\pi_0 f_{\mathbf{Y}|H=0}(\mathbf{y}_{\mathrm{obs}}) + \pi_1 f_{\mathbf{Y}|H=1}(\mathbf{y}_{\mathrm{obs}}) > 0 \tag{20.75}$$

*are satisfied. Here (20.74) is computed for $H$ having the prior $(\pi_0, \pi_1)$ and for the conditional densities $f_{\mathbf{Y}|H=0}(\cdot)$ and $f_{\mathbf{Y}|H=1}(\cdot)$.*

**Proof.** We begin by proving that (a) implies (b). That is, we assume that $T(\cdot)$ forms a sufficient statistic and proceed to prove the existence of the set $\mathcal{Y}_0$ and of the functions $\psi_0(\cdot), \psi_1(\cdot)$. Let $\mathcal{Y}_0$ and $\zeta \colon \mathbb{R}^{d'} \to [0, \infty]$ be as guaranteed by the definition of sufficient statistics (Definition 20.12.2) so

$$\frac{f_{\mathbf{Y}|H=0}(\mathbf{y}_{\mathrm{obs}})}{f_{\mathbf{Y}|H=1}(\mathbf{y}_{\mathrm{obs}})} = \zeta(T(\mathbf{y}_{\mathrm{obs}})), \tag{20.76}$$

whenever $\mathbf{y}_{\mathrm{obs}}$ satisfies (20.70). We next show how to construct for every pair $(\pi_0, \pi_1)$ the functions $\psi_0(\cdot), \psi_1(\cdot)$. We consider three cases separately: the case

$\pi_0 = 1 - \pi_1 = 1$, the case $\pi_0 = 1 - \pi_1 = 0$, and the case where both $\pi_0$ and $\pi_1$ are strictly positive.

In the first case $H$ is deterministically zero, and the functions $\psi_0(1,0,\mathbf{t}) = 1$ and $\psi_1(1,0,\mathbf{t}) = 0$ meet our requirements. In the second case $H$ is deterministically one, and the functions $\psi_0(0,1,\mathbf{t}) = 1 - \psi_1(0,1,\mathbf{t}) = 0$ meet our requirements.

It remains to treat the case where $\pi_0, \pi_1 > 0$. We shall show that in this case the functions

$$\psi_0(\pi_0, \pi_1, \mathbf{t}) \triangleq \frac{\pi_0 \zeta(\mathbf{t})}{\pi_0 \zeta(\mathbf{t}) + \pi_1}, \quad \psi_1(\pi_0, \pi_1, \mathbf{t}) \triangleq 1 - \psi_0(\pi_0, \pi_1, \mathbf{t}), \quad (20.77)$$

(where $\infty/(\infty + a)$ is defined as one for all finite $a$) meet our requirements. To that end we first note that $\psi_0(\pi_0, \pi_1, \mathbf{t})$ and $\psi_1(\pi_0, \pi_1, \mathbf{t})$ are nonnegative and sum to one. We next note that, for $\pi_0, \pi_1 > 0$, the condition (20.75) implies that $f_{\mathbf{Y}|H=0}(\mathbf{y}_{\mathrm{obs}})$ and $f_{\mathbf{Y}|H=1}(\mathbf{y}_{\mathrm{obs}})$ are not both zero. Consequently, if $\mathbf{y}_{\mathrm{obs}}$ satisfies (20.75) and also $\mathbf{y}_{\mathrm{obs}} \notin \mathcal{Y}_0$, then it satisfies (20.70) and $\mathrm{LR}(\mathbf{y}_{\mathrm{obs}}) = \zeta(T(\mathbf{y}_{\mathrm{obs}}))$. Thus, in the case $\pi_0, \pi_1 > 0$, we have that, whenever (20.75) and $\mathbf{y}_{\mathrm{obs}} \notin \mathcal{Y}_0$ hold,

$$
\begin{aligned}
\psi_0(\pi_0, \pi_1, T(\mathbf{y}_{\mathrm{obs}})) &= \frac{\pi_0 \zeta(T(\mathbf{y}_{\mathrm{obs}}))}{\pi_0 \zeta(T(\mathbf{y}_{\mathrm{obs}})) + \pi_1} \\
&= \frac{\pi_0 \, \mathrm{LR}(\mathbf{y}_{\mathrm{obs}})}{\pi_0 \, \mathrm{LR}(\mathbf{y}_{\mathrm{obs}}) + \pi_1} \\
&= \frac{\pi_0 f_{\mathbf{Y}|H=0}(\mathbf{y}_{\mathrm{obs}})/f_{\mathbf{Y}|H=1}(\mathbf{y}_{\mathrm{obs}})}{\pi_0 f_{\mathbf{Y}|H=0}(\mathbf{y}_{\mathrm{obs}})/f_{\mathbf{Y}|H=1}(\mathbf{y}_{\mathrm{obs}}) + \pi_1} \\
&= \frac{\pi_0 f_{\mathbf{Y}|H=0}(\mathbf{y}_{\mathrm{obs}})}{\pi_0 f_{\mathbf{Y}|H=0}(\mathbf{y}_{\mathrm{obs}}) + \pi_1 f_{\mathbf{Y}|H=1}(\mathbf{y}_{\mathrm{obs}})} \\
&= \Pr[H = 0 \,|\, \mathbf{Y} = \mathbf{y}_{\mathrm{obs}}]
\end{aligned}
$$

as required. This implies that, whenever (20.75) and $\mathbf{y}_{\mathrm{obs}} \notin \mathcal{Y}_0$ hold, we also have $\psi_1(\pi_0, \pi_1, T(\mathbf{y}_{\mathrm{obs}})) = \Pr[H = 1 \,|\, \mathbf{Y} = \mathbf{y}_{\mathrm{obs}}]$, since $\psi_1(\pi_0, \pi_1, \mathbf{t}) = 1 - \psi_0(\pi_0, \pi_1, \mathbf{t})$ and since $\Pr[H = 1 \,|\, \mathbf{Y} = \mathbf{y}_{\mathrm{obs}}] = 1 - \Pr[H = 0 \,|\, \mathbf{Y} = \mathbf{y}_{\mathrm{obs}}]$; see (20.10).

We now prove that (b) implies (a), i.e., that the existence of the set $\mathcal{Y}_0$ and of the functions $\psi_0(\cdot), \psi_1(\cdot)$ imply the existence of the function $\zeta(\cdot)$. In fact, we shall prove a stronger statement that if for *some* nondegenerate prior the a *posteriori* distribution of $H$ given $\mathbf{Y} = \mathbf{y}_{\mathrm{obs}}$ is computable from $T(\mathbf{y}_{\mathrm{obs}})$ (whenever (20.75) and $\mathbf{y}_{\mathrm{obs}} \notin \mathcal{Y}_0$ hold), then there exists some function $\zeta \colon \mathbb{R}^{d'} \to [0, \infty]$ such that $\mathrm{LR}(\mathbf{y}_{\mathrm{obs}}) = \zeta(T(\mathbf{y}_{\mathrm{obs}}))$, whenever $\mathbf{y}_{\mathrm{obs}}$ satisfies (20.70).

To construct $\zeta(\cdot)$ from $\psi_0(\cdot)$ and $\psi_1(\cdot)$, pick some arbitrary strictly positive $\tilde{\pi}_0, \tilde{\pi}_1$ summing to one (e.g., $\tilde{\pi}_0, \tilde{\pi}_1 = 1/2$), and define $\zeta(\cdot)$ by

$$\zeta(T(\mathbf{y}_{\mathrm{obs}})) = \frac{\tilde{\pi}_1 \psi_0(\tilde{\pi}_0, \tilde{\pi}_1, T(\mathbf{y}_{\mathrm{obs}}))}{\tilde{\pi}_0 \psi_1(\tilde{\pi}_0, \tilde{\pi}_1, T(\mathbf{y}_{\mathrm{obs}}))}, \quad (20.78)$$

using the convention that $a/0 = \infty$ for all $a > 0$; see (20.39).

We next verify that if $\mathbf{y}_{\mathrm{obs}}$ satisfies (20.70) then $\zeta(T(\mathbf{y}_{\mathrm{obs}})) = \mathrm{LR}(\mathbf{y}_{\mathrm{obs}})$. To this end, define $H$ to have the law $\Pr[H = 0] = \tilde{\pi}_0$ and $\Pr[H = 1] = \tilde{\pi}_1$, and let the conditional law of $\mathbf{Y}$ given $H$ be as specified by the given densities. Since $\tilde{\pi}_0$ and $\tilde{\pi}_1$ are strictly positive, it follows that whenever $f_{\mathbf{Y}|H=0}(\mathbf{y}_{\text{obs}})$ and $f_{\mathbf{Y}|H=1}(\mathbf{y}_{\text{obs}})$ are not both zero, we also have $\tilde{\pi}_0 f_{\mathbf{Y}|H=0}(\mathbf{y}_{\text{obs}}) + \tilde{\pi}_1 f_{\mathbf{Y}|H=1}(\mathbf{y}_{\text{obs}}) > 0$. Consequently, for strictly positive $\tilde{\pi}_0, \tilde{\pi}_1$ we have that (20.70) implies that $\mathbf{y}_{\text{obs}} \notin \mathcal{Y}_0$ and $\tilde{\pi}_0 f_{\mathbf{Y}|H=0}(\mathbf{y}_{\text{obs}}) + \tilde{\pi}_1 f_{\mathbf{Y}|H=1}(\mathbf{y}_{\text{obs}}) > 0$ and thus, for observations $\mathbf{y}_{\text{obs}}$ satisfying (20.70),

$$
\begin{aligned}
\zeta\big(T(\mathbf{y}_{\text{obs}})\big) &= \frac{\tilde{\pi}_1 \psi_0\big(\tilde{\pi}_0, \tilde{\pi}_1, T(\mathbf{y}_{\text{obs}})\big)}{\tilde{\pi}_0 \psi_1\big(\tilde{\pi}_0, \tilde{\pi}_1, T(\mathbf{y}_{\text{obs}})\big)} \\
&= \frac{\Pr[H=1]\Pr[H=0\,|\,\mathbf{Y}=\mathbf{y}_{\text{obs}}]}{\Pr[H=0]\Pr[H=1\,|\,\mathbf{Y}=\mathbf{y}_{\text{obs}}]} \\
&= \text{LR}(\mathbf{y}_{\text{obs}}),
\end{aligned}
$$

where the last equality follows by dividing the equation

$$
\Pr[H=0\,|\,\mathbf{Y}=\mathbf{y}_{\text{obs}}] = \frac{\Pr[H=0]f_{\mathbf{Y}|H=0}(\mathbf{y}_{\text{obs}})}{\Pr[H=0]f_{\mathbf{Y}|H=0}(\mathbf{y}_{\text{obs}}) + \Pr[H=1]f_{\mathbf{Y}|H=1}(\mathbf{y}_{\text{obs}})}
$$

(which is a restatement of (20.9a) for our case) by

$$
\Pr[H=1\,|\,\mathbf{Y}=\mathbf{y}_{\text{obs}}] = \frac{\Pr[H=1]f_{\mathbf{Y}|H=1}(\mathbf{y}_{\text{obs}})}{\Pr[H=0]f_{\mathbf{Y}|H=0}(\mathbf{y}_{\text{obs}}) + \Pr[H=1]f_{\mathbf{Y}|H=1}(\mathbf{y}_{\text{obs}})}
$$

(which is a restatement of (20.9b) for our case). $\qquad\square$

Once we have identified a sufficient statistic $T(\mathbf{Y})$, we can proceed to derive an optimal guessing rule using two methods that we describe next. Again, we focus on nondegenerate priors.

**Method 1:** We ignore the fact that $T(\mathbf{Y})$ forms a sufficient statistic and simply use the MAP rule (20.40):

$$
\phi_{\text{MAP}}(\mathbf{y}_{\text{obs}}) = \begin{cases} 0 & \text{if } \text{LR}(\mathbf{y}_{\text{obs}}) > \frac{\pi_1}{\pi_0}, \\ 1 & \text{if } \text{LR}(\mathbf{y}_{\text{obs}}) < \frac{\pi_1}{\pi_0}, \\ \mathcal{U}(\{0,1\}) & \text{if } \text{LR}(\mathbf{y}_{\text{obs}}) = \frac{\pi_1}{\pi_0}. \end{cases} \tag{20.79}
$$

(Because $T(\mathbf{Y})$ is a sufficient statistic, the likelihood-ratio function $\text{LR}(\mathbf{y}_{\text{obs}})$ will be computable from $T(\mathbf{y}_{\text{obs}})$ whenever $\text{LR}(\mathbf{y}_{\text{obs}})$ does not have the pathological form $0/0$ and does not lie in the exception set $\mathcal{Y}_0$. Such pathological observations occur with probability zero (20.12), so we need not worry about them.)

**Method 2:** By Proposition 20.12.3, there is no loss in optimality in forming our guess based on $T(\mathbf{Y})$. So we can use any optimal rule, e.g., the MAP rule, for guessing $H$ based on the new $d'$-dimensional observations $\mathbf{t}_{\text{obs}} = T(\mathbf{y}_{\text{obs}})$. This method requires computing the conditional distribution of the random $d'$-vector

$\mathbf{T} = T(\mathbf{Y})$ conditional on $H = 0$ and conditional on $H = 1$ and deciding according to the rule:

$$\phi_{\text{Guess}}(T(\mathbf{y}_{\text{obs}})) = \begin{cases} 0 & \text{if } \pi_0\, f_{\mathbf{T}|H=0}(T(\mathbf{y}_{\text{obs}})) > \pi_1\, f_{\mathbf{T}|H=1}(T(\mathbf{y}_{\text{obs}})), \\ 1 & \text{if } \pi_0\, f_{\mathbf{T}|H=0}(T(\mathbf{y}_{\text{obs}})) < \pi_1\, f_{\mathbf{T}|H=1}(T(\mathbf{y}_{\text{obs}})), \end{cases} \quad (20.80)$$

with ties being resolved at random.

Why would we want to use Method 2 when we have already computed the likelihood-ratio function to establish the sufficiency of the statistic? The answer is that sometimes one can demonstrate that $T(\mathbf{Y})$ forms a sufficient statistic by methods that are not based on the computation of the likelihood-ratio. In such cases, Method 2 may be advantageous. Also, sometimes the analysis of the probability of error in Method 2 is easier. The choice is ours.

Returning to the example of (20.66), we demonstrate Method 2 by calculating the law of the sufficient statistic $T = Y_1^2 + Y_2^2$ under each of the hypotheses. Recalling that the sum of the squares of two IID zero-mean Gaussians is exponential (Note 19.8.1) we obtain:

$$f_{T|H=0}(t) = \frac{1}{2\sigma_0^2} \exp\left(-\frac{t}{2\sigma_0^2}\right), \quad t \geq 0, \quad (20.81a)$$

$$f_{T|H=1}(t) = \frac{1}{2\sigma_1^2} \exp\left(-\frac{t}{2\sigma_1^2}\right), \quad t \geq 0. \quad (20.81b)$$

Consequently, the likelihood-ratio is given by

$$\frac{f_{T|H=0}(t)}{f_{T|H=1}(t)} = \frac{\sigma_1^2}{\sigma_0^2} \exp\left(t\left(\frac{1}{2\sigma_1^2} - \frac{1}{2\sigma_0^2}\right)\right), \quad t \geq 0,$$

and the log likelihood-ratio by

$$\ln \frac{f_{T|H=0}(t)}{f_{T|H=1}(t)} = \ln \frac{\sigma_1^2}{\sigma_0^2} + t\left(\frac{1}{2\sigma_1^2} - \frac{1}{2\sigma_0^2}\right), \quad t \geq 0.$$

We thus guess "$H = 0$" if the log likelihood-ratio is positive,

$$t \geq \frac{2\sigma_0^2\sigma_1^2}{\sigma_0^2 - \sigma_1^2} \ln \frac{\sigma_0^2}{\sigma_1^2},$$

i.e., if

$$y_1^2 + y_2^2 \geq \frac{2\sigma_0^2\sigma_1^2}{\sigma_0^2 - \sigma_1^2} \ln \frac{\sigma_0^2}{\sigma_1^2}.$$

We similarly guess "$H = 1$" if the log likelihood-ratio is negative, and flip a coin if it is zero. This is the same law we obtained in (20.68) based on Method 1.

## 20.13 Consequences of Optimality

Consider the problem of guessing an *a priori* uniformly distributed binary random variable $H$ based on the observable $Y$ whose conditional law given $H = 0$

is $\mathcal{N}(0, \sigma^2)$ and whose conditional distribution given $H = 1$ is $\mathcal{N}(1, \sigma^2)$. To derive an optimal guessing rule we could derive the MAP rule by computing the likelihood-ratio function as we did in Section 20.10. But having already carried out the calculations in Section 20.10 for testing whether an observation was drawn $\mathcal{N}(A, \sigma^2)$ or $\mathcal{N}(-A, \sigma^2)$, there is a better way. Let

$$T = Y - \frac{1}{2}. \tag{20.82}$$

Because there is a one-to-one relationship between $Y$ and $T$, there is no loss in optimality in subtracting $1/2$ from $Y$ to obtain $T$ and in then applying an optimal decision rule to $T$. Indeed, since $Y = T + 1/2$, it follows that $Y$ is the result of processing $T$ with respect to $H$, so no decision rule based on $Y$ can outperform an optimal decision rule based on $T$ (Theorem 20.11.5). (Of course, no decision rule based on $T$ can outperform an optimal one based on $Y$, because $T$ is the result of processing $Y$ with respect to $H$.) In fact, using the terminology of Section 20.12, $T: y \mapsto y - 1/2$ forms a sufficient statistic for guessing $H$ based on $Y$, because the likelihood-ratio function $\mathrm{LR}(y_{\mathrm{obs}}) = f_{Y|H=0}(y_{\mathrm{obs}})/f_{Y|H=1}(y_{\mathrm{obs}})$ can be expressed as $\zeta(T(y_{\mathrm{obs}}))$ for the mapping $\zeta: t \mapsto \mathrm{LR}(t + 1/2)$. Consequently, our assertion that there is no loss in optimality in forming our guess based on $T(Y)$ is just a consequence of Proposition 20.12.3.

Conditional on $H = 0$, the random variable $T(Y)$ is $\mathcal{N}(-0.5, \sigma^2)$, and, conditional on $H = 1$, it is $\mathcal{N}(+0.5, \sigma^2)$. Consequently, using the results of Section 20.10 (with the substitution of $1/2$ for $A$), we obtain that an optimal rule based on $T$ is to guess "$H = 0$" if $T$ is negative, and to guess "$H = 1$" if $T$ is positive. To summarize, the decision rule we derived is to guess "$H = 0$" if $Y - 1/2 < 0$ and to guess "$H = 1$" if $Y - 1/2 > 0$.

In the terminology of Section 20.12, we used the fact that the transformation in (20.82) is one-to-one to conclude that $T(\cdot)$ forms a sufficient statistic, and we then used Method 2 from that section to derive an optimal decision rule.

## 20.14 Multi-Dimensional Binary Gaussian Hypothesis Testing

We now come closer to the receiver front end. The kind of problem we would eventually like to address is the hypothesis testing problem in which, conditional on $H = 0$, the observable is a continuous-time waveform of the form $s_0(t) + N(t)$ whereas, conditional on $H = 1$, it is of the form $s_1(t) + N(t)$, where $(N(t), t \in \mathbb{R})$ is some continuous-time stochastic process modeling the noise. This problem will be addressed in Chapter 26. For now we only address the discrete time version of this problem.

### 20.14.1 The Setup

We consider the problem of guessing the random variable $H$ that takes on the values 0 and 1 with positive probabilities $\pi_0$ and $\pi_1$. The observable $\mathbf{Y} \in \mathbb{R}^J$ is a random vector with $J$ components $Y^{(1)}, \ldots, Y^{(J)}$.[8] Conditional on $H = 0$, the components of $\mathbf{Y}$ are independent Gaussians with $Y^{(j)} \sim \mathcal{N}\left(s_0^{(j)}, \sigma^2\right)$, where $\mathbf{s}_0$ is some deterministic vector of $J$ components $s_0^{(1)}, \ldots, s_0^{(J)}$, and where $\sigma^2 > 0$. Conditional on $H = 1$, the components of $\mathbf{Y}$ are independent with $Y^{(j)} \sim \mathcal{N}\left(s_1^{(j)}, \sigma^2\right)$, for some other deterministic vector $\mathbf{s}_1$ of $J$ components $s_1^{(1)}, \ldots, s_1^{(J)}$. We assume that $\mathbf{s}_0$ and $\mathbf{s}_1$ differ in at least one coordinate. The setup can be described as

$$H = 0 : Y^{(j)} = s_0^{(j)} + Z^{(j)}, \quad j = 1, 2, \ldots, J,$$

$$H = 1 : Y^{(j)} = s_1^{(j)} + Z^{(j)}, \quad j = 1, 2, \ldots, J,$$

where $Z^{(1)}, Z^{(2)}, \ldots, Z^{(J)}$ are IID $\mathcal{N}\left(0, \sigma^2\right)$.

For typographical reasons, instead of denoting the observed vector by $\mathbf{y}_{\mathrm{obs}}$, we now denote it by $\mathbf{y}$ and its $J$ components by $y^{(1)}, \ldots, y^{(J)}$.

## 20.14.2 An Optimal Decision Rule

To find an optimal guessing rule we compute the likelihood-ratio function:

$$
\begin{aligned}
\mathrm{LR}(\mathbf{y}) &= \frac{f_{\mathbf{Y}|H=0}(\mathbf{y})}{f_{\mathbf{Y}|H=1}(\mathbf{y})} \\
&= \frac{\prod_{j=1}^{J}\left(\frac{1}{\sqrt{2\pi\sigma^2}} \exp\left(-\frac{\left(y^{(j)} - s_0^{(j)}\right)^2}{2\sigma^2}\right)\right)}{\prod_{j=1}^{J}\left(\frac{1}{\sqrt{2\pi\sigma^2}} \exp\left(-\frac{\left(y^{(j)} - s_1^{(j)}\right)^2}{2\sigma^2}\right)\right)} \\
&= \prod_{j=1}^{J}\left(\exp\left(-\frac{\left(y^{(j)} - s_0^{(j)}\right)^2}{2\sigma^2} + \frac{\left(y^{(j)} - s_1^{(j)}\right)^2}{2\sigma^2}\right)\right), \quad \mathbf{y} \in \mathbb{R}^J.
\end{aligned}
$$

The log likelihood-ratio function is thus given by

$$
\begin{aligned}
\mathrm{LLR}(\mathbf{y}) &= \ln \mathrm{LR}(\mathbf{y}) \\
&= \frac{1}{2\sigma^2} \sum_{j=1}^{J}\left(\left(y^{(j)} - s_1^{(j)}\right)^2 - \left(y^{(j)} - s_0^{(j)}\right)^2\right) \\
&= \frac{1}{\sigma^2}\left(\langle \mathbf{y}, \mathbf{s}_0 - \mathbf{s}_1\rangle_{\mathrm{E}} + \frac{\|\mathbf{s}_1\|^2 - \|\mathbf{s}_0\|^2}{2}\right) \\
&= \frac{1}{\sigma^2}\left(\langle \mathbf{y}, \mathbf{s}_0 - \mathbf{s}_1\rangle_{\mathrm{E}} - \frac{\langle \mathbf{s}_0, \mathbf{s}_0 - \mathbf{s}_1\rangle_{\mathrm{E}} + \langle \mathbf{s}_1, \mathbf{s}_0 - \mathbf{s}_1\rangle_{\mathrm{E}}}{2}\right) \\
&= \frac{\|\mathbf{s}_0 - \mathbf{s}_1\|}{\sigma^2}\left(\left\langle \mathbf{y}, \frac{\mathbf{s}_0 - \mathbf{s}_1}{\|\mathbf{s}_0 - \mathbf{s}_1\|}\right\rangle_{\mathrm{E}} - \frac{\left\langle \mathbf{s}_0, \frac{\mathbf{s}_0 - \mathbf{s}_1}{\|\mathbf{s}_0 - \mathbf{s}_1\|}\right\rangle_{\mathrm{E}} + \left\langle \mathbf{s}_1, \frac{\mathbf{s}_0 - \mathbf{s}_1}{\|\mathbf{s}_0 - \mathbf{s}_1\|}\right\rangle_{\mathrm{E}}}{2}\right) \\
&= \frac{\|\mathbf{s}_0 - \mathbf{s}_1\|}{\sigma^2}\left(\langle \mathbf{y}, \boldsymbol{\phi}\rangle_{\mathrm{E}} - \frac{1}{2}\left(\langle \mathbf{s}_0, \boldsymbol{\phi}\rangle_{\mathrm{E}} + \langle \mathbf{s}_1, \boldsymbol{\phi}\rangle_{\mathrm{E}}\right)\right), \quad \mathbf{y} \in \mathbb{R}^J, \qquad (20.83)
\end{aligned}
$$

---

[8]We use $J$ rather than $d$ in order to comply with the notation of Section 21.6 ahead.

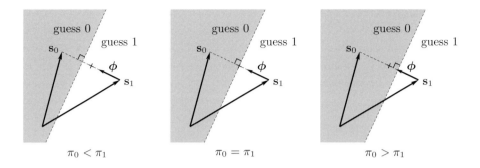

**Figure 20.5:** Effect of the ratio $\pi_0/\pi_1$ on the decision rule.

where for real vectors $\mathbf{u} = (u^{(1)}, \ldots, u^{(J)})^\mathsf{T}$ and $\mathbf{v} = (v^{(1)}, \ldots, v^{(J)})^\mathsf{T}$ taking value in $\mathbb{R}^J$ we define[9]

$$\langle \mathbf{u}, \mathbf{v} \rangle_\mathrm{E} \triangleq \sum_{j=1}^{J} u^{(j)} v^{(j)}, \tag{20.84}$$

$$\|\mathbf{u}\| \triangleq \sqrt{\langle \mathbf{u}, \mathbf{u} \rangle_\mathrm{E}} = \sqrt{\sum_{j=1}^{J} \left(u^{(j)}\right)^2}, \tag{20.85}$$

and where

$$\phi = \frac{\mathbf{s}_0 - \mathbf{s}_1}{\|\mathbf{s}_0 - \mathbf{s}_1\|} \tag{20.86}$$

is a unit-norm vector pointing from $\mathbf{s}_1$ to $\mathbf{s}_0$.

An optimal decision rule is to guess "$H = 0$" when $\mathrm{LLR}(\mathbf{y}) \geq \ln \frac{\pi_1}{\pi_0}$, i.e.,

$$\text{Guess "}H = 0\text{" if} \quad \langle \mathbf{y}, \phi \rangle_\mathrm{E} \geq \frac{\langle \mathbf{s}_0, \phi \rangle_\mathrm{E} + \langle \mathbf{s}_1, \phi \rangle_\mathrm{E}}{2} + \frac{\sigma^2}{\|\mathbf{s}_0 - \mathbf{s}_1\|} \ln \frac{\pi_1}{\pi_0}, \tag{20.87}$$

and to guess "$H = 1$" otherwise. This decision rule is illustrated in Figure 20.5. Depicted are the cases where $\pi_1/\pi_0$ is smaller than one, equal to one, and larger than one.

It is interesting to note that the projection $\langle \mathbf{y}, \phi \rangle_\mathrm{E} \, \phi$ of $\mathbf{y}$ onto the normalized vector $\phi = (\mathbf{s}_0 - \mathbf{s}_1)/\|\mathbf{s}_0 - \mathbf{s}_1\|$ forms a sufficient statistic for this problem. Indeed, by (20.83), the log likelihood-ratio (and hence the likelihood-ratio) function is computable from $\langle \mathbf{y}, \phi \rangle_\mathrm{E}$. The projection is depicted in Figure 20.6.

The rule (20.87) simplifies if $H$ has a uniform prior. In this case the rule is

$$\text{Guess "}H = 0\text{" if} \quad \langle \mathbf{y}, \phi \rangle_\mathrm{E} \geq \frac{\langle \mathbf{s}_0, \phi \rangle_\mathrm{E} + \langle \mathbf{s}_1, \phi \rangle_\mathrm{E}}{2}. \tag{20.88}$$

Note that in this case the guessing rule can be implemented even if $\sigma^2$ is unknown.

---

[9]This is sometimes called the standard inner product on $\mathbb{R}^J$ or the inner product between J-tuples. The subscript "E" stands here for "Euclidean."

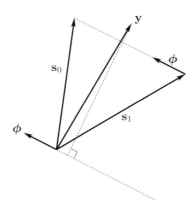

**Figure 20.6:** The projection of $\mathbf{y}$ onto the normalized vector $\phi = (\mathbf{s}_0 - \mathbf{s}_1)/\|\mathbf{s}_0 - \mathbf{s}_1\|$.

### 20.14.3 Error Probability Analysis

We next find the error probability associated with our guessing rule. We denote the conditional probabilities of error associated with our guessing rule by $p_{\mathrm{MAP}}(\mathrm{error}|H = 0)$ and $p_{\mathrm{MAP}}(\mathrm{error}|H = 1)$. Since our rule is optimal, its unconditional probability of error is $p^*(\mathrm{error})$ and is given by

$$p^*(\mathrm{error}) = \pi_0\, p_{\mathrm{MAP}}(\mathrm{error}|H = 0) + \pi_1\, p_{\mathrm{MAP}}(\mathrm{error}|H = 1). \tag{20.89}$$

Because in (20.87) we resolved ties by guessing "$H = 0$", it follows that to evaluate $p_{\mathrm{MAP}}(\mathrm{error}|H = 0)$ we need to evaluate the probability that a random vector $\mathbf{Y}$ drawn according to the density $f_{\mathbf{Y}|H=0}(\cdot)$ is such that the *a posteriori* probability of $H = 0$ is strictly smaller than the *a posteriori* probability of $H = 1$. Thus, if ties in the *a posteriori* distribution of $H$ are resolved in favor of guessing "$H = 0$", then

$$p_{\mathrm{MAP}}(\mathrm{error}|H = 0) = \Pr\!\left[\pi_0 f_{\mathbf{Y}|H=0}(\mathbf{Y}) < \pi_1 f_{\mathbf{Y}|H=1}(\mathbf{Y}) \,\middle|\, H = 0\right]. \tag{20.90}$$

This may seem self-referential, but it is not. Another way to state this is

$$p_{\mathrm{MAP}}(\mathrm{error}|H = 0) = \int_{\mathbf{y} \notin \mathcal{B}_{1,0}} f_{\mathbf{Y}|H=0}(\mathbf{y})\,\mathrm{d}\mathbf{y}, \tag{20.91}$$

where

$$\mathcal{B}_{1,0} = \left\{\mathbf{y} \in \mathbb{R}^{\mathsf{J}} : \pi_0 f_{\mathbf{Y}|H=0}(\mathbf{y}) \geq \pi_1 f_{\mathbf{Y}|H=1}(\mathbf{y})\right\}. \tag{20.92}$$

To compute this probability we need the following lemma:

**Lemma 20.14.1.** *Let $\pi_0$ and $\pi_1$ be strictly positive but not necessarily sum to one. Let the vectors $\mathbf{s}_0, \mathbf{s}_1 \in \mathbb{R}^{\mathsf{J}}$ differ in at least one component, i.e., $\|\mathbf{s}_0 - \mathbf{s}_1\| > 0$. Let*

$$f_0(\mathbf{y}) = \left(\frac{1}{\sqrt{2\pi\sigma^2}}\right)^{\mathsf{J}} \exp\left(-\frac{1}{2\sigma^2}\sum_{j=1}^{\mathsf{J}}\left(y^{(j)} - s_0^{(j)}\right)^2\right), \quad \mathbf{y} \in \mathbb{R}^{\mathsf{J}},$$

$$f_1(\mathbf{y}) = \left(\frac{1}{\sqrt{2\pi\sigma^2}}\right)^{\mathsf{J}} \exp\left(-\frac{1}{2\sigma^2}\sum_{j=1}^{\mathsf{J}}\left(y^{(j)} - s_1^{(j)}\right)^2\right), \quad \mathbf{y} \in \mathbb{R}^{\mathsf{J}},$$

*where* $\sigma^2 > 0$. *Define*

$$\mathcal{B}_{1,0} \triangleq \left\{\mathbf{y} \in \mathbb{R}^{\mathsf{J}} : \pi_0 f_0(\mathbf{y}) \geq \pi_1 f_1(\mathbf{y})\right\}.$$

*Then*

$$\int_{\mathbf{y}\notin\mathcal{B}_{1,0}} f_0(\mathbf{y})\,\mathrm{d}\mathbf{y} = \mathcal{Q}\left(\frac{\|\mathbf{s}_0 - \mathbf{s}_1\|}{2\sigma} + \frac{\sigma}{\|\mathbf{s}_0 - \mathbf{s}_1\|}\ln\frac{\pi_0}{\pi_1}\right). \tag{20.93}$$

*This equality continues to hold if we replace the weak inequality* $(\geq)$ *in the definition of* $\mathcal{B}_{1,0}$ *with a strict inequality* $(>)$.

**Proof.** Using a calculation identical to the one leading to (20.83) we obtain that the set $\mathcal{B}_{1,0}$ can also be expressed as

$$\mathcal{B}_{1,0} = \left\{\mathbf{y} \in \mathbb{R}^{\mathsf{J}} : \langle\mathbf{y}, \boldsymbol{\phi}\rangle_{\mathrm{E}} \geq \frac{\langle\mathbf{s}_0, \boldsymbol{\phi}\rangle_{\mathrm{E}} + \langle\mathbf{s}_1, \boldsymbol{\phi}\rangle_{\mathrm{E}}}{2} + \frac{\sigma^2}{\|\mathbf{s}_0 - \mathbf{s}_1\|}\ln\frac{\pi_1}{\pi_0}\right\}, \tag{20.94}$$

where $\boldsymbol{\phi}$ is defined in (20.86).

The density $f_0(\cdot)$ is the same as the density of the vector $\mathbf{s}_0 + \mathbf{Z}$, where the components $Z^{(1)}, \ldots, Z^{(\mathsf{J})}$ of $\mathbf{Z}$ are IID $\mathcal{N}(0, \sigma^2)$. Thus, the LHS of (20.93) can be expressed as

$$
\begin{aligned}
\int_{\mathbf{y}\notin\mathcal{B}_{1,0}} f_0(\mathbf{y})\,\mathrm{d}\mathbf{y} &= \Pr\left[\langle\mathbf{s}_0 + \mathbf{Z}, \boldsymbol{\phi}\rangle_{\mathrm{E}} < \frac{\langle\mathbf{s}_0, \boldsymbol{\phi}\rangle_{\mathrm{E}} + \langle\mathbf{s}_1, \boldsymbol{\phi}\rangle_{\mathrm{E}}}{2} + \frac{\sigma^2}{\|\mathbf{s}_0 - \mathbf{s}_1\|}\ln\frac{\pi_1}{\pi_0}\right] \\
&= \Pr\left[\langle\mathbf{Z}, \boldsymbol{\phi}\rangle_{\mathrm{E}} < \frac{\langle\mathbf{s}_1, \boldsymbol{\phi}\rangle_{\mathrm{E}} - \langle\mathbf{s}_0, \boldsymbol{\phi}\rangle_{\mathrm{E}}}{2} + \frac{\sigma^2}{\|\mathbf{s}_0 - \mathbf{s}_1\|}\ln\frac{\pi_1}{\pi_0}\right] \\
&= \Pr\left[-\langle\mathbf{Z}, \boldsymbol{\phi}\rangle_{\mathrm{E}} > \frac{\langle\mathbf{s}_0, \boldsymbol{\phi}\rangle_{\mathrm{E}} - \langle\mathbf{s}_1, \boldsymbol{\phi}\rangle_{\mathrm{E}}}{2} + \frac{\sigma^2}{\|\mathbf{s}_0 - \mathbf{s}_1\|}\ln\frac{\pi_0}{\pi_1}\right] \\
&= \Pr\left[-\langle\mathbf{Z}, \boldsymbol{\phi}\rangle_{\mathrm{E}} > \frac{\langle\mathbf{s}_0 - \mathbf{s}_1, \boldsymbol{\phi}\rangle_{\mathrm{E}}}{2} + \frac{\sigma^2}{\|\mathbf{s}_0 - \mathbf{s}_1\|}\ln\frac{\pi_0}{\pi_1}\right] \\
&= \Pr\left[\langle\mathbf{Z}, -\boldsymbol{\phi}\rangle_{\mathrm{E}} > \frac{\|\mathbf{s}_0 - \mathbf{s}_1\|}{2} + \frac{\sigma^2}{\|\mathbf{s}_0 - \mathbf{s}_1\|}\ln\frac{\pi_0}{\pi_1}\right] \\
&= \mathcal{Q}\left(\frac{\|\mathbf{s}_0 - \mathbf{s}_1\|}{2\sigma} + \frac{\sigma}{\|\mathbf{s}_0 - \mathbf{s}_1\|}\ln\frac{\pi_0}{\pi_1}\right),
\end{aligned}
$$

where the first equality follows from (20.94) and from the observation that the density $f_0(\cdot)$ is the density of $\mathbf{s}_0 + \mathbf{Z}$; the second because $\langle\cdot,\cdot\rangle_{\mathrm{E}}$ in linear in the first argument, so $\langle\mathbf{s}_0 + \mathbf{Z}, \boldsymbol{\phi}\rangle_{\mathrm{E}} = \langle\mathbf{s}_0, \boldsymbol{\phi}\rangle_{\mathrm{E}} + \langle\mathbf{Z}, \boldsymbol{\phi}\rangle_{\mathrm{E}}$; the third by noting that multiplying both sides of an inequality by $(-1)$ requires changing the direction of the inequality; the fourth by the linear relationship $\langle\mathbf{s}_1, \boldsymbol{\phi}\rangle_{\mathrm{E}} - \langle\mathbf{s}_0, \boldsymbol{\phi}\rangle_{\mathrm{E}} = \langle\mathbf{s}_1 - \mathbf{s}_0, \boldsymbol{\phi}\rangle_{\mathrm{E}}$; the fifth by (20.86); and the final equality because, as we next argue, $\langle\mathbf{Z}, -\boldsymbol{\phi}\rangle_{\mathrm{E}} \sim \mathcal{N}(0, \sigma^2)$, so we can employ (19.12a). To see that $\langle\mathbf{Z}, -\boldsymbol{\phi}\rangle_{\mathrm{E}} \sim \mathcal{N}(0, \sigma^2)$, note that, by (20.86), $\|-\boldsymbol{\phi}\| = 1$ and then employ Proposition 19.7.3.

This establishes the first part of the lemma. The result where the weak inequality is replaced with a strict inequality follows by replacing all the weak inequalities in the proof with the corresponding strict inequalities and vice versa. (If $X$ has a density, then $\Pr[X < \xi] = \Pr[X \leq \xi]$.) $\qquad\square$

By applying Lemma 20.14.1 to our problem we obtain

$$p_{\mathrm{MAP}}(\mathrm{error}|H = 0) = Q\left(\frac{\|\mathbf{s}_0 - \mathbf{s}_1\|}{2\sigma} + \frac{\sigma}{\|\mathbf{s}_0 - \mathbf{s}_1\|} \ln \frac{\pi_0}{\pi_1}\right). \qquad (20.95)$$

Similarly, one can show that

$$p_{\mathrm{MAP}}(\mathrm{error}|H = 1) = Q\left(\frac{\|\mathbf{s}_0 - \mathbf{s}_1\|}{2\sigma} + \frac{\sigma}{\|\mathbf{s}_0 - \mathbf{s}_1\|} \ln \frac{\pi_1}{\pi_0}\right). \qquad (20.96)$$

Consequently, by (20.89)

$$p^*(\mathrm{error}) = \pi_0 \, Q\left(\frac{\|\mathbf{s}_0 - \mathbf{s}_1\|}{2\sigma} + \frac{\sigma}{\|\mathbf{s}_0 - \mathbf{s}_1\|} \ln \frac{\pi_0}{\pi_1}\right)$$
$$+ \pi_1 \, Q\left(\frac{\|\mathbf{s}_0 - \mathbf{s}_1\|}{2\sigma} + \frac{\sigma}{\|\mathbf{s}_0 - \mathbf{s}_1\|} \ln \frac{\pi_1}{\pi_0}\right). \qquad (20.97)$$

In the special case where the prior is uniform we obtain from (20.95), (20.96), and (20.97)

$$p^*(\mathrm{error}) = p_{\mathrm{MAP}}(\mathrm{error}|H = 0) = p_{\mathrm{MAP}}(\mathrm{error}|H = 1) = Q\left(\frac{\|\mathbf{s}_0 - \mathbf{s}_1\|}{2\sigma}\right). \qquad (20.98)$$

This has a nice geometric interpretation. It is the probability that a $\mathcal{N}(0, \sigma^2)$ RV exceeds half the distance between the vectors $\mathbf{s}_0$ and $\mathbf{s}_1$. Stated differently, since $\|\mathbf{s}_0 - \mathbf{s}_1\| / \sigma$ is the number of standard deviations that separate $\mathbf{s}_0$ and $\mathbf{s}_1$, we can express the probability of error as the probability that a standard Gaussian exceeds half the distance between the vectors as measured in standard deviations of the noise.

### 20.14.4   The Bhattacharyya Bound

Finally, we compute the Bhattacharyya Bound for this problem. From (20.50) we obtain that, irrespective of the values of $\pi_0, \pi_1$,

$$p^*(\mathrm{error})$$
$$\leq \frac{1}{2} \int_{\mathbf{y} \in \mathbb{R}^J} \sqrt{f_{\mathbf{y}|H=0}(\mathbf{y}) f_{\mathbf{y}|H=1}(\mathbf{y})} \, \mathrm{d}\mathbf{y}$$
$$= \frac{1}{2} \int_{\mathbf{y}} \sqrt{\prod_{j=1}^{J}\left(\frac{1}{\sqrt{2\pi\sigma^2}} e^{-\frac{\left(y^{(j)} - s_0^{(j)}\right)^2}{2\sigma^2}}\right) \prod_{j=1}^{J}\left(\frac{1}{\sqrt{2\pi\sigma^2}} e^{-\frac{\left(y^{(j)} - s_1^{(j)}\right)^2}{2\sigma^2}}\right)} \, \mathrm{d}\mathbf{y}$$
$$= \frac{1}{2} \int_{\mathbf{y}} \sqrt{\prod_{j=1}^{J}\left(\frac{1}{\sqrt{2\pi\sigma^2}} e^{-\frac{\left(y^{(j)} - s_0^{(j)}\right)^2}{2\sigma^2}} \frac{1}{\sqrt{2\pi\sigma^2}} e^{-\frac{\left(y^{(j)} - s_1^{(j)}\right)^2}{2\sigma^2}}\right)} \, \mathrm{d}\mathbf{y}$$

$$
= \frac{1}{2} \int_{\mathbf{y}} \prod_{j=1}^{J} \left( \frac{1}{\sqrt{2\pi\sigma^2}} \exp\left( -\frac{\left(y^{(j)} - s_0^{(j)}\right)^2 + \left(y^{(j)} - s_1^{(j)}\right)^2}{4\sigma^2} \right) \right) d\mathbf{y}
$$

$$
= \frac{1}{2} \prod_{j=1}^{J} \int_{y^{(j)} \in \mathbb{R}} \frac{1}{\sqrt{2\pi\sigma^2}} e^{-\frac{2\left(y^{(j)}\right)^2 - 2y^{(j)}\left(s_0^{(j)} + s_1^{(j)}\right) + \left(s_0^{(j)}\right)^2 + \left(s_1^{(j)}\right)^2}{4\sigma^2}} dy^{(j)}
$$

$$
= \frac{1}{2} \prod_{j=1}^{J} \int_{-\infty}^{\infty} \frac{1}{\sqrt{2\pi\sigma^2}} \exp\left( -\frac{y^2 - y\left(s_0^{(j)} + s_1^{(j)}\right) + \frac{1}{2}\left(\left(s_0^{(j)}\right)^2 + \left(s_1^{(j)}\right)^2\right)}{2\sigma^2} \right) dy
$$

$$
= \frac{1}{2} \prod_{j=1}^{J} \int_{-\infty}^{\infty} \frac{1}{\sqrt{2\pi\sigma^2}} \exp\left( -\frac{\left(y - \frac{s_0^{(j)} + s_1^{(j)}}{2}\right)^2 + \frac{\left(s_0^{(j)} - s_1^{(j)}\right)^2}{4}}{2\sigma^2} \right) dy
$$

$$
= \frac{1}{2} \prod_{j=1}^{J} \exp\left( -\frac{\left(s_0^{(j)} - s_1^{(j)}\right)^2}{8\sigma^2} \right)
$$

$$
= \frac{1}{2} \exp\left( -\frac{1}{8\sigma^2} \sum_{j=1}^{J} \left(s_0^{(j)} - s_1^{(j)}\right)^2 \right)
$$

$$
= \frac{1}{2} \exp\left( -\frac{\|\mathbf{s}_0 - \mathbf{s}_1\|^2}{8\sigma^2} \right), \tag{20.99}
$$

where the last integral is evaluated using (19.22).

## 20.15 Guessing in the Presence of a Random Parameter

We now consider the guessing problem when the distribution of the observable $\mathbf{Y}$ depends not only on the hypothesis $H$ but also on a random parameter $\Theta$, which is independent of $H$. Based on the conditional densities $f_{\mathbf{Y}|\Theta,H=0}(\cdot), f_{\mathbf{Y}|\Theta,H=1}(\cdot)$, the nondegenerate prior $\pi_0, \pi_1 > 0$, and on the law of $\Theta$, we seek an optimal rule for guessing $H$. We distinguish between two cases depending on whether we must base our guess on the observed value $\mathbf{y}_{\text{obs}}$ of $\mathbf{Y}$ alone—**random parameter not observed**—or whether we also observe the value $\theta_{\text{obs}}$ of $\Theta$—**random parameter observed**. The analysis of both cases is conceptually straightforward.

### 20.15.1 Random Parameter Not Observed

The guessing problem when the random parameter is not observed is sometimes called "testing in the presence of a nuisance parameter." Conceptually, the situation is quite simple. We have only one observation, $\mathbf{Y} = \mathbf{y}_{\text{obs}}$, and an optimal decision rule is the MAP rule (Theorem 20.7.1). The MAP rule entails computing the likelihood-ratio function

$$
\text{LR}(\mathbf{y}_{\text{obs}}) = \frac{f_{\mathbf{Y}|H=0}(\mathbf{y}_{\text{obs}})}{f_{\mathbf{Y}|H=1}(\mathbf{y}_{\text{obs}})}, \tag{20.100}
$$

and comparing the result to the threshold $\pi_1/\pi_0$; see (20.40).

Often, however, the densities $f_{\mathbf{Y}|H=0}(\mathbf{y}_{\text{obs}})$ and $f_{\mathbf{Y}|H=1}(\mathbf{y}_{\text{obs}})$ appearing in (20.100) are not given directly. Instead we are given the density of $\Theta$ and the conditional density of $\mathbf{Y}$ given $(H, \Theta)$. (We shall encounter such a situation in Chapter 27 when we discuss noncoherent communications.) In such cases we can compute the conditional density $f_{\mathbf{Y}|H=0}(\mathbf{y}_{\text{obs}})$ as follows:

$$
\begin{aligned}
f_{\mathbf{Y}|H=0}(\mathbf{y}_{\text{obs}}) &= \int_\theta f_{\mathbf{Y},\Theta|H=0}(\mathbf{y}_{\text{obs}}, \theta) \, d\theta \\
&= \int_\theta f_{\mathbf{Y}|\Theta=\theta,H=0}(\mathbf{y}_{\text{obs}}) f_{\Theta|H=0}(\theta) \, d\theta \\
&= \int_\theta f_{\mathbf{Y}|\Theta=\theta,H=0}(\mathbf{y}_{\text{obs}}) f_{\Theta}(\theta) \, d\theta,
\end{aligned}
\tag{20.101}
$$

where the first equality follows because from the joint density one obtains the marginal density by integrating out the variable in which we are not interested; the second by the definition of the conditional density; and the final equality from our assumption that $\Theta$ and $H$ are independent. (In computations such as these it is best to think about the conditioning on $H = 0$ as defining a new law on $(\mathbf{Y}, \Theta)$—a new law to which all the regular probabilistic manipulations, such as marginalization and computation of conditional densities, continue to apply. We thus simply think of the conditioning on $H = 0$ as specifying the joint law of $(\mathbf{Y}, \Theta)$ that we have in mind.)

Repeating the calculation under $H = 1$ we obtain that the likelihood-ratio function is given by

$$
\boxed{
\text{LR}(\mathbf{y}_{\text{obs}}) = \frac{\int_\theta f_{\mathbf{Y}|\Theta=\theta,H=0}(\mathbf{y}_{\text{obs}}) f_{\Theta}(\theta) \, d\theta}{\int_\theta f_{\mathbf{Y}|\Theta=\theta,H=1}(\mathbf{y}_{\text{obs}}) f_{\Theta}(\theta) \, d\theta}.
}
\tag{20.102}
$$

The case where $\Theta$ is discrete can be similarly addressed. An optimal decision rule can now be derived based on this expression for the likelihood-ratio function and on the MAP rule (20.40).

## 20.15.2   Random Parameter Observed

When the random parameter is observed to be $\Theta = \theta_{\text{obs}}$, we merely view the problem as a standard hypothesis testing problem with the observation consisting of $\mathbf{Y}$ and $\Theta$. That is, we base our decision on the likelihood-ratio function

$$
\text{LR}(\mathbf{y}_{\text{obs}}, \theta_{\text{obs}}) = \frac{f_{\mathbf{Y},\Theta|H=0}(\mathbf{y}_{\text{obs}}, \theta_{\text{obs}})}{f_{\mathbf{Y},\Theta|H=1}(\mathbf{y}_{\text{obs}}, \theta_{\text{obs}})}.
\tag{20.103}
$$

The additional twist is that because $\Theta$ is independent of $H$ we have

$$
\begin{aligned}
f_{\mathbf{Y},\Theta|H=0}(\mathbf{y}_{\text{obs}}, \theta_{\text{obs}}) &= f_{\Theta|H=0}(\theta_{\text{obs}}) f_{\mathbf{Y}|\Theta=\theta_{\text{obs}},H=0}(\mathbf{y}_{\text{obs}}) \\
&= f_{\Theta}(\theta_{\text{obs}}) f_{\mathbf{Y}|\Theta=\theta_{\text{obs}},H=0}(\mathbf{y}_{\text{obs}}),
\end{aligned}
\tag{20.104}
$$

where the second equality follows from the independence of $\Theta$ and $H$. Repeating for the conditional law of the pair $(\mathbf{Y}, \Theta)$ given $H = 1$ we have

$$f_{\mathbf{Y},\Theta|H=1}(\mathbf{y}_{\text{obs}}, \theta_{\text{obs}}) = f_{\Theta}(\theta_{\text{obs}}) f_{\mathbf{Y}|\Theta=\theta_{\text{obs}},H=1}(\mathbf{y}_{\text{obs}}). \qquad (20.105)$$

Consequently, by (20.103), (20.104), and (20.105), we obtain that for $\theta_{\text{obs}}$ satisfying $f_{\Theta}(\theta_{\text{obs}}) \neq 0$

$$\boxed{\text{LR}(\mathbf{y}_{\text{obs}}, \theta_{\text{obs}}) = \frac{f_{\mathbf{Y}|H=0,\Theta=\theta_{\text{obs}}}(\mathbf{y}_{\text{obs}})}{f_{\mathbf{Y}|H=1,\Theta=\theta_{\text{obs}}}(\mathbf{y}_{\text{obs}})}.} \qquad (20.106)$$

An optimal decision rule can be again derived based on this expression for the likelihood-ratio and on the MAP rule (20.40).

## 20.16 Mathematical Notes

A standard reference on hypothesis testing is (Lehmann and Romano, 2005). It also contains a measure-theoretic treatment of the subject. For a precise mathematical definition of the condition $X$—o—$Y$—o—$Z$ we refer the reader to (Loève, 1963, Section 25.3). For a measure-theoretic treatment of sufficient statistic see (Loève, 1963, Section 24.4), (Billingsley, 1995, Section 34), (Romano and Siegel, 1986, pp. 154–156), and (Halmos and Savage, 1949). For a measure-theoretic treatment of the notion of conditional distribution see, for example, (Billingsley, 1995, Chapter 6), (Williams, 1991, Chapter 9), or (Lehmann and Romano, 2005, Chapter 2).

## 20.17 Exercises

**Exercise 20.1 (Hypothesis Testing).** Let $H$ take on the values 0 and 1 equiprobably. Conditional on $H = 0$, the observable $Y$ is equal to $a + Z$, where $Z$ is a Laplace RV, i.e., is of density

$$f_Z(z) = \frac{1}{2} e^{-|z|}, \ z \in \mathbb{R},$$

and $a > 0$ is a given constant. Conditional on $H = 1$, the observable $Y$ is given by $-a + Z$.

  (i) Find and draw the densities $f_{Y|H=0}(\cdot)$ and $f_{Y|H=1}(\cdot)$.

  (ii) Find an optimal rule for guessing $H$ based on $Y$.

  (iii) Compute the optimal probability of error.

  (iv) Compute the Bhattacharyya Bound.

**Exercise 20.2 (A Discrete Multi-Dimensional Problem).** Let $H$ take on the values 0 and 1 according to the prior $(\pi_0, \pi_1)$. Let the observation $\mathbf{Y} = (Y_1, \ldots, Y_n)^\mathsf{T}$ be an $n$-dimensional binary vector. Conditional on $H = 0$, the components of the vector $\mathbf{Y}$ are IID with

$$\Pr[Y_\ell = 1 \mid H = 0] = 1 - \Pr[Y_\ell = 0 \mid H = 0] = 0.25, \quad \ell = 1, \ldots, n.$$

Conditional on $H = 1$, the components are IID with

$$\Pr[Y_\ell = 1 \mid H = 1] = 1 - \Pr[Y_\ell = 0 \mid H = 1] = 0.75, \quad \ell = 1, \ldots, n.$$

(i) Find an optimal rule for guessing $H$ based on $\mathbf{Y}$.

(ii) Compute the optimal probability of error.

(iii) Compute the Bhattacharyya Bound.

*Hint: You may need to treat the cases of $n$ even and $n$ odd separately.*

**Exercise 20.3 (A Multi-Antenna Receiver).** Let $H$ take on the values 0 and 1 equiprobably. We wish to guess $H$ based on the random variables $Y_1$ and $Y_2$. Conditional on $H = 0$,

$$Y_1 = A + Z_1, \qquad Y_2 = A + Z_2,$$

and conditional on $H = 1$,

$$Y_1 = -A + Z_1, \qquad Y_2 = -A + Z_2.$$

Here $A$ is a positive constant, and $Z_1 \sim \mathcal{N}(0, \sigma_1^2)$, $Z_2 \sim \mathcal{N}(0, \sigma_2^2)$, and $H$ are independent.

(i) Find an optimal rule for guessing $H$ based on $(Y_1, Y_2)$.

(ii) Draw the decision regions in the $(Y_1, Y_2)$-plane for the special case where $\sigma_1 = 2\sigma_2$.

(iii) Returning to the general case, find a one-dimensional sufficient statistic.

(iv) Find the optimal probability of error in terms of $\sigma_1^2$, $\sigma_2^2$, and $A$.

(v) Consider a suboptimal receiver that declares "$H = 0$" if $Y_1 + Y_2 > 0$, and otherwise declares "$H = 1$." Evaluate the probability of error for this decoder as a function of $\sigma_1^2$, $\sigma_2^2$, and $A$.

**Exercise 20.4 (Binary Hypothesis Testing with General Costs).** Let $H$ take on the values 0 and 1 according to the prior $(\pi_0, \pi_1)$. The observable $\mathbf{Y}$ has conditional densities $f_{\mathbf{Y}|H=0}(\cdot)$ and $f_{\mathbf{Y}|H=1}(\cdot)$. Based on $\mathbf{Y}$, we wish to guess the value of $H$. Let the guess associated with $\mathbf{Y} = \mathbf{y}_{\mathrm{obs}}$ be denoted by $\phi_{\mathrm{Guess}}(\mathbf{y}_{\mathrm{obs}})$. Guessing "$H = \eta$" when $H = \nu$ costs $c(\eta, \nu)$, where $c(\cdot, \cdot)$ is a given function from $\{0, 1\} \times \{0, 1\}$ to the nonnegative reals. Find a decision rule that minimizes the expected cost

$$\mathsf{E}\Big[c\big(\phi_{\mathrm{Guess}}(\mathbf{Y}), H\big)\Big] = \sum_{\nu=0}^{1} \pi_\nu \sum_{\eta=0}^{1} c(\eta, \nu) \Pr\big[\phi_{\mathrm{Guess}}(\mathbf{Y}) = \eta \,\big|\, H = \nu\big].$$

**Exercise 20.5 (Binary Hypothesis Testing).** Let $H$ take on the values 0 and 1 according to the prior $(\pi_0, \pi_1)$, and let the observation consist of the RV $Y$. Conditional on $H$, the densities of $Y$ are given for every $y \in \mathbb{R}$ by

$$f_{Y|H=0}(y) = e^{-y}\, \mathrm{I}\{y \geq 0\}, \qquad f_{Y|H=1}(y) = \beta\, e^{-\frac{y^2}{2}}\, \mathrm{I}\{y \geq 0\},$$

where $\beta > 0$ is some constant.

(i) Determine $\beta$.

(ii) Find a decision rule that minimizes the probability of error.

(iii) For the rule that you have found, compute $\Pr(\mathrm{error}|H = 0)$.

*Hint: Different priors can lead to dramatically different decision rules.*

**Exercise 20.6 (Bhattacharyya Bound).**

(i) Show that the Bhattacharyya Bound never exceeds $1/2$.

(ii) When is it equal to $1/2$?

*Hint: You may find the Cauchy-Schwarz Inequality useful.*

**Exercise 20.7 (The Bhattacharyya Bound for Conditionally IID Observations).** Consider a binary hypothesis testing problem where, conditional on $H = 0$, the J components of the observed random vector $\mathbf{Y}$ are IID with each component of density $f_0(\cdot)$. Conditional on $H = 1$ the components of $\mathbf{Y}$ are IID with each component of density $f_1(\cdot)$. Express the Bhattacharyya Bound in terms of J and

$$\int_{\mathbb{R}} \sqrt{f_0(y)\, f_1(y)}\, \mathrm{d}y.$$

**Exercise 20.8 (Error Probability and $\mathcal{L}_1$-Distance).** Consider the setting of Theorem 20.5.2 when $H$ has a uniform prior. Show that in this case (20.26) can also be written as

$$\Pr[\phi_{\text{Guess}}^*(\mathbf{Y}) \neq H] = \frac{1}{2} - \frac{1}{4} \int_{\mathbb{R}^d} \left| f_{Y|H=0}(\mathbf{y}) - f_{Y|H=1}(\mathbf{y}) \right| \mathrm{d}\mathbf{y}.$$

**Exercise 20.9 (Conditionally Poisson Observations).** A RV $X$ is said to have a Poisson distribution of parameter ("intensity") $\lambda$, where $\lambda$ is some nonnegative real number, if $X$ takes value in the nonnegative integers and

$$\Pr[X = n] = e^{-\lambda} \frac{\lambda^n}{n!}, \quad n = 0, 1, 2, \ldots$$

(i) Find the Moment Generating Function of a Poisson RV of intensity $\lambda$.

(ii) Show that if $X$ and $Y$ are independent Poisson random variables of intensities $\lambda_x$ and $\lambda_y$, then their sum $X + Y$ is Poisson with parameter $\lambda_x + \lambda_y$.

(iii) Let $H$ take on the values 0 and 1 according to the prior $(\pi_0, \pi_1)$. We wish to guess $H$ based on the RV $Y$. Conditional on $H = 0$, the observation $Y$ is Poisson of intensity $\alpha + \lambda$, whereas conditional on $H = 1$ it is Poisson of intensity $\beta + \lambda$. Here $\alpha, \beta, \lambda$ are known non-negative constants. Show that the optimal probability of error is monotonically non-decreasing in $\lambda$.

*Hint: For Part (iii) recall Part (ii) and that no randomized decision rule can outperform an optimal deterministic rule.*

**Exercise 20.10 (Optical Communication).** Consider an optical communication system that uses binary on/off keying at a rate of $10^8$ bits per second. At the beginning of each time interval of duration $10^{-8}$ seconds a new data bit $D$ enters the transmitter. If $D = 0$, the laser is turned off for the duration of the interval; otherwise, if $D = 1$, the laser is turned on. The receiver counts the number $Y$ of photons received during the interval. Assume that, conditional on $D$, the observation $Y$ is a Poisson RV whose conditional PMF is

$$\Pr[Y = y \mid D = 0] = \frac{e^{-\mu} \mu^y}{y!}, \quad y = 0, 1, 2, \ldots, \tag{20.107}$$

$$\Pr[Y = y \mid D = 1] = \frac{e^{-\lambda} \lambda^y}{y!}, \quad y = 0, 1, 2, \ldots, \tag{20.108}$$

where $\lambda > \mu \geq 0$. Further assume that $\Pr[D = 0] = \Pr[D = 1] = 1/2$.

(i) Find an optimal guessing rule for guessing $D$ based on $Y$.

(ii) Compute the optimal probability of error. (Not necessarily in closed-form.)

(iii) Suppose that we now transmit each data bit over two time intervals, each of duration $10^{-8}$ seconds. (The system now supports a data rate of $0.5 \times 10^8$ bits per second.) The receiver produces the photon counts $Y_1$ and $Y_2$ over the two intervals. Assume that, conditional on $D = 0$, the counts $Y_1$ & $Y_2$ are IID with the PMF (20.107) and that, conditional on $D = 1$, they are IID with the PMF (20.108). Find a one-dimensional sufficient statistic for the problem and use it to find an optimal decision rule.

*Hint: For Part (iii), recall Part (ii) of Exercise 20.9.*

**Exercise 20.11 (Monotone Likelihood Ratio and Log-Concavity).** Let $H$ take on the values 0 and 1 according to the nondegenerate prior $(\pi_0, \pi_1)$. Conditional on $H = 0$, the observation $Y$ is given by

$$Y = \xi_0 + Z,$$

where $\xi_0 \in \mathbb{R}$ is some deterministic number and $Z$ is a RV of PDF $f_Z(\cdot)$. Conditional on $H = 1$, the observation $Y$ is given by

$$Y = \xi_1 + Z,$$

where $\xi_1 > \xi_0$.

(i) Show that if the PDF $f_Z(\cdot)$ is positive and is such that

$$f_Z(y_1 - \xi_0)\, f_Z(y_0 - \xi_1) \leq f_Z(y_1 - \xi_1)\, f_Z(y_0 - \xi_0), \quad \Big(y_1 > y_0, \ \xi_1 > \xi_0\Big), \quad (20.109)$$

then an optimal decision rule is to guess "$H = 0$" if $Y \leq y^\star$ and to guess "$H = 1$" if $Y > y^\star$ for some real number $y^\star$.

(ii) Show that if $z \mapsto \log f_Z(z)$ is a concave function, then (20.109) is satisfied.

Mathematicians state this result by saying that if $\mathbf{g} \colon \mathbb{R} \to \mathbb{R}$ is positive, then the mapping $(x, y) \mapsto g(x - y)$ has the Total Positivity property of Order 2 if, and only if, $\mathbf{g}$ is log-concave (Marshall and Olkin, 1979, Chapter 18, Section A, Example A.10). Statisticians state this result by saying that a location family generated by a positive PDF $f(\cdot)$ has monotone likelihood ratios if, and only if, $f(\cdot)$ is log-concave. For more on distributions with monotone likelihood ratios see (Lehmann and Romano, 2005, Chapter 3, Section 3.4).

*Hint: For Part (ii) recall that a function $\mathbf{g} \colon \mathbb{R} \mapsto \mathbb{R}$ is concave if for any $a < b$ and $0 < \alpha < 1$ we have $g(\alpha a + (1 - \alpha)b) \geq \alpha\, g(a) + (1 - \alpha)\, g(b)$. You may like to proceed as follows. Show that if $\mathbf{g}$ is concave then*

$$g(a - \Delta_2) + g(a + \Delta_2) \leq g(a - \Delta_1) + g(a + \Delta_1), \quad |\Delta_1| \leq |\Delta_2|.$$

*Defining $g(z) = \log f_Z(z)$, show that the logarithm of the LHS of (20.109) can be written as*

$$g\Big(\bar{y} - \bar{\xi} + \frac{1}{2}\Delta_y + \frac{1}{2}\Delta_\xi\Big) + g\Big(\bar{y} - \bar{\xi} - \frac{1}{2}\Delta_y - \frac{1}{2}\Delta_\xi\Big),$$

*where*

$$\bar{y} = (y_0 + y_1)/2, \quad \bar{\xi} = (\xi_0 + \xi_1)/2, \quad \Delta_y = y_1 - y_0, \quad \Delta_\xi = \xi_1 - \xi_0.$$

*Show that the logarithm of the RHS of (20.109) is given by*

$$g\Big(\bar{y} - \bar{\xi} + \frac{1}{2}\Delta_y - \frac{1}{2}\Delta_\xi\Big) + g\Big(\bar{y} - \bar{\xi} + \frac{1}{2}\Delta_\xi - \frac{1}{2}\Delta_y\Big).$$

**Exercise 20.12 (Is a Uniform Prior the Worst Prior?).** Based on an observation $Y$, we wish to guess the value of a RV $H$ taking on the values 0 and 1 according to the prior $(\pi_0, \pi_1)$. Conditional on $H = 0$, the observation $Y$ is uniform over the interval $[0, 1]$, and, conditional on $H = 1$, it is uniform over the interval $[0, 1/2]$.

(i) Find an optimal rule for guessing $H$ based on the observation $Y$. Note that the rule may depend on $\pi_0$.

(ii) Let $p^*(\text{error}; \pi_0)$ denote the optimal probability of error. Find $p^*(\text{error}; \pi_0)$ and plot it as a function of $\pi_0$ in the range $0 \leq \pi_0 \leq 1$.

(iii) Which value of $\pi_0$ maximizes $p^*(\text{error}; \pi_0)$?

Consider now the general problem where the RV $Y$ is of conditional densities $f_{Y|H=0}(\cdot)$, $f_{Y|H=1}(\cdot)$, and $H$ is of prior $(\pi_0, \pi_1)$. Let $p^*(\text{error}; \pi_0)$ denote the optimal probability of error for guessing $H$ based on $Y$.

(iv) Prove that

$$p^*\left(\text{error}; \frac{1}{2}\right) \geq \frac{1}{2} p^*(\text{error}; \pi_0) + \frac{1}{2} p^*(\text{error}; 1 - \pi_0), \quad \pi_0 \in [0, 1]. \qquad (20.110\text{a})$$

(v) Show that if the densities $f_{Y|H=0}(\cdot)$ and $f_{Y|H=1}(\cdot)$ satisfy

$$f_{Y|H=0}(y) = f_{Y|H=1}(-y), \quad y \in \mathbb{R}, \qquad (20.110\text{b})$$

then
$$p^*(\text{error}; \pi_0) = p^*(\text{error}; 1 - \pi_0), \quad \pi_0 \in [0, 1]. \qquad (20.110\text{c})$$

(vi) Show that if (20.110b) holds, then the uniform prior is the worst prior:

$$p^*(\text{error}; \pi_0) \leq p^*(\text{error}; 1/2), \quad \pi_0 \in [0, 1]. \qquad (20.110\text{d})$$

*Hint: For Part (iv) you might like to consider a new setup. In the new setup $\tilde{H} = M \oplus S$, where $\oplus$ denotes the exclusive-or operation and where the binary random variables $M$ and $S$ are independent with $S$ taking value in $\{0, 1\}$ equiprobably and with $\Pr[M = 0] = 1 - \Pr[M = 1] = \pi_0$. Assume that in the new setup $(M, S) - \!\!\circ\!\!- \tilde{H} - \!\!\circ\!\!- \tilde{Y}$ and that the conditional density of $\tilde{Y}$ given $\tilde{H} = 0$ is $f_{Y|H=0}(\cdot)$ and given $\tilde{H} = 1$ it is $f_{Y|H=1}(\cdot)$. Compare now the performance of an optimal decision rule for guessing $\tilde{H}$ based on $\tilde{Y}$ with the performance of an optimal decision rule for guessing $\tilde{H}$ based on the pair $(\tilde{Y}, S)$. Express these probabilities of error in terms of the parameters of the original problem.*

**Exercise 20.13 (Hypothesis Testing with a Random Parameter).** Let $Y = X + AZ$, where $X$, $A$, and $Z$ are independent random variables with $X$ taking on the values $\pm 1$ equiprobably, $A$ taking on the values 2 and 3 equiprobably, and $Z \sim \mathcal{N}(0, \sigma^2)$.

(i) Find an optimal rule for guessing $X$ based on the pair $(Y, A)$.

(ii) Repeat when you observe only $Y$.

**Exercise 20.14 (Bounding the Conditional Probability of Error).** Show that when the prior is uniform

$$p_{\text{MAP}}(\text{error}|H = 0) \leq \int \sqrt{f_{\mathbf{Y}|H=0}(\mathbf{y}) \, f_{\mathbf{Y}|H=1}(\mathbf{y})} \, d\mathbf{y}.$$

**Exercise 20.15 (Upper Bounds on the Conditional Probability of Error).**

(i) Let $H$ take on the values $0$ and $1$ according to the nondegenerate prior $(\pi_0, \pi_1)$. Let the observation $\mathbf{Y}$ have the conditional densities $f_{\mathbf{Y}|H=0}(\cdot)$ and $f_{\mathbf{Y}|H=1}(\cdot)$. Show that for every $\rho > 0$

$$p_{\mathrm{MAP}}(\mathrm{error}|H=0) \leq \left(\frac{\pi_1}{\pi_0}\right)^\rho \int f^\rho_{\mathbf{Y}|H=1}(\mathbf{y}) \, f^{1-\rho}_{\mathbf{Y}|H=0}(\mathbf{y}) \, \mathrm{d}\mathbf{y}.$$

(ii) A suboptimal decoder guesses "$H = 0$" if $q_0(\mathbf{y}) > q_1(\mathbf{y})$; guesses "$H = 1$" if $q_0(\mathbf{y}) < q_1(\mathbf{y})$; and otherwise tosses a coin. Here $q_0(\cdot)$ and $q_1(\cdot)$ are arbitrary positive functions. Show that for this decoder

$$p(\mathrm{error}|H=0) \leq \int \left(\frac{q_1(\mathbf{y})}{q_0(\mathbf{y})}\right)^\rho f_{\mathbf{Y}|H=0}(\mathbf{y}) \, \mathrm{d}\mathbf{y}, \quad \rho > 0.$$

*Hint: In Part (i) show that you can upper-bound* $\mathrm{I}\{\pi_1 \, f_{\mathbf{Y}|H=1}(\mathbf{y})/(\pi_0 \, f_{\mathbf{Y}|H=0}(\mathbf{y})) \geq 1\}$ *by* $\left(\pi_1 \, f_{\mathbf{Y}|H=1}(\mathbf{y})/(\pi_0 \, f_{\mathbf{Y}|H=0}(\mathbf{y}))\right)^\rho$.

**Exercise 20.16 (The Hellinger Distance).** The Hellinger distance between the densities $f(\cdot)$ and $g(\cdot)$ is defined as the square root of

$$\frac{1}{2} \int \left(\sqrt{f(\xi)} - \sqrt{g(\xi)}\right)^2 \mathrm{d}\xi$$

(though some authors drop the one-half).

(i) Show that the Hellinger distance between $f(\cdot)$ and $h(\cdot)$ is upper-bounded by the sum of the Hellinger distances between $f(\cdot)$ and $g(\cdot)$ and between $g(\cdot)$ and $h(\cdot)$.

(ii) Relate the Hellinger distance to the Bhattacharyya Bound.

(iii) Show that the Hellinger distance is upper-bounded by one.

**Exercise 20.17 (Artifacts of Suboptimality).** Let $H$ take on the values $0$ and $1$ equiprobably. Conditional on $H = 0$, the observation $Y$ is $\mathcal{N}(1, \sigma^2)$, and, conditional on $H = 1$, it is $\mathcal{N}(-1, \sigma^2)$. Alice guesses "$H = 0$" if $Y > 2$ and guesses "$H = 1$" otherwise.

(i) Compute the probability that Alice errs as a function of $\sigma^2$.

(ii) Show that this probability is not monotonically nondecreasing in $\sigma^2$.

(iii) Does her guessing rule minimize the probability of error?

(iv) Show that if you are obliged to use her rule, then adding noise to $Y$ prior to feeding it to her detector may be beneficial.

**Exercise 20.18 (The Bhattacharyya Bound and a Random Parameter).** Let $\Theta$ be independent of $H$ and of density $f_\Theta(\cdot)$. Express the Bhattacharyya Bound on the probability of guessing $H$ incorrectly in terms of $f_\Theta(\cdot)$, $f_{\mathbf{Y}|\Theta=\theta, H=0}(\cdot)$ and $f_{\mathbf{Y}|\Theta=\theta, H=1}(\cdot)$. Treat the case where $\Theta$ is not observed and the case where it is observed separately. Show that the Bhattacharyya Bound in the former case is always at least as large as in the latter case.

# Chapter 21

# Multi-Hypothesis Testing

## 21.1  Introduction

In Chapter 20 we discussed how to guess the outcome of a binary random variable. We now extend the discussion to random variables that take on more than two—but still a finite—number of values. Statisticians call this problem "multi-hypothesis testing" to indicate that there may be more than two hypotheses. Rather than using $H$, we now denote the random variable whose outcome we wish to guess by $M$. (In Chapter 20 we used $H$ for "hypothesis;" now we use $M$ for "message.") We denote the number of possible values that $M$ can take by $\mathsf{M}$ and assume that $\mathsf{M} \geq 2$. (The case $\mathsf{M} = 2$ corresponds to binary hypothesis testing.) As before the "labels" are not important and there is no loss in generality in assuming that $M$ takes value in the set $\mathcal{M} = \{1, \ldots, \mathsf{M}\}$. (In the binary case we used the traditional labels 0 and 1 but now we prefer $1, 2, \ldots, \mathsf{M}$.)

## 21.2  The Setup

A random variable $M$ takes value in the set $\mathcal{M} = \{1, \ldots, \mathsf{M}\}$, where $\mathsf{M} \geq 2$ according to the **prior**

$$\pi_m = \Pr[M = m], \quad m \in \mathcal{M}, \tag{21.1}$$

where

$$\pi_m \geq 0, \quad m \in \mathcal{M}, \tag{21.2}$$

and where

$$\sum_{m \in \mathcal{M}} \pi_m = 1. \tag{21.3}$$

We say that the prior is **nondegenerate** if

$$\pi_m > 0, \quad m \in \mathcal{M}, \tag{21.4}$$

with the inequalities being strict, so $M$ can take on any value in $\mathcal{M}$ with positive probability. We say that the prior is **uniform** if

$$\pi_1 = \cdots = \pi_\mathsf{M} = \frac{1}{\mathsf{M}}. \tag{21.5}$$

The **observation** is a random vector $\mathbf{Y}$ taking value in $\mathbb{R}^d$. We assume that for each $m \in \mathcal{M}$ the distribution of $\mathbf{Y}$ conditional on $M = m$ has the density[1]

$$f_{\mathbf{Y}|M=m}(\cdot), \quad m \in \mathcal{M}, \tag{21.6}$$

where $f_{\mathbf{Y}|M=m}(\cdot)$ is a nonnegative Borel measurable function that integrates to one over $\mathbb{R}^d$.

A **guessing rule** is a Borel measurable function $\phi_{\mathrm{Guess}} \colon \mathbb{R}^d \to \mathcal{M}$ from the space of possible observations $\mathbb{R}^d$ to the set of possible messages $\mathcal{M}$. We think about $\phi_{\mathrm{Guess}}(\mathbf{y}_{\mathrm{obs}})$ as the guess we form after observing that $\mathbf{Y} = \mathbf{y}_{\mathrm{obs}}$. The **error probability** associated with the guessing rule $\phi_{\mathrm{Guess}}(\cdot)$ is given by

$$\Pr\left[\phi_{\mathrm{Guess}}(\mathbf{Y}) \neq M\right]. \tag{21.7}$$

Note that two sources of randomness determine whether we err or not: the realization of $M$ and the generation of $\mathbf{Y}$ conditional on that realization. A guessing rule is said to be **optimal** if no other guessing rule achieves a lower probability of error.[2] The **optimal error probability** $p^*(\mathrm{error})$ is the probability of error associated with an optimal decision rule. In this chapter we shall derive optimal decision rules and study the optimal probability of error.

## 21.3    Optimal Guessing

Having observed that $\mathbf{Y} = \mathbf{y}_{\mathrm{obs}}$, we would like to guess $M$. An optimal guessing rule can be derived, as in the binary case, by first considering the scenario where there are no observables. Its extension to the more interesting case where we observe $\mathbf{Y}$ is straightforward.

### 21.3.1    Guessing in the Absence of Observables

In this scenario there are only $\mathsf{M}$ deterministic decision rules to choose from: the decision rule "guess 1", the decision rule "guess 2", etc. If we employ the "guess 1" rule, then we are correct if $M$ is indeed equal to 1 and thus with probability of success $\pi_1$ and corresponding probability of error of $1 - \pi_1$. In general, if we employ the "guess $m$" rule for some $m \in \mathcal{M}$, then our probability of success is $\pi_m$. Thus, of the $\mathsf{M}$ different rules at our disposal, the one that has the highest probability of success is the "guess $\tilde{m}$" rule, where $\tilde{m}$ is the outcome that is *a priori* the most likely. If this $\tilde{m}$ is not unique, then guessing any one of the outcomes that have the highest *a priori* probability is optimal.

---

[1]We feel no remorse for limiting ourselves to conditional distributions possessing a density. The reason is that, while the reader is encouraged to assume that the densities are with respect to the Lebesgue measure, this assumption is never used in the text. And using the Radon-Nikodym Theorem (Billingsley, 1995, Section 32), one can show that even in the most general case there exists a measure on $\mathbb{R}^d$ with respect to which the conditional laws of $\mathbf{Y}$ conditional on each of the possible values of $M$ are absolutely continuous. That measure can be taken, for example, as the sum of the conditional laws corresponding to each of the possible values that $M$ can take.

[2]As in the case of binary hypothesis testing, an optimal guessing rule always exists.

We conclude that in the absence of observables, the guessing rule "guess $\tilde{m}$" is optimal if, and only if,

$$\pi_{\tilde{m}} = \max_{m' \in \mathcal{M}} \pi_{m'}. \tag{21.8}$$

For an optimal guessing rule the probability of success is

$$p^*(\text{correct}) = \max_{m' \in \mathcal{M}} \{\pi_{m'}\}, \tag{21.9}$$

and the optimal error probability is thus

$$p^*(\text{error}) = 1 - \max_{m' \in \mathcal{M}} \{\pi_{m'}\}. \tag{21.10}$$

### 21.3.2 The Joint Law of $M$ and $\mathbf{Y}$

Using the prior $\{\pi_m\}$ and the conditional densities $\{f_{\mathbf{Y}|M=m}(\cdot)\}$, we can express the unconditional density of $\mathbf{Y}$ as

$$f_{\mathbf{Y}}(\mathbf{y}) = \sum_{m \in \mathcal{M}} \pi_m f_{\mathbf{Y}|M=m}(\mathbf{y}), \quad \mathbf{y} \in \mathbb{R}^d. \tag{21.11}$$

As in Section 20.4, we define for every $m \in \mathcal{M}$ and for every $\mathbf{y}_{\text{obs}} \in \mathbb{R}^d$ the conditional probability that $M = m$ conditional on $\mathbf{Y} = \mathbf{y}_{\text{obs}}$ by

$$\Pr[M = m \mid \mathbf{Y} = \mathbf{y}_{\text{obs}}] \triangleq \begin{cases} \dfrac{\pi_m f_{\mathbf{Y}|M=m}(\mathbf{y}_{\text{obs}})}{f_{\mathbf{Y}}(\mathbf{y}_{\text{obs}})} & \text{if } f_{\mathbf{Y}}(\mathbf{y}_{\text{obs}}) > 0, \\ \dfrac{1}{M} & \text{otherwise.} \end{cases} \tag{21.12}$$

By an argument similar to the one proving (20.12) we have

$$\Pr\left[\mathbf{Y} \in \{\tilde{\mathbf{y}} \in \mathbb{R}^d : f_{\mathbf{Y}}(\tilde{\mathbf{y}}) = 0\}\right] = 0, \tag{21.13}$$

which can also be written as

$$\Pr\left[f_{\mathbf{Y}}(\mathbf{Y}) = 0\right] = 0.$$

### 21.3.3 Guessing in the Presence of Observables

The problem of guessing in the presence of an observable is very similar to the one without observables. The intuition is that after observing that $\mathbf{Y} = \mathbf{y}_{\text{obs}}$, we associate with each $m \in \mathcal{M}$ the *a posteriori* probability $\Pr[M = m \mid \mathbf{Y} = \mathbf{y}_{\text{obs}}]$ and then guess $M$ as though there were no observables. Thus, rather than choosing the message that has the highest *a priori* probability as we do in the absence of observables, we should now choose the message that has the highest *a posteriori* probability.

After having observed that $\mathbf{Y} = \mathbf{y}_{\text{obs}}$ we should thus guess "$\tilde{m}$" where $\tilde{m}$ is the outcome in $\mathcal{M}$ that has the highest *a posteriori* probability. If more than one outcome attains the highest *a posteriori* probability, then we say that a **tie** has occurred and we need to resolve this tie by picking one (it does not matter which) of the outcomes that attains the maximum *a posteriori* probability. We thus guess "$\tilde{m}$," in analogy to (21.8), only if

$$\Pr[M = \tilde{m} \mid \mathbf{Y} = \mathbf{y}_{\text{obs}}] = \max_{m' \in \mathcal{M}} \left\{ \Pr[M = m' \mid \mathbf{Y} = \mathbf{y}_{\text{obs}}] \right\}.$$

(We shall later define the Maximum A Posteriori guessing rule as a randomized decision rule that picks uniformly at random from the outcomes that have the highest *a posteriori* probability; see Definition 21.3.2 ahead.)

In analogy with (21.9) we have that for this optimal rule

$$p^*(\text{correct} \mid \mathbf{Y} = \mathbf{y}_{\text{obs}}) = \max_{m' \in \mathcal{M}} \left\{ \Pr[M = m' \mid \mathbf{Y} = \mathbf{y}_{\text{obs}}] \right\},$$

and in analogy with (21.10),

$$p^*(\text{error} \mid \mathbf{Y} = \mathbf{y}_{\text{obs}}) = 1 - \max_{m' \in \mathcal{M}} \left\{ \Pr[M = m' \mid \mathbf{Y} = \mathbf{y}_{\text{obs}}] \right\}.$$

Consequently, the unconditional optimal probability of error can be expressed as

$$p^*(\text{error}) = \int_{\mathbb{R}^d} \left( 1 - \max_{m' \in \mathcal{M}} \left\{ \Pr[M = m' \mid \mathbf{Y} = \mathbf{y}] \right\} \right) f_{\mathbf{Y}}(\mathbf{y}) \, \mathrm{d}\mathbf{y},$$

where $f_{\mathbf{Y}}(\cdot)$ is the unconditional density function of $\mathbf{Y}$ and is given in (21.11).

We next proceed to make the above intuitive discussion more rigorous. We begin by defining for every possible observation $\mathbf{y}_{\text{obs}} \in \mathbb{R}^d$ the set of outcomes of maximal *a posteriori* probability:

$$\tilde{\mathcal{M}}(\mathbf{y}_{\text{obs}}) \triangleq \left\{ \tilde{m} \in \mathcal{M} : \Pr[M = \tilde{m} \mid \mathbf{Y} = \mathbf{y}_{\text{obs}}] = \max_{m' \in \mathcal{M}} \Pr[M = m' \mid \mathbf{Y} = \mathbf{y}_{\text{obs}}] \right\}. \tag{21.14}$$

As we next argue, this set can also be expressed as

$$\tilde{\mathcal{M}}(\mathbf{y}_{\text{obs}}) = \left\{ \tilde{m} \in \mathcal{M} : \pi_{\tilde{m}} \, f_{\mathbf{Y} \mid M = \tilde{m}}(\mathbf{y}_{\text{obs}}) = \max_{m' \in \mathcal{M}} \pi_{m'} \, f_{\mathbf{Y} \mid M = m'}(\mathbf{y}_{\text{obs}}) \right\}. \tag{21.15}$$

This can be shown by treating the case $f_{\mathbf{Y}}(\mathbf{y}_{\text{obs}}) > 0$ and the case $f_{\mathbf{Y}}(\mathbf{y}_{\text{obs}}) = 0$ separately. In the former case, (21.15) is verified by noting that in this case we have, by (21.12), that $\Pr[M = m' \mid \mathbf{Y} = \mathbf{y}_{\text{obs}}] = \pi_{m'} \, f_{\mathbf{Y} \mid M = m'}(\mathbf{y}_{\text{obs}}) / f_{\mathbf{Y}}(\mathbf{y}_{\text{obs}})$, so the result follows because scaling the scores of all the elements of a set by a positive number that is common to them all ($1/f_{\mathbf{Y}}(\mathbf{y}_{\text{obs}})$) does not change the subset of the elements with the highest score. In the latter case we note that, by (21.12), we have for all $m' \in \mathcal{M}$ that $\Pr[M = m' \mid \mathbf{Y} = \mathbf{y}_{\text{obs}}] = 1/M$, so the RHS of (21.14) is $\mathcal{M}$ and we also have by (21.11) for all $m' \in \mathcal{M}$ that $\pi_{m'} \, f_{\mathbf{Y} \mid M = m'}(\mathbf{y}_{\text{obs}}) = 0$ so the RHS of (21.15) is also $\mathcal{M}$.

Using the above definition of $\tilde{\mathcal{M}}(\mathbf{y}_{\text{obs}})$ we can now state the main theorem regarding optimal guessing rules.

**Theorem 21.3.1 (Optimal Multi-Hypothesis Testing).** *Let $M$ take value in the set $\mathcal{M} = \{1, \ldots, M\}$ with the prior (21.1), and let the observation $\mathbf{Y}$ be a random vector taking value in $\mathbb{R}^d$ with conditional densities $f_{\mathbf{Y} \mid M = 1}(\cdot), \ldots, f_{\mathbf{Y} \mid M = M}(\cdot)$. Any guessing rule $\phi^*_{\text{Guess}} : \mathbb{R}^d \to \mathcal{M}$ that satisfies*

$$\phi^*_{\text{Guess}}(\mathbf{y}_{\text{obs}}) \in \tilde{\mathcal{M}}(\mathbf{y}_{\text{obs}}), \quad \mathbf{y}_{\text{obs}} \in \mathbb{R}^d \tag{21.16}$$

*is optimal. Here $\tilde{\mathcal{M}}(\mathbf{y}_{\text{obs}})$ is the set defined in (21.14) or (21.15).*

**Proof.** Every (deterministic) guessing rule induces a partitioning of the space of possible outcomes $\mathbb{R}^d$ into M disjoint sets $\mathcal{D}_1, \dots, \mathcal{D}_M$:

$$\bigcup_{m=1}^{M} \mathcal{D}_m = \mathbb{R}^d, \tag{21.17a}$$

$$\mathcal{D}_m \cap \mathcal{D}_{m'} = \emptyset, \quad m \neq m', \tag{21.17b}$$

where $\mathcal{D}_m$ is the set of observations that result in the guessing rule producing the guess "$M = m$." Conversely, every partition $\mathcal{D}_1, \dots, \mathcal{D}_M$ of $\mathbb{R}^d$ corresponds to some deterministic guessing rule that guesses "$M = m$" whenever $\mathbf{y}_{\text{obs}} \in \mathcal{D}_m$. Searching for an optimal decision rule is thus equivalent to searching for an optimal way to partition $\mathbb{R}^d$. For every partition $\mathcal{D}_1, \dots, \mathcal{D}_M$ the probability of success of the guessing rule associated with it is given by

$$\Pr(\text{correct}) = \sum_{m \in \mathcal{M}} \pi_m \Pr(\text{correct} \mid M = m)$$

$$= \sum_{m \in \mathcal{M}} \pi_m \int_{\mathcal{D}_m} f_{\mathbf{Y}|M=m}(\mathbf{y}) \, \mathrm{d}\mathbf{y}$$

$$= \sum_{m \in \mathcal{M}} \pi_m \int_{\mathbb{R}^d} f_{\mathbf{Y}|M=m}(\mathbf{y}) \, \mathrm{I}\{\mathbf{y} \in \mathcal{D}_m\} \, \mathrm{d}\mathbf{y}$$

$$= \int_{\mathbb{R}^d} \left( \sum_{m \in \mathcal{M}} \pi_m f_{\mathbf{Y}|M=m}(\mathbf{y}) \, \mathrm{I}\{\mathbf{y} \in \mathcal{D}_m\} \right) \mathrm{d}\mathbf{y}.$$

To minimize the probability of error we maximize the probability of correct decision. We thus need to find a partition $\mathcal{D}_1, \dots, \mathcal{D}_M$ that maximizes the last integral.

To maximize the integral we shall maximize the integrand

$$\sum_{m \in \mathcal{M}} \pi_m f_{\mathbf{Y}|M=m}(\mathbf{y}) \, \mathrm{I}\{\mathbf{y} \in \mathcal{D}_m\}.$$

For a fixed value of $\mathbf{y}$, the value of the integrand depends on the set to which we have assigned $\mathbf{y}$. If $\mathbf{y}$ was assigned to $\mathcal{D}_1$ (i.e., if $\mathbf{y} \in \mathcal{D}_1$), then all the terms in the sum except for the first are zero, and the value of the integrand is $\pi_1 f_{\mathbf{Y}|M=1}(\mathbf{y})$. More generally, if $\mathbf{y}$ was assigned to $\mathcal{D}_m$, then all the terms in the sum except for the $m$-th term are zero, and the value of the integrand is $\pi_m f_{\mathbf{Y}|M=m}(\mathbf{y})$. For a fixed value of $\mathbf{y}$, the integrand will thus be maximized if we assign $\mathbf{y}$ to the set $\mathcal{D}_{\tilde{m}}$ (and correspondingly guess $\tilde{m}$), only if

$$\pi_{\tilde{m}} f_{\mathbf{Y}|M=\tilde{m}}(\mathbf{y}) = \max_{m' \in \mathcal{M}} \left\{ \pi_{m'} f_{\mathbf{Y}|M=m'}(\mathbf{y}) \right\}.$$

Thus, if $\phi^*_{\text{Guess}}(\cdot)$ satisfies the theorem's hypotheses, then it maximizes the integrand for every $\mathbf{y} \in \mathbb{R}^d$ and thus also maximizes the probability of guessing correctly. $\square$

### 21.3.4 The MAP and ML Rules

As in the binary hypothesis testing case, we can also consider randomized decision rules. Extending the definition of a randomized decision rule to our setting, one can show using arguments very similar to those of Section 20.6 that randomization does not help: no randomized decision rule can yield a smaller probability of error than an optimal deterministic rule. But randomized decision rules can yield more symmetric or more "fair" rules. Indeed, we shall define the **MAP** rule as the randomized rule that resolves ties by choosing one of the messages that achieves the highest *a posteriori* probability uniformly at random:

**Definition 21.3.2 (The M-ary MAP Decision Rule).** *The **Maximum A Posteriori decision rule** is the guessing rule that, after observing that $\mathbf{Y} = \mathbf{y}_{\mathrm{obs}}$, forms a guess by picking uniformly at random an element of the set $\tilde{\mathcal{M}}(\mathbf{y}_{\mathrm{obs}})$, which is defined in (21.14) or (21.15).*

**Theorem 21.3.3 (The MAP Rule Is Optimal).** *For the setting of Theorem 21.3.1 the MAP decision rule is optimal in the sense that it achieves the smallest probability of error among all deterministic or randomized decision rules. Thus,*

$$p^*(\text{error}) = \sum_{m \in \mathcal{M}} \pi_m \, p_{\mathrm{MAP}}(\text{error}|M = m), \qquad (21.18)$$

*where $p^*(\text{error})$ denotes the optimal probability of error and $p_{\mathrm{MAP}}(\text{error}|M = m)$ denotes the conditional probability of error of the MAP rule.*

**Proof.** Irrespective of the realization of the randomization that is used to pick an element of $\tilde{\mathcal{M}}(\mathbf{y}_{\mathrm{obs}})$, the resulting decision rule is optimal (Theorem 21.3.1). Consequently, the average probability of error that results when we average over this source of randomness must also be optimal. $\qquad\square$

The **Maximum-Likelihood** (ML) rule ignores the prior. It is identical to the MAP rule when the prior is uniform. Having observed that $\mathbf{Y} = \mathbf{y}_{\mathrm{obs}}$, the ML decoder produces as its guess a member of the set

$$\left\{ \tilde{m} \in \mathcal{M} : f_{\mathbf{Y}|M=\tilde{m}}(\mathbf{y}_{\mathrm{obs}}) = \max_{m' \in \mathcal{M}} f_{\mathbf{Y}|M=m'}(\mathbf{y}_{\mathrm{obs}}) \right\}$$

that is drawn uniformly at random.

The ML decoder thus guesses "$M = \tilde{m}$" only if

$$f_{\mathbf{Y}|M=\tilde{m}}(\mathbf{y}_{\mathrm{obs}}) = \max_{m' \in \mathcal{M}} f_{\mathbf{Y}|M=m'}(\mathbf{y}_{\mathrm{obs}}). \qquad (21.19)$$

(If more than one outcome achieves this maximum, it chooses uniformly at random one of the outcomes that achieves the maximum.)

### 21.3.5 Processing

As in Section 20.11, we say that $\mathbf{Z}$ is the result of processing $\mathbf{Y}$ with respect to $M$ if

$$M \!-\!\!\circ\!\!-\! \mathbf{Y} \!-\!\!\circ\!\!-\! \mathbf{Z}$$

forms a Markov chain. In analogy to Theorem 20.11.5, one can prove that if $\mathbf{Z}$ is the result of processing $\mathbf{Y}$ with respect to $M$, then no decision rule based on $\mathbf{Z}$ can outperform an optimal decision rule based on $\mathbf{Y}$.

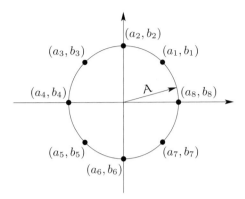

**Figure 21.1:** Eight equiprobable hypotheses; the situation corresponds to 8-PSK.

## 21.4   Example: Multi-Hypothesis Testing for 2D Signals

### 21.4.1   The Setup

Consider the case where $M$ is uniformly distributed over the set $\mathcal{M} = \{1, \ldots, \mathsf{M}\}$ and where we would like to guess the outcome of $M$ based on an observation consisting of a two-dimensional random vector $\mathbf{Y}$ of components $Y^{(1)}$ and $Y^{(2)}$. Conditional on $M = m$, the random variables $Y^{(1)}$ and $Y^{(2)}$ are independent with $Y^{(1)} \sim \mathcal{N}(a_m, \sigma^2)$ and $Y^{(2)} \sim \mathcal{N}(b_m, \sigma^2)$. We assume that $\sigma^2 > 0$, so the conditional densities can be written for every $m \in \mathcal{M}$ and every $y^{(1)}, y^{(2)} \in \mathbb{R}$ as

$$f_{Y^{(1)}, Y^{(2)}|M=m}\big(y^{(1)}, y^{(2)}\big) = \frac{1}{2\pi\sigma^2} \exp\left(-\frac{(y^{(1)} - a_m)^2 + (y^{(2)} - b_m)^2}{2\sigma^2}\right). \quad (21.20)$$

This hypothesis testing problem is related to QAM communication over an additive white Gaussian noise channel with a pulse shape that is orthogonal to its time shifts by integer multiples of the baud period. The setup is demonstrated in Figure 21.1 for the special case of $\mathsf{M} = 8$ with

$$a_m = \mathsf{A}\cos\left(\frac{2\pi m}{8}\right), \quad b_m = \mathsf{A}\sin\left(\frac{2\pi m}{8}\right), \quad m = 1, \ldots, 8. \quad (21.21)$$

This special case is related to 8-PSK communication, where M-PSK stands for M-ary Phase Shift Keying.

### 21.4.2   The "Nearest-Neighbor" Decoding Rule

We shall next derive an optimal decision rule. For typographical reasons we shall use $\mathbf{y}$ rather than $\mathbf{y}_{\text{obs}}$ to denote the observed vector. To find an optimal decoding rule we note that, since $M$ has a uniform prior, the Maximum-Likelihood rule

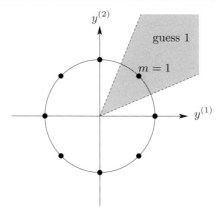

**Figure 21.2:** Shaded region corresponds to observations leading the ML rule to guess "$M = 1$."

(21.19) is optimal. Now $\tilde{m}$ maximizes the likelihood function if, and only if,

$$\left( f_{Y^{(1)},Y^{(2)}|M=\tilde{m}}(y^{(1)}, y^{(2)}) = \max_{m' \in \mathcal{M}} \left\{ f_{Y^{(1)},Y^{(2)}|M=m'}(y^{(1)}, y^{(2)}) \right\} \right)$$

$$\Leftrightarrow \left( \frac{1}{2\pi\sigma^2} e^{-\frac{\left(y^{(1)} - a_{\tilde{m}}\right)^2 + \left(y^{(2)} - b_{\tilde{m}}\right)^2}{2\sigma^2}} = \max_{m' \in \mathcal{M}} \left\{ \frac{1}{2\pi\sigma^2} e^{-\frac{\left(y^{(1)} - a_{m'}\right)^2 + \left(y^{(2)} - b_{m'}\right)^2}{2\sigma^2}} \right\} \right)$$

$$\Leftrightarrow \left( e^{-\frac{\left(y^{(1)} - a_{\tilde{m}}\right)^2 + \left(y^{(2)} - b_{\tilde{m}}\right)^2}{2\sigma^2}} = \max_{m' \in \mathcal{M}} \left\{ e^{-\frac{\left(y^{(1)} - a_{m'}\right)^2 + \left(y^{(2)} - b_{m'}\right)^2}{2\sigma^2}} \right\} \right)$$

$$\Leftrightarrow \left( -\frac{\left(y^{(1)} - a_{\tilde{m}}\right)^2 + \left(y^{(2)} - b_{\tilde{m}}\right)^2}{2\sigma^2} = \max_{m' \in \mathcal{M}} \left\{ -\frac{\left(y^{(1)} - a_{m'}\right)^2 + \left(y^{(2)} - b_{m'}\right)^2}{2\sigma^2} \right\} \right)$$

$$\Leftrightarrow \left( \frac{\left(y^{(1)} - a_{\tilde{m}}\right)^2 + \left(y^{(2)} - b_{\tilde{m}}\right)^2}{2\sigma^2} = \min_{m' \in \mathcal{M}} \left\{ \frac{\left(y^{(1)} - a_{m'}\right)^2 + \left(y^{(2)} - b_{m'}\right)^2}{2\sigma^2} \right\} \right)$$

$$\Leftrightarrow \left( \left(y^{(1)} - a_{\tilde{m}}\right)^2 + \left(y^{(2)} - b_{\tilde{m}}\right)^2 = \min_{m' \in \mathcal{M}} \left\{ \left(y^{(1)} - a_{m'}\right)^2 + \left(y^{(2)} - b_{m'}\right)^2 \right\} \right)$$

$$\Leftrightarrow \left( \|\mathbf{y} - \mathbf{s}_{\tilde{m}}\| = \min_{m' \in \mathcal{M}} \left\{ \|\mathbf{y} - \mathbf{s}_{m'}\| \right\} \right),$$

where $\mathbf{y} = (y^{(1)}, y^{(2)})^\mathsf{T}$, $\mathbf{s}_m \triangleq (a_m, b_m)^\mathsf{T}$ for $m \in \mathcal{M}$, and $\|\cdot\|$ denotes the Euclidean distance (23.4). It is thus seen that the ML rule (which is equivalent to the MAP rule because the prior is uniform) is equivalent to a "nearest-neighbor" decoding rule, which chooses the hypothesis under which the mean vector is closest to the observed vector (with ties being resolved at random). Figure 21.2 depicts the nearest-neighbor decoding rule for 8-PSK. The shaded region corresponds to the set of observables that result in the guess "$M = 1$," i.e., the set of points that are nearest to $\left( A \cos(2\pi/8), A \sin(2\pi/8) \right)$.

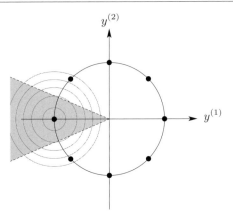

**Figure 21.3:** Contour lines of the density $f_{Y_1,Y_2|M=4}(\cdot)$. Shaded region corresponds to guessing "$M = 4$".

### 21.4.3   Exact Error Analysis for 8-PSK

The analysis of the probability of error can be a bit tricky. Here we only present the analysis for 8-PSK. If nothing else, it will motivate us to seek more easily computable bounds.

We shall compute the probability of error conditional on $M = 4$. But there is nothing special about this choice; the rotational symmetry of the problem implies that the probability of error does not depend on the hypothesis.

Conditional on $M = 4$, the observables $(Y^{(1)}, Y^{(2)})^\mathsf{T}$ can be expressed as

$$\left(Y^{(1)}, Y^{(2)}\right)^\mathsf{T} = (-\mathsf{A}, 0)^\mathsf{T} + \left(Z^{(1)}, Z^{(2)}\right)^\mathsf{T},$$

where $Z^{(1)}$ and $Z^{(2)}$ are independent $\mathcal{N}\left(0, \sigma^2\right)$ random variables:

$$f_{Z^{(1)},Z^{(2)}}\left(z^{(1)}, z^{(2)}\right) = \frac{1}{2\pi\sigma^2} \exp\left(-\frac{(z^{(1)})^2 + (z^{(2)})^2}{2\sigma^2}\right), \quad z^{(1)}, z^{(2)} \in \mathbb{R}.$$

Figure 21.3 depicts the contour lines of the density $f_{Y^{(1)},Y^{(2)}|M=4}(\cdot)$, which are centered on the mean $(a_4, b_4) = (-\mathsf{A}, 0)$. Note that $f_{Y^{(1)},Y^{(2)}|M=4}(\cdot)$ is symmetric about the horizontal axis:

$$f_{Y^{(1)},Y^{(2)}|M=4}\left(y^{(1)}, -y^{(2)}\right) = f_{Y^{(1)},Y^{(2)}|M=4}\left(y^{(1)}, y^{(2)}\right), \quad y^{(1)}, y^{(2)} \in \mathbb{R}. \quad (21.22)$$

The shaded region in the figure is the set of pairs $(y^{(1)}, y^{(2)})$ that cause the nearest-neighbor decoder to guess "$M = 4$".[3] Conditional on $M = 4$ an error results if $(Y^{(1)}, Y^{(2)})$ is outside the shaded region.

Referring now to Figure 21.4 we need to compute the probability that the noise $(Z^{(1)}, Z^{(2)})$ causes the received signal to lie in the union of the shaded areas. The

---

[3]It can be shown that the probability that the observation lies exactly on the boundary of the region is zero; see Proposition 21.6.2 ahead. We shall thus ignore this possibility.

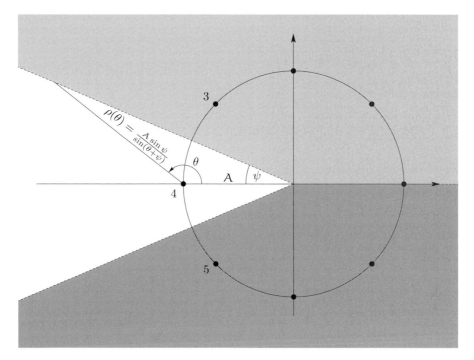

**Figure 21.4:** Error analysis for 8-PSK.

symmetry of $f_{Y^{(1)},Y^{(2)}|M=4}(\cdot)$ about the horizontal axis (21.22) implies that the probability that the received vector lies in the darkly-shaded region is the same as the probability that it lies in the lightly-shaded region. We shall thus compute the probability of the latter and double the result.

Let $\psi = \pi/8$ denote half the angle between the constellation points. To carry out the integration we shall use polar coordinates $(r, \theta)$ centered on the constellation point $(-A, 0)$ corresponding to Message 4:

$$p_{\mathrm{MAP}}(\text{error}|M=4) = 2 \int_0^{\pi-\psi} \int_{\rho(\theta)}^{\infty} \frac{1}{2\pi\sigma^2} e^{-\frac{r^2}{2\sigma^2}} r \, dr \, d\theta$$

$$= \frac{1}{\pi} \int_0^{\pi-\psi} \int_{\rho^2(\theta)/(2\sigma^2)}^{\infty} e^{-u} \, du \, d\theta$$

$$= \frac{1}{\pi} \int_0^{\pi-\psi} e^{-\frac{\rho^2(\theta)}{2\sigma^2}} \, d\theta, \tag{21.23}$$

where $\rho(\theta)$ is the distance we travel from the point $(-A, 0)$ at angle $\theta$ until we reach the lightly-shaded region, and where the second equality follows using the substitution $u \triangleq r^2/(2\sigma^2)$. Using the law of sines we have

$$\rho(\theta) = \frac{A \sin\psi}{\sin(\theta + \psi)}. \tag{21.24}$$

Since the symmetry of the problem implies that the conditional probability of error conditioned on $M = m$ does not depend on $m$, it follows from (21.23), (21.24), and

(21.18) that

$$p^*(\text{error}) = \frac{1}{\pi} \int_0^{\pi-\psi} e^{-\frac{A^2 \sin^2 \psi}{2 \sin^2(\theta+\psi)\sigma^2}} \, d\theta, \quad \psi = \frac{\pi}{8}. \tag{21.25}$$

## 21.5 The Union-of-Events Bound

Although simple, the **Union-of-Events Bound**, or **Union Bound** for short, is an extremely powerful and useful bound.[4] To derive it, recall that one of the axioms of probability is that the probability of the union of two *disjoint* events is the sum of their probabilities.[5] Given two *not necessarily disjoint* events $\mathcal{V}$ and $\mathcal{W}$, we can express the set $\mathcal{V}$ as in Figure 21.5 as the union of those elements of $\mathcal{V}$ that are not in $\mathcal{W}$ and those that are both in $\mathcal{V}$ and in $\mathcal{W}$:

$$\mathcal{V} = (\mathcal{V} \setminus \mathcal{W}) \cup (\mathcal{V} \cap \mathcal{W}). \tag{21.26}$$

Because the sets $\mathcal{V} \setminus \mathcal{W}$ and $\mathcal{V} \cap \mathcal{W}$ are disjoint, and because their union is $\mathcal{V}$, it follows that $\Pr(\mathcal{V}) = \Pr(\mathcal{V} \setminus \mathcal{W}) + \Pr(\mathcal{V} \cap \mathcal{W})$, which can also be written as

$$\Pr(\mathcal{V} \setminus \mathcal{W}) = \Pr(\mathcal{V}) - \Pr(\mathcal{V} \cap \mathcal{W}). \tag{21.27}$$

Writing the union $\mathcal{V} \cup \mathcal{W}$ as the union of two disjoint sets

$$\mathcal{V} \cup \mathcal{W} = \mathcal{W} \cup (\mathcal{V} \setminus \mathcal{W}) \tag{21.28}$$

as in Figure 21.6, we conclude that

$$\Pr(\mathcal{V} \cup \mathcal{W}) = \Pr(\mathcal{W}) + \Pr(\mathcal{V} \setminus \mathcal{W}), \tag{21.29}$$

which combines with (21.27) to prove that

$$\Pr(\mathcal{V} \cup \mathcal{W}) = \Pr(\mathcal{V}) + \Pr(\mathcal{W}) - \Pr(\mathcal{V} \cap \mathcal{W}). \tag{21.30}$$

Since probabilities are nonnegative, it follows from (21.30) that

$$\Pr(\mathcal{V} \cup \mathcal{W}) \leq \Pr(\mathcal{V}) + \Pr(\mathcal{W}), \tag{21.31}$$

which is the Union Bound. This bound can also be extended to derive an upper bound on the union of more sets. For example, we can show that for three events $\mathcal{U}, \mathcal{V}, \mathcal{W}$ we have $\Pr(\mathcal{U} \cup \mathcal{V} \cup \mathcal{W}) \leq \Pr(\mathcal{U}) + \Pr(\mathcal{V}) + \Pr(\mathcal{W})$. Indeed, by first applying the claim to the two sets $\mathcal{U}$ and $(\mathcal{V} \cup \mathcal{W})$ we obtain

$$\begin{aligned}
\Pr(\mathcal{U} \cup \mathcal{V} \cup \mathcal{W}) &= \Pr\big(\mathcal{U} \cup (\mathcal{V} \cup \mathcal{W})\big) \\
&\leq \Pr(\mathcal{U}) + \Pr(\mathcal{V} \cup \mathcal{W}) \\
&\leq \Pr(\mathcal{U}) + \Pr(\mathcal{V}) + \Pr(\mathcal{W}),
\end{aligned}$$

---

[4]It is also sometimes called Boole's Inequality.

[5]Actually the axiom is stronger; it states that this holds also for a countably infinite number of sets.

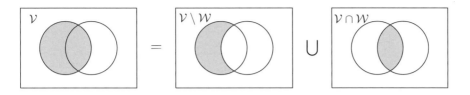

**Figure 21.5:** Diagram of two nondisjoint sets.

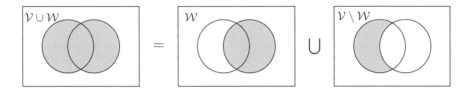

**Figure 21.6:** Diagram of the union of two nondisjoint sets.

where the last inequality follows by applying the inequality to the two sets $\mathcal{V}$ and $\mathcal{W}$. One can continue the argument by induction for a finite[6] collection of events to obtain:

**Theorem 21.5.1 (Union-of-Events Bound).** *If $\mathcal{V}_1, \mathcal{V}_2, \ldots$, is a finite or countably infinite collection of events then*

$$\Pr\left(\bigcup_j \mathcal{V}_j\right) \le \sum_j \Pr(\mathcal{V}_j). \tag{21.32}$$

We can think about the LHS of (21.32) as the probability that at least one of the events $\mathcal{V}_1, \mathcal{V}_2, \ldots$ occurs and of its RHS as the expected number of events that occur. Indeed, if for each $j$ we define the random variables $X_j(\omega) = \mathrm{I}\{\omega \in \mathcal{V}_j\}$ for all $\omega \in \Omega$, then the LHS of (21.32) is equal to $\Pr\left[\sum_j X_j > 0\right]$, and the RHS is $\sum_j \mathsf{E}[X_j]$, which can also be expressed as $\mathsf{E}\left[\sum_j X_j\right]$.

After the trivial bound that the probability of any event cannot exceed one, the Union Bound is probably the most important bound in Probability Theory. What makes it so useful is the fact that the RHS of (21.32) can be computed without regard to any dependencies between the events.

**Corollary 21.5.2.**

*(i) If each of a finite (or countably infinite) collection of events occurs with probability zero, then their union also occurs with probability zero.*

---

[6]In fact, this claim holds for a countably infinite number of events.

*(ii) If each of a finite (or countably infinite) collection of events occurs with probability one, then their intersection also occurs with probability one.*

**Proof.** To prove Part (i) we assume that each of the events $\mathcal{V}_1, \mathcal{V}_2, \ldots$ is of zero probability and compute

$$\Pr\left(\bigcup_j \mathcal{V}_j\right) \leq \sum_j \Pr(\mathcal{V}_j)$$

$$= \sum_j 0$$

$$= 0,$$

where the first inequality follows from the Union Bound, and where the subsequent equality follows from our assumption that $\Pr(\mathcal{V}_j) = 0$, for all $j$.

To prove Part (ii) we assume that each of the events $\mathcal{W}_1, \mathcal{W}_2, \ldots$ occurs with probability one and apply Part (i) to the sets $\mathcal{V}_1, \mathcal{V}_2, \ldots$, where $\mathcal{V}_j$ is the set-complement of $\mathcal{W}_j$, i.e., $\mathcal{V}_j = \Omega \setminus \mathcal{W}_j$:

$$\Pr\left(\bigcap_j \mathcal{W}_j\right) = 1 - \Pr\left(\left(\bigcap_j \mathcal{W}_j\right)^c\right)$$

$$= 1 - \Pr\left(\bigcup_j \mathcal{V}_j\right)$$

$$= 1,$$

where the first equality follows because the probabilities of an event and its complement sum to one; the second because the complement of an intersection is the union of the complements; and the final equality follows from Part (i) because the events $\mathcal{W}_j$ are, by assumption, of probability one so their complements are of probability zero. □

### 21.5.1 Applications to Hypothesis Testing

We shall now use the Union Bound to derive an upper bound on the conditional probability of error $p_{\mathrm{MAP}}(\mathrm{error}|M = m)$ of the MAP decoding rule. The bound we derive is applicable to any decision rule that satisfies the hypothesis of Theorem 21.3.1 as expressed in (21.16).

Define for every $m' \neq m$ the set $\mathcal{B}_{m,m'} \subset \mathbb{R}^d$ by

$$\mathcal{B}_{m,m'} = \left\{ \mathbf{y} \in \mathbb{R}^d : \pi_{m'} f_{\mathbf{Y}|M=m'}(\mathbf{y}) \geq \pi_m f_{\mathbf{Y}|M=m}(\mathbf{y}) \right\}. \tag{21.33}$$

Notice that $\mathbf{y} \in \mathcal{B}_{m,m'}$ does not imply that the MAP rule will guess $m'$: there may be a third hypothesis that is *a posteriori* even more likely than either $m$ or $m'$. Also, since the inequality in (21.33) is not strict, $\mathbf{y} \in \mathcal{B}_{m,m'}$ does not imply that the MAP rule will not guess $m$: there may be a tie, which may be resolved in favor of $m$. As we next argue, what is true is that *if $m$ was not guessed by the MAP rule,* *then some $m'$ which is not equal to $m$ must have had an* a posteriori *probability that is at least as high as that of $m$:*

$$\boxed{\big(m \text{ was not guessed}\big) \Rightarrow \Big(\mathbf{Y} \in \bigcup_{m' \neq m} \mathcal{B}_{m,m'}\Big).} \tag{21.34}$$

Indeed, if $m$ was not guessed by the MAP rule, then some other message was. Denoting that other message by $m'$, we note that $\pi_{m'} f_{\mathbf{Y}|M=m'}(\mathbf{y})$ must be at least as large as $\pi_m f_{\mathbf{Y}|M=m}(\mathbf{y})$ (because otherwise $m'$ would not have been guessed), so $\mathbf{y} \in \mathcal{B}_{m,m'}$.

Continuing from (21.34), we note that if the occurrence of an event $\mathcal{E}_1$ implies the occurrence of an event $\mathcal{E}_2$, then $\Pr(\mathcal{E}_1) \leq \Pr(\mathcal{E}_2)$. Consequently, by (21.34),

$$
\begin{aligned}
p_{\text{MAP}}(\text{error}|M=m) &\leq \Pr\Big[\mathbf{Y} \in \bigcup_{m' \neq m} \mathcal{B}_{m,m'} \,\Big|\, M = m\Big] \\
&= \Pr\Big(\bigcup_{m' \neq m} \big\{\omega \in \Omega : \mathbf{Y}(\omega) \in \mathcal{B}_{m,m'}\big\} \,\Big|\, M = m\Big) \\
&\leq \sum_{m' \neq m} \Pr\big(\{\omega \in \Omega : \mathbf{Y}(\omega) \in \mathcal{B}_{m,m'}\} \,\big|\, M = m\big) \\
&= \sum_{m' \neq m} \Pr\big[\mathbf{Y} \in \mathcal{B}_{m,m'} \,\big|\, M = m\big] \\
&= \sum_{m' \neq m} \int_{\mathcal{B}_{m,m'}} f_{\mathbf{Y}|M=m}(\mathbf{y}) \, d\mathbf{y}.
\end{aligned}
$$

We have thus derived:

**Proposition 21.5.3.** *For the setup of Theorem 21.3.1 let $p_{\text{MAP}}(\text{error}|M = m)$ denote the conditional probability of error conditional on $M = m$ of the MAP rule for guessing $M$ based on $\mathbf{Y}$. Then,*

$$p_{\text{MAP}}(\text{error}|M=m) \leq \sum_{m' \neq m} \Pr\big[\mathbf{Y} \in \mathcal{B}_{m,m'} \,\big|\, M = m\big] \tag{21.35}$$

$$= \sum_{m' \neq m} \int_{\mathcal{B}_{m,m'}} f_{\mathbf{Y}|M=m}(\mathbf{y}) \, d\mathbf{y}, \tag{21.36}$$

*where*

$$\mathcal{B}_{m,m'} = \Big\{\mathbf{y} \in \mathbb{R}^d : \pi_{m'} f_{\mathbf{Y}|M=m'}(\mathbf{y}) \geq \pi_m f_{\mathbf{Y}|M=m}(\mathbf{y})\Big\}. \tag{21.37}$$

*This bound is applicable to any decision rule satisfying the hypothesis of Theorem 21.3.1 as expressed in (21.16).*

The term $\Pr(\mathbf{Y} \in \mathcal{B}_{m,m'}|M = m)$ has an interesting interpretation. If ties occur with probability zero, then it corresponds to the conditional probability of error (given that $M = m$) incurred by a MAP decoder designed for the binary hypothesis

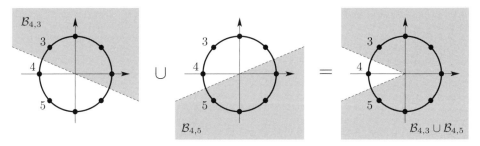

**Figure 21.7:** Error events for 8-PSK conditional on $M = 4$.

testing problem of guessing whether $M = m$ or $M = m'$ when the prior probability that $M = m$ is $\pi_m/(\pi_m + \pi_{m'})$ and that $M = m'$ is $\pi_{m'}/(\pi_m + \pi_{m'})$.

Alternatively, we can write (21.35) as

$$p_{\mathrm{MAP}}(\mathrm{error}|M = m) \leq \sum_{m' \neq m} \Pr\big[\pi_{m'} f_{\mathbf{Y}|M=m'}(\mathbf{Y}) \geq \pi_m f_{\mathbf{Y}|M=m}(\mathbf{Y}) \,\big|\, M = m\big].$$

$$(21.38)$$

## 21.5.2   Example: The Union Bound for 8-PSK

We next apply the Union Bound to upper-bound the probability of error associated with maximum-likelihood decoding of 8-PSK. For concreteness we focus on the conditional probability of error, conditional on $M = 4$. We shall see that in this case the RHS of (21.35) is still an upper bound on the probability of error even if we do not sum over all $m'$ that differ from $m$. Indeed, as we next argue, in upper-bounding the conditional probability of error of the ML decoder given $M = 4$, it suffices to sum over $m' \in \{3, 5\}$ only.

To show this we first note that for this problem the set $\mathcal{B}_{m,m'}$ of (21.33) corresponds to the set of vectors that are at least as close to $(a_{m'}, b_{m'})$ as to $(a_m, b_m)$:

$$\mathcal{B}_{m,m'} = \Big\{ \mathbf{y} \in \mathbb{R}^2 : \big(y^{(1)} - a_{m'}\big)^2 + \big(y^{(2)} - b_{m'}\big)^2 \leq \big(y^{(1)} - a_m\big)^2 + \big(y^{(2)} - b_m\big)^2 \Big\}.$$

As seen in Figure 21.7, given $M = 4$, an error will occur only if the observed vector $\mathbf{Y}$ is at least as close to $(a_3, b_3)$ as to $(a_4, b_4)$, or if it is at least as close to $(a_5, b_5)$ as to $(a_4, b_4)$. Thus, conditional on $M = 4$, an error can occur only if $\mathbf{Y} \in \mathcal{B}_{4,3} \cup \mathcal{B}_{4,5}$. (If $\mathbf{Y} \notin \mathcal{B}_{4,3} \cup \mathcal{B}_{4,5}$, then an error will certainly not occur. If $\mathbf{Y} \in \mathcal{B}_{4,3} \cup \mathcal{B}_{4,5}$, then an error may or may not occur. It will not occur in the case of a tie—corresponding to $\mathbf{Y}$ being on the boundary of $\mathcal{B}_{4,3} \cup \mathcal{B}_{4,5}$—provided that the tie is resolved in favor of $M = 4$.)

Note that the events $\mathbf{Y} \in \mathcal{B}_{4,5}$ and $\mathbf{Y} \in \mathcal{B}_{4,3}$ are not mutually exclusive, but, nevertheless, by the Union-of-Events Bound

$$p_{\mathrm{MAP}}(\mathrm{error}|M = 4) \leq \Pr[\mathbf{Y} \in \mathcal{B}_{4,3} \cup \mathcal{B}_{4,5} | M = 4]$$
$$\leq \Pr[\mathbf{Y} \in \mathcal{B}_{4,3} | M = 4] + \Pr[\mathbf{Y} \in \mathcal{B}_{4,5} | M = 4], \quad (21.39)$$

where the first inequality follows because, conditional on $M = 4$, an error can occur only if $\mathbf{y} \in \mathcal{B}_{4,3} \cup \mathcal{B}_{4,5}$; and where the second inequality follows from the Union-of-Events Bound. In fact, the first inequality holds with equality because, for this problem, the probability of a tie is zero; see Proposition 21.6.2 ahead.

From our analysis of multi-dimensional binary hypothesis testing (Lemma 20.14.1) we obtain that

$$\Pr\left[\mathbf{Y} \in \mathcal{B}_{4,3} \mid M = 4\right] = \mathcal{Q}\left(\frac{\sqrt{(a_4 - a_3)^2 + (b_4 - b_3)^2}}{2\sigma}\right)$$
$$= \mathcal{Q}\left(\frac{\mathsf{A}}{\sigma}\sin\left(\frac{\pi}{8}\right)\right) \tag{21.40}$$

and

$$\Pr\left[\mathbf{Y} \in \mathcal{B}_{4,5} \mid M = 4\right] = \mathcal{Q}\left(\frac{\sqrt{(a_4 - a_5)^2 + (b_4 - b_5)^2}}{2\sigma}\right)$$
$$= \mathcal{Q}\left(\frac{\mathsf{A}}{\sigma}\sin\left(\frac{\pi}{8}\right)\right). \tag{21.41}$$

Combining (21.39), (21.40), and (21.41) we obtain

$$p_{\mathrm{MAP}}(\mathrm{error}|M = 4) \leq 2\mathcal{Q}\left(\frac{\mathsf{A}}{\sigma}\sin\left(\frac{\pi}{8}\right)\right). \tag{21.42}$$

This is only an upper bound and not the exact error probability because the sets $\mathcal{B}_{4,3}$ and $\mathcal{B}_{4,5}$ are not disjoint so the events $\mathbf{Y} \in \mathcal{B}_{4,3}$ and $\mathbf{Y} \in \mathcal{B}_{4,5}$ are not disjoint and the Union-Bound is not tight; see Figure 21.7.

For this symmetric problem the conditional probability of error conditional on $M = m$ does not depend on the message $m$, and we thus also have by (21.18)

$$p^*(\mathrm{error}) \leq 2\mathcal{Q}\left(\frac{\mathsf{A}}{\sigma}\sin\left(\frac{\pi}{8}\right)\right). \tag{21.43}$$

### 21.5.3 Union-Bhattacharyya Bound

We next derive a bound which is looser than the Union Bound but which is often easier to evaluate in non-Gaussian settings. It is the multi-hypothesis testing version of the Bhattacharyya Bound (20.50).

Recall that, by Theorem 21.3.1, any guessing rule whose guess after observing that $\mathbf{Y} = \mathbf{y}_{\mathrm{obs}}$ is in the set

$$\tilde{\mathcal{M}}(\mathbf{y}_{\mathrm{obs}}) = \left\{\tilde{m} \in \mathcal{M} : \pi_{\tilde{m}}\, f_{\mathbf{Y}|M=\tilde{m}}(\mathbf{y}_{\mathrm{obs}}) = \max_{m'}\left\{\pi_{m'}\, f_{\mathbf{Y}|M=m'}(\mathbf{y}_{\mathrm{obs}})\right\}\right\}$$

is optimal. To analyze the optimal probability of error $p^*(\mathrm{error})$, we shall analyze one particular optimal decision rule. This rule is not the MAP rule, but it differs from the MAP rule only in the way it resolves ties. Rather than resolving ties at random, this rule resolves ties according to the index of the hypothesis: it chooses the message in $\tilde{\mathcal{M}}(\mathbf{y}_{\text{obs}})$ of smallest index. For example, if the messages of highest *a posteriori* probability are Messages 7, 9, and 17, i.e., if $\tilde{\mathcal{M}}(\mathbf{y}_{\text{obs}}) = \{7, 9, 17\}$, then it guesses "7." This decision rule may not appeal to the reader's sense of fairness but, by Theorem 21.3.1, it is nonetheless optimal. Consequently, if we denote the conditional probability of error of this decoder by $p(\text{error}|M = m)$, then

$$p^*(\text{error}) = \sum_{m \in \mathcal{M}} \pi_m \, p(\text{error}|M = m). \tag{21.44}$$

We next analyze the performance of this decision rule. For every $m' \neq m$ let

$$\mathcal{D}_{m,m'} = \begin{cases} \{\mathbf{y} \in \mathbb{R}^d : \pi_{m'} \, f_{\mathbf{Y}|M=m'}(\mathbf{y}) \geq \pi_m \, f_{\mathbf{Y}|M=m}(\mathbf{y})\} & \text{if } m' < m, \\ \{\mathbf{y} \in \mathbb{R}^d : \pi_{m'} \, f_{\mathbf{Y}|M=m'}(\mathbf{y}) > \pi_m \, f_{\mathbf{Y}|M=m}(\mathbf{y})\} & \text{if } m' > m. \end{cases} \tag{21.45}$$

Notice that

$$\mathcal{D}_{m,m'} = \mathcal{D}^{\text{c}}_{m',m}, \quad m \neq m'. \tag{21.46}$$

Conditional on $M = m$, our detector will err if, and only if, $\mathbf{y}_{\text{obs}} \in \cup_{m' \neq m} \mathcal{D}_{m,m'}$. Thus

$$\Pr(\text{error}|M = m) = \Pr\left[\mathbf{Y} \in \bigcup_{m' \neq m} \mathcal{D}_{m,m'} \,\middle|\, M = m\right]$$

$$= \Pr\left(\bigcup_{m' \neq m} \{\omega \in \Omega : \mathbf{Y}(\omega) \in \mathcal{D}_{m,m'}\} \,\middle|\, M = m\right)$$

$$\leq \sum_{m' \neq m} \Pr\left(\{\omega \in \Omega : \mathbf{Y}(\omega) \in \mathcal{D}_{m,m'}\} \,\middle|\, M = m\right)$$

$$= \sum_{m' \neq m} \Pr\left[\mathbf{Y} \in \mathcal{D}_{m,m'} \,\middle|\, M = m\right]$$

$$= \sum_{m' \neq m} \int_{\mathcal{D}_{m,m'}} f_{\mathbf{Y}|M=m}(\mathbf{y}) \, d\mathbf{y}, \tag{21.47}$$

where the inequality follows from the Union Bound. To upper-bound $p^*(\text{error})$ we use (21.44) and (21.47) to obtain

$$p^*(\text{error}) = \sum_{m=1}^{M} \pi_m \, \Pr(\text{error}|M = m)$$

$$\leq \sum_{m=1}^{M} \pi_m \sum_{m' \neq m} \int_{\mathcal{D}_{m,m'}} f_{\mathbf{Y}|M=m}(\mathbf{y}) \, d\mathbf{y}$$

$$= \sum_{m=1}^{M} \sum_{m' > m} \left(\pi_m \int_{\mathcal{D}_{m,m'}} f_{\mathbf{Y}|M=m}(\mathbf{y}) \, d\mathbf{y} + \pi_{m'} \int_{\mathcal{D}_{m',m}} f_{\mathbf{Y}|M=m'}(\mathbf{y}) \, d\mathbf{y}\right)$$

$$= \sum_{m=1}^{M} \sum_{m' > m} \left(\int_{\mathcal{D}_{m,m'}} \pi_m \, f_{\mathbf{Y}|M=m}(\mathbf{y}) \, d\mathbf{y} + \int_{\mathcal{D}^{\text{c}}_{m,m'}} \pi_{m'} \, f_{\mathbf{Y}|M=m'}(\mathbf{y}) \, d\mathbf{y}\right)$$

$$= \sum_{m=1}^{M} \sum_{m'>m} \int_{\mathbb{R}^d} \min \left\{ \pi_m \, f_{\mathbf{Y}|M=m}(\mathbf{y}), \, \pi_{m'} \, f_{\mathbf{Y}|M=m'}(\mathbf{y}) \right\} \mathrm{d}\mathbf{y}$$

$$\leq \sum_{m=1}^{M} \sum_{m'>m} \sqrt{\pi_m \pi_{m'}} \int_{\mathbb{R}^d} \sqrt{f_{\mathbf{Y}|M=m}(\mathbf{y}) f_{\mathbf{Y}|M=m'}(\mathbf{y})} \, \mathrm{d}\mathbf{y}$$

$$\leq \sum_{m=1}^{M} \sum_{m'>m} \frac{\pi_m + \pi_{m'}}{2} \int_{\mathbb{R}^d} \sqrt{f_{\mathbf{Y}|M=m}(\mathbf{y}) f_{\mathbf{Y}|M=m'}(\mathbf{y})} \, \mathrm{d}\mathbf{y}$$

$$= \frac{1}{2} \sum_{m \in \mathcal{M}} \sum_{m' \neq m} \frac{\pi_m + \pi_{m'}}{2} \int_{\mathbb{R}^d} \sqrt{f_{\mathbf{Y}|M=m}(\mathbf{y}) f_{\mathbf{Y}|M=m'}(\mathbf{y})} \, \mathrm{d}\mathbf{y},$$

where the equality in the first line follows from (21.44); the inequality in the second line from (21.47); the equality in the third line by rearranging the sum; the equality in the fourth line from (21.46); the equality in the fifth line from the definition of the set $\mathcal{D}_{m,m'}$; the inequality in the sixth line from the inequality $\min\{a,b\} \leq \sqrt{ab}$, which holds for all nonnegative $a, b \in \mathbb{R}$ (see (20.48)); the inequality in the seventh line from the Arithmetic-Geometric Inequality $\sqrt{cd} \leq (c+d)/2$, which holds for all $c, d \geq 0$ (see (20.49)); and the final equality by the symmetry of the summand. We have thus obtained the **Union-Bhattacharyya Bound**:

$$p^*(\text{error}) \leq \sum_{m \in \mathcal{M}} \sum_{m' \neq m} \frac{\pi_m + \pi_{m'}}{4} \int_{\mathbb{R}^d} \sqrt{f_{\mathbf{Y}|M=m}(\mathbf{y}) f_{\mathbf{Y}|M=m'}(\mathbf{y})} \, \mathrm{d}\mathbf{y}. \qquad (21.48)$$

For *a priori* equally likely hypotheses it takes the form

$$p^*(\text{error}) \leq \frac{1}{2M} \sum_{m \in \mathcal{M}} \sum_{m' \neq m} \int \sqrt{f_{\mathbf{Y}|M=m}(\mathbf{y}) f_{\mathbf{Y}|M=m'}(\mathbf{y})} \, \mathrm{d}\mathbf{y}, \qquad (21.49)$$

which is the Union-Bhattacharyya Bound for M-ary hypothesis testing with a uniform prior.

## 21.6 Multi-Dimensional M-ary Gaussian Hypothesis Testing

We next use Theorem 21.3.3 to study the multi-hypothesis testing version of the problem we addressed in Section 20.14. We begin with the problem setup and then proceed to derive the MAP decision rule. We then assess the performance of this rule by deriving an upper bound and a lower bound on its probability of error.

### 21.6.1 Problem Setup

A random variable $M$ takes value in the set $\mathcal{M} = \{1, \ldots, M\}$ with a nondegenerate prior (21.4). We wish to guess $M$ based on an observation consisting of a random column-vector $\mathbf{Y}$ taking value in $\mathbb{R}^J$ whose components are given by $Y^{(1)}, \ldots, Y^{(J)}$.[7]

---

[7] Our observation now takes value in $\mathbb{R}^J$ and not as before in $\mathbb{R}^d$. My excuse for using J instead of $d$ is that later, when we refer to this section, $d$ will have a different meaning and choosing J here reduces the chance of confusion later on.

For typographical reasons we denote the observed realization of $\mathbf{Y}$ by $\mathbf{y}$, instead of $\mathbf{y}_{\text{obs}}$. For every $m \in \mathcal{M}$ we have that, conditional on $M = m$, the components of $\mathbf{Y}$ are independent Gaussians, with $Y^{(j)} \sim \mathcal{N}(s_m^{(j)}, \sigma^2)$, where $\mathbf{s}_m$ is some deterministic vector of J components $s_m^{(1)}, \ldots, s_m^{(J)}$, and where $\sigma^2 > 0$. Recalling the density of the univariate Gaussian distribution (19.6) and using the conditional independence of the components of $\mathbf{Y}$ given $M = m$, we can express the conditional density $f_{\mathbf{Y}|M=m}(\mathbf{y})$ of the vector $\mathbf{Y}$ at every point $\mathbf{y} = (y^{(1)}, \ldots, y^{(J)})^{\mathsf{T}}$ in $\mathbb{R}^J$ as

$$f_{\mathbf{Y}|M=m}(\mathbf{y}) = \prod_{j=1}^{J} \left( \frac{1}{\sqrt{2\pi\sigma^2}} \exp\left( -\frac{(y^{(j)} - s_m^{(j)})^2}{2\sigma^2} \right) \right). \tag{21.50}$$

### 21.6.2 Optimal Guessing Rule

Using Theorem 21.3.3 we obtain that, having observed $\mathbf{y} = (y^{(1)}, \ldots, y^{(J)})^{\mathsf{T}} \in \mathbb{R}^J$, an optimal decision rule is the MAP rule, which picks uniformly at random an element from the set

$$\begin{aligned}
\tilde{\mathcal{M}}(\mathbf{y}) &= \left\{ \tilde{m} \in \mathcal{M} : \pi_{\tilde{m}} f_{\mathbf{Y}|M=\tilde{m}}(\mathbf{y}) = \max_{m' \in \mathcal{M}} \left\{ \pi_{m'} f_{\mathbf{Y}|M=m'}(\mathbf{y}) \right\} \right\} \\
&= \left\{ \tilde{m} \in \mathcal{M} : \ln\left( \pi_{\tilde{m}} f_{\mathbf{Y}|M=\tilde{m}}(\mathbf{y}) \right) = \max_{m' \in \mathcal{M}} \left\{ \ln\left( \pi_{m'} f_{\mathbf{Y}|M=m'}(\mathbf{y}) \right) \right\} \right\},
\end{aligned} \tag{21.51}$$

where the second equality follows from the strict monotonicity of the logarithm. We next obtain a more explicit description of $\tilde{\mathcal{M}}(\mathbf{y})$ for our setup. By (21.50),

$$\ln\left( \pi_m f_{\mathbf{Y}|M=m}(\mathbf{y}) \right) = \ln \pi_m - \frac{J}{2} \ln(2\pi\sigma^2) - \frac{1}{2\sigma^2} \sum_{j=1}^{J} (y^{(j)} - s_m^{(j)})^2. \tag{21.52}$$

The term $(J/2) \ln(2\pi\sigma^2)$ is a constant term that does not depend on the hypothesis. Consequently, it does not influence the set of messages that attain the highest score. (The tallest student in the class is the same irrespective of whether the height of all the students is measured when they are barefoot or when they are all wearing the one-inch heel school uniform shoes. The heel can only make a difference if different students wear shoes of different heel height.) Thus,

$$\tilde{\mathcal{M}}(\mathbf{y}) = \left\{ \tilde{m} \in \mathcal{M} : \ln \pi_{\tilde{m}} - \sum_{j=1}^{J} \frac{(y^{(j)} - s_{\tilde{m}}^{(j)})^2}{2\sigma^2} = \max_{m' \in \mathcal{M}} \left\{ \ln \pi_{m'} - \sum_{j=1}^{J} \frac{(y^{(j)} - s_{m'}^{(j)})^2}{2\sigma^2} \right\} \right\}.$$

The expression for $\tilde{\mathcal{M}}(\mathbf{y})$ can be further simplified if $M$ is *a priori* uniformly distributed. In this case we have

$$\begin{aligned}
\tilde{\mathcal{M}}(\mathbf{y}) &= \left\{ \tilde{m} \in \mathcal{M} : -\sum_{j=1}^{J} \frac{(y^{(j)} - s_{\tilde{m}}^{(j)})^2}{2\sigma^2} = \max_{m' \in \mathcal{M}} \left\{ -\sum_{j=1}^{J} \frac{(y^{(j)} - s_{m'}^{(j)})^2}{2\sigma^2} \right\} \right\} \\
&= \left\{ \tilde{m} \in \mathcal{M} : \sum_{j=1}^{J} (y^{(j)} - s_{\tilde{m}}^{(j)})^2 = \min_{m' \in \mathcal{M}} \left\{ \sum_{j=1}^{J} (y^{(j)} - s_{m'}^{(j)})^2 \right\} \right\}, \quad M \text{ uniform,}
\end{aligned}$$

where the first equality follows because when $M$ is uniform the additive term $\ln \pi_m$ is given by $\ln(1/M)$ and hence does not depend on the hypothesis; and where the second equality follows because changing the sign of all the elements of a set changes the largest ones to the smallest ones, and by noting that scaling the score by $2\sigma^2$ does not change the highest scoring messages (because we assumed that $\sigma^2 > 0$).

If we interpret the quantity

$$\|\mathbf{y} - \mathbf{s}_m\| = \sqrt{\sum_{j=1}^{J} \left(y^{(j)} - s_m^{(j)}\right)^2}$$

as the Euclidean distance between the vector $\mathbf{y}$ and the vector $\mathbf{s}_m$, then we see that, for a uniform prior on $M$, it is optimal to guess the message $m$ whose corresponding mean vector $\mathbf{s}_m$ is closest to the observed vector $\mathbf{y}$. Notice that to implement this "nearest-neighbor" decision rule we do not need to know the value of $\sigma^2$.

We next show that if, in addition to assuming a uniform prior on $M$, we also assume that the vectors $\mathbf{s}_1, \ldots, \mathbf{s}_M$ all have the same norm, i.e.,

$$\|\mathbf{s}_1\| = \|\mathbf{s}_2\| = \cdots = \|\mathbf{s}_M\|, \tag{21.53}$$

then

$$\tilde{\mathcal{M}}(\mathbf{y}) = \left\{ \tilde{m} \in \mathcal{M} : \sum_{j=1}^{J} y^{(j)} s_{\tilde{m}}^{(j)} = \max_{m' \in \mathcal{M}} \left\{ \sum_{j=1}^{J} y^{(j)} s_{m'}^{(j)} \right\} \right\},$$

so the MAP decision rules guesses the message $m$ whose mean vector $\mathbf{s}_m$ has the "highest correlation" with the received vector $\mathbf{y}$. To see this, we note that because $M$ has a uniform prior the "nearest-neighbor" decoding rule is optimal, and we then expand

$$\|\mathbf{y} - \mathbf{s}_m\|^2 = \sum_{j=1}^{J} \left(y^{(j)} - s_m^{(j)}\right)^2$$

$$= \sum_{j=1}^{J} \left(y^{(j)}\right)^2 - 2 \sum_{j=1}^{J} y^{(j)} s_m^{(j)} + \sum_{j=1}^{J} \left(s_m^{(j)}\right)^2,$$

where the first term does not depend on the hypothesis and where, by (21.53), the third term also does not depend on the hypothesis.

We summarize our findings in the following proposition.

**Proposition 21.6.1.** *Consider the problem described in Section 21.6.1 of guessing $M$ based on the observation $\mathbf{y}$.*

*(i) It is optimal to form the guess based on $\mathbf{y} = (y^{(1)}, \ldots, y^{(J)})^{\mathsf{T}}$ by choosing uniformly at random from the set*

$$\left\{ \tilde{m} \in \mathcal{M} : \ln \pi_{\tilde{m}} - \sum_{j=1}^{J} \frac{\left(y^{(j)} - s_{\tilde{m}}^{(j)}\right)^2}{2\sigma^2} = \max_{m' \in \mathcal{M}} \left\{ \ln \pi_{m'} - \sum_{j=1}^{J} \frac{\left(y^{(j)} - s_{m'}^{(j)}\right)^2}{2\sigma^2} \right\} \right\}.$$

(ii) *If $M$ is uniformly distributed, then this rule is equivalent to the "nearest-neighbor" decoding rule of picking uniformly at random an element of the set*

$$\left\{\tilde{m} \in \mathcal{M} : \|\mathbf{y} - \mathbf{s}_{\tilde{m}}\| = \min_{m' \in \mathcal{M}} \left\{ \|\mathbf{y} - \mathbf{s}_{m'}\| \right\} \right\}.$$

(iii) *If, in addition to $M$ being uniform, we also assume that the mean vectors satisfy (21.53), then this rule is equivalent to the "maximum-correlation" rule of picking at random an element of the set*

$$\left\{\tilde{m} \in \mathcal{M} : \sum_{j=1}^{J} y^{(j)} s_{\tilde{m}}^{(j)} = \max_{m' \in \mathcal{M}} \left\{ \sum_{j=1}^{J} y^{(j)} s_{m'}^{(j)} \right\} \right\}.$$

We next show that if the mean vectors $\mathbf{s}_1, \ldots, \mathbf{s}_M$ are distinct in the sense that for every pair $m' \neq m''$ in $\mathcal{M}$ there exists at least one component where the vectors $\mathbf{s}_{m'}$ and $\mathbf{s}_{m''}$ differ, i.e.,

$$\|\mathbf{s}_{m'} - \mathbf{s}_{m''}\| > 0, \quad m' \neq m'',$$

then the probability of ties is zero. That is, we will show that the probability of observing a vector $\mathbf{y}$ for which the set $\tilde{\mathcal{M}}(\mathbf{y})$ (21.51) has more than one element is zero. Stated in yet another way, the probability that the observable $\mathbf{Y}$ will be such that the MAP will require randomization is zero. Stated one last time:

**Proposition 21.6.2.** *If the mean vectors $\mathbf{s}_1, \ldots, \mathbf{s}_M$ in our setup are distinct, then with probability one the observed vector $\mathbf{y}$ is such that there is a unique message of highest* a posteriori *probability.*

**Proof.** Conditional on $\mathbf{Y} = \mathbf{y}$, associate with each message $m \in \mathcal{M}$ the score $\ln\left(\pi_m f_{\mathbf{Y}|M=m}(\mathbf{y})\right)$. We need to show that the probability of the observation $\mathbf{y}$ being such that at least two messages attain the highest score is zero. Instead, we shall prove the stronger statement that the probability of two messages attaining the same score (be it maximal or not) is zero.

We first show that it suffices to prove that for every $m \in \mathcal{M}$ and for every pair of messages $m' \neq m''$, we have that, conditional on $M = m$, the probability that $m'$ and $m''$ attain the same score is zero, i.e.,

$$\Pr\left(\text{score of Message } m' = \text{score of Message } m'' \,\middle|\, M = m\right) = 0, \quad m' \neq m''. \tag{21.54}$$

Indeed, once we show (21.54), it will follow that the *unconditional* probability that Message $m'$ attains the same score as Message $m''$ is zero, i.e.,

$$\Pr\left(\text{score of Message } m' = \text{score of Message } m''\right) = 0, \quad m' \neq m'', \tag{21.55}$$

because

$$\Pr\left(\text{score of Message } m' = \text{score of Message } m''\right)$$
$$= \sum_{m \in \mathcal{M}} \pi_m \Pr\left(\text{score of Message } m' = \text{score of Message } m'' \,\middle|\, M = m\right).$$

But (21.55) implies that the probability that any two or more messages attain the highest score is zero because

Pr(two or more messages attain the highest score)

$$
= \Pr\left( \bigcup_{\substack{m',m'' \in \mathcal{M} \\ m' \neq m''}} \{m' \text{ and } m'' \text{ attain the highest score}\} \right)
$$

$$
\leq \sum_{\substack{m',m'' \in \mathcal{M} \\ m' \neq m''}} \Pr\big(m' \text{ and } m'' \text{ attain the highest score}\big)
$$

$$
\leq \sum_{\substack{m',m'' \in \mathcal{M} \\ m' \neq m''}} \Pr\big(m' \text{ and } m'' \text{ attain the same score}\big),
$$

where the first equality follows because more than one message attains the highest score if, and only if, there exist two distinct messages $m'$ and $m''$ that attain the highest score; the subsequent inequality follows from the Union Bound (Theorem 21.5.1); and the final inequality by noting that if $m'$ and $m''$ both attain the highest score, then they both achieve the same score.

Having established that in order to complete the proof it suffices to establish (21.54), we proceed to do so. By (21.52) we obtain, upon opening the square, that the observation $\mathbf{Y}$ results in Messages $m'$ and $m''$ obtaining the same score if, and only if,

$$
\frac{1}{\sigma^2} \sum_{j=1}^{J} Y^{(j)} \big(s_{m'}^{(j)} - s_{m''}^{(j)}\big) = \ln \frac{\pi_{m''}}{\pi_{m'}} + \frac{1}{2\sigma^2} \big(\|\mathbf{s}_{m'}\|^2 - \|\mathbf{s}_{m''}\|^2\big). \tag{21.56}
$$

We next show that, conditional on $M = m$, the probability that $\mathbf{Y}$ satisfies (21.56) is zero. To that end we note that, conditional on $M = m$, the random variables $Y^{(1)}, \ldots, Y^{(J)}$ are independent random variables with $Y^{(j)}$ being Gaussian with mean $s_m^{(j)}$ and variance $\sigma^2$; see (21.50). Consequently, by Proposition 19.7.3, we have that, conditional on $M = m$, the LHS of (21.56) is a Gaussian random variable of variance

$$
\frac{1}{\sigma^2} \|\mathbf{s}_{m'} - \mathbf{s}_{m''}\|^2,
$$

which is positive because $m' \neq m''$ and because we assumed that the mean vectors are distinct. It follows that, conditional on $M = m$, the LHS of (21.56) is a Gaussian random variable of positive variance, and hence has zero probability of being equal to the deterministic number on the RHS of (21.56). This proves (21.54), and hence concludes the proof. $\qquad\square$

### 21.6.3 The Union Bound

We next use the Union Bound to upper-bound the optimal probability of error $p^*(\text{error})$. By (21.38)

$$
p_{\mathrm{MAP}}(\text{error}|M = m) \leq \sum_{m' \neq m} \Pr\big[\pi_{m'} f_{\mathbf{Y}|M=m'}(\mathbf{Y}) \geq \pi_m f_{\mathbf{Y}|M=m}(\mathbf{Y}) \,\big|\, M = m\big]
$$

$$= \sum_{m' \neq m} \mathcal{Q}\left( \frac{\|\mathbf{s}_m - \mathbf{s}_{m'}\|}{2\sigma} + \frac{\sigma}{\|\mathbf{s}_m - \mathbf{s}_{m'}\|} \ln \frac{\pi_m}{\pi_{m'}} \right), \qquad (21.57)$$

where the equality follows from Lemma 20.14.1. From this and from the optimality of the MAP rule (21.18) we thus obtain

$$p^*(\text{error}) \leq \sum_{m \in \mathcal{M}} \sum_{m' \neq m} \pi_m \mathcal{Q}\left( \frac{\|\mathbf{s}_m - \mathbf{s}_{m'}\|}{2\sigma} + \frac{\sigma}{\|\mathbf{s}_m - \mathbf{s}_{m'}\|} \ln \frac{\pi_m}{\pi_{m'}} \right). \qquad (21.58)$$

If $M$ is uniform, these bounds simplify to:

$$p_{\text{MAP}}(\text{error}|M = m) \leq \sum_{m' \neq m} \mathcal{Q}\left( \frac{\|\mathbf{s}_m - \mathbf{s}_{m'}\|}{2\sigma} \right), \quad M \text{ uniform}, \qquad (21.59)$$

$$p^*(\text{error}) \leq \frac{1}{\mathsf{M}} \sum_{m \in \mathcal{M}} \sum_{m' \neq m} \mathcal{Q}\left( \frac{\|\mathbf{s}_m - \mathbf{s}_{m'}\|}{2\sigma} \right), \quad M \text{ uniform}. \qquad (21.60)$$

### 21.6.4   A Lower Bound

We next derive a lower bound on the optimal error probability $p^*(\text{error})$. We do so by lower-bounding the conditional probability of error $p_{\text{MAP}}(\text{error}|M = m)$ of the MAP rule and by then using this lower bound to derive a lower bound on $p^*(\text{error})$ via (21.18).

We note that if Message $m'$ attains a score that is strictly higher than the one attained by Message $m$, then the MAP decoder will surely not guess "$M = m$." (The MAP may or may not guess "$M = m'$" depending on the score associated with messages other than $m$ and $m'$.) Thus, for each message $m' \neq m$ we have

$$p_{\text{MAP}}(\text{error}|M = m) \geq \Pr\left[ \pi_{m'} f_{\mathbf{Y}|M=m'}(\mathbf{Y}) > \pi_m f_{\mathbf{Y}|M=m}(\mathbf{Y}) \,\middle|\, M = m \right] \quad (21.61)$$

$$= \mathcal{Q}\left( \frac{\|\mathbf{s}_m - \mathbf{s}_{m'}\|}{2\sigma} + \frac{\sigma}{\|\mathbf{s}_m - \mathbf{s}_{m'}\|} \ln \frac{\pi_m}{\pi_{m'}} \right), \qquad (21.62)$$

where the equality follows from Lemma 20.14.1.

Noting that (21.62) holds for all $m' \neq m$, we can choose $m'$ to get the tightest bound. This yields the lower bound

$$p_{\text{MAP}}(\text{error}|M = m) \geq \max_{m' \in \mathcal{M} \setminus \{m\}} \mathcal{Q}\left( \frac{\|\mathbf{s}_m - \mathbf{s}_{m'}\|}{2\sigma} + \frac{\sigma}{\|\mathbf{s}_m - \mathbf{s}_{m'}\|} \ln \frac{\pi_m}{\pi_{m'}} \right) \quad (21.63)$$

and hence, by (21.18),

$$p^*(\text{error}) \geq \sum_{m \in \mathcal{M}} \pi_m \max_{m' \in \mathcal{M} \setminus \{m\}} \mathcal{Q}\left( \frac{\|\mathbf{s}_m - \mathbf{s}_{m'}\|}{2\sigma} + \frac{\sigma}{\|\mathbf{s}_m - \mathbf{s}_{m'}\|} \ln \frac{\pi_m}{\pi_{m'}} \right). \quad (21.64)$$

For uniform $M$ this expression can be simplified by noting that the $\mathcal{Q}$-function is strictly decreasing:

$$p_{\text{MAP}}(\text{error}|M = m) \geq \mathcal{Q}\left( \min_{m' \in \mathcal{M} \setminus \{m\}} \frac{\|\mathbf{s}_m - \mathbf{s}_{m'}\|}{2\sigma} \right), \quad M \text{ uniform}, \qquad (21.65)$$

$$p^*(\text{error}) \geq \frac{1}{M} \sum_{m \in \mathcal{M}} \mathcal{Q}\left( \min_{m' \in \mathcal{M} \setminus \{m\}} \frac{\|\mathbf{s}_m - \mathbf{s}_{m'}\|}{2\sigma} \right), \quad M \text{ uniform.} \qquad (21.66)$$

## 21.7   Additional Reading

For additional reading on multi-hypothesis testing see the recommended reading for Chapter 20. The problem of assessing the optimal probability of error for the multi-dimensional M-ary Gaussian hypothesis testing problem of Section 21.6 has received extensive attention in the coding literature. For a survey of these results see (Sason and Shamai, 2006).

## 21.8   Exercises

**Exercise 21.1 (Ternary Gaussian Detection).** Consider the following special case of the problem discussed in Section 21.6. Here $M$ is uniformly distributed over the set $\{1, 2, 3\}$, and the mean vectors $\mathbf{s}_1, \mathbf{s}_2, \mathbf{s}_3$ are given by

$$\mathbf{s}_1 = \mathbf{0}, \quad \mathbf{s}_2 = \mathbf{s}, \quad \mathbf{s}_3 = -\mathbf{s},$$

where $\mathbf{s}$ is some deterministic nonzero vector in $\mathbb{R}^{\mathsf{J}}$. Find the conditional probability of error of the MAP rule conditional on each hypothesis.

**Exercise 21.2 (4-PSK Detection).** Consider the setup of Section 21.4 with $M = 4$ and

$$(a_1, b_1) = (0, A), \quad (a_2, b_2) = (-A, 0), \quad (a_3, b_3) = (0, -A), \quad (a_4, b_4) = (A, 0).$$

(i) Sketch the decision regions of the MAP decision rule.

(ii) Using the $\mathcal{Q}$-function, express the conditional probabilities of error of this rule conditional on each hypothesis.

(iii) Compute an upper bound on $p_{\text{MAP}}(\text{error}|M = 1)$ using Propsition 21.5.3. Indicate on the figure which events are summed two or three times. Can you improve the bound by summing only over a subset of the alternative hypotheses?

*Hint: In Part (ii) first find the probability of correct detection.*

**Exercise 21.3 (A 7-ary QAM problem).** Consider the problem addressed in Section 21.4 in the special case where $M = 7$ and where

$$a_m = A \cos\left(\frac{2\pi m}{6}\right), \quad b_m = A \sin\left(\frac{2\pi m}{6}\right), \quad m = 1, \ldots, 6,$$

$$a_7 = 0, \quad b_7 = 0.$$

(i) Illustrate the decision regions of the MAP (nearest-neighbor) guessing rule.

(ii) Let $\mathbf{Z} = (Z^{(1)}, Z^{(2)})^{\mathsf{T}}$ be a random vector whose components are IID $\mathcal{N}(0, \sigma^2)$. Show that for every message $m \in \{1, \ldots, 7\}$ the conditional probability of error $p_{\text{MAP}}(\text{error}|M = m)$ can be upper-bounded by the probability that the Euclidean norm of $\mathbf{Z}$ exceeds $A/2$. Calculate this probability.

(iii) What is the upper bound on $p_{\mathrm{MAP}}(\mathrm{error}|M = m)$ that Proposition 21.5.3 yields in this case? Can you improve it by including fewer terms?

(iv) Compare the different bounds.

See (Viterbi and Omura, 1979, Chapter 2, Problem 2.2).

**Exercise 21.4 (Orthogonal Mean Vectors).** Let $M$ be uniformly distributed over the set $\mathcal{M} = \{1, \ldots, M\}$. Let the observable $\mathbf{Y}$ be a random J-vector. Conditional on $M = m$, the observable $\mathbf{Y}$ is given by

$$\mathbf{Y} = \sqrt{\mathsf{E}_{\mathrm{s}}}\,\boldsymbol{\phi}_m + \mathbf{Z},$$

where $\mathbf{Z}$ is a random J-vector whose components are IID $\mathcal{N}(0, \sigma^2)$, and where $\boldsymbol{\phi}_1, \ldots, \boldsymbol{\phi}_M$ are orthonormal in the sense that

$$\langle \boldsymbol{\phi}_{m'}, \boldsymbol{\phi}_{m''}\rangle_{\mathrm{E}} = \mathrm{I}\{m' = m''\}, \quad m', m'' \in \mathcal{M}.$$

Show that

$$p_{\mathrm{MAP}}(\mathrm{error}|M = m) = 1 - \frac{1}{\sqrt{2\pi}} \int_{-\infty}^{\infty} \left(1 - \mathcal{Q}(\xi)\right)^{M-1} e^{-\frac{(\xi - \alpha)^2}{2}}\, d\xi, \tag{21.67}$$

where $\alpha = \sqrt{\mathsf{E}_{\mathrm{s}}}/\sigma$.

**Exercise 21.5 (Equi-Energy Constellations).** Consider the setup of Section 21.6.1 with a uniform prior and with $\|\mathbf{s}_1\|^2 = \cdots = \|\mathbf{s}_M\|^2 = \mathsf{E}_{\mathrm{s}}$. Show that the optimal probability of correct decoding is given by

$$p^*(\mathrm{correct}) = \frac{1}{M} \exp\left(-\frac{\mathsf{E}_{\mathrm{s}}}{2\sigma^2}\right) \mathsf{E}\left[\exp\left(\frac{1}{\sigma^2} \max_m \langle \mathbf{V}, \mathbf{s}_m\rangle_{\mathrm{E}}\right)\right], \tag{21.68}$$

where $\mathbf{V}$ is a random J-vector whose components are IID $\mathcal{N}(0, \sigma^2)$. We recommend the following approach. Let $\mathcal{D}_1, \ldots, \mathcal{D}_M$ be a partition of $\mathbb{R}^{\mathsf{J}}$ such that for every $m \in \mathcal{M}$,

$$\mathbf{y} \in \mathcal{D}_m \Rightarrow \langle \mathbf{y}, \mathbf{s}_m\rangle_{\mathrm{E}} = \max_{m'} \langle \mathbf{y}, \mathbf{s}_{m'}\rangle_{\mathrm{E}}.$$

(i) Show that

$$p^*(\mathrm{correct}) = \frac{1}{M} \sum_{m \in \mathcal{M}} \Pr\left[\mathbf{Y} \in \mathcal{D}_m \mid M = m\right].$$

(ii) Show that the RHS of the above can be written as

$$\frac{1}{M} \exp\left(-\frac{\mathsf{E}_{\mathrm{s}}}{2\sigma^2}\right)$$

$$\int_{\mathbb{R}^{\mathsf{J}}} \frac{1}{(2\pi\sigma^2)^{\mathsf{J}/2}} \exp\left(-\frac{\|\mathbf{y}\|^2}{2\sigma^2}\right) \left(\sum_{m \in \mathcal{M}} \mathrm{I}\{\mathbf{y} \in \mathcal{D}_m\} \exp\left(\frac{1}{\sigma^2} \langle \mathbf{y}, \mathbf{s}_m\rangle_{\mathrm{E}}\right)\right) d\mathbf{y}.$$

(iii) Finally show that

$$\sum_{m \in \mathcal{M}} \mathrm{I}\{\mathbf{y} \in \mathcal{D}_m\} \exp\left(\frac{1}{\sigma^2} \langle \mathbf{y}, \mathbf{s}_m\rangle_{\mathrm{E}}\right) = \exp\left(\frac{1}{\sigma^2} \max_m \langle \mathbf{y}, \mathbf{s}_m\rangle_{\mathrm{E}}\right), \quad \mathbf{y} \in \mathbb{R}^{\mathsf{J}}.$$

See also Problem 23.7.

**Exercise 21.6 (When Is the Union Bound Tight?).** Under what conditions on the events $\mathcal{V}_1, \mathcal{V}_2, \ldots$ is the Union Bound (21.32) tight?

**Exercise 21.7 (The Union of Independent Events).** Show that if the events $\mathcal{V}_1, \mathcal{V}_2, \ldots, \mathcal{V}_n$ are independent then

$$\Pr\left(\bigcup_{j=1}^{n} \mathcal{V}_j\right) = 1 - \prod_{j=1}^{n}\left(1 - \Pr(\mathcal{V}_j)\right).$$

**Exercise 21.8 (A Lower Bound on the Probability of a Union).** Show that the probability of the union of $n$ events $\mathcal{V}_1, \ldots, \mathcal{V}_n$ can be lower-bounded by

$$\Pr\left(\bigcup_{j=1}^{n} \mathcal{V}_j\right) \geq \sum_{j=1}^{n} \Pr(\mathcal{V}_j) - \sum_{j=1}^{n-1} \sum_{\ell=j+1}^{n} \Pr(\mathcal{V}_j \cap \mathcal{V}_\ell).$$

Inequalities of this nature are sometimes called Bonferroni Inequalities.

**Exercise 21.9 (de Caen's Inequality).** Let $X$ be a RV taking value in the finite set $\mathcal{X}$, and let $\{\mathcal{A}_i\}_{i \in \mathcal{I}}$ be a finite family of subsets (not necessarily disjoint) of $\mathcal{X}$:

$$\mathcal{A}_i \subseteq \mathcal{X}, \quad i \in \mathcal{I}.$$

Define

$$\Pr(\mathcal{A}_i) \triangleq \Pr[X \in \mathcal{A}_i], \quad i \in \mathcal{I},$$

$$\deg(x) \triangleq \#\{i \in \mathcal{I} : x \in \mathcal{A}_i\}, \quad x \in \mathcal{X},$$

where $\#\mathcal{B}$ denotes the cardinality of a set $\mathcal{B}$.

(i) Show that

$$\Pr\left(\bigcup_{i \in \mathcal{I}} \mathcal{A}_i\right) = \sum_{i \in \mathcal{I}} \sum_{x \in \mathcal{A}_i} \frac{\Pr[X = x]}{\deg(x)}.$$

(ii) Use the Cauchy-Schwarz Inequality to show that for every $i \in \mathcal{I}$,

$$\left(\sum_{x \in \mathcal{A}_i} \frac{\Pr[X = x]}{\deg(x)}\right)\left(\sum_{x \in \mathcal{A}_i} \Pr[X = x]\deg(x)\right) \geq \left(\sum_{x \in \mathcal{A}_i} \Pr[X = x]\right)^2.$$

(iii) Use Parts (i) and (ii) to show that

$$\Pr\left(\bigcup_{i \in \mathcal{I}} \mathcal{A}_i\right) \geq \sum_{i \in \mathcal{I}} \frac{\left(\sum_{x \in \mathcal{A}_i} \Pr[X = x]\right)^2}{\sum_{j \in \mathcal{I}} \sum_{x' \in \mathcal{A}_i \cap \mathcal{A}_j} \Pr[X = x']}.$$

(iv) Conclude that

$$\Pr\left(\bigcup_{i \in \mathcal{I}} \mathcal{A}_i\right) \geq \sum_{i \in \mathcal{I}} \frac{\Pr(\mathcal{A}_i)^2}{\sum_{j \in \mathcal{I}} \Pr(\mathcal{A}_i \cap \mathcal{A}_j)}.$$

This is de Caen's Bound (de Caen, 1997).

**Exercise 21.10 (Asymptotic Tightness of the Union Bound).** Consider the hypothesis testing problem of Section 21.6 when the prior is uniform and the mean vectors $\mathbf{s}_1, \ldots, \mathbf{s}_M$ are distinct. Show that the Union Bound of (21.59) is asymptotically tight in the sense that the limiting ratio of the RHS of (21.59) to the LHS tends to one as $\sigma$ tends to zero.

*Hint: Use Exercise 21.8.*

# Chapter 22

# Sufficient Statistics

## 22.1 Introduction

In layman's terms, a sufficient statistic for guessing $M$ based on the observable $\mathbf{Y}$ is a random variable or a collection of random variables that contains all the information in $\mathbf{Y}$ that is relevant for guessing $M$. This is a particularly useful concept when the sufficient statistic is more concise than the observables. For example, if we observe the results of a thousand coin tosses $Y_1, \ldots, Y_{1000}$ and we wish to test whether the coin is fair or has a bias of $1/4$, then a sufficient statistic turns out to be the number of "heads" among the outcomes $Y_1, \ldots, Y_{1000}$.[1] Another example was encountered in Section 20.12. There the observable was a two-dimensional random vector, and the sufficient statistic summarized the information that was relevant for guessing $H$ in a scalar random variable; see (20.69).

In this chapter we provide a formal definition of sufficient statistics in the multi-hypothesis setting and explore the concept in some detail. We shall see that our definition is compatible with Definition 20.12.2, which we gave for the binary case. We only address the case where the observations take value in the $d$-dimensional Euclidean space $\mathbb{R}^d$. Extensions to observations consisting of a stochastic process are discussed in Section 26.3. Also, we only treat the case of guessing among a finite number of alternatives. We thus consider a finite set of messages

$$\mathcal{M} = \{1, \ldots, \mathsf{M}\}, \tag{22.1}$$

where $\mathsf{M} \geq 2$, and we assume that associated with each message $m \in \mathcal{M}$ is a density $f_{\mathbf{Y}|M=m}(\cdot)$ on $\mathbb{R}^d$, i.e., a nonnegative Borel measurable function that integrates to one.

The concept of sufficient statistics is defined for the family of densities

$$f_{\mathbf{Y}|M=m}(\cdot), \quad m \in \mathcal{M}; \tag{22.2}$$

it is unrelated to a prior. But when we wish to use it in the context of hypothesis testing we need to introduce a probabilistic setting. If, in addition to the family

---

[1]Testing whether a coin is fair or not is a more complicated hypothesis testing problem of a kind that we shall not address. It falls under the category of "composite hypothesis testing."

$\{f_{\mathbf{Y}|M=m}(\cdot)\}_{m\in\mathcal{M}}$, we introduce a prior $\{\pi_m\}_{m\in\mathcal{M}}$, then we can discuss the pair $(M, \mathbf{Y})$, where $\Pr[M = m] = \pi_m$, and where, conditionally on $M = m$, the distribution of $\mathbf{Y}$ is of density $f_{\mathbf{Y}|M=m}(\cdot)$. Thus, once we have introduced a prior $\{\pi_m\}_{m\in\mathcal{M}}$ we can, for example, discuss the density $f_{\mathbf{Y}}(\cdot)$ of $\mathbf{Y}$ as in (21.11)

$$f_{\mathbf{Y}}(\mathbf{y}) = \sum_{m\in\mathcal{M}} \pi_m\, f_{\mathbf{Y}|M=m}(\mathbf{y}), \quad \mathbf{y} \in \mathbb{R}^d, \tag{22.3}$$

and the conditional distribution of $M$ conditional on $\mathbf{Y} = \mathbf{y}$ as in (21.12)

$$\Pr[M = m \,|\, \mathbf{Y} = \mathbf{y}] \triangleq \begin{cases} \dfrac{\pi_m\, f_{\mathbf{Y}|M=m}(\mathbf{y})}{f_{\mathbf{Y}}(\mathbf{y})} & \text{if } f_{\mathbf{Y}}(\mathbf{y}) > 0, \\ \dfrac{1}{\mathsf{M}} & \text{otherwise,} \end{cases} \quad m \in \mathcal{M},\ \mathbf{y} \in \mathbb{R}^d. \tag{22.4}$$

## 22.2 Definition and Main Consequence

In this section we shall define sufficient statistics for a family of densities (22.2). We shall then state the main result about this notion, namely, that there is no loss in optimality in basing one's guess on a sufficient statistic.

Very roughly, $T(\cdot)$ (or sometimes $T(\mathbf{Y})$) forms a sufficient statistic for guessing $M$ based on $\mathbf{Y}$ if there exists a black box that, when fed $T(\mathbf{y}_{\text{obs}})$ (but not $\mathbf{y}_{\text{obs}}$) and any prior $\{\pi_m\}$ on $\mathcal{M}$ produces the *a posteriori* distribution of $M$ given $\mathbf{Y} = \mathbf{y}_{\text{obs}}$.

For technical reasons we make two exceptions. While the black box must always produce a probability vector, we only require that this vector be the *a posteriori* distribution of $M$ given $\mathbf{Y} = \mathbf{y}_{\text{obs}}$ for observations $\mathbf{y}_{\text{obs}}$ that satisfy

$$\sum_{m\in\mathcal{M}} \pi_m\, f_{\mathbf{Y}|M=m}(\mathbf{y}_{\text{obs}}) > 0 \tag{22.5}$$

and that lie outside some prespecified set $\mathcal{Y}_0 \subset \mathbb{R}^d$ of Lebesgue measure zero. Thus, if $\mathbf{y}_{\text{obs}}$ is in $\mathcal{Y}_0$ or if (22.5) is violated, then the output of the black box can be any probability vector. The exception set $\mathcal{Y}_0$ is not allowed to depend on $\{\pi_m\}$. Since it is of Lebesgue measure zero, the conditional probability that the observation $\mathbf{Y}$ lies in $\mathcal{Y}_0$ is zero:

$$\Pr[\mathbf{Y} \in \mathcal{Y}_0 \,|\, M = m] = 0, \quad m \in \mathcal{M}. \tag{22.6}$$

Note that the black box need not indicate whether $\mathbf{y}_{\text{obs}}$ is in $\mathcal{Y}_0$ and/or whether (22.5) holds. Figure 22.1 depicts such a black box.

**Definition 22.2.1 (Sufficient Statistics for M Densities).** *We say that a mapping* $T\colon \mathbb{R}^d \to \mathbb{R}^{d'}$ *forms a sufficient statistic for the densities* $f_{\mathbf{Y}|M=1}(\cdot), \ldots, f_{\mathbf{Y}|M=\mathsf{M}}(\cdot)$ *on* $\mathbb{R}^d$ *if it is Borel measurable and if for some* $\mathcal{Y}_0 \subset \mathbb{R}^d$ *of Lebesgue measure zero we have that for every prior* $\{\pi_m\}$ *there exist* $\mathsf{M}$ *Borel measurable functions from* $\mathbb{R}^{d'}$ *to* $[0, 1]$

$$T(\mathbf{y}_{\text{obs}}) \mapsto \psi_m\big(\{\pi_m\}, T(\mathbf{y}_{\text{obs}})\big), \quad m \in \mathcal{M},$$

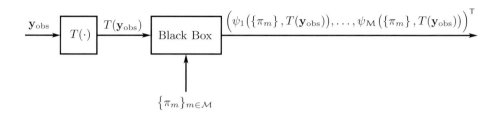

**Figure 22.1:** A black box that when fed any prior $\{\pi_m\}$ and $T(\mathbf{y}_{\mathrm{obs}})$ (but not the observation $\mathbf{y}_{\mathrm{obs}}$ directly) produces a probability vector that is equal to $(\Pr[M = 1 \mid \mathbf{Y} = \mathbf{y}_{\mathrm{obs}}], \ldots, \Pr[M = \mathsf{M} \mid \mathbf{Y} = \mathbf{y}_{\mathrm{obs}}])^{\mathsf{T}}$ whenever both the condition $\sum_{m \in \mathcal{M}} \pi_m f_{Y \mid M = m}(\mathbf{y}_{\mathrm{obs}}) > 0$ and the condition $\mathbf{y}_{\mathrm{obs}} \notin \mathcal{Y}_0$ are satisfied.

*such that the vector*

$$\left( \psi_1\big(\{\pi_m\}, T(\mathbf{y}_{\mathrm{obs}})\big), \ldots, \psi_{\mathsf{M}}\big(\{\pi_m\}, T(\mathbf{y}_{\mathrm{obs}})\big) \right)^{\mathsf{T}}$$

*is a probability vector and such that this probability vector is equal to*

$$\left( \Pr[M = 1 \mid \mathbf{Y} = \mathbf{y}_{\mathrm{obs}}], \ldots, \Pr[M = \mathsf{M} \mid \mathbf{Y} = \mathbf{y}_{\mathrm{obs}}] \right)^{\mathsf{T}} \qquad (22.7)$$

*whenever both the condition* $\mathbf{y}_{\mathrm{obs}} \notin \mathcal{Y}_0$ *and the condition*

$$\sum_{m=1}^{\mathsf{M}} \pi_m f_{\mathbf{Y} \mid M = m}(\mathbf{y}_{\mathrm{obs}}) > 0 \qquad (22.8)$$

*are satisfied. Here* (22.7) *is computed for* $M$ *having the prior* $\{\pi_m\}$ *and for the conditional law of* $\mathbf{Y}$ *given* $M$ *corresponding to the given densities.*

The main result regarding sufficient statistics is that if $T(\cdot)$ forms a sufficient statistic, then—even if the transformation $T(\cdot)$ is not reversible—there is no loss in optimality in basing one's guess on $T(\mathbf{Y})$.

**Proposition 22.2.2 (Guessing Based on $T(\mathbf{Y})$ Is Optimal).** *If* $T \colon \mathbb{R}^d \to \mathbb{R}^{d'}$ *is a sufficient statistic for the* $\mathsf{M}$ *densities* $\{f_{\mathbf{Y} \mid M = m}(\cdot)\}_{m \in \mathcal{M}}$, *then, given any prior* $\{\pi_m\}$, *there exists an optimal decision rule that bases its decision on* $T(\mathbf{Y})$.

**Proof.** To prove the proposition we shall exhibit a decision rule that is based on $T(\mathbf{Y})$ and that mimics the MAP rule based on $\mathbf{Y}$. Since the latter is optimal (Theorem 21.3.3), our proposed rule must also be optimal. Let $\{\psi_m(\cdot)\}$ be as in Definition 22.2.1. Given $\mathbf{Y} = \mathbf{y}_{\mathrm{obs}}$, the proposed decoder considers the set of all messages $\tilde{m}$ satisfying

$$\psi_{\tilde{m}}\big(\{\pi_m\}, T(\mathbf{y}_{\mathrm{obs}})\big) = \max_{m' \in \mathcal{M}} \psi_{m'}\big(\{\pi_m\}, T(\mathbf{y}_{\mathrm{obs}})\big) \qquad (22.9)$$

and picks uniformly at random from this set.

We next argue that this decision rule is optimal. To that end we shall show that, with probability one, this guessing rule is the same as the MAP rule for guessing $M$ based on $\mathbf{Y}$. Indeed, the guess produced by this rule is identical to the one produced by the MAP rule whenever $\mathbf{y}_{\text{obs}}$ satisfies (22.8) and lies outside $\mathcal{Y}_0$. Since the probability that $\mathbf{Y}$ satisfies (22.8) is, by (21.13), one, and since the probability that $\mathbf{Y}$ is outside $\mathcal{Y}_0$ is, by (22.6), also one, it follows from Corollary 21.5.2 that the probability that $\mathbf{Y}$ satisfies both (22.8) and the condition $\mathbf{Y} \notin \mathcal{Y}_0$ is also one. Thus, the proposed guessing rule, which bases its decision only on $T(\mathbf{y}_{\text{obs}})$ and on the prior has the same performance as the (optimal) MAP decision rule for guessing $M$ based on $\mathbf{Y}$. $\qquad\qquad\square$

## 22.3   Equivalent Conditions

In this section we derive a number of important equivalent definitions for sufficient statistics. These will further clarify the concept and will also be useful in identifying sufficient statistics. We shall try to state the theorems rigorously, but our proofs will be mostly heuristic. Rigorous proofs require some Measure Theory that we do not wish to assume. For a rigorous measure-theoretic treatment of this topic see (Halmos and Savage, 1949), (Lehmann and Romano, 2005, Section 2.6), or (Billingsley, 1995, Section 34).[2]

### 22.3.1   The Factorization Theorem

The following characterization is useful because it is purely algebraic. It explores the form that the densities $\{f_{\mathbf{Y}|M=m}(\cdot)\}$ must have for $T(\mathbf{Y})$ to form a sufficient statistic. Roughly speaking, $T(\cdot)$ is sufficient if the densities in the family all have the form of a product of two functions, where the first function depends on the message and on $T(\mathbf{y})$, and where the second function does not depend on the message but may depend on $\mathbf{y}$. We allow, however, an exception set $\mathcal{Y}_0 \subset \mathbb{R}^d$ of Lebesgue measure zero, so we only require that for every $m \in \mathcal{M}$

$$f_{\mathbf{Y}|M=m}(\mathbf{y}) = g_m\big(T(\mathbf{y})\big)\, h(\mathbf{y}), \quad \mathbf{y} \notin \mathcal{Y}_0. \tag{22.10}$$

Note that if such a factorization exists, then it also exists with the additional requirement that the functions be nonnegative. Indeed, if (22.10) holds, then by the nonnegativity of the densities

$$
\begin{aligned}
f_{\mathbf{Y}|M=m}(\mathbf{y}) &= \big|f_{\mathbf{Y}|M=m}(\mathbf{y})\big| \\
&= \big|g_m\big(T(\mathbf{y})\big)\, h(\mathbf{y})\big|, \quad \mathbf{y} \notin \mathcal{Y}_0 \\
&= \big|g_m\big(T(\mathbf{y})\big)\big|\, |h(\mathbf{y})|, \quad \mathbf{y} \notin \mathcal{Y}_0,
\end{aligned}
$$

thus yielding a factorization with the nonnegative functions

$$\big\{\mathbf{y} \mapsto \big|g_m\big(T(\mathbf{y})\big)\big|\big\}_{m \in \mathcal{M}} \quad \text{and} \quad \mathbf{y} \mapsto |h(\mathbf{y})|.$$

---

[2]Our setting is technically easier because we only consider the case where $\mathcal{M}$ is *finite* and because we restrict the observation space to $\mathbb{R}^d$.

Limiting ourselves to nonnegative factorizations, as we henceforth shall, is helpful in manipulating inequalities where multiplication by negative numbers requires changing the direction of the inequality. For our setting the Factorization Theorem can be stated as follows.[3]

**Theorem 22.3.1 (The Factorization Theorem).** *A Borel measurable function* $T \colon \mathbb{R}^d \to \mathbb{R}^{d'}$ *forms a sufficient statistic for the* $\mathsf{M}$ *densities* $\{f_{\mathbf{Y}|M=m}(\cdot)\}_{m \in \mathcal{M}}$ *on* $\mathbb{R}^d$ *if, and only if, there exists a set* $\mathcal{Y}_0 \subset \mathbb{R}^d$ *of Lebesgue measure zero and nonnegative Borel measurable functions* $\mathbf{g}_1, \ldots, \mathbf{g}_\mathsf{M} \colon \mathbb{R}^{d'} \to [0, \infty)$ *and* $\mathbf{h} \colon \mathbb{R}^d \to [0, \infty)$ *such that for every* $m \in \mathcal{M}$

$$f_{\mathbf{Y}|M=m}(\mathbf{y}) = g_m\big(T(\mathbf{y})\big)\, h(\mathbf{y}), \quad \mathbf{y} \in \mathbb{R}^d \setminus \mathcal{Y}_0. \tag{22.11}$$

**Proof.** We begin by showing that if $T(\cdot)$ is a sufficient statistic then there exists a factorization of the form (22.11). Let the set $\mathcal{Y}_0$ and the functions $\{\psi_m(\cdot)\}$ be as in Definition 22.2.1. Pick some $\tilde{\pi}_1, \ldots, \tilde{\pi}_\mathsf{M} > 0$ that sum to one, e.g., $\tilde{\pi}_m = 1/\mathsf{M}$ for all $m \in \mathcal{M}$, and let $M$ be of the prior $\{\tilde{\pi}_m\}$, so $\Pr[M = m] = \tilde{\pi}_m$ for all $m \in \mathcal{M}$. Let the conditional law of $\mathbf{Y}$ given $M$ be as specified by the given densities so, in particular,

$$f_{\mathbf{Y}}(\mathbf{y}) = \sum_{m \in \mathcal{M}} \tilde{\pi}_m\, f_{\mathbf{Y}|M=m}(\mathbf{y}), \quad \mathbf{y} \in \mathbb{R}^d. \tag{22.12}$$

Since $\{\tilde{\pi}_m\}$ are strictly positive, it follows from (22.12) that

$$\Big(f_{\mathbf{Y}}(\mathbf{y}) = 0\Big) \Rightarrow \Big(f_{\mathbf{Y}|M=m}(\mathbf{y}) = 0, \quad m \in \mathcal{M}\Big). \tag{22.13}$$

(The only way the sum of nonnegative numbers can be zero is if they are all zero. Thus, $f_{\mathbf{Y}}(\mathbf{y}) = 0$ always implies that all the terms $\{\tilde{\pi}_m\, f_{\mathbf{Y}|M=m}(\mathbf{y})\}$ are zero. But if $\{\tilde{\pi}_m\}$ are strictly positive, then this implies that all the terms $\{f_{\mathbf{Y}|M=m}(\mathbf{y})\}$ are zero.)

By the definition of the functions $\{\psi_m(\cdot)\}$ and of the conditional probability (22.4), we have for every $m \in \mathcal{M}$

$$\psi_m\big(\tilde{\pi}_1, \ldots, \tilde{\pi}_\mathsf{M}, T(\mathbf{y}_{\mathrm{obs}})\big) = \frac{\tilde{\pi}_m\, f_{\mathbf{Y}|M=m}(\mathbf{y}_{\mathrm{obs}})}{f_{\mathbf{Y}}(\mathbf{y}_{\mathrm{obs}})}, \quad \Big(\mathbf{y}_{\mathrm{obs}} \notin \mathcal{Y}_0 \text{ and } f_{\mathbf{Y}}(\mathbf{y}_{\mathrm{obs}}) > 0\Big). \tag{22.14}$$

We next argue that the densities factorize as

$$f_{\mathbf{Y}|M=m}(\mathbf{y}) = \underbrace{\frac{1}{\tilde{\pi}_m} \psi_m\big(\tilde{\pi}_1, \ldots, \tilde{\pi}_\mathsf{M}, T(\mathbf{y})\big)}_{g_m(T(\mathbf{y}))}\, \underbrace{f_{\mathbf{Y}}(\mathbf{y})}_{h(\mathbf{y})}, \quad \mathbf{y} \in \mathbb{R}^d \setminus \mathcal{Y}_0. \tag{22.15}$$

---

[3]A different, perhaps more elegant, way to state the theorem is in terms of probability distributions. Let $P_m$ be the probability distribution on $\mathbb{R}^d$ corresponding to $M = m$, where $m$ is in the finite set $\mathcal{M}$. Assume that $\{P_m\}$ are dominated by the $\sigma$-finite measure $\mu$. Then the Borel measurable mapping $T \colon \mathbb{R}^d \to \mathbb{R}^{d'}$ forms a sufficient statistic for the family $\{P_m\}$ if, and only if, there exists a Borel measurable nonnegative function $h(\cdot)$ from $\mathbb{R}^d$ to $\mathbb{R}$, and $\mathsf{M}$ nonnegative, Borel measurable functions $g_m(\cdot)$ from $\mathbb{R}^{d'}$ to $\mathbb{R}$ such that for each $m \in \mathcal{M}$ the function $\mathbf{y} \mapsto g_m(T(\mathbf{y}))\, h(\mathbf{y})$ is a version of the Radon-Nikodym derivative $\mathrm{d}P_m / \mathrm{d}\mu$ of $P_m$ with respect to $\mu$; see (Billingsley, 1995, Theorem 34.6) and (Lehmann and Romano, 2005, Corollary 2.6.1).

This can be argued as follows. If $f_{\mathbf{Y}}(\mathbf{y})$ is greater than zero, then (22.15) follows directly from (22.14). And if $f_{\mathbf{Y}}(\mathbf{y})$ is equal to zero, then RHS of (22.15) is equal to zero and, by (22.13), the LHS is also equal to zero.

We next prove that if the densities factorize as in (22.11), then $T(\cdot)$ forms a sufficient statistic. That is, we show how using the factorization (22.11) we can design the desired black box. The inputs to the black box are the prior $\{\pi_m\}$ and $T(\mathbf{y})$. The black box considers the vector

$$\left( \pi_1\, g_1\big(T(\mathbf{y})\big), \ldots, \pi_{\mathsf{M}}\, g_{\mathsf{M}}\big(T(\mathbf{y})\big) \right)^{\mathsf{T}}. \tag{22.16}$$

If all its components are zero, then the black box produces the uniform distribution (or any other distribution of the reader's choice). Otherwise, it produces the above vector but normalized to sum to one. Thus, if we denote by $\psi_m(\pi_1, \ldots, \pi_{\mathsf{M}}, T(\mathbf{y}))$ the probability that the black box assigns to $m$ when fed $\pi_1, \ldots, \pi_{\mathsf{M}}$ and $T(\mathbf{y})$, then

$$\psi_m(\pi_1, \ldots, \pi_{\mathsf{M}}, T(\mathbf{y})) \triangleq \begin{cases} \dfrac{1}{\mathsf{M}} & \text{if } \sum_{m'=1}^{\mathsf{M}} \pi_{m'}\, g_{m'}\big(T(\mathbf{y})\big) = 0, \\[2ex] \dfrac{\pi_m\, g_m\big(T(\mathbf{y})\big)}{\sum_{m' \in \mathcal{M}} \pi_{m'}\, g_{m'}\big(T(\mathbf{y})\big)} & \text{otherwise.} \end{cases} \tag{22.17}$$

To verify that $\psi_m(\pi_1, \ldots, \pi_{\mathsf{M}}, T(\mathbf{y})) = \Pr[M = m \,|\, \mathbf{Y} = \mathbf{y}]$ whenever $\mathbf{y}$ is such that $\mathbf{y} \notin \mathcal{Y}_0$ and (22.8) holds, we first note that, by the factorization (22.11),

$$\left( f_{\mathbf{Y}}(\mathbf{y}) > 0 \text{ and } \mathbf{y} \notin \mathcal{Y}_0 \right) \Rightarrow \left( h(\mathbf{y}) \sum_{m'=1}^{\mathsf{M}} \pi_{m'}\, g_{m'}\big(T(\mathbf{y})\big) > 0 \right),$$

so

$$\left( f_{\mathbf{Y}}(\mathbf{y}) > 0 \text{ and } \mathbf{y} \notin \mathcal{Y}_0 \right) \Rightarrow \left( h(\mathbf{y}) > 0 \quad \text{and} \quad \sum_{m'=1}^{\mathsf{M}} \pi_{m'}\, g_{m'}\big(T(\mathbf{y})\big) > 0 \right). \tag{22.18}$$

Consequently, if $\mathbf{y} \notin \mathcal{Y}_0$ and if (22.8) holds, then by (22.18) & (22.17)

$$\left( \psi_1\big(\pi_1, \ldots, \pi_{\mathsf{M}}, T(\mathbf{y})\big), \ldots, \psi_{\mathsf{M}}\big(\pi_1, \ldots, \pi_{\mathsf{M}}, T(\mathbf{y})\big) \right)^{\mathsf{T}}$$

is equal to the vector in (22.16) but scaled so that its components add to one. But the *a posteriori* probability vector is also a scaled version of (22.16) (scaled by $h(\mathbf{y})/f_{\mathbf{Y}}(\mathbf{y})$) that sums to one. Thus, if $\mathbf{y} \notin \mathcal{Y}_0$ and (22.8) holds, then the vector produced by the black box is identical to the *a posteriori* distribution vector. $\square$

### 22.3.2 Pairwise sufficiency

We next clarify the connection between sufficient statistics for binary hypothesis testing and for multi-hypothesis testing. We show that $T(\mathbf{Y})$ forms a sufficient statistic for the family of densities $\{f_{\mathbf{Y}|M=m}(\cdot)\}_{m \in \mathcal{M}}$ if, and only if, for every pair of messages $m' \neq m''$ in $\mathcal{M}$ we have that $T(\mathbf{Y})$ forms a sufficient statistic for the densities $f_{\mathbf{Y}|M=m'}(\cdot)$ and $f_{\mathbf{Y}|M=m''}(\cdot)$.

One part of this statement is trivial, namely, that if $T(\cdot)$ is sufficient for the family $\{f_{\mathbf{Y}|M=m}(\cdot)\}_{m \in \mathcal{M}}$ then it is also sufficient for any pair. Indeed, by the Factorization Theorem (Theorem 22.3.1), the sufficiency of $T(\cdot)$ for the family implies the existence of a set of Lebesgue measure zero $\mathcal{Y}_0 \subset \mathbb{R}^d$ and functions $\{g_m\}_{m \in \mathcal{M}}$, $\mathbf{h}$ such that for all $\mathbf{y} \in \mathbb{R}^d \setminus \mathcal{Y}_0$

$$f_{\mathbf{Y}|M=m}(\mathbf{y}) = g_m\big(T(\mathbf{y})\big)\, h(\mathbf{y}), \quad m \in \mathcal{M}. \tag{22.19}$$

In particular, if we limit ourselves to $m', m'' \in \mathcal{M}$ then for $\mathbf{y} \notin \mathcal{Y}_0$

$$f_{\mathbf{Y}|M=m'}(\mathbf{y}) = g_{m'}\big(T(\mathbf{y})\big)\, h(\mathbf{y}),$$

$$f_{\mathbf{Y}|M=m''}(\mathbf{y}) = g_{m''}\big(T(\mathbf{y})\big)\, h(\mathbf{y}),$$

which, by the Factorization Theorem, implies the sufficiency of $T(\cdot)$ for the pair of densities $f_{\mathbf{Y}|M=m'}(\cdot), f_{\mathbf{Y}|M=m''}(\cdot)$.

The nontrivial part of the proposition is that pairwise sufficiency implies sufficiency. Even this is quite easy when the densities are all strictly positive. It is a bit more tricky without this assumption.[4]

**Proposition 22.3.2 (Pairwise Sufficiency Implies Sufficiency).** *Consider* M *densities* $\{f_{\mathbf{Y}|M=m}(\cdot)\}_{m \in \mathcal{M}}$ *on* $\mathbb{R}^d$, *and assume that* $T \colon \mathbb{R}^d \to \mathbb{R}^{d'}$ *forms a sufficient statistic for every pair of densities* $f_{\mathbf{Y}|M=m'}(\cdot), f_{\mathbf{Y}|M=m''}(\cdot)$, *where* $m' \neq m''$ *are both in* $\mathcal{M}$. *Then* $T(\cdot)$ *is a sufficient statistic for the* M *densities* $\{f_{\mathbf{Y}|M=m}(\cdot)\}_{m \in \mathcal{M}}$.

**Proof.** To prove that $T(\cdot)$ forms a sufficient statistic for $\{f_{\mathbf{Y}|M=m}(\cdot)\}_{m=1}^{\mathrm{M}}$ we shall describe an algorithm (black box) that when fed any prior $\{\pi_m\}$ and $T(\mathbf{y}_{\mathrm{obs}})$ (but not $\mathbf{y}_{\mathrm{obs}}$) produces an M-dimensional probability vector that is equal to the *a posteriori* probability distribution vector

$$\Big(\mathrm{Pr}\big[M = 1 \mid \mathbf{Y} = \mathbf{y}_{\mathrm{obs}}\big], \ldots, \mathrm{Pr}\big[M = \mathrm{M} \mid \mathbf{Y} = \mathbf{y}_{\mathrm{obs}}\big]\Big)^{\mathsf{T}} \tag{22.20}$$

whenever $\mathbf{y}_{\mathrm{obs}} \in \mathbb{R}^d$ is such that

$$\mathbf{y}_{\mathrm{obs}} \notin \mathcal{Y}_0 \quad \text{and} \quad \sum_{m=1}^{\mathrm{M}} \pi_m f_{\mathbf{Y}|M=m}(\mathbf{y}_{\mathrm{obs}}) > 0, \tag{22.21}$$

where $\mathcal{Y}_0$ is a subset of $\mathbb{R}^d$ that does not depend on the prior $\{\pi_m\}$ and that is of Lebesgue measure zero.

To describe the algorithm we first use the Factorization Theorem (Theorem 22.3.1) to recast the proposition's hypothesis as saying that for every pair $m' \neq m''$ in $\mathcal{M}$ there exists a set $\mathcal{Y}_0^{(m',m'')} \subset \mathbb{R}^d$ of Lebesgue measure zero and there exist nonnegative functions $\mathbf{g}_{m'}^{(m',m'')}, \mathbf{g}_{m''}^{(m',m'')} \colon \mathbb{R}^{d'} \to \mathbb{R}$ and $\mathbf{h}^{(m',m'')} \colon \mathbb{R}^d \to \mathbb{R}$ such that

$$f_{\mathbf{Y}|M=m'}(\mathbf{y}) = g_{m'}^{(m',m'')}\big(T(\mathbf{y})\big)\, h^{(m',m'')}(\mathbf{y}), \quad \mathbf{y} \in \mathbb{R}^d \setminus \mathcal{Y}_0^{(m',m'')}, \tag{22.22a}$$

---

[4]This result does not extend to the case where the random variable $M$ can take on infinitely many values.

$$f_{\mathbf{Y}|M=m''}(\mathbf{y}) = g_{m''}^{(m',m'')}\left(T(\mathbf{y})\right) h^{(m',m'')}(\mathbf{y}), \quad \mathbf{y} \in \mathbb{R}^d \setminus \mathcal{Y}_0^{(m',m'')}. \qquad (22.22\mathrm{b})$$

Let

$$\mathcal{Y}_0 = \bigcup_{\substack{m',m'' \in \mathcal{M} \\ m' \neq m''}} \mathcal{Y}_0^{(m',m'')}, \qquad (22.23)$$

and note that, being the union of a finite number of sets of Lebesgue measure zero, $\mathcal{Y}_0$ is of Lebesgue measure zero.

We now use the above functions $\mathbf{g}_{m'}^{(m',m'')}, \mathbf{g}_{m''}^{(m',m'')}$ to describe the algorithm. Note that $\mathbf{y}_{\mathrm{obs}}$ is never fed directly to the algorithm; only $T(\mathbf{y}_{\mathrm{obs}})$ is used. Let the prior

$$\pi_m = \Pr[M = m], \quad m \in \mathcal{M} \qquad (22.24)$$

be given, and assume without loss of generality that it is nondegenerate in the sense that

$$\pi_m > 0, \quad m \in \mathcal{M}. \qquad (22.25)$$

(If that is not the case, we can set the black box to produce 0 in the coordinates of the output vector corresponding to messages of prior probability zero and then proceed to ignore such messages.) Let $\mathbf{y}_{\mathrm{obs}} \in \mathbb{R}^d$ be arbitrary.

There are two phases to the algorithm. In the first phase the algorithm produces some $m^* \in \mathcal{M}$ whose *a posteriori* probability is guaranteed to be positive whenever (22.21) holds. In fact, if (22.21) holds, then no message has an *a posteriori* probability higher than that of $m^*$ (but this is immaterial to us because we are not content with showing that from $T(\mathbf{y}_{\mathrm{obs}})$ we can compute the message that *a posteriori* has the highest probability; we want to be able to compute the entire *a posteriori* probability vector). In the second phase the algorithm uses $m^*$ to compute the desired *a posteriori* probability vector.

The first phase of the algorithm runs in $\mathsf{M}$ steps. In Step 1 we set $m[1] = 1$. In Step 2 we set

$$m[2] = \begin{cases} 1 & \text{if } \dfrac{\pi_1 g_1^{(1,2)}\left(T(\mathbf{y}_{\mathrm{obs}})\right)}{\pi_2 g_2^{(1,2)}\left(T(\mathbf{y}_{\mathrm{obs}})\right)} > 1, \\ 2 & \text{otherwise.} \end{cases}$$

And in Step $\nu$ for $\nu \in \{2, \dots, \mathsf{M}\}$ we set

$$m[\nu] = \begin{cases} m[\nu-1] & \text{if } \dfrac{\pi_{m[\nu-1]}\, g_{m[\nu-1]}^{(m[\nu-1],\nu)}\left(T(\mathbf{y}_{\mathrm{obs}})\right)}{\pi_\nu\, g_\nu^{(m[\nu-1],\nu)}\left(T(\mathbf{y}_{\mathrm{obs}})\right)} > 1, \\ \nu & \text{otherwise.} \end{cases} \qquad (22.26)$$

Here we use the convention that $a/0 = +\infty$ whenever $a > 0$ and that $0/0 = 1$. We complete the first phase by setting

$$m^* = m[\mathsf{M}]. \qquad (22.27)$$

In the second phase we compute the vector

$$\alpha[m] = \frac{\pi_m\, g_m^{(m,m^*)}\left(T(\mathbf{y}_{\mathrm{obs}})\right)}{\pi_{m^*}\, g_{m^*}^{(m,m^*)}\left(T(\mathbf{y}_{\mathrm{obs}})\right)}, \quad m \in \mathcal{M}. \qquad (22.28)$$

If at least one of the components of $\alpha[\cdot]$ is $+\infty$, then we produce as the algorithm's output the uniform distribution on $\mathcal{M}$. (The output corresponding to this case is immaterial because it will turn out that this case is only possible if $\mathbf{y}_{\text{obs}}$ is such that either $\mathbf{y}_{\text{obs}} \in \mathcal{Y}_0$ or $\sum_m \pi_m f_{\mathbf{Y}|M=m}(\mathbf{y}_{\text{obs}}) = 0$, in which case the algorithm's output is not required to be equal to the *a posteriori* distribution.) Otherwise, the algorithm's output is the vector

$$\left( \frac{\alpha[1]}{\sum_{\nu=1}^{M} \alpha[\nu]}, \ldots, \frac{\alpha[M]}{\sum_{\nu=1}^{M} \alpha[\nu]} \right)^{\mathsf{T}}. \tag{22.29}$$

Having described the algorithm, we now proceed to prove that it produces the *a posteriori* probability vector whenever (22.21) holds. We need to show that if (22.21) holds then

$$\Pr[M = m \,|\, \mathbf{Y} = \mathbf{y}_{\text{obs}}] = \frac{\alpha[m]}{\sum_{\nu=1}^{M} \alpha[\nu]}, \quad m \in \mathcal{M}. \tag{22.30}$$

Since there is nothing to prove if (22.21) does not hold, we shall henceforth assume for the rest of the proof that it does. In this case we have by (22.4)

$$\Pr[M = m \,|\, \mathbf{Y} = \mathbf{y}_{\text{obs}}] = \frac{\pi_m \, f_{\mathbf{Y}|M=m}(\mathbf{y}_{\text{obs}})}{f_{\mathbf{Y}}(\mathbf{y}_{\text{obs}})}, \quad m \in \mathcal{M}. \tag{22.31}$$

We shall prove (22.30) in two steps. In the first step we show that the result $m^*$ of the algorithm's first phase satisfies

$$\Pr[M = m^* \,|\, \mathbf{Y} = \mathbf{y}_{\text{obs}}] > 0. \tag{22.32}$$

To establish (22.32) we shall prove the stronger statement that

$$\Pr\big[M = m^* \,\big|\, \mathbf{Y} = \mathbf{y}_{\text{obs}}\big] = \max_{m \in \mathcal{M}} \Pr\big[M = m \,\big|\, \mathbf{Y} = \mathbf{y}_{\text{obs}}\big]. \tag{22.33}$$

This latter statement follows from the more general claim that for any $\nu \in \mathcal{M}$ (and not only for $\nu = M$) we have, subject to (22.21),

$$\Pr\big[M = m[\nu] \,\big|\, \mathbf{Y} = \mathbf{y}_{\text{obs}}\big] = \max_{1 \le m \le \nu} \Pr\big[M = m \,\big|\, \mathbf{Y} = \mathbf{y}_{\text{obs}}\big]. \tag{22.34}$$

For $\nu = 1$, Statement (22.34) is trivial. For $2 \le \nu \le M$, (22.34) follows from

$$\Pr\big[M = m[\nu] \,\big|\, \mathbf{Y} = \mathbf{y}_{\text{obs}}\big] =$$
$$\max\Big\{ \Pr\big[M = \nu \,\big|\, \mathbf{Y} = \mathbf{y}_{\text{obs}}\big], \Pr\big[M = m[\nu - 1] \,\big|\, \mathbf{Y} = \mathbf{y}_{\text{obs}}\big] \Big\}, \tag{22.35}$$

which we now prove. We prove (22.35) by considering two cases separately depending on whether $\Pr[M = \nu \,|\, \mathbf{Y} = \mathbf{y}_{\text{obs}}]$ and $\Pr[M = m[\nu-1] \,|\, \mathbf{Y} = \mathbf{y}_{\text{obs}}]$ are both zero or not. In the former case there is nothing to prove because (22.35) holds irrespective of whether (22.26) results in $m[\nu]$ being set to $\nu$ or to $m[\nu - 1]$. In the latter case we have by (22.31) and (22.25) that $f_{\mathbf{Y}|M=\nu}(\mathbf{y}_{\text{obs}})$ and $f_{\mathbf{Y}|M=m[\nu-1]}(\mathbf{y}_{\text{obs}})$

are not both zero. Consequently, by (22.22), in this case $h^{(m[\nu-1],\nu)}(\mathbf{y}_{\mathrm{obs}})$ is not only nonnegative but strictly positive. It follows that the choice (22.26) guarantees (22.35) because

$$
\frac{\pi_{m[\nu-1]}\, g_{m[\nu-1]}^{(m[\nu-1],\nu)}\big(T(\mathbf{y}_{\mathrm{obs}})\big)}{\pi_\nu\, g_\nu^{(m[\nu-1],\nu)}\big(T(\mathbf{y}_{\mathrm{obs}})\big)}
$$

$$
= \frac{\pi_{m[\nu-1]}\, g_{m[\nu-1]}^{(m[\nu-1],\nu)}\big(T(\mathbf{y}_{\mathrm{obs}})\big)\, h^{(m[\nu-1],\nu)}(\mathbf{y}_{\mathrm{obs}})}{\pi_\nu\, g_\nu^{(m[\nu-1],\nu)}\big(T(\mathbf{y}_{\mathrm{obs}})\big)\, h^{(m[\nu-1],\nu)}(\mathbf{y}_{\mathrm{obs}})}
$$

$$
= \frac{\pi_{m[\nu-1]}\, g_{m[\nu-1]}^{(m[\nu-1],\nu)}\big(T(\mathbf{y}_{\mathrm{obs}})\big)\, h^{(m[\nu-1],\nu)}(\mathbf{y}_{\mathrm{obs}})/f_{\mathbf{Y}}(\mathbf{y}_{\mathrm{obs}})}{\pi_\nu\, g_\nu^{(m[\nu-1],\nu)}\big(T(\mathbf{y}_{\mathrm{obs}})\big)\, h^{(m[\nu-1],\nu)}(\mathbf{y}_{\mathrm{obs}})/f_{\mathbf{Y}}(\mathbf{y}_{\mathrm{obs}})}
$$

$$
= \frac{\Pr\big[M = m[\nu-1]\,\big|\,\mathbf{Y} = \mathbf{y}_{\mathrm{obs}}\big]}{\Pr[M = \nu]\,|\,\mathbf{Y} = \mathbf{y}_{\mathrm{obs}}]},
$$

where the first equality follows because $h^{(m[\nu-1],\nu)}(\mathbf{y}_{\mathrm{obs}})$ is strictly positive; the second because in this part of the proof we are assuming (22.21); and where the last equality follows from (22.22) and (22.31). This establishes (22.35), which implies (22.34), which in turn implies (22.33), which in turn implies (22.32), and thus concludes the proof of the first step.

In the second step of the proof we use (22.32) to establish (22.30). This is straightforward because, in view of (22.31), we have that (22.32) implies that $f_{\mathbf{Y}|M=m^*}(\mathbf{y}_{\mathrm{obs}}) > 0$ so, by (22.22b), we have that

$$
h^{(m,m^*)}(\mathbf{y}_{\mathrm{obs}}) > 0, \quad m \in \mathcal{M},
$$

$$
g_{m^*}^{(m,m^*)}(\mathbf{y}_{\mathrm{obs}}) > 0, \quad m \in \mathcal{M}.
$$

Consequently

$$
\alpha[m] = \frac{\pi_m\, g_m^{(m,m^*)}\big(T(\mathbf{y}_{\mathrm{obs}})\big)}{\pi_{m^*}\, g_{m^*}^{(m,m^*)}\big(T(\mathbf{y}_{\mathrm{obs}})\big)}
$$

$$
= \frac{\pi_m\, g_m^{(m,m^*)}\big(T(\mathbf{y}_{\mathrm{obs}})\big)\, h^{(m,m^*)}(\mathbf{y}_{\mathrm{obs}})}{\pi_{m^*}\, g_{m^*}^{(m,m^*)}\big(T(\mathbf{y}_{\mathrm{obs}})\big)\, h^{(m,m^*)}(\mathbf{y}_{\mathrm{obs}})}
$$

$$
= \frac{\pi_m\, g_m^{(m,m^*)}\big(T(\mathbf{y}_{\mathrm{obs}})\big)\, h^{(m,m^*)}(\mathbf{y}_{\mathrm{obs}})/f_{\mathbf{Y}}(\mathbf{y}_{\mathrm{obs}})}{\pi_{m^*}\, g_{m^*}^{(m,m^*)}\big(T(\mathbf{y}_{\mathrm{obs}})\big)\, h^{(m,m^*)}(\mathbf{y}_{\mathrm{obs}})/f_{\mathbf{Y}}(\mathbf{y}_{\mathrm{obs}})}
$$

$$
= \frac{\Pr[M = m\,|\,\mathbf{Y} = \mathbf{y}_{\mathrm{obs}}]}{\Pr[M = m^*\,|\,\mathbf{Y} = \mathbf{y}_{\mathrm{obs}}]},
$$

from which (22.30) follows by (22.32). $\qquad\square$

### 22.3.3 Markov Condition

We now characterize sufficient statistics using Markov chains and conditional independence. These concepts were introduced in Section 20.11. The key result we ask the reader to recall is Theorem 20.11.3. We rephrase it for our present setting as follows.

**Proposition 22.3.3.** *The statement that* $M \!-\!\!\circ\!-T(\mathbf{Y}) \!-\!\!\circ\!-\mathbf{Y}$ *forms a Markov chain is equivalent to each of the following statements:*

(a) *The conditional distribution of* $M$ *given* $\big(T(\mathbf{Y}), \mathbf{Y}\big)$ *is the same as given* $T(\mathbf{Y})$.

(b) $M$ *and* $\mathbf{Y}$ *are conditionally independent given* $T(\mathbf{Y})$.

(c) *The conditional distribution of* $\mathbf{Y}$ *given* $\big(M, T(\mathbf{Y})\big)$ *is the same as given* $T(\mathbf{Y})$.

Statement (a) can also be written as:

(a') *The conditional distribution of* $M$ *given* $\mathbf{Y}$ *is the same as given* $T(\mathbf{Y})$.

Indeed, the conditional distribution of any random variable—in particular $M$—given $\big(T(\mathbf{Y}), \mathbf{Y}\big)$ is the same as given $\mathbf{Y}$ only, because $T(\mathbf{Y})$ carries no information that is not in $\mathbf{Y}$.

Statement (a') can be rephrased as saying that the conditional distribution of $M$ given $\mathbf{Y}$ can be computed from $T(\mathbf{Y})$. Since this is the key requirement of sufficient statistics, we obtain:

**Proposition 22.3.4.** *A Borel measurable function* $T \colon \mathbb{R}^d \to \mathbb{R}^{d'}$ *forms a sufficient statistic for the* M *densities* $\{f_{\mathbf{Y}|M=m}(\cdot)\}_{m \in \mathcal{M}}$ *if, and only if, for any prior* $\{\pi_m\}$

$$M \!-\!\!\circ\!-T(\mathbf{Y}) \!-\!\!\circ\!-\mathbf{Y} \qquad\qquad (22.36)$$

*forms a Markov chain.*

**Proof.** The proof of this proposition is omitted. It is not difficult, but it requires some measure-theoretic tools.[5]                                                                                   $\square$

Using Proposition 22.3.4 and Proposition 22.3.3 (*cf.* (b)) we obtain that a Borel measurable function $T(\cdot)$ forms a sufficient statistic for guessing $M$ based on $\mathbf{Y}$ if, and only if, for any prior $\{\pi_m\}$ on $M$, the message $M$ and the observation $\mathbf{Y}$ are conditionally independent given $T(\mathbf{Y})$.

We next explore the implications of Proposition 22.3.4 and the equivalence of the Markovity $M \!-\!\!\circ\!-T(\mathbf{Y}) \!-\!\!\circ\!-\mathbf{Y}$ and Statement (c) in Proposition 22.3.3. These imply that a Borel measurable function $T(\cdot)$ forms a sufficient statistic if, and only if, the conditional distribution of $\mathbf{Y}$ given $\big(T(\mathbf{Y}), M = m\big)$ is the same for all $m \in \mathcal{M}$. Or, in other words, a Borel measurable function $T(\cdot)$ forms a sufficient statistic if, and only if, the conditional distribution of $\mathbf{Y}$ given $T(\mathbf{Y})$ does not depend on which of the densities in $\{f_{\mathbf{Y}|M=m}(\cdot)\}$ governs the law of $\mathbf{Y}$. This characterization has interesting implications regarding the possibility of simulating observables. These implications are explored next.

---

[5] If $T(\cdot)$ forms a sufficient statistic, then by Definition 22.2.1 $\psi_m\big(\{\pi_m\}, T(\mathbf{Y})\big)$ is a version of the conditional probability that $M = m$ conditional on the $\sigma$-algebra generated by $\mathbf{Y}$, and it is also measurable with respect to the $\sigma$-algebra generated by $T(\mathbf{Y})$. The reverse direction follows from (Lehmann and Romano, 2005, Lemma 2.3.1).

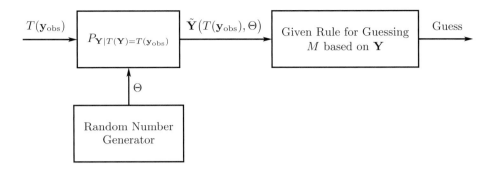

**Figure 22.2:** If $T(\mathbf{Y})$ forms a sufficient statistic for guessing $M$ based on $\mathbf{Y}$, then—even though $\mathbf{Y}$ cannot typically be recovered from $T(\mathbf{Y})$—the performance of any given detector based on $\mathbf{Y}$ can be achieved based on $T(\mathbf{Y})$ and a local random number generator as follows. Using $T(\mathbf{y}_{\mathrm{obs}})$ and local randomness $\Theta$, one produces a $\tilde{\mathbf{Y}}$ whose conditional law given $M = m$ is the same as that of $\mathbf{Y}$, for each $m \in \mathcal{M}$. One then feeds $\tilde{\mathbf{Y}}$ to the given detector.

### 22.3.4 Simulating Observables

For $T(\mathbf{Y})$ to form a sufficient statistic, we do not require that $T(\cdot)$ be invertible, i.e., that $\mathbf{Y}$ be recoverable from $T(\mathbf{Y})$. Indeed, the notion of sufficient statistics is most useful when this transformation is not invertible, in which case $T(\cdot)$ "summarizes" the information in the observation $\mathbf{Y}$ that is needed for guessing $M$. Nevertheless, as we shall next show, if $T(\mathbf{Y})$ forms a sufficient statistic, then from $T(\mathbf{Y})$ we can produce (using a local random number generator) a vector $\tilde{\mathbf{Y}}$ that *appears statistically* like $\mathbf{Y}$ in the sense that the conditional law of $\tilde{\mathbf{Y}}$ given $M$ is identical to the conditional law of $\mathbf{Y}$ given $M$.

To expand on this, we first explain what we mean by "we can produce ... $\tilde{\mathbf{Y}}$" and then elaborate on the consequences of the vector $\tilde{\mathbf{Y}}$ having the same conditional law given $M = m$ as $\mathbf{Y}$. By "producing" $\tilde{\mathbf{Y}}$ from $T(\mathbf{Y})$ we mean that $\tilde{\mathbf{Y}}$ is the result of processing $T(\mathbf{Y})$ with respect to $M$. Stated differently, for every $\mathbf{t} \in \mathbb{R}^{d'}$ there corresponds a probability distribution $P_{\tilde{\mathbf{Y}}|\mathbf{t}}$ (not dependent on $m$) that can be used to generate $\tilde{\mathbf{Y}}$ as follows: having observed $T(\mathbf{y}_{\mathrm{obs}})$, we use a local random number generator to generate the vector $\tilde{\mathbf{Y}}$ according to the distribution $P_{\tilde{\mathbf{Y}}|\mathbf{t}}$, where $\mathbf{t} = T(\mathbf{y}_{\mathrm{obs}})$; see Figure 22.2.

By $\tilde{\mathbf{Y}}$ appearing statistically the same as $\mathbf{Y}$ we mean that the conditional law of $\tilde{\mathbf{Y}}$ given $M = m$ is the same as that of $\mathbf{Y}$, i.e., is of density $f_{\mathbf{Y}|M=m}(\cdot)$. Consequently, anything that can be learned about $M$ from $\mathbf{Y}$ can also be learned about $M$ from $\tilde{\mathbf{Y}}$. Also, any guessing device that was designed to guess $M$ based on the input $\mathbf{Y}$ will yield the same probability of error when, instead of being fed $\mathbf{Y}$, it is fed $\tilde{\mathbf{Y}}$. Thus, if $p(\mathrm{error}|M = m)$ is the conditional error probability associated with a guessing device that is fed $\mathbf{Y}$, then it is also the conditional probability of error that will be incurred by this device if, rather than $\mathbf{Y}$, it is fed $\tilde{\mathbf{Y}}$; see Figure 22.2.

Before stating this as a theorem, let us consider the following simple example. Suppose that our observation consists of $d$ random variables $Y_1, \ldots, Y_d$ and that, conditional on $H = 0$, these random variables are IID Bernoulli($p_0$), i.e., they each take on the value 1 with probability $p_0$ and the value 0 with probability $1 - p_0$. Conditional on $H = 1$, these $d$ random variables are IID Bernoulli($p_1$). Here $0 < p_0, p_1 < 1$ and $p_0 \neq p_1$. Consequently, the conditional probability mass functions are

$$P_{Y_1, \ldots, Y_d | H=0}(y_1, \ldots, y_d) = \prod_{j=1}^{d} \left( p_0^{y_j} (1 - p_0)^{1-y_j} \right)$$

$$= p_0^{\sum_{j=1}^{d} y_j} (1 - p_0)^{d - \sum_{j=1}^{d} y_j},$$

and

$$P_{Y_1, \ldots, Y_d | H=1}(y_1, \ldots, y_d) = p_1^{\sum_{j=1}^{d} y_j} (1 - p_1)^{d - \sum_{j=1}^{d} y_j},$$

so $T(Y_1, \ldots, Y_d) \triangleq \sum_{j=1}^{d} Y_j$ forms a sufficient statistic by the Factorization Theorem.[6] From $T(y_1, \ldots, y_d)$ one cannot recover the sequence $y_1, \ldots, y_d$. Indeed, specifying that $T(y_1, \ldots, y_d) = t$ does not determine which of the random variables is one; it only determines how many of them are one. There are thus $\binom{d}{t}$ possible outcomes $(y_1, \ldots y_d)$ that are consistent with $T(y_1, \ldots, y_d)$ being equal to $t$. We leave it to the reader to verify that if we use a local random number generator to pick one of these outcomes uniformly at random then the result $(\tilde{Y}_1, \ldots \tilde{Y}_d)$ will have the same conditional law given $H$ as $(Y_1, \ldots, Y_d)$. We do not, of course, guarantee that $(\tilde{Y}_1, \ldots \tilde{Y}_d)$ be identical to $(Y_1, \ldots, Y_d)$. (The transformation $T(\cdot)$ is, after all, not reversible.)

For additional insight let us consider our example of (20.66). For $T(y_1, y_2) = y_1^2 + y_2^2$ we can generate $\tilde{\mathbf{Y}}$ from a uniform random variable $\Theta \sim \mathcal{U}([0, 1))$ as

$$\tilde{Y}_1 = \sqrt{T(\mathbf{Y})} \cos(2\pi\Theta)$$

$$\tilde{Y}_2 = \sqrt{T(\mathbf{Y})} \sin(2\pi\Theta).$$

That is, after observing $T(\mathbf{y}_{\text{obs}}) = t$, we generate $(\tilde{Y}_1, \tilde{Y}_2)$ uniformly over the tuples that are at radius $\sqrt{t}$ from the origin.

This last example also demonstrates the difficulty of stating the result. The random vector $\mathbf{Y}$ in this example has a density, both when conditioned on $H = 0$ and when conditioned on $H = 1$. The same applies to the random variable $T(\mathbf{Y})$. However, the distribution that is used to generate $\tilde{\mathbf{Y}}$ from $T(\mathbf{Y})$ is neither discrete nor has a density. All its mass is concentrated on the circle of radius $\sqrt{t}$, so it cannot have a density, and it is uniformly distributed over that circle, so it cannot be discrete.

**Theorem 22.3.5 (Simulating the Observables from the Sufficient Statistic).** *Let* $T \colon \mathbb{R}^d \to \mathbb{R}^{d'}$ *be Borel measurable and let* $f_{\mathbf{Y}|M=1}(\cdot), \ldots, f_{\mathbf{Y}|M=\mathsf{M}}(\cdot)$ *be* $\mathsf{M}$ *densities on* $\mathbb{R}^d$. *Then the following two statements are equivalent:*

*(a)* $T(\cdot)$ *forms a sufficient statistic for the given densities.*

---

[6] For illustration purposes we are extending the discussion here to discrete distributions.

(b) *To every* $\mathbf{t}$ *in* $\mathbb{R}^{d'}$ *there corresponds a distribution on* $\mathbb{R}^d$ *such that the following holds: for every* $m \in \{1, \ldots, \mathsf{M}\}$, *if* $\mathbf{Y} = \mathbf{y}_{\mathrm{obs}}$ *is generated according to the density* $f_{\mathbf{Y}|M=m}(\cdot)$ *and if the random vector* $\tilde{\mathbf{Y}}$ *is then generated according to the distribution corresponding to* $\mathbf{t}$, *where* $\mathbf{t} = T(\mathbf{y}_{\mathrm{obs}})$, *then* $\tilde{\mathbf{Y}}$ *is of density* $f_{\mathbf{Y}|M=m}(\cdot)$.

**Proof.** For a measure-theoretic statement and proof see (Lehmann and Romano, 2005, Theorem 2.6.1). Here we only present some intuition. Ignoring some of the technical details, the proof is very simple. The sufficiency of $T(\cdot)$ is equivalent to $M \!\!-\!\!\circ\!\!-\!\! T(\mathbf{Y}) \!\!-\!\!\circ\!\!-\!\! \mathbf{Y}$ forming a Markov chain for every prior on $M$. This latter condition is equivalent by Proposition 22.3.3 (*cf.* (c)) to the conditional distribution of $\mathbf{Y}$ given $\big(T(\mathbf{Y}), M\big)$ being the same as given $T(\mathbf{Y})$ only. This latter condition is equivalent to the conditional distribution of $\mathbf{Y}$ given $T(\mathbf{Y})$ not depending on which density in the family $\{f_{\mathbf{Y}|M=m}(\cdot)\}_{m \in \mathcal{M}}$ was used to generate $\mathbf{Y}$, i.e., to the existence of a conditional distribution of $\mathbf{Y}$ given $T(\mathbf{Y})$ that does not depend on $m \in \mathcal{M}$. $\qquad\square$

## 22.4   Identifying Sufficient Statistics

Often a sufficient statistic can be identified without having to compute and factorize the conditional densities of the observation. A number of such cases are described in this section.

### 22.4.1   Invertible Transformation

We begin by showing that, ignoring some technical details, any invertible transformation forms a sufficient statistic. It may not be a particularly helpful sufficient statistic because it does not "summarize" the observation, but it is a sufficient statistic nonetheless.

**Proposition 22.4.1 (Reversible Transformations Yield Sufficient Statistics).** *If* $T \colon \mathbb{R}^d \to \mathbb{R}^{d'}$ *is Borel measurable with a Borel measurable inverse, then* $T(\cdot)$ *forms a sufficient statistic for guessing* $M$ *based on* $\mathbf{Y}$.

**Proof.** We provide two proofs. The first uses the definition. We need to verify that from $T(\mathbf{y}_{\mathrm{obs}})$ one can compute the conditional distribution of $M$ given $\mathbf{Y} = \mathbf{y}_{\mathrm{obs}}$. This is obvious because if $\mathbf{t} = T(\mathbf{y}_{\mathrm{obs}})$, then one can compute $\Pr[M = m \,|\, \mathbf{Y} = \mathbf{y}_{\mathrm{obs}}]$ from $\mathbf{t}$ by first applying the inverse $T^{-1}(\mathbf{t})$ to recover $\mathbf{y}_{\mathrm{obs}}$ and by then substituting the result in the expression for $\Pr[M = m \,|\, \mathbf{Y} = \mathbf{y}_{\mathrm{obs}}]$ (21.12).

A second proof can be based on Proposition 22.3.4. We need to verify that for any prior $\{\pi_m\}$

$$M \!\!-\!\!\circ\!\!-\!\! T(\mathbf{Y}) \!\!-\!\!\circ\!\!-\!\! \mathbf{Y}$$

forms a Markov chain. To this end we note that, by Theorem 20.11.3, it suffices to verify that $M$ and $\mathbf{Y}$ are conditionally independent given $T(\mathbf{Y})$. This is clear because the invertibility of $T(\cdot)$ guarantees that, conditional on $T(\mathbf{Y})$, the random vector $\mathbf{Y}$ is deterministic and hence independent of any random variable and *a fortiori* of $M$.                                                                    □

## 22.4.2  A Sufficient Statistic Is Computable from the Statistic

Intuitively, we think about $T(\cdot)$ as forming a sufficient statistic if $T(\mathbf{Y})$ contains all the information about $\mathbf{Y}$ that is relevant to guessing $M$. For this intuition to make sense it had better be the case that if $T(\cdot)$ forms a sufficient statistic for guessing $M$ based on $\mathbf{Y}$, and if $T(\mathbf{Y})$ is computable from $S(\mathbf{Y})$, then $S(\cdot)$ also forms a sufficient statistic. Fortunately, this is so:

**Proposition 22.4.2.** *Suppose that a Borel measurable mapping $T\colon \mathbb{R}^d \to \mathbb{R}^{d'}$ forms a sufficient statistic for the $\mathsf{M}$ densities $\{f_{\mathbf{Y}|M=m}(\cdot)\}_{m\in\mathcal{M}}$ on $\mathbb{R}^d$. Let the mapping $S\colon \mathbb{R}^d \to \mathbb{R}^{d''}$ be Borel measurable. If $T(\cdot)$ can be written as the composition $\psi \circ S$ of $S$ with some Borel measurable function $\psi\colon \mathbb{R}^{d''} \to \mathbb{R}^{d'}$, then $S(\cdot)$ also forms a sufficient statistic for these densities.*

**Proof.** We need to show that $\Pr[M = m \,|\, \mathbf{Y} = \mathbf{y}_{\mathrm{obs}}]$ is computable from $S(\mathbf{y}_{\mathrm{obs}})$. This follows because, by assumption, $T(\mathbf{y}_{\mathrm{obs}})$ is computable from $S(\mathbf{y}_{\mathrm{obs}})$ and because the sufficiency of $T(\cdot)$ implies that $\Pr[M = m \,|\, \mathbf{Y} = \mathbf{y}_{\mathrm{obs}}]$ is computable from $T(\mathbf{y}_{\mathrm{obs}})$.                                                                    □

## 22.4.3  Establishing Sufficiency in Two Steps

It is sometimes convenient to establish sufficiency in two steps: in the first step we establish that $T(\mathbf{Y})$ is sufficient for guessing $M$ based on $\mathbf{Y}$, and in the second step we establish that $S(\mathbf{T})$ is sufficient for guessing $M$ based on $T(\mathbf{Y})$. The next proposition demonstrates that it then follows that $S(T(\mathbf{Y}))$ forms a sufficient statistic for guessing $M$ based on $\mathbf{Y}$.

**Proposition 22.4.3.** *If $T\colon \mathbb{R}^d \to \mathbb{R}^{d'}$ forms a sufficient statistic for the $\mathsf{M}$ densities $\{f_{\mathbf{Y}|M=m}(\cdot)\}_{m\in\mathcal{M}}$ and if $S\colon \mathbb{R}^{d'} \to \mathbb{R}^{d''}$ forms a sufficient statistic for the corresponding family of densities of $T(\mathbf{Y})$, then the composition $S \circ T$ forms a sufficient statistic for the densities $\{f_{\mathbf{Y}|M=m}(\cdot)\}_{m\in\mathcal{M}}$.*

**Proof.** We shall establish the sufficiency of $S \circ T$ by proving that for any prior $\{\pi_m\}$

$$M \!-\!\circ\!- S\big(T(\mathbf{Y})\big) \!-\!\circ\!- \mathbf{Y}.$$

This follows because for every $m \in \mathcal{M}$ and every $\mathbf{y}_{\mathrm{obs}} \in \mathbb{R}^d$

$$\Pr\big[M = m \,\big|\, S\big(T(\mathbf{Y})\big) = S\big(T(\mathbf{y}_{\mathrm{obs}})\big)\big] = \Pr\big[M = m \,\big|\, T(\mathbf{Y}) = T(\mathbf{y}_{\mathrm{obs}})\big]$$
$$= \Pr\big[M = m \,\big|\, \mathbf{Y} = \mathbf{y}_{\mathrm{obs}}\big],$$

where the first equality follows from the sufficiency of $S(T(\mathbf{Y}))$ for guessing $M$ based on $T(\mathbf{Y})$, and where the second equality follows from the sufficiency of $T(\mathbf{Y})$ for guessing $M$ based on $\mathbf{Y}$.                                                                    □

### 22.4.4   Guessing whether $M$ Lies in a Given Subset of $\mathcal{M}$

We motivate the next result with the following example, which arises in the detection of PAM signals in white Gaussian noise (Section 28.3 ahead). Suppose that the distribution of the observable $\mathbf{Y}$ is determined by the value of a $k$-tuple of bits $(D_1, \ldots, D_k)$. Thus, to each of the $2^k$ values that the $k$-tuple $(D_1, \ldots, D_k)$ can take, there corresponds a distribution on $\mathbf{Y}$ of some given density $f_{\mathbf{Y}|D_1=d_1,\ldots,D_k=d_k}(\cdot)$. Suppose now that $T(\cdot)$ forms a sufficient statistic for this family of $\mathsf{M} = 2^k$ densities. The result we next describe guarantees that $T(\cdot)$ is also sufficient for the binary hypothesis testing problem of guessing whether a specific bit $D_j$ is zero or one. More precisely, we shall show that if $\{\pi_{(d_1,\ldots,d_k)}\}$ is any nondegenerate prior on the $2^k$ different $k$-tuples of bits, then $T(\cdot)$ forms a sufficient statistic for the two densities

$$\sum_{\substack{(d_1,\ldots,d_k) \\ d_j=0}} \pi_{(d_1,\ldots,d_k)} f_{\mathbf{Y}|D_1=d_1,\ldots,D_k=d_k}(\mathbf{y}), \quad \sum_{\substack{(d_1,\ldots,d_k) \\ d_j=1}} \pi_{(d_1,\ldots,d_k)} f_{\mathbf{Y}|D_1=d_1,\ldots,D_k=d_k}(\mathbf{y}).$$

**Proposition 22.4.4 (Guessing whether $M$ Is in $\mathcal{K}$).** *Let* $T\colon \mathbb{R}^d \to \mathbb{R}^{d'}$ *form a sufficient statistic for the* $\mathsf{M}$ *densities* $\{f_{\mathbf{Y}|M=m}(\cdot)\}_{m\in\mathcal{M}}$. *Let the set* $\mathcal{K} \subset \mathcal{M}$ *be a nonempty strict subset of* $\mathcal{M}$. *Let* $\{\pi_m\}$ *be a prior on* $\mathcal{M}$ *satisfying*

$$0 < \sum_{m\in\mathcal{K}} \pi_m < 1.$$

*Then $T(\cdot)$ forms a sufficient statistic for the two densities*

$$\mathbf{y} \mapsto \sum_{m\in\mathcal{K}} \pi_m f_{\mathbf{Y}|M=m}(\mathbf{y}) \quad and \quad \mathbf{y} \mapsto \sum_{m\notin\mathcal{K}} \pi_m f_{\mathbf{Y}|M=m}(\mathbf{y}). \tag{22.37}$$

**Proof.** By the Factorization Theorem it follows that the sufficiency of $T(\cdot)$ for the family $\{f_{\mathbf{Y}|M=m}(\cdot)\}_{m\in\mathcal{M}}$ is equivalent to the condition that for every $m \in \mathcal{M}$ and for every $\mathbf{y} \notin \mathcal{Y}_0$

$$f_{\mathbf{Y}|M=m}(\mathbf{y}) = g_m\big(T(\mathbf{y})\big) h(\mathbf{y}), \tag{22.38}$$

where the set $\mathcal{Y}_0 \subset \mathbb{R}^d$ is of Lebesgue measure zero; where $\{g_m(\cdot)\}_{m\in\mathcal{M}}$ are nonnegative Borel measurable functions from $\mathbb{R}^{d'}$; and where $h(\cdot)$ is a nonnegative Borel measurable function from $\mathbb{R}^d$. Consequently,

$$\sum_{m\in\mathcal{K}} \pi_m f_{\mathbf{Y}|M=m}(\mathbf{y}) = \sum_{m\in\mathcal{K}} \pi_m g_m\big(T(\mathbf{y})\big) h(\mathbf{y})$$

$$= \bigg(\sum_{m\in\mathcal{K}} \pi_m g_m\big(T(\mathbf{y})\big)\bigg) h(\mathbf{y}), \quad \mathbf{y} \notin \mathcal{Y}_0, \tag{22.39a}$$

and

$$\sum_{m\notin\mathcal{K}} \pi_m f_{\mathbf{Y}|M=m}(\mathbf{y}) = \sum_{m\notin\mathcal{K}} \pi_m g_m\big(T(\mathbf{y})\big) h(\mathbf{y})$$

$$= \bigg(\sum_{m\notin\mathcal{K}} \pi_m g_m\big(T(\mathbf{y})\big)\bigg) h(\mathbf{y}), \quad \mathbf{y} \notin \mathcal{Y}_0. \tag{22.39b}$$

The factorization (22.39) of the densities in (22.37) proves that $T(\cdot)$ is also sufficient for these two densities. $\qquad\qquad\qquad\qquad\qquad\qquad\qquad\qquad\qquad\qquad\qquad\square$

**Note 22.4.5.** The proposition also extends to more general partitions as follows. Suppose that $T(\cdot)$ is sufficient for the family $\{f_{\mathbf{Y}|M=m}(\cdot)\}_{m\in\mathcal{M}}$. Let $\mathcal{K}_1,\ldots,\mathcal{K}_\kappa$ be disjoint nonempty subsets of $\mathcal{M}$ whose union is equal to $\mathcal{M}$, and let the prior $\{\pi_m\}$ be such that

$$\sum_{m\in\mathcal{K}_j} \pi_m > 0, \quad j \in \{1,\ldots,\kappa\}.$$

Then $T(\cdot)$ is sufficient for the $\kappa$ densities

$$\mathbf{y} \mapsto \sum_{m\in\mathcal{K}_1} \pi_m\, f_{\mathbf{Y}|M=m}(\mathbf{y}),\ldots,\mathbf{y} \mapsto \sum_{m\in\mathcal{K}_\kappa} \pi_m\, f_{\mathbf{Y}|M=m}(\mathbf{y}).$$

### 22.4.5 Conditionally Independent Observations

Our next result deals with a situation where we need to guess $M$ based on two observations: $\mathbf{Y}_1$ and $\mathbf{Y}_2$. We assume that $T_1(\mathbf{Y}_1)$ forms a sufficient statistic for guessing $M$ when only $\mathbf{Y}_1$ is observed, and that $T_2(\mathbf{Y}_2)$ forms a sufficient statistic for guessing $M$ when only $\mathbf{Y}_2$ is observed. It is tempting to conjecture that in this case the pair $(T_1(\mathbf{Y}_1), T_2(\mathbf{Y}_2))$ must form a sufficient statistic for guessing $M$ when both $\mathbf{Y}_1$ and $\mathbf{Y}_2$ are observed. But, without additional assumptions, this is not the case. An example where this fails can be constructed as follows. Let $M$ and $Z$ be independent with $M$ taking on the values 0 and 1 equiprobably and with $Z \sim \mathcal{N}(0,1)$. Suppose that $Y_1 = M+Z$ and that $Y_2 = Z$. In this case the invertible mapping $T_1(Y_1) = Y_1$ forms a sufficient statistic for guessing $M$ based on $Y_1$ alone, and the mapping $T_2(Y_2) = 17$ forms a sufficient statistic for guessing $M$ based on $Y_2$ alone (because $M$ and $Z$ are independent). Nevertheless, the pair $(Y_1, 17)$ is not sufficient for guessing $M$ based on the pair $(Y_1, Y_2)$. Basing one's guess of $M$ on $(Y_1, 17)$ is not as good as basing it on the pair $(Y_1, Y_2)$. (The reader is encouraged to verify that $Y_1 - Y_2$ is sufficient for guessing $M$ based on $(Y_1, Y_2)$ and that $M$ can be guessed error-free from $Y_1 - Y_2$.)

The additional assumption we need is that $Y_1$ and $Y_2$ be conditionally independent given $M$. (It would make no sense to assume that they are independent, because they are presumably both related to $M$.) This assumption is valid in many applications. For example, it occurs when a signal is received at two different antennas with the additive noises in the two antennas being independent.

**Proposition 22.4.6 (Conditionally Independent Observations).** *Let the mapping* $T_1 \colon \mathbb{R}^{d_1} \to \mathbb{R}^{d_1'}$ *form a sufficient statistic for guessing $M$ based on the observation* $\mathbf{Y}_1 \in \mathbb{R}^{d_1}$, *and let* $T_2 \colon \mathbb{R}^{d_2} \to \mathbb{R}^{d_2'}$ *form a sufficient statistic for guessing $M$ based on the observation* $\mathbf{Y}_2 \in \mathbb{R}^{d_2}$. *If* $\mathbf{Y}_1$ *and* $\mathbf{Y}_2$ *are conditionally independent given* $M$, *then the pair* $(T_1(\mathbf{Y}_1), T_2(\mathbf{Y}_2))$ *forms a sufficient statistic for guessing $M$ based on the pair* $(\mathbf{Y}_1, \mathbf{Y}_2)$.

**Proof.** The proof we offer is based on the Factorization Theorem. The hypothesis that $T_1 \colon \mathbb{R}^{d_1} \to \mathbb{R}^{d_1'}$ forms a sufficient statistic for guessing $M$ based on the observation $\mathbf{Y}_1$ implies the existence of nonnegative functions $\{\mathbf{g}_m^{(1)}\}_{m\in\mathcal{M}}$ and $\mathbf{h}^{(1)}$ and a subset $\mathcal{Y}_0^{(1)} \subset \mathbb{R}^{d_1}$ of Lebesgue measure zero such that

$$f_{\mathbf{Y}_1 \mid M=m}(\mathbf{y}_1) = g_m^{(1)}\big(T_1(\mathbf{y}_1)\big)\, h^{(1)}(\mathbf{y}_1), \quad m \in \mathcal{M}, \quad \mathbf{y}_1 \notin \mathcal{Y}_0^{(1)}. \tag{22.40}$$

Similarly, the hypothesis that $T_2(\cdot)$ is sufficient for guessing $M$ based on $\mathbf{Y}_2$ implies the existence of nonnegative functions $\big\{\mathbf{g}_m^{(2)}\big\}_{m \in \mathcal{M}}$ and $h^{(2)}$ and a subset of Lebesgue measure zero $\mathcal{Y}_0^{(2)} \subset \mathbb{R}^{d_2}$ such that

$$f_{\mathbf{Y}_2 \mid M=m}(\mathbf{y}_2) = g_m^{(2)}\big(T_2(\mathbf{y}_2)\big)\, h^{(2)}(\mathbf{y}_2), \quad m \in \mathcal{M}, \quad \mathbf{y}_2 \notin \mathcal{Y}_0^{(2)}. \tag{22.41}$$

The conditional independence of $\mathbf{Y}_1$ and $\mathbf{Y}_2$ given $M$ implies[7]

$$f_{\mathbf{Y}_1, \mathbf{Y}_2 \mid M=m}(\mathbf{y}_1, \mathbf{y}_2) = f_{\mathbf{Y}_1 \mid M=m}(\mathbf{y}_1)\, f_{\mathbf{Y}_2 \mid M=m}(\mathbf{y}_2),$$
$$m \in \mathcal{M},\ \mathbf{y}_1 \in \mathbb{R}^{d_1},\ \mathbf{y}_2 \in \mathbb{R}^{d_2}. \tag{22.42}$$

Combining (22.40), (22.41), and (22.42), we obtain

$$f_{\mathbf{Y}_1, \mathbf{Y}_2 \mid M=m}(\mathbf{y}_1, \mathbf{y}_2) = \underbrace{g_m^{(1)}\big(T_1(\mathbf{y}_1)\big)\, g_m^{(2)}\big(T_2(\mathbf{y}_2)\big)}_{g_m\big(T_1(\mathbf{y}_1), T_2(\mathbf{y}_2)\big)}\, \underbrace{h^{(1)}(\mathbf{y}_1)\, h^{(2)}(\mathbf{y}_2)}_{h(\mathbf{y}_1, \mathbf{y}_2)},$$
$$m \in \mathcal{M},\ \mathbf{y}_1 \notin \mathcal{Y}_0^{(1)},\ \mathbf{y}_2 \notin \mathcal{Y}_0^{(2)}. \tag{22.43}$$

The set of pairs $(\mathbf{y}_1, \mathbf{y}_2) \in \mathbb{R}^{d_1} \times \mathbb{R}^{d_2}$ for which $\mathbf{y}_1$ is in $\mathcal{Y}_0^{(1)}$ and/or $\mathbf{y}_2$ is in $\mathcal{Y}_0^{(2)}$ is of Lebesgue measure zero, and consequently, the factorization (22.43) implies that the pair $\big(T_1(\mathbf{Y}_1), T_2(\mathbf{Y}_2)\big)$ forms a sufficient statistic for guessing $M$ based on $(\mathbf{Y}_1, \mathbf{Y}_2)$. □

## 22.5 Irrelevant Data

Closely related to the notion of sufficient statistics is the notion of irrelevant data. This notion is particularly useful when we think about the data as consisting of two parts. Heuristically speaking, we say that the second part of the data is irrelevant for guessing $M$ given the first, if it adds no information about $M$ that is not already contained in the first part. In such cases the second part of the data can be ignored. It should be emphasized that the question whether a part of the observation is irrelevant depends not only on its dependence on the random variable to be guessed but also on the other part of the observation.

**Definition 22.5.1 (Irrelevant Data).** *We say that $R$ is **irrelevant** for guessing $M$ given $\mathbf{Y}$, if $\mathbf{Y}$ forms a sufficient statistic for guessing $M$ based on $(\mathbf{Y}, R)$.*

Equivalently, $R$ is irrelevant for guessing $M$ given $\mathbf{Y}$, if for any prior $\{\pi_m\}$ on $M$

$$M \multimap \mathbf{Y} \multimap (\mathbf{Y}, R), \tag{22.44}$$

i.e.,

$$M \multimap \mathbf{Y} \multimap R. \tag{22.45}$$

---

[7]Technically speaking, this must only hold outside a set of Lebesgue measure zero, but we do not want to make things even more cumbersome.

**Example 22.5.2.** Let $H$ take on the values 0 and 1, and assume that, conditional on $H = 0$, the observation $Y$ is $\mathcal{N}(0, \sigma_0^2)$ and that, conditional on $H = 1$, it is $\mathcal{N}(0, \sigma_1^2)$. Rather than thinking of this problem as a decision problem with a single observation, let us think of it as a decision problem with two observations $(Y_1, Y_2)$, where $Y_1$ is the absolute value of $Y$, and where $Y_2$ is the sign of $Y$. Thus $Y = Y_1 Y_2$, where $Y_1 \geq 0$ and $Y_2 \in \{+1, -1\}$. (The probability that $Y = 0$ is zero under each hypothesis, so we need not define the sign of zero.) We now show that $Y_2$ (= the sign of $Y$) is irrelevant data for guessing $H$ given $Y_1$ (= the magnitude of $Y$). Or, in other words, the magnitude of $Y$ is a sufficient statistic for guessing $H$ based on $(Y_1, Y_2)$. Indeed the likelihood-ratio function

$$
\begin{aligned}
\mathrm{LR}(y_1, y_2) &= \frac{f_{Y_1, Y_2 | H = 0}(y_1, y_2)}{f_{Y_1, Y_2 | H = 1}(y_1, y_2)} \\
&= \frac{\frac{1}{\sqrt{2\pi\sigma_0^2}} \exp\left(-\frac{(y_1 y_2)^2}{2\sigma_0^2}\right)}{\frac{1}{\sqrt{2\pi\sigma_1^2}} \exp\left(-\frac{(y_1 y_2)^2}{2\sigma_1^2}\right)} \\
&= \frac{\sigma_1}{\sigma_0} \exp\left(\frac{y_1^2}{2\sigma_1^2} - \frac{y_1^2}{2\sigma_0^2}\right)
\end{aligned}
$$

can be computed from the magnitude $y_1$ only, so $Y_1$ is a sufficient statistic for guessing $H$ based on $(Y_1, Y_2)$.

The following two notes clarify that the notion of irrelevance is different from that of statistical independence. Neither implies the other.

**Note 22.5.3.** A RV can be independent of the RV that we wish to guess and yet not be irrelevant.

**Proof.** We provide an example of a RV $R$ that is independent of the RV $H$ that we wish to guess and that is nonetheless not irrelevant. Suppose that $H$ takes on the values 0 and 1, and assume that under both hypotheses $Y \sim \text{Bernoulli}(1/2)$:

$$
\Pr[Y = 1 \mid H = 0] = \Pr[Y = 1 \mid H = 1] = \frac{1}{2}.
$$

Further assume that under $H = 0$ the RV $R$ is given by $0 \oplus Y = Y$, whereas under $H = 1$ it is given by $1 \oplus Y$. Here $\oplus$ denotes the exclusive-or operation or mod-2 addition.

The distribution of $R$ does not depend on the hypothesis; it is Bernoulli$(1/2)$ both conditional on $H = 0$ and conditional on $H = 1$. But $R$ is not irrelevant for guessing $H$ given $Y$. In fact, if we had to guess $H$ based on $Y$ only, our probability of error would be $1/2$. But if we base our decision on $Y$ and $R$, then our probability of error is zero because

$$
H = Y \oplus R. \qquad \square
$$

**Note 22.5.4.** A RV can be irrelevant even if it is statistically dependent on the RV that we wish to guess.

**Proof.** As an example, consider the case where $R$ is equal to $Y$ with probability one and that $Y$ (and hence also $R$) is statistically dependent on the RV $M$ that we wish to guess. Since $R$ is deterministically equal to $Y$, it follows that, conditional on $Y$, the random variable $R$ is deterministic. Consequently, since a deterministic RV is independent of every RV, it follows that $M$ and $R$ are conditionally independent given $Y$, i.e., that (22.45) holds. Thus, even though in this example $R$ is statistically dependent on $M$, it is irrelevant for guessing $M$ given $Y$. The intuitive explanation is that, in this example, $R$ is irrelevant for guessing $M$ given $Y$ not because it conveys no information about $M$ (it does!) but because it conveys no information about $M$ that is not already conveyed by $Y$. $\qquad\square$

Condition (22.44) is often difficult to establish directly, especially when the distribution of the pair $(R, \mathbf{Y})$ is specified in terms of its conditional density given $M$, because in this case the conditional law of $(M, R)$ given $\mathbf{Y}$ can be unwieldy. In some cases the following proposition can be used to establish that $R$ is irrelevant.

**Proposition 22.5.5 (A Condition that Implies Irrelevance).** *Suppose that the conditional law of $R$ given $M = m$ does not depend on $m$ and that, for each $m \in \mathcal{M}$, we have that, conditionally on $M = m$, the observations $\mathbf{Y}$ and $R$ are independent. Then $R$ is irrelevant for guessing $M$ given $\mathbf{Y}$.*

**Proof.** We provide the proof for the case where the pair $(\mathbf{Y}, R)$ has a conditional density given $M$. The discrete case or the mixed case (where one has a conditional density and the other a conditional PMF) can be treated with the same approach. To prove this proposition we shall demonstrate that $\mathbf{Y}$ is a sufficient statistic for guessing $H$ based on $(\mathbf{Y}, R)$ using the Factorization Theorem. To that end, we express the conditional density of $(\mathbf{Y}, R)$ as

$$
\begin{aligned}
f_{\mathbf{Y}, R | M=m}(\mathbf{y}, r) &= f_{\mathbf{Y} | M=m}(\mathbf{y})\, f_{R | M=m}(r) \\
&= f_{\mathbf{Y} | M=m}(\mathbf{y})\, f_R(r) \\
&= g_m(\mathbf{y})\, h(\mathbf{y}, r), \qquad\qquad (22.46)
\end{aligned}
$$

where the first equality follows from the conditional independence of $\mathbf{Y}$ and $R$ given $M$; the second from the hypothesis that the conditional density of $R$ given $M = m$ does not depend on $m$ and by denoting this density by $f_R(\cdot)$; and the final equality follows by defining $g_m(\mathbf{y}) \triangleq f_{\mathbf{Y} | M=m}(\mathbf{y})$ and $h(\mathbf{y}, r) \triangleq f_R(r)$. The factorization (22.46) demonstrates that $\mathbf{Y}$ forms a sufficient statistic for guessing $M$ based on $(\mathbf{Y}, R)$, i.e., that $R$ is irrelevant for guessing $M$ given $\mathbf{Y}$. $\qquad\square$

## 22.6 Testing with Random Parameters

The notions of sufficient statistics and irrelevance also apply when testing in the presence of a random parameter. If the random parameter $\Theta$ is *not observed*, then $T(\mathbf{Y})$ is sufficient if, and only if, for any prior $\{\pi_m\}$ on $M$

$$
M \,\text{—}\!\circ\!\text{—}\, T(\mathbf{Y}) \,\text{—}\!\circ\!\text{—}\, \mathbf{Y}. \qquad\qquad (22.47)
$$

If $\Theta$ is of density $f_\Theta(\cdot)$ and independent of $M$, then, as in (20.101), we can express the conditional density of $\mathbf{Y}$ given $M = m$ as

$$f_{\mathbf{Y}|M=m}(\mathbf{y}) = \int_\theta f_{\mathbf{Y}|\Theta=\theta, M=m}(\mathbf{y}) f_\Theta(\theta) \, d\theta,$$

so $T(\cdot)$ forms a sufficient statistic if, and only if, it forms a sufficient statistic for the M densities

$$\left\{ \mathbf{y} \mapsto \int_\theta f_{\mathbf{Y}|\Theta=\theta, M=m}(\mathbf{y}) f_\Theta(\theta) \, d\theta \right\}_{m \in \mathcal{M}}.$$

Similarly, $R$ is irrelevant for guessing $M$ given $\mathbf{Y}$ if, and only if,

$$M \multimap \mathbf{Y} \multimap R$$

forms a Markov chain for every prior $\{\pi_m\}$ on $\mathcal{M}$.

If the parameter $\Theta$ is *observed*, then $T(\mathbf{Y}, \Theta)$ is a sufficient statistic if, and only if, for any prior $\{\pi_m\}$ on $\mathcal{M}$

$$M \multimap T(\mathbf{Y}, \Theta) \multimap (\mathbf{Y}, \Theta).$$

If $\Theta$ is independent of $M$ and of density $f_\Theta(\cdot)$, then the density $f_{\mathbf{Y}, \Theta|M=m}(\cdot)$ can be expressed, as in (20.104), as

$$f_{\mathbf{Y}, \Theta|M=m}(\mathbf{y}, \theta) = f_\Theta(\theta) f_{\mathbf{Y}|\Theta=\theta, M=m}(\mathbf{y}),$$

so $T(\cdot)$ forms a sufficient statistic if, and only if, it forms a sufficient statistic for the M densities

$$\left\{ (\theta, \mathbf{y}) \mapsto f_\Theta(\theta) f_{\mathbf{Y}|\Theta=\theta, M=m}(\mathbf{y}) \right\}_{m \in \mathcal{M}}.$$

Similarly, $R$ is irrelevant for guessing $M$ given $(\mathbf{Y}, \Theta)$ if, and only if,

$$M \multimap (\mathbf{Y}, \Theta) \multimap R.$$

The following lemma provides an easily-verifiable condition that guarantees that $R$ is irrelevant for guessing $H$ based on $\mathbf{Y}$, irrespective of whether the random parameter is observed or not.

**Lemma 22.6.1.** *If for any prior $\{\pi_m\}$ on $M$ we have that $R$ is independent of the triplet $(M, \Theta, \mathbf{Y})$,[8] then $R$ is irrelevant for guessing $M$ given $(\Theta, \mathbf{Y})$ and also for guessing $M$ given $\mathbf{Y}$.*

**Proof.** To prove the lemma when $\Theta$ is observed, we need to show that the independence of $R$ and the triplet $(M, \Theta, \mathbf{Y})$ implies

$$M \multimap (\mathbf{Y}, \Theta) \multimap R,$$

---

[8]Note that being independent of the triplet is a stronger condition than being independent of each of the members of the triplet!

i.e., that the conditional distribution of $R$ given $(\mathbf{Y}, \Theta)$ is the same as given $(M, \mathbf{Y}, \Theta)$. This is indeed the case because $R$ is independent of $(M, \mathbf{Y}, \Theta)$ so the two conditional distributions are equal to the unconditional distribution of $R$.

To prove the lemma in the case where $\Theta$ is unobserved, we need to show that the independence of $R$ and the triplet $(M, \Theta, \mathbf{Y})$ implies that

$$M \text{---}\circ\text{---} \mathbf{Y} \text{---}\circ\text{---} R.$$

Again, one can do so by noting that the conditional distribution of $R$ given $\mathbf{Y}$ is equal to the conditional distribution of $R$ given $(\mathbf{Y}, M)$ because both are equal to the unconditional distribution of $R$. $\qquad\square$

## 22.7   Additional Reading

The classical definition of sufficient statistic as a mapping $T(\cdot)$ such that the distribution of $\mathbf{Y}$ given $\big(T(\mathbf{Y}), M = m\big)$ does not depend on $m$ is due to R. A. Fisher. A. N. Kolmogorov defined $T(\cdot)$ to be sufficient if for every prior $\{\pi_m\}$ the *a posteriori* distribution of $M$ given $\mathbf{Y}$ can be computed from $T(\mathbf{Y})$. In our setting where $M$ takes on a finite number of values the two definitions are equivalent. For an example where the definitions differ, see (Blackwell and Ramamoorthi, 1982).

For a discussion of pairwise sufficiency and its relation to sufficiency, see (Halmos and Savage, 1949).

## 22.8   Exercises

**Exercise 22.1 (Another Proof of Proposition 22.4.6).** Give an alternative proof of Proposition 22.4.6 using Theorem 22.3.5.

**Exercise 22.2 (Hypothesis Testing with Two Observations).** Let $H$ take on the values $0$ and $1$ equiprobably. Let $\mathbf{Y}_1$ be a random vector taking value in $\mathbb{R}^2$, and let $Y_2$ be a random variable. Conditional on $H = 0$,

$$\mathbf{Y}_1 = \boldsymbol{\mu} + \mathbf{Z}_1, \quad Y_2 = \alpha + Z_2,$$

and, conditional on $H = 1$,

$$\mathbf{Y}_1 = -\boldsymbol{\mu} + \mathbf{Z}_1, \quad Y_2 = -\alpha + Z_2.$$

Here $H$, $\mathbf{Z}_1$, and $Z_2$ are independent with the components of $\mathbf{Z}_1$ being IID $\mathcal{N}(0,1)$, with $Z_2$ being a mean-one exponential, and with $\boldsymbol{\mu} \in \mathbb{R}^2$ and $\alpha \in \mathbb{R}$ being deterministic.

(i) Find an optimal rule for guessing $H$ based on $\mathbf{Y}_1$. Find a one-dimensional sufficient statistic.

(ii) Find an optimal rule for guessing $H$ based on $Y_2$.

(iii) Find a two-dimensional sufficient statistic $(T_1, T_2)$ for guessing $H$ based on $(\mathbf{Y}_1, Y_2)$.

(iv) Find an optimal rule for guessing $H$ based on the pair $(T_1, T_2)$.

**Exercise 22.3 (Sufficient Statistics and the Bhattacharyya Bound).** Show that if the mapping $T : \mathbb{R}^d \to \mathbb{R}^{d'}$ is a sufficient statistic for the densities $f_{\mathbf{Y}|H=0}(\cdot)$ & $f_{\mathbf{Y}|H=1}(\cdot)$, and if $\mathbf{T} = T(\mathbf{Y})$ is of conditional densities $f_{\mathbf{T}|H=0}(\cdot)$ and $f_{\mathbf{T}|H=1}(\cdot)$, then

$$\frac{1}{2} \int_{\mathbb{R}^d} \sqrt{f_{\mathbf{Y}|H=0}(\mathbf{y}) \, f_{\mathbf{Y}|H=1}(\mathbf{y})} \, \mathrm{d}\mathbf{y} = \frac{1}{2} \int_{\mathbb{R}^{d'}} \sqrt{f_{\mathbf{T}|H=0}(\mathbf{t}) \, f_{\mathbf{T}|H=1}(\mathbf{t})} \, \mathrm{d}\mathbf{t}.$$

*Hint: You may want to first derive the identity*

$$\int_{\mathbb{R}^d} \sqrt{f_{\mathbf{Y}|H=0}(\mathbf{y}) \, f_{\mathbf{Y}|H=1}(\mathbf{y})} \, \mathrm{d}\mathbf{y} = \mathsf{E}\left[ \left( \frac{f_{\mathbf{Y}|H=0}(\mathbf{Y})}{f_{\mathbf{Y}|H=1}(\mathbf{Y})} \right)^{1/2} \,\middle|\, H = 1 \right].$$

**Exercise 22.4 (Sufficient Statistics and Irrelevant Data).**

(i) Show that if the hypotheses of Proposition 22.5.5 are satisfied, then the random variables $\mathbf{Y}$ and $R$ must be independent also when one does not condition on $M$.

(ii) Show that the conditions for irrelevance in that proposition are not necessary.

**Exercise 22.5 (Two More Characterizations of Sufficient Statistics).** Let $P_{Y|H=0}(\cdot)$ and $P_{Y|H=1}(\cdot)$ be probability mass functions on the finite set $\mathcal{Y}$. We say that $T(Y)$ forms a sufficient statistic for guessing $H$ based on $Y$ if $H \!-\!\circ\!-\! T(Y) \!-\!\circ\!-\! Y$ for every prior on $H$. Show that each of the following conditions is equivalent to $T(Y)$ forming a sufficient statistic for guessing $H$ based on $Y$:

(a) For every $y \in \mathcal{Y}$ satisfying $P_{Y|H=0}(y) + P_{Y|H=1}(y) > 0$ we have

$$\frac{P_{Y|H=0}(y)}{P_{Y|H=1}(y)} = \frac{P_{T|H=0}(T(y))}{P_{T|H=1}(T(y))},$$

where we adopt the convention (20.39).

(b) For every prior $(\pi_0, \pi_1)$ on $H$ there exists a decision rule that bases its decision on $\pi_0$, $\pi_1$, and $T(Y)$ and that is optimal for guessing $H$ based on $Y$.

**Exercise 22.6 (Pairwise Sufficiency Implies Sufficiency).** Prove Proposition 22.3.2 in the case where the conditional densities of the observable given each of the hypotheses are positive.

**Exercise 22.7 (Simulating the Observable).** In all the examples we gave in Section 22.3.4 the random vector $\tilde{\mathbf{Y}}$ was generated from $T(\mathbf{y}_{\mathrm{obs}})$ uniformly over the set of vectors $\boldsymbol{\xi}$ in $\mathbb{R}^d$ satisfying $T(\boldsymbol{\xi}) = T(\mathbf{y}_{\mathrm{obs}})$. Provide an example where this is not the case.

*Hint: The setup of Proposition 22.5.5 might be useful.*

**Exercise 22.8 (Densities with Zeros).** Conditional on $H = 0$, the $d$ components of $\mathbf{Y}$ are IID and uniformly distributed over the interval $[\alpha_0, \beta_0]$. Conditional on $H = 1$, they are IID and uniformly distributed over the interval $[\alpha_1, \beta_1]$. Show that the tuple

$$\left( \max\{Y^{(1)}, \dots, Y^{(d)}\}, \ \min\{Y^{(1)}, \dots, Y^{(d)}\} \right)$$

forms a sufficient statistic for guessing $H$ based on $\mathbf{Y}$.

**Exercise 22.9 (Optimality Does Not Imply Sufficiency).** Let $H$ take value in the set $\{0,1\}$, and let $d = 2$. Suppose that

$$Y_j = (1 - 2H) + \Theta Z_j, \quad j = 1, \ldots, d,$$

where $H, \Theta, Z_1, \ldots, Z_d$ are independent with $\Theta$ taking on the distinct positive values $\sigma_0$ and $\sigma_1$ with probability $\rho_0$ and $\rho_1$ respectively, and with $Z_1, \ldots, Z_d$ being IID $\mathcal{N}(0,1)$. Let $T = \sum_j Y_j$.

(i) Show that $T$ forms a sufficient statistic for guessing $H$ based on $Y_1, \ldots, Y_d$ when $\Theta$ *is* observed.

(ii) Show that $T$ does *not* form a sufficient statistic for guessing $H$ based on $Y_1, \ldots, Y_d$ when $\Theta$ *is not* observed.

(iii) Show that notwithstanding Part (ii), if $H$ has a uniform prior, then the decision rule that guesses "$H = 0$" whenever $T \geq 0$ is optimal both when $\Theta$ is observed and when it is not observed.

**Exercise 22.10 (Markovity Implies Markovity).** Suppose that for every prior on $M$

$$(M, \mathbf{A})\!-\!\!\circ\!\!-T(\mathbf{Y})\!-\!\!\circ\!\!-\mathbf{Y}$$

forms a Markov chain, where $M$ takes value in the set $\mathcal{M} = \{1, \ldots, \mathsf{M}\}$, where $\mathbf{A}$ and $\mathbf{Y}$ are random vectors, and where $T(\cdot)$ is Borel measurable. Does this imply that $T(\cdot)$ forms a sufficient statistic for guessing $M$ based on $\mathbf{Y}$?

# Chapter 23

# The Multivariate Gaussian Distribution

## 23.1 Introduction

The **multivariate Gaussian distribution** is arguably the most important multivariate distribution in Digital Communications. It is the extension of the univariate Gaussian distribution from scalars to vectors. A random vector of this distribution is said to be a **Gaussian vector**, and its components are said to be **jointly Gaussian**. In this chapter we shall define this distribution, provide some useful characterizations, and study some of its key properties. To emphasize its connection to the univariate distribution, we shall derive it along the same lines we followed in deriving the univariate Gaussian distribution in Chapter 19.

There are a number of equivalent ways to define the multivariate Gaussian distribution, and authors typically pick one definition and then proceed over the course of numerous pages to derive alternate characterizations. We shall also proceed in this way, but to satisfy the impatient reader's curiosity we shall state the various equivalent definitions in this section. The proof of their equivalence will be spread over the whole chapter.

In the following definition we use the notation introduced in Section 17.2. In particular, all vectors are column vectors, and we denote the components of the vector $\mathbf{a} \in \mathbb{R}^n$ by $a^{(1)}, \dots, a^{(n)}$.

**Definition 23.1.1 (Standard Gaussians, Centered Gaussians, and Gaussians).**

(i) *A random vector $\mathbf{W}$ taking value in $\mathbb{R}^n$ is said to be a **standard Gaussian** if its $n$ components $W^{(1)}, \dots, W^{(n)}$ are independent and each is a zero-mean unit-variance univariate Gaussian.*

(ii) *A random vector $\mathbf{X}$ taking value in $\mathbb{R}^n$ is said to be a **centered Gaussian** if there exists some deterministic $n \times m$ matrix $\mathsf{A}$ such that the distribution of $\mathbf{X}$ is the same as the distribution of $\mathsf{A}\mathbf{W}$, i.e.,*

$$\mathbf{X} \overset{\mathscr{L}}{=} \mathsf{A}\mathbf{W}, \tag{23.1}$$

*where $\mathbf{W}$ is a standard Gaussian with $m$ components.*

454

*(iii) A random vector $\mathbf{X}$ taking value in $\mathbb{R}^n$ is said to be **Gaussian** if there exists some deterministic $n \times m$ matrix $\mathsf{A}$ and some deterministic vector $\boldsymbol{\mu} \in \mathbb{R}^n$ such that the distribution of $\mathbf{X}$ is equal to the distribution of $\mathsf{A}\mathbf{W} + \boldsymbol{\mu}$, i.e., if*

$$\mathbf{X} \overset{\mathscr{L}}{=} \mathsf{A}\mathbf{W} + \boldsymbol{\mu}, \tag{23.2}$$

*where $\mathbf{W}$ is a standard Gaussian with $m$ components.*

The random vectors $\mathsf{A}\mathbf{W} + \boldsymbol{\mu}$ and $\mathbf{X}$ can have identical laws only if they have identical mean vectors. As we shall see, the linearity of expectation and the fact that a standard Gaussian is of zero mean imply that the mean vector of $\mathsf{A}\mathbf{W} + \boldsymbol{\mu}$ is equal to $\boldsymbol{\mu}$. Thus, $\mathsf{A}\mathbf{W} + \boldsymbol{\mu}$ and $\mathbf{X}$ can have identical laws only if $\boldsymbol{\mu} = \mathsf{E}[\mathbf{X}]$. Consequently, $\mathbf{X}$ is a Gaussian random vector if, and only if, for some $\mathsf{A}$ and $\mathbf{W}$ as above $\mathbf{X} \overset{\mathscr{L}}{=} \mathsf{A}\mathbf{W} + \mathsf{E}[\mathbf{X}]$. Stated differently, $\mathbf{X}$ is a Gaussian random vector if, and only if, $\mathbf{X} - \mathsf{E}[\mathbf{X}]$ is a centered Gaussian.

While Definition 23.1.1 allows for the matrix $\mathsf{A}$ to be rectangular, we shall see in Corollary 23.6.13 that every centered Gaussian can be generated from a standard Gaussian by multiplication by a *square* matrix. That is, if $\mathbf{X}$ is an $n$-dimensional centered Gaussian, then there exists an $n \times n$ square matrix $\mathsf{A}$ such that $\mathbf{X} \overset{\mathscr{L}}{=} \mathsf{A}\mathbf{W}$, where $\mathbf{W}$ is a standard Gaussian.

In fact, we shall see in Theorem 23.6.14 that we can even limit ourselves to square matrices that are the product of an orthogonal matrix by a diagonal matrix. Since multiplying $\mathbf{W}$ by a diagonal matrix merely scales its components while leaving them independent and Gaussian, it follows that $\mathbf{X}$ is a centered Gaussian if, and only if, its law is the same as the law of the result of applying an orthogonal transformation to a random vector whose components are independent zero-mean univariate Gaussians (not necessarily of equal variance).

In view of Definition 23.1.1, it is not surprising that applying a linear transformation to a Gaussian vector results in a Gaussian vector ((23.43) ahead). The reverse is perhaps more surprising: $\mathbf{X}$ is a Gaussian vector if, and only if, the result of applying any deterministic linear functional to $\mathbf{X}$ has a univariate Gaussian distribution (Theorem 23.6.17 ahead).

We conclude this section with the following pact with the reader.

(i) Unless preceded by the word "random" or "Gaussian," all scalars, vectors, and matrices in this chapter are **deterministic**.

(ii) Unless preceded by the word "complex," all scalars, vectors, and matrices in this chapter are **real**.

But, without violating this pact, we shall sometimes get excited and throw in the words "real" and "deterministic" even when unnecessary.

## 23.2 Notation and Preliminaries

Our notation in this chapter expands upon the one introduced in Section 17.2. To minimize page flipping, we repeat here parts of that section.

Deterministic vectors are denoted by boldface lowercase letters such as $\mathbf{w}$, whereas random vectors are denoted by boldface uppercase letters such as $\mathbf{W}$. When we deal with deterministic matrices we make an exception to our rule of trying to denote deterministic quantities by lowercase letters.[1] Thus, deterministic matrices are denoted by uppercase letters. But to make it clear that we are dealing with a deterministic matrix and not a scalar random variable, we use special fonts to distinguish the two. Thus $\mathsf{A}$ denotes a deterministic matrix, whereas $A$ denotes a random variable. Random matrices, which only appear briefly in this book, are denoted by uppercase letters of yet another font, e.g., $\mathbb{H}$.

An $n \times m$ deterministic real matrix $\mathsf{A}$ is an array of real numbers having $n$ rows and $m$ columns

$$
\mathsf{A} = \begin{pmatrix} a^{(1,1)} & a^{(1,2)} & \cdots & a^{(1,m)} \\ a^{(2,1)} & a^{(2,2)} & \cdots & a^{(2,m)} \\ \vdots & \vdots & \ddots & \vdots \\ a^{(n,1)} & a^{(n,2)} & \cdots & a^{(n,m)} \end{pmatrix}.
$$

The Row-$j$ Column-$\ell$ element of the matrix $\mathsf{A}$ is denoted

$$
a^{(j,\ell)} \quad \text{or} \quad [\mathsf{A}]_{j,\ell}.
$$

The transpose of an $n \times m$ matrix $\mathsf{A}$ is the $m \times n$ matrix $\mathsf{A}^{\mathsf{T}}$ whose Row-$j$ Column-$\ell$ entry is equal to the Row-$\ell$ Column-$j$ entry of $\mathsf{A}$:

$$
[\mathsf{A}^{\mathsf{T}}]_{j,\ell} = [\mathsf{A}]_{\ell,j}, \quad j \in \{1, \ldots, m\}, \ \ell \in \{1, \ldots, n\}.
$$

We shall repeatedly use the fact that if the matrix-product $\mathsf{AB}$ is defined (i.e., if the number of columns of $\mathsf{A}$ is the same as the number of rows of $\mathsf{B}$), then the transpose of the product is the product of the transposes in reverse order

$$
(\mathsf{AB})^{\mathsf{T}} = \mathsf{B}^{\mathsf{T}} \mathsf{A}^{\mathsf{T}}. \tag{23.3}
$$

The $n \times n$ **identity matrix** whose diagonal elements are all 1 and whose off-diagonal elements are all 0 is denoted $\mathsf{I}_n$. The **all-zero matrix** whose components are all zero is denoted $\mathsf{0}$.

An $n \times 1$ matrix is an $n$-vector, or a vector for short. Thus, unless otherwise specified, all the vectors we shall encounter are column vectors.[2] The components of an $n$-vector $\mathbf{a}$ are denoted by $a^{(1)}, \ldots, a^{(n)}$ so

$$
\mathbf{a} = \begin{pmatrix} a^{(1)} \\ \vdots \\ a^{(n)} \end{pmatrix},
$$

or, in a typographically more efficient form,

$$
\mathbf{a} = (a^{(1)}, \ldots, a^{(n)})^{\mathsf{T}}.
$$

---

[1]We have already made some exceptions to this rule when we dealt with deterministic constants that are by convention always denoted using uppercase letters, e.g., bandwidth $\mathcal{W}$, amplitude $\mathsf{A}$, baud period $\mathsf{T}_{\mathrm{s}}$, etc.

[2]An exception to this rule is in our treatment of linear codes where the tradition of using row vectors is too strong to change.

The vector whose components are all zero is denoted by $\mathbf{0}$. The square root of the sum of the squares of the components of a real $n$-vector $\mathbf{a}$ is denoted by $\|\mathbf{a}\|$:

$$\|\mathbf{a}\| = \sqrt{\sum_{\ell=1}^{n} \left(a^{(\ell)}\right)^2}, \quad \mathbf{a} \in \mathbb{R}^n. \tag{23.4}$$

If $\mathbf{a} = (a^{(1)}, \ldots, a^{(n)})^\mathsf{T}$ and $\mathbf{b} = (b^{(1)}, \ldots, b^{(n)})^\mathsf{T}$, then[3]

$$\mathbf{a}^\mathsf{T}\mathbf{b} = \sum_{\ell=1}^{n} a^{(\ell)} b^{(\ell)}$$
$$= \mathbf{b}^\mathsf{T}\mathbf{a}.$$

In particular,

$$\|\mathbf{a}\|^2 = \sum_{\ell=1}^{n} \left(a^{(\ell)}\right)^2$$
$$= \mathbf{a}^\mathsf{T}\mathbf{a}. \tag{23.5}$$

Note the difference between $\mathbf{a}^\mathsf{T}\mathbf{a}$ and $\mathbf{a}\mathbf{a}^\mathsf{T}$: the former is the scalar $\|\mathbf{a}\|^2$ whereas the latter is the $n \times n$ matrix whose Row-$j$ Column-$\ell$ element is $a^{(j)}a^{(\ell)}$.

The determinant of a square matrix $\mathsf{A}$ is denoted by $\det \mathsf{A}$. We note that a matrix and its transpose have equal determinants

$$\det \left(\mathsf{A}^\mathsf{T}\right) = \det \mathsf{A}, \tag{23.6}$$

and that the determinant of the product of two square matrices is the product of the determinants

$$\det (\mathsf{A}\mathsf{B}) = \det (\mathsf{A}) \det (\mathsf{B}). \tag{23.7}$$

We say that a square $n \times n$ matrix $\mathsf{A}$ is **singular** if its determinant is zero or, equivalently, if its columns are linearly dependent or, equivalently, if its rows are linearly dependent or, equivalently, if there exists some nonzero vector $\boldsymbol{\alpha} \in \mathbb{R}^n$ such that $\mathsf{A}\boldsymbol{\alpha} = \mathbf{0}$.

## 23.3   Some Results on Matrices

We next survey some of the results from Matrix Theory that we shall be using. Particularly important to us are results on positive semidefinite matrices, because, as we shall see in Proposition 23.6.1, every covariance matrix is positive semidefinite, and every positive semidefinite matrix is the covariance matrix of some random vector.

---

[3] In (20.84) we denoted $\mathbf{a}^\mathsf{T}\mathbf{b}$ by $\langle \mathbf{a}, \mathbf{b} \rangle_\mathrm{E}$.

## 23.3.1   Orthogonal Matrices

**Definition 23.3.1 (Orthogonal Matrices).** *An $n \times n$ real matrix $\mathsf{U}$ is said to be **orthogonal** if*

$$\mathsf{U}\mathsf{U}^{\mathsf{T}} = \mathsf{I}_n. \tag{23.8}$$

As proved in (Axler, 1997, Chapter 7, Theorem 7.36), the condition (23.8) is equivalent to the condition

$$\mathsf{U}^{\mathsf{T}}\mathsf{U} = \mathsf{I}_n. \tag{23.9}$$

Thus, a real matrix is orthogonal if, and only if, its transpose is orthogonal. From (23.8) and (23.9) we also obtain:

**Note 23.3.2.** The inverse of an orthogonal matrix is its transpose.

If we write an $n \times n$ matrix $\mathsf{U}$ in terms of its columns as

$$\mathsf{U} = \begin{pmatrix} \uparrow & \cdots & \uparrow \\ & \cdots & \\ \boldsymbol{\psi}_1 & \cdots & \boldsymbol{\psi}_n \\ & \cdots & \\ \downarrow & \cdots & \downarrow \end{pmatrix},$$

then (23.9) can be expressed as

$$\mathsf{I}_n = \mathsf{U}^{\mathsf{T}}\mathsf{U}$$

$$= \begin{pmatrix} \leftarrow & \boldsymbol{\psi}_1^{\mathsf{T}} & \rightarrow \\ \cdots & \cdots & \cdots \\ \cdots & \cdots & \cdots \\ \leftarrow & \boldsymbol{\psi}_n^{\mathsf{T}} & \rightarrow \end{pmatrix} \begin{pmatrix} \uparrow & \cdots & \uparrow \\ & \cdots & \\ \boldsymbol{\psi}_1 & \cdots & \boldsymbol{\psi}_n \\ & \cdots & \\ \downarrow & \cdots & \downarrow \end{pmatrix}$$

$$= \begin{pmatrix} \boldsymbol{\psi}_1^{\mathsf{T}}\boldsymbol{\psi}_1 & \boldsymbol{\psi}_1^{\mathsf{T}}\boldsymbol{\psi}_2 & \cdots & \boldsymbol{\psi}_1^{\mathsf{T}}\boldsymbol{\psi}_n \\ \boldsymbol{\psi}_2^{\mathsf{T}}\boldsymbol{\psi}_1 & \boldsymbol{\psi}_2^{\mathsf{T}}\boldsymbol{\psi}_2 & \cdots & \boldsymbol{\psi}_2^{\mathsf{T}}\boldsymbol{\psi}_n \\ \vdots & \vdots & \ddots & \vdots \\ \boldsymbol{\psi}_n^{\mathsf{T}}\boldsymbol{\psi}_1 & \boldsymbol{\psi}_n^{\mathsf{T}}\boldsymbol{\psi}_2 & \cdots & \boldsymbol{\psi}_n^{\mathsf{T}}\boldsymbol{\psi}_n \end{pmatrix},$$

thus showing that a real $n \times n$ matrix $\mathsf{U}$ is orthogonal if, and only if, its $n$ columns $\boldsymbol{\psi}_1, \ldots, \boldsymbol{\psi}_n$ satisfy

$$\boldsymbol{\psi}_\nu^{\mathsf{T}}\boldsymbol{\psi}_{\nu'} = \mathrm{I}\{\nu = \nu'\}, \quad \nu, \nu' \in \{1, \ldots, n\}. \tag{23.10}$$

Using the same argument but starting with (23.8) we can prove a similar result about the rows of an orthogonal matrix: if the rows of a real $n \times n$ matrix $\mathsf{U}$ are denoted by $\boldsymbol{\phi}_1^{\mathsf{T}}, \ldots, \boldsymbol{\phi}_n^{\mathsf{T}}$, i.e.,

$$\mathsf{U} = \begin{pmatrix} \leftarrow & \boldsymbol{\phi}_1^{\mathsf{T}} & \rightarrow \\ \cdots & \cdots & \cdots \\ \cdots & \cdots & \cdots \\ \leftarrow & \boldsymbol{\phi}_n^{\mathsf{T}} & \rightarrow \end{pmatrix},$$

then $\mathsf{U}$ is orthogonal if, and only if,

$$\phi_\nu^\mathsf{T} \phi_{\nu'} = \mathrm{I}\{\nu = \nu'\}, \quad \nu, \nu' \in \{1, \dots, n\}. \tag{23.11}$$

Recalling that the determinant of a product of square matrices is the product of the determinants and that the determinant of a matrix is equal to the determinant of its transpose, we obtain that for every square matrix $\mathsf{U}$

$$\det\left(\mathsf{U}\mathsf{U}^\mathsf{T}\right) = \left(\det \mathsf{U}\right)^2. \tag{23.12}$$

Consequently, by taking the determinant of both sides of (23.8) we obtain that the determinant of an orthogonal matrix must be either $+1$ or $-1$. It should, however, be noted that there are numerous examples of matrices of unit determinant that are not orthogonal.

We leave it to the reader to verify that a $2 \times 2$ matrix is orthogonal if, and only if, it is equal to one of the following matrices for some choice of $-\pi \leq \theta < \pi$

$$\begin{pmatrix} \cos\theta & -\sin\theta \\ \sin\theta & \cos\theta \end{pmatrix}, \begin{pmatrix} \cos\theta & \sin\theta \\ \sin\theta & -\cos\theta \end{pmatrix}. \tag{23.13}$$

The former matrix corresponds to a rotation by $\theta$ and has determinant $+1$, and the latter to a reflection followed by a rotation

$$\begin{pmatrix} \cos\theta & \sin\theta \\ \sin\theta & -\cos\theta \end{pmatrix} = \begin{pmatrix} \cos\theta & -\sin\theta \\ \sin\theta & \cos\theta \end{pmatrix} \begin{pmatrix} 1 & 0 \\ 0 & -1 \end{pmatrix}$$

and has determinant $-1$.

### 23.3.2 Symmetric Matrices

A matrix $\mathsf{A}$ is said to be **symmetric** if it is equal to its transpose:

$$\mathsf{A}^\mathsf{T} = \mathsf{A}.$$

Only square matrices can be symmetric. A vector $\boldsymbol{\psi} \in \mathbb{R}^n$ is said to be an **eigenvector** of the matrix $\mathsf{A}$ corresponding to the real **eigenvalue** $\lambda \in \mathbb{R}$ if $\boldsymbol{\psi}$ is nonzero and if $\mathsf{A}\boldsymbol{\psi} = \lambda\boldsymbol{\psi}$. The following is a key result about the eigenvectors of symmetric real matrices.

**Proposition 23.3.3 (Eigenvectors and Eigenvalues of Symmetric Real Matrices).**
*If $\mathsf{A}$ is a symmetric real $n \times n$ matrix, then $\mathsf{A}$ has $n$ (not necessarily distinct) real eigenvalues $\lambda_1, \dots, \lambda_n \in \mathbb{R}$ with corresponding eigenvectors $\boldsymbol{\psi}_1, \dots, \boldsymbol{\psi}_n \in \mathbb{R}^n$ satisfying*

$$\boldsymbol{\psi}_\nu^\mathsf{T} \boldsymbol{\psi}_{\nu'} = \mathrm{I}\{\nu = \nu'\}, \quad \nu, \nu' \in \{1, \dots, n\}. \tag{23.14}$$

**Proof.** See, for example, (Axler, 1997, Chapter 7, Theorem 7.13, p. 136), or (Herstein, 2001, Section 6.10, pp. 346–348), or (Horn and Johnson, 1985, Chapter 4, Section 1, Theorem 4.5.1). ☐

The vectors $\boldsymbol{\psi}_1, \ldots, \boldsymbol{\psi}_n$ are eigenvectors of the matrix $\mathsf{A}$ corresponding to the eigenvalues $\lambda_1, \ldots, \lambda_n$ if

$$\mathsf{A}\boldsymbol{\psi}_\nu = \lambda_\nu \boldsymbol{\psi}_\nu, \quad \nu \in \{1, \ldots, n\}. \tag{23.15}$$

We next express this in an alternative way. We begin by noting that

$$\mathsf{A} \begin{pmatrix} \uparrow & \cdots & \uparrow \\ & \cdots & \\ \boldsymbol{\psi}_1 & \cdots & \boldsymbol{\psi}_n \\ & \cdots & \\ \downarrow & \cdots & \downarrow \end{pmatrix} = \begin{pmatrix} \uparrow & \cdots & \uparrow \\ & \cdots & \\ \mathsf{A}\boldsymbol{\psi}_1 & \cdots & \mathsf{A}\boldsymbol{\psi}_n \\ & \cdots & \\ \downarrow & \cdots & \downarrow \end{pmatrix}$$

and that

$$\begin{pmatrix} \uparrow & \cdots & \uparrow \\ & \cdots & \\ \boldsymbol{\psi}_1 & \cdots & \boldsymbol{\psi}_n \\ & \cdots & \\ \downarrow & \cdots & \downarrow \end{pmatrix} \begin{pmatrix} \lambda_1 & 0 & \cdots & 0 \\ 0 & \lambda_2 & \ddots & \vdots \\ \vdots & \ddots & \ddots & 0 \\ 0 & \cdots & 0 & \lambda_n \end{pmatrix} = \begin{pmatrix} \uparrow & \cdots & \uparrow \\ & \cdots & \\ \lambda_1 \boldsymbol{\psi}_1 & \cdots & \lambda_n \boldsymbol{\psi}_n \\ & \cdots & \\ \downarrow & \cdots & \downarrow \end{pmatrix}.$$

Consequently, Condition (23.15) can be written as

$$\mathsf{A}\mathsf{U} = \mathsf{U}\Lambda, \tag{23.16}$$

where

$$\mathsf{U} = \begin{pmatrix} \uparrow & \cdots & \uparrow \\ & \cdots & \\ \boldsymbol{\psi}_1 & \cdots & \boldsymbol{\psi}_n \\ & \cdots & \\ \downarrow & \cdots & \downarrow \end{pmatrix} \quad \text{and} \quad \Lambda = \begin{pmatrix} \lambda_1 & 0 & \cdots & 0 \\ 0 & \lambda_2 & \ddots & \vdots \\ \vdots & \ddots & \ddots & 0 \\ 0 & \cdots & 0 & \lambda_n \end{pmatrix}. \tag{23.17}$$

Condition (23.14) is equivalent to the condition that the above matrix $\mathsf{U}$ is orthogonal. By multiplying (23.16) from the right by the inverse of $\mathsf{U}$ (which, because $\mathsf{U}$ is orthogonal and by (23.8), is $\mathsf{U}^\mathsf{T}$) we obtain the equivalent form $\mathsf{A} = \mathsf{U}\Lambda\mathsf{U}^\mathsf{T}$. Consequently, an equivalent statement of Proposition 23.3.3 is:

**Proposition 23.3.4 (Spectral Theorem for Real Symmetric Matrices).** *A symmetric real $n \times n$ matrix $\mathsf{A}$ can be written in the form*

$$\mathsf{A} = \mathsf{U}\Lambda\mathsf{U}^\mathsf{T}$$

*where, as in (23.17), $\Lambda$ is a diagonal real $n \times n$ matrix whose diagonal elements are the eigenvalues of $\mathsf{A}$, and where $\mathsf{U}$ is a real $n \times n$ orthogonal matrix whose $\nu$-th column is an eigenvector of $\mathsf{A}$ corresponding to the eigenvalue in the $\nu$-th position on the diagonal of $\Lambda$.*

The reverse is also true: if $\mathsf{A} = \mathsf{U}\Lambda\mathsf{U}^\mathsf{T}$ for a real diagonal matrix $\Lambda$ and for a real orthogonal matrix $\mathsf{U}$, then $\mathsf{A}$ is symmetric, its eigenvalues are the diagonal elements of $\Lambda$, and the $\nu$-th column of $\mathsf{U}$ is an eigenvector of the matrix $\mathsf{A}$ corresponding to the eigenvalue in the $\nu$-th position on the diagonal of $\Lambda$.

### 23.3.3   Positive Semidefinite Matrices

**Definition 23.3.5 (Positive Semidefinite and Positive Definite Matrices).**

(i) *We say that the $n \times n$ real matrix* $\mathsf{K}$ *is* ***positive semidefinite*** *or* ***nonnegative definite*** *and write*

$$\mathsf{K} \succeq 0$$

*if* $\mathsf{K}$ *is symmetric and*

$$\boldsymbol{\alpha}^\mathsf{T} \mathsf{K} \boldsymbol{\alpha} \geq 0, \quad \boldsymbol{\alpha} \in \mathbb{R}^n.$$

(ii) *We say that the $n \times n$ real matrix* $\mathsf{K}$ *is* ***positive definite*** *and write*

$$\mathsf{K} \succ 0$$

*if* $\mathsf{K}$ *is symmetric and*

$$\boldsymbol{\alpha}^\mathsf{T} \mathsf{K} \boldsymbol{\alpha} > 0, \quad \left( \boldsymbol{\alpha} \neq \mathbf{0}, \; \boldsymbol{\alpha} \in \mathbb{R}^n \right).$$

The following two propositions characterize positive semidefinite and positive definite matrices. For proofs, see (Axler, 1997, Chapter 7, Theorem 7.27).

**Proposition 23.3.6 (Characterizing Positive Semidefinite Matrices).** *Let $\mathsf{K}$ be a real $n \times n$ matrix. Then the statement that $\mathsf{K}$ is positive semidefinite is equivalent to each of the following statements:*

(a) *The matrix $\mathsf{K}$ can be written in the form*

$$\mathsf{K} = \mathsf{S}^\mathsf{T} \mathsf{S} \tag{23.18}$$

*for some real $n \times n$ matrix $\mathsf{S}$.[4]*

(b) *The matrix $\mathsf{K}$ is symmetric and all its eigenvalues are nonnegative.*

(c) *The matrix $\mathsf{K}$ can be written in the form*

$$\mathsf{K} = \mathsf{U} \Lambda \mathsf{U}^\mathsf{T}, \tag{23.19}$$

*where $\Lambda$ is a real $n \times n$ diagonal matrix with nonnegative entries on the diagonal and where $\mathsf{U}$ is a real $n \times n$ orthogonal matrix.*

**Proposition 23.3.7 (Characterizing Positive Definite Matrices).** *Let $\mathsf{K}$ be a real $n \times n$ matrix. Then the statement that $\mathsf{K}$ is positive definite is equivalent to each of the following statements.*

(a) *The matrix $\mathsf{K}$ can be written in the form $\mathsf{K} = \mathsf{S}^\mathsf{T} \mathsf{S}$ for some real $n \times n$ nonsingular matrix $\mathsf{S}$.*

(b) *The matrix $\mathsf{K}$ is symmetric and all its eigenvalues are positive.*

---

[4]Even if $\mathsf{S}$ is not a square matrix, $\mathsf{S}^\mathsf{T}\mathsf{S} \succeq 0$.

(c) *The matrix* $\mathsf{K}$ *can be written in the form*

$$\mathsf{K} = \mathsf{U}\mathsf{\Lambda}\mathsf{U}^\mathsf{T},$$

where $\mathsf{\Lambda}$ *is a real* $n \times n$ *diagonal matrix with positive entries on the diagonal and where* $\mathsf{U}$ *is a real* $n \times n$ *orthogonal matrix.*

Given a positive semidefinite matrix $\mathsf{K}$, how can we find a matrix $\mathsf{S}$ satisfying $\mathsf{K} = \mathsf{S}^\mathsf{T}\mathsf{S}$? In general, there can be many such matrices. For example, if $\mathsf{K}$ is the identity matrix, then $\mathsf{S}$ can be any orthogonal matrix. We mention here two useful choices. Being symmetric, the matrix $\mathsf{K}$ can be written in the form

$$\mathsf{K} = \mathsf{U}\mathsf{\Lambda}\mathsf{U}^\mathsf{T}, \tag{23.20}$$

where $\mathsf{U}$ and $\mathsf{\Lambda}$ are as in (23.17). Since $\mathsf{K}$ is positive semidefinite, the diagonal elements of $\mathsf{\Lambda}$ (which are the eigenvalues of $\mathsf{K}$) are nonnegative. Consequently, we can define the matrix

$$\mathsf{\Lambda}^{1/2} = \begin{pmatrix} \sqrt{\lambda_1} & 0 & \cdots & 0 \\ 0 & \sqrt{\lambda_2} & \ddots & \vdots \\ \vdots & \ddots & \ddots & 0 \\ 0 & \cdots & 0 & \sqrt{\lambda_n} \end{pmatrix}.$$

One choice of the matrix $\mathsf{S}$ is

$$\mathsf{S} = \mathsf{\Lambda}^{1/2}\mathsf{U}^\mathsf{T}. \tag{23.21}$$

Indeed, with this definition of $\mathsf{S}$ we have

$$\begin{aligned} \mathsf{S}^\mathsf{T}\mathsf{S} &= \left(\mathsf{\Lambda}^{1/2}\mathsf{U}^\mathsf{T}\right)^\mathsf{T}\mathsf{\Lambda}^{1/2}\mathsf{U}^\mathsf{T} \\ &= \mathsf{U}\mathsf{\Lambda}^{1/2}\mathsf{\Lambda}^{1/2}\mathsf{U}^\mathsf{T} \\ &= \mathsf{U}\mathsf{\Lambda}\mathsf{U}^\mathsf{T} \\ &= \mathsf{K}, \end{aligned}$$

where the first equality follows from the definition of $\mathsf{S}$; the second from the rule $(\mathsf{A}\mathsf{B})^\mathsf{T} = \mathsf{B}^\mathsf{T}\mathsf{A}^\mathsf{T}$ and from the symmetry of the diagonal matrix $\mathsf{\Lambda}^{1/2}$; the third from the definition of $\mathsf{\Lambda}^{1/2}$; and where the final equality follows from (23.20).

A different choice for $\mathsf{S}$, which will be less useful to us in this chapter, is[5]

$$\mathsf{U}\mathsf{\Lambda}^{1/2}\mathsf{U}^\mathsf{T}.$$

The following lemmas will be used in Section 23.4.3 when we study random vectors of singular covariance matrices.

**Lemma 23.3.8.** *Let* $\mathsf{K}$ *be a real* $n \times n$ *positive semidefinite matrix, and let* $\boldsymbol{\alpha}$ *be a vector in* $\mathbb{R}^n$. *Then* $\boldsymbol{\alpha}^\mathsf{T}\mathsf{K}\boldsymbol{\alpha} = 0$ *if, and only if,* $\mathsf{K}\boldsymbol{\alpha} = \mathbf{0}$.

---

[5]This is the only choice for $\mathsf{S}$ that is positive semidefinite (Axler, 1997, Chapter 7, Proposition 7.28), (Horn and Johnson, 1985, Chapter 7, Section 7.2, Theorem 7.2.6).

**Proof.** One direction is trivial and does not require that $K$ be positive semidefinite: if $K\boldsymbol{\alpha} = \mathbf{0}$, then $\boldsymbol{\alpha}^{\mathsf{T}}K\boldsymbol{\alpha}$ must also be equal to zero. Indeed, in this case we have by the associativity of matrix multiplication $\boldsymbol{\alpha}^{\mathsf{T}}K\boldsymbol{\alpha} = \boldsymbol{\alpha}^{\mathsf{T}}(K\boldsymbol{\alpha}) = \boldsymbol{\alpha}^{\mathsf{T}}\mathbf{0} = 0$.

To prove the other direction, we first note that, since $K$ is positive semidefinite, there exists some $n \times n$ matrix $S$ such that $K = S^{\mathsf{T}}S$. Hence,

$$\boldsymbol{\alpha}^{\mathsf{T}}K\boldsymbol{\alpha} = \boldsymbol{\alpha}^{\mathsf{T}}S^{\mathsf{T}}S\boldsymbol{\alpha}$$
$$= (S\boldsymbol{\alpha})^{\mathsf{T}}(S\boldsymbol{\alpha})$$
$$= \|S\boldsymbol{\alpha}\|^2, \quad \boldsymbol{\alpha} \in \mathbb{R}^n,$$

where the second equality follows from the rule for transposing a product (23.3). and where the third equality follows from (23.5). Consequently, if $\boldsymbol{\alpha}^{\mathsf{T}}K\boldsymbol{\alpha} = 0$, then $\|S\boldsymbol{\alpha}\|^2 = 0$, so $S\boldsymbol{\alpha} = \mathbf{0}$, and hence $S^{\mathsf{T}}S\boldsymbol{\alpha} = \mathbf{0}$, i.e., $K\boldsymbol{\alpha} = \mathbf{0}$. $\square$

**Lemma 23.3.9.** *If $K$ is a real $n \times n$ positive definite matrix, then $\boldsymbol{\alpha}^{\mathsf{T}}K\boldsymbol{\alpha} = 0$ if, and only if, $\boldsymbol{\alpha} = \mathbf{0}$.*

**Proof.** Follows directly from Definition 23.3.5 of positive semidefinite matrices. $\square$

## 23.4 Random Vectors

### 23.4.1 Definitions

Recall that an $n$-**dimensional random vector** or a **random $n$-vector** $\mathbf{X}$ defined over the probability space $(\Omega, \mathcal{F}, P)$ is a (measurable) mapping from the set of experiment outcomes $\Omega$ to the $n$-dimensional Euclidean space $\mathbb{R}^n$. A random vector $\mathbf{X}$ is very much like a random variable, except that rather than taking value in the real line $\mathbb{R}$, it takes value in $\mathbb{R}^n$. In fact, an $n$-dimensional random vector can be viewed as an array of $n$ random variables.[6]

The density of a random vector is the joint density of its components. The density of a random $n$-vector is thus a nonnegative (Borel measurable) function from $\mathbb{R}^n$ to the nonnegative reals that integrates to one.

Similarly, an $n \times m$ **random matrix** $\mathbb{H}$ is an $n \times m$ array of random variables defined over a common probability space.

### 23.4.2 Expectations and Covariance Matrices

The **expectation** $\mathsf{E}[\mathbf{X}]$ of a random $n$-vector $\mathbf{X} = (X^{(1)}, \ldots, X^{(n)})^{\mathsf{T}}$ is a vector whose components are the expectations of the corresponding components of $\mathbf{X}$:[7]

$$\mathsf{E}[\mathbf{X}] \triangleq \left(\mathsf{E}[X^{(1)}], \ldots, \mathsf{E}[X^{(n)}]\right)^{\mathsf{T}}. \tag{23.22}$$

---

[6]In dealing with random vectors one often abandons the "coordinate free" approach and views vectors in a particular coordinate system. This allows one to speak of the covariance *matrix* in more familiar terms.

[7]The expectation of a random vector is only defined if the expectation of each of its components is defined.

The $j$-th element of $\mathsf{E}[\mathbf{X}]$ is thus the expectation of the $j$-th component of $\mathbf{X}$, namely, $\mathsf{E}\left[X^{(j)}\right]$. Similarly, the expectation of a random matrix is the matrix of expectations.

If all the components of a random $n$-vector $\mathbf{X}$ are of finite variance, then we say that $\mathbf{X}$ is of **finite variance**. We then define its $n \times n$ **covariance matrix** $\mathsf{K_{XX}}$ as

$$\mathsf{K_{XX}} \triangleq \mathsf{E}\left[(\mathbf{X} - \mathsf{E}[\mathbf{X}])\,(\mathbf{X} - \mathsf{E}[\mathbf{X}])^{\mathsf{T}}\right]. \tag{23.23}$$

That is,

$$
\begin{aligned}
\mathsf{K_{XX}} &= \mathsf{E}\left[ \begin{pmatrix} X^{(1)} - \mathsf{E}\left[X^{(1)}\right] \\ \vdots \\ \vdots \\ X^{(n)} - \mathsf{E}\left[X^{(n)}\right] \end{pmatrix} \begin{pmatrix} X^{(1)} - \mathsf{E}\left[X^{(1)}\right] & \cdots & \cdots & X^{(n)} - \mathsf{E}\left[X^{(n)}\right] \end{pmatrix} \right] \\
&= \begin{pmatrix} \mathsf{Var}\left[X^{(1)}\right] & \mathsf{Cov}\left[X^{(1)}, X^{(2)}\right] & \cdots & \mathsf{Cov}\left[X^{(1)}, X^{(n)}\right] \\ \mathsf{Cov}\left[X^{(2)}, X^{(1)}\right] & \mathsf{Var}\left[X^{(2)}\right] & \cdots & \mathsf{Cov}\left[X^{(2)}, X^{(n)}\right] \\ \vdots & \vdots & \ddots & \vdots \\ \mathsf{Cov}\left[X^{(n)}, X^{(1)}\right] & \mathsf{Cov}\left[X^{(n)}, X^{(2)}\right] & \cdots & \mathsf{Var}\left[X^{(n)}\right] \end{pmatrix}. \tag{23.24}
\end{aligned}
$$

If $n = 1$ and the $n$-dimensional random vector $\mathbf{X}$ hence a scalar, then the covariance matrix $\mathsf{K}_{XX}$ is a $1 \times 1$ matrix whose sole component is the variance of the sole component of $\mathbf{X}$.

Note that from the $n \times n$ covariance matrix $\mathsf{K_{XX}}$ of a random $n$-vector $\mathbf{X}$ it is easy to compute the covariance matrix of a subset of $\mathbf{X}$'s components. For example, if we are only interested in the $2 \times 2$ covariance matrix of $(X^{(1)}, X^{(2)})^{\mathsf{T}}$, then we just pick the first two columns and the first two rows of $\mathsf{K_{XX}}$. More generally, the $r \times r$ covariance matrix of $(X^{(j_1)}, X^{(j_2)}, \ldots, X^{(j_r)})^{\mathsf{T}}$ for $1 \leq j_1 < j_2 < \cdots < j_r \leq n$ is obtained from $\mathsf{K_{XX}}$ by picking Rows and Columns $j_1, \ldots, j_r$. For example, if

$$\mathsf{K_{XX}} = \begin{pmatrix} 30 & 31 & 9 & 7 \\ 31 & 39 & 11 & 13 \\ 9 & 11 & 9 & 12 \\ 7 & 13 & 12 & 26 \end{pmatrix},$$

then the covariance matrix of $(X^{(2)}, X^{(4)})^{\mathsf{T}}$ is $\begin{pmatrix} 39 & 13 \\ 13 & 26 \end{pmatrix}$.

We next explore the behavior of the mean vector and the covariance matrix of a random vector when it is multiplied by a deterministic matrix. Regarding the mean, we shall show that since matrix multiplication is a linear transformation, it commutes with the expectation operation. Consequently, if $\mathbb{H}$ is a random $n \times m$ matrix and $\mathsf{A}$ is a deterministic $\nu \times n$ matrix, then

$$\mathsf{E}[\mathsf{A}\mathbb{H}] = \mathsf{A}\mathsf{E}[\mathbb{H}], \tag{23.25a}$$

and similarly if $\mathsf{B}$ is a deterministic $m \times \nu$ matrix, then

$$\mathsf{E}[\mathbb{H}\mathsf{B}] = \mathsf{E}[\mathbb{H}]\,\mathsf{B}. \tag{23.25b}$$

To prove (23.25a) we write out the Row-$j$ Column-$\ell$ element of the $\nu \times m$ matrix $\mathsf{E}[\mathsf{A}\mathbb{H}]$ and use the linearity of expectation to relate it to the Row-$j$ Column-$\ell$ element of the matrix $\mathsf{A}\mathsf{E}[\mathbb{H}]$:

$$
\begin{aligned}
\big[\mathsf{E}[\mathsf{A}\mathbb{H}]\big]_{j,\ell} &= \mathsf{E}\Big[\sum_{\kappa=1}^{n}[\mathsf{A}]_{j,\kappa}[\mathbb{H}]_{\kappa,\ell}\Big] \\
&= \sum_{\kappa=1}^{n}\mathsf{E}\Big[[\mathsf{A}]_{j,\kappa}[\mathbb{H}]_{\kappa,\ell}\Big] \\
&= \sum_{\kappa=1}^{n}[\mathsf{A}]_{j,\kappa}\mathsf{E}\big[[\mathbb{H}]_{\kappa,\ell}\big] \\
&= \big[\mathsf{A}\mathsf{E}[\mathbb{H}]\big]_{j,\ell}, \quad j \in \{1,\ldots,\nu\},\ \ell \in \{1,\ldots,m\}.
\end{aligned}
$$

The proof of (23.25b) is almost identical and is omitted.

The transpose operation also commutes with expectation: if $\mathbb{H}$ is a random matrix then

$$
\mathsf{E}\big[\mathbb{H}^{\mathsf{T}}\big] = \big(\mathsf{E}[\mathbb{H}]\big)^{\mathsf{T}}. \tag{23.26}
$$

As to the covariance matrix, we next show that if $\mathsf{A}$ is a deterministic matrix and if $\mathbf{X}$ is a random vector, then the covariance matrix $\mathsf{K}_{\mathbf{YY}}$ of the random vector $\mathbf{Y} = \mathsf{A}\mathbf{X}$ can be expressed in terms of the covariance matrix $\mathsf{K}_{\mathbf{XX}}$ of $\mathbf{X}$ as

$$
\mathsf{K}_{\mathbf{YY}} = \mathsf{A}\,\mathsf{K}_{\mathbf{XX}}\,\mathsf{A}^{\mathsf{T}}, \quad \mathbf{Y} = \mathsf{A}\mathbf{X}. \tag{23.27}
$$

Indeed,

$$
\begin{aligned}
\mathsf{K}_{\mathbf{YY}} &\triangleq \mathsf{E}\big[(\mathbf{Y}-\mathsf{E}[\mathbf{Y}])(\mathbf{Y}-\mathsf{E}[\mathbf{Y}])^{\mathsf{T}}\big] \\
&= \mathsf{E}\big[(\mathsf{A}\mathbf{X}-\mathsf{E}[\mathsf{A}\mathbf{X}])(\mathsf{A}\mathbf{X}-\mathsf{E}[\mathsf{A}\mathbf{X}])^{\mathsf{T}}\big] \\
&= \mathsf{E}\big[\mathsf{A}(\mathbf{X}-\mathsf{E}[\mathbf{X}])(\mathsf{A}(\mathbf{X}-\mathsf{E}[\mathbf{X}]))^{\mathsf{T}}\big] \\
&= \mathsf{E}\big[\mathsf{A}(\mathbf{X}-\mathsf{E}[\mathbf{X}])(\mathbf{X}-\mathsf{E}[\mathbf{X}])^{\mathsf{T}}\mathsf{A}^{\mathsf{T}}\big] \\
&= \mathsf{A}\mathsf{E}\big[(\mathbf{X}-\mathsf{E}[\mathbf{X}])(\mathbf{X}-\mathsf{E}[\mathbf{X}])^{\mathsf{T}}\mathsf{A}^{\mathsf{T}}\big] \\
&= \mathsf{A}\mathsf{E}\big[(\mathbf{X}-\mathsf{E}[\mathbf{X}])(\mathbf{X}-\mathsf{E}[\mathbf{X}])^{\mathsf{T}}\big]\mathsf{A}^{\mathsf{T}} \\
&= \mathsf{A}\,\mathsf{K}_{\mathbf{XX}}\,\mathsf{A}^{\mathsf{T}}.
\end{aligned}
$$

A key property of covariance matrices is that, as we shall next show, they are all positive semidefinite. That is, the covariance matrix $\mathsf{K}_{\mathbf{XX}}$ of any random vector $\mathbf{X}$ is a symmetric matrix satisfying

$$
\boldsymbol{\alpha}^{\mathsf{T}}\mathsf{K}_{\mathbf{XX}}\,\boldsymbol{\alpha} \geq 0, \quad \boldsymbol{\alpha} \in \mathbb{R}^{n}. \tag{23.28}
$$

(In Proposition 23.6.1 we shall see that this property fully characterizes covariance matrices: every positive semidefinite matrix is the covariance matrix of some random vector.)

To prove (23.28) it suffices to consider the case where $\mathbf{X}$ is of zero mean because the covariance matrix of $\mathbf{X}$ is the same as the covariance matrix of $\mathbf{X}-\mathsf{E}[\mathbf{X}]$. The symmetry of $\mathsf{K_{XX}}$ follows from the definition of the covariance matrix (23.23); from the fact that expectation and transposition commute (23.26); and from the formula for the transpose of a product of matrices (23.3):

$$
\begin{aligned}
\mathsf{K_{XX}^T} &= \left(\mathsf{E}[\mathbf{X}\mathbf{X}^\mathsf{T}]\right)^\mathsf{T} \\
&= \mathsf{E}\left[\left(\mathbf{X}\mathbf{X}^\mathsf{T}\right)^\mathsf{T}\right] \\
&= \mathsf{E}[\mathbf{X}\mathbf{X}^\mathsf{T}] \\
&= \mathsf{K_{XX}}.
\end{aligned}
\tag{23.29}
$$

The nonnegativity of $\boldsymbol{\alpha}^\mathsf{T}\mathsf{K_{XX}}\boldsymbol{\alpha}$ for any deterministic $\boldsymbol{\alpha} \in \mathbb{R}^n$ follows by noting that by (23.27) (applied with $\mathsf{A} = \boldsymbol{\alpha}^\mathsf{T}$) the term $\boldsymbol{\alpha}^\mathsf{T}\mathsf{K_{XX}}\boldsymbol{\alpha}$ is the variance of the scalar random variable $\boldsymbol{\alpha}^\mathsf{T}\mathbf{X}$, i.e.,

$$
\boldsymbol{\alpha}^\mathsf{T}\mathsf{K_{XX}}\boldsymbol{\alpha} = \mathsf{Var}[\boldsymbol{\alpha}^\mathsf{T}\mathbf{X}]
\tag{23.30}
$$

and, as such, is nonnegative.

### 23.4.3   Singular Covariance Matrices

A random vector having a singular covariance matrix can be unwieldy because it cannot have a density function. Indeed, as we shall see in Corollary 23.4.2, any such random vector has at least one component that is determined (with probability one) by the other components. In this section we shall propose a way of manipulating such vectors. Roughly speaking, the idea is that if $\mathbf{X}$ has a singular covariance matrix, then we choose a subset of its components so that the covariance matrix of the chosen subset be nonsingular and so that each component that was not chosen be equal (with probability one) to a deterministic affine function of the chosen components. We then manipulate only the chosen components and, with some deterministic bookkeeping "on the side," take care of the components that were not chosen. This idea is made precise in Corollary 23.4.3.

To illustrate the idea, suppose that $\mathbf{X}$ is a zero-mean random vector of covariance matrix

$$
\mathsf{K_{XX}} = \begin{pmatrix} 3 & 5 & 7 \\ 5 & 9 & 13 \\ 7 & 13 & 19 \end{pmatrix}.
$$

An application of Proposition 23.4.1 ahead will show that because the three columns of $\mathsf{K_{XX}}$ satisfy the linear relationship

$$
- \begin{pmatrix} 3 \\ 5 \\ 7 \end{pmatrix} + 2 \begin{pmatrix} 5 \\ 9 \\ 13 \end{pmatrix} - \begin{pmatrix} 7 \\ 13 \\ 19 \end{pmatrix} = \mathbf{0},
$$

it follows that

$$
-X^{(1)} + 2X^{(2)} - X^{(3)} = 0, \quad \text{with probability one.}
$$

Consequently, in manipulating $\mathbf{X}$ we can pick the two components $X^{(2)}, X^{(3)}$, which are of nonsingular covariance matrix $\left(\begin{smallmatrix} 9 & 13 \\ 13 & 19 \end{smallmatrix}\right)$ (obtained by picking the last two rows and the last two columns of $\mathsf{K_{XX}}$), and keep track "on the side" of the fact that $X^{(1)}$ is equal, with probability one, to $2X^{(2)} - X^{(3)}$. We could, of course, also pick the components $X^{(1)}, X^{(2)}$ of nonsingular covariance matrix $\left(\begin{smallmatrix} 3 & 5 \\ 5 & 9 \end{smallmatrix}\right)$ and keep track "on the side" of the relationship $X^{(3)} = 2X^{(2)} - X^{(1)}$.

To avoid cumbersome language, for the remainder of this section we shall take all equalities between random variables to stand for equalities with probability one. Thus, if we write $X^{(1)} = 2X^{(2)} - X^{(3)}$ we mean that the probability that $X^{(1)}$ is equal to $2X^{(2)} - X^{(3)}$ is one.

The justification of the procedure is in the following proposition and its two corollaries.

**Proposition 23.4.1.** *Let $\mathbf{X}$ be a zero-mean random $n$-vector of covariance matrix $\mathsf{K_{XX}}$. Then its $\ell$-th component $X^{(\ell)}$ is a deterministic linear combination of $X^{(\ell_1)}, \ldots, X^{(\ell_\eta)}$ if, and only if, the $\ell$-th column of $\mathsf{K_{XX}}$ is a linear combination of Columns $\ell_1, \ldots, \ell_\eta$. Here $\ell, \eta, \ell_1, \ldots, \ell_\eta, \in \{1, \ldots, n\}$ are arbitrary.*

**Proof.** If $\ell \in \{\ell_1, \ldots, \ell_\eta\}$, then the result is trivial. We shall therefore present a proof only for the case where $\ell \notin \{\ell_1, \ldots, \ell_\eta\}$. In this case, the $\ell$-th component of the random $n$-vector $\mathbf{X}$ is a linear combination of the $\eta$ components $X^{(\ell_1)}, \ldots, X^{(\ell_\eta)}$ if, and only if, there exists a vector $\boldsymbol{\alpha} \in \mathbb{R}^n$ satisfying

$$\alpha^{(\ell)} = -1, \tag{23.31a}$$

$$\alpha^{(\kappa)} = 0, \quad \kappa \notin \{\ell, \ell_1, \ldots, \ell_\eta\}, \tag{23.31b}$$

and

$$\boldsymbol{\alpha}^{\mathsf{T}}\mathbf{X} = 0. \tag{23.31c}$$

Since $\mathbf{X}$ is of zero mean, the condition $\boldsymbol{\alpha}^{\mathsf{T}}\mathbf{X} = 0$ is equivalent to the condition $\mathsf{Var}[\boldsymbol{\alpha}^{\mathsf{T}}\mathbf{X}] = 0$. By (23.30) and Lemma 23.3.8 this latter condition is equivalent to the condition $\mathsf{K_{XX}}\,\boldsymbol{\alpha} = \mathbf{0}$. Now $\mathsf{K_{XX}}\,\boldsymbol{\alpha}$ is a linear combination of the columns of $\mathsf{K_{XX}}$ where the first column is multiplied by $\alpha^{(1)}$, the second by $\alpha^{(2)}$, etc. Consequently, the condition that $\mathsf{K_{XX}}\,\boldsymbol{\alpha} = \mathbf{0}$ for some $\boldsymbol{\alpha} \in \mathbb{R}^n$ satisfying (23.31a) & (23.31b) is equivalent to the condition that the $\ell$-th column of $\mathsf{K_{XX}}$ is a linear combination of Columns $\ell_1, \ldots, \ell_\eta$. $\qquad \square$

**Corollary 23.4.2.** *The covariance matrix of a zero-mean random $n$-vector $\mathbf{X}$ is singular if, and only if, some component of $\mathbf{X}$ is a linear combination of the other components.*

**Proof.** Follows from Proposition 23.4.1 by noting that a square matrix is singular if, and only if, its columns are linearly dependent. $\qquad \square$

**Corollary 23.4.3.** *Let $\mathbf{X}$ be a zero-mean random $n$-vector of covariance matrix $\mathsf{K_{XX}}$. If Columns $\ell_1, \ldots, \ell_d$ of $\mathsf{K_{XX}}$ form a basis for the subspace of $\mathbb{R}^n$ spanned by the columns of $\mathsf{K_{XX}}$, then every component of $\mathbf{X}$ can be written as a linear combination of the components $X^{(\ell_1)}, \ldots, X^{(\ell_d)}$, and the random $d$-vector $\left(X^{(\ell_1)}, \ldots, X^{(\ell_d)}\right)^{\mathsf{T}}$ has a nonsingular $d \times d$ covariance matrix.*

**Proof.** Since Columns $\ell_1, \ldots, \ell_d$ form a basis for the subspace spanned by the columns of $\mathsf{K_{XX}}$, every column $\ell$ can be written as a linear combination of these columns. Consequently, by Proposition 23.4.1, every component of $\mathbf{X}$ can be written as a linear combination of $X^{(\ell_1)}, \ldots, X^{(\ell_d)}$. To prove that the $d \times d$ covariance matrix $\mathsf{K_{\tilde{X}\tilde{X}}}$ of the random $d$-vector $\tilde{\mathbf{X}} = \left(X^{(\ell_1)}, \ldots, X^{(\ell_d)}\right)^{\mathsf{T}}$ is nonsingular, we note that if this were not the case, then by Corollary 23.4.2 applied to $\tilde{\mathbf{X}}$ it would follow that one of the components of $\tilde{\mathbf{X}}$ is a linear combination of the other $d-1$ components. But by Proposition 23.4.1 applied to $\mathbf{X}$, this would imply that the columns $\ell_1, \ldots, \ell_d$ of $\mathsf{K_{XX}}$ are not linearly independent, in contradiction to the corollary's hypothesis that they form a basis. $\square$

### 23.4.4 The Characteristic Function

If $\mathbf{X}$ is a random $n$-vector, then its characteristic function $\Phi_{\mathbf{X}}(\cdot)$ is a mapping from $\mathbb{R}^n$ to $\mathbb{C}$ that maps each vector $\boldsymbol{\varpi} = (\varpi^{(1)}, \ldots \varpi^{(n)})^{\mathsf{T}}$ in $\mathbb{R}^n$ to $\Phi_{\mathbf{X}}(\boldsymbol{\varpi})$, where

$$
\Phi_{\mathbf{X}}(\boldsymbol{\varpi}) \triangleq \mathsf{E}\left[e^{i\boldsymbol{\varpi}^{\mathsf{T}}\mathbf{X}}\right]
$$

$$
= \mathsf{E}\left[\exp\left(i\sum_{\ell=1}^{n} \varpi^{(\ell)} X^{(\ell)}\right)\right], \quad \boldsymbol{\varpi} \in \mathbb{R}^n.
$$

If $\mathbf{X}$ has the density $f_{\mathbf{X}}(\cdot)$, then

$$
\Phi_{\mathbf{X}}(\boldsymbol{\varpi}) = \int_{-\infty}^{\infty} \cdots \int_{-\infty}^{\infty} f_{\mathbf{X}}(\mathbf{x})\, e^{i\sum_{\ell=1}^{n} \varpi^{(\ell)} x^{(\ell)}} \, \mathrm{d}x^{(1)} \cdots \mathrm{d}x^{(n)},
$$

which is reminiscent of the multi-dimensional Fourier Transform of $f_{\mathbf{X}}(\cdot)$ (ignoring $2\pi$'s and the sign of i).

**Proposition 23.4.4 (Identical Distributions and Characteristic Functions).** *Two random $n$-vectors $\mathbf{X}, \mathbf{Y}$ are of the same distribution if, and only if, they have identical characteristic functions:*

$$
\left(\mathbf{X} \overset{\mathscr{L}}{=} \mathbf{Y}\right) \Leftrightarrow \left(\Phi_{\mathbf{X}}(\boldsymbol{\varpi}) = \Phi_{\mathbf{Y}}(\boldsymbol{\varpi}), \quad \boldsymbol{\varpi} \in \mathbb{R}^n\right). \tag{23.32}
$$

**Proof.** See (Dudley, 2003, Chapter 9, Section 5, Theorem 9.5.1). $\square$

This proposition is extremely useful. We shall demonstrate its power by using it to show that two random variables $X$ and $Y$ are independent if, and only if,

$$
\mathsf{E}\left[e^{i(\varpi_1 X + \varpi_2 Y)}\right] = \mathsf{E}\left[e^{i\varpi_1 X}\right]\mathsf{E}\left[e^{i\varpi_2 Y}\right], \quad \varpi_1, \varpi_2 \in \mathbb{R}. \tag{23.33}
$$

One direction is straightforward. If $X$ and $Y$ are independent, then for any Borel measurable functions $g(\cdot)$ and $h(\cdot)$ the random variables $g(X)$ and $h(Y)$ are also independent. Thus, the independence of $X$ and $Y$ implies the independence of the random variables $e^{i\varpi_1 X}$ and $e^{i\varpi_2 Y}$ and hence implies that the expectation of their product is the product of their expectations:

$$
\begin{aligned}
\mathsf{E}\left[e^{i(\varpi_1 X + \varpi_2 Y)}\right] &= \mathsf{E}\left[e^{i\varpi_1 X} e^{i\varpi_2 Y}\right] \\
&= \mathsf{E}\left[e^{i\varpi_1 X}\right] \mathsf{E}\left[e^{i\varpi_2 Y}\right], \quad \varpi_1, \varpi_2 \in \mathbb{R}.
\end{aligned}
$$

As to the other direction, suppose that $X'$ has the same law as $X$, that $Y'$ has the same law as $Y$, and that $X'$ and $Y'$ are independent. Since $X'$ has the same law as $X$, it follows that

$$
\mathsf{E}\left[e^{i\varpi_1 X'}\right] = \mathsf{E}\left[e^{i\varpi_1 X}\right], \quad \varpi_1 \in \mathbb{R}, \tag{23.34}
$$

and similarly for $Y'$

$$
\mathsf{E}\left[e^{i\varpi_2 Y'}\right] = \mathsf{E}\left[e^{i\varpi_2 Y}\right], \quad \varpi_2 \in \mathbb{R}. \tag{23.35}
$$

Consequently, since $X'$ and $Y'$ are independent

$$
\begin{aligned}
\mathsf{E}\left[e^{i(\varpi_1 X' + \varpi_2 Y')}\right] &= \mathsf{E}\left[e^{i\varpi_1 X'} e^{i\varpi_2 Y'}\right] \\
&= \mathsf{E}\left[e^{i\varpi_1 X'}\right] \mathsf{E}\left[e^{i\varpi_2 Y'}\right] \\
&= \mathsf{E}\left[e^{i\varpi_1 X}\right] \mathsf{E}\left[e^{i\varpi_2 Y}\right], \quad \varpi_1, \varpi_2 \in \mathbb{R},
\end{aligned}
$$

where the third equality follows from (23.34) and (23.35).

We thus see that if (23.33) holds, then the characteristic function of the vector $(X, Y)^\mathsf{T}$ is identical to the characteristic function of the vector $(X', Y')^\mathsf{T}$. By Proposition 23.4.4 the joint distribution of $(X, Y)$ must then be the same as the joint distribution of $(X', Y')$. Since according to the latter distribution the two components are independent, it follows that the same must be true according to the former, i.e., $X$ and $Y$ must be independent.

## 23.5 A Standard Gaussian Vector

Recall Definition 23.1.1 that a random $n$-vector $\mathbf{W}$ is a standard Gaussian if its $n$ components are independent zero-mean unit-variance Gaussian random variables. Its density $f_{\mathbf{W}}(\cdot)$ is then given by

$$
\begin{aligned}
f_{\mathbf{W}}(\mathbf{w}) &= \prod_{\ell=1}^{n}\left(\frac{1}{\sqrt{2\pi}} \exp\left(-\frac{\left(w^{(\ell)}\right)^2}{2}\right)\right) \\
&= \frac{1}{(2\pi)^{n/2}} \exp\left(-\frac{1}{2} \sum_{\ell=1}^{n}\left(w^{(\ell)}\right)^2\right) \\
&= (2\pi)^{-n/2} e^{-\frac{1}{2}\|\mathbf{w}\|^2}, \quad \mathbf{w} \in \mathbb{R}^n. \tag{23.36}
\end{aligned}
$$

The definition of a standard Gaussian random vector is an extension of the definition of a standard Gaussian random variable: the sole component of a standard one-dimensional Gaussian vector is a scalar $\mathcal{N}(0,1)$ random variable. Conversely, every $\mathcal{N}(0,1)$ random variable can be viewed as a one-dimensional standard Gaussian.

If $\mathbf{W}$ is a standard Gaussian random $n$-vector then, as we next show, its mean vector and covariance matrix are given by

$$\mathsf{E}[\mathbf{W}] = \mathbf{0}, \quad \text{and} \quad \mathsf{K}_{\mathbf{WW}} = \mathsf{I}_n. \tag{23.37}$$

Indeed, the mean of a random vector is the vector of the means (23.22), so the fact that $\mathsf{E}[\mathbf{W}] = \mathbf{0}$ is a consequence of all the components of $\mathbf{W}$ having zero mean. And using (23.24) it can be easily shown that the covariance matrix of $\mathbf{W}$ is the identity matrix because the components of $\mathbf{W}$ are independent and hence, *a fortiori* uncorrelated, and because they are each of unit variance.

## 23.6   Gaussian Random Vectors

Recall Definition 23.1.1 that a random $n$-vector $\mathbf{X}$ is said to be Gaussian if for some positive integer $m$ there exists an $n \times m$ matrix $\mathsf{A}$; a standard Gaussian random $m$-vector $\mathbf{W}$; and a deterministic vector $\boldsymbol{\mu} \in \mathbb{R}^n$ such that

$$\mathbf{X} \stackrel{\mathscr{L}}{=} \mathsf{A}\mathbf{W} + \boldsymbol{\mu}. \tag{23.38}$$

From (23.38), from the second order properties of standard Gaussians (23.37), and from the behavior of the mean vector and covariance matrix under linear transformation (23.25a) & (23.27) we obtain

$$\left( \mathbf{X} \stackrel{\mathscr{L}}{=} \mathsf{A}\mathbf{W} + \boldsymbol{\mu} \text{ and } \mathbf{W} \text{ standard} \right) \Rightarrow \left( \mathsf{E}[\mathbf{X}] = \boldsymbol{\mu} \text{ and } \mathsf{K}_{\mathbf{XX}} = \mathsf{A}\mathsf{A}^\mathsf{T} \right). \tag{23.39}$$

Recall also that $\mathbf{X}$ is a centered Gaussian if $\mathbf{X} \stackrel{\mathscr{L}}{=} \mathsf{A}\mathbf{W}$ for $\mathsf{A}$ and $\mathbf{W}$ as above.

*Every standard Gaussian vector is a centered Gaussian* because every standard Gaussian $n$-vector $\mathbf{W}$ is equal to $\mathsf{A}\mathbf{W}$ when $\mathsf{A}$ is the $n \times n$ identity matrix $\mathsf{I}_n$. The reverse is not true: not every centered Gaussian is a standard Gaussian. Indeed, standard Gaussians have the identity covariance matrix (23.37), whereas the centered Gaussian vector $\mathsf{A}\mathbf{W}$ has, by (23.39), the covariance matrix $\mathsf{A}\mathsf{A}^\mathsf{T}$, which need not be the identity matrix.

Also, $\mathbf{X}$ *is a Gaussian vector if, and only if,* $\mathbf{X} - \mathsf{E}[\mathbf{X}]$ *is a centered Gaussian* because, by (23.39),

$$\left( \mathbf{X} \stackrel{\mathscr{L}}{=} \mathsf{A}\mathbf{W} + \boldsymbol{\mu} \text{ for some } \boldsymbol{\mu} \in \mathbb{R}^n \text{ and } \mathbf{W} \text{ standard Gaussian} \right)$$

$$\Leftrightarrow \left( \mathbf{X} \stackrel{\mathscr{L}}{=} \mathsf{A}\mathbf{W} + \mathsf{E}[\mathbf{X}] \text{ and } \mathbf{W} \text{ standard Gaussian} \right)$$

$$\Leftrightarrow \left( \mathbf{X} - \mathsf{E}[\mathbf{X}] \stackrel{\mathscr{L}}{=} \mathsf{A}\mathbf{W} \text{ and } \mathbf{W} \text{ standard Gaussian} \right). \tag{23.40}$$

From (23.40) it also follows that *the centered Gaussians are the Gaussian vectors of zero mean.*[8]

---

[8]Thus, the name "centered Gaussian," which we gave in Definition 23.1.1 was not misleading. A vector is a "centered Gaussian" if, and only if, it is Gaussian and centered.

Using the definition of a centered Gaussian and using (23.39) we can readily show that every positive semidefinite matrix is the covariance matrix of some centered Gaussian. In fact, more is true:

**Proposition 23.6.1 (Covariance Matrices and Positive Semidefinite Matrices).**
*The covariance matrix of every finite-variance random vector is positive semidefinite, and every positive semidefinite matrix is the covariance matrix of some centered Gaussian random vector.*

**Proof.** The covariance matrix of every random vector is positive semidefinite because every covariance matrix is symmetric (23.29) and satisfies (23.28). We next establish the reverse. Given an $n \times n$ positive semidefinite matrix $\mathsf{K}$ we shall construct a centered Gaussian $\mathbf{X}$ whose covariance matrix $\mathsf{K_{XX}}$ is equal to $\mathsf{K}$. We begin by noting that, since $\mathsf{K}$ is positive semidefinite, it follows from Proposition 23.3.6 that there exists some $n \times n$ matrix $\mathsf{S}$ such that $\mathsf{S^T S} = \mathsf{K}$. Let $\mathbf{W}$ be a standard Gaussian $n$-vector and consider the vector $\mathbf{X} = \mathsf{S^T W}$. Being the result of a linear transformation of the standard Gaussian $\mathbf{W}$, this vector is a centered Gaussian. We complete the proof by showing that its covariance matrix $\mathsf{K_{XX}}$ is the prespecified matrix $\mathsf{K}$. This follows from the calculation

$$
\begin{aligned}
\mathsf{K_{XX}} &= \mathsf{S^T S} \\
&= \mathsf{K},
\end{aligned}
$$

where the first equality follows from (23.39) (by substituting $\mathsf{S^T}$ for $\mathsf{A}$ and $\mathbf{0}$ for $\boldsymbol{\mu}$) and the second from our choice of $\mathsf{S}$ as satisfying $\mathsf{S^T S} = \mathsf{K}$. $\qquad\square$

## 23.6.1 Examples and Basic Properties

In this section we provide some examples of Gaussian vectors and some simple properties that follow from their definition.

(i) *Every univariate $\mathcal{N}(\mu, \sigma^2)$ random variable, when viewed as a one dimensional random vector, is a Gaussian random vector.*

   *Proof:* Such a univariate random variable has the same law as $\sigma W + \mu$, when $W$ is a standard univariate Gaussian.

(ii) *Any deterministic vector is a Gaussian vector.*

   *Proof:* Choose the matrix $\mathsf{A}$ as the all-zero matrix $\mathsf{0}$.

(iii) *If the components of $\mathbf{X}$ are independent univariate Gaussians (not necessarily of equal variance), then $\mathbf{X}$ is a Gaussian vector.*

   *Proof:* Choose $\mathsf{A}$ to be an appropriate diagonal matrix.

For the purposes of stating the next proposition we remind the reader that the random vectors $\mathbf{X} = \left(X^{(1)}, \ldots, X^{(n_x)}\right)^\mathsf{T}$ and $\mathbf{Y} = \left(Y^{(1)}, \ldots, Y^{(n_y)}\right)^\mathsf{T}$ are independent if, for every choice of $\xi_1, \ldots, \xi_{n_x} \in \mathbb{R}$ and $\eta_1, \ldots, \eta_{n_y} \in \mathbb{R}$,

$$\Pr\left[X^{(1)} \leq \xi_1, \ldots, X^{(n_x)} \leq \xi_{n_x}, Y^{(1)} \leq \eta_1, \ldots, Y^{(n_y)} \leq \eta_{n_y}\right]$$
$$= \Pr\left[X^{(1)} \leq \xi_1, \ldots, X^{(n_x)} \leq \xi_{n_x}\right] \Pr\left[Y^{(1)} \leq \eta_1, \ldots, Y^{(n_y)} \leq \eta_{n_y}\right].$$

The following proposition is a consequence of the fact that if $\mathbf{X}_1$ & $\mathbf{X}_2$ are independent, $\mathbf{X}_1 \overset{\mathscr{L}}{=} \mathbf{X}_1'$, $\mathbf{X}_2 \overset{\mathscr{L}}{=} \mathbf{X}_2'$, and $\mathbf{X}_1'$ & $\mathbf{X}_2'$ are independent, then

$$\begin{pmatrix} \mathbf{X}_1 \\ \mathbf{X}_2 \end{pmatrix} \overset{\mathscr{L}}{=} \begin{pmatrix} \mathbf{X}_1' \\ \mathbf{X}_2' \end{pmatrix}.$$

**Proposition 23.6.2 (Stacking Independent Gaussian Vectors).** *Stacking two independent Gaussian vectors one on top of the other results in a Gaussian vector.*

**Proof.** Let the random $n_1$-vector $\mathbf{X}_1 = (X_1^{(1)}, \ldots, X_1^{(n_1)})^\mathsf{T}$ be Gaussian, and let the random $n_2$-vector $\mathbf{X}_2 = (X_2^{(1)}, \ldots, X_2^{(n_2)})^\mathsf{T}$ be Gaussian and independent of $\mathbf{X}_1$. We need to show that the $(n_1 + n_2)$-vector

$$\left(X_1^{(1)}, \ldots, X_1^{(n_1)}, X_2^{(1)}, \ldots, X_2^{(n_2)}\right)^\mathsf{T} \tag{23.41}$$

is Gaussian.

Let the pair $(\mathsf{A}_1, \boldsymbol{\mu}_1)$ represent $\mathbf{X}_1$ in the sense that $\mathbf{X}_1 \overset{\mathscr{L}}{=} \mathsf{A}_1 \mathbf{W}_1 + \boldsymbol{\mu}_1$, where $\mathsf{A}_1$ is $n_1 \times m_1$, $\boldsymbol{\mu}_1 \in \mathbb{R}^{n_1}$, and $\mathbf{W}_1$ is a standard Gaussian $m_1$-vector. Similarly, let the pair $(\mathsf{A}_2, \boldsymbol{\mu}_2)$ represent $\mathbf{X}_2$, where $\mathsf{A}_2$ is $n_2 \times m_2$ and $\boldsymbol{\mu}_2 \in \mathbb{R}^{n_2}$. We next show that the vector (23.41) can be represented using the $(n_1 + n_2) \times (m_1 + m_2)$ block-diagonal matrix $\mathsf{A}$ of diagonal components $\mathsf{A}_1$ and $\mathsf{A}_2$, and using the vector $\boldsymbol{\mu} \in \mathbb{R}^{n_1 + n_2}$ that results when the vector $\boldsymbol{\mu}_1$ is stacked on top of the vector $\boldsymbol{\mu}_2$:

$$\mathsf{A} = \begin{pmatrix} \mathsf{A}_1 & 0 \\ 0 & \mathsf{A}_2 \end{pmatrix} \qquad \boldsymbol{\mu} = \begin{pmatrix} \boldsymbol{\mu}_1 \\ \boldsymbol{\mu}_2 \end{pmatrix}. \tag{23.42}$$

Indeed, if $\mathbf{W}$ is a standard Gaussian $(n_1 + n_2)$-vector and if we denote by $\mathbf{W}_1$ its first $n_1$ components and by $\mathbf{W}_2$ its last $n_2$ components, then the random vectors $\mathbf{W}_1$ and $\mathbf{W}_2$ are independent, and each is a standard Gaussian. Consequently,

$$\mathsf{A}\mathbf{W} + \boldsymbol{\mu} = \begin{pmatrix} \mathsf{A}_1 & 0 \\ 0 & \mathsf{A}_2 \end{pmatrix} \begin{pmatrix} \mathbf{W}_1 \\ \mathbf{W}_2 \end{pmatrix} + \begin{pmatrix} \boldsymbol{\mu}_1 \\ \boldsymbol{\mu}_2 \end{pmatrix}$$
$$= \begin{pmatrix} \mathsf{A}_1 \mathbf{W}_1 + \boldsymbol{\mu}_1 \\ \mathsf{A}_2 \mathbf{W}_2 + \boldsymbol{\mu}_2 \end{pmatrix}$$
$$\overset{\mathscr{L}}{=} \begin{pmatrix} \mathbf{X}_1 \\ \mathbf{X}_2 \end{pmatrix},$$

where the first equality follows from the definition of $\mathsf{A}$ and $\boldsymbol{\mu}$ in (23.42); the second equality by computing the matrix product in blocks; and where the equality in distribution follows because the fact that $\mathbf{W}_1$ is a standard Gaussian implies that $\mathbf{X}_1 \overset{\mathscr{L}}{=} \mathsf{A}_1 \mathbf{W}_1 + \boldsymbol{\mu}_1$, the fact that $\mathbf{W}_2$ is a standard Gaussian implies that $\mathbf{X}_2 \overset{\mathscr{L}}{=} \mathsf{A}_2 \mathbf{W}_2 + \boldsymbol{\mu}_2$, and the fact that $\mathbf{W}_1$ and $\mathbf{W}_2$ are independent implies that $\mathsf{A}_1 \mathbf{W}_1 + \boldsymbol{\mu}_1$ and $\mathsf{A}_2 \mathbf{W}_2 + \boldsymbol{\mu}_2$ are independent. $\qquad \square$

**Proposition 23.6.3 (An Affine Transformation of a Gaussian Is a Gaussian).**
*Let* $\mathbf{X}$ *be a Gaussian n-vector. If* $\mathsf{C}$ *is a* $\nu \times n$ *matrix and if* $\mathbf{d} \in \mathbb{R}^\nu$, *then the random $\nu$-vector* $\mathsf{C}\mathbf{X} + \mathbf{d}$ *is Gaussian.*

**Proof.** If $\mathbf{X} \stackrel{\mathscr{L}}{=} \mathsf{A}\mathbf{W} + \boldsymbol{\mu}$, where $\mathsf{A}$ is a deterministic $n \times m$ matrix, $\mathbf{W}$ is a standard Gaussian $m$-vector, and $\boldsymbol{\mu} \in \mathbb{R}^n$, then

$$\mathsf{C}\mathbf{X} + \mathbf{d} \stackrel{\mathscr{L}}{=} \mathsf{C}(\mathsf{A}\mathbf{W} + \boldsymbol{\mu}) + \mathbf{d}$$
$$= (\mathsf{C}\mathsf{A})\mathbf{W} + (\mathsf{C}\boldsymbol{\mu} + \mathbf{d}), \tag{23.43}$$

which demonstrates that $\mathsf{C}\mathbf{X} + \mathbf{d}$ is Gaussian, because $(\mathsf{C}\mathsf{A})$ is a deterministic $\nu \times m$ matrix, $\mathbf{W}$ is a standard Gaussian $m$-vector, and $\mathsf{C}\boldsymbol{\mu} + \mathbf{d}$ is a deterministic vector in $\mathbb{R}^\nu$. $\qquad\square$

This proposition has some important consequences. The first is that if we permute the components of a Gaussian vector then the resulting vector is also Gaussian. This explains why we sometimes say of random variables that they are jointly Gaussian without specifying an order. Indeed, by the following corollary, the Gaussianity of $(X, Y, Z)^\mathsf{T}$ is equivalent to the Gaussianity of $(Y, X, Z)^\mathsf{T}$, etc.

**Corollary 23.6.4.** *Permuting the components of a Gaussian vector results in a Gaussian vector.*

**Proof.** Follows from Proposition 23.6.3 by choosing $\mathsf{C}$ to be the appropriate permutation matrix, i.e., the matrix that results from permuting the columns of the identity matrix. For example,

$$\begin{pmatrix} X^{(3)} \\ X^{(1)} \\ X^{(2)} \end{pmatrix} = \begin{pmatrix} 0 & 0 & 1 \\ 1 & 0 & 0 \\ 0 & 1 & 0 \end{pmatrix} \begin{pmatrix} X^{(1)} \\ X^{(2)} \\ X^{(3)} \end{pmatrix}. \qquad\square$$

**Corollary 23.6.5 (Subsets of Jointly Gaussians Are Jointly Gaussian).** *Constructing a random p-vector from a Gaussian n-vector by picking p of its components (allowing for repetition) yields a Gaussian vector.*

**Proof.** Let $\mathbf{X}$ be a Gaussian $n$-vector. For any choice of $j_1, \ldots, j_p \in \{1, \ldots, n\}$, we can express the random $p$-vector $(X^{(j_1)}, \ldots, X^{(j_p)})^\mathsf{T}$ as $\mathsf{C}\mathbf{X}$, where $\mathsf{C}$ is a deterministic $p \times n$ matrix whose Row-$\nu$ Column-$\ell$ component is given by

$$[\mathsf{C}]_{\nu,\ell} = \mathrm{I}\{\ell = j_\nu\}.$$

For example

$$\begin{pmatrix} X^{(3)} \\ X^{(1)} \end{pmatrix} = \begin{pmatrix} 0 & 0 & 1 \\ 1 & 0 & 0 \end{pmatrix} \begin{pmatrix} X^{(1)} \\ X^{(2)} \\ X^{(3)} \end{pmatrix}.$$

The result thus follows from Proposition 23.6.3. $\qquad\square$

**Proposition 23.6.6.** *Each component of a Gaussian vector is a univariate Gaussian.*

**Proof.** Let $\mathbf{X}$ be a Gaussian $n$-vector, and let $j \in \{1, \ldots, n\}$ be arbitrary. We need to show that $X^{(j)}$ is Gaussian. Since $\mathbf{X}$ is Gaussian, there exist an $n \times m$ matrix $\mathsf{A}$, a vector $\boldsymbol{\mu} \in \mathbb{R}^n$, and a standard Gaussian $\mathbf{W}$ such that the vector $\mathbf{X}$ has the same law as the random vector $\mathsf{A}\mathbf{W} + \boldsymbol{\mu}$ (Definition 23.1.1). In particular, the $j$-th component of $\mathbf{X}$ has the same law as the $j$-th component of $\mathsf{A}\mathbf{W} + \boldsymbol{\mu}$, i.e.,

$$X^{(j)} \overset{\mathscr{L}}{=} \sum_{\ell=1}^{m} a^{(j,\ell)} W^{(\ell)} + \mu^{(j)}, \quad j \in \{1, \ldots, n\}.$$

The sum on the RHS is a linear combination of the independent univariate Gaussians $W^{(1)}, \ldots, W^{(m)}$ and is thus, by Proposition 19.7.3, Gaussian. The result of adding $\mu^{(j)}$ is still Gaussian. $\qquad\square$

We caution the reader that while each component of a Gaussian vector has a univariate Gaussian distribution, there exist random vectors that are not Gaussian and that yet have Gaussian components.

## 23.6.2 The Mean and Covariance Determine the Law of a Gaussian

From (23.39) it follows that if $\mathbf{X} \overset{\mathscr{L}}{=} \mathsf{A}\mathbf{W} + \boldsymbol{\mu}$, where $\mathbf{W}$ is a standard Gaussian, then $\boldsymbol{\mu}$ must be equal to $\mathsf{E}[\mathbf{X}]$. Thus, the mean of $\mathbf{X}$ fully determines the vector $\boldsymbol{\mu}$. The matrix $\mathsf{A}$, however, is not determined by the covariance of $\mathbf{X}$. Indeed, by (23.39), the covariance matrix $\mathsf{K}_{\mathbf{XX}}$ of $\mathbf{X}$ is equal to $\mathsf{A}\mathsf{A}^\mathsf{T}$, so $\mathsf{K}_{\mathbf{XX}}$ only determines the product $\mathsf{A}\mathsf{A}^\mathsf{T}$. Since there are many different ways to express $\mathsf{K}_{\mathbf{XX}}$ as the product of a matrix by its transpose, there are many choices of $\mathsf{A}$ (even of different dimensions) that result in $\mathsf{A}\mathbf{X} + \boldsymbol{\mu}$ having the given covariance matrix. *Prima facie*, one might think that these different choices for $\mathsf{A}$ yield different Gaussian distributions. But this is not the case. In this section we shall show that, while the choice of $\mathsf{A}$ is not unique, all choices that result in $\mathsf{A}\mathsf{A}^\mathsf{T}$ having the given covariance matrix $\mathsf{K}_{\mathbf{XX}}$ give rise to the same distribution.

We shall derive this result by computing the characteristic function $\Phi_{\mathbf{X}}(\cdot)$ of a random $n$-vector $\mathbf{X}$ whose law is equal to the law of $\mathsf{A}\mathbf{W} + \boldsymbol{\mu}$, where $\mathbf{W}$, $\mathsf{A}$, and $\boldsymbol{\mu}$ are as above and by then showing that $\Phi_{\mathbf{X}}(\cdot)$ depends on $\mathsf{A}$ only via $\mathsf{A}\mathsf{A}^\mathsf{T}$, i.e., that $\Phi_{\mathbf{X}}(\boldsymbol{\varpi})$ can be computed for every $\boldsymbol{\varpi} \in \mathbb{R}^n$ from $\boldsymbol{\varpi}$, $\mathsf{A}\mathsf{A}^\mathsf{T}$, and $\boldsymbol{\mu}$. Since, by (23.39), $\mathsf{A}\mathsf{A}^\mathsf{T}$ is equal to the covariance matrix $\mathsf{K}_{\mathbf{XX}}$ of $\mathbf{X}$, it will follow that the characteristic functions of all Gaussian vectors of a given mean vector and a given covariance matrix are identical. Since random vectors of identical characteristic functions must have identical distributions (Proposition 23.4.4), it will follow that all Gaussian vectors of a given mean vector and a given covariance matrix have identical distributions.

We thus proceed to compute the characteristic function of a random $n$-vector $\mathbf{X}$ whose law is the law of $\mathsf{A}\mathbf{W} + \boldsymbol{\mu}$, where $\mathbf{W}$ is a standard Gaussian $m$-vector, $\mathsf{A}$ is $n \times m$, and $\boldsymbol{\mu} \in \mathbb{R}^n$. By (23.39) it follows that $\mathsf{K}_{\mathbf{XX}} = \mathsf{A}\mathsf{A}^\mathsf{T}$. To that end we need to compute $\mathsf{E}[e^{i\boldsymbol{\varpi}^\mathsf{T}\mathbf{X}}]$ for every $\boldsymbol{\varpi} \in \mathbb{R}^n$. From Proposition 23.6.3 (with the substitution of the $1 \times n$ matrix $\boldsymbol{\varpi}^\mathsf{T}$ for $\mathsf{C}$ and of the scalar zero for $\mathbf{d}$), it follows that $\boldsymbol{\varpi}^\mathsf{T}\mathbf{X}$ is a Gaussian vector with only one component. By Proposition 23.6.6, this sole component is a univariate Gaussian. Its mean is, by (23.25a), $\boldsymbol{\varpi}^\mathsf{T}\boldsymbol{\mu}$ and its variance is, by (23.30), $\boldsymbol{\varpi}^\mathsf{T}\mathsf{K}_{\mathbf{XX}}\boldsymbol{\varpi}$. Thus,

$$\boldsymbol{\varpi}^\mathsf{T}\mathbf{X} \sim \mathcal{N}\!\left(\boldsymbol{\varpi}^\mathsf{T}\boldsymbol{\mu},\ \boldsymbol{\varpi}^\mathsf{T}\mathsf{K}_{\mathbf{XX}}\boldsymbol{\varpi}\right), \quad \boldsymbol{\varpi}\in\mathbb{R}^n. \tag{23.44}$$

Using the expression (19.29) for the characteristic function of the univariate Gaussian distribution (with the substitution $\boldsymbol{\varpi}^\mathsf{T}\boldsymbol{\mu}$ for $\mu$, the substitution $\boldsymbol{\varpi}^\mathsf{T}\mathsf{K}_{\mathbf{XX}}\boldsymbol{\varpi}$ for $\sigma^2$, and the substitution 1 for $\varpi$), we obtain that the characteristic function $\Phi_{\mathbf{X}}(\cdot)$, which is defined as $\mathsf{E}\!\left[e^{i\boldsymbol{\varpi}^\mathsf{T}\mathbf{X}}\right]$, is given by

$$\Phi_{\mathbf{X}}(\boldsymbol{\varpi}) = e^{-\frac{1}{2}\boldsymbol{\varpi}^\mathsf{T}\mathsf{K}_{\mathbf{XX}}\boldsymbol{\varpi}+i\boldsymbol{\varpi}^\mathsf{T}\boldsymbol{\mu}}, \quad \boldsymbol{\varpi}\in\mathbb{R}^n. \tag{23.45}$$

Since this characteristic function is fully determined by the mean vector and the covariance matrix of $\mathbf{X}$, it follows that the distribution is also determined by the mean and covariance. We have thus proved:

**Theorem 23.6.7 (The Mean and Covariance of a Gaussian Determine its Law).** *Two Gaussian vectors of equal mean vectors and of equal covariance matrices have identical distributions.*

**Note 23.6.8.** Theorem 23.6.7 and Proposition 23.6.1 combine to prove that for every $\boldsymbol{\mu}\in\mathbb{R}^n$ and every $n\times n$ positive semidefinite matrix $\mathsf{K}$ there exists one, and only one, Gaussian distribution of mean $\boldsymbol{\mu}$ and covariance matrix $\mathsf{K}$. We denote this Gaussian distribution by $\mathcal{N}(\boldsymbol{\mu},\mathsf{K})$.

By (23.45) it follows that if $\mathbf{X}\sim\mathcal{N}(\boldsymbol{\mu},\mathsf{K})$ then

$$\boxed{\Phi_{\mathbf{X}}(\boldsymbol{\varpi}) = e^{-\frac{1}{2}\boldsymbol{\varpi}^\mathsf{T}\mathsf{K}\boldsymbol{\varpi}+i\boldsymbol{\varpi}^\mathsf{T}\boldsymbol{\mu}}, \quad \boldsymbol{\varpi}\in\mathbb{R}^n.} \tag{23.46}$$

Theorem 23.6.7 has important consequences, one of which has to do with the properties of independence and uncorrelatedness. Recall that any two independent random variables (of finite mean) are also uncorrelated. The reverse is not in general true. But for jointly Gaussians it is: if $X$ and $Y$ are jointly Gaussian, then $X$ and $Y$ are independent if, and only if, they are uncorrelated. More generally:

**Corollary 23.6.9.** *Let $\mathbf{X}$ be a centered Gaussian $(n_1+n_2)$-vector. Let the random $n_1$-vector $\mathbf{X}_1 = (X^{(1)},\ldots,X^{(n_1)})^\mathsf{T}$ correspond to its first $n_1$ components, and let $\mathbf{X}_2 = (X^{(n_1+1)},\ldots,X^{(n_1+n_2)})^\mathsf{T}$ correspond to the rest of its components. Then the vectors $\mathbf{X}_1$ and $\mathbf{X}_2$ are independent if, and only if, they are uncorrelated, i.e., if, and only if,*

$$\mathsf{E}\!\left[\mathbf{X}_1\mathbf{X}_2^\mathsf{T}\right] = 0. \tag{23.47}$$

**Proof.** The easy direction, which has nothing to do with Gaussianity, is that if $\mathbf{X}_1$ and $\mathbf{X}_2$ are centered and independent, then (23.47) holds. Indeed, by the independence and the fact that the vectors are of zero mean we have

$$\begin{aligned}
\mathsf{E}\!\left[\mathbf{X}_1\mathbf{X}_2^\mathsf{T}\right] &= \mathsf{E}[\mathbf{X}_1]\,\mathsf{E}\!\left[\mathbf{X}_2^\mathsf{T}\right] \\
&= \mathsf{E}[\mathbf{X}_1]\,(\mathsf{E}[\mathbf{X}_2])^\mathsf{T} \\
&= \mathbf{0}\mathbf{0}^\mathsf{T} \\
&= 0.
\end{aligned}$$

We now prove the reverse using the Gaussianity. We begin by expressing the covariance matrix of $\mathbf{X}$ in terms of the covariance matrices of $\mathbf{X}_1$ and $\mathbf{X}_2$ as

$$
\mathsf{K}_{\mathbf{XX}} = \begin{pmatrix} \mathsf{E}\left[\mathbf{X}_1\mathbf{X}_1^\mathsf{T}\right] & \mathsf{E}\left[\mathbf{X}_1\mathbf{X}_2^\mathsf{T}\right] \\ \mathsf{E}\left[\mathbf{X}_2\mathbf{X}_1^\mathsf{T}\right] & \mathsf{E}\left[\mathbf{X}_2\mathbf{X}_2^\mathsf{T}\right] \end{pmatrix}
$$

$$
= \begin{pmatrix} \mathsf{K}_{\mathbf{X}_1\mathbf{X}_1} & 0 \\ 0 & \mathsf{K}_{\mathbf{X}_2\mathbf{X}_2} \end{pmatrix}, \tag{23.48}
$$

where the second equality follows from (23.47).

Next, let $\mathbf{X}_1'$ and $\mathbf{X}_2'$ be independent random vectors such that $\mathbf{X}_1' \overset{\mathscr{L}}{=} \mathbf{X}_1$ and $\mathbf{X}_2' \overset{\mathscr{L}}{=} \mathbf{X}_2$. Let $\mathbf{X}'$ be the $(n_1 + n_2)$-vector that results from stacking $\mathbf{X}_1'$ on top of $\mathbf{X}_2'$. Since $\mathbf{X}$ is Gaussian, it follows from Corollary 23.6.5 that $\mathbf{X}_1$ must also be Gaussian, and since $\mathbf{X}_1'$ has the same law as $\mathbf{X}_1$, it too is Gaussian. Similarly, $\mathbf{X}_2'$ is also Gaussian. And since $\mathbf{X}_1'$ and $\mathbf{X}_2'$ are, by construction, independent, it follows from Proposition 23.6.2 that $\mathbf{X}'$ is a centered Gaussian.

Having established that $\mathbf{X}'$ is Gaussian, we next compute its covariance matrix. Since, by construction, $\mathbf{X}_1'$ and $\mathbf{X}_2'$ are independent and centered,

$$
\mathsf{K}_{\mathbf{X}'\mathbf{X}'} = \begin{pmatrix} \mathsf{K}_{\mathbf{X}_1'\mathbf{X}_1'} & 0 \\ 0 & \mathsf{K}_{\mathbf{X}_2'\mathbf{X}_2'} \end{pmatrix}
$$

$$
= \begin{pmatrix} \mathsf{K}_{\mathbf{X}_1\mathbf{X}_1} & 0 \\ 0 & \mathsf{K}_{\mathbf{X}_2\mathbf{X}_2} \end{pmatrix}, \tag{23.49}
$$

where the second equality follows because the equality in law between $\mathbf{X}_1'$ and $\mathbf{X}_1$ implies that $\mathsf{K}_{\mathbf{X}_1'\mathbf{X}_1'} = \mathsf{K}_{\mathbf{X}_1\mathbf{X}_1}$ and similarly for $\mathbf{X}_2'$.

Comparing (23.49) and (23.48) we conclude that $\mathbf{X}$ and $\mathbf{X}'$ are centered Gaussians of identical covariance matrices. Consequently, by Theorem 23.6.7, $\mathbf{X}' \overset{\mathscr{L}}{=} \mathbf{X}$. And since the first $n_1$ components of $\mathbf{X}'$ are independent of its last $n_2$ components, the same must also be true for $\mathbf{X}$. ∎

**Corollary 23.6.10.** *If the components of the Gaussian random vector $\mathbf{X}$ are uncorrelated and the matrix $\mathsf{K}_{\mathbf{XX}}$ is therefore diagonal, then the components of $\mathbf{X}$ are independent.*

**Proof.** By repeated application of Corollary 23.6.9. ∎

Another consequence of the fact that there is only one multivariate Gaussian distribution of a given mean vector and of a given covariance matrix has to do with pairwise independence and independence. Recall that the random variables $X_1, \ldots, X_n$ are **pairwise independent** if for each pair of distinct indices $\nu', \nu'' \in \{1, \ldots, n\}$ the random variables $X_{\nu'}$ and $X_{\nu''}$ are independent, i.e., if for all such $\nu', \nu''$ and all $\xi_{\nu'}, \xi_{\nu''} \in \mathbb{R}$

$$
\Pr\left[X_{\nu'} \leq \xi_{\nu'},\, X_{\nu''} \leq \xi_{\nu''}\right] = \Pr\left[X_{\nu'} \leq \xi_{\nu'}\right] \Pr\left[X_{\nu''} \leq \xi_{\nu''}\right]. \tag{23.50}
$$

The random variables $X_1, \ldots, X_n$ are **independent** if for all $\xi_1, \ldots, \xi_n$ in $\mathbb{R}$

$$
\Pr\left[X_j \leq \xi_j,\, \text{for all } j \in \{1, \ldots, n\}\right] = \prod_{j=1}^{n} \Pr\left[X_j \leq \xi_j\right]. \tag{23.51}
$$

Independence implies pairwise independence, but the two are not equivalent. One can find triplets of random variables that are pairwise independent but not independent.[9] But if $X_1, \ldots, X_n$ are jointly Gaussian, then pairwise independence is equivalent to independence:

**Corollary 23.6.11.** *If the components of a Gaussian random vector are pairwise independent, then they are independent.*

**Proof.** If the components of the Gaussian $n$-vector $\mathbf{X}$ are pairwise independent, then they are pairwise uncorrelated and the covariance matrix $\mathsf{K_{XX}}$ must be diagonal. Denote the diagonal elements by $\lambda_1, \ldots, \lambda_n$. Let $\boldsymbol{\mu}$ be the mean vector of $\mathbf{X}$. Another Gaussian vector of this mean and of this covariance matrix is the Gaussian vector whose components are independent $\mathcal{N}\big(\mu^{(j)}, \lambda_j\big)$. Since the mean and covariance determine the distribution of Gaussian vectors, it follows that the two vectors, in fact, have identical laws so the components of $\mathbf{X}$ are also independent. $\quad\square$

**Corollary 23.6.12.** *If $\mathbf{W}$ is a standard Gaussian $n$-vector, and if $\mathsf{U}$ is an $n \times n$ orthogonal matrix, then $\mathsf{U}\mathbf{W}$ is also a standard Gaussian vector.*

**Proof.** By Definition 23.1.1 it follows that the random vector $\mathsf{U}\mathbf{W}$ is a centered Gaussian. By (23.39) we obtain that the orthogonality of the matrix $\mathsf{U}$ implies that the covariance matrix of this centered Gaussian is the identity matrix, which is also the covariance matrix of $\mathbf{W}$; see (23.37). Consequently, $\mathsf{U}\mathbf{W}$ and $\mathbf{W}$ are two centered Gaussian vectors of identical covariance matrices and hence, by Theorem 23.6.7, of equal law. Since $\mathbf{W}$ is standard, this implies that $\mathsf{U}\mathbf{W}$ must also be standard. $\quad\square$

The next corollary shows that if $\mathbf{X}$ is a centered Gaussian $n$-vector, then $\mathbf{X} \overset{\mathscr{L}}{=} \mathsf{A}\mathbf{W}$ for a standard Gaussian $n$-vector $\mathbf{W}$ and some *square* matrix $\mathsf{A}$. That is, if the law of an $n$-vector $\mathbf{X}$ is equal to the law of $\tilde{\mathsf{A}}\tilde{\mathbf{W}}$ where $\tilde{\mathsf{A}}$ is an $n \times m$ matrix and where $\tilde{\mathbf{W}}$ is a standard Gaussian $m$-vector, then the law of $\mathbf{X}$ is also identical to the law of $\mathsf{A}\mathbf{W}$, where $\mathsf{A}$ is some $n \times n$ matrix and where $\mathbf{W}$ is a standard Gaussian $n$-vector. Consequently, we could have required in Definition 23.1.1 that the matrix $\mathsf{A}$ be square without changing the set of distributions that we define as Gaussian.

**Corollary 23.6.13.** *If $\mathbf{X}$ is a centered Gaussian $n$-vector, then there exists a deterministic square $n \times n$ matrix $\mathsf{A}$ such that $\mathbf{X} \overset{\mathscr{L}}{=} \mathsf{A}\mathbf{W}$, where $\mathbf{W}$ is a standard Gaussian $n$-vector.*

**Proof.** Let $\mathsf{K_{XX}}$ denote the covariance matrix of $\mathbf{X}$. Being a covariance matrix, $\mathsf{K_{XX}}$ must be positive semidefinite (Proposition 23.6.1). Consequently, by Proposition 23.3.7, there exists some $n \times n$ matrix $\mathsf{S}$ such that

$$\mathsf{K_{XX}} = \mathsf{S}^\mathsf{T}\mathsf{S}. \tag{23.52}$$

Consider now the centered Gaussian $\mathsf{S}^\mathsf{T}\mathbf{W}$, where $\mathbf{W}$ is a standard Gaussian $n$-vector. By (23.39), the covariance matrix of $\mathsf{S}^\mathsf{T}\mathbf{W}$ is $\mathsf{S}^\mathsf{T}\mathsf{S}$, which by (23.52) is

---

[9]A classical example is the triple $X, Y, Z$ where $X$ and $Y$ are IID each taking on the values $\pm 1$ equiprobably and where $Z$ is their product.

equal to $\mathsf{K_{XX}}$. Thus $\mathbf{X}$ and $\mathsf{S}^\mathsf{T}\mathbf{W}$ are centered Gaussians of the same covariance, and so they must be of the same law. We have thus established that the law of $\mathbf{X}$ is the same as the law of the product of a square matrix ($\mathsf{S}^\mathsf{T}$) by a standard Gaussian ($\mathbf{W}$). $\qquad\square$

### 23.6.3 A Canonical Representation of a Centered Gaussian

The representation of a centered Gaussian vector as the result of the multiplication of a deterministic matrix by a standard Gaussian vector is not unique. Indeed, whenever the $n \times m$ matrix $\mathsf{A}$ satisfies $\mathsf{AA}^\mathsf{T} = \mathsf{K}$ it follows that if $\mathbf{W}$ is a standard Gaussian $m$-vector, then $\mathsf{A}\mathbf{W} \sim \mathcal{N}(\mathbf{0}, \mathsf{K})$. (This follows because $\mathsf{A}\mathbf{W}$ is a random $n$-vector of covariance matrix $\mathsf{AA}^\mathsf{T}$ (23.39); it is, by Definition 23.1.1, a centered Gaussian; and all centered Gaussians of a given covariance matrix have the same law.) We saw in Corollary 23.6.13 that $\mathsf{A}$ can always be chosen as a square matrix. Thus, to every $\mathsf{K} \succeq \mathbf{0}$ there exists a square matrix $\mathsf{A}$ such that $\mathsf{A}\mathbf{W} \sim \mathcal{N}(\mathbf{0}, \mathsf{K})$. In this section we shall focus on a particular choice of the matrix $\mathsf{A}$ that is useful in the analysis of Gaussian vectors. In this representation $\mathsf{A}$ is a square matrix that can be written as the product of an orthogonal matrix by a diagonal matrix. The diagonal matrix acts on $\mathbf{W}$ by stretching and shrinking its components, and the orthogonal matrix then rotates (and possibly reflects) the result.

**Theorem 23.6.14 (A Canonical Representation of a Gaussian Vector).** *Let $\mathbf{X}$ be a centered Gaussian $n$-vector of covariance matrix $\mathsf{K_{XX}}$. Then*

$$\mathbf{X} \overset{\mathscr{L}}{=} \mathsf{U}\Lambda^{1/2}\mathbf{W},$$

*where $\mathbf{W}$ is a standard Gaussian $n$-vector; the $n \times n$ matrix $\mathsf{U}$ is orthogonal; the $n \times n$ matrix $\Lambda$ is diagonal; the diagonal elements of $\Lambda$ are the eigenvalues of $\mathsf{K_{XX}}$; and the $j$-th column of $\mathsf{U}$ is an eigenvector corresponding to the eigenvalue of $\mathsf{K_{XX}}$ that is equal to the $j$-th diagonal element of $\Lambda$.*

**Proof.** By Proposition 23.6.1, $\mathsf{K_{XX}}$ is positive semidefinite and *a fortiori* symmetric. Consequently, by Proposition 23.3.6, there exists a diagonal matrix $\Lambda$ whose diagonal elements are the eigenvalues of $\mathsf{K_{XX}}$ and there exists an orthogonal matrix $\mathsf{U}$ such that $\mathsf{K_{XX}} \mathsf{U} = \mathsf{U}\Lambda$, so the $j$-th column of $\mathsf{U}$ is an eigenvector corresponding to the eigenvalue given by the $j$-th diagonal element of $\Lambda$. Since, $\mathsf{K_{XX}} \succeq \mathbf{0}$, it follows that all its eigenvalues are nonnegative, and we can define the matrix $\Lambda^{1/2}$ as the matrix whose components are the componentwise nonnegative square roots of the matrix $\Lambda$. As in (23.21), choose $\mathsf{S} = \Lambda^{1/2}\mathsf{U}^\mathsf{T}$. We then have that $\mathsf{K_{XX}} = \mathsf{S}^\mathsf{T}\mathsf{S}$. If $\mathbf{W}$ is a standard Gaussian, then $\mathsf{S}^\mathsf{T}\mathbf{W}$ is a centered Gaussian of zero mean and covariance $\mathsf{S}^\mathsf{T}\mathsf{S}$. Since $\mathsf{S}^\mathsf{T}\mathsf{S} = \mathsf{K_{XX}}$ and since there is only one centered multivariate Gaussian distribution of a given covariance matrix, it follows that the law of $\mathsf{S}^\mathsf{T}\mathbf{W}$ ($= \mathsf{U}\Lambda^{1/2}\mathbf{W}$) is the same as the law of $\mathbf{X}$. $\qquad\square$

**Corollary 23.6.15.** *A centered Gaussian vector can be expressed as the result of an orthogonal transformation applied to a random vector whose components are independent centered univariate Gaussians of different variances. These variances are the eigenvalues of the covariance matrix.*

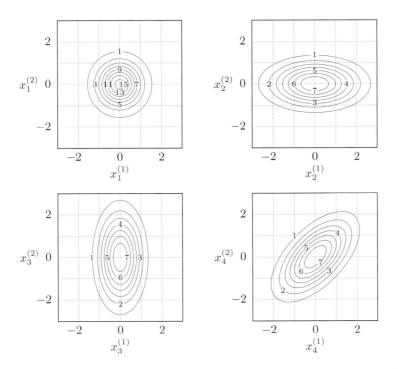

**Figure 23.1:** Contour plot of the density of four different two dimensional Gaussian random variables: from left to right and top to bottom $\mathbf{X}_1, \ldots, \mathbf{X}_4$.

**Proof.** Because the matrix $\Lambda$ in the theorem is diagonal, we can write $\Lambda^{1/2}\mathbf{W}$ as

$$\Lambda^{1/2}\mathbf{W} = \begin{pmatrix} \sqrt{\lambda_1}W^{(1)} \\ \vdots \\ \sqrt{\lambda_n}W^{(n)} \end{pmatrix},$$

where $\lambda_1, \ldots, \lambda_n$ are the diagonal elements of $\Lambda$, i.e., the eigenvalues of $\mathsf{K_{XX}}$. Thus, the random vector $\Lambda^{1/2}\mathbf{W}$ has independent components with the $\nu$-th component being $\mathcal{N}(0, \lambda_\nu)$. $\qquad\square$

Figures 23.1 and 23.2 demonstrate this canonical representation. They depict the contour lines and mesh plots of the density functions of the following four two-dimensional Gaussian vectors:

$$\mathbf{X}_1 = \begin{pmatrix} 1 & 0 \\ 0 & 1 \end{pmatrix} \mathbf{W}, \qquad \mathsf{K_{X_1X_1}} = \mathsf{I}_2,$$

$$\mathbf{X}_2 = \begin{pmatrix} 2 & 0 \\ 0 & 1 \end{pmatrix} \mathbf{W}, \qquad \mathsf{K_{X_2X_2}} = \begin{pmatrix} 4 & 0 \\ 0 & 1 \end{pmatrix},$$

$$\mathbf{X}_3 = \begin{pmatrix} 1 & 0 \\ 0 & 2 \end{pmatrix} \mathbf{W}, \qquad \mathsf{K_{X_3X_3}} = \begin{pmatrix} 1 & 0 \\ 0 & 4 \end{pmatrix},$$

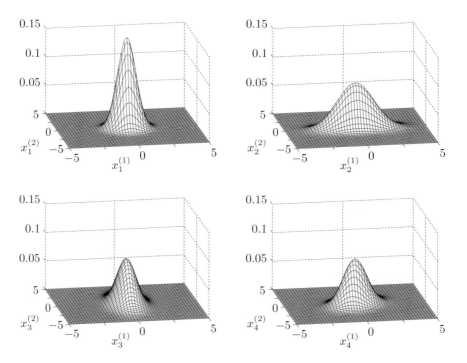

**Figure 23.2:** Mesh plots of the density functions of Gaussian random vectors: from left to right and top to down $\mathbf{X}_1, \dots, \mathbf{X}_4$.

$$\mathbf{X}_4 = \frac{1}{\sqrt{2}} \begin{pmatrix} 1 & 1 \\ -1 & 1 \end{pmatrix} \begin{pmatrix} 1 & 0 \\ 0 & 2 \end{pmatrix} \mathbf{W}, \qquad \mathsf{K}_{\mathbf{X}_4\mathbf{X}_4} = \frac{1}{2} \begin{pmatrix} 5 & 3 \\ 3 & 5 \end{pmatrix},$$

where $\mathbf{W}$ is a standard Gaussian vector with two components.

Theorem 23.6.14 can be used to find a linear transformation that transforms a given Gaussian vector to a standard Gaussian. The following is the multivariate version of the univariate result showing that if $X \sim \mathcal{N}(\mu, \sigma^2)$, where $\sigma^2 > 0$, then $(X - \mu)/\sigma$ has a $\mathcal{N}(0,1)$ distribution (19.8).

**Proposition 23.6.16 (From Gaussians to Standard Gaussians).** *Let the random $n$-vector $\mathbf{X}$ be $\mathcal{N}(\boldsymbol{\mu}, \mathsf{K})$, where $\mathsf{K} \succ 0$ and $\boldsymbol{\mu} \in \mathbb{R}^n$. Let the $n \times n$ matrices $\Lambda$ and $\mathsf{U}$ be such that $\Lambda$ is diagonal, $\mathsf{U}$ is orthogonal, and $\mathsf{KU} = \mathsf{U}\Lambda$. Then*

$$\Lambda^{-1/2}\mathsf{U}^\mathsf{T}(\mathbf{X} - \boldsymbol{\mu}) \sim \mathcal{N}(\mathbf{0}, \mathsf{I}_n),$$

*where $\Lambda^{-1/2}$ is the diagonal matrix whose diagonal entries are the reciprocals of the square roots of the diagonal elements of $\Lambda$.*

**Proof.** Since an affine transformation of a Gaussian vector is Gaussian (Proposition 23.6.3), it follows that $\Lambda^{-1/2}\mathsf{U}^\mathsf{T}(\mathbf{X} - \boldsymbol{\mu})$ is a Gaussian vector. And since the mean and covariance of a Gaussian vector fully specify its law (Theorem 23.6.7), the result will follow once we show that the mean of $\Lambda^{-1/2}\mathsf{U}^\mathsf{T}(\mathbf{X} - \boldsymbol{\mu})$ is the zero vector and its covariance matrix is the identity matrix. This can be readily verified using (23.39). □

## 23.6.4 The Density of a Gaussian Vector

As we saw in Corollary 23.4.2, if the covariance matrix of a centered vector is singular, then at least one of its components can be expressed as a deterministic linear combination of its other components. Consequently, random vectors with singular covariance matrices cannot have a density. If the covariance matrix is nonsingular, then the vector may or may not have a density. If it is Gaussian, then it does. In this section we shall derive the density of the multivariate Gaussian distribution when the covariance matrix is nonsingular.

We begin with the centered case. To derive the density of a centered Gaussian $n$-vector of positive definite covariance matrix $\mathsf{K}$ we shall use Theorem 23.6.14 to represent the $\mathcal{N}(\mathbf{0}, \mathsf{K})$ distribution as the distribution of $\mathsf{U}\Lambda^{1/2}\mathbf{W}$ where $\mathsf{U}$ is an orthogonal matrix and $\Lambda$ is a diagonal matrix satisfying $\mathsf{K}\mathsf{U} = \mathsf{U}\Lambda$. Note that $\Lambda$ is nonsingular because its diagonal elements are the eigenvalues of $\mathsf{K}$, which we assume to be positive definite.

Let

$$\mathsf{B} = \mathsf{U}\Lambda^{1/2}, \tag{23.53}$$

so the density we are after is the density of $\mathsf{B}\mathbf{W}$. Note that, by (23.53),

$$\mathsf{B}\mathsf{B}^{\mathsf{T}} = \mathsf{U}\Lambda^{1/2}\Lambda^{1/2}\mathsf{U}^{\mathsf{T}}$$
$$= \mathsf{U}\Lambda\mathsf{U}^{\mathsf{T}}$$
$$= \mathsf{K}. \tag{23.54}$$

Also, by (23.54),

$$|\det(\mathsf{B})| = \sqrt{\det(\mathsf{B})\det(\mathsf{B})}$$
$$= \sqrt{\det(\mathsf{B})\det(\mathsf{B}^{\mathsf{T}})}$$
$$= \sqrt{\det(\mathsf{B}\mathsf{B}^{\mathsf{T}})}$$
$$= \sqrt{\det(\mathsf{K})}, \tag{23.55}$$

where the first equality follows by expressing $|x|$ as $\sqrt{x^2}$; the second follows because a square matrix and its transpose have the same determinant; the third because the determinant of the product of square matrices is the product of the determinants; and where the last equality follows from (23.54).

Using the formula for computing the density of $\mathsf{B}\mathbf{W}$ from that of $\mathbf{W}$ (Theorem 17.3.4), we have that if $\mathbf{X} \overset{\mathscr{L}}{=} \mathsf{B}\mathbf{W}$, then

$$f_{\mathbf{X}}(\mathbf{x}) = \frac{f_{\mathbf{W}}(\mathsf{B}^{-1}\mathbf{x})}{|\det(\mathsf{B})|}$$

$$= \frac{\exp\left(-\frac{1}{2}\left(\mathsf{B}^{-1}\mathbf{x}\right)^{\mathsf{T}}\left(\mathsf{B}^{-1}\mathbf{x}\right)\right)}{(2\pi)^{n/2}|\det(\mathsf{B})|}$$

$$= \frac{\exp\left(-\frac{1}{2}\mathbf{x}^{\mathsf{T}}\left(\mathsf{B}^{-1}\right)^{\mathsf{T}}\left(\mathsf{B}^{-1}\mathbf{x}\right)\right)}{(2\pi)^{n/2}|\det(\mathsf{B})|}$$

$$= \frac{\exp\left(-\frac{1}{2}\mathbf{x}^{\mathsf{T}}\left(\mathsf{B}\mathsf{B}^{\mathsf{T}}\right)^{-1}\mathbf{x}\right)}{(2\pi)^{n/2}|\det(\mathsf{B})|}$$

$$= \frac{\exp\left(-\frac{1}{2}\mathbf{x}^{\mathsf{T}}\mathsf{K}^{-1}\mathbf{x}\right)}{(2\pi)^{n/2}|\det(\mathsf{B})|}$$

$$= \frac{\exp\left(-\frac{1}{2}\mathbf{x}^{\mathsf{T}}\mathsf{K}^{-1}\mathbf{x}\right)}{(2\pi)^{n/2}\sqrt{\det(\mathsf{K})}},$$

where the second equality follows from the density of the standard Gaussian (23.36); the third from the rule for the transpose of the product of matrices (23.3); the fourth from the representation of the inverse of the product of matrices as the product of the inverses in reverse order $(\mathsf{AB})^{-1} = \mathsf{B}^{-1}\mathsf{A}^{-1}$ and because transposition and inversion commute; the fifth from (23.54); and the sixth from (23.55). It follows that if $X \sim \mathcal{N}(\mathbf{0}, \mathsf{K})$ where $\mathsf{K}$ is nonsingular, then

$$f_{\mathbf{X}}(\mathbf{x}) = \frac{\exp\left(-\frac{1}{2}\mathbf{x}^{\mathsf{T}}\mathsf{K}^{-1}\mathbf{x}\right)}{\sqrt{(2\pi)^n\det(\mathsf{K})}}, \quad \mathbf{x} \in \mathbb{R}^n.$$

Accounting for the mean, we have that if $X \sim \mathcal{N}(\boldsymbol{\mu}, \mathsf{K})$ where $\mathsf{K}$ is nonsingular, then

$$f_{\mathbf{X}}(\mathbf{x}) = \frac{\exp\left(-\frac{1}{2}(\mathbf{x} - \boldsymbol{\mu})^{\mathsf{T}}\mathsf{K}^{-1}(\mathbf{x} - \boldsymbol{\mu})\right)}{\sqrt{(2\pi)^n\det(\mathsf{K})}}, \quad \mathbf{x} \in \mathbb{R}^n. \tag{23.56}$$

### 23.6.5  Linear Functionals of Gaussian Vectors

A **linear functional** on $\mathbb{R}^n$ is a linear mapping from $\mathbb{R}^n$ to $\mathbb{R}$. For example, if $\boldsymbol{\alpha}$ is any fixed vector in $\mathbb{R}^n$, then the mapping

$$\mathbf{x} \mapsto \boldsymbol{\alpha}^{\mathsf{T}}\mathbf{x} \tag{23.57}$$

is a linear functional on $\mathbb{R}^n$. In fact, as we next show, every linear functional on $\mathbb{R}^n$ has this form. This can be proved by using linearity to verify that we can choose the $j$-th component of $\boldsymbol{\alpha}$ to equal the result of applying the linear functional to the vector $\mathbf{e}_j$ whose components are all zero except for its $j$-th component which is equal to one.

If $\mathbf{X}$ is a Gaussian $n$-vector and if $\boldsymbol{\alpha} \in \mathbb{R}^n$, then, by Proposition 23.6.3 (applied with the substitution of the $1 \times n$ matrix $\boldsymbol{\alpha}^{\mathsf{T}}$ for $\mathsf{C}$), it follows that $\boldsymbol{\alpha}^{\mathsf{T}}\mathbf{X}$ is a Gaussian vector with only one component. By Proposition 23.6.6, this sole component must have a univariate Gaussian distribution. We thus conclude that the result of applying a linear functional to a Gaussian vector is a Gaussian random variable.

We next show that the reverse is also true: if $\mathbf{X}$ is of mean $\boldsymbol{\mu}$ and of covariance matrix K and if the result of applying every linear functional to $\mathbf{X}$ has a univariate Gaussian distribution, then $\mathbf{X} \sim \mathcal{N}(\boldsymbol{\mu}, \mathsf{K})$.[10] To prove this result we compute the characteristic function of $\mathbf{X}$. For every $\boldsymbol{\varpi} \in \mathbb{R}^n$ the mapping $\mathbf{x} \mapsto \boldsymbol{\varpi}^\mathsf{T} \mathbf{x}$ is a linear functional on $\mathbb{R}^n$. Consequently, our assumption that the result of the application of every linear functional to $\mathbf{X}$ has a univariate Gaussian distribution implies (23.44). From here we can follow the steps leading to (23.46) to conclude that the characteristic function of $\mathbf{X}$ must be given by the RHS of (23.46). Since this is also the characteristic function of a $\mathcal{N}(\boldsymbol{\mu}, \mathsf{K})$ random vector, it follows that $\mathbf{X} \sim \mathcal{N}(\boldsymbol{\mu}, \mathsf{K})$, because random vectors of identical characteristic functions must have identical distributions (Proposition 23.4.4). We have thus proved:

**Theorem 23.6.17 (Gaussian Vectors and Linear Functionals).** *A random vector* $\mathbf{X}$ *is Gaussian if, and only if, every linear functional of* $\mathbf{X}$ *has a univariate Gaussian distribution.*

## 23.7 Jointly Gaussian Vectors

Three miracles occur when we compute the conditional distribution of $\mathbf{X}$ given $\mathbf{Y} = \mathbf{y}$ for jointly Gaussian random vectors $\mathbf{X}$ and $\mathbf{Y}$. Before describing these miracles we need to define jointly Gaussian vectors.

**Definition 23.7.1 (Jointly Gaussian Vectors).** *Two random vectors are said to be* ***jointly Gaussian*** *if the vector that results when one is stacked on top of the other is Gaussian.*

That is, the random $n_x$-vector $\mathbf{X} = (X^{(1)}, \dots, X^{(n_x)})^\mathsf{T}$ and the random $n_y$-vector $\mathbf{Y} = (Y^{(1)}, \dots, Y^{(n_y)})^\mathsf{T}$ are jointly Gaussian if the random $(n_x + n_y)$-vector

$$\left( X^{(1)}, \dots, X^{(n_x)}, Y^{(1)}, \dots, Y^{(n_y)} \right)^\mathsf{T}$$

is Gaussian.

By Corollary 23.6.5, the random vectors $\mathbf{X}$ and $\mathbf{Y}$ can only be jointly Gaussian if each is Gaussian. But this is not enough: both $\mathbf{X}$ and $\mathbf{Y}$ can be Gaussian without them being jointly Gaussian. However, if $\mathbf{X}$ and $\mathbf{Y}$ are *independent* Gaussian vectors, then, by Proposition 23.6.2, they are jointly Gaussian.

**Proposition 23.7.2.** *Independent Gaussian vectors are jointly Gaussian.*

By Corollary 23.6.9 we have:

**Proposition 23.7.3.** *If two jointly Gaussian random vectors are uncorrelated, then they are independent.*

---

[10]It is not difficult to show that the assumption that $\mathbf{X}$ is of finite variance is not necessary. If every linear functional of $\mathbf{X}$ is of finite variance, then $\mathbf{X}$ must be of finite variance. Thus, we could have stated the result as follows: if a random vector is such that the result of applying every linear functional to it is a univariate Gaussian, then it is a multivariate Gaussian.

Having defined jointly Gaussian random vectors we next turn to the main result of this section. Loosely speaking, it states that if $\mathbf{X}$ and $\mathbf{Y}$ are jointly Gaussian, then in computing the conditional distribution of $\mathbf{X}$ given $\mathbf{Y} = \mathbf{y}$ three miracles occur:

(i) the conditional distribution is a multivariate Gaussian;

(ii) its mean vector is an affine function of $\mathbf{y}$;

(iii) and its covariance matrix does not depend on $\mathbf{y}$.

Before stating this more formally, we justify two simplifying assumptions. The first assumption is that the covariance matrix of $\mathbf{Y}$ is nonsingular, so

$$\mathsf{K_{YY}} \succ 0.$$

The reason is that if the covariance matrix of $\mathbf{Y}$ is singular, then, by Corollary 23.4.2, some of its components are with probability one affine functions of the others, and we then have to consider two cases. If the realization $\mathbf{y}$ satisfies these affine relations, then we can just pick a subset of the components of $\mathbf{Y}$ that determine all the other components and that have a nonsingular covariance matrix as in Section 23.4.3 and ignore the other components of $\mathbf{y}$; the ignored components do not alter the conditional distribution of $\mathbf{X}$ given $\mathbf{Y} = \mathbf{y}$. The other case where the realization $\mathbf{y}$ does not satisfy the relations that $\mathbf{Y}$ satisfies with probability one can be ignored because it occurs with probability zero.

The second assumption we make is that both $\mathbf{X}$ and $\mathbf{Y}$ are centered. There is no loss in generality in making this assumption for the following reason. Conditioning on $\mathbf{Y} = \mathbf{y}$ when $\mathbf{Y}$ has mean $\boldsymbol{\mu}_y$ is equivalent to conditioning on $\mathbf{Y} - \boldsymbol{\mu}_y = \mathbf{y} - \boldsymbol{\mu}_y$. And if $\mathbf{X}$ has mean $\boldsymbol{\mu}_x$, then we can compute the conditional distribution of $\mathbf{X}$ by computing the conditional distribution of $\mathbf{X} - \boldsymbol{\mu}_x$ and by then shifting the resulting distribution by $\boldsymbol{\mu}_x$. Thus, the conditional density $f_{\mathbf{X}|\mathbf{Y}=\mathbf{y}}(\cdot)$ is given by

$$f_{\mathbf{X}|\mathbf{Y}=\mathbf{y}}(\mathbf{x}) = f_{\mathbf{X}-\boldsymbol{\mu}_x|\mathbf{Y}-\boldsymbol{\mu}_y=\mathbf{y}-\boldsymbol{\mu}_y}(\mathbf{x} - \boldsymbol{\mu}_x), \tag{23.58}$$

where $\mathbf{X} - \boldsymbol{\mu}_x$ & $\mathbf{Y} - \boldsymbol{\mu}_y$ are jointly Gaussian and centered whenever $\mathbf{X}$ & $\mathbf{Y}$ are jointly Gaussian. It is now straightforward to verify that if the miracles hold for the centered case

$$\mathbf{x} \mapsto f_{\mathbf{X}-\boldsymbol{\mu}_x|\mathbf{Y}-\boldsymbol{\mu}_y=\mathbf{y}-\boldsymbol{\mu}_y}(\mathbf{x})$$

then they also hold for the general case

$$\mathbf{x} \mapsto f_{\mathbf{X}-\boldsymbol{\mu}_x|\mathbf{Y}-\boldsymbol{\mu}_y=\mathbf{y}-\boldsymbol{\mu}_y}(\mathbf{x} - \boldsymbol{\mu}_x).$$

**Theorem 23.7.4.** *Let $\mathbf{X}$ and $\mathbf{Y}$ be centered and jointly Gaussian with covariance matrices $\mathsf{K_{XX}}$ and $\mathsf{K_{YY}}$. Assume that $\mathsf{K_{YY}} \succ 0$. Then the conditional distribution of $\mathbf{X}$ conditional on $\mathbf{Y} = \mathbf{y}$ is a multivariate Gaussian of mean*

$$\mathsf{E}\!\left[\mathbf{X}\mathbf{Y}^{\mathsf{T}}\right] \mathsf{K_{YY}^{-1}}\, \mathbf{y} \tag{23.59}$$

*and covariance matrix*

$$\mathsf{K_{XX}} - \mathsf{E}\!\left[\mathbf{X}\mathbf{Y}^{\mathsf{T}}\right] \mathsf{K_{YY}^{-1}}\, \mathsf{E}\!\left[\mathbf{Y}\mathbf{X}^{\mathsf{T}}\right]. \tag{23.60}$$

**Proof.** Let $n_x$ and $n_y$ denote the number of components of $\mathbf{X}$ and $\mathbf{Y}$. Let $\mathsf{D}$ be any deterministic real $n_x \times n_y$ matrix. Then clearly

$$\mathbf{X} = \mathsf{D}\mathbf{Y} + (\mathbf{X} - \mathsf{D}\mathbf{Y}). \tag{23.61}$$

Since $\mathbf{X}$ and $\mathbf{Y}$ are jointly Gaussian, the vector $\left(\mathbf{X}^{\mathsf{T}}, \mathbf{Y}^{\mathsf{T}}\right)^{\mathsf{T}}$ is Gaussian. Consequently, since

$$\begin{pmatrix} \mathbf{X} - \mathsf{D}\mathbf{Y} \\ \mathbf{Y} \end{pmatrix} = \begin{pmatrix} \mathsf{I}_{n_x} & -\mathsf{D} \\ 0 & \mathsf{I}_{n_y} \end{pmatrix} \begin{pmatrix} \mathbf{X} \\ \mathbf{Y} \end{pmatrix},$$

it follows from Proposition 23.6.3 that

$$(\mathbf{X} - \mathsf{D}\mathbf{Y}) \text{ and } \mathbf{Y} \text{ are centered and jointly Gaussian.} \tag{23.62}$$

Suppose now that the matrix $\mathsf{D}$ is chosen so that $(\mathbf{X} - \mathsf{D}\mathbf{Y})$ and $\mathbf{Y}$ be uncorrelated:

$$\mathsf{E}\left[(\mathbf{X} - \mathsf{D}\mathbf{Y})\mathbf{Y}^{\mathsf{T}}\right] = 0. \tag{23.63}$$

By (23.62) and Proposition 23.7.3 it then follows that the random vector $(\mathbf{X} - \mathsf{D}\mathbf{Y})$ is independent of $\mathbf{Y}$. Consequently, with this choice of $\mathsf{D}$ we have that (23.61) expresses $\mathbf{X}$ as the sum of two terms where the first, $\mathsf{D}\mathbf{Y}$, is fully determined by $\mathbf{Y}$ and where the second, $(\mathbf{X} - \mathsf{D}\mathbf{Y})$, is independent of $\mathbf{Y}$. It follows that the conditional distribution of $\mathbf{X}$ given $\mathbf{Y} = \mathbf{y}$ is the same as the distribution of $(\mathbf{X} - \mathsf{D}\mathbf{Y})$ but shifted by $\mathsf{D}\mathbf{y}$. By (23.62) and Corollary 23.6.5, $(\mathbf{X} - \mathsf{D}\mathbf{Y})$ is a centered Gaussian, so the conditional distribution of $\mathbf{X}$ given $\mathbf{Y} = \mathbf{y}$ is that of the centered Gaussian $(\mathbf{X} - \mathsf{D}\mathbf{Y})$ shifted by the vector $\mathsf{D}\mathbf{y}$. This already establishes the three "miracles" we discussed before: the conditional distribution of $\mathbf{X}$ given $\mathbf{Y} = \mathbf{y}$ is Gaussian; its mean $\mathsf{D}\mathbf{y}$ is a linear function of $\mathbf{Y}$; and its covariance matrix, which is the covariance matrix of $(\mathbf{X} - \mathsf{D}\mathbf{Y})$, does not depend on the realization $\mathbf{y}$ of $\mathbf{Y}$.

The remaining claims, namely that the mean of the conditional distribution is as given in (23.59) and that the covariance matrix is as given in (23.60) now follow from straightforward calculations. Indeed, by solving (23.63) for $\mathsf{D}$ we obtain

$$\mathsf{D} = \mathsf{E}\left[\mathbf{X}\mathbf{Y}^{\mathsf{T}}\right] \mathsf{K}_{\mathbf{Y}\mathbf{Y}}^{-1}, \tag{23.64}$$

so $\mathsf{D}\mathbf{y}$ is given by (23.59). To show that the covariance of the conditional law of $\mathbf{X}$ given $\mathbf{Y} = \mathbf{y}$ is as given in (23.60), we note that this covariance is the covariance of $(\mathbf{X} - \mathsf{D}\mathbf{Y})$, which is given by

$$\begin{aligned}
\mathsf{E}\left[(\mathbf{X} - \mathsf{D}\mathbf{Y})(\mathbf{X} - \mathsf{D}\mathbf{Y})^{\mathsf{T}}\right] &= \mathsf{E}\left[(\mathbf{X} - \mathsf{D}\mathbf{Y})\mathbf{X}^{\mathsf{T}}\right] - \mathsf{E}\left[(\mathbf{X} - \mathsf{D}\mathbf{Y})(\mathsf{D}\mathbf{Y})^{\mathsf{T}}\right] \\
&= \mathsf{E}\left[(\mathbf{X} - \mathsf{D}\mathbf{Y})\mathbf{X}^{\mathsf{T}}\right] - \mathsf{E}\left[(\mathbf{X} - \mathsf{D}\mathbf{Y})\mathbf{Y}^{\mathsf{T}}\right] \mathsf{D}^{\mathsf{T}} \\
&= \mathsf{E}\left[(\mathbf{X} - \mathsf{D}\mathbf{Y})\mathbf{X}^{\mathsf{T}}\right] \\
&= \mathsf{K}_{\mathbf{X}\mathbf{X}} - \mathsf{D}\mathsf{E}\left[\mathbf{Y}\mathbf{X}^{\mathsf{T}}\right] \\
&= \mathsf{K}_{\mathbf{X}\mathbf{X}} - \mathsf{E}\left[\mathbf{X}\mathbf{Y}^{\mathsf{T}}\right] \mathsf{K}_{\mathbf{Y}\mathbf{Y}}^{-1} \mathsf{E}\left[\mathbf{Y}\mathbf{X}^{\mathsf{T}}\right],
\end{aligned}$$

where the first equality follows by opening the second set of parentheses; the second by (23.3) and (23.25b); the third by (23.63); the fourth by opening the parentheses and using the linearity of the expectation; and the final equality by (23.64). $\qquad \square$

Theorem 23.7.4 has important consequences in Estimation Theory. A key result in Estimation Theory is that if after observing that $\mathbf{Y} = \mathbf{y}$ for some $\mathbf{y} \in \mathbb{R}^{n_y}$ we would like to estimate the random $n_x$-vector $\mathbf{X}$ using a (Borel measurable) function $\mathbf{g} \colon \mathbb{R}^{n_y} \to \mathbb{R}^{n_x}$ so as to minimize the estimation error

$$\mathsf{E}\big[\|\mathbf{X} - g(\mathbf{Y})\|^2\big], \tag{23.65}$$

then an optimal choice for $g(\cdot)$ is the conditional expectation

$$g(\mathbf{y}) = \mathsf{E}\big[\mathbf{X} \,\big|\, \mathbf{Y} = \mathbf{y}\big], \quad \mathbf{y} \in \mathbb{R}^{n_y}. \tag{23.66}$$

Theorem (23.7.4) demonstrates that if $\mathbf{X}$ and $\mathbf{Y}$ are jointly Gaussian and centered, then $\mathsf{E}[\mathbf{X} \,|\, \mathbf{Y} = \mathbf{y}]$ is a linear function of $\mathbf{y}$ and is explicitly given by (23.59). Thus, for jointly Gaussian centered random vectors, there is no loss in optimality in limiting ourselves to linear estimators.

The optimality of choosing $g(\cdot)$ as in (23.66) has a simple intuitive explanation. We first note that it suffices to establish the result when $n_x = 1$, i.e., when estimating a random variable rather than a random vector. Indeed, the squared-norm error in estimating a random vector $\mathbf{X}$ with $n_x$ components is the sum of the squared errors in estimating its components. To minimize the sum, one should therefore minimize each of the terms. And the problem of estimating the $j$-th component of $\mathbf{X}$ based on the observation $\mathbf{Y} = \mathbf{y}$ is a problem of estimating a random variable. Stated differently, to estimate $\mathbf{X}$ so as to minimize the error (23.65) we should separately estimate each of its components.

Having established that it suffices to prove the optimality of (23.66) when $n_x = 1$, we now assume that $n_x = 1$ and denote the random variable to be estimated by $X$. To study how to estimate $X$ after observing that $\mathbf{Y} = \mathbf{y}$, we first consider the case where there is no observation. In this case, the estimate is a constant, and by Lemma 14.4.1 the optimal choice of that constant is the mean $\mathsf{E}[X]$. We now view the general case where we observe $\mathbf{Y} = \mathbf{y}$ as though there were no observables but $X$ had the *a posteriori* distribution given $\mathbf{Y} = \mathbf{y}$. Utilizing the result for the case where there are no observables yields that estimating $X$ by $\mathsf{E}[X \,|\, \mathbf{Y} = \mathbf{y}]$ is optimal.

## 23.8   Moments and Wick's Formula

We next describe without proof a technique for computing moments of centered Gaussian vectors. A sketch of a proof can be found in (Zvonkin, 1997).

**Theorem 23.8.1 (Wick's Formula).** *Let $\mathbf{X}$ be a centered Gaussian $n$-vector and let $\mathbf{g}_1, \ldots, \mathbf{g}_{2k} \colon \mathbb{R}^n \to \mathbb{R}$ be an even number of (not necessarily different) linear functionals on $\mathbb{R}^n$. Then*

$$\mathsf{E}\big[g_1(\mathbf{X})\, g_2(\mathbf{X}) \cdots g_{2k}(\mathbf{X})\big]$$
$$= \sum \mathsf{E}\big[g_{p_1}(\mathbf{X})\, g_{q_1}(\mathbf{X})\big] \mathsf{E}\big[g_{p_2}(\mathbf{X})\, g_{q_2}(\mathbf{X})\big] \cdots \mathsf{E}\big[g_{p_k}(\mathbf{X})\, g_{q_k}(\mathbf{X})\big], \tag{23.67}$$

*where the summation is over all permutations $p_1, q_1, p_2, q_2, \ldots, p_k, q_k$ of $1, 2, \ldots, 2k$ such that*

$$p_1 < p_2 < \cdots < p_k \tag{23.68a}$$

*and*

$$p_1 < q_1, \quad p_2 < q_2, \quad \cdots, \quad p_k < q_k. \tag{23.68b}$$

The number of terms on the RHS of (23.67) is $1 \times 3 \times 5 \times \cdots \times (2k-1)$.

**Example 23.8.2.** Suppose that $n = 1$, so $X$ is a centered univariate Gaussian. Let $\sigma^2$ be its variance, and suppose we wish to compute $\mathsf{E}\left[X^4\right]$. We can express this in the form of Theorem 23.8.1 with $k = 2$ and $g_1(x) = g_2(x) = g_3(x) = g_4(x) = x$. By Wick's Formula

$$\begin{aligned}
\mathsf{E}\left[X^4\right] &= \mathsf{E}[g_1(X)\,g_2(X)]\mathsf{E}[g_3(X)\,g_4(X)] + \mathsf{E}[g_1(X)\,g_3(X)]\mathsf{E}[g_2(X)\,g_4(X)] \\
&\quad + \mathsf{E}[g_1(X)\,g_4(X)]\mathsf{E}[g_2(X)\,g_3(X)] \\
&= 3\sigma^4,
\end{aligned}$$

which is in agreement with (19.31).

**Example 23.8.3.** Suppose that $\mathbf{X}$ is a bivariate centered Gaussian whose components are of unit variance and of correlation coefficient $\rho \in [-1, 1]$. We compute $\mathsf{E}\left[(X^{(1)})^2(X^{(2)})^2\right]$ using Theorem 23.8.1 by setting $k = 2$ and by defining $g_1(\mathbf{x}) = g_2(\mathbf{x}) = x^{(1)}$ and $g_3(\mathbf{x}) = g_4(\mathbf{x}) = x^{(2)}$. By Wick's Formula

$$\begin{aligned}
&\mathsf{E}\left[\left(X^{(1)}\right)^2\left(X^{(2)}\right)^2\right] \\
&= \mathsf{E}[g_1(\mathbf{X})\,g_2(\mathbf{X})]\mathsf{E}[g_3(\mathbf{X})\,g_4(\mathbf{X})] + \mathsf{E}[g_1(\mathbf{X})\,g_3(\mathbf{X})]\mathsf{E}[g_2(\mathbf{X})\,g_4(\mathbf{X})] \\
&\quad + \mathsf{E}[g_1(\mathbf{X})\,g_4(\mathbf{X})]\mathsf{E}[g_2(\mathbf{X})\,g_3(\mathbf{X})] \\
&= \mathsf{E}\left[\left(X^{(1)}\right)^2\right]\mathsf{E}\left[\left(X^{(2)}\right)^2\right] + \mathsf{E}\left[X^{(1)}X^{(2)}\right]\mathsf{E}\left[X^{(1)}X^{(2)}\right] \\
&\quad + \mathsf{E}\left[X^{(1)}X^{(2)}\right]\mathsf{E}\left[X^{(1)}X^{(2)}\right] \\
&= 1 + 2\rho^2. \tag{23.69}
\end{aligned}$$

Similarly,

$$\mathsf{E}\left[\left(X^{(1)}\right)^3 X^{(2)}\right] = 3\rho. \tag{23.70}$$

## 23.9 The Limit of Gaussian Vectors Is a Gaussian Vector

The results of Section 19.9 on limits of Gaussian random variables extend to Gaussian vectors. In this setting we consider random vectors $\mathbf{X}, \mathbf{X}_1, \mathbf{X}_2, \ldots$ defined over the probability space $(\Omega, \mathcal{F}, P)$. We say that the sequence of random vectors $\mathbf{X}_1, \mathbf{X}_2, \ldots$ converges to the random vector $\mathbf{X}$ **with probability one** or **almost surely** if

$$\Pr\left(\left\{\omega \in \Omega : \lim_{n\to\infty} \mathbf{X}_n(\omega) = \mathbf{X}(\omega)\right\}\right) = 1. \tag{23.71}$$

The sequence $\mathbf{X}_1, \mathbf{X}_2, \ldots$ converges to the random vector $\mathbf{X}$ **in probability** if

$$\lim_{n\to\infty} \Pr\left[\|\mathbf{X}_n - \mathbf{X}\| \geq \epsilon\right] = 0, \quad \epsilon > 0. \tag{23.72}$$

The sequence $\mathbf{X}_1, \mathbf{X}_2, \ldots$ converges to the random vector $\mathbf{X}$ **in mean square** if

$$\lim_{n\to\infty} \mathsf{E}\left[\|\mathbf{X}_n - \mathbf{X}\|^2\right] = 0. \tag{23.73}$$

Finally, the sequence of random vectors $\mathbf{X}_1, \mathbf{X}_2, \ldots$ taking value in $\mathbb{R}^d$ converges to the random vector $\mathbf{X}$ **weakly** or **in distribution** if

$$\lim_{n \to \infty} \Pr\left[X_n^{(1)} \leq \xi^{(1)}, \ldots, X_n^{(d)} \leq \xi^{(d)}\right] = \Pr\left[X^{(1)} \leq \xi^{(1)}, \ldots, X^{(d)} \leq \xi^{(d)}\right] \quad (23.74)$$

for every vector $\boldsymbol{\xi} \in \mathbb{R}^d$ such that

$$\lim_{\epsilon \downarrow 0} \Pr\left[X^{(1)} \leq \xi^{(1)} - \epsilon, \ldots, X^{(d)} \leq \xi^{(d)} - \epsilon\right]$$
$$= \Pr\left[X^{(1)} \leq \xi^{(1)}, \ldots, X^{(d)} \leq \xi^{(d)}\right]. \quad (23.75)$$

In analogy to Theorem 19.9.1 we next show that, irrespective of which of the above forms of convergence we consider, if a sequence of Gaussian vectors converges to some random vector $\mathbf{X}$, then $\mathbf{X}$ must be Gaussian.

**Theorem 23.9.1.** *Let the random d-vectors* $\mathbf{X}, \mathbf{X}_1, \mathbf{X}_2, \ldots$ *be defined over a common probability space. Let* $\mathbf{X}_1, \mathbf{X}_2, \ldots$ *each be Gaussian (with possibly different mean vectors and covariance matrices). If the sequence* $\mathbf{X}_1, \mathbf{X}_2, \ldots$ *converges to* $\mathbf{X}$ *in the sense of (23.71) or (23.72) or (23.73), then* $\mathbf{X}$ *must be Gaussian.*

**Proof.** The proof is based on Theorem 23.6.17, which demonstrates that it suffices to consider linear functionals of the vectors in the sequence and on the analogous result for scalars (Theorem 19.9.1). We demonstrate the idea by considering the case where the convergence is almost sure. If $\mathbf{X}_1, \mathbf{X}_2, \ldots$ converges almost surely to $\mathbf{X}$, then for every $\boldsymbol{\alpha} \in \mathbb{R}^d$ the sequence $\boldsymbol{\alpha}^\mathsf{T}\mathbf{X}_1, \boldsymbol{\alpha}^\mathsf{T}\mathbf{X}_2, \ldots$ converges almost surely to $\boldsymbol{\alpha}^\mathsf{T}\mathbf{X}$. Since, by Theorem 23.6.17, linear functionals of Gaussian vectors are univariate Gaussians, it follows that the sequence $\boldsymbol{\alpha}^\mathsf{T}\mathbf{X}_1, \boldsymbol{\alpha}^\mathsf{T}\mathbf{X}_2, \ldots$ is a sequence of Gaussian random variables. And since it converges almost surely to $\boldsymbol{\alpha}^\mathsf{T}\mathbf{X}$, it follows from Theorem 19.9.1 that $\boldsymbol{\alpha}^\mathsf{T}\mathbf{X}$ must be Gaussian. Since this is true for every $\boldsymbol{\alpha}$ in $\mathbb{R}^d$, it follows from Theorem 23.6.17 that $\mathbf{X}$ must be a Gaussian vector. $\quad\square$

In analogy to Theorem 19.9.2 we have the following result on weakly converging Gaussian vectors.

**Theorem 23.9.2 (Weakly Converging Gaussian Vectors).** *Let the sequence of random d-vectors* $\mathbf{X}_1, \mathbf{X}_2, \ldots$ *be such that* $\mathbf{X}_n \sim \mathcal{N}(\boldsymbol{\mu}_n, \mathsf{K}_n)$ *for* $n = 1, 2, \ldots$ *Then the sequence converges in distribution to some limiting distribution, if, and only if, there exist some* $\boldsymbol{\mu} \in \mathbb{R}^d$ *and some* $d \times d$ *matrix* $\mathsf{K}$ *such that*

$$\boldsymbol{\mu}_n \to \boldsymbol{\mu} \text{ and } \mathsf{K}_n \to \mathsf{K}. \quad (23.76)$$

*And if the sequence does converge in distribution, then it converges to the multivariate Gaussian distribution of mean vector* $\boldsymbol{\mu}$ *and covariance matrix* $\mathsf{K}$.

**Proof.** See (Gikhman and Skorokhod, 1996, Chapter I, Section 3, Theorem 4). $\quad\square$

## 23.10    Additional Reading

There are numerous books on Matrix Theory that discuss orthogonal matrices and positive semidefinite matrices. We mention here (Zhang, 1999, Section 5.2), (Herstein, 2001, Chapter 6, Section 6.10), and (Axler, 1997, Chapter 7) on orthogonal matrices, and (Zhang, 1999, Chapter 6), (Axler, 1997, Chapter 7), and (Horn and Johnson, 1985, Chapter 7) on positive semidefinite matrices. Much more on the multivariate Gaussian can be found in (Tong, 1990) and (Johnson and Kotz, 1972, Chapter 35). For more on estimation and linear estimation, see Poor (1994) and (Kailath, Sayed, and Hassibi, 2000).

## 23.11    Exercises

**Exercise 23.1 (Covariance Matrices).** Which of the following matrices cannot be a covariance matrix of some real random vector?

$$A = \begin{pmatrix} 5 & 0 \\ 0 & -1 \end{pmatrix}, \qquad B = \begin{pmatrix} 5 & 1 \\ 2 & 2 \end{pmatrix}, \qquad C = \begin{pmatrix} 2 & 10 \\ 10 & 1 \end{pmatrix}, \qquad D = \begin{pmatrix} 1 & -1 \\ -1 & 1 \end{pmatrix}.$$

**Exercise 23.2 (An Orthogonal Matrix of Determinant 1).** Show that in Theorem 23.6.14 the orthogonal matrix $U$ can be chosen to have determinant $+1$.

**Exercise 23.3 (A Mixture of Gaussians).** Let $X \sim \mathcal{N}(\mu_x, \sigma_x^2)$ and $Y \sim \mathcal{N}(\mu_y, \sigma_y^2)$ be Gaussian random variables. Let $E$ take on the values 0 and 1 equiprobably and independently of $(X, Y)$. Define the mixture RV

$$Z = \begin{cases} X & \text{if } E = 0, \\ Y & \text{if } E = 1. \end{cases}$$

Must $Z$ be Gaussian? Can $Z$ be Gaussian? Compute $Z$'s characteristic function.

**Exercise 23.4 (Multivariate Gaussians).** Show that if $Z$ is a univariate Gaussian, then the random vector $(Z, Z)^{\mathsf{T}}$ is a Gaussian vector. What is its canonical representation?

**Exercise 23.5 (Manipulating Gaussians).** Let $W_1, W_2, \ldots, W_5$ be IID $\mathcal{N}(0, 1)$. Define $Y = 3W_1 + 4W_2 - 2W_3 + W_4 - W_5$ and $Z = W_1 - 4W_2 - 2W_3 + 3W_4 - W_5$. What is the joint distribution of $(Y, Z)$?

**Exercise 23.6 (Largest Eigenvalue).** Let $\mathbf{X}$ be a zero-mean Gaussian $n$-vector of covariance matrix $K \succeq 0$, and let $\lambda_{\max}$ denote the maximal eigenvalue of $K$. Show that for some random $n$-vector $\mathbf{Z}$ independent of $\mathbf{X}$

$$\mathbf{X} + \mathbf{Z} \sim \mathcal{N}(\mathbf{0}, \lambda_{\max} I_n),$$

where $I_n$ denotes the $n \times n$ identity matrix.

**Exercise 23.7 (The Error Probability Revisited).** Show that $p^*(\text{correct})$ of (21.68) in Problem 21.5 can be rewritten as

$$p^*(\text{correct}) = \frac{1}{M} \exp\left(-\frac{E_s}{2\sigma^2}\right) \mathsf{E}\left[\exp\left(\frac{1}{\sigma} \max_m\{\Xi^{(m)}\}\right)\right],$$

where $(\Xi^{(1)}, \ldots, \Xi^{(M)})^\mathsf{T}$ is a centered Gaussian with a covariance matrix whose Row-$j$ Column-$\ell$ entry is $\langle \mathbf{s}_j, \mathbf{s}_\ell \rangle_{\text{E}}$.

**Exercise 23.8 (Gaussian Marginals).** Let $X$ and $Z$ be IID $\mathcal{N}(0,1)$. Let $Y = |Z|\,\text{sgn}(X)$, where $\text{sgn}(X)$ is 1 if $X \geq 0$ and is $-1$ otherwise. Show that $X$ is Gaussian, that $Y$ is Gaussian, but that they are not jointly Gaussian. Sketch the contour lines of their joint probability density function.

**Exercise 23.9 (Characteristic Function of a Random Vector).** Let $\mathbf{X}$ be a random vector with two components whose characteristic function is $\Phi_{\mathbf{X}}(\cdot)$. Express the characteristic function of the sum of its components in terms of $\Phi_{\mathbf{X}}(\cdot)$.

**Exercise 23.10 (The Distribution of Linear Functionals).** Let $\mathbf{X}$ and $\mathbf{Y}$ be random $n$-vectors of components $X^{(1)}, \ldots, X^{(n)}$ and $Y^{(1)}, \ldots, Y^{(n)}$. Assume that for all deterministic coefficients $\alpha_1, \ldots, \alpha_n \in \mathbb{R}$ the random variables $\sum_{\nu=1}^n \alpha_\nu X^{(\nu)}$ and $\sum_{\nu=1}^n \alpha_\nu Y^{(\nu)}$ have the same distribution, i.e.,

$$\left(\sum_{j=1}^n \alpha_j X^{(j)} \overset{\mathscr{L}}{=} \sum_{j=1}^n \alpha_j Y^{(j)}\right), \quad \left(\alpha_1, \ldots, \alpha_n \in \mathbb{R}\right).$$

  (i) Show that the characteristic function of $\mathbf{X}$ must be equal to that of $\mathbf{Y}$.

  (ii) Show that $\mathbf{X}$ and $\mathbf{Y}$ must have the same distribution.

**Exercise 23.11 (Independence, Uncorrelatedness and Gaussianity).** Let the random variables $X$ and $H$ be independent with $X \sim \mathcal{N}(0,1)$ and with $H$ taking on the values $\pm 1$ equiprobably. Let $Y = HX$ denote their product.

  (i) Find the density of $Y$.

  (ii) Are $X$ and $Y$ correlated?

  (iii) Compute $\Pr[|X| \geq 1]$ and $\Pr[|Y| \geq 1]$.

  (iv) Compute the probability that both $|X|$ and $|Y|$ exceed 1.

  (v) Are $X$ and $Y$ independent?

  (vi) Is the vector $(X, Y)^\mathsf{T}$ a Gaussian vector?

**Exercise 23.12 (Expected Maximum of Jointly Gaussians).**

  (i) Let $(X_1, X_2, \ldots, X_n, Y)$ have an arbitrary joint distribution with $\mathsf{E}[Y] = 0$. Here $Y$ need not be independent of $(X_1, X_2, \ldots, X_n)$. Prove that

$$\mathsf{E}\left[\max_{1 \leq j \leq n}\{X_j + Y\}\right] = \mathsf{E}\left[\max_{1 \leq j \leq n}\{X_j\}\right].$$

  (ii) Use Part (i) to prove that if $(U, V)$ are jointly Gaussian and of zero mean, then

$$\mathsf{E}[\max\{U, V\}] = \sqrt{\frac{\mathsf{E}[(U - V)^2]}{2\pi}}.$$

**Exercise 23.13 (The Density of a Bivariate Gaussian).** Let $X$ and $Y$ be jointly Gaussian with means $\mu_x$ and $\mu_y$ and with positive variances $\sigma_x^2$ and $\sigma_y^2$. Let

$$\rho = \frac{\mathsf{Cov}[X, Y]}{\sigma_x \, \sigma_y}$$

be their correlation coefficient. Assume $|\rho| < 1$.

(i) Find the joint density of $X$ and $Y$.

(ii) Find the conditional density of $X$ given $Y = y$.

**Exercise 23.14 (A Training Symbol).** Conditional on $(X_1, X_2) = (x_1, x_2)$, the observable $(Y_1, Y_2)$ is given by

$$Y_\nu = A x_\nu + Z_\nu, \quad \nu = 1, 2,$$

where $Z_1$, $Z_2$, and $A$ are independent with $Z_1, Z_2 \sim$ IID $\mathcal{N}(0, \sigma^2)$ and $A \sim \mathcal{N}(0, 1)$. Suppose that $X_1 = 1$ (deterministically) and that $X_2$ takes on the values $\pm 1$ equiprobably.

(i) Derive an optimal rule for guessing $X_2$ based on $(Y_1, Y_2)$.

(ii) Consider a decoder that operates in two stages. In the first stage the decoder estimates $A$ from $Y_1$ with an estimator that minimizes the mean squared-error. In the second stage it uses the ML decoding rule for guessing $X_2$ based on $Y_2$ by pretending that $A$ is given by its estimate from the first stage. Compute the probability of error of this decoder. Is it optimal?

**Exercise 23.15 (On Wick's Formula).** Let $\mathbf{X}$ be a centered Gaussian $n$-vector, and let $\mathbf{g}_1, \ldots, \mathbf{g}_{2k+1} \colon \mathbb{R}^n \to \mathbb{R}$ be an odd number of (not necessarily different) linear functionals from $\mathbb{R}^n$ to $\mathbb{R}$. Show that

$$\mathsf{E}\big[g_1(\mathbf{X})\, g_2(\mathbf{X}) \cdots g_{2k+1}(\mathbf{X})\big] = 0.$$

**Exercise 23.16 (Jointly Gaussians with Positive Correlation).** Let $X$ and $Y$ be jointly Gaussian with means $\mu_x$ and $\mu_y$; positive variances $\sigma_x^2$ and $\sigma_y^2$; and correlation coefficient $\rho$ as in Exercise 23.13 satisfying $|\rho| < 1$.

(i) Show that, conditional on $Y = y$, the distribution of $X$ is Gaussian with mean $\mu_x + \rho \frac{\sigma_x}{\sigma_y}(y - \mu_y)$ and variance $\sigma_x^2(1 - \rho^2)$.

(ii) Show that if $\rho \geq 0$, then the family $f_{X|Y}(x|y)$ has the **monotone likelihood ratio** property that the mapping

$$x \mapsto \frac{f_{X|Y}(x|y)}{f_{X|Y}(x|y')}$$

is nondecreasing whenever $y' \leq y$. Here $f_{X|Y}(\cdot|y)$ is the conditional density of $X$ given $Y = y$.

(iii) Show that if $\rho \geq 0$, then the joint density $f_{X,Y}(\cdot)$ has the **Total Positivity of Order 2** (TP$_2$) property, i.e.,

$$f_{X,Y}(x', y)\, f_{X,Y}(x, y') \leq f_{X,Y}(x, y)\, f_{X,Y}(x', y'), \quad \big(x' < x, \ y' < y\big).$$

See (Tong, 1990, Chapter 4, Section 4.3.1, Fact 4.3.1 and Theorem 4.3.1).

**Exercise 23.17 (Price's Theorem).** Let $\mathbf{X}$ be a centered Gaussian $n$-vector of covariance matrix $\Lambda$. Let $\lambda^{(j,\ell)} = \mathsf{E}\big[X^{(j)}X^{(\ell)}\big]$ be the Row-$j$ Column-$\ell$ entry of $\Lambda$. Let $f_{\mathbf{X}}(\mathbf{x};\Lambda)$ denote the density of $\mathbf{X}$ (when $\Lambda$ is nonsingular).

(i) Expressing the FT of the partial derivative of a function in terms of the FT of the original function and using the characteristic function of a Gaussian (23.46), derive **Plackett's Identities**

$$\frac{\partial f_{\mathbf{X}}(\mathbf{x};\Lambda)}{\partial \lambda^{(j,j)}} = \frac{1}{2}\frac{\partial^2 f_{\mathbf{X}}(\mathbf{x};\Lambda)}{\partial (x^{(j)})^2}, \qquad \frac{\partial f_{\mathbf{X}}(\mathbf{x};\Lambda)}{\partial \lambda^{(j,\ell)}} = \frac{\partial^2 f_{\mathbf{X}}(\mathbf{x};\Lambda)}{\partial x^{(j)}\partial x^{(\ell)}}, \quad j \neq \ell.$$

(ii) Using integration by parts, derive **Price's Theorem**: if $\mathbf{h}\colon \mathbb{R}^n \to \mathbb{R}$ is twice continuously differentiable with $\mathbf{h}$ and its first and second derivatives growing at most polynomially in $\|\mathbf{x}\|$ as $\|\mathbf{x}\| \to \infty$, then

$$\frac{\partial \mathsf{E}[h(\mathbf{X})]}{\partial \lambda^{(j,\ell)}} = \int_{\mathbb{R}^n} \frac{\partial^2 h(\mathbf{x})}{\partial x^{(j)}\partial x^{(\ell)}} f_{\mathbf{X}}(\mathbf{x};\Lambda)\, d\mathbf{x}, \quad j \neq \ell.$$

(See (Adler, 1990, Chapter 2, Section 2.2) for the case where $\Lambda$ is singular.)

(iii) Show that if in addition to the assumptions of Part (ii) we also assume that for some $j \neq \ell$

$$\frac{\partial^2 h(\mathbf{x})}{\partial x^{(j)}\partial x^{(\ell)}} \geq 0, \quad \mathbf{x} \in \mathbb{R}^n, \tag{23.77}$$

then $\mathsf{E}[h(\mathbf{X})]$ is a nondecreasing function of $\lambda^{(j,\ell)}$.

(iv) Conclude that if $h(\mathbf{x}) = \prod_{\nu=1}^n g_\nu(x^{(\nu)})$, where for each $\nu \in \{1,\dots,n\}$ the function $\mathbf{g}_\nu\colon \mathbb{R} \to \mathbb{R}$ is nonnegative, nondecreasing, twice continuously differentiable, and satisfying the growth conditions of $\mathbf{h}$ in Part (ii), then

$$\mathsf{E}\left[\prod_{\nu=1}^n g_\nu\big(X^{(\nu)}\big)\right]$$

is monotonically nondecreasing in $\lambda^{(j,\ell)}$ whenever $j \neq \ell$.

(v) By choosing $g_\nu(\cdot)$ to approximate the step function $\alpha \mapsto \mathrm{I}\{\alpha \geq \zeta^{(\nu)}\}$ for properly chosen $\zeta^{(\nu)}$, prove **Slepian's Inequality**: if $\mathbf{X} \sim \mathcal{N}(\boldsymbol{\mu},\Lambda)$, then for every choice of $\xi^{(1)},\dots,\xi^{(n)} \in \mathbb{R}$
$$\Pr\Big[X^{(1)} \geq \xi^{(1)}, \dots, X^{(n)} \geq \xi^{(n)}\Big]$$

is monotonically nondecreasing in $\lambda^{(j,\ell)}$ whenever $j \neq \ell$. See (Tong, 1990, Chapter 5, Section 5.1.4, Theorem 5.1.7).

(vi) Modify the arguments in Parts (iv) and (v) to show that if $\mathbf{X} \sim \mathcal{N}(\boldsymbol{\mu},\Lambda)$, then for every choice of $\xi^{(1)},\dots,\xi^{(n)} \in \mathbb{R}$
$$\Pr\Big[X^{(1)} \leq \xi^{(1)}, \dots, X^{(n)} \leq \xi^{(n)}\Big]$$

is monotonically nondecreasing in $\lambda^{(j,\ell)}$ whenever $j \neq \ell$. See (Adler, 1990, Chapter 2, Section 2.2, Corollary 2.4).

**Exercise 23.18 (Jointly Gaussians of Equal Sign).** Let $X$ and $Y$ be jointly Gaussian and centered with positive variances and correlation coefficient $\rho$. Prove that

$$\Pr[XY > 0] = \frac{1}{2} + \frac{\phi}{\pi},$$

where $-\pi/2 \leq \phi \leq \pi/2$ is such that $\sin\phi = \rho$. We propose the following approach.

(i) Show that it suffices to prove the result when $X$ and $Y$ are of unit variance.

(ii) Show that, for such $X$ and $Y$, if we define

$$W = \frac{1}{\sqrt{1-\rho^2}} X - \frac{\rho}{\sqrt{1-\rho^2}} Y, \qquad Z = Y,$$

then $W$ and $Z$ are IID $\mathcal{N}(0,1)$.

(iii) Show that $X$ and $Y$ can be expressed as

$$X = R\sin(\Theta + \phi), \qquad Y = R\cos\Theta,$$

where $\phi$ is as defined before, $\Theta$ is uniformly distributed over the interval $[-\pi, \pi)$, $R$ is independent of $\Theta$, and $f_R(r) = r\,e^{-r^2/2}\,\mathrm{I}\{r > 0\}$.

(iv) Justify the calculation

$$\begin{aligned}
\Pr[XY > 0] &= 2\Pr[X > 0,\, Y > 0] \\
&= 2\Pr[\sin(\Theta + \phi) > 0,\, \cos\Theta > 0] \\
&= \frac{1}{2} + \frac{\phi}{\pi}.
\end{aligned}$$

*Hint: Exercise 19.7 may be useful for Part (iii).*

# Chapter 24

# Complex Gaussians and Circular Symmetry

## 24.1 Introduction

This chapter introduces the complex Gaussian distribution and the circular symmetry property. We start with the scalar case and then extend these notions to random vectors. We rely heavily on Chapter 17 for the basic properties of complex random variables and on Chapter 23 for the properties of the multivariate Gaussian distribution.

## 24.2 Scalars

### 24.2.1 Standard Complex Gaussians

**Definition 24.2.1 (Standard Complex Gaussian).** *A **standard complex Gaussian** is a complex random variable whose real and imaginary parts are independent* $\mathcal{N}(0, 1/2)$ *random variables.*

If $W$ is a standard complex Gaussian, then its density is given by

$$f_W(w) = \frac{1}{\pi} e^{-|w|^2}, \quad w \in \mathbb{C}, \tag{24.1}$$

because

$$\begin{aligned}
f_W(w) &= f_{\mathrm{Re}(W),\mathrm{Im}(W)}\big(\mathrm{Re}(w), \mathrm{Im}(w)\big) \\
&= f_{\mathrm{Re}(W)}\big(\mathrm{Re}(w)\big) f_{\mathrm{Im}(W)}\big(\mathrm{Im}(w)\big) \\
&= \frac{1}{\sqrt{\pi}} e^{-\mathrm{Re}(w)^2} \frac{1}{\sqrt{\pi}} e^{-\mathrm{Im}(w)^2} \\
&= \frac{1}{\pi} e^{-|w|^2}, \quad w \in \mathbb{C},
\end{aligned}$$

where the first equality follows from the definition of the density $f_W(w)$ of a CRV $W$ at $w \in \mathbb{C}$ as the joint density $f_{\mathrm{Re}(W),\mathrm{Im}(W)}$ of its real and imaginary parts $\big(\mathrm{Re}(W), \mathrm{Im}(W)\big)$ evaluated at $\big(\mathrm{Re}(w), \mathrm{Im}(w)\big)$ (Section 17.3.1); the second because the real and imaginary parts of a standard complex Gaussian are independent; the third because the real and imaginary parts of a standard Gaussian are zero-mean variance-$1/2$ real Gaussians whose density can thus be computed from (19.6) (by substituting $1/2$ for $\sigma^2$); and where the final equality follows because for any complex number $w$ we have $\mathrm{Re}(w)^2 + \mathrm{Im}(w)^2 = |w|^2$.

Because the real and imaginary parts of a standard complex Gaussian $W$ are of zero mean, it follows that

$$\mathsf{E}[W] = \mathsf{E}[\mathrm{Re}(W)] + \mathrm{i}\,\mathsf{E}[\mathrm{Im}(W)]$$
$$= 0.$$

And because they are each of variance $1/2$, it follows from (17.14c) that a standard complex Gaussian $W$ has unit-variance

$$\mathsf{Var}[W] = \mathsf{E}\big[|W|^2\big] = 1. \tag{24.2}$$

Moreover, since a standard complex Gaussian is of zero mean and since its real and imaginary parts are of equal variance and uncorrelated, a standard Gaussian is proper (Definition 17.3.1 and Proposition 17.3.2), i.e.,

$$\mathsf{E}[W] = 0 \quad \text{and} \quad \mathsf{E}\big[W^2\big] = 0. \tag{24.3}$$

Finally note that, by (24.1), the density $f_W(\cdot)$ of a standard complex Gaussian is **radially-symmetric**, i.e., its value at $w \in \mathbb{C}$ depends on $w$ only via its modulus $|w|$. A CRV whose density is radially-symmetric is said to be **circularly-symmetric**, but the definition of circular symmetry applies also to complex random variables that do not have a density. This is the topic of the next section.

## 24.2.2 Circular Symmetry

**Definition 24.2.2 (Circularly-Symmetric CRV).** *A CRV $Z$ is said to be **circularly-symmetric** if for any deterministic $\phi \in [-\pi, \pi)$ the distribution of $e^{\mathrm{i}\phi}Z$ is identical to the distribution of $Z$:*

$$e^{\mathrm{i}\phi}Z \overset{\mathscr{L}}{=} Z, \quad \phi \in [-\pi, \pi). \tag{24.4}$$

**Note 24.2.3.** If the expectation of a circularly-symmetric CRV is defined, then it must be zero.

**Proof.** Let $Z$ be circularly-symmetric. It then follows from (24.4) that $e^{\mathrm{i}\phi}Z$ and $Z$ are of equal expectation, so

$$\mathsf{E}[Z] = \mathsf{E}\big[e^{\mathrm{i}\phi}Z\big]$$
$$= e^{\mathrm{i}\phi}\mathsf{E}[Z], \quad \phi \in [-\pi, \pi),$$

which, by considering a $\phi$ for which $e^{\mathrm{i}\phi} \neq 1$, implies that $\mathsf{E}[Z]$ must be zero. $\qquad\square$

To shed some light on the definition of circular symmetry we shall need Proposition 24.2.5 ahead, which is highly intuitive but a bit cumbersome to state. Before stating it we provide its discrete counterpart, which is a bit easier to state: it makes formal the intuition that if after giving the wheel-of-fortune an arbitrary spin, you give it another fair spin, then the combined result is a fair spin that does not depend on the initial spin. The case $\eta = 2$ is critical in cryptography. It shows that taking the mod-2 sum of a binary source sequence with a sequence of IID random bits results in a sequence that is independent of the source sequence.

**Proposition 24.2.4.** *Fix a positive integer $\eta$, and define the set $\mathcal{A} = \{0, \ldots, \eta - 1\}$. Let $N$ be a RV taking value in the set $\mathcal{A}$. Then the following statements are equivalent:*

(a) *The RV $N$ is uniformly distributed over the set $\mathcal{A}$.*

(b) *For any integer-valued RV $K$ that is independent of $N$, the RV $(N+K)$ mod $\eta$ is independent of $K$ and uniformly distributed over $\mathcal{A}$.*[1]

**Proof.** We first show (b) $\Rightarrow$ (a). To this end, define $K$ to be a RV that takes on the value zero deterministically. Being deterministic, it is independent of every RV, and in particular of $N$. Statement (b) thus guarantees that $(N + 0)$ mod $\eta$ is uniformly distributed over $\mathcal{A}$. Since we have assumed from the outset that $N$ takes value in $\mathcal{A}$, it follows that $(N + 0)$ mod $\eta = N$, so the uniformity of $(N + 0)$ mod $\eta$ over $\mathcal{A}$ implies the uniformity of $N$ over $\mathcal{A}$.

We next show (a) $\Rightarrow$ (b). To this end, we need to show that if $N$ is uniformly distributed over $\mathcal{A}$ and if $K$ is independent of $N$, then[2]

$$\Pr\left[((N + K) \bmod \eta) = a \,\middle|\, K = k\right] = \frac{1}{\eta}, \quad \left(k \in \mathbb{Z},\, a \in \mathcal{A}\right). \tag{24.5}$$

By the independence of $N$ and $K$ it follows that

$$\Pr\left[((N+K) \bmod \eta) = a \,\middle|\, K = k\right] = \Pr\left[((N+k) \bmod \eta) = a\right], \quad \left(k \in \mathbb{Z},\, a \in \mathcal{A}\right),$$

so to prove (24.5) it suffices to prove

$$\Pr\left[((N + k) \bmod \eta) = a\right] = \frac{1}{\eta}, \quad \left(k \in \mathbb{Z},\, a \in \mathcal{A}\right). \tag{24.6}$$

This can be proved as follows. Because $N$ is uniformly distributed over $\mathcal{A}$, it follows that $N+k$ is uniformly distributed over the set $\{k, k+1, \ldots, k+\eta-1\}$. And, because the mapping $m \mapsto (m \bmod \eta)$ is a one-to-one mapping from $\{k, k+1, \ldots, k+\eta-1\}$ onto $\mathcal{A}$, this implies that $(N + k)$ mod $\eta$ is also uniformly distributed over $\mathcal{A}$, thus establishing (24.6). $\qquad\qquad\square$

**Proposition 24.2.5.** *Let $\Theta$ be a RV taking value in $[-\pi, \pi)$. Then the following statements are equivalent:*

---

[1]Here $m$ mod $\eta$ is the remainder of dividing $m$ by $\eta$, i.e., the unique $\nu \in \mathcal{A}$ such that $m - \nu$ is an integer multiple of $\eta$. E.g. 17 mod 8 = 1.

[2]Recall that the random variables $X$ and $Y$ are independent if, and only if, the conditional distribution of $X$ given $Y$ is equal to the marginal distribution of $X$.

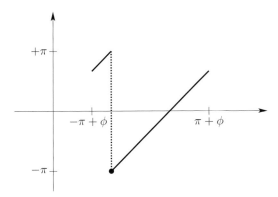

**Figure 24.1:** The function $\xi \mapsto \left(\xi \bmod [-\pi, +\pi)\right)$ plotted for $\xi \in [-\pi + \phi, \pi + \phi)$.

(a) The RV $\Theta$ is uniformly distributed over $[-\pi, \pi)$.

(b) For any real RV $\Phi$ that is independent of $\Theta$, the RV $(\Theta + \Phi) \bmod [-\pi, \pi)$ is independent of $\Phi$ and uniformly distributed over the interval $[-\pi, \pi)$.[3]

**Proof.** The proof is similar to the proof of Proposition 24.2.4 but with an added twist. The twist is needed because if $X$ has a uniform density and if a function $\mathbf{g}$ is one-to-one (injective) and onto (surjective), then $g(X)$ need not be uniformly distributed. (For example, if $X \sim \mathcal{U}([0,1])$ and if $\mathbf{g}\colon [0,1] \to [0,1]$ maps $\xi$ to $\xi^2$, then $g(X)$ is not uniform.)

To prove that (b) implies (a) we simply apply (b) to the deterministic RV $\Phi = 0$.

We next prove that (a) implies (b). As in the discrete case, it suffices to show that if $\Theta$ is uniformly distributed over $[-\pi, \pi)$, then for any deterministic $\phi \in \mathbb{R}$ the distribution of $(\Theta + \phi) \bmod [-\pi, \pi)$ is uniform over $[-\pi, \pi)$, irrespective of $\phi$. To this end we first note that because $\Theta$ is uniform over $[-\pi, \pi)$ it follows that $\Theta + \phi$ is uniform over $[\phi - \pi, \phi + \pi)$. Consider now the mapping $\mathbf{g}\colon [\phi - \pi, \phi + \pi) \to [-\pi, \pi)$ defined by $\mathbf{g}\colon \xi \mapsto \left(\xi \bmod [-\pi, \pi)\right)$. This function is a one-to-one mapping onto $[-\pi, \pi)$ and is differentiable except at the point $\xi^* \in [\phi - \pi, \phi + \pi)$ satisfying $\xi^* \bmod [-\pi, \pi) = \pi$, i.e., the point $\xi^* \in [\phi - \pi, \phi + \pi)$ of the form $\xi^* = 2\pi m + \pi$ for some integer $m$. At all other points its derivative is 1; see Figure 24.1. (Incidentally, $-\pi + \phi$ is mapped to a negative number if $\phi < \xi^*$ and to a positive number if $\phi > \xi^*$. In Figure 24.1 we assume the latter.) Applying the formula for computing the density of $g(X)$ from the density of $X$ (Theorem 17.3.4) we find that if $\Theta + \phi$ is uniform over $[\phi - \pi, \phi + \pi)$, then $g(\phi + \Theta)$ is uniform over $[-\pi, \pi)$. $\qquad\square$

With the aid of Proposition 24.2.5 we can now give alternative characterizations of circular symmetry.

---

[3]Here $x \bmod [-\pi, \pi)$ is the unique $\xi \in [-\pi, \pi)$ such that $x - \xi$ is an integer multiple of $2\pi$.

**Proposition 24.2.6 (Characterizing Circular Symmetry).** *Let $Z$ be a CRV with a density. Then each of the following statements is equivalent to the statement that $Z$ is circularly-symmetric:*

(a) *The distribution of $e^{i\phi}Z$ is identical to the distribution of $Z$, for any deterministic $\phi \in [-\pi, \pi)$.*

(b) *The CRV $Z$ has a radially-symmetric density function, i.e., a density $f_Z(\cdot)$ whose value at $z$ depends on $z$ only via its modulus $|z|$.*

(c) *The CRV $Z$ can be written as $Z = R\,e^{i\Theta}$, where $R \geq 0$ and $\Theta$ are independent real random variables and $\Theta \sim \mathcal{U}([-\pi, \pi))$.*

**Proof.** Statement (a) is the definition of circular symmetry (Definition 24.2.2).

The proof of (a) $\Rightarrow$ (b) is slightly obtuse because the density of a CRV is not unique.[4] We begin by noting that if $Z$ is of density $f_Z(\cdot)$, then by (17.34) the CRV $e^{i\phi}Z$ is of density $w \mapsto f_Z(e^{-i\phi}w)$. Thus, if $Z \stackrel{\mathscr{L}}{=} e^{i\phi}Z$ and if $Z$ is of density $f_Z(\cdot)$, then $Z$ is also of density $w \mapsto f_Z(e^{-i\phi}w)$. Consequently, if $Z$ is circularly-symmetric, then for every $\phi \in [-\pi, \pi)$ the mapping $w \mapsto f_Z(e^{-i\phi}w)$ is a density for $Z$. We can therefore conclude that the mapping

$$ w \mapsto \frac{1}{2\pi} \int_{-\pi}^{\pi} f_Z(e^{-i\phi}w)\,d\phi $$

is also a density for $Z$, and this function is radially-symmetric.

The fact that (b) $\Rightarrow$ (c) follows because if we define $R$ to be the magnitude of $Z$ and $\Theta$ to be its argument, then $Z = R\,e^{i\Theta}$, and

$$ \begin{aligned} f_{R,\Theta}(r, \theta) &= r f_Z(r\,e^{i\theta}) \\ &= r f_Z(r) \\ &= \left(2\pi r f_Z(r)\right)\frac{1}{2\pi}, \end{aligned} $$

where the first equality follows from (17.29) and the second from our assumption that $f_Z(z)$ depends on $z$ only via its modulus $|z|$. The joint density of $R, \Theta$ is thus of a product form, thereby indicating that $R$ and $\Theta$ are independent. And it does not depend on $\theta$, thus indicating that its marginal $\Theta$ is uniformly distributed.

We finally show that (c) $\Rightarrow$ (a). To that end we assume that $R \geq 0$ and $\Theta$ are independent with $\Theta$ being uniformly distributed over $[-\pi, \pi)$ and proceed to show that $R\,e^{i\Theta}$ is circularly-symmetric, i.e., that

$$ R\,e^{i\Theta} \stackrel{\mathscr{L}}{=} R\,e^{i(\Theta + \phi)}, \quad \phi \in [-\pi, \pi). \tag{24.7} $$

To prove (24.7) we note that

$$ \begin{aligned} e^{i(\Theta + \phi)} &= e^{i\left((\Theta + \phi) \bmod [-\pi, \pi)\right)} \\ &\stackrel{\mathscr{L}}{=} e^{i\Theta}, \end{aligned} \tag{24.8} $$

---

[4]And not all the functions that are densities for a given circularly-symmetric CRV $Z$ are radially-symmetric. The radial symmetry can be broken on a set of Lebesgue measure zero. We can therefore only claim that there exists "a" radially-symmetric density function for $Z$.

where the first equality follows from the periodicity of the complex exponentials, and where the equality in distribution follows from Proposition 24.2.5 because $\Theta \sim \mathcal{U}\left([-\pi, \pi)\right)$. The proof is now completed by noting that (24.7) follows from (24.8) and from the independence of $R$ and $\Theta$. (If $X$ is independent of $Y$, if $X$ is independent of $Z$, and if $Y \stackrel{\mathscr{L}}{=} Z$, then $(X, Y) \stackrel{\mathscr{L}}{=} (X, Z)$ and hence $XY \stackrel{\mathscr{L}}{=} XZ$.) $\square$

**Example 24.2.7.** Let the CRV $Z$ be given by $Z = e^{i\Phi}$, where $\Phi \sim \mathcal{U}\left([-\pi, \pi)\right)$. Then $Z$ is uniformly distributed over the unit circle $\{z : |z| = 1\}$ and is circularly-symmetric. It does not have a density.

### 24.2.3 Properness and Circular Symmetry

**Proposition 24.2.8.** *Every finite-variance circularly-symmetric CRV is proper.*

**Proof.** Let $Z$ be a finite-variance circularly-symmetric CRV. By Note 24.2.3 it follows that $\mathsf{E}[Z] = 0$. To conclude the proof it remains to show that $\mathsf{E}\left[Z^2\right] = 0$. To this end we note that

$$
\begin{aligned}
\mathsf{E}\left[Z^2\right] &= e^{-i2\phi}\mathsf{E}\left[\left(e^{i\phi}Z\right)^2\right] \\
&= e^{-i2\phi}\mathsf{E}\left[Z^2\right], \quad \phi \in [-\pi, \pi),
\end{aligned} \tag{24.9}
$$

where the first equality follows by rewriting $Z^2$ as $e^{-i2\phi}\left(e^{i\phi}Z\right)^2$, and where the second equality follows because the circular symmetry of $Z$ guarantees that $Z$ and $e^{i\phi}Z$ have the same law, so the expectation of their squares must be equal. But (24.9) cannot be satisfied for all $\phi \in [-\pi, \pi)$ (or for that matter for any $\phi$ such that $e^{i2\phi} \neq 1$) unless $\mathsf{E}\left[Z^2\right] = 0$. $\square$

**Note 24.2.9.** Not every proper CRV is circularly-symmetric.

**Proof.** Consider the CRV $Z$ that takes on the four values $1 + i$, $1 - i$, $-1 + i$, and $-1 - i$ equiprobably. Its real and imaginary parts are independent, each taking on the values $\pm 1$ equiprobably. Computing $\mathsf{E}[Z]$ and $\mathsf{E}\left[Z^2\right]$ we find that they are both zero, so $Z$ is proper. To see that $Z$ is not circularly-symmetric consider the random variable $e^{i\pi/4}Z$. Its distribution is different from the distribution of $Z$ because $Z$ takes value in the set $\{1 + i, -1 + i, 1 - i, -1 - i\}$, and $e^{i\pi/4}Z$ takes value in the rotated set $\{\sqrt{2}, -\sqrt{2}, \sqrt{2}i, -\sqrt{2}i\}$. $\square$

The fact that not every proper CRV is circularly-symmetric is not surprising because whether a CRV is proper or not is determined solely by its mean and by the covariance matrix of its real and imaginary parts, whereas circular symmetry has to do with the entire distribution.

### 24.2.4 Complex Gaussians

The definition of a complex Gaussian builds on the definition of a real Gaussian vector (Definition 23.1.1).

**Definition 24.2.10 (Complex Gaussian).** *A complex Gaussian is a CRV whose real and imaginary parts are jointly Gaussian real random variables. A **centered complex Gaussian** is a complex Gaussian of zero mean.*

An example of a complex Gaussian is the standard complex Gaussian, which we encountered in Section 24.2.1.

The class of complex Gaussians is closed under multiplication by deterministic complex numbers. Thus, if $Z$ is a complex Gaussian and if $\alpha \in \mathbb{C}$ is deterministic, then $\alpha Z$ is also a complex Gaussian. Indeed,

$$\begin{pmatrix} \mathrm{Re}(\alpha Z) \\ \mathrm{Im}(\alpha Z) \end{pmatrix} = \begin{pmatrix} \mathrm{Re}(\alpha) & -\mathrm{Im}(\alpha) \\ \mathrm{Im}(\alpha) & \mathrm{Re}(\alpha) \end{pmatrix} \begin{pmatrix} \mathrm{Re}(Z) \\ \mathrm{Im}(Z) \end{pmatrix},$$

so the claim follows from the fact that multiplying a real Gaussian vector by a deterministic real matrix results in a real Gaussian vector (Proposition 23.6.3). We leave it to the reader to verify that, more generally, if $Z$ is a complex Gaussian and if $\alpha, \beta \in \mathbb{C}$ are deterministic, then $\alpha Z + \beta Z^*$ is also a complex Gaussian. (This is a special case of Proposition 24.3.9 ahead.)

Not every centered complex Gaussian can be expressed as the scaling of a standard complex Gaussian by some complex number. But the following result characterizes those that can:

**Proposition 24.2.11.**

(i) *For every centered complex Gaussian $Z$ we can find coefficients $\alpha, \beta \in \mathbb{C}$ so that*

$$Z \overset{\mathscr{L}}{=} \alpha W + \beta W^*, \tag{24.10}$$

*where $W$ is a standard complex Gaussian.*

(ii) *A centered complex Gaussian $Z$ is proper if, and only if, there exists some $\alpha \in \mathbb{C}$ such that $Z \overset{\mathscr{L}}{=} \alpha W$, where $W$ is a standard complex Gaussian.*

**Proof.** We begin with Part (i). First note that since $Z$ is a complex Gaussian, its real and imaginary parts are jointly Gaussian, and it follows from Corollary 23.6.13 that there exist deterministic real numbers $a^{(1,1)}, a^{(1,2)}, a^{(2,1)}, a^{(2,2)}$ such that

$$\begin{pmatrix} \mathrm{Re}(Z) \\ \mathrm{Im}(Z) \end{pmatrix} \overset{\mathscr{L}}{=} \begin{pmatrix} a^{(1,1)} & a^{(1,2)} \\ a^{(2,1)} & a^{(2,2)} \end{pmatrix} \begin{pmatrix} W_1 \\ W_2 \end{pmatrix}, \tag{24.11}$$

where $W_1$ and $W_2$ are independent real standard Gaussians. Next note that by direct computation

$$\begin{pmatrix} \mathrm{Re}(\alpha W + \beta W^*) \\ \mathrm{Im}(\alpha W + \beta W^*) \end{pmatrix} = \begin{pmatrix} \frac{\mathrm{Re}(\alpha)+\mathrm{Re}(\beta)}{\sqrt{2}} & \frac{\mathrm{Im}(\beta)-\mathrm{Im}(\alpha)}{\sqrt{2}} \\ \frac{\mathrm{Im}(\beta)+\mathrm{Im}(\alpha)}{\sqrt{2}} & \frac{\mathrm{Re}(\alpha)-\mathrm{Re}(\beta)}{\sqrt{2}} \end{pmatrix} \begin{pmatrix} \sqrt{2}\,\mathrm{Re}(W) \\ \sqrt{2}\,\mathrm{Im}(W) \end{pmatrix}. \tag{24.12}$$

Since, by the definition of a standard complex Gaussian $W$,

$$\begin{pmatrix} W_1 \\ W_2 \end{pmatrix} \overset{\mathscr{L}}{=} \begin{pmatrix} \sqrt{2}\,\mathrm{Re}(W) \\ \sqrt{2}\,\mathrm{Im}(W) \end{pmatrix} \tag{24.13}$$

it follows from (24.11), (24.12), and (24.13) that if $\alpha$ and $\beta$ are chosen so that

$$\begin{pmatrix} \frac{\mathrm{Re}(\alpha)+\mathrm{Re}(\beta)}{\sqrt{2}} & \frac{\mathrm{Im}(\beta)-\mathrm{Im}(\alpha)}{\sqrt{2}} \\ \frac{\mathrm{Im}(\beta)+\mathrm{Im}(\alpha)}{\sqrt{2}} & \frac{\mathrm{Re}(\alpha)-\mathrm{Re}(\beta)}{\sqrt{2}} \end{pmatrix} = \begin{pmatrix} a^{(1,1)} & a^{(1,2)} \\ a^{(2,1)} & a^{(2,2)} \end{pmatrix},$$

i.e., if

$$\alpha = \frac{1}{\sqrt{2}}\left(\left(a^{(1,1)}+a^{(2,2)}\right)+\mathrm{i}\left(a^{(2,1)}-a^{(1,2)}\right)\right),$$

$$\beta = \frac{1}{\sqrt{2}}\left(\left(a^{(1,1)}-a^{(2,2)}\right)+\mathrm{i}\left(a^{(2,1)}+a^{(1,2)}\right)\right),$$

then

$$\begin{pmatrix} \mathrm{Re}(Z) \\ \mathrm{Im}(Z) \end{pmatrix} \overset{\mathscr{L}}{=} \begin{pmatrix} \mathrm{Re}(\alpha W + \beta W^*) \\ \mathrm{Im}(\alpha W + \beta W^*) \end{pmatrix},$$

and (24.10) is satisfied.

We next turn to Part (ii). One direction is straightforward: if $Z \overset{\mathscr{L}}{=} \alpha W$, then $Z$ must be proper because from (24.3) it follows that $\mathsf{E}[\alpha W] = \alpha \mathsf{E}[W] = 0$ and $\mathsf{E}\left[(\alpha W)^2\right] = \alpha^2 \mathsf{E}\left[W^2\right] = 0$.

We next prove the other direction that if $Z$ is a proper complex Gaussian, then $Z \overset{\mathscr{L}}{=} \alpha W$ for some $\alpha \in \mathbb{C}$ and some standard complex Gaussian $W$. Let $Z$ be a proper complex Gaussian. By Part (i) it follows that there exist $\alpha, \beta \in \mathbb{C}$ such that (24.10) is satisfied. Consequently, for this choice of $\alpha$ and $\beta$ we have

$$\begin{aligned} 0 &= \mathsf{E}\left[Z^2\right] \\ &= \mathsf{E}\left[(\alpha W + \beta W^*)^2\right] \\ &= \alpha^2 \mathsf{E}\left[W^2\right] + 2\alpha\beta \mathsf{E}[WW^*] + \beta^2 \mathsf{E}\left[(W^*)^2\right] \\ &= 2\alpha\beta, \end{aligned}$$

where the first equality follows because $Z$ is proper; the second because $\alpha$ and $\beta$ have been chosen so that (24.10) holds; the third by opening the brackets and using the linearity of expectation; and the fourth by (24.3) and (24.2). It follows that either $\alpha$ or $\beta$ must be zero. Since $W \overset{\mathscr{L}}{=} W^*$, there is no loss in generality in assuming that $\beta = 0$, thus establishing the existence of $\alpha \in \mathbb{C}$ such that $Z \overset{\mathscr{L}}{=} \alpha W$. $\square$

By Proposition 24.2.11 (ii) we conclude that if $Z$ is a proper complex Gaussian, then $Z \overset{\mathscr{L}}{=} \alpha W$ for some $\alpha \in \mathbb{C}$ and some standard complex Gaussian $W$. Consequently, the density of such a CRV $Z$ (that is not deterministically zero) is given by

$$\begin{aligned} f_Z(z) &= \frac{f_W(z/\alpha)}{|\alpha|^2} \\ &= \frac{1}{\pi|\alpha|^2} e^{-\frac{|z|^2}{|\alpha|^2}}, \quad z \in \mathbb{C}, \end{aligned}$$

where the first equality follows from the way the density of a CRV behaves under linear transformations (Theorem 17.3.7 or Lemma 17.4.6), and where the second equality follows from (24.1). We thus conclude that if $Z$ is a proper complex Gaussian, then its density is radially-symmetric, and $Z$ must be circularly-symmetric. The reverse is also true: since every complex Gaussian is of finite variance, and since every finite-variance circularly-symmetric CRV is also proper (Proposition 24.2.8), we conclude that every circularly-symmetric complex Gaussian is proper. Thus:

**Proposition 24.2.12.** *A complex Gaussian is circularly-symmetric if, and only if, it is proper.*

The picture that thus emerges is the following.

  (i) Every finite-variance circularly-symmetric CRV is proper.

 (ii) Some proper CRVs are not circularly symmetric.

(iii) A Gaussian CRV is circularly-symmetric, if and only if, it is proper.

We shall soon see that these observations extend to vectors too. In fact, the reader is encouraged to consult Figure 24.2 on Page 508, which holds also for CRVs.

## 24.3   Vectors

### 24.3.1   Standard Complex Gaussian Vectors

**Definition 24.3.1 (Standard Complex Gaussian Vector).** *A **standard complex Gaussian vector** is a complex random vector whose components are IID and each of them is a standard complex Gaussian random variable.*

If $\mathbf{W}$ is a standard complex Gaussian $n$-vector, then, by the independence of its $n$ components and by (24.1), its density is given by

$$f_{\mathbf{W}}(\mathbf{w}) = \frac{1}{\pi^n} \, e^{-\mathbf{w}^\dagger \mathbf{w}} \, . \quad \mathbf{w} \in \mathbb{C}^n. \tag{24.14}$$

By the independence of its components and by (24.3)

$$\mathsf{E}[\mathbf{W}] = \mathbf{0} \quad \text{and} \quad \mathsf{E}[\mathbf{W}\mathbf{W}^\mathsf{T}] = 0. \tag{24.15}$$

Thus, every standard complex Gaussian vector is proper (Section 17.4.2). By the independence of the components and by (24.2) it also follows that

$$\mathsf{E}[\mathbf{W}\mathbf{W}^\dagger] = \mathsf{I}_n, \tag{24.16}$$

where we remind the reader that $\mathsf{I}_n$ denotes the $n \times n$ identity matrix.

### 24.3.2   Circularly-Symmetric Complex Random Vectors

**Definition 24.3.2 (Circularly-Symmetric Complex Random Vectors).** *We say that the complex random vector $\mathbf{Z}$ is **circularly-symmetric** if for every $\phi \in [-\pi, \pi)$ the law of $e^{i\phi}\mathbf{Z}$ is identical to the law of $\mathbf{Z}$.*

An equivalent definition can be given in terms of linear functionals:

**Proposition 24.3.3 (Circular Symmetry and Linear Functionals).** *Each of the following statements is equivalent to the statement that the complex random $n$-vector $\mathbf{Z}$ is circularly-symmetric.*

(a) *For every $\phi \in [-\pi, \pi)$ the law of the complex random vector $e^{i\phi}\mathbf{Z}$ is the same as the law of $\mathbf{Z}$:*

$$e^{i\phi}\mathbf{Z} \stackrel{\mathscr{L}}{=} \mathbf{Z}, \quad \phi \in [-\pi, \pi). \tag{24.17}$$

(b) *For every deterministic vector $\boldsymbol{\alpha} \in \mathbb{C}^n$, the CRV $\boldsymbol{\alpha}^{\mathsf{T}}\mathbf{Z}$ is circularly-symmetric:*

$$e^{i\phi}\boldsymbol{\alpha}^{\mathsf{T}}\mathbf{Z} \stackrel{\mathscr{L}}{=} \boldsymbol{\alpha}^{\mathsf{T}}\mathbf{Z}, \quad \left(\boldsymbol{\alpha} \in \mathbb{C}^n, \ \phi \in [-\pi, \pi)\right). \tag{24.18}$$

**Proof.** Statement (a) is just the definition of circular symmetry. We next show that the two statements (a) and (b) are equivalent. We begin by proving that (a) implies (b). This is the easy part because applying the same linear functional to two random vectors that have the same law results in random variables that have the same law. Consequently, (24.17) implies (24.18).

We now prove that (b) implies (a). We thus assume (24.18) and set out to prove (24.17). By Theorem 17.4.4 it follows that to establish (24.17) it suffices to show that the random vectors on the RHS and LHS of (24.17) have the same characteristic function, i.e., that

$$\mathsf{E}\left[e^{i\,\mathrm{Re}\left(\boldsymbol{\varpi}^{\dagger} e^{i\phi}\,\mathbf{Z}\right)}\right] = \mathsf{E}\left[e^{i\,\mathrm{Re}(\boldsymbol{\varpi}^{\dagger}\mathbf{Z})}\right], \quad \boldsymbol{\varpi} \in \mathbb{C}^n. \tag{24.19}$$

But this readily follows from (24.18) because upon substituting $\boldsymbol{\varpi}^{\dagger}$ for $\boldsymbol{\alpha}^{\mathsf{T}}$ in (24.18) we obtain that

$$\boldsymbol{\varpi}^{\dagger}\mathbf{Z} \stackrel{\mathscr{L}}{=} \boldsymbol{\varpi}^{\dagger} e^{i\phi}\,\mathbf{Z}, \quad \boldsymbol{\varpi} \in \mathbb{C}^n,$$

and this implies (24.19), because if $Z_1 \stackrel{\mathscr{L}}{=} Z_2$, then $\mathsf{E}[g(Z_1)] = \mathsf{E}[g(Z_2)]$ for any measurable function $\mathbf{g}$ and, in particular, for the function $\mathbf{g}\colon \xi \mapsto e^{i\,\mathrm{Re}(\xi)}$. $\qquad\square$

The following proposition demonstrates that circular symmetry is preserved by linear transformations.

**Proposition 24.3.4 (Circular Symmetry and Linear Transformations).** *Let $\mathbf{Z}$ be a circularly-symmetric complex random $n$-vector and let $\mathsf{A}$ be a deterministic complex $m \times n$ matrix. Then the complex random $m$-vector $\mathsf{A}\mathbf{Z}$ is also circularly-symmetric.*

**Proof.** By Proposition 24.3.3 it follows that to establish that $\mathsf{A}\mathbf{Z}$ is circularly-symmetric it suffices to show that for every deterministic $\boldsymbol{\alpha} \in \mathbb{C}^m$ the random variable $\boldsymbol{\alpha}^{\mathsf{T}}\mathsf{A}\mathbf{Z}$ is circularly-symmetric. To show this, fix some arbitrary $\boldsymbol{\alpha} \in \mathbb{C}^m$. Because $\mathbf{Z}$ is circularly-symmetric, it follows from Proposition 24.3.3 that for every deterministic vector $\boldsymbol{\beta} \in \mathbb{C}^n$, the random variable $\boldsymbol{\beta}^{\mathsf{T}}\mathbf{Z}$ is circularly-symmetric. Choosing $\boldsymbol{\beta} = \mathsf{A}^{\mathsf{T}}\boldsymbol{\alpha}$ establishes that $\boldsymbol{\alpha}^{\mathsf{T}}\mathsf{A}\mathbf{Z}$ is circularly-symmetric. $\qquad\square$

### 24.3.3   Proper vs. Circularly-Symmetric Vectors

We now extend the relationship between properness and circular symmetry to vectors:

**Proposition 24.3.5 (Circular Symmetry Implies Properness).**

   (i) *Every finite-variance circularly-symmetric random vector is proper.*

   (ii) *Some proper random vectors are not circularly-symmetric.*

**Proof.** Part (ii) requires no proof because a CRV can be viewed as a complex random vector taking value in $\mathbb{C}^1$, and we have already seen in Section 24.2.3 an example of a CRV which is proper but not circularly-symmetric (Note 24.2.9).

We now prove Part (i). Let $\mathbf{Z}$ be a finite-variance circularly-symmetric random $n$-vector. To establish that $\mathbf{Z}$ is proper we will show that for every $\boldsymbol{\alpha} \in \mathbb{C}^n$ the CRV $\boldsymbol{\alpha}^\mathsf{T}\mathbf{Z}$ is proper (Proposition 17.4.2). To this end, fix an arbitrary $\boldsymbol{\alpha} \in \mathbb{C}^n$. By Proposition 24.3.3 it follows that the CRV $\boldsymbol{\alpha}^\mathsf{T}\mathbf{Z}$ is circularly-symmetric. And because $\mathbf{Z}$ is of finite variance, so is $\boldsymbol{\alpha}^\mathsf{T}\mathbf{Z}$. Being a circularly-symmetric CRV of finite variance, it follows from Section 24.2.3 that $\boldsymbol{\alpha}^\mathsf{T}\mathbf{Z}$ must be proper.     □

### 24.3.4   Complex Gaussian Vectors

**Definition 24.3.6 (Complex Gaussian Vectors).** *A complex random $n$-vector $\mathbf{Z}$ is said to be a **complex Gaussian vector** if the real random $2n$-vector*

$$\left(\operatorname{Re}\!\left(Z^{(1)}\right), \ldots, \operatorname{Re}\!\left(Z^{(n)}\right), \operatorname{Im}\!\left(Z^{(1)}\right), \ldots, \operatorname{Im}\!\left(Z^{(n)}\right)\right)^\mathsf{T} \tag{24.20}$$

*consisting of the real and imaginary parts of its components is a real Gaussian vector. A **centered complex Gaussian vector** is a zero-mean complex Gaussian vector.*

Note that, Theorem 23.6.7 notwithstanding, the distribution of a centered complex Gaussian vector is *not* uniquely specified by its covariance matrix. It is uniquely specified by the covariance matrix if the Gaussian vector is additionally known to be proper. This is a direct consequence of the following proposition.

**Proposition 24.3.7.** *The distribution of a centered complex Gaussian vector $\mathbf{Z}$ is uniquely specified by the matrices*

$$\mathsf{K} = \mathsf{E}\!\left[\mathbf{Z}\mathbf{Z}^\dagger\right] \quad and \quad \mathsf{L} = \mathsf{E}\!\left[\mathbf{Z}\mathbf{Z}^\mathsf{T}\right].$$

**Proof.** Let $\mathbf{R}$ be the real $2n$-vector that results from stacking the real part of $\mathbf{Z}$ on top of its imaginary part as in (24.20). We will prove the proposition by showing that the matrices $\mathsf{K}$ and $\mathsf{L}$ uniquely specify the distribution of $\mathbf{R}$.

Since $\mathbf{Z}$ is a complex Gaussian $n$-vector, $\mathbf{R}$ is a real Gaussian $2n$-vector. Since $\mathbf{Z}$ is of zero mean, so is $\mathbf{R}$. Consequently, the distribution of $\mathbf{R}$ is fully characterized by its covariance matrix $\mathsf{E}\!\left[\mathbf{R}\mathbf{R}^\mathsf{T}\right]$ (Theorem 23.6.7). The proof will thus be concluded once we show that the matrices $\mathsf{L}$ and $\mathsf{K}$ determine the covariance matrix of $\mathbf{R}$. Indeed, as we next verify,

$$\mathsf{E}\left[\mathbf{R}\mathbf{R}^{\mathsf{T}}\right] = \frac{1}{2}\begin{pmatrix} \operatorname{Re}(\mathsf{K}) + \operatorname{Re}(\mathsf{L}) & \operatorname{Im}(\mathsf{L}) - \operatorname{Im}(\mathsf{K}) \\ \operatorname{Im}(\mathsf{L}) + \operatorname{Im}(\mathsf{K}) & \operatorname{Re}(\mathsf{K}) - \operatorname{Re}(\mathsf{L}) \end{pmatrix}. \tag{24.21}$$

To verify (24.21) one needs to compute each of the block entries separately. We shall see how this is done by computing the top-right entry. The rest of the entries are left for the reader to verify.

$$\begin{aligned} \mathsf{E}\left[\operatorname{Re}(\mathbf{Z})\operatorname{Im}(\mathbf{Z})^{\mathsf{T}}\right] &= \mathsf{E}\left[\left(\frac{\mathbf{Z}+\mathbf{Z}^*}{2}\right)\left(\frac{\mathbf{Z}-\mathbf{Z}^*}{2i}\right)^{\mathsf{T}}\right] \\ &= \mathsf{E}\left[\left(\frac{\mathbf{Z}+\mathbf{Z}^*}{2}\right)\left(\frac{\mathbf{Z}^{\mathsf{T}}-\mathbf{Z}^{\dagger}}{2i}\right)\right] \\ &= \frac{1}{2}\left(\frac{\mathsf{E}\left[\mathbf{Z}\mathbf{Z}^{\mathsf{T}}\right]-\mathsf{E}\left[\mathbf{Z}^*\mathbf{Z}^{\dagger}\right]}{2i} - \frac{\mathsf{E}\left[\mathbf{Z}\mathbf{Z}^{\dagger}\right]-\mathsf{E}\left[\mathbf{Z}^*\mathbf{Z}^{\mathsf{T}}\right]}{2i}\right) \\ &= \frac{1}{2}\big(\operatorname{Im}(\mathsf{L}) - \operatorname{Im}(\mathsf{K})\big). \qquad \square \end{aligned}$$

**Corollary 24.3.8.** *The distribution of a proper complex Gaussian vector is uniquely specified by its covariance matrix.*

**Proof.** Follows from Proposition 24.3.7 by noting that by specifying that a complex Gaussian is proper we are specifying that the matrix $\mathsf{L}$ is zero (Definition 17.4.1).
$\square$

**Proposition 24.3.9 (Linear Transformations of Complex Gaussians).** *If $\mathbf{Z}$ is a complex Gaussian $n$-vector and if $\mathsf{A}$ and $\mathsf{B}$ are deterministic $m \times n$ complex matrices, then the $m$-vector*

$$\mathsf{A}\mathbf{Z} + \mathsf{B}\mathbf{Z}^*$$

*is a complex Gaussian.*

**Proof.** Define the complex random $m$-vector $\mathbf{C} \triangleq \mathsf{A}\mathbf{Z} + \mathsf{B}\mathbf{Z}^*$. To prove that $\mathbf{C}$ is Gaussian we recall that linearly transforming a real Gaussian vector yields a real Gaussian vector (Proposition 23.6.3), and we note that the real random $2m$-vector whose components are the real and imaginary parts of $\mathbf{C}$ can be expressed as the result of applying a linear transformation to the real Gaussian $2n$-vector whose components are the real and imaginary parts of the components of $\mathbf{Z}$:

$$\begin{pmatrix} \operatorname{Re}(\mathbf{C}) \\ \operatorname{Im}(\mathbf{C}) \end{pmatrix} = \begin{pmatrix} \operatorname{Re}(\mathsf{A}) + \operatorname{Re}(\mathsf{B}) & \operatorname{Im}(\mathsf{B}) - \operatorname{Im}(\mathsf{A}) \\ \operatorname{Im}(\mathsf{A}) + \operatorname{Im}(\mathsf{B}) & \operatorname{Re}(\mathsf{A}) - \operatorname{Re}(\mathsf{B}) \end{pmatrix} \begin{pmatrix} \operatorname{Re}(\mathbf{Z}) \\ \operatorname{Im}(\mathbf{Z}) \end{pmatrix}. \qquad \square$$

**Proposition 24.3.10 (Characterizing Complex Gaussian Vectors).** *Each of the following statements is equivalent to the statement that $\mathbf{Z}$ is a complex Gaussian $n$-vector.*

*(a) The real random vector whose $2n$ components correspond to the real and imaginary parts of $\mathbf{Z}$ is a real Gaussian vector.*

(b) *For every deterministic vector $\boldsymbol{\alpha} \in \mathbb{C}^n$, the CRV $\boldsymbol{\alpha}^{\mathsf{T}}\mathbf{Z}$ is a complex Gaussian random variable.*

(c) *There exist complex $n \times m$ matrices $\mathsf{A}$ and $\mathsf{B}$ and a vector $\boldsymbol{\mu} \in \mathbb{C}^n$ such that*

$$\mathbf{Z} \stackrel{\mathscr{L}}{=} \mathsf{A}\mathbf{W} + \mathsf{B}\mathbf{W}^* + \boldsymbol{\mu}$$

*for some standard complex Gaussian random $m$-vector $\mathbf{W}$.*

**Proof.** Statement (a) is just the definition of a Gaussian complex random vector.

We next prove the equivalence of (a) and (b). That (a) implies (b) follows from Proposition 24.3.9 (by substituting $\boldsymbol{\alpha}^{\mathsf{T}}$ for $\mathsf{A}$ and $0$ for $\mathsf{B}$).

To prove that (b) $\Rightarrow$ (a) it suffices (by Definition 24.3.6 and Theorem 23.6.17) to show that (b) implies that any real linear functional of the real random $2n$-vector comprising the real and imaginary parts of $\mathbf{Z}$ is a real Gaussian random variable, i.e., that for every choice of the real constants $\alpha^{(1)}, \ldots, \alpha^{(n)}$ and $\beta^{(1)}, \ldots, \beta^{(n)}$ the random variable

$$\sum_{j=1}^{n} \alpha^{(j)} \operatorname{Re}\left(Z^{(j)}\right) + \sum_{j=1}^{n} \beta^{(j)} \operatorname{Im}\left(Z^{(j)}\right) \tag{24.22}$$

is a Gaussian real random variable. To that end we rewrite (24.22) as

$$\sum_{j=1}^{n} \alpha^{(j)} \operatorname{Re}\left(Z^{(j)}\right) + \sum_{j=1}^{n} \beta^{(j)} \operatorname{Im}\left(Z^{(j)}\right) = \boldsymbol{\alpha}^{\mathsf{T}} \operatorname{Re}(\mathbf{Z}) + \boldsymbol{\beta}^{\mathsf{T}} \operatorname{Im}(\mathbf{Z}) \tag{24.23}$$

$$= \operatorname{Re}\left((\boldsymbol{\alpha} - \mathrm{i}\boldsymbol{\beta})^{\mathsf{T}}\mathbf{Z}\right), \tag{24.24}$$

where we define the real vectors $\boldsymbol{\alpha}$ and $\boldsymbol{\beta}$ as $\boldsymbol{\alpha} \triangleq (\alpha^{(1)}, \ldots, \alpha^{(n)})^{\mathsf{T}} \in \mathbb{R}^n$ and $\boldsymbol{\beta} \triangleq (\beta^{(1)}, \ldots, \beta^{(n)})^{\mathsf{T}} \in \mathbb{R}^n$. Now (b) implies that $(\boldsymbol{\alpha} - \mathrm{i}\boldsymbol{\beta})^{\mathsf{T}}\mathbf{Z}$ is a Gaussian complex random variable, so its real part $\operatorname{Re}((\boldsymbol{\alpha} - \mathrm{i}\boldsymbol{\beta})^{\mathsf{T}}\mathbf{Z})$ must be a real Gaussian random variable (Definition 24.2.10 and Proposition 23.6.6), thus establishing that (b) implies that (24.22) is a real Gaussian random variable.

We next turn to proving the equivalence of (a) and (c). That (c) implies (a) follows directly from Proposition 24.3.9 applied to the Gaussian vector $\mathbf{W}$. The proof of the implication (a) $\Rightarrow$ (c) is very similar to the proof of its scalar version (24.10). We first note that since we can choose $\boldsymbol{\mu} = \mathsf{E}[\mathbf{Z}]$, it suffices to prove the result for the centered case. Now (a) implies that there exist $n \times n$ matrices $\mathsf{D}, \mathsf{E}, \mathsf{F}, \mathsf{G}$ such that

$$\begin{pmatrix} \operatorname{Re}(\mathbf{Z}) \\ \operatorname{Im}(\mathbf{Z}) \end{pmatrix} \stackrel{\mathscr{L}}{=} \begin{pmatrix} \mathsf{D} & \mathsf{E} \\ \mathsf{F} & \mathsf{G} \end{pmatrix} \begin{pmatrix} \mathbf{W}_1 \\ \mathbf{W}_2 \end{pmatrix}, \tag{24.25}$$

where $\mathbf{W}_1$ and $\mathbf{W}_2$ are independent real standard Gaussian $n$-vectors (Definition 23.1.1). On the other hand

$$\begin{pmatrix} \operatorname{Re}(\mathsf{A}\mathbf{W} + \mathsf{B}\mathbf{W}^*) \\ \operatorname{Im}(\mathsf{A}\mathbf{W} + \mathsf{B}\mathbf{W}^*) \end{pmatrix} = \begin{pmatrix} \frac{\operatorname{Re}(\mathsf{A}) + \operatorname{Re}(\mathsf{B})}{\sqrt{2}} & \frac{\operatorname{Im}(\mathsf{B}) - \operatorname{Im}(\mathsf{A})}{\sqrt{2}} \\ \frac{\operatorname{Im}(\mathsf{B}) + \operatorname{Im}(\mathsf{A})}{\sqrt{2}} & \frac{\operatorname{Re}(\mathsf{A}) - \operatorname{Re}(\mathsf{B})}{\sqrt{2}} \end{pmatrix} \begin{pmatrix} \sqrt{2}\operatorname{Re}(\mathbf{W}) \\ \sqrt{2}\operatorname{Im}(\mathbf{W}) \end{pmatrix}. \tag{24.26}$$

If $\mathbf{W}$ is a standard complex Gaussian, then

$$\begin{pmatrix} \sqrt{2}\operatorname{Re}(\mathbf{W}) \\ \sqrt{2}\operatorname{Im}(\mathbf{W}) \end{pmatrix} \stackrel{\mathscr{L}}{=} \begin{pmatrix} \mathbf{W}_1 \\ \mathbf{W}_2 \end{pmatrix},$$

where $\mathbf{W}_1$ and $\mathbf{W}_2$ are as above. Consequently, the representations (24.25) and (24.26) agree if

$$\begin{pmatrix} \mathsf{D} & \mathsf{E} \\ \mathsf{F} & \mathsf{G} \end{pmatrix} = \begin{pmatrix} \frac{\mathrm{Re(A)+Re(B)}}{\sqrt{2}} & \frac{\mathrm{Im(B)-Im(A)}}{\sqrt{2}} \\ \frac{\mathrm{Im(B)+Im(A)}}{\sqrt{2}} & \frac{\mathrm{Re(A)-Re(B)}}{\sqrt{2}} \end{pmatrix},$$

i.e., if we set

$$\mathsf{A} = \frac{1}{\sqrt{2}}\big((\mathsf{D} + \mathsf{G}) + \mathrm{i}(\mathsf{F} - \mathsf{E})\big),$$

$$\mathsf{B} = \frac{1}{\sqrt{2}}\big((\mathsf{D} - \mathsf{G}) + \mathrm{i}(\mathsf{F} + \mathsf{E})\big). \qquad \square$$

### 24.3.5  Proper Complex Gaussian Vectors

A proper complex Gaussian vector is a complex Gaussian vector that is also proper (Definition 17.4.1). Thus, $\mathbf{Z}$ is a proper complex Gaussian vector if it is a centered complex Gaussian vector satisfying $\mathsf{E}\big[\mathbf{ZZ}^\mathsf{T}\big] = 0$.

Recall that, by Proposition 24.3.5, every finite-variance circularly-symmetric complex random vector is also proper, but that some random vectors are proper and not circularly-symmetric. We next show that for Gaussian vectors, circular symmetry is equivalent to properness.

**Proposition 24.3.11 (For Complex Gaussians, Proper = Circularly-Symmetric).**
*A complex Gaussian vector is proper if, and only if, it is circularly-symmetric.*

**Proof.** Every circularly-symmetric complex Gaussian is proper, because every complex Gaussian is of finite-variance, and every finite-variance circularly-symmetric complex random vector is proper (Proposition 24.3.5).

We now turn to the reverse implication, i.e., that if a complex Gaussian vector is proper, then it is circularly-symmetric. Assume that $\mathbf{Z}$ is a proper Gaussian $n$-vector. We will prove that $\mathbf{Z}$ is circularly-symmetric using Proposition 24.3.3 by showing that for every deterministic vector $\boldsymbol{\alpha} \in \mathbb{C}^n$ the random variable $\boldsymbol{\alpha}^\mathsf{T}\mathbf{Z}$ is circularly-symmetric.

To that end, fix some arbitrary $\boldsymbol{\alpha} \in \mathbb{C}^n$. Since $\mathbf{Z}$ is a Gaussian vector, it follows that $\boldsymbol{\alpha}^\mathsf{T}\mathbf{Z}$ is a Gaussian CRV (Proposition 24.3.9 with the substitution of $\boldsymbol{\alpha}^\mathsf{T}$ for A and 0 for B). Moreover, since $\mathbf{Z}$ is proper, so is $\boldsymbol{\alpha}^\mathsf{T}\mathbf{Z}$ (Proposition 17.4.2). We have thus established that $\boldsymbol{\alpha}^\mathsf{T}\mathbf{Z}$ is a proper Gaussian CRV and hence, by Proposition 24.2.12, also circularly-symmetric. $\qquad \square$

The relationship between circular symmetry, properness, and Gaussianity is illustrated in Figure 24.2.

We next address the existence of a proper complex Gaussian of a given covariance matrix. We first recall that we say that a complex $n \times n$ matrix $\mathsf{K}$ is **complex positive semidefinite** and write $\mathsf{K} \succeq 0$ if $\boldsymbol{\alpha}^\dagger \mathsf{K}\boldsymbol{\alpha}$ is a nonnegative real number for every $\boldsymbol{\alpha} \in \mathbb{C}^n$. Recall also that an $n \times n$ complex matrix $\mathsf{K}$ is a complex positive definite matrix if, and only if, there exists a complex $n \times n$ matrix $\mathsf{S}$ such that $\mathsf{K} = \mathsf{SS}^\dagger$; see (Axler, 1997, Chapter 7, Theorem 7.27).

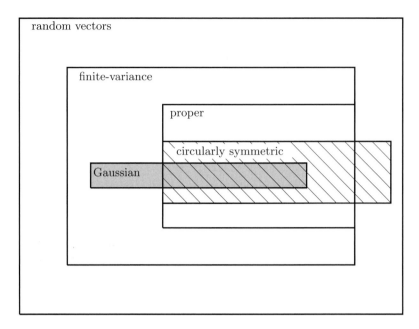

**Figure 24.2:** The relationship between circular symmetry, Gaussianity, and properness. The outer region corresponds to all complex random vectors. Within that is the set of all vectors whose components are of finite variance. Within it is the family of all proper random vectors. The slanted lines indicate the circularly-symmetric vectors, and the gray area corresponds to the Gaussian vectors. The same relations hold for scalars and for stochastic processes.

**Proposition 24.3.12.**

(i) *Given any $n \times n$ complex positive semidefinite matrix $\mathsf{K}$, there exists a proper complex Gaussian $n$-vector whose covariance matrix is $\mathsf{K}$.*

(ii) *The distribution of a proper Gaussian complex vector is fully specified by its covariance matrix.*

(iii) *If $\mathbf{Z}$ is a proper complex Gaussian $n$-vector of nonsingular covariance matrix $\mathsf{K}$, then its density is given by:*

$$f_{\mathbf{Z}}(\mathbf{z}) = \frac{1}{\pi^n \det \mathsf{K}} e^{-\mathbf{z}^\dagger \mathsf{K}^{-1} \mathbf{z}}, \quad \mathbf{z} \in \mathbb{C}^n. \tag{24.27}$$

**Note 24.3.13.** We denote the distribution of a proper Gaussian complex vector of covariance matrix $\mathsf{K}$ by

$$\mathcal{N}_{\mathbb{C}}(\mathbf{0}, \mathsf{K}).$$

**Proof.** To prove (i) we note that since $\mathsf{K}$ is positive semidefinite, it follows that there exists an $n \times n$ matrix $\mathsf{S}$ such that

$$\mathsf{K} = \mathsf{S}\mathsf{S}^\dagger. \tag{24.28}$$

Consider now the vector

$$\mathbf{Z} = \mathsf{S}\mathbf{W}, \tag{24.29}$$

where $\mathbf{W}$ is a standard complex Gaussian $n$-vector. We will show that $\mathbf{Z}$ has the desired properties. First, it must be Gaussian because it is the result of applying a deterministic linear mapping to the Gaussian vector $\mathbf{W}$ (Proposition 24.3.9). It is centered because $\mathbf{W}$ is centered (24.15) and because $\mathsf{E}[\mathsf{S}\mathbf{W}] = \mathsf{S}\mathsf{E}[\mathbf{W}]$. It is proper because it is the result of linearly transforming the proper complex random vector $\mathbf{W}$ (Proposition 17.4.3 and (24.15)). Finally, its covariance matrix is

$$
\begin{aligned}
\mathsf{E}\big[(\mathsf{S}\mathbf{W})(\mathsf{S}\mathbf{W})^{\dagger}\big] &= \mathsf{E}\big[\mathsf{S}\mathbf{W}\mathbf{W}^{\dagger}\mathsf{S}^{\dagger}\big] \\
&= \mathsf{S}\mathsf{E}\big[\mathbf{W}\mathbf{W}^{\dagger}\big]\mathsf{S}^{\dagger} \\
&= \mathsf{S}\mathsf{I}_n\mathsf{S}^{\dagger} \\
&= \mathsf{K}.
\end{aligned}
$$

Part (ii) was proved in Corollary 24.3.8.

To prove (iii) we use (24.29) & (24.28) along with the change of variables formula (Lemma 17.4.6) and the density of a standard Gaussian complex random vector (24.14) to obtain

$$
\begin{aligned}
f_{\mathbf{Z}}(\mathbf{z}) &= \frac{1}{|\det \mathsf{S}|^2} f_{\mathbf{W}}(\mathsf{S}^{-1}\mathbf{z}) \\
&= \frac{1}{\pi^n \det(\mathsf{S}\mathsf{S}^{\dagger})} e^{-(\mathsf{S}^{-1}\mathbf{z})^{\dagger}\mathsf{S}^{-1}\mathbf{z}} \\
&= \frac{1}{\pi^n \det \mathsf{K}_{ZZ}} e^{-\mathbf{z}^{\dagger}\mathsf{K}^{-1}\mathbf{z}}, \quad \mathbf{z} \in \mathbb{C}^n. \qquad \square
\end{aligned}
$$

## 24.4 Exercises

**Exercise 24.1 (The Complex Conjugate of a Circularly-Symmetric CRV).** Must the complex conjugate of a circularly-symmetric CRV be circularly-symmetric?

**Exercise 24.2 (Scaled Circularly-Symmetric CRV).** Show that if $Z$ is circularly-symmetric and if $\alpha \in \mathbb{C}$ is deterministic, then the distribution of $\alpha Z$ depends on $\alpha$ only via its magnitude $|\alpha|$.

**Exercise 24.3 (The $n$-th Power of a Circularly-Symmetric CRV).** Show that if $Z$ is a circularly-symmetric CRV and if $n$ is a positive integer, then $Z^n$ is circularly-symmetric.

**Exercise 24.4 (The Characteristic Function of Circularly-Symmetric CRVs).** Show that a CRV $Z$ is circularly-symmetric if, and only if, its characteristic function $\Phi_Z(\cdot)$ is radially-symmetric in the sense that $\Phi_Z(\varpi)$ depends on $\varpi$ only via its magnitude $|\varpi|$.

**Exercise 24.5 (Multiplying Independent CRVs).** Show that the product of two independent complex random variables is circularly-symmetric whenever (at least) one of them is circularly-symmetric.

**Exercise 24.6 (The Complex Conjugate of a Gaussian CRV).** Must the complex conjugate of a Gaussian CRV be Gaussian?

**Exercise 24.7 (Independent Components).** Show that if the complex random variables $W$ and $Z$ are circularly-symmetric and independent, then the random vector $(W, Z)^\mathsf{T}$ is circularly-symmetric.

**Exercise 24.8 (The Characteristic Function of a Proper Complex Gaussian Vector).** Compute the characteristic function of a proper complex Gaussian vector of covariance matrix $\mathsf{K}$.

**Exercise 24.9 (Jointly Circularly-Symmetric Complex Gaussians).** As in Definition 23.7.1, we can also define jointly complex Gaussians and jointly circularly-symmetric complex Gaussians. Extend the results of Section 23.7 by showing:

(i) Two centered jointly complex Gaussian vectors $\mathbf{Z}_1$ and $\mathbf{Z}_2$ are independent if, and only if, they satisfy

$$\mathsf{E}\big[\mathbf{Z}_1 \mathbf{Z}_2^\dagger\big] = 0 \text{ and } \mathsf{E}\big[\mathbf{Z}_1 \mathbf{Z}_2^\mathsf{T}\big] = 0.$$

(ii) Two jointly circularly-symmetric complex Gaussian vectors $\mathbf{Z}_1$ and $\mathbf{Z}_2$ are independent if, and only if, they satisfy

$$\mathsf{E}\big[\mathbf{Z}_1 \mathbf{Z}_2^\dagger\big] = 0.$$

(iii) If $\mathbf{Z}_1, \mathbf{Z}_2$ are centered jointly complex Gaussians, then, conditional on $\mathbf{Z}_2 = \mathbf{z}_2$, the complex random vector $\mathbf{Z}_1$ is a complex Gaussian such that

$$\mathsf{E}\Big[\big(\mathbf{Z}_1 - \mathsf{E}[\mathbf{Z}_1 \,|\, \mathbf{Z}_2 = \mathbf{z}_2]\big)\big(\mathbf{Z}_1 - \mathsf{E}[\mathbf{Z}_1 \,|\, \mathbf{Z}_2 = \mathbf{z}_2]\big)^\dagger \,\Big|\, \mathbf{Z}_2 = \mathbf{z}_2\Big]$$

and

$$\mathsf{E}\Big[\big(\mathbf{Z}_1 - \mathsf{E}[\mathbf{Z}_1 \,|\, \mathbf{Z}_2 = \mathbf{z}_2]\big)\big(\mathbf{Z}_1 - \mathsf{E}[\mathbf{Z}_1 \,|\, \mathbf{Z}_2 = \mathbf{z}_2]\big)^\mathsf{T} \,\Big|\, \mathbf{Z}_2 = \mathbf{z}_2\Big]$$

do not depend on $\mathbf{z}_2$ and such that the conditional mean $\mathsf{E}[\mathbf{Z}_1 \,|\, \mathbf{Z}_2 = \mathbf{z}_2]$ can be expressed as $\mathsf{A}\mathbf{z}_2 + \mathsf{B}\mathbf{z}_2^*$ for some matrices $\mathsf{A}$ and $\mathsf{B}$ that do not depend on $\mathbf{z}_2$.

(iv) If $\mathbf{Z}_1, \mathbf{Z}_2$ are jointly circularly-symmetric complex Gaussians, then, conditional on $\mathbf{Z}_2 = \mathbf{z}_2$, the complex random vector $\mathbf{Z}_1$ is a circularly-symmetric complex Gaussian of a covariance matrix that does not depend on $\mathbf{z}_2$ and of a mean that can be expressed as $\mathsf{A}\mathbf{z}_2$ for some matrix $\mathsf{A}$ that does not depend on $\mathbf{z}_2$.

**Exercise 24.10 (Limits of Complex Gaussians).** Extend the definition of almost-sure convergence (23.71) to complex random vectors, and show that if the complex Gaussian $d$-vectors $\mathbf{Z}_1, \mathbf{Z}_2, \ldots$ converge to $\mathbf{Z}$ almost surely, then $\mathbf{Z}$ must be a complex Gaussian.

**Exercise 24.11 (Limits of Circularly-Symmetric Complex Random Variables).** Consider a sequence $Z_1, Z_2, \ldots$ of circularly-symmetric complex random variables that converges almost surely to the CRV $Z$. Show that $Z$ must be circularly-symmetric. Extend this result to complex random vectors.

*Hint: Consider the characteristic functions of $Z, Z_1, Z_2, \ldots$, and recall the proof of Theorem 19.9.1.*

**Exercise 24.12 (Limits of Circularly-Symmetric Complex Gaussians).** Let $Z_1, Z_2, \ldots$ be a sequence of circularly-symmetric complex Gaussians that converges almost surely to the CRV $Z$. Show that $Z$ must be a circularly-symmetric Gaussian. Extend to complex random vectors.

*Hint: Either combine Exercises 24.10 & 24.11 or prove directly using the characteristic function as in the proof of Theorem 19.9.1.*

# Chapter 25

# Continuous-Time Stochastic Processes

## 25.1 Notation

Recall from Section 12.2 that a continuous-time stochastic process $\big(X(t),\ t \in \mathbb{R}\big)$ is a family of random variables that are defined on a common probability space $(\Omega, \mathcal{F}, P)$ and that are indexed by the real line (time). We denote by $X(t)$ the **time-$t$ sample** of $\big(X(t),\ t \in \mathbb{R}\big)$, i.e., the random variable to which $t$ is mapped (the RV indexed by $t$). This RV is sometimes also called **the state at time** $t$. Rather than writing $\big(X(t),\ t \in \mathbb{R}\big)$, we sometimes denote the SP by $\big(X(t)\big)$ or by $\mathbf{X}$. Perhaps the clearest way to denote the process is as a mapping:

$$\mathbf{X}\colon \Omega \times \mathbb{R} \to \mathbb{R}, \qquad (\omega, t) \mapsto X(\omega, t).$$

For a fixed $t \in \mathbb{R}$, the time-$t$ sample $X(t)$ is the mapping $X(\cdot, t)$ from $\Omega$ to the real line, i.e., the RV $\omega \mapsto X(\omega, t)$ indexed by $t$. If we fix $\omega \in \Omega$ and view $X(\omega, \cdot)$ as a mapping $t \mapsto X(\omega, t)$, then we obtain a function of time. This function is called a **trajectory**, **sample-path**, **path**, **sample-function**, or **realization**.

$$\begin{array}{lll} \omega \mapsto X(\omega, t) & \text{time-}t \text{ sample for a fixed } t \in \mathbb{R} & \text{(random variable)} \\ t \mapsto X(\omega, t) & \text{trajectory for a fixed } \omega \in \Omega & \text{(function of time)} \end{array}$$

Recall also from Section 12.2 that the process is **centered** if for every $t \in \mathbb{R}$ the RV $X(t)$ is of zero mean. It is **of finite variance** if for every $t \in \mathbb{R}$ the RV $X(t)$ is of finite variance.

## 25.2 The Finite-Dimensional Distributions

The finite-dimensional distributions (FDDs) of a continuous-time SP is the family of all joint distributions of $n$-tuples of the form $(X(t_1), \ldots, X(t_n))$, where $n$ can be any positive integer and $t_1, \ldots, t_n \in \mathbb{R}$ are arbitrary epochs. To specify the FDDs of a SP $\big(X(t)\big)$ one must thus specify for every $n \in \mathbb{N}$ and for every choice of the epochs $t_1, \ldots, t_n \in \mathbb{R}$ the distribution of the $n$-tuple $\big(X(t_1), \ldots, X(t_n)\big)$. This is a conceptually clear if formidable task. We denote the cumulative distribution function of the $n$-tuple $(X(t_1), \ldots, X(t_n))$ by

$$F_n\big(\xi_1, \ldots, \xi_n; t_1, \ldots, t_n\big) \triangleq \Pr\big[X(t_1) \leq \xi_1, \ldots, X(t_n) \leq \xi_n\big].$$

We next show that the FDDs of every SP $\big(X(t)\big)$ must satisfy two key properties: the symmetry property and the consistency property. The **symmetry property** is that $F_n(\cdot; \cdot)$ is unaltered when we simultaneously permute its right arguments (the $t$'s) and its left arguments (the $\xi$'s) by the same permutation. That is, for every $n \in \mathbb{N}$; every choice of the epochs $t_1, \ldots, t_n \in \mathbb{R}$; every $\xi_1, \ldots, \xi_n \in \mathbb{R}$; and every permutation $\pi$ on $\{1, \ldots, n\}$

$$F_n\big(\xi_{\pi(1)}, \ldots, \xi_{\pi(n)}; t_{\pi(1)}, \ldots, t_{\pi(n)}\big) = F_n\big(\xi_1, \ldots, \xi_n; t_1, \ldots, t_n\big). \tag{25.1}$$

This property is a generalization to $n$-tuples of the obvious fact that if $X$ and $Y$ are random variables, then $\Pr[X \leq x, Y \leq y] = \Pr[Y \leq y, X \leq x]$ for every $x, y \in \mathbb{R}$.

The **consistency property** is that whenever $n \in \mathbb{N}$ and $t_1, \ldots, t_n, \xi_1, \ldots, \xi_n \in \mathbb{R}$,

$$\lim_{\xi_n \to \infty} F_n\big(\xi_1, \ldots, \xi_{n-1}, \xi_n; t_1, \ldots, t_{n-1}, t_n\big) = F_{n-1}\big(\xi_1, \ldots, \xi_{n-1}; t_1, \ldots, t_{n-1}\big). \tag{25.2}$$

This property is a consequence of the fact that the set

$$\big\{\omega \in \Omega : X(\omega, t_1) \leq \xi_1, \ldots, X(\omega, t_{n-1}) \leq \xi_{n-1}, X(\omega, t_n) \leq \xi_n\big\}$$

is increasing in $\xi_n$ and converges as $\xi_n$ tends to infinity to the set

$$\big\{\omega \in \Omega : X(\omega, t_1) \leq \xi_1, \ldots, X(\omega, t_{n-1}) \leq \xi_{n-1}\big\}.$$

The key result on the existence of stochastic processes of given FDDs is Kolmogorov's Existence Theorem, which states that the symmetry and consistency properties suffice for a family of finite-dimensional distributions to correspond to the FDDs of some SP.

**Theorem 25.2.1 (Kolmogorov's Existence Theorem).** *Let* $G_1(\cdot; \cdot)$, $G_2(\cdot; \cdot)$, $\ldots$ *be a sequence of functions* $G_n \colon \mathbb{R}^n \times \mathbb{R}^n \to [0, 1]$ *satisfying*

*1) that for every* $n \geq 1$ *and every* $t_1, \ldots, t_n \in \mathbb{R}$ *the function* $G_n(\cdot; t_1, \ldots, t_n)$ *is a valid joint distribution function;[1]*

*2) the symmetry property*

$$G_n\big(\xi_{\pi(1)}, \ldots, \xi_{\pi(n)}; t_{\pi(1)}, \ldots, t_{\pi(n)}\big) = G_n\big(\xi_1, \ldots, \xi_n; t_1, \ldots, t_n\big),$$
$$t_1, \ldots, t_n, \xi_1, \ldots, \xi_n \in \mathbb{R}, \ \pi \ \text{a permutation on } \{1, \ldots, n\}; \tag{25.3}$$

---

[1] A function $F \colon \mathbb{R}^n \to [0, 1]$ is a valid joint distribution function if there exist random variables $X_1, \ldots, X_n$ whose joint distribution function is $F(\cdot)$, i.e.,

$$\Pr[X_1 \leq \xi_1, \ldots, X_n \leq \xi_n] = F(\xi_1, \ldots, \xi_n), \quad \xi_1, \ldots, \xi_n \in \mathbb{R}.$$

Not every function $F \colon \mathbb{R}^n \to [0, 1]$ is a valid joint distribution function. For example, a valid joint distribution function must be monotonic in each variable. See, for example, (Billingsley, 1995, Theorem 12.5) for a characterization of joint distribution functions.

*3) and the consistency property*

$$\lim_{\xi_n \to \infty} G_n(\xi_1, \ldots, \xi_{n-1}, \xi_n; t_1, \ldots, t_{n-1}, t_n)$$

$$= G_{n-1}(\xi_1, \ldots, \xi_{n-1}; t_1, \ldots, t_{n-1}),$$

$$t_1, \ldots, t_n, \xi_1, \ldots, \xi_n \in \mathbb{R}. \quad (25.4)$$

*Then there exists a SP $(X(t))$ whose FDDs are given by $\{G_n(\cdot; \cdot)\}$ in the sense that*

$$\Pr[X(t_1) \le \xi_1, \ldots, X(t_n) \le \xi_n] = G_n(\xi_1, \ldots, \xi_n; t_1, \ldots, t_n)$$

*for every $n \in \mathbb{N}$, all $t_1, \ldots, t_n \in \mathbb{R}$, and all $\xi_1, \ldots, \xi_n \in \mathbb{R}$.*

**Proof.** See, for example, (Billingsley, 1995, Chapter 7, Section 36), (Cramér and Leadbetter, 2004, Section 3.3), (Grimmett and Stirzaker, 2001, Section 8.6), or (Doob, 1990, Chapter I § 5). $\quad\square$

In the study of $n$-tuples of random variables we can use the joint distribution function to answer, at least in principle, most of our probability questions. When it comes to stochastic processes, however, there are interesting questions that cannot be answered using the FDDs. For example, it can be shown that the probability of the event that the SP $(X(t))$ produces a sample-path that is continuous at time zero cannot be computed from the FDDs. This is not due to our limited analytic capabilities but rather because there exist two stochastic processes of identical FDDs where for one process this event is of zero probability whereas for the other it is of probability one (Cramér and Leadbetter, 2004, Section 3.6). Fortunately, most of the questions of interest to us in Digital Communications can be answered based on the FDDs.

An exception is a very subtle point related to measurability. From the FDDs alone one cannot determine whether the trajectories are measurable functions of time, i.e., whether it makes sense to talk about integrals of the form $\int_{-\infty}^{\infty} x(\omega, t)\, \mathrm{d}t$. This issue will be revisited in Section 25.9.

The above discussion motivates us to define the set of events whose probability can be determined from the FDDs using the axioms of probability, i.e., using the rules that the probability of the set of all possible outcomes $\Omega$ is one and that the probability of a countable union of disjoint events is the infinite sum of the probabilities of the events. In the mathematical literature what we are defining is called the **$\sigma$-algebra generated by** $(X(t),\ t \in \mathbb{R})$ or the **$\sigma$-algebra generated by the cylindrical sets of** $(X(t),\ t \in \mathbb{R})$.[2] For the classical definition see, for example, (Billingsley, 1995, Section 36).

**Definition 25.2.2 ($\sigma$-Algebra Generated by a SP).** *The $\sigma$-algebra generated by a SP $(X(t),\ t \in \mathbb{R})$ which is defined over the probability space $(\Omega, \mathcal{F}, P)$ is the set of events (i.e., elements of $\mathcal{F}$) whose probability can be computed from the FDDs of $(X(t))$ using only the axioms of probability.*

---

[2]It is the smallest $\sigma$-algebra with respect to which all the random variables $(X(t),\ t \in \mathbb{R})$ are measurable.

We now rephrase our previous statement about continuity as saying that the set of $\omega \in \Omega$ for which the function $t \mapsto X(\omega, t)$ is continuous at $t = 0$ is not in the $\sigma$-algebra generated by $(X(t))$. The probability of such sets cannot be inferred from the FDDs alone. If such sets are assigned a probability it must be based on some additional information that is not captured by the FDDs.

The FDDs provide a natural way to define independence between stochastic processes.

**Definition 25.2.3 (Independent Stochastic Processes).** *Two stochastic processes* $(X(t))$ *and* $(Y(t))$ *defined on the same probability space* $(\Omega, \mathcal{F}, P)$ *are said to be* **independent stochastic processes** *if for every* $n \in \mathbb{N}$ *and any choice of the epochs* $t_1, \ldots, t_n \in \mathbb{R}$, *the n-tuples* $(X(t_1), \ldots, X(t_n))$ *and* $(Y(t_1), \ldots, Y(t_n))$ *are independent.*

## 25.3 Definition of a Gaussian SP

By far the most important processes for modeling noise in Digital Communications are the Gaussian processes. Fortunately, these processes are among the mathematically most tractable. The definition of a Gaussian SP builds on that of a Gaussian vector (Definition 23.1.1).

**Definition 25.3.1 (Gaussian Stochastic Processes).** *A SP* $(X(t))$ *is said to be a* **Gaussian stochastic process** *if for every* $n \in \mathbb{N}$ *and every choice of the epochs* $t_1, \ldots, t_n \in \mathbb{R}$, *the random vector* $(X(t_1), \ldots, X(t_n))^\mathsf{T}$ *is Gaussian.*

**Note 25.3.2.** Gaussian stochastic processes are of finite variance.

**Proof.** If $(X(t))$ is a Gaussian process, then *a fortiori* at each epoch $t \in \mathbb{R}$, the random variable $X(t)$ is a univariate Gaussian (choose $n = 1$ in the above definition) and hence, by the definition of the univariate distribution (Definition 19.3.1), of finite variance. $\square$

One of the things that make Gaussian processes tractable is the ease with which their FDDs can be specified.

**Proposition 25.3.3 (The FDDs of a Gaussian SP).** *If* $(X(t))$ *is a centered Gaussian SP, then all its FDDs are determined by the mapping that specifies the covariance between any two of its samples:*

$$(t_1, t_2) \mapsto \mathsf{Cov}\big[X(t_1), X(t_2)\big], \quad t_1, t_2 \in \mathbb{R}. \tag{25.5}$$

**Proof.** Let $(X(t))$ be a centered Gaussian SP. We shall show that for any choice of the epochs $t_1, \ldots, t_n \in \mathbb{R}$ we can compute the joint distribution of $X(t_1), \ldots X(t_n)$ from the mapping (25.5). To this end we note that since $(X(t))$ is a Gaussian SP, the random vector $(X(t_1), \ldots X(t_n))^\mathsf{T}$ is Gaussian (Definition 25.3.1). Consequently, its distribution is fully specified by its mean vector and covariance matrix (Theorem 23.6.7). Its mean vector is zero, because we assumed that $(X(t))$ is centered. To conclude the proof we thus only need to show that the covariance matrix of $(X(t_1), \ldots X(t_n))^\mathsf{T}$ is determined by the mapping (25.5). But this is obvious because the covariance matrix of $(X(t_1), \ldots X(t_n))^\mathsf{T}$ is the $n \times n$ matrix

$$
\begin{pmatrix}
\mathsf{Cov}[X(t_1), X(t_1)] & \mathsf{Cov}[X(t_1), X(t_2)] & \cdots & \mathsf{Cov}[X(t_1), X(t_n)] \\
\cdots & \cdots & \cdots & \cdots \\
\cdots & \cdots & \cdots & \cdots \\
\cdots & \cdots & \cdots & \cdots \\
\mathsf{Cov}[X(t_n), X(t_1)] & \mathsf{Cov}[X(t_n), X(t_2)] & \cdots & \mathsf{Cov}[X(t_n), X(t_n)]
\end{pmatrix}, \qquad (25.6)
$$

and each of the entries in this matrix is specified by the mapping (25.5). $\qquad \square$

Things become even simpler if the Gaussian process is **wide-sense stationary** (Definition 25.4.2 ahead). In this case the RHS of (25.5) is determined by $t_1 - t_2$, so the mapping (25.5) (and hence all the FDDs) is determined by the mapping $\tau \mapsto \mathsf{Cov}[X(t), X(t + \tau)]$. But before discussing wide-sense stationary Gaussian stochastic processes in Section 25.5, we first define stationarity and wide-sense stationarity for general processes that are not necessarily Gaussian.

## 25.4 Stationary Continuous-Time Processes

Our treatment of stationary continuous-time processes is similar to the treatment of their discrete-time counterparts (Chapter 13). The following is the continuous-time analogue of Definition 13.2.1.

**Definition 25.4.1 (Stationary Continuous-Time SP).** *We say that a continuous-time SP* $(X(t))$ *is* ***stationary*** *(or* ***strict sense stationary***, *or* ***strongly stationary***) *if for every* $n \in \mathbb{N}$, *any epochs* $t_1, \ldots, t_n \in \mathbb{R}$, *and every* $\tau \in \mathbb{R}$,

$$
\big(X(t_1 + \tau), \ldots, X(t_n + \tau)\big) \stackrel{\mathscr{L}}{=} \big(X(t_1), \ldots, X(t_n)\big). \qquad (25.7)
$$

By considering the case where $n = 1$ we obtain that if $(X(t))$ is stationary, then all its samples have the same distribution

$$
X(t) \stackrel{\mathscr{L}}{=} X(t + \tau), \quad t, \tau \in \mathbb{R}. \qquad (25.8)
$$

That is, the distribution of the random variable $X(t)$ does not depend on $t$. By considering $n = 2$ we obtain that if $(X(t))$ is stationary, then the joint distribution of any two of its samples depends on how far apart they are and not on the absolute time at which they are taken

$$
\big(X(t_1), X(t_2)\big) \stackrel{\mathscr{L}}{=} \big(X(t_1 + \tau), X(t_2 + \tau)\big), \quad t_1, t_2, \tau \in \mathbb{R}. \qquad (25.9)
$$

That is, the joint distribution of $\big(X(t_1), X(t_2)\big)$ can be computed from $t_2 - t_1$.

As we did for discrete-time processes (Definition 13.3.1), we can also define wide-sense stationarity of continuous-time processes. Recall that a process $(X(t))$ is said to be of finite variance if at every time $t \in \mathbb{R}$ the random variable $X(t)$ is of finite variance.

**Definition 25.4.2 (Wide-Sense Stationary Continuous-Time SP).** *A continuous-time SP* $(X(t))$ *is said to be* ***wide-sense stationary*** *(or* ***weakly stationary*** *or* ***second-order stationary****) if the following three conditions are met:*

1) *It is of finite variance.*

2) *Its mean is constant*

$$\mathsf{E}[X(t)] = \mathsf{E}[X(t+\tau)], \quad t, \tau \in \mathbb{R}. \tag{25.10}$$

3) *The covariance between its samples satisfies*

$$\mathsf{Cov}[X(t_1), X(t_2)] = \mathsf{Cov}[X(t_1+\tau), X(t_2+\tau)], \quad t_1, t_2, \tau \in \mathbb{R}. \tag{25.11}$$

By considering the case where $t_1 = t_2$ in (25.11), we obtain that all the samples of a WSS SP have the same variance:

$$\mathsf{Var}[X(t)] = \mathsf{Var}[X(0)], \quad t \in \mathbb{R}. \tag{25.12}$$

**Note 25.4.3.** Every finite-variance stationary SP is WSS.

**Proof.** This follows because (25.8) implies (25.10), and because (25.9) implies (25.11). $\qquad\square$

The reverse is not true: some WSS processes are not stationary. (Wide-sense stationarity concerns only means and covariances, whereas stationarity has to do with distributions.)

The following definition of the autocovariance function of a continuous-time WSS SP is the analogue of Definition 13.5.1.

**Definition 25.4.4 (Autocovariance Function).** *The* ***autocovariance function*** $\mathsf{K}_{XX} \colon \mathbb{R} \to \mathbb{R}$ *of a WSS continuous-time SP* $(X(t))$ *is defined for every* $\tau \in \mathbb{R}$ *by*

$$\mathsf{K}_{XX}(\tau) \triangleq \mathsf{Cov}[X(t+\tau), X(t)], \tag{25.13}$$

*where the RHS does not depend on t because* $(X(t))$ *is assumed to be WSS.*

By evaluating (25.13) at $\tau = 0$ and using (25.12), we can express the variance of $X(t)$ in terms of the autocovariance function $\mathsf{K}_{XX}$ as

$$\mathsf{Var}[X(t)] = \mathsf{K}_{XX}(0), \quad t \in \mathbb{R}. \tag{25.14}$$

We end this section with a few simple inequalities related to WSS stochastic processes and their autocovariance functions.

**Lemma 25.4.5.** *Let* $(X(t))$ *be a WSS SP of autocovariance function* $\mathsf{K}_{XX}$. *Then*

$$|\mathsf{K}_{XX}(\tau)| \leq \mathsf{K}_{XX}(0), \quad \tau \in \mathbb{R}, \tag{25.15}$$

$$\mathsf{E}[|X(t)|] \leq \sqrt{\mathsf{K}_{XX}(0) + \mathsf{E}[X(0)]^2}, \quad t \in \mathbb{R}, \tag{25.16}$$

*and*

$$\mathsf{E}[|X(t)\,X(t')|] \leq \mathsf{K}_{XX}(0) + \mathsf{E}[X(0)]^2, \quad t, t' \in \mathbb{R}. \tag{25.17}$$

**Proof.** Inequality (25.15) follows from the Covariance Inequality (Corollary 3.5.2):

$$|\mathsf{K}_{XX}(\tau)| = |\mathsf{Cov}[X(t+\tau), X(t)]|$$
$$\leq \sqrt{\mathsf{Var}[X(t+\tau)]}\sqrt{\mathsf{Var}[X(t)]}$$
$$= \mathsf{K}_{XX}(0),$$

where the last equality follows from (25.14).

Inequality (25.16) follows from the nonnegativity of the variance of $|X(t)|$ and the assumption that $(X(t))$ is WSS:

$$0 \leq \mathsf{Var}[|X(t)|]$$
$$= \mathsf{E}[X^2(t)] - (\mathsf{E}[|X(t)|])^2$$
$$= \mathsf{Var}[X(t)] + (\mathsf{E}[X(t)])^2 - (\mathsf{E}[|X(t)|])^2$$
$$= \mathsf{K}_{XX}(0) + (\mathsf{E}[X(0)])^2 - (\mathsf{E}[|X(t)|])^2.$$

Finally, Inequality (25.17) follows from the Cauchy-Schwarz Inequality for random variables (Theorem 3.5.1)

$$|\mathsf{E}[UV]| \leq \sqrt{\mathsf{E}[U^2]\,\mathsf{E}[V^2]}$$

by substituting $|X(t)|$ for $U$ and $|X(t')|$ for $V$ and by noting that

$$\mathsf{E}[|X(t)|^2] = \mathsf{E}[X^2(t)]$$
$$= \mathsf{Var}[X(t)] + (\mathsf{E}[X(t)])^2$$
$$= \mathsf{K}_{XX}(0) + (\mathsf{E}[X(0)])^2, \quad t \in \mathbb{R}. \qquad \square$$

## 25.5  Stationary Gaussian Stochastic Processes

For Gaussian stochastic processes we do not distinguish between stationarity and wide-sense stationarity. The reason is that, while for general processes the two concepts are different (in that every finite-variance stationary SP is WSS, but not every WSS SP is stationary), for Gaussian stochastic processes the two concepts are equivalent. These relationships between stationarity and wide-sense stationarity for general stochastic processes and for Gaussian stochastic processes are illustrated in Figure 25.1.

**Proposition 25.5.1 (Stationary Gaussian Stochastic Processes).**

(i) *A Gaussian SP is stationary if, and only if, it is WSS.*

(ii) *The FDDs of a centered stationary Gaussian SP are fully specified by its autocovariance function.*

**Proof.** We begin by proving (i). One direction has only little to do with Gaussianity. Since every Gaussian SP is of finite variance (Note 25.3.2), and since every

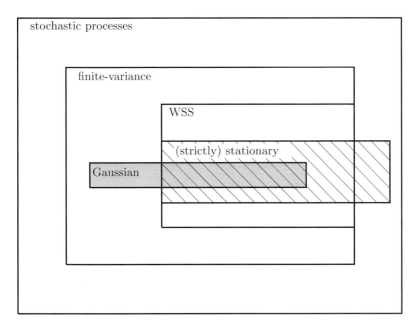

**Figure 25.1:** The relationship between wide-sense stationarity, Gaussianity, and strict-sense stationarity. The outer region corresponds to all stochastic processes. Within it is the set of all finite-variance processes and within that the set of all wide-sense stationary processes. The slanted lines indicate the strict-sense stationary processes, and the gray area corresponds to the Gaussian stochastic processes.

finite-variance stationary SP is WSS (Note 25.4.3), it follows that every stationary Gaussian SP is WSS.

Gaussianity plays a much more important role in the proof of the reverse direction, namely, that every WSS Gaussian SP is stationary. We prove this by showing that if $\big(X(t)\big)$ is Gaussian and WSS, then for every $n \in \mathbb{N}$ and any $t_1,\dots,t_n,\tau \in \mathbb{R}$ the joint distribution of $X(t_1),\dots,X(t_n)$ is identical to the joint distribution of $X(t_1 + \tau),\dots,X(t_n + \tau)$. To this end, let $n \in \mathbb{N}$ and $t_1,\dots,t_n,\tau \in \mathbb{R}$ be fixed.

Because $\big(X(t)\big)$ is Gaussian, $(X(t_1),\dots,X(t_n))^\mathsf{T}$ and $(X(t_1 + \tau),\dots,X(t_n + \tau))^\mathsf{T}$ are both Gaussian vectors (Definition 25.3.1). And since $\big(X(t)\big)$ is WSS, the two are of the same mean vector (see (25.10)). The former's covariance matrix is

$$
\begin{pmatrix}
\mathsf{Cov}[X(t_1), X(t_1)] & \cdots & \mathsf{Cov}[X(t_1), X(t_n)] \\
\cdots & \cdots & \cdots \\
\cdots & \cdots & \cdots \\
\cdots & \cdots & \cdots \\
\mathsf{Cov}[X(t_n), X(t_1)] & \cdots & \mathsf{Cov}[X(t_n), X(t_n)]
\end{pmatrix}
$$

and the latter's is

$$
\begin{pmatrix}
\mathsf{Cov}[X(t_1 + \tau), X(t_1 + \tau)] & \cdots & \mathsf{Cov}[X(t_1 + \tau), X(t_n + \tau)] \\
\cdots & \cdots & \cdots \\
\cdots & \cdots & \cdots \\
\cdots & \cdots & \cdots \\
\mathsf{Cov}[X(t_n + \tau), X(t_1 + \tau)] & \cdots & \mathsf{Cov}[X(t_n + \tau), X(t_n + \tau)]
\end{pmatrix}.
$$

Since $(X(t))$ is WSS, the two covariance matrices are identical (see (25.11)). But two Gaussian vectors of equal mean vectors and of equal covariance matrices have identical distributions (Theorem 23.6.7), so the distribution of $(X(t_1), \ldots, X(t_n))^{\mathsf{T}}$ is identical to that of $(X(t_1 + \tau), \ldots, X(t_n + \tau))^{\mathsf{T}}$. Since this has been established for all choices of $n \in \mathbb{N}$ and all choices of $t_1, \ldots, t_n, \tau \in \mathbb{R}$, the SP $(X(t))$ is stationary.

Part (ii) follows from Proposition 25.3.3 and the definition of wide-sense stationarity. Indeed, by Proposition 25.3.3, all the FDDs of a centered Gaussian SP $(X(t))$ are determined by the mapping (25.5). If $(X(t))$ is additionally WSS, then the RHS of (25.5) can be computed from $t_1 - t_2$ and is given by $\mathsf{K}_{XX}(t_1 - t_2)$, so the mapping (25.5) is fully specified by the autocovariance function $\mathsf{K}_{XX}$.     $\square$

## 25.6    Properties of the Autocovariance Function

Many of the definitions and results on continuous-time WSS stochastic processes have analogous discrete-time counterparts. But some technical issues are encountered only in continuous time. For example, most results on continuous-time WSS stochastic processes require that the autocovariance function of the process be continuous at the origin, i.e., satisfy

$$
\lim_{\delta \to 0} \mathsf{K}_{XX}(\delta) = \mathsf{K}_{XX}(0), \tag{25.18}
$$

and this condition has no discrete-time counterpart. As we next show, this condition is equivalent to the condition

$$
\lim_{\delta \to 0} \mathsf{E}\left[\left(X(t + \delta) - X(t)\right)^2\right] = 0, \quad t \in \mathbb{R}. \tag{25.19}
$$

This equivalence follows from the identity

$$
\mathsf{E}\left[\left(X(t) - X(t + \delta)\right)^2\right] = 2\left(\mathsf{K}_{XX}(0) - \mathsf{K}_{XX}(\delta)\right), \quad t, \delta \in \mathbb{R}, \tag{25.20}
$$

which can be proved as follows. We first note that it suffices to prove it for centered processes, and for such processes we then compute:

$$
\begin{aligned}
\mathsf{E}\left[\left(X(t) - X(t + \delta)\right)^2\right] &= \mathsf{E}\left[X^2(t) - 2X(t)X(t + \delta) + X^2(t + \delta)\right] \\
&= \mathsf{E}\left[X^2(t)\right] - 2\mathsf{E}\left[X(t)X(t + \delta)\right] + \mathsf{E}\left[X^2(t + \delta)\right] \\
&= \mathsf{K}_{XX}(0) - 2\mathsf{K}_{XX}(\delta) + \mathsf{K}_{XX}(0) \\
&= 2\left(\mathsf{K}_{XX}(0) - \mathsf{K}_{XX}(\delta)\right),
\end{aligned}
$$

where the first equality follows by opening the square; the second by the linearity of expectation; the third by the definition of $K_{XX}$; and the final equality by collecting terms.

We note here that if the autocovariance function of a WSS process is continuous at the origin, then it is continuous everywhere. In fact, it is uniformly continuous:

**Lemma 25.6.1.** *If the autocovariance function of a WSS continuous-time SP is continuous at the origin, then it is a uniformly continuous function.*

**Proof.** We first note that it suffices to prove the lemma for centered processes. Let $\bigl(X(t)\bigr)$ be such a process. For every $\tau, \delta \in \mathbb{R}$ we then have

$$
\begin{aligned}
\bigl| K_{XX}(\tau + \delta) - K_{XX}(\tau) \bigr| &= \bigl| \mathsf{E}[X(\tau + \delta)\, X(0)] - \mathsf{E}[X(\tau)\, X(0)] \bigr| \\
&= \bigl| \mathsf{E}\bigl[ \bigl( X(\tau + \delta) - X(\tau) \bigr) X(0) \bigr] \bigr| \\
&= \bigl| \mathsf{Cov}\bigl[ X(\tau + \delta) - X(\tau), X(0) \bigr] \bigr| \\
&\leq \sqrt{ \mathsf{E}\bigl[ \bigl( X(\tau + \delta) - X(\tau) \bigr)^2 \bigr] } \sqrt{ \mathsf{E}[X^2(0)] } \\
&= \sqrt{ 2\bigl( K_{XX}(0) - K_{XX}(\delta) \bigr) } \sqrt{ K_{XX}(0) } \\
&= \sqrt{ 2\, K_{XX}(0) \bigl( K_{XX}(0) - K_{XX}(\delta) \bigr) }, \qquad (25.21)
\end{aligned}
$$

where the equality in the first line follows from the definition of the autocovariance function because $\bigl(X(t)\bigr)$ is centered; the equality in the second line by the linearity of expectation; the equality in the third line by the definition of the covariance between two zero-mean random variables; the inequality in the fourth line by the Covariance Inequality (Corollary 3.5.2); the equality in the fifth line by (25.20); and the final equality by trivial algebra. The uniform continuity of $K_{XX}$ now follows from (25.21) by noting that its RHS does not depend on $\tau$ and that, by our assumption about the continuity of $K_{XX}$ at zero, it tends to zero as $\delta \to 0$. $\qquad\square$

We next derive two important properties of autocovariance functions and then demonstrate in Theorem 25.6.2 that these properties characterize those functions that can arise as the autocovariance functions of a WSS SP. These properties are the continuous-time analogues of (13.12) & (13.13), and the proofs are almost identical. We first state the properties and then proceed to prove them.

The first property is that the autocovariance function $K_{XX}$ of any continuous-time WSS process $\bigl(X(t)\bigr)$ is a **symmetric function**

$$
K_{XX}(-\tau) = K_{XX}(\tau), \quad \tau \in \mathbb{R}. \qquad (25.22)
$$

The second is that it is a **positive definite function** in the sense that for every $n \in \mathbb{N}$, and for every choice of the coefficients $\alpha_1, \ldots, \alpha_n \in \mathbb{R}$ and of the epochs $t_1, \ldots, t_n \in \mathbb{R}$

$$
\sum_{\nu=1}^{n} \sum_{\nu'=1}^{n} \alpha_\nu \alpha_{\nu'}\, K_{XX}(t_\nu - t_{\nu'}) \geq 0. \qquad (25.23)
$$

To prove (25.22) we calculate

$$
\begin{aligned}
\mathsf{K}_{XX}(\tau) &= \mathsf{Cov}\big[X(t+\tau), X(t)\big] \\
&= \mathsf{Cov}\big[X(t'), X(t'-\tau)\big] \\
&= \mathsf{Cov}\big[X(t'-\tau), X(t')\big] \\
&= \mathsf{K}_{XX}(-\tau), \quad \tau \in \mathbb{R},
\end{aligned}
$$

where the first equality follows from the definition of $\mathsf{K}_{XX}(\tau)$ (25.13); the second by defining $t' \triangleq t + \tau$; the third because $\mathsf{Cov}[X, Y] = \mathsf{Cov}[Y, X]$ (for real random variables); and the final equality by the definition of $\mathsf{K}_{XX}(-\tau)$ (25.13).

To prove (25.23) we compute

$$
\begin{aligned}
\sum_{\nu=1}^{n}\sum_{\nu'=1}^{n}\alpha_\nu\alpha_{\nu'}\,\mathsf{K}_{XX}(t_\nu - t_{\nu'}) &= \sum_{\nu=1}^{n}\sum_{\nu'=1}^{n}\alpha_\nu\alpha_{\nu'}\,\mathsf{Cov}[X(t_\nu), X(t_{\nu'})] \\
&= \mathsf{Cov}\left[\sum_{\nu=1}^{n}\alpha_\nu X(t_\nu), \sum_{\nu'=1}^{n}\alpha_{\nu'} X(t_{\nu'})\right] \\
&= \mathsf{Var}\left[\sum_{\nu=1}^{n}\alpha_\nu X(t_\nu)\right] \quad\quad (25.24) \\
&\geq 0.
\end{aligned}
$$

The next theorem demonstrates that Properties (25.22) and (25.23) characterize the autocovariance functions of WSS stochastic processes (*cf.* Theorem 13.5.2).

**Theorem 25.6.2.** *Every symmetric positive definite function is the autocovariance function of some stationary Gaussian SP.*

**Proof.** The proof is based on Kolmogorov's Existence Theorem (Theorem 25.2.1) and is only sketched here. Let $\mathsf{K}(\cdot)$ be a symmetric and positive definite function from $\mathbb{R}$ to $\mathbb{R}$. The idea is to consider for every $n \in \mathbb{N}$ and for every choice of the epochs $t_1, \ldots, t_n \in \mathbb{R}$ the joint distribution function $G_n(\cdot; t_1, \ldots, t_n)$ corresponding to the centered multivariate Gaussian distribution of covariance matrix

$$
\begin{pmatrix}
\mathsf{K}(t_1 - t_1) & \mathsf{K}(t_1 - t_2) & \cdots & \mathsf{K}(t_1 - t_n) \\
\cdots & \cdots & \cdots & \cdots \\
\cdots & \cdots & \cdots & \cdots \\
\cdots & \cdots & \cdots & \cdots \\
\mathsf{K}(t_n - t_1) & \mathsf{K}(t_n - t_2) & \cdots & \mathsf{K}(t_n - t_n)
\end{pmatrix}
$$

and to verify that the sequence $\{G_n(\cdot; \cdot)\}$ satisfies the symmetry and consistency requirements of Kolmogorov's Existence Theorem. The details, which can be found in (Doob, 1990, Chapter II, Section § 3, Theorem 3.1), are omitted. $\square$

## 25.7 The Power Spectral Density of a Continuous-Time SP

Under suitable conditions, engineers usually define the power spectral density of a WSS SP as the Fourier Transform of its autocovariance function. There is nothing wrong with this definition, and we encourage the reader to think about the PSD in this way.[3] We, however, prefer a slightly more general definition that allows us also to consider discontinuous spectra and, more importantly, allows us to infer that any integrable, nonnegative, symmetric function is the PSD of some Gaussian SP (Proposition 25.7.3). Fortunately, the two definitions agree whenever the autocovariance function is continuous and integrable.

Before defining the PSD, we pause to discuss the Fourier Transform of the autocovariance. If the autocovariance function $\mathsf{K}_{XX}$ of a WSS SP $\big(X(t)\big)$ is integrable, i.e., if

$$\int_{-\infty}^{\infty} \big|\mathsf{K}_{XX}(\tau)\big|\,\mathrm{d}\tau < \infty, \tag{25.25}$$

then we can discuss its FT $\hat{\mathsf{K}}_{XX}$. The following proposition summarizes the main properties of the FT of continuous integrable autocovariance functions.

**Proposition 25.7.1.** *If the autocovariance function $\mathsf{K}_{XX}$ is continuous at the origin and integrable, then its Fourier Transform $\hat{\mathsf{K}}_{XX}$ is nonnegative*

$$\hat{\mathsf{K}}_{XX}(f) \geq 0, \quad f \in \mathbb{R} \tag{25.26}$$

*and symmetric*

$$\hat{\mathsf{K}}_{XX}(-f) = \hat{\mathsf{K}}_{XX}(f), \quad f \in \mathbb{R}. \tag{25.27}$$

*Moreover, the Inverse Fourier Transform recovers $\mathsf{K}_{XX}$ in the sense that[4]*

$$\mathsf{K}_{XX}(\tau) = \int_{-\infty}^{\infty} \hat{\mathsf{K}}_{XX}(f)\,e^{i2\pi f\tau}\,\mathrm{d}f, \quad \tau \in \mathbb{R}. \tag{25.28}$$

**Proof.** This result can be deduced from three results in (Feller, 1971, Chapter XIX): the theorem in Section 3, Bochner's Theorem in Section 2, and Lemma 2 in Section 2. □

**Definition 25.7.2 (The PSD of a Continuous-Time WSS SP).** *We say that the WSS continuous-time SP $\big(X(t)\big)$ is of **power spectral density** (PSD) $\mathsf{S}_{XX}$ if $\mathsf{S}_{XX}$ is a nonnegative, symmetric, integrable function from $\mathbb{R}$ to $\mathbb{R}$ whose Inverse Fourier Transform is the autocovariance function $\mathsf{K}_{XX}$ of $\big(X(t)\big)$:*

$$\mathsf{K}_{XX}(\tau) = \int_{-\infty}^{\infty} \mathsf{S}_{XX}(f)\,e^{i2\pi f\tau}\,\mathrm{d}f, \quad \tau \in \mathbb{R}. \tag{25.29}$$

A few remarks regarding this definition:

---

[3]Engineers can, however, be a bit sloppy in that they sometimes speak of a SP whose PSD is discontinuous, e.g., the Brickwall function $f \mapsto \mathrm{I}\{|f| \leq \mathsf{W}\}$. This is inconsistent with their definition because the FT of an integrable function must be continuous (Theorem 6.2.11), and consequently if the autocovariance function is integrable then its FT cannot be discontinuous. Our more general definition does not suffer from this problem and allows for discontinuous PSDs.

[4]Recall that without additional assumptions one is not guaranteed that the Inverse Fourier Transform of the Fourier Transform of a function will be identical to the original function. Here we need not make any additional assumptions because we already assumed that the autocovariance function is continuous and because autocovariance functions are positive definite.

(i) By the uniqueness of the IFT (the analogue of Theorem 6.2.12 for the IFT) it follows that if two functions are PSDs of the same WSS SP, then they must be equal except on a set of frequencies of Lebesgue measure zero. Consequently, we shall often speak of "the" PSD as though it were unique.

(ii) By Proposition 25.7.1, if $\mathsf{K}_{XX}$ is continuous and integrable, then $\big(X(t)\big)$ has a PSD in the sense of Definition 25.7.2, and this PSD is the FT of $\mathsf{K}_{XX}$. There are, however, autocovariance functions that are not integrable and that nonetheless have a PSD in the sense of Definition 25.7.2. For example, $\tau \mapsto \operatorname{sinc}(\tau)$.

Thus, every continuous autocovariance function that has a PSD in the engineers' sense (i.e., that is integrable) also has the same PSD according to our definition, but our definition is more general in that some autocovariance functions that have a PSD according to our definition are not integrable and therefore do not have a PSD in the engineers' sense.

(iii) By substituting $\tau = 0$ in (25.29) and using (25.14) we can express the variance of $X(t)$ in terms of the PSD $\mathsf{S}_{XX}$ as

$$\operatorname{Var}\big[X(t)\big] = \mathsf{K}_{XX}(0) = \int_{-\infty}^{\infty} \mathsf{S}_{XX}(f)\,\mathrm{d}f, \quad t \in \mathbb{R}. \tag{25.30}$$

(iv) Only processes with continuous autocovariance functions have PSDs, because the RHS of (25.29), being the IFT of an integrable function, must be continuous (Theorem 6.2.11 (ii)).

(v) It can be shown that if the autocovariance function can be written as the IFT of some integrable function, then this latter function must be nonnegative (except on a set of frequencies of Lebesgue measure zero). This is the continuous-time analogue of Proposition 13.6.3.

The nonnegativity, symmetry, and integrability conditions characterize PSDs in the following sense:

**Proposition 25.7.3.** *Every nonnegative, symmetric, integrable function is the PSD of some stationary Gaussian SP whose autocovariance function is continuous.*

**Proof.** Let $\mathsf{S}(\cdot)$ be some integrable, nonnegative, and symmetric function from $\mathbb{R}$ to the nonnegative reals. Define $\mathsf{K}(\cdot)$ to be its IFT

$$\mathsf{K}(\tau) = \int_{-\infty}^{\infty} \mathsf{S}(f)\,e^{i2\pi f\tau}\,\mathrm{d}f, \quad \tau \in \mathbb{R}. \tag{25.31}$$

We shall verify that $\mathsf{K}(\cdot)$ satisfies the hypotheses of Theorem 25.6.2, namely, that it is symmetric and positive definite. It will then follow from Theorem 25.6.2 that there exists a stationary Gaussian SP $\big(X(t)\big)$ whose autocovariance function $\mathsf{K}_{XX}$ is equal to $\mathsf{K}(\cdot)$ and is thus given by

$$\mathsf{K}_{XX}(\tau) = \int_{-\infty}^{\infty} \mathsf{S}(f)\,e^{i2\pi f\tau}\,\mathrm{d}f, \quad \tau \in \mathbb{R}. \tag{25.32}$$

This will establish that $\big(X(t)\big)$ is of PSD $\mathsf{S}(\cdot)$. The continuity of $\mathsf{K}_{XX}$ will follow from the continuity of the IFT of integrable functions (Theorem 6.2.11).

To conclude the proof we need to show that the function $\mathsf{K}(\cdot)$ defined in (25.31) is symmetric and positive definite. The symmetry follows from our assumption that $\mathsf{S}(\cdot)$ is symmetric:

$$
\begin{aligned}
\mathsf{K}(-\tau) &= \int_{-\infty}^{\infty} \mathsf{S}(f)\, e^{\mathrm{i}2\pi f(-\tau)}\, \mathrm{d}f \\
&= \int_{-\infty}^{\infty} \mathsf{S}(-\tilde{f})\, e^{\mathrm{i}2\pi \tilde{f}\tau}\, \mathrm{d}\tilde{f} \\
&= \int_{-\infty}^{\infty} \mathsf{S}(\tilde{f})\, e^{\mathrm{i}2\pi \tilde{f}\tau}\, \mathrm{d}\tilde{f} \\
&= \mathsf{K}(\tau), \quad \tau \in \mathbb{R},
\end{aligned}
$$

where the first equality follows from (25.31); the second from the change of variable $\tilde{f} \triangleq -f$; the third by the symmetry of $\mathsf{S}(\cdot)$; and the final equality again by (25.31).

We next prove that $\mathsf{K}(\cdot)$ is positive definite. To that end we fix some $n \in \mathbb{N}$, some constants $\alpha_1, \dots, \alpha_n \in \mathbb{R}$, and some epochs $t_1, \dots, t_n \in \mathbb{R}$ and compute:

$$
\begin{aligned}
\sum_{\nu=1}^{n}\sum_{\nu'=1}^{n} \alpha_\nu \alpha_{\nu'} \mathsf{K}(t_\nu - t_{\nu'}) &= \sum_{\nu=1}^{n}\sum_{\nu'=1}^{n} \alpha_\nu \alpha_{\nu'} \int_{-\infty}^{\infty} \mathsf{S}(f)\, e^{\mathrm{i}2\pi f(t_\nu - t_{\nu'})}\, \mathrm{d}f \\
&= \int_{-\infty}^{\infty} \mathsf{S}(f) \left( \sum_{\nu=1}^{n}\sum_{\nu'=1}^{n} \alpha_\nu \alpha_{\nu'}\, e^{\mathrm{i}2\pi f(t_\nu - t_{\nu'})} \right) \mathrm{d}f \\
&= \int_{-\infty}^{\infty} \mathsf{S}(f) \left( \sum_{\nu=1}^{n}\sum_{\nu'=1}^{n} \alpha_\nu\, e^{\mathrm{i}2\pi f t_\nu}\, \alpha_{\nu'}\, e^{-\mathrm{i}2\pi f t_{\nu'}} \right) \mathrm{d}f \\
&= \int_{-\infty}^{\infty} \mathsf{S}(f) \left( \sum_{\nu=1}^{n} \alpha_\nu\, e^{\mathrm{i}2\pi f t_\nu} \right) \left( \sum_{\nu'=1}^{n} \alpha_{\nu'}\, e^{\mathrm{i}2\pi f t_{\nu'}} \right)^{*} \mathrm{d}f \\
&= \int_{-\infty}^{\infty} \mathsf{S}(f) \left| \sum_{\nu=1}^{n} \alpha_\nu\, e^{\mathrm{i}2\pi f t_\nu} \right|^{2} \mathrm{d}f \\
&\geq 0,
\end{aligned}
$$

where the first equality follows from (25.31); the subsequent equalities by simple algebra; and the last inequality from our assumption that $\mathsf{S}(\cdot)$ is nonnegative. $\square$

## 25.8 The Spectral Distribution Function

In this section we shall state without proof Bochner's Theorem on continuous positive definite functions and discuss its application to continuous autocovariance functions. We shall then define the spectral distribution function of WSS stochastic processes. The concept of a spectral distribution function is more general than that of a PSD, because every WSS with a continuous autocovariance function has a spectral distribution function, but only some have a PSD. Nevertheless, for our purposes, the notion of PSD will suffice, and the results of this section will not be used in the rest of the book.

Recall that the characteristic function $\Phi_X(\cdot)$ of a RV $X$ is the mapping from $\mathbb{R}$ to $\mathbb{C}$ defined by

$$\varpi \mapsto \mathsf{E}\big[e^{\mathrm{i}\varpi X}\big], \quad \varpi \in \mathbb{R}. \tag{25.33}$$

If $X$ is symmetric (i.e., has a symmetric distribution) in the sense that

$$\Pr[X \geq x] = \Pr[X \leq -x], \quad x \in \mathbb{R}, \tag{25.34}$$

then $\Phi_X(\cdot)$ only takes on real values and is a symmetric function, as the following argument shows. The symmetry of the distribution of $X$ implies that $X$ and $-X$ have the same distribution, which implies that their exponentiations have the same law

$$e^{\mathrm{i}\varpi X} \overset{\mathscr{L}}{=} e^{-\mathrm{i}\varpi X}, \quad \varpi \in \mathbb{R}, \tag{25.35}$$

and *a fortiori* that the expectation of the two exponentials are equal

$$\mathsf{E}\big[e^{\mathrm{i}\varpi X}\big] = \mathsf{E}\big[e^{-\mathrm{i}\varpi X}\big], \quad \varpi \in \mathbb{R}. \tag{25.36}$$

The LHS of (25.36) is $\Phi_X(\varpi)$, and the RHS is $\Phi_X(-\varpi)$, thus demonstrating the symmetry of $\Phi_X(\cdot)$. To establish that (25.34) also implies that $\Phi_X(\cdot)$ is real, we note that, by (25.36),

$$\begin{aligned}
\Phi_X(\varpi) &= \mathsf{E}\big[e^{\mathrm{i}\varpi X}\big] \\
&= \frac{1}{2}\Big(\mathsf{E}\big[e^{\mathrm{i}\varpi X}\big] + \mathsf{E}\big[e^{-\mathrm{i}\varpi X}\big]\Big) \\
&= \mathsf{E}\left[\frac{e^{\mathrm{i}\varpi X} + e^{-\mathrm{i}\varpi X}}{2}\right] \\
&= \mathsf{E}\big[\cos(\varpi X)\big], \quad \varpi \in \mathbb{R},
\end{aligned}$$

which is real. Here the first equality follows from (25.33); the second from (25.36); and the third from the linearity of expectation.

Bochner's Theorem establishes a correspondence between continuous, symmetric, positive definite functions and characteristic functions.

**Theorem 25.8.1 (Bochner's Theorem).** *Let the mapping $\Phi(\cdot)$ from $\mathbb{R}$ to $\mathbb{R}$ be continuous. Then the following two conditions are equivalent:*

    *a) $\Phi(\cdot)$ is the characteristic function of some RV having a symmetric distribution.*

    *b) $\Phi(\cdot)$ is a symmetric positive definite function satisfying $\Phi(0) = 1$.*

**Proof.** See (Feller, 1971, Chapter XIX, Section 2) or (Loève, 1963, Chapter IV, Section 14) or (Katznelson, 1976, Chapter VI, Section 2.8). $\square$

Bochner's Theorem is the key to understanding autocovariance functions:

**Proposition 25.8.2.** *Let $\big(X(t)\big)$ be a WSS SP whose autocovariance function $\mathsf{K}_{XX}$ is continuous. Then:*

*(i) There exists a symmetric RV $S$ such that*

$$\mathsf{K}_{XX}(\tau) = \mathsf{K}_{XX}(0)\,\mathsf{E}\big[e^{\mathrm{i}2\pi\tau S}\big], \quad \tau \in \mathbb{R}. \tag{25.37}$$

*(ii) If $\mathsf{K}_{XX}(0) > 0$, then the distribution of $S$ in (25.37) is uniquely determined by $\mathsf{K}_{XX}$, and $\big(X(t)\big)$ has a PSD if, and only if, $S$ has a density.*

**Proof.** If $\mathsf{K}_{XX}(0) = 0$, then $\big(X(t)\big)$ is deterministic in the sense that for every epoch $t \in \mathbb{R}$ the variance of $X(t)$ is zero. By the inequality $|\mathsf{K}_{XX}(\tau)| \le \mathsf{K}_{XX}(0)$ (Lemma 25.4.5, (25.15)) it follows that if $\mathsf{K}_{XX}(0) = 0$ then $\mathsf{K}_{XX}(\tau) = 0$ for all $\tau \in \mathbb{R}$, and (25.37) holds in this case for any choice of $S$ and there is nothing else to prove.

Consider now the case $\mathsf{K}_{XX}(0) > 0$. To prove Part (i) we note that because $\mathsf{K}_{XX}$ is by assumption continuous, and because all autocovariance functions are symmetric and positive definite (see (25.22) and (25.23)), it follows that the mapping

$$\tau \mapsto \frac{\mathsf{K}_{XX}(\tau)}{\mathsf{K}_{XX}(0)}, \quad \tau \in \mathbb{R}$$

is a continuous, symmetric, positive definite mapping that takes on the value one at $\tau = 0$. Consequently, by Bochner's Theorem, there exists a RV $R$ of a symmetric distribution such that

$$\frac{\mathsf{K}_{XX}(\tau)}{\mathsf{K}_{XX}(0)} = \mathsf{E}\big[e^{\mathrm{i}\tau R}\big], \quad \tau \in \mathbb{R}.$$

It follows that if we define $S$ as $R/(2\pi)$ then (25.37) will hold, and Part (i) is thus also established for the case where $\mathsf{K}_{XX}(0) > 0$.

We now conclude the treatment of the case $\mathsf{K}_{XX}(0) > 0$ by proving Part (ii) for this case. That the distribution of $S$ is unique follows because (25.37) implies that

$$\mathsf{E}\big[e^{\mathrm{i}\varpi S}\big] = \frac{\mathsf{K}_{XX}(\varpi/(2\pi))}{\mathsf{K}_{XX}(0)}, \quad \varpi \in \mathbb{R},$$

so $\mathsf{K}_{XX}$ determines the characteristic function of $S$ and hence also its distribution (Theorem 17.4.4).

Because the distribution of $S$ is symmetric, if $S$ has a density then it also has a symmetric density. Denote by $f_S(\cdot)$ a symmetric density function for $S$. In terms of $f_S(\cdot)$ we can rewrite (25.37) as

$$\mathsf{K}_{XX}(\tau) = \int_{-\infty}^{\infty} \mathsf{K}_{XX}(0)\,f_S(s)\,e^{\mathrm{i}2\pi s\tau}\,\mathrm{d}s, \quad \tau \in \mathbb{R},$$

so the nonnegative symmetric function $\mathsf{K}_{XX}(0)\,f_S(\cdot)$ is a PSD of $\big(X(t)\big)$. Conversely, if $\big(X(t)\big)$ has PSD $\mathsf{S}_{XX}$, then

$$\mathsf{K}_{XX}(\tau) = \int_{-\infty}^{\infty} \mathsf{S}_{XX}(f)\,e^{\mathrm{i}2\pi f\tau}\,\mathrm{d}f, \quad \tau \in \mathbb{R}, \tag{25.38}$$

and (25.37) holds with $S$ having the density

$$f_S(s) = \frac{\mathsf{S}_{XX}(s)}{\mathsf{K}_{XX}(0)}, \quad s \in \mathbb{R}. \tag{25.39}$$

(The RHS of (25.39) is symmetric, nonnegative, and integrates to 1 by (25.30).)   □

Proposition 25.8.2 motivates us to define the spectral distribution function of a continuous autocovariance function (or of a WSS SP having such an autocovariance function) as follows.

**Definition 25.8.3 (Spectral Distribution Function).** *The **spectral distribution function** of a continuous autocovariance function* $\mathsf{K}_{XX}$ *is the mapping*

$$\xi \mapsto \mathsf{K}_{XX}(0)\,\Pr[S \le \xi], \tag{25.40}$$

*where $S$ is a random variable for which (25.37) holds.*

## 25.9   The Average Power

We next address the average power in the sample-paths of a SP. We would like to better understand formal expressions of the form

$$\frac{1}{\mathsf{T}} \int_{-\mathsf{T}/2}^{\mathsf{T}/2} X^2(\omega, t)\,\mathrm{d}t$$

for a SP $\big(X(t)\big)$ defined on the probability space $(\Omega, \mathcal{F}, P)$. Recalling that if we fix $\omega \in \Omega$ then we can view the trajectory $t \mapsto X(\omega, t)$ as a function of time, we would like to think about the integral above as the time-integral of the square of the trajectory $t \mapsto X(\omega, t)$. Since the result of this integral is a (nonnegative) number that depends on $\omega$, we would like to view this result as a nonnegative RV

$$\omega \mapsto \frac{1}{\mathsf{T}} \int_{-\mathsf{T}/2}^{\mathsf{T}/2} X^2(\omega, t)\,\mathrm{d}t, \quad \omega \in \Omega.$$

Mathematicians, however, would object to our naive approach on two grounds. The first is that it is *prima facie* unclear whether for every fixed $\omega \in \Omega$ the mapping $t \mapsto X^2(\omega, t)$ is sufficiently well-behaved to allow us to discuss its integral. (It may not be Lebesgue measurable.) The second is that, even if this integral could be carried out for every $\omega \in \Omega$, it is *prima facie* unclear that the result would be a RV. While it would certainly be a mapping from $\Omega$ to the extended reals (allowing for $+\infty$), it is not clear that it would satisfy the technical measurability conditions that random variables must meet.[5]

---

[5]By "$X$ is a random variable possibly taking on the value $+\infty$" we mean that $X$ is a mapping from $\Omega$ to $\mathbb{R} \cup \{+\infty\}$ with the set $\{\omega \in \Omega : X(\omega) \le \xi\}$ being an event for every $\xi \in \mathbb{R}$ and with the set $\{\omega \in \Omega : X(\omega) = +\infty\}$ also being an event.

To address these objections we shall assume that $\big(X(t)\big)$ is a "measurable stochastic process." This is a technical condition that will be foreign to most readers and that will be inessential to the rest of this book. We mention it here because, in order to be mathematically honest, we shall have to slip this attribute into some of the theorems that we shall later state. Nothing will be lost on readers who replace "measurable stochastic process" with "stochastic process satisfying a mild technical condition."

Fortunately, this technical condition is, indeed, very mild. For example, Proposition 25.7.3 still holds if we slip in the attribute "measurable" before the words "Gaussian process." Similarly, in Theorem 25.6.2, if we add the hypothesis that the given function is continuous at the origin, then we can slip in the attribute "measurable" before the words "stationary Gaussian stochastic process."[6]

For the benefit of readers who are familiar with Measure Theory, we provide the following definition.

**Definition 25.9.1 (Measurable SP).** *Let $\big(X(t),\ t \in \mathbb{R}\big)$ be a SP defined over the probability space $(\Omega, \mathcal{F}, P)$. We say that the process is a **measurable stochastic process** if the mapping $(\omega, t) \mapsto X(\omega, t)$ is a measurable mapping from $\Omega \times \mathbb{R}$ to $\mathbb{R}$ when the range $\mathbb{R}$ is endowed with the Borel $\sigma$-algebra on $\mathbb{R}$ and when the domain $\Omega \times \mathbb{R}$ is endowed with the $\sigma$-algebra defined by the product of $\mathcal{F}$ on $\Omega$ by the Borel $\sigma$-algebra on $\mathbb{R}$.*

The nice thing about measurable stochastic processes is that if $\big(X(t)\big)$ is a measurable SP, then for every $\omega \in \Omega$ the trajectory $t \mapsto X(\omega, t)$ is a Borel (and hence also Lebesgue) measurable function of time; see (Halmos, 1950, Chapter 7, Section 34, Theorem B) or (Billingsley, 1995, Chapter 3, Section 18, Theorem 18.1 (ii)). Moreover, for such processes we can sometimes use Fubini's Theorem to swap the order in which we compute time-integrals and expectations; see (Halmos, 1950, Chapter 7, Section 36) or (Billingsley, 1995, Chapter 3, Section 18, Theorem 18.3 (ii)).

We can now state the main result of this section regarding the average power in a WSS SP.

**Proposition 25.9.2 (Power in a Centered WSS SP).** *If $\big(X(t)\big)$ is a measurable, centered, WSS SP defined over the probability space $(\Omega, \mathcal{F}, P)$ and having the autocovariance function $\mathsf{K}_{XX}$, then for every $a, b \in \mathbb{R}$ satisfying $a < b$ the mapping*

$$\omega \mapsto \frac{1}{b-a} \int_a^b X^2(\omega, t)\, \mathrm{d}t \tag{25.41}$$

*defines a RV (possibly taking on the value $+\infty$) satisfying*

$$\frac{1}{b-a}\, \mathsf{E}\!\left[ \int_a^b X^2(t)\, \mathrm{d}t \right] = \mathsf{K}_{XX}(0). \tag{25.42}$$

---

[6]These are but very special cases of a much more general result that states that given FDDs corresponding to a WSS SP of an autocovariance that is continuous at the origin, there exists a SP of the given FDDs that is also measurable. See, for example, (Doob, 1990, Chapter II, Section § 2, Theorem 2.6). (Replacing the values $\pm\infty$ with zero may ruin the separability but not the measurability.)

**Proof.** The proof of (25.42) is straightforward and merely requires swapping the order of integration and expectation. This swap can be justified using Fubini's Theorem. Heuristically, the swapping of expectation and integration can be justified by thinking about the integral as being a Riemann integral that can be approximated by finite sums and by then recalling the linearity of expectation that guarantees that the expectation of a finite sum is the sum of the expectations. We then have

$$\mathsf{E}\left[\int_a^b X^2(t)\,\mathrm{d}t\right] = \int_a^b \mathsf{E}\big[X^2(t)\big]\,\mathrm{d}t$$

$$= \int_a^b \mathsf{K}_{XX}(0)\,\mathrm{d}t$$

$$= (b-a)\,\mathsf{K}_{XX}(0),$$

where the first equality follows by swapping the integration with the expectation; the second because our assumption that $\big(X(t)\big)$ is centered implies that for every $t \in \mathbb{R}$ the RV $X(t)$ is centered and by (25.13); and the final equality because the integrand is constant.

That (25.41) is a RV (possibly taking on the value $+\infty$) follows from Fubini's Theorem.          □

Recalling Definition 14.6.1 of the power in a SP as

$$\lim_{\mathsf{T}\to\infty} \mathsf{E}\left[\frac{1}{\mathsf{T}}\int_{-\mathsf{T}/2}^{\mathsf{T}/2} X^2(t)\,\mathrm{d}t\right],$$

we conclude:

**Corollary 25.9.3.** *The power in a centered, measurable, WSS SP $\big(X(t)\big)$ of auto-covariance function $\mathsf{K}_{XX}$ is equal to $\mathsf{K}_{XX}(0)$.*


## 25.10   Linear Functionals

For the problem of detecting continuous-time signals corrupted by noise, we shall be interested in stochastic integrals of the form

$$\int_{-\infty}^{\infty} X(t)\,s(t)\,\mathrm{d}t \tag{25.43}$$

for WSS stochastic processes $\big(X(t)\big)$ defined over a probability space $(\Omega, \mathcal{F}, P)$ and for properly well-behaved deterministic functions $s(\cdot)$. We would like to think about the result of such an integral as defining a RV

$$\omega \mapsto \int_{-\infty}^{\infty} X(\omega, t)\,s(t)\,\mathrm{d}t \tag{25.44}$$

that maps each $\omega \in \Omega$ to the real number that is the result of the integration over time of the product of the trajectory $t \mapsto X(\omega, t)$ corresponding to $\omega$ by the deterministic function $t \mapsto s(t)$. That is, each $\omega$ is mapped to the inner product between its trajectory $t \mapsto X(\omega, t)$ and the function $s(\cdot)$.

This is an excellent way of thinking about such integrals, but we do run into some mathematical objections similar to those we encountered in Section 25.9. For example, it is not obvious that for each $\omega \in \Omega$ the mapping $t \mapsto X(\omega, t) s(t)$ is a sufficiently well-behaved function for the time-integral to be defined. As we shall see, for this reason we must impose certain restrictions on $s(\cdot)$, and we will not claim that $t \mapsto X(\omega, t) s(t)$ is integrable for *every* $\omega \in \Omega$ but only for $\omega$'s in some subset of $\Omega$ having probability one. Also, even if this issue is addressed, it is unclear that the mapping of $\omega$ to the result of the integration is a RV. While it is clearly a mapping from $\Omega$ to the reals, it is unclear that it satisfies the additional mathematical requirement of measurability, i.e., that for every $\xi \in \mathbb{R}$ the set

$$\left\{ \omega \in \Omega : \int_{-\infty}^{\infty} X(\omega, t) s(t) \, \mathrm{d}t \le \xi \right\}$$

be an event, i.e., an element of $\mathcal{F}$.

We ask the reader to take it on faith that these issues can be resolved and to focus on the relatively straightforward computation of the mean and variance of (25.44). The resolution of the measurability issues is provided in Proposition 25.10.1, whose proof is recommended only to readers with background in Measure Theory.

We shall assume throughout that $(X(t))$ is WSS and that the deterministic function $\mathbf{s} \colon \mathbb{R} \to \mathbb{R}$ is integrable. We begin by heuristically deriving the mean:

$$\begin{aligned}
\mathsf{E}\left[ \int_{-\infty}^{\infty} X(t) s(t) \, \mathrm{d}t \right] &= \int_{-\infty}^{\infty} \mathsf{E}\big[ X(t) s(t) \big] \, \mathrm{d}t \\
&= \int_{-\infty}^{\infty} \mathsf{E}\big[ X(t) \big] s(t) \, \mathrm{d}t \\
&= \mathsf{E}\big[ X(0) \big] \int_{-\infty}^{\infty} s(t) \, \mathrm{d}t,
\end{aligned} \qquad (25.45)$$

with the following heuristic justification. The first equality follows by swapping the expectation with the time-integration; the second because $s(\cdot)$ is deterministic; and the last equality from our assumption that $(X(t))$ is WSS, which implies that $(X(t))$ is of constant mean: $\mathsf{E}[X(t)] = \mathsf{E}[X(0)]$ for all $t \in \mathbb{R}$.

We next heuristically derive the variance of the integral in terms of the autocovariance function $\mathsf{K}_{XX}$ of the process $(X(t))$. We begin by considering the case where $(X(t))$ is of zero mean. In this case we have

$$\begin{aligned}
\mathsf{Var}\left[ \int_{-\infty}^{\infty} X(t) s(t) \, \mathrm{d}t \right] &= \mathsf{E}\left[ \left( \int_{-\infty}^{\infty} X(t) s(t) \, \mathrm{d}t \right)^2 \right] \\
&= \mathsf{E}\left[ \left( \int_{-\infty}^{\infty} X(t) s(t) \, \mathrm{d}t \right) \left( \int_{-\infty}^{\infty} X(\tau) s(\tau) \, \mathrm{d}\tau \right) \right] \\
&= \mathsf{E}\left[ \int_{-\infty}^{\infty} \int_{-\infty}^{\infty} X(t) s(t) X(\tau) s(\tau) \, \mathrm{d}t \, \mathrm{d}\tau \right]
\end{aligned}$$

$$= \int_{-\infty}^{\infty} \int_{-\infty}^{\infty} s(t)\, s(\tau)\, \mathsf{E}\big[X(t)\, X(\tau)\big]\, \mathrm{d}t\, \mathrm{d}\tau$$

$$= \int_{-\infty}^{\infty} \int_{-\infty}^{\infty} s(t)\, \mathsf{K}_{XX}(t-\tau)\, s(\tau)\, \mathrm{d}t\, \mathrm{d}\tau, \qquad (25.46)$$

where the first equality follows because (25.45) and our assumption that $\big(X(t)\big)$ is centered combine to guarantee that $\int X(t)\, s(t)\, \mathrm{d}t$ is of zero mean; the second by writing $a^2$ as $a$ times $a$; the third by writing the product of integrals over $\mathbb{R}$ as a double integral (i.e., as an integral over $\mathbb{R}^2$); the fourth by swapping the double-integral with the expectation; and the final equality by the definition of the autocovariance function (Definition 25.4.4) and because $\big(X(t)\big)$ is centered.

There are two equivalent ways of writing the RHS of (25.46) that we wish to point out. The first is obtained from (25.46) by changing the integration variables from $(t,\tau)$ to $(\sigma,\tau)$, where $\sigma \triangleq t-\tau$ and by performing the integration first over $\tau$ and then over $\sigma$:

$$\mathsf{Var}\left[\int_{-\infty}^{\infty} X(t)\, s(t)\, \mathrm{d}t\right] = \int_{-\infty}^{\infty} \int_{-\infty}^{\infty} s(t)\, \mathsf{K}_{XX}(t-\tau)\, s(\tau)\, \mathrm{d}t\, \mathrm{d}\tau$$

$$= \int_{-\infty}^{\infty} \int_{-\infty}^{\infty} s(\sigma+\tau)\, \mathsf{K}_{XX}(\sigma)\, s(\tau)\, \mathrm{d}\sigma\, \mathrm{d}\tau$$

$$= \int_{-\infty}^{\infty} \mathsf{K}_{XX}(\sigma) \int_{-\infty}^{\infty} s(\sigma+\tau)\, s(\tau)\, \mathrm{d}\tau\, \mathrm{d}\sigma$$

$$= \int_{-\infty}^{\infty} \mathsf{K}_{XX}(\sigma)\, \mathsf{R}_{\mathrm{ss}}(\sigma)\, \mathrm{d}\sigma, \qquad (25.47)$$

where $\mathsf{R}_{\mathrm{ss}}$ is the self-similarity function of $\mathbf{s}$ (Definition 11.2.1 and Section 11.4).

The second equivalent way of writing (25.46) can be derived from (25.47) when $\big(X(t)\big)$ is of PSD $\mathsf{S}_{XX}$. Since (25.47) has the form of an inner product, we can use Proposition 6.2.4 to write this inner product in the frequency domain by noting that the FT of $\mathsf{R}_{\mathrm{ss}}$ is $f \mapsto |\hat{s}(f)|^2$ (see (11.35)) and that $\mathsf{K}_{XX}$ is the IFT of its PSD $\mathsf{S}_{XX}$. The result is that

$$\mathsf{Var}\left[\int_{-\infty}^{\infty} X(t)\, s(t)\, \mathrm{d}t\right] = \int_{-\infty}^{\infty} \mathsf{S}_{XX}(f)\, \big|\hat{s}(f)\big|^2\, \mathrm{d}f. \qquad (25.48)$$

We next show that (25.46) (and hence also (25.47) & (25.48), which are equivalent ways of writing (25.46)) remains valid also when $\big(X(t)\big)$ is of mean $\mu$ (not necessarily zero). To see this we can consider the zero-mean SP $\big(\tilde{X}(t)\big)$ defined at every epoch $t \in \mathbb{R}$ by $\tilde{X}(t) = X(t) - \mu$ and formally compute

$$\mathsf{Var}\left[\int_{-\infty}^{\infty} X(t)\, s(t)\, \mathrm{d}t\right] = \mathsf{Var}\left[\int_{-\infty}^{\infty} \big(\tilde{X}(t) + \mu\big)\, s(t)\, \mathrm{d}t\right]$$

$$= \mathsf{Var}\left[\int_{-\infty}^{\infty} \tilde{X}(t)\, s(t)\, \mathrm{d}t + \mu \int_{-\infty}^{\infty} s(t)\, \mathrm{d}t\right]$$

$$= \mathsf{Var}\left[\int_{-\infty}^{\infty} \tilde{X}(t)\, s(t)\, \mathrm{d}t\right]$$

$$= \int_{-\infty}^{\infty} \int_{-\infty}^{\infty} s(t) \, \mathsf{K}_{\tilde{X}\tilde{X}}(t - \tau) \, s(\tau) \, \mathrm{d}t \, \mathrm{d}\tau$$

$$= \int_{-\infty}^{\infty} \int_{-\infty}^{\infty} s(t) \, \mathsf{K}_{XX}(t - \tau) \, s(\tau) \, \mathrm{d}t \, \mathrm{d}\tau, \qquad (25.49)$$

where the first equality follows from the definition of $\tilde{X}(t)$ as $X(t) - \mu$; the second by the linearity of integration; the third because adding a deterministic quantity to a RV does not change its covariance; the fourth by (25.46) applied to the zero-mean process $\big(\tilde{X}(t)\big)$; and the final equality because the autocovariance function of $\big(\tilde{X}(t)\big)$ is the same as the autocovariance function of $\big(X(t)\big)$ (Definition 25.4.4).

As above, once a result is proved for centered stochastic processes, its extension to WSS stochastic processes with a mean can be straightforward. Consequently, we shall often derive our results for centered WSS stochastic processes and leave it to the reader to extend them to mean-$\mu$ stochastic processes by expressing such stochastic processes as the sum of a zero-mean SP and the deterministic constant $\mu$.

As promised, we now state the results about the mean and variance of (25.44) in a mathematically defensible proposition.

**Proposition 25.10.1 (Mean and Variance of Linear Functionals of a WSS SP).**
*Let $\big(X(t)\big)$ be a measurable WSS SP defined over the probability space $(\Omega, \mathcal{F}, P)$ and having the autocovariance function $\mathsf{K}_{XX}$. Let $\mathbf{s} \colon \mathbb{R} \to \mathbb{R}$ be some deterministic integrable function. Then:*

(i) *For every $\omega \in \Omega$ the mapping $t \mapsto X(\omega, t) \, s(t)$ is Lebesgue measurable.*

(ii) *The set*

$$\mathcal{N} \triangleq \left\{ \omega \in \Omega : \int_{-\infty}^{\infty} \big| X(\omega, t) \, s(t) \big| \, \mathrm{d}t = \infty \right\} \qquad (25.50)$$

*is an event and is of probability zero.*

(iii) *The mapping from $\Omega \setminus \mathcal{N}$ to $\mathbb{R}$ defined by*

$$\omega \mapsto \int_{-\infty}^{\infty} X(\omega, t) \, s(t) \, \mathrm{d}t \qquad (25.51)$$

*is measurable with respect to $\mathcal{F}$.*

(iv) *The mapping from $\Omega$ to $\mathbb{R}$ defined by*

$$\omega \mapsto \begin{cases} \displaystyle \int_{-\infty}^{\infty} X(\omega, t) \, s(t) \, \mathrm{d}t & \text{if } \omega \notin \mathcal{N}, \\ 0 & \text{otherwise,} \end{cases} \qquad (25.52)$$

*defines a random variable.*

(v) *The mean of this RV is*

$$\mathsf{E}\big[X(0)\big] \int_{-\infty}^{\infty} s(t) \, \mathrm{d}t.$$

*(vi) Its variance is*

$$\int_{-\infty}^{\infty} \int_{-\infty}^{\infty} s(t)\, \mathsf{K}_{XX}(t-\tau)\, s(\tau)\, \mathrm{d}\tau\, \mathrm{d}t, \tag{25.53}$$

*which can also be expressed as*

$$\int_{-\infty}^{\infty} \mathsf{K}_{XX}(\sigma)\, \mathsf{R}_{\mathbf{ss}}(\sigma)\, \mathrm{d}\sigma, \tag{25.54}$$

*where $\mathsf{R}_{\mathbf{ss}}$ is the self-similarity function of $\mathbf{s}$.*

*(vii) If $\big(X(t)\big)$ is of PSD $\mathsf{S}_{XX}$, then the variance of this RV can be expressed as*

$$\int_{-\infty}^{\infty} \mathsf{S}_{XX}(f)\, \big|\hat{s}(f)\big|^2\, \mathrm{d}f. \tag{25.55}$$

**Proof.** Part (i) follows because the measurability of the process $\big(X(t)\big)$ guarantees that for every $\omega \in \Omega$ the mapping $t \mapsto X(\omega, t)$ is Borel measurable and hence *a fortiori* Lebesgue measurable; see (Billingsley, 1995, Chapter 3, Section 18, Theorem 18.1 (ii)).

If $\mathbf{s}$ happens to be Borel measurable, then Parts (ii)–(v) follow directly by Fubini's Theorem (Billingsley, 1995, Chapter 3, Section 18, Theorem 18.3) because in this case the mapping $(\omega, t) \mapsto X(\omega, t)\, s(t)$ is measurable (with respect to the product of $\mathcal{F}$ by the Borel $\sigma$-algebra on the real line) and because

$$\int_{-\infty}^{\infty} \mathsf{E}\big[\big|X(t)\, s(t)\big|\big]\, \mathrm{d}t = \int_{-\infty}^{\infty} \mathsf{E}\big[|X(t)|\big]\, |s(t)|\, \mathrm{d}t$$

$$\leq \sqrt{\mathsf{E}[X^2(0)]} \int_{-\infty}^{\infty} |s(t)|\, \mathrm{d}t$$

$$< \infty,$$

where the first inequality follows from (25.16), and where the second inequality follows from our assumption that $\mathbf{s}$ is integrable.

To prove Parts (i)–(v) for the case where $\mathbf{s}$ is Lebesgue measurable but not Borel measurable, recall that every Lebesgue measurable function is equal (except on a set of Lebesgue measure zero) to a Borel measurable function (Rudin, 1974, Chapter 7, Lemma 1), and note that the RHS of (25.50) and the mappings in (25.51) and (25.52) are unaltered when $\mathbf{s}$ is replaced with a function that is identical to it outside a set of Lebesgue measure zero.

We next prove Part (vi) under the assumption that $\big(X(t)\big)$ is centered. The more general case then follows from the argument leading to (25.49). To prove Part (vi) we need to justify the steps leading to (25.46). For the reader's convenience we repeat these steps here and then proceed to justify them.

$$\mathsf{Var}\left[\int_{-\infty}^{\infty} X(t)\, s(t)\, \mathrm{d}t\right] = \mathsf{E}\left[\left(\int_{-\infty}^{\infty} X(t)\, s(t)\, \mathrm{d}t\right)^2\right]$$

$$= \mathsf{E}\left[\left(\int_{-\infty}^{\infty} X(t)\, s(t)\, \mathrm{d}t\right)\left(\int_{-\infty}^{\infty} X(\tau)\, s(\tau)\, \mathrm{d}\tau\right)\right]$$

$$= \mathsf{E}\left[\int_{-\infty}^{\infty}\int_{-\infty}^{\infty} X(t)\,s(t)X(\tau)\,s(\tau)\,\mathrm{d}t\,\mathrm{d}\tau\right]$$

$$= \int_{-\infty}^{\infty}\int_{-\infty}^{\infty} s(t)\,s(\tau)\,\mathsf{E}[X(t)X(\tau)]\,\mathrm{d}t\,\mathrm{d}\tau$$

$$= \int_{-\infty}^{\infty}\int_{-\infty}^{\infty} s(t)\,\mathsf{K}_{XX}(t-\tau)\,s(\tau)\,\mathrm{d}t\,\mathrm{d}\tau.$$

The first equality holds because for centered processes, by Part (v), the RV on the LHS is of zero mean; the second follows by writing $a^2$ as $a$ times $a$; the third follows because for $\omega$'s satisfying $\int |X(\omega,t)\,s(t)|\,\mathrm{d}t < \infty$ we can use Fubini's Theorem to replace the iterated integrals with a double integral and because other $\omega$'s occur with zero probability and therefore do not influence the expectation; the fourth equality entails swapping the expectation with the integration over $\mathbb{R}^2$ and can be justified by Fubini's Theorem because, by (25.17),

$$\int_{-\infty}^{\infty}\int_{-\infty}^{\infty} |s(t)\,s(\tau)|\,\mathsf{E}\big[|X(t)X(\tau)|\big]\,\mathrm{d}t\,\mathrm{d}\tau \leq \mathsf{K}_{XX}(0)\int_{-\infty}^{\infty}\int_{-\infty}^{\infty} |s(t)|\,|s(\tau)|\,\mathrm{d}t\,\mathrm{d}\tau$$

$$= \mathsf{K}_{XX}(0)\,\|\mathbf{s}\|_1^2$$

$$< \infty;$$

and the final equality follows from the definition of the autocovariance function (Definition 25.4.4).

Having derived (25.53) we can derive (25.54) by following the steps leading to (25.47). The only issue that needs clarification is the justification for replacing the integral over $\mathbb{R}^2$ with the iterated integrals. This is justified using Fubini's Theorem by noting that, by (25.15), $|\mathsf{K}_{XX}(\sigma)| \leq \mathsf{K}_{XX}(0)$ and that $\mathbf{s}$ is integrable:

$$\int_{-\infty}^{\infty} |s(\tau)| \int_{-\infty}^{\infty} |s(\sigma+\tau)\,\mathsf{K}_{XX}(\sigma)|\,\mathrm{d}\sigma\,\mathrm{d}\tau \leq \mathsf{K}_{XX}(0)\int_{-\infty}^{\infty} |s(\tau)|\int_{-\infty}^{\infty} |s(\sigma+\tau)|\,\mathrm{d}\sigma\,\mathrm{d}\tau$$

$$= \mathsf{K}_{XX}(0)\,\|\mathbf{s}\|_1^2$$

$$< \infty.$$

Finally, Part (vii) follows from (25.54) and from Proposition 6.2.4 by noting that, by (11.34) & (11.35), $\mathsf{R}_{\mathsf{ss}}$ is integrable and of FT

$$\hat{\mathsf{R}}_{\mathsf{ss}}(f) = |\hat{s}(f)|^2, \quad f \in \mathbb{R},$$

and that, by Definition 25.7.2, if $\mathsf{S}_{XX}$ is the PSD of $\big(X(t)\big)$, then $\mathsf{S}_{XX}$ is integrable and its IFT is $\mathsf{K}_{XX}$, i.e.,

$$\mathsf{K}_{XX}(\sigma) = \int_{-\infty}^{\infty} \mathsf{S}_{XX}(f)\,e^{\mathrm{i}2\pi f\sigma}\,\mathrm{d}f. \qquad \square$$

**Note 25.10.2.**

(i) In the future we shall sometimes write

$$\int_{-\infty}^{\infty} X(t)\,s(t)\,\mathrm{d}t$$

instead of the mathematically more explicit (25.52) & (25.50). Sometimes, however, we shall make the argument $\omega \in \Omega$ more explicit:

$$\left(\int_{-\infty}^{\infty} X(t)\, s(t)\, \mathrm{d}t\right)(\omega) = \begin{cases} \int_{-\infty}^{\infty} X(\omega, t)\, s(t)\, \mathrm{d}t & \text{if } \int_{-\infty}^{\infty} |X(\omega, t)\, s(t)|\, \mathrm{d}t < \infty, \\ 0 & \text{otherwise.} \end{cases}$$

(ii) If $\mathbf{s}_1$ and $\mathbf{s}_2$ are indistinguishable integrable real signals (Definition 2.5.2), then the random variables $\int_{-\infty}^{\infty} X(t) s_1(t)\, \mathrm{d}t$ and $\int_{-\infty}^{\infty} X(t) s_2(t)\, \mathrm{d}t$ are identical.

(iii) For every $\alpha \in \mathbb{R}$

$$\int_{-\infty}^{\infty} X(t)\bigl(\alpha\, s(t)\bigr)\, \mathrm{d}t = \alpha \int_{-\infty}^{\infty} X(t)\, s(t)\, \mathrm{d}t. \tag{25.56}$$

(iv) We caution the very careful readers that if $\mathbf{s}_1$ and $\mathbf{s}_2$ are integrable functions, then there may be some $\omega$'s in $\Omega$ for which the stochastic integral $\left(\int_{-\infty}^{\infty} X(t)\, (s_1(t) + s_2(t))\, \mathrm{d}t\right)(\omega)$ is not equal to the sum of the stochastic integrals $\left(\int_{-\infty}^{\infty} X(t)\, s_1(t)\, \mathrm{d}t\right)(\omega)$ and $\left(\int_{-\infty}^{\infty} X(t)\, s_2(t)\, \mathrm{d}t\right)(\omega)$. This can happen, for example, if the trajectory $t \mapsto X(\omega, t)$ corresponding to $\omega$ is such that either $\int |X(\omega, t)\, s_1(t)|\, \mathrm{d}t$ or $\int |X(\omega, t)\, s_2(t)|\, \mathrm{d}t$ is infinite, but not both. Fortunately, as we shall see in Lemma 25.10.3, such $\omega$'s occur with zero probability.

(v) The value that we have chosen to assign to the integral in (25.52) when $\omega$ is in $\mathcal{N}$ is immaterial. Such $\omega$'s occur with zero probability, so this value does not influence the distribution of the integral.[7]

**Lemma 25.10.3 ("Almost" Linearity of Stochastic Integration).** *Let $\bigl(X(t)\bigr)$ be a measurable WSS SP, let $\mathbf{s}_1, \ldots, \mathbf{s}_m \colon \mathbb{R} \to \mathbb{R}$ be integrable, and let $\gamma_1, \ldots, \gamma_m$ be real. Then the random variables*

$$\omega \mapsto \left(\int_{-\infty}^{\infty} X(t)\left(\sum_{j=1}^{m} \gamma_j\, s_j(t)\right) \mathrm{d}t\right)(\omega) \tag{25.57}$$

*and*

$$\omega \mapsto \sum_{j=1}^{m} \gamma_j\left(\left(\int_{-\infty}^{\infty} X(t)\, s_j(t)\, \mathrm{d}t\right)(\omega)\right) \tag{25.58}$$

*differ on at most a set of $\omega$'s of probability zero. In particular, the two random variables have the same distribution.*

**Note 25.10.4.** In view of this lemma we shall write, somewhat imprecisely,

$$\int_{-\infty}^{\infty} X(t)\bigl(\alpha_1\, s_1(t) + \alpha_2\, s_2(t)\bigr)\, \mathrm{d}t = \alpha_1 \int_{-\infty}^{\infty} X(t)\, s_1(t)\, \mathrm{d}t + \alpha_2 \int_{-\infty}^{\infty} X(t)\, s_2(t)\, \mathrm{d}t.$$

---

[7]The value zero is convenient because it guarantees that (25.56) holds even for $\omega$'s for which the mapping $t \mapsto X(\omega, t)\, s(t)$ is not integrable.

**Proof of Lemma 25.10.3.** Let $(\Omega, \mathcal{F}, P)$ be the probability space over which the SP $(X(t))$ is defined. Define the function

$$\mathbf{s}_0 \colon t \mapsto \sum_{j=1}^{m} \gamma_j \, s_j(t) \tag{25.59}$$

and the sets

$$\mathcal{N}_j = \left\{ \omega \in \Omega : \int_{-\infty}^{\infty} |X(\omega, t) \, s_j(t)| \, \mathrm{d}t = \infty \right\}, \quad j = 0, 1, \ldots, m.$$

By (25.59) and the Triangle Inequality (2.12)

$$|X(\omega, t) \, s_0(t)| \leq \sum_{j=1}^{m} |\gamma_j| \, |X(\omega, t) \, s_j(t)|, \quad \omega \in \Omega, \ t \in \mathbb{R},$$

which implies that

$$\mathcal{N}_0 \subseteq \bigcup_{j=1}^{m} \mathcal{N}_j.$$

By the Union Bound (or more specifically by Corollary 21.5.2 (i)), the set on the RHS is of probability zero. The proof is concluded by noting that, outside this set, the random variables (25.57) and (25.58) are identical. This follows because, for $\omega$'s outside this set, all the integrals are finite so linearity holds. □

## 25.11 Linear Functionals of Gaussian Processes

We continue our discussion of integrals of the form $\int X(t) \, s(t) \, \mathrm{d}t$, but this time with the additional assumption that $(X(t))$ is Gaussian. The main result of this section is Proposition 25.11.1, which states that, subject to some technical conditions, the result of this integral is a Gaussian RV. In fact, Proposition 25.11.1 is a bit more general and addresses expressions of the form

$$\int_{-\infty}^{\infty} X(t) \, s(t) \, \mathrm{d}t + \sum_{\nu=1}^{n} \alpha_\nu X(t_\nu), \tag{25.60}$$

where $(X(t))$ is a stationary Gaussian process, $\mathbf{s} \colon \mathbb{R} \to \mathbb{R}$ is integrable, $n$ is an arbitrary nonnegative integer, and the coefficients $\alpha_1, \ldots, \alpha_n \in \mathbb{R}$ and the epochs $t_1, \ldots, t_n \in \mathbb{R}$ are arbitrary. It shows that, subject to the additional technical condition that $(X(t))$ is measurable, the result of (25.60) is a Gaussian RV. Consequently, its distribution is fully specified by its mean and variance, which, as we shall see, can be easily computed from the autocovariance function $\mathsf{K}_{XX}$.

The proof of the Gaussianity of (25.60) (Proposition 25.11.1 ahead) is technical, so we encourage the reader to focus on the following heuristic argument. Suppose that the integral is a Riemann integral and that we can therefore approximate it with a finite sum

$$\int_{-\infty}^{\infty} X(t) \, s(t) \, \mathrm{d}t \approx \sum_{k=-\mathsf{K}}^{\mathsf{K}} \delta \, X(\delta k) \, s(\delta k)$$

for some large enough K and small enough $\delta > 0$. (Do not bother trying to sort out the exact sense in which this approximation holds. This is, after all, a heuristic argument.) Consequently, we can approximate (25.60) by

$$\int_{-\infty}^{\infty} X(t)\, s(t)\, \mathrm{d}t + \sum_{\nu=1}^{n} \alpha_\nu X(t_\nu) \approx \sum_{k=-K}^{K} \delta\, s(\delta k)\, X(\delta k) + \sum_{\nu=1}^{n} \alpha_\nu X(t_\nu). \qquad (25.61)$$

But the RHS of the above is just a linear combination of the random variables

$$X(-\mathsf{K}\delta), \ldots, X(\mathsf{K}\delta), X(t_1), \ldots, X(t_\nu),$$

which are jointly Gaussian because $\big(X(t)\big)$ is a Gaussian SP. Since a linear functional of jointly Gaussian random variables is Gaussian (Theorem 23.6.17), the RHS of (25.61) is Gaussian, thus making it plausible that its LHS is also Gaussian.

Before stating the main result of this section in a mathematically defensible way, we now proceed to compute the mean and variance of (25.60). We assume that $s(\cdot)$ is integrable and that $\big(X(t)\big)$ is measurable and WSS. (Gaussianity is inessential for the computation of the mean and variance.) The computation is very similar to the one leading to (25.45) and (25.46). For the mean we have:

$$\mathsf{E}\!\left[\int_{-\infty}^{\infty} X(t)\, s(t)\, \mathrm{d}t + \sum_{\nu=1}^{n} \alpha_\nu X(t_\nu)\right] = \mathsf{E}\!\left[\int_{-\infty}^{\infty} X(t)\, s(t)\, \mathrm{d}t\right] + \sum_{\nu=1}^{n} \alpha_\nu \mathsf{E}\big[X(t_\nu)\big]$$

$$= \mathsf{E}[X(0)]\left(\int_{-\infty}^{\infty} s(t)\, \mathrm{d}t + \sum_{\nu=1}^{n} \alpha_\nu\right), \qquad (25.62)$$

where the first equality follows from the linearity of expectation and where the second equality follows from (25.45) and from the wide-sense stationarity of $\big(X(t)\big)$, which implies that $\mathsf{E}[X(t)] = \mathsf{E}[X(0)]$, for all $t \in \mathbb{R}$.

For the purpose of computing the variance of (25.60), we assume that $\big(X(t)\big)$ is centered. The result continues to hold if $\big(X(t)\big)$ has a nonzero mean, because the mean of $\big(X(t)\big)$ does not influence the variance of (25.60). We begin by expanding the variance as

$$\mathsf{Var}\!\left[\int_{-\infty}^{\infty} X(t)\, s(t)\, \mathrm{d}t + \sum_{\nu=1}^{n} \alpha_\nu X(t_\nu)\right] = \mathsf{Var}\!\left[\int_{-\infty}^{\infty} X(t)\, s(t)\, \mathrm{d}t\right]$$

$$+ \mathsf{Var}\!\left[\sum_{\nu=1}^{n} \alpha_\nu X(t_\nu)\right] + 2\sum_{\nu=1}^{n} \alpha_\nu \mathsf{Cov}\!\left[\int_{-\infty}^{\infty} X(t)\, s(t)\, \mathrm{d}t, X(t_\nu)\right] \qquad (25.63)$$

and by noting that, by (25.47),

$$\mathsf{Var}\!\left[\int_{-\infty}^{\infty} X(t)\, s(t)\, \mathrm{d}t\right] = \int_{-\infty}^{\infty} \mathsf{K}_{XX}(\sigma)\, \mathsf{R}_{ss}(\sigma)\, \mathrm{d}\sigma \qquad (25.64)$$

and that, by (25.24),

$$\mathsf{Var}\!\left[\sum_{\nu=1}^{n} \alpha_\nu X(t_\nu)\right] = \sum_{\nu=1}^{n} \sum_{\nu'=1}^{n} \alpha_\nu \alpha_{\nu'}\, \mathsf{K}_{XX}(t_\nu - t_{\nu'}). \qquad (25.65)$$

To complete the computation of the variance of (25.60) it remains to compute the covariance in the last term in (25.63):

$$\mathsf{E}\left[X(t_\nu)\int_{-\infty}^{\infty} X(t)\,s(t)\,\mathrm{d}t\right] = \mathsf{E}\left[\int_{-\infty}^{\infty} X(t)X(t_\nu)\,s(t)\,\mathrm{d}t\right]$$

$$= \int_{-\infty}^{\infty} s(t)\,\mathsf{E}[X(t)X(t_\nu)]\,\mathrm{d}t$$

$$= \int_{-\infty}^{\infty} s(t)\,\mathsf{K}_{XX}(t - t_\nu)\,\mathrm{d}t. \qquad (25.66)$$

Combining (25.63) with (25.64)–(25.66) we obtain

$$\mathsf{Var}\left[\int_{-\infty}^{\infty} X(t)\,s(t)\,\mathrm{d}t + \sum_{\nu=1}^{n}\alpha_\nu X(t_\nu)\right] = \int_{-\infty}^{\infty}\mathsf{K}_{XX}(\sigma)\,\mathsf{R}_{\mathsf{ss}}(\sigma)\,\mathrm{d}\sigma$$

$$+ \sum_{\nu=1}^{n}\sum_{\nu'=1}^{n}\alpha_\nu\alpha_{\nu'}\,\mathsf{K}_{XX}(t_\nu - t_{\nu'}) + 2\sum_{\nu=1}^{n}\alpha_\nu\int_{-\infty}^{\infty} s(t)\,\mathsf{K}_{XX}(t - t_\nu)\,\mathrm{d}t. \qquad (25.67)$$

We now state the main result about linear functionals of Gaussian stochastic processes. The proof is recommended for mathematically-inclined readers only.

**Proposition 25.11.1 (Linear Functional of Stationary Gaussian Processes).** *Consider the setup of Proposition 25.10.1 with the additional assumption that the process $\big(X(t)\big)$ is Gaussian. Additionally introduce the coefficients $\alpha_1,\ldots,\alpha_n \in \mathbb{R}$ and the epochs $t_1,\ldots,t_n \in \mathbb{R}$ for some $n \in \mathbb{N}$. Then there exists an event $\mathcal{N} \in \mathcal{F}$ of zero probability such that for all $\omega \notin \mathcal{N}$ the mapping $t \mapsto X(\omega,t)\,s(t)$ is a Lebesgue integrable function:*

$$\Big(\text{the mapping } t \mapsto X(\omega,t)\,s(t) \text{ is in } \mathcal{L}_1\Big), \quad \omega \notin \mathcal{N}, \qquad (25.68\mathrm{a})$$

*and the mapping from $\Omega$ to $\mathbb{R}$*

$$\omega \mapsto \begin{cases} \displaystyle\int_{-\infty}^{\infty} X(\omega,t)\,s(t)\,\mathrm{d}t + \sum_{\nu=1}^{n}\alpha_\nu X(\omega,t_\nu) & \text{if } \omega \notin \mathcal{N}, \\ 0 & \text{otherwise} \end{cases} \qquad (25.68\mathrm{b})$$

*is a Gaussian RV whose mean and variance are given in (25.62) and (25.67).*

**Proof.** We prove this result when $\big(X(t)\big)$ is centered. The extension to the more general case follows by noting that adding a deterministic constant to a zero-mean Gaussian results in a Gaussian. We also assume that $s(\cdot)$ is Borel measurable, because once the theorem is established for this case it immediately also extends to the case where $s(\cdot)$ is only Lebesgue measurable by noting that every Lebesgue measurable function is equal almost everywhere to a Borel measurable function.

The existence of the event $\mathcal{N}$ and the fact that the mapping (25.68b) is a RV follow from Proposition 25.10.1. We next show that the RV

$$Y(\omega) \triangleq \begin{cases} \displaystyle\int_{-\infty}^{\infty} X(\omega,t)\,s(t)\,\mathrm{d}t + \sum_{\nu=1}^{n}\alpha_\nu X(\omega,t_\nu) & \text{if } \omega \notin \mathcal{N}, \\ 0 & \text{otherwise}, \end{cases} \qquad (25.69)$$

is Gaussian.

To that end, define for every $k \in \mathbb{N}$ the function

$$s_k(t) = \begin{cases} s(t) & \text{if } |t| \le k \text{ and } |s(t)| \le \sqrt{k}, \\ 0 & \text{otherwise.} \end{cases} \quad t \in \mathbb{R}. \tag{25.70}$$

Note that for every $\omega \in \Omega$

$$\lim_{k \to \infty} X(\omega, t) \, s_k(t) = X(\omega, t) \, s(t), \quad t \in \mathbb{R},$$

and

$$\left| X(\omega, t) \, s_k(t) \right| \le \left| X(\omega, t) \, s(t) \right|, \quad t \in \mathbb{R},$$

so, by the Dominated Convergence Theorem and (25.68a),

$$\lim_{k \to \infty} \int_{-\infty}^{\infty} X(\omega, t) \, s_k(t) \, dt = \int_{-\infty}^{\infty} X(\omega, t) \, s(t) \, dt, \quad \omega \notin \mathcal{N}. \tag{25.71}$$

Define now for every $k \in \mathbb{N}$ the RV

$$Y_k(\omega) = \begin{cases} \displaystyle\int_{-\infty}^{\infty} X(\omega, t) \, s_k(t) \, dt + \sum_{\nu=1}^{n} \alpha_\nu X(\omega, t_\nu) & \text{if } \omega \notin \mathcal{N}, \\ 0 & \text{otherwise.} \end{cases} \tag{25.72}$$

It follows from (25.71) that the sequence $Y_1, Y_2, \ldots$ converges almost surely to $Y$. To prove that $Y$ is Gaussian, it thus suffices to prove that for every $k \in \mathbb{N}$ the RV $Y_k$ is Gaussian (Theorem 19.9.1).

To prove that $Y_k$ is Gaussian, we begin by showing that it is of finite variance. To that end, it suffices to show that the RV

$$\tilde{Y}_k(\omega) \triangleq \begin{cases} \int_{-\infty}^{\infty} X(\omega, t) \, s_k(t) \, dt & \text{if } \omega \notin \mathcal{N}, \\ 0 & \text{otherwise} \end{cases} \tag{25.73}$$

is of finite variance. We prove this by using the definition of $s_k(\cdot)$ (25.70) and by using the Cauchy-Schwarz Inequality to show that for every $\omega \notin \mathcal{N}$

$$\tilde{Y}_k^2(\omega) = \left( \int_{-\infty}^{\infty} X(\omega, t) \, s_k(t) \, dt \right)^2$$

$$= \left( \int_{-k}^{k} X(\omega, t) \, s_k(t) \, dt \right)^2$$

$$\le \int_{-k}^{k} X^2(\omega, t) \, dt \int_{-k}^{k} s_k^2(t) \, dt$$

$$\le \left( \int_{-k}^{k} X^2(\omega, t) \, dt \right) 2k^2,$$

where the equality in the first line follows from the definition of $\tilde{Y}_k$ (25.73); the equality in the second line from the definition of $s_k(\cdot)$ (25.70); the inequality in the third line from the Cauchy-Schwarz Inequality; and the final inequality again by (25.70). Since $\mathcal{N}$ is an event of probability zero, it follows from this inequality that

$$\mathsf{E}\left[\tilde{Y}_k^2\right] \leq 4k^3 \, \mathsf{K}_{XX}(0) < \infty,$$

thus establishing that $\tilde{Y}_k$, and hence also $Y_k$, is of finite variance.

To prove that $Y_k$ is Gaussian we shall use some results about the Hilbert space $L^2(\Omega, \mathcal{F}, P)$ of (the equivalence classes of) the random variables that are defined over $(\Omega, \mathcal{F}, P)$ and that have a finite second moment; see, for example, (Shiryaev, 1996, Chapter II, Section 11). Let $\mathcal{G}$ denote the closed linear subspace of $L^2(\Omega, \mathcal{F}, P)$ that is generated by the random variables $\big(X(t), \ t \in \mathbb{R}\big)$. Thus, $\mathcal{G}$ contains all finite linear combinations of the random variables $\big(X(t), \ t \in \mathbb{R}\big)$ as well as the mean-square limits of such linear combinations. Since the process $\big(X(t), \ t \in \mathbb{R}\big)$ is Gaussian, it follows that all such linear combinations are Gaussian. And since mean-square limits of Gaussian random variables are Gaussian (Theorem 19.9.1), it follows that $\mathcal{G}$ contains only random variables that have a Gaussian distribution (Shiryaev, 1996, Chapter II, Section 13, Paragraph 6). To prove that $Y_k$ is Gaussian it thus suffices to show that it is an element of $\mathcal{G}$.

To prove that $Y_k$ is an element of $\mathcal{G}$, decompose $Y_k$ as

$$Y_k = Y_k^{\mathcal{G}} + Y_k^{\perp}, \tag{25.74}$$

where $Y_k^{\mathcal{G}}$ is the projection of $Y_k$ onto $\mathcal{G}$ and where $Y_k^{\perp}$ is consequently perpendicular to every element of $\mathcal{G}$ and *a fortiori* to all the random variables $\big(X(t), \ t \in \mathbb{R}\big)$:

$$\mathsf{E}\left[X(t)Y_k^{\perp}\right] = 0, \quad t \in \mathbb{R}. \tag{25.75}$$

Since $Y_k$ is of finite variance, this decomposition is possible and

$$\mathsf{E}\left[\left(Y_k^{\mathcal{G}}\right)^2\right], \mathsf{E}\left[\left(Y_k^{\perp}\right)^2\right] < \infty. \tag{25.76}$$

To prove that $Y_k$ is an element of $\mathcal{G}$ we shall next show that $\mathsf{E}\left[\left(Y_k^{\perp}\right)^2\right] = 0$ or, equivalently (in view of (25.74)), that

$$\mathsf{E}\left[Y_k Y_k^{\perp}\right] = 0. \tag{25.77}$$

To establish (25.77) we evaluate its LHS as follows:

$$\mathsf{E}[Y_k Y_k^{\perp}] = \mathsf{E}\left[\left(\int_{-\infty}^{\infty} X(t)\,s_k(t)\,\mathrm{d}t + \sum_{\nu=1}^{n} \alpha_{\nu} X(t_{\nu})\right) Y_k^{\perp}\right]$$

$$= \mathsf{E}\left[\left(\int_{-\infty}^{\infty} X(t)\,s_k(t)\,\mathrm{d}t\right) Y_k^{\perp}\right] + \sum_{\nu=1}^{n} \alpha_{\nu} \underbrace{\mathsf{E}\left[X(t_{\nu})Y_k^{\perp}\right]}_{0}$$

$$= \mathsf{E}\left[\left(\int_{-\infty}^{\infty} X(t)\,s_k(t)\,\mathrm{d}t\right) Y_k^{\perp}\right]$$

$$= \int_{-\infty}^{\infty} \mathsf{E}\left[X(t)\,s_k(t)Y_k^{\perp}\right]\,\mathrm{d}t$$

$$= \int_{-\infty}^{\infty} \underbrace{\mathsf{E}\left[X(t)Y_k^{\perp}\right]}_{0} s_k(t)\, \mathrm{d}t$$

$$= 0,$$

where the first equality follows from the definition of $Y_k$ (25.72); the second from the linearity of expectation; the third from the orthogonality (25.75); the fourth by an application of Fubini's Theorem that we shall justify shortly; the fifth because $s_k(\cdot)$ is a deterministic function; and the final equality again by (25.75). This establishes (25.77) subject to a verification that the conditions of Fubini's Theorem are satisfied, a verification we conduct now. That $(\omega, t) \mapsto X(\omega, t)Y_k^{\perp}(\omega)\, s_k(t)$ is measurable follows because $\bigl(X(t),\ t \in \mathbb{R}\bigr)$ is a measurable SP; $Y_k^{\perp}$, being a RV, is measurable with respect to $\mathcal{F}$; and because the Borel measurability of $s(\cdot)$ also implies the Borel measurability of $s_k(\cdot)$. The integrability of this function follows from the Cauchy-Schwarz Inequality for random variables

$$\int_{-\infty}^{\infty} \mathsf{E}\left[\left|X(t)Y_k^{\perp}\right|\right]|s_k(t)|\, \mathrm{d}t \leq \int_{-\infty}^{\infty} \sqrt{\mathsf{E}[X^2(t)]}\sqrt{\mathsf{E}\left[\left(Y_k^{\perp}\right)^2\right]}|s_k(t)|\, \mathrm{d}t$$

$$\leq \sqrt{\mathsf{K}_{XX}(0)}\sqrt{\mathsf{E}\left[\left(Y_k^{\perp}\right)^2\right]}\, 2k\, \sqrt{k}$$

$$< \infty,$$

where the second inequality follows from the definition of $s_k(\cdot)$ (25.70), and where the third inequality follows from (25.76). This justifies the use of Fubini's Theorem in the proof of (25.77). We have thus demonstrated that $Y_k$ is in $\mathcal{G}$, and hence, like all elements of $\mathcal{G}$, is Gaussian. This concludes the proof of the Gaussianity of $Y_k$ for every $k \in \mathbb{N}$ and hence the Gaussianity of $Y$.

It only remains to verify that the mean and variance of $Y$ are as stated in the theorem. The only part of the derivation of (25.67) that we have not yet justified is the derivation of (25.66) and, in particular, the swapping of the expectation and integration. But this is easily justified using Fubini's Theorem because, by (25.17),

$$\int_{-\infty}^{\infty} \mathsf{E}\left[|X(t_\nu)X(t)|\right]|s(t)|\, \mathrm{d}t \leq \bigl(\mathsf{K}_{XX}(0) + \mathsf{E}[X(0)]^2\bigr)\, \|\mathsf{s}\|_1 < \infty. \tag{25.78}$$

$\square$

Proposition 25.11.1 is extremely powerful because it allows us to determine the distribution of a linear functional of a Gaussian SP from its mean and variance. In the next section we shall extend this result and show that any finite number of linear functionals of a Gaussian SP are jointly Gaussian. Their joint distribution is thus fully determined by the mean vector and the covariance matrix, which, as we shall see, can be readily computed from the autocovariance function.

## 25.12 The Joint Distribution of Linear Functionals

Let us now shift our focus from the distribution of a single linear functional to the joint distribution of a collection of such functionals. Specifically, we consider $m$

functionals

$$\int_{-\infty}^{\infty} X(t)\, s_j(t)\, \mathrm{d}t + \sum_{\nu=1}^{n_j} \alpha_{j,\nu} X\big(t_{j,\nu}\big), \quad j = 1,\ldots,m \tag{25.79}$$

of the measurable, stationary Gaussian SP $\big(X(t)\big)$. Here the $m$ real-valued signals $\mathbf{s}_1,\ldots,\mathbf{s}_m$ are integrable, $n_1,\ldots,n_m$ are in $\mathbb{N}$, and $\alpha_{j,\nu}, t_{j,\nu}$ are deterministic constants for all $\nu \in \{1,\ldots,n_j\}$.

The main result of this section is that if $\big(X(t)\big)$ is a Gaussian SP, then the random variables in (25.79) are jointly Gaussian.

**Theorem 25.12.1 (Linear Functionals of a Gaussian SP Are Jointly Gaussian).** *The $m$ linear functionals*

$$\int_{-\infty}^{\infty} X(t)\, s_j(t)\, \mathrm{d}t + \sum_{\nu=1}^{n_j} \alpha_{j,\nu} X\big(t_{j,\nu}\big), \quad j = 1,\ldots,m$$

*of a measurable, stationary, Gaussian SP $\big(X(t)\big)$ are jointly Gaussian, whenever $m \in \mathbb{N}$; the $m$ functions $\{\mathbf{s}_j\}_{j=1}^m$ are integrable functions from $\mathbb{R}$ to $\mathbb{R}$; the integers $\{n_j\}$ are nonnegative; and the coefficients $\{\alpha_{j,\nu}\}$ and the epochs $\{t_{j,\nu}\}$ are deterministic real numbers for all $j \in \{1,\ldots,m\}$ and all $\nu \in \{1,\ldots,n_j\}$.*

**Proof.** It suffices to show that any linear combination of these linear functionals has a univariate Gaussian distribution (Theorem 23.6.17). This follows from Proposition 25.11.1 and Lemma 25.10.3 because, by Lemma 25.10.3, for any choice of the coefficients $\gamma_1,\ldots,\gamma_m \in \mathbb{R}$ the linear combination

$$\gamma_1\left(\int_{-\infty}^{\infty} X(t)\, s_1(t)\, \mathrm{d}t + \sum_{\nu=1}^{n_1} \alpha_{1,\nu} X\big(t_{1,\nu}\big)\right) + \cdots$$

$$+ \gamma_m\left(\int_{-\infty}^{\infty} X(t)\, s_m(t)\, \mathrm{d}t + \sum_{\nu=1}^{n_m} \alpha_{m,\nu} X\big(t_{m,\nu}\big)\right)$$

has the same distribution as the linear functional

$$\int_{-\infty}^{\infty} X(t)\left(\sum_{j=1}^m \gamma_j\, s_j(t)\right) \mathrm{d}t + \sum_{j=1}^m \sum_{\nu=1}^{n_j} \gamma_j \alpha_{j,\nu} X\big(t_{j,\nu}\big),$$

which, by Proposition 25.11.1, has a univariate Gaussian distribution. $\quad\square$

It follows from Theorem 25.12.1 that if $\big(X(t)\big)$ is a measurable, stationary, Gaussian SP, then the joint distribution of the random variables in (25.79) is fully specified by their means and their covariance matrix. If $\big(X(t)\big)$ is centered, then by (25.62) these random variables are centered, so their joint distribution is determined by their covariance matrix. We next show how this covariance matrix can be computed from the autocovariance function $\mathsf{K}_{XX}$. To this end we assume that $\big(X(t)\big)$ is centered, and expand the covariance between any two such functionals as follows:

$$
\mathsf{Cov}\left[\int_{-\infty}^{\infty} X(t)\,s_j(t)\,\mathrm{d}t + \sum_{\nu=1}^{n_j} \alpha_{j,\nu} X(t_{j,\nu}),\ \int_{-\infty}^{\infty} X(t)\,s_k(t)\,\mathrm{d}t + \sum_{\nu'=1}^{n_k} \alpha_{k,\nu'} X(t_{k,\nu'})\right]
$$

$$
= \mathsf{Cov}\left[\int_{-\infty}^{\infty} X(t)\,s_j(t)\,\mathrm{d}t,\ \int_{-\infty}^{\infty} X(t)\,s_k(t)\,\mathrm{d}t\right]
$$

$$
+ \sum_{\nu=1}^{n_j} \alpha_{j,\nu} \mathsf{Cov}\left[X(t_{j,\nu}),\ \int_{-\infty}^{\infty} X(t)\,s_k(t)\,\mathrm{d}t\right]
$$

$$
+ \sum_{\nu'=1}^{n_k} \alpha_{k,\nu'} \mathsf{Cov}\left[X(t_{k,\nu'}),\ \int_{-\infty}^{\infty} X(t)\,s_j(t)\,\mathrm{d}t\right]
$$

$$
+ \sum_{\nu=1}^{n_j} \sum_{\nu'=1}^{n_k} \alpha_{j,\nu}\alpha_{k,\nu'} \mathsf{Cov}\left[X(t_{j,\nu}), X(t_{k,\nu'})\right], \quad j,k \in \{1,\dots,m\}. \quad (25.80)
$$

The second and third terms on the RHS can be computed from the autocovariance function $\mathsf{K}_{XX}$ using (25.66). The fourth term can be computed from $\mathsf{K}_{XX}$ by noting that $\mathsf{Cov}[X(t_{j,\nu}), X(t_{k,\nu'})] = \mathsf{K}_{XX}(t_{j,\nu} - t_{k,\nu'})$ (Definition 25.4.4). We now evaluate the first term:

$$
\mathsf{Cov}\left[\int_{-\infty}^{\infty} X(t)\,s_j(t)\,\mathrm{d}t,\ \int_{-\infty}^{\infty} X(t)\,s_k(t)\,\mathrm{d}t\right]
$$

$$
= \mathsf{E}\left[\int_{-\infty}^{\infty} X(t)\,s_j(t)\,\mathrm{d}t \int_{-\infty}^{\infty} X(\tau)\,s_k(\tau)\,\mathrm{d}\tau\right]
$$

$$
= \mathsf{E}\left[\int_{-\infty}^{\infty}\int_{-\infty}^{\infty} X(t)\,s_j(t)\,X(\tau)\,s_k(\tau)\,\mathrm{d}t\,\mathrm{d}\tau\right]
$$

$$
= \int_{-\infty}^{\infty}\int_{-\infty}^{\infty} \mathsf{E}[X(t)\,X(\tau)]\,s_j(t)\,s_k(\tau)\,\mathrm{d}t\,\mathrm{d}\tau
$$

$$
= \int_{-\infty}^{\infty}\int_{-\infty}^{\infty} \mathsf{K}_{XX}(t-\tau)\,s_j(t)\,s_k(\tau)\,\mathrm{d}t\,\mathrm{d}\tau, \quad\quad (25.81)
$$

which is the generalization of (25.53). By changing variables from $(t,\tau)$ to $(t,\sigma)$, where $\sigma \triangleq t-\tau$, we can obtain the generalization of (25.54). Starting from (25.81)

$$
\mathsf{Cov}\left[\int_{-\infty}^{\infty} X(t)\,s_j(t)\,\mathrm{d}t,\ \int_{-\infty}^{\infty} X(t)\,s_k(t)\,\mathrm{d}t\right] = \int_{-\infty}^{\infty}\int_{-\infty}^{\infty} \mathsf{K}_{XX}(t-\tau)\,s_j(t)\,s_k(\tau)\,\mathrm{d}t\,\mathrm{d}\tau
$$

$$
= \int_{-\infty}^{\infty} \mathsf{K}_{XX}(\sigma) \int_{-\infty}^{\infty} s_j(t)\,s_k(t-\sigma)\,\mathrm{d}t\,\mathrm{d}\sigma
$$

$$
= \int_{-\infty}^{\infty} \mathsf{K}_{XX}(\sigma) \int_{-\infty}^{\infty} s_j(t)\,\overleftarrow{s}_k(\sigma - t)\,\mathrm{d}t\,\mathrm{d}\sigma
$$

$$
= \int_{-\infty}^{\infty} \mathsf{K}_{XX}(\sigma)\,\big(\mathbf{s}_j \star \overleftarrow{\mathbf{s}}_k\big)(\sigma)\,\mathrm{d}\sigma. \quad (25.82)
$$

If $\big(X(t)\big)$ is of PSD $\mathsf{S}_{XX}$, then we can rewrite (25.82) in the frequency domain using Proposition 6.2.4 in much the same way that we rewrote (25.46) in the form (25.48):

$$
\mathsf{Cov}\left[\int_{-\infty}^{\infty} X(t)\,s_j(t)\,\mathrm{d}t,\ \int_{-\infty}^{\infty} X(t)\,s_k(t)\,\mathrm{d}t\right] = \int_{-\infty}^{\infty} \mathsf{S}_{XX}(f)\,\hat{s}_j(f)\,\hat{s}_k^*(f)\,\mathrm{d}f, \quad (25.83)
$$

where we have used the fact that the FT of $\mathbf{s}_j \star \bar{\mathbf{s}}_k$ is the product of the FT of $\mathbf{s}_j$ and the FT of $\bar{\mathbf{s}}_k$, and that the FT of $\bar{\mathbf{s}}_k$ is $f \mapsto \hat{s}_k(-f)$, which, because $\mathbf{s}_k$ is real, is also given by $f \mapsto \hat{s}_k^*(f)$.

The key second-order properties of linear functionals of measurable WSS stochastic processes are summarized in the following theorem. Using these properties and (25.80) we can compute the covariance matrix of the linear functionals in (25.79), a matrix which fully specifies their joint distribution whenever $(X(t))$ is a centered Gaussian SP.

**Theorem 25.12.2 (Covariance Properties of Linear Functionals of a WSS SP).**
*Let $(X(t))$ be a measurable WSS SP.*

*(i) If the real signal $\mathbf{s}$ is integrable, then*

$$\mathsf{Var}\left[\int_{-\infty}^{\infty} X(t)\, s(t)\, \mathrm{d}t\right] = \int_{-\infty}^{\infty} \mathsf{K}_{XX}(\sigma)\, \mathsf{R}_{\mathbf{ss}}(\sigma)\, \mathrm{d}\sigma, \qquad (25.84)$$

*where $\mathsf{R}_{\mathbf{ss}}$ is the self-similarity function of $\mathbf{s}$. Furthermore, for every fixed epoch $\tau \in \mathbb{R}$*

$$\mathsf{Cov}\left[\int_{-\infty}^{\infty} X(t)\, s(t)\, \mathrm{d}t, X(\tau)\right] = \int_{-\infty}^{\infty} s(t)\, \mathsf{K}_{XX}(\tau - t)\, \mathrm{d}t, \quad \tau \in \mathbb{R}. \quad (25.85)$$

*If $\mathbf{s}_1, \mathbf{s}_2$ are real-valued integrable signals, then*

$$\mathsf{Cov}\left[\int_{-\infty}^{\infty} X(t)\, s_1(t)\, \mathrm{d}t, \int_{-\infty}^{\infty} X(t)\, s_2(t)\, \mathrm{d}t\right] = \int_{-\infty}^{\infty} \mathsf{K}_{XX}(\sigma)\, (\mathbf{s}_1 \star \bar{\mathbf{s}}_2)(\sigma)\, \mathrm{d}\sigma. \quad (25.86)$$

*(ii) If $(X(t))$ is of PSD $\mathsf{S}_{XX}$, then for $\mathbf{s}$, $\mathbf{s}_1$, $\mathbf{s}_2$, and $\tau$ as above*

$$\mathsf{Var}\left[\int_{-\infty}^{\infty} X(t)\, s(t)\, \mathrm{d}t\right] = \int_{-\infty}^{\infty} \mathsf{S}_{XX}(f)\, |\hat{s}(f)|^2\, \mathrm{d}f, \qquad (25.87)$$

$$\mathsf{Cov}\left[\int_{-\infty}^{\infty} X(t)\, s(t)\, \mathrm{d}t, X(\tau)\right] = \int_{-\infty}^{\infty} \mathsf{S}_{XX}(f)\, \hat{s}(f)\, e^{\mathrm{i}2\pi f \tau}\, \mathrm{d}f, \qquad (25.88)$$

*and*

$$\mathsf{Cov}\left[\int_{-\infty}^{\infty} X(t)\, s_1(t)\, \mathrm{d}t, \int_{-\infty}^{\infty} X(t)\, s_2(t)\, \mathrm{d}t\right] = \int_{-\infty}^{\infty} \mathsf{S}_{XX}(f)\, \hat{s}_1(f)\, \hat{s}_2^*(f)\, \mathrm{d}f. \qquad (25.89)$$

**Proof.** Most of these claims have already been proved. Indeed, (25.84) was proved in Proposition 25.10.1 (vi), and (25.85) was proved in Proposition 25.11.1 using Fubini's Theorem and (25.78). However, (25.86) was only derived heuristically in (25.81) and (25.82). To rigorously justify this derivation one can use Fubini's Theorem, or use the relation

$$\mathsf{Cov}[X, Y] = \frac{1}{2}\left(\mathsf{Var}[X + Y] - \mathsf{Var}[X] - \mathsf{Var}[Y]\right)$$

and the result for the variance, namely, (25.84).

All the results in Part (ii) of this theorem follow from the corresponding results in Part (i) using the definition of the PSD and Proposition 6.2.4. $\qquad\square$

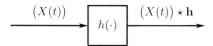

**Figure 25.2:** Passing a SP $(X(t))$ through a stable filter of impulse response $h(\cdot)$. If $(X(t))$ is measurable and WSS, then so is the output $(X(t)) \star \mathbf{h}$. If, additionally, $(X(t))$ is of PSD $\mathsf{S}_{XX}$, then the output is of PSD $f \mapsto \mathsf{S}_{XX}(f) |\hat{h}(f)|^2$. If $(X(t))$ is additionally Gaussian, then so is the output.

## 25.13 Filtering WSS Processes

We next discuss the result of passing a WSS SP through a stable filter, i.e., the convolution of a SP with a deterministic integrable function. Our main result is that, subject to some technical conditions, the following hold:

(i) Passing a WSS SP through a stable filter produces a WSS SP.

(ii) If the input to the filter is of PSD $\mathsf{S}_{XX}$, then the output of the filter is of PSD $f \mapsto \mathsf{S}_{XX}(f) |\hat{h}(f)|^2$, where $\hat{h}(\cdot)$ is the filter's frequency response.

(iii) If the input to the filter is a Gaussian SP, then so is the output.

We state this result in Theorem 25.13.2. But first we must define the convolution of a SP with an integrable deterministic signal. Our approach is to build on our definition of linear functionals of WSS stochastic processes (Section 25.10) and to define the convolution of $(X(t))$ with $h(\cdot)$ as the SP that maps every epoch $t \in \mathbb{R}$ to the RV

$$\int_{-\infty}^{\infty} X(\sigma) h(t - \sigma) \, d\sigma,$$

where the above integral is the linear functional

$$\int_{-\infty}^{\infty} X(\sigma) s(\sigma) \, d\sigma \quad \text{with} \quad \mathbf{s} \colon \sigma \mapsto h(t - \sigma).$$

With this approach the key results will follow by applying Theorem 25.12.2 with the proper substitutions.

**Definition 25.13.1 (Filtering a Stochastic Process).** *The convolution of a measurable, WSS SP $(X(t))$ with an integrable function $\mathbf{h} \colon \mathbb{R} \to \mathbb{R}$ is denoted by*

$$(X(t)) \star \mathbf{h}$$

*and is defined as the SP that maps every $t \in \mathbb{R}$ to the RV*

$$\int_{-\infty}^{\infty} X(\sigma) h(t - \sigma) \, d\sigma, \tag{25.90}$$

*where the stochastic integral in (25.90) is the stochastic integral that was defined in Note 25.10.2*

$$\big(X(t)\big)\star\mathbf{h}\colon (\omega, t) \mapsto \begin{cases} \displaystyle\int_{-\infty}^{\infty} X(\omega, \sigma)\, h(t-\sigma)\, \mathrm{d}\sigma & \text{if } \int_{-\infty}^{\infty} \big|X(\omega, \sigma)\, h(t-\sigma)\big|\, \mathrm{d}\sigma < \infty, \\ 0 & \text{otherwise.} \end{cases}$$

**Theorem 25.13.2.** *Let $\big(Y(t)\big)$ be the result of convolving the measurable, centered, WSS SP $\big(X(t)\big)$ of autocovariance function $\mathsf{K}_{XX}$ with the integrable function $\mathbf{h}\colon \mathbb{R} \to \mathbb{R}$.*

(i) *The SP $\big(Y(t)\big)$ is centered, measurable, and WSS with autocovariance function*

$$\mathsf{K}_{YY} = \mathsf{K}_{XX} \star \mathsf{R}_{\mathbf{hh}}, \tag{25.91}$$

*where $\mathsf{R}_{\mathbf{hh}}$ is the self-similarity function of $\mathbf{h}$ (Section 11.4).*

(ii) *If $\big(X(t)\big)$ is of PSD $\mathsf{S}_{XX}$, then $\big(Y(t)\big)$ is of PSD*

$$\mathsf{S}_{YY}(f) = \big|\hat{h}(f)\big|^2 \mathsf{S}_{XX}(f), \quad f \in \mathbb{R}. \tag{25.92}$$

(iii) *For every $t, \tau \in \mathbb{R}$,*

$$\mathsf{E}\big[X(t)\, Y(t+\tau)\big] = \big(\mathsf{K}_{XX} \star \mathbf{h}\big)(\tau), \tag{25.93}$$

*where the RHS does not depend on $t$.[8]*

(iv) *If $\big(X(t)\big)$ is Gaussian, then so is $\big(Y(t)\big)$. Moreover, for every choice of $n, m \in \mathbb{N}$ and for every choice of the epochs $t_1, \ldots, t_n, t_{n+1}, \ldots, t_{n+m} \in \mathbb{R}$, the random variables*

$$X(t_1), \ldots, X(t_n), Y(t_{n+1}), \ldots, Y(t_{n+m}) \tag{25.94}$$

*are jointly Gaussian.[9]*

**Proof.** For fixed $t, \tau \in \mathbb{R}$ we use Definition 25.13.1 to express $Y(t)$ and $Y(t+\tau)$ as

$$Y(t) = \int_{-\infty}^{\infty} X(\sigma)\, s_1(\sigma)\, \mathrm{d}\sigma, \tag{25.95}$$

and

$$Y(t+\tau) = \int_{-\infty}^{\infty} X(\sigma)\, s_2(\sigma)\, \mathrm{d}\sigma, \tag{25.96}$$

where

$$s_1\colon \sigma \mapsto h(t-\sigma), \tag{25.97}$$

$$s_2\colon \sigma \mapsto h(t+\tau-\sigma). \tag{25.98}$$

---

[8]Two stochastic processes $\big(X(t)\big)$ and $\big(Y(t)\big)$ are said to be **jointly wide-sense stationary** if each is WSS and if $\mathsf{E}[X(t)Y(t+\tau)]$ does not depend on $t$.

[9]That is, $\big(X(t)\big)$ and $\big(Y(t)\big)$ are **jointly Gaussian processes**.

We are now ready to prove Part (i). That $(Y(t))$ is centered follows from the representation of $Y(t)$ in (25.95) & (25.97) as a linear functional of $(X(t))$ and from the hypothesis that $(X(t))$ is centered (Proposition 25.10.1).

To establish that $(Y(t))$ is WSS we use the representations (25.95)–(25.98) and Theorem 25.12.2 regarding the covariance between two linear functionals as follows.

$$\mathsf{Cov}\big[Y(t+\tau),Y(t)\big] = \mathsf{Cov}\left[\int_{-\infty}^{\infty} X(\sigma)\,s_2(\sigma)\,\mathrm{d}\sigma,\ \int_{-\infty}^{\infty} X(\sigma)\,s_1(\sigma)\,\mathrm{d}\sigma\right]$$

$$= \int_{-\infty}^{\infty} \mathsf{K}_{XX}(\sigma)\,\big(\mathbf{s}_2 \star \bar{\mathbf{s}}_1\big)(\sigma)\,\mathrm{d}\sigma, \tag{25.99}$$

where the convolution can be evaluated as

$$\big(\mathbf{s}_2 \star \bar{\mathbf{s}}_1\big)(\sigma) = \int_{-\infty}^{\infty} s_2(\mu)\,\bar{s}_1(\sigma - \mu)\,\mathrm{d}\mu$$

$$= \int_{-\infty}^{\infty} h(t+\tau-\mu)\,h(t+\sigma-\mu)\,\mathrm{d}\mu$$

$$= \int_{-\infty}^{\infty} h(\tilde{\mu}+\tau-\sigma)\,h(\tilde{\mu})\,\mathrm{d}\tilde{\mu}$$

$$= \mathsf{R}_{\mathsf{hh}}(\tau-\sigma), \tag{25.100}$$

where $\tilde{\mu} \triangleq t+\sigma-\mu$. Combining (25.99) with (25.100) yields

$$\mathsf{Cov}\big[Y(t+\tau),Y(t)\big] = \big(\mathsf{K}_{XX} \star \mathsf{R}_{\mathsf{hh}}\big)(\tau), \quad t,\tau \in \mathbb{R}, \tag{25.101}$$

where the RHS does not depend on $t$. This establishes that $(Y(t))$ is WSS and proves (25.91).[10]

To conclude the proof of Part (i) we now show that $(Y(t))$ is measurable. The proof is technical and requires background in Measure Theory. Readers are encouraged to skip it and move on to the proof of Part (ii).

We first note that, as in the proof of Proposition 25.10.1, it suffices to prove the result for impulse response functions $\mathbf{h}$ that are Borel measurable; the extension to Lebesgue measurable functions will then follow by approximating $\mathbf{h}$ by a Borel measurable function that differs from it on a set of Lebesgue measure zero (Rudin, 1974, Chapter 7, Lemma 1) and by then applying Part (ii) of Note 25.10.2. We hence now assume that $\mathbf{h}$ is Borel measurable.

We shall prove that $(Y(t))$ is measurable by proving that the (nonstationary) process $(\omega,t) \mapsto Y(\omega,t)/(1+t^2)$ is measurable. This we shall prove using Fubini's Theorem applied to the function from $(\Omega \times \mathbb{R}) \times \mathbb{R}$ to $\mathbb{R}$ defined by

$$\big((\omega,t),\sigma\big) \mapsto \frac{X(\omega,\sigma)\,h(t-\sigma)}{1+t^2}, \quad \big((\omega,t) \in \Omega \times \mathbb{R},\ \sigma \in \mathbb{R}\big). \tag{25.102}$$

This function is measurable because, by assumption, $(X(t))$ is measurable and because the measurability of the function $h(\cdot)$ implies the measurability of the

---

[10]That $(Y(t))$ is of finite variance follows from (25.101) by setting $\tau = 0$ and noting that the convolution on the RHS of (25.101) is between a bounded function ($\mathsf{K}_{XX}$) and an integrable function ($\mathsf{R}_{\mathsf{hh}}$) and is thus defined and finite at every $\tau \in \mathbb{R}$ and a fortiori at $\tau = 0$.

function $(t, \sigma) \mapsto h(t - \sigma)$ (as proved, for example, in (Rudin, 1974, p. 157)). We next verify that this function is integrable. To that end, we first integrate its absolute value over $(\omega, t)$ and then over $\sigma$. The integral over $(\omega, t)$ is given by

$$\int_{t=-\infty}^{\infty} \frac{\mathsf{E}[|X(\sigma)|] \, |h(t - \sigma)|}{1 + t^2} \, \mathrm{d}t \leq \sqrt{\mathsf{K}_{XX}(0)} \int_{t=-\infty}^{\infty} \frac{|h(t - \sigma)|}{1 + t^2} \, \mathrm{d}t,$$

where the inequality follows from (25.16) and from our assumption that $\big(X(t)\big)$ is centered. We next need to integrate the RHS over $\sigma$. Invoking Fubini's Theorem to exchange the order of integration over $t$ and $\sigma$ we obtain that the integral of the absolute value of the function defined in (25.102) is upper-bounded by

$$\int_{\sigma=-\infty}^{\infty} \sqrt{\mathsf{K}_{XX}(0)} \int_{t=-\infty}^{\infty} \frac{|h(t - \sigma)|}{1 + t^2} \, \mathrm{d}t \, \mathrm{d}\sigma = \sqrt{\mathsf{K}_{XX}(0)} \int_{t=-\infty}^{\infty} \int_{\sigma=-\infty}^{\infty} \frac{|h(t - \sigma)|}{1 + t^2} \, \mathrm{d}\sigma \, \mathrm{d}t$$

$$= \sqrt{\mathsf{K}_{XX}(0)} \int_{t=-\infty}^{\infty} \frac{\|\mathbf{h}\|_1}{1 + t^2} \, \mathrm{d}t$$

$$= \pi \sqrt{\mathsf{K}_{XX}(0)} \, \|\mathbf{h}\|_1$$

$$< \infty.$$

Having established that the function in (25.102) is measurable and integrable, we can now use Fubini's Theorem to deduce that its integral over $\sigma$ is measurable as a mapping of $(\omega, t)$, i.e., that the mapping

$$(\omega, t) \mapsto \int_{\sigma=-\infty}^{\infty} \frac{X(\omega, \sigma) \, h(t - \sigma)}{1 + t^2} \, \mathrm{d}\sigma \qquad (25.103)$$

is measurable. Since the RHS of (25.103) is $Y(\omega, t)/(1 + t^2)$, we conclude that the mapping $(\omega, t) \mapsto Y(\omega, t)/(1 + t^2)$ is measurable and hence also $(\omega, t) \mapsto Y(\omega, t)$.

We next prove Part (ii) using (25.91) and Proposition 6.2.5. Because $\mathbf{h}$ is integrable, its self-similarity function $\mathsf{R_{hh}}$ is integrable and of FT

$$\hat{\mathsf{R}}_{\mathbf{hh}}(f) = \big|\hat{h}(f)\big|^2, \quad f \in \mathbb{R} \qquad (25.104)$$

(Section 11.4). And since, by assumption, $\big(X(t)\big)$ is of PSD $\mathsf{S}_{XX}$, it follows that $\mathsf{S}_{XX}$ is integrable and that its IFT is $\mathsf{K}_{XX}$:

$$\mathsf{K}_{XX}(\tau) = \int_{-\infty}^{\infty} \mathsf{S}_{XX}(f) \, e^{\mathrm{i}2\pi f \tau} \, \mathrm{d}f, \quad \tau \in \mathbb{R}. \qquad (25.105)$$

Consequently, by Proposition 6.2.5,

$$\big(\mathsf{K}_{XX} \star \mathsf{R_{hh}}\big)(\tau) = \int_{-\infty}^{\infty} \big|\hat{h}(f)\big|^2 \mathsf{S}_{XX}(f) \, e^{\mathrm{i}2\pi f \tau} \, \mathrm{d}f, \quad \tau \in \mathbb{R}.$$

Combining this with (25.91) yields

$$\mathsf{K}_{YY}(\tau) = \int_{-\infty}^{\infty} \big|\hat{h}(f)\big|^2 \mathsf{S}_{XX}(f) \, e^{\mathrm{i}2\pi f \tau} \, \mathrm{d}f, \quad \tau \in \mathbb{R},$$

and thus establishes that the PSD of $\big(Y(t)\big)$ is as given in (25.92).

We next turn to Part (iii). To establish (25.93) we use the representation (25.96) & (25.98) and Theorem 25.12.2:

$$
\begin{aligned}
\mathsf{E}\big[X(t)Y(t+\tau)\big] &= \mathsf{Cov}\left[X(t), \int_{-\infty}^{\infty} X(\sigma)\, s_2(\sigma)\, \mathrm{d}\sigma\right] \\
&= \int_{-\infty}^{\infty} s_2(\sigma)\, \mathsf{K}_{XX}(t-\sigma)\, \mathrm{d}\sigma \\
&= \int_{-\infty}^{\infty} h(t+\tau-\sigma)\, \mathsf{K}_{XX}(t-\sigma)\, \mathrm{d}\sigma \\
&= \int_{-\infty}^{\infty} \mathsf{K}_{XX}(-\mu)\, h(\tau-\mu)\, \mathrm{d}\mu \\
&= \int_{-\infty}^{\infty} \mathsf{K}_{XX}(\mu)\, h(\tau-\mu)\, \mathrm{d}\mu \\
&= \big(\mathsf{K}_{XX} \star \mathbf{h}\big)(\tau), \quad \tau \in \mathbb{R},
\end{aligned}
$$

where $\mu \triangleq \sigma - t$, and where we have used the symmetry of the autocovariance function.

Finally, we prove Part (iv). The proof is a simple application of Theorem 25.12.1. To prove that $\big(Y(t)\big)$ is a Gaussian process we need to show that, for every positive integer $n$ and for every choice of the epochs $t_1, \ldots, t_n$, the random variables $Y(t_1), \ldots, Y(t_n)$ are jointly Gaussian. This follows directly from Theorem 25.12.1 because $Y(t_\nu)$ can be expressed as

$$
\begin{aligned}
Y(t_\nu) &= \int_{-\infty}^{\infty} X(\sigma)\, h(t_\nu - \sigma)\, \mathrm{d}\sigma \\
&= \int_{-\infty}^{\infty} X(\sigma)\, s_\nu(\sigma)\, \mathrm{d}\sigma, \quad \nu = 1, \ldots, n,
\end{aligned}
$$

where

$$
\mathbf{s}_\nu \colon \sigma \mapsto h(t_\nu - \sigma), \quad \nu = 1, \ldots, n
$$

are all integrable.

The joint Gaussianity of the random variables in (25.94) can also be deduced from Theorem 25.12.1. Indeed, $X(t_\nu)$ can be trivially expressed as the functional

$$
X(t_\nu) = \int_{-\infty}^{\infty} X(\sigma)\, s_\nu(\sigma)\, \mathrm{d}\sigma + \alpha_\nu X(t_\nu), \quad \nu = 1, \ldots, n
$$

when $\mathbf{s}_\nu$ is chosen to be the zero function and when $\alpha_\nu$ is chosen as 1, and $Y(t_\nu)$ can be similarly expressed as

$$
Y(t_\nu) = \int_{-\infty}^{\infty} X(\sigma)\, s_\nu(\sigma)\, \mathrm{d}\sigma + \alpha_\nu X(t_\nu), \quad \nu = n+1, \ldots, n+m
$$

when $\mathbf{s}_\nu \colon \sigma \mapsto h(t_\nu - \sigma)$ and $\alpha_\nu = 0$. $\qquad\square$

The mathematically astute reader may have noted that, in defining the result of passing a WSS SP through a stable filter of impulse response **h**, we did not preclude the possibility that for every $\omega$ there may be some epochs $t$ for which the mapping $\sigma \mapsto X(\omega, \sigma)\, h(t - \sigma)$ is not integrable. So far, we have only established that for every epoch $t$ the set $\mathcal{N}_t$ of $\omega$'s for which this mapping is not integrable is of probability zero.

We next show that if **h** is well-behaved in the sense that it is not only integrable but also satisfies

$$\int_{-\infty}^{\infty} h^2(t)\,(1 + t^2)\,\mathrm{d}t < \infty, \tag{25.106}$$

then whenever $\omega$ is outside some set $\mathcal{N} \subset \Omega$ of probability zero, the mapping $\sigma \mapsto X(\omega, \sigma)\, h(t - \sigma)$ is integrable *for all* $t \in \mathbb{R}$. Thus, for $\omega$'s outside this set of probability zero, we can think of the response of the filter as being the convolution of the trajectory $t \mapsto X(\omega, t)$ and the impulse response $t \mapsto h(t)$. For such $\omega$'s this convolution never blows up.

We show this in two steps. In the first step we note that if **h** satisfies (25.106) and if the trajectory $t \mapsto X(\omega, t)$ satisfies

$$\int_{-\infty}^{\infty} \frac{X^2(\omega, t)}{1 + t^2}\,\mathrm{d}t < \infty, \tag{25.107}$$

then the function $\sigma \mapsto X(\omega, \sigma)\, h(t - \sigma)$ is integrable for every $t \in \mathbb{R}$ (Proposition 3.4.4).

In the second step we show that outside a set of $\omega$'s of probability zero, all the trajectories $t \mapsto X(\omega, t)$ satisfy (25.107):

**Lemma 25.13.3.** *Let $\big(X(t)\big)$ be a WSS measurable SP defined over the probability space $(\Omega, \mathcal{F}, P)$. Then*

$$\mathsf{E}\left[\int_{-\infty}^{\infty} \frac{X^2(t)}{1 + t^2}\,\mathrm{d}t\right] < \infty, \tag{25.108}$$

*and the set*

$$\left\{\omega \in \Omega : \int_{-\infty}^{\infty} \frac{X^2(\omega, t)}{1 + t^2}\,\mathrm{d}t < \infty\right\} \tag{25.109}$$

*is an event of probability one.*

**Proof.** Since $\big(X(t)\big)$ is measurable, the mapping

$$(\omega, t) \mapsto \frac{X^2(\omega, t)}{1 + t^2} \tag{25.110}$$

is nonnegative and measurable. By Fubini's Theorem it follows that if we define

$$W(\omega) \triangleq \int_{-\infty}^{\infty} \frac{X^2(\omega, t)}{1 + t^2}\,\mathrm{d}t, \quad \omega \in \Omega, \tag{25.111}$$

then $W$ is a nonnegative RV taking value in the interval $[0, \infty]$. Consequently, the set $\{\omega \in \Omega : W(\omega) < \infty\}$ is measurable. Moreover, by Fubini's Theorem,

$$
\begin{aligned}
\mathsf{E}[W] &= \int_{-\infty}^{\infty} \mathsf{E}\left[\frac{X^2(t)}{1+t^2}\right] \, \mathrm{d}t \\
&= \int_{-\infty}^{\infty} \frac{\mathsf{E}\left[X^2(t)\right]}{1+t^2} \, \mathrm{d}t \\
&= \mathsf{E}\left[X^2(0)\right] \int_{-\infty}^{\infty} \frac{1}{1+t^2} \, \mathrm{d}t \\
&= \pi \, \mathsf{E}\left[X^2(0)\right] \\
&< \infty.
\end{aligned}
$$

Thus, $W$ is a RV taking value in the interval $[0, \infty]$ and having finite expectation, so the event $\{\omega \in \Omega : W(\omega) < \infty\}$ must be of probability one.  □


## 25.14   The PSD Revisited

Theorem 25.13.2 describes the PSD of the output of a stable filter that is fed a WSS SP $\big(X(t)\big)$. By integrating this PSD, we obtain the value at the origin of the autocovariance function of the filter's output (see (25.30)). Since the latter is the power of the filter's output (Corollary 25.9.3), we have:

**Theorem 25.14.1 (Wiener-Khinchin).** *If a measurable, centered, WSS SP $\big(X(t)\big)$ of autocovariance function $\mathsf{K}_{XX}$ is passed through a stable filter of impulse response $\mathbf{h} \colon \mathbb{R} \to \mathbb{R}$, then the average power of the filter's output is given by*

$$
\text{Power of } \mathbf{X} \star \mathbf{h} = \langle \mathsf{K}_{XX}, \mathsf{R}_{\mathbf{hh}} \rangle. \tag{25.112}
$$

*If, additionally, $\big(X(t)\big)$ is of PSD $\mathsf{S}_{XX}$, then this power is given by*

$$
\text{Power of } \mathbf{X} \star \mathbf{h} = \int_{-\infty}^{\infty} \mathsf{S}_{XX}(f) \left|\hat{h}(f)\right|^2 \, \mathrm{d}f. \tag{25.113}
$$

**Proof.** To prove (25.112), we note that by (25.91) the autocovariance function of the filtered process is $\mathsf{K}_{XX} \star \mathsf{R}_{\mathbf{hh}}$, which evaluates at the origin to (25.112). The result thus follows from Proposition 25.9.2, which shows that the power in the filtered process is given by its autocovariance function evaluated at the origin.

To prove (25.113), we note that $\mathsf{K}_{XX}$ is the IFT of $\mathsf{S}_{XX}$ and that, by (11.35), $\hat{\mathsf{R}}_{\mathbf{hh}}(f) = |\hat{h}(f)|^2$, so the RHS of (25.113) is equal to the RHS of (25.112) by Proposition 6.2.4.  □

We next show that for WSS stochastic processes, the operational PSD (Definition 15.3.1) and the PSD (Definition 25.7.2) are equivalent. That is, a WSS SP has an operational PSD if, and only if, it has a PSD, and if the two exist, then they are equal (outside a set of frequencies of Lebesgue measure zero). Before stating this as a theorem, we present a lemma that will be needed in the proof. It is very much in the spirit of Lemma 15.3.2.

**Lemma 25.14.2.** *Let* $\mathbf{g}\colon \mathbb{R} \to \mathbb{R}$ *be a symmetric continuous function satisfying the condition that for every integrable real signal* $\mathbf{h}\colon \mathbb{R} \to \mathbb{R}$

$$\int_{-\infty}^{\infty} g(t)\, \mathsf{R_{hh}}(t)\, \mathrm{d}t = 0. \tag{25.114}$$

*Then* $\mathbf{g}$ *is the all-zero function.*

**Proof.** For every $a > 0$ consider the function

$$h(t) = \frac{1}{\sqrt{a}}\, \mathrm{I}\{|t| \le a/2\}, \quad t \in \mathbb{R}$$

whose self-similarity function is

$$\mathsf{R_{hh}}(t) = \left(1 - \frac{|t|}{a}\right) \mathrm{I}\{|t| \le a\}, \quad t \in \mathbb{R}. \tag{25.115}$$

Since $\mathbf{h}$ is integrable, it follows from (25.114) that

$$
\begin{aligned}
0 &= \int_{-\infty}^{\infty} g(t)\, \mathsf{R_{hh}}(t)\, \mathrm{d}t \\
&= 2 \int_{0}^{\infty} g(t)\, \mathsf{R_{hh}}(t)\, \mathrm{d}t \\
&= 2 \int_{0}^{a} g(t) \left(1 - \frac{t}{a}\right) \mathrm{d}t, \quad a > 0,
\end{aligned}
\tag{25.116}
$$

where the second equality follows from the hypothesis that $g(\cdot)$ is symmetric and from the symmetry of $\mathsf{R_{hh}}$, and where the third equality follows from (25.115). Defining

$$\mathsf{G}(t) = \int_{0}^{t} g(\xi)\, \mathrm{d}\xi, \quad t \ge 0, \tag{25.117}$$

and using integration by parts, we obtain from (25.116) that

$$0 = \mathsf{G}(\xi) \left(1 - \frac{\xi}{a}\right)\Big|_{0}^{a} + \frac{1}{a} \int_{0}^{a} \mathsf{G}(\xi)\, \mathrm{d}\xi, \quad a > 0,$$

from which we obtain

$$a\mathsf{G}(0) = \int_{0}^{a} \mathsf{G}(\xi)\, \mathrm{d}\xi, \quad a > 0.$$

Differentiating with respect to $a$ yields

$$\mathsf{G}(0) = \mathsf{G}(a), \quad a \ge 0,$$

which combines with (25.117) to yield

$$\int_{0}^{a} g(t)\, \mathrm{d}t = 0, \quad a \ge 0. \tag{25.118}$$

Differentiating with respect to $a$ and using the continuity of $\mathbf{g}$ (Rudin, 1976, Chapter 6, Theorem 6.20) yields that $g(a)$ is zero for all $a \ge 0$ and hence, by its symmetry, for all $a \in \mathbb{R}$. $\qquad\square$

**Theorem 25.14.3 (The PSD and Operational PSD of a WSS SP).** *Let $(X(t))$ be a measurable, centered, WSS SP of a continuous autocovariance function $\mathsf{K}_{XX}$. Let $\mathsf{S}(\cdot)$ be a nonnegative, symmetric, integrable function. Then the following two conditions are equivalent:*

*(a) $\mathsf{K}_{XX}$ is the Inverse Fourier Transform of $\mathsf{S}(\cdot)$.*

*(b) For every integrable $\mathbf{h}\colon \mathbb{R} \to \mathbb{R}$, the power in $\mathbf{X} \star \mathbf{h}$ is given by*

$$\text{Power of } \mathbf{X} \star \mathbf{h} = \int_{-\infty}^{\infty} \mathsf{S}(f)\, |\hat{h}(f)|^2 \, \mathrm{d}f. \qquad (25.119)$$

**Proof.** That (a) implies (b) follows from the Wiener-Khinchin Theorem because (a) implies that $(X(t))$ is of PSD $\mathsf{S}(\cdot)$. It remains to prove that (b) implies (a). To this end we now assume that Condition (b) is satisfied and proceed to prove that $\mathsf{K}_{XX}$ must then be equal to the IFT of $\mathsf{S}(\cdot)$. By Theorem 25.14.1, the power in $\mathbf{X} \star \mathbf{h}$ is given by (25.112). Consequently, Condition (b) implies that

$$\int_{-\infty}^{\infty} \mathsf{S}(f)\, |\hat{h}(f)|^2 \, \mathrm{d}f = \int_{-\infty}^{\infty} \mathsf{K}_{XX}(\tau)\, \mathsf{R}_{\mathbf{hh}}(\tau) \, \mathrm{d}\tau, \qquad (25.120)$$

for every integrable $\mathbf{h}\colon \mathbb{R} \to \mathbb{R}$.

If $\mathbf{h}$ is integrable, then the FT of $\mathsf{R}_{\mathbf{hh}}$ is the mapping $f \mapsto |\hat{h}(f)|^2$ (see (11.35)). If, in addition, $\mathbf{h}$ is a real signal, then $\mathsf{R}_{\mathbf{hh}}$ is a symmetric function, and its IFT is thus identical to its FT (Proposition 6.2.3 (ii)). Thus, if $\mathbf{h}$ is real and integrable, then the IFT of $\mathsf{R}_{\mathbf{hh}}$ is the mapping $f \mapsto |\hat{h}(f)|^2$. (Using the dummy variable $f$ for the IFT is unusual but legitimate.) Consequently, by Proposition 6.2.4 (applied with the substitution of $\mathsf{S}(\cdot)$ for $\mathbf{x}$ and of $\mathsf{R}_{\mathbf{hh}}$ for $\mathbf{g}$),

$$\int_{-\infty}^{\infty} \mathsf{S}(f)\, |\hat{h}(f)|^2 \, \mathrm{d}f = \int_{-\infty}^{\infty} \hat{\mathsf{S}}(\tau)\, \mathsf{R}_{\mathbf{hh}}(\tau) \, \mathrm{d}\tau. \qquad (25.121)$$

By (25.120) & (25.121) and by the symmetry of $\mathsf{S}(\cdot)$ (which implies that $\hat{\mathsf{S}} = \check{\mathsf{S}}$) we obtain that

$$\int_{-\infty}^{\infty} \left(\check{\mathsf{S}}(\tau) - \mathsf{K}_{XX}(\tau)\right) \mathsf{R}_{\mathbf{hh}}(\tau) \, \mathrm{d}\tau = 0, \quad \mathbf{h} \in \mathcal{L}_1. \qquad (25.122)$$

It thus follows from Lemma 15.3.2 that the mapping $\tau \mapsto \check{\mathsf{S}}(\tau) - \mathsf{K}_{XX}(\tau)$ is the all-zero function, and Condition (a) is established. $\qquad \square$

## 25.15 White Gaussian Noise

The most important continuous-time SP in Digital Communications is **white Gaussian noise**, which is often used to model the additive noise in communication systems. In this section we define this process and study its key properties. Our definition differs from the one in most textbooks, most notably in that we define white Gaussian noise only with respect to some given bandwidth $\mathsf{W}$. We give our reasons and comment on the implications in Section 25.15.2 after providing our definition and deriving the key results.

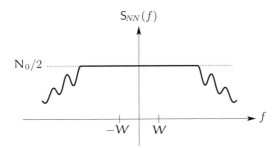

**Figure 25.3:** The PSD of a SP $(N(t))$ which is of double-sided spectral density $N_0/2$ with respect to the bandwidth $W$.

### 25.15.1   Definition and Main Properties

The parameters defining white Gaussian noise are the bandwidth $W$ with respect to which the process is white and the double-sided spectral density $N_0/2$.

**Definition 25.15.1 (White Gaussian Noise).** *We say that $(N(t))$ is **white Gaussian noise of double-sided spectral density** $N_0/2$ **with respect to the bandwidth** $W$ if $(N(t))$ is a measurable, stationary, centered, Gaussian SP that has a PSD $S_{NN}$ satisfying*

$$S_{NN}(f) = \frac{N_0}{2}, \quad f \in [-W, W]. \tag{25.123}$$

An example of the PSD of white Gaussian noise of double-sided spectral density $N_0/2$ with respect to the bandwidth $W$ is depicted in Figure 25.3. Note that our definition of white Gaussian noise only specifies the PSD for frequencies $f$ satisfying $|f| \leq W$. We leave the value of the PSD at other frequencies unspecified. But the PSD should, of course, be a valid PSD, i.e., it must be nonnegative, symmetric, and integrable (Definition 25.7.2). Recall also that by Proposition 25.7.3 every nonnegative, symmetric, integrable function is the PSD of some measurable stationary Gaussian SP.[11]

The following proposition summarizes the key properties of white Gaussian noise. The reader is encouraged to recall the definition of an integrable function that is bandlimited to $W$ Hz (Definition 6.4.9); the definition of the inner product between two energy-limited real signals (3.1); the definition of $\|s\|_2$ as $\sqrt{\langle s, s \rangle}$; and the definition of orthonormality of the functions $\phi_1, \ldots, \phi_m$ (Definition 4.6.1).

**Proposition 25.15.2 (Key Properties of White Gaussian Noise).** *Let $(N(t))$ be white Gaussian noise of double-sided spectral density $N_0/2$ with respect to the bandwidth $W$.*

---

[11] As we have noted in the paragraph preceding Definition 25.9.1, Proposition 25.7.3 can be strengthened to also guarantee measurability. Every nonnegative, symmetric, and integrable function is the PSD of some measurable, stationary, and Gaussian SP whose autocovariance function is continuous.

(i) *If* $\mathbf{s}$ *is any integrable function that is bandlimited to* $\mathsf{W}$ *Hz, then*

$$\int_{-\infty}^{\infty} N(t)\, s(t)\, \mathrm{d}t \sim \mathcal{N}\left(0,\ \frac{\mathsf{N}_0}{2}\, \|\mathbf{s}\|_2^2\right).$$

(ii) *If* $\mathbf{s}_1, \ldots, \mathbf{s}_m$ *are integrable functions that are bandlimited to* $\mathsf{W}$ *Hz, then the* $m$ *random variables*

$$\int_{-\infty}^{\infty} N(t)\, s_1(t)\, \mathrm{d}t, \ldots, \int_{-\infty}^{\infty} N(t)\, s_m(t)\, \mathrm{d}t$$

*are jointly Gaussian centered random variables of covariance matrix*

$$\frac{\mathsf{N}_0}{2} \begin{pmatrix} \langle \mathbf{s}_1, \mathbf{s}_1 \rangle & \langle \mathbf{s}_1, \mathbf{s}_2 \rangle & \cdots & \langle \mathbf{s}_1, \mathbf{s}_m \rangle \\ \langle \mathbf{s}_2, \mathbf{s}_1 \rangle & \langle \mathbf{s}_2, \mathbf{s}_2 \rangle & \cdots & \langle \mathbf{s}_2, \mathbf{s}_m \rangle \\ \vdots & \vdots & \ddots & \vdots \\ \langle \mathbf{s}_m, \mathbf{s}_1 \rangle & \langle \mathbf{s}_m, \mathbf{s}_2 \rangle & \cdots & \langle \mathbf{s}_m, \mathbf{s}_m \rangle \end{pmatrix}.$$

(iii) *If* $\boldsymbol{\phi}_1, \ldots, \boldsymbol{\phi}_m$ *are integrable functions that are bandlimited to* $\mathsf{W}$ *Hz and are orthonormal, then the random variables*

$$\int_{-\infty}^{\infty} N(t)\, \phi_1(t)\, \mathrm{d}t, \ldots, \int_{-\infty}^{\infty} N(t)\, \phi_m(t)\, \mathrm{d}t$$

*are IID* $\mathcal{N}(0, \mathsf{N}_0/2)$.

(iv) *If* $\mathbf{s}$ *is any integrable function that is bandlimited to* $\mathsf{W}$ *Hz, and if* $\mathsf{K}_{NN}$ *is the autocovariance function of* $(N(t))$, *then*

$$\mathsf{K}_{NN} \star \mathbf{s} = \frac{\mathsf{N}_0}{2}\, \mathbf{s}. \tag{25.124}$$

(v) *If* $\mathbf{s}$ *is an integrable function that is bandlimited to* $\mathsf{W}$ *Hz, then for every epoch* $t \in \mathbb{R}$

$$\mathrm{Cov}\left[\int_{-\infty}^{\infty} N(\sigma)\, s(\sigma)\, \mathrm{d}\sigma,\, N(t)\right] = \frac{\mathsf{N}_0}{2}\, s(t). \tag{25.125}$$

**Proof.** Parts (i) and (iii) are special cases of Part (ii), so it suffices to prove Parts (ii), (iv), and (v). We begin with Part (ii). We first note that since $\{\mathbf{s}_j\}$ are assumed to be integrable and bandlimited to $\mathsf{W}$ Hz, and since Note 6.4.12 guarantees that every bandlimited integrable signal is also of finite energy, it follows that the functions $\{\mathbf{s}_j\}$ are energy-limited and the inner products $\langle \mathbf{s}_j, \mathbf{s}_k \rangle$ are well-defined. By (25.89)

$$\begin{aligned} \mathrm{Cov}\left[\int_{-\infty}^{\infty} N(t)\, s_j(t)\, \mathrm{d}t,\, \int_{-\infty}^{\infty} N(t)\, s_k(t)\, \mathrm{d}t\right] &= \int_{-\infty}^{\infty} \mathsf{S}_{NN}(f)\, \hat{s}_j(f)\, \hat{s}_k^*(f)\, \mathrm{d}f \\ &= \int_{-\mathsf{W}}^{\mathsf{W}} \mathsf{S}_{NN}(f)\, \hat{s}_j(f)\, \hat{s}_k^*(f)\, \mathrm{d}f \\ &= \frac{\mathsf{N}_0}{2} \int_{-\mathsf{W}}^{\mathsf{W}} \hat{s}_j(f)\, \hat{s}_k^*(f)\, \mathrm{d}f \\ &= \frac{\mathsf{N}_0}{2}\, \langle \mathbf{s}_j, \mathbf{s}_k \rangle, \quad j, k \in \{1, \ldots, m\}, \end{aligned}$$

where the second equality follows because $\mathbf{s}_j$ and $\mathbf{s}_k$ are bandlimited to $W$ Hz; the third from (25.123); and the final equality from Parseval's Theorem.

To prove Part (iv), we start with the definition of the convolution and compute

$$
\begin{aligned}
\left(\mathsf{K}_{NN} \star \mathbf{s}\right)(t) &= \int_{-\infty}^{\infty} s(\tau)\, \mathsf{K}_{NN}(t - \tau)\, \mathrm{d}\tau \\
&= \int_{-\infty}^{\infty} s(\tau) \int_{-\infty}^{\infty} \mathsf{S}_{NN}(f)\, e^{\mathrm{i}2\pi f(t-\tau)}\, \mathrm{d}f\, \mathrm{d}\tau \\
&= \int_{-\infty}^{\infty} \mathsf{S}_{NN}(f)\, \hat{s}(f)\, e^{\mathrm{i}2\pi ft}\, \mathrm{d}f \\
&= \int_{-W}^{W} \mathsf{S}_{NN}(f)\, \hat{s}(f)\, e^{\mathrm{i}2\pi ft}\, \mathrm{d}f \\
&= \frac{\mathsf{N}_0}{2} \int_{-W}^{W} \hat{s}(f)\, e^{\mathrm{i}2\pi ft}\, \mathrm{d}f \\
&= \frac{\mathsf{N}_0}{2}\, s(t), \quad t \in \mathbb{R},
\end{aligned}
$$

where the second equality follows from the definition of the PSD of $\left(N(t)\right)$ (Definition 25.7.2); the third by Proposition 6.2.5; the fourth because $\mathbf{s}$ is, by assumption, bandlimited to $W$ Hz (Proposition 6.4.10 cf. (c)); the fifth from our assumption that $\left(N(t)\right)$ is white with respect to the bandwidth $W$ (25.123); and the final equality from Proposition 6.4.10 (cf. (b)).

Part (v) now follows from (25.85) and Part (iv). Alternatively, it can be proved using (25.88) and (25.123) as follows:

$$
\begin{aligned}
\mathsf{Cov}\left[\int_{-\infty}^{\infty} N(\sigma)\, s(\sigma)\, \mathrm{d}\sigma,\, N(t)\right] &= \int_{-\infty}^{\infty} \mathsf{S}_{NN}(f)\, \hat{s}(f)\, e^{\mathrm{i}2\pi ft}\, \mathrm{d}f \\
&= \int_{-W}^{W} \mathsf{S}_{NN}(f)\, \hat{s}(f)\, e^{\mathrm{i}2\pi ft}\, \mathrm{d}f \\
&= \frac{\mathsf{N}_0}{2} \int_{-W}^{W} \hat{s}(f)\, e^{\mathrm{i}2\pi ft}\, \mathrm{d}f \\
&= \frac{\mathsf{N}_0}{2}\, s(t), \quad t \in \mathbb{R},
\end{aligned}
$$

where the first equality follows from (25.88); the second because $\mathbf{s}$ is bandlimited to $W$ Hz (Proposition 6.4.10 cf. (c)); the third from (25.123); and the last from Proposition 6.4.10 (cf. (b)). $\qquad\square$

## 25.15.2 Other Definitions

As we noted earlier, our definition of white Gaussian noise is different from the one given in most textbooks on Digital Communications. The key difference is that we define whiteness *with respect to a certain bandwidth* $W$, whereas most textbooks do not add this qualifier. Thus, while we require that the PSD $\mathsf{S}_{NN}(f)$ be equal to $\mathsf{N}_0/2$ only for frequencies $f$ satisfying $|f| \leq W$ (leaving $\mathsf{S}_{NN}(f)$ unspecified at other frequencies), other textbooks require that $S_{NN}(f)$ be equal to $N_0/2$ *for all* frequencies $f \in \mathbb{R}$. With our definition of white noise we can only prove that (25.124) holds for integrable signals that are bandlimited to $W$ Hz, whereas with the other textbooks' definition one could presumably derive this relationship for all integrable functions.

We prefer our definition because there does not exist a Gaussian SP $(N(t))$ whose PSD is equal to $N_0/2$ at all frequencies. Indeed, the function of frequency that is equal to $N_0/2$ at all frequencies is not integrable and therefore does not qualify as a PSD (Definition 25.7.2). Were such a PSD to exist, we would obtain from (25.30) that such a process would have infinite variance and thus be neither WSS (Definition 25.4.2) nor Gaussian (Note 25.3.2).

Requiring that (25.124) hold for all integrable (continuous) signals would require that $K_{NN}$ be given by the product of $N_0/2$ and Dirac's delta, which opens a whole can of worms. Nevertheless, the reader should be aware that in some books white noise is defined as a centered, stationary Gaussian noise whose autocovariance function is given by Dirac's Delta scaled by $N_0/2$ or, equivalently, whose PSD is equal to $N_0/2$ at all frequencies.

### 25.15.3 White Noise in Passband

**Definition 25.15.3 (White Gaussian Noise in Passband).** *We say that $(N(t))$ is white Gaussian noise of double-sided power spectral density* $N_0/2$ *with respect to the bandwidth* $W$ *around the carrier frequency* $f_c$ *if $(N(t))$ is a centered, measurable, stationary, Gaussian process that has a PSD* $S_{NN}$ *satisfying*

$$S_{NN}(f) = \frac{N_0}{2}, \quad \bigl||f| - f_c\bigr| \le \frac{W}{2}, \tag{25.126}$$

*and if* $f_c > W/2$.

**Note 25.15.4.** For white Gaussian noise with respect to the bandwidth $W$ around the carrier frequency $f_c$, all the claims of Proposition 25.15.2 hold provided that we replace the requirement that the functions $\mathbf{s}$, $\{\mathbf{s}_j\}$, and $\{\phi_j\}$ be integrable functions that are bandlimited to $W$ Hz with the requirement that they be integrable functions that are bandlimited to $W$ Hz around the carrier frequency $f_c$.

## 25.16  Exercises

**Exercise 25.1 (Constructing a SP from a RV).** Let $W$ be a standard Gaussian RV. Define the continuous-time SP $(X(t))$ by

$$X(t) = e^{-|t|} W, \quad t \in \mathbb{R}.$$

(i)  Is $(X(t))$ a stationary SP?

(ii)  Is $(X(t))$ a Gaussian SP?

**Exercise 25.2 (Delaying and Adding).** Let $\big(X(t)\big)$ be a stationary Gaussian SP of mean $\mu_x$ and autocovariance function $\mathsf{K}_{XX}$. Define

$$Y(t) = X(t) + X(t - t_\mathrm{D}), \quad t \in \mathbb{R},$$

where $t_\mathrm{D} \in \mathbb{R}$ is deterministic.

  (i) Is $\big(Y(t)\big)$ a Gaussian SP?

  (ii) Compute the mean and the autocovariance function of $\big(Y(t)\big)$.

  (iii) Is $\big(Y(t)\big)$ stationary?

**Exercise 25.3 (Random Variables and Stochastic Processes).** Let the random variables $X$ and $Y$ be IID $\mathcal{N}(0, \sigma^2)$, and let

$$Z(t) = X \cos(2\pi t) + Y \sin(2\pi t), \quad t \in \mathbb{R}.$$

  (i) Is $Z(0.2)$ Gaussian?

  (ii) Is $\big(Z(t)\big)$ a Gaussian SP?

  (iii) Is it stationary?

**Exercise 25.4 (Stochastic Processes through Nonlinearities).**

  (i) Let $\big(X(t)\big)$ be a stationary SP and let

$$Y(t) = g(X(t)), \quad t \in \mathbb{R},$$

  where $\mathbf{g} \colon \mathbb{R} \to \mathbb{R}$ is some (Borel measurable) deterministic function. Show that the SP $\big(Y(t)\big)$ is stationary. Under what conditions is $\big(Y(t)\big)$ WSS?

  (ii) Let $\big(X(t)\big)$ be a centered stationary Gaussian SP of autocovariance function $\mathsf{K}_{XX}$. Let $Y(t) = \mathrm{sgn}(X(t))$, where $\mathrm{sgn}(\xi)$ is equal to $+1$ whenever $\xi \geq 0$ and is equal to $-1$ otherwise. Is $\big(Y(t)\big)$ centered? Is it WSS? If so, what is its autocovariance function?

*Hint: For Part (ii) recall Exercise 23.18.*

**Exercise 25.5 (WSS Stochastic Processes).** Let $A$ and $B$ be IID random variables taking on the values $\pm 1$ equiprobably. Define the SP $\big(Z(t)\big)$ as

$$Z(t) = A \cos(2\pi t) + B \sin(2\pi t), \quad t \in \mathbb{R}.$$

  (i) Is the SP $\big(Z(t)\big)$ WSS?

  (ii) Define the SP $\big(W(t)\big)$ by $W(t) = Z^2(t)$. Is $\big(W(t)\big)$ WSS?

**Exercise 25.6 (Valid Autocovariance Functions).** Let $\mathsf{K}_{XX}$ and $\mathsf{K}_{YY}$ be the autocovariance functions of some WSS stochastic processes $\big(X(t)\big)$ and $\big(Y(t)\big)$.

  (i) Show that $\mathsf{K}_{XX} + \mathsf{K}_{YY}$ is an autocovariance function of some WSS SP.

  (ii) Repeat for $\tau \mapsto \mathsf{K}_{XX}(\tau) \, \mathsf{K}_{YY}(\tau)$.

**Exercise 25.7 (Time Reversal).** Let $\mathsf{K}_{XX}$ be the autocovariance function of some WSS SP $\big(X(t), \; t \in \mathbb{R}\big)$. Is the time-reversed SP $(\omega, t) \mapsto X(\omega, -t)$ WSS? If so, express its autocovariance function in terms of $\mathsf{K}_{XX}$.

**Exercise 25.8 (Classifying Stochastic Processes).** Let $(X(t))$ and $(Y(t))$ be independent centered stationary Gaussian stochastic processes of unit variance and autocovariance functions $\mathsf{K}_{XX}$ and $\mathsf{K}_{YY}$. Define the stochastic processes $(S(t)), (T(t)), (U(t)), (V(t))$, and $(W(t))$ at every $t \in \mathbb{R}$ as:

$$
\begin{aligned}
S(t) &= X(t) + Y(t + \tau_1), \quad T(t) = X(t)\,Y(t + \tau_2), \\
U(t) &= X(t) + X(t + \tau_3), \quad V(t) = X(t)\,X(t + \tau_4), \\
W(t) &= X(t) + X(-t),
\end{aligned}
$$

where $\tau_1, \tau_2, \tau_3, \tau_4 \in \mathbb{R}$ are deterministic. Which of these stochastic processes is Gaussian? Which is WSS? Which is stationary?

**Exercise 25.9 (A Linear Functional of a Gaussian SP).** Let $(X(t),\ t \in \mathbb{R})$ be a measurable stationary Gaussian SP of mean 2 and of autocovariance function $\mathsf{K}_{XX} \colon \tau \mapsto \exp(-|\tau|)$. Compute

$$
\Pr\left[\int_0^2 X(t)\,dt \geq 2\right].
$$

**Exercise 25.10 (Two Filters).** Let $(X(t))$ be a centered stationary Gaussian SP of autocovariance function $\mathsf{K}_{XX}$ and PSD $\mathsf{S}_{XX}$. Define

$$
(Y(t)) = (X(t)) \star \mathbf{h}_y, \quad (Z(t)) = (X(t)) \star \mathbf{h}_z,
$$

where $\mathbf{h}_y, \mathbf{h}_z \in \mathcal{L}_1$. Thus, $(Y(t))$ is the result of passing $(X(t))$ through a stable filter of impulse response $\mathbf{h}_y$ and similarly $(Z(t))$.

(i) What is the joint distribution of $Y(t_1)$ and $Z(t_2)$ for given epochs $t_1, t_2 \in \mathbb{R}$?

(ii) Give a necessary and sufficient condition on $\hat{\mathbf{h}}_y$, $\hat{\mathbf{h}}_z$, and $\mathsf{S}_{XX}$ for $Y(17)$ to be independent of $Z(17)$.

(iii) Give a necessary and sufficient condition on $\hat{\mathbf{h}}_y$, $\hat{\mathbf{h}}_z$, and $\mathsf{S}_{XX}$ for $(Z(t))$ to be independent of $(Y(t))$.

**Exercise 25.11 (Linear Functionals of White Gaussian Noise).** Find the distribution of

$$
\int_0^{\mathsf{T}_s} N(t)\,dt \quad \text{and of} \quad \int_0^\infty e^{-t} N(t)\,dt
$$

when $(N(t),\ t \in \mathbb{R})$ is white Gaussian noise of double-sided PSD $\mathsf{N}_0/2$ with respect to the bandwidth of interest. (Ignore the fact that the mappings $t \mapsto \mathrm{I}\{0 \leq t \leq \mathsf{T}_s\}$ and $t \mapsto e^{-t}\,\mathrm{I}\{t \geq 0\}$ are not bandlimited.)

**Exercise 25.12 (Approximately White SP).** Let $(X(t),\ t \in \mathbb{R})$ be a measurable, centered, stationary, Gaussian SP of autocovariance function

$$
\mathsf{K}_{XX}(\tau) = \frac{\mathsf{B}\mathsf{N}_0}{4}\, e^{-\mathsf{B}|\tau|}, \quad \tau \in \mathbb{R},
$$

where $\mathsf{N}_0, \mathsf{B} > 0$ are given constants. Throughout this problem $\mathsf{N}_0$ is fixed.

(i) Plot $\mathsf{K}_{XX}$ for several values of $\mathsf{B}$. What does $\mathsf{K}_{XX}$ look like when $\mathsf{B} \gg 1$? Show that $\mathsf{K}_{XX}(\tau) > 0$ for all $\tau \in \mathbb{R}$; that

$$
\int_{-\infty}^\infty \mathsf{K}_{XX}(\tau)\,d\tau = \frac{\mathsf{N}_0}{2};
$$

and that for every $\delta > 0$,

$$\lim_{B \to \infty} \int_{-\delta}^{\delta} \mathsf{K}_{XX}(\tau) \, d\tau = \frac{\mathsf{N}_0}{2}.$$

(In this sense, $\mathsf{K}_{XX}$ approximates Dirac's Delta scaled by $\mathsf{N}_0/2$ when $\mathsf{B}$ is large.)

(ii) Compute $\mathsf{E}[X(t)^2]$. Plot this as a function of $\mathsf{B}$, with $\mathsf{N}_0$ held fixed. What happens when $\mathsf{B} \gg 1$?

(iii) Compute the PSD $\mathsf{S}_{XX}$. Plot it for several values of $\mathsf{B}$. What does it look like when $\mathsf{B} \gg 1$?

(iv) For the orthonormal signals defined for every $t \in \mathbb{R}$ by

$$\phi_1(t) = \begin{cases} 1 & \text{if } 0 \le t \le 1, \\ 0 & \text{otherwise,} \end{cases} \qquad \phi_2(t) = \begin{cases} 1 & \text{if } 0 \le t \le \frac{1}{2}, \\ -1 & \text{if } \frac{1}{2} < t \le 1, \\ 0 & \text{otherwise} \end{cases}$$

compute $\mathsf{E}[\langle \mathbf{X}, \phi_1 \rangle \langle \mathbf{X}, \phi_2 \rangle]$. What happens to this expression when $\mathsf{B} \gg 1$?

# Chapter 26

# Detection in White Gaussian Noise

## 26.1 Introduction

In this chapter we finally address the detection problem in continuous time. The setup is described in Section 26.2. The key result of this chapter is that—even though in this setup the observation consists of a stochastic process (i.e., a continuum of random variables)—the problem can be reduced without loss of optimality to a finite-dimensional problem where the observation consists of a random vector. Before stating this result precisely in Section 26.4, we shall take a detour in Section 26.3 to discuss the definition of sufficient statistics when the observation consists of a continuous-time SP. The proof of the main result is delayed until Section 26.8. In Section 26.5 we analyze the conditional law of the sufficient statistic vector under each of the hypotheses. This analysis enables us in Section 26.6 to derive an optimal guessing rule and in Section 26.7 to analyze its performance. Section 26.9 addresses the front-end filter, which is a critical element of any practical implementation of the decision rule. Extensions to passband detection are then described in Section 26.10, followed by some examples in Section 26.11. Section 26.12 treats the problem of detection in "colored" noise, and the chapter concludes with a discussion of the detection problem for mean signals that are not bandlimited.

## 26.2 Setup

A discrete random variable $M$ ("message") takes value in the set $\mathcal{M} = \{1, \ldots, \mathsf{M}\}$, where $\mathsf{M} \geq 2$, according to the *a priori* probabilities

$$\pi_m = \Pr[M = m], \quad m \in \mathcal{M}, \tag{26.1}$$

where $\pi_1, \ldots, \pi_\mathsf{M}$ are positive[1]

$$\pi_m > 0, \quad m \in \mathcal{M} \tag{26.2}$$

---

[1]There is no loss in generality in addressing the detection problem only for strictly positive priors. Hypotheses that have a zero prior can be ignored at the receiver without loss in optimality.

and sum to one

$$\sum_{m \in \mathcal{M}} \pi_m = 1. \tag{26.3}$$

The observation consists of the continuous-time SP $\big(Y(t),\ t \in \mathbb{R}\big)$, which, conditional on $M = m$, can be expressed as

$$Y(t) = s_m(t) + N(t), \quad t \in \mathbb{R}, \tag{26.4}$$

where the "mean signals" $\mathbf{s}_1, \ldots, \mathbf{s}_M$ are real, deterministic, integrable signals that are bandlimited to $\mathsf{W}$ Hz (Definition 6.4.9), and where the "noise" $\big(N(t)\big)$ is independent of $M$ and is white Gaussian noise of double-sided spectral density $\mathsf{N}_0/2$ with respect to the bandwidth $\mathsf{W}$ (Definition 25.15.1). Based on the observation $\big(Y(t)\big)$ we wish to guess $M$ with the smallest possible probability of error.[2]

## 26.3 Sufficient Statistics when Observing a SP

The definition of sufficient statistics for the infinite-dimensional hypothesis testing problem where the observation consists of a SP is conceptually very similar to the definition in the finite-dimensional case where the observation consists of a random vector (Definition 22.2.1). But some new technical difficulties do arise. Foremost is that we cannot speak of the probability density function (in the usual sense) of the observation given each of the hypotheses.[3] Consequently, we need a new definition that does not involve such densities.

### 26.3.1 Definition of Sufficient Statistics

Loosely speaking, a sufficient statistic for guessing a RV $M$ taking value in the finite set $\mathcal{M}$ based on an observation consisting of a SP $\big(Y(t)\big)$ is a random vector $\mathbf{T} = (T^{(1)}, \ldots, T^{(d')})^\mathsf{T}$ that satisfies two conditions. The first is that it can be computed from the observed SP, and the second is that—once we are given $\mathbf{T}$—any finite number of samples $\eta \in \mathbb{N}$ of the observations $Y(t_1), \ldots, Y(t_\eta)$ are irrelevant for guessing $M$. Thus, once $\mathbf{T}$ has been revealed to us, our optimal guess for $M$ will not be improved if we are additionally given the values of $\big(Y(t)\big)$ at any finite number of (deterministic) epochs.

Recall the definition of the $\sigma$-algebra generated by the SP $\big(Y(t)\big)$ (Definition 25.2.2) and the definition of irrelevant data (Definition 22.5.1).

**Definition 26.3.1 (Sufficient Statistic: Observable SP).** *We say that the random vector $\mathbf{T}$ forms a sufficient statistic for guessing the RV $M$ taking value in the finite set $\mathcal{M}$ based on the observed SP $\big(Y(t)\big)$ if the following two conditions hold:*

---

[2]In mathematical terms we are looking for a mapping from the set of all sample-paths of $\big(Y(t)\big)$ to $\mathcal{M}$ that is measurable with respect to the $\sigma$-algebra generated by $\big(Y(t)\big)$ and that minimizes the probability of error among all such functions.

[3]One could, instead, speak of the Radon-Nikodym derivative with respect to a reference measure, but we prefer not to pursue this approach.

1) $\mathbf{T}$ *is measurable with respect to the $\sigma$-algebra generated by $\big(Y(t)\big)$.*

2) *For every $\eta \in \mathbb{N}$ and every choice of the epochs $t_1, \ldots, t_\eta \in \mathbb{R}$, the $\eta$-tuple $\big(Y(t_1), \ldots, Y(t_\eta)\big)$ is irrelevant for guessing $M$ based on $\mathbf{T}$.*

Condition 2) is equivalent to

$$M \multimap \mathbf{T} \multimap \big(Y(t_1), \ldots, Y(t_\eta)\big) \tag{26.5}$$

forming a Markov chain for any prior on $M$.

As we shall see in Section 26.4, such a sufficient statistic can always be found for the setup described in Section 26.2.

## 26.3.2   Consequences of Sufficiency

It would have been nice if, in analogy with Proposition 22.2.2, we could have said that if $\mathbf{T}$ forms a sufficient statistic for guessing $M$ based on the observed SP $\big(Y(t)\big)$, then the best performance in guessing $M$ based on $\big(Y(t)\big)$ can be achieved by a decision rule that bases its decision on $\mathbf{T}$. This statement is almost correct, but it requires a qualification.

A pathological example that demonstrates the need for a qualification is the following. Suppose that $M$ takes on the values 1 and 2 equiprobably and that $R$ is a RV that is independent of $M$ and that has a density. For example, $R$ could be a mean-one exponential. Suppose further that, conditional on $M = 1$, the observed SP $\big(Y(t)\big)$ is deterministically zero, and that, conditional on $M = 2$, the observed SP is zero at all times $t \in \mathbb{R}$ except at time $R$ when it takes on the value 1. In this case the conditional law of $\big(Y(t_1), \ldots, Y(t_\eta)\big)$ does not depend on whether the conditioning is on $M = 1$ or on $M = 2$. Thus, if we define the RV $T$ to equal 17 deterministically, then $T$ forms a sufficient statistic for guessing $M$ based on $\big(Y(t)\big)$. The smallest probability of a guessing error based on $T$ is $1/2$.[4] Nevertheless, a detector that guesses "$M = 1$" if the observed trajectory is the all-zero function and "$M = 2$" if the observed trajectory is discontinuous is correct with probability one.

It is interesting to note that the latter guessing rule is not measurable with respect to the $\sigma$-algebra generated by $\big(Y(t)\big)$. As the next theorem demonstrates, the qualifier that we need to add is that we only consider guessing rules that are measurable with respect to the $\sigma$-algebra generated by $\big(Y(t)\big)$. Barring this qualifier, if $\mathbf{T}$ is sufficient, then there is no loss in optimality in basing our guess on $\mathbf{T}$ only.

**Theorem 26.3.2.** *Consider the multi-hypothesis testing problem of guessing a RV $M$ taking value in the set $\mathcal{M} = \{1, \ldots, \mathsf{M}\}$ based on an observation consisting of a SP $\big(Y(t), \ t \in \mathbb{R}\big)$. Let $\mathbf{T}$ be a random vector that forms a sufficient statistic for guessing $M$ based on $\big(Y(t)\big)$. Then no decision rule that is measurable with respect to the $\sigma$-algebra generated by $\big(Y(t)\big)$ can have a lower probability of error than an optimal rule for guessing $M$ based on $\mathbf{T}$.*

---

[4]This is also the smallest probability of error in guessing $M$ based on $\big(Y(t_1), \ldots, Y(t_\eta)\big)$, irrespective of the (finite) value of the positive integer $\eta$ and of the (deterministic) choice of the epochs $t_1, \ldots, t_\eta$.

**Proof.** Let $\phi(\cdot)$ be any decision rule that is measurable with respect to the $\sigma$-algebra generated by $(Y(t))$, i.e., a decision rule whose disjoint decision sets

$$\mathcal{D}_m \triangleq \phi^{-1}(m), \quad m \in \mathcal{M}$$

are all measurable with respect to this $\sigma$-algebra. The conditional probability that the rule $\phi(\cdot)$ guesses correctly is

$$\Pr\big(\phi(\cdot) \text{ is correct} \,\big|\, M = m\big) = \Pr\big[(Y(t)) \in \mathcal{D}_m \,\big|\, M = m\big], \quad m \in \mathcal{M}. \quad (26.6)$$

We shall show that $\phi(\cdot)$ can be approximated by a decision rule $\tilde{\phi}(\cdot)$ that bases its decision on a finite number of samples $Y(t_1), \ldots, Y(t_\eta)$, where $\eta \in \mathbb{N}$ and where $t_1, \ldots, t_\eta \in \mathbb{R}$ are deterministic epochs. The approximation is in the sense that, conditional on each $m \in \mathcal{M}$, the probability of success of $\tilde{\phi}(\cdot)$ is within $\epsilon$ of that of $\phi(\cdot)$. We shall then show that the best decision rule based on $\mathbf{T}$ is at least as good as $\tilde{\phi}(\cdot)$ and is thus also within $\epsilon$ of $\phi(\cdot)$. Since these steps will be performed for an arbitrary $\epsilon > 0$, and since the performance of the best decoder based on $\mathbf{T}$ does not depend on $\epsilon$, this will demonstrate that $\phi(\cdot)$ is no better than the best decision rule based on $\mathbf{T}$. And since $\phi(\cdot)$ here is an arbitrary measurable decision rule, it will follow that no measurable decision rule can outperform an optimal rule based on $\mathbf{T}$ and the theorem will be proved.

To follow this outline we first need some basic set-theoretic notation. Given two sets $\mathcal{A}$ and $\mathcal{B}$ we denote by $\mathcal{A} \setminus \mathcal{B}$ the set consisting of those elements of $\mathcal{A}$ that are not in $\mathcal{B}$. We denote by $\mathcal{A} \triangle \mathcal{B}$ the symmetric set difference between $\mathcal{A}$ and $\mathcal{B}$ consisting of those elements that are in one of the sets but not in the other. Thus, $\mathcal{A} \triangle \mathcal{B} = (\mathcal{A} \setminus \mathcal{B}) \cup (\mathcal{B} \setminus \mathcal{A}).$[5]

A standard result from Measure Theory (Halmos, 1950, Exercise (8), Section 14) guarantees that for every $\epsilon > 0$ there exist epochs $t_1, \ldots, t_\eta \in \mathbb{R}$ and sets $\hat{\mathcal{D}}_1, \ldots \hat{\mathcal{D}}_\mathsf{M}$ (not necessarily disjoint) that are all measurable with respect to the $\sigma$-algebra generated by $Y(t_1), \ldots, Y(t_\eta)$ and such that

$$\Pr\big[(Y(t)) \in \mathcal{D}_{m'} \triangle \hat{\mathcal{D}}_{m'} \,\big|\, M = m\big] < \frac{\epsilon}{\mathsf{M}}, \quad m, m' \in \mathcal{M}. \quad (26.7)$$

Define now the disjoint sets $\tilde{\mathcal{D}}_1, \ldots, \tilde{\mathcal{D}}_\mathsf{M}$ inductively by defining $\tilde{\mathcal{D}}_1 = \hat{\mathcal{D}}_1$ and

$$\tilde{\mathcal{D}}_m = \hat{\mathcal{D}}_m \setminus \bigcup_{m' < m} \tilde{\mathcal{D}}_{m'}, \quad m \in \{2, \ldots, \mathsf{M}\}. \quad (26.8)$$

By construction, these sets are disjoint. And because $\hat{\mathcal{D}}_1, \ldots, \hat{\mathcal{D}}_\mathsf{M}$ are measurable with respect to the $\sigma$-algebra generated by $Y(t_1), \ldots, Y(t_\eta)$, so are $\tilde{\mathcal{D}}_1, \ldots, \tilde{\mathcal{D}}_\mathsf{M}$.

---

[5]As an aside we mention that the indicator functions of the sets $\mathcal{A}$, $\mathcal{B}$ and $\mathcal{A} \triangle \mathcal{B}$ are related via the relation

$$\mathrm{I}\{x \in \mathcal{A} \triangle \mathcal{B}\} = \mathrm{I}\{x \in \mathcal{A}\} \oplus \mathrm{I}\{x \in \mathcal{B}\},$$

where $\oplus$ denotes exclusive-or, i.e., mod-2 addition ($0 \oplus 0 = 0$, $0 \oplus 1 = 1$, $1 \oplus 0 = 1$, and $1 \oplus 1 = 0$). This relationship simplifies the proof of some of the key properties of the symmetric set difference, especially when combined with the analogous relation for intersection

$$\mathrm{I}\{x \in \mathcal{A} \cap \mathcal{B}\} = \mathrm{I}\{x \in \mathcal{A}\} \, \mathrm{I}\{x \in \mathcal{B}\},$$

where on the RHS of the above we use mod-2 multiplication.

We next consider a decoder $\tilde{\phi}(\cdot)$ that guesses "$M = m$" whenever the sample-path of $(Y(t))$ is in the set $\tilde{\mathcal{D}}_m$. (If the sample-path of $(Y(t))$ does not fall in any of the sets $\{\tilde{\mathcal{D}}_m\}$, then the decoder produces an error flag.) This decoder bases its guess only on $Y(t_1), \ldots, Y(t_n)$ and yet, as we shall next show, succeeds with probability that is at least within $\epsilon$ of the probability of success of $\phi(\cdot)$, i.e.,

$$\Pr\big(\tilde{\phi}(\cdot) \text{ is correct} \,\big|\, M = m\big) \geq \Pr\big(\phi(\cdot) \text{ is correct} \,\big|\, M = m\big) - \epsilon, \quad m \in \mathcal{M}. \quad (26.9)$$

This will imply, in particular, that when averaged over $M$

$$\Pr\big(\tilde{\phi}(\cdot) \text{ is correct}\big) \geq \Pr\big(\phi(\cdot) \text{ is correct}\big) - \epsilon, \quad (26.10)$$

irrespective of the prior on $M$. But by (26.5) an optimal decision rule based on $\mathbf{T}$ is at least as good as $\tilde{\phi}(\cdot)$ and is thus also within $\epsilon$ of $\phi(\cdot)$.[6] Since $\epsilon$ is arbitrary, it follows that an optimal decision rule based on $\mathbf{T}$ is at least as good as $\phi(\cdot)$, thus proving the theorem.

To complete the proof it thus remains to prove (26.9). To that end we note that, since for any sets $\mathcal{A}$ and $\mathcal{B}$ we have $\mathcal{A} \supseteq \mathcal{B} \cap \mathcal{A}$, it follows that $\hat{\mathcal{D}}_m \supseteq \mathcal{D}_m \cap \hat{\mathcal{D}}_m$ and hence, by (26.8),

$$\tilde{\mathcal{D}}_m \supseteq \big(\mathcal{D}_m \cap \hat{\mathcal{D}}_m\big) \setminus \bigcup_{m' < m} \tilde{\mathcal{D}}_{m'}$$

$$\supseteq \big(\mathcal{D}_m \cap \hat{\mathcal{D}}_m\big) \setminus \bigcup_{m' < m} \Big(\mathcal{D}_{m'} \cup \big(\hat{\mathcal{D}}_{m'} \setminus \mathcal{D}_{m'}\big)\Big) \quad (26.11)$$

$$= \big(\mathcal{D}_m \cap \hat{\mathcal{D}}_m\big) \setminus \Big( \bigcup_{m' < m} \mathcal{D}_{m'} \cup \bigcup_{m' < m} \big(\hat{\mathcal{D}}_{m'} \setminus \mathcal{D}_{m'}\big) \Big)$$

$$= \Big( \big(\mathcal{D}_m \cap \hat{\mathcal{D}}_m\big) \setminus \bigcup_{m' < m} \mathcal{D}_{m'} \Big) \setminus \Big( \bigcup_{m' < m} \big(\hat{\mathcal{D}}_{m'} \setminus \mathcal{D}_{m'}\big) \Big)$$

$$= \big(\mathcal{D}_m \cap \hat{\mathcal{D}}_m\big) \setminus \Big( \bigcup_{m' < m} \big(\hat{\mathcal{D}}_{m'} \setminus \mathcal{D}_{m'}\big) \Big) \quad (26.12)$$

$$= \Big(\mathcal{D}_m \setminus \big(\mathcal{D}_m \setminus \hat{\mathcal{D}}_m\big)\Big) \setminus \Big( \bigcup_{m' < m} \big(\hat{\mathcal{D}}_{m'} \setminus \mathcal{D}_{m'}\big) \Big)$$

$$= \mathcal{D}_m \setminus \Big( \big(\mathcal{D}_m \setminus \hat{\mathcal{D}}_m\big) \cup \bigcup_{m' < m} \big(\hat{\mathcal{D}}_{m'} \setminus \mathcal{D}_{m'}\big) \Big), \quad (26.13)$$

where (26.11) follows because $\mathcal{D}_{m'} \cup \big(\hat{\mathcal{D}}_{m'} \setminus \mathcal{D}_{m'}\big) = \mathcal{D}_{m'} \cup \hat{\mathcal{D}}_{m'} \supseteq \hat{\mathcal{D}}_{m'}$, and because, by construction, $\hat{\mathcal{D}}_{m'}$ contains $\mathcal{D}_{m'}$ (see (26.8)); where the equality (26.12) follows because the sets $\{\mathcal{D}_m\}$ are disjoint; and where the other equalities follow by standard set-theoretic identities. It follows from (26.13) that

$$\mathcal{D}_m \subseteq \tilde{\mathcal{D}}_m \cup \big(\mathcal{D}_m \setminus \hat{\mathcal{D}}_m\big) \cup \bigcup_{m' < m} \big(\hat{\mathcal{D}}_{m'} \setminus \mathcal{D}_{m'}\big), \quad (26.14)$$

---

[6]An optimal decision rule based on $\mathbf{T}$ is the Maximum A Posteriori rule that computes the conditional distribution of $M$ given $\mathbf{T}$. But the Markov condition (26.5) implies that the conditional law of $M$ given $\mathbf{T}$ is the same as the conditional law given $\mathbf{T}$ & $(Y(t_1), \ldots, Y(t_\eta))$, so an optimal decision rule based on $\mathbf{T}$ is as good as an optimal decision rule given $\mathbf{T}$ *and* $(Y(t_1), \ldots, Y(t_\eta))$ and is therefore at least as good as $\tilde{\phi}(\cdot)$, which is based on $(Y(t_1), \ldots, Y(t_\eta))$ alone.

because for arbitrary sets $\mathcal{A}, \mathcal{B}, \mathcal{C}$, the relation $\mathcal{A} \supseteq \mathcal{B} \setminus \mathcal{C}$ implies that $\mathcal{B} \subseteq \mathcal{A} \cup \mathcal{C}$. From (26.14) and the Union-of-Events Bound (Theorem 21.5.1) we obtain

$$
\begin{aligned}
\Pr\big(\phi(\cdot) \text{ is correct} \,\big|\, M = m\big) \\
&= \Pr\big[(Y(t)) \in \mathcal{D}_m \,\big|\, M = m\big] \\
&\leq \Pr\big[(Y(t)) \in \tilde{\mathcal{D}}_m \,\big|\, M = m\big] + \Pr\big[(Y(t)) \in (\mathcal{D}_m \setminus \hat{\mathcal{D}}_m) \,\big|\, M = m\big] \\
&\quad + \sum_{m' < m} \Pr\big[(Y(t)) \in \hat{\mathcal{D}}_{m'} \setminus \mathcal{D}_{m'} \,\big|\, M = m\big] \\
&\leq \Pr\big[(Y(t)) \in \tilde{\mathcal{D}}_m \,\big|\, M = m\big] + \sum_{m' \leq m} \Pr\big[(Y(t)) \in \hat{\mathcal{D}}_{m'} \triangle \mathcal{D}_{m'} \,\big|\, M = m\big] \\
&\leq \Pr\big(\tilde{\phi}(\cdot) \text{ is correct} \,\big|\, M = m\big) + m \frac{\epsilon}{\mathsf{M}} \\
&\leq \Pr\big(\tilde{\phi}(\cdot) \text{ is correct} \,\big|\, M = m\big) + \epsilon, \quad m \in \mathcal{M},
\end{aligned}
$$

where the first inequality follows from (26.14) and the Union-of-Events bound; the second inequality follows because for any two sets $\mathcal{A}$ and $\mathcal{B}$ we have $\mathcal{A} \setminus \mathcal{B} \subseteq \mathcal{A} \triangle \mathcal{B}$ and also $\mathcal{B} \setminus \mathcal{A} \subseteq \mathcal{A} \triangle \mathcal{B}$; the third inequality from (26.7); and the final inequality because $m \in \{1, \ldots, \mathsf{M}\}$. This concludes the proof by establishing (26.9). □

## 26.4 Main Result

The main result of this chapter is Theorem 26.4.1, which provides a sufficient statistic for the setup of Section 26.2. A more general version (Theorem 27.3.1) will be proved in Chapter 27. Nevertheless, we have chosen to provide a separate proof of Theorem 26.4.1 in Section 26.8 because the proof of this case is simpler.

**Theorem 26.4.1 (Inner Products with the Mean Signals Suffice).** *In the setup of Section 26.2, the random vector*

$$
\left( \int_{-\infty}^{\infty} Y(t)\, s_1(t) \, \mathrm{d}t, \ldots, \int_{-\infty}^{\infty} Y(t)\, s_{\mathsf{M}}(t) \, \mathrm{d}t \right)^{\mathsf{T}} \tag{26.15}
$$

*forms a sufficient statistic for guessing $M$ based on $(Y(t))$.*

**Proof.** See Section 26.8. □

Because the RV $\int Y(t)\, s_m(t)\, \mathrm{d}t$ can be viewed as a mapping that maps each $\omega \in \Omega$ to the inner product between its trajectory $t \mapsto Y(\omega, t)$ and the signal $t \mapsto s_m(t)$, we denote this random variable by $\langle \mathbf{Y}, \mathbf{s}_m \rangle$.[7] With this notation, the main result is that the $\mathsf{M}$ inner products

$$
\langle \mathbf{Y}, \mathbf{s}_1 \rangle, \ldots, \langle \mathbf{Y}, \mathbf{s}_{\mathsf{M}} \rangle \tag{26.16}
$$

form a sufficient statistic for guessing $M$ based on $(Y(t))$.

---

[7]Here, as throughout, $(\Omega, \mathcal{F}, P)$ denotes the probability space over which all the random variables and stochastic processes in the setup are defined.

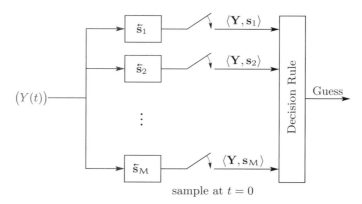

sample at $t = 0$

**Figure 26.1:** Computing the inner product between the observed SP and each of the mean signals and then basing one's decision on these inner products.

This theorem is extremely useful because in combination with Theorem 26.3.2 it demonstrates that, without loss of optimality, we can limit ourselves to guessing rules that use the observation to compute the M inner products (26.16) and that then base their decision on these inner products. Figure 26.1 illustrates such decision rules. The theorem thus helps us to convert the guessing problem from one with a continuous-time observation $(Y(t))$ to a problem of the kind we addressed in Section 21.3, where the observable is a finite-dimensional random vector (the inner products vector, which takes value in in $\mathbb{R}^M$).

We can generalize Theorem 26.4.1 using the linearity of the stochastic integral (Lemma 25.10.3). This generalization allows us to further reduce the dimension of the sufficient statistic vector from the number of messages M to the dimension $d$ of the linear subspace $\mathrm{span}(\mathbf{s}_1, \ldots, \mathbf{s}_M)$ spanned by the mean signals $\mathbf{s}_1, \ldots, \mathbf{s}_M$:

$$d \triangleq \mathrm{Dim}\big(\mathrm{span}(\mathbf{s}_1, \ldots, \mathbf{s}_M)\big). \tag{26.17}$$

**Corollary 26.4.2.** *Let $\tilde{\mathbf{s}}_1, \ldots, \tilde{\mathbf{s}}_n$ be integrable signals that are bandlimited to W Hz.*[8] *If every mean signal can be written as a linear combination of $(\tilde{\mathbf{s}}_1, \ldots, \tilde{\mathbf{s}}_n)$, then the random n-vector*

$$\left( \int_{-\infty}^{\infty} Y(t)\, \tilde{s}_1(t)\, \mathrm{d}t, \ldots, \int_{-\infty}^{\infty} Y(t)\, \tilde{s}_n(t)\, \mathrm{d}t \right)^{\mathsf{T}} \tag{26.18}$$

*forms a sufficient statistic for guessing M based on $(Y(t))$.*

**Proof of the corollary.** By the corollary's hypothesis, every mean signal $\mathbf{s}_m$ can be written as a linear combination of the signals $\{\tilde{\mathbf{s}}_j\}_{j=1}^n$. Thus, to each $m \in \mathcal{M}$ there correspond $n$ coefficients (not necessarily unique) $\alpha_m^{(1)}, \ldots, \alpha_m^{(n)} \in \mathbb{R}$ such that

$$\mathbf{s}_m = \sum_{j=1}^{n} \alpha_m^{(j)}\, \tilde{\mathbf{s}}_j. \tag{26.19}$$

---

[8]The result also holds if the signals are not bandlimited, but we prefer to assume that they are.

Consequently, by the linearity of integration (Lemma 25.10.3), we can compute the integrals appearing in (26.15) from the random $n$-vector (26.18) using the relation

$$\int_{-\infty}^{\infty} Y(t)\, s_m(t)\, \mathrm{d}t = \sum_{j=1}^{n} \alpha_m^{(j)} \int_{-\infty}^{\infty} Y(t)\, \tilde{s}_j(t)\, \mathrm{d}t, \quad m \in \mathcal{M}.$$

From the vector in (26.18) we can thus compute the vector in (26.15), and since the latter forms a sufficient statistic (Theorem 26.4.1) it follows that the former must also form a sufficient statistic (Proposition 22.4.2).[9] $\qquad\qquad\square$

We note that Corollary 26.4.2 does, indeed, generalize the theorem because, by choosing $n = \mathsf{M}$ with $\tilde{\mathbf{s}}_m = \mathbf{s}_m$ for all $m \in \mathcal{M}$, we recover the theorem from the corollary. More interesting is the case where $(\tilde{\mathbf{s}}_1, \ldots, \tilde{\mathbf{s}}_n)$ forms a basis for $\mathrm{span}(\mathbf{s}_1, \ldots, \mathbf{s}_\mathsf{M})$. In this case the corollary provides a sufficient statistic consisting of a random $d$-vector, where $d$ is the dimension of $\mathrm{span}(\mathbf{s}_1, \ldots, \mathbf{s}_\mathsf{M})$. This reduces the number of inner products needed to implement the receiver from $\mathsf{M}$ to $d$. As we shall see, it is particularly convenient to choose $(\tilde{\mathbf{s}}_1, \ldots, \tilde{\mathbf{s}}_n)$ as an orthonormal basis for $\mathrm{span}(\mathbf{s}_1, \ldots, \mathbf{s}_\mathsf{M})$. In this case we shall prefer to refer to $\{\tilde{\mathbf{s}}_j\}$ as $\{\boldsymbol{\phi}_\ell\}_{\ell=1}^{d}$, where, as before, $d$ is the dimension of $\mathrm{span}(\mathbf{s}_1, \ldots, \mathbf{s}_\mathsf{M})$.

## 26.5 Analyzing the Sufficient Statistic

### 26.5.1 The Conditional Law of the Sufficient Statistic

Having reduced the guessing problem from one where the observation is a SP to one where it is a random vector, we can proceed to derive an optimal decision rule based on this vector. To derive such a rule we need the conditional distribution of this vector conditional on each of the hypotheses. Fortunately, this is easy for the problem at hand, because the Gaussianity of the noise $(N(t))$ implies that, conditional on each of the hypotheses, the vectors in (26.15) and (26.18) are Gaussian (Theorem 25.12.1). Their conditional distributions are thus fully specified by their mean vectors and covariance matrices.

The calculation of the mean vectors is straightforward. Indeed, by linearity and by Proposition 25.10.1,

$$\mathsf{E}\left[\int_{-\infty}^{\infty} Y(t)\, s_j(t)\, \mathrm{d}t \,\middle|\, M = m\right] = \mathsf{E}\left[\int_{-\infty}^{\infty} \bigl(s_m(t) + N(t)\bigr)\, s_j(t)\, \mathrm{d}t\right]$$

$$= \langle \mathbf{s}_m, \mathbf{s}_j \rangle + \mathsf{E}\left[\int_{-\infty}^{\infty} N(t)\, s_j(t)\, \mathrm{d}t\right]$$

$$= \langle \mathbf{s}_m, \mathbf{s}_j \rangle, \quad j, m \in \mathcal{M}.$$

Thus, for every $m \in \mathcal{M}$, the conditional mean of the vector in (26.15), conditional on $M = m$, is the vector

$$\bigl(\langle \mathbf{s}_m, \mathbf{s}_1 \rangle, \ldots, \langle \mathbf{s}_m, \mathbf{s}_\mathsf{M} \rangle\bigr)^{\mathsf{T}}. \tag{26.20}$$

---

[9]For the pedantic reader one should add that, by Proposition 25.10.1, the vector in (26.18) is measurable with respect to the $\sigma$-algebra generated by $(Y(t))$.

The calculation of the conditional covariance matrices requires a simple application of Proposition 25.15.2. It yields that the covariance matrix of the vector in (26.15), conditional on $M = m$, is given by the $\mathsf{M} \times \mathsf{M}$ matrix

$$\frac{\mathsf{N}_0}{2} \begin{pmatrix} \langle \mathbf{s}_1, \mathbf{s}_1 \rangle & \langle \mathbf{s}_1, \mathbf{s}_2 \rangle & \cdots & \langle \mathbf{s}_1, \mathbf{s}_\mathsf{M} \rangle \\ \langle \mathbf{s}_2, \mathbf{s}_1 \rangle & \langle \mathbf{s}_2, \mathbf{s}_2 \rangle & \cdots & \langle \mathbf{s}_2, \mathbf{s}_\mathsf{M} \rangle \\ \cdots & \cdots & \ddots & \cdots \\ \langle \mathbf{s}_\mathsf{M}, \mathbf{s}_1 \rangle & \langle \mathbf{s}_\mathsf{M}, \mathbf{s}_2 \rangle & \cdots & \langle \mathbf{s}_\mathsf{M}, \mathbf{s}_\mathsf{M} \rangle \end{pmatrix}. \tag{26.21}$$

Note that the conditional covariance matrix does not depend on the hypothesis $m$ on which we are conditioning, because this hypothesis only influences the mean of $\big( Y(t) \big)$.

More generally, for the sufficient statistic vector in (26.18) we obtain that for every $m \in \mathcal{M}$ the conditional distribution of this vector, conditional on $M = m$, is Gaussian with the $n$-dimensional mean vector

$$\big( \langle \mathbf{s}_m, \tilde{\mathbf{s}}_1 \rangle, \ldots, \langle \mathbf{s}_m, \tilde{\mathbf{s}}_n \rangle \big)^{\mathsf{T}} \tag{26.22}$$

and the $n \times n$ covariance matrix

$$\frac{\mathsf{N}_0}{2} \begin{pmatrix} \langle \tilde{\mathbf{s}}_1, \tilde{\mathbf{s}}_1 \rangle & \langle \tilde{\mathbf{s}}_1, \tilde{\mathbf{s}}_2 \rangle & \cdots & \langle \tilde{\mathbf{s}}_1, \tilde{\mathbf{s}}_n \rangle \\ \langle \tilde{\mathbf{s}}_2, \tilde{\mathbf{s}}_1 \rangle & \langle \tilde{\mathbf{s}}_2, \tilde{\mathbf{s}}_2 \rangle & \cdots & \langle \tilde{\mathbf{s}}_2, \tilde{\mathbf{s}}_n \rangle \\ \cdots & \cdots & \ddots & \cdots \\ \langle \tilde{\mathbf{s}}_n, \tilde{\mathbf{s}}_1 \rangle & \langle \tilde{\mathbf{s}}_n, \tilde{\mathbf{s}}_2 \rangle & \cdots & \langle \tilde{\mathbf{s}}_n, \tilde{\mathbf{s}}_n \rangle \end{pmatrix}. \tag{26.23}$$

(The assumption that the signals $\{\tilde{\mathbf{s}}_j\}$ are bandlimited to $\mathsf{W}$ Hz is not needed in Corollary 26.4.2, but it is needed for the above conditional law to hold.)

### 26.5.2 It Is all in the Geometry!

It is interesting to note that the conditional mean vector in (26.20) and the conditional covariance matrix in (26.21) are fully determined by $\mathsf{N}_0/2$ and by the inner products

$$\big\{ \langle \mathbf{s}_{m'}, \mathbf{s}_{m''} \rangle \big\}_{m', m'' \in \mathcal{M}}; \tag{26.24}$$

the PSD of the noise $\big( N(t) \big)$ outside the band $f \in [-\mathsf{W}, \mathsf{W}]$ is immaterial. Similarly, except in determining the pairwise inner products, the exact waveforms of the mean signals are immaterial. Since the conditional distribution of the sufficient statistic vector (26.15) is Gaussian, and since the distribution of a Gaussian vector is fully determined by its mean vector and its covariance matrices (Theorem 23.6.7), we can conclude:

**Note 26.5.1.** The conditional distribution of the sufficient statistic vector (26.15) given each of the hypotheses is determined by $\mathsf{N}_0$ and by the inner products in (26.24). The PSD of the noise at frequencies outside the band $[-\mathsf{W}, \mathsf{W}]$ is immaterial.

Note, however, that the calculation of the sufficient statistic from the observation $\big( Y(t) \big)$ requires more than just knowledge of the inner products in (26.24); the calculation of the vector (26.15) requires knowledge of the waveforms $\mathbf{s}_1, \ldots, \mathbf{s}_\mathsf{M}$.

Since an optimal decision rule for guessing $M$ based on $\bigl(Y(t)\bigr)$ can be based on the sufficient statistic (Theorem 26.3.2), and since the conditional distribution of the sufficient statistic given each of the hypotheses depends only on $\mathsf{N}_0$ and the inner products (26.24), it follows that:

**Proposition 26.5.2.** *For the setup of Section 26.2, the minimal probability of error that can be achieved in guessing $M$ based on $\bigl(Y(t)\bigr)$ is determined by $\mathsf{N}_0$, by the inner products (26.24), and by the prior $\{\pi_m\}_{m\in\mathcal{M}}$.*

### 26.5.3   Orthonormal Bases

The conditional distribution of the sufficient statistic given each of the hypotheses is easier to manipulate if we choose the functions $\{\tilde{\mathsf{s}}_j\}$ in (26.18) to form an orthonormal basis for the linear subspace spanned by the mean signals. In this case we denote the basis functions by $\phi_1,\ldots,\phi_d$ so

$$\operatorname{span}(\phi_1,\ldots,\phi_d) = \operatorname{span}(\mathsf{s}_1,\ldots,\mathsf{s}_M), \tag{26.25a}$$

$$\langle\phi_{\ell'},\phi_{\ell''}\rangle = \mathrm{I}\{\ell' = \ell''\}, \quad \ell',\ell'' \in \{1,\ldots,d\}, \tag{26.25b}$$

where $d$ is the dimension of the linear subspace spanned by the mean signals (26.17). Such functions $\phi_1,\ldots,\phi_d$ can be found, for example, using the Gram-Schmidt procedure (Section 4.6.6).[10] We denote the sufficient statistic vector (26.18) by $\mathbf{T} = (T^{(1)},\ldots,T^{(d)})^{\mathsf{T}}$:

$$T^{(\ell)} = \int_{-\infty}^{\infty} Y(t)\,\phi_\ell(t)\,\mathrm{d}t$$

$$= \langle\mathbf{Y},\phi_\ell\rangle, \quad \ell = 1,\ldots,d. \tag{26.26}$$

Figure 26.2 depicts a block diagram of a circuit that computes the inner products of the received waveform with each of the basis signals.

By (26.22) and (26.23) we obtain that for every $m \in \mathcal{M}$ the conditional distribution of $\mathbf{T}$ given that $M = m$ is Gaussian with mean

$$\mathsf{E}\bigl[\mathbf{T} \mid M = m\bigr] = \bigl(\langle\mathsf{s}_m,\phi_1\rangle,\ldots,\langle\mathsf{s}_m,\phi_d\rangle\bigr)^{\mathsf{T}} \tag{26.27}$$

and covariance matrix $(\mathsf{N}_0/2)\,\mathsf{I}_d$, where $\mathsf{I}_d$ denotes the $d \times d$ identity matrix. The components of $\mathbf{T}$ are thus conditionally independent and of equal variance $\mathsf{N}_0/2$ (but not of equal mean). Consequently, we can express the conditional density of $\mathbf{T}$, conditional on $M = m$, at every point $\mathbf{t} = (t^{(1)},\ldots,t^{(d)})^{\mathsf{T}} \in \mathbb{R}^d$ using this conditional independence and the explicit form of the univariate Gaussian density (19.6) as

$$f_{\mathbf{T}\mid M=m}(\mathbf{t}) = \prod_{\ell=1}^{d} \frac{1}{\sqrt{2\pi\mathsf{N}_0/2}} \exp\left(-\frac{\bigl(t^{(\ell)} - \langle\mathsf{s}_m,\phi_\ell\rangle\bigr)^2}{2\mathsf{N}_0/2}\right)$$

$$= \frac{1}{(\pi\mathsf{N}_0)^{d/2}} \exp\left(-\frac{1}{\mathsf{N}_0}\sum_{\ell=1}^{d}\bigl(t^{(\ell)} - \langle\mathsf{s}_m,\phi_\ell\rangle\bigr)^2\right), \quad \mathbf{t} \in \mathbb{R}^d. \tag{26.28}$$

---

[10]Since the mean signals are bandlimited, the only zero-energy element of $\operatorname{span}(\mathsf{s}_1,\ldots,\mathsf{s}_M)$ is the all-zero signal (Note 6.4.2). Consequently, $\operatorname{span}(\mathsf{s}_1,\ldots,\mathsf{s}_M)$ has an orthonormal basis (Proposition 4.6.10), which can be found using the Gram-Schmidt procedure (Section 4.6.6).

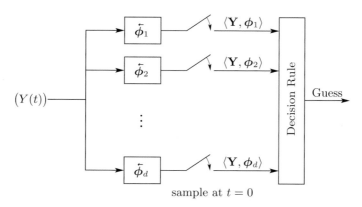

**Figure 26.2:** Computing the inner products $T^{(\ell)} = \langle \mathbf{Y}, \phi_\ell \rangle$ for $\ell = 1, \ldots, d$ from the received waveform.

Note that with proper translation (Table 26.1) the conditional distribution of $\mathbf{T}$ is very similar to the one we addressed in Section 21.6; see (21.50). In fact, it is a special case of the distribution studied in Section 21.6: $\mathbf{Y}$ there corresponds to $\mathbf{T}$ here; J there corresponds to $d$ here; $\sigma^2$ there corresponds to $N_0/2$ here; and the mean vector $\mathbf{s}_m$ associated with Message $m$ corresponds to the vector

$$\left( \langle \mathbf{s}_m, \phi_1 \rangle, \ldots, \langle \mathbf{s}_m, \phi_d \rangle \right)^{\mathsf{T}} \tag{26.29}$$

here. Consequently, we can use the results from Section 21.6 and, more specifically, Proposition 21.6.1, to derive an optimal decision rule for guessing $M$ based on $\mathbf{T}$. We adopt this approach when we next derive an optimal decision rule for our setup.

| | In Section 21.6 | Here |
|---|---|---|
| number of components of observed vector | J | $d$ |
| variance of noise added to each component | $\sigma^2$ | $N_0/2$ |
| number of hypotheses | M | M |
| conditional mean of observation given $M = m$ | $\left(s_m^{(1)}, \ldots, s_m^{(J)}\right)^{\mathsf{T}}$ | $\left( \langle \mathbf{s}_m, \phi_1 \rangle, \ldots, \langle \mathbf{s}_m, \phi_d \rangle \right)^{\mathsf{T}}$ |
| sum of squared components of mean vector | $\sum_{j=1}^{J} \left(s_m^{(j)}\right)^2$ | $\sum_{\ell=1}^{d} \left( \langle \mathbf{s}_m, \phi_\ell \rangle \right)^2 = \int_{-\infty}^{\infty} s_m^2(t) \, \mathrm{d}t$ |

**Table 26.1:** The setup in Section 21.6 and here.

## 26.6 Optimal Guessing Rule

We are finally ready to derive an optimal guessing rule for our setup. Recall that, by Corollary 26.4.2, if $(\phi_1, \ldots, \phi_d)$ is an orthonormal basis for the linear space spanned by the mean signals, then the random vector $\mathbf{T}$ defined in (26.26) forms a sufficient statistic for guessing $M$ based on $\big(Y(t)\big)$. Consequently, by Theorem 26.3.2, it is optimal to use the observation $\big(Y(t)\big)$ to compute the vector $\mathbf{T}$ and to then use the MAP rule to guess $M$ based on $\mathbf{T}$ (Theorem 21.3.3). We shall do just that. We first present the resulting rule in terms of the orthonormal basis $(\phi_1, \ldots, \phi_d)$ and then show that the rule does not depend on the specific choice of the orthonormal basis.

In deriving the decision rule we shall repeatedly use the fact that if $(\phi_1, \ldots, \phi_d)$ is an orthonormal basis for $\mathrm{span}(\mathbf{s}_1, \ldots, \mathbf{s}_M)$, then, by Proposition 4.6.4,

$$\mathbf{s}_m = \sum_{\ell=1}^{d} \langle \mathbf{s}_m, \phi_\ell \rangle \phi_\ell, \quad m \in \mathcal{M}, \tag{26.30}$$

and, by Proposition 4.6.9,

$$\|\mathbf{s}_m\|_2^2 = \sum_{\ell=1}^{d} \langle \mathbf{s}_m, \phi_\ell \rangle^2, \quad m \in \mathcal{M}. \tag{26.31}$$

### 26.6.1 The Decision Rule in Terms of $(\phi_1, \ldots, \phi_d)$

As we have noted, the conditional density $f_{\mathbf{T}|M=m}(\cdot)$ in (26.28) is of the form we discussed in Section 21.6 (Table 26.1). By Proposition 21.6.1 we thus obtain:

**Theorem 26.6.1.** *Let $M$, $\mathbf{s}_1, \ldots, \mathbf{s}_M$, and $\big(Y(t)\big)$ be as in our setup, and let the $d$-tuple $(\phi_1, \ldots, \phi_d)$ be an orthonormal basis for $\mathrm{span}(\mathbf{s}_1, \ldots, \mathbf{s}_M)$.*

*(i) The decision rule that guesses uniformly at random from among all the messages $\tilde{m} \in \mathcal{M}$ for which*

$$\ln \pi_{\tilde{m}} - \frac{\sum_{\ell=1}^{d} \big(\langle \mathbf{Y}, \phi_\ell \rangle - \langle \mathbf{s}_{\tilde{m}}, \phi_\ell \rangle\big)^2}{\mathsf{N}_0}$$
$$= \max_{m' \in \mathcal{M}} \left\{ \ln \pi_{m'} - \frac{\sum_{\ell=1}^{d} \big(\langle \mathbf{Y}, \phi_\ell \rangle - \langle \mathbf{s}_{m'}, \phi_\ell \rangle\big)^2}{\mathsf{N}_0} \right\} \tag{26.32}$$

*minimizes the probability of a guessing error.*

*(ii) If $M$ has a uniform distribution, then this rule does not depend on the value of $\mathsf{N}_0$. It chooses uniformly at random from among all the messages $\tilde{m} \in \mathcal{M}$ for which*

$$\sum_{\ell=1}^{d} \big(\langle \mathbf{Y}, \phi_\ell \rangle - \langle \mathbf{s}_{\tilde{m}}, \phi_\ell \rangle\big)^2 = \min_{m' \in \mathcal{M}} \left\{ \sum_{\ell=1}^{d} \big(\langle \mathbf{Y}, \phi_\ell \rangle - \langle \mathbf{s}_{m'}, \phi_\ell \rangle\big)^2 \right\}. \tag{26.33}$$

*(iii) If $M$ has a uniform distribution and, in addition, the mean signals are of equal energy, i.e.,*

$$\|\mathbf{s}_1\|_2 = \|\mathbf{s}_2\|_2 = \cdots = \|\mathbf{s}_M\|_2, \tag{26.34}$$

*then these decision rules are equivalent to the maximum-correlation rule that guesses uniformly from among all the messages $\tilde{m} \in \mathcal{M}$ for which*

$$\sum_{\ell=1}^{d} \langle \mathbf{s}_{\tilde{m}}, \boldsymbol{\phi}_\ell \rangle \langle \mathbf{Y}, \boldsymbol{\phi}_\ell \rangle = \max_{m' \in \mathcal{M}} \sum_{\ell=1}^{d} \langle \mathbf{s}_{m'}, \boldsymbol{\phi}_\ell \rangle \langle \mathbf{Y}, \boldsymbol{\phi}_\ell \rangle. \tag{26.35}$$

**Proof.** The theorem follows directly from Proposition 21.6.1. For Part (iii) we need to note that, by (26.31), Condition (26.34) is equivalent to the condition

$$\sum_{\ell=1}^{d} \langle \mathbf{s}_1, \boldsymbol{\phi}_\ell \rangle^2 = \sum_{\ell=1}^{d} \langle \mathbf{s}_2, \boldsymbol{\phi}_\ell \rangle^2 = \cdots = \sum_{\ell=1}^{d} \langle \mathbf{s}_\mathsf{M}, \boldsymbol{\phi}_\ell \rangle^2, \tag{26.36}$$

which is the condition needed in Proposition 21.6.1. □

Note that, because $(\boldsymbol{\phi}_1, \ldots, \boldsymbol{\phi}_d)$ is an orthonormal basis for $\text{span}(\mathbf{s}_1, \ldots, \mathbf{s}_\mathsf{M})$, the signals $\mathbf{s}_{m'}$ and $\mathbf{s}_{m''}$ differ, if, and only if, the vectors $(\langle \mathbf{s}_{m'}, \boldsymbol{\phi}_1 \rangle, \ldots, \langle \mathbf{s}_{m'}, \boldsymbol{\phi}_d \rangle)^\mathsf{T}$ and $(\langle \mathbf{s}_{m''}, \boldsymbol{\phi}_1 \rangle, \ldots, \langle \mathbf{s}_{m''}, \boldsymbol{\phi}_d \rangle)^\mathsf{T}$ in $\mathbb{R}^d$ differ. Consequently, by Proposition 21.6.2:

**Note 26.6.2.** If the mean signals $\mathbf{s}_1, \ldots, \mathbf{s}_\mathsf{M}$ are distinct, then the probability of a tie, i.e., that more than one message $\tilde{m} \in \mathcal{M}$ satisfies (26.32), is zero.

## 26.6.2 The Decision Rule without Reference to a Basis

We next derive a representation of our decision rule without reference to a specific orthonormal basis.

**Theorem 26.6.3.** *Consider the problem of guessing $M$ based on $\big(Y(t)\big)$ in our setup.*

*(i) The decision rule that guesses uniformly at random from among all the messages $\tilde{m} \in \mathcal{M}$ for which*

$$\ln \pi_{\tilde{m}} + \frac{2}{\mathsf{N}_0} \left( \int_{-\infty}^{\infty} Y(t) \, s_{\tilde{m}}(t) \, \mathrm{d}t - \frac{1}{2} \int_{-\infty}^{\infty} s_{\tilde{m}}^2(t) \, \mathrm{d}t \right)$$
$$= \max_{m' \in \mathcal{M}} \left\{ \ln \pi_{m'} + \frac{2}{\mathsf{N}_0} \left( \int_{-\infty}^{\infty} Y(t) \, s_{m'}(t) \, \mathrm{d}t - \frac{1}{2} \int_{-\infty}^{\infty} s_{m'}^2(t) \, \mathrm{d}t \right) \right\} \tag{26.37}$$

*minimizes the probability of error.*

*(ii) If $M$ has a uniform distribution, then this rule does not depend on the value of $\mathsf{N}_0$. It chooses uniformly at random from among all the messages $\tilde{m} \in \mathcal{M}$ for which*

$$\int_{-\infty}^{\infty} Y(t) \, s_{\tilde{m}}(t) \, \mathrm{d}t - \frac{1}{2} \int_{-\infty}^{\infty} s_{\tilde{m}}^2(t) \, \mathrm{d}t$$
$$= \max_{m' \in \mathcal{M}} \left\{ \int_{-\infty}^{\infty} Y(t) \, s_{m'}(t) \, \mathrm{d}t - \frac{1}{2} \int_{-\infty}^{\infty} s_{m'}^2(t) \, \mathrm{d}t \right\}. \tag{26.38}$$

*(iii) If $M$ has a uniform distribution and, in addition, the mean signals are of equal energy, i.e.,*

$$\|\mathbf{s}_1\|_2 = \|\mathbf{s}_2\|_2 = \cdots = \|\mathbf{s}_M\|_2\,,$$

*then these decision rules are equivalent to the maximum-correlation rule that guesses uniformly from among all the messages $\tilde{m} \in \mathcal{M}$ for which*

$$\int_{-\infty}^{\infty} Y(t)\, s_{\tilde{m}}(t)\, \mathrm{d}t = \max_{m' \in \mathcal{M}} \left\{ \int_{-\infty}^{\infty} Y(t)\, s_{m'}(t)\, \mathrm{d}t \right\}. \tag{26.39}$$

**Proof.** We shall prove Part (i) using Theorem 26.6.1 (i). To this end we begin by noting that

$$\ln \pi_{m'} - \frac{\sum_{\ell=1}^{d} \big( \langle \mathbf{Y}, \boldsymbol{\phi}_\ell \rangle - \langle \mathbf{s}_{m'}, \boldsymbol{\phi}_\ell \rangle \big)^2}{\mathsf{N}_0}$$

can be expressed by opening the square as

$$\ln \pi_{m'} - \frac{1}{\mathsf{N}_0} \sum_{\ell=1}^{d} \langle \mathbf{Y}, \boldsymbol{\phi}_\ell \rangle^2 + \frac{2}{\mathsf{N}_0} \sum_{\ell=1}^{d} \langle \mathbf{Y}, \boldsymbol{\phi}_\ell \rangle \langle \mathbf{s}_{m'}, \boldsymbol{\phi}_\ell \rangle - \frac{1}{\mathsf{N}_0} \sum_{\ell=1}^{d} \langle \mathbf{s}_{m'}, \boldsymbol{\phi}_\ell \rangle^2.$$

Since the term

$$-\frac{1}{\mathsf{N}_0} \sum_{\ell=1}^{d} \langle \mathbf{Y}, \boldsymbol{\phi}_\ell \rangle^2$$

does not depend on the hypothesis, it is optimal to choose a message at random from among all the message $\tilde{m}$ satisfying

$$\ln \pi_{\tilde{m}} + \frac{2}{\mathsf{N}_0} \sum_{\ell=1}^{d} \langle \mathbf{Y}, \boldsymbol{\phi}_\ell \rangle \langle \mathbf{s}_{\tilde{m}}, \boldsymbol{\phi}_\ell \rangle - \frac{1}{\mathsf{N}_0} \sum_{\ell=1}^{d} \langle \mathbf{s}_{\tilde{m}}, \boldsymbol{\phi}_\ell \rangle^2$$

$$= \max_{m' \in \mathcal{M}} \left\{ \ln \pi_{m'} + \frac{2}{\mathsf{N}_0} \sum_{\ell=1}^{d} \langle \mathbf{Y}, \boldsymbol{\phi}_\ell \rangle \langle \mathbf{s}_{m'}, \boldsymbol{\phi}_\ell \rangle - \frac{1}{\mathsf{N}_0} \sum_{\ell=1}^{d} \langle \mathbf{s}_{m'}, \boldsymbol{\phi}_\ell \rangle^2 \right\}.$$

Part (i) of the theorem now follows from this rule using (26.31) and by noting that

$$\sum_{\ell=1}^{d} \langle \mathbf{Y}, \boldsymbol{\phi}_\ell \rangle \langle \mathbf{s}_m, \boldsymbol{\phi}_\ell \rangle = \left\langle \mathbf{Y}, \sum_{\ell=1}^{d} \langle \mathbf{s}_m, \boldsymbol{\phi}_\ell \rangle \boldsymbol{\phi}_\ell \right\rangle$$

$$= \langle \mathbf{Y}, \mathbf{s}_m \rangle, \quad m \in \mathcal{M},$$

where the first equality follows by linearity (Lemma 25.10.3) and the second from (26.30).

Part (ii) follows by noting that if $M$ is uniform, then $\ln \pi_m$ does not depend on the hypothesis $m$.

Part (iii) follows from Part (ii) because if all the mean signals are of equal energy, then the term

$$\int_{-\infty}^{\infty} s_m^2(t)\, \mathrm{d}t$$

does not depend on the hypothesis. $\qquad\square$

By Note 26.6.2 we have:

**Note 26.6.4.** If the mean signals are distinct, then the probability of a tie is zero.

## 26.7   Performance Analysis

The decision rule we derived in Section 26.6.1 uses the observed SP $(Y(t))$ to compute the vector $\mathbf{T}$ of inner products with an orthonormal basis $(\boldsymbol{\phi}_1, \ldots, \boldsymbol{\phi}_d)$ via (26.26), with the result that the vector $\mathbf{T}$ has the conditional law specified in (26.28). Our decision rule then performs MAP decoding of $M$ based on $\mathbf{T}$. Consequently, the performance of our decoding rule is identical to the performance of the MAP rule for guessing $M$ based on a vector $\mathbf{T}$ having the conditional law (26.28). The performance of this latter decoding rule was studied in Section 21.6. All that remains is to translate the results from that section in order to obtain performance bounds on our decoder.

To translate the results from Section 21.6 we need to substitute $\mathsf{N}_0/2$ for $\sigma^2$ there; $d$ for $\mathsf{J}$ there; and (26.29) for the mean vectors there. But there is one more translation we need: the bounds in Section 21.6 are expressed in terms of the Euclidean distance between the mean vectors, and here we prefer to express the bounds in terms of the distance between the mean signals. Fortunately, as we next show, the translation is straightforward. Because $(\boldsymbol{\phi}_1, \ldots, \boldsymbol{\phi}_d)$ is an orthonormal basis for $\mathrm{span}(\mathbf{s}_1, \ldots, \mathbf{s}_\mathsf{M})$, it follows from Proposition 4.6.9 that

$$\sum_{\ell=1}^{d} \langle \mathbf{v}, \boldsymbol{\phi}_\ell \rangle^2 = \|\mathbf{v}\|_2^2, \quad \mathbf{v} \in \mathrm{span}(\mathbf{s}_1, \ldots, \mathbf{s}_\mathsf{M}). \tag{26.40}$$

Substituting $\mathbf{s}_{m'} - \mathbf{s}_{m''}$ for $\mathbf{v}$ in this identity yields

$$\sum_{\ell=1}^{d} \left( \langle \mathbf{s}_{m'}, \boldsymbol{\phi}_\ell \rangle - \langle \mathbf{s}_{m''}, \boldsymbol{\phi}_\ell \rangle \right)^2 = \|\mathbf{s}_{m'} - \mathbf{s}_{m''}\|_2^2$$

$$= \int_{-\infty}^{\infty} \left( s_{m'}(t) - s_{m''}(t) \right)^2 \mathrm{d}t,$$

where we have also used the fact that for $\mathbf{v} = \mathbf{s}_{m'} - \mathbf{s}_{m''}$ we have, by the linearity of the inner product in its left argument, $\langle \mathbf{v}, \boldsymbol{\phi}_\ell \rangle = \langle \mathbf{s}_{m'}, \boldsymbol{\phi}_\ell \rangle - \langle \mathbf{s}_{m''}, \boldsymbol{\phi}_\ell \rangle$. Thus, the squared Euclidean distance between two mean vectors in Section 21.6 is equal to the energy in the difference between the corresponding mean signals in our setup.

Denoting by $p_{\mathrm{MAP}}(\mathrm{error}|M = m)$ the conditional probability of error of our decoder conditional on $M = m$, and denoting by $p^*(\mathrm{error})$ its unconditioned probability of error (which is the optimal probability of error)

$$p^*(\mathrm{error}) = \sum_{m \in \mathcal{M}} \pi_m \, p_{\mathrm{MAP}}(\mathrm{error}|M = m), \tag{26.41}$$

we obtain from (21.57)

$$p_{\mathrm{MAP}}(\mathrm{error}|M = m) \leq \sum_{m' \neq m} \mathcal{Q}\left( \frac{\|\mathbf{s}_m - \mathbf{s}_{m'}\|_2}{\sqrt{2\mathsf{N}_0}} + \frac{\sqrt{\mathsf{N}_0/2}}{\|\mathbf{s}_m - \mathbf{s}_{m'}\|_2} \ln \frac{\pi_m}{\pi_{m'}} \right) \tag{26.42}$$

and hence by, (26.41),

$$p^*(\text{error}) \leq \sum_{m \in \mathcal{M}} \pi_m \sum_{m' \neq m} \mathcal{Q}\left( \frac{\|\mathbf{s}_m - \mathbf{s}_{m'}\|_2}{\sqrt{2N_0}} + \frac{\sqrt{N_0/2}}{\|\mathbf{s}_m - \mathbf{s}_{m'}\|_2} \ln \frac{\pi_m}{\pi_{m'}} \right). \quad (26.43)$$

When $M$ is uniform these bounds simplify to

$$\boxed{p_{\text{MAP}}(\text{error}|M=m) \leq \sum_{m' \neq m} \mathcal{Q}\left( \sqrt{\frac{\|\mathbf{s}_m - \mathbf{s}_{m'}\|_2^2}{2N_0}} \right), \quad M \text{ uniform}} \quad (26.44)$$

and

$$p^*(\text{error}) \leq \frac{1}{M} \sum_{m \in \mathcal{M}} \sum_{m' \neq m} \mathcal{Q}\left( \sqrt{\frac{\|\mathbf{s}_m - \mathbf{s}_{m'}\|_2^2}{2N_0}} \right), \quad M \text{ uniform.} \quad (26.45)$$

Similarly, we can use the results from Section 21.6 to lower-bound the probability of a guessing error. Indeed, using (21.63) we obtain

$$p_{\text{MAP}}(\text{error}|M=m) \geq \max_{m' \neq m} \mathcal{Q}\left( \frac{\|\mathbf{s}_m - \mathbf{s}_{m'}\|_2}{\sqrt{2N_0}} + \frac{\sqrt{N_0/2}}{\|\mathbf{s}_m - \mathbf{s}_{m'}\|_2} \ln \frac{\pi_m}{\pi_{m'}} \right), \quad (26.46)$$

$$p^*(\text{error}) \geq \sum_{m \in \mathcal{M}} \pi_m \max_{m' \neq m} \mathcal{Q}\left( \frac{\|\mathbf{s}_m - \mathbf{s}_{m'}\|_2}{\sqrt{2N_0}} + \frac{\sqrt{N_0/2}}{\|\mathbf{s}_m - \mathbf{s}_{m'}\|_2} \ln \frac{\pi_m}{\pi_{m'}} \right). \quad (26.47)$$

For a uniform prior these bounds simplify to

$$\boxed{p_{\text{MAP}}(\text{error}|M=m) \geq \max_{m' \neq m} \mathcal{Q}\left( \sqrt{\frac{\|\mathbf{s}_m - \mathbf{s}_{m'}\|_2^2}{2N_0}} \right), \quad M \text{ uniform,}} \quad (26.48)$$

$$p^*(\text{error}) \geq \frac{1}{M} \sum_{m \in \mathcal{M}} \max_{m' \neq m} \mathcal{Q}\left( \sqrt{\frac{\|\mathbf{s}_m - \mathbf{s}_{m'}\|_2^2}{2N_0}} \right), \quad M \text{ uniform.} \quad (26.49)$$

## 26.8 Proof of Theorem 26.4.1

### 26.8.1 A Lemma

We begin with a lemma regarding sufficient statistics in testing whether a random vector $\mathbf{Y}$ was drawn $\mathcal{N}(\boldsymbol{\mu}, \Lambda)$ or $\mathcal{N}(-\boldsymbol{\mu}, \Lambda)$.

**Lemma 26.8.1.** *Let $H$ be a binary RV, and let the random vector $\mathbf{Y}$ be $\mathcal{N}(\boldsymbol{\mu}, \Lambda)$ conditional on $H = 0$ and $\mathcal{N}(-\boldsymbol{\mu}, \Lambda)$ conditional on $H = 1$. If $\boldsymbol{\mu}$ is a scalar multiple of the last column of $\Lambda$, then the last component of $\mathbf{Y}$ forms a sufficient statistic for guessing $H$ based on $\mathbf{Y}$.*

**Proof.** Let $n$ denote the number of components of the vectors $\mathbf{Y}$ and $\boldsymbol{\mu}$, so $\Lambda$ is $n \times n$. To show that $Y^{(n)}$ is a sufficient statistic we shall calculate the log likelihood-ratio function and then show that it is computable from $Y^{(n)}$. This approach, while straightforward, does not prove the lemma in its fullest generality because it only covers the case where $\mathbf{Y}$ has a density, i.e., when the covariance matrix $\Lambda$ is nonsingular. Referring the reader to Section 26.14 on Page 606 for a somewhat less intuitive proof that covers all cases, we proceed here to address the case where $\Lambda$ is nonsingular.

The condition that $\boldsymbol{\mu}$ is a scalar multiple of the last column of $\Lambda$ is equivalent to the existence of some $\alpha \in \mathbb{R}$ such that

$$\boldsymbol{\mu} = \Lambda \begin{pmatrix} 0 \\ \vdots \\ 0 \\ \alpha \end{pmatrix}. \tag{26.50}$$

When $\Lambda$ is nonsingular we can use the explicit form of the density of the multivariate Gaussian distribution (23.56) to express the log likelihood-ratio as

$$
\begin{aligned}
\ln \frac{f_{\mathbf{Y}|H=0}(\mathbf{y})}{f_{\mathbf{Y}|H=1}(\mathbf{y})} &= \ln \frac{\frac{1}{\sqrt{(2\pi)^n \det \Lambda}} e^{-\frac{1}{2}(\mathbf{y}-\boldsymbol{\mu})^\mathsf{T}\Lambda^{-1}(\mathbf{y}-\boldsymbol{\mu})}}{\frac{1}{\sqrt{(2\pi)^n \det \Lambda}} e^{-\frac{1}{2}(\mathbf{y}+\boldsymbol{\mu})^\mathsf{T}\Lambda^{-1}(\mathbf{y}+\boldsymbol{\mu})}} \\
&= \frac{1}{2}(\mathbf{y}+\boldsymbol{\mu})^\mathsf{T}\Lambda^{-1}(\mathbf{y}+\boldsymbol{\mu}) - \frac{1}{2}(\mathbf{y}-\boldsymbol{\mu})^\mathsf{T}\Lambda^{-1}(\mathbf{y}-\boldsymbol{\mu}) \\
&= \mathbf{y}^\mathsf{T}\Lambda^{-1}\boldsymbol{\mu} + \boldsymbol{\mu}^\mathsf{T}\Lambda^{-1}\mathbf{y} \\
&= 2\mathbf{y}^\mathsf{T}\Lambda^{-1}\boldsymbol{\mu} \tag{26.51} \\
&= 2\mathbf{y}^\mathsf{T}\Lambda^{-1}\Lambda(0,\dots,0,\alpha)^\mathsf{T} \tag{26.52} \\
&= 2\alpha y^{(n)}, \quad \mathbf{y} \in \mathbb{R}^n, \tag{26.53}
\end{aligned}
$$

where (26.51) holds because the scalar $\boldsymbol{\mu}^\mathsf{T}\Lambda^{-1}\mathbf{y}$ is equal to its transpose, and the latter—by the transposition law $(\mathsf{AB})^\mathsf{T} = \mathsf{B}^\mathsf{T}\mathsf{A}^\mathsf{T}$—is given by $\mathbf{y}^\mathsf{T}(\Lambda^{-1})^\mathsf{T}\boldsymbol{\mu}$, which by the symmetry of $\Lambda$ (and hence also of its inverse), is equal to $\mathbf{y}^\mathsf{T}\Lambda^{-1}\boldsymbol{\mu}$; and where (26.52) follows from (26.50).

It follows from (26.53) that the likelihood-ratio is computable from the last component of $\mathbf{Y}$, thus establishing that this component forms a sufficient statistic (Definition 20.12.2). $\qquad\square$

### 26.8.2 The Binary Antipodal Case

We begin the proof of Theorem 26.4.1 by considering the special case of binary hypothesis testing ($\mathsf{M} = 2$) where the mean signals are antipodal to each other, i.e., when their sum is the all-zero signal. Since we are now treating the binary hypothesis testing setting, we denote the RV we wish to guess by $H$ and assume that it takes value in the set $\{0, 1\}$. We denote the mean signal corresponding to $H = 0$ by $\mathbf{s}$, so the mean signal corresponding to $H = 1$ is $-\mathbf{s}$. We assume that $\mathbf{s}$

is an integrable signal that is bandlimited to $W$ Hz. Conditional on $H = 0$, the received signal $(Y(t))$ is given at each $t \in \mathbb{R}$ by $s(t) + N(t)$, where $(N(t))$ is white Gaussian noise of PSD $N_0/2$ with respect to the bandwidth $W$. Conditional on $H = 1$ the time-$t$ received signal is $-s(t) + N(t)$.

Recall from Definition 26.3.1 that to show that $\langle \mathbf{Y}, \mathbf{s} \rangle$ forms a sufficient statistic for guessing $H$ based on the observation $(Y(t))$ we need to show that for every positive integer $\eta$ and every choice of the epochs $t_1, \ldots, t_\eta \in \mathbb{R}$ the RV $\langle \mathbf{Y}, \mathbf{s} \rangle$ forms a sufficient statistic for guessing $H$ based on the observation consisting of the random vector[11]

$$\big(Y(t_1), \ldots, Y(t_\eta), \langle \mathbf{Y}, \mathbf{s} \rangle\big)^{\mathsf{T}}. \tag{26.54}$$

This we prove by showing that this vector satisfies the assumptions of Lemma 26.8.1.

Denoting the conditional mean of this vector, conditional on $H = 0$, by $\boldsymbol{\mu}$, we have

$$\boldsymbol{\mu} = \big(s(t_1), \ldots, s(t_\eta), \|\mathbf{s}\|_2^2\big)^{\mathsf{T}}, \tag{26.55}$$

because

$$\begin{aligned}
\mathsf{E}\big[Y(t_\nu) \,\big|\, H = 0\big] &= \mathsf{E}\big[s(t_\nu) + N(t_\nu) \,\big|\, H = 0\big] \\
&= s(t_\nu) + \mathsf{E}[N(t_\nu)] \\
&= s(t_\nu), \quad \nu = 1, \ldots, \eta,
\end{aligned}$$

and

$$\begin{aligned}
\mathsf{E}\big[\langle \mathbf{Y}, \mathbf{s} \rangle \,\big|\, H = 0\big] &= \mathsf{E}\big[\langle \mathbf{s} + \mathbf{N}, \mathbf{s} \rangle\big] \\
&= \|\mathbf{s}\|_2^2 + \mathsf{E}[\langle \mathbf{N}, \mathbf{s} \rangle] \\
&= \|\mathbf{s}\|_2^2
\end{aligned}$$

(Theorem 25.12.2). The conditional covariance matrix $\Lambda$ of the vector in (26.54) conditional on $H = 0$ is given by the $(\eta + 1) \times (\eta + 1)$ matrix

$$\Lambda = \begin{pmatrix}
\mathsf{K}_{NN}(0) & \mathsf{K}_{NN}(t_1 - t_2) & \ldots & \mathsf{K}_{NN}(t_1 - t_\eta) & s(t_1)\mathsf{N}_0/2 \\
\mathsf{K}_{NN}(t_2 - t_1) & \mathsf{K}_{NN}(0) & \ldots & \mathsf{K}_{NN}(t_2 - t_\eta) & s(t_2)\mathsf{N}_0/2 \\
\ldots & \ldots & \ldots & \ldots & \ldots \\
\ldots & \ldots & \ldots & \ldots & \ldots \\
\ldots & \ldots & \ldots & \ldots & \ldots \\
\mathsf{K}_{NN}(t_\eta - t_1) & \mathsf{K}_{NN}(t_\eta - t_2) & \ldots & \mathsf{K}_{NN}(0) & s(t_\eta)\mathsf{N}_0/2 \\
s(t_1)\mathsf{N}_0/2 & s(t_2)\mathsf{N}_0/2 & \ldots & s(t_\eta)\mathsf{N}_0/2 & \|\mathbf{s}\|_2^2 \mathsf{N}_0/2
\end{pmatrix} \tag{26.56}$$

(Proposition 25.15.2). Conditional on $H = 1$, the mean of the vector in (26.54) is $-\boldsymbol{\mu}$ and the covariance matrix is also $\Lambda$. From Proposition 25.11.1 regarding linear functionals of Gaussian stochastic processes, it follows that, conditional on $H = 0$, the vector in (26.54) is Gaussian. Likewise conditional on $H = 1$. And by (26.55) & (26.56) the mean vector $\boldsymbol{\mu}$ is equal to the last column of the covariance matrix (26.56) scaled by $2/\mathsf{N}_0$.

---

[11]The measurability of $\langle \mathbf{Y}, \mathbf{s} \rangle$ with respect to the $\sigma$-algebra generated by $(Y(t))$ follows from Proposition 25.10.1.

Having established that the random vector (26.54) satisfies the hypotheses of Lemma 26.8.1, we can infer that its last component, namely $\langle \mathbf{Y}, \mathbf{s} \rangle$, forms a sufficient statistic for guessing $H$ based on the vector in (26.54). Since $\eta \in \mathbb{N}$ and $t_1, \ldots, t_\eta \in \mathbb{R}$ are here arbitrary, this proves that $\langle \mathbf{Y}, \mathbf{s} \rangle$ is sufficient for guessing $H$ based on $(Y(t))$, thus proving Theorem 26.4.1 for the two-hypotheses case with antipodal mean signals.

### 26.8.3   The General Binary Case

We next prove Theorem 26.4.1 in the more general binary hypothesis testing setting where the mean signals are not necessarily antipodal. We denote the mean signals corresponding to $H = 0$ and $H = 1$ by $\mathbf{s}_0$ and $\mathbf{s}_1$, and we assume that both are integrable signals that are bandlimited to $W$ Hz. We need to show that the vector

$$\left( \langle \mathbf{Y}, \mathbf{s}_0 \rangle, \langle \mathbf{Y}, \mathbf{s}_1 \rangle \right)^{\mathsf{T}} \tag{26.57}$$

forms a sufficient statistic.

Before giving a formal proof, we provide some intuition. Based on the observation $(Y(t))$, the receiver can compute the waveform

$$\tilde{Y}(t) = Y(t) - \frac{s_0(t) + s_1(t)}{2}, \quad t \in \mathbb{R}.$$

Since the transformation from $(Y(t))$ to $(\tilde{Y}(t))$ is reversible, there is no loss in optimality in basing one's decision on $(\tilde{Y}(t))$. Conditional on $H = 0$ the SP $(\tilde{Y}(t))$ is of the form

$$\tilde{Y}(t) = \underbrace{s_0(t) + N(t)}_{Y(t)} - \frac{s_0(t) + s_1(t)}{2}$$

$$= \frac{s_0(t) - s_1(t)}{2} + N(t), \quad t \in \mathbb{R},$$

whereas conditional on $H = 1$ it is of the form

$$\tilde{Y}(t) = \underbrace{s_1(t) + N(t)}_{Y(t)} - \frac{s_0(t) + s_1(t)}{2}$$

$$= -\frac{s_0(t) - s_1(t)}{2} + N(t), \quad t \in \mathbb{R}.$$

Consequently, the problem of guessing $H$ based on $(\tilde{Y}(t))$ is the antipodal problem we addressed before with the received waveform being $(\tilde{Y}(t))$ and with the mean signals corresponding to $H = 0$ and $H = 1$ being $(\mathbf{s}_0 - \mathbf{s}_1)/2$ and $-(\mathbf{s}_0 - \mathbf{s}_1)/2$ respectively. From our treatment of the antipodal case, we know that for this problem $\langle \tilde{\mathbf{Y}}, (\mathbf{s}_0 - \mathbf{s}_1)/2 \rangle$ forms a sufficient statistic. This sufficient statistic can be written more explicitly as

$$\left\langle \tilde{\mathbf{Y}}, \frac{\mathbf{s}_0 - \mathbf{s}_1}{2} \right\rangle = \left\langle \mathbf{Y} - \frac{\mathbf{s}_0 + \mathbf{s}_1}{2}, \frac{\mathbf{s}_0 - \mathbf{s}_1}{2} \right\rangle$$

$$= \frac{1}{2} \langle \mathbf{Y}, \mathbf{s}_0 \rangle - \frac{1}{2} \langle \mathbf{Y}, \mathbf{s}_1 \rangle - \frac{1}{4} \left( \|\mathbf{s}_0\|_2^2 - \|\mathbf{s}_1\|_2^2 \right),$$

thus demonstrating that this sufficient statistic is computable from the vector in (26.57).[12]

For readers who prefer a more formal proof we offer the following. Define

$$\mathbf{s} = \frac{\mathbf{s}_0 - \mathbf{s}_1}{2}. \tag{26.58}$$

Since $\langle \mathbf{Y} - (\mathbf{s}_0 + \mathbf{s}_1)/2, \mathbf{s} \rangle$ is computable from the vector in (26.57), it follows from Proposition 22.4.2 that to prove that the vector in (26.57) is sufficient it is enough to establish that $\langle \mathbf{Y} - (\mathbf{s}_0 + \mathbf{s}_1)/2, \mathbf{s} \rangle$ is sufficient. We thus need to show that for every $\eta \in \mathbb{N}$ and for every choice of the epochs $t_1, \ldots, t_\eta \in \mathbb{R}$, the RV $\langle \mathbf{Y} - (\mathbf{s}_0 + \mathbf{s}_1)/2, \mathbf{s} \rangle$ forms a sufficient statistic for the hypothesis testing problem of guessing $H$ based on $Y(t_1), \ldots, Y(t_\eta)$, $\langle \mathbf{Y} - (\mathbf{s}_0 + \mathbf{s}_1)/2, \mathbf{s} \rangle$, i.e., that for every prior on $H$,

$$H \mathrel{-\!\circ\!-} \left\langle \mathbf{Y} - \frac{\mathbf{s}_0 + \mathbf{s}_1}{2}, \mathbf{s} \right\rangle \mathrel{-\!\circ\!-} \Big( Y(t_1), \ldots, Y(t_\eta) \Big).$$

Equivalently, since subtracting deterministic quantities does not alter conditional independence, it suffices to show that

$$H \mathrel{-\!\circ\!-} \left\langle \mathbf{Y} - \frac{\mathbf{s}_0 + \mathbf{s}_1}{2}, \mathbf{s} \right\rangle \mathrel{-\!\circ\!-} \left( Y(t_1) - \frac{s_0(t_1) + s_1(t_1)}{2}, \ldots, Y(t_\eta) - \frac{s_0(t_\eta) + s_1(t_\eta)}{2} \right).$$

This can be proved by applying Lemma 26.8.1 to the vector

$$\left( Y(t_1) - \frac{s_0(t_1) + s_1(t_1)}{2}, \ldots, Y(t_\eta) - \frac{s_0(t_\eta) + s_1(t_\eta)}{2}, \left\langle \mathbf{Y} - \frac{\mathbf{s}_0 + \mathbf{s}_1}{2}, \mathbf{s} \right\rangle \right)^{\mathsf{T}}$$

which, conditional on $H = 0$, is Gaussian, with the covariance matrix in (26.56) and with the mean vector being the RHS of (26.55) (with $\mathbf{s}$ defined in (26.58)) and which, conditional on $H = 1$, is Gaussian with the same covariance matrix (26.56) but with the conditional mean being antipodal to the RHS of (26.55).

### 26.8.4 The General Case

We now prove the general (not necessarily binary) case of Theorem 26.4.1. There is surprisingly little left to do. The key is Proposition 22.3.2, which demonstrates that if a function of the observation is sufficient for testing between any two of the hypotheses, then it is sufficient for the multi-hypothesis testing problem.

To prove that the vector (26.15) of inner products forms a sufficient statistic we need to show that for every $\eta \in \mathbb{N}$ and for any choice of the epochs $t_1, \ldots, t_\eta \in \mathbb{R}$ the inner products vector (26.15) forms a sufficient statistic for guessing $M$ based on the observation consisting of $Y(t_1), \ldots, Y(t_\eta)$ and of the inner products vector (Definition 26.3.1). By Proposition 22.3.2, it is enough to show this when testing between any two fixed distinct messages $m', m'' \in \mathcal{M}$. But in this case the sufficiency of the inner products vector (26.15) follows directly from Section 26.8.3 and

---

[12]This is only a heuristic argument because it only shows that it is optimal to guess $H$ based on the vector (26.57). It does not prove that this vector forms a sufficient statistic.

Proposition 22.4.2 because, by the general binary hypothesis testing case treated in Section 26.8.3, the two inner products $\langle \mathbf{Y}, \mathbf{s}_{m'} \rangle$ & $\langle \mathbf{Y}, \mathbf{s}_{m''} \rangle$ suffice for this problem, and these two inner products are obviously computable from the inner products vector (26.15) (simply by ignoring its other components). This completes the proof of Theorem 26.4.1.

## 26.9 The Front-End Filter

Receivers in practice rarely have the structure depicted in Figure 26.1 because—although mathematically optimal—its hardware implementation is challenging. The difficulty is related to the "dynamic range" problem in implementing the matched filter: it is very difficult to design a perfectly-linear system to exact specification. Linearity is usually only guaranteed for a certain range of input amplitudes. Once the amplitude of the signal exceeds a certain level, the circuit often "clips" the input waveform and no longer behaves linearly. Similarly, input signals that are too small might be below the sensitivity of the circuit and might therefore produce no output, thus violating linearity. This is certainly the case with circuits that employ analog-to-digital conversion followed by digital processing, because analog-to-digital converters can only represent the input using a fixed number of bits. The problem with the structure depicted in Figure 26.1 is that the noise $(N(t))$ is typically much larger than the mean signal, so it becomes very difficult to design a circuit to exact specifications that will be linear enough to guarantee that its action on the received waveform (consisting of the weak transmitted waveform and the strong additive noise) be the sum of the required responses to the mean signal and to the noise-signal. (That the noise is typically much larger than the mean signals can be seen from the heuristic plot of its PSD; see Figure 25.3. White Gaussian noise is often of PSD $N_0/2$ over frequency bands that are much larger than the band $[-W, W]$ so, by (25.30), the variance of the noise can be extremely large.)

The engineering solution to the dynamic range problem is to pass the received waveform through a "front-end filter" and to then feed this filter's output to the matched filter; see Figure 26.3. Except for a few very stringent requirements, the specifications of the front-end filter are relatively lax. The first specification is that the filter be linear over a very large range of input levels. This is usually accomplished by using only passive elements to design the filter. The second requirement is that the front-end filter's frequency response be of unit-gain over the mean signals' frequency band $[-W, W]$ so that it will not distort the mean signals.[13] Additionally, we require that the filter be stable and that its frequency response decay to zero sharply for frequencies outside the band $[-W, W]$. This latter condition guarantees that the filter's response to the noise be of small variance so that the dynamic range of the signal at the filter's output be moderate. If we denote the front-end filter's impulse response by $\mathbf{h}_{\mathrm{FE}}$, then the key mathematical requirements are linearity; stability, i.e.,

$$\int_{-\infty}^{\infty} \left| h_{\mathrm{FE}}(t) \right| \, \mathrm{d}t < \infty; \tag{26.59}$$

---

[13]Imprecisions here can often be corrected using signal processing.

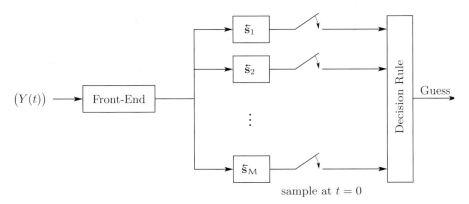

Figure 26.3: Feeding the signal to a front-end filter and then computing the inner products with the mean signals.

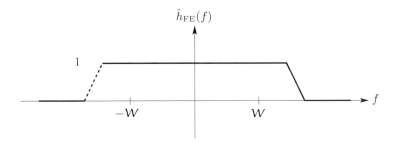

Figure 26.4: An example of the frequency response of a front-end filter.

and the unit-gain requirement

$$\hat{h}_{\mathrm{FE}}(f) = 1, \quad |f| \leq \mathsf{W}. \tag{26.60}$$

An example of the frequency response of a front-end filter is depicted in Figure 26.4.

In the rest of this section we shall prove that, as long as these assumptions are met, there is no loss in optimality in introducing the front-end filter as in Figure 26.3. (In the ideal mathematical world there is, of course, nothing to be gained from this filter, because the structure we introduced in Figure 26.1 is optimal.)

The crux of the proof is in showing that—like $(Y(t))$—the front-end filter's output is the sum of the transmitted signal and white Gaussian noise of PSD $\mathsf{N}_0/2$ with respect to the bandwidth $\mathsf{W}$. Once this is established, the result follows by recalling that the conditional joint distribution of the matched filters' outputs does not depend on the PSD of the noise outside the band $[-\mathsf{W}, \mathsf{W}]$ (Note 26.5.1).

We thus proceed to analyze the front-end filter's output, which we denote by $(\tilde{Y}(t))$:

$$(\tilde{Y}(t)) = (Y(t)) \star \mathbf{h}_{\mathrm{FE}}. \tag{26.61}$$

We first note that (26.60) and the assumption that $\mathbf{s}_m$ is an integrable signal that is bandlimited to $W$ Hz guarantee that

$$\mathbf{s}_m \star \mathbf{h}_{\mathrm{FE}} = \mathbf{s}_m, \quad m \in \mathcal{M} \tag{26.62}$$

(Proposition 6.5.2 and Proposition 6.4.10 *cf.* (b)). By (26.62) and by the linearity of the filter we can thus express the filter's output (conditional on $M = m$) as

$$\begin{aligned}
\left(\tilde{Y}(t)\right) &= \left(Y(t)\right) \star \mathbf{h}_{\mathrm{FE}} \\
&= \mathbf{s}_m \star \mathbf{h}_{\mathrm{FE}} + \left(N(t)\right) \star \mathbf{h}_{\mathrm{FE}} \\
&= \mathbf{s}_m + \left(N(t)\right) \star \mathbf{h}_{\mathrm{FE}}.
\end{aligned} \tag{26.63}$$

We next show that the SP $\left(N(t)\right) \star \mathbf{h}_{\mathrm{FE}}$ on the RHS of (26.63) is white Gaussian noise of PSD $\mathsf{N}_0/2$ with respect to the bandwidth $W$. This follows from Theorem 25.13.2. Indeed, being the result of passing a measurable stationary Gaussian SP through a stable filter, it is a measurable stationary Gaussian SP. And its PSD is

$$f \mapsto \mathsf{S}_{NN}(f) \left|\hat{h}_{\mathrm{FE}}(f)\right|^2, \tag{26.64}$$

which is equal to $\mathsf{N}_0/2$ for all frequencies $f \in [-W, W]$, because for these frequencies $\mathsf{S}_{NN}(f)$ is equal to $\mathsf{N}_0/2$ and $\hat{h}_{\mathrm{FE}}(f)$ is equal to one. Note that at frequencies outside the band $[-W, W]$ the PSD of $\left(N(t)\right) \star \mathbf{h}_{\mathrm{FE}}$ may differ from that of $\left(N(t)\right)$.

We thus conclude that the front-end filter's output, like its input, can be expressed as the transmitted signal corrupted by white Gaussian noise of PSD $\mathsf{N}_0/2$ with respect to the bandwidth $W$. Note 26.5.1 now guarantees that for every $m \in \mathcal{M}$ we have that, conditional on $M = m$, the distribution of

$$\left(\int_{-\infty}^{\infty} \tilde{Y}(t) \, s_1(t) \, \mathrm{d}t, \ldots, \int_{-\infty}^{\infty} \tilde{Y}(t) \, s_{\mathsf{M}}(t) \, \mathrm{d}t\right)^{\mathsf{T}}$$

is identical to the conditional distribution of the random vector in (26.15).

The advantage of the front-end filter becomes apparent when we re-examine the PSD of the noise at its output. If the front-end filter's frequency response decays very sharply to zero for frequencies outside the band $[-W, W]$, then, by (26.64), this PSD will be nearly zero outside this band. Consequently, the variance of the noise at the front-end filter's output—which is the integral of this PSD—will be greatly reduced. This will guarantee that the dynamic range at the filter's output be much smaller than at its input, thus simplifying the implementation of the matched filters.

## 26.10    Detection in Passband

The detection problem in passband is very similar to the one in baseband. The difference is that the mean signals $\{\mathbf{s}_m\}$ are now assumed to be integrable signals that are bandlimited to $W$ Hz around the carrier frequency $f_{\mathrm{c}}$ (Definition 7.2.1) and that the noise is now assumed to be white Gaussian noise of PSD $\mathsf{N}_0/2$ with respect to the bandwidth $W$ around $f_{\mathrm{c}}$ (Definition 25.15.3).

Here too, the inner products in (26.16) form a sufficient statistic. So do those in (26.18) whenever the signals $\{\tilde{\mathbf{s}}_j\}$ satisfy

$$\mathbf{s}_m \in \text{span}(\tilde{\mathbf{s}}_1, \ldots, \tilde{\mathbf{s}}_n), \quad m \in \mathcal{M}$$

and are integrable signals that are bandlimited to $W$ Hz around $f_c$.

For every $m \in \mathcal{M}$ the conditional distribution of the vector of inner products in (26.18), conditional on $M = m$, is Gaussian with mean vector (26.22) and covariance matrix (26.23). The latter covariance matrix can also be written in terms of the baseband representation of the mean signals using the relation

$$\langle \tilde{\mathbf{s}}_{j'}, \tilde{\mathbf{s}}_{j''} \rangle = 2\,\text{Re}\big(\langle \tilde{\mathbf{s}}_{j',\text{BB}}, \tilde{\mathbf{s}}_{j'',\text{BB}} \rangle\big), \tag{26.65}$$

where $\tilde{\mathbf{s}}_{j',\text{BB}}$ and $\tilde{\mathbf{s}}_{j'',\text{BB}}$ are the baseband representations of $\tilde{\mathbf{s}}_{j'}$ and $\tilde{\mathbf{s}}_{j''}$ (Theorem 7.6.10).

The computation of the inner products (26.18) can be performed in passband by feeding the signal $\mathbf{Y}$ directly to filters that are matched to the passband signals $\{\tilde{\mathbf{s}}_j\}$, or in baseband by expressing the inner product $\langle \mathbf{Y}, \tilde{\mathbf{s}}_j \rangle$ in terms of the baseband representation $\tilde{\mathbf{s}}_{j,\text{BB}}$ of $\tilde{\mathbf{s}}_j$ as follows:

$$\begin{aligned}\langle \mathbf{Y}, \tilde{\mathbf{s}}_j \rangle &= \big\langle \mathbf{Y}, t \mapsto 2\,\text{Re}\big(\tilde{s}_{j,\text{BB}}(t)\, e^{i2\pi f_c t}\big)\big\rangle \\ &= 2 \int_{-\infty}^{\infty} \big(Y(t)\cos(2\pi f_c t)\big)\,\text{Re}\big(\tilde{s}_{j,\text{BB}}(t)\big)\,\text{d}t \\ &\quad - 2 \int_{-\infty}^{\infty} \big(Y(t)\sin(2\pi f_c t)\big)\,\text{Im}\big(\tilde{s}_{j,\text{BB}}(t)\big)\,\text{d}t.\end{aligned}$$

This expression suggests computing the inner product $\langle \mathbf{Y}, \tilde{\mathbf{s}}_j \rangle$ using two baseband matched filters: one that is matched to $\text{Re}(\tilde{\mathbf{s}}_{j,\text{BB}})$ and that is fed the product of $\big(Y(t)\big)$ and $\cos(2\pi f_c t)$, and one that is matched to $\text{Im}(\tilde{\mathbf{s}}_{j,\text{BB}})$ and that is fed the product of $\big(Y(t)\big)$ and $\sin(2\pi f_c t)$.[14]

As discussed in Section 26.9, in practice one typically first feeds the received signal $\big(Y(t)\big)$ to a stable highly-linear bandpass filter of frequency response $\hat{h}_{\text{PB-FE}}(\cdot)$ satisfying

$$\hat{h}_{\text{PB-FE}}(f) = 1, \quad \big||f| - f_c\big| \le W/2, \tag{26.66}$$

with the frequency response decaying drastically at other frequencies to guarantee that the filter's output be of small dynamic range.

---

[14]Since the baseband representation of an integrable passband signal that is bandlimited to $W$ Hz around the carrier frequency $f_c$ is integrable (Proposition 7.6.2), it follows that our assumption that $\tilde{\mathbf{s}}_j$ is an integrable function that is bandlimited to $W$ Hz around the carrier frequency $f_c$ guarantees that both $t \mapsto \cos(2\pi f_c t)\,\text{Re}\big(\tilde{s}_{j,\text{BB}}(t)\big)$ and $t \mapsto \sin(2\pi f_c t)\,\text{Im}\big(\tilde{s}_{j,\text{BB}}(t)\big)$ are integrable. Hence, with probability one, both the integrals $\int_{-\infty}^{\infty} \big(Y(t)\cos(2\pi f_c t)\big)\,\text{Re}\big(\tilde{s}_{j,\text{BB}}(t)\big)\,\text{d}t$ and $\int_{-\infty}^{\infty} \big(Y(t)\sin(2\pi f_c t)\big)\,\text{Im}\big(\tilde{s}_{j,\text{BB}}(t)\big)\,\text{d}t$ exist.

## 26.11    Some Examples

### 26.11.1    Binary Hypothesis Testing

Before treating the general binary hypothesis testing problem we begin with the case of antipodal signaling with a uniform prior. In this case

$$\mathbf{s}_0 = -\mathbf{s}_1 = \mathbf{s}, \tag{26.67}$$

where $\mathbf{s}$ is some integrable signal that is bandlimited to $W$ Hz. We denote its energy by $\mathsf{E}_\mathrm{s}$, i.e.,

$$\mathsf{E}_\mathrm{s} = \|\mathbf{s}\|_2^2 \tag{26.68}$$

and assume that it is strictly positive. In this case the dimension of the linear subspace spanned by the mean signals is one, and this subspace is spanned by the unit-norm signal

$$\phi = \frac{\mathbf{s}}{\|\mathbf{s}\|_2}. \tag{26.69}$$

Depending on the outcome of a fair coin toss, either $\mathbf{s}$ or $-\mathbf{s}$ is sent over the channel. We observe the SP $(Y(t))$ given by the sum of the transmitted signal and white Gaussian noise of PSD $\mathsf{N}_0/2$ with respect to the bandwidth $W$, and we wish to guess which signal was sent. How should we form our guess?

By Theorem 26.4.1 a sufficient statistic for this guessing problem is $T = \langle \mathbf{Y}, \boldsymbol{\phi} \rangle$. Conditional on $H = 0$, we have $T \sim \mathcal{N}(\sqrt{\mathsf{E}_\mathrm{s}}, \mathsf{N}_0/2)$, whereas, conditional on $H = 1$, we have $T \sim \mathcal{N}(-\sqrt{\mathsf{E}_\mathrm{s}}, \mathsf{N}_0/2)$. How to guess $H$ based on $T$ is the problem we addressed in Section 20.10. There we showed that it is optimal to guess "$H = 0$" if $T \geq 0$ and to guess "$H = 1$" if $T < 0$. (The case $T = 0$ occurs with probability zero, so we need not worry about it.) An optimal decision rule for guessing $H$ based on $(Y(t))$ is thus:

$$\boxed{\text{Guess ``}H = 0\text{'' if } \int_{-\infty}^{\infty} Y(t)\, s(t)\, \mathrm{d}t \geq 0.} \tag{26.70}$$

Let $p_{\mathrm{MAP}}(\text{error}|\mathbf{s})$ denote the conditional probability of error of this decision rule given that $\mathbf{s}$ was sent; let $p_{\mathrm{MAP}}(\text{error}|-\mathbf{s})$ be similarly defined; and let $p^*(\text{error})$ denote the optimal probability of error of this problem. By the optimality of our rule,

$$p^*(\text{error}) = \frac{1}{2}\big(p_{\mathrm{MAP}}(\text{error}|\mathbf{s}) + p_{\mathrm{MAP}}(\text{error}|-\mathbf{s})\big).$$

Using the expression for the error probability derived in Section 20.10 we obtain

$$p^*(\text{error}) = Q\left(\sqrt{\frac{2\,\|\mathbf{s}\|_2^2}{\mathsf{N}_0}}\right), \tag{26.71}$$

which, in view of (26.68), can also be written as

$$\boxed{p^*(\text{error}) = Q\left(\sqrt{\frac{2\mathsf{E}_\mathrm{s}}{\mathsf{N}_0}}\right).} \tag{26.72}$$

Note that, as expected from Section 26.5.2 and in particular from Proposition 26.5.2, the probability of error is determined by the "geometry" of the problem, which in this case is summarized by the energy in $\mathbf{s}$.

There is also a nice geometric interpretation to (26.72). The distance between the mean signals $\mathbf{s}$ and $-\mathbf{s}$ is $\|\mathbf{s} - (-\mathbf{s})\|_2 = 2\sqrt{\mathsf{E_s}}$. Half the distance is $\sqrt{\mathsf{E_s}}$. The inner product between the noise and the unit-length vector $\phi$ pointing from $-\mathbf{s}$ to $\mathbf{s}$ is $\mathcal{N}(0, \mathsf{N_0}/2)$. Half the distance thus corresponds to $\sqrt{\mathsf{E_s}}/\sqrt{\mathsf{N_0}/2}$ standard deviations of this inner product. The probability of error is thus the probability that a standard Gaussian is greater than half the distance between the signals as measured by standard deviations of the inner product between the noise and the unit-length vector pointing from $-\mathbf{s}$ towards $\mathbf{s}$.

Consider now the more general binary hypothesis testing problem where both hypotheses are still equally likely, but where now the mean signals $\mathbf{s}_0$ and $\mathbf{s}_1$ are not antipodal, i.e., they do not sum to zero. Our approach to this problem is to reduce it to the antipodal case we already treated. We begin by forming the signal $(\tilde{Y}(t))$ by subtracting $(\mathbf{s}_0 + \mathbf{s}_1)/2$ from the received signal, so

$$\tilde{Y}(t) = Y(t) - \frac{1}{2}\big(s_0(t) + s_1(t)\big), \quad t \in \mathbb{R}. \tag{26.73}$$

Since $Y(t)$ can be recovered from $\tilde{Y}(t)$ by adding $\big(s_0(t) + s_1(t)\big)/2$, the smallest probability of a guessing error that can be achieved based on $(\tilde{Y}(t))$ is no larger than that which can be achieved based on $(Y(t))$. (The two are, in fact, the same because $(\tilde{Y}(t))$ can be computed from $(Y(t))$.)

The advantage of using $(\tilde{Y}(t))$ becomes apparent once we compute its conditional law given $H$. Conditional on $H = 0$, we have $\tilde{Y}(t) = (s_0(t) - s_1(t))/2 + N(t)$, whereas conditional on $H = 1$, we have $\tilde{Y}(t) = -(s_0(t) - s_1(t))/2 + N(t)$. Thus, the guessing problem given $(\tilde{Y}(t))$ is exactly the problem we addressed in the antipodal case with $(\mathbf{s}_0 - \mathbf{s}_1)/2$ playing the role of $\mathbf{s}$. We thus obtain that an optimal decision rule is to guess "$H = 0$" if $\int \tilde{Y}(t)\big(s_0(t) - s_1(t)\big)/2 \, dt$ is nonnegative. Or stated in terms of $(Y(t))$ using (26.73):

$$\boxed{\text{Guess ``}H = 0\text{'' if } \int_{-\infty}^{\infty} \left(Y(t) - \frac{s_0(t) + s_1(t)}{2}\right) \frac{s_0(t) - s_1(t)}{2} \, dt \geq 0.} \tag{26.74}$$

The error probability associated with this decision rule is obtained from (26.71) by substituting $(\mathbf{s}_0 - \mathbf{s}_1)/2$ for $\mathbf{s}$:

$$\boxed{p^*(\text{error}) = Q\left(\sqrt{\frac{\|\mathbf{s}_0 - \mathbf{s}_1\|_2^2}{2\mathsf{N_0}}}\right).} \tag{26.75}$$

This expression too has a nice geometric interpretation. The inner product between the noise and the unit-norm signal that is pointing from $\mathbf{s}_1$ to $\mathbf{s}_0$ is $\mathcal{N}(0, \mathsf{N_0}/2)$. The "distance" between the signals is $\|\mathbf{s}_0 - \mathbf{s}_1\|_2$. Half the distance is $\|\mathbf{s}_0 - \mathbf{s}_1\|_2/2$, which corresponds to

$$\frac{\|\mathbf{s}_0 - \mathbf{s}_1\|_2/2}{\sqrt{\mathsf{N_0}/2}}$$

standard deviations of a $\mathcal{N}(0, \mathsf{N}_0/2)$ random variable. The probability of error (26.75) is thus the probability that the inner product between the noise and the unit-norm signal that is pointing from $\mathbf{s}_1$ to $\mathbf{s}_0$ exceeds half the distance between the signals.

### 26.11.2   8-PSK

We next present an example of detection in passband. For concreteness we consider 8-PSK, which stands for "8-ary Phase Shift Keying." Here the number of hypotheses is eight, so $\mathcal{M} = \{1, 2, \ldots, 8\}$ and $\mathsf{M} = 8$. We assume that $M$ is uniformly distributed over $\mathcal{M}$. Conditional on $M = m$, the received signal is given by

$$Y(t) = s_m(t) + N(t), \quad t \in \mathbb{R}, \tag{26.76}$$

where

$$s_m(t) = 2 \operatorname{Re}\!\left(c_m s_{\mathrm{BB}}(t)\, e^{\mathrm{i} 2\pi f_c t}\right), \quad t \in \mathbb{R}; \tag{26.77}$$

$$c_m = \alpha\, e^{\mathrm{i} m \frac{2\pi}{8}} \tag{26.78}$$

for some positive real $\alpha$; the baseband signal $\mathbf{s}_{\mathrm{BB}}$ is an integrable complex signal that is bandlimited to $W/2$ Hz and of unit energy

$$\|\mathbf{s}_{\mathrm{BB}}\|_2 = 1; \tag{26.79}$$

the carrier frequency $f_c$ satisfies $f_c > W/2$; and $\bigl(N(t)\bigr)$ is white Gaussian noise of PSD $\mathsf{N}_0/2$ with respect to the bandwidth $W$ around the carrier frequency $f_c$ (Definition 25.15.3). Irrespective of $M$, the transmitted energy $\mathsf{E}_s$ is given by

$$
\begin{aligned}
\mathsf{E}_s &= \|\mathbf{s}_m\|_2^2 \\
&= \int_{-\infty}^{\infty} \left(2 \operatorname{Re}\!\left(c_m s_{\mathrm{BB}}(t)\, e^{\mathrm{i} 2\pi f_c t}\right)\right)^2 \mathrm{d}t \\
&= 2\alpha^2,
\end{aligned}
\tag{26.80}
$$

as can be verified using the relationship between energy in passband and baseband (Theorem 7.6.10) and using (26.79).

The transmitted waveform $\mathbf{s}_m$ can also be written in a form that is highly suggestive of a choice of an orthonormal basis for $\operatorname{span}(\mathbf{s}_1, \ldots, \mathbf{s}_\mathsf{M})$:

$$
s_m(t) = \sqrt{2}\operatorname{Re}(c_m)\,\underbrace{\sqrt{2}\operatorname{Re}\!\left(s_{\mathrm{BB}}(t)\, e^{\mathrm{i} 2\pi f_c t}\right)}_{\phi_1(t)} + \sqrt{2}\operatorname{Im}(c_m)\,\underbrace{\sqrt{2}\operatorname{Re}\!\left(\mathrm{i}\, s_{\mathrm{BB}}(t)\, e^{\mathrm{i} 2\pi f_c t}\right)}_{\phi_2(t)}
$$

$$
= \sqrt{2}\operatorname{Re}(c_m)\,\phi_1(t) + \sqrt{2}\operatorname{Im}(c_m)\,\phi_2(t),
$$

where

$$\phi_1(t) \triangleq \sqrt{2}\operatorname{Re}\!\left(s_{\mathrm{BB}}(t)\, e^{\mathrm{i} 2\pi f_c t}\right), \quad t \in \mathbb{R},$$

$$\phi_2(t) \triangleq \sqrt{2}\operatorname{Re}\!\left(\mathrm{i}\, s_{\mathrm{BB}}(t)\, e^{\mathrm{i} 2\pi f_c t}\right), \quad t \in \mathbb{R}.$$

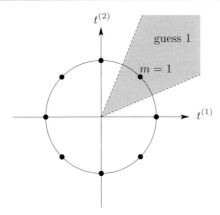

**Figure 26.5:** Region of points $(t^{(1)}, t^{(2)})$ resulting in guessing "$M = 1$."

From Theorem 7.6.10 on inner products in passband and baseband, it follows that $\phi_1$ and $\phi_2$ are orthogonal. Also, from that theorem and from (26.79), it follows that they are of unit energy. Thus, the tuple $(\phi_1, \phi_2)$ is an orthonormal basis for $\text{span}(\mathbf{s}_1, \ldots, \mathbf{s}_M)$. Consequently, the vector $\mathbf{T} = (\langle \mathbf{Y}, \phi_1 \rangle, \langle \mathbf{Y}, \phi_2 \rangle)^{\mathsf{T}}$ forms a sufficient statistic for guessing $M$ based on $(Y(t))$, and, conditional on $M = m$, the components of $\mathbf{T}$ are independent with $T^{(1)} \sim \mathcal{N}(\sqrt{2}\alpha \cos(2\pi m/8), \mathsf{N}_0/2)$ and with $T^{(2)} \sim \mathcal{N}(\sqrt{2}\alpha \sin(2\pi m/8), \mathsf{N}_0/2)$. We have thus reduced the guessing problem to that of guessing $M$ based on a two-dimensional vector $\mathbf{T}$.

The problem of guessing $M$ based on $\mathbf{T}$ was studied in Section 21.4. To lift the results from that section, we need to substitute $\sqrt{2}\alpha$ for A and to substitute $\mathsf{N}_0/2$ for $\sigma^2$. For example, the region where we guess "$M = 1$" is given in Figure 26.5.

For the scenario we described, some engineers prefer to use complex random variables (Chapter 17). Rather than viewing $\mathbf{T}$ as a two-dimensional real random vector, they prefer to view it as a (scalar) complex random variable whose real part is $\langle \mathbf{Y}, \phi_1 \rangle$ and whose imaginary part is $\langle \mathbf{Y}, \phi_2 \rangle$. Conditional on $M = m$, this CRV has the form

$$\sqrt{2}c_m + Z, \quad Z \sim \mathcal{N}_{\mathbb{C}}(0, \mathsf{N}_0), \tag{26.81}$$

where $\mathcal{N}_{\mathbb{C}}(0, \mathsf{N}_0)$ denotes the circularly-symmetric variance-$\mathsf{N}_0$ complex Gaussian distribution (Note 24.3.13).

The expression for the probability of error of our detector can also be lifted from Section 21.4. Substituting, as above, $\sqrt{2}\alpha$ for A and $\mathsf{N}_0/2$ for $\sigma^2$, we obtain from (21.25) that the conditional probability of error $p_{\text{MAP}}(\text{error}|M = m)$ of our proposed decision rule is given for every $m \in \mathcal{M}$ by

$$p_{\text{MAP}}(\text{error}|M = m) = \frac{1}{\pi} \int_0^{\pi - \psi} e^{-\frac{2\alpha^2 \sin^2 \psi}{\mathsf{N}_0 \sin^2(\theta + \psi)}} \, \mathrm{d}\theta, \quad \psi = \frac{\pi}{8}.$$

The conditional probability of error can also be expressed in terms of the transmitted energy $\mathsf{E}_\mathsf{s}$ using (26.80). Doing that and recalling that the conditional probability of error does not depend on the transmitted message, we obtain that the average probability of error $p^*(\text{error})$ is given by

$$p^*(\text{error}) = \frac{1}{\pi} \int_0^{\pi-\psi} e^{-\frac{\text{E}_\text{s} \sin^2 \psi}{\text{N}_0 \sin^2(\theta+\psi)}} \, d\theta, \quad \psi = \frac{\pi}{8}. \tag{26.82}$$

**Note 26.11.1.** The expression (26.82) continues to hold also for M-PSK where $c_m = \alpha \, e^{\text{i}2\pi m/\text{M}}$ for $\text{M} \geq 2$ not necessarily equal to 8, provided that we define $\psi = \pi/\text{M}$ in (26.82).

## 26.11.3 Orthogonal Keying

We next consider M-ary orthogonal keying. We assume that the RV $M$ that we wish to guess is uniformly distributed over the set $\mathcal{M} = \{1, \ldots, \text{M}\}$, where $\text{M} \geq 2$. The mean signals are assumed to be orthogonal and of equal (strictly) positive energy $\text{E}_\text{s}$:

$$\langle \mathbf{s}_{m'}, \mathbf{s}_{m''} \rangle = \text{E}_\text{s} \, \text{I}\{m' = m''\}, \quad m', m'' \in \mathcal{M}. \tag{26.83}$$

Since $M$ is uniform, and since the mean signals are of equal energy, it follows from Theorem 26.6.3 that to minimize the probability of guessing incorrectly, it is optimal to correlate the received waveform $(Y(t))$ with each of the mean signals and to pick the message whose mean signal gives the highest correlation:

$$\text{Guess ``}m\text{'' if } \langle \mathbf{Y}, \mathbf{s}_m \rangle = \max_{m' \in \mathcal{M}} \langle \mathbf{Y}, \mathbf{s}_{m'} \rangle \tag{26.84}$$

with ties (which occur with probability zero) being resolved by picking a random message among those that attain the highest correlation.

We next address the probability of error of this optimal decision rule. We first define the vector $(T^{(1)}, \ldots, T^{(\text{M})})^\mathsf{T}$ by

$$T^{(\ell)} = \int_{-\infty}^{\infty} Y(t) \frac{s_\ell(t)}{\sqrt{\text{E}_\text{s}}} \, dt, \quad \ell \in \{1, \ldots, \text{M}\}$$

and recast the decision rule as guessing "$M = m$" if $T^{(m)} = \max_{m' \in \mathcal{M}} T^{(m')}$, with ties being resolved at random among the components of $\mathbf{T}$ that are maximal.

Let $p_{\text{MAP}}(\text{error}|M = m)$ denote the conditional probability of error of this decoding rule, conditional on $M = m$. Conditional on $M = m$, an error occurs in two cases: when $m$ does not attain the highest score or when $m$ attains the highest score but this score is also attained by some other message and the tie is not resolved in $m$'s favor. Since the probability of a tie is zero (Note 26.6.4), we may ignore the second case and only compute the probability that an incorrect message is assigned a score that is (strictly) higher than the one associated with $m$. Thus,

$$p_{\text{MAP}}(\text{error}|M = m)$$
$$= \Pr\left[\max\{T^{(1)}, \ldots, T^{(m-1)}, T^{(m+1)}, \ldots, T^{(\text{M})}\} > T^{(m)} \,\big|\, M = m\right]. \tag{26.85}$$

From (26.28) and the orthogonality of the signals (26.83) we have that, conditional on $M = m$, the random vector $\mathbf{T}$ is Gaussian with the mean of its $m$-th component being $\sqrt{\mathsf{E_s}}$, the mean of its other components being zero, and with all the components being of variance $\mathsf{N_0}/2$ and independent of each other. Thus, the conditional probability of error given $M = m$ is "the probability that at least one of $\mathsf{M} - 1$ IID $\mathcal{N}(0, \mathsf{N_0}/2)$ random variables exceeds the value of a $\mathcal{N}(\sqrt{\mathsf{E_s}}, \mathsf{N_0}/2)$ random variable that is independent of them." Having recast the probability of error conditional on $M = m$ in a way that does not involve $m$ (the clause in quotes makes no reference to $m$), we conclude that the conditional probability of error given that $M = m$ does not depend on $m$:

$$p_{\mathrm{MAP}}(\mathrm{error}|M = m) = p_{\mathrm{MAP}}(\mathrm{error}|M = 1), \quad m \in \mathcal{M}. \tag{26.86}$$

This conditional probability of error can be computed starting from (26.85) as:

$$
\begin{aligned}
&p_{\mathrm{MAP}}(\mathrm{error}|M = 1)\\
&= \Pr\big[\max\{T^{(2)}, \dots, T^{(\mathsf{M})}\} > T^{(1)} \,\big|\, M = 1\big]\\
&= 1 - \Pr\big[\max\{T^{(2)}, \dots, T^{(\mathsf{M})}\} \le T^{(1)} \,\big|\, M = 1\big]\\
&= 1 - \int_{-\infty}^{\infty} f_{T^{(1)}|M=1}(t) \Pr\big[\max\{T^{(2)}, \dots, T^{(\mathsf{M})}\} \le t \,\big|\, M = 1, T^{(1)} = t\big]\, \mathrm{d}t\\
&= 1 - \int_{-\infty}^{\infty} f_{T^{(1)}|M=1}(t) \Pr\big[\max\{T^{(2)}, \dots, T^{(\mathsf{M})}\} \le t \,\big|\, M = 1\big]\, \mathrm{d}t\\
&= 1 - \int_{-\infty}^{\infty} f_{T^{(1)}|M=1}(t) \Pr\big[T^{(2)} \le t, \dots, T^{(\mathsf{M})} \le t \,\big|\, M = 1\big]\, \mathrm{d}t\\
&= 1 - \int_{-\infty}^{\infty} f_{T^{(1)}|M=1}(t) \Big(\Pr\big[T^{(2)} \le t \,\big|\, M = 1\big]\Big)^{\mathsf{M}-1}\, \mathrm{d}t\\
&= 1 - \int_{-\infty}^{\infty} f_{T^{(1)}|M=1}(t) \left(1 - \mathcal{Q}\Big(\frac{t}{\sqrt{\mathsf{N_0}/2}}\Big)\right)^{\mathsf{M}-1}\, \mathrm{d}t\\
&= 1 - \int_{-\infty}^{\infty} \frac{1}{\sqrt{\pi \mathsf{N_0}}} e^{-\frac{(t-\sqrt{\mathsf{E_s}})^2}{\mathsf{N_0}}} \left(1 - \mathcal{Q}\Big(\frac{t}{\sqrt{\mathsf{N_0}/2}}\Big)\right)^{\mathsf{M}-1}\, \mathrm{d}t\\
&= 1 - \frac{1}{\sqrt{2\pi}} \int_{-\infty}^{\infty} e^{-\tau^2/2} \left(1 - \mathcal{Q}\Big(\tau + \sqrt{\frac{2\mathsf{E_s}}{\mathsf{N_0}}}\Big)\right)^{\mathsf{M}-1}\, \mathrm{d}\tau, \tag{26.87}
\end{aligned}
$$

where the first equality follows from (26.85); the second because the conditional probability of an event and its complement add to one; the third by conditioning on $T^{(1)} = t$ and integrating it out, i.e., by noting that for any random variable $X$ of density $f_X(\cdot)$ and for any random variable $Y$,

$$\Pr\big[Y \le X\big] = \int_{-\infty}^{\infty} f_X(x) \Pr\big[Y \le x \,\big|\, X = x\big]\, \mathrm{d}x, \tag{26.88}$$

with $X$ here being equal to $T^{(1)}$ and with $Y$ here being $\max\{T^{(2)}, \dots, T^{(\mathsf{M})}\}$; the fourth from the conditional independence of $T^{(1)}$ and $(T^{(2)}, \dots, T^{(\mathsf{M})})$ given $M = 1$, which implies the conditional independence of $T^{(1)}$ and $\max\{T^{(2)}, \dots, T^{(\mathsf{M})}\}$ given $M = 1$; the fifth because the maximum of random variables does not exceed $t$ if, and only if, none of them exceeds $t$

$$\Big(\max\{T^{(2)}, \dots, T^{(\mathsf{M})}\} \le t\Big) \Leftrightarrow \Big(T^{(2)} \le t, \dots, T^{(\mathsf{M})} \le t\Big);$$

the sixth because, conditional on $M = 1$, the random variables $T^{(2)}, \dots, T^{(\mathsf{M})}$ are IID so

$$\Pr\big[T^{(2)} \le t, \dots, T^{(\mathsf{M})} \le t \,\big|\, M = 1\big] = \Big(\Pr\big[T^{(2)} \le t \,\big|\, M = 1\big]\Big)^{\mathsf{M}-1};$$

the seventh because, conditional on $M = 1$, we have $T^{(2)} \sim \mathcal{N}(0, \mathsf{N}_0/2)$ and using (19.12b); the eighth because, conditional on $M = 1$, we have $T^{(1)} \sim \mathcal{N}(\sqrt{\mathsf{E}_{\mathrm{s}}}, \mathsf{N}_0/2)$ so its conditional density can be written explicitly using (19.6); and the final equality using the change of variable

$$\tau \triangleq \frac{t - \sqrt{\mathsf{E}_{\mathrm{s}}}}{\sqrt{\mathsf{N}_0/2}}. \tag{26.89}$$

Using (26.86) and (26.87) we obtain that if $p^*(\text{error})$ denotes the unconditional probability of error, then $p^*(\text{error}) = p_{\mathrm{MAP}}(\text{error}|M = 1)$ and

$$p^*(\text{error}) = 1 - \frac{1}{\sqrt{2\pi}} \int_{-\infty}^{\infty} e^{-\tau^2/2} \left(1 - \mathcal{Q}\left(\tau + \sqrt{\frac{2\mathsf{E}_{\mathrm{s}}}{\mathsf{N}_0}}\right)\right)^{\mathsf{M}-1} \mathrm{d}\tau. \tag{26.90}$$

An alternative expression for the probability of error can be derived using the Binomial Expansion

$$(a + b)^n = \sum_{j=0}^{n} \binom{n}{j} a^{n-j} b^j, \quad \Big(n \in \mathbb{N},\ a, b \in \mathbb{R}\Big). \tag{26.91}$$

Substituting

$$a = 1, \qquad b = -\mathcal{Q}\left(\tau + \sqrt{\frac{2\mathsf{E}_{\mathrm{s}}}{\mathsf{N}_0}}\right), \qquad n = \mathsf{M} - 1,$$

in (26.91) yields

$$\left(1 - \mathcal{Q}\left(\tau + \sqrt{\frac{2\mathsf{E}_{\mathrm{s}}}{\mathsf{N}_0}}\right)\right)^{\mathsf{M}-1} = \sum_{j=0}^{\mathsf{M}-1} (-1)^j \binom{\mathsf{M}-1}{j} \left(\mathcal{Q}\left(\tau + \sqrt{\frac{2\mathsf{E}_{\mathrm{s}}}{\mathsf{N}_0}}\right)\right)^j$$

$$= 1 + \sum_{j=1}^{\mathsf{M}-1} (-1)^j \binom{\mathsf{M}-1}{j} \left(\mathcal{Q}\left(\tau + \sqrt{\frac{2\mathsf{E}_{\mathrm{s}}}{\mathsf{N}_0}}\right)\right)^j,$$

from which we obtain from (26.90) (using the linearity of integration and the fact that the Gaussian density integrates to one)

$$p^*(\text{error}) = \sum_{j=1}^{\mathsf{M}-1} (-1)^{j+1} \binom{\mathsf{M}-1}{j} \int_{-\infty}^{\infty} \frac{1}{\sqrt{2\pi}} e^{-\tau^2/2} \left(\mathcal{Q}\left(\tau + \sqrt{\frac{2\mathsf{E}_{\mathrm{s}}}{\mathsf{N}_0}}\right)\right)^j \mathrm{d}\tau. \tag{26.92}$$

For the case where $M = 2$ the expression (26.90) for the probability of error can be simplified to

$$p^*(\text{error}) = \mathcal{Q}\left(\sqrt{\frac{\mathsf{E_s}}{\mathsf{N_0}}}\right), \quad M = 2, \tag{26.93}$$

as we proceed to show in two different ways. The first way is to note that for $M = 2$ the probability of error can be expressed, using (26.85) and (26.86), as

$$p_{\text{MAP}}(\text{error}|M = 1) = \Pr\left[T^{(2)} > T^{(1)} \mid M = 1\right]$$
$$= \Pr\left[T^{(2)} - T^{(1)} > 0 \mid M = 1\right]$$
$$= \mathcal{Q}\left(\sqrt{\frac{\mathsf{E_s}}{\mathsf{N_0}}}\right),$$

where the last equality follows because, conditional on $M = 1$, the random variables $T^{(1)}$ and $T^{(2)}$ are independent Gaussians of variance $\mathsf{N_0}/2$ with the first having mean $\sqrt{\mathsf{E_s}}$ and the second having zero mean, so their difference $T^{(2)} - T^{(1)}$ is $\mathcal{N}\left(-\sqrt{\mathsf{E_s}}, \mathsf{N_0}\right)$. (The probability that a $\mathcal{N}\left(-\sqrt{\mathsf{E_s}}, \mathsf{N_0}\right)$ RV exceeds zero can be computed using (19.12a).) The second way of showing (26.93) it to use (26.75) and to note that the orthogonality of $\mathbf{s}_1$ and $\mathbf{s}_2$ implies $\|\mathbf{s}_1 - \mathbf{s}_2\|_2^2 = \|\mathbf{s}_1\|_2^2 + \|\mathbf{s}_2\|_2^2 = 2\mathsf{E_s}$.

### 26.11.4 The $M$-ary Simplex

We next describe a detection problem that is intimately related to the problem we addressed in Section 26.11.3. To motivate the problem we first note:

**Proposition 26.11.2.** *Consider the setup described in Section 26.2. If $\mathbf{s}$ is any integrable signal that is bandlimited to $\mathsf{W}$ Hz, then the probability of error associated with the mean signals $\{\mathbf{s}_1, \ldots, \mathbf{s}_M\}$ and the prior $\{\pi_m\}$ is the same as with the mean signals $\{\mathbf{s}_1 - \mathbf{s}, \ldots, \mathbf{s}_M - \mathbf{s}\}$ and the same prior.*

**Proof.** We have essentially given a proof of this result in Section 14.3 and also in Section 26.11.1 in our analysis of nonantipodal signaling. The idea is that, by subtracting the signal $\mathbf{s}$ from the received waveform, the receiver can make the problem with mean signals $\{\mathbf{s}_1, \ldots, \mathbf{s}_M\}$ appear as though it were the problem with mean signals $\{\mathbf{s}_1 - \mathbf{s}, \ldots, \mathbf{s}_M - \mathbf{s}\}$. Conversely, by adding $\mathbf{s}$, the receiver can make the latter appear as though it were the former. Consequently, the best performance achievable in the two settings must be identical. $\square$

The expected transmitted energy when employing the mean signals $\{\mathbf{s}_1, \ldots, \mathbf{s}_M\}$ may be different than when employing the mean signals $\{\mathbf{s}_1 - \mathbf{s}, \ldots, \mathbf{s}_M - \mathbf{s}\}$. In subtracting the signal $\mathbf{s}$ one can change the average transmitted energy for better or worse. As we argued in Section 14.3, to minimize the expected transmitted energy, one should choose $\mathbf{s}$ to correspond to the "center of gravity" of the mean signals:

**Proposition 26.11.3.** *Let the prior $\{\pi_m\}$ and mean signals $\{\mathbf{s}_m\}$ be given. Let*

$$\mathbf{s}_* = \sum_{m \in \mathcal{M}} \pi_m \, \mathbf{s}_m . \tag{26.94}$$

*Then, for any energy-limited signal $\mathbf{s}$*

$$\sum_{m \in \mathcal{M}} \pi_m \, \|\mathbf{s}_m - \mathbf{s}_*\|_2^2 \le \sum_{m \in \mathcal{M}} \pi_m \, \|\mathbf{s}_m - \mathbf{s}\|_2^2 , \tag{26.95}$$

*with equality if, and only if, $\mathbf{s}$ is indistinguishable from $\mathbf{s}_*$.*

**Proof.** Writing $\mathbf{s}_m - \mathbf{s}$ as $(\mathbf{s}_m - \mathbf{s}_*) + (\mathbf{s}_* - \mathbf{s})$ we have

$$\sum_{m \in \mathcal{M}} \pi_m \, \|\mathbf{s}_m - \mathbf{s}\|_2^2$$

$$= \sum_{m \in \mathcal{M}} \pi_m \, \|(\mathbf{s}_m - \mathbf{s}_*) + (\mathbf{s}_* - \mathbf{s})\|_2^2$$

$$= \sum_{m \in \mathcal{M}} \pi_m \, \|\mathbf{s}_m - \mathbf{s}_*\|_2^2 + \sum_{m \in \mathcal{M}} \pi_m \, \|\mathbf{s}_* - \mathbf{s}\|_2^2 + 2 \sum_{m \in \mathcal{M}} \pi_m \, \langle \mathbf{s}_m - \mathbf{s}_*, \mathbf{s}_* - \mathbf{s} \rangle$$

$$= \sum_{m \in \mathcal{M}} \pi_m \, \|\mathbf{s}_m - \mathbf{s}_*\|_2^2 + \|\mathbf{s}_* - \mathbf{s}\|_2^2 + 2 \left\langle \sum_{m \in \mathcal{M}} \pi_m (\mathbf{s}_m - \mathbf{s}_*), \mathbf{s}_* - \mathbf{s} \right\rangle$$

$$= \sum_{m \in \mathcal{M}} \pi_m \, \|\mathbf{s}_m - \mathbf{s}_*\|_2^2 + \|\mathbf{s}_* - \mathbf{s}\|_2^2 + 2 \, \langle \mathbf{s}_* - \mathbf{s}_*, \mathbf{s}_* - \mathbf{s} \rangle$$

$$= \sum_{m \in \mathcal{M}} \pi_m \, \|\mathbf{s}_m - \mathbf{s}_*\|_2^2 + \|\mathbf{s}_* - \mathbf{s}\|_2^2$$

$$\ge \sum_{m \in \mathcal{M}} \pi_m \, \|\mathbf{s}_m - \mathbf{s}_*\|_2^2 ,$$

with the inequality being an equality if, and only if, $\|\mathbf{s}_* - \mathbf{s}\|_2^2 = 0$.    □

We can now construct the simplex signals as follows. We start with $\mathsf{M}$ orthonormal waveforms $\boldsymbol{\phi}_1, \ldots, \boldsymbol{\phi}_\mathsf{M}$

$$\langle \boldsymbol{\phi}_{m'}, \boldsymbol{\phi}_{m''} \rangle = \mathrm{I}\{m' = m''\}, \quad m', m'' \in \mathcal{M} \tag{26.96}$$

that are integrable and bandlimited to $W$ Hz. We set $\bar{\boldsymbol{\phi}}$ to be their "center of gravity" with respect to the uniform prior

$$\bar{\boldsymbol{\phi}} = \frac{1}{\mathsf{M}} \sum_{m \in \mathcal{M}} \boldsymbol{\phi}_m. \tag{26.97}$$

Using (26.96), (26.97), and the basic properties of the inner product (3.6)–(3.10) it is easily verified that

$$\left\langle \boldsymbol{\phi}_{m'} - \bar{\boldsymbol{\phi}}, \boldsymbol{\phi}_{m''} - \bar{\boldsymbol{\phi}} \right\rangle = \mathrm{I}\left\{m' = m''\right\} - \frac{1}{\mathsf{M}}, \quad m', m'' \in \mathcal{M}. \tag{26.98}$$

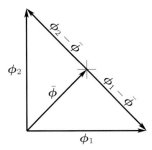

**Figure 26.6:** Starting with two orthonormal signals and subtracting the "center of gravity" from each we obtain two antipodal signals. Scaling these antipodal signals results in the simplex constellation with two signals.

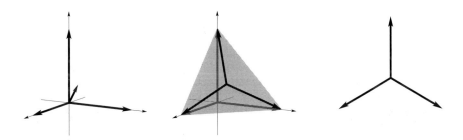

**Figure 26.7:** Constructing the simplex constellation with three points from three orthonormal signals. Left figure depicts the orthonormal constellation and its center of gravity; middle figure depicts the result of subtracting the center of gravity, and the right figure depicts the result of scaling (from a different perspective).

We now define the M-ary simplex constellation with energy $E_s$ by

$$\mathbf{s}_m = \sqrt{E_s} \sqrt{\frac{M}{M-1}} \left( \boldsymbol{\phi}_m - \bar{\boldsymbol{\phi}} \right), \quad m \in \mathcal{M}. \tag{26.99}$$

The construction for the case where $M = 2$ is depicted in Figure 26.6. It yields the binary antipodal signaling scheme. The construction for $M = 3$ is depicted in Figure 26.7.

From (26.99) and (26.98) we obtain for distinct $m', m'' \in \mathcal{M}$

$$\|\mathbf{s}_m\|_2^2 = E_s \quad \text{and} \quad \langle \mathbf{s}_{m'}, \mathbf{s}_{m''} \rangle = -\frac{E_s}{M-1}. \tag{26.100}$$

Also, from (26.99) we see that $\{\mathbf{s}_m\}$ can be viewed as the result of subtracting the center of gravity from orthogonal signals of energy $E_s M/(M-1)$. Consequently, the least error probability that can be achieved in detecting simplex signals of

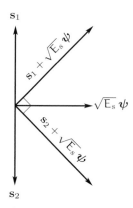

**Figure 26.8:** Adding a properly scaled signal $\psi$ that is orthogonal to all the elements of a simplex constellation results in an orthogonal constellation.

energy $E_s$ is the same as the least error probability that can be achieved in detecting orthogonal signals of energy

$$\frac{M}{M-1} E_s \tag{26.101}$$

(Proposition 26.11.2). From the expression for the error probability in orthogonal signaling (26.90) we obtain for the simplex signals with a uniform prior

$$\boxed{p^*(\text{error}) = 1 - \frac{1}{\sqrt{2\pi}} \int_{-\infty}^{\infty} e^{-\tau^2/2} \left(1 - Q\left(\tau + \sqrt{\frac{M}{M-1} \frac{2E_s}{N_0}}\right)\right)^{M-1} d\tau.}$$

$$\tag{26.102}$$

The decision rule for the simplex constellation can also be derived by exploiting the relationship to orthogonal keying. For example, if $\psi$ is a unit-energy integrable signal that is bandlimited to $W$ Hz and that is orthogonal to the signals $\{s_1, \ldots, s_M\}$, then, by (26.100), the waveforms

$$\left\{s_m + \frac{1}{\sqrt{M-1}} \sqrt{E_s}\, \psi\right\}_{m \in \mathcal{M}} \tag{26.103}$$

are orthogonal, each of energy $E_s M/(M-1)$. (See Figure 26.8 for a demonstration of the process of obtaining an orthogonal constellation with $M = 2$ signals by adding a signal $\psi$ to each of the signals in a binary simplex constellation.) Consequently, in order to decode the simplex signals contaminated by white Gaussian noise with respect to the bandwidth $W$, we can add $\frac{1}{\sqrt{M-1}} \sqrt{E_s}\, \psi$ to the received waveform and then feed the result to an optimal detector for orthogonal keying.

## 26.11.5  Bi-Orthogonal Keying

Starting with an orthogonal constellation, we can double the number of signals without reducing the minimum distance. This construction, which results in the

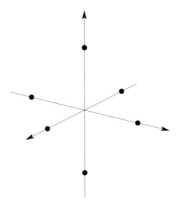

**Figure 26.9:** A bi-orthogonal constellation with six signals.

"bi-orthogonal signal set" is the topic of this section. To construct the bi-orthogonal signal set with $2\kappa$ signals, we start with $\kappa \geq 1$ orthonormal signals $(\phi_1, \ldots, \phi_\kappa)$ and define the $2\kappa$ bi-orthogonal signal set $\{\mathbf{s}_{1,\mathrm{u}}, \mathbf{s}_{1,\mathrm{d}}, \ldots, \mathbf{s}_{\kappa,\mathrm{u}}, \mathbf{s}_{\kappa,\mathrm{d}}\}$ by

$$\mathbf{s}_{\nu,\mathrm{u}} = +\sqrt{\mathsf{E}_\mathrm{s}}\,\phi_\nu \quad \text{and} \quad \mathbf{s}_{\nu,\mathrm{d}} = -\sqrt{\mathsf{E}_\mathrm{s}}\,\phi_\nu, \quad \nu \in \{1, \ldots, \kappa\}. \tag{26.104}$$

We can think of "u" as standing for "up" and of "d" as standing for "down," so to each signal $\phi_\nu$ there correspond two signals in the bi-orthogonal constellation: the "up signal" that corresponds to multiplying $\sqrt{\mathsf{E}_\mathrm{s}}\phi_\nu$ by $+1$ and the "down signal" that corresponds to multiplying $\sqrt{\mathsf{E}_\mathrm{s}}\phi_\nu$ by $-1$. Only bi-orthogonal signal sets with an even number of signals are defined. The constructed signals are all of energy $\mathsf{E}_\mathrm{s}$:

$$\|\mathbf{s}_{\nu,\mathrm{u}}\|_2 = \|\mathbf{s}_{\nu,\mathrm{d}}\|_2 = \sqrt{\mathsf{E}_\mathrm{s}}, \quad \nu \in \{1, \ldots, \kappa\}. \tag{26.105}$$

A bi-orthogonal constellation with six points ($\kappa = 3$) is depicted in Figure 26.9. Suppose that each of the signals $\phi_1, \ldots, \phi_\kappa$ is an integrable signal that is band-limited to $\mathsf{W}$ Hz and that, consequently, so are all the signals in the constructed bi-orthogonal signal set. A signal is picked uniformly at random from the signal set and is sent over a channel. We observe the stochastic process $(Y(t))$ given by the sum of the transmitted signal and white Gaussian noise of PSD $\mathsf{N}_0/2$ with respect to the bandwidth $\mathsf{W}$. How should we guess which signal was sent?

Since the signal was chosen equiprobably, and since all the signals in the signal set are of the same energy, it is optimal to consider the inner products

$$\langle \mathbf{Y}, \mathbf{s}_{1,\mathrm{u}} \rangle, \langle \mathbf{Y}, \mathbf{s}_{1,\mathrm{d}} \rangle, \ldots, \langle \mathbf{Y}, \mathbf{s}_{\kappa,\mathrm{u}} \rangle, \langle \mathbf{Y}, \mathbf{s}_{\kappa,\mathrm{d}} \rangle \tag{26.106}$$

and to pick the signal in the signal set corresponding to the largest of these inner products. By (26.104) we have for every $\nu \in \{1, \ldots, \kappa\}$ that $\mathbf{s}_{\nu,\mathrm{u}} = -\mathbf{s}_{\nu,\mathrm{d}}$ so $\langle \mathbf{Y}, \mathbf{s}_{\nu,\mathrm{u}} \rangle = -\langle \mathbf{Y}, \mathbf{s}_{\nu,\mathrm{d}} \rangle$ and hence

$$\max\{\langle \mathbf{Y}, \mathbf{s}_{\nu,\mathrm{u}} \rangle, \langle \mathbf{Y}, \mathbf{s}_{\nu,\mathrm{d}} \rangle\} = |\langle \mathbf{Y}, \mathbf{s}_{\nu,\mathrm{u}} \rangle|, \quad \nu \in \{1, \ldots, \kappa\}. \tag{26.107}$$

Equivalently, by (26.104),

$$\max\{\langle \mathbf{Y}, \mathbf{s}_{\nu,\mathrm{u}} \rangle, \langle \mathbf{Y}, \mathbf{s}_{\nu,\mathrm{d}} \rangle\} = \sqrt{\mathsf{E}_\mathrm{s}}\,|\langle \mathbf{Y}, \boldsymbol{\phi}_\nu \rangle|, \quad \nu \in \{1, \ldots, \kappa\}.$$

To find the maximum of the $2\kappa$ terms in (26.106) we can first compute for each $\nu \in \{1, \ldots, \kappa\}$ the maximum between $\langle \mathbf{Y}, \mathbf{s}_{\nu,\mathrm{u}} \rangle$ and $\langle \mathbf{Y}, \mathbf{s}_{\nu,\mathrm{d}} \rangle$ and then compute the maximum of the $\kappa$ results:

$$\max\Big\{\langle \mathbf{Y}, \mathbf{s}_{1,\mathrm{u}} \rangle, \langle \mathbf{Y}, \mathbf{s}_{1,\mathrm{d}} \rangle, \ldots, \langle \mathbf{Y}, \mathbf{s}_{\kappa,\mathrm{u}} \rangle, \langle \mathbf{Y}, \mathbf{s}_{\kappa,\mathrm{d}} \rangle\Big\}$$
$$= \max\Big\{\max\{\langle \mathbf{Y}, \mathbf{s}_{1,\mathrm{u}} \rangle, \langle \mathbf{Y}, \mathbf{s}_{1,\mathrm{d}} \rangle\}, \ldots, \max\{\langle \mathbf{Y}, \mathbf{s}_{\kappa,\mathrm{u}} \rangle, \langle \mathbf{Y}, \mathbf{s}_{\kappa,\mathrm{d}} \rangle\}\Big\}.$$

Using this approach, we obtain from (26.107) the following optimal two-step procedure: first find which $\nu^*$ in $\{1, \ldots, \kappa\}$ attains the maximum of the absolute values of the inner products

$$\max_{\nu \in \{1, \ldots, \kappa\}}\big\{|\langle \mathbf{Y}, \boldsymbol{\phi}_\nu \rangle|\big\}$$

and then, after you have found $\nu^*$, guess "$\mathbf{s}_{\nu^*,\mathrm{u}}$" if $\langle \mathbf{Y}, \boldsymbol{\phi}_{\nu^*} \rangle > 0$ and guess "$\mathbf{s}_{\nu^*,\mathrm{d}}$" if $\langle \mathbf{Y}, \boldsymbol{\phi}_{\nu^*} \rangle \leq 0$.

We next compute the probability of error of this optimal guessing rule. It is not difficult to see that the conditional probability of error does not depend on the message we condition on. For concreteness, we shall analyze the probability of error associated with the message corresponding to the signal $\mathbf{s}_{1,\mathrm{u}}$, a probability that we denote by $p_{\mathrm{MAP}}(\mathrm{error}|\mathbf{s}_{1,\mathrm{u}})$, with the corresponding conditional probability of correct decoding $p_{\mathrm{MAP}}(\mathrm{correct}|\mathbf{s}_{1,\mathrm{u}}) = 1 - p_{\mathrm{MAP}}(\mathrm{error}|\mathbf{s}_{1,\mathrm{u}})$. To simplify the typesetting, we shall denote the conditional probability of the event $\mathcal{A}$ given that $\mathbf{s}_{1,\mathrm{u}}$ is sent by $\Pr(\mathcal{A}|\mathbf{s}_{1,\mathrm{u}})$.

Since the probability of ties in the likelihood function is zero (Note 26.6.4)

$$p_{\mathrm{MAP}}(\mathrm{correct}|\mathbf{s}_{1,\mathrm{u}})$$
$$= \Pr\Big(-\langle \mathbf{Y}, \boldsymbol{\phi}_1 \rangle \leq \langle \mathbf{Y}, \boldsymbol{\phi}_1 \rangle \text{ and } \max_{2 \leq \nu \leq \kappa}\big\{|\langle \mathbf{Y}, \boldsymbol{\phi}_\nu \rangle|\big\} \leq \langle \mathbf{Y}, \boldsymbol{\phi}_1 \rangle \,\Big|\, \mathbf{s}_{1,\mathrm{u}}\Big)$$
$$= \Pr\Big(\langle \mathbf{Y}, \boldsymbol{\phi}_1 \rangle \geq 0 \text{ and } \max_{2 \leq \nu \leq \kappa}\big\{|\langle \mathbf{Y}, \boldsymbol{\phi}_\nu \rangle|\big\} \leq \langle \mathbf{Y}, \boldsymbol{\phi}_1 \rangle \,\Big|\, \mathbf{s}_{1,\mathrm{u}}\Big)$$
$$= \int_0^\infty f_{\langle \mathbf{Y}, \boldsymbol{\phi}_1 \rangle | \mathbf{s}_{1,\mathrm{u}}}(t) \Pr\Big[\max_{2 \leq \nu \leq \kappa}\big\{|\langle \mathbf{Y}, \boldsymbol{\phi}_\nu \rangle|\big\} \leq t \,\Big|\, \mathbf{s}_{1,\mathrm{u}}, \langle \mathbf{Y}, \boldsymbol{\phi}_1 \rangle = t\Big]\, \mathrm{d}t$$
$$= \int_0^\infty f_{\langle \mathbf{Y}, \boldsymbol{\phi}_1 \rangle | \mathbf{s}_{1,\mathrm{u}}}(t) \Pr\Big[\max_{2 \leq \nu \leq \kappa}\big\{|\langle \mathbf{Y}, \boldsymbol{\phi}_\nu \rangle|\big\} \leq t \,\Big|\, \mathbf{s}_{1,\mathrm{u}}\Big]\, \mathrm{d}t$$
$$= \int_0^\infty f_{\langle \mathbf{Y}, \boldsymbol{\phi}_1 \rangle | \mathbf{s}_{1,\mathrm{u}}}(t) \Big(\Pr\big[|\langle \mathbf{Y}, \boldsymbol{\phi}_2 \rangle| \leq t \,\big|\, \mathbf{s}_{1,\mathrm{u}}\big]\Big)^{\kappa-1}\, \mathrm{d}t$$
$$= \int_0^\infty \frac{1}{\sqrt{\pi \mathsf{N}_0}} e^{-\frac{(t-\sqrt{\mathsf{E}_\mathrm{s}})^2}{\mathsf{N}_0}} \left(1 - 2Q\left(\frac{t}{\sqrt{\mathsf{N}_0/2}}\right)\right)^{\kappa-1}\, \mathrm{d}t$$
$$= \frac{1}{\sqrt{2\pi}} \int_{-\sqrt{\frac{2\mathsf{E}_\mathrm{s}}{\mathsf{N}_0}}}^\infty e^{-\tau^2/2} \left(1 - 2Q\left(\tau + \sqrt{\frac{2\mathsf{E}_\mathrm{s}}{\mathsf{N}_0}}\right)\right)^{\kappa-1}\, \mathrm{d}\tau, \tag{26.108}$$

with the following justification. The first equality follows from the definition of our optimal decoder and from the fact that ties occur with probability zero. The second equality follows by trivial algebra ($-\xi \leq \xi$ if, and only if, $\xi \geq 0$). The third equality follows by conditioning on $\langle \mathbf{Y}, \mathbf{s}_{1,\mathrm{u}} \rangle$ being equal to $t$ and integrating $t$ out while noting that a correct decision can only be made if $t \geq 0$, in which case the condition $\langle \mathbf{Y}, \boldsymbol{\phi}_1 \rangle \geq 0$ is satisfied automatically. The fourth equality follows because, conditional on the signal $\mathbf{s}_{1,\mathrm{u}}$ being sent, the random variable $\langle \mathbf{Y}, \mathbf{s}_{1,\mathrm{u}} \rangle$ is independent of the random variables $\{|\langle \mathbf{Y}, \boldsymbol{\phi}_\nu \rangle|\}_{2 \leq \nu \leq \kappa}$. The fifth equality follows because, conditional on $\mathbf{s}_{1,\mathrm{u}}$ being sent, the random variables $\{|\langle \mathbf{Y}, \boldsymbol{\phi}_\nu \rangle|\}_{2 \leq \nu \leq \kappa}$ are IID. The sixth equality follows because, conditional on $\mathbf{s}_{1,\mathrm{u}}$ being sent, we have $\langle \mathbf{Y}, \boldsymbol{\phi}_1 \rangle \sim \mathcal{N}(\sqrt{\mathsf{E_s}}, \mathsf{N}_0/2)$ and $\langle \mathbf{Y}, \boldsymbol{\phi}_2 \rangle \sim \mathcal{N}(0, \mathsf{N}_0/2)$, so

$$
\Pr\left[ |\langle \mathbf{Y}, \boldsymbol{\phi}_2 \rangle| \leq t \,\middle|\, \mathbf{s}_{1,\mathrm{u}} \right]
$$
$$
= \Pr\left[ \frac{|\langle \mathbf{Y}, \boldsymbol{\phi}_2 \rangle|}{\sqrt{\mathsf{N}_0/2}} \leq \frac{t}{\sqrt{\mathsf{N}_0/2}} \,\middle|\, \mathbf{s}_{1,\mathrm{u}} \right]
$$
$$
= 1 - \Pr\left[ \frac{|\langle \mathbf{Y}, \boldsymbol{\phi}_2 \rangle|}{\sqrt{\mathsf{N}_0/2}} \geq \frac{t}{\sqrt{\mathsf{N}_0/2}} \,\middle|\, \mathbf{s}_{1,\mathrm{u}} \right]
$$
$$
= 1 - \Pr\left[ \frac{\langle \mathbf{Y}, \boldsymbol{\phi}_2 \rangle}{\sqrt{\mathsf{N}_0/2}} \geq \frac{t}{\sqrt{\mathsf{N}_0/2}} \,\middle|\, \mathbf{s}_{1,\mathrm{u}} \right] - \Pr\left[ \frac{\langle \mathbf{Y}, \boldsymbol{\phi}_2 \rangle}{\sqrt{\mathsf{N}_0/2}} \leq \frac{-t}{\sqrt{\mathsf{N}_0/2}} \,\middle|\, \mathbf{s}_{1,\mathrm{u}} \right]
$$
$$
= 1 - 2\mathcal{Q}\left( \frac{t}{\sqrt{\mathsf{N}_0/2}} \right).
$$

Finally, (26.108) follows from the substitution $\tau \triangleq (t - \sqrt{\mathsf{E_s}})/\sqrt{\mathsf{N}_0/2}$ as in (26.89).

Since the conditional probability of error does not depend on the message, it follows that all conditional probabilities of error are equal to the average probability of error $p^*(\mathrm{error})$ and

$$
\boxed{p^*(\mathrm{error}) = 1 - \frac{1}{\sqrt{2\pi}} \int_{-\sqrt{\frac{2\mathsf{E_s}}{\mathsf{N}_0}}}^{\infty} e^{-\tau^2/2} \left( 1 - 2\mathcal{Q}\left( \tau + \sqrt{\frac{2\mathsf{E_s}}{\mathsf{N}_0}} \right) \right)^{\kappa-1} \mathrm{d}\tau,} \tag{26.109}
$$

or, using the Binomial Expansion (26.91) with the substitution of $-\mathcal{Q}\left( \tau + \sqrt{\frac{2\mathsf{E_s}}{\mathsf{N}_0}} \right)$ for $b$ and of 1 for $a$,

$$
p^*(\mathrm{error}) = \sum_{j=1}^{\kappa-1} (-1)^{j+1} 2^j \binom{\kappa-1}{j} \frac{1}{\sqrt{2\pi}} \int_{-\sqrt{\frac{2\mathsf{E_s}}{\mathsf{N}_0}}}^{\infty} e^{-\tau^2/2} \left( \mathcal{Q}\left( \tau + \sqrt{\frac{2\mathsf{E_s}}{\mathsf{N}_0}} \right) \right)^j \mathrm{d}\tau.
$$
$$
\tag{26.110}
$$

The probability of error associated with an orthogonal constellation with $\kappa$ signals is better than that of the bi-orthogonal constellation with $2\kappa$ signals and equal average energy. But the comparison is not quite fair because the bi-orthogonal constellation is richer.

## 26.12 Detection in Colored Noise

Our focus throughout has been on the detection problem when the noise is "white" in the sense that its PSD is flat over the frequency band to which the mean signals are limited. We now extend the discussion to "colored" noise, i.e., to noise whose PSD is not constant over the bandwidth of interest. We continue to assume that the mean signals $\{\mathbf{s}_m\}_{m \in \mathcal{M}}$ are integrable signals that are bandlimited to $\mathsf{W}$ Hz and that the noise $\big(N(t)\big)$ is independent of the message $M$ and is a measurable, stationary, Gaussian SP. Its PSD $\mathsf{S}_{NN}$, however, is now an arbitrary nonnegative, symmetric, integrable function that is not necessarily constant over the band $[-\mathsf{W}, \mathsf{W}]$. Conditional on $M = m$, the received waveform $\big(Y(t)\big)$ is given at time $t$ by $s_m(t) + N(t)$.

Our approach is based on "whitening the noise" and is only applicable when the noise can be whitened with respect to the bandwidth $\mathsf{W}$, i.e., when there exists a whitening filter for the noise with respect to $\mathsf{W}$:

**Definition 26.12.1 (Whitening Filter for $\mathsf{S}_{NN}$ with respect to $\mathsf{W}$).** *A filter of impulse response* $\mathbf{h} \colon \mathbb{R} \to \mathbb{R}$ *is said to be a **whitening filter for** $\mathsf{S}_{NN}$ (or for $\big(N(t)\big)$) **with respect to the bandwidth** $\mathsf{W}$ if it is stable and its frequency response* $\hat{\mathbf{h}}$ *satisfies*

$$\mathsf{S}_{NN}(f)\,|\hat{h}(f)|^2 = 1, \quad |f| \le \mathsf{W}. \tag{26.111}$$

Only the magnitude of the frequency response of the whitening filter is specified in (26.111) and only for frequencies in the band $[-\mathsf{W}, \mathsf{W}]$. The response is unspecified outside this band. Consequently:

**Note 26.12.2.** There may be many different whitening filters for $\mathsf{S}_{NN}$ with respect to the bandwidth $\mathsf{W}$.

If $\mathsf{S}_{NN}$ is zero at some frequencies in $[-\mathsf{W}, \mathsf{W}]$, then there is no whitening filter for $\mathsf{S}_{NN}$ with respect to $\mathsf{W}$. Likewise, a whitening filter for $\mathsf{S}_{NN}$ does not exist if $\mathsf{S}_{NN}$ is not continuous in $[-\mathsf{W}, \mathsf{W}]$ (because the frequency response of a stable filter must be continuous (Theorem 6.2.11), and if $\mathsf{S}_{NN}$ is discontinuous, then so is $f \mapsto 1/\sqrt{|\mathsf{S}_{NN}(f)|}$). Thus:

**Note 26.12.3.** There does not always exist a whitening filter for $\mathsf{S}_{NN}$ with respect to $\mathsf{W}$.

We shall see, however, in Proposition 26.12.8 that a whitening filter exists whenever throughout the interval $[-\mathsf{W}, \mathsf{W}]$ the PSD $\mathsf{S}_{NN}$ is strictly positive and is twice continuously differentiable.

The filter is called "whitening" because, by Theorem 25.13.2, we have:

**Proposition 26.12.4.** *If* $\big(N(t),\ t \in \mathbb{R}\big)$ *is a measurable, stationary, Gaussian SP of PSD* $\mathsf{S}_{NN}$, *and if* $\mathbf{h}$ *is the impulse response of a whitening filter for* $\mathsf{S}_{NN}$ *with respect to* $\mathsf{W}$, *then* $\big(N(t)\big) \star \mathbf{h}$ *is white Gaussian noise of PSD* 1 *with respect to the bandwidth* $\mathsf{W}$.

Assuming that the noise can be whitened with respect to the bandwidth $\mathsf{W}$, we pick some whitening filter of impulse response $\mathbf{h}$ and denote by $\big(\tilde{Y}(t)\big)$ the result of feeding the observed SP $\big(Y(t)\big)$ to this filter:

$$\big(\tilde{Y}(t)\big) = \big(Y(t)\big) \star \mathbf{h}. \tag{26.112}$$

Conditional on $M = m$, the output of the whitening filter is given by

$$\tilde{Y}(t) = \left(\mathbf{s}_m + \left(N(t)\right)\right) \star \mathbf{h}$$
$$= \tilde{\mathbf{s}}_m + \tilde{N}(t), \quad t \in \mathbb{R}, \tag{26.113}$$

where

$$\tilde{\mathbf{s}}_m = \mathbf{s}_m \star \mathbf{h}, \quad m \in \mathcal{M}, \tag{26.114}$$

and

$$\left(\tilde{N}(t)\right) = \left(N(t)\right) \star \mathbf{h}. \tag{26.115}$$

By Proposition 6.5.2, $\tilde{\mathbf{s}}_m$ is an integrable signal that is bandlimited to $W$ Hz and

$$\tilde{s}_m(t) = \int_{-W}^{W} \hat{s}_m(f)\,\hat{h}(f)\,e^{\mathrm{i}2\pi ft}\,\mathrm{d}f, \quad t \in \mathbb{R}. \tag{26.116}$$

And, by Proposition 26.12.4, $\left(\tilde{N}(t)\right)$ is white Gaussian noise of PSD 1 with respect to the bandwidth $W$.

Loosely speaking, the main result of this section is that there is no loss in optimality in guessing $M$ based on the whitening filter's output $\left(\tilde{Y}(t)\right)$. This is not very surprising for the following reason. While passing $\left(Y(t)\right)$ through the whitening filter is not necessarily an invertible operation, it "almost" is, in the sense that we can recover the original observation inside the band $[-W, W]$. Since the transmitted signals are bandlimited to $W$ Hz, we do not expect that the observation outside this band will influence our guess.

Once this result is proved, the detection problem is reduced to detecting known signals (the signals $\{\tilde{\mathbf{s}}_m\}$) in white Gaussian noise (the SP $\left(\tilde{N}(t)\right)$). Employing Theorem 26.4.1, we obtain that if guessing $M$ based on the whitening filter's output is optimal, then so is basing one's guess on the inner products vector

$$\left(\langle \tilde{\mathbf{Y}}, \tilde{\mathbf{s}}_1 \rangle, \dots, \langle \tilde{\mathbf{Y}}, \tilde{\mathbf{s}}_\mathsf{M} \rangle \right)^\mathsf{T}, \tag{26.117}$$

thus reducing the continuous-time detection problem to one where the observation is a random vector taking value in $\mathbb{R}^\mathsf{M}$.

We next describe the sufficient statistic for our problem more carefully. Rather than expressing the sufficient statistic as in (26.117), we prefer to express it directly in terms of the observed signal $\left(Y(t)\right)$ as the vector

$$\left(\langle \mathbf{Y}, \overleftarrow{\mathbf{h}} \star \tilde{\mathbf{s}}_1 \rangle, \dots, \langle \mathbf{Y}, \overleftarrow{\mathbf{h}} \star \tilde{\mathbf{s}}_\mathsf{M} \rangle \right)^\mathsf{T}, \tag{26.118}$$

where the equivalence of the two forms can be formally derived as follows:

$$\langle \tilde{\mathbf{Y}}, \tilde{\mathbf{s}}_m \rangle = \int_{-\infty}^{\infty} \left(\int_{-\infty}^{\infty} Y(\sigma)\,h(t-\sigma)\,\mathrm{d}\sigma\right) \tilde{s}_m(t)\,\mathrm{d}t$$
$$= \int_{-\infty}^{\infty} \int_{-\infty}^{\infty} Y(\sigma)\,h(t-\sigma)\,\tilde{s}_m(t)\,\mathrm{d}t\,\mathrm{d}\sigma$$
$$= \int_{-\infty}^{\infty} Y(\sigma) \left(\int_{-\infty}^{\infty} h(t-\sigma)\,\tilde{s}_m(t)\,\mathrm{d}t\right) \mathrm{d}\sigma$$

$$= \int_{-\infty}^{\infty} Y(\sigma) \left(\overleftarrow{\mathbf{h}} \star \tilde{\mathbf{s}}_m\right)(\sigma) \, \mathrm{d}\sigma$$

$$= \left\langle \mathbf{Y}, \overleftarrow{\mathbf{h}} \star \tilde{\mathbf{s}}_m \right\rangle.$$

Note that for each $m \in \mathcal{M}$ the convolution $\overleftarrow{\mathbf{h}} \star \tilde{\mathbf{s}}_m$ is the result of passing the signal $\tilde{\mathbf{s}}_m$, which is an integrable signal that is bandlimited to $W$ Hz, through the stable filter of impulse response $\overleftarrow{\mathbf{h}}$, so $\overleftarrow{\mathbf{h}} \star \tilde{\mathbf{s}}_m$ is an integrable signal that is bandlimited to $W$ Hz (Proposition 6.5.2). This integrability guarantees that the inner products in (26.118) are well-defined (Proposition 25.10.1).

We can now state the main result of this section:

**Theorem 26.12.5 (Detecting Known Signals in Colored Noise).** *Let $M$ take value in the finite set $\mathcal{M} = \{1, \ldots, \mathsf{M}\}$, and let the signals $\mathbf{s}_1, \ldots, \mathbf{s}_\mathsf{M}$ be integrable signals that are bandlimited to $W$ Hz. Let the conditional law of $\left(Y(t)\right)$ given $M = m$ be that of $s_m(t) + N(t)$, where $\left(N(t)\right)$ is a stationary, measurable, Gaussian SP of PSD $\mathsf{S}_{NN}$ that can be whitened with respect to the bandwidth $W$. Let $\mathbf{h}$ be the impulse response of a whitening filter for $\left(N(t)\right)$. Then:*

(i) *The inner-products vector (26.118) forms a sufficient statistic for guessing $M$ based on the observation $\left(Y(t)\right)$.*

(ii) *Conditional on $M = m$, this vector is Gaussian with mean*

$$\left(\langle \tilde{\mathbf{s}}_m, \tilde{\mathbf{s}}_1 \rangle, \ldots, \langle \tilde{\mathbf{s}}_m, \tilde{\mathbf{s}}_\mathsf{M} \rangle\right)^\mathsf{T} \tag{26.119}$$

*and $\mathsf{M} \times \mathsf{M}$ covariance matrix*

$$\begin{pmatrix} \langle \tilde{\mathbf{s}}_1, \tilde{\mathbf{s}}_1 \rangle & \langle \tilde{\mathbf{s}}_1, \tilde{\mathbf{s}}_2 \rangle & \cdots & \langle \tilde{\mathbf{s}}_1, \tilde{\mathbf{s}}_\mathsf{M} \rangle \\ \langle \tilde{\mathbf{s}}_2, \tilde{\mathbf{s}}_1 \rangle & \langle \tilde{\mathbf{s}}_2, \tilde{\mathbf{s}}_2 \rangle & \cdots & \langle \tilde{\mathbf{s}}_2, \tilde{\mathbf{s}}_\mathsf{M} \rangle \\ \vdots & \vdots & \ddots & \vdots \\ \langle \tilde{\mathbf{s}}_\mathsf{M}, \tilde{\mathbf{s}}_1 \rangle & \langle \tilde{\mathbf{s}}_\mathsf{M}, \tilde{\mathbf{s}}_2 \rangle & \cdots & \langle \tilde{\mathbf{s}}_\mathsf{M}, \tilde{\mathbf{s}}_\mathsf{M} \rangle \end{pmatrix}, \tag{26.120}$$

*where*

$$\tilde{\mathbf{s}}_j = \mathbf{s}_j \star \mathbf{h}, \quad j \in \mathcal{M}, \tag{26.121}$$

*and where the inner product $\langle \tilde{\mathbf{s}}_{m'}, \tilde{\mathbf{s}}_{m''} \rangle$ can also be expressed as*

$$\langle \tilde{\mathbf{s}}_{m'}, \tilde{\mathbf{s}}_{m''} \rangle = \int_{-W}^{W} \hat{s}_{m'}(f) \, \hat{s}_{m''}^*(f) \, \frac{1}{\mathsf{S}_{NN}(f)} \, \mathrm{d}f, \quad m', m'' \in \mathcal{M}. \tag{26.122}$$

(iii) *If $(\boldsymbol{\phi}_1, \ldots, \boldsymbol{\phi}_{d'})$ is an orthonormal $d'$-tuple of integrable signals that are bandlimited to $W$ Hz, and if*

$$\tilde{\mathbf{s}}_m \in \mathrm{span}(\boldsymbol{\phi}_1, \ldots, \boldsymbol{\phi}_{d'}), \quad m \in \mathcal{M}, \tag{26.123}$$

*then the inner products vector*

$$\left(\langle \mathbf{Y}, \overleftarrow{\mathbf{h}} \star \boldsymbol{\phi}_1 \rangle, \ldots, \langle \mathbf{Y}, \overleftarrow{\mathbf{h}} \star \boldsymbol{\phi}_{d'} \rangle\right)^\mathsf{T}, \tag{26.124}$$

*forms a sufficient statistic for guessing $M$ based on $\big(Y(t)\big)$ and, conditional on $M = m$, is a multivariate Gaussian of covariance matrix $\mathsf{I}_{d'}$ and of mean vector*

$$\big(\langle \tilde{\mathbf{s}}_m, \boldsymbol{\phi}_1\rangle, \ldots, \langle \tilde{\mathbf{s}}_m, \boldsymbol{\phi}_{d'}\rangle\big)^{\mathsf{T}}. \tag{26.125}$$

**Proof.** The sufficiency of the vector (26.118) can be established by first proving the result in the binary antipodal case, and by then generalizing the result as we did in the proof of Theorem 26.4.1.

In the binary antipodal case we denote the RV to be guessed by $H$ and assume that it takes value in $\{0, 1\}$. We assume that, conditional on $H = 0$, the time-$t$ received waveform is $s(t) + N(t)$ whereas, conditional on $H = 1$, it is $-s(t) + N(t)$. We show that for every $\eta \in \mathbb{N}$ and any choice of the epochs $t_1, \ldots, t_\eta \in \mathbb{R}$, the inner product $\langle \mathbf{Y}, \overset{\leftarrow}{\mathbf{h}} \star \mathbf{s}\rangle$ forms a sufficient statistic for guessing $H$ based on the vector

$$\Big(Y(t_1), \ldots, Y(t_\eta), \langle \mathbf{Y}, \overset{\leftarrow}{\mathbf{h}} \star \mathbf{s}\rangle\Big)^{\mathsf{T}}.$$

As in the proof of Theorem 26.4.1, this can be established using Lemma 26.8.1 as follows. One first notes that, conditional on $H$, this vector is Gaussian (Proposition 25.11.1). One then notes that the conditional covariance matrix of this vector conditional on $H = 0$ is the same as conditional on $H = 1$ and that this covariance matrix can be computed using Theorem 25.12.2. Finally one shows that the vector's conditional mean vector, conditional on $H = 0$, is antipodal to its conditional mean vector, conditional on $H = 1$, and that both are scaled versions of the last column of the conditional covariance matrix.

Once the sufficiency of the vector (26.118) has been established, the computation of its conditional law is straightforward: by Proposition 25.11.1 it is conditionally Gaussian, and its conditional mean (26.119) and conditional covariance (26.120) are readily derived using Theorem 25.12.2. The derivation of (26.122) follows from (26.116) using the Mini Parseval Theorem (Proposition 6.2.6 (i)) and (26.111).

An alternative way of deriving the conditional distribution is to note that the vector (26.118) can also be expressed as the vector (26.117) and to then use the result from Section 26.5.1 by substituting 1 for $\mathsf{N}_0/2$ and $\tilde{\mathbf{s}}_m$ for $\mathbf{s}_m$ for all $m \in \mathcal{M}$.

Part (iii) follows directly from Parts (i) and (ii). $\qquad\square$

Since the inner products $\langle \tilde{\mathbf{s}}_{m'}, \tilde{\mathbf{s}}_{m''}\rangle$ for $m', m'' \in \mathcal{M}$ determine the conditional law of the sufficient statistic (see (26.119) & (26.120)), and since, by (26.122), the inner product $\langle \tilde{\mathbf{s}}_{m'}, \tilde{\mathbf{s}}_{m''}\rangle$ does not depend on the choice of the whitening filter we obtain:

**Note 26.12.6.** Neither the conditional distribution of the sufficient statistic vector in (26.118) nor the optimal proability of error depends on the choice of the whitening filter.

Using Theorem 26.12.5 we can now derive an optimal rule for guessing $M$. Indeed, in analogy to Theorem 26.6.3 we have:

**Theorem 26.12.7.** *Consider the setting of Theorem 26.12.5 with $M$ of prior $\{\pi_m\}$. The decision rule that guesses uniformly at random from among all the messages* *m̃ ∈ M for which*

$$\ln \pi_{\tilde{m}} - \frac{1}{2} \sum_{\ell=1}^{d} \left( \langle \mathbf{Y}, \overset{\leftarrow}{\mathbf{h}} \star \phi_\ell \rangle - \langle \mathbf{s}_{\tilde{m}}, \overset{\leftarrow}{\mathbf{h}} \star \phi_\ell \rangle \right)^2$$

$$= \max_{m' \in M} \left\{ \ln \pi_{m'} - \frac{1}{2} \sum_{\ell=1}^{d} \left( \langle \mathbf{Y}, \overset{\leftarrow}{\mathbf{h}} \star \phi_\ell \rangle - \langle \mathbf{s}_{m'}, \overset{\leftarrow}{\mathbf{h}} \star \phi_\ell \rangle \right)^2 \right\} \quad (26.126)$$

*minimizes the probability of error whenever* **h** *is a whitening filter and the tuple* $(\phi_1, \ldots, \phi_d)$ *forms an orthonormal basis for* $\mathrm{span}(\mathbf{s}_1 \star \mathbf{h}, \ldots, \mathbf{s}_M \star \mathbf{h})$.

Before concluding our discussion of detection in the presence of colored noise we derive here a sufficient condition for the existence of a whitening filter.

**Proposition 26.12.8 (Existence of a Whitening Filter).** *Let* $\mathsf{W} > 0$ *be fixed. If throughout the interval* $[-\mathsf{W}, \mathsf{W}]$ *the PSD* $\mathsf{S}_{NN}$ *is strictly positive and twice continuously differentiable, then there exists a whitening filter for* $\mathsf{S}_{NN}$ *with respect to the bandwidth* $\mathsf{W}$.

**Proof.** The proof hinges on the following basic result from harmonic analysis (Katznelson, 1976, Chapter VI, Section 1, Exercise 7): if a function $f \mapsto g(f)$ is twice continuously differentiable and is zero outside some interval $[-\Delta, \Delta]$, then it is the FT of some integrable function.

To prove the proposition using this result we begin by picking some $\Delta > \mathsf{W}$. We now define a function $\mathbf{g} \colon \mathbb{R} \to \mathbb{R}$ as follows. For $f \geq \Delta$, we define $g(f) = 0$. For $f$ in the interval $[0, \mathsf{W}]$, we define $g(f) = 1/\sqrt{\mathsf{S}_{NN}(f)}$. And for $f \in (\mathsf{W}, \Delta)$, we define $g(f)$ so that $\mathbf{g}$ be twice continuously differentiable in $[0, \infty)$. We can thus think of $\mathbf{g}$ in $[\mathsf{W}, \Delta]$ as an interpolation function whose values and first two derivatives are specified at the endpoints of the interval. Finally, for $f < 0$, we define $g(f)$ as $g(-f)$. Figure 26.10 depicts $\mathsf{S}_{NN}$, $\mathbf{g}$, $\mathsf{W}$, and $\Delta$.

A whitening filter for $\mathsf{S}_{NN}$ with respect to the bandwidth $\mathsf{W}$ is the integrable function whose FT is $\mathbf{g}$ and whose existence is guaranteed by the quoted result. $\square$

## 26.13 Detecting Signals of Infinite Bandwidth

So far we have only dealt with the detection problem when the mean signals are bandlimited. What if the mean signals are not bandlimited? The difficulty in this case is that we cannot assume that the noise PSD is constant over the bandwidth occupied by the mean signals, or that the noise can be whitened with respect to this bandwidth.

We can address this issue in three different ways. In the first we can try to find the optimal detector by studying this more complicated hypothesis testing problem. It will no longer be the case that the inner products vector (26.15) forms a sufficient statistic. It will turn out that the optimal detector greatly depends on the relationship between the rate of decay of the PSD of the noise as the frequency tends

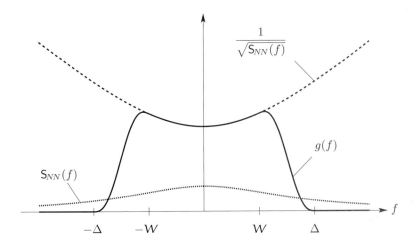

**Figure 26.10:** The frequency response of a whitening filter for the PSD $S_{NN}$ with respect to the bandwidth $W$.

to $\pm\infty$ and the rate of decay of the FT of the mean signals. This approach will often lead to bad designs, because the structure of the receiver will depend greatly on how we model the noise, and inaccuracies in our modeling of the noise PSD at ultra-high frequencies might lead us completely astray in our design.

A more level-headed approach that is valid if the noise PSD is "essentially flat over the bandwidth of interest" is to ignore the fact that the mean signals are not bandlimited and to base our decision on the inner products vector, even if this is not fully justified mathematically. This approach leads to robust designs that are insensitive to inaccuracies in our modeling of the noise process. If the PSD is not essentially flat, we can whiten it with respect to a sufficiently large band $[-W, W]$ that contains most of the energy of the mean signals.

The third approach is to use very complicated mathematical machinery involving the Itô Calculus (Karatzas and Shreve, 1991) to model the noise in a way that will result in the inner products forming a sufficient statistic. We have chosen not to pursue this approach because it requires modeling the noise as a process of infinite power, which is physically unappealing. This approach just shifts the burden of proof from one place to another. Indeed, the Itô Calculus can now prove for us that the inner products vector is sufficient, but we need a leap of faith in modeling the noise as a process of infinite power.

In the future, in dealing with mean signals that are not bandlimited, we shall refer to the "white noise paradigm" as the paradigm under which the receiver forms its decision based on the inner products vector (26.15) and under which these inner products have the conditional law derived in Section 26.5.1.

## 26.14   A Proof of Lemma 26.8.1

We next present a proof of Lemma 26.8.1 that is also valid when the matrix $\Lambda$ is singular. We denote the Row-$j$ Column-$k$ component of this matrix by $\lambda^{(j,k)}$.

**Proof.** We first treat the case where the variance $\lambda^{(n,n)}$ of the last component $Y^{(n)}$ of the random $n$-vector $\mathbf{Y}$ is zero. By the Covariance Inequality it follows that for every $j \in \{1, \ldots, n\}$

$$\left|\lambda^{(j,n)}\right| = \left|\mathsf{Cov}\big[Y^{(j)}, Y^{(n)}\big]\right| \leq \sqrt{\mathsf{Var}\big[Y^{(j)}\big]}\sqrt{\mathsf{Var}\big[Y^{(n)}\big]} = \sqrt{\lambda^{(j,j)}}\sqrt{\lambda^{(n,n)}},$$

so in this case the $n$-th column of $\Lambda$ is zero. Consequently, since the mean vector $\boldsymbol{\mu}$ is by assumption proportional to the last column of $\Lambda$, it follows that in this case $\boldsymbol{\mu} = \mathbf{0}$. But for $\boldsymbol{\mu} = \mathbf{0}$ the conditional law of $\mathbf{Y}$ given $H = 0$ is the same as given $H = 1$, so $\mathbf{Y}$ is useless for guessing $H$, and any measurable function of $Y$, and *a fortiori* its last component, forms a sufficient statistic (albeit also useless).

We next turn to the more interesting case where

$$\lambda^{(n,n)} > 0. \tag{26.127}$$

In this case we can write the assumption that $\boldsymbol{\mu}$ is a scaled version of the last column of $\Lambda$ as

$$\mu^{(j)} = \mu^{(n)} \frac{\lambda^{(j,n)}}{\lambda^{(n,n)}}, \quad j \in \{1, \ldots, n\}. \tag{26.128}$$

We need to show that, irrespective of the prior on $H$, (26.128) implies that

$$H -\!\circ\!- Y^{(n)} -\!\circ\!- \mathbf{Y} \tag{26.129}$$

forms a Markov chain or, equivalently, that

$$H -\!\circ\!- Y^{(n)} -\!\circ\!- \mathbf{R} \tag{26.130}$$

forms a Markov chain, where $\mathbf{R}$ is the random $(n-1)$-vector of components

$$R^{(j)} = Y^{(j)} - \frac{\lambda^{(j,n)}}{\lambda^{(n,n)}} Y^{(n)}, \quad j \in \{1, \ldots, n-1\}. \tag{26.131}$$

(Conditional on $Y^{(n)}$, we have that $Y^{(n)}$ is deterministic, so $R^{(j)}$ and $Y^{(j)}$ only differ by a deterministic constant.) Thus, we need to show that $\mathbf{R}$ is irrelevant for guessing $H$ based on $Y^{(n)}$. This we prove using Proposition 22.5.5 by showing that $Y^{(n)}$ and $\mathbf{R}$ are conditionally independent given $H$

$$Y^{(n)} -\!\circ\!- H -\!\circ\!- \mathbf{R} \tag{26.132}$$

and that $H$ and $\mathbf{R}$ are independent.

We begin by proving (26.132). We first note that, conditional on $H = 0$, the vector

$$\left(R^{(1)}, \ldots, R^{(n-1)}, Y^{(n)}\right)^{\mathsf{T}}$$

is Gaussian, because it is the result of linearly transforming the vector $\mathbf{Y}$, which, conditional on $H = 0$, is Gaussian (Proposition 23.6.3). Also, conditional on $H = 0$, we have that $Y^{(n)}$ is uncorrelated with the components of $\mathbf{R}$ because

$$
\begin{aligned}
\mathsf{Cov}&\left[Y^{(n)}, R^{(j)}\,\middle|\, H = 0\right] \\
&= \mathsf{Cov}\left[Y^{(n)}, Y^{(j)} - \frac{\lambda^{(j,n)}}{\lambda^{(n,n)}}Y^{(n)}\,\middle|\, H = 0\right] \\
&= \mathsf{Cov}\left[Y^{(n)}, Y^{(j)}\,\middle|\, H = 0\right] - \frac{\lambda^{(j,n)}}{\lambda^{(n,n)}}\mathsf{Cov}\left[Y^{(n)}, Y^{(n)}\,\middle|\, H = 0\right] \\
&= \lambda^{(j,n)} - \frac{\lambda^{(j,n)}}{\lambda^{(n,n)}}\lambda^{(n,n)} \\
&= 0, \quad j \in \{1, \ldots, n-1\}.
\end{aligned}
$$

By Corollary 23.6.9, we conclude that, conditional on $H = 0$, we have that $Y^{(n)}$ is independent of $\mathbf{R}$. Repeating this argument for the case where the conditioning is on $H = 1$ proves (26.132).

We next verify that $\mathbf{R}$ and $H$ are independent. We do so by showing that the conditional distribution of $\mathbf{R}$ given $H = 0$ is identical to its conditional distribution given $H = 1$. Since under both conditionings $\mathbf{R}$ is Gaussian, it suffices to show that the conditional covariance of $\mathbf{R}$ given $H = 0$ is the same as given $H = 1$ and similarly for the mean. To show that the covariances are the same is easy, because the conditional covariance of $\mathbf{R}$ is determined by $\Lambda$, which is the same under the two hypotheses. As to the mean we have

$$
\begin{aligned}
\mathsf{E}\left[R^{(j)}\,\middle|\, H = 0\right] &= \mathsf{E}\left[Y^{(j)} - \frac{\lambda^{(j,n)}}{\lambda^{(n,n)}}Y^{(n)}\,\middle|\, H = 0\right] \\
&= \mathsf{E}\left[Y^{(j)}\,\middle|\, H = 0\right] - \frac{\lambda^{(j,n)}}{\lambda^{(n,n)}}\mathsf{E}\left[Y^{(n)}\,\middle|\, H = 0\right] \\
&= \mu^{(j)} - \frac{\lambda^{(j,n)}}{\lambda^{(n,n)}}\mu^{(n)} \\
&= 0, \quad j \in \{1, \ldots, n-1\},
\end{aligned}
$$

where the last equality follows from (26.128). Similarly, under $H = 1$ we have

$$
\begin{aligned}
\mathsf{E}\left[R^{(j)}\,\middle|\, H = 1\right] &= \mathsf{E}\left[Y^{(j)} - \frac{\lambda^{(j,n)}}{\lambda^{(n,n)}}Y^{(n)}\,\middle|\, H = 1\right] \\
&= \mathsf{E}\left[Y^{(j)}\,\middle|\, H = 1\right] - \frac{\lambda^{(j,n)}}{\lambda^{(n,n)}}\mathsf{E}\left[Y^{(n)}\,\middle|\, H = 1\right] \\
&= -\mu^{(j)} - \frac{\lambda^{(j,n)}}{\lambda^{(n,n)}}\left(-\mu^{(n)}\right) \\
&= 0, \quad j \in \{1, \ldots, n-1\},
\end{aligned}
$$

thus establishing that the mean of $\mathbf{R}$ does not depend on $H$ either.

Having established that $\mathbf{R}$ and $H$ are independent, it now follows from the conditional independence of $Y^{(n)}$ and $\mathbf{R}$ given $H$ (26.132) that $\mathbf{R}$ is irrelevant for guessing $H$ based on $Y^{(n)}$ (Proposition 22.5.5). $\qquad\square$

## 26.15   Exercises

**Exercise 26.1 (Reducing the Number of Matched Filters).** We saw in Section 26.4 how to obtain a $d$-dimensional sufficient statistics vector, where $d$ is the dimension of the linear subspace spanned by the mean signals (26.17). Show that, given any integrable signal $s_0$ that is bandlimited to $W$ Hz, we can find a $d'$-dimensional sufficient statistics vector, where

$$d' = \text{Dim}\big(\text{span}(\mathbf{s}_1 - \mathbf{s}_0, \dots, \mathbf{s}_M - \mathbf{s}_0)\big).$$

Show that $d'$ is sometimes smaller than $d$.

**Exercise 26.2 (Nearest-Neighbor Decoding Revisited).** The form of the decoder in Theorem 26.6.3 (ii) is different from the nearest-neighbor rule of Proposition 21.6.1 (ii). Why does minimizing $\|\mathbf{Y} - \mathbf{s}_m\|_2$ not make mathematical sense in the setting of Theorem 26.6.3?

**Exercise 26.3 (Proving Sufficiency).** In Section 26.8.3 we sketched an argument for the sufficiency of the vector in (26.57). Fill in the details.

**Exercise 26.4 (Minimum Shift Keying).** Let the signals $s_0, s_1$ be given at every $t \in \mathbb{R}$ by

$$s_0(t) = \sqrt{\frac{2E_s}{T_s}} \cos(2\pi f_0 t)\, \text{I}\{0 \leq t \leq T_s\}, \quad s_1(t) = \sqrt{\frac{2E_s}{T_s}} \cos(2\pi f_1 t)\, \text{I}\{0 \leq t \leq T_s\}.$$

(i) Compute the energies $\|\mathbf{s}_0\|_2^2$, $\|\mathbf{s}_1\|_2^2$. You may assume that $f_1 T_s \gg 1$ and $f_2 T_s \gg 1$.

(ii) Under what conditions on $f_0$, $f_1$, and $T_s$ are $\mathbf{s}_0$ and $\mathbf{s}_1$ orthogonal?

(iii) Assume that the parameters are chosen as in Part (ii). Let $H$ take on the values 0 and 1 equiprobably, and assume that, conditional on $H = \nu$, the time-$t$ received waveform is $s_\nu(t) + N(t)$ where $\big(N(t)\big)$ is white Gaussian noise of double-sided PSD $N_0/2$ with respect to the bandwidth of interest, and $\nu \in \{0, 1\}$. Find an optimal rule for guessing $H$ based on the received waveform.

(iv) Compute the optimal probability of error.

**Exercise 26.5 (Signaling in White Gaussian Noise).** Let the RV $M$ take value in the set $\mathcal{M} = \{1, 2, 3, 4\}$ uniformly. Conditional on $M = m$, the observed waveform $\big(Y(t)\big)$ is given at every time $t \in \mathbb{R}$ by $s_m(t) + N(t)$, where the signals $\mathbf{s}_1, \mathbf{s}_2, \mathbf{s}_3, \mathbf{s}_4$ are given by

$$s_1(t) = A\, \text{I}\{0 \leq t \leq T\}, \qquad s_2(t) = A\, \text{I}\{0 \leq t \leq T/2\} - A\, \text{I}\{T/2 < t \leq T\},$$
$$s_3(t) = 2A\, \text{I}\{0 \leq t \leq T/2\}, \quad s_4(t) = -A\, \text{I}\{0 \leq t \leq T/2\} + A\, \text{I}\{T/2 < t \leq T\},$$

and where $\big(N(t)\big)$ is white Gaussian noise of PSD $N_0/2$ over the bandwidth of interest. (Ignore the fact that the signals are not bandlimited.)

(i) Derive the MAP rule for guessing $M$ based on $\big(Y(t)\big)$.

(ii) Use the Union-of-Events Bound to upper bound $p_{\text{MAP}}(\text{error}|M = 3)$. Are all the terms in the bound needed?

(iii) Compute $p_{\text{MAP}}(\text{error}|M = 3)$ exactly.

(iv) Show that by subtracting a waveform $\mathbf{s}_*$ from each of the signals $\mathbf{s}_1, \mathbf{s}_2, \mathbf{s}_3, \mathbf{s}_4$, we can reduce the average transmitted energy without degrading performance. What waveform $\mathbf{s}_*$ should be subtracted to minimize the transmitted energy?

**Exercise 26.6 (QPSK).** Let the IID random bits $D_1$ and $D_2$ be mapped to the symbols $X_1$, $X_2$ according to the rule

$$(0,0) \mapsto (1,0), \quad (0,1) \mapsto (-1,0), \quad (1,0) \mapsto (0,1), \quad (1,1) \mapsto (0,-1).$$

The received waveform $(Y(t))$ is given by

$$Y(t) = A X_1 \, \phi_1(t) + A X_2 \, \phi_2(t) + N(t), \quad t \in \mathbb{R},$$

where $A > 0$, the signals $\phi_1, \phi_2$ are orthonormal integrable signals that are bandlimited to $W$ Hz, and the SP $(N(t))$ is independent of $(D_1, D_2)$ and is white Gaussian noise of PSD $N_0/2$ with respect to the bandwidth $W$.

(i) Find an optimal rule for guessing $(D_1, D_2)$ based on $(Y(t))$.

(ii) Find an optimal rule for guessing $D_1$ based on $(Y(t))$.

(iii) Compare the rule that you have found in Part (ii) with the rule that guesses that $D_1$ is the first component of the tuple produced by the decoder that you have found in Part (i). Evaluate the probability of error for both rules.

(iv) Repeat when $(D_1, D_2)$ are mapped to $(X_1, X_2)$ according to the rule

$$(0,0) \mapsto (1,0), \quad (0,1) \mapsto (0,1), \quad (1,0) \mapsto (-1,0), \quad (1,1) \mapsto (0,-1).$$

**Exercise 26.7 (Mismatched Decoding of Antipodal Signaling).** Let the received waveform $(Y(t))$ be given at every $t \in \mathbb{R}$ by $(1 - 2H) \, s(t) + N(t)$, where $\mathbf{s}$ is an integrable signal that is bandlimited to $W$ Hz, $(N(t))$ is white Gaussian noise of PSD $N_0/2$ with respect to the bandwidth $W$, and $H$ takes on the values 0 and 1 equiprobably and independently of $(N(t))$. Let $\mathbf{s}'$ be an integrable signal that is bandlimited to $W$ Hz. A suboptimal detector feeds the received waveform to a matched filter for $\mathbf{s}'$ and guesses according to the filter's time-0 output: if it is positive, it guesses "$H = 0$," and if it is negative, it guesses "$H = 1$." Express this detector's probability of error in terms of $\mathbf{s}$, $\mathbf{s}'$, and $N_0$.

**Exercise 26.8 (Imperfect Automatic Gain Control).** Let the received signal $(Y(t))$ be given by

$$Y(t) = A X \, s(t) + N(t), \quad t \in \mathbb{R},$$

where $A > 0$ is some deterministic positive constant, $X$ is a RV that takes value in the set $\{-3, -1, +1, +3\}$ uniformly, $\mathbf{s}$ is an integrable signal that is bandlimited to $W$ Hz, and $(N(t))$ is white Gaussian noise of double-sided PSD $N_0/2$ with respect to the bandwidth $W$.

(i) Find an optimal rule for guessing $X$ based on $(Y(t))$.

(ii) Using the $\mathcal{Q}$-function compute the optimal probability of error.

(iii) Suppose you use the rule you have found in Part (i), but the received signal is

$$Y(t) = \frac{3}{4} A X \, s(t) + N(t), \quad t \in \mathbb{R}.$$

(You were misinformed about the amplitude of the signal.) What is the probability of error now?

**Exercise 26.9 (Positive Semidefinite Matrices).**

(i) Let $\mathbf{s}_1, \ldots, \mathbf{s}_M$ be of finite energy. Show that the $M \times M$ matrix whose Row-$j$ Column-$\ell$ entry is $\langle \mathbf{s}_j, \mathbf{s}_\ell \rangle$ is positive semidefinite.

(ii) Show that any $M \times M$ positive semidefinite matrix can be expressed in this form with a proper choice of the signals $\mathbf{s}_1, \ldots, \mathbf{s}_M$.

**Exercise 26.10 (A Lower Bound on the Minimum Distance).** Let $\mathbf{s}_1, \ldots, \mathbf{s}_M$ be equi-energy signals of energy $E_s$. Let

$$\bar{d}^2 \triangleq \frac{1}{M(M-1)} \sum_{m'} \sum_{m'' \neq m'} \| \mathbf{s}_{m'} - \mathbf{s}_{m''} \|_2^2$$

denote the average squared-distance between the signals.

(i) Justify the following bound on $\bar{d}$:

$$\bar{d}^2 = \frac{1}{M(M-1)} \sum_{m'=1}^{M} \sum_{m''=1}^{M} \| \mathbf{s}_{m'} - \mathbf{s}_{m''} \|_2^2$$

$$= \frac{2M}{M-1} E_s - \frac{2M}{M-1} \frac{1}{M^2} \sum_{m'=1}^{M} \sum_{m''=1}^{M} \langle \mathbf{s}_{m'}, \mathbf{s}_{m''} \rangle$$

$$= \frac{2M}{M-1} E_s - \frac{2M}{M-1} \left\| \frac{1}{M} \sum_{m=1}^{M} \mathbf{s}_m \right\|_2^2$$

$$\leq \frac{2M}{M-1} E_s.$$

(ii) Show that if, in addition, $\langle \mathbf{s}_{m'}, \mathbf{s}_{m''} \rangle = \rho E_s$ for all $m' \neq m''$ in $\{1, \ldots, M\}$, then

$$-\frac{1}{M-1} \leq \rho \leq 1.$$

(iii) Are equalities possible in the above bounds?

**Exercise 26.11 (Generalizations of the Simplex).** Let $p^*(\text{error}; E_s; \rho; M; N_0)$ denote the optimal probability of error for the setup of Section 26.2 for the case where the prior on $M$ is uniform and where

$$\langle \mathbf{s}_{m'}, \mathbf{s}_{m''} \rangle = \begin{cases} E_s & \text{if } m' = m'', \\ \rho E_s & \text{otherwise,} \end{cases} \quad m', m'' \in \{1, \ldots, M\}.$$

Show that

$$p^*(\text{error}; E_s; \rho; M; N_0) = p^*(\text{error}; E_s(1-\rho); 0; M; N_0), \quad -\frac{1}{M-1} \leq \rho \leq 1.$$

*Hint: You may need a different proof depending on the sign of $\rho$.*

**Exercise 26.12 (Decoding the Simplex without Gain Control).** Let the simplex constellation $\mathbf{s}_1, \ldots, \mathbf{s}_M$ be constructed from the orthonormal signals $\phi_1, \ldots, \phi_M$ as in Section 26.11.4. In that section we proposed to decode by adding

$$\frac{1}{\sqrt{M-1}} \sqrt{E_s} \psi$$

to the received signal $\mathbf{Y}$ and then feeding the result to a decoder that was designed for the orthogonal signals

$$\mathbf{s}_1 + \frac{1}{\sqrt{M-1}}\sqrt{E_s}\boldsymbol{\psi}, \ldots, \mathbf{s}_M + \frac{1}{\sqrt{M-1}}\sqrt{E_s}\boldsymbol{\psi}.$$

Here $\boldsymbol{\psi}$ is any signal that is orthogonal to the signals $\{\mathbf{s}_1, \ldots, \mathbf{s}_M\}$. Show that feeding the signal $\mathbf{Y} + \alpha\boldsymbol{\psi}$ to the above orthogonal-keying decoder also results in an optimal decoding rule, irrespective of the value of $\alpha \in \mathbb{R}$.

**Exercise 26.13 (Pretending the Noise Is White).** Let $H$ take on the values 0 and 1 equiprobably, and let the received waveform $(Y(t))$ be given at time $t$ by

$$Y(t) = (1 - 2H)\,s(t) + N(t),$$

where $\mathbf{s} \colon t \mapsto \mathrm{I}\{0 \le t \le 1\}$, and where the SP $(N(t))$ is independent of $H$ and is a measureable, centered, stationary, Gaussian SP of autocovariance function

$$\mathsf{K}_{NN}(\tau) = \frac{1}{4\alpha}e^{-|\tau|/\alpha}, \quad \tau \in \mathbb{R},$$

where $0 < \alpha < \infty$ is some deterministic real parameter. Compute the probability of error of a detector that guesses "$H = 0$" whenever

$$\int_0^1 Y(t)\,\mathrm{d}t \ge 0.$$

To what does this probability of error converge when $\alpha$ tends to zero?

**Exercise 26.14 (Antipodal Signaling in Colored Noise).** Let $\mathbf{s}$ be an integrable signal that is bandlimited to $W$ Hz, and let $H$ take on the values 0 and 1 equiprobably. Let the time-$t$ value of the received signal $(Y(t))$ be given by $(1 - 2H)\,s(t) + N(t)$, where $(N(t))$ is a measurable, centered, stationary, Gaussian SP of autocovariance function $\mathsf{K}_{NN}$. Assume that $H$ and $(N(t))$ are independent, and that $\mathsf{K}_{NN}$ can be whitened with respect to the bandwidth $W$. Find the optimal probability of error in guessing $H$ based on $(Y(t))$.

**Exercise 26.15 (Modeling Artifacts).** Let $H$ take on the values 0 and 1 equiprobably, and let the received signal $(Y(t))$ be given by

$$Y(t) = (1 - 2H)\,s(t) + N(t), \quad t \in \mathbb{R},$$

where $\mathbf{s} \colon t \mapsto \mathrm{I}\{0 \le t \le 1\}$ and the SP $(N(t))$ is independent of $H$ and is a measurable, centered, stationary, Gaussian SP of autocovariance function

$$\mathsf{K}_{NN}(\tau) = \alpha\,e^{-\tau^2/\beta}, \tau \in \mathbb{R},$$

for some $\alpha, \beta > 0$.

Argue heuristically that—irrespective of the values of $\alpha$ and $\beta$—for any $\epsilon > 0$ we can find a rule for guessing $H$ based on $(Y(t))$ whose probability of error is smaller than $\epsilon$.

*Hint: Study $\hat{s}(f)$ and $\mathsf{S}_{NN}(f)$ at high frequencies $f$.*

**Exercise 26.16 (Measurability in Theorem 26.3.2).**

(i) Let $\big(N(t)\big)$ be white Gaussian noise of double-sided PSD $N_0/2$ with respect to the bandwidth $W$. Let $R$ be a unit-mean exponential RV that is independent of $\big(N(t)\big)$. Define the SP

$$\tilde{N}(t) = N(t)\,\mathrm{I}\{t \neq R\}, \quad t \in \mathbb{R}.$$

Show that $\big(\tilde{N}(t)\big)$ is white Gaussian noise of double-sided PSD $N_0/2$ with respect to the bandwidth $W$.

(ii) Let $\mathbf{s}$ be a nonzero integrable signal that is bandlimited to $W$ Hz. To be concrete,

$$s(t) = \mathrm{sinc}^2(Wt), \quad t \in \mathbb{R}.$$

Suppose that the SP $\big(N(t)\big)$ is as above and that for every $\omega \in \Omega$ the sample-path $t \mapsto N(\omega, t)$ is continuous. Construct $\big(\tilde{N}(t)\big)$ as above. Suppose you wish to test whether you are observing $\mathbf{s}$ or $-\mathbf{s}$ in the additive noise $\big(\tilde{N}(t)\big)$. Show that you can guess with zero probability of error by finding an epoch where the observed SP is discontinuous and by comparing the value of the received signal at that epoch to the value of $\mathbf{s}$. (This does not violate Theorem 26.3.2 because this decision rule is not measurable with respect to the Borel $\sigma$-algebra generated by the observed SP.)

# Chapter 27

# Noncoherent Detection and Nuisance Parameters

## 27.1 Introduction and Motivation

In this chapter we discuss a problem that arises in noncoherent detection. To motivate the problem, consider a setup where a transmitter sends one of two different passband waveforms

$$t \mapsto 2\,\mathrm{Re}\big(s_{0,\mathrm{BB}}(t)\,e^{\mathrm{i}2\pi f_\mathrm{c} t}\big) \quad \text{or} \quad t \mapsto 2\,\mathrm{Re}\big(s_{1,\mathrm{BB}}(t)\,e^{\mathrm{i}2\pi f_\mathrm{c} t}\big),$$

where $s_{0,\mathrm{BB}}$ and $s_{1,\mathrm{BB}}$ are integrable baseband signals that are bandlimited to $W/2$ Hz, and where the carrier frequency $f_\mathrm{c}$ satisfies $f_\mathrm{c} > W/2$. To motivate our problem it is instructive to consider the case where

$$f_\mathrm{c} \gg W. \tag{27.1}$$

(In wireless communications it is common for $f_\mathrm{c}$ to be three orders of magnitude larger than $W$.) Let $X(t)$ denote the transmitted waveform at time $t$. Suppose that the received waveform $\big(Y(t)\big)$ is a delayed version of the transmitted waveform corrupted by white Gaussian noise of PSD $N_0/2$ with respect to the bandwidth $W$ around the carrier frequency $f_\mathrm{c}$ (Definition 25.15.3):

$$Y(t) = X(t - t_\mathrm{D}) + N(t), \quad t \in \mathbb{R},$$

where $t_\mathrm{D}$ denotes the delay (typically proportional to the distance between the transmitter and the receiver) and $\big(N(t)\big)$ is the additive noise. Suppose further that the receiver estimates the delay to be $t'_\mathrm{D}$ and moves its clock back by defining

$$t' \triangleq t - t'_\mathrm{D}. \tag{27.2}$$

If $\tilde{Y}(t')$ is what the receiver receives when its clock shows $t'$, then by (27.2)

$$\begin{aligned}
\tilde{Y}(t') &= Y(t' + t'_\mathrm{D}) \\
&= X(t' + t'_\mathrm{D} - t_\mathrm{D}) + N(t' + t'_\mathrm{D}) \\
&= X(t' + t'_\mathrm{D} - t_\mathrm{D}) + \tilde{N}(t'), \quad t' \in \mathbb{R},
\end{aligned}$$

where $\tilde{N}(t') \triangleq N(t' + t'_{\mathrm{D}})$ and is thus, by the stationarity of $(N(t))$, also white Gaussian noise of PSD $\mathsf{N}_0/2$ with respect to the bandwidth $W$ around $f_{\mathrm{c}}$. The term $X(t' + t'_{\mathrm{D}} - t_{\mathrm{D}})$ can be more explicitly written for every $t' \in \mathbb{R}$ as

$$X(t' + t'_{\mathrm{D}} - t_{\mathrm{D}}) = 2\,\mathrm{Re}\Big(s_{\nu,\mathrm{BB}}(t' + t'_{\mathrm{D}} - t_{\mathrm{D}})\,e^{\mathrm{i}2\pi f_{\mathrm{c}}(t' + t'_{\mathrm{D}} - t_{\mathrm{D}})}\Big), \tag{27.3}$$

where $\nu$ is either zero or one, depending on which waveform is sent.

We next argue that if

$$\left| t'_{\mathrm{D}} - t_{\mathrm{D}} \right| \ll \frac{1}{W}, \tag{27.4}$$

then

$$s_{\nu,\mathrm{BB}}(t' + t'_{\mathrm{D}} - t_{\mathrm{D}}) \approx s_{\nu,\mathrm{BB}}(t'), \quad t' \in \mathbb{R}. \tag{27.5}$$

This can be seen by considering a Taylor Series expansion for $s_{\nu,\mathrm{BB}}(\cdot)$ around $t'$

$$s_{\nu,\mathrm{BB}}(t' + t'_{\mathrm{D}} - t_{\mathrm{D}}) \approx s_{\nu,\mathrm{BB}}(t') + \left.\frac{\mathrm{d}s_{\nu,\mathrm{BB}}(\tau)}{\mathrm{d}\tau}\right|_{\tau=t'} (t'_{\mathrm{D}} - t_{\mathrm{D}})$$

and by then using Bernstein's Inequality (Theorem 6.7.1) to heuristically argue that the derivative of the baseband signal is of order of magnitude $W$, so its product by the timing error is, by (27.4), negligible.

From (27.3) and (27.5) we obtain that, as long as (27.4) holds,

$$X(t' + t'_{\mathrm{D}} - t_{\mathrm{D}}) \approx 2\,\mathrm{Re}\Big(s_{\nu,\mathrm{BB}}(t')\,e^{\mathrm{i}2\pi f_{\mathrm{c}}(t' + t'_{\mathrm{D}} - t_{\mathrm{D}})}\Big)$$

$$= 2\,\mathrm{Re}\Big(s_{\nu,\mathrm{BB}}(t')\,e^{\mathrm{i}(2\pi f_{\mathrm{c}}t' + \theta)}\Big), \quad t' \in \mathbb{R}, \tag{27.6a}$$

where

$$\theta = 2\pi f_{\mathrm{c}}(t'_{\mathrm{D}} - t_{\mathrm{D}}) \quad \mathrm{mod}\ [-\pi, \pi). \tag{27.6b}$$

(Recall that $\xi \ \mathrm{mod}\ [-\pi, \pi)$ is the element in the interval $[-\pi, \pi)$ that differs from $\xi$ by an integer multiple of $2\pi$.) Note that even if (27.4) holds, the term $2\pi f_{\mathrm{c}}(t'_{\mathrm{D}} - t_{\mathrm{D}})$ may be much larger than 1 when $f_{\mathrm{c}} \gg W$.

We conclude that if the error in estimating the delay is negligible compared to the reciprocal of the signal bandwidth but significantly larger than the reciprocal of the carrier frequency, then the received waveform can be modeled as

$$\tilde{Y}(t') = 2\,\mathrm{Re}\Big(s_{\nu,\mathrm{BB}}(t')\,e^{\mathrm{i}(2\pi f_{\mathrm{c}}t' + \theta)}\Big) + \tilde{N}(t'), \quad t' \in \mathbb{R}, \tag{27.7}$$

where the receiver needs to determine whether $\nu$ is equal to zero or one; $(\tilde{N}(t'))$ is additive white Gaussian noise of PSD $\mathsf{N}_0/2$ with respect to the bandwidth $W$ around $f_{\mathrm{c}}$; and where the phase $\theta$ is unknown to the receiver. Since the phase is unknown to the receiver, the detection is said to be **noncoherent**. In the statistics literature an unknown parameter such as $\theta$ is called a **nuisance parameter**.

It would make engineering sense to ask for a decision rule for guessing $\nu$ based on $(\tilde{Y}(t'))$ that would work well irrespective of the value of $\theta$, but this is not the question we shall ask. This question is related to "composite hypothesis testing," which is not treated in this book.[1] Instead we shall adopt a probabilistic approach. We shall assume that $\theta$ is a random variable—and therefore henceforth denote it by $\Theta$ and its realization by $\theta$—that is uniformly distributed over the interval $[-\pi, \pi)$ independently of the noise and the message, and we shall seek a decision rule that has the smallest *average* probability of error. Thus, if we denote the probability of error conditional on $\Theta = \theta$ by $p(\text{error}|\theta)$, then we seek a decision rule based on $(\tilde{Y}(t))$ that minimizes

$$\frac{1}{2\pi} \int_{-\pi}^{\pi} p(\text{error}|\theta) \, d\theta. \tag{27.8}$$

The conservative reader may prefer to minimize the probability of error on the "worst case $\theta$"

$$\sup_{\theta \in [-\pi, \pi)} p(\text{error}|\theta) \tag{27.9}$$

but, miraculously, it will turn out that the decoder we shall derive to minimize (27.8) has a conditional probability of error $p(\text{error}|\theta)$ that does not depend on the realization $\theta$ so, as we shall see in Section 27.7, our decoder also minimizes (27.9).

## 27.2 The Setup

We next define our hypothesis testing problem. We denote time by $t$ and the received waveform by $(Y(t))$ (even though in the scenario we described in Section 27.1 these correspond to $t'$ and $(\tilde{Y}(t'))$, i.e., to the time coordinate and to the corresponding signal at the receiver). We denote the RV we wish to guess by $H$ and assume a uniform prior:

$$\Pr[H = 0] = \Pr[H = 1] = \frac{1}{2}. \tag{27.10}$$

For each $\nu \in \{0, 1\}$ the observation $(Y(t))$ is, conditionally on $H = \nu$, a SP of the form

$$Y(t) = S_\nu(t) + N(t), \quad t \in \mathbb{R}, \tag{27.11}$$

where $(N(t))$ is white Gaussian noise of positive PSD $\mathsf{N}_0/2$ with respect to the bandwidth $W$ around the carrier frequency $f_c$ (Definition 25.15.3), and where $S_\nu(t)$ can be described as

$$\begin{aligned}
S_\nu(t) &= 2\operatorname{Re}\left(s_{\nu,\text{BB}}(t)\, e^{\mathrm{i}(2\pi f_c t + \Theta)}\right) \\
&= 2\operatorname{Re}\left(s_{\nu,\text{BB}}(t)\, e^{\mathrm{i}2\pi f_c t}\right) \cos\Theta - 2\operatorname{Im}\left(s_{\nu,\text{BB}}(t)\, e^{\mathrm{i}2\pi f_c t}\right) \sin\Theta \\
&= 2\operatorname{Re}\left(s_{\nu,\text{BB}}(t)\, e^{\mathrm{i}2\pi f_c t}\right) \cos\Theta + 2\operatorname{Re}\left(\mathrm{i}\, s_{\nu,\text{BB}}(t)\, e^{\mathrm{i}2\pi f_c t}\right) \sin\Theta \\
&= s_{\nu,\text{c}}(t) \cos\Theta + s_{\nu,\text{s}}(t) \sin\Theta, \quad t \in \mathbb{R}, \tag{27.12}
\end{aligned}$$

where $\Theta$ is a RV that is uniformly distributed over the interval $[-\pi, \pi)$ independently of $(H, (N(t)))$, and where we define for $\nu \in \{0, 1\}$

$$s_{\nu,\text{c}}(t) \triangleq 2\operatorname{Re}\left(s_{\nu,\text{BB}}(t)\, e^{\mathrm{i}2\pi f_c t}\right), \quad t \in \mathbb{R}, \tag{27.13a}$$

$$s_{\nu,\text{s}}(t) \triangleq 2\operatorname{Re}\left(\mathrm{i}\, s_{\nu,\text{BB}}(t)\, e^{\mathrm{i}2\pi f_c t}\right), \quad t \in \mathbb{R}. \tag{27.13b}$$

---

[1] See, for example, (Lehmann and Romano, 2005, Chapter 3).

Notice that by (27.13) and by the relationship between inner products in baseband and passband (Theorem 7.6.10),

$$\langle \mathbf{s}_{\nu,\mathrm{c}}, \mathbf{s}_{\nu,\mathrm{s}} \rangle = 0, \quad \nu = 0, 1. \tag{27.14}$$

We assume that the baseband signals $\mathbf{s}_{0,\mathrm{BB}}, \mathbf{s}_{1,\mathrm{BB}}$ are integrable complex signals that are bandlimited to $W/2$ Hz and that they are orthogonal:

$$\langle \mathbf{s}_{0,\mathrm{BB}}, \mathbf{s}_{1,\mathrm{BB}} \rangle = 0. \tag{27.15}$$

Consequently, by (27.13) and Theorem 7.6.10,

$$\langle \mathbf{s}_{0,\mathrm{c}}, \mathbf{s}_{1,\mathrm{c}} \rangle = \langle \mathbf{s}_{0,\mathrm{s}}, \mathbf{s}_{1,\mathrm{c}} \rangle = \langle \mathbf{s}_{0,\mathrm{c}}, \mathbf{s}_{1,\mathrm{s}} \rangle = \langle \mathbf{s}_{0,\mathrm{s}}, \mathbf{s}_{1,\mathrm{s}} \rangle = 0. \tag{27.16}$$

We finally assume that the baseband signals $\mathbf{s}_{0,\mathrm{BB}}$ and $\mathbf{s}_{1,\mathrm{BB}}$ are of equal positive energy:

$$\|\mathbf{s}_{0,\mathrm{BB}}\|_2^2 = \|\mathbf{s}_{1,\mathrm{BB}}\|_2^2 > 0. \tag{27.17}$$

Defining[2]

$$\mathsf{E}_{\mathrm{s}} = 2 \|\mathbf{s}_{0,\mathrm{BB}}\|_2^2 \tag{27.18}$$

we have by the relationship between energy in baseband and passband (Theorem 7.6.10)

$$\mathsf{E}_{\mathrm{s}} = \|\mathbf{S}_0\|_2^2 = \|\mathbf{S}_1\|_2^2 = \|\mathbf{s}_{0,\mathrm{s}}\|_2^2 = \|\mathbf{s}_{0,\mathrm{c}}\|_2^2 = \|\mathbf{s}_{1,\mathrm{s}}\|_2^2 = \|\mathbf{s}_{1,\mathrm{c}}\|_2^2 . \tag{27.19}$$

By (27.14), (27.16), and (27.18)

$$\frac{1}{\sqrt{\mathsf{E}_{\mathrm{s}}}} \big( \mathbf{s}_{0,\mathrm{c}}, \mathbf{s}_{0,\mathrm{s}}, \mathbf{s}_{1,\mathrm{c}}, \mathbf{s}_{1,\mathrm{s}} \big) \quad \text{is an orthonormal 4-tuple.} \tag{27.20}$$

Our problem is to guess $H$ based on the observation $\big( Y(t) \big)$.

## 27.3    A Sufficient Statistic

To derive an optimal guessing rule, we begin by deriving a sufficient statistic vector. This vector takes value in $\mathbb{R}^4$ and enables us to simplify the guessing problem from one where the observation consists of a SP to one where it consists of a random 4-vector. We shall later find an even more concise sufficient statistic vector with only two components. We denote the sufficient statistic vector by $\mathbf{T}$ and its four components by $T_{0,\mathrm{c}}, T_{0,\mathrm{s}}, T_{1,\mathrm{c}},$ and $T_{1,\mathrm{s}}$:

$$\mathbf{T} = \big( T_{0,\mathrm{c}}, T_{0,\mathrm{s}}, T_{1,\mathrm{c}}, T_{1,\mathrm{s}} \big)^{\mathsf{T}}.$$

We denote its realization by $\mathbf{t}$ with corresponding components

$$\mathbf{t} = \big( t_{0,\mathrm{c}}, t_{0,\mathrm{s}}, t_{1,\mathrm{c}}, t_{1,\mathrm{s}} \big)^{\mathsf{T}}.$$

---

[2]The "s" in $\mathsf{E}_{\mathrm{s}}$ stands for "signal," whereas the "s" in $\mathbf{s}_{0,\mathrm{s}}$ and $\mathbf{s}_{1,\mathrm{s}}$ stands for "sine."

The vector $\mathbf{T}$ is defined by

$$\mathbf{T} \triangleq \left( \left\langle \mathbf{Y}, \frac{\mathbf{s}_{0,\mathrm{c}}}{\sqrt{\mathsf{E}_\mathrm{s}}} \right\rangle, \left\langle \mathbf{Y}, \frac{\mathbf{s}_{0,\mathrm{s}}}{\sqrt{\mathsf{E}_\mathrm{s}}} \right\rangle, \left\langle \mathbf{Y}, \frac{\mathbf{s}_{1,\mathrm{c}}}{\sqrt{\mathsf{E}_\mathrm{s}}} \right\rangle, \left\langle \mathbf{Y}, \frac{\mathbf{s}_{1,\mathrm{s}}}{\sqrt{\mathsf{E}_\mathrm{s}}} \right\rangle \right)^{\mathsf{T}} \tag{27.21}$$

$$= \frac{1}{\sqrt{\mathsf{E}_\mathrm{s}}} \left( \int_{-\infty}^{\infty} Y(t)\, s_{0,\mathrm{c}}(t)\, \mathrm{d}t, \ldots, \int_{-\infty}^{\infty} Y(t)\, s_{1,\mathrm{s}}(t)\, \mathrm{d}t \right)^{\mathsf{T}}. \tag{27.22}$$

We now prove that it forms a sufficient statistic for guessing $H$ based on the observation $\big(Y(t)\big)$. It is interesting to note that this sufficiency also holds for $\Theta$ of arbitrary distribution (not necessarily uniform) provided that the pair $(H, \Theta)$ is independent of the additive noise. Moreover, it holds even if the baseband signals $\mathbf{s}_{0,\mathrm{BB}}$ and $\mathbf{s}_{1,\mathrm{BB}}$ are not orthogonal.

Before proving the sufficiency of $\mathbf{T}$ we give a plausibility argument. To that end we consider a new (hypothetical) scenario where $\Theta$, rather than being uniform, now takes value in a finite set $\{\theta_1, \ldots, \theta_\kappa\}$ according to some arbitrary distribution. Suppose further that rather than just being interested in $H$ we also wish to guess the value of $\Theta$. Thus, rather than just guessing $H$ we wish to guess the pair $(H, \Theta)$, which takes value in the set

$$\big\{ (0, \theta_1), (1, \theta_1), (0, \theta_2), (1, \theta_2), \ldots, (0, \theta_\kappa), (1, \theta_\kappa) \big\}.$$

In this new scenario we have for every $\nu \in \{0, 1\}$ and every $\eta \in \{1, \ldots, \kappa\}$ that, conditional on $(H, \Theta) = (\nu, \theta_\eta)$, the observation $\big(Y(t)\big)$ consists of the signal $t \mapsto s_{\nu,\mathrm{c}}(t) \cos \theta_\eta + s_{\nu,\mathrm{s}}(t) \sin \theta_\eta$ corrupted by additive Gaussian noise $\big(N(t)\big)$. Since for every such $\nu$ and $\eta$ the signal $t \mapsto s_{\nu,\mathrm{c}}(t) \cos \theta_\eta + s_{\nu,\mathrm{s}}(t) \sin \theta_\eta$ can be written as a linear combination of the signals $\mathbf{s}_{0,\mathrm{c}}$, $\mathbf{s}_{0,\mathrm{s}}$, $\mathbf{s}_{1,\mathrm{c}}$, and $\mathbf{s}_{1,\mathrm{s}}$, it follows from Theorem 26.4.1 that in this new scenario $\mathbf{T}$ forms a sufficient statistic for guessing the pair $(H, \Theta)$ based on $\big(Y(t)\big)$. But what if we are only interested in guessing $H$? Guessing $H$ in this scenario reduces to guessing whether the pair $(H, \Theta)$ is in the set $\{(0, \theta_1), (0, \theta_2), \ldots, (0, \theta_\kappa)\}$ or in the set $\{(1, \theta_1), (1, \theta_2), \ldots, (1, \theta_\kappa)\}$. Consequently, by Proposition 22.4.4, in the new scenario $\mathbf{T}$ is also sufficient for guessing $H$. Since $\kappa$ in this argument can be as large as we want, it is plausible that $\mathbf{T}$ is also a sufficient statistic for guessing $H$ in our original problem where $\Theta$ is uniform over $[-\pi, \pi)$.

The key to the above heuristic argument is that, irrespective of the realization of $\Theta$ and of the value of $\nu$, the signal $\mathbf{S}_\nu$ lies in the four dimensional subspace spanned by the signals $\mathbf{s}_{0,\mathrm{c}}$, $\mathbf{s}_{0,\mathrm{s}}$, $\mathbf{s}_{1,\mathrm{c}}$, and $\mathbf{s}_{1,\mathrm{s}}$. The sufficiency thus follows from a more general theorem that we state next.

**Theorem 27.3.1 (White Gaussian Noise with Nuisance Parameters).** *Let $\mathcal{V}$ be a $d$-dimensional subspace of the set of all integrable signals that are bandlimited to $\mathsf{W}$ Hz, and let $(\boldsymbol{\phi}_1, \ldots, \boldsymbol{\phi}_d)$ be an orthonormal basis for $\mathcal{V}$. Let the RV $M$ take value in a finite set $\mathcal{M}$. Suppose that, conditional on $M = m$, the SP $\big(Y(t)\big)$ is given by*

$$Y(t) = \sum_{\ell=1}^{d} A^{(\ell)} \phi_\ell(t) + N(t), \tag{27.23}$$

*where* $\mathbf{A} = (A^{(1)}, \ldots, A^{(d)})^{\mathsf{T}}$ *is a random d-vector whose law typically depends on m, where the SP* $(N(t))$ *is white Gaussian noise with respect to the bandwidth* $\mathsf{W}$*, and where* $(N(t))$ *is independent of the pair* $(M, \mathbf{A})$*. Then the vector*

$$\mathbf{T} = (\langle \mathbf{Y}, \boldsymbol{\phi}_1 \rangle, \ldots, \langle \mathbf{Y}, \boldsymbol{\phi}_d \rangle)^{\mathsf{T}} \tag{27.24}$$

*forms a sufficient statistic for guessing M based on* $(Y(t))$*.*

*The theorem also holds in passband, i.e., if* $\mathcal{V}$ *is a d-dimensional subspace of the set of all integrable signals that are bandlimited to* $\mathsf{W}$ Hz *around the carrier frequency* $f_c$ *and if* $(N(t))$ *is white with respect to the bandwidth* $\mathsf{W}$ *around* $f_c$*.*

**Note 27.3.2.** Theorem 27.3.1 continues to hold even if $(\boldsymbol{\phi}_1, \ldots, \boldsymbol{\phi}_d)$ are not orthonormal; it suffices that they form a basis for $\mathcal{V}$.

**Proof of Note 27.3.2.** This follows from Proposition 22.4.2 and from the observation that if $(\mathbf{u}_1, \ldots, \mathbf{u}_d)$ forms a basis for $\mathcal{V}$ and if $(\mathbf{v}_1, \ldots, \mathbf{v}_d)$ forms another basis for $\mathcal{V}$, then the inner products $\{\langle \mathbf{Y}, \mathbf{v}_\ell \rangle\}_{\ell=1}^d$ are computable from the inner products $\{\langle \mathbf{Y}, \mathbf{u}_\ell \rangle\}_{\ell=1}^d$ (Lemma 25.10.3). $\qquad \square$

Before presenting the proof of Theorem 27.3.1 we give two examples of its application. The first is a simple case where, conditional on $M$, the vector $\mathbf{A}$ is deterministic. This corresponds to the problem of detecting a known signal corrupted by additive white Gaussian noise. This case was treated in Theorem 26.4.1 and slightly generalized in Corollary 26.4.2. We thus see that Theorem 27.3.1 is a generalization of Theorem 26.4.1 & Corollary 26.4.2.[3]

The second example of the application of this theorem is for the noncoherent detection problem at hand. Here $d = 4$ and

$$\mathcal{V} = \text{span}(\mathbf{s}_{0,c}, \mathbf{s}_{0,s}, \mathbf{s}_{1,c}, \mathbf{s}_{1,s}), \tag{27.25}$$

with $\boldsymbol{\phi}_1 \triangleq \mathbf{s}_{0,c}/\sqrt{\mathsf{E_s}}$, $\boldsymbol{\phi}_2 \triangleq \mathbf{s}_{0,s}/\sqrt{\mathsf{E_s}}$, $\boldsymbol{\phi}_3 \triangleq \mathbf{s}_{1,c}/\sqrt{\mathsf{E_s}}$, and $\boldsymbol{\phi}_4 \triangleq \mathbf{s}_{1,s}/\sqrt{\mathsf{E_s}}$. We note that, conditional on $H = 0$, the received waveform $(Y(t))$ can be written in the form (27.23) where $A^{(3)}$ & $A^{(4)}$ are deterministically zero and the pair $(A^{(1)}, A^{(2)})$ is uniformly distributed over the unit circle:

$$(A^{(1)})^2 + (A^{(2)})^2 = 1.$$

Similarly, conditional on $H = 1$, the random variables $A^{(1)}$ and $A^{(2)}$ are deterministically zero and the pair $(A^{(3)}, A^{(4)})$ is uniformly distributed over the unit circle. Thus, once we prove Theorem 27.3.1, it will follow that the vector in (27.22) forms a sufficient statistic.

**Proof of Theorem 27.3.1.** To derive the sufficiency of $\mathbf{T}$ we need to show that for every $\eta \in \mathbb{N}$ and any choice of the epochs $t_1, \ldots, t_\eta \in \mathbb{R}$ the random vector $\mathbf{T}$ forms

---

[3]The setup of Corollary 26.4.2 may appear slightly more general than our setting because the signals $\tilde{\mathbf{s}}_1, \ldots, \tilde{\mathbf{s}}_n$ are not assumed to be orthonormal. But, using the linearity of the inner product (Lemma 25.10.3), it is readily seen that from the inner products (27.24) one can compute the inner products $\{\langle \mathbf{Y}, \tilde{\mathbf{s}}_j \rangle\}_{j=1}^n$ and vice versa.

a sufficient statistic for guessing $M$ based on $(Y(t_1), \ldots, Y(t_\eta), \mathbf{T})$. That is, we need to show that, irrespective of the prior distribution of $M$,

$$M \; \text{---}\circ\text{---} \; \mathbf{T} \; \text{---}\circ\text{---} \; (Y(t_1), \ldots, Y(t_\eta)). \tag{27.26}$$

Define the random variables

$$\tilde{Y}(t_\kappa) \triangleq Y(t_\kappa) - \sum_{\ell=1}^{d} \phi_\ell(t_\kappa) \langle \mathbf{Y}, \phi_\ell \rangle \tag{27.27}$$

$$= Y(t_\kappa) - \sum_{\ell=1}^{d} \phi_\ell(t_\kappa) T^{(\ell)}, \quad \kappa = 1, \ldots, \eta \tag{27.28}$$

and stack them in a vector $\tilde{\mathbf{Y}} \triangleq (\tilde{Y}(t_1), \ldots, \tilde{Y}(t_\eta))^\mathsf{T}$. Since, conditional on $\mathbf{T}$, the random variables $Y(t_\kappa)$ and $\tilde{Y}(t_\kappa)$ only differ by a constant (which depends on $\mathbf{T}$), it follows that to prove (27.26) it suffices to prove

$$M \; \text{---}\circ\text{---} \; \mathbf{T} \; \text{---}\circ\text{---} \; \tilde{\mathbf{Y}}. \tag{27.29}$$

Instead of proving (27.29), we shall prove

$$(M, \mathbf{A}) \; \text{---}\circ\text{---} \; \mathbf{T} \; \text{---}\circ\text{---} \; \tilde{\mathbf{Y}}, \tag{27.30}$$

which implies (27.29). (If the pair $(X, Y)$ is independent of $Z$, then $X$ is independent of $Z$. Likewise if we condition on $T$: if conditional on $T$ the pair $(X, Y)$ is independent of $Z$, then conditional on $T$ we also have that $X$ is independent of $Z$.)

By Proposition 22.5.5 it follows that to establish (27.30) it suffices to show that

$$\tilde{\mathbf{Y}} \text{ is independent of } (M, \mathbf{A}) \tag{27.31}$$

and

$$\mathbf{T} \; \text{---}\circ\text{---} \; (M, \mathbf{A}) \; \text{---}\circ\text{---} \; \tilde{\mathbf{Y}}. \tag{27.32}$$

We first prove (27.31) by showing that conditional on $(M, \mathbf{A}) = (m, \mathbf{a})$ the random vector $\tilde{\mathbf{Y}}$ is Gaussian with a mean vector and a covariance matrix that do not depend on $m$ and $\mathbf{a}$. That conditional on $(M, \mathbf{A}) = (m, \mathbf{a})$ the random vector $\tilde{\mathbf{Y}}$ is Gaussian follows because under this conditioning $\mathbf{T}$ and $Y(t_1), \ldots Y(t_\eta)$ are jointly Gaussian (Theorem 25.12.1) so the result of linearly transforming them to form $\tilde{\mathbf{Y}}$ must also be Gaussian (Proposition 23.6.3). For the mean we have from (27.27)

$$\mathsf{E}\big[\tilde{Y}(t_\kappa) \,\big|\, (M, \mathbf{A}) = (m, \mathbf{a})\big]$$

$$= \mathsf{E}\bigg[ Y(t_\kappa) - \sum_{\ell=1}^{d} \phi_\ell(t_\kappa) \langle \mathbf{Y}, \phi_\ell \rangle \,\bigg|\, (M, \mathbf{A}) = (m, \mathbf{a}) \bigg]$$

$$= \mathsf{E}\big[ Y(t_\kappa) \,\big|\, (M, \mathbf{A}) = (m, \mathbf{a}) \big] - \sum_{\ell=1}^{d} \phi_\ell(t_\kappa) \, \mathsf{E}\big[ \langle \mathbf{Y}, \phi_\ell \rangle \,\big|\, (M, \mathbf{A}) = (m, \mathbf{a}) \big]$$

$$= \sum_{\ell=1}^{d} a^{(\ell)} \phi_\ell(t_\kappa) - \sum_{\ell=1}^{d} \phi_\ell(t_\kappa) \bigg\langle \sum_{\ell'=1}^{d} a^{(\ell')} \phi_{\ell'}, \phi_\ell \bigg\rangle$$

$$= \sum_{\ell=1}^{d} a^{(\ell)} \phi_\ell(t_\kappa) - \sum_{\ell=1}^{d} \phi_\ell(t_\kappa) a^{(\ell)}$$
$$= 0, \quad \kappa \in \{1, \ldots, \eta\},$$

where the first equality follows from the definition of $\tilde{Y}(t_\kappa)$; the second from the linearity of conditional expectation; the third because $(N(t))$ is of zero mean; and the fourth from the orthonormality of $(\phi_1, \ldots, \phi_d)$. We thus conclude that for every $m \in \mathcal{M}$ and every $\mathbf{a} \in \mathbb{R}^d$,

$$\mathsf{E}\big[\tilde{\mathbf{Y}} \mid (M, \mathbf{A}) = (m, \mathbf{a})\big] = \mathbf{0}. \tag{27.33}$$

Likewise, the conditional covariance matrix of $\tilde{\mathbf{Y}}$ given $(M, \mathbf{A}) = (m, \mathbf{a})$ does not depend on the value of $m$ and $\mathbf{a}$: it is the covariance matrix of $(N(t_1), \ldots, N(t_\eta))^{\mathsf{T}}$. By establishing that, conditional on $(M, \mathbf{A}) = (m, \mathbf{a})$, the vector $\tilde{\mathbf{Y}}$ has a multivariate Gaussian distribution whose mean vector and covariance matrix do not depend on $(m, \mathbf{a})$ we have established (27.31).

We next prove (27.32). By Theorem 25.12.1, we have that, conditional on $(M, \mathbf{A})$, the random vectors $\mathbf{T}$ and $\tilde{\mathbf{Y}}$ are jointly Gaussian. To establish that they are conditionally independent given $(M, \mathbf{A})$ it thus suffices to establish that they are conditionally uncorrelated (Proposition 23.7.3). We now proceed to compute their conditional covariance and show that it is zero. Since the conditional mean of $\tilde{\mathbf{Y}}$ is zero (27.33), it follows that we need to show that

$$\mathsf{E}\bigg[\Big(T^{(\ell)} - \mathsf{E}\big[T^{(\ell)} \mid (M, \mathbf{A}) = (m, \mathbf{a})\big]\Big) \tilde{Y}(t_\kappa) \,\bigg|\, (M, \mathbf{A}) = (m, \mathbf{a})\bigg] = 0,$$
$$m \in \mathcal{M}, \ \mathbf{a} \in \mathbb{R}^d, \ \ell \in \{1, \ldots, d\}, \ \kappa \in \{1, \ldots, \eta\}. \tag{27.34}$$

Before embarking on this calculation, we make two preliminary algebraic manipulations. The first entails using (27.23), (27.24), and the orthonormality of $(\phi_1, \ldots, \phi_d)$ to express $T^{(\ell)}$ as

$$T^{(\ell)} = A^{(\ell)} + \langle \mathbf{N}, \phi_\ell \rangle, \quad \ell = 1, \ldots, d. \tag{27.35}$$

This representation makes it clear that

$$T^{(\ell)} - \mathsf{E}\big[T^{(\ell)} \mid (M, \mathbf{A}) = (m, \mathbf{a})\big] = \langle \mathbf{N}, \phi_\ell \rangle, \quad \ell = 1, \ldots, d. \tag{27.36}$$

The second manipulation involves rewriting $\tilde{Y}(t_\kappa)$ using (27.23) and (27.27) as:

$$\tilde{Y}(t_\kappa) = Y(t_\kappa) - \sum_{\ell'=1}^{d} \phi_{\ell'}(t_\kappa) T^{(\ell')}$$

$$= \sum_{\ell'=1}^{d} A^{(\ell')} \phi_{\ell'}(t_\kappa) + N(t_\kappa) - \sum_{\ell'=1}^{d} \phi_{\ell'}(t_\kappa) T^{(\ell')}$$

$$= N(t_\kappa) - \sum_{\ell'=1}^{d} \big(T^{(\ell')} - A^{(\ell')}\big) \phi_{\ell'}(t_\kappa)$$

$$= N(t_\kappa) - \sum_{\ell'=1}^{d} \langle \mathbf{N}, \phi_{\ell'} \rangle \phi_{\ell'}(t_\kappa), \quad \kappa \in \{1, \ldots, \eta\}, \tag{27.37}$$

where the first equality follows from the definition of $\tilde{Y}(t_\kappa)$ (27.27); the second from (27.23); the third by rearranging terms; and the final equality from (27.35).

It follows from (27.36) and (27.37) that to establish (27.34) it suffices to show that for every $\ell \in \{1, \ldots, d\}$ and $\kappa \in \{1, \ldots, \eta\}$

$$\mathsf{E}\left[\langle \mathbf{N}, \boldsymbol{\phi}_\ell \rangle \left( N(t_\kappa) - \sum_{\ell'=1}^{d} \langle \mathbf{N}, \boldsymbol{\phi}_{\ell'} \rangle \, \phi_{\ell'}(t_\kappa) \right)\right] = 0. \tag{27.38}$$

This follows from Proposition 25.15.2 and the orthonormality of $(\boldsymbol{\phi}_1, \ldots, \boldsymbol{\phi}_d)$:

$$\mathsf{E}\left[\langle \mathbf{N}, \boldsymbol{\phi}_\ell \rangle \left( N(t_\kappa) - \sum_{\ell'=1}^{d} \langle \mathbf{N}, \boldsymbol{\phi}_{\ell'} \rangle \, \phi_{\ell'}(t_\kappa) \right)\right]$$

$$= \mathsf{E}\left[\langle \mathbf{N}, \boldsymbol{\phi}_\ell \rangle \, N(t_\kappa)\right] - \sum_{\ell'=1}^{d} \phi_{\ell'}(t_\kappa) \mathsf{E}\left[\langle \mathbf{N}, \boldsymbol{\phi}_\ell \rangle \, \langle \mathbf{N}, \boldsymbol{\phi}_{\ell'} \rangle\right]$$

$$= \frac{\mathsf{N}_0}{2} \phi_\ell(t_\kappa) - \sum_{\ell'=1}^{d} \phi_{\ell'}(t_\kappa) \frac{\mathsf{N}_0}{2} \mathrm{I}\{\ell = \ell'\}$$

$$= \frac{\mathsf{N}_0}{2} \phi_\ell(t_\kappa) - \frac{\mathsf{N}_0}{2} \phi_\ell(t_\kappa)$$

$$= 0.$$

Combining (27.38) with (27.36) and (27.37) establishes (27.34), i.e., that for every $m \in \mathcal{M}$, $\mathbf{a} \in \mathbb{R}^d$, $\ell \in \{1, \ldots, d\}$, and $\kappa \in \{1, \ldots, \eta\}$

$$\mathsf{Cov}\left[T^{(\ell)}, \tilde{Y}(t_\kappa) \,\middle|\, (M, \mathbf{A}) = (m, \mathbf{a})\right] = 0. \tag{27.39}$$

This combines with the conditional joint Gaussianity of vectors $\mathbf{T}$ and $\tilde{\mathbf{Y}}$ given $(M, \mathbf{A})$ to establish (27.32). The combination of (27.32) and (27.31) implies (27.30), which implies (27.29). Since (27.29) is equivalent to (27.26), this establishes the theorem for baseband signals.

For passband signals the proof is almost identical except that in deriving (27.38) we use Note 25.15.4 instead of Proposition 25.15.2. $\qquad\square$

## 27.4 The Conditional Law of the Sufficient Statistic

Having established in the previous section that the vector $\mathbf{T}$ defined in (27.21) forms a sufficient statistic for guessing $H$ based on $(Y(t))$, we next proceed to calculate its conditional distribution given $H$. This will allow us to compute the likelihood-ratio $f_{\mathbf{T}|H=0}(\mathbf{t})/f_{\mathbf{T}|H=1}(\mathbf{t})$ and to thus obtain an optimal guessing rule.

Rather than computing the conditional distribution directly, we begin with the simpler conditional distribution of $\mathbf{T}$ given $(H, \Theta)$. Conditional on $(H, \Theta)$, the vector $\mathbf{T}$ is Gaussian (Theorem 25.12.1). Consequently, to compute its conditional distribution we only need to compute its conditional mean vector and covariance matrix, which we proceed to do. Conditional on $(H, \Theta) = (\nu, \theta)$, the observed process $(Y(t))$ can be expressed as

$$Y(t) = s_{\nu,c}(t) \cos \theta + s_{\nu,s}(t) \sin \theta + N(t), \quad t \in \mathbb{R}. \tag{27.40}$$

Hence, since $(N(t))$ is of zero mean, we have from (27.22) and (27.20)

$$\mathsf{E}\big[\mathbf{T} \,\big|\, (H, \Theta) = (0, \theta)\big] = \sqrt{\mathsf{E}_s} \big(\cos \theta, \sin \theta, 0, 0\big)^{\mathsf{T}}, \tag{27.41a}$$

$$\mathsf{E}\big[\mathbf{T} \,\big|\, (H, \Theta) = (1, \theta)\big] = \sqrt{\mathsf{E}_s} \big(0, 0, \cos \theta, \sin \theta\big)^{\mathsf{T}}, \tag{27.41b}$$

as we next calculate. The calculation is a bit tedious because we need to compute the conditional mean of each of four random variables conditional on each of two hypotheses, thus requiring eight calculations, which are all very similar but not identical. We shall carry out only one calculation:

$$
\begin{aligned}
\mathsf{E}\big[T_{0,c} \,\big|\, (H, \Theta) = (0, \theta)\big] &= \frac{1}{\sqrt{\mathsf{E}_s}} \Big(\langle \mathbf{s}_{0,c} \cos \theta + \mathbf{s}_{0,s} \sin \theta, \mathbf{s}_{0,c}\rangle + \mathsf{E}[\langle \mathbf{N}, \mathbf{s}_{0,c}\rangle]\Big) \\
&= \frac{1}{\sqrt{\mathsf{E}_s}} \langle \mathbf{s}_{0,c} \cos \theta + \mathbf{s}_{0,s} \sin \theta, \mathbf{s}_{0,c}\rangle \\
&= \frac{1}{\sqrt{\mathsf{E}_s}} \Big(\|\mathbf{s}_{0,c}\|_2^2 \cos \theta + \langle \mathbf{s}_{0,s}, \mathbf{s}_{0,c}\rangle \sin \theta\Big) \\
&= \sqrt{\mathsf{E}_s} \cos \theta,
\end{aligned}
$$

where the first equality follows from (27.40); the second because $(N(t))$ is of zero mean (Proposition 25.10.1); the third from the linearity of the inner product and by writing $\langle \mathbf{s}_{0,c}, \mathbf{s}_{0,c}\rangle$ as $\|\mathbf{s}_{0,c}\|_2^2$; and the final equality from (27.20).

We next compute the conditional covariance matrix of $\mathbf{T}$ given $(H, \Theta) = (\nu, \theta)$. By the orthonormality (27.20) and the whiteness of the noise (Proposition 25.15.2) we have that, irrespective of $\nu$ and $\theta$, this conditional covariance matrix is given by the $4 \times 4$ matrix $(\mathsf{N}_0/2)\mathsf{I}_4$, where $\mathsf{I}_4$ is the $4 \times 4$ identity matrix.

Using the explicit form of the Gaussian distribution (19.6) and defining

$$\sigma^2 \triangleq \frac{\mathsf{N}_0}{2}, \tag{27.42}$$

we can thus write the conditional density as

$$
\begin{aligned}
f_{\mathbf{T} \mid H=0, \Theta=\theta}&(\mathbf{t}) \\
&= \frac{1}{(2\pi\sigma^2)^2} \exp\Big(-\frac{1}{2\sigma^2}\Big(\big(t_{0,c} - \sqrt{\mathsf{E}_s} \cos \theta\big)^2 + \big(t_{0,s} - \sqrt{\mathsf{E}_s} \sin \theta\big)^2 + t_{1,c}^2 + t_{1,s}^2\Big)\Big) \\
&= \frac{1}{(2\pi\sigma^2)^2} \exp\Big(-\frac{\mathsf{E}_s}{2\sigma^2} - \frac{t_0 + t_1}{2}\Big) \\
&\quad \times \exp\Big(\frac{1}{\sigma^2} \sqrt{\mathsf{E}_s}\, t_{0,c} \cos \theta + \frac{1}{\sigma^2} \sqrt{\mathsf{E}_s}\, t_{0,s} \sin \theta\Big), \quad \mathbf{t} \in \mathbb{R}^4, \tag{27.43}
\end{aligned}
$$

where the second equality follows by opening the squares, by using the identity $\cos^2\theta + \sin^2\theta = 1$, and by defining

$$T_0 \triangleq \frac{T_{0,c}^2 + T_{0,s}^2}{\sigma^2}, \qquad\qquad t_0 \triangleq \frac{t_{0,c}^2 + t_{0,s}^2}{\sigma^2}, \qquad (27.44\text{a})$$

$$T_1 \triangleq \frac{T_{1,c}^2 + T_{1,s}^2}{\sigma^2}, \qquad\qquad t_1 \triangleq \frac{t_{1,c}^2 + t_{1,s}^2}{\sigma^2}. \qquad (27.44\text{b})$$

(We define $T_0$ and $T_1$ not only to simplify the typesetting but also for ulterior motives that have to do with the further reduction of the sufficient statistic from a random vector of four components to one with only two, namely, the vector $(T_0, T_1)^\mathsf{T}$.)

To derive $f_{\mathbf{T}|H=0}(\mathbf{t})$ (unconditioned on $\Theta$) we can integrate out $\Theta$. Thus, for every $\mathbf{t} = (t_{0,c}, t_{0,s}, t_{1,c}, t_{1,s})^\mathsf{T}$ in $\mathbb{R}^4$

$$f_{\mathbf{T}|H=0}(\mathbf{t}) = \int_{-\pi}^{\pi} f_{\Theta|H=0}(\theta) \, f_{\mathbf{T}|H=0,\Theta=\theta}(\mathbf{t}) \, \mathrm{d}\theta$$

$$= \int_{-\pi}^{\pi} f_\Theta(\theta) \, f_{\mathbf{T}|H=0,\Theta=\theta}(\mathbf{t}) \, \mathrm{d}\theta$$

$$= \frac{1}{2\pi} \int_{-\pi}^{\pi} f_{\mathbf{T}|H=0,\Theta=\theta}(\mathbf{t}) \, \mathrm{d}\theta$$

$$= \frac{1}{(2\pi\sigma^2)^2} \, e^{-\mathsf{E}_s/(2\sigma^2)} \, e^{-t_1/2} \, e^{-t_0/2}$$

$$\times \frac{1}{2\pi} \int_{-\pi}^{\pi} \exp\left(\frac{1}{\sigma^2}\sqrt{\mathsf{E}_s}\, t_{0,c}\cos\theta + \frac{1}{\sigma^2}\sqrt{\mathsf{E}_s}\, t_{0,s}\sin\theta\right) \mathrm{d}\theta$$

$$= \frac{1}{(2\pi\sigma^2)^2} \, e^{-\mathsf{E}_s/(2\sigma^2)} \, e^{-(t_0+t_1)/2}$$

$$\times \frac{1}{2\pi} \int_{-\pi}^{\pi} \exp\left(\sqrt{\frac{\mathsf{E}_s}{\sigma^2}}\sqrt{t_0}\cos\left(\theta - \tan^{-1}(t_{0,s}/t_{0,c})\right)\right) \mathrm{d}\theta$$

$$= \frac{1}{(2\pi\sigma^2)^2} \, e^{-\mathsf{E}_s/(2\sigma^2)} \, e^{-(t_0+t_1)/2}$$

$$\times \frac{1}{2\pi} \int_{-\pi-\tan^{-1}(t_{0,s}/t_{0,c})}^{\pi-\tan^{-1}(t_{0,s}/t_{0,c})} \exp\left(\sqrt{\frac{\mathsf{E}_s}{\sigma^2}}\sqrt{t_0}\cos\psi\right) \mathrm{d}\psi$$

$$= \frac{1}{(2\pi\sigma^2)^2} \, e^{-\mathsf{E}_s/(2\sigma^2)} \, e^{-(t_0+t_1)/2} \frac{1}{2\pi}\int_{-\pi}^{\pi} \exp\left(\sqrt{\frac{\mathsf{E}_s}{\sigma^2}}\sqrt{t_0}\cos\psi\right) \mathrm{d}\psi$$

$$= \frac{1}{(2\pi\sigma^2)^2} \, e^{-\mathsf{E}_s/(2\sigma^2)} \, e^{-(t_1+t_0)/2} \, \mathsf{I}_0\left(\sqrt{\frac{\mathsf{E}_s}{\sigma^2}}\sqrt{t_0}\right), \qquad (27.45)$$

where the first equality follows by averaging out $\Theta$; the second because $\Theta$ and $H$ are independent; the third because $\Theta$ is uniform; the fourth by the explicit form of $f_{\mathbf{T}|H=0,\Theta=\theta}(\mathbf{t})$ (27.43); the fifth by the trigonometric identity

$$\alpha\cos\theta + \beta\sin\theta = \sqrt{\alpha^2+\beta^2}\cos\left(\theta - \tan^{-1}(\beta/\alpha)\right); \qquad (27.46)$$

the sixth by the change of variable $\psi \triangleq \theta - \tan^{-1}(t_{0,s}/t_{0,c})$; the seventh from the periodicity of the cosine function; and the final equality by recalling that the zeroth-order modified Bessel function $I_0(\cdot)$ is defined by

$$I_0(\xi) \triangleq \frac{1}{2\pi} \int_{-\pi}^{\pi} e^{\xi \cos \phi} \, d\phi \tag{27.47}$$

$$= \frac{1}{\pi} \int_0^{\pi} e^{\xi \cos \phi} \, d\phi$$

$$= \frac{1}{\pi} \int_0^{\pi/2} \left( e^{\xi \cos \phi} + e^{-\xi \cos \phi} \right) d\phi, \quad \xi \in \mathbb{R}. \tag{27.48}$$

By symmetry,

$$f_{\mathbf{T}|H=1}(\mathbf{t}) = \frac{1}{(2\pi\sigma^2)^2} \, e^{-\mathsf{E}_s/(2\sigma^2)} \, e^{-(t_0+t_1)/2} \, I_0\left( \sqrt{\frac{\mathsf{E}_s}{\sigma^2}} \sqrt{t_1} \right), \quad \mathbf{t} \in \mathbb{R}^4. \tag{27.49}$$

## 27.5  An Optimal Detector

By (27.45) and (27.49), the likelihood-ratio is given by

$$\frac{f_{\mathbf{T}|H=0}(\mathbf{t})}{f_{\mathbf{T}|H=1}(\mathbf{t})} = \frac{I_0\left( \sqrt{\frac{\mathsf{E}_s}{\sigma^2}} \sqrt{t_0} \right)}{I_0\left( \sqrt{\frac{\mathsf{E}_s}{\sigma^2}} \sqrt{t_1} \right)}, \quad \mathbf{t} \in \mathbb{R}^4, \tag{27.50}$$

which is computable from $t_0$ and $t_1$. This proves that the pair $(T_0, T_1)$ defined in (27.44) forms a sufficient statistic for guessing $H$ based on $\mathbf{T}$ (Definition 20.12.2). Having identified $(T_0, T_1)$ as a sufficient statistic, we now proceed to derive an optimal decision rule using two different methods. The first method, which is summarized in (20.79), ignores the fact that $(T_0, T_1)$ is sufficient and proceeds to base the decision on the likelihood-ratio of $\mathbf{T}$ (27.50). The second method, which is summarized in (20.80), bases the decision on the likelihood-ratio of the pair $(T_0, T_1)$.

**Method 1:**  Since we assumed a uniform prior (27.10), an optimal decision rule is to guess "$H = 0$" whenever $f_{\mathbf{T}|H=0}(\mathbf{t})/f_{\mathbf{T}|H=1}(\mathbf{t}) \geq 1$, which, by (27.50) is equivalent to

$$\text{Guess "}H = 0\text{" if } \quad I_0\left( \sqrt{\frac{\mathsf{E}_s}{\sigma^2}} \sqrt{t_0} \right) \geq I_0\left( \sqrt{\frac{\mathsf{E}_s}{\sigma^2}} \sqrt{t_1} \right). \tag{27.51}$$

This rule can be further simplified by noting that $I_0(\xi)$ is (strictly) increasing in $\xi$ for $\xi \geq 0$. (This can be verified by computing the derivative from (27.48)

$$\frac{d\, I_0(\xi)}{d\xi} = \frac{1}{\pi} \int_0^{\pi/2} \cos \phi \left( e^{\xi \cos \phi} - e^{-\xi \cos \phi} \right) d\phi$$

and by noting that for $\xi > 0$ the integrand is positive for all $\phi \in (0, \pi/2)$.) Consequently, the function $\xi \mapsto I_0(\sqrt{\xi})$ is also (strictly) increasing and the guessing rule (27.51) is thus equivalent to the rule

$$\boxed{\text{Guess ``}H = 0\text{'' if } t_0 \geq t_1.} \tag{27.52}$$

In terms of the observable $(Y(t))$ this can be paraphrased using (27.44) and (27.22) as guessing "$H = 0$" whenever

$$\left( \int_{-\infty}^{\infty} Y(t) \operatorname{Re}\big(s_{0,\mathrm{BB}}(t)\, e^{\mathrm{i}2\pi f_c t}\big)\, \mathrm{d}t \right)^2 + \left( \int_{-\infty}^{\infty} Y(t) \operatorname{Re}\big(\mathrm{i}\, s_{0,\mathrm{BB}}(t)\, e^{\mathrm{i}2\pi f_c t}\big)\, \mathrm{d}t \right)^2$$
$$\geq \left( \int_{-\infty}^{\infty} Y(t) \operatorname{Re}\big(s_{1,\mathrm{BB}}(t)\, e^{\mathrm{i}2\pi f_c t}\big)\, \mathrm{d}t \right)^2 + \left( \int_{-\infty}^{\infty} Y(t) \operatorname{Re}\big(\mathrm{i}\, s_{1,\mathrm{BB}}(t)\, e^{\mathrm{i}2\pi f_c t}\big)\, \mathrm{d}t \right)^2.$$

**Method 2:** We next obtain the same result by considering the likelihood-ratio function of the sufficient statistic $(T_0, T_1)$

$$\frac{f_{T_0, T_1 | H=0}(t_0, t_1)}{f_{T_0, T_1 | H=1}(t_0, t_1)}.$$

We begin by arguing that, conditional on $H = 0$, the random variables $T_0$, $T_1$, and $\Theta$ are independent with

$$f_{T_0, T_1, \Theta | H=0}(t_0, t_1, \theta) = \frac{1}{2\pi}\, f_{\chi^2_{2, \lambda_1}}(t_0)\, f_{\chi^2_{2, \lambda_0}}(t_1), \tag{27.53}$$

where $f_{\chi^2_{n, \lambda}}(x)$ denotes the density at $x$ of the noncentral $\chi^2$ distribution with $n$ degrees of freedom and noncentrality parameter $\lambda$ (Section 19.8.2), and where

$$\lambda_0 = 0 \quad \text{and} \quad \lambda_1 = \frac{\mathsf{E_s}}{\sigma^2}. \tag{27.54}$$

To prove (27.53) we compute for every $t_0, t_1 \in \mathbb{R}$ and $\theta \in [-\pi, \pi)$

$$f_{T_0, T_1, \Theta | H=0}(t_0, t_1, \theta) = f_{\Theta | H=0}(\theta)\, f_{T_0, T_1 | H=0, \Theta=\theta}(t_0, t_1)$$
$$= \frac{1}{2\pi}\, f_{T_0, T_1 | H=0, \Theta=\theta}(t_0, t_1)$$
$$= \frac{1}{2\pi}\, f_{T_0 | H=0, \Theta=\theta}(t_0)\, f_{T_1 | H=0, \Theta=\theta}(t_1)$$
$$= \frac{1}{2\pi}\, f_{\chi^2_{2, \lambda_1}}(t_0)\, f_{\chi^2_{2, \lambda_0}}(t_1),$$

where the first equality follows from the definition of the conditional density; the second because $\Theta$ is independent of $H$ and is uniformly distributed over the interval $[-\pi, \pi)$; the third because, conditional on $(H, \Theta) = (0, \theta)$, the random variables $T_{0,c}, T_{0,s}, T_{1,c}, T_{1,s}$ are independent (Section 27.4), and because $T_0$ is a function of $(T_{0,c}, T_{0,s})$ whereas $T_1$ is a function of $(T_{1,c}, T_{1,s})$ (see (27.44)); and the final equality follows because, conditional on $(H, \Theta) = (0, \theta)$, the random variables $T_{0,c}, T_{0,s}, T_{1,c}, T_{1,s}$ are variance-$\sigma^2$ Gaussians with means specified in (27.41a) (Section 19.8.2).

Integrating out $\theta$ in (27.53) we obtain that, conditional on $H$, the random variables $T_0$ and $T_1$ are independent with

$$f_{T_0, T_1 | H = 0}(t_0, t_1) = f_{\chi^2_{2, \lambda_1}}(t_0)\, f_{\chi^2_{2, \lambda_0}}(t_1) \tag{27.55a}$$

$$f_{T_0, T_1 | H = 1}(t_0, t_1) = f_{\chi^2_{2, \lambda_0}}(t_0)\, f_{\chi^2_{2, \lambda_1}}(t_1), \tag{27.55b}$$

where the expression for $f_{T_0, T_1 | H = 1}(t_0, t_1)$ is obtained using analogous steps.

Since $H$ has a uniform prior, an optimal decision rule is thus to guess "$H = 0$" whenever

$$f_{\chi^2_{2, \lambda_1}}(t_0)\, f_{\chi^2_{2, \lambda_0}}(t_1) \geq f_{\chi^2_{2, \lambda_0}}(t_0)\, f_{\chi^2_{2, \lambda_1}}(t_1).$$

Since $\lambda_1 > \lambda_0$, this will hold, by Proposition 19.8.3, whenever $t_0 \geq t_1$. And by the same proposition the inequality

$$f_{\chi^2_{2, \lambda_1}}(t_0)\, f_{\chi^2_{2, \lambda_0}}(t_1) \leq f_{\chi^2_{2, \lambda_0}}(t_0)\, f_{\chi^2_{2, \lambda_1}}(t_1)$$

will hold whenever $t_0 \leq t_1$. It is thus optimal to guess "$H = 0$" whenever $t_0 \geq t_1$ and to guess "$H = 1$" whenever $t_0 < t_1$. (It does not matter how we guess when $t_0 = t_1$.) The decision rule (27.52) has thus been recovered.

## 27.6 The Probability of Error

In this section we compute the probability of error for the optimal guessing rule (27.52). Since the probability of a tie (i.e., of $T_0 = T_1$) is zero both conditional on $H = 0$ and conditional on $H = 1$, we shall analyze a slightly simpler guessing rule that guesses "$H = 0$" if $T_0 > T_1$, and guesses "$H = 1$" if $T_1 > T_0$.

We begin with the conditional probability of error given that $H = 0$, i.e., with $\Pr[T_1 \geq T_0 \,|\, H = 0]$. Conditional on $H = 0$, the question of whether our decoder errs depends *prima facie* not only on the realization of the additive noise $(N(t))$ but also on the realization of $\Theta$. But this is not the case because, conditionally on $H = 0$, the pair $(T_0, T_1)$ is independent of $\Theta$ (see (27.53)), so the realization of $\Theta$ does not play a role in the sense that for every $\theta \in [-\pi, \pi)$

$$\Pr[T_1 \geq T_0 \,|\, H = 0, \Theta = \theta] = \Pr[T_1 \geq T_0 \,|\, H = 0, \Theta = 0]. \tag{27.56}$$

Conditional on $(H, \Theta) = (0, \theta)$ we have by (27.53) that $T_0$ and $T_1$ are independent with $T_0 \sim \chi^2_{2, \lambda_1}$ and with $T_1 \sim \chi^2_{2, \lambda_0}$, i.e., with $T_1$ having a mean-2 exponential distribution (Note 19.8.1)

$$f_{T_1 | H = 0, \Theta = \theta}(t_1) = \frac{1}{2} e^{-\frac{t_1}{2}}, \quad t_1 \geq 0.$$

Consequently, for every $\theta \in [-\pi, \pi)$ and $\xi \geq 0$,

$$\Pr[T_1 \geq \xi \,|\, H = 0, \Theta = \theta] = \int_\xi^\infty \frac{1}{2} e^{-t/2} \, \mathrm{d}t = e^{-\xi/2}. \tag{27.57}$$

Starting with (27.56) we now have for every $\theta \in [-\pi, \pi)$

$$
\begin{aligned}
\Pr\big[T_1 \geq T_0 \,\big|\, H = 0, \Theta = \theta\big] \\
= \Pr\big[T_1 \geq T_0 \,\big|\, H = 0, \Theta = 0\big] \\
= \int_0^\infty f_{T_0 | H = 0, \Theta = 0}(t_0) \, \Pr\big[T_1 \geq t_0 \,\big|\, H = 0, \Theta = 0, T_0 = t_0\big] \, \mathrm{d}t_0 \\
= \int_0^\infty f_{T_0 | H = 0, \Theta = 0}(t_0) \, \Pr\big[T_1 \geq t_0 \,\big|\, H = 0, \Theta = 0\big] \, \mathrm{d}t_0 \\
= \int_0^\infty f_{T_0 | H = 0, \Theta = 0}(t_0) \, e^{-t_0/2} \, \mathrm{d}t_0 \\
= \mathsf{E}\Big[ e^{sT_0} \,\Big|\, H = 0, \Theta = 0 \Big]\Big|_{s = -1/2} \\
= M_{\chi^2_{2, \mathsf{E}_\mathrm{s}/\sigma^2}}(s)\Big|_{s = -1/2} \\
= \frac{1}{2}\, e^{-\frac{\mathsf{E}_\mathrm{s}}{4\sigma^2}},
\end{aligned}
\tag{27.58}
$$

where the first equality follows from (27.56); the second from (26.88); the third because conditional on $H = 0$ (and $\Theta = 0$) the random variables $T_0$ and $T_1$ are independent; the fourth from (27.57); the fifth by expressing $\int f_Z(z)\, g(z)\, \mathrm{d}z$ as $\mathsf{E}[g(Z)]$ (with $g(\cdot)$ the exponential function); the sixth by the definition of the MGF (19.23) and because, conditional on $H = 0$ and $\Theta = 0$, we have that $T_0 \sim \chi^2_{2, \mathsf{E}_\mathrm{s}/\sigma^2}$; and the final equality from the explicit expression for the MGF of a $\chi^2_{2, \mathsf{E}_\mathrm{s}/\sigma^2}$ RV, i.e., from (19.45) with the substitution $n = 2$ for the number of degrees of freedom, $\lambda = \mathsf{E}_\mathrm{s}/\sigma^2$ for the noncentrality parameter, and $s = -1/2$.

By symmetry we also have for every $\theta \in [-\pi, \pi)$

$$
\Pr\big[T_0 \geq T_1 \,\big|\, H = 1, \Theta = \theta\big] = \frac{1}{2}\, e^{-\frac{\mathsf{E}_\mathrm{s}}{4\sigma^2}}.
\tag{27.59}
$$

Thus, if we denote by $p_{\mathrm{MAP}}(\mathrm{error}|\Theta = \theta)$ the conditional probability of error of our decoder conditional on $\Theta = \theta$, then by the uniformity of the prior (27.10) and by (27.58) & (27.59)

$$
\begin{aligned}
p_{\mathrm{MAP}}(\mathrm{error}|\Theta = \theta) \\
= \Pr[H = 0]\, p_{\mathrm{MAP}}(\mathrm{error}|H = 0, \Theta = \theta) + \Pr[H = 1]\, p_{\mathrm{MAP}}(\mathrm{error}|H = 1, \Theta = \theta) \\
= \frac{1}{2} \Pr\big[T_1 \geq T_0 \,\big|\, H = 0, \Theta = \theta\big] + \frac{1}{2} \Pr\big[T_0 \geq T_1 \,\big|\, H = 1, \Theta = \theta\big] \\
= \frac{1}{2}\, e^{-\frac{\mathsf{E}_\mathrm{s}}{4\sigma^2}}, \quad \theta \in [-\pi, \pi).
\end{aligned}
\tag{27.60}
$$

Integrating (27.60) over $\theta$ yields the optimal unconditional probability of error

$$
\boxed{\; p^*(\mathrm{error}) = \frac{1}{2}\, e^{-\frac{\mathsf{E}_\mathrm{s}}{4\sigma^2}}. \;}
\tag{27.61}
$$

Using (27.42), this can also be expressed as

$$
p^*(\mathrm{error}) = \frac{1}{2}\, e^{-\frac{\mathsf{E}_\mathrm{s}}{2\mathsf{N}_0}}.
\tag{27.62}
$$

## 27.7    Discussion

The detector we derived has the property that its error probability does not depend on the realization of the nuisance parameter $\Theta$; see (27.60). This property makes the detector robust with respect to the distribution of $\Theta$: since the conditional probability of error does not depend on the realization of $\Theta$, neither does the average performance depend on the distribution of $\Theta$. (Of course, if $\Theta$ is not uniform, then our decoder need not be optimal.)

We next show that our guessing rule is also conservative in the sense that it minimizes the worst-case performance:

$$\sup_{\theta \in [-\pi, \pi)} p(\mathrm{error}|\Theta = \theta).$$

That is, for any guessing rule of conditional error probability $p'(\mathrm{error}|\Theta = \theta)$

$$\sup_{\theta \in [-\pi, \pi)} p'(\mathrm{error}|\Theta = \theta) \geq \sup_{\theta \in [-\pi, \pi)} p_{\mathrm{MAP}}(\mathrm{error}|\Theta = \theta) = \frac{1}{2} e^{-\frac{E_s}{4\sigma^2}}. \qquad (27.63)$$

Thus, while other decoders may outperform our decoder for *some* realizations of $\Theta$, for other realizations their probability of error will be at least as high. Indeed, if $p'(\mathrm{error}|\Theta = \theta)$ is the conditional probability of error associated with any guessing rule, then

$$\sup_{\theta \in [-\pi, \pi)} p'(\mathrm{error}|\Theta = \theta) \geq \frac{1}{2\pi} \int_{-\pi}^{\pi} p'(\mathrm{error}|\Theta = \theta) \, \mathrm{d}\theta$$

$$\geq \frac{1}{2\pi} \int_{-\pi}^{\pi} p_{\mathrm{MAP}}(\mathrm{error}|\Theta = \theta) \, \mathrm{d}\theta$$

$$= \sup_{\theta \in [-\pi, \pi)} p_{\mathrm{MAP}}(\mathrm{error}|\Theta = \theta) \, \mathrm{d}\theta$$

$$= e^{-\frac{E_s}{4\sigma^2}},$$

where the first inequality follows because the average (over $\theta$) can never exceed the supremum; the second inequality because the decoder we designed minimizes the unconditional probability of error; and the last two equalities follow from (27.60), i.e., from the fact that the conditional probability of error $p_{\mathrm{MAP}}(\mathrm{error}|\Theta = \theta)$ of our decoder does not depend on $\theta$ and is equal to the RHS of (27.60).

It is interesting to assess the degradation in performance due to our ignorance of $\Theta$. To that end we now compare the performance of our detector with that of the "coherent detector." The coherent decoder is an optimal decoder for the setting where the realization of $\Theta$ is known to the receiver, i.e., when the receiver can form its guess based on both $(Y(t))$ and $\Theta$. If the receiver knows $\Theta = \theta$, then it can compute $\mathbf{S}_0$ and $\mathbf{S}_1$, and the problem reduces to the problem of deciding which of two equi-energy orthogonal waveforms $\mathbf{S}_0$ and $\mathbf{S}_1$ is being observed in white Gaussian noise (the binary version of the problem we discussed in Section 26.11.3). An optimal decision rule would be

$$\text{guess ``}H = 0\text{'' if } \int_{-\infty}^{\infty} Y(t) \, S_0(t) \, \mathrm{d}t > \int_{-\infty}^{\infty} Y(t) \, S_1(t) \, \mathrm{d}t$$

with resulting probability of error (see (26.93))

$$p^*_{\text{coherent}}(\text{error}|\Theta = \theta) = Q\left(\frac{\|\mathbf{S}_0 - \mathbf{S}_1\|_2 / 2}{\sigma}\right)$$

$$= Q\left(\sqrt{\frac{\mathsf{E}_s}{2\sigma^2}}\right)$$

$$\approx \frac{1}{\sqrt{\pi \mathsf{E}_s / \sigma^2}} \exp\left(-\frac{\mathsf{E}_s}{4\sigma^2}\right), \quad \frac{\mathsf{E}_s}{\sigma^2} \gg 1, \qquad (27.64)$$

where the approximation follows from (19.18). Integrating over $\theta$ we obtain

$$p^*_{\text{coherent}}(\text{error}) \approx \frac{1}{\sqrt{\pi \mathsf{E}_s / \sigma^2}} \exp\left(-\frac{\mathsf{E}_s}{4\sigma^2}\right), \quad \frac{\mathsf{E}_s}{\sigma^2} \gg 1. \qquad (27.65)$$

Comparing (27.65) with (27.61) we see that if $\mathsf{E}_s/\sigma^2$ is large, then we pay only a small penalty for not knowing the phase.[4] Of course, if the phase were known precisely we mights have used antipodal signaling with the resulting probability of error being lower; see (26.72).[5]

## 27.8 Extension to M ≥ 2 Signals

We next briefly address the M-ary version of the problem of noncoherent detection of orthogonal signals. We now denote the RV to be guessed by $M$ and replace (27.10) with the assumption that $M$ is uniformly distributed over the set $\mathcal{M} = \{1, \ldots, \mathsf{M}\}$, where $\mathsf{M} \geq 2$. We wish to guess the value of $M$ based on the observation $\big(Y(t)\big)$ (27.11), where $\nu$ now takes value in $\mathcal{M}$ and where the orthogonality conditions (27.15) & (27.18) are now written as

$$\langle \mathbf{s}_{\nu',\text{BB}}, \mathbf{s}_{\nu'',\text{BB}} \rangle = \frac{1}{2} \mathsf{E}_s \, \mathrm{I}\{\nu' = \nu''\}, \quad \nu', \nu'' \in \mathcal{M}. \qquad (27.66)$$

We first argue that the vector

$$\big(T_1, \ldots, T_\mathsf{M}\big)^\mathsf{T} \qquad (27.67)$$

forms a sufficient statistic, where, in analogy to (27.44), we define

$$T_\nu = \frac{T_{\nu,\text{c}}^2 + T_{\nu,\text{s}}^2}{\sigma^2}, \quad \nu \in \mathcal{M},$$

and where

$$T_{\nu,\text{c}} = \left\langle \mathbf{Y}, \frac{s_{\nu,\text{c}}}{\sqrt{\mathsf{E}_s}} \right\rangle \quad \text{and} \quad T_{\nu,\text{s}} = \left\langle \mathbf{Y}, \frac{s_{\nu,\text{s}}}{\sqrt{\mathsf{E}_s}} \right\rangle, \quad \nu \in \mathcal{M}.$$

To this end, we first note that it is enough that we show pairwise sufficiency (Proposition 22.3.2). Pairwise sufficiency can be proved using Proposition 22.4.2

---

[4] Although $p^*(\text{error})/p^*_{\text{coherent}}(\text{error})$ tends to infinity, it does so only subexponentially.

[5] Comparing (26.93) and (26.72) we see that, to achieve the same probability of error, binary orthogonal keying requires twice as much energy as antipodal signaling.

because for every $m' \neq m''$ in $\mathcal{M}$ our analysis of the binary problem shows that the tuple $(T_{m'}, T_{m''})$ forms a sufficient statistic for testing between $m'$ and $m''$, and this tuple is computable from the vector in (27.67).

Our analysis of the binary case shows that, after observing $(Y(t))$, the *a posteriori* probability of the event $M = m$ is larger than the *a posteriori* distribution of the event $M = m'$ whenever $T_m > T_{m'}$. Consequently, Message $m$ has the highest *a posteriori* probability if $T_m = \max_{m' \in \mathcal{M}} T_{m'}$. Thus, the decision rule

$$
\text{Guess ``}M = m\text{'' if } T_m = \max_{m' \in \mathcal{M}} T_{m'} \tag{27.68}
$$

is optimal. The probability of a tie is zero, so it does not matter how ties are resolved.

We next turn to the analysis of the probability of error. We shall assume that a tie results in an error, so, conditional on $M = m$, an error occurs whenever $\max\{T_1, \ldots, T_{m-1}, T_{m+1}, \ldots, T_M\} \geq T_m$. We first show that, as in the binary case, the probability of error associated with this guessing rule depends neither on the realization of $\Theta$ nor on the message, i.e., that for every $m \in \mathcal{M}$ and $\theta \in [-\pi, \pi)$

$$
p_{\text{MAP}}(\text{error}|M = m, \Theta = \theta) = p_{\text{MAP}}(\text{error}|M = 1, \Theta = 0). \tag{27.69}
$$

To see this note that, conditional on $(M, \Theta) = (m, \theta)$, the components of the vector (27.67) are independent, with the $m$-th component being $\chi^2_{2, \text{E}_s/\sigma^2}$ and with the other components being $\chi^2_{2,0}$. Consequently, irrespective of $\theta$ and $m$, the conditional probability of error is the probability that a $\chi^2_{2, \text{E}_s/\sigma^2}$ RV is exceeded by, or is equal to, at least one of $M - 1$ IID $\chi^2_{2,0}$ random variables that are independent of it. In the analysis of the probability of error we shall thus assume that $M = 1$ and that $\theta = 0$.

The probability that the maximum among the random variables $T_2, \ldots, T_M$ exceeds or is equal to $\xi$ is given for every $\xi \geq 0$ by

$$
\begin{aligned}
\Pr&\big[\max\{T_2, \ldots, T_M\} \geq \xi \,\big|\, M = 1, \Theta = 0\big] \\
&= 1 - \Pr\big[\max\{T_2, \ldots, T_M\} < \xi \,\big|\, M = 1, \Theta = 0\big] \\
&= 1 - \Pr\big[T_2 < \xi, \ldots, T_M < \xi \,\big|\, M = 1, \Theta = 0\big] \\
&= 1 - \big(\Pr\big[T_2 < \xi \,\big|\, M = 1, \Theta = 0\big]\big)^{M-1} \\
&= 1 - \big(1 - e^{-\xi/2}\big)^{M-1} \\
&= 1 - \sum_{j=0}^{M-1} (-1)^j \binom{M-1}{j} e^{-j\xi/2}, \tag{27.70}
\end{aligned}
$$

where the first equality follows because the probabilities of an event and of its complement sum to one; the second because the maximum is smaller than $\xi$ if, and only if, all the random variables are smaller than $\xi$; the third because, conditionally on $M = 1$ and $\Theta = 0$, the random variables $T_2, \ldots, T_M$ are IID; the fourth because conditional on $M = 1$ and $\Theta = 0$, the RV $T_2$ is a mean-2 exponential (Note 19.8.1); and the final equality follows from the binomial formula (26.91) with the substitution $a = 1$, $b = -e^{-\xi/2}$, and $n = M - 1$. The probability of error is thus:

$$
\begin{aligned}
\Pr&\big[\max\{T_2, \ldots, T_M\} \geq T_1 \,\big|\, M = 1, \Theta = \theta\big] \\
&= \Pr\big[\max\{T_2, \ldots, T_M\} \geq T_1 \,\big|\, M = 1, \Theta = 0\big] \\
&= \int_0^\infty f_{T_1 | M=1, \Theta=0}(t_1) \Pr\big[\max\{T_2, \ldots, T_M\} \geq t_1 \,\big|\, M = 1, \Theta = 0, T_1 = t_1\big] \, dt_1 \\
&= \int_0^\infty f_{T_1 | M=1, \Theta=0}(t_1) \Pr\big[\max\{T_2, \ldots, T_M\} \geq t_1 \,\big|\, M = 1, \Theta = 0\big] \, dt_1 \\
&= \int_0^\infty f_{T_1 | M=1, \Theta=0}(t_1) \left( 1 - \sum_{j=0}^{M-1} (-1)^j \binom{M-1}{j} e^{-j t_1 / 2} \right) dt_1 \\
&= 1 - \sum_{j=0}^{M-1} (-1)^j \binom{M-1}{j} \int_0^\infty f_{T_1 | M=1, \Theta=0}(t_1) \, e^{-j t_1 / 2} \, dt_1 \\
&= 1 - \sum_{j=0}^{M-1} (-1)^j \binom{M-1}{j} \mathsf{E}\Big[ e^{s T_1} \,\big|\, M = 1, \Theta = 0 \Big]\Big|_{s = -j/2} \\
&= 1 - \sum_{j=0}^{M-1} (-1)^j \binom{M-1}{j} M_{\chi^2_{2, \mathsf{E_s}/\sigma^2}}(s)\Big|_{s = -j/2} \\
&= 1 - \sum_{j=0}^{M-1} (-1)^j \binom{M-1}{j} \frac{1}{j+1} e^{-\frac{j}{j+1} \frac{\mathsf{E_s}}{2\sigma^2}},
\end{aligned}
$$

where the justifications are very similar to the justifications of (27.58) except that we use (27.70) instead of (27.57). Denoting the probability of error by $p^*(\text{error})$ and noting that for $j = 0$ the summand is 1, we have

$$
p^*(\text{error}) = \sum_{j=1}^{M-1} (-1)^{j+1} \binom{M-1}{j} \frac{1}{j+1} e^{-\frac{j}{j+1} \frac{\mathsf{E_s}}{2\sigma^2}}, \tag{27.71}
$$

or, upon recalling that $\sigma^2$ was defined in (27.42) as $\mathsf{N}_0/2$,

$$
\boxed{p^*(\text{error}) = \sum_{j=1}^{M-1} (-1)^{j+1} \binom{M-1}{j} \frac{1}{j+1} e^{-\frac{j}{j+1} \frac{\mathsf{E_s}}{\mathsf{N}_0}}.} \tag{27.72}
$$

## 27.9 Exercises

**Exercise 27.1 (The Conditional Law of the Sufficient Statistic).** Conditional on $M = m$, are the components of the random vector $\mathbf{T}$ in Theorem 27.3.1 independent? What about conditional on $(M, \mathbf{A}) = (m, \mathbf{a})$ for $m \in \mathcal{M}$ and $\mathbf{a} \in \mathbb{R}^d$?

**Exercise 27.2 (A Silly Design Criterion).** Let $\tilde{p}(\text{error} | \Theta = \theta)$ denote the conditional probability of error given $\Theta = \theta$ of some decision rule for the setup of Section 27.2. Show that

$$\inf_{-\pi \leq \theta < \pi} \tilde{p}(\text{error}|\Theta = \theta) \geq \mathcal{Q}\left(\sqrt{\frac{\mathsf{E_s}}{\mathsf{N_0}}}\right).$$

Can you think of a detector that achieves this bound with equality? Would you recommend using it?

**Exercise 27.3 (A Coherent Detector for an Incoherent Channel).** Alice designs a coherent detector for the setup of Section 27.2 by pretending that $\Theta$ is deterministically equal to zero and by then using the results on the detection of known signals in white Gaussian noise. Show that if her detector is used over our channel where $\Theta \sim \mathcal{U}\big([-\pi, \pi)\big)$, then the resulting average probability of error (averaged over $\Theta$) is $1/2$.

**Exercise 27.4 (Noncoherent Antipodal Signaling).** Show that if in the setup of Section 27.2 the baseband signals $\mathbf{s}_{0,\mathrm{BB}}$ and $\mathbf{s}_{1,\mathrm{BB}}$—rather than orthogonal—are antipodal in the sense that $\mathbf{s}_{0,\mathrm{BB}} = -\mathbf{s}_{1,\mathrm{BB}}$, then the optimal probability of error is $1/2$.

**Exercise 27.5 (A Fading Scenario).** Consider the setup of Section 27.2 but with (27.11) replaced by $Y(t) = AS_\nu(t) + N(t)$, where $A$ is a Rayleigh RV that is independent of $\big(H, \Theta, (N(t))\big)$. Find an optimal detector and the associated probability of error when $A$ is observed by the receiver. Repeat when $A$ is unobserved.

**Exercise 27.6 (Uniform Phase Noise Is the Worst Phase Noise).** Consider the setup of Section 27.2 but with $\Theta$ not necessarily uniformly distributed over $[-\pi, \pi)$. Show that the optimal probability of error is upper-bounded by the optimal probability of error corresponding to the case where $\Theta \sim \mathcal{U}\big([-\pi, \pi)\big)$.

**Exercise 27.7 (Unknown Frequency-Selective Channel).** Let $H$ take on the values $0$ and $1$ equiprobably, and let $\mathbf{s}$ be an integrable signal that is bandlimited to $W$ Hz. When $H = 0$ the transmitted signal is $\mathbf{s}$, and when $H = 1$ it is $-\mathbf{s}$. Let $U$ take on the values $\{\text{up}, \text{down}\}$ equiprobably and independently of $H$. When $U = \text{up}$ the transmitted signal is passed through a stable filter of impulse response $\mathbf{h}_{\mathrm{u}}$; when $U = \text{down}$ it is passed through a stable filter of impulse response $\mathbf{h}_{\mathrm{d}}$. At the receiver, white Gaussian noise $\big(N(t)\big)$ of PSD $\mathsf{N_0}/2$ over the bandwidth $W$ is added to the received signal. The noise is independent of $(H, U)$. Based on the received waveform $\big(Y(t)\big)$, the receiver wishes to guess $H$. The receiver has no knowledge of the realization of the switch $U$.

(i) Find a two-dimensional sufficient statistic vector $(T_1, T_2)^\mathsf{T}$ for this problem.

(ii) Find a decision rule that minimizes the probability of error. Express your rule using the function $\phi(x, y; \sigma_x^2, \sigma_y^2, \rho)$, which is the value at the point $(x, y)$ of the joint density of the zero-mean jointly Gaussian random variables $X$, $Y$ of variances $\sigma_x^2$ and $\sigma_y^2$ and covariance $\mathsf{E}[XY] = \sigma_x \sigma_y \rho$.

**Exercise 27.8 (Noncoherent Detection with Two Antennas).** Consider the setup of Section 27.2 but with the signal now received at two antennas. Denote the received signals by $\big(Y_1(t)\big)$ and $\big(Y_2(t)\big)$

$$Y_1(t) = 2\,\mathrm{Re}\Big(s_{\nu,\mathrm{BB}}(t)\,e^{i(2\pi f_c t + \Theta_1)}\Big) + N_1(t), \quad t \in \mathbb{R},$$

$$Y_2(t) = 2\,\mathrm{Re}\Big(s_{\nu,\mathrm{BB}}(t)\,e^{i(2\pi f_c t + \Theta_2)}\Big) + N_2(t), \quad t \in \mathbb{R},$$

where the additive white noises $\big(N_1(t)\big)$ and $\big(N_2(t)\big)$ at the two antennas are independent.

(i) Suppose that the random phase at the two antennas $\Theta_1$ and $\Theta_2$ are unknown but identical. Find an optimal detector and the optimal probability of error.

(ii) Assume now that $\Theta_1$ and $\Theta_2$ are independent. Find an optimal guessing rule for $H$.

**Exercise 27.9 (Unknown Polarity).** Consider the setup of Section 27.2 but with $\Theta$ now taking on the values $-\pi$ and $0$ equiprobably.

(i) Find an optimal decision rule for guessing $H$.

(ii) Bob suggests accounting for the random phase as follows. Pretend that the transmitted signal is drawn uniformly from the set $\{\pm\mathbf{s}_{0,c}, \pm\mathbf{s}_{1,c}\}$ and that it is observed in white Gaussian noise. Feed the received signal to an optimal receiver for guessing which of these four signals is being observed in white Gaussian noise, and if the receiver produces the guess "$\mathbf{s}_{0,c}$" or "$-\mathbf{s}_{0,c}$", declare "$H = 0$"; otherwise declare "$H = 1$". Is Bob's receiver optimal?

**Exercise 27.10 (Additional Channel Randomness).** Consider the setup of Section 27.2 but when the observed SP $\big(Y(t),\ t \in \mathbb{R}\big)$, rather than being given by (27.11), is now given by

$$Y(t) = S_\nu(t) + AN(t), \quad t \in \mathbb{R},$$

where $A$ is a positive RV that is independent of $\big(H, \Theta, (N(t))\big)$. Find an optimal decision rule when $A$ is observed. Repeat when $A$ is not observed.

**Exercise 27.11 (Mismatched Noncoherent Detection).** Suppose that the signal fed to the detector of Section 27.5 is

$$2\operatorname{Re}\Big(u_{\mathrm{BB}}(t)\, e^{\mathrm{i}(2\pi f_c t + \Theta)}\Big) + N(t), \quad t \in \mathbb{R},$$

where $\mathbf{u}_{\mathrm{BB}}$ is an integrable signal that is bandlimited to $W/2$ Hz and that is orthogonal to $\mathbf{s}_{0,\mathrm{BB}}$, and where the other quantities are as defined in Section 27.2. Compute the probability that the detector produces the guess "$H = 0$." Express your answer in terms of the inner product $\langle \mathbf{u}_{\mathrm{BB}}, \mathbf{s}_{1,\mathrm{BB}} \rangle$, the energy in $\mathbf{u}_{\mathrm{BB}}$, and $\mathsf{N}_0$.

# Chapter 28

# Detecting PAM and QAM Signals in White Gaussian Noise

## 28.1 Introduction and Setup

In Chapter 26 we addressed the problem of detecting one of $M$ bandwidth-$W$ signals corrupted by additive Gaussian noise that is white with respect to the bandwidth $W$. Except for assuming that the mean signals are integrable signals that are bandlimited to $W$ Hz, we made no assumptions about their structure. In this chapter we study the implication of the results of Chapter 26 for Pulse Amplitude Modulation, where the mean signals correspond to different possible outputs of a PAM modulator. The conclusions we shall draw are extremely important to the design of receivers for systems employing PAM.

The most important result of this chapter is that, loosely speaking, for PAM signals contaminated by additive white Gaussian noise, the inner products between the received waveform and the time shifts of the pulse shape by integer multiples of the baud period $T_s$ form a sufficient statistic. Thus, if we feed the received waveform to a matched filter that is matched to the pulse shape defining the PAM signals, then the matched filter's outputs sampled at integer multiples of the baud period $T_s$ form a sufficient statistic (Theorem 5.8.2). Using this result we can reduce the guessing problem from one with an observation consisting of a continuous-time stochastic process to one with an observation consisting of a discrete-time SP. In fact, since we shall only consider the problem of detecting a finite number of data bits, the reduction will be to a finite number of random variables. This will justify the canonical structure of a PAM receiver where the received continuous-time waveform is fed to a matched filter whose sampled output is then used by the decision circuitry to produce its guess. We shall derive the results first for PAM and then briefly describe their extension to QAM in Section 28.5.

The setup we study is one where $k$ data bits $D_1, \ldots, D_k$ are mapped by an encoder $\varphi \colon \{0,1\}^k \to \mathbb{R}^n$ to the real symbols $X_1, \ldots, X_n$, which are then used to produce the transmitted waveform

$$X(t) = A \sum_{\ell=1}^{n} X_\ell \, g(t - \ell T_s), \quad t \in \mathbb{R}, \tag{28.1}$$

where $A > 0$ is a scaling constant; $T_s > 0$ is the baud period; and $g(\cdot)$ is the pulse shape, which is assumed to be a real integrable signal that is bandlimited to $W$ Hz. The received waveform $(Y(t))$ is given by

$$Y(t) = X(t) + N(t)$$

$$= A \sum_{\ell=1}^{n} X_\ell \, g(t - \ell T_s) + N(t), \quad t \in \mathbb{R}, \tag{28.2}$$

where $(N(t))$ is white Gaussian noise of PSD $N_0/2$ with respect to the bandwidth $W$ and is independent of the data bits $D_1, \ldots, D_k$ and hence also of $(X(t))$. Based on the received waveform $(Y(t))$ we wish to guess the data bits $D_1, \ldots, D_k$. To simplify the typesetting we shall stack the $k$ data bits $D_1, \ldots, D_k$ in a vector

$$\mathbf{D} = (D_1, \ldots, D_k)^\mathsf{T}, \tag{28.3}$$

stack the $n$ symbols $X_1, \ldots, X_n$ in a vector

$$\mathbf{X} = (X_1, \ldots, X_n)^\mathsf{T}, \tag{28.4}$$

and write

$$\mathbf{X} = \varphi(\mathbf{D}). \tag{28.5}$$

We denote the transmitted waveform corresponding to the realization $\mathbf{D} = \mathbf{d}$ by

$$x(t; \mathbf{d}) = A \sum_{\ell=1}^{n} x_\ell \, g(t - \ell T_s), \quad t \in \mathbb{R}, \tag{28.6}$$

where $(x_1, \ldots, x_n)^\mathsf{T} = \varphi(\mathbf{d})$ is the real $n$-vector to which $\mathbf{d}$ is mapped by $\varphi(\cdot)$. Thus, conditional on $\mathbf{D} = \mathbf{d}$,

$$Y(t) = x(t; \mathbf{d}) + N(t), \quad t \in \mathbb{R}. \tag{28.7}$$

## 28.2    Sufficient Statistic and Its Conditional Law

We can view the vector $\mathbf{D} = (D_1, \ldots, D_k)^\mathsf{T}$ as a message and view the $2^k$ different values it can take as the set of messages. To promote this view we define

$$\mathcal{D} \triangleq \{0, 1\}^k \tag{28.8}$$

to be the set of all $2^k$ binary $k$-tuples and view $\mathcal{D}$ as the set of possible messages. While in Chapter 21 on multi-hypothesis testing we always denoted the set of messages by $\mathcal{M}$ and assumed that its elements are the integers $1, \ldots, \mathsf{M}$, we never attached a meaning to the "labels" we associated with the messages. So there is no harm in now labeling the messages by the binary $k$-tuples. Associated with every message $\mathbf{d} \in \mathcal{D}$ is its prior $\pi_\mathbf{d}$

$$\pi_\mathbf{d} = \Pr[\mathbf{D} = \mathbf{d}]$$

$$= \Pr[D_1 = d_1, \ldots, D_k = d_k], \quad \mathbf{d} \in \mathcal{D}. \tag{28.9}$$

If we assume that the data bits are IID random bits (Definition 14.5.1), then $\pi_{\mathbf{d}} = 2^{-k}$ for every $k$-tuple $\mathbf{d} \in \mathcal{D}$, but this assumption is inessential to our derivation of the sufficient statistic. (Recall that sufficiency is defined for a family of conditional distributions; the prior plays no role.)

Conditional on $\mathbf{D} = \mathbf{d}$, the transmitted waveform is given by $x(\cdot; \mathbf{d})$; see (28.6). Thus, the problem of guessing $\mathbf{D}$ is equivalent to guessing which of the $2^k$ signals

$$\left\{ t \mapsto x(t; \mathbf{d}) \right\}_{\mathbf{d} \in \mathcal{D}} \tag{28.10}$$

is being observed in white Gaussian noise of PSD $N_0/2$ with respect to the bandwidth $W$. From (28.6) it follows that for every message $\mathbf{d} \in \mathcal{D}$ the transmitted waveform $t \mapsto x(t; \mathbf{d})$ is a (deterministic) linear combination of the $n$ functions $\{t \mapsto g(t - \ell\mathsf{T_s})\}_{\ell=1}^n$. Moreover, if the pulse shape $g(\cdot)$ is an integrable function that is bandlimited to $W$ Hz, then so is each waveform $t \mapsto x(t; \mathbf{d})$. Consequently, from Corollary 26.4.2 and from (26.23) we obtain:

**Proposition 28.2.1 (Sufficient Statistic for PAM in White Noise).** *Let the conditional law of $\big(Y(t)\big)$ given $\mathbf{D} = \mathbf{d}$ be given by (28.5), (28.6), and (28.7), where the pulse shape $\mathbf{g}$ is a real integrable signal that is bandlimited to $W$ Hz, and where $\big(N(t)\big)$ is white Gaussian noise of PSD $N_0/2$ with respect to the bandwidth $W$. Then the $n$ inner products*

$$T^{(\ell)} = \int_{-\infty}^{\infty} Y(t)\, g(t - \ell\mathsf{T_s})\, \mathrm{d}t, \quad \ell \in \{1, \ldots, n\} \tag{28.11}$$

*form a sufficient statistic for guessing $\mathbf{D}$ based on $\big(Y(t)\big)$.*

*Moreover, conditional on $\mathbf{D} = \mathbf{d}$, the vector $\mathbf{T} = (T^{(1)}, \ldots, T^{(n)})^{\mathsf{T}}$ is a Gaussian $n$-vector whose $\ell$-th component $T^{(\ell)}$ is of conditional mean*

$$\mathsf{E}\!\left[ T^{(\ell)} \,\middle|\, \mathbf{D} = \mathbf{d} \right] = \mathsf{A} \sum_{\ell'=1}^{n} x_{\ell'}\, \mathsf{R_{gg}}\big((\ell - \ell')\mathsf{T_s}\big), \quad \ell \in \{1, \ldots, n\} \tag{28.12}$$

*and whose conditional covariance matrix is*

$$\frac{\mathsf{N_0}}{2} \begin{pmatrix} \mathsf{R_{gg}}(0) & \mathsf{R_{gg}}(\mathsf{T_s}) & \cdots & \mathsf{R_{gg}}\big((n-1)\mathsf{T_s}\big) \\ \mathsf{R_{gg}}(\mathsf{T_s}) & \mathsf{R_{gg}}(0) & \cdots & \mathsf{R_{gg}}\big((n-2)\mathsf{T_s}\big) \\ \cdots & \cdots & \cdots & \cdots \\ \cdots & \cdots & \cdots & \cdots \\ \mathsf{R_{gg}}\big((n-1)\mathsf{T_s}\big) & \mathsf{R_{gg}}\big((n-2)\mathsf{T_s}\big) & \cdots & \mathsf{R_{gg}}(0) \end{pmatrix}, \tag{28.13}$$

*i.e.,*

$$\mathsf{Cov}\!\left[ T^{(\ell')}, T^{(\ell'')} \,\middle|\, \mathbf{D} = \mathbf{d} \right] = \frac{\mathsf{N_0}}{2}\, \mathsf{R_{gg}}\big((\ell' - \ell'')\mathsf{T_s}\big), \quad \ell', \ell'' \in \{1, \ldots, n\}. \tag{28.14}$$

*Here $\mathsf{R_{gg}}$ is the self-similarity function of the real pulse shape $\mathbf{g}$ (Definition 11.2.1), and $(x_1, \ldots, x_n)^{\mathsf{T}} = \varphi(\mathbf{d})$ is the real $n$-tuple to which $\mathbf{d}$ is encoded.*

**Proof.** This follows directly from Corollary 26.4.2 and from (26.23) upon substituting the mapping $t \mapsto g(t - \ell\mathsf{T_s})$ for $\tilde{\mathsf{s}}_j$ and upon computing the inner product

$$\big\langle t \mapsto g(t - \ell\mathsf{T_s}),\, t \mapsto g(t - \ell'\mathsf{T_s}) \big\rangle = \mathsf{R_{gg}}\big((\ell - \ell')\mathsf{T_s}\big), \quad \ell, \ell' \in \mathbb{Z}. \qquad \square$$

## 28.3  Consequences of Sufficiency and Other Optimality Criteria

The sufficiency of the random vector $\mathbf{T} = (T^{(1)}, \ldots, T^{(n)})^{\mathsf{T}}$ and Theorem 26.3.2 guarantee that if our design objective is to minimize the probability of a message error, then there is no loss in optimality in basing our guess on $\mathbf{T}$. We shall next consider other design criteria and show that, for these too, there is no loss in optimality in basing our guess on $\mathbf{T}$.

We first elaborate on what a **message error** is. If we denote our guess by

$$\tilde{\mathbf{d}} = (\tilde{d}_1, \ldots, \tilde{d}_k)^{\mathsf{T}},$$

then a message error occurs if our guess differs from the message $\mathbf{d}$ in at least one component, i.e., if $\tilde{d}_\ell \neq d_\ell$ for some $\ell \in \{1, \ldots, n\}$. The probability of a message error is thus

$$\Pr[\tilde{\mathbf{D}} \neq \mathbf{D}]. \tag{28.15}$$

Designing the receiver to minimize the probability of a message error is reasonable, for example, when the $k$ data bits constitute a computer file, and we wish to minimize the probability that the file is corrupted. In such applications the user is often only interested in knowing whether the file was successfully received (no error occurred) or if the file was corrupted (at least one error occurred). Minimizing the probability of a message error corresponds to minimizing the probability that the file is corrupted.

In other applications, engineers are more interested in the **average probability of a bit error** or **bit error rate** (BER). That is, they may wish to minimize

$$\frac{1}{k} \sum_{j=1}^{k} \Pr[\tilde{D}_j \neq D_j]. \tag{28.16}$$

To better appreciate the difference between the average probability of a bit error (28.16) and the probability of a message error (28.15), define the RV

$$E_j = \mathrm{I}\{\tilde{D}_j \neq D_j\}, \quad j \in \{1, \ldots, k\},$$

which indicates whether the $j$-th bit was incorrectly decoded. Minimizing the probability of a message error minimizes

$$\Pr\left[\sum_{j=1}^{k} E_j > 0\right],$$

whereas minimizing the average probability of a bit error minimizes

$$\frac{1}{k} \mathsf{E}\left[\sum_{j=1}^{k} E_j\right]. \tag{28.17}$$

Thus, minimizing the probability of a message error is equivalent to minimizing the probability that one or more of the data bits is corrupted, whereas minimizing the average probability of a bit error is equivalent to minimizing the expected number of data bits that are decoded erroneously.

We next argue that there is no loss in optimality in basing our guess on $\mathbf{T}$ also when designing to minimize the average probability of a bit error (28.16). We first note that to minimize (28.16) we should choose for each $j \in \{1, \ldots, k\}$ our guess $\tilde{D}_j$ to minimize

$$\Pr\left[\tilde{D}_j \neq D_j\right].$$

That is, we should consider the binary hypothesis testing problem of guessing whether $D_j$ is equal to zero or one, and we should guess $\tilde{D}_j$ to minimize the probability of error associated with this problem. To conclude our argument we next show that for the purpose of minimizing $\Pr\left[\tilde{D}_j \neq D_j\right]$, there is no loss in optimality in basing our decision on $\mathbf{T}$. To show this, it suffices, by the binary version of Theorem 26.3.2, to establish that $\mathbf{T}$ also forms a sufficient statistic for guessing $D_j$ based on $\big(Y(t)\big)$. That is, we need to show that for every $\eta \in \mathbb{N}$ and any choice of the epochs $t_1, \ldots, t_\eta \in \mathbb{R}$, the vector $\mathbf{T}$ forms a sufficient statistic for guessing $D_j$ based on $\big(Y(t_1), \ldots, Y(t_\eta), \mathbf{T}\big)$. This follows from the sufficiency of $\mathbf{T}$ for guessing $\mathbf{D}$ based on $\big(Y(t_1), \ldots, Y(t_\eta), \mathbf{T}\big)$ and from Proposition 22.4.4, which shows that the sufficiency of $\mathbf{T}$ for guessing $\mathbf{D}$ also implies its sufficiency for guessing whether $\mathbf{D}$ is in the set of $k$-tuples whose $j$-th component is zero or in its complement set of $k$-tuples whose $j$-th component is one.

More generally we have:

**Proposition 28.3.1.** *Consider the setup of Proposition 28.2.1. Let $\psi \colon \mathbf{d} \mapsto \psi(\mathbf{d})$ be any function of the data bits, and let $\mathbf{D}$ have an arbitrary prior. Then no guessing rule for guessing $\psi(\mathbf{D})$ based on $\big(Y(t)\big)$ can outperform an optimal rule for guessing $\psi(\mathbf{D})$ based on $T^{(1)}, \ldots, T^{(n)}$.*

**Proof.** Any function from $\{0,1\}^k$ can take on at most $2^k$ different values. Let $q$ denote the number of different values that $\psi(\cdot)$ takes, i.e.,

$$q = \#\left\{\psi(\mathbf{d}) : \mathbf{d} \in \{0,1\}^k\right\},$$

where $\#\mathcal{A}$ denotes the number of elements in the set $\mathcal{A}$. Denote these different values by $\gamma_1, \ldots, \gamma_q$. The $q$ subsets of $\mathcal{D}$

$$\left\{\mathbf{d} \in \{0,1\}^k : \psi(\mathbf{d}) = \gamma_\kappa\right\}, \quad \kappa \in \{1, \ldots, q\}$$

are disjoint sets whose union is $\{0,1\}^k$. That is, they form a partition of $\{0,1\}^k$. Guessing $\psi(\mathbf{D})$ is equivalent to guessing which subset in this partition contains $\mathbf{D}$. For this we know that $(T^{(1)}, \ldots, T^{(n)})$ forms a sufficient statistic because it forms a sufficient statistic for guessing $\mathbf{D}$ and hence, by Note 22.4.5, it also forms a sufficient statistic for guessing which subset in the partition contains $\mathbf{D}$. The result now follows from Theorem 26.3.2. □

The examples we have seen so far correspond to the case where $\boldsymbol{\psi} \colon \mathbf{d} \mapsto \mathbf{d}$ (with the probability of guessing $\psi(\mathbf{D})$ incorrectly corresponding to a message error) and the case $\boldsymbol{\psi} \colon \mathbf{d} \mapsto d_j$ (with the probability of guessing $\psi(\mathbf{D})$ incorrectly corresponding

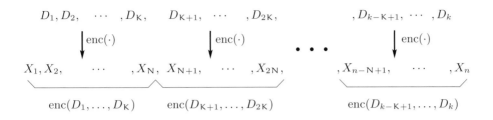

**Figure 28.1:** Block-mode encoding.

to the probability that the $j$-th bit $D_j$ is incorrectly decoded). Another useful example is when $\boldsymbol{\psi}\colon \mathbf{d} \mapsto \left(d_\nu, \ldots, d_{\nu'}\right)$ for some given $\nu, \nu' \in \mathbb{N}$ satisfying $\nu' \geq \nu$. This situation corresponds to the case where $\left(D_\nu, \ldots, D_{\nu'}\right)$ constitutes a **packet** and we are interested in the probability that the packet is erroneously decoded.

Yet another example arises in block-mode transmission—which is described in Section 10.4 and which is depicted in Figure 28.1—where the data bits $D_1, \ldots, D_k$ are mapped to the symbols $X_1, \ldots, X_n$ using a $(\mathsf{K}, \mathsf{N})$ binary-to-reals block encoder

$$\mathbf{enc}\colon \{0,1\}^{\mathsf{K}} \to \mathbb{R}^{\mathsf{N}}.$$

Here we assume that $k$ is divisible by $\mathsf{K}$ and that $n = \mathsf{N}\, k/\mathsf{K}$.

If we wish to guess the $\mathsf{K}$-tuple $\left(D_{(\nu-1)\mathsf{K}+1}, \ldots, D_{(\nu-1)\mathsf{K}+\mathsf{K}}\right)$ with the smallest probability of error, then there is no loss in optimality in basing our guess on $T^{(1)}, \ldots, T^{(n)}$. This follows by applying Proposition 28.3.1 with the function $\psi(\mathbf{d}) = \left(d_{(\nu-1)\mathsf{K}+1}, \ldots, d_{(\nu-1)\mathsf{K}+\mathsf{K}}\right)$.

## 28.4   Consequences of Orthonormality

The conditional distribution of the inner products in (28.11) becomes simpler when the time shifts of the pulse shape by integer multiples of $\mathsf{T_s}$ are orthonormal. In this case we denote the pulse shape by $\phi(\cdot)$ and state the orthonormality condition as

$$\int_{-\infty}^{\infty} \phi(t - \ell \mathsf{T_s})\, \phi(t - \ell' \mathsf{T_s})\, \mathrm{d}t = \mathrm{I}\{\ell = \ell'\}, \quad \ell, \ell' \in \mathbb{Z}, \qquad (28.18)$$

or, equivalently, as

$$\mathsf{R}_{\phi\phi}(\ell \mathsf{T_s}) = \begin{cases} 1 & \text{if } \ell = 0, \\ 0 & \text{if } \ell \neq 0, \end{cases} \quad \ell \in \mathbb{Z}. \qquad (28.19)$$

### 28.4.1   The Conditional Law of the Sufficient Statistic

From Proposition 28.2.1 we obtain a key result on PAM communication in white Gaussian noise:

**Corollary 28.4.1.** *Consider PAM where data bits $D_1, \ldots, D_k$ are mapped by an encoder to the real symbols $X_1, \ldots, X_n$, which are then mapped to the waveform*

$$X(t) = \mathsf{A} \sum_{\ell=1}^{n} X_\ell \, \phi(t - \ell \mathsf{T}_{\mathrm{s}}), \quad t \in \mathbb{R}, \tag{28.20}$$

*where the pulse shape $\phi(\cdot)$ is an integrable signal that is bandlimited to $\mathsf{W}$ Hz and whose time shifts by integer multiples of the baud period $\mathsf{T}_{\mathrm{s}}$ are orthonormal. Let the observed waveform $\bigl(Y(t)\bigr)$ be given by*

$$Y(t) = X(t) + N(t), \quad t \in \mathbb{R},$$

*where $\bigl(N(t), \, t \in \mathbb{R}\bigr)$ is independent of the data bits and is white Gaussian noise of PSD $\mathsf{N}_0/2$ with respect to the bandwidth $\mathsf{W}$.*

*(i) The $n$ inner products*

$$T^{(\ell)} = \int_{-\infty}^{\infty} Y(t) \, \phi(t - \ell \mathsf{T}_{\mathrm{s}}) \, \mathrm{d}t, \quad \ell \in \{1, \ldots, n\} \tag{28.21}$$

*form a sufficient statistic for guessing $(D_1, \ldots, D_k)$ based on $\bigl(Y(t)\bigr)$.*

*(ii) Conditional on $\mathbf{D} = \mathbf{d}$ with corresponding encoder outputs $(X_1, \ldots, X_n) = (x_1, \ldots, x_n)$, the inner products (28.21) are independent with*

$$T^{(\ell)} \sim \mathcal{N}\left(\mathsf{A} x_\ell, \frac{\mathsf{N}_0}{2}\right), \quad \ell \in \{1, \ldots, n\}. \tag{28.22}$$

*(iii) The conditional distribution of these inner products can also be expressed as*

$$T^{(\ell)} = \mathsf{A} x_\ell + Z_\ell, \quad \ell \in \{1, \ldots, n\}, \tag{28.23a}$$

*where*

$$Z_1, \ldots, Z_n \sim \mathrm{IID} \, \mathcal{N}\left(0, \frac{\mathsf{N}_0}{2}\right). \tag{28.23b}$$

From Proposition 28.3.1 we obtain that $T^{(1)}, \ldots, T^{(n)}$ also form a sufficient statistic for guessing the value of any function of the data bits $D_1, \ldots, D_k$.

## 28.4.2 A Further Reduction in the Sufficient Statistic

We next show a further reduction (from $n$ to $\mathsf{N}$ random variables) of the sufficient statistic in block-mode transmission (with the pulse shape $\phi(\cdot)$ still satisfying (28.19)). For this reduction to hold we need to assume that the data bits are independent or that the $k/\mathsf{K}$ tuples

$$\bigl(D_1, \ldots, D_\mathsf{K}\bigr), \bigl(D_{\mathsf{K}+1}, \ldots, D_{2\mathsf{K}}\bigr), \ldots, \bigl(D_{k-\mathsf{K}+1}, \ldots, D_k\bigr) \tag{28.24}$$

are independent.

**Proposition 28.4.2.** *In addition to the assumptions of Corollary 28.4.1, assume that $X_1, \ldots, X_n$ are generated from $D_1, \ldots, D_k$ in block-mode using a $(\mathsf{K}, \mathsf{N})$ binary-to-reals block encoder. Further assume that the $\mathsf{K}$-tuples in (28.24) are independent. Then for every $\nu \in \{1, \ldots, k/\mathsf{K}\}$, the $\mathsf{N}$-tuple*

$$\left( T^{((\nu-1)\mathsf{N}+1)}, \ldots, T^{(\nu\mathsf{N})} \right) \tag{28.25}$$

*forms a sufficient statistic for guessing the $\mathsf{K}$-tuple*

$$\left( D_{(\nu-1)\mathsf{K}+1}, \ldots, D_{\nu\mathsf{K}} \right) \tag{28.26}$$

*or any function thereof.*

**Proof.** Fix some $\nu \in \{1, \ldots, k/\mathsf{K}\}$. For every choice of $\eta \in \mathbb{N}$ and of the epochs $t_1, \ldots, t_\eta \in \mathbb{R}$, the $n$-tuple of matched filter outputs $(T^{(1)}, \ldots, T^{(n)})$ forms a sufficient statistic for guessing $D_1, \ldots, D_k$ based on $\left( Y(t_1), \ldots, Y(t_\eta), \mathbf{T} \right)$ (Proposition 28.2.1). Consequently, by Note 22.4.5, this $n$-tuple is also sufficient for guessing the $\mathsf{K}$-tuple (28.26). We shall next show that the $\mathsf{N}$-tuple (28.25) is sufficient for guessing the $\mathsf{K}$-tuple (28.26) based on the $n$-tuple $(T^{(1)}, \ldots, T^{(n)})$. It will then follow from Proposition 22.4.3 that the $\mathsf{N}$-tuple (28.25) is also sufficient for guessing the $\mathsf{K}$-tuple (28.26) based on $\left( Y(t_1), \ldots, Y(t_\eta), \mathbf{T} \right)$, thus establishing the proposition.

That the $\mathsf{N}$-tuple (28.25) is sufficient for guessing the $\mathsf{K}$-tuple (28.26) based on the $n$-tuple $(T^{(1)}, \ldots, T^{(n)})$ is equivalent to the irrelevancy of

$$\mathbf{R} \triangleq \left( \left( T^{(1)}, \ldots, T^{(\mathsf{N})} \right), \ldots, \left( T^{((\nu-2)\mathsf{N}+1)}, \ldots, T^{((\nu-1)\mathsf{N})} \right), \right.$$
$$\left. \left( T^{(\nu\mathsf{N}+1)}, \ldots, T^{((\nu+1)\mathsf{N})} \right), \ldots, \left( T^{(n-\mathsf{N}+1)}, \ldots, T^{(n)} \right) \right)^{\mathsf{T}}$$

for guessing the $\mathsf{K}$-tuple (28.26) based on the $\mathsf{N}$-tuple (28.25). To prove this irrelevancy, it suffices to prove two claims: that $\mathbf{R}$ is independent of the $\mathsf{K}$-tuple (28.26) and that, conditionally on this $\mathsf{K}$-tuple, $\mathbf{R}$ is independent of the $\mathsf{N}$-tuple (28.25) (Proposition 22.5.5). These claims follow from three observations: that, by the orthonormaility assumption (28.19), $\mathbf{R}$ is determined by the data bits

$$D_1, \ldots, D_{(\nu-1)\mathsf{K}}, D_{\nu\mathsf{K}+1}, \ldots, D_k \tag{28.27}$$

and by the random variables

$$Z_1, \ldots, Z_{(\nu-1)\mathsf{N}}, Z_{\nu\mathsf{N}+1}, \ldots, Z_n; \tag{28.28}$$

that the $\mathsf{N}$-tuple (28.25) is determined by the $\mathsf{K}$-tuple (28.26) and by the random variables

$$Z_{(\nu-1)\mathsf{N}+1}, \ldots, Z_{\nu\mathsf{N}}; \tag{28.29}$$

and that the tuples in (28.26), (28.27), (28.28), and (28.29) are independent.

Having established that the $\mathsf{N}$-tuple (28.25) forms a sufficient statistic for guessing the $\mathsf{K}$-tuple (28.26), it now follows, using arguments very similar to those employed in proving Proposition 28.3.1, that the $\mathsf{N}$-tuple (28.25) is also sufficient for guessing the value of any function $\psi(\cdot)$ of the $\mathsf{K}$-tuple (28.26). $\qquad\square$

### 28.4.3 The Discrete-Time Single-Block Model

Proposition 28.4.2 is the starting point of much of the literature on block codes, upon which we shall touch in Chapter 29. In Coding Theory $\mathsf{N}$ is usually called the **blocklength**, and $\mathsf{K}/\mathsf{N}$ is called the **rate** in bits per dimension. Coding theorists envision that the function $\mathrm{enc}(\cdot)$ is used to map $k$ bits to $n$ real numbers using the block-encoding rule of Figure 10.1 (with $k$ being divisible by $\mathsf{K}$) and that the resulting real symbols are then transmitted over a white Gaussian noise channel using PAM with a pulse shape satisfying the orthogonality condition (28.19). Assuming that the data tuples are independent, and by then resorting to Proposition 28.4.2, coding theorists focus on the problem of decoding the $\mathsf{K}$-tuple (28.26) from the $\mathsf{N}$ matched filter outputs (28.25).

In this problem the index $\nu$ of the block is immaterial, and coding theorists re-label the data bits of the $\mathsf{K}$ tuple (28.26) as $D_1, \ldots, D_{\mathsf{K}}$; they re-label the symbols to which they are mapped as $X_1, \ldots, X_{\mathsf{N}}$; and they re-label the corresponding observations as $Y_1, \ldots, Y_{\mathsf{N}}$. The resulting model is the **discrete-time single-block model** where

$$\big(X_1, \ldots, X_{\mathsf{N}}\big) = \mathrm{enc}\big(D_1, \ldots, D_{\mathsf{K}}\big), \tag{28.30a}$$

$$Y_\eta = \mathsf{A} X_\eta + Z_\eta, \quad \eta \in \{1, \ldots, \mathsf{N}\}, \tag{28.30b}$$

$$Z_\eta \sim \mathcal{N}\left(0, \frac{\mathsf{N}_0}{2}\right), \quad \eta \in \{1, \ldots, \mathsf{N}\}, \tag{28.30c}$$

where $Z_1, \ldots, Z_{\mathsf{N}}$ are IID and independent of $D_1, \ldots, D_{\mathsf{K}}$. We recall that this model is appropriate when the pulse shape $\phi$ satisfies the orthonormality condition (28.18); the data bits are "block IID" in the sense that the $k/\mathsf{K}$ tuples in (28.24) are independent; and the additive noise is white Gaussian noise of double-sided spectral density $\mathsf{N}_0/2$ with respect to the bandwidth occupied by the pulse shape $\phi$. It is customary to additionally assume that $D_1, \ldots, D_{\mathsf{K}}$ are IID random bits (Definition 14.5.1). This is a good assumption if, prior to transmission, the data bits are compressed using an efficient data compression algorithm.

## 28.5 Extension to QAM Communications

### 28.5.1 Introduction and Setup

We next extend our discussion to the detection of QAM signals. We assume that an encoding function

$$\varphi \colon \{0, 1\}^k \to \mathbb{C}^n$$

is used to map the $k$ data bits $\mathbf{D} = (D_1, \ldots, D_k)^{\mathsf{T}}$ to the $n$ complex symbols $\mathbf{C} = (C_1, \ldots, C_n)^{\mathsf{T}}$ and that the resulting complex symbols are then mapped to the passband signal $\big(X_{\mathrm{PB}}(t)\big)$, which is given by

$$X_{\mathrm{PB}}(t) = 2\,\mathrm{Re}\big(X_{\mathrm{BB}}(t)\,e^{\mathrm{i}2\pi f_c t}\big), \quad t \in \mathbb{R},$$

where

$$X_{\mathrm{BB}}(t) = A \sum_{\ell=1}^{n} C_\ell\, g(t - \ell T_\mathrm{s}), \quad t \in \mathbb{R};$$

the pulse shape $g(\cdot)$ is a complex integrable signal that is bandlimited to $W/2$ Hz; $A > 0$ is a real constant; and $f_\mathrm{c} > W/2$. Conditionally on $\mathbf{D} = \mathbf{d}$, we denote the transmitted signal by

$$x(t; \mathbf{d}) = 2A \operatorname{Re}\left( \sum_{\ell=1}^{n} c_\ell\, g(t - \ell T_\mathrm{s})\, e^{\mathrm{i}2\pi f_\mathrm{c} t} \right) \tag{28.31}$$

$$= \sqrt{2}A \sum_{\ell=1}^{n} \operatorname{Re}(c_\ell) \overbrace{2 \operatorname{Re}\left( \underbrace{\frac{1}{\sqrt{2}}\, g(t - \ell T_\mathrm{s})\, e^{\mathrm{i}2\pi f_\mathrm{c} t}}_{g_{\mathrm{I},\ell,\mathrm{BB}}(t)} \right)}^{g_{\mathrm{I},\ell}(t)}$$

$$+ \sqrt{2}A \sum_{\ell=1}^{n} \operatorname{Im}(c_\ell) \overbrace{2 \operatorname{Re}\left( \underbrace{\mathrm{i}\frac{1}{\sqrt{2}}\, g(t - \ell T_\mathrm{s})\, e^{\mathrm{i}2\pi f_\mathrm{c} t}}_{g_{\mathrm{Q},\ell,\mathrm{BB}}(t)} \right)}^{g_{\mathrm{Q},\ell}(t)}, \quad t \in \mathbb{R}, \tag{28.32}$$

where $\mathbf{c} = \varphi(\mathbf{d})$ is the result of encoding the data bits $\mathbf{d}$; where (28.32) follows from (16.7); and where $\{g_{\mathrm{I},\ell}\}$, $\{g_{\mathrm{Q},\ell}\}$, $\{g_{\mathrm{I},\ell,\mathrm{BB}}\}$, $\{g_{\mathrm{Q},\ell,\mathrm{BB}}\}$ are as indicated in (28.32) and as defined in (16.8) and (16.9).

We consider the case where, conditional on $\mathbf{D} = \mathbf{d}$, the received waveform $\big(Y(t)\big)$ is given by

$$Y(t) = x(t; \mathbf{d}) + N(t), \quad t \in \mathbb{R}, \tag{28.33}$$

where $\big(N(t)\big)$ is white Gaussian noise of PSD $N_0/2$ with respect to the bandwidth $W$ around the carrier frequency $f_\mathrm{c}$ (Definition 25.15.3).

### 28.5.2 Real Sufficient Statistics

The representation (28.32) makes it clear that for every $\mathbf{d} \in \{0,1\}^k$ the signal $t \mapsto x(t; \mathbf{d})$ can be expressed as a linear combination of the $2n$ real-valued signals

$$\{g_{\mathrm{I},\ell}\}_{\ell=1}^{n}, \quad \{g_{\mathrm{Q},\ell}\}_{\ell=1}^{n}. \tag{28.34}$$

Since these signals are integrable signals that are bandlimited to $W$ Hz around the carrier frequency $f_\mathrm{c}$, it follows that the $2n$ inner products

$$T_{\mathrm{I}}^{(\ell)} = \int_{-\infty}^{\infty} Y(t)\, g_{\mathrm{I},\ell}(t)\, \mathrm{d}t, \quad \ell \in \{1, \ldots, n\}, \tag{28.35a}$$

$$T_{\mathrm{Q}}^{(\ell)} = \int_{-\infty}^{\infty} Y(t)\, g_{\mathrm{Q},\ell}(t)\, \mathrm{d}t, \quad \ell \in \{1, \ldots, n\} \tag{28.35b}$$

form a real sufficient statistic for guessing $\mathbf{D}$ based on $\big(Y(t)\big)$ (Section 26.10).

To describe the distribution of the sufficient statistic conditional on each of the hypotheses, we next express the inner products between the functions in (28.34) in terms of the self-similarity function $\mathsf{R_{gg}}$ of the complex pulse shape $\mathbf{g}$

$$\mathsf{R_{gg}}(\tau) = \int_{-\infty}^{\infty} g(t+\tau)\, g^*(t)\, \mathrm{d}t, \quad \tau \in \mathbb{R} \tag{28.36}$$

(Definition 11.2.1). Key to these calculations is the relationship between the inner product between real passband signals and the inner product between their complex baseband representations (Theorem 7.6.10). Thus,

$$
\begin{aligned}
\langle \mathsf{g}_{\mathrm{I},\ell'}, \mathsf{g}_{\mathrm{I},\ell} \rangle &= 2\,\mathrm{Re}\big(\langle \mathsf{g}_{\mathrm{I},\ell',\mathrm{BB}}, \mathsf{g}_{\mathrm{I},\ell,\mathrm{BB}} \rangle\big) \\
&= \mathrm{Re}\big(\langle t \mapsto g(t-\ell'\mathsf{T_s}), t \mapsto g(t-\ell\mathsf{T_s}) \rangle\big) \\
&= \mathrm{Re}\left(\int_{-\infty}^{\infty} g(t-\ell'\mathsf{T_s})\, g^*(t-\ell\mathsf{T_s})\, \mathrm{d}t\right) \\
&= \mathrm{Re}\big(\mathsf{R_{gg}}((\ell-\ell')\mathsf{T_s})\big), \quad \ell,\ell' \in \mathbb{Z},
\end{aligned} \tag{28.37a}
$$

where the first equality follows by relating the inner product in passband to the inner product in baseband; the second from the expressions for the corresponding baseband representations (16.9a); the third from the definition of the inner product for complex-valued signals (3.4); and the final equality from the definition of the self-similarity function (28.36). Similarly,

$$
\begin{aligned}
\langle \mathsf{g}_{\mathrm{Q},\ell'}, \mathsf{g}_{\mathrm{Q},\ell} \rangle &= 2\,\mathrm{Re}\big(\langle \mathsf{g}_{\mathrm{Q},\ell',\mathrm{BB}}, \mathsf{g}_{\mathrm{Q},\ell,\mathrm{BB}} \rangle\big) \\
&= \mathrm{Re}\big(\langle t \mapsto \mathrm{i}g(t-\ell'\mathsf{T_s}), t \mapsto \mathrm{i}g(t-\ell\mathsf{T_s}) \rangle\big) \\
&= \mathrm{Re}\left(\int_{-\infty}^{\infty} \mathrm{i}g(t-\ell'\mathsf{T_s})\,(-\mathrm{i})\, g^*(t-\ell\mathsf{T_s})\, \mathrm{d}t\right) \\
&= \mathrm{Re}\left(\int_{-\infty}^{\infty} g(t-\ell'\mathsf{T_s})\, g^*(t-\ell\mathsf{T_s})\, \mathrm{d}t\right) \\
&= \mathrm{Re}\big(\mathsf{R_{gg}}((\ell-\ell')\mathsf{T_s})\big), \quad \ell,\ell' \in \mathbb{Z},
\end{aligned} \tag{28.37b}
$$

and

$$
\begin{aligned}
\langle \mathsf{g}_{\mathrm{Q},\ell'}, \mathsf{g}_{\mathrm{I},\ell} \rangle &= 2\,\mathrm{Re}\big(\langle \mathsf{g}_{\mathrm{Q},\ell',\mathrm{BB}}, \mathsf{g}_{\mathrm{I},\ell,\mathrm{BB}} \rangle\big) \\
&= \mathrm{Re}\big(\langle t \mapsto \mathrm{i}g(t-\ell'\mathsf{T_s}), t \mapsto g(t-\ell\mathsf{T_s}) \rangle\big) \\
&= \mathrm{Re}\left(\mathrm{i}\int_{-\infty}^{\infty} g(t-\ell'\mathsf{T_s})\, g^*(t-\ell\mathsf{T_s})\, \mathrm{d}t\right) \\
&= -\mathrm{Im}\left(\int_{-\infty}^{\infty} g(t-\ell'\mathsf{T_s})\, g^*(t-\ell\mathsf{T_s})\, \mathrm{d}t\right) \\
&= -\mathrm{Im}\big(\mathsf{R_{gg}}((\ell-\ell')\mathsf{T_s})\big), \quad \ell,\ell' \in \mathbb{Z},
\end{aligned} \tag{28.37c}
$$

where the first equality leading to (28.37c) follows from the relationship between the inner product between real passband signals and the inner product between their baseband representations (Theorem 7.6.10); the second from the expressions for the corresponding baseband representations (16.9); the third from the definition of the inner product between complex signals (3.4); the fourth from the identity $\mathrm{Re}(\mathrm{i}z) = -\mathrm{Im}(z)$; and where the last equality follows from the definition of the self-similarity function of complex signals (28.36).

We are now ready to compute the conditional law of the sufficient statistic given each of the hypotheses. Conditional on $\mathbf{D} = \mathbf{d}$ with corresponding $\mathbf{c} = \varphi(\mathbf{d})$, the $2n$ random variables $\{T_\mathrm{I}^{(\ell)}, T_\mathrm{Q}^{(\ell)}\}_{\ell=1}^n$ are jointly Gaussian (Section 26.10). Their conditional law is thus fully specified by the conditional mean vector and by the conditional covariance matrix. We begin with the computation of the former:

$$
\begin{aligned}
\mathsf{E}&\left[T_\mathrm{I}^{(\ell)} \,\middle|\, \mathbf{D} = \mathbf{d}\right] \\
&= \left\langle t \mapsto x(\mathbf{d}; t), \mathbf{g}_{\mathrm{I},\ell} \right\rangle \\
&= \left\langle \sqrt{2}\mathsf{A} \sum_{\ell'=1}^n \mathrm{Re}(c_{\ell'})\, \mathbf{g}_{\mathrm{I},\ell'} + \sqrt{2}\mathsf{A} \sum_{\ell'=1}^n \mathrm{Im}(c_{\ell'})\, \mathbf{g}_{\mathrm{Q},\ell'}, \mathbf{g}_{\mathrm{I},\ell} \right\rangle \\
&= \sqrt{2}\mathsf{A} \sum_{\ell'=1}^n \Big(\mathrm{Re}(c_{\ell'})\, \langle \mathbf{g}_{\mathrm{I},\ell'}, \mathbf{g}_{\mathrm{I},\ell} \rangle + \mathrm{Im}(c_{\ell'})\, \langle \mathbf{g}_{\mathrm{Q},\ell'}, \mathbf{g}_{\mathrm{I},\ell} \rangle\Big) \\
&= \sqrt{2}\mathsf{A} \sum_{\ell'=1}^n \Big(\mathrm{Re}(c_{\ell'})\, \mathrm{Re}\big(\mathsf{R}_{\mathbf{gg}}((\ell - \ell')\mathsf{T}_\mathrm{s})\big) - \mathrm{Im}(c_{\ell'})\, \mathrm{Im}\big(\mathsf{R}_{\mathbf{gg}}((\ell - \ell')\mathsf{T}_\mathrm{s})\big)\Big) \\
&= \sqrt{2}\mathsf{A} \sum_{\ell'=1}^n \mathrm{Re}\Big(c_{\ell'}\, \mathsf{R}_{\mathbf{gg}}((\ell - \ell')\mathsf{T}_\mathrm{s})\Big) \\
&= \sqrt{2}\mathsf{A}\, \mathrm{Re}\left(\sum_{\ell'=1}^n c_{\ell'}\, \mathsf{R}_{\mathbf{gg}}((\ell - \ell')\mathsf{T}_\mathrm{s})\right),
\end{aligned}
\tag{28.38a}
$$

where the first equality follows from the definition of $T_\mathrm{I}^{(\ell)}$ (28.35a), from (28.33), and from our assumption that the noise $(N(t))$ is of zero mean; the second from (28.32); the third from the linearity of the inner product; the fourth by expressing the inner products using the self-similarity function, i.e., using (28.37a) and (28.37c); the fifth by the complex-numbers identity $\mathrm{Re}(wz) = \mathrm{Re}(w)\,\mathrm{Re}(z) - \mathrm{Im}(w)\,\mathrm{Im}(z)$; and the final equality because the sum of the real parts is the real part of the sum. Similarly,

$$
\begin{aligned}
\mathsf{E}&\left[T_\mathrm{Q}^{(\ell)} \,\middle|\, \mathbf{D} = \mathbf{d}\right] \\
&= \left\langle t \mapsto x(\mathbf{d}; t), \mathbf{g}_{\mathrm{Q},\ell} \right\rangle \\
&= \left\langle \sqrt{2}\mathsf{A} \sum_{\ell'=1}^n \mathrm{Re}(c_{\ell'})\, \mathbf{g}_{\mathrm{I},\ell'} + \sqrt{2}\mathsf{A} \sum_{\ell'=1}^n \mathrm{Im}(c_{\ell'})\, \mathbf{g}_{\mathrm{Q},\ell'}, \mathbf{g}_{\mathrm{Q},\ell} \right\rangle \\
&= \sqrt{2}\mathsf{A} \sum_{\ell'=1}^n \Big(\mathrm{Re}(c_{\ell'})\, \langle \mathbf{g}_{\mathrm{I},\ell'}, \mathbf{g}_{\mathrm{Q},\ell} \rangle + \mathrm{Im}(c_{\ell'})\, \langle \mathbf{g}_{\mathrm{Q},\ell'}, \mathbf{g}_{\mathrm{Q},\ell} \rangle\Big) \\
&= \sqrt{2}\mathsf{A} \sum_{\ell'=1}^n \left(\mathrm{Re}(c_{\ell'})\Big(-\mathrm{Im}\big(\mathsf{R}_{\mathbf{gg}}((\ell' - \ell)\mathsf{T}_\mathrm{s})\big)\Big) + \mathrm{Im}(c_{\ell'})\, \mathrm{Re}\big(\mathsf{R}_{\mathbf{gg}}((\ell - \ell')\mathsf{T}_\mathrm{s})\big)\right) \\
&= \sqrt{2}\mathsf{A} \sum_{\ell'=1}^n \left(\mathrm{Re}(c_{\ell'})\, \mathrm{Im}\big(\mathsf{R}_{\mathbf{gg}}((\ell - \ell')\mathsf{T}_\mathrm{s})\big) + \mathrm{Im}(c_{\ell'})\, \mathrm{Re}\big(\mathsf{R}_{\mathbf{gg}}((\ell - \ell')\mathsf{T}_\mathrm{s})\big)\right)
\end{aligned}
$$

$$= \sqrt{2}A \sum_{\ell'=1}^{n} \text{Im}\Big( c_{\ell'} \, \mathsf{R}_{\mathbf{gg}}\big((\ell - \ell')\mathsf{T}_{\mathrm{s}}\big)\Big)$$

$$= \sqrt{2}A \, \text{Im}\Big( \sum_{\ell'=1}^{n} c_{\ell'} \, \mathsf{R}_{\mathbf{gg}}\big((\ell - \ell')\mathsf{T}_{\mathrm{s}}\big)\Big), \tag{28.38b}$$

where the first equality follows from the definition of $T_{\mathrm{Q}}^{(\ell)}$ (28.35b), from (28.33), and from our assumption that the noise $(N(t))$ is of zero mean; the second from (28.32); the third from the linearity of the inner product; the fourth by expressing the inner products using the self-similarity function, i.e., using (28.37c) and (28.37b); the fifth by the conjugate symmetry of the self-similarity function (Proposition 11.2.2); the sixth by the complex-numbers identity $\text{Im}(wz) = \text{Re}(w)\,\text{Im}(z) + \text{Im}(w)\,\text{Re}(z)$; and the final equality by noting that the sum of the imaginary parts is equal to the imaginary part of the sum.

The conditional covariances are easily computed using Note 25.15.4. Using the inner products expressions (28.37), we obtain:

$$\text{Cov}\Big[T_{\mathrm{I}}^{(\ell')}, T_{\mathrm{I}}^{(\ell'')} \,\Big|\, \mathbf{D} = \mathbf{d}\Big] = \frac{\mathsf{N}_0}{2}\big\langle \mathbf{g}_{\mathrm{I},\ell'}, \mathbf{g}_{\mathrm{I},\ell''}\big\rangle$$

$$= \frac{\mathsf{N}_0}{2}\,\text{Re}\Big(\mathsf{R}_{\mathbf{gg}}\big((\ell' - \ell'')\mathsf{T}_{\mathrm{s}}\big)\Big), \tag{28.39a}$$

$$\text{Cov}\Big[T_{\mathrm{Q}}^{(\ell')}, T_{\mathrm{Q}}^{(\ell'')} \,\Big|\, \mathbf{D} = \mathbf{d}\Big] = \frac{\mathsf{N}_0}{2}\big\langle \mathbf{g}_{\mathrm{Q},\ell'}, \mathbf{g}_{\mathrm{Q},\ell''}\big\rangle$$

$$= \frac{\mathsf{N}_0}{2}\,\text{Re}\Big(\mathsf{R}_{\mathbf{gg}}\big((\ell' - \ell'')\mathsf{T}_{\mathrm{s}}\big)\Big), \tag{28.39b}$$

and

$$\text{Cov}\Big[T_{\mathrm{I}}^{(\ell')}, T_{\mathrm{Q}}^{(\ell'')} \,\Big|\, \mathbf{D} = \mathbf{d}\Big] = \frac{\mathsf{N}_0}{2}\big\langle \mathbf{g}_{\mathrm{I},\ell'}, \mathbf{g}_{\mathrm{Q},\ell''}\big\rangle$$

$$= -\frac{\mathsf{N}_0}{2}\,\text{Im}\Big(\mathsf{R}_{\mathbf{gg}}\big((\ell' - \ell'')\mathsf{T}_{\mathrm{s}}\big)\Big). \tag{28.39c}$$

We summarize our results on QAM detection in white Gaussian noise as follows.

**Proposition 28.5.1 (QAM Detection in White Noise: Real Sufficient Statistics).**
*Let a QAM signal (28.32) of an integrable pulse shape $g(\cdot)$ that is bandlimited to $\mathsf{W}/2$ Hz be observed in white Gaussian noise of PSD $\mathsf{N}_0/2$ with respect to the bandwidth $\mathsf{W}$ around the carrier frequency $f_{\mathrm{c}}$. Then:*

*(i) The $2n$ inner products*

$$T_{\mathrm{I}}^{(\ell)} = \int_{-\infty}^{\infty} Y(t)\, g_{\mathrm{I},\ell}(t)\,\mathrm{d}t, \quad \ell \in \{1,\dots,n\}, \tag{28.40a}$$

$$T_{\mathrm{Q}}^{(\ell)} = \int_{-\infty}^{\infty} Y(t)\, g_{\mathrm{Q},\ell}(t)\,\mathrm{d}t, \quad \ell \in \{1,\dots,n\} \tag{28.40b}$$

*form a sufficient statistic for guessing* $\mathbf{D}$ *based on* $(Y(t))$, *where*

$$g_{\mathrm{I},\ell}(t) = 2\,\mathrm{Re}\Big(\frac{1}{\sqrt{2}}\,g(t - \ell\mathsf{T_s})\,e^{\mathrm{i}2\pi f_c}\Big), \quad t \in \mathbb{R},$$

$$g_{\mathrm{Q},\ell}(t) = 2\,\mathrm{Re}\Big(\frac{1}{\sqrt{2}}\mathrm{i}\,g(t - \ell\mathsf{T_s})\,e^{\mathrm{i}2\pi f_c}\Big), \quad t \in \mathbb{R}.$$

*(ii) Conditional on* $\mathbf{D} = \mathbf{d}$ *with corresponding transmitted symbols* $\mathbf{c} = \varphi(\mathbf{d})$, *these $2n$ real random variables are jointly Gaussian with conditional means as specified by (28.38) and with conditional covariances as specified by (28.39).*

## 28.5.3 Complex Sufficient Statistics

The notation is simpler if we introduce the $n$ *complex* random variables

$$T^{(\ell)} \triangleq T_{\mathrm{I}}^{(\ell)} + \mathrm{i}\,T_{\mathrm{Q}}^{(\ell)}$$

$$= \int_{-\infty}^{\infty} Y(t)\,g_{\mathrm{I},\ell}(t)\,\mathrm{d}t + \mathrm{i}\int_{-\infty}^{\infty} Y(t)\,g_{\mathrm{Q},\ell}(t)\,\mathrm{d}t, \quad \ell \in \{1,\ldots,n\}. \qquad (28.41)$$

These $n$ complex random variables form a sufficient statistic in the sense that their real and imaginary parts form a sufficient statistic. Using (28.38) we obtain

$$\mathsf{E}\Big[T^{(\ell)}\,\Big|\,\mathbf{D} = \mathbf{d}\Big]$$

$$= \mathsf{E}\Big[T_{\mathrm{I}}^{(\ell)}\,\Big|\,\mathbf{D} = \mathbf{d}\Big] + \mathrm{i}\,\mathsf{E}\Big[T_{\mathrm{Q}}^{(\ell)}\,\Big|\,\mathbf{D} = \mathbf{d}\Big]$$

$$= \sqrt{2}A\,\mathrm{Re}\Big(\sum_{\ell'=1}^{n} c_{\ell'}\,\mathsf{R_{gg}}\big((\ell - \ell')\mathsf{T_s}\big)\Big) + \mathrm{i}\sqrt{2}A\,\mathrm{Im}\Big(\sum_{\ell'=1}^{n} c_{\ell'}\,\mathsf{R_{gg}}\big((\ell - \ell')\mathsf{T_s}\big)\Big)$$

$$= \sqrt{2}A\sum_{\ell'=1}^{n} c_{\ell'}\,\mathsf{R_{gg}}\big((\ell - \ell')\mathsf{T_s}\big), \quad \ell \in \{1,\ldots,n\}. \qquad (28.42)$$

The advantage of the complex notation is that—as we shall see in Proposition 28.5.2 ahead—conditional on $\mathbf{D} = \mathbf{d}$, the random vector $\mathbf{T} - \mathsf{E}[\mathbf{T}|\mathbf{D} = \mathbf{d}]$ is proper (Definition 17.4.1). And since conditionally on $\mathbf{D} = \mathbf{d}$ it is also Gaussian, it follows from Proposition 24.3.11 that, conditional on $\mathbf{D} = \mathbf{d}$, the random vector $\mathbf{T} - \mathsf{E}[\mathbf{T}|\mathbf{D} = \mathbf{d}]$ is a circularly-symmetric complex Gaussian (Definition 24.3.2). Its conditional law is thus determined by its conditional covariance matrix (Corollary 24.3.8). This covariance matrix is an $n \times n$ (complex) matrix, whereas the covariance matrix for the $2n$ real variables in Proposition 28.5.1 is a $(2n) \times (2n)$ (real) matrix.

We summarize our results for QAM detection with complex sufficient statistics in the following.

**Proposition 28.5.2 (QAM in White Noise: Complex Sufficient Statistics).** *Consider the setup of Proposition 28.5.1.*

*(i) The complex random vector* $\mathbf{T} = (T^{(1)}, \ldots, T^{(n)})^{\mathsf{T}}$ *defined by*

$$T^{(\ell)} = \int_{-\infty}^{\infty} Y(t)\,g_{\mathrm{I},\ell}(t)\,\mathrm{d}t + \mathrm{i}\int_{-\infty}^{\infty} Y(t)\,g_{\mathrm{Q},\ell}(t)\,\mathrm{d}t, \quad \ell \in \{1,\ldots,n\},$$

*forms a sufficient statistic for guessing* $\mathbf{D}$ *based on* $(Y(t))$.

(ii) *The $\ell$-th component of* $\mathbf{T}$ *can be expressed as*

$$T^{(\ell)} = \sqrt{2}A \sum_{\ell'=1}^{n} C_{\ell'} \, \mathsf{R}_{\mathbf{gg}} \big( (\ell - \ell') \mathsf{T}_{\mathsf{s}} \big) + Z^{(\ell)}, \quad \ell \in \{1, \ldots, n\},$$

*where* $\mathsf{R}_{\mathbf{gg}}$ *is the self-similarity function of the pulse shape* $g(\cdot)$ *(28.36), and where the random vector* $\mathbf{Z} = (Z^{(1)}, \ldots, Z^{(n)})^{\mathsf{T}}$ *is independent of* $\mathbf{D}$ *and is a circularly-symmetric complex Gaussian of covariance*

$$\mathsf{Cov}\Big[ Z^{(\ell')}, Z^{(\ell'')} \Big] = \mathsf{E}\Big[ Z^{(\ell')} \big( Z^{(\ell'')} \big)^* \Big]$$

$$= \mathsf{N}_0 \, \mathsf{R}_{\mathbf{gg}} \big( (\ell' - \ell'') \mathsf{T}_{\mathsf{s}} \big), \quad \ell', \ell'' \in \{1, \ldots, n\}. \tag{28.43}$$

(iii) *If the time shifts of the pulse shape by integer multiples of* $\mathsf{T}_{\mathsf{s}}$ *are orthonormal, then*

$$T^{(\ell)} = \sqrt{2}AC_\ell + Z^{(\ell)}, \quad \ell \in \{1, \ldots, n, \}, \tag{28.44}$$

*where the complex random variables* $\{Z^{(\ell)}\}$ *are independent of* $\{D_j\}$ *and are IID circularly-symmetric complex Gaussians of variance* $\mathsf{N}_0$.

**Proof.** Part (i) follows directly from Proposition 28.5.1 because, by definition, the sufficiency of $\mathbf{T}$ is equivalent to the sufficiency of its real and imaginary parts.

To prove Part (ii) define

$$Z^{(\ell)} \triangleq T^{(\ell)} - \sqrt{2}A \sum_{\ell'=1}^{n} C_{\ell'} \, \mathsf{R}_{\mathbf{gg}} \big( (\ell - \ell') \mathsf{T}_{\mathsf{s}} \big), \quad \ell' \in \{1, \ldots, n\}, \tag{28.45}$$

and note that by (28.42) the conditional distribution of $\mathbf{Z}$ given $\mathbf{D} = \mathbf{d}$ is of zero mean. Moreover, from Proposition 28.5.1 and from the definition of a complex Gaussian random vector as one whose real and imaginary parts are jointly Gaussian (Definition 24.3.6), it follows that, conditional on $\mathbf{D} = \mathbf{d}$, the vector $\mathbf{Z}$ is Gaussian. To prove that it is proper we compute

$$\mathsf{E}\Big[ Z^{(\ell')} Z^{(\ell'')} \,\Big|\, \mathbf{D} = \mathbf{d} \Big]$$

$$= \mathsf{E}\Big[ \mathrm{Re}\big( Z^{(\ell')} \big) \, \mathrm{Re}\big( Z^{(\ell'')} \big) - \mathrm{Im}\big( Z^{(\ell')} \big) \, \mathrm{Im}\big( Z^{(\ell'')} \big) \,\Big|\, \mathbf{D} = \mathbf{d} \Big]$$

$$\quad + i\, \mathsf{E}\Big[ \mathrm{Re}\big( Z^{(\ell')} \big) \, \mathrm{Im}\big( Z^{(\ell'')} \big) + \mathrm{Im}\big( Z^{(\ell')} \big) \, \mathrm{Re}\big( Z^{(\ell'')} \big) \,\Big|\, \mathbf{D} = \mathbf{d} \Big]$$

$$= \mathsf{Cov}\Big[ T_{\mathrm{I}}^{(\ell')}, T_{\mathrm{I}}^{(\ell'')} \,\Big|\, \mathbf{D} = \mathbf{d} \Big] - \mathsf{Cov}\Big[ T_{\mathrm{Q}}^{(\ell')}, T_{\mathrm{Q}}^{(\ell'')} \,\Big|\, \mathbf{D} = \mathbf{d} \Big]$$

$$\quad + i\, \Big( \mathsf{Cov}\Big[ T_{\mathrm{I}}^{(\ell')}, T_{\mathrm{Q}}^{(\ell'')} \,\Big|\, \mathbf{D} = \mathbf{d} \Big] + \mathsf{Cov}\Big[ T_{\mathrm{Q}}^{(\ell')}, T_{\mathrm{I}}^{(\ell'')} \,\Big|\, \mathbf{D} = \mathbf{d} \Big] \Big)$$

$$= 0, \quad \ell', \ell'' \in \{1, \ldots, n\},$$

where the second equality follows from (28.45) and the last equality from (28.39).

The calculation of the conditional covariance matrix is very similar except that $Z^{(\ell'')}$ is now conjugated:

$$
\begin{aligned}
\mathsf{Cov}&\Big[ Z^{(\ell')}, Z^{(\ell'')}\,\Big|\, \mathbf{D} = \mathbf{d}\Big] \\
&= \mathsf{E}\Big[ \mathrm{Re}\big(Z^{(\ell')}\big)\,\mathrm{Re}\big(Z^{(\ell'')}\big) + \mathrm{Im}\big(Z^{(\ell')}\big)\,\mathrm{Im}\big(Z^{(\ell'')}\big)\,\Big|\, \mathbf{D} = \mathbf{d}\Big] \\
&\quad + i\,\mathsf{E}\Big[ -\mathrm{Re}\big(Z^{(\ell')}\big)\,\mathrm{Im}\big(Z^{(\ell'')}\big) + \mathrm{Im}\big(Z^{(\ell')}\big)\,\mathrm{Re}\big(Z^{(\ell'')}\big)\,\Big|\, \mathbf{D} = \mathbf{d}\Big] \\
&= \mathsf{Cov}\Big[ T_{\mathrm{I}}^{(\ell')}, T_{\mathrm{I}}^{(\ell'')}\,\Big|\, \mathbf{D} = \mathbf{d}\Big] + \mathsf{Cov}\Big[ T_{\mathrm{Q}}^{(\ell')}, T_{\mathrm{Q}}^{(\ell'')}\,\Big|\, \mathbf{D} = \mathbf{d}\Big] \\
&\quad + i\left( -\mathsf{Cov}\Big[ T_{\mathrm{I}}^{(\ell')}, T_{\mathrm{Q}}^{(\ell'')}\,\Big|\, \mathbf{D} = \mathbf{d}\Big] + \mathsf{Cov}\Big[ T_{\mathrm{Q}}^{(\ell')}, T_{\mathrm{I}}^{(\ell'')}\,\Big|\, \mathbf{D} = \mathbf{d}\Big]\right) \\
&= \frac{\mathsf{N}_0}{2}\,\mathrm{Re}\Big( \mathsf{R}_{\mathbf{gg}}\big((\ell' - \ell'')\mathsf{T}_{\mathrm{s}}\big)\Big) + \frac{\mathsf{N}_0}{2}\,\mathrm{Re}\Big( \mathsf{R}_{\mathbf{gg}}\big((\ell' - \ell'')\mathsf{T}_{\mathrm{s}}\big)\Big) \\
&\quad + i\left( \frac{\mathsf{N}_0}{2}\,\mathrm{Im}\Big( \mathsf{R}_{\mathbf{gg}}\big((\ell' - \ell'')\mathsf{T}_{\mathrm{s}}\big)\Big) - \frac{\mathsf{N}_0}{2}\,\mathrm{Im}\Big( \mathsf{R}_{\mathbf{gg}}\big((\ell'' - \ell')\mathsf{T}_{\mathrm{s}}\big)\Big)\right) \\
&= \mathsf{N}_0\,\mathsf{R}_{\mathbf{gg}}\big((\ell' - \ell'')\mathsf{T}_{\mathrm{s}}\big), \quad \ell', \ell'' \in \{1, \dots, n\},
\end{aligned}
\tag{28.46}
$$

where the first equality follows from the definition of the covariance between complex random variables (17.17); the second by (28.45); the third by (28.39); and the last equality by the conjugate-symmetry of the self-similarity function (Proposition 11.2.2 (iii)).

Conditional on $\mathbf{D} = \mathbf{d}$, the complex $n$-vector $\mathbf{Z}$ is thus a proper Gaussian, and its conditional law is thus fully specified by its conditional covariance matrix (Corollary 24.3.8). By (28.46), this conditional covariance matrix does not depend on $\mathbf{d}$, and we thus conclude that the conditional law of $\mathbf{Z}$ conditional on $\mathbf{D} = \mathbf{d}$ does not depend on $\mathbf{d}$, i.e., that $\mathbf{Z}$ is independent of $\mathbf{D}$.

Part (iii) follows from Part (ii). $\qquad\qquad\qquad\qquad\qquad\qquad\qquad\qquad\square$

## 28.6   Additional Reading

Proposition 28.2.1 and Proposition 28.5.2 are the starting points of much of the literature on equalization and on the use of the Viterbi Algorithm for channels with inter-symbol interference (ISI). See, for example, (Proakis, 2000, Chapter 10), (Viterbi and Omura, 1979, Chapter 4, Section 4.9), and (Barry, Lee, and Messerschmitt, 2004, Chapter 8).

## 28.7   Exercises

**Exercise 28.1 (A Dispersive Channel).** Let the transmitted signal $\big(X(t)\big)$ be as in (28.1), and let the received signal $\big(Y(t)\big)$ be given by

$$
Y(t) = \big(\mathbf{X} \star \mathbf{h}\big)(t) + N(t), \quad t \in \mathbb{R},
$$

where $\big(N(t)\big)$ is white Gaussian noise of PSD $\mathsf{N}_0/2$ with respect to the bandwidth $W$, and where $\mathbf{h}$ is the impulse response of some stable real filter.

(i) Show that the $n$ inner products

$$\int_{-\infty}^{\infty} Y(t) \, (\mathbf{g} \star \mathbf{h})(t - \ell \mathsf{T}_\mathsf{s}) \, dt, \quad \ell \in \{1, \ldots, n\}$$

form a sufficient statistic for guessing $D_1, \ldots D_k$ based on $\big(Y(t)\big)$.

(ii) Compute their conditional law.

**Exercise 28.2 (PAM in Colored Noise).** Let the transmitted signal $\big(X(t)\big)$ be as in (28.1), and let the received signal $\big(Y(t)\big)$ be given by

$$Y(t) = X(t) + N(t), \quad t \in \mathbb{R},$$

where $\big(N(t)\big)$ is a centered, stationary, measurable, Gaussian SP of PSD $\mathsf{S}_{NN}$ that can be whitened with respect to the bandwidth $\mathsf{W}$. Let $\mathbf{h}$ be the impulse response of a whitening filter for $\big(N(t)\big)$ with respect to $\mathsf{W}$.

(i) Show that the $n$ inner products

$$\int_{-\infty}^{\infty} Y(t) \, (\mathbf{g} \star \mathbf{h} \star \tilde{\mathbf{h}})(t - \ell \mathsf{T}_\mathsf{s}) \, dt, \quad \ell \in \{1, \ldots, n\}$$

form a sufficient statistic for guessing $D_1, \ldots D_k$ based on $\big(Y(t)\big)$.

(ii) Compute their conditional law.

**Exercise 28.3 (A Channel with an Echo).** Data bits $D_1, \ldots, D_k$ are mapped to real symbols $X_1, \ldots, X_k$ using the antipodal mapping, so $X_\ell = 1 - 2D_\ell$, for every $\ell \in \{1, \ldots, k\}$. The transmitted signal $\big(X(t)\big)$ is given by $X(t) = \mathsf{A} \sum_\ell X_\ell \, \phi(t - \ell \mathsf{T}_\mathsf{s})$, where $\phi$ is an integrable signal that is bandlimited to $\mathsf{W}$ Hz and that satisfies the orthonormality condition (28.18). The received signal $\big(Y(t)\big)$ is

$$Y(t) = X(t) + \alpha X(t - \mathsf{T}_\mathsf{s}) + N(t), \quad t \in \mathbb{R},$$

where $\big(N(t)\big)$ is white Gaussian noise of PSD $\mathsf{N}_0/2$ with respect to the bandwidth $\mathsf{W}$, and $\alpha$ is a real constant. Let $Y_\ell$ be the time-$\ell \mathsf{T}_\mathsf{s}$ output of a filter that is matched to $\phi$ and that is fed $\big(Y(t)\big)$.

(i) Do $Y_1, \ldots, Y_{k+1}$ form a sufficient statistic for guessing $(D_1, \ldots, D_k)$?

(ii) Consider a suboptimal rule that guesses "$D_j = 0$" if $Y_j \geq 0$, and otherwise guesses "$D_j = 1$." Express the probability that this rule guesses $D_j$ incorrectly in terms of $j$, $\alpha$, $\mathsf{A}$, and $\mathsf{N}_0$. To what does this probability of error converge when $\mathsf{N}_0$ tends to zero?

**Exercise 28.4 (Another Channel with an Echo).** Consider the setup of Exercise 28.3 but where the echo is delayed by a noninteger multiple of the baud period. Thus,

$$Y(t) = X(t) + \alpha X(t - \tau) + N(t), \quad t \in \mathbb{R},$$

where $0 < \tau < \mathsf{T}_\mathsf{s}$. Show that the $2k$ inner products

$$\int_{-\infty}^{\infty} Y(t) \, \phi(t - \ell \mathsf{T}_\mathsf{s}) \, dt, \quad \int_{-\infty}^{\infty} Y(t) \, \phi(t - \ell \mathsf{T}_\mathsf{s} - \tau) \, dt, \quad \ell \in \{1, \ldots, k\}$$

form a sufficient statistic for guessing $(D_1, \ldots, D_k)$ based on $\big(Y(t)\big)$.

**Exercise 28.5 (A Multiple-Access Scenario).** Two transmitters communicate with a single receiver. The receiver observes the signal

$$Y(t) = A_1 X_1\, \phi_1(t) + A_2 X_2\, \phi_2(t) + N(t), \quad t \in \mathbb{R},$$

where $A_1, A_2 > 0$; $\phi_1$ and $\phi_2$ are orthonormal integrable signals that are bandlimited to $W$ Hz; the pair $(X_1, X_2)$ takes value in the set $\{(+1, +1), (+1, -1), (-1, +1), (-1, -1)\}$ equiprobably; and where $(N(t))$ is white Gaussian noise of PSD $N_0/2$ with respect to the bandwidth $W$.

  (i) Can you recover $(X_1, X_2)$ from $A_1 X_1 \phi_1 + A_2 X_2 \phi_2$?

  (ii) Find an optimal receiver for guessing $(X_1, X_2)$ based on $(Y(t))$.

  (iii) Compute the optimal probability of error for guessing $(X_1, X_2)$ based on $(Y(t))$.

  (iv) Suppose that a genie informs the receiver of the value of $X_2$. How should the receiver then guess $X_1$ based on $(Y(t))$ and the information provided by the genie?

  (v) A receiver guesses "$X_1 = +1$" if $\langle \mathbf{Y}, \phi_1 \rangle > 0$ and guesses "$X_1 = -1$" otherwise. Is this receiver optimal for guessing $X_1$?

**Exercise 28.6 (Two Receiver Antennas).** Consider the setup of (28.1). We observe two signals $(Y_1(t))$, $(Y_2(t))$ that are given at every epoch $t \in \mathbb{R}$ by

$$Y_1(t) = (\mathbf{X} \star \mathbf{h}_1)(t) + N_1(t), \quad Y_2(t) = (\mathbf{X} \star \mathbf{h}_2)(t) + N_2(t),$$

where $\mathbf{h}_1$ and $\mathbf{h}_2$ are the impulse responses of two real stable filters, and where the stochastic processes $(N_1(t))$ and $(N_2(t))$ are independent white Gaussian noise processes of PSD $N_0/2$ with respect to the bandwidth $W$.

  (i) Extend Definition 26.3.1 to the case where the observation consists of two stochastic processes.

  (ii) Show that the $2n$ inner products

$$\int_{-\infty}^{\infty} Y_1(t)\, (\mathbf{g} \star \mathbf{h}_1)(t - \ell T_s)\, dt, \quad \int_{-\infty}^{\infty} Y_2(t)\, (\mathbf{g} \star \mathbf{h}_2)(t - \ell T_s)\, dt, \quad \ell \in \{1, \ldots, n\}$$

form a sufficient statistic for guessing $D_1, \ldots D_k$ based on $(Y_1(t))$ and $(Y_2(t))$.

**Exercise 28.7 (Bits of Unequal Importance).** Consider the setup of Section 28.3 but where some data bits are more important than others. We therefore wish to minimize the weighted average

$$\sum_{j=1}^{k} \alpha_j \Pr\!\left[\hat{D}_j \neq D_j\right],$$

for some positive $\alpha_1, \ldots, \alpha_k$ that sum to one.

  (i) Is it still optimal to base our guess of $D_1, \ldots, D_k$ on the inner products in (28.11)?

  (ii) Does this criterion lead to a different receiver design than the bit error rate?

**Exercise 28.8 (Sandwiching the Probability of a Message Error).** In the notation of Section 28.3, show that

$$\frac{1}{k}\sum_{j=1}^{k} \Pr[\tilde{D}_j \neq D_j] \leq \max_{1 \leq j \leq k} \left\{ \Pr[\tilde{D}_j \neq D_j] \right\} \leq \Pr\!\left[\tilde{\mathbf{D}} \neq \mathbf{D}\right] \leq \sum_{j=1}^{k} \Pr[\tilde{D}_j \neq D_j].$$

**Exercise 28.9 (Sandwiching the Bit Error Rate).** In the notation of Section 28.3, show that

$$\frac{1}{k}\Pr[\tilde{\mathbf{D}} \neq \mathbf{D}] \leq \frac{1}{k}\sum_{j=1}^{k}\Pr[\tilde{D}_j \neq D_j] \leq \Pr[\tilde{\mathbf{D}} \neq \mathbf{D}].$$

**Exercise 28.10 (Transmission via an Unknown Dispersive Channel).** A random switch that is outside our control and whose realization is not observed determines whether the observed output $(Y(t))$ is given by

$$\mathbf{X} \star \mathbf{h}_1 + \mathbf{N} \quad \text{or} \quad \mathbf{X} \star \mathbf{h}_2 + \mathbf{N},$$

where $(X(t))$ is the transmitted signal of (28.1); $(N(t))$ is white Gaussian noise of PSD $\mathsf{N}_0/2$ with respect to the bandwidth $\mathcal{W}$; and $\mathbf{h}_1$ & $\mathbf{h}_2$ are the impulse responses of two stable real filters. Show that the $2n$ inner products

$$\int_{-\infty}^{\infty} Y(t)\,(\mathbf{g}\star\mathbf{h}_1)(t-\ell\mathsf{T}_\mathrm{s})\,\mathrm{d}t, \quad \int_{-\infty}^{\infty} Y(t)\,(\mathbf{g}\star\mathbf{h}_2)(t-\ell\mathsf{T}_\mathrm{s})\,\mathrm{d}t, \quad \ell \in \{1,\ldots,n\}$$

form a sufficient statistic for guessing $D_1,\ldots D_k$ based on $(Y(t))$.

# Chapter 29

# Linear Binary Block Codes with Antipodal Signaling

## 29.1  Introduction and Setup

We have thus far said very little about the design of good encoders. We mentioned block encoders but, apart from defining and studying some of their basic properties (such as rate and energy per symbol), we have said very little about how to design such encoders. The design of block encoders falls under the heading of "Coding Theory" and is the subject of numerous books such as (MacWilliams and Sloane, 1977), (van Lint, 1998), (Blahut, 2002), (Roth, 2006) and (Richardson and Urbanke, 2008). Here we provide only a glimpse of this theory for one class of such encoders: the class of binary linear block encoders with antipodal pulse amplitude modulation. Such encoders map the data bits $D_1, \ldots, D_K$ to the real symbols $X_1, \ldots, X_N$ by first applying a one-to-one linear mapping of binary K-tuples to binary N-tuples and by then applying the antipodal mapping

$$0 \mapsto +1$$
$$1 \mapsto -1$$

to each component of the binary N-tuple to produce the $\{\pm 1\}$-valued symbols $X_1, \ldots, X_N$.

Our emphasis in this chapter is not on the design of such encoders, but on how their properties influence the performance of communication systems that employ them in combination with Pulse Amplitude Modulation. We thus assume that the transmitted waveform is given by

$$A \sum_\ell X_\ell \, \phi(t - \ell T_s), \quad t \in \mathbb{R}, \tag{29.1}$$

where $A > 0$ is a scaling factor, $T_s > 0$ is the baud period, $\phi(\cdot)$ is a real integrable signal that is bandlimited to $W$ Hz, and where the time shifts of $\phi(\cdot)$ by integer multiples of $T_s$ are orthonormal

$$\int_{-\infty}^{\infty} \phi(t - \ell T_s) \, \phi(t - \ell' T_s) \, dt = I\{\ell = \ell'\}, \quad \ell, \ell' \in \mathbb{Z}. \tag{29.2}$$

The summation in (29.1) can be finite, as in the block-mode that we discussed in Section 10.4, or infinite, as in the bi-infinite block-mode that we discussed in Section 14.5.2. We shall further assume that the PAM signal is transmitted over an additive noise channel where the transmitted signal is corrupted by Gaussian noise that is white with respect to the bandwidth $W$. We also assume that the data are IID random bits (Definition 14.5.1).

In Section 29.2 we briefly discuss the binary field $\mathbb{F}_2$ and discuss some of the basic properties of the set of all binary $\kappa$-tuples when it is viewed as a vector space over this field. This allows us in Section 29.3 to define linear binary encoders and codes. Section 29.4 introduces binary encoders with antipodal signaling, and Section 29.5 discusses the power and power spectral density when they are employed in conjunction with PAM. Section 29.6 begins the study of decoding with a discussion of two performance criteria: the probability of a **block error** (also called **message error**) and the probability of a **bit error**. It also recalls the discrete-time single-block channel model. Section 29.7 contains the design and performance analysis of the guessing rule that minimizes the probability of a block error, and Section 29.8 contains a similar analysis for the guessing rule that minimizes the probability of a bit error. Section 29.9 explains why performance analysis and simulation is often done under the assumption that the transmitted data is the all-zero data. Section 29.10 discusses how the encoder and the PAM parameters influence the overall system performance. The chapter concludes with a discussion of the (suboptimal) Hard Decision decoding rule in Section 29.11 and of bounds on the minimum distance of a code in Section 29.12.

## 29.2    The Binary Field $\mathbb{F}_2$ and the Vector Space $\mathbb{F}_2^\kappa$

### 29.2.1    The Binary Field $\mathbb{F}_2$

The binary field $\mathbb{F}_2$ consists of two elements that we denote by 0 and 1. An operation that we denote by $\oplus$ is defined between any two elements of $\mathbb{F}_2$ through the relation

$$0 \oplus 0 = 0, \quad 0 \oplus 1 = 1, \quad 1 \oplus 0 = 1, \quad 1 \oplus 1 = 0. \tag{29.3}$$

This operation is sometimes called "mod 2 addition" or "exclusive-or" or "GF(2) addition." (Here GF(2) stands for the Galois Field of two elements after the French mathematician Évariste Galois (1811–1832) who did ground-breaking work on finite fields and groups.) Another operation—"GF(2) multiplication"—is denoted by a dot and is defined via the relation

$$0 \cdot 0 = 0, \quad 0 \cdot 1 = 0, \quad 1 \cdot 0 = 0, \quad 1 \cdot 1 = 1. \tag{29.4}$$

Combined with these operations, the set $\mathbb{F}_2$ forms a **field**, which is sometimes called the Galois Field of size two. We leave it to the reader to verify that the $\oplus$ operation satisfies

$$a \oplus b = b \oplus a, \quad a, b \in \mathbb{F}_2,$$

$$(a \oplus b) \oplus c = a \oplus (b \oplus c), \quad a, b, c \in \mathbb{F}_2,$$

$$a \oplus 0 = 0 \oplus a = a, \quad a \in \mathbb{F}_2,$$

$$a \oplus a = 0, \quad a \in \mathbb{F}_2;$$

and that the operations $\oplus$ and $\cdot$ satisfy the distributive law

$$(a \oplus b) \cdot c = (a \cdot c) \oplus (b \cdot c), \quad a, b, c \in \mathbb{F}_2.$$

## 29.2.2 The Vector Field $\mathbb{F}_2^\kappa$

We denote the set of all binary $\kappa$-tuples by $\mathbb{F}_2^\kappa$ and define the componentwise-$\oplus$ operation between $\kappa$-tuples $\mathbf{u} = (u_1, \ldots, u_\kappa) \in \mathbb{F}_2^\kappa$ and $\mathbf{v} = (v_1, \ldots, v_\kappa) \in \mathbb{F}_2^\kappa$ as

$$\mathbf{u} \oplus \mathbf{v} \triangleq (u_1 \oplus v_1, \ldots, u_\kappa \oplus v_\kappa), \quad \mathbf{u}, \mathbf{v} \in \mathbb{F}_2^\kappa. \tag{29.5}$$

We define the product between a scalar $\alpha \in \mathbb{F}_2$ and a $\kappa$-tuple $\mathbf{u} = (u_1, \ldots, u_\kappa) \in \mathbb{F}_2^\kappa$ by

$$\alpha \cdot \mathbf{u} \triangleq (\alpha \cdot u_1, \ldots, \alpha \cdot u_\kappa). \tag{29.6}$$

With these operations the set $\mathbb{F}_2^\kappa$ forms a vector space over the field $\mathbb{F}_2$. The all-zero $\kappa$-tuple is denoted by $\mathbf{0}$.

## 29.2.3 Linear Mappings

A mapping $\mathsf{T} \colon \mathbb{F}_2^\kappa \to \mathbb{F}_2^\eta$ is said to be **linear** if

$$\mathsf{T}(\alpha \cdot \mathbf{u} \oplus \beta \cdot \mathbf{v}) = \alpha \cdot \mathsf{T}(\mathbf{u}) \oplus \beta \cdot \mathsf{T}(\mathbf{v}), \quad \left(\alpha, \beta \in \mathbb{F}_2, \, \mathbf{u}, \mathbf{v} \in \mathbb{F}_2^\kappa\right). \tag{29.7}$$

The **kernel** of a linear mapping $\mathsf{T} \colon \mathbb{F}_2^\kappa \to \mathbb{F}_2^\eta$ is denoted by $\mathrm{Ker}(\mathsf{T})$ and is the set of all $\kappa$-tuples in $\mathbb{F}_2^\kappa$ that are mapped by $\mathsf{T}(\cdot)$ to the all-zero $\eta$-tuple $\mathbf{0}$:

$$\mathrm{Ker}(\mathsf{T}) = \left\{ \mathbf{u} \in \mathbb{F}_2^\kappa : \mathsf{T}(\mathbf{u}) = \mathbf{0} \right\}. \tag{29.8}$$

The kernel of every linear mapping contains the all-zero tuple $\mathbf{0}$.

The **image** of $\mathsf{T} \colon \mathbb{F}_2^\kappa \to \mathbb{F}_2^\eta$ is denoted by $\mathrm{Image}(\mathsf{T})$ and consists of those elements of $\mathbb{F}_2^\eta$ to which some element of $\mathbb{F}_2^\kappa$ is mapped by $\mathsf{T}(\cdot)$:

$$\mathrm{Image}(\mathsf{T}) = \left\{ \mathsf{T}(\mathbf{u}) : \mathbf{u} \in \mathbb{F}_2^\kappa \right\}. \tag{29.9}$$

The key results from Linear Algebra that we need are summarized in the following proposition.

**Proposition 29.2.1.** *Let $\mathsf{T} \colon \mathbb{F}_2^\kappa \to \mathbb{F}_2^\eta$ be linear.*

(i) *The kernel of $\mathsf{T}(\cdot)$ is a linear subspace of $\mathbb{F}_2^\kappa$.*

(ii) *The mapping $\mathsf{T}(\cdot)$ is one-to-one if, and only if, $\mathrm{Ker}(\mathsf{T}) = \{\mathbf{0}\}$.*

(iii) *The image of $\mathsf{T}(\cdot)$ is a linear subspace of $\mathbb{F}_2^\eta$.*

(iv) *The sum of the dimension of the kernel and the dimension of the image space is equal to the dimension of the domain:*

$$\mathrm{Dim}\big(\mathrm{Ker}(\mathsf{T})\big) + \mathrm{Dim}\big(\mathrm{Image}(\mathsf{T})\big) = \kappa. \tag{29.10}$$

(v) *If $\mathcal{U}$ is a linear subspace of $\mathbb{F}_2^\eta$ of dimension $\kappa$, then there exists a one-to-one linear mapping from $\mathbb{F}_2^\kappa$ to $\mathbb{F}_2^\eta$ whose image is $\mathcal{U}$.*

### 29.2.4 Hamming Distance and Hamming Weight

The **Hamming distance** $d_H(\mathbf{u}, \mathbf{v})$ between two binary $\kappa$-tuples $\mathbf{u}$ and $\mathbf{v}$ is defined as the number of components in which they differ. For example, the Hamming distance between the tuples $(1, 0, 1, 0)$ and $(0, 0, 1, 1)$ is two. It is easy to prove that for $\mathbf{u}, \mathbf{v}, \mathbf{w} \in \mathbb{F}_2^\kappa$:

$$d_H(\mathbf{u}, \mathbf{v}) \geq 0 \text{ with equality if, and only if, } \mathbf{u} = \mathbf{v}; \tag{29.11a}$$

$$d_H(\mathbf{u}, \mathbf{v}) = d_H(\mathbf{v}, \mathbf{u}); \tag{29.11b}$$

$$d_H(\mathbf{u}, \mathbf{w}) \leq d_H(\mathbf{u}, \mathbf{v}) + d_H(\mathbf{v}, \mathbf{w}). \tag{29.11c}$$

The **Hamming weight** $w_H(\mathbf{u})$ of a binary $\kappa$-tuple $\mathbf{u}$ is defined as the number of its nonzero components. Thus,

$$w_H(\mathbf{u}) = d_H(\mathbf{u}, \mathbf{0}), \quad \mathbf{u} \in \mathbb{F}_2^\kappa, \tag{29.12}$$

and

$$d_H(\mathbf{u}, \mathbf{v}) = w_H(\mathbf{u} \oplus \mathbf{v}), \quad \mathbf{u}, \mathbf{v} \in \mathbb{F}_2^\kappa. \tag{29.13}$$

### 29.2.5 The Componentwise Antipodal Mapping

The **antipodal mapping** $\Upsilon \colon \mathbb{F}_2 \to \{-1, +1\}$ maps the zero element of $\mathbb{F}_2$ to the real number $+1$ and the unit element of $\mathbb{F}_2$ to $-1$:

$$\Upsilon(0) = +1, \quad \Upsilon(1) = -1. \tag{29.14}$$

This rule is not as arbitrary as it may seem. Although one might be somewhat surprised that we do not map $1 \in \mathbb{F}_2$ to $+1$, we have our reasons. We prefer the mapping (29.14) because it maps mod-2 sums to real products. Thus,

$$\Upsilon(a \oplus b) = \Upsilon(a)\Upsilon(b), \quad a, b \in \mathbb{F}_2, \tag{29.15}$$

where the operation on the RHS between $\Upsilon(a)$ and $\Upsilon(b)$ is the regular real-numbers multiplication. This extends by induction to any finite number of elements of $\mathbb{F}_2$:

$$\Upsilon(c_1 \oplus c_2 \oplus \cdots \oplus c_\nu) = \prod_{\ell=1}^{\nu} \Upsilon(c_\ell), \quad c_1, \ldots, c_\nu \in \mathbb{F}_2. \tag{29.16}$$

The **componentwise antipodal mapping** $\Upsilon_\eta \colon \mathbb{F}_2^\eta \to \{-1, +1\}^\eta$ maps elements of $\mathbb{F}_2^\eta$ to elements of $\{-1, +1\}^\eta$ by applying the mapping (29.14) to each component:

$$\Upsilon_\eta \colon (c_1, \ldots, c_\eta) \mapsto (\Upsilon(c_1), \ldots, \Upsilon(c_\eta)). \tag{29.17}$$

For example, $\Upsilon_3$ maps the triplet $(0, 0, 1)$ to $(+1, +1, -1)$.

### 29.2.6 Hamming Distance and Euclidean Distance

We next relate the Hamming distance $d_H(\mathbf{u}, \mathbf{v})$ between any two binary $\eta$-tuples $\mathbf{u} = (u_1, \ldots, u_\eta)$ and $\mathbf{v} = (v_1, \ldots, v_\eta)$ to the squared Euclidean distance between the results of applying the componentwise antipodal mapping $\Upsilon_\eta$ to them. We argue that

$$d_E^2(\Upsilon_\eta(\mathbf{u}), \Upsilon_\eta(\mathbf{v})) = 4\, d_H(\mathbf{u}, \mathbf{v}), \quad \mathbf{u}, \mathbf{v} \in \mathbb{F}_2^\eta, \tag{29.18}$$

where $d_E(\cdot, \cdot)$ denotes the Euclidean distance, so

$$d_E^2(\Upsilon_\eta(\mathbf{u}), \Upsilon_\eta(\mathbf{v})) = \sum_{\nu=1}^\eta \big(\Upsilon(u_\nu) - \Upsilon(v_\nu)\big)^2. \tag{29.19}$$

To prove this relationship it suffices to consider the case where $\eta = 1$, because the Hamming distance is the sum of the Hamming distances between the respective components, and likewise for the squared Euclidean distance. To prove this result for $\eta = 1$ we note that if the Hamming distance is zero, then $u$ and $v$ are identical and hence so are $\Upsilon(u)$ and $\Upsilon(v)$, so the Euclidean distance between them must be zero. And if the Hamming distance is one, then $u \neq v$, and hence $\Upsilon(u)$ and $\Upsilon(v)$ are of opposite sign but of equal unit magnitude, so the squared Euclidean distance between them is four.

## 29.3 Binary Linear Encoders and Codes

**Definition 29.3.1 (Linear $(K, N)$ $\mathbb{F}_2$ Encoder and Code).** *Let $N$ and $K$ be positive integers.*

*(i) A **linear** $(K, N)$ $\mathbb{F}_2$ **encoder** is a one-to-one linear mapping from $\mathbb{F}_2^K$ to $\mathbb{F}_2^N$.*

*(ii) A **linear** $(K, N)$ $\mathbb{F}_2$ **code** is a linear subspace of $\mathbb{F}_2^N$ of dimension $K$.[1]*

*In both definitions $N$ is called the **blocklength** and $K$ is called the **dimension**.*

For example, the $(K, K+1)$ **systematic single parity check encoder** is the mapping

$$(d_1, \ldots, d_K) \mapsto (d_1, \ldots, d_K, d_1 \oplus d_2 \oplus \cdots \oplus d_K). \tag{29.20}$$

It appends to the data tuple a single bit that is chosen so that the resulting $(K+1)$-tuple be of even Hamming weight. The $(K, K+1)$ **single parity check code** is the subset of $\mathbb{F}_2^{K+1}$ consisting of those binary $(K+1)$-tuples whose Hamming weight is even.

Recall that the image of a mapping $\mathbf{g}\colon \mathcal{A} \to \mathcal{B}$ is the subset of $\mathcal{B}$ comprising those elements $y \in \mathcal{B}$ to which there corresponds some $x \in \mathcal{A}$ such that $g(x) = y$.

---

[1]The terminology here is not standard. In the Coding Theory literature a linear $(K, N)$ $\mathbb{F}_2$ code is often called a "binary linear $[N, K]$ code."

**Proposition 29.3.2 ($\mathbb{F}_2$ Encoders and Codes).**

(i) *If* $\mathsf{T} \colon \mathbb{F}_2^\mathsf{K} \to \mathbb{F}_2^\mathsf{N}$ *is a linear* $(\mathsf{K}, \mathsf{N})$ $\mathbb{F}_2$ *encoder, then its image is a linear* $(\mathsf{K}, \mathsf{N})$ $\mathbb{F}_2$ *code.*

(ii) *Every linear* $(\mathsf{K}, \mathsf{N})$ $\mathbb{F}_2$ *code is the image of some (nonunique) linear* $(\mathsf{K}, \mathsf{N})$ $\mathbb{F}_2$ *encoder.*

**Proof.** We begin with Part (i). Let $\mathsf{T} \colon \mathbb{F}_2^\mathsf{K} \to \mathbb{F}_2^\mathsf{N}$ be a linear $(\mathsf{K}, \mathsf{N})$ $\mathbb{F}_2$ encoder. That its image is a linear subspace of $\mathbb{F}_2^\mathsf{N}$ follows from Proposition 29.2.1 (iii). That its dimension must be $\mathsf{K}$ follows from Proposition 29.2.1 (iv) (see (29.10)) because the fact that $\mathsf{T}(\cdot)$ is one-to-one implies, by Proposition 29.2.1 (ii), that $\mathrm{Ker}(\mathsf{T}) = \{\mathbf{0}\}$ so $\mathrm{Dim}\big(\mathrm{Ker}(\mathsf{T})\big) = 0$.

To prove Part (ii) we note that $\mathbb{F}_2^\mathsf{K}$ is of dimension $\mathsf{K}$ and that, by definition, every linear $(\mathsf{K}, \mathsf{N})$ $\mathbb{F}_2$ code is also of dimension $\mathsf{K}$. The result now follows by noting that there exists a one-to-one linear mapping between any two subspaces of equal dimensions over the same field (Proposition 29.2.1 (v)). □

Any linear transformation from a finite-dimensional space to a finite-dimensional space can be represented as matrix multiplication. A linear $(\mathsf{K}, \mathsf{N})$ $\mathbb{F}_2$ encoder is no exception. What is perhaps unusual is that coding theorists use row vectors to denote the data $\mathsf{K}$-tuples and the $\mathsf{N}$-tuples to which they are mapped. They consequently use matrix multiplication from the left. This tradition is so ingrained that we shall begrudgingly adopt it.

**Definition 29.3.3 (Matrix Representation of an Encoder).** *We say that the linear* $(\mathsf{K}, \mathsf{N})$ $\mathbb{F}_2$ *encoder* $\mathsf{T} \colon \mathbb{F}_2^\mathsf{K} \to \mathbb{F}_2^\mathsf{N}$ *is* ***represented by the matrix*** $\mathsf{G}$ *if* $\mathsf{G}$ *is a* $\mathsf{K} \times \mathsf{N}$ *matrix whose elements are in* $\mathbb{F}_2$ *and*

$$\mathsf{T}(\mathbf{d}) = \mathbf{d}\mathsf{G}, \quad \mathbf{d} \in \mathbb{F}_2^\mathsf{K}. \tag{29.21}$$

Note that in the matrix multiplication in (29.21) we use $\mathbb{F}_2$ arithmetic, so the $\eta$-th component of $\mathbf{d}\mathsf{G}$ is given by $d^{(1)} \cdot g^{(1,\eta)} \oplus \cdots \oplus d^{(\mathsf{K})} \cdot g^{(\mathsf{K},\eta)}$, where $g^{(\kappa,\eta)}$ is the Row-$\kappa$ Column-$\eta$ component of the matrix $\mathsf{G}$, and where $d^{(\kappa)}$ is the $\kappa$-th component of $\mathbf{d}$.

For example, the $(\mathsf{K}, \mathsf{K} + 1)$ $\mathbb{F}_2$ systematic single parity check encoder (29.20) is represented by the $\mathsf{K} \times (\mathsf{K} + 1)$ matrix

$$\begin{pmatrix} 1 & 0 & 0 & \cdots & 0 & 1 \\ 0 & 1 & 0 & \cdots & 0 & 1 \\ 0 & 0 & 1 & \cdots & 0 & 1 \\ \vdots & \vdots & \vdots & \ddots & 0 & 1 \\ 0 & 0 & 0 & \cdots & 1 & 1 \end{pmatrix}. \tag{29.22}$$

The matrix $\mathsf{G}$ in (29.21) is uniquely specified by the linear transformation $\mathsf{T}(\cdot)$: its $\eta$-th row is the result of applying $\mathsf{T}(\cdot)$ to the $\mathsf{K}$-tuple $(0, \ldots, 0, 1, 0, \ldots, 0)$ (the $\mathsf{K}$-tuple whose components are all zero except for the $\eta$-th, which is one).

Moreover, every $\mathsf{K} \times \mathsf{N}$ binary matrix $\mathsf{G}$ defines a linear transformation $\mathsf{T}(\cdot)$ via (29.21), but this linear transformation need not be one-to-one. It is one-to-one if, and only if, the subspace of $\mathbb{F}_2^\mathsf{N}$ spanned by the rows of $\mathsf{G}$ is of dimension $\mathsf{K}$.

**Definition 29.3.4 (Generator Matrix).** *A matrix* $\mathsf{G}$ *is a **generator matrix** for a given linear* $(\mathsf{K},\mathsf{N})$ $\mathbb{F}_2$ *code if* $\mathsf{G}$ *is a binary* $\mathsf{K} \times \mathsf{N}$ *matrix such that the image of the mapping* $\mathbf{d} \mapsto \mathbf{d}\mathsf{G}$ *is the given code.*

Note that there may be numerous generator matrices for a given code. For example, the matrix (29.22) is a generator matrix for the single parity check code. But there are others. Indeed, replacing any row of the above matrix by the sum of that row and another different row results in another generator matrix for this code.

Coding theorists like to distinguish between a **code property** and an **encoder property**. Code properties are properties that are common to all encoders of the same image. Encoder properties are specific to an encoder. Examples of code properties are the blocklength and dimension. We shall soon encounter more. An example of an encoder property is the property of being systematic:

**Definition 29.3.5 (Systematic Encoder).** *A linear* $(\mathsf{K},\mathsf{N})$ $\mathbb{F}_2$ *encoder* $\mathsf{T}\colon \mathbb{F}_2^{\mathsf{K}} \to \mathbb{F}_2^{\mathsf{N}}$ *is said to be **systematic** (or **strictly systematic**) if, for every* $\mathsf{K}$-*tuple* $\big(d_1,\ldots,d_{\mathsf{K}}\big)$ *in* $\mathbb{F}_2^{\mathsf{K}}$, *the first* $\mathsf{K}$ *components of* $\mathsf{T}\big((d_1,\ldots,d_{\mathsf{K}})\big)$ *are equal to* $d_1,\ldots,d_{\mathsf{K}}$.

For example, the encoder (29.20) is systematic. An encoder whose image is the single-parity check code and which is not systematic is the encoder

$$\big(d_1,\ldots,d_{\mathsf{K}}\big) \mapsto \big(d_1, d_1 \oplus d_2, d_2 \oplus d_3, \ldots, d_{\mathsf{K}-1} \oplus d_{\mathsf{K}}, d_{\mathsf{K}}\big). \tag{29.23}$$

The reader is encouraged to verify that if a linear $(\mathsf{K},\mathsf{N})$ $\mathbb{F}_2$ encoder $\mathsf{T}\colon \mathbb{F}_2^{\mathsf{K}} \to \mathbb{F}_2^{\mathsf{N}}$ is represented by the matrix $\mathsf{G}$, then $\mathsf{T}(\cdot)$ is systematic if, and only if, the $\mathsf{K} \times \mathsf{K}$ matrix that results from deleting the last $\mathsf{N} - \mathsf{K}$ columns of $\mathsf{G}$ is the $\mathsf{K} \times \mathsf{K}$ identity matrix.

**Definition 29.3.6 (Parity-Check Matrix).** *A **parity-check matrix** for a given linear* $(\mathsf{K},\mathsf{N})$ $\mathbb{F}_2$ *code is a* $\mathsf{K} \times \mathsf{N}$ *matrix* $\mathsf{H}$ *such that a (row)* $\mathsf{N}$-*tuple* $\mathbf{c}$ *is in the code if, and only if,* $\mathbf{c}\mathsf{H}^{\mathsf{T}}$ *is the all-zero (row) vector.*

For example, a parity-check matrix for the $(\mathsf{K},\mathsf{K}+1)$ single-parity check code is the $1 \times (\mathsf{K}+1)$ matrix

$$\mathsf{H} = (1,1,\ldots,1).$$

(Codes typically have numerous different parity-check matrices, but the single-parity check code is an exception.)

## 29.4 Binary Encoders with Antipodal Signaling

**Definition 29.4.1.**

(i) *We say that a* $(\mathsf{K},\mathsf{N})$ *binary-to-reals block encoder* **enc**$\colon \{0,1\}^{\mathsf{K}} \to \mathbb{R}^{\mathsf{N}}$ *is a **linear binary** $(\mathsf{K},\mathsf{N})$ **block encoder with antipodal signaling** if*

$$\text{enc}(\mathbf{d}) = \Upsilon_{\mathsf{N}}\big(\mathsf{T}(\mathbf{d})\big), \quad \mathbf{d} \in \mathbb{F}_2^{\mathsf{K}}, \tag{29.24}$$

*where* $\mathsf{T}\colon \mathbb{F}_2^\mathsf{K} \to \mathbb{F}_2^\mathsf{N}$ *is a linear* $(\mathsf{K}, \mathsf{N})$ $\mathbb{F}_2$ *encoder, and where* $\Upsilon_\mathsf{N}(\cdot)$ *is the componentwise antipodal mapping* (29.17). *Thus, if* $(X_1, \ldots, X_\mathsf{N})$ *denotes the* $\mathsf{N}$-*tuple produced by* $\mathrm{enc}(\cdot)$ *when fed the data* $\mathsf{K}$-*tuple* $(D_1, \ldots, D_\mathsf{K})$, *then*

$$X_\eta = \begin{cases} +1 & \text{if the } \eta\text{-th components of } \mathsf{T}\big((D_1, \ldots, D_\mathsf{K})\big) \text{ is zero,} \\ -1 & \text{otherwise.} \end{cases} \qquad (29.25)$$

*(ii) A **linear binary** $(\mathsf{K}, \mathsf{N})$ **block code with antipodal signaling** is the image of some linear binary $(\mathsf{K}, \mathsf{N})$ block encoder with antipodal signaling.*

In analogy to Proposition 29.3.2, the image of every linear binary $(\mathsf{K}, \mathsf{N})$ block encoder with antipodal signaling is a linear binary $(\mathsf{K}, \mathsf{N})$ block code with antipodal signaling.

If $\mathrm{enc}(\cdot)$ can be represented by the application of $\mathsf{T}(\cdot)$ to the data $\mathsf{K}$-tuple followed by the application of the componentwise antipodal mapping $\Upsilon_\mathsf{N}$, then we shall write

$$\mathbf{enc} = \Upsilon_\mathsf{N} \circ \mathsf{T}. \qquad (29.26)$$

Since $\Upsilon_\mathsf{N}$ is invertible, there is a one-to-one correspondence between $\mathsf{T}$ and $\mathbf{enc}$.

An important code property is the distribution of the result of applying an encoder to IID random bits.

**Proposition 29.4.2.** *Let* $\mathsf{T}\colon \mathbb{F}_2^\mathsf{K} \to \mathbb{F}_2^\mathsf{N}$ *be a linear* $(\mathsf{K}, \mathsf{N})$ $\mathbb{F}_2$ *encoder.*

*(i) Applying $\mathsf{T}$ to a $\mathsf{K}$-tuple of IID random bits results in a random $\mathsf{N}$-tuple that is uniformly distributed over $\mathrm{Image}(\mathsf{T})$.*

*(ii) Applying $\Upsilon_\mathsf{N} \circ \mathsf{T}$ to IID random bits produces an $\mathsf{N}$-tuple that is uniformly distributed over the image of $\mathrm{Image}(\mathsf{T})$ under the componentwise antipodal mapping $\Upsilon_\mathsf{N}$.*

**Proof.** Part (i) follows from the fact that the mapping $\mathsf{T}(\cdot)$ is one-to-one. Part (ii) follows from Part (i) and from the fact that $\Upsilon_\mathsf{N}(\cdot)$ is one-to-one. $\qquad \square$

For example, it follows from Proposition 29.4.2 (ii) and from (29.16) that if we feed IID random bits to any encoder (be it systematic or not) whose image is the $(\mathsf{K}, \mathsf{K} + 1)$ single parity check code and then employ the componentwise antipodal mapping $\Upsilon_\mathsf{N}(\cdot)$, then the resulting random $(\mathsf{K} + 1)$-tuple $(X_1, \ldots, X_{\mathsf{K}+1})$ will be uniformly distributed over the set

$$\left\{ (\xi_1, \ldots, \xi_{\mathsf{K}+1}) \in \{-1, +1\}^{\mathsf{K}+1} : \prod_{\eta=1}^{\mathsf{K}+1} \xi_\eta = +1 \right\}.$$

**Corollary 29.4.3.** *Any property that is determined by the joint distribution of the result of applying the encoder to IID random bits is a code property.*

Examples of such properties are the power and operational power spectral density, which are discussed next.

## 29.5   Power and Operational Power Spectral Density

To discuss the transmitted power and the operational power spectral density we shall consider bi-infinite block encoding (Section 14.5.2). We shall then use the results of Section 14.5.2 and Section 15.4.3 to compute the power and operational PSD of the transmitted signal in this mode.

The impatient reader who is only interested in the transmitted power for pulse shapes satisfying the orthogonality condition (29.2) can apply the results of Section 14.5.3 directly to obtain that, subject to the decay condition (14.46), the transmitted power P is given by

$$
\boxed{\mathsf{P} = \frac{\mathsf{A}^2}{\mathsf{T_s}}.}
\tag{29.27}
$$

We next extend the discussion to general pulse shapes and to the operational PSD. To remind the reader that we no longer assume the orthogonality condition (29.2), we shall now denote the pulse shape by $g(\cdot)$ and assume that it is bandlimited to $\mathsf{W}$ Hz and that it satisfies the decay condition (14.17). Before proceeding with the analysis of the power and PSD, we wish to characterize linear binary $(\mathsf{K}, \mathsf{N})$ block encoders with antipodal signaling that map IID random bits to zero-mean N-tuples. Note that by Corollary 29.4.3 this is, in fact, a code property. Thus, if $\mathbf{enc} = \mathbf{\Upsilon}_{\mathsf{N}} \circ \mathsf{T}$, then the question of whether $\mathrm{enc}(\cdot)$ maps IID random bits to zero-mean N-tuples depends only on the image of $\mathsf{T}$. Aiding us in this characterization is the following lemma on linear functionals. A linear functional on $\mathbb{F}_2^\kappa$ is a linear mapping from $\mathbb{F}_2^\kappa$ to $\mathbb{F}_2$. The zero functional maps every $\kappa$-tuple in $\mathbb{F}_2^\kappa$ to zero.

**Lemma 29.5.1.** *Let $\mathsf{L}\colon \mathbb{F}_2^\mathsf{K} \to \mathbb{F}_2$ be a linear functional that is not the zero functional. Then the RV $X$ defined by*

$$
X = \begin{cases} +1 & \text{if } \mathsf{L}\big((D_1, \ldots, D_\mathsf{K})\big) = 0, \\ -1 & \text{if } \mathsf{L}\big((D_1, \ldots, D_\mathsf{K})\big) = 1 \end{cases}
\tag{29.28}
$$

*is of zero mean whenever $D_1, \ldots, D_\mathsf{K}$ are IID random bits.*

**Proof.** We begin by expressing the expectation of $X$ as

$$
\begin{aligned}
\mathsf{E}[X] &= \sum_{\mathbf{d} \in \mathbb{F}_2^\mathsf{K}} \Pr[\mathbf{D} = \mathbf{d}] \, \mathbf{\Upsilon}\big(\mathsf{L}(\mathbf{d})\big) \\
&= 2^{-\mathsf{K}} \sum_{\mathbf{d} \in \mathbb{F}_2^\mathsf{K}} \mathbf{\Upsilon}\big(\mathsf{L}(\mathbf{d})\big) \\
&= 2^{-\mathsf{K}} \sum_{\mathbf{d} \in \mathbb{F}_2^\mathsf{K} : \mathsf{L}(\mathbf{d}) = 0} (+1) + 2^{-\mathsf{K}} \sum_{\mathbf{d} \in \mathbb{F}_2^\mathsf{K} : \mathsf{L}(\mathbf{d}) = 1} (-1) \\
&= 2^{-\mathsf{K}} \Big( \# \mathsf{L}^{-1}(0) - \# \mathsf{L}^{-1}(1) \Big),
\end{aligned}
$$

where

$$
\mathsf{L}^{-1}(0) = \big\{ \mathbf{d} \in \mathbb{F}_2^\mathsf{K} : \mathsf{L}(\mathbf{d}) = 0 \big\}
$$

is the set of all K-tuples in $\mathbb{F}_2^K$ that are mapped by $\mathsf{L}(\cdot)$ to 0, where $\mathsf{L}^{-1}(1)$ is analogously defined, and where $\#\mathcal{A}$ denotes the number of elements in the set $\mathcal{A}$. It follows that to prove that $\mathsf{E}[X] = 0$ it suffices to show that if $\mathsf{L}(\cdot)$ is not deterministically zero, then the sets $\mathsf{L}^{-1}(0)$ and $\mathsf{L}^{-1}(1)$ have the same number of elements. We prove this by exhibiting a one-to-one mapping from $\mathsf{L}^{-1}(0)$ onto $\mathsf{L}^{-1}(1)$. (If there is a one-to-one mapping from a finite set $\mathcal{A}$ onto a finite set $\mathcal{B}$, then $\mathcal{A}$ and $\mathcal{B}$ must have the same number of elements.) To exhibit this mapping, note that the assumption that $\mathsf{L}(\cdot)$ is not the zero transformation implies that the set $\mathsf{L}^{-1}(1)$ is not empty. Let $\mathbf{d}_*$ be an element of this set, so

$$\mathsf{L}(\mathbf{d}_*) = 1. \tag{29.29}$$

The required mapping maps each $\mathbf{d}_0 \in \mathsf{L}^{-1}(0)$ to $\mathbf{d}_0 \oplus \mathbf{d}_*$:

$$\mathsf{L}^{-1}(0) \ni \mathbf{d}_0 \mapsto \mathbf{d}_0 \oplus \mathbf{d}_*. \tag{29.30}$$

We next verify that it is a one-to-one mapping from $\mathsf{L}^{-1}(0)$ onto $\mathsf{L}^{-1}(1)$. That it is one-to-one follows because if $\mathbf{d}_0 \oplus \mathbf{d}_* = \mathbf{d}_0' \oplus \mathbf{d}_*$ then by adding $\mathbf{d}_*$ to both sides we obtain $\mathbf{d}_0 \oplus \mathbf{d}_* \oplus \mathbf{d}_* = \mathbf{d}_0' \oplus \mathbf{d}_* \oplus \mathbf{d}_*$, i.e., that $\mathbf{d}_0 = \mathbf{d}_0'$ (because $\mathbf{d}_* \oplus \mathbf{d}_* = 0$). That this mapping maps each element of $\mathsf{L}^{-1}(0)$ to an element of $\mathsf{L}^{-1}(1)$ follows because, as we next show, if $\mathbf{d}_0 \in \mathsf{L}^{-1}(0)$, then $\mathsf{L}(\mathbf{d}_0 \oplus \mathbf{d}_*) = 1$. Indeed, if $\mathbf{d}_0 \in \mathsf{L}^{-1}(0)$, then

$$\mathsf{L}(\mathbf{d}_0) = 0, \tag{29.31}$$

and consequently,

$$\begin{aligned}
\mathsf{L}(\mathbf{d}_0 \oplus \mathbf{d}_*) &= \mathsf{L}(\mathbf{d}_0) \oplus \mathsf{L}(\mathbf{d}_*) \\
&= 0 \oplus 1 \\
&= 1,
\end{aligned}$$

where the first equality follows from the linearity of $\mathsf{L}(\cdot)$, and where the second equality follows from (29.29) and (29.31). That the mapping is onto follows by noting that if $\mathbf{d}_1$ is any element of $\mathsf{L}^{-1}(1)$, then $\mathbf{d}_1 \oplus \mathbf{d}_*$ is in $\mathsf{L}^{-1}(0)$ and it is mapped by this mapping to $\mathbf{d}_1$. $\qquad\square$

Using this lemma we can show:

**Proposition 29.5.2.** *Let* $(X_1, \ldots, X_N)$ *be the result of applying a linear binary* $(K, N)$ *block encoder with antipodal signaling to a binary* K*-tuple comprising IID random bits.*

(i) *For every* $\eta \in \{1, \ldots, N\}$, *the RV* $X_\eta$ *is either deterministically equal to* $+1$, *or else of zero mean.*

(ii) *For every* $\eta, \eta' \in \{1, \ldots, N\}$, *the random variables* $X_\eta$ *and* $X_{\eta'}$ *are either deterministically equal to each other or else* $\mathsf{E}[X_\eta X_{\eta'}] = 0$.

**Proof.** Let the linear binary $(K, N)$ block encoder with antipodal signaling $\mathbf{enc}(\cdot)$ be given by $\mathbf{enc} = \mathbf{\Upsilon}_N \circ \mathsf{T}$, where $\mathsf{T} \colon \mathbb{F}_2^K \to \mathbb{F}_2^N$ is one-to-one and linear. Let

$(X_1, \ldots, X_N)$ be the result of applying **enc** to the K-tuple $\mathbf{D} = (D_1, \ldots, D_K)$, where $D_1, \ldots, D_K$ are IID random bits.

To prove Part (i), fix some $\eta \in \{1, \ldots, N\}$, and let $\mathsf{L}(\cdot)$ be the linear functional that maps $\mathbf{d}$ to the $\eta$-th component of $\mathsf{T}(\mathbf{d})$, so $X_\eta = \Upsilon(\mathsf{L}(\mathbf{D}))$, where $\mathbf{D}$ denotes the row vector comprising the K IID random bits. If $\mathsf{L}(\cdot)$ maps all data K-tuples to zero, then $X_\eta$ is deterministically equal to $+1$. Otherwise, $\mathsf{E}[X_\eta] = 0$ by Lemma 29.5.1.

To prove Part (ii), let the matrix $\mathsf{G}$ represent the mapping $\mathsf{T}(\cdot)$, so $X_\eta = \Upsilon(\mathbf{D}\mathsf{G}^{(\cdot, \eta)})$, where $\mathsf{G}^{(\cdot, \eta)}$ denotes the $\eta$-th column of $\mathsf{G}$. Expressing $X_{\eta'}$ in a similar way, we obtain from (29.15)

$$
\begin{aligned}
X_\eta X_{\eta'} &= \Upsilon\left(\mathbf{D}\mathsf{G}^{(\cdot, \eta)}\right) \Upsilon\left(\mathbf{D}\mathsf{G}^{(\cdot, \eta')}\right) \\
&= \Upsilon\left(\mathbf{D}\mathsf{G}^{(\cdot, \eta)} \oplus \mathbf{D}\mathsf{G}^{(\cdot, \eta')}\right) \\
&= \Upsilon\left(\mathbf{D}\left(\mathsf{G}^{(\cdot, \eta)} \oplus \mathsf{G}^{(\cdot, \eta')}\right)\right).
\end{aligned}
\tag{29.32}
$$

Consequently, if we define the linear functional $\mathsf{L} \colon \mathbf{d} \mapsto \mathbf{d}\left(\mathsf{G}^{(\cdot, \eta)} \oplus \mathsf{G}^{(\cdot, \eta')}\right)$, then $X_\eta X_{\eta'} = \Upsilon(\mathsf{L}(\mathbf{D}))$. This linear functional is the zero functional if the $\eta$-th column of $\mathsf{G}$ is identical to its $\eta'$-th column, i.e., if $X_\eta$ is deterministically equal to $X_{\eta'}$. Otherwise, it is not the zero functional, and $\mathsf{E}[X_\eta X_{\eta'}]$ $\left(= \mathsf{E}\left[\Upsilon(\mathsf{L}(\mathbf{D}))\right]\right)$ must be zero (Lemma 29.5.1). $\qquad \square$

**Proposition 29.5.3 (Producing Zero-Mean Uncorrelated Symbols).** *A linear binary* (K, N) *block encoder with antipodal signaling* **enc** $= \Upsilon_N \circ \mathsf{T}$ *produces zero-mean uncorrelated symbols when fed IID random bits if, and only if, the columns of the matrix* $\mathsf{G}$ *representing* $\mathsf{T}(\cdot)$ *are distinct and neither of these columns is the all-zero column.*

**Proof.** The $\eta$-th symbol $X_\eta$ produced by **enc** $= \Upsilon_N \circ \mathsf{T}$ when fed the K-tuple of IID random bits $\mathbf{D} = (D_1, \ldots, D_K)$ is given by

$$
\begin{aligned}
X_\eta &= \Upsilon(\mathbf{D}\mathsf{G}^{(\cdot, \eta)}) \\
&= \Upsilon(D_1 \cdot \mathsf{G}^{(1, \eta)} \oplus \cdots \oplus D_K \cdot \mathsf{G}^{(K, \eta)})
\end{aligned}
$$

where $\mathsf{G}^{(\cdot, \eta)}$ is the $\eta$-th column of the $\mathsf{K} \times \mathsf{N}$ generator matrix of $\mathsf{T}(\cdot)$. Since the linear functional

$$
\mathbf{d} \mapsto d_1 \cdot \mathsf{G}^{(1, \eta)} \oplus \cdots \oplus d_K \cdot \mathsf{G}^{(K, \eta)}
$$

is the zero functional if, and only if,

$$
\mathsf{G}^{(1, \eta)} = \cdots = \mathsf{G}^{(K, \eta)} = 0,
\tag{29.33}
$$

it follows that $X_\eta$ is deterministically zero if, and only if, the $\eta$-th column of $\mathsf{G}$ is zero. From this and Lemma 29.5.1 it follows that all the symbols produced by **enc** are of zero mean if, and only if, none of the columns of $\mathsf{G}$ is zero.

A similar argument shows that the product $X_\eta X_{\eta'}$, which by (29.32) is given by

$$
\Upsilon\left(\mathbf{D}\left(\mathsf{G}^{(\cdot, \eta)} \oplus \mathsf{G}^{(\cdot, \eta')}\right)\right),
$$

is deterministically zero if, and only if, the functional

$$\mathbf{d} \mapsto d_1 \cdot (\mathsf{G}^{(1,\eta)} \oplus \mathsf{G}^{(1,\eta')}) \oplus \cdots \oplus d_{\mathsf{K}} \cdot (\mathsf{G}^{(\mathsf{K},\eta)} \oplus \mathsf{G}^{(\mathsf{K},\eta')})$$

is zero, i.e., if, and only if, the $\eta$-th and $\eta'$-th columns of $\mathsf{G}$ are equal. Otherwise, by Lemma 29.5.1, we have $\mathsf{E}[X_\eta X_{\eta'}] = 0$.      $\square$

**Note 29.5.4.** By Corollary 29.4.3 the property of producing zero-mean uncorrelated symbols is a code property.

**Proposition 29.5.5 (Power and PSD).** *Let the linear binary* $(\mathsf{K}, \mathsf{N})$ *block encoder with antipodal signaling* $\mathrm{enc} = \Upsilon_\mathsf{N} \circ \mathsf{T}$ *produce zero-mean uncorrelated symbols when fed IID random bits, and let the pulse shape* $\mathbf{g}$ *satisfy the decay condition* (14.17). *Then the transmitted power* $\mathsf{P}$ *in bi-infinite block-encoding mode is given by*

$$\boxed{\mathsf{P} = \frac{\mathsf{A}^2}{\mathsf{T_s}} \|\mathbf{g}\|_2^2} \tag{29.34}$$

*and the operational PSD is*

$$\boxed{\mathsf{S}_{XX}(f) = \frac{\mathsf{A}^2}{\mathsf{T_s}} |\hat{g}(f)|^2, \quad f \in \mathbb{R}.} \tag{29.35}$$

**Proof.** The expression (29.34) for the power follows either from (14.33) or (14.38). The expression for the operational PSD follows either from (15.20) or from (15.23).      $\square$

Engineers rarely check whether an encoder produces uncorrelated symbols when fed IID random bits. The reason may be that they usually deal with pulse shapes $\phi$ satisfying the orthogonality condition (29.2) and the decay condition (14.46). For such pulse shapes the power is given by (29.27) without any additional assumptions. Also, by Theorem 15.4.1, the bandwidth of the PAM signal is typically equal to the bandwidth of the pulse shape. In fact, by that theorem, for linear binary $(\mathsf{K}, \mathsf{N})$ block encoders with antipodal signaling

$$\boxed{\text{bandwidth of PAM signal} = \text{bandwidth of pulse shape,}} \tag{29.36}$$

whenever $\mathsf{A} \neq 0$; the pulse shape $\mathbf{g}$ is a Borel measurable function satisfying the decay condition (14.17) for some $\alpha, \beta > 0$; and the encoder produces zero-mean symbols when fed IID random bits. Thus, if one is not interested in the exact form of the operational PSD but only in its support, then one need not check whether the encoder produces uncorrelated symbols when fed IID random bits.

## 29.6 Performance Criteria

Designing an optimal decoder for linear binary block encoders with antipodal signaling is conceptually very simple but algorithmically very difficult. The structure of the decoder depends on what we mean by "optimal." In this chapter we focus on two notions of optimality: minimizing the probability of a **block error**—also called **message error**—and minimizing the probability of a **bit error**. Referring to Figure 28.1, we say that a block error occurred in decoding the $\nu$-th block if at least one of the data bits $\left(D_{(\nu-1)K+1}, \ldots, D_{(\nu-1)K+K}\right)$ was incorrectly decoded. We say that a bit error occurred in decoding the $j$-th bit if $D_j$ was incorrectly decoded.

We consider the case where IID random bits are transmitted in block-mode and where the transmitted waveform is corrupted by additive Gaussian noise that is white with respect to the bandwidth $W$ of the pulse shape. The pulse shape is assumed to satisfy the orthonormality condition (29.2) and the decay condition (14.17). From Proposition 28.3.1 it follows that for both optimality criteria, there is no loss in optimality in feeding the received waveform to a matched filter for $\phi$ and in basing the decision on the filter's output sampled at integer multiples of $T_s$. Moreover, for the purposes of decoding a given message it suffices to consider only the samples corresponding to the symbols that were produced when the encoder encoded the given message (Proposition 28.4.2). Similarly, for decoding a given data bit it suffices to consider only the samples corresponding to the symbols that were produced when the encoder encoded the message of which the given bit is part. These observations lead us (as in Section 28.4.3) to the discrete-time single-block model (28.30). For convenience, we repeat this model here (with the additional assumption that the data are IID random bits):

$$\left(X_1, \ldots, X_N\right) = \mathrm{enc}\left(D_1, \ldots, D_K\right); \tag{29.37a}$$

$$Y_\eta = A X_\eta + Z_\eta, \quad \eta \in \{1, \ldots, N\}; \tag{29.37b}$$

$$Z_1, \ldots, Z_N \sim \mathrm{IID}\ \mathcal{N}\left(0, \frac{N_0}{2}\right); \tag{29.37c}$$

$$D_1, \ldots, D_K \sim \mathrm{IID}\ \mathcal{U}\left(\{0, 1\}\right), \tag{29.37d}$$

where $(Z_1, \ldots, Z_N)$ are independent of $(D_1, \ldots, D_K)$. We also introduce some additional notation. We use $x_\eta(\mathbf{d})$ for the $\eta$-th component of the N-tuple to which the binary K-tuple $\mathbf{d}$ is mapped by $\mathrm{enc}(\cdot)$:

$$x_\eta(\mathbf{d}) \triangleq \eta\text{-th component of } \mathrm{enc}(\mathbf{d}), \quad \left(\eta \in \{1, \ldots, N\},\ \mathbf{d} \in \mathbb{F}_2^K\right). \tag{29.38}$$

Denoting the conditional density of $(Y_1, \ldots, Y_N)$ given $(X_1, \ldots, X_N)$ by $f_{\mathbf{Y}|\mathbf{X}}(\cdot)$, we have for every $\mathbf{y} \in \mathbb{R}^N$ of components $y_1, \ldots, y_N$ and for every $\mathbf{x} \in \{-1, +1\}^N$ of components $x_1, \ldots, x_N$

$$f_{\mathbf{Y}|\mathbf{X}=\mathbf{x}}(\mathbf{y}) = (\pi N_0)^{-N/2} \prod_{\eta=1}^N \exp\left(-\frac{(y_\eta - A x_\eta)^2}{N_0}\right). \tag{29.39}$$

| Parameter | In Section 21.6 | In Section 29.7 |
|---|---|---|
| number of observations | J | N |
| number of hypotheses | M | $2^K$ |
| set of hypotheses | $\{1, \ldots, M\}$ | $\mathbb{F}_2^K$ |
| dummy hypothesis variable | $m$ | $\mathbf{d}$ |
| prior | $\{\pi_m\}$ | uniform |
| conditional mean tuple | $\left(s_m^{(1)}, \ldots, s_m^{(J)}\right)$ | $\left(Ax_1(\mathbf{d}), \ldots, Ax_N(\mathbf{d})\right)$ |
| conditional variance | $\sigma^2$ | $N_0/2$ |

**Table 29.1:** A conversion table for the setups of Section 21.6 and of Section 29.7.

Likewise, for every $\mathbf{y} \in \mathbb{R}^N$ and every data tuple $\mathbf{d} \in \mathbb{F}_2^K$,

$$f_{\mathbf{Y}|\mathbf{D}=\mathbf{d}}(\mathbf{y}) = (\pi N_0)^{-N/2} \prod_{\eta=1}^{N} \exp\left(-\frac{\left(y_\eta - Ax_\eta(\mathbf{d})\right)^2}{N_0}\right). \tag{29.40}$$

## 29.7   Minimizing the Block Error Rate

### 29.7.1   Optimal Decoding

To minimize the probability of a block error, we need to use the random N-vector $\mathbf{Y} = (Y_1, \ldots, Y_N)$ to guess the K-tuple $\mathbf{D} = (D_1, \ldots, D_K)$. This is the type of problem we addressed in Section 21.6. The translation between the setup of that section and our current setup is summarized in Table 29.1: the number of observations, which was given there by J, is here N; the number of hypotheses, which was given there by M, is here $2^K$; the set of possible messages, which was given there by $\mathcal{M} = \{1, \ldots, M\}$, is here the set of binary K-tuples $\mathbb{F}_2^K$; the dummy variable for a generic message, which was given there by $m$, is here the binary K-tuple $\mathbf{d}$; the prior, which was denoted there by $\{\pi_m\}$, is here uniform; the mean tuple corresponding to the $m$-th message, which was given there by $\left(s_m^{(1)}, \ldots, s_m^{(J)}\right)$ is here $\left(Ax_1(\mathbf{d}), \ldots, Ax_N(\mathbf{d})\right)$ (see (29.38)); and the conditional variance of each observation, which was given there by $\sigma^2$, is here $N_0/2$.

Because all the symbols produced by the encoder take value in $\{-1, +1\}$, it follows that

$$\sum_{\eta=1}^{N} \left(Ax_\eta(\mathbf{d})\right)^2 = A^2 N, \quad \mathbf{d} \in \mathbb{F}_2^K,$$

so all the mean tuples are of equal Euclidean norm. From Proposition 21.6.1 (iii) we thus obtain that, to minimize the probability of a block error, our guess should be the K-tuple $\mathbf{d}^*$ that satisfies

$$\sum_{\eta=1}^{N} x_\eta(\mathbf{d}^*) Y_\eta = \max_{\mathbf{d} \in \mathbb{F}_2^K} \sum_{\eta=1}^{N} x_\eta(\mathbf{d}) Y_\eta, \tag{29.41}$$

with ties being resolved uniformly at random among the data tuples that achieve the maximum. Our guess should thus be the data sequence that when fed to the encoder produces the $\{\pm 1\}$-valued N-tuple of highest correlation with the observed tuple $\mathbf{Y}$. Note that, by definition, all block encoders are one-to-one mappings and thus the mean tuples are distinct. Consequently, by Proposition 21.6.2, the probability that more than one tuple $\mathbf{d}^*$ satisfies (29.41) is zero.

Since guessing the data tuple is equivalent to guessing the N-tuple to which it is mapped, we can also describe the optimal decision rule in terms of the encoder's output.

**Proposition 29.7.1 (The Max-Correlation Decision Rule).** *Consider the problem of guessing* $\mathbf{D}$ *based on* $\mathbf{Y}$ *for the setup of Section 29.6.*

(i) *Picking at random a message from the set*

$$\left\{ \tilde{\mathbf{d}} \in \mathbb{F}_2^K : \sum_{\eta=1}^N x_\eta(\tilde{\mathbf{d}}) \, Y_\eta = \max_{\mathbf{d} \in \mathbb{F}_2^K} \sum_{\eta=1}^N x_\eta(\mathbf{d}) \, Y_\eta \right\} \tag{29.42}$$

*minimizes the probability of incorrectly guessing* $\mathbf{D}$.

(ii) *The probability that the above set contains more than one element is zero.*

(iii) *For the problem of guessing the encoder's output, picking at random an N-tuple from the set*

$$\left\{ \tilde{\mathbf{x}} \in \mathrm{Image}(\mathbf{enc}) : \sum_{\eta=1}^N \tilde{x}_\eta \, Y_\eta = \max_{\mathbf{x} \in \mathrm{Image}(\mathbf{enc})} \sum_{\eta=1}^N x_\eta \, Y_\eta \right\} \tag{29.43}$$

*minimizes the probability of error. This set contains more than one element with probability zero.*

Conceptually, the problem of finding an N-tuple that has the highest correlation with $(Y_1, \ldots, Y_N)$ among all the N-tuples in the image of $\mathrm{enc}(\cdot)$ is very simple: one goes over the list of all the $2^K$ N-tuples that are in the image of $\mathrm{enc}(\cdot)$ and picks the one that has the highest correlation with $(Y_1, \ldots, Y_N)$. But algorithmically this is very difficult because $2^K$ is in most applications a huge number. It is one of the challenges of Coding Theory to come up with encoders for which the decoding does not require an exhaustive search over all $2^K$ tuples. As we shall see, the single parity check code is an example of such a code. But the performance of this encoder is, alas, not stellar.

## 29.7.2 Wagner's Rule

For the $(K, K+1)$ systematic single parity check encoder (29.20), the decoding can be performed very efficiently using a decision algorithm that is called **Wagner's Rule** in honor of C.A. Wagner. Unlike the brute-force approach that considers all possible data tuples and which thus has a complexity which is *exponential* in $K$, the complexity of Wagner's Rule is *linear* in $K$.

Wagner's Rule can be summarized as follows. *Consider the* $(K + 1)$ *tuple*

$$\xi_\eta \triangleq \begin{cases} +1 & \text{if } Y_\eta \geq 0, \\ -1 & \text{otherwise,} \end{cases} \qquad \eta = 1, \ldots, K + 1. \tag{29.44}$$

*If this tuple has an even number of negative components, then guess that the encoder's output is* $(\xi_1, \ldots, \xi_{K+1})$ *and that the data sequence is thus the inverse of* $(\xi_1, \ldots, \xi_K)$ *under the componentwise antipodal mapping* $\Upsilon_K$*, i.e., that the data tuple is* $(1 - \xi_1)/2, \ldots, (1 - \xi_K)/2$*. Otherwise, flip the sign of* $\xi_{\eta_*}$ *corresponding to the* $Y_{\eta_*}$ *of smallest magnitude. I.e., guess that the encoder's output is*

$$\xi_1, \ldots, \xi_{\eta_* - 1}, -\xi_{\eta_*}, \xi_{\eta_* + 1} \cdots, \xi_{K+1}, \tag{29.45}$$

*and that the data bits are*

$$\frac{1 - \xi_1}{2}, \ldots, \frac{1 - \xi_{\eta_* - 1}}{2}, \frac{1 + \xi_{\eta_*}}{2}, \frac{1 - \xi_{\eta_* + 1}}{2} \cdots, \frac{1 - \xi_K}{2}, \tag{29.46}$$

*where* $\eta_*$ *is the element of* $\{1, \ldots, K + 1\}$ *satisfying*

$$|Y_{\eta_*}| = \min_{1 \leq \eta \leq K+1} |Y_\eta|. \tag{29.47}$$

**Proof that Wagner's Rule is Optimal.** Recall that the $(K, K + 1)$ single parity check code with antipodal signaling consists of all $\pm 1$-valued $(K + 1)$-tuples having an even number of $-1$'s. We seek to find the tuple that among all such tuples maximizes the correlation with the received tuple $(Y_1, \ldots, Y_{K+1})$. The tuple defined in (29.44) is the tuple that among *all* tuples in $\{-1, +1\}^{K+1}$ has the highest correlation with $(Y_1, \ldots, Y_{K+1})$. Since flipping the sign of $\xi_\eta$ reduces the correlation by $2|Y_\eta|$, the tuple (29.45) has the second-highest correlation among all the tuples in $\{-1, +1\}^{K+1}$. Since the tuples (29.44) and (29.45) differ in one component, exactly one of them has an even number of negative components. That tuple thus maximizes the correlation among all tuples in $\{-1, +1\}^{K+1}$ that have an even number of negative components and is thus the tuple we are after.

Since the encoder is systematic, the data tuple that generates a given encoder output is easily found by considering the first $K$ components of the encoder output and by then applying the mapping $+1 \mapsto 0$ and $-1 \mapsto 1$, i.e., $\xi \mapsto (1 - \xi)/2$. $\square$

## 29.7.3 The Probability of a Block Error

We next address the performance of the detector that we designed in Section 29.7.1 when we sought to minimize the probability of a block error. We continue to assume that the encoder is a linear binary $(K, N)$ block encoder with antipodal signaling, so the encoder function $\text{enc}(\cdot)$ can be written as $\mathbf{enc} = \Upsilon_N \circ T$ where $T \colon \mathbb{F}_2^K \to \mathbb{F}_2^N$ is a linear one-to-one mapping and $\Upsilon_N(\cdot)$ is the componentwise antipodal mapping.

### An Upper Bound

It is usually very difficult to precisely evaluate the probability of a block error. A very useful bound is the Union Bound, which we encountered in Section 21.6.3. Denoting by $p_{\mathrm{MAP}}(\mathrm{error}|\mathbf{D} = \mathbf{d})$ the probability of error of our guessing rule conditional on the binary K-tuple $\mathbf{D} = \mathbf{d}$ being fed to the encoder, we can use (21.59), Table 29.1, and (29.18) to obtain

$$p_{\mathrm{MAP}}(\mathrm{error}|\mathbf{D} = \mathbf{d}) \leq \sum_{\mathbf{d}' \in \mathbb{F}_2^K \setminus \{\mathbf{d}\}} \mathcal{Q}\left(\sqrt{\frac{2A^2 d_{\mathrm{H}}(\mathsf{T}(\mathbf{d}'), \mathsf{T}(\mathbf{d}))}{N_0}}\right). \tag{29.48}$$

It is customary to group all the equal terms on the RHS of (29.48) and to write the bound in the equivalent form

$$p_{\mathrm{MAP}}(\mathrm{error}|\mathbf{D} = \mathbf{d}) \leq \sum_{\nu=1}^{N} \#\left\{\mathbf{d}' \in \mathbb{F}_2^K : d_{\mathrm{H}}(\mathsf{T}(\mathbf{d}'), \mathsf{T}(\mathbf{d})) = \nu\right\} \mathcal{Q}\left(\sqrt{\frac{2A^2 \nu}{N_0}}\right), \tag{29.49}$$

where

$$\#\left\{\mathbf{d}' \in \mathbb{F}_2^K : d_{\mathrm{H}}(\mathsf{T}(\mathbf{d}'), \mathsf{T}(\mathbf{d})) = \nu\right\} \tag{29.50}$$

is the number of data tuples that are mapped by $\mathsf{T}(\cdot)$ to a binary N-tuple that is at Hamming distance $\nu$ from $\mathsf{T}(\mathbf{d})$, and where the sum excludes $\nu = 0$ because the fact that $\mathsf{T}(\cdot)$ is one-to-one implies that if $\mathbf{d}' \neq \mathbf{d}$ then the Hamming distance between $\mathsf{T}(\mathbf{d}')$ and $\mathsf{T}(\mathbf{d})$ must be at least one.

We next show that the linearity of $\mathsf{T}(\cdot)$ implies that the RHS of (29.49) does not depend on $\mathbf{d}$. (In Section 29.9 we show that this is also true of the LHS.) To this end we show that for every $\nu \in \{1, \ldots, N\}$ and for every $\mathbf{d} \in \mathbb{F}_2^K$,

$$\#\left\{\mathbf{d}' \in \mathbb{F}_2^K : d_{\mathrm{H}}(\mathsf{T}(\mathbf{d}'), \mathsf{T}(\mathbf{d})) = \nu\right\} = \#\left\{\tilde{\mathbf{d}} \in \mathbb{F}_2^K : w_{\mathrm{H}}(\mathsf{T}(\tilde{\mathbf{d}})) = \nu\right\} \tag{29.51}$$

$$= \#\left\{\mathbf{c} \in \mathrm{Image}(\mathsf{T}) : w_{\mathrm{H}}(\mathbf{c}) = \nu\right\}, \tag{29.52}$$

where the RHS of (29.51) is the evaluation of the LHS at $\mathbf{d} = \mathbf{0}$. To prove (29.51) we note that the mapping $\mathbf{d}' \mapsto \mathbf{d}' \oplus \mathbf{d}$ is a one-to-one mapping from the set whose cardinality is written on the LHS to the set whose cardinality is written on the RHS, because

$$\left(d_{\mathrm{H}}(\mathsf{T}(\mathbf{d}'), \mathsf{T}(\mathbf{d})) = \nu\right) \Leftrightarrow \left(w_{\mathrm{H}}(\mathsf{T}(\mathbf{d}) \oplus \mathsf{T}(\mathbf{d}')) = \nu\right)$$

$$\Leftrightarrow \left(w_{\mathrm{H}}(\mathsf{T}(\mathbf{d} \oplus \mathbf{d}')) = \nu\right),$$

where the first equivalence follows from (29.13), and where the second equivalence follows from the linearity of $\mathsf{T}(\cdot)$. To prove (29.52) we merely substitute $\mathbf{c}$ for $\mathsf{T}(\tilde{\mathbf{d}})$ in (29.51) and use the fact that $\mathsf{T}(\cdot)$ is one-to-one.

Combining (29.49) with (29.52) we obtain the bound

$$p_{\mathrm{MAP}}(\mathrm{error}|\mathbf{D} = \mathbf{d}) \leq \sum_{\nu=1}^{N} \#\left\{\mathbf{c} \in \mathrm{Image}(\mathsf{T}) : w_{\mathrm{H}}(\mathbf{c}) = \nu\right\} \mathcal{Q}\left(\sqrt{\frac{2A^2 \nu}{N_0}}\right). \tag{29.53}$$

The list of $N + 1$ nonnegative integers

$$\left( \#\{\mathbf{c} \in \mathrm{Image}(\mathsf{T}) : \mathrm{w_H}(\mathbf{c}) = 0\}, \ldots, \#\{\mathbf{c} \in \mathrm{Image}(\mathsf{T}) : \mathrm{w_H}(\mathbf{c}) = N\} \right)$$

(whose first term is equal to one and whose terms sum to $2^K$) is called the **weight enumerator** of the code.

For example, for the $(K, K+1)$ single parity check code

$$\#\left\{ \tilde{\mathbf{d}} \in \mathbb{F}_2^K : \mathrm{w_H}\left(\mathsf{T}(\tilde{\mathbf{d}})\right) = \nu \right\} = \begin{cases} 0 & \text{if } \nu \text{ is odd,} \\ \binom{K+1}{\nu} & \text{if } \nu \text{ is even,} \end{cases} \quad \nu = 0, \ldots, K+1$$

because this code consists of all $(K+1)$-tuples of even Hamming weight. Consequently, this code's weight enumerator is

$$\left( 1, 0, \binom{K+1}{2}, 0, \binom{K+1}{4}, 0, \ldots, 0, \binom{K+1}{K+1} \right), \quad \text{if } K \text{ is odd;}$$

$$\left( 1, 0, \binom{K+1}{2}, 0, \binom{K+1}{4}, 0, \ldots, \binom{K+1}{K}, 0 \right), \quad \text{if } K \text{ is even.}$$

The **minimum Hamming distance** $\mathrm{d_{min,H}}$ of a linear $(K, N)$ $\mathbb{F}_2$ code is the smallest Hamming distance between distinct elements of the code. (If $K = 0$, i.e., if the only codeword is the all-zero codeword, then, by convention, the minimum distance is said to be infinite.) By (29.52) it follows that (for $K > 0$) the minimum Hamming distance of a code is also the smallest weight that a nonzero codeword can have

$$\mathrm{d_{min,H}} = \min_{\mathbf{c} \in \mathrm{Image}(\mathsf{T}) \setminus \{\mathbf{0}\}} \mathrm{w_H}(\mathbf{c}). \tag{29.54}$$

With this definition we can rewrite (29.53) as

$$p_{\mathrm{MAP}}(\text{error}|\mathbf{D} = \mathbf{d}) \leq \sum_{\nu = \mathrm{d_{min,H}}}^{N} \#\{\mathbf{c} \in \mathrm{Image}(\mathsf{T}) : \mathrm{w_H}(\mathbf{c}) = \nu\} \, Q\left( \sqrt{\frac{2A^2\nu}{N_0}} \right). \tag{29.55}$$

Engineers sometimes approximate the RHS of (29.55) by its first term:

$$\#\{\mathbf{c} \in \mathrm{Image}(\mathsf{T}) : \mathrm{w_H}(\mathbf{c}) = \mathrm{d_{min,H}}\} \, Q\left( \sqrt{\frac{2A^2\mathrm{d_{min,H}}}{N_0}} \right). \tag{29.56}$$

This is reasonable when $A^2/N_0 \gg 1$ because the $Q(\cdot)$ function decays very rapidly; see (19.18).

The term

$$\#\{\mathbf{c} \in \mathrm{Image}(\mathsf{T}) : \mathrm{w_H}(\mathbf{c}) = \mathrm{d_{min,H}}\} \tag{29.57}$$

is sometimes called the **number of nearset neighbors**.

**A Lower Bound**

Using the results of Section 21.6.4, we can obtain a lower bound on the probability of a block error. Indeed, by (21.65), Table 29.1, (29.18), the monotonicity of $\mathcal{Q}(\cdot)$, and the definition of $d_{\min,H}$

$$p_{\mathrm{MAP}}(\mathrm{error}|\mathbf{D} = \mathbf{d}) \geq \mathcal{Q}\left(\sqrt{\frac{2A^2 d_{\min,H}}{N_0}}\right). \tag{29.58}$$

## 29.8 Minimizing the Bit Error Rate

In some applications we want to minimize the number of data bits that are incorrectly decoded. This performance criterion leads to a different guessing rule, which we derive and analyze in this section.

### 29.8.1 Optimal Decoding

We next derive the guessing rule that minimizes the average probability of a bit error, or the Bit Error Rate. Conceptually, this is simple. For each $\kappa \in \{1, \ldots, K\}$ our guess of the $\kappa$-th data bit $D_\kappa$ should minimize the probability of error. This problem falls under the category of binary hypothesis testing, and, since $D_\kappa$ is *a priori* equally likely to be 0 or 1, the Maximum-Likelihood rule of Section 20.8 is optimal. To compute the likelihood-ratio function, we treat the other data bits $D_1, \ldots, D_{\kappa-1}, D_{\kappa+1}, \ldots, D_K$ as unobserved random parameters (Section 20.15.1). Thus, using (20.101) with the random parameter $\Theta$ now corresponding to the tuple $(D_1, \ldots, D_{\kappa-1}, D_{\kappa+1}, \ldots, D_K)$ we obtain[2]

$$
\begin{aligned}
f_{\mathbf{Y}|D_\kappa = 0}&(y_1, \ldots, y_N) \\
&= 2^{-(K-1)} \sum_{\mathbf{d} \in \mathcal{A}_{\kappa,0}} f_{Y_1, \ldots, Y_N | \mathbf{D} = \mathbf{d}}(y_1, \ldots, y_N) \tag{29.59}
\end{aligned}
$$

$$
= 2^{-(K-1)} (\pi N_0)^{-N/2} \sum_{\mathbf{d} \in \mathcal{A}_{\kappa,0}} \prod_{\eta=1}^{N} \exp\left(-\frac{(y_\eta - A x_\eta(\mathbf{d}))^2}{N_0}\right), \tag{29.60}
$$

where the set $\mathcal{A}_{\kappa,0}$ consists of those tuples in $\mathbb{F}_2^K$ whose $\kappa$-th component is zero

$$\mathcal{A}_{\kappa,0} = \left\{(d_1, \ldots, d_K) \in \mathbb{F}_2^K : d_\kappa = 0\right\}. \tag{29.61}$$

Likewise,

$$
\begin{aligned}
f_{\mathbf{Y}|D_\kappa = 1}&(y_1, \ldots, y_N) \\
&= 2^{-(K-1)} \sum_{\mathbf{d} \in \mathcal{A}_{\kappa,1}} f_{Y_1, \ldots, Y_N | \mathbf{D} = \mathbf{d}}(y_1, \ldots, y_N) \tag{29.62}
\end{aligned}
$$

---

[2]Our assumption that the data are IID random bits guarantees that the random parameter $\Theta \triangleq (D_1, \ldots, D_{\kappa-1}, D_{\kappa+1}, \ldots, D_K)$ is independent of the RV $D_\kappa$ that we wish to guess.

$$= 2^{-(\mathsf{K}-1)} (\pi \mathsf{N}_0)^{-\mathsf{N}/2} \sum_{\mathbf{d} \in \mathcal{A}_{\kappa,1}} \prod_{\eta=1}^{\mathsf{N}} \exp\left(-\frac{\left(y_\eta - \mathsf{A} x_\eta(\mathbf{d})\right)^2}{\mathsf{N}_0}\right), \quad (29.63)$$

where we similarly define

$$\mathcal{A}_{\kappa,1} = \left\{ (d_1, \ldots, d_\mathsf{K}) \in \mathbb{F}_2^\mathsf{K} : d_\kappa = 1 \right\}. \quad (29.64)$$

Using Theorem 20.7.1 and (29.60) & (29.63) we obtain the following.

**Proposition 29.8.1 (Minimizing the BER).** *Consider the problem of guessing $D_\kappa$ based on $\mathbf{Y}$ for the setup of Section 29.6, where $\kappa \in \{1, \ldots, \mathsf{K}\}$. The decision rule that guesses "$D_\kappa = 0$" if*

$$\sum_{\mathbf{d} \in \mathcal{A}_{\kappa,0}} \prod_{\eta=1}^{\mathsf{N}} \exp\left(-\frac{\left(y_\eta - \mathsf{A} x_\eta(\mathbf{d})\right)^2}{\mathsf{N}_0}\right) > \sum_{\mathbf{d} \in \mathcal{A}_{\kappa,1}} \prod_{\eta=1}^{\mathsf{N}} \exp\left(-\frac{\left(y_\eta - \mathsf{A} x_\eta(\mathbf{d})\right)^2}{\mathsf{N}_0}\right);$$

*that guesses "$D_\kappa = 1$" if*

$$\sum_{\mathbf{d} \in \mathcal{A}_{\kappa,0}} \prod_{\eta=1}^{\mathsf{N}} \exp\left(-\frac{\left(y_\eta - \mathsf{A} x_\eta(\mathbf{d})\right)^2}{\mathsf{N}_0}\right) < \sum_{\mathbf{d} \in \mathcal{A}_{\kappa,1}} \prod_{\eta=1}^{\mathsf{N}} \exp\left(-\frac{\left(y_\eta - \mathsf{A} x_\eta(\mathbf{d})\right)^2}{\mathsf{N}_0}\right);$$

*and that guesses at random in case of equality minimizes the probability of guessing the data bit $D_\kappa$ incorrectly.*

The difficulty in implementing this decision rule is that, unless we exploit some algebraic structure, the computation of the sums above has exponential complexity because the number of terms in each sum is $2^{\mathsf{K}-1}$.

It is interesting to note that, unlike the decision rule that minimizes the probability of a block error, the above decision rule depends on the value of $\mathsf{N}_0/2$.

### 29.8.2  The Probability of a Bit Error

We next obtain bounds on the probability that the detector of Proposition 29.8.1 errs in guessing the $\kappa$-th data bit $D_\kappa$. We denote this probability by $p_\kappa^*$.

**An Upper Bound**

Since the detector of Proposition 29.8.1 is optimal, the probability that it errs in decoding the $\kappa$-th data bit $D_\kappa$ cannot exceed the probability of error of the suboptimal rule whose guess for $D_\kappa$ is the $\kappa$-th bit of the message produced by the detector of Section 29.7. Thus, if $\phi_{\mathrm{MAP}}(\cdot)$ denotes the decision rule of Section 29.7, then

$$p_\kappa^* \leq \Pr\left[\mathbf{D} \oplus \phi_{\mathrm{MAP}}(\mathbf{Y}) \in \mathcal{A}_{\kappa,1}\right], \quad \kappa \in \{1, \ldots, \mathsf{K}\}, \quad (29.65)$$

where the set $\mathcal{A}_{\kappa,1}$ was defined in (29.64) as the set of messages whose $\kappa$-th component is equal to one, and where $\mathbf{Y}$ is the observed N-tuple whose components are given in (29.37b).

Since the data are IID random bits, we can rewrite (29.65) as

$$p_\kappa^* \leq \frac{1}{2^\mathsf{K}} \sum_{\mathbf{d} \in \mathbb{F}_2^\mathsf{K}} \sum_{\tilde{\mathbf{d}} \in \mathcal{A}_{\kappa,1}} \Pr\left[\phi_{\mathrm{MAP}}(\mathbf{Y}) = \mathbf{d} \oplus \tilde{\mathbf{d}} \,\middle|\, \mathbf{D} = \mathbf{d}\right], \quad \kappa \in \{1, \ldots, \mathsf{K}\}. \quad (29.66)$$

Since $\phi_{\mathrm{MAP}}(\mathbf{Y})$ can only equal $\mathbf{d} \oplus \tilde{\mathbf{d}}$ if $\mathbf{Y}$ is at least as close in Euclidean distance to $\mathrm{enc}(\mathbf{d} \oplus \tilde{\mathbf{d}})$ as it is to $\mathrm{enc}(\mathbf{d})$, it follows from Lemma 20.14.1, Table 29.1, and (29.18) that

$$\Pr\left[\phi_{\mathrm{MAP}}(\mathbf{Y}) = \mathbf{d} \oplus \tilde{\mathbf{d}} \,\middle|\, \mathbf{D} = \mathbf{d}\right] \leq \mathcal{Q}\left(\frac{A d_\mathrm{E}\left(\Upsilon_\mathsf{N}(\mathsf{T}(\mathbf{d} \oplus \tilde{\mathbf{d}})), \Upsilon_\mathsf{N}(\mathsf{T}(\mathbf{d}))\right)}{2\sqrt{\frac{N_0}{2}}}\right)$$

$$= \mathcal{Q}\left(\sqrt{\frac{A^2 d_\mathrm{E}^2\left(\Upsilon_\mathsf{N}(\mathsf{T}(\mathbf{d} \oplus \tilde{\mathbf{d}})), \Upsilon_\mathsf{N}(\mathsf{T}(\mathbf{d}))\right)}{2N_0}}\right)$$

$$= \mathcal{Q}\left(\sqrt{\frac{2A^2 d_\mathrm{H}\left(\mathsf{T}(\mathbf{d} \oplus \tilde{\mathbf{d}}), \mathsf{T}(\mathbf{d})\right)}{N_0}}\right)$$

$$= \mathcal{Q}\left(\sqrt{\frac{2A^2 w_\mathrm{H}\left(\mathsf{T}(\mathbf{d} \oplus \tilde{\mathbf{d}}) \oplus \mathsf{T}(\mathbf{d})\right)}{N_0}}\right)$$

$$= \mathcal{Q}\left(\sqrt{\frac{2A^2 w_\mathrm{H}\left(\mathsf{T}(\tilde{\mathbf{d}})\right)}{N_0}}\right). \quad (29.67)$$

It follows from (29.66) and (29.67) upon noting that RHS of (29.67) does not depend on the transmitted message $\mathbf{d}$ that

$$p_\kappa^* \leq \sum_{\tilde{\mathbf{d}} \in \mathcal{A}_{\kappa,1}} \mathcal{Q}\left(\sqrt{\frac{2A^2 w_\mathrm{H}\left(\mathsf{T}(\tilde{\mathbf{d}})\right)}{N_0}}\right), \quad \kappa \in \{1, \ldots, \mathsf{K}\}. \quad (29.68)$$

This bound is sometimes written as

$$\boxed{p_\kappa^* \leq \sum_{\nu = \mathrm{d_{min,H}}}^\mathsf{N} \gamma(\nu, \kappa)\, \mathcal{Q}\left(\sqrt{\frac{2A^2 \nu}{N_0}}\right), \quad \kappa \in \{1, \ldots, \mathsf{K}\},} \quad (29.69a)$$

where $\gamma(\nu, \kappa)$ denotes the number of elements $\tilde{\mathbf{d}}$ of $\mathbb{F}_2^\mathsf{K}$ whose $\kappa$-th component is equal to one and for which $\mathsf{T}(\tilde{\mathbf{d}})$ is of Hamming weight $\nu$, i.e.,

$$\gamma(\nu, \kappa) = \#\left\{\tilde{\mathbf{d}} \in \mathcal{A}_{\kappa,1} : w_\mathrm{H}\left(\mathsf{T}(\tilde{\mathbf{d}})\right) = \nu\right\}, \quad (29.69b)$$

and where the minimum Hamming distance $\mathrm{d_{min,H}}$ is defined in (29.54).

Sometimes one is more interested in the arithmetic average of $p_\kappa^*$

$$\frac{1}{\mathsf{K}} \sum_{\kappa=1}^\mathsf{K} p_\kappa^*, \quad (29.70)$$

which is the **optimal bit error rate**. We next show that (29.68) leads to the upper bound

$$
\frac{1}{K} \sum_{\kappa=1}^{K} p_\kappa^* \leq \frac{1}{K} \sum_{d \in \mathbb{F}_2^K} w_H(d) \, Q\left(\sqrt{\frac{2A^2 w_H(T(d))}{N_0}}\right).
\tag{29.71}
$$

This follows from the calculation

$$
\begin{aligned}
\sum_{\kappa=1}^{K} p_\kappa^* &\leq \sum_{\kappa=1}^{K} \sum_{d \in \mathcal{A}_{\kappa,1}} Q\left(\sqrt{\frac{2A^2 w_H(T(d))}{N_0}}\right) \\
&= \sum_{\kappa=1}^{K} \sum_{d \in \mathbb{F}_2^K} Q\left(\sqrt{\frac{2A^2 w_H(T(d))}{N_0}}\right) I\{d \in \mathcal{A}_{\kappa,1}\} \\
&= \sum_{d \in \mathbb{F}_2^K} Q\left(\sqrt{\frac{2A^2 w_H(T(d))}{N_0}}\right) \sum_{\kappa=1}^{K} I\{d \in \mathcal{A}_{\kappa,1}\} \\
&= \sum_{d \in \mathbb{F}_2^K} Q\left(\sqrt{\frac{2A^2 w_H(T(d))}{N_0}}\right) w_H(d),
\end{aligned}
$$

where the inequality in the first line follows from (29.68); the equality in the second by introducing the indicator function for the set $\mathcal{A}_{\kappa,1}$ and extending the summation; the equality in the third line by changing the order of summation; and the equality in the last line by noting that every $d \in \mathbb{F}_2^K$ is in exactly $w_H(d)$ of the sets $\mathcal{A}_{1,1}, \ldots, \mathcal{A}_{K,1}$.

## A Lower Bound

We next show that, for every $\kappa \in \{1, \ldots, K\}$, the probability $p_\kappa^*$ that the optimal detector for guessing the $\kappa$-th data bit errs is lower-bounded by

$$
p_\kappa^* \geq \max_{d \in \mathcal{A}_{\kappa,1}} Q\left(\sqrt{\frac{2A^2 w_H(T(d))}{N_0}}\right),
\tag{29.72}
$$

where $\mathcal{A}_{\kappa,1}$ denotes the set of binary K-tuples whose $\kappa$-th component is equal to one (29.64). To derive (29.72), fix some $d \in \mathcal{A}_{\kappa,1}$ and note that for every $d' \in \mathbb{F}_2^K$

$$
(d' \in \mathcal{A}_{\kappa,0}) \Leftrightarrow (d' \oplus d \in \mathcal{A}_{\kappa,1}).
\tag{29.73}
$$

This allows us to express $f_{Y|D_\kappa=1}(y)$ for every $y \in \mathbb{R}^N$ as

$$
\begin{aligned}
f_{Y|D_\kappa=1}(y) &= 2^{-(K-1)} \sum_{\tilde{d} \in \mathcal{A}_{\kappa,1}} f_{Y|D=\tilde{d}}(y) \\
&= 2^{-(K-1)} \sum_{d' \in \mathcal{A}_{\kappa,0}} f_{Y|D=d \oplus d'}(y),
\end{aligned}
\tag{29.74}
$$

where the first equality follows from (29.62) and the second from (29.73).

Using the exact expression for the probability of error in binary hypothesis testing (20.20) we have:

$$
\begin{aligned}
p_\kappa^* &= \frac{1}{2} \int_{\mathbf{y} \in \mathbb{R}^N} \min\left\{ f_{\mathbf{Y}|D_\kappa=0}(\mathbf{y}), f_{\mathbf{Y}|D_\kappa=1}(\mathbf{y}) \right\} d\mathbf{y} \\
&= \frac{1}{2} \int \min\left\{ 2^{-(K-1)} \sum_{\mathbf{d}' \in \mathcal{A}_{\kappa,0}} f_{\mathbf{Y}|\mathbf{D}=\mathbf{d}'}(\mathbf{y}), \; 2^{-(K-1)} \sum_{\mathbf{d}' \in \mathcal{A}_{\kappa,0}} f_{\mathbf{Y}|\mathbf{D}=\mathbf{d}\oplus\mathbf{d}'}(\mathbf{y}) \right\} d\mathbf{y} \\
&= 2^{-(K-1)} \frac{1}{2} \int \min\left\{ \sum_{\mathbf{d}' \in \mathcal{A}_{\kappa,0}} f_{\mathbf{Y}|\mathbf{D}=\mathbf{d}'}(\mathbf{y}), \; \sum_{\mathbf{d}' \in \mathcal{A}_{\kappa,0}} f_{\mathbf{Y}|\mathbf{D}=\mathbf{d}\oplus\mathbf{d}'}(\mathbf{y}) \right\} d\mathbf{y} \\
&\geq 2^{-(K-1)} \frac{1}{2} \int \sum_{\mathbf{d}' \in \mathcal{A}_{\kappa,0}} \min\left\{ f_{\mathbf{Y}|\mathbf{D}=\mathbf{d}'}(\mathbf{y}), \; f_{\mathbf{Y}|\mathbf{D}=\mathbf{d}\oplus\mathbf{d}'}(\mathbf{y}) \right\} d\mathbf{y} \\
&= 2^{-(K-1)} \sum_{\mathbf{d}' \in \mathcal{A}_{\kappa,0}} \int \frac{1}{2} \min\left\{ f_{\mathbf{Y}|\mathbf{D}=\mathbf{d}'}(\mathbf{y}), \; f_{\mathbf{Y}|\mathbf{D}=\mathbf{d}\oplus\mathbf{d}'}(\mathbf{y}) \right\} d\mathbf{y} \\
&= 2^{-(K-1)} \sum_{\mathbf{d}' \in \mathcal{A}_{\kappa,0}} Q\left( \sqrt{\frac{2A^2 d_H\left(\mathsf{T}(\mathbf{d}'), \mathsf{T}(\mathbf{d}' \oplus \mathbf{d})\right)}{N_0}} \right) \\
&= 2^{-(K-1)} \sum_{\mathbf{d}' \in \mathcal{A}_{\kappa,0}} Q\left( \sqrt{\frac{2A^2 w_H\left(\mathsf{T}(\mathbf{d})\right)}{N_0}} \right) \\
&= Q\left( \sqrt{\frac{2A^2 w_H\left(\mathsf{T}(\mathbf{d})\right)}{N_0}} \right), \quad \mathbf{d} \in \mathcal{A}_{\kappa,1},
\end{aligned}
$$

where the first line follows from (20.20); the second by the explicit forms (29.59) & (29.74) of the conditional densities $f_{\mathbf{Y}|D_\kappa=0}(\cdot)$ and $f_{\mathbf{Y}|D_\kappa=1}(\cdot)$; the third by pulling the common term $2^{-(K-1)}$ outside the minimum; the fourth because the minimum between two sums with an equal number of terms is lower-bounded by the sum of the minima between the corresponding terms; the fifth by swapping the summation and integration; the sixth by Expression (20.20) for the optimal probability of error for the binary hypothesis testing between $\mathbf{D} = \mathbf{d}'$ and $\mathbf{D} = \mathbf{d} \oplus \mathbf{d}'$; the seventh by the linearity of $\mathsf{T}(\cdot)$; and the final line because the cardinality of $\mathcal{A}_{\kappa,0}$ is $2^{(K-1)}$. Since the above derivation holds for every $\mathbf{d} \in \mathcal{A}_{\kappa,1}$, we may choose $\mathbf{d}$ to yield the tightest bound, thus establishing (29.72).

## 29.9 Assuming the All-Zero Codeword

When simulating linear binary block encoders with antipodal signaling over the Gaussian channel we rarely simulate the data as IID random bits. Instead we assume that the message that is fed to the encoder is the all-zero message and that the encoder's output is hence the N-tuple whose components are all +1. In this section we shall explain why it is correct to do so. More specifically, we shall show that $p_{\mathrm{MAP}}(\text{error}|\mathbf{D} = \mathbf{d})$ does not depend on the message $\mathbf{d}$ and is thus equal to

$p_{\mathrm{MAP}}(\mathrm{error}|\mathbf{D} = \mathbf{0})$. We shall also prove an analogous result for the decoder that minimizes the probability of a bit error. The proofs are based on two features of our setup: the encoder is linear and the Gaussian channel with antipodal inputs is symmetric in the sense that

$$f_{Y|X=-1}(y) = f_{Y|X=+1}(-y), \quad y \in \mathbb{R}. \tag{29.75}$$

Indeed, by (29.37b),

$$f_{Y|X=-1}(y) = \frac{1}{\sqrt{\pi \mathrm{N}_0}} e^{-\frac{(y+A)^2}{\mathrm{N}_0}}$$

$$= \frac{1}{\sqrt{\pi \mathrm{N}_0}} e^{-\frac{(-y-A)^2}{\mathrm{N}_0}}$$

$$= f_{Y|X=+1}(-y), \quad y \in \mathbb{R}.$$

**Definition 29.9.1 (Memoryless Binary-Input/Output-Symmetric Channel).** *We say that the conditional distribution of* $\mathbf{Y} = (Y_1, \ldots, Y_{\mathrm{N}})$ *conditional on* $\mathbf{X} = (X_1, \ldots, X_{\mathrm{N}})$ *corresponds to a* ***memoryless binary-input/output-symmetric channel*** *if*

$$f_{\mathbf{Y}|\mathbf{X}=\mathbf{x}}(\mathbf{y}) = \prod_{\eta=1}^{\mathrm{N}} f_{Y|X=x_\eta}(y_\eta), \quad \mathbf{x} \in \{-1, +1\}^{\mathrm{N}}, \tag{29.76a}$$

*where*

$$f_{Y|X=-1}(y) = f_{Y|X=+1}(-y), \quad y \in \mathbb{R}. \tag{29.76b}$$

For every $\mathbf{d} \in \mathbb{F}_2^{\mathrm{K}}$ define the mapping $\psi_{\mathbf{d}} \colon \mathbb{R}^{\mathrm{N}} \to \mathbb{R}^{\mathrm{N}}$ as

$$\psi_{\mathbf{d}} \colon (y_1, \ldots, y_{\mathrm{N}}) \mapsto (y_1 x_1(\mathbf{d}), \ldots, y_{\mathrm{N}} x_{\mathrm{N}}(\mathbf{d})). \tag{29.77}$$

The function $\psi_{\mathbf{d}}(\cdot)$ thus changes the sign of those components of its argument that correspond to the negative components of $\mathrm{enc}(\mathbf{d})$. The key properties of this mapping are summarized in the following lemma.

**Lemma 29.9.2.** *As in (29.38), let* $x_\eta(\mathbf{d})$ *denote the result of applying the antipodal mapping* $\Upsilon$ *to the* $\eta$-*th component of* $\mathsf{T}(\mathbf{d})$, *where* $\mathsf{T} \colon \mathbb{F}_2^{\mathrm{K}} \to \mathbb{F}_2^{\mathrm{N}}$ *is some one-to-one linear mapping. Let the conditional law of* $(Y_1, \ldots, Y_{\mathrm{N}})$ *given* $\mathbf{D} = \mathbf{d}$ *be given by* $\prod_{\eta=1}^{\mathrm{N}} f_{Y|X=x_\eta(\mathbf{d})}(y_\eta)$, *where* $f_{Y|X}(\cdot)$ *satisfies the symmetry property (29.75). Let* $\psi_{\mathbf{d}}(\cdot)$ *be defined as in (29.77). Then*

*(i)* $\psi_{\mathbf{0}}(\cdot)$ *maps each* $\mathbf{y} \in \mathbb{R}^{\mathrm{N}}$ *to itself.*

*(ii) For any* $\mathbf{d}, \mathbf{d}' \in \mathbb{F}_2^{\mathrm{K}}$ *the composition of* $\psi_{\mathbf{d}'}$ *with* $\psi_{\mathbf{d}}$ *is given by* $\psi_{\mathbf{d} \oplus \mathbf{d}'}$:

$$\psi_{\mathbf{d}} \circ \psi_{\mathbf{d}'} = \psi_{\mathbf{d} \oplus \mathbf{d}'}. \tag{29.78}$$

*(iii)* $\psi_{\mathbf{d}}$ *is equal to its inverse*

$$\psi_{\mathbf{d}}(\psi_{\mathbf{d}}(\mathbf{y})) = \mathbf{y}, \quad \mathbf{y} \in \mathbb{R}^{\mathrm{N}}. \tag{29.79}$$

*(iv) For every* $\mathbf{d} \in \mathbb{F}_2^{\mathrm{K}}$ *the Jacobian of the mapping* $\psi_{\mathbf{d}}(\cdot)$ *is one.*

(v) *For every* $\mathbf{d} \in \mathbb{F}_2^{\mathsf{K}}$ *and every* $\mathbf{y} \in \mathbb{R}^{\mathsf{N}}$,

$$f_{\mathbf{Y}|\mathbf{D}=\mathbf{d}}(\mathbf{y}) = f_{\mathbf{Y}|\mathbf{D}=\mathbf{0}}\big(\psi_{\mathbf{d}}(\mathbf{y})\big). \tag{29.80}$$

(vi) *For any* $\mathbf{d}, \mathbf{d}' \in \mathbb{F}_2^{\mathsf{K}}$ *and every* $\mathbf{y} \in \mathbb{R}^{\mathsf{N}}$,

$$f_{\mathbf{Y}|\mathbf{D}=\mathbf{d}'}\big(\psi_{\mathbf{d}}(\mathbf{y})\big) = f_{\mathbf{Y}|\mathbf{D}=\mathbf{d}'\oplus\mathbf{d}}(\mathbf{y}). \tag{29.81}$$

**Proof.** Part (i) follows from the definition (29.77) because the linearity of $\mathsf{T}(\cdot)$ and the definition of $\Upsilon_{\mathsf{N}}$ guarantee that $x_\eta(\mathbf{0}) = +1$, for all $\eta \in \{1, \ldots, \mathsf{N}\}$. Part (ii) follows by linearity and from (29.15):

$$\begin{aligned}
(\psi_{\mathbf{d}} \circ \psi_{\mathbf{d}'})(y_1, \ldots, y_{\mathsf{N}}) &= \psi_{\mathbf{d}}\big(y_1 x_1(\mathbf{d}'), \ldots, y_{\mathsf{N}} x_{\mathsf{N}}(\mathbf{d}')\big) \\
&= \big(y_1 x_1(\mathbf{d}') x_1(\mathbf{d}), \ldots, y_{\mathsf{N}} x_{\mathsf{N}}(\mathbf{d}') x_{\mathsf{N}}(\mathbf{d})\big) \\
&= \big(y_1 x_1(\mathbf{d}' \oplus \mathbf{d}), \ldots, y_{\mathsf{N}} x_{\mathsf{N}}(\mathbf{d}' \oplus \mathbf{d})\big) \\
&= \psi_{\mathbf{d}\oplus\mathbf{d}'}(y_1, \ldots, y_{\mathsf{N}}),
\end{aligned}$$

where in the third equality we used (29.15) and the linearity of the encoder. Part (iii) follows from Parts (i) and (ii). Part (iv) follows from Part (iii) or directly by computing the partial derivative matrix and noting that it is diagonal with the diagonal elements being $\pm 1$ only. Part (v) follows from (29.75). To prove Part (vi) we substitute $\mathbf{d}'$ for $\mathbf{d}$ and $\psi_{\mathbf{d}}(\mathbf{y})$ for $\mathbf{y}$ in Part (v) to obtain

$$\begin{aligned}
f_{\mathbf{Y}|\mathbf{D}=\mathbf{d}'}\big(\psi_{\mathbf{d}}(\mathbf{y})\big) &= f_{\mathbf{Y}|\mathbf{D}=\mathbf{0}}\Big(\psi_{\mathbf{d}'}\big(\psi_{\mathbf{d}}(\mathbf{y})\big)\Big) \\
&= f_{\mathbf{Y}|\mathbf{D}=\mathbf{0}}\big(\psi_{\mathbf{d}\oplus\mathbf{d}'}(\mathbf{y})\big) \\
&= f_{\mathbf{Y}|\mathbf{D}=\mathbf{d}\oplus\mathbf{d}'}(\mathbf{y}),
\end{aligned}$$

where the second equality follows from Part (ii), and where the third equality follows from Part (v). $\qquad\square$

With the aid of this lemma we can now justify the all-zero assumption in the analysis of the probability of a block error. We shall state the result not only for the Gaussian setup but also for the more general case where the conditional density $f_{\mathbf{Y}|\mathbf{X}}(\cdot)$ corresponds to a memoryless binary-input/output-symmetric channel.

**Theorem 29.9.3.** *Consider the setup of Section 29.6 with the conditional density* $f_{\mathbf{Y}|\mathbf{X}}(\cdot)$ *corresponding to a memoryless binary-input/output-symmetric channel. Let* $p_{\mathrm{MAP}}(\mathrm{error}|\mathbf{D} = \mathbf{d})$ *denote the conditional probability of a block error for the detector of Proposition 29.7.1, conditional on the data tuple being* $\mathbf{d}$. *Then,*

$$p_{\mathrm{MAP}}(\mathrm{error}|\mathbf{D} = \mathbf{d}) = p_{\mathrm{MAP}}(\mathrm{error}|\mathbf{D} = \mathbf{0}), \quad \mathbf{d} \in \mathbb{F}_2^{\mathsf{K}}. \tag{29.82}$$

**Proof.** The proof of this result is not very difficult, but there is a slight technicality that arises from the way ties are resolved. Since on the Gaussian channel ties occur with probability zero (Proposition 21.6.2), this issue could be ignored. But we prefer not to ignore it because we would like the proof to apply also to channels satisfying (29.76) that are not necessarily Gaussian. To address ties, we shall assume that they are resolved at random as in Proposition 29.7.1 (i.e., as in the Definition 21.3.2 of the MAP rule).

For every $\mathbf{d} \in \mathbb{F}_2^{\mathsf{K}}$ and every $\nu \in \{1, \ldots, 2^{\mathsf{K}}\}$, define the set $\mathcal{D}_{\mathbf{d},\nu} \subset \mathbb{R}^{\mathsf{N}}$ to contain those $\mathbf{y} \in \mathbb{R}^{\mathsf{N}}$ for which the following two conditions hold:

$$f_{\mathbf{Y}|\mathbf{D}=\mathbf{d}}(\mathbf{y}) = \max_{\mathbf{d}' \in \mathbb{F}_2^{\mathsf{K}}} f_{\mathbf{Y}|\mathbf{D}=\mathbf{d}'}(\mathbf{y}), \tag{29.83a}$$

$$\#\left\{\tilde{\mathbf{d}} \in \mathbb{F}_2^{\mathsf{K}} : f_{\mathbf{Y}|\mathbf{D}=\tilde{\mathbf{d}}}(\mathbf{y}) = f_{\mathbf{Y}|\mathbf{D}=\mathbf{d}}(\mathbf{y})\right\} = \nu. \tag{29.83b}$$

Whenever $\mathbf{y} \in \mathcal{D}_{\mathbf{d},\nu}$, the MAP rule guesses "$\mathbf{D} = \mathbf{d}$" with probability $1/\nu$. Thus,

$$p_{\mathrm{MAP}}(\mathrm{error}|\mathbf{D} = \mathbf{d}) = 1 - \sum_{\nu=1}^{2^{\mathsf{K}}} \frac{1}{\nu} \int_{\mathbf{y} \in \mathbb{R}^{\mathsf{N}}} \mathrm{I}\{\mathbf{y} \in \mathcal{D}_{\mathbf{d},\nu}\} f_{\mathbf{Y}|\mathbf{D}=\mathbf{d}}(\mathbf{y}) \, \mathrm{d}\mathbf{y}. \tag{29.84}$$

The key is to note that, by Lemma 29.9.2 (v), for every $\mathbf{d} \in \mathbb{F}_2^{\mathsf{K}}$ and $\nu \in \{1, \ldots, 2^{\mathsf{K}}\}$

$$\left(\mathbf{y} \in \mathcal{D}_{\mathbf{d},\nu}\right) \Leftrightarrow \left(\psi_{\mathbf{d}}(\mathbf{y}) \in \mathcal{D}_{\mathbf{0},\nu}\right). \tag{29.85}$$

(Please pause to verify this.) Consequently, by (29.84),

$$p_{\mathrm{MAP}}(\mathrm{error}|\mathbf{D} = \mathbf{d}) = 1 - \sum_{\nu=1}^{2^{\mathsf{K}}} \frac{1}{\nu} \int_{\mathbf{y} \in \mathbb{R}^{\mathsf{N}}} \mathrm{I}\{\mathbf{y} \in \mathcal{D}_{\mathbf{d},\nu}\} f_{\mathbf{Y}|\mathbf{D}=\mathbf{d}}(\mathbf{y}) \, \mathrm{d}\mathbf{y}$$

$$= 1 - \sum_{\nu=1}^{2^{\mathsf{K}}} \frac{1}{\nu} \int_{\mathbf{y} \in \mathbb{R}^{\mathsf{N}}} \mathrm{I}\{\mathbf{y} \in \mathcal{D}_{\mathbf{d},\nu}\} f_{\mathbf{Y}|\mathbf{D}=\mathbf{0}}(\psi_{\mathbf{d}}(\mathbf{y})) \, \mathrm{d}\mathbf{y}$$

$$= 1 - \sum_{\nu=1}^{2^{\mathsf{K}}} \frac{1}{\nu} \int_{\tilde{\mathbf{y}} \in \mathbb{R}^{\mathsf{N}}} \mathrm{I}\{\psi_{\mathbf{d}}(\tilde{\mathbf{y}}) \in \mathcal{D}_{\mathbf{d},\nu}\} f_{\mathbf{Y}|\mathbf{D}=\mathbf{0}}(\tilde{\mathbf{y}}) \, \mathrm{d}\tilde{\mathbf{y}}$$

$$= 1 - \sum_{\nu=1}^{2^{\mathsf{K}}} \frac{1}{\nu} \int_{\tilde{\mathbf{y}} \in \mathbb{R}^{\mathsf{N}}} \mathrm{I}\{\tilde{\mathbf{y}} \in \mathcal{D}_{\mathbf{0},\nu}\} f_{\mathbf{Y}|\mathbf{D}=\mathbf{0}}(\tilde{\mathbf{y}}) \, \mathrm{d}\tilde{\mathbf{y}}$$

$$= p_{\mathrm{MAP}}(\mathrm{error}|\mathbf{D} = \mathbf{0}),$$

where the first equality follows from (29.84); the second by Lemma 29.9.2 (v); the third by defining $\tilde{\mathbf{y}} \triangleq \psi_{\mathbf{d}}(\mathbf{y})$ and using Parts (iv) and (iii) of Lemma 29.9.2; the fourth by (29.85); and the final equality by (29.84). $\qquad \square$

We now formulate a similar result for the detector of Proposition 29.8.1. Let $p_{\kappa}^*(\mathrm{error}|\mathbf{D} = \mathbf{d})$ denote the conditional probability that the decoder of Proposition 29.8.1 incorrectly decodes the $\kappa$-th data bit, conditional on the tuple $\mathbf{d}$ being fed to the encoder. Since the data are IID random bits,

$$p_{\kappa}^* = 2^{-\mathsf{K}} \sum_{\mathbf{d} \in \mathbb{F}_2^{\mathsf{K}}} p_{\kappa}^*(\mathrm{error}|\mathbf{D} = \mathbf{d}). \tag{29.86}$$

Since ties are resolved at random

$$p_\kappa^*(\text{error}|\mathbf{D} = \mathbf{d})$$
$$= \Pr\left[\sum_{\mathbf{d}' \in \mathcal{A}_{\kappa,0}} f_{\mathbf{Y}|\mathbf{D}=\mathbf{d}'}(\mathbf{Y}) > \sum_{\mathbf{d}' \in \mathcal{A}_{\kappa,1}} f_{\mathbf{Y}|\mathbf{D}=\mathbf{d}'}(\mathbf{Y}) \,\middle|\, \mathbf{D} = \mathbf{d}\right]$$
$$+ \frac{1}{2} \Pr\left[\sum_{\mathbf{d}' \in \mathcal{A}_{\kappa,0}} f_{\mathbf{Y}|\mathbf{D}=\mathbf{d}'}(\mathbf{Y}) = \sum_{\mathbf{d}' \in \mathcal{A}_{\kappa,1}} f_{\mathbf{Y}|\mathbf{D}=\mathbf{d}'}(\mathbf{Y}) \,\middle|\, \mathbf{D} = \mathbf{d}\right], \quad \mathbf{d} \in \mathcal{A}_{\kappa,1}, \quad (29.87)$$

and

$$p_\kappa^*(\text{error}|\mathbf{D} = \mathbf{d})$$
$$= \Pr\left[\sum_{\mathbf{d}' \in \mathcal{A}_{\kappa,0}} f_{\mathbf{Y}|\mathbf{D}=\mathbf{d}'}(\mathbf{Y}) < \sum_{\mathbf{d}' \in \mathcal{A}_{\kappa,1}} f_{\mathbf{Y}|\mathbf{D}=\mathbf{d}'}(\mathbf{Y}) \,\middle|\, \mathbf{D} = \mathbf{d}\right]$$
$$+ \frac{1}{2} \Pr\left[\sum_{\mathbf{d}' \in \mathcal{A}_{\kappa,0}} f_{\mathbf{Y}|\mathbf{D}=\mathbf{d}'}(\mathbf{Y}) = \sum_{\mathbf{d}' \in \mathcal{A}_{\kappa,1}} f_{\mathbf{Y}|\mathbf{D}=\mathbf{d}'}(\mathbf{Y}) \,\middle|\, \mathbf{D} = \mathbf{d}\right], \quad \mathbf{d} \in \mathcal{A}_{\kappa,0}. \quad (29.88)$$

**Theorem 29.9.4.** *Under the assumptions of Theorem 29.9.3, we have for every $\kappa \in \{1, \ldots, \mathsf{K}\}$*

$$p_\kappa^*(\text{error}|\mathbf{D} = \mathbf{d}) = p_\kappa^*(\text{error}|\mathbf{D} = \mathbf{0}), \quad \mathbf{d} \in \mathbb{F}_2^{\mathsf{K}}, \quad (29.89)$$

*and consequently*

$$p_\kappa^* = p_\kappa^*(\text{error}|\mathbf{D} = \mathbf{0}). \quad (29.90)$$

**Proof.** It suffices to prove (29.89) because (29.90) will then follow by (29.86). To prove (29.89) we begin by defining $e(\mathbf{d})$ for $\mathbf{d} \in \mathbb{F}_2^{\mathsf{K}}$ as follows. If $\mathbf{d}$ is in $\mathcal{A}_{\kappa,1}$, then we define $e(\mathbf{d})$ as

$$e(\mathbf{d}) \triangleq \Pr\left[\sum_{\mathbf{d}' \in \mathcal{A}_{\kappa,0}} f_{\mathbf{Y}|\mathbf{D}=\mathbf{d}'}(\mathbf{Y}) > \sum_{\mathbf{d}' \in \mathcal{A}_{\kappa,1}} f_{\mathbf{Y}|\mathbf{D}=\mathbf{d}'}(\mathbf{Y}) \,\middle|\, \mathbf{D} = \mathbf{d}\right], \quad \mathbf{d} \in \mathcal{A}_{\kappa,1}.$$

Otherwise, if $\mathbf{d}$ is in $\mathcal{A}_{\kappa,0}$, then we define $e(\mathbf{d})$ as

$$e(\mathbf{d}) \triangleq \Pr\left[\sum_{\mathbf{d}' \in \mathcal{A}_{\kappa,0}} f_{\mathbf{Y}|\mathbf{D}=\mathbf{d}'}(\mathbf{Y}) < \sum_{\mathbf{d}' \in \mathcal{A}_{\kappa,1}} f_{\mathbf{Y}|\mathbf{D}=\mathbf{d}'}(\mathbf{Y}) \,\middle|\, \mathbf{D} = \mathbf{d}\right], \quad \mathbf{d} \in \mathcal{A}_{\kappa,0}.$$

We shall prove (29.89) for the case where

$$\mathbf{d} \in \mathcal{A}_{\kappa,1}. \quad (29.91)$$

The proof for the case where $\mathbf{d} \in \mathcal{A}_{\kappa,0}$ is almost identical and is omitted. For $\mathbf{d}$ satisfying (29.91) we shall prove that $e(\mathbf{d})$ does not depend on $\mathbf{d}$. The second term in (29.87) which accounts for the random resolution of ties can be treated very similarly. To show that $e(\mathbf{d})$ does not depend on $\mathbf{d}$ we compute:

$$
\begin{aligned}
e(\mathbf{d}) \\
&= \int_{\mathbf{y}\in\mathbb{R}^N} \mathrm{I}\Big\{ \sum_{\mathbf{d}'\in\mathcal{A}_{\kappa,0}} f_{\mathbf{Y}|\mathbf{D}=\mathbf{d}'}(\mathbf{y}) > \sum_{\mathbf{d}'\in\mathcal{A}_{\kappa,1}} f_{\mathbf{Y}|\mathbf{D}=\mathbf{d}'}(\mathbf{y}) \Big\} f_{\mathbf{Y}|\mathbf{D}=\mathbf{d}}(\mathbf{y}) \, d\mathbf{y} \\
&= \int_{\mathbf{y}\in\mathbb{R}^N} \mathrm{I}\Big\{ \sum_{\mathbf{d}'\in\mathcal{A}_{\kappa,0}} f_{\mathbf{Y}|\mathbf{D}=\mathbf{d}'}(\mathbf{y}) > \sum_{\mathbf{d}'\in\mathcal{A}_{\kappa,1}} f_{\mathbf{Y}|\mathbf{D}=\mathbf{d}'}(\mathbf{y}) \Big\} f_{\mathbf{Y}|\mathbf{D}=\mathbf{0}}\big(\psi_{\mathbf{d}}(\mathbf{y})\big) \, d\mathbf{y} \\
&= \int_{\tilde{\mathbf{y}}\in\mathbb{R}^N} \mathrm{I}\Big\{ \sum_{\mathbf{d}'\in\mathcal{A}_{\kappa,0}} f_{\mathbf{Y}|\mathbf{D}=\mathbf{d}'}\big(\psi_{\mathbf{d}}(\tilde{\mathbf{y}})\big) > \sum_{\mathbf{d}'\in\mathcal{A}_{\kappa,1}} f_{\mathbf{Y}|\mathbf{D}=\mathbf{d}'}\big(\psi_{\mathbf{d}}(\tilde{\mathbf{y}})\big) \Big\} f_{\mathbf{Y}|\mathbf{D}=\mathbf{0}}(\tilde{\mathbf{y}}) \, d\tilde{\mathbf{y}} \\
&= \int_{\tilde{\mathbf{y}}\in\mathbb{R}^N} \mathrm{I}\Big\{ \sum_{\mathbf{d}'\in\mathcal{A}_{\kappa,0}} f_{\mathbf{Y}|\mathbf{D}=\mathbf{d}'\oplus\mathbf{d}}(\tilde{\mathbf{y}}) > \sum_{\mathbf{d}'\in\mathcal{A}_{\kappa,1}} f_{\mathbf{Y}|\mathbf{D}=\mathbf{d}'\oplus\mathbf{d}}(\tilde{\mathbf{y}}) \Big\} f_{\mathbf{Y}|\mathbf{D}=\mathbf{0}}(\tilde{\mathbf{y}}) \, d\tilde{\mathbf{y}} \\
&= \int_{\tilde{\mathbf{y}}\in\mathbb{R}^N} \mathrm{I}\Big\{ \sum_{\tilde{\mathbf{d}}\in\mathcal{A}_{\kappa,1}} f_{\mathbf{Y}|\mathbf{D}=\tilde{\mathbf{d}}}(\tilde{\mathbf{y}}) > \sum_{\tilde{\mathbf{d}}\in\mathcal{A}_{\kappa,0}} f_{\mathbf{Y}|\mathbf{D}=\tilde{\mathbf{d}}}(\tilde{\mathbf{y}}) \Big\} f_{\mathbf{Y}|\mathbf{D}=\mathbf{0}}(\tilde{\mathbf{y}}) \, d\tilde{\mathbf{y}} \\
&= e(\mathbf{0}),
\end{aligned}
$$

where the second equality follows from Lemma 29.9.2 (v); the third by defining the vector $\tilde{\mathbf{y}}$ as $\tilde{\mathbf{y}} \triangleq \psi_{\mathbf{d}}(\mathbf{y})$ and by Parts (iv) and (iii) of Lemma 29.9.2; the fourth by Lemma 29.9.2 (vi); and the fifth equality by defining $\tilde{\mathbf{d}} \triangleq \mathbf{d} \oplus \mathbf{d}'$ and using (29.73). $\qquad\square$

## 29.10    System Parameters

We next summarize how the system parameters such as power, bandwidth, and block error rate are related to the parameters of the encoder. We only address the case where the pulse shape $\phi$ satisfies the orthonormality condition (29.2). As we next show, in this case the bandwidth $W$ in Hz of the pulse shape can be expressed as

$$
\boxed{\; W = \frac{1}{2}\, \mathsf{R}_{\mathrm{b}}\, \frac{\mathsf{N}}{\mathsf{K}}\, (1 + \text{excess bandwidth}), \;}
\tag{29.92}
$$

where $\mathsf{R}_{\mathrm{b}}$ is the bit rate at which the data are fed to the modem in bits per second, and where the excess bandwidth, which is defined in Definition 11.3.6, is nonnegative. To verify (29.92) note that if the data arrive at the encoder at the rate of $\mathsf{R}_{\mathrm{b}}$ bits per second and if the encoder produces $\mathsf{N}$ real symbols for every $\mathsf{K}$ bits that are fed to it, then the encoder produces real symbols at a rate

$$
\mathsf{R}_{\mathrm{s}} = \frac{\mathsf{N}}{\mathsf{K}}\, \mathsf{R}_{\mathrm{b}} \left[\frac{\text{real symbol}}{\text{second}}\right],
\tag{29.93}
$$

so the baud period must be

$$
\mathsf{T}_{\mathrm{s}} = \frac{\mathsf{K}}{\mathsf{N}}\, \frac{1}{\mathsf{R}_{\mathrm{b}}}.
\tag{29.94}
$$

It then follows from Definition 11.3.6 that the bandwidth of $\phi$ is given by (29.92) with the excess bandwidth being nonnegative by Corollary 11.3.5.

As to the transmitted power P, by (29.27) and (29.94) it is given by

$$
\boxed{\mathsf{P} = \mathsf{E_b}\,\mathsf{R_b},}
\tag{29.95}
$$

where $\mathsf{E_b}$ denotes the energy per data bit and is given by

$$
\mathsf{E_b} = \frac{\mathsf{N}}{\mathsf{K}}\,\mathsf{A}^2.
\tag{29.96}
$$

It is customary to describe the error probability by which one measures performance as a function of the energy-per-bit $\mathsf{E_b}$.[3] Thus, for example, one typically writes the upper bound (29.55) on the probability of a block error using (29.96) as

$$
\boxed{
\begin{aligned}
&p_{\text{MAP}}(\text{error}|\mathbf{D} = \mathbf{d}) \\
&\leq \sum_{\nu=\mathsf{d}_{\min,\mathrm{H}}}^{\mathsf{N}} \#\big\{\mathbf{c} \in \text{Image}(\mathsf{T}) : \mathrm{w_H}(\mathbf{c}) = \nu\big\}\, \mathcal{Q}\left(\sqrt{\frac{2\mathsf{E_b}(\mathsf{K}/\mathsf{N})\nu}{\mathsf{N}_0}}\right).
\end{aligned}}
\tag{29.97}
$$

## 29.11 Hard vs. Soft Decisions

In Section 29.7 we derived the decision rule that minimizes the probability of a block error. We saw that, in general, its complexity is exponential in the dimension K of the code because a brute-force implementation of this rule requires correlating the N-tuple $\mathbf{Y}$ with each of the $2^{\mathsf{K}}$ tuples in Image(**enc**). For the single parity check rule we found a much simpler implementation of this rule, but for general codes the decoding problem can be very difficult.

A suboptimal decoding rule that is sometimes implemented is the **Hard Decision** decoding rule, which has two steps. In the first step one uses the observed real-valued N-tuple $(Y_1, \ldots, Y_N)$ to form the binary tuple $(\hat{C}_1, \ldots, \hat{C}_N)$ according to the rule

$$
\hat{C}_\eta = \begin{cases} 0 & \text{if } Y_\eta \geq 0, \\ 1 & \text{if } Y_\eta < 0, \end{cases} \quad \eta = 1, \ldots, \mathsf{N},
$$

and in the second step one searches for the message $\mathbf{d}$ for which $T(\mathbf{d})$ is closest in Hamming distance to $(\hat{C}_1, \ldots, \hat{C}_N)$. The advantage of this decoding rule is that the first step is very simple and that the second step can be often performed very efficiently if the code has a strong algebraic structure.

## 29.12 The Varshamov and Singleton Bounds

Motivated by the approximation (29.56) and by (29.58), a fair bit of effort in Coding Theory has been invested in finding $(\mathsf{K}, \mathsf{N})$ codes that have a large minimum

---

[3] The terms "energy-per-bit," "energy-per-data-bit," and "energy-per-information-bit" are used interchangeably.

Hamming weight and reasonable decoding complexity. One of the key existence results in this area is the Varshamov Bound. We state here a special case of this bound pertaining to our binary setting.

**Theorem 29.12.1 (The Varshamov Bound).** *Let* $\mathsf{K}$ *and* $\mathsf{N}$ *be positive integers, and let* $d$ *be an integer in the range* $2 \leq d \leq \mathsf{N} - \mathsf{K} + 1$. *If*

$$\sum_{\nu=0}^{d-2} \binom{\mathsf{N}-1}{\nu} < 2^{\mathsf{N}-\mathsf{K}}, \tag{29.98}$$

*then there exists a linear* $(\mathsf{K}, \mathsf{N})$ $\mathbb{F}_2$ *code whose minimum distance* $d_{\min,\mathrm{H}}$ *satisfies* $d_{\min,\mathrm{H}} \geq d$.

**Proof.** See, for example, (MacWilliams and Sloane, 1977, Chapter 1, Section 10, Theorem 12) or (Blahut, 2002, Chapter 12, Section 3, Theorem 12.3.3). $\qquad \square$

A key upper bound on $d_{\min,\mathrm{H}}$ is given by the Singleton Bound.

**Theorem 29.12.2 (The Singleton Bound).** *If* $\mathsf{N}$ *and* $\mathsf{K}$ *are positive integers, then the minimum Hamming distance* $d_{\min,\mathrm{H}}$ *of any linear* $(\mathsf{K}, \mathsf{N})$ $\mathbb{F}_2$ *code must satisfy*

$$d_{\min,\mathrm{H}} \leq \mathsf{N} - \mathsf{K} + 1. \tag{29.99}$$

**Proof.** See, for example, (Blahut, 2002, Chapter 3, Section 3, Theorem 3.2.6) or (van Lint, 1998, Chapter 5, Section 2, Corollary 5.2.2) or Exercise 29.10. $\qquad \square$

## 29.13 Additional Reading

We have only had a glimpse of Coding Theory. A good starting point for the literature on Algebraic Coding Theory is (Roth, 2006). For more on the modern coding techniques such as low-density parity-check codes (LDPC) and turbo-codes, see (Richardson and Urbanke, 2008).

The degredation resulting from hard decsions is addressed, e.g., in (Viterbi and Omura, 1979, Chapter 3, Section 3.4).

The results of Section 29.9 can be extended also to non-binary codes with other mappings. See, for example, (Loeliger, 1991) and (Forney, 1991).

For some of the literature on the minimum distance and its asymptotic behavior in the block length, see, for example, (Roth, 2006, Chapter 4)

For more on the decoding complexity see the notes on Section 2.4 in Chapter 2 of (Roth, 2006).

## 29.14 Exercises

**Exercise 29.1 (Orthogonality of Signals).** Recall that, given a binary K-tuple $\mathbf{d} \in \mathbb{F}_2^{\mathsf{K}}$ and a linear $(\mathsf{K}, \mathsf{N})$ $\mathbb{F}_2$ encoder $\mathsf{T}(\cdot)$, we use $x_\eta(\mathbf{d})$ to denote the result of applying the antipodal mapping $\Upsilon(\cdot)$ to the $\eta$-th component of $\mathbf{T}(\mathbf{d})$. Let the pulse shape $\phi$ be such that its time shifts by integer multiples of the baud period $\mathsf{T_s}$ are orthonormal. Show that

$$\left\langle t \mapsto \sum_{\eta=1}^{\mathsf{N}} x_\eta(\mathbf{d})\,\phi(t - \eta\mathsf{T_s}),\; t \mapsto \sum_{\eta=1}^{\mathsf{N}} x_\eta(\mathbf{d'})\,\phi(t - \eta\mathsf{T_s}) \right\rangle = 0,$$

if, and only if, $d_{\mathrm{H}}\big(\mathbf{T}(\mathbf{d}), \mathbf{T}(\mathbf{d'})\big) = \mathsf{N}/2$.

**Exercise 29.2 (How Many Encoders Does a Code Have?).** Let the linear $(\mathsf{K}, \mathsf{N})$ $\mathbb{F}_2$ encoder $\mathbf{T} \colon \mathbb{F}_2^{\mathsf{K}} \to \mathbb{F}_2^{\mathsf{N}}$ be represented by the $\mathsf{K} \times \mathsf{N}$ matrix $\mathsf{G}$. Show that any linear $(\mathsf{K}, \mathsf{N})$ $\mathbb{F}_2$ encoder whose image is equal to the image of $\mathbf{T}$ can be written in the form

$$\mathbf{d} \mapsto \mathbf{d}\mathsf{AG},$$

where $\mathsf{A}$ is a $\mathsf{K} \times \mathsf{K}$ invertible matrix whose entries are in $\mathbb{F}_2$. How many such matrices $\mathsf{A}$ are there?

**Exercise 29.3 (The (4,7) Hamming Code).** A systematic encoder for the linear $(4, 7)$ $\mathbb{F}_2$ **Hamming code** maps the four data bits $d_1, d_2, d_3, d_4$ to the 7-tuple

$$\big(d_1, d_2, d_3, d_4, d_1 \oplus d_3 \oplus d_4, d_1 \oplus d_2 \oplus d_4, d_2 \oplus d_3 \oplus d_4\big).$$

Suppose that this encoder is used in conjunction with the componentwise antipodal mapping $\Upsilon_7(\cdot)$ over the white Gaussian noise channel with PAM of pulse shape whose time shifts by integer multiples of the baud period are orthonormal.

  (i) Write out the 16 binary codewords and compute the code's weight enumerator.

  (ii) Assuming that the codewords are equally likely and that the decoding minimizes the probability of a message error, use the Union Bound to upper-bound the probability of codeword error. Express your bound using the transmitted energy per bit $\mathsf{E_b}$.

  (iii) Find a lower bound on the probability that the first bit $D_1$ is incorrectly decoded. Express your bound in terms of the energy per bit. Compare with the exact expression in uncoded communication.

**Exercise 29.4 (The Repetition Code).** Consider the linear $(1, \mathsf{N})$ $\mathbb{F}_2$ **repetition code** consisting of the all-zero and all-one $\mathsf{N}$-tuples $(0, \ldots, 0)$ and $(1, \ldots, 1)$.

  (i) Find its weight enumerator.

  (ii) Find an optimal decoder for a system employing this code with the componentwise antipodal mapping $\Upsilon_{\mathsf{N}}(\cdot)$ over the white Gaussian noise channel in conjunction with PAM with a pulse shape whose times shifts by integer multiples of the baud period are orthonormal.

  (iii) Find the optimal probability of error. Express your answer using the energy per bit $\mathsf{E_b}$. Compare with uncoded antipodal signaling.

  (iv) Describe the hard decision rule for this setup. Find its performance in terms of $\mathsf{E_b}$.

**Exercise 29.5 (The Dual Code).** We say that two binary $\kappa$-tuples $\mathbf{u} = (u_1, \ldots, u_\kappa)$ and $\mathbf{v} = (v_1, \ldots, v_\kappa)$ are orthogonal if

$$u_1 \cdot v_1 \oplus u_2 \cdot v_2 \oplus \cdots \oplus u_\kappa \cdot v_\kappa = 0.$$

Consider the set of all $\mathsf{N}$-tuples that are orthogonal to every codeword of some given linear $(\mathsf{K}, \mathsf{N})$ $\mathbb{F}_2$ code. Show that this set is a linear $(\mathsf{N} - \mathsf{K}, \mathsf{N})$ $\mathbb{F}_2$ code. This code is called the **dual code**. What is the dual code of the $(\mathsf{K}, \mathsf{K} + 1)$ single parity check code?

**Exercise 29.6 (Hadamard Code).** For a positive integer $N$ which is a power of two, define the $N \times N$ binary matrix $H_N$ recursively as

$$H_2 = \begin{pmatrix} 0 & 0 \\ 0 & 1 \end{pmatrix}, \quad H_N = \begin{pmatrix} H_{N/2} & H_{N/2} \\ H_{N/2} & \bar{H}_{N/2} \end{pmatrix}, \quad N = 4, 8, 16, \ldots, \tag{29.100}$$

where $\bar{H}$ denotes the componentwise negation of the matrix $H$, that is, the matrix whose Row-$j$ Column-$\ell$ element is given by $1 \oplus [H]_{j,\ell}$, where $[H]_{j,\ell}$ is the Row-$j$ Column-$\ell$ element of $H$. Consider the set of all rows of $H_N$.

(i) Show that this collection of $N$ binary $N$-tuples forms a linear $(\log_2 N, N)$ $\mathbb{F}_2$ code. This code is called the **Hadamard code**. Find this code's weight enumerator.

(ii) Suppose that, as in Section 29.6, this code is used in conjunction with PAM over the white Gaussian noise channel and that $Y_1, \ldots, Y_N$ are as defined there. Show that the following rule minimizes the probability of a message error: compute the vector

$$\tilde{H}_N \begin{pmatrix} Y_1 \\ Y_2 \\ \vdots \\ Y_N \end{pmatrix} \tag{29.101}$$

and guess that the $m$-th message was sent if the $m$-th component of this vector is largest. Here $\tilde{H}_N$ is the $N \times N$ matrix whose Row-$j$ Column-$\ell$ entry is the result of applying $\Upsilon(\cdot)$ to the Row-$j$ Column-$\ell$ entry of $H_N$.

(iii) A brute-force computation of the vector in (29.101) requires $N^2$ additions, which translates to $N^2 / \log_2 N$ additions per information bit. Use the structure of $H_N$ that is given in (29.100) to show that this can be done with $N \log_2 N$ additions (or $N$ additions per information bit).

*Hint: For Part (iii) provide an algorithm for which $c(N) = 2c(N/2) + N$, where $c(n)$ denotes the number of additions needed to compute this vector when the matrix is $n \times n$. Show that the solution to this recursion for $c(2) = 2$ is $c(n) = n \log_2 n$.*

**Exercise 29.7 (Bi-Orthogonal Code).** Referring to the notation introduced in Exercise 29.6, consider the $2N \times N$ matrix

$$\begin{pmatrix} H_N \\ \bar{H}_N \end{pmatrix},$$

where $N$ is some positive power of two.

(i) Show that the rows of this matrix form a linear $(\log_2(2N), N)$ $\mathbb{F}_2$ code.

(ii) Compute the code's weight enumerator.

(iii) Explain why we chose the title "Bi-Orthogonal Code" for this exercise.

(iv) Find an efficient decoding algorithm for the setup of Section 29.6.

**Exercise 29.8 (Non-IID Data).** How would you modify the decision rule of Section 29.8 if the data bits $(D_1, \ldots, D_K)$ are not necessarily IID but have the general joint probability mass function $P_{\mathbf{D}}(\cdot)$?

**Exercise 29.9 (Asymmetric Channels).** Show that Theorem 29.9.3 will no longer hold if we drop the hypothesis that the channel is symmetric.

**Exercise 29.10 (A Proof of the Singleton Bound).** Use the following steps to prove the Singleton Bound.

(i) Consider a linear $(K, N)$ $\mathbb{F}_2$ code. Let $\pi \colon \mathbb{F}_2^N \to \mathbb{F}_2^{K-1}$ map each N-tuple to the $(K - 1)$-tuple consisting of its first $K - 1$ components. By comparing the number of codewords with the cardinality of the range of $\pi$, argue that there must exist two codewords whose first $K - 1$ components are identical.

(ii) Show that these two codewords are at Hamming distance of at most $N - K + 1$.

(iii) Show that the minimum Hamming distance of the code is at most $N - K + 1$.

(iv) Does linearity play a role in the proof?

**Exercise 29.11 (Binary MDS Codes).** Codes that satisfy the Singleton Bound with equality are called Maximum Distance Separable (MDS). Show that the linear $(K, K + 1)$ $\mathbb{F}_2$ single parity check code is MDS. Can you think of other binary MDS codes?

**Exercise 29.12 (Existence via the Varshamov Bound).** Can the existence of a linear $(4, 7)$ $\mathbb{F}_2$ code of minimum Hamming distance 3 be deduced from the Varshamov Bound?

# Appendix A

# On the Fourier Series

## A.1 Introduction and Preliminaries

We survey here some of the results on the Fourier Series that are used in the book. The Fourier Series has numerous other applications that we do not touch upon. For those we refer the reader to (Katznelson, 1976), (Dym and McKean, 1972), and (Körner, 1988).

To simplify typography, we denote the half-open interval $[-1/2, 1/2)$ by $\mathbb{I}$:

$$\mathbb{I} \triangleq \left\{ \theta \in \mathbb{R} : -\frac{1}{2} \leq \theta < \frac{1}{2} \right\}. \tag{A.1}$$

**Definition A.1.1 (Fourier Series Coefficient).** *The $\eta$-th **Fourier Series Coefficient** of an integrable function* $\mathbf{g} \colon \mathbb{I} \to \mathbb{C}$ *is denoted by $\hat{g}(\eta)$ and is defined for every integer $\eta$ by*

$$\hat{g}(\eta) \triangleq \int_{\mathbb{I}} g(\theta) \, e^{-\mathrm{i}2\pi\eta\theta} \, \mathrm{d}\theta. \tag{A.2}$$

The **periodic extension** of the function $\mathbf{g} \colon \mathbb{I} \to \mathbb{C}$ is denoted by $\mathbf{g}_\mathrm{P} \colon \mathbb{R} \to \mathbb{C}$ and is defined as

$$g_\mathrm{P}(n + \theta) = g(\theta), \quad \left( n \in \mathbb{Z}, \ \theta \in \mathbb{I} \right). \tag{A.3}$$

We say that $\mathbf{g} \colon \mathbb{I} \to \mathbb{C}$ is **periodically continuous** if its periodic extension $\mathbf{g}_\mathrm{P}$ is continuous, i.e., if $g(\cdot)$ is continuous in $\mathbb{I}$ and if, additionally,

$$\lim_{\theta\uparrow 1/2} g(\theta) = g(-1/2). \tag{A.4}$$

A **degree-$n$ trigonometric polynomial** is a function of the form

$$\theta \mapsto \sum_{\eta=-n}^{n} a_\eta \, e^{\mathrm{i}2\pi\eta\theta}, \quad \theta \in \mathbb{R}, \tag{A.5}$$

where $a_n$ and $a_{-n}$ are not both zero. Note that if $p(\cdot)$ is a trigonometric polynomial, then $p(\theta + 1) = p(\theta)$ for all $\theta \in \mathbb{R}$.

If $\mathbf{g} \colon \mathbb{I} \to \mathbb{C}$ is integrable, and if $p(\cdot)$ is a trigonometric polynomial, then we define the convolution $\mathbf{g} \star \mathbf{p}$ at every $\theta \in \mathbb{R}$ as

$$(\mathbf{g} \star \mathbf{p})(\theta) = \int_{\mathbb{I}} g(\vartheta)\, p(\theta - \vartheta)\, \mathrm{d}\vartheta \tag{A.6}$$

$$= \int_{\mathbb{I}} p(\vartheta)\, g_{\mathrm{P}}(\theta - \vartheta)\, \mathrm{d}\vartheta. \tag{A.7}$$

**Lemma A.1.2 (Convolution with a Trigonometric Polynomial).** *The convolution of an integrable function* $\mathbf{g} \colon \mathbb{I} \to \mathbb{C}$ *with the trigonometric polynomial*

$$\theta \mapsto \sum_{\eta=-n}^{n} a_\eta\, e^{\mathrm{i}2\pi\eta\theta} \tag{A.8}$$

*is the trigonometric polynomial*

$$\theta \mapsto \sum_{\eta=-n}^{n} \hat{g}(\eta)\, a_\eta\, e^{\mathrm{i}2\pi\eta\theta}, \quad \theta \in \mathbb{R}. \tag{A.9}$$

**Proof.** Denote the trigonometric polynomial in (A.8) by $p(\cdot)$. By swapping summation and integration we obtain

$$(\mathbf{g} \star \mathbf{p})(\theta) = \int_{\mathbb{I}} g(\vartheta)\, p(\theta - \vartheta)\, \mathrm{d}\vartheta$$

$$= \int_{\mathbb{I}} g(\vartheta) \sum_{\eta=-n}^{n} a_\eta\, e^{\mathrm{i}2\pi\eta(\theta-\vartheta)}\, \mathrm{d}\vartheta$$

$$= \sum_{\eta=-n}^{n} \int_{\mathbb{I}} g(\vartheta)\, a_\eta\, e^{\mathrm{i}2\pi\eta(\theta-\vartheta)}\, \mathrm{d}\vartheta$$

$$= \sum_{\eta=-n}^{n} a_\eta\, e^{\mathrm{i}2\pi\eta\theta} \int_{\mathbb{I}} g(\vartheta)\, e^{-\mathrm{i}2\pi\eta\vartheta}\, \mathrm{d}\vartheta$$

$$= \sum_{\eta=-n}^{n} a_\eta\, e^{\mathrm{i}2\pi\eta\theta}\, \hat{g}(\eta), \quad \theta \in \mathbb{R}. \qquad \square$$

**Definition A.1.3 (Fejér's Kernel).** *Fejér's degree-$n$ kernel* $\mathbf{k}_n$ *is the trigonometric polynomial*

$$k_n(\theta) = \sum_{\eta=-n}^{n} \left(1 - \frac{|\eta|}{n+1}\right) e^{\mathrm{i}2\pi\eta\theta} \tag{A.10a}$$

$$= \begin{cases} n+1 & \text{if } \theta \in \mathbb{Z}, \\ \dfrac{1}{n+1}\left(\dfrac{\sin\big((n+1)\pi\theta\big)}{\sin(\pi\theta)}\right)^2 & \text{if } \theta \in \mathbb{R} \setminus \mathbb{Z}. \end{cases} \tag{A.10b}$$

The key properties of Fejér's kernel are that it is nonnegative

$$k_n(\theta) \geq 0, \quad \theta \in \mathbb{R}; \tag{A.11a}$$

that it integrates over $\mathbb{I}$ to one

$$\int_{\mathbb{I}} k_n(\theta)\, d\theta = 1; \tag{A.11b}$$

and that for every fixed $0 < \delta < 1/2$

$$\lim_{n\to\infty} \int_{\delta < |\theta| < \frac{1}{2}} k_n(\theta)\, d\theta = 0. \tag{A.11c}$$

Here (A.11a) follows from (A.10b); (A.11b) follows from (A.10a) by term-by-term integration over $\mathbb{I}$; and (A.11c) follows from the inequality

$$k_n(\theta) \le \frac{1}{n+1}\left(\frac{1}{\sin \pi\delta}\right)^2, \quad \delta \le |\theta| \le \frac{1}{2},$$

which follows from (A.10b) by upper-bounding the numerator by 1 and by using the monotonicity of $\sin^2(\pi\theta)$ in $|\theta| \in [0, 1/2]$.

For an integrable function $\mathbf{g}\colon \mathbb{I} \to \mathbb{C}$, we define for every $n \in \mathbb{N}$ and $\theta \in \mathbb{R}$

$$\sigma_n(\mathbf{g}, \theta) \triangleq (\mathbf{g} \star \mathbf{k}_n)(\theta) \tag{A.12}$$

$$= \sum_{\eta=-n}^{n} \left(1 - \frac{|\eta|}{n+1}\right) \hat{g}(\eta)\, e^{i2\pi\eta\theta}, \tag{A.13}$$

where the second equality follows from (A.10a) and Lemma A.1.2. We also define for $\mathbf{g}\colon \mathbb{R} \to \mathbb{C}$ or $\mathbf{g}\colon \mathbb{I} \to \mathbb{C}$

$$\|\mathbf{g}\|_{\mathbb{I}, 1} = \int_{\mathbb{I}} |g(\theta)|\, d\theta. \tag{A.14}$$

Finally, for every function $\mathbf{h}\colon \mathbb{I} \to \mathbb{C}$ and $\vartheta \in \mathbb{R}$ we define the mapping $\mathbf{h}_\vartheta\colon \mathbb{R} \to \mathbb{C}$ as

$$\mathbf{h}_\vartheta\colon \theta \mapsto h_{\mathrm{P}}(\theta - \vartheta). \tag{A.15}$$

## A.2   Reconstruction in $\mathcal{L}_1$

**Lemma A.2.1.** *If* $\mathbf{g}\colon \mathbb{R} \to \mathbb{C}$ *is integrable over* $\mathbb{I}$ *and* $g(\theta + 1) = g(\theta)$ *for every* $\theta \in \mathbb{R}$, *then*

$$\lim_{\vartheta\to 0} \int_{\mathbb{I}} |g(\theta) - g(\theta - \vartheta)|\, d\theta = 0. \tag{A.16}$$

**Proof.** This is easy to see if $\mathbf{g}$ is continuous, because in this case $\mathbf{g}$ is uniformly continuous. The general result follows from this case by picking a periodic continuous function $\mathbf{h}$ that approximates $\mathbf{g}$ in the sense that $\|\mathbf{g} - \mathbf{h}\|_{\mathbb{I}, 1} < \epsilon/2$; by computing

$$\begin{aligned}
\|\mathbf{g} - \mathbf{g}_\vartheta\|_{\mathbb{I}, 1} &= \|\mathbf{g} - \mathbf{h} + \mathbf{h} - \mathbf{g}_\vartheta\|_{\mathbb{I}, 1} \\
&= \|\mathbf{g} - \mathbf{h} + \mathbf{h} - \mathbf{h}_\vartheta + \mathbf{h}_\vartheta - \mathbf{g}_\vartheta\|_{\mathbb{I}, 1} \\
&\le \|\mathbf{g} - \mathbf{h}\|_{\mathbb{I}, 1} + \|\mathbf{h} - \mathbf{h}_\vartheta\|_{\mathbb{I}, 1} + \|\mathbf{h}_\vartheta - \mathbf{g}_\vartheta\|_{\mathbb{I}, 1} \\
&= \|\mathbf{g} - \mathbf{h}\|_{\mathbb{I}, 1} + \|\mathbf{h} - \mathbf{h}_\vartheta\|_{\mathbb{I}, 1} + \|\mathbf{h} - \mathbf{g}\|_{\mathbb{I}, 1} \\
&\le \epsilon + \|\mathbf{h} - \mathbf{h}_\vartheta\|_{\mathbb{I}, 1};
\end{aligned} \tag{A.17}$$

and by then applying the result to $\mathbf{h}$, which is continuous. $\qquad\qquad\qquad\square$

**Theorem A.2.2 (Reconstruction in $\mathcal{L}_1$).** *If* $\mathbf{g}\colon \mathbb{I} \to \mathbb{C}$ *is integrable, then*

$$\lim_{n\to\infty} \int_{\mathbb{I}} \big|g(\theta) - \sigma_n(\mathbf{g},\theta)\big|\, \mathrm{d}\theta = 0. \tag{A.18}$$

**Proof.** Let $\mathbf{g}_{\mathrm{P}}$ be the periodic extension of $\mathbf{g}$. Then for every $\delta \in (0,1/2)$,

$$\int_{\mathbb{I}} \big|g(\theta) - \sigma_n(\mathbf{g},\theta)\big|\, \mathrm{d}\theta = \int_{\mathbb{I}} \left| g_{\mathrm{P}}(\theta) - \int_{\mathbb{I}} g_{\mathrm{P}}(\theta - \vartheta)\, k_n(\vartheta)\, \mathrm{d}\vartheta \right| \mathrm{d}\theta$$

$$= \int_{\mathbb{I}} \left| \int_{\mathbb{I}} k_n(\vartheta)\big(g_{\mathrm{P}}(\theta) - g_{\mathrm{P}}(\theta - \vartheta)\big)\, \mathrm{d}\vartheta \right| \mathrm{d}\theta,$$

$$= \int_{\mathbb{I}} \left| \int_{-\delta}^{\delta} + \int_{\delta < |\vartheta| < \frac{1}{2}} k_n(\vartheta)\big(g_{\mathrm{P}}(\theta) - g_{\mathrm{P}}(\theta - \vartheta)\big)\, \mathrm{d}\vartheta \right| \mathrm{d}\theta, \tag{A.19}$$

where the first equality follows from the definition of $\sigma_n(\mathbf{g},\theta)$ (A.12), and where the second equality follows from (A.11b). We now bound the two integrals in (A.19) separately:

$$\int_{\mathbb{I}} \left| \int_{-\delta}^{\delta} k_n(\vartheta)\big(g_{\mathrm{P}}(\theta) - g_{\mathrm{P}}(\theta - \vartheta)\big)\, \mathrm{d}\vartheta \right| \mathrm{d}\theta$$

$$\leq \int_{\mathbb{I}} \int_{-\delta}^{\delta} k_n(\vartheta)\big|g_{\mathrm{P}}(\theta) - g_{\mathrm{P}}(\theta - \vartheta)\big|\, \mathrm{d}\vartheta\, \mathrm{d}\theta \tag{A.20}$$

$$= \int_{-\delta}^{\delta} \int_{\mathbb{I}} k_n(\vartheta)\big|g_{\mathrm{P}}(\theta) - g_{\mathrm{P}}(\theta - \vartheta)\big|\, \mathrm{d}\theta\, \mathrm{d}\vartheta$$

$$= \int_{-\delta}^{\delta} k_n(\vartheta) \int_{\mathbb{I}} \big|g_{\mathrm{P}}(\theta) - g_{\mathrm{P}}(\theta - \vartheta)\big|\, \mathrm{d}\theta\, \mathrm{d}\vartheta$$

$$\leq \int_{-\delta}^{\delta} k_n(\vartheta) \max_{|\vartheta'|\leq\delta} \left\{ \int_{\mathbb{I}} \big|g_{\mathrm{P}}(\theta) - g_{\mathrm{P}}(\theta - \vartheta')\big|\, \mathrm{d}\theta \right\} \mathrm{d}\vartheta$$

$$\leq \int_{\mathbb{I}} k_n(\vartheta) \max_{|\vartheta'|\leq\delta} \left\{ \int_{\mathbb{I}} \big|g_{\mathrm{P}}(\theta) - g_{\mathrm{P}}(\theta - \vartheta')\big|\, \mathrm{d}\theta \right\} \mathrm{d}\vartheta$$

$$= \max_{|\vartheta|\leq\delta} \int_{\mathbb{I}} \big|g_{\mathrm{P}}(\theta) - g_{\mathrm{P}}(\theta - \vartheta)\big|\, \mathrm{d}\theta, \tag{A.21}$$

where the first inequality follows from the Triangle Inequality for Integrals (Proposition 2.4.1) and the nonnegativity of $k_n(\cdot)$ (A.11a), and where the last equality follows because $k_n(\cdot)$ integrates to one (A.11b).

The second integral in (A.19) is bounded as follows:

$$\int_{\mathbb{I}} \int_{\delta < |\vartheta| < \frac{1}{2}} k_n(\vartheta)\big|g_{\mathrm{P}}(\theta) - g_{\mathrm{P}}(\theta - \vartheta)\big|\, \mathrm{d}\vartheta\, \mathrm{d}\theta$$

$$= \int_{\delta < |\vartheta| < \frac{1}{2}} k_n(\vartheta) \int_{\mathbb{I}} \big|g_{\mathrm{P}}(\theta) - g_{\mathrm{P}}(\theta - \vartheta)\big|\, \mathrm{d}\theta\, \mathrm{d}\vartheta$$

$$\leq \max_{\vartheta' \in \mathbb{I}} \left\{ \int_{\mathbb{I}} \big|g_{\mathrm{P}}(\theta) - g_{\mathrm{P}}(\theta - \vartheta')\big|\, \mathrm{d}\theta \right\} \int_{\delta < |\vartheta| < \frac{1}{2}} k_n(\vartheta)\, \mathrm{d}\vartheta$$

$$\leq 2\, \|\mathbf{g}\|_{\mathbb{I},1} \int_{\delta < |\vartheta| < \frac{1}{2}} k_n(\vartheta)\, \mathrm{d}\vartheta. \tag{A.22}$$

From (A.19), (A.21), and (A.22) we obtain

$$\int_{\mathbb{I}} \left| g(\theta) - \sigma_n(\mathbf{g}, \theta) \right| d\theta \leq \max_{|\vartheta| \leq \delta} \int_{\mathbb{I}} \left| g_{\mathrm{P}}(\theta - \vartheta) - g_{\mathrm{P}}(\theta) \right| d\theta$$

$$+ 2 \left\| \mathbf{g} \right\|_{\mathbb{I}, 1} \int_{\delta < |\vartheta| < \frac{1}{2}} k_n(\vartheta) \, d\vartheta. \tag{A.23}$$

Inequality (A.23) establishes the theorem as follows. For every $\epsilon > 0$ we can find by Lemma A.2.1 some $\delta > 0$ such that

$$\max_{|\vartheta| \leq \delta} \int_{\mathbb{I}} \left| g_{\mathrm{P}}(\theta - \vartheta) - g_{\mathrm{P}}(\theta) \right| d\theta < \epsilon, \tag{A.24}$$

and keeping this $\delta > 0$ fixed we have by (A.11c)

$$\lim_{n \to \infty} 2 \left\| \mathbf{g} \right\|_{\mathbb{I}, 1} \int_{\delta < |\vartheta| < \frac{1}{2}} k_n(\vartheta) \, d\vartheta = 0. \tag{A.25}$$

It thus follows from (A.23), (A.24), and (A.25) that

$$\overline{\lim_{n \to \infty}} \int_{\mathbb{I}} \left| g(\theta) - \sigma_n(\mathbf{g}, \theta) \right| d\theta < \epsilon, \tag{A.26}$$

from which the theorem follows because $\epsilon > 0$ was arbitrary. $\qquad\square$

From Theorem A.2.2 we obtain:

**Theorem A.2.3 (Uniqueness Theorem).** *Let* $\mathbf{g}_1, \mathbf{g}_2 \colon \mathbb{I} \to \mathbb{C}$ *be integrable. If*

$$\hat{g}_1(\eta) = \hat{g}_2(\eta), \quad \eta \in \mathbb{Z}, \tag{A.27}$$

*then* $\mathbf{g}_1$ *and* $\mathbf{g}_2$ *are equal except on a set of Lebesgue measure zero.*

**Proof.** Let $\mathbf{g} = \mathbf{g}_1 - \mathbf{g}_2$. By (A.27)

$$\hat{g}(\eta) = 0, \quad \eta \in \mathbb{Z}, \tag{A.28}$$

and consequently, by (A.13), $\sigma_n(\mathbf{g}, \theta) = 0$ for every $n \in \mathbb{N}$ and $\theta \in \mathbb{I}$. By Theorem A.2.2

$$\lim_{n \to \infty} \int_{\mathbb{I}} \left| g(\theta) - \sigma_n(\mathbf{g}, \theta) \right| d\theta = 0, \tag{A.29}$$

which combines with (A.28) to establish that

$$\int_{\mathbb{I}} \left| g(\theta) \right| d\theta = 0.$$

Thus, $\mathbf{g}$ is zero except on a set of Lebesgue measure zero (Proposition 2.5.3 (i)), and the result follows by recalling that $\mathbf{g} = \mathbf{g}_1 - \mathbf{g}_2$. $\qquad\square$

**Theorem A.2.4 (Riemann-Lebesgue Lemma).** *If* $\mathbf{g} \colon \mathbb{I} \to \mathbb{C}$ *is integrable, then*

$$\lim_{|\eta| \to \infty} \hat{g}(\eta) = 0. \tag{A.30}$$

**Proof.** Given any $\epsilon > 0$, let $\mathbf{p}$ be a degree-$n$ trigonometric polynomial satisfying

$$\int_{\mathbb{I}} \left| g(\theta) - p(\theta) \right| \mathrm{d}\theta < \epsilon. \tag{A.31}$$

(Such a trigonometric polynomial exists for some $n \in \mathbb{N}$ by Theorem A.2.2). Expressing $\mathbf{g}$ as $(\mathbf{g} - \mathbf{p}) + \mathbf{p}$ and using the linearity of the Fourier Series Coefficients we obtain for every integer $\eta$ whose magnitude exceeds the degree $n$ of $\mathbf{p}$

$$\begin{aligned}
\left| \hat{g}(\eta) \right| &= \left| \widehat{(g - p)}(\eta) + \hat{p}(\eta) \right| \\
&= \left| \widehat{(g - p)}(\eta) \right| \\
&\leq \int_{\mathbb{I}} \left| g(\theta) - p(\theta) \right| \mathrm{d}\theta \\
&< \epsilon,
\end{aligned} \tag{A.32}$$

where the equality in the first line follows from the linearity of the Fourier Series Coefficient; the equality in the second line because $|\eta|$ is larger than the degree $n$ of $\mathbf{p}$; the inequality in the third line because for every integrable $\mathbf{h} \colon \mathbb{I} \to \mathbb{C}$ we have

$$\begin{aligned}
\left| \hat{h}(\eta) \right| &= \left| \int_{\mathbb{I}} h(\theta) \, e^{-\mathrm{i}2\pi\eta\theta} \, \mathrm{d}\theta \right| \\
&\leq \int_{\mathbb{I}} \left| h(\theta) \, e^{-\mathrm{i}2\pi\eta\theta} \right| \mathrm{d}\theta \\
&= \int_{\mathbb{I}} \left| h(\theta) \right| \mathrm{d}\theta, \quad \eta \in \mathbb{Z};
\end{aligned}$$

and where the inequality in the last line of (A.32) follows from (A.31). $\qquad\square$

## A.3 Geometric Considerations

Every square-integrable function that is zero outside the interval $[-1/2, 1/2]$ is also integrable (Proposition 3.4.3). For such functions we can discuss the inner product and some of the related geometry. The main result is the following.

**Theorem A.3.1 (Complete Orthonormal System).** *The bi-infinite sequence of functions $\ldots, \boldsymbol{\phi}_{-1}, \boldsymbol{\phi}_0, \boldsymbol{\phi}_1, \ldots$ defined for every $\eta \in \mathbb{Z}$ by*

$$\phi_\eta(\theta) = e^{\mathrm{i}2\pi\eta\theta} \, \mathrm{I}\{\theta \in \mathbb{I}\}, \quad \theta \in \mathbb{R}$$

*forms a complete orthonormal system for the subspace of $\mathcal{L}_2$ consisting of those energy-limited functions that are zero outside the interval $\mathbb{I}$.*

**Proof.** The orthonormality follows by direct calculation

$$\int_{\mathbb{I}} e^{\mathrm{i}2\pi\eta\theta} \, e^{-\mathrm{i}2\pi\eta'\theta} \, \mathrm{d}\theta = \mathrm{I}\{\eta = \eta'\}, \quad \eta, \eta' \in \mathbb{Z}. \tag{A.33}$$

To show completeness it suffices by Proposition 8.5.5 (ii) to show that a square-integrable function $\mathbf{g} \colon \mathbb{I} \to \mathbb{C}$ that satisfies

$$\langle \mathbf{g}, \boldsymbol{\phi}_\eta \rangle = 0, \quad \eta \in \mathbb{Z} \tag{A.34}$$

must be equal to the all-zero function except on a subset of $\mathbb{I}$ of Lebesgue measure zero. To show this, we note that

$$\langle \mathbf{g}, \boldsymbol{\phi}_\eta \rangle = \hat{g}(\eta), \quad \eta \in \mathbb{Z}, \tag{A.35}$$

so (A.34) is equivalent to

$$\hat{g}(\eta) = 0, \quad \eta \in \mathbb{Z}, \tag{A.36}$$

and hence, by Theorem A.2.3, $\mathbf{g}$ must be zero except on a set of Lebesgue measure zero. $\qquad\qquad\qquad\qquad\qquad\qquad\qquad\qquad\qquad\qquad\qquad\qquad\qquad\qquad\qquad\quad\square$

Recalling Definition 8.2.1 and Proposition 8.2.2 (d) we obtain that, because the functions $\dots, \boldsymbol{\phi}_{-1}, \boldsymbol{\phi}_0, \boldsymbol{\phi}_1, \dots$ form a CONS and because $\langle \mathbf{g}, \boldsymbol{\phi}_\eta \rangle = \hat{g}(\eta)$, we have:

**Theorem A.3.2.** *Let* $\mathbf{g}, \mathbf{h} \colon \mathbb{I} \to \mathbb{C}$ *be square integrable. Then*

$$\int_{\mathbb{I}} \left| g(\theta) \right|^2 \mathrm{d}\theta = \sum_{\eta=-\infty}^{\infty} \left| \hat{g}(\eta) \right|^2 \tag{A.37}$$

*and*

$$\int_{\mathbb{I}} g(\theta)\, h^*(\theta)\, \mathrm{d}\theta = \sum_{\eta=-\infty}^{\infty} \hat{g}(\eta)\, \hat{h}^*(\eta). \tag{A.38}$$

There is nothing special about the interval $\mathbb{I}$, and, indeed, by scaling we obtain:

**Theorem A.3.3.** *Let* $S$ *be nonnegative.*

(i) *The bi-infinite sequence of functions defined for every* $\eta \in \mathbb{Z}$ *by*

$$s \mapsto \frac{1}{\sqrt{S}}\, e^{i2\pi\eta s/S}\, \mathrm{I}\left\{ -\frac{S}{2} \leq s < \frac{S}{2} \right\}, \quad s \in \mathbb{R} \tag{A.39}$$

*forms a CONS for the class of square-integrable functions that are zero outside the interval* $[-S/2, S/2)$.

(ii) *If* $\mathbf{g}$ *is square integrable and zero outside the interval* $[-S/2, S/2)$, *then*

$$\int_{-S/2}^{S/2} \left| g(\xi) \right|^2 \mathrm{d}\xi = \sum_{\eta=-\infty}^{\infty} \left| \int_{-S/2}^{S/2} g(\xi) \frac{1}{\sqrt{S}}\, e^{-i2\pi\eta s/S}\, \mathrm{d}\xi \right|^2. \tag{A.40}$$

(iii) *If* $\mathbf{g}, \mathbf{h} \colon \mathbb{R} \to \mathbb{C}$ *are square integrable and zero outside the interval* $[-S/2, S/2)$, *then*

$$\int_{-S/2}^{S/2} g(\xi)\, h^*(\xi)\, \mathrm{d}\xi$$

$$= \sum_{\eta=-\infty}^{\infty} \left( \int_{-S/2}^{S/2} g(\xi) \frac{1}{\sqrt{S}}\, e^{-i2\pi\eta s/S}\, \mathrm{d}\xi \right) \left( \int_{-S/2}^{S/2} h(\xi) \frac{1}{\sqrt{S}}\, e^{-i2\pi\eta s/S}\, \mathrm{d}\xi \right)^*.$$

**Note A.3.4.** The theorem continues to hold if we replace the half-open interval with the open interval $(-S/2, S/2)$ or with the closed interval $[-S/2, S/2]$, because the integrals are insensitive to these replacements.

**Note A.3.5.** We refer to

$$\int_{-S/2}^{S/2} g(\xi) \frac{1}{\sqrt{S}} e^{-i2\pi\eta s/S} \, d\xi$$

as the $\eta$-th **Fourier Series Coefficient of g with respect to the interval** $[-S/2, S/2)$.

**Lemma A.3.6 (A Mini Parseval Theorem).**

(i) If

$$x(t) = \int_{-W}^{W} g(f) \, e^{i2\pi ft} \, df, \quad t \in \mathbb{R}, \tag{A.41}$$

where $\mathbf{g} \colon \mathbb{R} \to \mathbb{C}$ satisfies

$$\int_{-W}^{W} |g(f)|^2 \, df < \infty, \tag{A.42}$$

then

$$\int_{-\infty}^{\infty} |x(t)|^2 \, dt = \int_{-W}^{W} |g(f)|^2 \, df. \tag{A.43}$$

(ii) If for both $\nu = 1$ and $\nu = 2$

$$x_\nu(t) = \int_{-W}^{W} g_\nu(f) \, e^{i2\pi ft} \, df, \quad t \in \mathbb{R}, \tag{A.44}$$

where the functions $\mathbf{g}_1, \mathbf{g}_2 \colon \mathbb{R} \to \mathbb{C}$ satisfy

$$\int_{-W}^{W} |g_\nu(f)|^2 \, df < \infty, \quad \nu = 1, 2, \tag{A.45}$$

then

$$\int_{-\infty}^{\infty} x_1(t) \, x_2^*(t) \, dt = \int_{-W}^{W} g_1(f) \, g_2^*(f) \, df. \tag{A.46}$$

**Proof.** We first prove Part (i). We begin by expressing the energy in $\mathbf{x}$ in the form

$$\int_{-\infty}^{\infty} |x(t)|^2 \, dt = \sum_{\ell=-\infty}^{\infty} \int_{-\frac{\ell}{2W}}^{-\frac{\ell}{2W}+\frac{1}{2W}} |x(t)|^2 \, dt$$

$$= \sum_{\ell=-\infty}^{\infty} \int_0^{\frac{1}{2W}} \left| x\left(\alpha - \frac{\ell}{2W}\right) \right|^2 \, d\alpha$$

$$= \int_0^{\frac{1}{2W}} \sum_{\ell=-\infty}^{\infty} \left| x\left(\alpha - \frac{\ell}{2W}\right) \right|^2 \, d\alpha, \tag{A.47}$$

where in the second equality we changed the integration variable to $\alpha \triangleq t + \ell/(2W)$; and where the third equality follows from Fubini's Theorem and the nonnegativity of the integrand. The proof of Part (i) will follow from (A.47) once we show that for every $\alpha \in \mathbb{R}$

$$\sum_{\ell=-\infty}^{\infty} \left| x\left(\alpha - \frac{\ell}{2W}\right) \right|^2 = 2W \int_{-W}^{W} |g(f)|^2 \, \mathrm{d}f. \tag{A.48}$$

This can be shown by noting that by (A.41)

$$\frac{1}{\sqrt{2W}} x\left(\alpha - \frac{\ell}{2W}\right) = \int_{-W}^{W} \frac{1}{\sqrt{2W}} e^{-\mathrm{i}2\pi f \frac{\ell}{2W}} e^{\mathrm{i}2\pi f \alpha} g(f) \, \mathrm{d}f,$$

so $(2W)^{-1/2} x(\alpha - \ell/(2W))$ is the $\ell$-th Fourier Series Coefficient of the mapping $f \mapsto e^{\mathrm{i}2\pi f \alpha} g(f)$ with respect to the interval $[-W, W)$ and consequently

$$\sum_{\ell=-\infty}^{\infty} \left| \frac{1}{\sqrt{2W}} x\left(\alpha - \frac{\ell}{2W}\right) \right|^2 = \int_{-W}^{W} \left| e^{\mathrm{i}2\pi f \alpha} g(f) \right|^2 \, \mathrm{d}f$$

$$= \int_{-W}^{W} |g(f)|^2 \, \mathrm{d}f,$$

where the first equality follows from Theorem A.3.3 (ii) and the second because the magnitude of $e^{\mathrm{i}2\pi f \alpha}$ is one.

To prove Part (ii) we note that by opening the square and then applying Part (i) to the function $\beta \mathbf{x}_1 + \mathbf{x}_2$ we obtain for every $\beta \in \mathbb{C}$

$$|\beta|^2 \int_{-\infty}^{\infty} |x_1(t)|^2 \, \mathrm{d}t + \int_{-\infty}^{\infty} |x_2(t)|^2 \, \mathrm{d}t + 2\,\mathrm{Re}\left( \beta \int_{-\infty}^{\infty} x_1(t)\, x_2^*(t) \, \mathrm{d}t \right)$$

$$= \int_{-\infty}^{\infty} |\beta x_1(t) + x_2(t)|^2 \, \mathrm{d}t$$

$$= \int_{-W}^{W} |\beta g_1(f) + g_2(f)|^2 \, \mathrm{d}f$$

$$= |\beta|^2 \int_{-\infty}^{\infty} |g_1(f)|^2 \, \mathrm{d}f + \int_{-\infty}^{\infty} |g_2(f)|^2 \, \mathrm{d}f + 2\,\mathrm{Re}\left( \beta \int_{-\infty}^{\infty} g_1(f)\, g_2^*(f) \, \mathrm{d}f \right).$$

Consequently, upon applying Part (i) to $\mathbf{x}_1$ and to $\mathbf{x}_2$ we obtain

$$\mathrm{Re}\left( \beta \int_{-\infty}^{\infty} x_1(t)\, x_2^*(t) \, \mathrm{d}t \right) = \mathrm{Re}\left( \beta \int_{-W}^{W} g_1(f)\, g_2^*(f) \, \mathrm{d}f \right), \quad \beta \in \mathbb{C},$$

which implies

$$\int_{-\infty}^{\infty} x_1(t)\, x_2^*(t) \, \mathrm{d}t = \int_{-W}^{W} g_1(f)\, g_2^*(f) \, \mathrm{d}f. \qquad \square$$

**Corollary A.3.7.**

*(i) Let* $\mathbf{y}\colon \mathbb{R} \to \mathbb{C}$ *be of finite energy, and let* $\mathsf{T} > 0$ *be arbitrary. Let*

$$\tilde{g}(f) = \int_{-\mathsf{T}}^{\mathsf{T}} y(t)\, e^{-\mathrm{i}2\pi f t}\, \mathrm{d}t, \quad f \in \mathbb{R}.$$

*Then*

$$\int_{-\mathsf{T}}^{\mathsf{T}} |y(t)|^2\, \mathrm{d}t = \int_{-\infty}^{\infty} |\tilde{g}(f)|^2\, \mathrm{d}f.$$

*(ii) Let the signals* $\mathbf{x}_1, \mathbf{x}_2 \colon \mathbb{R} \to \mathbb{C}$ *be of finite energy, and let* $\mathsf{T} > 0$. *Define*

$$g_\nu(f) = \int_{-\mathsf{T}}^{\mathsf{T}} x_\nu(t)\, e^{-\mathrm{i}2\pi f t}\, \mathrm{d}t, \quad \left(\nu = 1, 2, \quad f \in \mathbb{R}\right).$$

*Then*

$$\int_{-\mathsf{T}}^{\mathsf{T}} x_1(t)\, x_2^*(t)\, \mathrm{d}t = \int_{-\infty}^{\infty} g_1(f)\, g_2^*(f)\, \mathrm{d}f.$$

**Proof.** Part (i) follows from Lemma A.3.6 (i) by substituting $\mathsf{T}$ for $\mathsf{W}$; by substituting $\tilde{\mathbf{y}}$ for $\mathbf{g}$; and by swapping the dummy variables $f$ and $t$. Part (ii) follows analogously. □

## A.4 Pointwise Reconstruction

If $\mathbf{g}\colon \mathbb{I} \to \mathbb{C}$ is periodically continuous, then we can reconstruct its value at every point from its Fourier Series Coefficients:

**Theorem A.4.1 (Reconstructing Periodically Continuous Functions).** *Let the function* $\mathbf{g}\colon \mathbb{I} \to \mathbb{C}$ *be periodically continuous. Then*

$$\lim_{n \to \infty} \max_{\theta \in \mathbb{I}} \left\{ |g(\theta) - \sigma_n(\mathbf{g}, \theta)| \right\} = 0. \tag{A.49}$$

**Proof.** Let $\mathbf{g}_\mathrm{P}$ denote the periodic extension of $\mathbf{g}$. Then for every $\theta \in \mathbb{I}$,

$$g(\theta) - \sigma_n(\mathbf{g}, \theta) = g(\theta) - \int_{\mathbb{I}} k_n(\vartheta)\, g_\mathrm{P}(\theta - \vartheta)\, \mathrm{d}\vartheta$$

$$= \int_{\mathbb{I}} k_n(\vartheta) \big( g_\mathrm{P}(\theta) - g_\mathrm{P}(\theta - \vartheta) \big)\, \mathrm{d}\vartheta, \tag{A.50}$$

where the first equality follows from the definition of $\sigma_n(\mathbf{g}, \theta)$ (A.12) and the second from (A.11b). Consequently, for every $\theta \in \mathbb{I}$,

$$\big| g(\theta) - \sigma_n(\mathbf{g}, \theta) \big|$$

$$\leq \int_{\mathbb{I}} k_n(\vartheta) \big| g_\mathrm{P}(\theta) - g_\mathrm{P}(\theta - \vartheta) \big|\, \mathrm{d}\vartheta$$

$$= \int_{-\delta}^{\delta} + \int_{\delta < |\vartheta| < \frac{1}{2}} k_n(\vartheta) \big| g_\mathrm{P}(\theta) - g_\mathrm{P}(\theta - \vartheta) \big|\, \mathrm{d}\vartheta, \quad 0 \leq \delta < \frac{1}{2}. \tag{A.51}$$

We next treat the two integrals separately. For the first we have for every $\theta \in \mathbb{I}$ and every $0 \leq \delta < 1/2$,

$$\int_{-\delta}^{\delta} k_n(\vartheta) |g_\mathrm{P}(\theta) - g_\mathrm{P}(\theta - \vartheta)| \, \mathrm{d}\vartheta \leq \max_{|\vartheta'| \leq \delta} \{ |g_\mathrm{P}(\theta) - g_\mathrm{P}(\theta - \vartheta')| \} \int_{-\delta}^{\delta} k_n(\vartheta) \, \mathrm{d}\vartheta$$

$$\leq \max_{|\vartheta| \leq \delta} \{ |g_\mathrm{P}(\theta) - g_\mathrm{P}(\theta - \vartheta)| \}, \tag{A.52}$$

where the first inequality follows from the the nonnegativity of $k_n(\cdot)$ (A.11a), and where the second inequality follows because $k_n(\cdot)$ is nonnegative and integrates over $\mathbb{I}$ to one (A.11b). For the second integral in (A.51) we have for every $\theta \in \mathbb{I}$ and every $0 \leq \delta < 1/2$,

$$\int_{\delta < |\vartheta| < \frac{1}{2}} k_n(\vartheta) |g_\mathrm{P}(\theta) - g_\mathrm{P}(\theta - \vartheta)| \, \mathrm{d}\vartheta \leq 2 \max_{\theta' \in \mathbb{I}} \{ |g(\theta')| \} \int_{\delta < |\vartheta| < \frac{1}{2}} k_n(\vartheta) \, \mathrm{d}\vartheta, \tag{A.53}$$

where the maximum on the RHS is finite because $\mathbf{g}$ is periodically continuous. Combining (A.51), (A.52), and (A.53) we obtain for every $0 \leq \delta < 1/2$

$$\max_{\theta \in \mathbb{I}} \{ |g(\theta) - \sigma_n(\mathbf{g}, \theta)| \}$$

$$\leq \max_{\theta \in \mathbb{I}} \max_{|\vartheta| \leq \delta} \{ |g_\mathrm{P}(\theta) - g_\mathrm{P}(\theta - \vartheta)| \} + 2 \max_{\theta' \in \mathbb{I}} \{ |g(\theta')| \} \int_{\delta < |\vartheta| < \frac{1}{2}} k_n(\vartheta) \, \mathrm{d}\vartheta. \tag{A.54}$$

Because $g(\cdot)$ is periodically continuous it follows that its periodic extension $\mathbf{g}_\mathrm{P}$ is uniformly continuous. Consequently, for every $\epsilon > 0$ we can find some $\delta > 0$ such that

$$\max_{|\vartheta| \leq \delta} |g_\mathrm{P}(\theta) - g_\mathrm{P}(\theta - \vartheta)| < \epsilon, \quad \theta \in \mathbb{I}. \tag{A.55}$$

By letting $n$ tend to infinity in (A.54) we obtain from (A.11c) and (A.55)

$$\varlimsup_{n \to \infty} \max_{\theta \in \mathbb{I}} \{ |g(\theta) - \sigma_n(\mathbf{g}, \theta)| \} < \epsilon,$$

which establishes the result because $\epsilon > 0$ was arbitrary. $\qquad\square$

As a corollary we obtain:

**Corollary A.4.2 (Weierstrass's Approximation Theorem).** *Every periodically continuous function from $\mathbb{I}$ to $\mathbb{C}$ can be approximated uniformly using trigonometric polynomials.*

# Bibliography

ADAMS, R. A., AND J. J. F. FOURNIER (2003): *Sobolev Spaces*. Elsevier B.V., Amsterdam, The Netherlands, second edn.

ADLER, R. J. (1990): *An Introduction to Continuity, Extrema, and Related Topics for General Gaussian Processes*. Institute of Mathematical Statistics, Hayward, CA.

AXLER, S. (1997): *Linear Algebra Done Right*. Springer-Verlag New York, Inc., New York, second edn.

BARRY, J. R., E. A. LEE, AND D. G. MESSERSCHMITT (2004): *Digital Communication*. Springer-Verlag New York, Inc., New York, third edn.

BILLINGSLEY, P. (1995): *Probability and Measure*. John Wiley & Sons, Inc., New York, third edn.

BLACKWELL, D., AND R. V. RAMAMOORTHI (1982): "A Bayes but not classically sufficient statistic," *The Annals of Statistics*, 10(3), 1025–1026.

BLAHUT, R. E. (2002): *Algebraic Codes for Data Transmission*. Cambridge University Press, New York.

BOAS, JR., R. P. (1954): *Entire Functions*. Academic Press Inc., New York.

BOGACHEV, V. I. (1998): *Gaussian Measures*. American Mathematical Society, Providence, RI.

BRYC, W. (1995): *The Normal Distribution: Characterizations with Applications*. Springer-Verlag New York, Inc., New York.

CHUNG, K. L. (2001): *A Course in Probability Theory*. Academic Press, San Diego, third edn.

CRAMÉR, H., AND M. R. LEADBETTER (2004): *Stationary and Related Stochastic Processes: Sample Function Properties and Their Applications*. Dover Publications, Inc., Mineola, NY.

DE CAEN, D. (1997): "A lower bound on the probability of a union," *Discrete Mathematics*, 169, 217–220.

DOOB, J. L. (1990): *Stochastic Processes*. John Wiley & Sons, Inc., New York.

DUDLEY, R. M. (2003): *Real Analysis and Probability.* Cambridge University Press, New York, second edn.

DYM, H., AND H. P. MCKEAN (1972): *Fourier Series and Integrals.* Academic Press, Inc., San Diego.

FAREBROTHER, R. W. (1988): *Linear Least Squares Computations.* Marcel Dekker, Inc., New York.

FELLER, W. (1971): *An Introduction to Probability Theory and Its Applications,* vol. II. John Wiley & Sons, Inc., New York, second edn.

FINNER, H., AND M. ROTERS (1997): "Log-concavity and inequalities for chi-square, F and beta distributions with applications in multiple comparisons," *Statistica Sinica,* 7(3), 771–787.

FORNEY, JR., G. D. (1991): "Geometrically uniform codes," *IEEE Transactions on Information Theory,* 37(5), 1241–1260.

GALLAGER, R. G. (2008): *Principles of Digital Communication.* Cambridge University Press, New York.

GIKHMAN, I. I., AND A. V. SKOROKHOD (1996): *Introduction to the Theory of Random Processes.* Dover Publications, Inc., Mineola, NY.

GOLUB, G. H., AND C. F. VAN LOAN (1996): *Matrix Computations.* The John Hopkins University Press, Baltimore, third edn.

GRIMMETT, G., AND D. STIRZAKER (2001): *Probability and Random Processes.* Oxford University Press, New York, third edn.

HALMOS, P. R. (1950): *Measure Theory.* D. Van Nostrand Company, Inc., Princeton, NJ.

HALMOS, P. R., AND L. J. SAVAGE (1949): "Application of the Radon-Nikodym theorem to the theory of sufficient statistics," *The Annals of Mathematical Statistics,* 20(2), 225–241.

HERSTEIN, I. N. (2001): *Topics in Algebra.* John Wiley & Sons, Inc., New York, second edn.

HORN, R. A., AND C. R. JOHNSON (1985): *Matrix Analysis.* Cambridge University Press, New York.

JOHNSON, N. L., AND S. KOTZ (1972): *Distributions in Statistics: Continuous Multivariate Distributions.* John Wiley & Sons, Inc., New York.

JOHNSON, N. L., S. KOTZ, AND N. BALAKRISHNAN (1994a): *Continuous Univariate Distributions,* vol. 1. John Wiley & Sons, Inc., New York, second edn.

——— (1994b): *Continuous Univariate Distributions,* vol. 2. John Wiley & Sons, Inc., New York, second edn.

KAILATH, T., A. H. SAYED, AND B. HASSIBI (2000): *Linear Estimation.* Prentice Hall, Inc., Upper Saddle River, NJ.

KARATZAS, I., AND S. E. SHREVE (1991): *Brownian Motion and Stochastic Calculus.* Springer-Verlag New York, Inc., New York, second edn.

KATZNELSON, Y. (1976): *An Introduction to Harmonic Analysis.* Dover Publications, Inc., Mineola, NY, second edn.

KÖRNER, T. W. (1988): *Fourier Analysis.* Cambridge University Press, New York.

KWAKERNAAK, H., AND R. SIVAN (1991): *Modern Signals and Systems.* Prentice Hall, Inc., Englewood Cliffs, NJ.

LEHMANN, E. L., AND J. P. ROMANO (2005): *Testing Statistical Hypotheses.* Springer-Verlag New York, Inc., New York, third edn.

LOELIGER, H.-A. (1991): "Signal sets matched to groups," *IEEE Transactions on Information Theory*, 37(6), 1675–1682.

LOÈVE, M. (1963): *Probability Theory.* D. Van Nostrand Company, Inc., Princeton, NJ, third edn.

LOGAN, JR., B. F. (1978): "Theory of analytic modulation systems," *The Bell System Technical Journal*, 57(3), 491–576.

MACWILLIAMS, F. J., AND N. J. A. SLOANE (1977): *The Theory of Error-Correcting Codes.* North-Holland, Amsterdam, The Netherlands.

MARSHALL, A. W., AND I. OLKIN (1979): *Inequalities: Theory of Majorization and Its Applications.* Academic Press, Inc., San Diego.

NEHARI, Z. (1975): *Conformal Mapping.* Dover Publications, Inc., Mineola, NY.

OPPENHEIM, A. V., AND A. S. WILLSKY (1997): *Signals & Systems.* Prentice Hall, Inc., Upper Saddle River, NJ, second edn.

PINSKY, M. A. (2002): *Introduction to Fourier Analysis and Wavelets.* Brooks/Cole, Pacific Grove, CA.

POOR, H. V. (1994): *An Introduction to Signal Detection and Estimation.* Springer-Verlag New York, Inc., New York, second edn.

PORAT, B. (2008): *Digital Processing of Random Signals: Theory and Methods.* Dover Publications, Inc., Mineola, NY.

POURAHMADI, M. (2001): *Foundations of Time Series Analysis and Prediction Theory.* John Wiley & Sons, Inc., New York.

PROAKIS, J. G. (2000): *Digital Communications.* McGraw-Hill, Inc., New York, fourth edn.

REQUICHA, A. A. G. (1980): "The zeros of entire functions: theory and engineering applications," *Proceedings of the IEEE*, 68(3), 308–328.

RICHARDSON, T., AND R. URBANKE (2008): *Modern Coding Theory.* Cambridge University Press, New York.

RIESZ, F., AND B. SZ.-NAGY (1990): *Functional Analysis.* Dover Publications, Inc., Mineola, NY.

ROMANO, J. P., AND A. F. SIEGEL (1986): *Counterexamples in Probability and Statistics.* Chapman & Hall, New York.

ROTH, R. M. (2006): *Introduction to Coding Theory.* Cambridge University Press, New York.

ROYDEN, H. L. (1988): *Real Analysis.* Prentice Hall, Inc., Upper Saddle River, NJ, third edn.

RUDIN, W. (1974): *Real & Complex Analysis.* McGraw-Hill, Inc., New York, second edn.

———— (1976): *Principles of Mathematical Analysis.* McGraw-Hill, Inc., New York, third edn.

SASON, I., AND S. SHAMAI (2006): "Performance analysis of linear codes under maximum-likelihood decoding: a tutorial," *Foundations and Trends in Communications and Information Theory,* 3(1/2), 1–222.

SHIRYAEV, A. N. (1996): *Probability.* Springer-Verlag New York, Inc., New York, second edn.

SIMON, B. (2005): *Orthogonal Polynomials on the Unit Circle. Part 1: Classical Theory.* American Mathematical Society, Providence RI.

SIMON, M. K. (2002): *Probability Distributions Involving Gaussian Random Variables: A Handbook for Engineers and Scientists.* Kluwer Academic Publishers, Norwell, MA.

SLEPIAN, D. (1976): "On bandwidth," *Proceedings of the IEEE,* 64(3), 292–300.

SLEPIAN, D., AND H. O. POLLAK (1961): "Prolate spheroidal wave functions, Fourier analysis and uncertainty—I," *The Bell System Technical Journal,* 40, 43–63.

STEELE, J. M. (2004): *The Cauchy-Schwarz Master Class: An Introduction to the Art of Mathematical Inequalities.* Cambridge University Press, New York.

STEIN, E. M., AND G. WEISS (1990): *Introduction to Fourier Analysis on Euclidean Spaces.* Princeton University Press, Princeton, NJ.

TONG, Y. L. (1990): *The Multivariate Normal Distribution.* Springer-Verlag New York, Inc., New York.

UNSER, M. (2000): "Sampling—50 years after Shannon," *Proceedings of the IEEE,* 88(4), 569–587.

VAN LINT, J. H. (1998): *Introduction to Coding Theory.* Springer-Verlag New York, Inc., New York, third edn.

VERDÚ, S. (1998): *Multiuser Detection.* Cambridge University Press, New York.

VITERBI, A. J., AND J. K. OMURA (1979): *Principles of Digital Communication and Coding.* McGraw-Hill, Inc., New York.

WILLIAMS, D. (1991): *Probability with Martingales.* Cambridge University Press, New York.

ZHANG, F. (1999): *Matrix Theory: Basic Results and Techniques.* Springer-Verlag New York, Inc., New York.

ZVONKIN, A. (1997): "Matrix integrals and map enumeration: an accessible introduction," *Mathematical and Computer Modelling,* 26(8–10), 281–304.

# Theorems Referenced by Name

# Abbreviations

## Abbreviations in Mathematics

| | |
|---|---|
| CDF | Cumulative Distribution Function |
| CONS | Complete Orthonormal System |
| CRV | Complex Random Variable |
| CSP | Complex Stochastic Process |
| FDD | Finite-Dimensional Distribution |
| FT | Fourier Transform |
| IFT | Inverse Fourier Transform |
| IID | Independent and Identically Distributed |
| LHS | Left-Hand Side |
| MGF | Moment Generating Function |
| PDF | Probability Density Function |
| PMF | Probability Mass Function |
| PSD | Power Spectral Density |
| RHS | Right-Hand Side |
| RV | Random Variable |
| SP | Stochastic Process |
| WSS | Wide-Sense Stationary |

## Abbreviations in Communications

| | |
|---|---|
| BER | Bit Error Rate |
| BPF | Bandpass Filter |
| LPF | Lowpass Filter |
| M-PSK | M-ary Phase Shift Keying |
| PAM | Pulse Amplitude Modulation |
| PSK | Phase Shift Keying |
| QAM | Quadrature Amplitude Modulation |
| QPSK | Quadrature Phase Keying |

# List of Symbols

## General

| | |
|---|---|
| $A \Rightarrow B$ | Statement $B$ is true whenever Statement $A$ is true. |
| $A \Leftrightarrow B$ | Statement $A$ is true if, and only if, Statement $B$ is true. |
| $\sum$ | Summation. |
| $\prod$ | Product. |
| $\triangleq$ | Equal by definition. |
| $\square$ | End of proof. |

## Sets

| | |
|---|---|
| $\emptyset$ | Empty set. |
| $\{- : -\}$ | The set of all objects described before the colon that satisfy the condition stated after the colon. |
| $\# \mathcal{A}$ | Number of elements of the set $\mathcal{A}$. |
| $a \in \mathcal{A}$ | Set membership: $a$ is an element of $\mathcal{A}$. |
| $a \notin \mathcal{A}$ | Exclusion: $a$ is not an element of $\mathcal{A}$. |
| $\mathcal{A} \subset \mathcal{B}$ | Proper subset: every element of $\mathcal{A}$ is an element of $\mathcal{B}$ but some elements of $\mathcal{B}$ are not elements of $\mathcal{A}$. |
| $\mathcal{A} \subseteq \mathcal{B}$ | Subset: every element of $\mathcal{A}$ is also an element of $\mathcal{B}$. |
| $\mathcal{B} \setminus \mathcal{A}$ | Setminus: $\{b \in \mathcal{B} : b \notin \mathcal{A}\}$. |
| $\mathcal{A}^c$ | Set-complement. |
| $\mathcal{A} \triangle \mathcal{B}$ | Symmetric Set Difference: $(\mathcal{A} \setminus \mathcal{B}) \cup (\mathcal{B} \setminus \mathcal{A})$. |
| $\mathcal{A} \times \mathcal{B}$ | Cartesian product: $\{(a, b) : a \in \mathcal{A}, b \in \mathcal{B}\}$. |
| $\mathcal{A}^n$ | $n$-fold Cartesian product: $\underbrace{\mathcal{A} \times \mathcal{A} \times \cdots \times \mathcal{A}}_{n \text{ times}}$. |
| $\mathcal{A} \cap \mathcal{B}$ | Intersection: $\{\xi \in \mathcal{A} : \xi \in \mathcal{B}\}$. |
| $\mathcal{A} \cup \mathcal{B}$ | Union: elements of $\mathcal{A}$ or $\mathcal{B}$. |

## Specific Sets

| | |
|---|---|
| $\mathbb{N}$ | Natural Numbers: $\{1, 2, \ldots\}$. |
| $\mathbb{Z}$ | Integers: $\{\ldots, -2, -1, 0, 1, 2, \ldots\}$. |
| $\mathbb{R}$ | Real Numbers. |
| $\mathbb{C}$ | Complex Numbers. |

$\mathbb{F}_2$        Binary Field (Section 29.2).

$\mathbb{I}$        Unit interval $[-1/2, 1/2)$; see (A.1).

## Intervals and Some Functions

| | |
|---|---|
| $\leq, <, \geq, >$ | Inequality signs. |
| $+\infty, -\infty, \infty$ | Infinities. |
| $[a, b]$ | Closed Interval: $\{\xi \in \mathbb{R} : a \leq \xi \leq b\}$. |
| $[a, b)$ | Interval open on the right: $\{\xi \in \mathbb{R} : a \leq \xi < b\}$. |
| $(a, b]$ | Interval open on the left: $\{\xi \in \mathbb{R} : a < \xi \leq b\}$. |
| $(a, b)$ | Open interval: $\{\xi \in \mathbb{R} : a < \xi < b\}$. |
| $[0, \infty]$ | Nonnegative reals including infinity: $\{\xi \in \mathbb{R} : \xi \geq 0\} \cup \{\infty\}$. |
| $\lfloor \xi \rfloor$ | Floor: the largest integer not larger than $\xi$. |
| $\lceil \xi \rceil$ | Ceiling: the smallest integer not smaller than $\xi$. |
| max | Maximum. |
| min | Minimum. |
| sup | Least upper bound. |
| inf | Greatest lower bound. |

## Complex Numbers

| | |
|---|---|
| $\mathbb{C}$ | Complex field. |
| i | $i = \sqrt{-1}$. |
| $\mathrm{Re}(z)$ | Real part of $z$. |
| $\mathrm{Im}(\cdot)$ | Imaginary part of $z$. |
| $\lvert z \rvert$ | Modulus of $z$. |
| $z^*$ | Complex conjugate of $z$. |
| $\mathcal{D}(z_0, r)$ | Open disc: $\{z \in \mathbb{C} : \lvert z - z_0 \rvert < r\}$. |

## Limits

| | |
|---|---|
| $a_n \to a$ | Convergence: the sequence $a_1, a_2, \ldots$ converges to $a$. |
| $\lim_{n \to \infty} a_n$ | Limit: the limit of $a_n$ as $n$ tends to infinity. |
| $\to$ | Converges to. |
| $\overline{\lim}_{n \to \infty} a_n$ | Upper limit (limit superior). |
| $\underline{\lim}_{n \to \infty} a_n$ | Lower limit (limit inferior). |

## Defining and Operating on Functions

| | |
|---|---|
| $\mathbf{g} \colon \mathcal{D} \to \mathcal{R}$ | Function of name $\mathbf{g}$, domain $\mathcal{D}$, and range $\mathcal{R}$. |
| $\mathbf{g} \colon t \mapsto t^2$ | Function of name $\mathbf{g}$ mapping $t$ to $t^2$. (Domain & range unspecified.) |
| $\mathbf{g} \circ \mathbf{h}$ | Composition: $\xi \mapsto g(h(\xi))$. |
| d | Differentiation operator. |
| $\frac{\partial g(\mathbf{x})}{\partial x^{(j)}}$ | Partial derivative of $g(\cdot)$ with respect to $x^{(j)}$. |

| | |
|---|---|
| $\frac{\partial g(\mathbf{x})}{\partial \mathbf{x}}$ | Jacobian matrix. |
| $\int_{\mathcal{D}}$ | Integral over the region $\mathcal{D}$. |
| $f(\xi)\big|_{\xi=a}$ | The evaluation of the function $\xi \mapsto g(\xi)$ at $a$. |
| $g(\xi)\big|_a^b$ | The evaluation $g(b) - g(a)$. |

## Function Norms, Relations, and Equivalence Classes

| | |
|---|---|
| $\|\mathbf{x}\|_1$ | See (2.6). |
| $\|\mathbf{x}\|_2$ | See (3.12). |
| $\|\mathbf{x}\|_{\mathbb{I},1}$ | See (A.14). |
| $\mathbf{x} \equiv \mathbf{y}$ | $\mathbf{x}$ and $\mathbf{y}$ are indistinguishable; see Definition 2.5.2. |
| $[\mathbf{u}]$ | The equivalence class of $\mathbf{x}$; see (4.60). |

## Function Spaces

| | |
|---|---|
| $\mathcal{L}_1$ | Integrable functions from $\mathbb{R}$ to $\mathbb{C}$ or $\mathbb{R}$ to $\mathbb{R}$ (depending on context); see Sections 2.2 and 2.3. |
| $\mathcal{L}_2$ | Square-integrable functions from $\mathbb{R}$ to $\mathbb{C}$ or $\mathbb{R}$ to $\mathbb{R}$ (depending on context); see Section 3.1. |
| $L_2$ | Collection of equivalence classes of square-integrable functions; see Section 4.7. |

## Special Functions

| | |
|---|---|
| $\mathrm{I}\{\text{statement}\}$ | Indicator function. Its value is 1 if the statement is true and 0 otherwise. |
| $\mathbf{0}$ | All-zero function: $t \mapsto 0$. |
| $n!$ | $n$ factorial: $1 \times 2 \times \cdots \times n$. |
| $\binom{n}{k}$ | Number of subsets of $\{1, \ldots, n\}$ containing $k$ (distinct) elements $(= n!/(k!(n-k)!))$. |
| $\sqrt{\xi}$ | Nonnegative square root of $\xi$. |
| $\cos(\cdot)$ | Cosine function (argument in radians). |
| $\sin(\cdot)$ | Sine function (argument in radians). |
| $\mathrm{sinc}(\cdot)$ | Sinc function; see (5.20). |
| $\tan^{-1}(\cdot)$ | Inverse tangent. |
| $\mathcal{Q}(\cdot)$ | $\mathcal{Q}$-function; see (19.9). |
| $\Gamma(\cdot)$ | Gamma function; see (19.39). |
| $\mathrm{I}_0(\cdot)$ | The zeroth-order modified Bessel function; see (27.47). |
| $\ln(\cdot)$ | Natural logarithm (base $e$). |
| $\exp(\cdot)$ | Exponential function: $\exp(\xi) = e^\xi$. |
| $\xi \mod [-\pi, \pi)$ | element of $[-\pi, \pi)$ that differs from $\xi$ by an integer multiple of $2\pi$. |

## Operations on Signals

| | |
|---|---|
| $\tilde{\mathbf{x}}$ | The mirror image of $\mathbf{x}$; see (5.1). |
| $\hat{\mathbf{x}}$ | The Fourier Transform of the signal $\mathbf{x}$; see (6.1). |
| $\check{\mathbf{x}}$ | Inverse Fourier Transform of $\mathbf{x}$; see (6.4). |
| $\langle \mathbf{x}, \mathbf{y} \rangle$ | Inner product between the signals $\mathbf{x}$ and $\mathbf{y}$; see (3.1) and (3.4). |
| $\mathbf{x} \star \mathbf{y}$ | Convolution of $\mathbf{x}$ with $\mathbf{y}$; see (5.2). |
| $\mathbf{x} + \mathbf{y}$ | The signal $t \mapsto x(t) + y(t)$. |
| $\alpha \mathbf{x}$ | The scaling of the signal $\mathbf{x}$ by complex or real number $\alpha$, i.e., the signal $t \mapsto \alpha x(t)$. |
| $\mathsf{R}_{\mathbf{xx}}$ | Self-similarity function of signal $\mathbf{x}$. |
| $\hat{g}(\eta)$ | The $\eta$-th Fourier Series Coefficient; see (A.2). |

## Filters

| | |
|---|---|
| $\widehat{\mathrm{LPF}}_{W_c}(\cdot)$ | Frequency response of a unit-gain lowpass filter of cutoff frequency $W_c$. That is, $\widehat{\mathrm{LPF}}_{W_c}(f) = \mathrm{I}\{|f| \le W_c\}$. |
| $\mathrm{LPF}_{W_c}(\cdot)$ | Impulse response of a unit-gain lowpass filter of cutoff frequency $W_c$. That is, $\mathrm{LPF}_{W_c}(t) = 2W_c \,\mathrm{sinc}(2W_c t)$. |
| $\widehat{\mathrm{BPF}}_{W,f_c}(\cdot)$ | Frequency response of a unit-gain bandpass filter of bandwidth $W$ around the carrier frequency $f_c$. That is, the mapping of $f$ to $\mathrm{I}\{||f| - f_c| \le W/2\}$. It is assumed that $f_c > W/2$. |
| $\mathrm{BPF}_{W,f_c}(\cdot)$ | Impulse response of a unit-gain bandpass filter of bandwidth $W$ around the carrier frequency $f_c$. That is, the mapping of $t$ to $2W \cos(2\pi f_c t)\,\mathrm{sinc}(Wt)$. It is assumed that $f_c > W/2$. |

## PAM Signaling

| | |
|---|---|
| $\mathbf{g}$ or $\phi$ | Pulse shape; see Section 10.7. |
| $\mathsf{T}_s$ | Baud period; see Section 10.7. |
| $1/\mathsf{T}_s$ | Baud rate. |
| $\mathcal{X}$ | Constellation; see Section 10.8. |
| $\delta$ | Minimum distance of a constellation; see Section 10.8. |
| $\mathrm{enc}(\cdot)$ | Block encoder; see Definition 10.4.1 and (18.3). |
| $x(t; \mathbf{d})$ | Transmitted signal at time $t$ when the data are $\mathbf{d}$; see (28.6). |

## QAM Signaling

| | |
|---|---|
| $\mathbf{g}$ or $\phi$ | Pulse shape; see Sections 16.3 & 16.5. |
| $\mathsf{T}_s$ | Baud period; see Section 16.3. |
| $1/\mathsf{T}_s$ | Baud rate. |
| $\mathcal{C}$ | Constellation; see Section 16.7. |
| $\delta$ | Minimum distance of a constellation; see Section 16.7. |
| $\mathrm{enc}(\cdot)$ | Block encoder; see (18.3). |
| $x(t; \mathbf{d})$ | The transmitted signal at time $t$ when the data are $\mathbf{d}$; see (28.31). |

## Matrices

| | |
|---|---|
| $n \times m$ matrix | A matrix with $n$ rows and $m$ columns. |
| $0$ | The all-zero matrix. |
| $\mathsf{I}_n$ | The $n \times n$ identity matrix. |
| $a^{(k,\ell)}$ | The Row-$k$ Column-$\ell$ component of the matrix $\mathsf{A}$. |
| $\mathsf{A}^*$ | Componentwise complex conjugate. |
| $\mathsf{A}^\mathsf{T}$ | Transpose of $\mathsf{A}$. |
| $\mathsf{A}^\dagger$ | Hermitian conjugate of $\mathsf{A}$. |
| $\mathrm{tr}(\mathsf{A})$ | Trace of $\mathsf{A}$. |
| $\det(\mathsf{A})$ | Determinant of $\mathsf{A}$. |
| $\mathrm{Re}(\mathsf{A})$ | Componentwise real part of $\mathsf{A}$. |
| $\mathrm{Im}(\mathsf{A})$ | Componentwise imaginary part of $\mathsf{A}$. |
| $\mathsf{A} \succeq 0$ | $\mathsf{A}$ is a positive semidefinite matrix. |
| $\mathsf{A} \succ 0$ | $\mathsf{A}$ is a positive definite matrix. |

## Vectors

| | |
|---|---|
| $\mathbb{R}^n$ | Set of column vectors of $n$ real components. |
| $\mathbb{C}^n$ | Set of column vectors of $n$ complex components. |
| $\mathbf{0}$ | The all-zero vector. |
| $a^{(j)}$ | The $j$-component of the column vector $\mathbf{a}$. |
| $\mathbf{a}^\mathsf{T}$ | The transpose of the vector $\mathbf{a}$. |
| $\|\mathbf{a}\|$ | Euclidean norm of $\mathbf{a}$; see (20.85). |
| $\langle \mathbf{a}, \mathbf{b} \rangle_\mathrm{E}$ | Euclidean inner product; see (20.84). |
| $\mathrm{d}_\mathrm{E}(\mathbf{a}, \mathbf{b})$ | Euclidean distance between $\mathbf{a}$ and $\mathbf{b}$, i.e., $\|\mathbf{a} - \mathbf{b}\|$. |

## Linear Algebra

| | |
|---|---|
| $\mathrm{span}(\mathbf{v}_1, \ldots, \mathbf{v}_n)$ | Linear subspace spanned by the $n$-tuple $(\mathbf{v}_1, \ldots, \mathbf{v}_n)$; see (4.8). |
| $\mathrm{Dim}(\mathcal{V})$ | Dimension of the subspace $\mathcal{V}$. |
| $\mathrm{Ker}(\mathsf{T})$ | Kernel of the linear mapping $\mathsf{T}(\cdot)$. |
| $\mathrm{Image}(\mathsf{T})$ | Image of the linear mapping $\mathsf{T}(\cdot)$. |

## Probability

| | |
|---|---|
| $(\Omega, \mathcal{F}, P)$ | Probability triplet; see Page 3. |
| $P_X(\cdot)$ | Probability Mass Function (PMF) of $X$. |
| $P_{X,Y}(\cdot, \cdot)$ | Joint PMF of $(X, Y)$. |
| $P_{X|Y}(\cdot|\cdot)$ | Conditional PMF of $X$ given $Y$. |
| $f_X(\cdot)$ | Probability density function of $X$. |
| $f_{X,Y}(\cdot, \cdot)$ | Joint PDF of $(X, Y)$. |
| $f_{X|Y}(\cdot|\cdot)$ | Conditional PDF of $X$ given $Y$. |
| $f_{X|\mathcal{A}}$ | Conditional PDF of $X$ given the event $\mathcal{A}$. |
| $F_X(\cdot)$ | Cumulative distribution function of $X$. |
| $\Phi_X(\cdot)$ | Characteristic function of $X$. |
| $M_X(\cdot)$ | Moment generating function of $X$; see (19.23). |
| $\mathsf{E}[X]$ | Expectation of $X$; see (17.9). |
| $\mathsf{Var}[X]$ | Variance of $X$; see (17.14a). |
| $\mathsf{Cov}[X, Y]$ | Covariance between $X$ and $Y$; see (17.17). |
| $\mathsf{E}[\cdot|\cdot]$ | Conditional expectation. |
| $\mathrm{Pr}(\cdot)$ | Probability of an event. |
| $\mathrm{Pr}(\cdot|\cdot)$ | Conditional probability of an event. |
| $\mathrm{Pr}[\cdot]$ | Probability that a RV satisfies some condition. |
| $\mathrm{Pr}[\cdot|\cdot]$ | Conditional version of $\mathrm{Pr}[\cdot]$. |
| $\overset{\mathscr{L}}{=}$ | Equal in law. |
| $X\!-\!\!\circ\!-\!Y\!-\!\!\circ\!-\!Z$ | $X$ and $Z$ are conditionally independent given $Y$. |
| $\{X_k\}$ | Sequence of random variables $\ldots, X_{-1}, X_0, X_1, \ldots$ |
| $X \sim \text{Distribution}$ | $X$ has the specified distribution. |
| $\chi^2_{n,\lambda}$ | Noncentral $\chi^2$ distribution with $n$ degrees of freedom and noncentrality parameter $\lambda$. |
| $\text{Bernoulli}(p)$ | Bernoulli distribution (takes on the values 0 and 1 with probabilities $p$ and $1 - p$). |
| $\mathcal{U}(\mathcal{A})$ | Uniform distribution over the set $\mathcal{A}$. |
| $\mathcal{N}_{\mathbb{C}}(\mathbf{0}, \mathsf{K})$ | Multivariate proper complex Gaussian distribution of covariance $\mathsf{K}$; see Note 24.3.13. |
| $\mathcal{N}(\boldsymbol{\mu}, \mathsf{K})$ | Multivariate real Gaussian distribution of mean $\boldsymbol{\mu}$ and covariance $\mathsf{K}$. |

## Stochastic Processes

| | |
|---|---|
| $\big(X(n)\big), \big(X_n, \; n \in \mathbb{Z}\big)$ | Discrete-time stochastic process. |
| $\big(X(t)\big), \big(X(t), \; t \in \mathbb{R}\big)$ | Continuous-time stochastic process. |
| $\mathsf{K}_{XX}$ | Autocovariance function. |
| $\mathsf{S}_{XX}$ | Power spectral density (PSD). |
| $\rho_{XX}(\cdot)$ | Correlation function. |

## Hypothesis Testing

| | |
|---|---|
| $\mathcal{B}_{m,m'}$ | The subset of $\mathbb{R}^d$ defined in (21.33). |
| $H$ | RV to be guessed in binary hypothesis testing. |
| $\mathrm{LLR}(\cdot)$ | Log likelihood-ratio function; see (20.41). |
| $\mathrm{LR}(\cdot)$ | Likelihood-ratio function; see (20.38). |
| $\mathsf{M}$ | Number of hypotheses in multi-hypothesis testing. |
| $\mathcal{M}$ | Set of hypotheses $\{1, \ldots, \mathsf{M}\}$. |
| $M$ | RV to be guessed in multi-hypothesis testing. |
| $\phi_{\mathrm{Guess}}$ | Generic guessing rule; see Sections 20.2 & 21.2. |
| $\phi_{\mathrm{Guess}}^*$ | Generic optimal guessing rule. |
| $\phi_{\mathrm{MAP}}$ | MAP Decision Rule. |
| $\phi_{\mathrm{ML}}$ | Maximum-Likelihood Rule. |
| $p^*(\mathrm{error})$ | Optimal probability of error. |
| $p_{\mathrm{MAP}}(\mathrm{error}|\cdot)$ | Conditional probability of error of MAP rule. |

## The Binary Field and Binary Tuples

| | |
|---|---|
| $\mathbb{F}_2$ | Binary field $\{0, 1\}$. |
| $\mathbb{F}_2^\kappa$ | The set of binary $\kappa$-tuples. |
| $\oplus$ | Addition in $\mathbb{F}_2$; see (29.3). |
| $\cdot$ | Multiplication in $\mathbb{F}_2$; see (29.4). |
| $\mathrm{d_H}(\mathbf{u}, \mathbf{v})$ | Hamming distance; see Section 29.2.4. |
| $\mathrm{w_H}(\mathbf{u})$ | Hamming weight; see Section 29.2.4. |
| $\Upsilon$ and $\Upsilon_\eta$ | Antipodal mappings (29.14) and (29.17). |

## Coding

| | |
|---|---|
| $\mathcal{A}_{\kappa,0}$ | Binary N-tuples whose $\kappa$-th component is zero; see (29.61). |
| $\mathcal{A}_{\kappa,1}$ | Binary N-tuples whose $\kappa$-th component is one; see (29.64). |
| $\mathbf{c}$ | Generic element of Image($\mathsf{T}$). |
| $\mathrm{d_{min,H}}$ | Minimum Hamming distance; see (29.54). |
| $\mathbf{enc}$ | Encoder. |
| $p_\kappa^*$ | Optimal probability of error in guessing the $\kappa$-th data bit. |
| $p_{\mathrm{MAP}}(\mathrm{error}|\mathbf{D} = \mathbf{d})$ | Conditional probability of error of the MAP rule designed to minimize block errors. |
| $\psi_\mathbf{d}(\cdot)$ | See (29.77). |
| $\mathbf{x}$ | Generic element of Image($\mathbf{enc}$). |
| $x_\eta(\mathbf{d})$ | The $\eta$-th symbol in the N-tuple enc($\mathbf{d}$). |

# Index